Lecture Notes in Networks and Systems

Volume 366

The series "Lecture Notes in Networks and Systems" publishes the latest developments in Networks and Systems—quickly, informally and with high quality. Original research reported in proceedings and post-proceedings represents the core of LNNS.

Volumes published in LNNS embrace all aspects and subfields of, as well as new challenges in, Networks and Systems.

The series contains proceedings and edited volumes in systems and networks, spanning the areas of Cyber-Physical Systems, Autonomous Systems, Sensor Networks, Control Systems, Energy Systems, Automotive Systems, Biological Systems, Vehicular Networking and Connected Vehicles, Aerospace Systems, Automation, Manufacturing, Smart Grids, Nonlinear Systems, Power Systems, Robotics, Social Systems, Economic Systems and other. Of particular value to both the contributors and the readership are the short publication timeframe and the world-wide distribution and exposure which enable both a wide and rapid dissemination of research output.

The series covers the theory, applications, and perspectives on the state of the art and future developments relevant to systems and networks, decision making, control, complex processes and related areas, as embedded in the fields of interdisciplinary and applied sciences, engineering, computer science, physics, economics, social, and life sciences, as well as the paradigms and methodologies behind them.

Indexed by SCOPUS, INSPEC, WTI Frankfurt eG, zbMATH, SCImago.

All books published in the series are submitted for consideration in Web of Science.

More information about this series at https://link.springer.com/bookseries/15179

Duy Cuong Nguyen · Ngoc Pi Vu ·
Banh Tien Long · Horst Puta ·
Kai-Uwe Sattler
Editors

Advances in Engineering Research and Application

Proceedings of the International Conference on Engineering Research and Applications, ICERA 2021

Springer

Editors
Duy Cuong Nguyen
Faculty of Electronic Engineering
Thai Nguyen University of Technology
Thai Nguyen, Vietnam

Ngoc Pi Vu
Faculty of Mechanical Engineering
Thai Nguyen University of Technology
Thai Nguyen, Vietnam

Banh Tien Long
Vietnam Association for Science Editing
Hanoi University of Science and Technology
Hanoi, Vietnam

Horst Puta
Institute for Automation
and Systems Engineering
Ilmenau University of Technology (IUT)
Ilmenau, Germany

Kai-Uwe Sattler
Department of Computer Science
and Automation
Ilmenau University of Technology (IUT)
Ilmenau, Germany

ISSN 2367-3370 ISSN 2367-3389 (electronic)
Lecture Notes in Networks and Systems
ISBN 978-3-030-92573-4 ISBN 978-3-030-92574-1 (eBook)
https://doi.org/10.1007/978-3-030-92574-1

This Springer imprint is published by the registered company Springer Nature Switzerland AG
The registered company address is: Gewerbestrasse 11, 6330 Cham, Switzerland

Preface

This book presents the up-to-date contributions from many researchers of different universities in the field of researches, theories, and practices in engineering research and applications. The topics of this volume are mechanical engineering, materials and mechanics of materials, mechatronics and micromechatronics, automotive engineering, electrical and electronics engineering, and information and communication technology.

The up-to-date contributions reported in this book were carefully reviewed by experts and also approved by editors for the last review. All 95 accepted papers were presented and discussed at the 4th International Conference on Engineering Research and Applications 2021 (ICERA 2021), held in Thai Nguyen City, Vietnam, on December 1–2, 2021. The total papers sent to this conference are 225 papers. As a result of the two-stage review process, only 95 excellent contributions were selected for the presentation at the conference and publication in this book.

ICERA 2021 was organized by Thai Nguyen University of Technology (TNUT), Vietnam, and cooperated with Ilmenau University of Technology, Germany. The first conference started in 2018. So far, the 4th conference has attracted a lot of contributions from researchers of many different universities around the world.

To contribute the proceedings, we are grateful to thank the authors for their contributions to the conference and this volume. We would like to deeply thank the reviewers for their recommendations as improvement guidance for the selected papers. We would specially thank our keynote speakers, Prof. Kai-Uwe Sattler (Ilmenau University of Technology, Germany), Prof. Trung Q. Duong (Queen's University Belfast, UK, and Research Chair of the Royal Academy of Engineering, UK), Prof. Le Anh Tuan (Hanoi University of Science and Technology, Vietnam), and Prof. Roger A. Sauer (RWTH Aachen University, Germany), for their valuable and inspiring contributions to scientists, researchers, and listeners. We also thank the members of the Organizing Committee of ICERA 2021 and Springer Publishers for their excellent technical and editorial support.

December 2021

Kai-Uwe Sattler
Duy Cuong Nguyen
Ngoc Pi Vu
Banh Tien Long
Horst Puta

Keynote Addresses

Accelerating Data Management Using Modern Hardware Architectures

Kai-Uwe Sattler

TU Ilmenau, Germany
kus@tu-ilmenau.de

Abstract. Over the past thirty years, database management systems have been established as one of the most successful software concepts. In today's business environment, they constitute the centerpiece of almost all critical IT systems. Furthermore, the application of database systems has moved beyond pure transaction-oriented scenarios. Instead, they are more and more utilized as data integration platforms to realize a unified access model to heterogeneous or even distributed data. In addition, database technology in a broader sense is exploited in pure analytical applications, e.g., driving data science and AI scenarios. To meet the challenges stemming from this trend, data management architectures must evolve. A successful and promising way is to exploit modern hardware technologies to accelerate database processing.

In this talk, we present our research results for the use of such hardware technologies. Using the example of graph processing, we discuss the design of data structures for memory hierarchies with persistent memory and the use of accelerators like GPU and FPGA for database operations.

Intelligent 6G Ultra-reliable and Low-Latency Communications for Industrial Automation

Trung Q. Duong

Chair of Telecommunications at Queen's University Belfast, UK
trung.q.duong@qub.ac.uk

Abstract. Consider these startling statistics (from CISCO), global mobile data traffic is increasing exponentially and it is expected that there will be 100 billion 'Internet of things' devices by the year 2025. Thus, the need for improved wireless reliability and latency is greater than ever. However, implementing algorithms to ensure low-latency communication for massive numbers of power-constrained mobile devices conflicts directly with the need for ultra-reliability. Recently, ultra-reliable and low latency communications (URLLC) has been considered as one of the promising technologies for future wireless networks. URLLC brings many new kinds of services and applications that require URLLC as a prerequisite. For instance, URLLC can be applied in smart healthcare systems with telesurgery applications or in entertainment with a wide range of augmented reality (AR) and virtual reality (VR) utilities. Importantly, URLLC opens opportunities to transform current industrial automation to the next generation, which entails wireless control and real-time monitoring for autonomous factories. Despite its huge potential in the Industrial Revolution, research of URLLC in industry automation is still in the early stage. There are many open research problems in URLLC, which are unresolved and, implicitly, challenge the viability of 6G. This talk will discuss 6G technologies in industrial automation that require high data rates with ultra-reliability at very low latency for which URLLC is a natural choice.

Short bio: Dr. Trung Q. Duong (Senior Member of IEEE) is Professor, Chair of Telecommunications at Queen's University Belfast, UK, and Research Chair of the Royal Academy of Engineering, UK. His current research interests include wireless communications, signal processing, and machine learning. He has published more than 350 papers with 12,500+ citations and h-index 61. He currently serves as Editor for the IEEE Transactions on Wireless Communications and Executive Editor for the IEEE Communications Letters. He was awarded the Best Paper Award at the IEEE Vehicular Technology Conference (VTC Spring) in 2013, IEEE International Conference on Communications (ICC) 2014, IEEE Global

Communications Conference (GLOBECOM) 2016, IEEE Digital Signal Processing Conference (DSP) 2017, and GLOBECOM 2019. He is the recipient of prestigious Royal Academy of Engineering Research Fellowship (2015–2020) and has won a prestigious Newton Prize 2017.

Isogeometric Finite Element Formulations: Motivation, Implementation, and Applications

Roger A. Sauer

Graduate School AICES, RWTH Aachen University
sauer@aices.rwth-aachen.de

Abstract. Isogeometric finite elements are a new class of finite elements that have been developed by combining computer-aided design (CAD) and finite element analysis [1]. They use spline-based shape functions that have several advantages over classical finite element shape functions. They can be embedded straightforwardly into existing finite element software using the Bezier extraction operator [2], although special attention has to be paid to boundary conditions, especially on curved surfaces. The isogeometric paradigm is particularly advantageous for solving high-order partial differential equations, as they appear, for example, in general Kirchhoff–Love shell theory [3]. This is illustrated by several application examples from science and engineering.

References

1. Hughes, T. J. R., Cottrell, J. A., Bazilevs, Y.: Isogeometric analysis: CAD, finite elements, NURBS, exact geometry and mesh refinement. Comput. Methods Appl. Mech. Eng., **194**, 4135–4195 (2005)
2. Borden, M. J., Scott, M. A., Evans, J. A., Hughes, T. J. R.: Isogeometric finite element data structures based on Bezier extraction of NURBS. Int. J. Numer. Meth. Eng., **87**, 15–47 (2011)
3. Duong, T. X., Roohbakhshan, F., Sauer, R. A.: A new rotation-free isogeometric thin shell formulation and a corresponding continuity constraint for patch boundaries. Comput. Methods Appl. Mech. Eng., **316**, 43–83 (2017)

Electric Vehicle Development in Vietnam: Status Update and Shortcoming Perspective

Le Anh Tuan[1] and Nguyen Thi Yen Lien[2]

[1] Hanoi University of Science and Technology, Vietnam
tuan.leanh@hust.edu.vn
[2] University of Transport and Communications, Vietnam

Abstract. Vietnam committed to cut off 9% GHG emissions in 2030, compared to the business as usual scenario using domestic resources in the nationally determined contribution (NDC); this number is increased to 27% when receiving international support. In Vietnam, the transport sector contributes about 18% of the country's total CO_2 emissions; therefore, the targets for carbon emissions in this sector must be set accordingly.

Among many low-carbon transport measures, the CO_2 emission reduction target in the transport sector could only be achieved clearly when electric vehicle is promoted and implemented at the different levels under a sustainable development roadmap.

In this lecture, the status and shortcoming perspective of electric vehicle in Vietnam were reviewed based on the data collected from recent research papers and reports. By this approach, the data related to the electric vehicle fleet characteristics, the investment and technology status, and the electric vehicle development-related policies at national level and city level were summarized and analyzed. The status of electric power sources and circuit is also mentioned to present the potential for electric vehicle development in a due course.

Keywords: Electric vehicle development, Electric vehicle fleet, EV development policy, Vietnam

Contents

A Robust Control Approach to Self-balancing Bicycles

Chung Nguyen Van, Duong Dinh Binh, Hien Nguyen Thi, Hieu Le Xuan, Mai Hoang Thi, Thu Nguyen Thanh, Van Anh Nguyen Thi, and Tung Lam Nguyen(✉)

Hanoi University of Science and Technology, Hanoi, Vietnam
lam.nguyentung@hust.edu.vn

Abstract. The paper presents a robust control approach to self-balancing bikes (SBB) against unknown system parameters and external disturbances. The system is mechanically balanced by the momentum force produced by a flywheel. Initially, sliding mode control is employed for balancing the bike in roll motion and steering. Based on the designed control, a high-gain observer is developed to estimate the bike roll inertia momentum which is difficult to derive in practice. The stability of the overall system including estimated inertia momentum is given. Comprehensive simulations are performed to demonstrate the proposed control.

Keywords: Self-balancing · Sliding mode control · High-gain observer · Lyapunov stability

1 Introduction

Bicycles and their variants have been used for centuries. In comparison to cars and motorbikes, bicycles offer great maneuverable and environmentally friendly abilities. Possessing complicated dynamics, bicycles attract many scientists in the quest of developing self-balancing and riderless bicycles [1–5]. Deng et al. propose a hardware design in [6] for bike balancing using PID in steering control. The paper's main concern is the design part of the system specifying necessary hardware and structures. Similarity, design-oriented approaches are introduced in [7, 8]. Based on a linearized mathematical model of the SBB, Guo et al. formulate a linear- quadratic optimal control for the system. However, due to the small roll and steering angles approximation, the control is valid in a limited range. Linear quadratic control is also investigated in [9] in comparison with model predictive control. Based on the assumption of full feedback and known parameters, a feedback linearization is suggested in [10]. To provide the system with disturbances rejection ability, a control with scissored-pair control moment gyroscope is employed in [11], the control shows promising performance against disturbances but the design might be problematic in practice due to its complicity. In [12], the authors propose a selective balancing control through flywheel and handlebar control. The method could cause a chattering phenomenon since control sources are switched via examining the balancing border. Another interesting research

D. C. Nguyen et al. (Eds.): ICERA 2021, LNNS 366, pp. 1–11, 2022.
https://doi.org/10.1007/978-3-030-92574-1_1

on SBB balancing control can be found in [13] where the authors successfully design a control that can maintain lateral stability regardless of speed changes. The closed-loop system is validated numerically and experimentally. Not only considering the balancing problem, but also path tracking and roll balancing are investigated simultaneously in [14].

The most challenging problem in controlling the SBB is accurately acquiring system parameters such as roll moment of inertia since roll inertia varies in accordance with the carried load. In the paper, we propose an estimating mechanism based on high-gain observer to tackle the problem of unknown roll inertia. It is shown that the system stability is maintained under sliding mode control action with the provided estimated information. The simulations are carried out based on a miniature SBB design to validate the control.

2 System Modelling

The principle of SBB is based on the concept of an inverted pendulum. To maintain the balance, a flywheel mounted on the bicycle is designed to rotate around its axis by using a DC motor. When the whole system is out of the equilibrium position, a torque occurs due to the gravity that makes the bicycle falling down. The flywheel will generate a reaction torque that is equal to the torque by gravity Fig. 1 and Fig. 2.

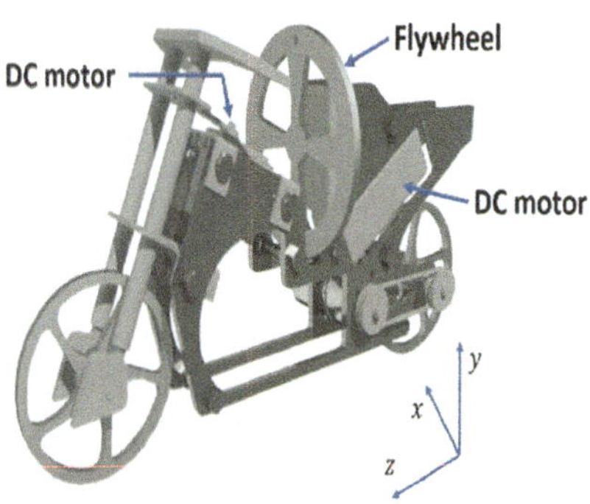

Fig. 1. A SBB system.

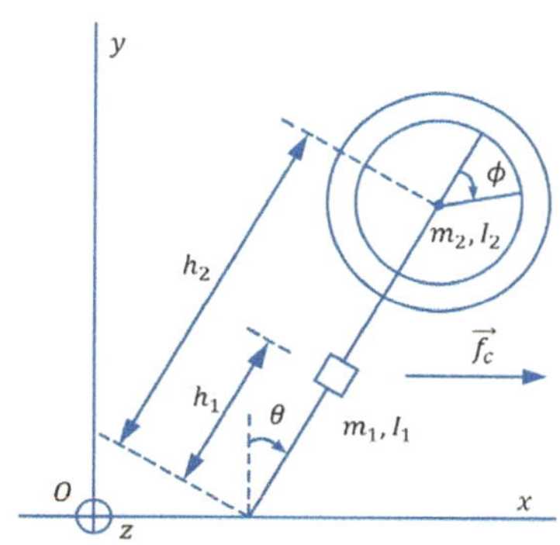

Fig. 2. System coordinate.

2.1 Motor Model

It is assumed that motor mechanical torque T, is proportional to armature current i_a by a motor constant K_m,

$$T = K_m \cdot i_a \tag{1}$$

The voltage applied to the motor armature can be expressed as:

$$V = R_a \cdot i_a + L_a \cdot \frac{di_a}{dt} + K_e \cdot \omega \tag{2}$$

where $e_b = K_e\omega$ is the motor electromotive back-emf, K_e is the motor back-emf constant, ω is the motor angular speed, L_a and R_a are the armature inductance and resistance, respectively. The motor inductance can be neglected since, in general, $L_a \ll R_a$. By combining Eq. 1 with Eq. 2, the motor model can be formulated as:

$$T = K_m\left(\frac{V - K_e \cdot \omega}{R_a}\right) \tag{3}$$

2.2 Self-balancing Bicycle

Balancing problem. The SBB balancing model is illustrated in Fig. 2, in which m_1 is the weight of the bicycle (excluding the flywheel), m_2 is the weight of the flywheel, h_1 is the length measured from the ground to the bicycle centre of gravity (excluding the flywheel), h_2 is the length from the ground to the flywheel's rotation axis, I_1 is the inertia moment of the bicycle around its centre of gravity and I_2 is the inertia moment of the flywheel around its centre of gravity. From Fig. 1, l is the distance between the two front and rear axles along the body length, α is the head angle of the steering axis. The vehicle is considered moving laterally along the x-axis while it is moving forward along the z-axis with steering. Thus, there are four degrees of freedom including $\theta-$ the tilt angle of the bicycle compared to the $y-$ axis, $\phi-$ the angle of rotation of the flywheel, $\delta-$ steering angle, $x-$ lateral coordinates of the bicycle. Suppose that the velocity v_z along the $z-$ axis is constant. The total kinetic energy of the system is:

$$\begin{aligned} K = {} & \tfrac{1}{2}I_1\dot{\theta}^2 + \tfrac{1}{2}I_2\left(\dot{\theta}+\dot{\phi}\right)^2 + \tfrac{1}{2}(m_1+m_2)v_z^2 + \tfrac{1}{2}m_1\left(h_1\sin\,\theta\cdot\dot{\theta}\right)^2 \\ & + \tfrac{1}{2}m_2\left(h_2\sin\theta\cdot\dot{\theta}\right)^2 + \tfrac{1}{2}m_1\left(\dot{x}+h_1\cos\theta\cdot\dot{\theta}\right)^2 + \tfrac{1}{2}m_2\left(\dot{x}+h_2\cos\theta\cdot\dot{\theta}\right)^2 \end{aligned} \tag{4}$$

The total potential energy of the system relative to the ground is:

$$P = (m_1h_1 + m_2h_2)\cdot g\cdot\cos\theta \tag{5}$$

Lagrange equation is represented as follows:

$$\frac{d}{dt}\left(\frac{\partial L}{\partial \dot{q}_i}\right) - \frac{\partial L}{\partial q_i} = Q_i \tag{6}$$

where $L = K - P$ is the Lagrange function, Q_i is the external torque (or force) and q_i is the generality coordinates. By taking $q_i = \theta, \phi, x$ respectively we get:

$$\begin{cases} (m_1h_1+m_2h_2)\ddot{x}\cos\theta+\left(m_1h_1^2+m_2h_2^2+I_1+I_2\right)\ddot{\theta} \\ \qquad +I_2\ddot{\phi}-(m_1h_1+m_2h_2)g\sin\theta=0 \\ I_2\left(\ddot{\theta}+\ddot{\phi}\right)=T_1=K_{m_1}\left(\frac{V_\phi-K_{e_1}\dot{\phi}}{R_1}\right) \\ (m_1+m_2)\ddot{x}+(m_1h_1+m_2h_2)\sin\theta \overset{=}{\ } f_c \end{cases} \tag{7}$$

where $f_c \approx \frac{(m_1+m_2)v_z^2\sin\alpha}{l\cos\theta}\delta$ is the centrifugal force generated while the vehicle is moving forward in curvature with centre O (Fig. 2), T_1 and V_ϕ are respectively the torque and input voltage of the DC motor attached to the shaft of the flywheel, R_1, K_{m_1} and K_{e_1} are its corresponding parameters as formulated in (3). Removing $\ddot{x}$ in Sys. 7 and linearizing (we assume $\sin\theta \approx \theta$ and $\cos\theta \approx 1$), the dynamic model of the system becomes:

$$\begin{cases} \frac{(m_1h_1+m_2h_2)v_z^2\sin\alpha}{l}\delta+\left[\frac{m_1m_2(h_1-h_2)^2}{m_1+m_2}+I_1+I_2\right]\ddot{\theta}+I_2\ddot{\phi}=(m_1h_1+m_2h_2)g\theta \\ I_2\left(\ddot{\theta}+\ddot{\phi}\right)=K_{m_1}\left(\frac{V_\phi-K_{e_1}\dot{\phi}}{R_1}\right) \end{cases} \tag{8}$$

Steering Model. Steering angle δ is controlled by a servo motor mounted on the steer with a transmission ratio 1:1. The model of the motor is formulated as:

$$T_2=K_{m_2}\left(\frac{V_\delta-K_{e_2}\dot{\delta}}{R_2}\right) \tag{9}$$

in which T_2, V_δ are the torque and armature voltage applied to the motor, R_2, K_{m_2} and K_{e_2} are its corresponding parameters. The relationship between the produced torque and steering angle is:

$$T_2=J_2\cdot\ddot{\delta}+D_2\cdot\dot{\delta} \tag{10}$$

where J_2 is the equivalent moment of inertia (including both the armature inertia and the load inertia), D_2 is the system equivalent viscous damping. By combining Eq. 9 with Eq. 10, it leads to:

$$\ddot{\delta}=-\left(\frac{D_2}{J_2}+\frac{K_{m_2}K_{e_2}}{J_2R_2}\right)\dot{\delta}+\frac{K_{m_2}}{J_2R_2}V_\delta \tag{11}$$

2.3 State-Space Representation

Let $\underline{x}=\left[\theta\dot{\theta}\dot{\phi}\delta\dot{\delta}\right]^T$, $\underline{u}=\left[V_\phi V_\delta\right]^T$. From Sys. 8 and Eq. 11, the state-space of the SBB can be expressed as:

$$\begin{cases} \dot{\underline{x}} = A\underline{x} + B\underline{u} \\ \underline{y} = \underline{x} \end{cases} \tag{12}$$

where:

$$A = \begin{pmatrix} 0 & 1 & 0 & 0 & 0 \\ a_{21} & 0 & a_{23} & a_{24} & 0 \\ a_{31} & 0 & a_{33} & a_{34} & 0 \\ 0 & 0 & 0 & 0 & 1 \\ 0 & 0 & 0 & 0 & a_{55} \end{pmatrix}, \quad B = \begin{pmatrix} 0 & 0 \\ b_{21} & 0 \\ b_{31} & 0 \\ 0 & 0 \\ 0 & b_{52} \end{pmatrix},$$

$$M = m_1 I_1 + m_2 I_2 + m_1 m_2 (h_1 - h_2)^2,$$

$$a_{21} = -a_{31} = \frac{g(m_1 + m_2)(m_1 h_1 + m_2 h_2)}{M}, a_{23} = \frac{K_{m_1} K_{e_1} (m_1 + m_2)}{MR_1},$$

$$a_{24} = -a_{34} = -\frac{(m_1 + m_2)(m_1 h_1 + m_2 h_2) v_z^2 \sin\alpha}{Ml},$$

$$a_{33} = -\frac{K_{m_1} K_{e_1}}{I_2 R_1}\left[1 + \frac{(m_1 + m_2) I_2}{M}\right], a_{55} = -\frac{D_2}{J_2} - \frac{K_{m_2} K_{e_2}}{J_2 R_2},$$

$$b_{21} = -\frac{K_{m_1}(m_1 + m_2)}{MR_1}, b_{31} = \frac{K_{m_1}}{I_2 R_1}\left[1 + \frac{(m_1 + m_2) I_2}{M}\right], b_{52} = \frac{K_{m_2}}{J_2 R_2}.$$

3 Sliding Mode Control

3.1 Balancing Control

Let $e_1 = \theta - \theta_r$, where θ_r is the reference tilting angle. The sliding surface is chosen as $\ddot{e}_1 + k_1 \dot{e}_1 + k_2 e_1 = 0$, where $k_1, k_2 > 0$. Since $s^2 + k_1 s + k_2$ is Hurwitz polynomial, it turns out that $e_1 \to 0$ when $t \to \infty$, i.e $\theta \to \theta_r$. We have:

$$\ddot{\theta} = \ddot{\theta}_r - k_1 \dot{e}_1 - k_2 e_1 \tag{13}$$

Substituting $\ddot{\theta} = a_{21}\theta + a_{23}\dot{\phi} + a_{24}\delta + b_{21} V_\phi$ from Sys. 12, the control law of tilt angle follows:

$$V_\phi = \frac{\ddot{\theta}_r - k_1 \dot{e}_1 - k_2 e_1 - a_{21}\theta - a_{23}\dot{\phi} - a_{24}\delta}{b_{21}} \tag{14}$$

3.2 Steering Control

Let $e_2 = \delta - \delta_r$, where δ_r is the reference steering angle. Similarly, we chose the sliding surface $\ddot{e}_2 + k_3\dot{e}_2 + k_4 e_2 = 0$ where $k_3, k_4 > 0$, which ensures that $\delta \to \delta_r (t \to \infty)$.

$$\ddot{\delta} = \ddot{\delta}_r - k_3\dot{e}_2 - k_4 e_2 \tag{15}$$

Substituting $\ddot{\delta} = a_{55}\dot{\delta} + b_{52}V_\delta$ from Sys. 12, the control law of steering angle is represented as follows:

$$V_\delta = \frac{\ddot{\delta}_r - k_3\dot{e}_2 - k_4 e_2 - a_{55}\dot{\delta}}{b_{52}} \tag{16}$$

With the controls given in (14) and (16), the system is stable. That means the vehicle can balance itself and it can control the movement of the system (steering and moving forward). However, in practice, the inertia of the vehicle might change while it is steering on a rough surface or containing objects with unknown mass. Thus, it requires a suitable observer to estimate uncertain parameters.

4 High-Gain Disturbance Observer

It is supposed that the inertia moment of the vehicle is $I_1 = I_1 + \Delta I_1$, where I_1 represents the known nominal value and ΔI_1, an (unknown) modelling error. Sys. 8 changes into:

$$\begin{cases} \frac{(m_1h_1 + m_2h_2)v_z^2 \sin\alpha}{l}\delta + \left[\frac{m_1m_2(h_1-h_2)^2}{m_1+m_2} + I_1 + I_2\right]\ddot{\theta} \\ \quad + I_2\ddot{\phi} + \Delta I_1\ddot{\theta} = (m_1h_1 + m_2h_2)g\theta \\ \quad I_2\left(\ddot{\theta} + \ddot{\phi}\right) = K_{m_1}\left(\frac{V_\phi - K_{e_1}\dot{\phi}}{R_1}\right) \end{cases} \tag{17}$$

Now, let's define the disturbance:

$$d = \frac{\Delta I_1\ddot{\theta}}{I_1 + \frac{m_1m_2(h_1-h_2)^2}{m_1+m_2}} \tag{18}$$

With this definition, the state-space can be represented as follows:

$$\begin{pmatrix} \dot{x}_1 \\ \dot{x}_2 \\ \dot{x}_3 \end{pmatrix} = \begin{pmatrix} 0 & 1 & 0 \\ a_{21} & 0 & a_{23} \\ a_{31} & 0 & a_{33} \end{pmatrix}\begin{pmatrix} x_1 \\ x_2 \\ x_3 \end{pmatrix} + \begin{pmatrix} 0 & 0 \\ b_{21} & a_{24} \\ b_{31} & a_{34} \end{pmatrix}\begin{pmatrix} V_\phi \\ \delta \end{pmatrix} + \begin{pmatrix} 0 \\ -1 \\ 1 \end{pmatrix} d \tag{19}$$

in which $\underline{x} = [\theta\dot{\theta}\dot{\phi}]^T, \underline{u} = [V_\phi \delta]^T$ and a_{ij}, b_{ij} are defined in (12).

4.1 High-Gain Disturbance Observer Design

We assume that all states are measurable. Let $f(\underline{x}, \underline{u}) = a_{21}x_1 + a_{23}x_3 + b_{21}V_\phi + a_{24}\delta$. From the second equation of Sys. 19, it follows:

$$\dot{x}_2 = f(\underline{x}, \underline{u}) - d \tag{20}$$

Since $\dot{x}_2$ is the derivative of signal and there may be considerable noise in the measurement, it is not implementable to estimate d directly from Eq. 20. Let $\widehat{d}$ be an estimation of d, with the estimation error $\tilde{d} = d - \widehat{d}$. A reference model is chosen as:

$$\dot{\tilde{d}} + k\tilde{d} = 0 \tag{21}$$

with $k > 0$, we have $\tilde{d} \to 0$, i.e $\widehat{d} \to d$. An observer follows as a result:

$$\dot{\widehat{d}} = k\left(f - \dot{x}_2 - \widehat{d}\right) + \dot{d} \tag{22}$$

It's easy to see that the above observer has the same drawback, measuring $\dot{d}$ is not practical to implement. We adjust the reference model (21) and the corresponding observer (22) as follows:

$$\dot{\tilde{d}} + k\tilde{d} = \dot{d} \tag{23}$$

$$\dot{\widehat{d}} = k\left(f - \dot{x}_2 - \widehat{d}\right) \tag{24}$$

To reduce the effect of $\dot{d}$ in Eq. 23 (with the constraints $|\dot{d}|$ is bounded), let $k = \frac{1}{\varepsilon}$ – the high-gain coefficient. By using an auxiliary state variable $z = \widehat{d} + \frac{x_2}{\varepsilon}$, the new obtained observer:

$$\dot{z} = \frac{1}{\varepsilon}\left(f + \frac{x_2}{\varepsilon} - z\right) \tag{25}$$

allows us to estimate $\widehat{d} \to d$ efficiently by measuring states and input signals.

4.2 Controller Design

Rewrite Eq. 20 as:

$$\ddot{\theta} = a_{21}x_1 + a_{23}x_3 + b_{21}V_\phi + a_{24}\delta - d \tag{26}$$

Let $h(\underline{x}, \underline{u}) = a_{21}x_1 + a_{23}x_3 + a_{24}\delta \Rightarrow \ddot{\theta} = h + b_{21}V_\phi - d$. Suppose that d is known and let $e = \theta - \theta_r$, we propose a reference model as follows:

$$\ddot{e} + (k_5 + k_6)\dot{e} + (1 + k_5k_6)e = 0 \tag{27}$$

where $k_5, k_6 > 0$. The control law of tilting angle θ is obtained consequently:

$$V_\phi = \frac{\ddot{\theta}_r - (k_5 + k_6)\dot{e} - (1 + k_5k_6)e - h + d}{b_{21}} \tag{28}$$

If d is unknown, replace d in Eq. 28 by $\widehat{d}$, the control law leads to:

$$V_\phi = \frac{\ddot{\theta}_r - (k_5 + k_6)\dot{e} - (1 + k_5k_6)e - h + \widehat{d}}{b_{21}} \tag{29}$$

4.3 System Stability Analysis

Let $e_1 = e, e_2 = \dot{e} + k_5e$, the Lyapunov candidate function is defined as $V = \frac{1}{2}e_1^2 + \frac{1}{2}e_2^2 + \frac{1}{2}\tilde{d}^2$. With the adjusting control law V_ϕ represented in (29), the sliding sufface (27) becomes:

$$\ddot{e} + (k_5 + k_6)\dot{e} + (1 + k_5k_6)e + \tilde{d} = 0 \tag{30}$$

$$\dot{e}_2 + k_6e_2 + e_1 + \tilde{d} = 0 \tag{31}$$

which gives us $\dot{e}_2 = -k_6e_2 - e_1 - \tilde{d}$. Thus,

$$\begin{aligned} \dot{V} &= e_1\dot{e}_1 + e_2\dot{e}_2 + \tilde{d}\dot{\tilde{d}} = e_1(e_2 - k_5e_1) + e_2(-k_6e_2 - e_1 - \tilde{d}) + \tilde{d}\left(-\tfrac{\tilde{d}}{\varepsilon} + \dot{d}\right) \\ &= -k_5e_1^2 - \tfrac{k_6}{2}e_2^2 - \beta\tilde{d}^2 - \tfrac{k_6}{2}\left(e_2 + \tfrac{\tilde{d}}{k_6}\right)^2 - \beta\left(\tilde{d} - \tfrac{\dot{d}}{2\beta}\right)^2 + \tfrac{\dot{d}^2}{4\beta} \end{aligned} \tag{32}$$

in which $\beta = \frac{1}{2\varepsilon} - \frac{1}{4k_6}$.

If we choose $\varepsilon < 2k_6$ i.e. $\beta > 0$ then $\dot{V} < -\lambda V + \frac{\dot{d}^2}{4\beta}$, where $\lambda = \min\{2k_5, k_6, 2\beta\} > 0$, the system is Input to State Stability (ISS).

5 Simulations Results

Simulations results and some parameters are shown in Table 1.

Table 1. The parameters of SBB.

Parameters	Value	Unit	Parameters	Value	Unit
g	9.8	m/s^2	v_z	0.1	m/s
l	0.253	m	R_1	0.5	Ω
α	$\pi/2$	rad	K_{m_1}	0.045	N.m/A
I_1	0.0135	kg.m^2	K_{e_1}	0.045	V.s/rad
I_2	$1.43 \cdot 10^{-6}$	kg.m^2	R_2	0.4	Ω
m_1	1.12	kg	K_{m_2}	0.035	N.m/A
m_2	0.08	kg	K_{e_2}	0.035	V.s/rad
h_1	0.097	m	J_2	0.01	kg.m^2
h_2	0.131	m	D_2	0.05	kg/s

Figure 3, Fig. 4 and Fig. 5 show the high-gain disturbance observer where disturbances acting on the systems are chosen as $d = e^{-t}, d = \sin(1.5t)$, respectively. In Fig. 6, the initial value of the tilt angle is set as 0.27 rad, the steering angle is designed to pursue the reference $\delta_r = 0.4\sin(1.5t)$. It takes about 2 s for the system to move itself back to the balance position. Illustrated in Fig. 7 and Fig. 8 are the velocity of flywheel and voltage inputs, in order to control the balance and the direction of the bicycle.

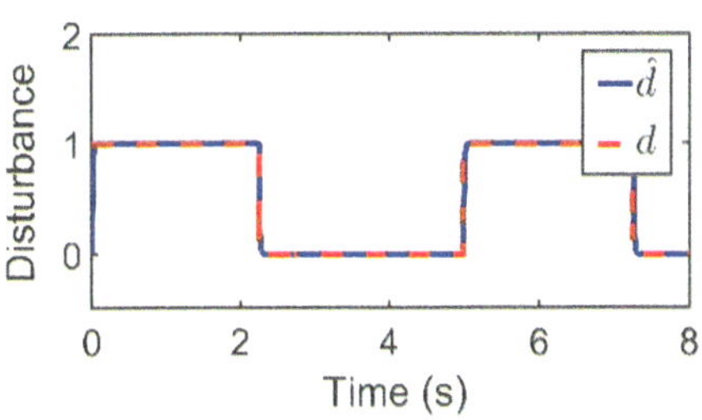

Fig. 3. Disturbance estimation (d is rectangular pulse)

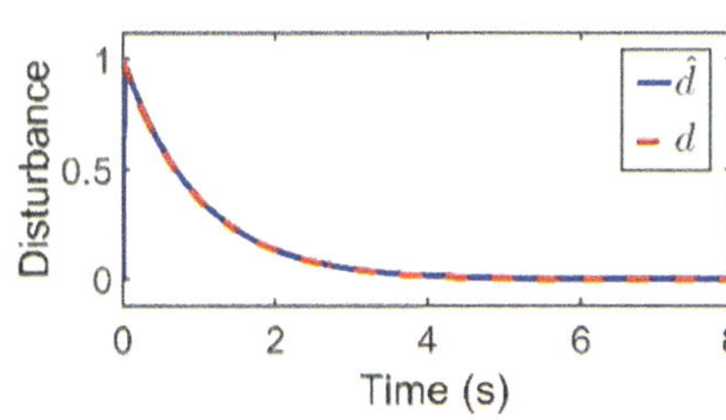

Fig. 4. Disturbance estimation ($d = e^{-t}$)

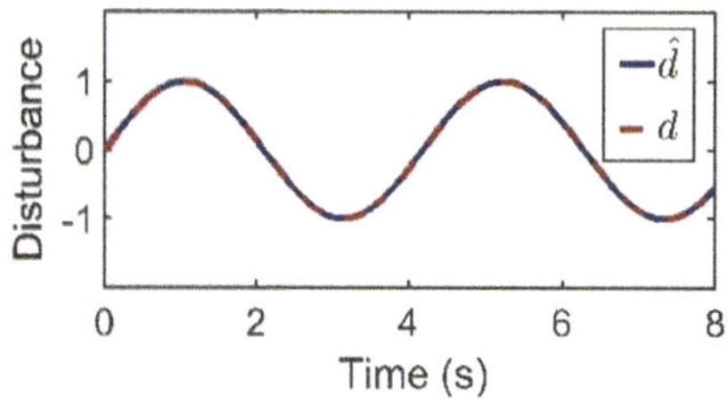

Fig. 5. Disturbance estimation ($d = \sin(1.5t)$)

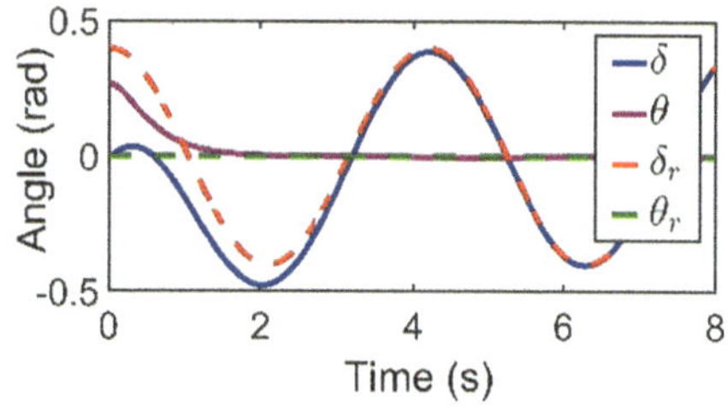

Fig. 6. The output signal of tilt angle and steering angle

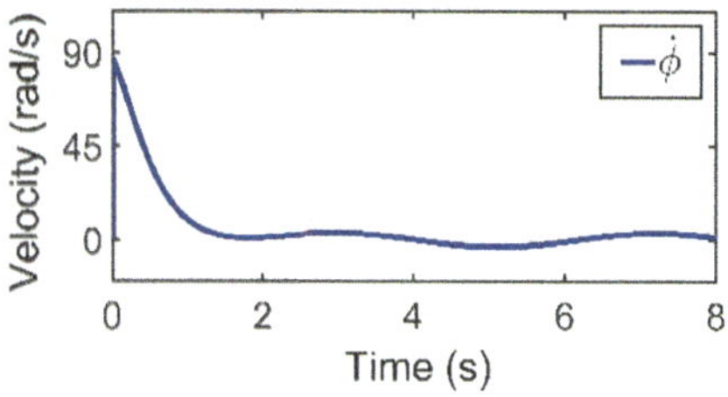

Fig. 7. Flywheel velocity

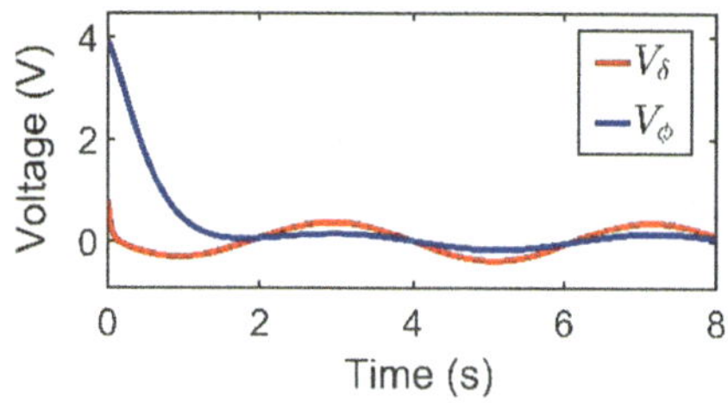

Fig. 8. Input voltage

6 Conclusion

This paper covered the design and control of a self-balancing bicycle using sliding mode control. Sliding mode control was the basis of stabilizing the system. An observer design based on high-gain techniques is presented to estimate unknown parameters for a nonlinear system. The results of the simulation show that the proposed control can keep the bicycle self-balanced stably and can resist disturbances. In the future, the control method will be tested under an experiment to ensure the efficient results of this paper.

Acknowledgement. This research is funded by Hanoi University of Science and Technology (HUST) under project number T2021-PC-001.

References

1. Astrom, K.J., Klein, R.E., Lennartsson, A.: Bicycle dynamics and control: adapted bicycles for education and research. IEEE Control Syst. **25**(4), 26–47 (2005)
2. Tamaldin, N., Yusof, H.I.M., Abdollah, M.F.B., Omar, G., Rosley, M.I.F.: Design self-balancing bicycle. In: Proceedings of Mechanical Engineering Research Day 2017 (Merd), (May). pp. 160–161 (2017)
3. Turnwald, A., Liu, S.: A nonlinear bike model for purposes of controller and observer design. IFAC-PapersOnLine **51**(2), 391–396 (2018)

4. Umur Erd: Modelling, Validation and Control of an Autonomous Bicycle (2019)
5. Shafiekhani, A., Mahjoob, M.J., Akraminia, M.: Design and implementation of an adaptive critic-based neuro-fuzzy controller on an unmanned bicycle. Mechatronics **28**, 115–123 (2015)
6. Deng, W., Moore, S., Bush, J., Mabey, M., Zhang, W.: Towards automated bicycles: achieving self-balance using steering control. ASME 2018 Dynamic Systems and Control Conference, vol. 2, pp. 1–9. DSCC 2018 (2018)
7. Aphiratsakun, N., Techakittiroj, K.: Autonomous AU bicycle: self balancing and tracking control (AUSB 2). In: 2013 IEEE International Conference on Robotics and Biomimetics. ROBIO 2013, (December), pp. 480–485 (2013)
8. Lam, P.Y., Sin, T.K.: Gyroscopic stabilization of a self-balancing robot bicycle. Int. J. Autom. Technol. **5**(6), 916–923 (2011)
9. Kanjanawanishkul, K.: LQR and MPC controller design and comparison for a stationary self-balancing bicycle robot with a reaction wheel. Kybernetika **54**(1), 173–191 (2015)
10. Guo, L., Liao, Q., Wei, S., Huang, Y.: A kind of bicycle robot dynamic modeling and nonlinear control. In: 2010 IEEE International Conference on Information and Automation. ICIA 2010, pp. 1613–1617 (2010)
11. Park, S.-H., Yi, S.-Y.: Active balancing control for unmanned bicycle using scissored-pair control moment gyroscope. Int. J. Control Autom. Syst. **18**(1), 217–224 (2019). https://doi.org/10.1007/s12555-018-0749-7
12. Cui, L., Wang, S., Lai, J., Chen, X., Yang, S., Zhang, Z., Jiang, Z.P.: Nonlinear balance control of an unmanned bicycle: Design and experiments. In: IEEE International Conference on Intelligent Robots and Systems. pp. 7279–7284 (2020)
13. Yeh, T.J., Lu, H.T., Tseng, P.H.: Balancing control of a self-driving bicycle. In: ICINCO 2019 – Proceedings of the 16th International Conference on Informatics in Control, Automation and Robotics, vol. 2(Icinco), pp. 34–41 (2019)
14. Shafiei, M.H., Emami, M.: Design of a robust path tracking controller for an unmanned bicycle with guaranteed stability of roll dynamics. Syst. Sci. Control Eng. **7**(1), 12–19 (2019)

A Case Study of Minimizing Cutting Force in Hard Milling JIS SKD61 Steel Under Nanofluid-MQL Condition

The-Vinh Do(✉)

Thai Nguyen University of Technology, Thai Nguyen City, Thai Nguyen Province, Vietnam
thevinh8880@tnut.edu.vn

Abstract. In the machining of hard material, the cutting force has a direct impact on the lifetime of the cutting tool, the vibration of the system, and more importantly the quality of the machined surface. On the other hand, the cutting force can be easily observed by using a three-component dynamometer. The goal of this research was to minimize the cutting force during milling JIS SKD61 steel under minimum quantity lubrication (MQL) aided SiO_2 nanoparticles. The resultant cutting force was determined by three components: the feed force (*Fx*), the thrust force (*Fy*), and the tangential force (*Fz*). Cutting speed, feed rate, and depth of cut together with the hardness of the work-piece were selected as input parameters. Experiments were carried out based on the G. Taguchi method with L27 orthogonal arrays was used to find the effects of input parameters on the resultant cutting force. Analysis of variance (ANOVA) showed that the depth of cut had the most influence on the cutting force.

Keywords: Hard milling · Cutting force · SiO_2 nanoparticle · JIS SKD61 · Taguchi method

1 Introduction

JIS SKD61 (also known as AISI H13) has an important role in dies and molds manufacturing. With the possession of great hardenability, high strength, and toughness, JIS SKD61 has been using widely for forging dies, extrusion dies, die-casting dies and other hot work dies. The hardness of this steel varies in different applications but ranges from 40 to 55 HRC as reported in [1–3]. There has been a great deal of research on JIS SKD61 to find proper machining conditions in order to reduce cutting force and gaining better surface quality.

Pardeep Kumar et al. [4] studied the influence of different cutting tools on cutting force and surface roughness of AISI H13 steel. The experiments in the turning process revealed that the tangential force increased as the work-piece hardness increased while the thrust force was 40–70% higher than the tangential force. Tongchao Ding et al. [5] built empirical models for cutting forces and surface roughness in milling AISI H13 based on four parameters: cutting speed, feed rate, radial depth of cut, and axial depth of cut. They stated that the linear model fitted for cutting forces while the quadratic

D. C. Nguyen et al. (Eds.): ICERA 2021, LNNS 366, pp. 12–18, 2022.
https://doi.org/10.1007/978-3-030-92574-1_2

model fitted for surface roughness. Moreover, the axial depth of cut and feed rate were two dominant parameters over cutting forces. In a study by Aouici et al. [6] An investigation of the effect of cutting parameters including cutting speed, feed rate, and depth of cut on cutting force in the milling of 60 HRC AISI D3 hardened steel. The authors found that the feed rate is the most influencing factor on the cutting force. In a study [7], T-V Do and L-N-A Vu have optimized the milling machining mode to reduce cutting forces. The authors concluded that the cutting process with higher cutting-speed, lower feed-rate, lower depth-of-cut, and lower workpiece hardness results in low cutting forces. The influence of cutting parameters on the cutting force has also been considered in many other studies. The authors all concluded that the cutting parameters have a strong impact on the cutting force [7–11].

In order to improve the efficiency of the machining process, new technology called "nanofluid" is applied. In the nanofluid, solid particles of size from 10 to 100 nm are added into cutting fluids [12–15]. Because of their ability to enhance thermal conductivity and tribological characteristics, nanofluids help improve the efficiency of cutting processes such as reducing cutting heat, reducing cutting forces, enhancing tool life, and improving machining roughness [13, 16–18].

In this study, the Taguchi method was used to minimize the cutting force during the milling process of hardened SKD 61 steel under the nanofluid machining condition. Cutting parameters (cutting speed, feed rate, and depth of cut, and the hardness of the work-piece) were investigated to determine their influence on the resultant cutting force determined by the three cutting force components including the feed force, the thrust force, and the tangential force.

2 Experimental Setup

Table 1. Cutting parameters

Cutting parameters	Levels		
	1	2	3
Cutting-speed *v* (m/min)	40	60	80
Feed-rate *f* (mm/tooth)	0.01	0.02	0.03
Depth-of-cut *d* (mm)	0.2	0.4	0.6
Hardness-of-workpiece *h* (HRC)	40	45	50

All experiments were conducted on a 3-axis vertical machining center. Workpiece material is JIS SKD61 steel hardened to reach from 40-50HRC. The cutting tool is ϕ10 TiAlN coated endmill that is substituted for each test to ensure the reliability of the test data. A piezo-electric three-component dynamometer of Kistler was used to measure the cutting force components. Nanofluid is applied under minimum quantity lubrication conditions (MQL) with 90 ml/h for the flow-rate, 3 kg/cm^2 for the air-pressure, and CT232 cutting oil for the lubricant. 100nm SiO_2 particles were added to the cutting oil at a concentration of 2%wt. Table 1 shows cutting parameters including the cutting-speed, the feed-rate, the depth-of-cut, and the hardness-of-workpiece with their levels. The input parameters were selected based on workpiece material, tool material, workpiece hardness, cutting tool company recommendations and the author's previous research results.

3 Results and Discussions

The S/N is calculated by Minitab software version 18. The purpose of this study is to reduce cutting force. Thus, the smaller – the – better type of S/N ratio was selected and calculated as the following equation.

$$S/N = -10Log_{10}\left(\frac{1}{n}\sum_{i=1}^{n} y_{ij}^2\right) \quad (1)$$

Where n is the number of repetitions at each experiment, y_{ij} is the observed data.

Four cutting parameters with three levels were investigated to determine their effects on the response. Therefore, L27 array of Taguchi methodology was used for the design of experiment. The resultant cutting force (*Fc*) on the cutter is given as the following equation [19]:

$$Fc = \sqrt{Fx^2 + Fy^2 + Fz^2} \quad (2)$$

Table 2 shows the results of the experiment and S/N for the resultant cutting force.

Table 2. The experimental results

Run	Input factors				Output factors				S/N for *Fc*
	v	*f*	*d*	*h*	*Fx* (N)	*Fy* (N)	*Fz* (N)	*Fc* (N)	
1	40	0.01	0.2	40	45.6	37.879	5.203	59.508	−35.4916
2	40	0.01	0.4	45	111.8	82.75	14.988	139.898	−42.9162
3	40	0.01	0.6	50	157.8	187.883	22.717	246.408	−47.8331
4	40	0.02	0.2	45	77.1	68.434	10.1	103.584	−40.3058
5	40	0.02	0.4	50	163.06	161.263	23.486	230.534	−47.2547
6	40	0.02	0.6	40	180.32	188.541	26.691	262.251	−48.3743
7	40	0.03	0.2	50	122.851	102.639	15.795	160.862	−44.1291
8	40	0.03	0.4	40	150.37	132.01	18.82	200.978	−46.063
9	40	0.03	0.6	45	233.45	230.275	32.49	329.516	−50,3575
10	60	0.01	0.2	45	40.319	29.211	4.81	50.020	−33.9829
11	60	0.01	0.4	50	116.188	95.182	15.523	150.997	−43.5794
12	60	0.01	0.6	40	108.261	112.044	15.457	156.567	−43.894
13	60	0.02	0.2	50	82.83	55.3	9.4	100.036	−40.0032
14	60	0.02	0.4	40	107.176	101.336	14.596	148.218	−43,418
15	60	0.02	0.6	45	140.191	172.734	23.389	223.691	−46.993

(*continued*)

Table 2. (*continued*)

Run	Input factors				Output factors				S/N for *Fc*
	v	*f*	*d*	*h*	*Fx* (N)	*Fy* (N)	*Fz* (N)	*Fc* (N)	
16	60	0.03	0.2	40	60.661	58.623	8.275	84.764	−38.5642
17	60	0.03	0.4	45	133.589	127.723	18.926	185.788	−45.3804
18	60	0.03	0.6	50	246.585	200.07	31.9	319.139	−50.0796
19	80	0.01	0.2	50	43.649	56.168	6.491	71.430	−37.0776
20	80	0.01	0.4	40	55.805	58.717	7.978	81.397	−38.2122
21	80	0.01	0.6	45	94.758	113.25	14.898	148.414	−43.4295
22	80	0.02	0.2	40	38.148	34.816	5.664	51.957	−34.3128
23	80	0.02	0.4	45	91.016	117.059	14.441	148.981	−43.4626
24	80	0.02	0.6	50	183.948	199.55	26.163	272.656	−48.7123
25	80	0.03	0.2	45	47.843	68.857	8.524	84.278	−38.5144
26	80	0.03	0.4	50	160.69	159.051	22.104	227.172	−47.1271
27	80	0.03	0.6	40	113.87	95.357	13.806	149.164	−43.4733

Table 3. Response table for signal to noise

Level	*v*	*f*	*d*	*h*
1	−44.75	**−40.71**	**−38.04**	**−41.31**
2	−42.88	−43.65	−44.16	−42.82
3	**−41.59**	−44.85	−47.02	−45.09
Delta	3.16	4.14	8.97	3.78
Rank	4	2	1	3

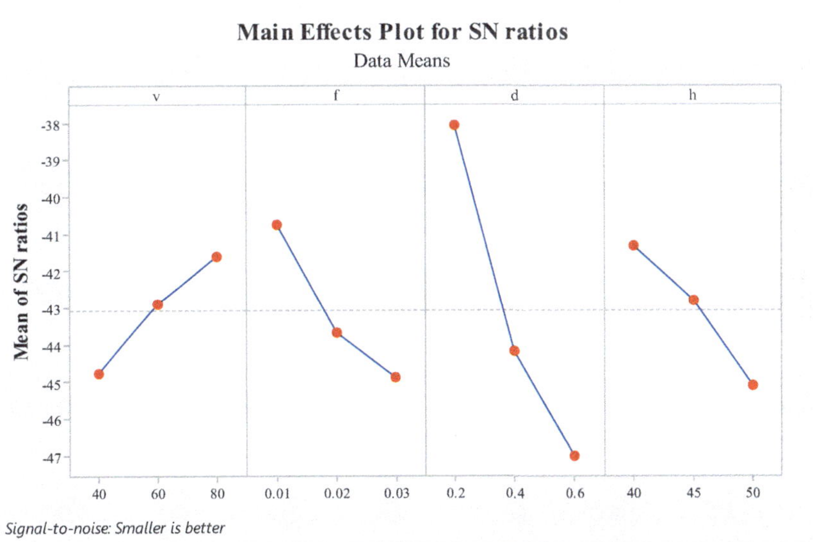

Fig. 1. Main effects plot for S/N of the cutting force

Table 4. The ANOVA for cutting force

Source	DF	Adj-SS	Adj-MS	F-Value	P-Value	C%
v	2	14099	7049.3	12.68	0.000	8.42
f	2	23589	11794.5	21.21	0.000	14.09
d	2	100396	50198.2	90.27	0.000	59.95
h	2	19369	9684.3	17.41	0.000	11.57
Error	18	10010	556.1	–	–	
Total	26	167462	–	–	–	
Rsq = 94.02%						

Table 3 shows the response table for S/N. The ranked highest of each level of sandblasting parameters is the third level for the cutting speed, the first level for the feed rate, the first level for the depth of cut, and the first level for the hardness. Consequently, the optimal condition of the hard milling process will be (3–1–1–1). The rank of the depth of cut is "1", which means that depth of cut is the most influential factor on the cutting force. The second factor affecting cutting force is the feed rate.

Figure 1 shows the main effects plot for S/N of the cutting force. It shows that the optimal condition of the hard milling process for minimum cutting force is the cutting speed of 80 m/min, the feed rate of 0.01 mm/tooth, the depth of cut of 0.2 mm, and hardness of 40HRC.

Table 4 indicates the ANOVA of S/N for the cutting force. As the result, the depth of cut had the most influence on the cutting force followed by the feed rate. The depth of cut and feed rate contribute 59.95% and 14.09% on total effect, respectively. P values of all parameters are less than 0.05. It means that the effect of the four parameters selected in the study has statistical significance. The coefficient of determination R-sq of 94.02% means that 94.02% of the variability in the cutting force can be explained by the parameters selected in this study.

A confirmation test with optimized cutting parameters was carried out to examine the results of the study. The result of the test is 39.8N for the feed force (Fx), 29.3N for the thrust force (Fy), and 4.33N for the tangential force (Fz). So, the resultant cutting force obtained from the confirmation test is 49.61N. This value is less than the smallest value of the cutting force shown in Table 2. It proves that the result of the study is of value.

4 Conclusion

In this study, the Taguchi method was used to minimize the cutting force during the milling process of hardened SKD 61 steel under the nanofluid machining condition. Cutting parameters were investigated to determine their influence on the cutting force. Some of the main conclusions can be drawn as the following:

- The optimal condition of the hard milling process for minimum cutting force is the cutting speed of 80 m/min, the feed rate of 0.01 mm/tooth, the depth of cut of 0.2 mm, and hardness of 40 HRC.
- Under nanofluid MQL condition, the depth of cut is the most influential factor on the cutting force followed by the feed rate. They contribute 59.95% and 14.09% on total effect, respectively.
- The coefficient of determination is high (94.02%) which means that the model is reliable.

Acknowledgements. The authors wish to thank Thai Nguyen University of Technology. This work was supported by Thai Nguyen University of Technology.

References

1. Do, T.-V., Hsu, Q.-C.: Optimization of minimum quantity lubricant conditions and cutting parameters in hard milling of AISI H13 steel. Appl. Sci. **6**, 83 (2016)
2. Do, T.V., Nguyen, Q.M., Pham, M.T.: Optimization of cutting parameters for improving surface roughness during hard milling of AISI H13 steel. Key Eng. Mater. 35–39 (2020)
3. Yan, H., Hua, J., Shivpuri, R.: Flow stress of AISI H13 die steel in hard machining. Mater. Des. **28**, 272–277 (2007)
4. Kumar, P., et al.: Influence of different grades of CBN inserts on cutting force and surface roughness of AISI H13 die tool steel during hard turning operation. Materials **12**, 177 (2019)
5. Ding, T., Zhang, S., Wang, Y., Zhu, X.: Empirical models and optimal cutting parameters for cutting forces and surface roughness in hard milling of AISI H13 steel. Int. J. Adv. Manuf. Technol. **51**, 45–55 (2010)
6. Aouici, H., Bouchelaghem, H., Yallese, M.A., Elbah, M., Fnides, B.: Machinability investigation in hard turning of AISI D3 cold work steel with ceramic tool using response surface methodology. Int. J. Adv. Manuf. Technol. **73**(9–12), 1775–1788 (2014). https://doi.org/10.1007/s00170-014-5950-0
7. Do, T.-V., Le, N.-A.-V.: Optimization of surface roughness and cutting force in MQL hard-milling of AISI H13 steel. In: Fujita, H., Nguyen, D.C., Vu, N.P., Banh, T.L., Puta, H.H. (eds.) ICERA 2018. LNNS, vol. 63, pp. 448–454. Springer, Cham (2019). https://doi.org/10.1007/978-3-030-04792-4_58
8. Nguyen, H.-T., Hsu, Q.-C.: Study on cutting forces and material removal rate in hard milling of SKD 61 alloy steel. J. Chin. Soc. Mech. Eng. **38**, 41–51 (2017)
9. Nguyen, N.-T., Kao, Y.-C., Bui, G.T., Nguyen, Q., Nguyen, Q.-M., Do, T.-V.: An experimental investigation of dynamic cutting forces in the stable milling processes. In: Fujita, H., Nguyen, D.C., Vu, N.P., Banh, T.L., Puta, H.H. (eds.) ICERA 2018. LNNS, vol. 63, pp. 158–166. Springer, Cham (2019). https://doi.org/10.1007/978-3-030-04792-4_22
10. Çalışkan, H., Kurbanoğlu, C., Panjan, P., Kramar, D.: Investigation of the performance of carbide cutting tools with hard coatings in hard milling based on the response surface methodology. Int. J. Adv. Manuf. Technol. **66**, 883–893 (2013)
11. Do, T.-V.: Effect of cutting parameters on cutting force of hard milling of SKD 61 steel under nanofluid-MQL condition. Int. J. Mech. Eng. Robot. Res. **9**, 1104–1109 (2020)
12. Lee, P.-H., Nam, J.S., Li, C., Lee, S.W.: An experimental study on micro-grinding process with nanofluid minimum quantity lubrication (MQL). Int. J. Precis. Eng. Manuf. **13**, 331–338 (2012)

13. Sayuti, M., Sarhan, A.A., Hamdi, M.: An investigation of optimum SiO 2 nanolubrication parameters in end milling of aerospace Al6061-T6 alloy. Int. J. Adv. Manu. Technol. **67**, 833–849 (2013)
14. Yu, W., Xie, H.: A review on nanofluids: preparation, stability mechanisms, and applications. J. Nanomater. **2012** (2012)
15. Phan, T.-D., Do, T.-V., Pham, T.-L., Duong, H.-L.: Optimization of cutting parameters and nanoparticle concentration in hard milling for surface roughness of JIS SKD61 steel using linear regression and taguchi method. In: Sattler, K.-U., Nguyen, D.C., Vu, N.P., Long, B.T., Puta, H. (eds.) ICERA 2020. LNNS, vol. 178, pp. 628–635. Springer, Cham (2021). https://doi.org/10.1007/978-3-030-64719-3_69
16. Sayuti, M., Sarhan, A.A., Salem, F.: Novel uses of SiO2 nano-lubrication system in hard turning process of hardened steel AISI4140 for less tool wear, surface roughness and oil consumption. J. Clean. Prod. **67**, 265–276 (2014)
17. Do, T.-V.: Empirical model for surface roughness in hard milling of AISI H13 steel under nanofluid-MQL condition based on analysis of cutting parameters. J. Mech. Eng. Res. Dev. **43**, 89–94 (2020)
18. Vu, N.-C., Dang, X.-P., Huang, S.-C.: Multi-objective optimization of hard milling process of AISI H13 in terms of productivity, quality, and cutting energy under nanofluid minimum quantity lubrication condition.Measurement and Control 0020294020919457 (2020)
19. Altintas, Y., Ber, A.: Manufacturing automation: metal cutting mechanics, machine tool vibrations, and CNC design. Appl. Mech. Rev. **54**, B84–B84 (2001)

A Comparative Analysis of Ride Performance of Double-Drum Vibratory Roller with Two Cab Mount Systems

Le Van Quynh[1(✉)], Le Anh Vu[2], Bui Van Cuong[1], Hoang Anh Tan[1], and Le Xuan Long[1]

[1] Faculty of Automotive and Power Machinery Engineering, Thai Nguyen University of Technology, Thai Nguyen, Vietnam
lequynh@tnut.edu.vn

[2] Faculty of Automobile Engineering, Hung Yen University of Technology and Education, Hung Yen, Vietnam

Abstract. The purpose of this paper is to compare the ride performance of a liquid-filled cab mount system (LCMs) with an annular orifice and an original rubber cab mount system (RCMs) under the operating conditions. In order to achieve these goals, a nonlinear dynamical model of LCM is set up to determine its vertical force which is connected with a half-vehicle ride dynamic model of a double-drum vibratory roller under two survey cases. The root-mean-square (RMS) and power spectral density (PSD) acceleration responses of the driver's seat and pitching cab angle are chosen as the objective functions. The ride performance of LCMs is verified and compared with RCMs through objective functions. The study results show that the values of the root mean square (RMS) acceleration responses with LCMs are respectively reduced in comparison with RCMs and the peak amplitude values of PSD acceleration responses are respectively reduced when compared with REMs in low frequency region from 0.5Hz to 60Hz when both the vehicle and drums operate under two cases. The performance of LCMs is better than that of RCMs in improving the ride comfort of a double-drum vibratory roller under two survey cases.

Keywords: Double-drum vibratory roller · Cab · Liquid-filled mount system · Rubber mount system · Dynamic model · Ride comfort

1 Introduction

In recent years, earthmoving equipment market has required increasingly not only working capacity but also ride comfort. Thus, in order to reduce the effect of vibration to operators, identification and elimination of vibration sources are the most important tasks to achieve optimum design. For vibratory roller, it often operates in harsh environments where the vibration sources are transmitted to the cabin and driver's body through the cabin isolation system. Therefore, the optimal solutions for reducing vibration transmitted to the cab during the operation of the machine have been an interesting topic for designers and researchers. An overview study on the development of cab isolation systems of off-road vibratory rollers was presented by reference [1].

D. C. Nguyen et al. (Eds.): ICERA 2021, LNNS 366, pp. 19–30, 2022.
https://doi.org/10.1007/978-3-030-92574-1_3

The dynamic characteristics of vibration isolation system of single-drum vibratory roller were analyzed by using both experiment and simulation [2]. The ride comfort of the vibratory roller with three cab's isolation mounts such as the traditional rubber mounts, the hydraulic mounts, and the pneumatic mounts was analyzed and evaluated by using a three-dimensional nonlinear dynamic model [3]. The stiffness and damping coefficients of the cab's isolation system was analyzed by using a 3D nonlinear dynamic model of a single drum vibratory roller [4]. The influences of different design parameters such as driver seat, cab, drum isolators, chassis and the tires on the whole body vibration responses was analyzed by using a 12-degrees-of-freedom in-plane ride dynamic model of a single-drum compactor [5]. A 3D nonlinear dynamic model of a single drum vibratory roller with the nonlinear geometric characteristics of wheel-deformation of soil ground contact was applied to analyze the effectiveness of ride comfort by both experiment and simulation methods [6]. Ride comfort evaluation of vibratory roller under different soil ground was analyzed by using a nonlinear dynamic model with the tire-deformation soil surface contact and elastic-plastic soil [7]. To improve the ride comfort of the vehicle cab, the cab's low frequency shaking in the direction of forward motion is solved by the optimal design methods for the main and auxiliary isolation system of cab [8, 9]. The optimal design parameters of drum's isolation systems for a double-drum vibrating roller were found out by using half-vehicle ride dynamic model and the genetic algorithm (GA) [10]. Similarly, the optimal design parameters of cab's isolation system for a single-drum vibratory roller were found out by using half-vehicle ride dynamic model and the genetic algorithm (GA) [11]. The optimal design parameters of cab's isolation system for vibratory roller were found out by using 3D nonlinear dynamic model and a multi-objective genetic algorithm based on the improved genetic algorithm NSGA-II [12, 14]. The damping coefficient of cab's is controlled by using a PID-Fuzzy controller [13]. A Magneto-rheological (MR) damper of drum's isolation system was applied to control by using fuzzy controller [15]. The hydraulic mounts are not only widely used in the field of vibration reduction of internal combustion engines, but also used in the field of vibration reduction of cab for the earth moving equipment. The ride performance of the hydraulic mounts and the rubber mounts of an earth-moving machinery cab was analyzed and evaluated by using six-degree of freedom (DOF) mathematical model of the cab [16]. The hydro pneumatic isolation (HPI) of the cab was analyzed by using an eleven degrees of freedom full vehicle model [17]. The passive hydraulic damping rubber mount (PHDRM) of an earth-moving machinery cab was analyzed by using ABAQUS combined simulation mode [18]. A novel liquid-filled mount with an annular orifice was investigated at low frequency by using experimental and analytical methods [19].

The main objective of this paper is to compare the ride performance of liquid-filled cab mount system (LCMs) with an annular orifice and an original rubber cab mount system (RCMs) of a double-drum vibratory roller under two survey cases such as vehicle moves into the workshop (Case 1) and vehicle operates in the workshop (Case 2). Firstly, a nonlinear dynamical model of LCM based on the reference [19] is set up to determine its vertical force. And then it is connected with a half-vehicle ride dynamic model of a double-drum vibratory roller under two survey cases. Finally, the ride performance of LCM is verified and compared with RCM through objective functions such as the root-mean-square (RMS) and power spectral density (PSD) acceleration responses of the driver's seat and pitching cab angle based on the International Standard ISO 2631–1: 1997 [23] under two survey cases.

2 Nonlinear Dynamical Model of Cab Mount

2.1 Rubber Cab Mount (RCM)

Rubber cab mount (RCM) is widely used on types of vibratory rollers. The model of rubber cab mount (RCM) of a double-drum vibratory roller is proposed in Fig. 1 where, the 3D model of RCM is shown Fig. 1(a), the structural diagram of RCM is shown Fig. 1(b) and the nonlinear dynamical model of RCM is shown in Fig. 1(c) where, m_f, and m_c are the masses of fame and cab, respectively, k_r is the elastic stiffness of rubber parts, F_{fmax} is the maximum friction force, b is the coefficient of the viscous force, and α is the order of the fractional derivative, z_f and z_c are the vertical displacements of frame and cab, respectively, F_{cr} is the vertical force of RCM.

From Fig. 1(c), the vertical force of RCM is defined below

$$F_{cr} = F_e + F_f + F_a \tag{1}$$

where, F_e is the elastic force of RCM which is defined by Eq. (2), F_f is the friction force of RCM which is defined by Eq. (3), Fa is the viscous force of RCM [20] which is defined by Eq. (4).

$$F_e = k_r(z_c - z_f) \tag{2}$$

$$F_f = F_{fmax} sign(\dot{z}_c - \dot{z}_f) \tag{3}$$

$$F_a = bD^\alpha(z_c - z_f) \tag{4}$$

where, D is the (non-dimensional) damping.

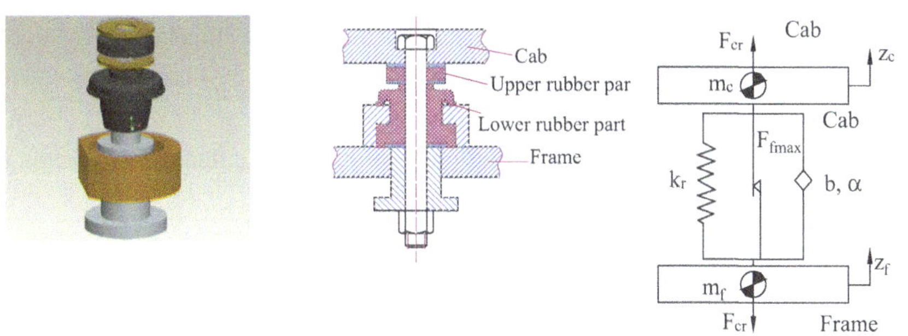

(a) 3D model of RCM (b) Structural diagram of RCM (c)Nonlinear dynamical model of RCM

Fig. 1. Proposed model of rubber cab mount (RCM)

2.2 Liquid-Filled Cab Mount (LCM) with an Annular Orifice

Liquid-filled cab mount (LCM) with an annular orifice is proposed based on the literature [19]. The model of liquid-filled cab mount (LCM) with an annular orifice is shown in Fig. 2 where, the cross-sectional view of LCM is shown Fig. 2(a), the structural diagram of LCM is shown Fig. 2(b) and the nonlinear dynamical model of LCM is shown in Fig. 2(c) where p_1 and p_2 are the pressure in chamber 1 and chamber 2, respectively, and F_{ch} is the vertical force of LCM.

From Fig. 2 (c), the vertical force of RLM is defined below

$$F_{ch} = F_{cre} + F_h \tag{5}$$

where, F_{cre} is the vertical force of LCM which is defined similarly to Eq. (1), F_h is the damping force of LCM which is generated by the liquid portion and it is defined by Eq. (6).

The damping force of LCM with the flow through the annular gap is determined by

$$F_h = A_d \Delta p = \Delta p(p_1 - p_2) \tag{6}$$

where, Δp is the pressure loss at the annular gap, A_d is the effective area of the damping plate.

The equation of continuity for the flow through the annular orifice can be written as

$$A_d \dot{z}_c = -A_a \dot{z}_a \tag{7}$$

where, A_c and A_a is the effective area of the chamber and the annular orifice of LCM, $\dot{z}_a$ is the average flow velocity in the annular orifice.

The pressure loss at the annular gap is described by two processes of the piecewise damping characteristic and they are defined below.

The Damping Plate Moves Down: The pressure loss of the liquid reciprocating in the gap is described by the mixed viscous damping model determined by

$$\Delta p = c_a^{11} \dot{z}_a + c_a^{12} |\dot{z}_a| \dot{z}_a \tag{8}$$

where, c_a^{11} and c_a^{12} are the positive constants which are functions of the geometric structure of the annular gap and the fluid properties, $\dot{z}_a$ is the average flow velocity in the annular gap.

The Damping Plate Moves Upward: The pressure loss of the liquid reciprocating in the gap [21] is determined by

$$\Delta p = c_a^{21} \dot{z}_a + c_a^{22} |\dot{z}_a| \dot{z}_a + \breve{c}_g \left(\frac{\dot{z}_a}{\delta} \right)^2 \dot{z}_a \tag{9}$$

where, c_a^{21}, c_a^{22} and $\breve{c}_g$ are the positive constants which are functions of the geometric structure of the annular gap and the fluid properties, δ is the gap between the damping plate and the bottom of the rubber when the damping plate is at a negative amplitude.

By substituting Eq. (7) into Eq. (8), the pressure loss at annular orifice can be achieved by

$$\Delta p = c_a^{11}\left(\frac{A_c}{A_a}\right)\dot{z} + c_a^{12}\left(\frac{A_c}{A_a}\right)^2 |\dot{z}|\dot{z} \text{ with } \dot{z} < 0 \tag{10}$$

By substituting Eq. (7) into Eq. (9), the pressure loss at annular orifice can be achieved by

$$\Delta p = c_a^{11}\left(\frac{A_c}{A_a}\right)\dot{z} + c_a^{12}\left(\frac{A_c}{A_a}\right)^2 |\dot{z}|\dot{z} + \breve{c}_g\left(\frac{\dot{A}_c}{A_a}\right)\left(\frac{\dot{z}}{\delta}\right)^2 \dot{z} \text{ with } \dot{z} \geq 0 \tag{11}$$

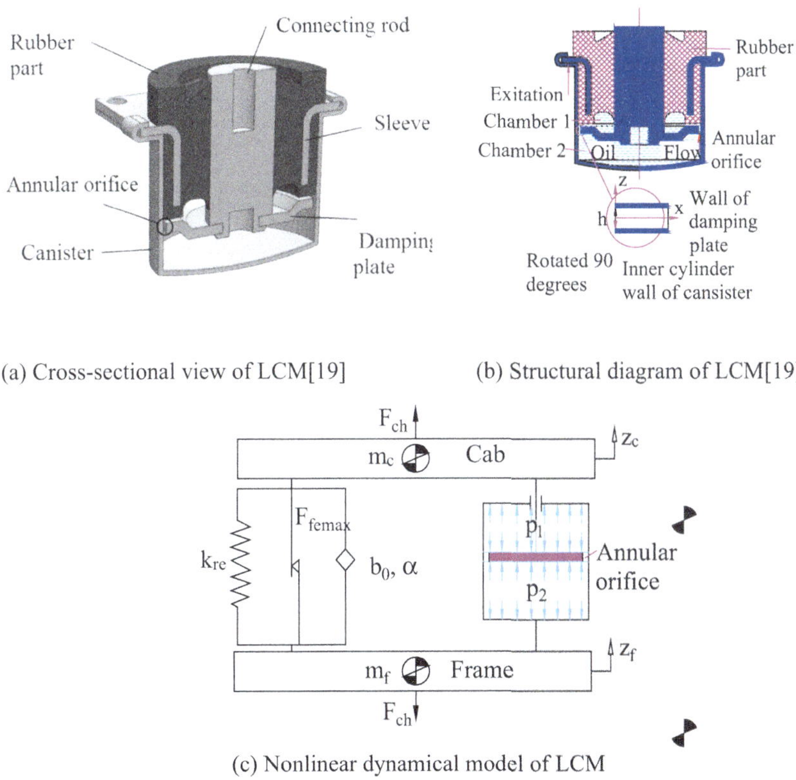

(a) Cross-sectional view of LCM[19]

(b) Structural diagram of LCM[19]

(c) Nonlinear dynamical model of LCM

Fig. 2. Proposed model of liquid-filled cab mount (LCM)

From Eq. (10), and Eq. (11), the pressure difference between the two chambers of LCM can be determined by

$$\Delta p = \begin{cases} c_a^{11}\left(\frac{A_c}{A_a}\right)\dot{z} + c_a^{12}\left(\frac{A_c}{A_a}\right)^2|\dot{z}|\dot{z} & with\, \dot{z} < 0 \\ c_a^{11}\left(\frac{A_c}{A_a}\right)\dot{z} + c_a^{12}\left(\frac{A_c}{A_a}\right)^2|\dot{z}|\dot{z} + \breve{c}_g\left(\frac{A_c}{A_a}\right)\left(\frac{\dot{z}}{\delta}\right)^2\dot{z} & with\, \dot{z} \geq 0 \end{cases} \tag{12}$$

By substituting Eq. (12) into Eq. (6), the vertical damping force of the hydraulic actuator with the flow through the orifice and the annular gap can be achieved by

$$F_h = \begin{cases} c_f^{11}\dot{z} + c_f^{12}|\dot{z}|\dot{z} & with\, \dot{z} < 0 \\ c_f^{11}\dot{z} + c_f^{12}|\dot{z}|\dot{z} + c_g\left(\frac{\dot{z}}{\delta}\right)^2\dot{z} & with\, \dot{z} \geq 0 \end{cases} \tag{13}$$

Where

$$c_f^{11} = c_a^{11}A_d\left(\frac{A_c}{A_a}\right); c_f^{12} = c_a^{12}A_d\left(\frac{A_c}{A_a}\right)^2; c_f^{21} = c_a^{21}A_d\left(\frac{A_c}{A_a}\right);$$

$$c_f^{22} = c_a^{22}A_d\left(\frac{A_c}{A_a}\right)^2; c_g = \breve{c}_g A_d\left(\frac{A_c}{A_a}\right)$$

3 Half-Vehicle Dynamic Model of a Double-Drum Vibratory Roller

A structure of a double-drum vibratory roller with four rubber cab mounts (see Fig. 1) are selected to investigate and compare the ride performance of a liquid-filled cab mount system (LCMs) in comparison with an original rubber cab mount system (RCMs). A half-vehicle ride dynamic model of a double-drum vibratory roller is shown in Fig. 3, where, m_{di}, m_b, m_c and m_s are the masses of the dynamic drums, frame, cab and driver's seat, respectively; I_f and I_c are the moment of inertia of the frame and cab, respectively; k_{si}, k_{di}, k_{si}, k_s and c_{csi}, c_{di}, c_{ci}, c_s are the stiffness and damping coefficients of elastic ground surfaces, front and rear mount systems of drums, cab mount system and driver's seat suspension system, respectively; z_{di}, z_f, z_c and z_s are the vertical displacements of the front and rear drums, frame, cab and driver's seat, respectively; φ_f and φ_c are the pitch angle displacements of frame and cab, respectively; qi are the excitation of ground surface roughness at drums, respectively; l_j are the distances; $F_{ei} = F_{0i}\sin(\omega_i t)$ are the force excitations of the vibrating drums; F_{0i} are the amplitude of force excitations; ω_i are the angular frequencies of the vibrators; and v is the vehicle speed ($i = 1 \div 2$, $j = 1 \div 6$).

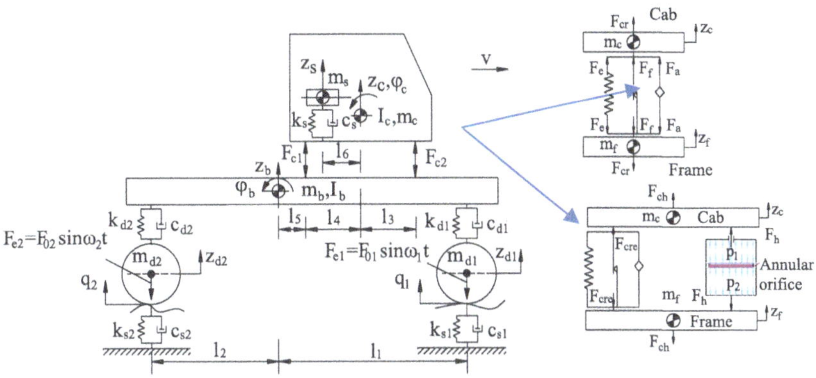

Fig. 3. Half-vehicle dynamic model of a double-drum vibratory roller

Equations of Motion: The equations of motion for a double-drum vibratory roller using Newton's second law of motion are written in two operating conditions below. From Fig. 3, the motion equations of vehicle mass are written as follows:

The equation of motion for driver's seat is written as follows

$$m_s\ddot{z}_s = -F_s \tag{14}$$

The equations of motion for cab are written as follows

$$m_c\ddot{z}_c = -F_s \sum\nolimits_{i=1}^{i=2} F_{ci} \tag{15}$$

$$I_c\ddot{\varphi}_c = F_s l_6 + \sum\nolimits_{i=1}^{i=2} (-1)^i F_{ci} l_{i+2} \tag{16}$$

The equations of motion for frame are written as follows

$$m_f\ddot{z}_f = \sum_{i=1}^{i=2} F_{ci} - \sum\nolimits_{i=1}^{i=2} F_{di} \tag{17}$$

$$I_f\ddot{\varphi}_f = F_{c2} l_5 + \sum_{i=1}^{i=2} (-1)^i F_{di} l_i + F_{ci} \sum\nolimits_{j=3}^{j=5} l_j \tag{18}$$

where, F_s is the vertical force of driver's seat which is determined by Eq. (18), F_{ci} are the vertical forces of cab mount system which are determined by Eq. (1) or Eq. (5) and F_{di} are the vertical forces of front and rear mount systems of drums which could be determined through two cases.

$$F_s = k_s(z_s - z_c + l_6\varphi_c) + c(\dot{z}_s - \dot{z}_c + l_6\dot{\varphi}_c) \tag{19}$$

Case 1: Vehicle Moves into the Workshop: The drums of vibratory roller in contact with the rigid road surface is the contact point which is considered in this study. The vertical forces of front and rear mount systems of drums are defined as

$$F_{d1} = k_{d1}\left(z_f + l_1\varphi_f - q_{d1}\right) + c_{d1}\left(\dot{z_f} + l_1\dot{\varphi}_f - \dot{q_{d1}}\right) \tag{20}$$

$$F_{d2} = k_{d2}\left(z_f - l_2\varphi_f - q_{d2}\right) + c_{d2}\left(\dot{z_f} - l_2\dot{\varphi}_f - \dot{q_{d2}}\right) \tag{21}$$

where, q_{di} are the excitations of road surface roughness at drums. The road surface roughness according to the International Standards Organization (ISO 8608) [22] is road excitation which is simulated in space domain and acts as an input to the vehicle-road model.

Case 2: Vehicle operates in the workshop: The equations of motion for the dynamic drums are written as follows

$$m_{d1}\ddot{z}_{d1} = F_{e1} + F_{d1} - k_{s1}z_{d1} - c_{s1}\dot{z_{d1}} \tag{22}$$

$$m_{d1}\ddot{z}_{d1} = F_{e2} + F_{d2} - k_{s2}z_{d2} - c_{s2}\dot{z_{d2}} \tag{23}$$

4 Results and Discussion

The equations of motion of Fig. 3 under two operating cases are solved using Matlab/simulink software with reference parameters of vehicle and two types of cab mounts [21]. The simulation results of the time domain and power spectral density (PSD) acceleration responses of the vertical driver's seat (a_s) and cab's pitch angle (a_{cphi}) when the vehicle moves on the ISO class D road surface (poor condition) at the vehicle speed of 10 km/h (Case1) are shown Fig. 4 in Case 1. From the results of Fig. 4 (a), (c), the values of root-mean-square (RMS) of acceleration responses of the vertical driver's seat (a_{ws}) and cab pitching angle (a_{wcphi}) are respectively determined through Eq. (23) as a_{ws} = 1.9858 m/s^2, a_{wcphi} = 3.4500 rad/s^2 with rubber cab mount system (RCMs) and a_{wa} = 1.6600 m/s^2, a_{wcphi} = 1.9442 rad/s^2 with liquid-filled cab mount system (LCMs). The obtained results show that the a_{ws} and a_{wcphi} values with LCMs respectively reduce by 19.63% and 77.45% in comparison with RCMs. As the study results, the ride comfort performance of LCMs is greater than that of RCMs when vehicle moves on ISO class D road surface at the driving speed of 10 km/h. From the results of Fig. 4 (b), (d), it indicates that the peak amplitude values of PSD acceleration responses of the vertical driver's seat and cab's pitch angle with LCMs are respectively reduce in comparison with RCMs. Especially, the peak amplitude value of PSD acceleration responses of the vertical driver's seat with LCMs reduces by 22.5 times in comparison with RCMs at frequency about 8 Hz (see Fig. 5 (b)) and the peak amplitude value of PSD acceleration responses of cab pitching angle reduce by 11.15%

at frequency about 2Hz (see Fig. 4 (d)). According to the International Standard ISO 2631[23], the resulting motions affect the ride quality perceived by the occupants and their comfort as at these frequencies humans are most sensitive to vertical vibrations from 4 Hz to10 Hz, and lateral and longitudinal from 0.5 Hz to 2 Hz vibrations. As the study results, the ride performance of LCMs has significantly reduced vibrations in comparison with RCMs at the frequency region from 0.5 Hz to 10 Hz.

The root mean square (RMS) acceleration responses according to ISO standard 2631–1:1997 [24] are chosen as the objective functions which are determined by the formula:

$$a_w = \left[\frac{1}{T} \int_0^T a^2(t)dt \right]^{\frac{1}{2}} \tag{24}$$

where a (t) is the weighted acceleration (translational and rotational) as a function of time in m/s^2; T is the duration of the measurements in s.

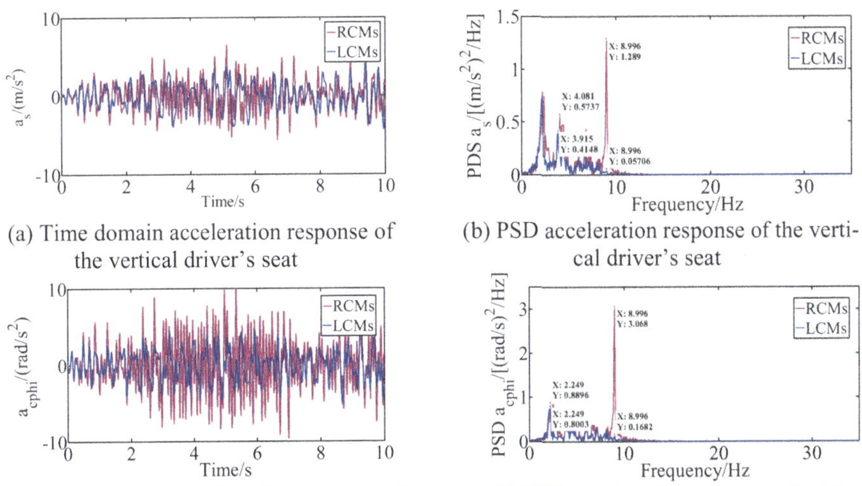

(a) Time domain acceleration response of the vertical driver's seat

(b) PSD acceleration response of the vertical driver's seat

(c) Time domain acceleration response of cab's pitch angle

(d) PSD acceleration response of cab's pitch angle

Fig. 4. The comparison of time domain and PSD acceleration responses of the vertical driver's seat and cab's pitch angle with RCMs and LCMs at Case 1

The simulation results of the time domain and power spectral density (PSD) acceleration responses of the vertical driver's seat (a_s) and cab's pitch angle (a_{cphi}) are shown in Fig. 5 when the vehicle compacts on the original place with the parameters of the elastic soil grounds as $k_{s1} = 10^7$ N/m, $c_{s1} = 2.1 \times 10^5$ (N.s/m) and $k_{s2} = 1.2$ 10^7 N/m, $c_{s2} = 2.8 \times 10^5$ (N.s/m) at the front and rear drums with the excitation forces as $F_{01} = 0.128 \times 10^6$ N, $f_{01} = 48$ Hz and $F_{02} = 0.96 \times 10^5$ N, $f_{01} = 54$ Hz.

From the results of Fig. 5 (a), (c), we can determine the a_{ws} and a_{wcphi} values through Eq. (23) are $a_{ws} = 0.0547$ m/s^2 and $a_{wcphi} = 0.0727$ rad/s^2 with LCMs and $a_{ws} = 0.0556$ m/s^2 and $a_{wcphi} = 0.0887$ rad/s^2 with RCMs. The results show that the a_{ws} and a_{wcphi} values with LCMs reduce by 1.64% and 22.00% in comparison with RCMs, which means that the performance of LCMs in improving the ride comfort is better than the RCMs. Especially, the ride performance of LCMs has significantly reduced vibrations in comparison with RCMs for cab's low-frequency shaking in the direction of forward motion when vibratory roller operates. From the results of Fig. 5 (b), (d), the peak amplitude value of PSD acceleration responses of the vertical driver's seat with LCMs reduces by 19.20% in comparison with RCMs at frequency about 5 Hz (see Fig. 5 (b)) and the peak amplitude value of PSD acceleration responses of cab pitching angle reduce by 43.32% at frequency about 4.8 Hz (see Fig. 5 (d)). From these results, we could see that it appeared to resonate at the excitation frequencies of front and rear drums ($f_{01} = 48$ Hz and $f_{02} = 54$ Hz). However, the ride performance of LCMs has significantly reduced vibrations in comparison with RCMs at the excitation frequencies of front and rear drums.

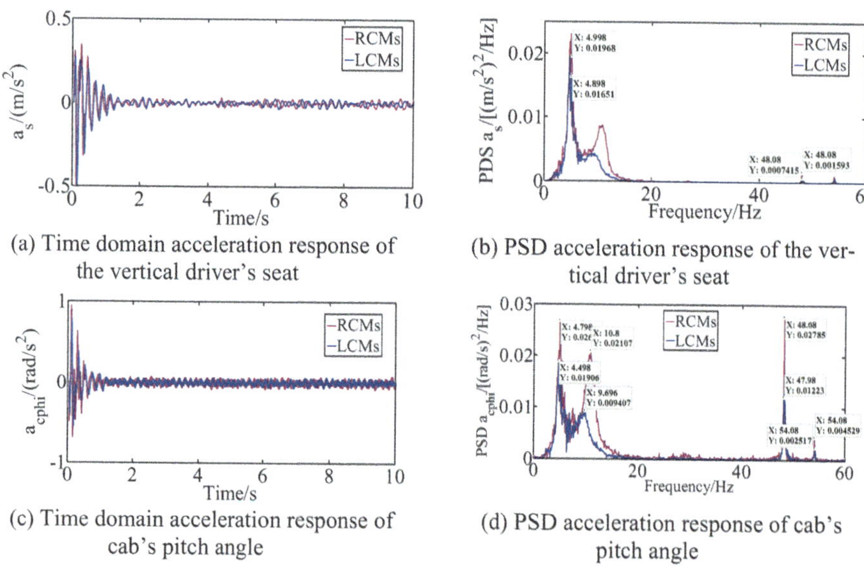

(a) Time domain acceleration response of the vertical driver's seat

(b) PSD acceleration response of the vertical driver's seat

(c) Time domain acceleration response of cab's pitch angle

(d) PSD acceleration response of cab's pitch angle

Fig. 5. The comparison of time domain and PSD acceleration responses of the vertical driver's seat and cab's pitch angle with RCMs and LCMs at Case 2

5 Conclusions

In this paper, a comparative analysis the ride performance of a liquid-filled cab mount system (LCMs) with an annular orifice and an original rubber cab mount system (RCMs) of a double-drum vibratory roller is proposed for the evaluations when vehicle operates under two survey cases. The major conclusions drawn from the comparative analysis results are: (1) The a_{ws} and a_{wcphi} values with LCMs respectively reduce by 19.63% and 77.45% in comparison with RCMs and the ride performance of LCMs has significantly reduced vibrations in comparison with RCMs at the frequency region from 0.5 Hz to 10 Hz when vehicle operates in Case 1; (2) The a_{ws} and a_{wcphi} values with LCMs reduce by 1.64% and 22.00% in comparison with RCMs and the peak amplitude value of PSD acceleration responses of the vertical driver's seat and cab's pitch angle with LCMs are respectively reduce in comparison with RCMs at the frequency region from 0.5 Hz to 60 Hz when vehicle operates in Case 2. In addition, the study results provide useful material for designers and manufacturers in the field of vibration reduction for the cab of construction machines.

Acknowledgment. The work described in this paper was supported by Thai Nguyen University of Technology for a scientific project.

References

1. Nguyen, V., Le, V.: Development of cab isolation systems of off-road vibratory rollers: review research. Math Models Eng **6**(2), 93–102 (2020)
2. Li, J., Zhang, Z., Xu, H., Feng, Z.: Dynamic characteristics of the vibratory roller test-bed vibration isolation system: Simulation and experiment. J Terrramech **56**, 139–156 (2014)
3. Liem, N.V., et al.: Vibration analysis and modeling of an off-road vibratory roller equipped with three different cab's isolation mounts. Shock Vib **2018**, 8527574 (2018)
4. Van Quynh, L., et al.: Influence of design parameters of cab's isolation system on vibratory roller ride comfort under the deformed ground surfaces. Int Res J Eng Technol (IRJET) **6**(6), 1974–1978 (2019)
5. Kordestani, A., et al.: Analysis of ride vibration environment of soil compactors. SAE Int J Commer Veh **3**(1), 259–272 (2010)
6. Le V (2017) Ride comfort analysis of vibratory roller via numerical simulation and experiment. DEStech Transactions on Engineering and Technology Research
7. Le VQ, Zhang JR et al (2013) Ride comfort evaluation of vibratory roller under different soil ground. Tran of the Chinese Society of Agricultural Engineering 29: 39–47
8. Quynh, L.V., Jianrun, Z., Liem, N.V., Cuong, B.V., Long, L.X., Phuong, D.T.: Experimental modal analysis and optimal design of cab's isolation system for a single drum vibratory roller. Vibroengineering PROCEDIA **31**, 52–56 (2020)
9. Quynh, L.V., et al.: Vibration analysis and optimal design for cab's isolation system of vibratory roller. Adv Mat Res **199**(200), 936–940 (2011)
10. Quynh, L.V., Thao, V.T.P., Phong, T.T.: Optimal design parameters of drum's isolation system for a double-drum vibratory roller. Vibroengineering PROCEDIA **31**, 74–79 (2020)
11. Van Quynh, L., Duy, N.T., Van Liem, N., Van Cuong, B., Long, L.X.: Optimal design of cab's isolation system for a single-drum vibratory roller. ICERA 2020. Lecture Notes in Networks and Systems **178**, 619–627 (2021)
12. Le, V.Q., Nguyen, K.T.: Optimal design parameters of cab's isolation system for vibratory roller using a multi-objective genetic algorithm. Appl Mech Mater **875**, 105–112 (2018)
13. Jiao, R., Nguyen, V., Le, V., Hoang, A.: Performance of PID-Fuzzy control for cab isolation mounts of soil compactors. Math Models Eng **5**(4), 137–145 (2019)
14. Le V (2013) Vibration study and control for cab of vibratory roller. Southeast University

15. Liu, S.N., Yan, S.R., Li, S.W., Zheng, Y.G.: Dynamic and control of vibratory road roller based on magneto-rheological semi-active damper. Adv Mat Res **479–481**, 1200–1204 (2012)
16. Sun, X., Zhang, J.: Performance of earth-moving machinery cab with hydraulic mounts in low frequency. J Vib Control **20**(5), 724–735 (2012)
17. Nguyen, V., Jiao, R., Le, V.: Ride comfort performance of hydro pneumatic isolation for soil compactors cab in low frequency region. J Vibroengineering **22**(5), 1174–1186 (2020)
18. Sun, X., Liao, X., Wang, H.: Modeling and dynamic analysis of hydraulic damping rubber mount for cab under larger amplitude excitation. J Vibroengineering **23**(3), 542–558 (2021)
19. Sun X, Zhang C, Fu Q, Zhang H, Dong H (2020) Measurement and modelling for harmonic dynamic characteristics of a liquid-filled isolator with a rubber element and high-viscosity silicone oil at low frequency. Mechanical Systems and Signal Processing, vol 140. 106659
20. Sjoberg, M.M., Kari, L.: Non-linear behavior of a rubber isolator system using fractional derivatives. Veh Syst Dyn **37**(3), 217–236 (2002)
21. Van Quynh, L.: Dynamic analysis and control of cab mount system for construction machine. Science research topic. Thai Nguyen University, Thai Nguyen, Vietnam (2021)
22. ISO 8608 (1995) Mechanical vibration-road surface profiles-reporting of measured data. International Organization for Standardization
23. ISO 2631-1 (1997) Mechanical vibration and shock-evaluation of human exposure to whole-body vibration, part i: general requirements. The International Organization for Standardization

A Comparison of Methods on Building Empirical Model of Milling Working Status Based on Vibration

Thanh-Dat Phan(✉) and The-Vinh Do

Thai Nguyen University of Technology, No. 666, 3/2 Street, Tich Luong Ward, Thai Nguyen City, Thai Nguyen Province 25000, Vietnam
leophanvn@tnut.edu.vn

Abstract. This research presents different approaches of data analyzing and modelling to detect the milling status (i.e., milling/idling) based on the vibration of 3 axes. Data harvested during experiments, which represented in time domain, was transformed into frequency domain by Fast Fourier Transform (FFT) method in order to get more useful information and reduce noise. Neural network (NN) and Support Vector Machine (SVM) were two supervised machine learning and classification models which were used to train the data. The structure of the two models were selected by using Genetic Algorithm (GA) and Non-Dominated Sorting Genetic Algorithm – II (NSGA-II). A preferable recognition rate of 99.6% was achieved by Neural network with Genetic algorithm implemented.

Keywords: Milling status · Neural network · Batch training · Genetic algorithm · LibSVM · NSGA-II · Fast fourier transform

1 Introduction

Milling is a machining process which is widely used to produce a custom-design part/product by removing materials from the work-piece. Many techniques have been developed to improve the efficiency of the milling.

However, disruptive events still happen during manufacturing due to machine breakdown, material absence or demand fluctuations [1]. The main reason of machine breakdown is from tool failure. Manufacturers and researchers can only reduce this issue by monitoring the working machines. As discussed in [2], there are two ways of monitoring: direct and indirect. While the direct methods use human labor, consume time, and deal with problems when they happen, the indirect methods use mathematical models, intelligent tools, and predict the problems beforehand. Cutting tool, surface roughness, machining parameters… are considered to be the subjects of monitoring system, and researches focus mostly on the condition of the cutting tool. Plastic deformation, mechanical breakage and gradual mechanical wear are main problems that occur during the working time [3]. These circumstances may lead the cutter to idle state although the machine is still in working state without the observation of the operators. Therefore, designing an automatic monitoring system is vital for the CNC

D. C. Nguyen et al. (Eds.): ICERA 2021, LNNS 366, pp. 31–41, 2022.
https://doi.org/10.1007/978-3-030-92574-1_4

milling machines. Most of researches did not approach to the idle/working state of the cutting tool. The state of the art for this approach can be found in [4], which is 93%.

Vibration, acoustic emission, cutting force, temperature, sound… are among means that used for condition monitoring [5]. In this research, vibration was used since it provided a similar periodic signal shape to that of the cutting force as suggested in [6]. It stores information in the form of time series data; thus, it is essential to analyze and extract useful information along with reducing noise to support the classifiers in recognizing the condition status.

Artificial neural networks (ANNs) which often refers to Neural Network (NN) by far is the most powerful computational intelligence tool. NN was introduced in early 1940s [7] with the interesting idea behind it: brain-like or brain-inspired computers. NN typically has three layers: input layer, hidden layer and output layer. The input layer receives untrained data for the network, while the hidden and output layers contain nodes. All of these layers and nodes are connected layer by layer. As mentioned in [8], neural network classifiers have been employed by many researchers due to its merits such as high fault tolerance and adaptability, noise suppression capability, and data-driven nature.

2 Methodology

The schematic of milling status detection is shown in Fig. 1:

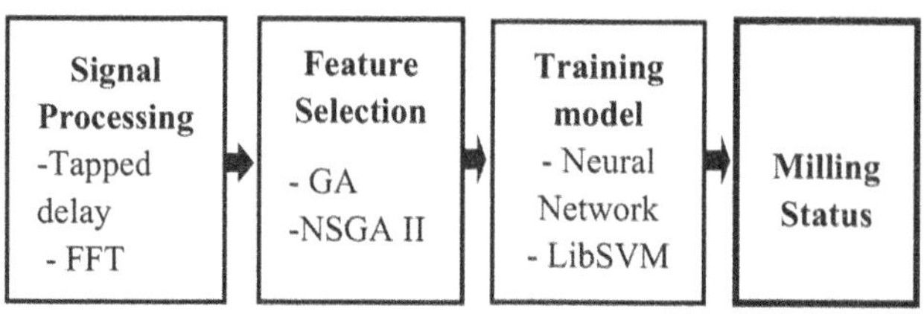

Fig. 1. Schematic of milling status detection

2.1 Signal Processing

2.1.1 Tapped Delay Parameter

As described in [6], the raw signal received from experiments need to be analyzed to extract significant information of the monitoring subject. Therefore, a tapped delay parameter N_d was added to increase the information contained in untrained data. It is straight forward that greater value of N_d will give more information, but it is also resulted in more noise and computational time. Therefore, choosing a suitable value for is critical to the effectiveness of the training model. If the data at time of the vibration signal can be presented as $[a_k^x, a_k^y, a_k^z]$, the data after added tapped delay parameter should be: $p_k = \left[a_k^x, a_{k-1}^x, \ldots, a_{k-d}^x, a_k^y, a_{k-1}^y, \ldots, a_{k-d}^y, a_k^z, a_{k-1}^z, \ldots, a_{k-d}^z\right]$.

2.2 Fast Fourier Transform (FFT)

Fourier transforms including Short-time Fourier transforms (STFT), discrete Fourier transforms (DFT) and fast Fourier transforms (FFT) help to extract the frequency information from the time series data. In [9], Kono et al. analyzed the frequency domain to find the corresponding frequency components to geometric errors. S. Palani and U. Natarajan used machine vision to analyze the surface of cutting area and evaluate the roughness. The computational time in their experiment for 256 x 256 Fourier transform was approximately 2s [10]. The measuring system (i.e., a self-organizing AAN) acquired the accuracy of 97.5% in prediction.

In this research, the data was processed through Fast Fourier Transform (FFT) algorithm which bases on the Discrete Fourier Transform (DFT) (1)

$$P_k = \sum_{n=0}^{N-1} p_n e^{-i2\pi k\frac{n}{N}} 0 \leq k \leq N-1 \tag{1}$$

where stands for N-point signal in time domain $(p_0, p_1 ..., p_{N-1})$, n and k are the discrete time and frequency indexes, belongs to N frequency domain signals $(P_0, P_1 ..., P_{N-1})$.

Because the length of the signal in frequency domain is equal to the signal in time domain, the number of FFT coefficient N_c should has a reasonable length.

2.3 Feature Selection

2.3.1 Genetic Algorithm (GA)

The genetic algorithm was applied to find the optimal parameters $[N_d, N_c, N_h]$ which are number of tapped delay parameter, number of FFT coefficient and number of hidden neurons of Neural network.

The Algorithm is described as follow:

Step 1. Initialization: Create a population of 30 chromosomes of 3 parameter $[N_d, N_c, N_h]$.

Step 2. Evaluation: Train the NN with each chromosome set and get the final test performance T_i.

Step 3. Start the loop from step 4 to step 7.

Step 4. Evaluating each chromosome's fitness value based on the result of objective function.

Compute the fitness F_i of each chromosome. The fittest chromosomes have higher probability P_i will be selected for next generation.

$$F_i = \frac{1}{1 + T_i} \tag{2}$$

$$C_i = \sum_{1}^{i} P_i \tag{3}$$

Step 5. Chromosomes selection: Use Roulette Wheel method

Step 6. Crossover: a crossover rate μ_c = 0.7 was applied
Step 7. Mutation: a mutation rate μ_m = 0.1 was applied
Step 8. Solution (Best Chromosomes): 30 generations were processed to find the best $[N_d, N_c, N_h]$.

2.3.2 Non-dominated Sorting Genetic Algorithm (NSGA-II):

The algorithm is explained as six following steps:

Step 1: Population Initialization:
Generate a random set of population.
Step 2: Non-Dominated sort:
The initialized population is sorted based on non-domination.
* For each individual p in the population P,
Initialize a set $S_p = \emptyset$ that will contain all dominated individual by p.
Initialize a set $n_p = \emptyset$ that will contain all individual dominated p.
* For each individual $n = n_q - 1$,
If p dominates q then $S_p = S_p \cup \{q\}$.
If q dominates p then $n_p = n_p + 1$.
* If n_p = 0, set rank of p to one (i.e., $p_{rank} = 1$). Update the first font set by adding p $F_1 = F_1 \cup \{p\}$.
* Then, initialize the front counter $i = 1$.
* While $F_i \neq \emptyset$:
A set $Q = \emptyset$ will be used to store the members of the next front (i.e., $(i+1)^{th}$).
* For each $p \in F_i$ do the following:
For each $q \in S_p$, decrease the domination count for q (i.e., $F_i = Q$). If n_q = 0, it means there are no members in subsequent fronts that dominate q. So set the rank for q and update the set Q = >$q_{rank} = i+1$ and $Q = Q \cup q$.
* Increase the front counter: $i = i + 1$.
* The next front is Q = >$F_i = Q$.
Step 3: Crowding Distance
After finishing the non-dominated sort, we assigned the crowding distance.
For each F_i, suppose we have m objective function and n members

- Member j^{th} will be set an initialize distance $F_i(d_j) = 0$.
- Sort the members based on objective function, $I = sort(F_i, m)$.
- Assign infinite distance to boundary values of each member in F_i. $I(d_n) = \infty$.
- for $k = 2$ to $(n - 1)$,

$$I(d_k) = I(d_k) + \frac{I_m^{k+1} - I_m^{k-1}}{f_m^{\max} - f_m^{\min}} \tag{4}$$

where I_m^k is the value of mth for the kth of I

The idea of crowding distance is to find Euclidian distance between each member of a front based on $m \times m$ dimensional hyper space.

Step 4: Selection

This step is carried out by using a crowded-comparison-operator $\prec_n$. We have already known that each member i in the population with two properties: the non-domination rank p_{rank} and the crow distance $F_i(d_j)$.

$$p \prec_n q \text{ if } p_{rank} < q_{rank} \text{ or } F_i(d_p) > F_i(d_q)(\text{if } p_{rank} = q_{rank}).$$

The member would be selected by binary tournament selection using $\prec_n$.

Step 5: Genetic Operators

In this step, we use two processes.

a. Simulated Binary Crossover process:

$$c_{1,k} = \frac{1}{2}\left[(1 - \beta_k)p_{1,k} + (1 + \beta_k)p_{2,k}\right] \tag{5}$$

$$c_{2,k} = \frac{1}{2}\left[(1 + \beta_k)p_{1,k} + (1 - \beta_k)p_{2,k}\right] \tag{6}$$

where $c_{i,k}$ is the ith child with k^{th} component, $p_{i,k}$ is the selected parent, β_k is the sample from a random number $(\beta_k \geq 0)$

$$p(\beta) = \frac{1}{2}(\eta_c + 1)\beta^{\eta_c} \quad 0 \leq \beta \leq 1. \tag{7}$$

$$p(\beta) = \frac{1}{2}(\eta_c + 1)\frac{1}{\beta^{\eta_c + 2}} \quad \beta > 1. \tag{8}$$

By a uniformly distributed random number u $(0<u<1)$, η_c is the distribution index for crossover process:

$$\beta(u) = (2u)^{\frac{1}{(\eta+1)}} \tag{9}$$

$$\beta(u) = \frac{1}{[2(1-u)]^{\frac{1}{(\eta+1)}}} \tag{10}$$

b. Polynomial Mutation process:

$$c_k = p_k + (p_k^u - p_k^l)\delta_k \tag{11}$$

where δ_k is small variation, p_k^u and p_k^l are upper bound and lower bound on parent component.

For r_k is a uniformly sampled random number between (0,1), and η_m is the mutation distribution index,

$$\delta_k = (2r_k)^{\frac{1}{\eta_m+1}} - 1 \quad r_k < 0.5 \tag{12}$$

$$\delta_k = 1 - [2(1 - r_k)]^{\frac{1}{\eta_m + 1}} \quad r_k \geq 0.5 \tag{13}$$

Step 6: Finishing

The process continues with subsequent generations when the offspring population is combined with the current generation. Hence, best generation will be selected based on previous selection steps.

2.4 Training Model

2.4.1 Neural Network

NN is indeed a powerful computational framework, which has been successfully applied in diverse applications including condition monitoring and tool breakage diagnosis. In [11] G. Zhang et al. showed that for training data, a hidden layer is sufficient. The structure of the network is presented as in Fig. 2:

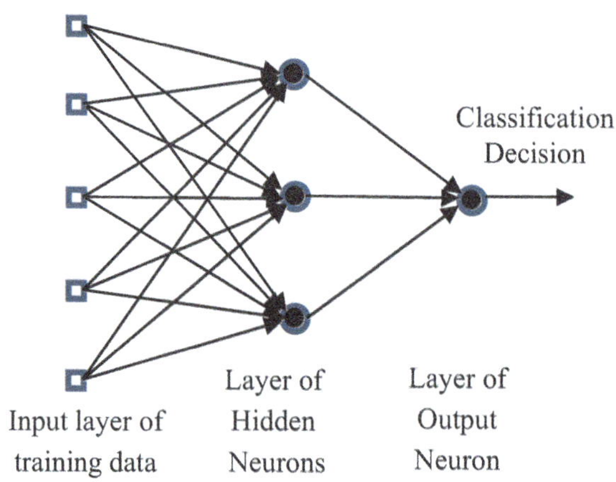

Fig. 2. A three-layer feedforward NN

Because batch training (i.e., offline) method was used at time k of the training, the input at time k of the NN can be presented as $p_k = [a_k^x, \ldots, a_{c+k-1}^x, a_k^y, \ldots, a_{c+k-1}^y, a_k^z, \ldots, a_{c+k-1}^z]$ where a_k^x is the coefficient extracted from FFT at time k of x-axis acceleration. Each component of the input p_k connects to the hidden neurons via the weights and biases. After applying the activation function to the hidden neurons, the output from the hidden neuron i^{th} can be described as:

$$u_k^i = v(\sum_{j=1}^{C} w_k^{ji} \cdot p_k^j + w_k^{jb}) \tag{14}$$

where u_k^i is the output at the ith hidden neuron at time k, $C = 3 \cdot c$ is the number of input sample, w_k^{ji}, w_k^{jb} are weights and biases of ith hidden neuron at time k, v is the sigmoid function.

The output of the NN at time k was calculated as follow:

$$y_k = s(\sum_{i=1}^{h} w_k^{oi} \cdot u_k^i + w_k^{ob}) \tag{15}$$

where y_k is the output of the NN at time k,u_k^i is the output at the ith, hidden neuron at time k. w_k^{oi}, w_k^{ob} are weights and bias of output layer at time k, s is the softmax function of the output layer.

The sigmoid and softmax function are described in Eqs. (16) and (17) respectively.

$$sigmoid(p) = \frac{1}{1 + e^{-p}} \tag{16}$$

$$softmax(p) = \frac{e^p}{\sum e^p} \tag{17}$$

GA and NSGA – II were applied to find the optimal values of $[N_d, N_c, N_h]$ parameters.

2.5 Support Vector Machine (SVM)

SVM is a new powerful and popular contemporary tool in machine learning problems. The LibSVM package Matlab code [12] is used in this research. The problem applying here is one-class SVM as described in [13]. It is supposed that the training vector $x_i \in R^n, i = 1, ..., l$, the problem can be described in SVM as:

$$\min_{\omega,\xi,\rho} \frac{1}{2}\omega^T\omega - \rho + \frac{1}{vl}\sum_{i=1}^{l}\xi_i \tag{18}$$

for $\omega^T\phi(x_i) \geq \rho - \xi_i,\ \xi_i \geq 0,\ i = 1, ..., l.$

where $\phi(x_i)$ is the kernel function and maps x_i into a higher dimensional space, ω is the vector variable, ρ is the optimal for the primal problem, $v \in (0, 1]$ is two boundaries of fraction of training errors and fraction of support vectors.

This dual problem is

$$\min_{\alpha} \frac{1}{2}\alpha^T Q\alpha \tag{19}$$

for $0 \leq \alpha_i \leq 1/(vl),\ i = 1, ..., l,\ e^T\alpha = 1.$

where $Q_{ij} = K(x_i, x_j) = \phi(x_i)^T\phi(x_j)$.

and decision function is

$$\mathrm{sgn}(\sum_{i=1}^{l} \alpha_i K(x_i, x) - \rho) \tag{20}$$

The LibSVM will then solve the problem in a scale version of (19) which subjects to:

$$e^T \alpha = v 0 \leq \alpha_i \leq 1,\ i = 1, \ldots, l, \tag{21}$$

The first step to apply the LibSVM is to choose some optional parameters for the training including: svm type (−s), kernel type (−t), degree (−d), gamma (−g), coef0 (−r), cost (−c), nu (−n), epsilon (−p), cache size (−m), epsilon (−e), shrinking (−h), probability estimates (−b), weight (−wi).

This step can be done by using a short Matlab code provided for the LibSVM. According to [11], we chose the initial parameter c = 8, g = 2 and h = 0 for more accurate and faster training.

The input in the NN training was also used in LibSVM. A ten folds cross-validation method was used to divide the training data. The result was the average training result of ten runs.

NSGA-II was applied to decide the N_d and N_c parameters to get the best input set for the LibSVM training.

2.6 Experiment Setup

2.6.1 System Setup for Data Collecting

An attachable electromagnetic energy harvester which is driven by wireless sensing system was designed as in [4] to collect the data. Total 59,946 samples of data for each sample contained signals from x-axis, y-axis and z-axis were used for experiments. The data value was in range from 0.8581 to 2.5484 according to the axis.

2.7 System to Execute the NN Program

Matlab R2016a program for NN training ran on an Intel Core i7 CPU 3.60 Ghz with 28G RAM and a NVIDIA GeForce GT 610 4GB display card. The parameters for Neural Network toolbox are shown in Table 1.

Table 1. Built in matlab neural network toolbox parameters

Data division	Random	**Gradient**	1.00e-06
Training	Trainlm	**Min MSE goal**	1.00e-09
Performance	Cross-entropy	**Max epochs**	50
Derivative	Default	**Validation Checks**	10

3 Experiment Result and Discussion

Table 2 shows the final result of three methods applied in this research to build predicting models of the milling status.

Table 2. Experiment results

Methods	Parameters			Result
	N_d	N_c	N_h	
NN + GA	71	48	53	99.6%
NN + NSGA II	30	17	30	92.6%
SVM + NSGA II	44	44	N/A	91.2%

The confusion matrix of the testing set of the three models are presented below (Fig. 3):

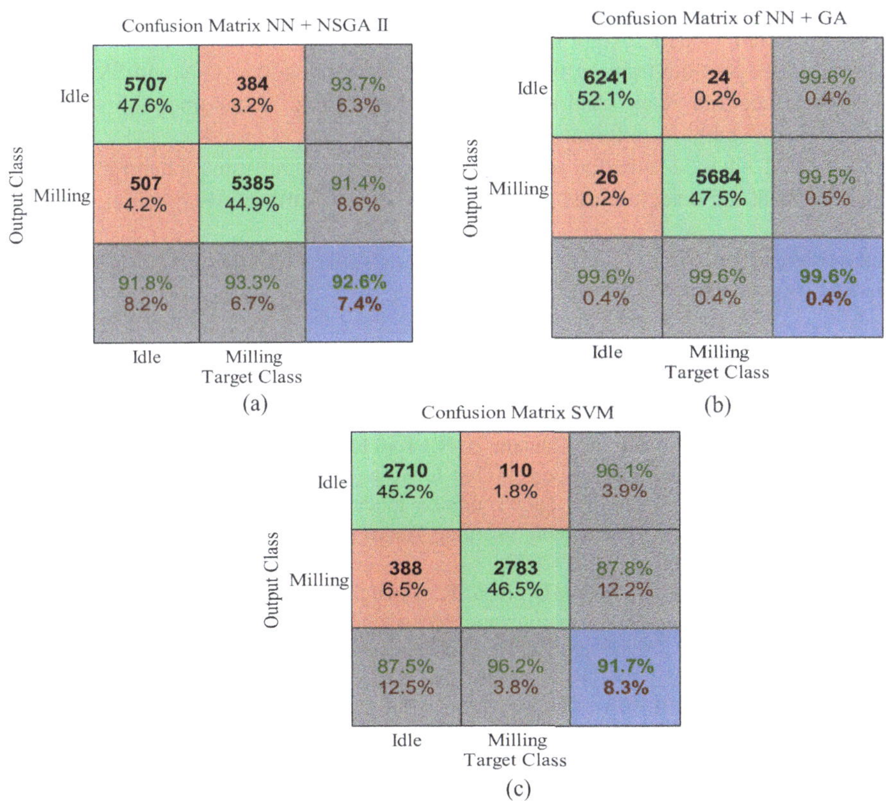

Fig. 3. (a): Confusion matrix of best NN with NSGA-II implemented, (b): Confusion matrix of best NN with GA implemented, (c): Confusion matrix of SVM with NSGA-II implemented

The best result achieved when neural network was applied with the help of Genetic Algorithm. However, NSGA – II implemented method optimized not only the classification accuracy but also the topology of the NN. Therefore, the result is a trade-off between the accuracy and the NN's topology, which affects the training time. There was also evidence in some runs which showed a better result when working with SVM (i.e., 99.7%). Therefore, further investigation should improve our achievement.

4 Conclusion

The research tried to implement multiple methods in classification of milling status during working. The high accuracy of 99.6% detecting the working and idling status of the machining process archived when using the Neural network and Genetic Algorithm. Other approaches were tried during the experiment such as "Exhaustive search" or "Extracting Texture Feature for Time Series classification" based on recurrent plot [14] but the result were not very satisfy. Some conclusion can be drawn from the final results:

Genetic Algorithm (GA) works perfectly to optimize the Neural network structure as well as other essential parameters.

Non sorted Genetic Algorithm – II (NSGA-II) method helps both finding the optimal target and the training time of the training models (i.e., NN and SVM)

Neural network is an ideal tool to detect the milling status since the results are excellent and stable in different trial runs.

LibSVM showed a better result (i.e., 99.7%) in some runs. Therefore, we are confident that the detection rate can be improved with further investigation.

Acknowledgement. This research is funded by Thai Nguyen University of Technology, Viet Nam under grant number T2020-B39.

References

1. Cho, S., Asfour, S., Onar, A., Kaundinya, N.: Tool breakage detection using support vector machine learning in a milling process. Int. J. Mach. Tools Manuf. **45**, 241–249 (2005)
2. Pai, S.P., Nagabhushana, T.: Tool condition monitoring using artificial neural network models. Handbook of Research on Emerging Trends and Applications of Machine Learning (ed). IGI Global, pp. 550–576 (2020)
3. Lauro, C., Brandão, L., Baldo, D., Reis, R., Davim, J.: Monitoring and processing signal applied in machining processes–a review. Measurement **58**, 73–86 (2014)
4. Chung, T.-K., et al.: An attachable electromagnetic energy harvester driven wireless sensing system demonstrating milling-processes and cutter-wear/breakage-condition monitoring. Sensors **16**, 269 (2016)
5. Botsaris, P.N., Tsanakas, J.A.: State-of-the-art in methods applied to tool condition monitoring (TCM) in unmanned machining operations: a review. In: Proceedings of the international conference of COMADEM, pp. 73–87 (2008)

6. Zhou, Y., Xue, W.: Review of tool condition monitoring methods in milling processes. Int. J. Adv. Manuf. Technol. **96**(5–8), 2509–2523 (2018). https://doi.org/10.1007/s00170-018-1768-5
7. Bishop, J., Mitchell, R.: Neural networks-an introduction. In: IEE Colloquium on Neural Networks for Systems: Principles and Applications, pp. 1/1–1/3 (1991)
8. Ambhore, N., Kamble, D., Chinchanikar, S., Wayal, V.: Tool condition monitoring system: a review. Mater Today: Proc **2**, 3419–3428 (2015)
9. Kono, D., Matsubara, A., Yamaji, I., Fujita, T.: High-precision machining by measurement and compensation of motion error. Int J Mach Tools Manuf **48**, 1103–1110 (2008)
10. Palani, S., Natarajan, U.: Prediction of surface roughness in CNC end milling by machine vision system using artificial neural network based on 2D Fourier transform. Int. J. Adv. Manuf. Technol. **54**, 1033–1042 (2011)
11. Zhang, G., Patuwo, B.E., Hu, M.Y.: Forecasting with artificial neural networks: The state of the art. Int. J. Forecast. **14**, 35–62 (1998)
12. Chang, C.-C., Lin, C.-J.: LIBSVM: A library for support vector machines. ACM Transactions on Intelligent Systems and Technology (TIST) **2**, 1–27 (2011)
13. Schölkopf, B., Smola, A.J., Bach, F.: Learning with Kernels: Support Vector Machines, Regularization, Optimization, and Beyond. MIT Press (2002)
14. Souza, V.M,. Silva, D.F., Batista, G.E.: Extracting texture features for time series classification. In: 2014 22nd International Conference on Pattern Recognition, pp. 1425–1430 (2014)

A Comprehensive Framework Integrating Attribute-Based Access Control and Privacy Protection Models

Anh Tuan Truong(✉)

Ho Chi Minh City University of Technology, VNU-HCM,
Ho Chi Minh City, Vietnam
anhtt@hcmut.edu.vn

Abstract. NoSQL databases have recently become increasingly popular as data platforms for big data and real-time web applications. Due to the simplicity in design but effectiveness in horizontal scaling and performance, NoSQL databases can be a better alternative approach in comparison with traditional relational databases. However, the lack of a fine-grained access control system together with a data privacy protection mechanism is one of the most important issues in NoSQL databases. In this paper, we investigate the attribute based access control model (ABAC) and use it as the main access control system in NoSQL databases. Moreover, we propose and implement a comprehensive framework for enforcing attribute-based security policies stored in JSON document together with a data privacy protection mechanism in the fine-grained level. We use Polish notation for modeling conditional expressions (i.e., the combination of subject, resource, and environment attributes) so that ABAC policies can be flexible, dynamic and fine grained. Moreover, for data privacy protection, privacy rules of the policies are constrained not only by access and intended purpose but also by subject, resource, and environment attributes as well as data disclosure level. The experiment is carried out to show the relationship between the processing time for access decision together with the privacy protection mechanism and the complexity of access and privacy policies.

Keywords: Hard milling · Cutting force · SiO2 nanoparticle · JIS SKD61 · Taguchi method

1 Introduction

Access Control is one of the essential and traditional security weapons of data protection. Due to the rapid development of large scale dynamic systems, traditional access control models such as Mandatory Access Control (MAC), Discretionary Access Control (DAC) [15], and Role-Based Access Control (RBAC) [14] have gradually revealed their disadvantages [11], for example, applied for an only closed system, role explosion, and inflexibility in specifying dynamic policies and contextual conditions. Recently, the Attribute-Based Access Control (ABAC) model [2, 3, 16–19] has been investigated to overcome such problems in traditional models. In ABAC,

D. C. Nguyen et al. (Eds.): ICERA 2021, LNNS 366, pp. 42–58, 2022.
https://doi.org/10.1007/978-3-030-92574-1_5

access requests are determined by evaluating access rules against the attributes of the subjects, objects, and environment relevant to the request. Ideally, the mentioned attributes can be any characteristic that describes the subjects, objects, and status of the system [8, 9].

Privacy is a major concern in any database management system since the databases may store personal and sensitive data of a user and such data can be disclosed without user permission and control, especially in mobile and ubiquitous computing applications and systems [12]. According to [20], privacy is defined as the claim of individuals, groups, or institutions to determine for themselves when, how, and to what extent information about them is communicated to others. Some research works have been addressed such privacy issues recently such as in [21–27] that consider constraints on the purpose of data usage for some purpose-based access control (PBAC) models.

Although PBAC has been integrated into some access control models such as RBAC [21, 27], to the best of our knowledge, no research has considered the integration of PBAC into ABAC. In this paper, we propose and implement a comprehensive framework for enforcing ABAC security policies together with a data privacy protection mechanism based on purpose-based access control. The main idea is to (1) use JSON to specify attribute-based policies that are leveraged by our work in [1], (2) integrate the PBAC model [5] into the ABAC model, and (3) develop the framework by Java and MongoDB database for policy evaluation mechanism.

The rest of the paper is organized as follows. Section 2 gives a brief survey of related works. Section 3 presents the overview of our approach. In Sect. 4, we introduce the structure of policies and main components in our proposed model. The experiment for evaluating the processing time is shown in Sect. 5. Concluding remarks and future works are discussed in Sect. 6.

2 Purpose-Based Frameworks

The authors in [7] have proposed a purpose-based framework for supporting privacy-preserving access control policies and mechanisms. They have also developed algorithms to help a system to detect and analyze the conflicts when adding new policies. However, they do not mention much about how to model the conditional expression and the proposed algorithms focus only on simple attributes and lack evaluating complex conditional expressions.

Research in [10] has presented an attribute-based protection model for JSON documents. Their approach allows a new attribute called "security-label" to JSON elements and specifies access control policies based on these values. The advantage of the separation of labeling and authorization policies is that they can be specified and administered independently possibly by different levels of administrators. However, the number of label assignments can be very large because it is calculated by the exponential function. Therefore, space storage is a potential problem when the system is expanded.

IBM has proposed a formal language called Enterprise Privacy Authorization Language [13] called EPAL for writing enterprise privacy policies to govern data handling practices in IT systems. An EPAL policy defines lists of hierarchies of data categories, user categories, and purpose. User-categories are the entities that use collected data, data categories define different categories of collected data, and purposes model the services for which data is intended to be used. An EPAL policy also defines sets of actions, obligations, and conditions. Actions model how the data is used and obligations define actions that must be taken by the environment EPAL. Lastly, conditions are Boolean expressions that evaluate the context. Privacy authorization rules are defined using these elements and each rule allows or denies actions on data categories by user-categories for certain purposes under certain conditions while mandating certain obligations.

Ji-Won Byun and Ninghui Li [4] have presented a comprehensive approach for privacy-preserving access control based on the notion of purpose. A key feature of their approach is that it allows multiple purposes to be associated with each data element and supports explicit prohibitions, thus allowing privacy officers to specify that some data should not be used for certain purposes.

3 Attribute-Based Access Control and Privacy Protection Models

We propose and implement a comprehensive framework for enforcing ABAC security policies together with a data privacy protection mechanism based on purpose-based access control. The architecture of our proposed approach is in Fig. 1. The framework includes two main modules: the authorization process and the privacy protection engine. Such main modules will be invoked by the Request Processor module that forwards user requests to engines. When a subject requests access to an object, the authorization process of ABAC and its privacy protection model will carry out through two stages (called 2-stage authorization) as follows (see Fig. 1):

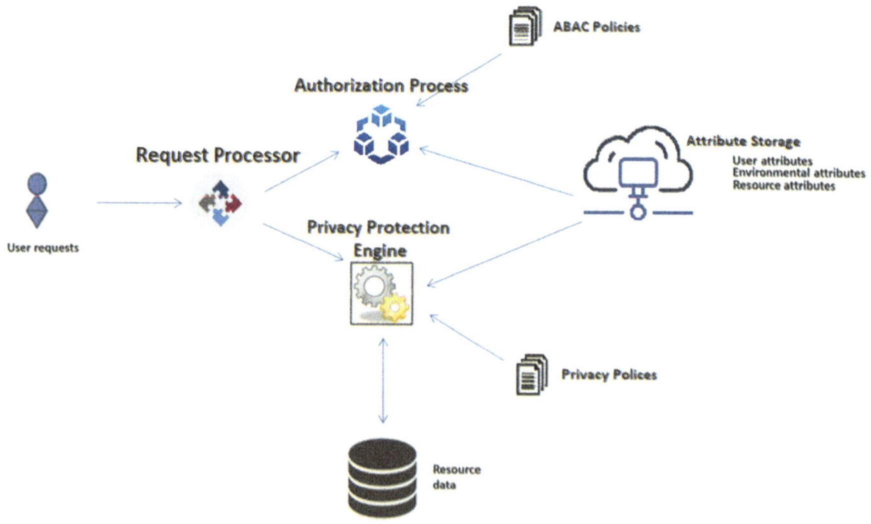

Fig. 1. Proposed Framework for integrating Attribute-Based Access Control and Privacy Protection Model

The First Stage (ABAC Model): Based on the pre-defined *access control policies*, the authorization process check if the request is legitimate with rights for the subject.

The Second Stage (Privacy Protection Model): the privacy protection process will verify the request by checking privacy compliance based on *privacy policies*.

Access Control Policies: As mentioned above, we integrate ABAC as our main access control system in NoSQL databases. According to the NIST standard [17], the ABAC authorization decision is based on rules that simultaneously specify a set of conditions on numerous attributes such as subjects, objects, actions, and environments for certain valid permission. To determine if an access request is granted or denied, the authorization process will retrieve the set of access control policies. In general, a policy contains rules which specify conditions for a user/subject can obtain access to an object. Then, the decision is made based on rules inside retrieved policies which are the Boolean expressions evaluated by the user's defined function, subject, resource, environment attribute. The relationship diagram between policies and rules is illustrated in Fig. 2. One access control policy may include many access control rules; one access control rule will be transformed into one function structure (see Sect. 4.1) and the access control policy is a set of functional structures.

One of the main problems in any access control system is the conflicts between policies and rules. The conflict may happen between policies or even inside a policy (i.e., between rules in the policy). To overcome this issue, we integrate algorithms for combining access control policies such as to permit override, deny override, etc. into the policy set and policies. Such algorithms have been proposed in XACML [6], a popular implementation of the ABAC model.

Privacy Policies: Such policies contain rules that specify whether access data should be shown, hidden, or generalized. At the second stage, the privacy protection engine will retrieve the privacy policies, and base on the Boolean expressions in rules, the engine evaluates the user's defined function, subject, resource, environment attributes to return how data/object can be returned to the user. We discuss more details on such privacy policies in Sect. 4.2.

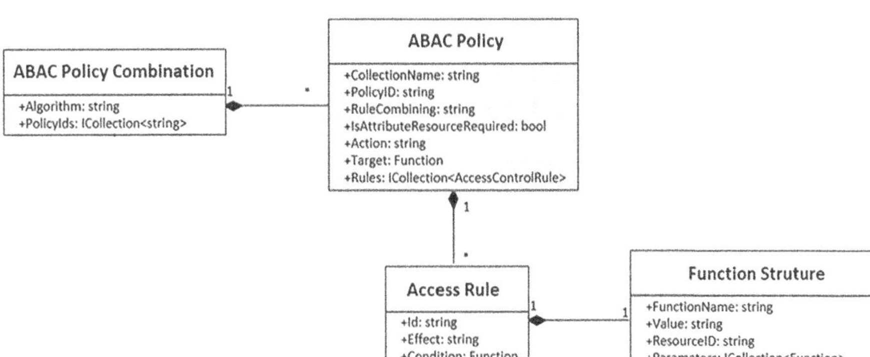

Fig. 2. Access Control Policy class diagram

Clearly, in order to assemble the ABAC model and Privacy protection model into NoSQL databases, one of the main works is to model the ABAC access control policies and Privacy policies. In the following, we describe our proposals to model such policies and integrate them into the access control system in NoSQL databases.

4 Modeling the ABAC Policies and Privacy Policies

As shown above, the policies contain access rules that specify Boolean expressions. Such expressions are the cores of the policies and based on them the authorization process and the privacy protection engine can determine the answer for an access request. In practice, modeling rule expressions in access control are one of the most challenging things because rules ensure that every access to a system and its resource is regulated and only accesses satisfying the rules are permitted to execute. In the following, we illustrate how to model the components of policies in JSON documents.

4.1 Modeling the ABAC Policies

In reality, an access control rule contains Boolean expressions. Similar to the Boolean logic formula, an expression in a rule will be formed from atomic expressions. Thus, the first work is to represent an atomic expression in a JSON document. Let consider an example of a rule:

> "A user with role doctor cannot view data of those users who has Area Number of SSN field which equals LA"

This rule includes two atomic expressions: (i) "A user with role doctor" and (ii) "those users who have Area Number of SSN field which equals LA".

To deal with such expression, we propose to model each atomic expression as a functional structure in JSON with its parameters can be a set of tuples (value, resource_id). The description for a function structure is defined as follows (Fig. 3):

Fields	Description
Function_name	The name of a function
Parameters	Parameters of each function can be a constant value, a value from environment, subject, resource or a value returned from another function
Value	If the value of this field is defined so it means that this function represents a constant value or the JSON path to get data from the resource (in this case "resource_id" field is also defined)
Resource_id	The identifier or name of the resource

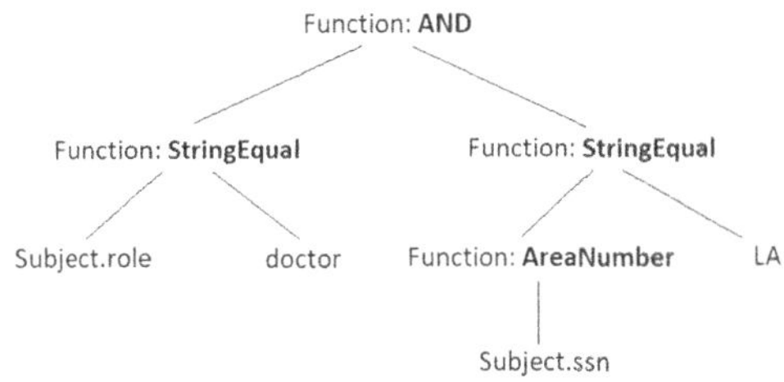

Fig. 3. A tree structure for expression E

Because each atomic expression is represented by a function structure, now, the complex expression can be visualized by a tree structure where each parent node is the name of the function and leaf node is the constant value or non-parameter function.

For example, consider again the rule:

"A user with role doctor cannot view data of those users who has Area Number of SSN field which equals LA"

The two atomic expressions in the rule can be transformed into an equivalent Boolean expression E as follows:

E = "StringEqual (Subject.role, 'doctor') AND StringEqual (AreaNumber (Subject.ssn), 'LA')"

where StringEqual is the function that compares two strings. The expression E now can be easily visualized using a tree structure as in Fig. 4.

Now, every expression in access rules can be transformed into tree structures by using the above approach. Algorithm 1 presents the pseudo-code of the module we write to take care of this transformation. It will take a raw expression as its input and output the corresponding tree structure in JSON formation. The main idea is to parse the expression into tokens (e.g., the keywords in the expressions; this can be done by using the function get tokens of any parsing tool) and then organize atomic expressions into functional structures. The function structure then will be combined to form the tree structure for the whole expression.

The output when we run Algorithm 1 on the expression in the rule mentioned above is the tree structure in Fig. 5.

Algorithm 1: Algorithm for parsing Boolean expression
Input: rawExpression: string
Output: function: Function class

```
Let listToken: List<string> ← getTokens(rawExpression); /* gets tokens from rawExpression
Let stackToken: Stack<string>   /* A stack stores names of functions */
Let queueToken: Queue<string>   /* A queue stores tokens in Reverse Polish Notation form */
Let queueFunction: Queue<Function>   /* A queue stores functions */
begin
     foreach token in listToken do
            if IsFunctionName( token ) Or IsLogicalOperator( token ) Or token == "(" then
                 stackToken.push (token)
           else if token == ")" then
                 while stackToken.length > 0 do
                        var temp := stackToken.pop()
                         if temp == "(" then
                               queueToken.enqueue(stackToken.pop())
                               break
                         else queueToken.enqueue(temp)
                  end while
           else queueToken.enqueue (token)
     end foreach
     while stackToken.length > 0 do queueToken.enqueue (stackToken.pop())
     while queueToken.length > 0 do
           token := queueToken.dequeue()
           if IsFunctionName( token ) do
                  var function :=  Function.CreateFunction (token)
                  for i := 1 to GetNumberParameters (function) do
                         function.Parameters.Add (queueFunction.dequeue())
                  end for
                  queueFunction.enqueue (function)
           else queueFunction.enqueue (Function.CreateConstantValue(token))
     end while
     return queueFunction.enqueue()
end
```

```
{
  "function_name" : "AND",
  "parameters" : [
      {
        "function_name" : "StringEqual ",
        "parameters" : [
            {   "value" : "role",    "resource_id" : "Subject"  },
            {   "value" : "doctor",  "resource_id" :   null     }
        ]
      },
      {
        "function_name" : "StringEqual",
        "parameters" :[
          {
            "function_name" : "AreaNumber ",
            "parameters" : [
                {   "value" : "ssn",    "resource_id" : "Subject"  }
            ]
          },
          {   "value" : "LA",    "resource_id" : null}
        ]
      }
  ]
}
```

Fig. 4. The JSON output of the Algorithm on the Boolean expression **E**

4.2 Modeling the Privacy Policy

As mentioned in Sect. 3, a privacy policy contains some privacy rules that specify whether access data should be shown, hidden, or generalized. The relationship between the privacy policy and rules is depicted in Fig. 6. In general, we adapt the approach to represent ABAC policies to privacy policies: One privacy policy includes some field rules (or privacy rules). Note that we organize the data/resources into data fields and each field rule controls the privacy protection mechanism for data fields. Since we have any data fields and each data field may require different levels of privacy protection, we proposed the attribute "FieldEffects" for each privacy rule that specifies data disclosure levels for each field of JSON data.

Each field rule contains a Boolean privacy expression like the access control rule. Again, the expression in a privacy rule will be formed from atomic privacy expressions and then such expression will be transformed into functional structures. Moreover, to integrate PBAC into the privacy protection model, access purpose is modeled as an attribute of the environment, and intended purpose is considered as an attribute of data objects in the resource content. Let consider the following privacy policy:

> "Privacy Policy 1: The body of any email documents must be processed by MaskSensitive function"

This policy has a single privacy expression E = "StringEqual (Document.type, 'emails')" and can be expressed using the structure below:

```
"PolicyID": "Privacy  Policy 1",
"Collection Name" : "emails"
"Target" : null,
"IsAttributeResourceRequired" : false,
"Rules" : [(
   "Id" : "Rule 1"
   "Condition" : null,
   "FieldEffects" : [
   {
      "name" : "body",
      "effect_function" : "MaskSensitive"
   }]
)]
```

Here, the data field considered is the body of any email, and privacy information in this body will be masked by invoking a specific function MaskSensitive.

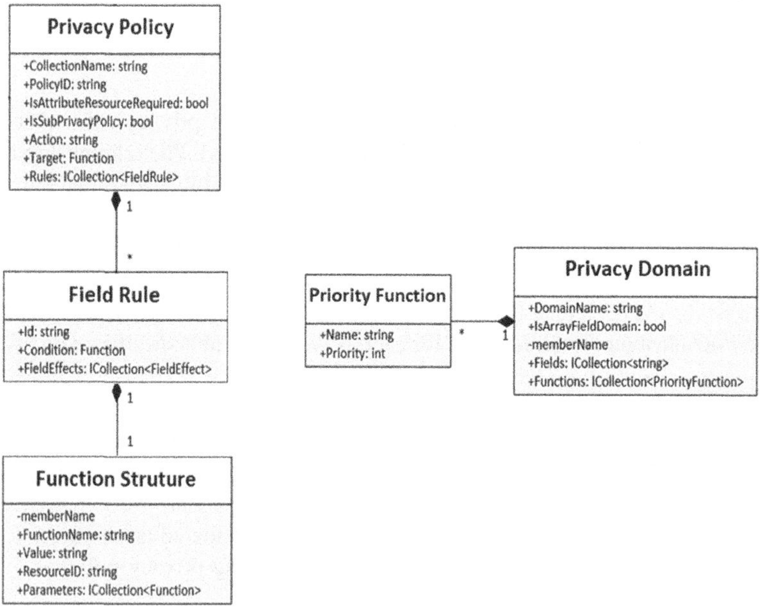

Fig. 5. Privacy Policy class diagram

Clearly, similar to ABAC policies, some conflicts may happen since some privacy rules may be applied to the same data fields. Thus, we propose a structure called "PrivacyDomain" for solving conflicts in privacy protection levels. In general, a privacy domain means that in which domain we are considering privacy protection.

Privacy domains contain privacy functions that execute some algorithms for privacy protection. By using privacy domains, now a privacy function can only affect the data fields but in a certain specific domain that the function is defined, thus, reduce the conflict between the privacy function with other private functions. Let consider the definition of the following privacy domains:

```
{"DomainName" : "DefaultDomain",
"fields: [],
"isArrayFieldDomain" : false,
"Functions": [ {"name" : "Hide", "priority" : 2},
     {"name" : "Show", "priority" : 1}]
}
{"DomainName" : "DateTimeDomain",
"fields: [Employee.personal_info.birth_date"],
"isArrayFieldDomain" : false,
"Functions": [ {"name" : "ShowYear", "priority" : 2},
     {"name" : "ShowMonthAndYear", "priority" : 1}]
}
```

Fig. 6. Privacy Domain example

We define two privacy domains: the DeaultDomain for generic privacy functions (e.g., Hide and Show functions) and DateTimeDomain for functions that regulate date and time (e.g., ShowYear and ShowMonthAndYear functions). Note that DefaultDomain is aiming at general purposes, thus, its attribute "fields" is set to [] (i.e., privacy functions defined in this domain can apply to any data field) while the functions in DateTimeDomain can only be applied to data field "birth_date". Since privacy functions in the same domain may conflict with other ones, we need to provide priority to such functions. The priority of the function helps to choose the suitable privacy function to apply to the data field when there exist conflicts (e.g., the priority of function DateTime.ShowYear is higher than one of DateTime.ShowYearAndMonth).

In our framework, we need the DefaultDomain because in case we want to apply privacy protection to the whole data (not a certain data field), the privacy functions defined in this domain can be invoked. Since the privacy functions in DefaultDomain can be applied to any data fields, this may lead to the case in which there is a function in other domains that may also be applied to the same data field as default privacy functions; this is considered as a conflict (Note that we only have conflicts between a function in DefaultDomain and functions in certain domains, the conflicts between functions in different domains do not happen since the domains cannot intersect to each other). To overcome this issue, we also propose a rule, named the CPF rule, to choose a unique privacy function that is applicable for the certain data field in case of conflict. The CPF rule is described as follows:

CPF: *P("Optional") < P("DefaultDomain.Show") < P(X.Y1) < ... < P(X.Yn) < P ("DefaultDomain.Hide")*

where P is the priority; X.Y means that the privacy function Y in domain X. The main idea of this rule is that we will select the privacy function with the highest priority. The lowest priority is "Optional" (i.e., equivalent to any function), then the privacy function "Show" of the DefaultDomain, then the priority of privacy functions as provided in their domain, and finally, the function "Hide" of the default domain. Let consider the following example: assume that we have some data fields and all the privacy functions that can apply to them as follows:

Fields	Conflict privacy functions
Name	Optional, DefaultDomain.Show
Personal_info.birth_date	DateTimeDomain.ShowMonthAndYear, DateTimeDomain.ShowYear, default domain.Show
Personal_info.ssn	SsnDomain.AreaNumber, SsnDomain.SerialNumber, Optional
Fields	Privacy function chosen
Name	Default domain.Show
Personal_info.birth_date	DateTimeDomain.ShowYear
Personal_info.ssn	SsnDomain.AreaNumber

Applying the CPF rule to the above conflict cases, the unique privacy function for each data field will be as below:

Let consider the data field personal_info.birth_date, 3 privacy functions applied to it: the priority of DefaultDomain.dShow is the lowest; for the two privacy functions DateTimeDomain.ShowMonthAndYear and DateTimeDomain.ShowYear, the priority of DateTimeDomain.ShowYear is bigger than the priority of DateTimeDomain.ShowMonthAndYear, then, the function DateTimeDomain.ShowYear is chosen as the unique privacy function for personal_info.birth_date.

Advanced Support for an Array of Embedded Subdocuments. JSON objects can contain not only a single-value field (string, number, Boolean value) but also contain an array of objects field (it's also known as an array of embedded subdocuments in MongoDB) which makes the JSON object more complex and expressive. To support privacy for the array of objects field, we propose to extend the attribute "effect_-function" in the Privacy Policy structure (see the example "Privacy Policy 1" above) so that its value can be an identifier of another privacy policy that is used for privacy protection of this array field. For example, let consider a complex JSON Object named "Project Bonus":

```
{
  "project_name" : "E-learning system",
  "bonuses": [      //this is array of embedded subdocuments
    { "name": "John", "bonus": 600 },
    { "name": "Bob", "bonus": 900  }
  ]
}
```

This object contains an array of fields (e.g., "John", "Bob", …). Assume that for privacy protection, we have a privacy rule: "Employee can only view his/her bonus for each project".

Such policy can be expressed as follows (the policy is named "Policy 1"):

```
{
  "policy_id": "Policy 1",
  "collection_name": "ProjectBonus",
  "rules": [
    {
      "rule_id": "rule 1",
      "field_effects": [
        {
          "name": "project_name",
          "effect_function": "DefaultDomain.Show"
        },
        {
          "name": "bonuses",
          "effect_function": " Policy3"
        }
      ],
      "condition": {}
    }
  ]
}
```

For the field effect of the rule inside Policy 1: the data filed "project_name" is protected using by using the function Show in the generic DefaultDomain while the privacy protection of the data filed "bonuses" (that is an array of object fields) is pointed to another privacy policy (e.g., Policy3). Then, inside policy 3, we can define privacy functions for data fields of bonuses:

```
{
  "policy_id": "Policy3",
  "collection_name" : "ProjectBonus.bonuses",
  ....
  "rules" : [
    {
      "rule_id" : "rule 1",
      "field_effects" : [
        { "name" : "name", "effect_function" : "DefaultDomain.Show" },
        { "name" : "bonus", "effect_function": "DefaultDomain.Hide" },
      ],
      "condition" : { }
    },
  ]
}
```

Now, if John requests to view the "ProjectBonus" collection, the data in such collection will be displayed as follows (due to Policy3):

```
{
  "project_name" : "E-learning system",
  "bonuses": [
    { "name": "John",  "bonus": 600 },
    { "name": "Bob",   "bonus": ""  }
  ]
}
```

5 Evaluation

We carried out experiments to show the relationship between the processing time of authorization and the complexity of resource documents. The system configuration for the experiments is Intel core i5–3230 M 2.60 GHz, 8 GB RAM. The proposed framework is implemented using C#,.NET Core and we use MongoDB for storing policies and data. For resource documents, we use the Employee Salaries dataset from https://data.world and classify into two (sub) data sets: (i) documents contain only single-value fields as described in Fig. 5 (data set 1); and (ii) complex documents which contain also arrays of fields as in Fig. 6 (data set 2). For each document in the sets, the number of records ranges from 2000 to 12000 records. Each record in the resource documents must be carried out through 2-stage authorization (i.e., ABAC authorization and Privacy protection). The attributes of subjects are shown in Fig. 7. All experiments include 5 ABAC access control polices and 5 privacy policies, each policy contains at least three rules, and each rule required at least three attributes to evaluate (Figs. 8, 9 and 10).

```
{
  "date_created" : "10/18/2014",
    "is_deleted" : false,
    "tax" : 89.4,
    "number_developers" : 43,
    "leader" : {
        "name" : "Felicle",
        "info" : {
            "ssn" : "341-11-9027",
            "date_of_birth" : "10/30/1998",
            "salary" : 542.3,
            "phone" : "355-(325)701-2694"
        }
    },
    "dept_no" : 706,
    "location" : "Albania",
    "name" : "Research and Development",
    "long_location" : "19.71611",
    "lat_location" : "41.06306",
    "level_domain" : "info"
}
```

Fig. 7. Data set 1: single-valued fields

```
{
    "date_created" : "3/3/2017",
    "is_deleted" : false,
    "tax" : 8.52,
    "number_developers" : 51,
    "leader" : {
        "name" : "Kirby",
        "info" : {
            "ssn" : "307-18-0580",
            "date_of_birth" : "10/13/1996",
            "salary" : 709.6,
            "phone" : "386-(522)612-8742"
        }
    },
    "dept_no" : 1269,
    "location" : "Slovenia",
    "name" : "Human Resources",
    "long_location" : "15.83139",
    "lat_location" : "46.57611",
    "level_domain" : "name",
    "developers" : [
        {
            "name" : "Katti",
            "info" : {
                "ssn" : "605-19-2690",
                "salary" : 2175,
                "phone" : "1-(752)629-0495"
            }
        }
    ]
}
```

Fig. 8. Data set 2: Complex documents

```
{
  "first_name": "Raina",
  "last_name": "Gillett",
  "email": "abc@gmail.com",
  "gender": "Female",
  "age": 35,
  "active": true,
  "job_title": "VP Product Management",
  "department_name": "Business Development",
  "location": "Kenya",
  "role": "admin"
}
```

Fig. 9. Figure 7. Subject structure

For each document, we run the frameworks 5 times with different user requests. Table 1 shows the experiment results when running test cases in Data set 1. Column 2 presents the number of records in the test case; Columns 3 to 7 report the processing times for 5 five different requests on the test case document in milliseconds (ms). Column 8 shows the average time to process the document. According to the result, the processing time increases (probably linear) when the complexity of the document increases.

Table 1. Processing time on the Data set 1

ID	Number records	T1 (ms)	T2 (ms)	T3 (ms)	T4 (ms)	T5 (ms)	Average (ms)
1	2000	2157	2274	2222	2189	2302	**2228**
2	4000	4501	4642	4644	4325	4494	**4521**
3	6000	6863	6690	6720	6761	6458	**6698**
4	8000	9022	8929	8908	9035	9010	**8980**
5	10000	11762	11619	11547	11273	11465	**11533**
6	12000	14290	13430	13643	13646	13980	**13797**

Table 2. Processing time of the second experiment.

ID	Number records	T1 (ms)	T2 (ms)	T3 (ms)	T4 (ms)	T5 (ms)	Average (ms)
1	2000	2641	2775	2708	2611	2507	**2648**
2	4000	5141	5270	5260	5265	5253	**5237**
3	6000	7879	7931	7692	8141	8176	**7963**
4	8000	10946	11226	10712	10768	10830	**10896**
5	10000	13700	13344	13581	13716	13526	**13573**
6	12000	16269	16483	16085	16219	16111	**16233**

Data set 2 includes resource documents that are more complex than one in Data set 1. Such documents contain an array of embedded documents. Each record has an array of embedded documents field which contains at most five elements inside it. Table 2 reports the experiment results when running test cases in Data set 2. Similar to Table 1, Column 2 presents the number of records in the test case; Column 3 to 7 shows the processing times for 5 five different requests on the test case document. Column 8 reports the average time to process the complex document. Similarly, according to the result, the processing time increases when the complexity of the document increases.

Such experiment results are demonstrated in charts as in Fig. 7. According to this, the processing time of our proposed framework increases linearity with the complexity of object data. In the first experiment, the average processing time for one record is approximately 1.1 ms, and in the second experiment which contains an array of embedded documents is about 1.3 ms per record.

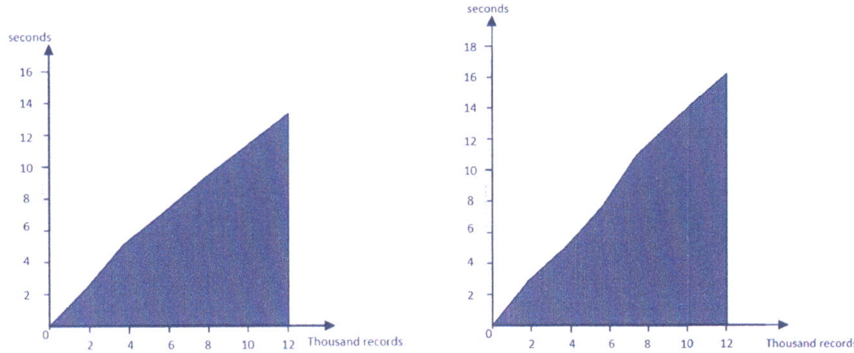

(a)

Fig. 10. Chart of processing time in the first and second experiments.

6 Conclusion

In this research, we have proposed a comprehensive framework for enforcing ABAC security policies stored in JSON documents and integrating a privacy protection model at the fine-grained level. The main idea is that the access control process must go through 2 stages: the authorization process based on the ABAC model, and the privacy protection process based on the PBAC model. We have used function structure and tree structure for modeling expressions which are the combination of subject, resource, and environment attributes so that the policies can be easy to transform into JSON formation. Specially, we have proposed a structure called "PrivacyDomain" for solving conflicts in privacy protection levels and provide a mechanism to process an array of embedded subdocuments.

The processing time of our proposed framework is mainly based on the evaluation process of ABAC and Privacy policies that contain expressions (e.g., propositional logic formula). Thus, some heuristics can be applied in order to speed up such an evaluation process. This investigation will be left as our near future work.

Acknowledgments. This work was funded by Vietnam National University-Ho Chi Minh City under the research project C2020–20-30.

References

1. Thi, Q.N.T., Dang, T.K., Van, H.L., Son, H.X.: Using JSON to specify privacy preserving-enabled attribute-based access control policies. In: Wang, G., Atiquzzaman, M., Yan, Z., Choo, K.-K. (eds.) SpaCCS 2017. LNCS, vol. 10656, pp. 561–570. Springer, Cham (2017). https://doi.org/10.1007/978-3-319-72389-1_44
2. Bertino, E., Ghinita, G., Kamra, A.: Access Control for Databases: Concepts and Systems
3. Hu, V.C., et al.: Now Publishers, Hanover (2011)
4. Sandlin, K., Miller, R., Scarfone, K.: Guide to attribute-based access control (ABAC) definition and considerations (draft). NIST Spec. Publ. **800**, 162 (2013)
5. Byun, J.W., Li, N.: Purpose-based access control for privacy protection in relational database systems. VLDB J. **17**(4), 603–619 (2008)

6. Colombo, P., Ferrari, E.: Enforcement of purpose-based access control within relational database management systems. IEEE Trans. Knowl. Data Eng. **26**(11), 2703–2716 (2014)
7. Rissanen, E.: eXtensible Access Control Markup Language (XACML) Version 3.0
8. (committee specification 01): Technical Report. OASIS (2010). http://docs.oasisopen.org/xacml/3.0/xacml-3.0-core-spec-cd-03-en.Pdf
9. Hua, L.S., Varadharajan, V.: Purpose-based access control policies and conflicting analysis. Security and Privacy – Silver Linings in the Cloud. 217–228 (2010)
10. Colombo, P., Ferrari, E.: Fine-grained access control within NoSQL document-oriented datastores. Data Sci. Eng. **1**(3), 127–138 (2016). https://doi.org/10.1007/s41019-016-0015-z
11. InterNational Committee for Information Technology Standards: Information technology – Next Generation Access Control – Generic Operations and Data Structures (2015)
12. Biswas, P., Sandhu, R., Krishnan, R.: An Attribute-Based Protection Model for JSON Documents. In: Chen, J., Piuri, V., Su, C., Yung, M. (eds.) NSS 2016. LNCS, vol. 9955, pp. 303–317. Springer, Cham (2016). https://doi.org/10.1007/978-3-319-46298-1_20
13. Gartner: Gartner says "By 2020, 70% of businesses will use attribute-based access control (ABAC) to protect critical assets". [Online] Available at https://www.avatier.com/products/identity-management/resources/gartner-iam-2020-predictions/. (2016) Accessed 3 March 2017
14. Information Commissioner's Office: Data Protection Principles. [Online] Available at https://ico.org.uk/for-organisations/guide-to-data-protection/data-protection-principles/
15. Enterprise Privacy Authorization Language: [Online] Available at https://www.w3.org/2003/p3p-ws/pp/ibm3.html
16. Sandhu, R., Coyne, E., Feinstein, H., Youmann, C.: Role-based access control models. IEEE Computer **2**(29), 38–47 (1996)
17. ISO/IEC DIS 10181-3: Information Technology, OSI Security Model, Security Frameworks, Part 3: Access Control (1993)
18. Hu, V.C., et al.: Guide to Attribute-Based Access Control (ABAC) Definition and Considerations (draft). NIST Special Publication. **800**(162) (2013)
19. Hu, V.C., Kuhn, D.R., Ferraiolo, D.F.: Attribute-based access control. IEEE Computer **48** (2), 85–88 (2015)
20. Huang, J., Nicol, D.M., Bobba, R., Huh, J.H.: A framework integrating attribute-based policies into role-based access control. In: Proceedings of the 17th ACM symposium on access control models and technologies, pp. 187–196. ACM (2012)
21. Jin, X., Krishnan, R., Sandhu, R.: A unified attribute-based access control model covering DAC, MAC, and RBAC. In: IFIP Annual Conference on Data and Applications Security and Privacy, pp. 41–55. Springer (2012)
22. Westin, A.F.: Privacy and Freedom, vol 7. Atheneum, New York (1967)
23. Byun, J.W., Bertino, E., Li, N.: Purpose based access control of complex data for privacy protection. In: Proceedings of the Tenth ACM Symposium on Access Control Models and Technologies, pp. 102–110. ACM (2005)
24. Byun, J.W., Li, N.: Purpose based access control for privacy protection in relational database systems. VLDB J. **17**(4), 603–619 (2008)
25. Kabir, M.E., Wang, H.: Conditional purpose based access control model for privacy protection. In: Proceedings of the twentieth australasian conference on australasian database, vol 92, pp 135–142. Australian Computer Society Inc. (2009)
26. Wang, H., Sun, L., Bertino, E.: Building access control policy model for privacy-preserving and testing policy connecting problems. J. Comput. Syst. Sci. **80**(8), 1493–1503 (2014)
27. Kabir, M.E., Wang, H., Bertino, E.: A role-involved conditional purpose-based access control model. In: E-Government, E-Services and Global Processes, pp. 167–180. Springer (2010)

A Force Model for Controlling the Destemming Process of the Fresh Chilli Fruit

Quoc-Khanh Huynh[1], Chi-Ngon Nguyen[1], Jeng-Rong Ho[2], Phuong Lan Tran-Nguyen[1], Huu-Cuong Nguyen[1], Hong-Phuc Vo-Nguyen[3], and Van-Cuong Nguyen[1](✉)

[1] College of Engineering Technology, Can Tho University, Can Tho City, Vietnam
{hqkhanh, ncngon, tnplan, nhcuong, nvcuong}@ctu.edu.vn

[2] Department of Mechanical Engineering, National Central University, Taoyuan City, Taiwan, R.O.C.
jrho@ncu.edu.tw

[3] School of Medicine, Can Tho University of Medicine and Pharmacy, Can Tho City, Vietnam
vnhphuc@ctump.edu.vn

Abstract. Stem removal is necessary for fresh chilli fruit processing, especially in the Mekong Delta region. In previous studies, the destemming method using clamping belts has been done which allowed to grip and completely separate the stems. However, the control of the gripping force has not yet been implemented, thus causing unable to clamp the stems, or clamping strongly to break the stems. This study focuses on building a force model to control the destemming process, helping to improve the success rate. The mathematical relationship between the gripping force and the measured force on the sensor has been shown. The experimental model was set up and run to evaluate the working ability. Tests were carried out on 1,000 chillies with a successful rate of up to 96.2%. In further studies, this model would be extended to design a destemming system.

Keywords: Destemming process · Force control model · Fresh chilli

1 Introduction

Chilli is a popular food and also an important agricultural product, especially in the Mekong Delta [1, 2]. Many methods have been used to harvest and process the fresh chillies, and the stem separation process is considered an important step [3, 4]. Several methods have been proposed in previous studies of the automatic destemming system [5–7]. In particular, the mechanism which used a timing belt to clamp and separate the stems offers an outstanding advantage over other methods. That is the ability to completely separate the stem from the fruit body [8].

In the working of the timing belt destemming system as in Fig. 1, when the pressure of the destemming timing belt on the destemming roller is low, there is not enough gripping force to separate the stems. Otherwise, the high pressure breaks and destroys the stems, resulting in loss of gripping ability. Therefore, it is necessary to control the destemming force.

D. C. Nguyen et al. (Eds.): ICERA 2021, LNNS 366, pp. 59–68, 2022.
https://doi.org/10.1007/978-3-030-92574-1_6

This study focuses on building the destemming force control model so that the gripping force caused by the belt is kept in a suitable range. A mathematical relationship between the destemming force generated by the timing belt and the force measured on the sensor at the tension adjustment has been established. An experimental model was fabricated to evaluate the control ability. The test results are recorded and discussed. In addition, limitations and future research directions are also pointed out.

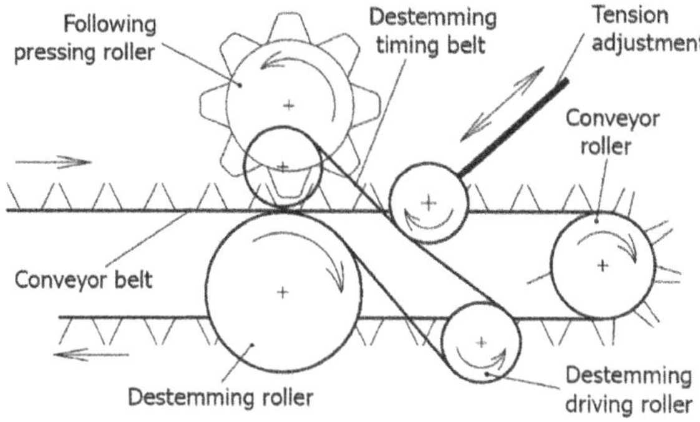

Fig. 1. Destemming working principle.

2 Proposed Destemming Force Model

2.1 The Working Principle

The destemming force model is proposed and presented in Fig. 2. In which, a sliding mechanism driven by a screw acts to create belt tension. Two ends of the screw are in contact with the belt tensioner roller and the loadcell, respectively. The destemming belt is a timing type belt, so the internal deformation is small and would be ignored in this calculation model.

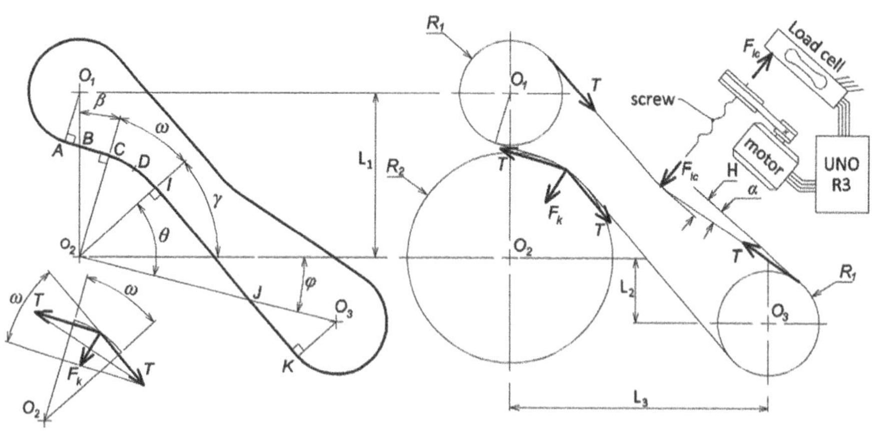

Fig. 2. Destemming force model.

Belt tension force is transmitted to every point of the belt. At the contacting position of the belt and the destemming roller, pressure will be generated to grip and pull the stem down away from the conveyor. While the fruit's body is kept on the conveyor by the following pressing roller and moved horizontally to the end. As a result, the stem is separated from the chilli's body.

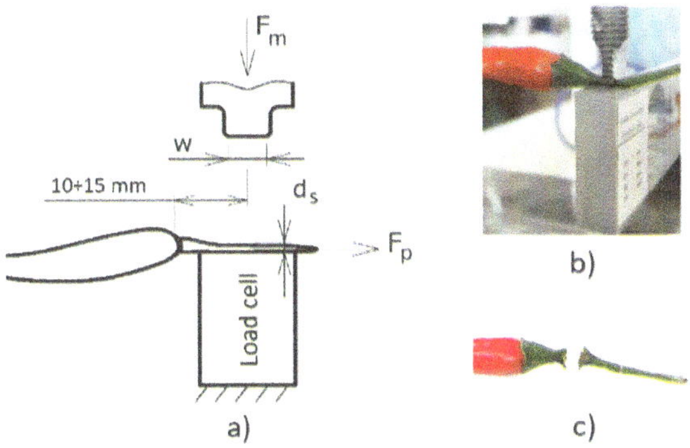

Fig. 3. Measurement of the destroying pressure. a) Measurement method, b) Measuring, c) Broken stem

2.2 The Measurement of Required Destemming Force and Destroying Pressure

In this study, the minimum destemming force $F_{d(min)}$ is defined as the smallest value of the force needed to remove the stem from the fruit's body. And the destroying pressure P is the pressure at which the stem is broken and could be divided into two parts by the $F_{d(min)}$ force.

The minimum required destemming force has been measured in [9] and is quoted as:

$$F_{d(\min)} = 4.7\,(\mathrm{N}) \tag{1}$$

The measurement of the destroying pressure is described in Fig. 3. As Fig. 3a, a pressing plate in the width w is used to apply the measuring force F_m on the stem with diameter d_s. While the force F_m is gradually increased, there is always a constant axial pulling force F_p that continuously pulls the stem away from the fruit. This value of F_p was chosen to be 5 (N), approximately equal to $F_{d(min)}$. The value of F_m was recorded at the time the stem was destroyed and separated from the body as in Fig. 3b–c.

The destroying pressure P (N/mm^2) is calculated as:

$$P = \frac{F_m}{d_s \times w} \tag{2}$$

An experiment was done on 150 fruits. The average destroying pressure is P = 128.8 ± 8.3 (N/mm^2).

2.3 The Destemming Force Model

As shown in Fig. 3, the relationship between tension force T and the measured force on the load cell F_{lc} is:

$$F_{lc} = T \times \sin(\alpha) = T \times \frac{H}{\sqrt{H^2 + \left(\frac{(L_1 + L_2)^2 + (L_3)^2}{4}\right)}} \tag{3}$$

The beginning contacting angle β is calculated as:

$$\beta = \cos^{-1}\left(\frac{(R_1 + R_2)}{L_1}\right) \tag{4}$$

The inclined angle of the destemming driving roller φ is:

$$\rho = \tan^{-1}\left(\frac{L_2}{L_3}\right) \tag{5}$$

The angle θ is obtained by:

$$\theta = \cos^{-1}\left(\frac{R_2}{\sqrt{\left((L_2)^2 + (L_3)^2\right) \times \left(\frac{R_2}{R_1 + R_2}\right)}}\right) \tag{6}$$

From Fig. 3, the end contacting angle γ is:

$$\gamma = \theta - \varphi \tag{7}$$

The pressing force F_k caused by the belt tension on the destemming roller is deduced as follows:

$$F_k = 2 \times T \times \sin\left(\frac{\pi/2 - \beta - \gamma}{2}\right) \tag{8}$$

The destemming force F caused by the pressing force Fk and the friction co-efficient f between the timing belt and the stem is:

$$F = f \times F_k \tag{9}$$

Combining Eq. (3), (8), and (9), the relationship between the destemming force F and the measured force on the loadcell F_{lc} is calculated as:

$$F = 2 \times f \times \frac{\sin\left(\pi/4 - \beta/2 - \gamma/2\right)}{\sin(\alpha)} \times F_{lc} \tag{10}$$

Finally, the above destemming force F and pressing force F_k must satisfy the following system of equations:

$$\begin{cases} F \geq F_{d(\min)} \\ F_k \leq P \times d_s \times w_b \end{cases} \tag{11}$$

Where: w_b is the width of the destemming belt.

3 Fabrication and Control Test

3.1 Force Control Individual Test

The device which is used to control the destemming force is fabricated and assembled into the system as in Fig. 4a. And the control flowchart is shown in Fig. 4b. The loadcell's force value is read in and compared with a pre-defined value F_0. Based on the comparison result, the motor is driven to rotate clockwise or counterclockwise to increase or decrease the belt tension. The value of F_0 is obtained from the force model built in Sect. 2. The process is repeated continuously to maintain the stability of destemming force.

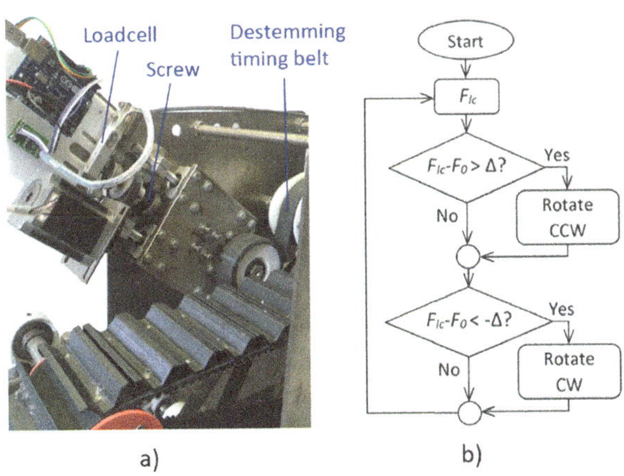

Fig. 4. Destemming force control flowchart.

The input of loadcell is updated continuously and there is a certain change in each updating time. This phenomenon drives the motor to change direction continuously, causing instability in the control process. Therefore, a stable value Δ is determined by recording the reading values of loadcell by time. The loadcell is calibrated first, then a sample weight of 150 (g) (14,709 (N)) is placed on it and the reading is recorded.

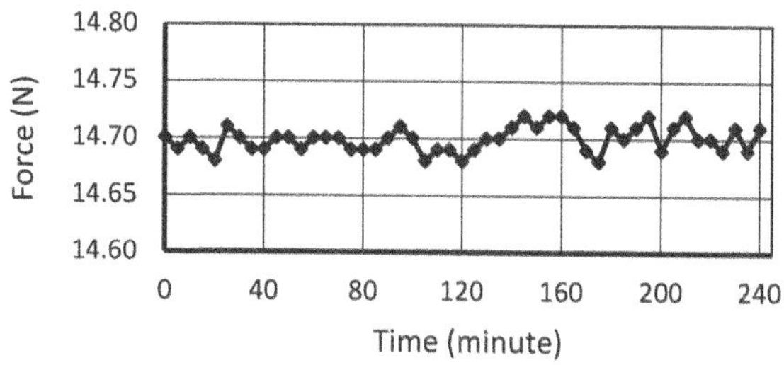

Fig. 5. The stabilization of loadcell.

The loadcell testing result is shown in Fig. 5. The evaluation of up to 240 min shows that the difference between the maximum and minimum reading values is 0.04 (N). Therefore, the stable value Δ is chosen as 0.04 (N). When the absolute difference between reading value F_{lc} and preset is less than Δ, the motor locks the screw. This is also considered as the tolerance value of the control process.

The microprocessor UNO R3 and module HX711 were used as the controller and weight scale signal converter. A short video describing the working of only the control device is presented at the link: https://youtu.be/0ZVCy2qS9uM. The working of this device in the system will be presented in Sect. 3.2.

3.2 Destemming Test

The force control device is installed on the destemming system as shown in Fig. 6. To avoid the effects of the feeding mechanism, the chillies are manually inputted to the groove so that the whole fruit is inside the conveyor belt. It is then moved to the position of the sorter which pushes the stem out of the conveyor. This is the ready state for destemming.

Then, the fruits with the stem out are moved into the destemming mechanism. Here, the fruit's body is kept on the conveyor by the following pressing roller, while the stem is gripped by the destemming timing belt and moved downward. The force sensor and controller are always activated to keep the destemming force within the appropriate value range.

The test was done in a group of 1,000 fruits. Their total length and maximum diameter are respectively in the range of 80 ± 6 (mm) and 8.5 ± 0.7 (mm). The running speed of the conveyor belt was set up to 60 (mm/s), this is equivalent to 1.5 fruits/second.

The average success rate was obtained up to 96.2%. the video of the brief test of 11 fruits is prepared and presented at the link: https://youtu.be/j1VbUdmiTGQ. To describe clearly the process when capturing the video, the experiment was run at a slow speed. The destemming force was also adjusted a little lower than the requirement to occur in three cases:

- Nine fruits have been successfully destemmed as in Fig. 6b–c. Their stems were all kept and pulled with an appropriate force;
- One fruit could not be destemmed as in Fig. 6d. Because the sorting mechanism did not work properly, so the stem was not pushed far enough from the conveyor belt to be gripped by the destemming belt;
- One unsuccessful case as described in Fig. 6e. This is because the gripping force on the stem was less than the required force $F_{d(min)}$.

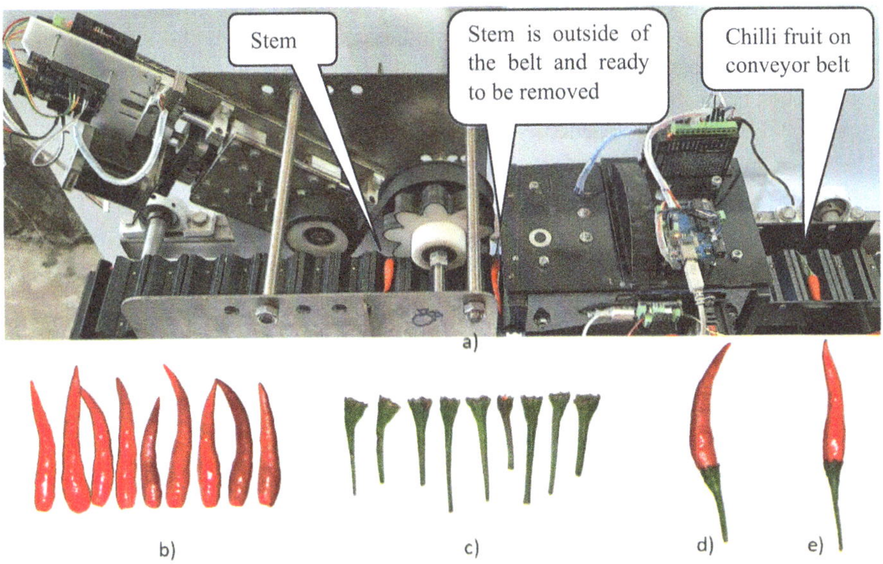

Fig. 6. The destemming test. a) On the system, b) Destemmed chilli, c) Removed stems, d, e) Non-destemmed fruits

4 Results and Discussions

The testing result achieved the average success rate of the system at 96.2%. The overall error rate of 3.8% is comprised of the arrangement and the destemming mechanism errors.

The destemming force was set up in the appropriate range during the test. There were only 2 in 1,000 fruits that the stem is not gripped tightly enough. And the stem broken has not been found in all cases. The remaining failure cases were mainly due to the error of the arrangement device which has not pushed the stem in the correct position for destemming.

Compared to the results in [6], no fruit's body broken errors appeared in this experiment. This is because the arrangement mechanism has been adjusted to better positioning ability and could avoid the over-far distance of the stem from the conveyor belt. The following pressing roller also worked properly to keep the fruit's body on the conveyor during the gripping process. In addition, the testing fruits have been chosen in a suitable range of sizes.

The running speed at 60 (mm/s) is relatively slow. It is not really meeting the industrial requirement. This is caused by the limit of the TCS3200 color sensors which was used in the arrangement device. So, the development in rising processing speed is an important further improvement. A better color sensor would be a solution to this problem.

This destemming mechanism works well when chilli is supplied individually one by one, so the fruit feeder is considered as a further research direction.

After destemming for a while, the liquid is extracted from the stem under the compression of the belt and roller. This affects the friction co-efficient, thus changes the gripping force. In this test, only 1,000 fruits were destemmed, so the change is not significant. However, this phenomenon should be evaluated in detail in coming studies.

5 Conclusion

A force model for fresh chilli destemming has been built in this study. The mathematical relationship is used in controlling the stem separation process by stabilizing the generated gripping force in an appropriate range. The experiment model was fabricated to test and achieved the success destemming rate at 96.2%. This result is useful in developing an automatic destemming system for fresh chili. The limitations on running speed and the change of the coefficient of friction are mentioned to suggest further research directions.

Nomenclature

Symbol	Explanation	Unit
A, B, C, D, I, J	Special points	
d_s	Diameter of fruit's stem	mm
f	Friction co-efficient	
F	Destemming force	N
$F_{d(min)}$	Minimum destemming force	N
F_k	The pressing force	N
F_{lc}	Measured force on loadcell	N
F_m	Measuring force for destroying pressure measurement	N
F_p	Pulling force in destroying pressure measurement	N
F_0	Preset force value	N
H	Slack of the destemming timing belt	mm
L_1, L_2, L_3	Distances of the roller centers	mm
O_1, O_2, O_3	Centers of the rollers	

(*continued*)

(*continued*)

Symbol	Explanation	Unit
P	Destroyed pressure of the stem	N/mm^2
R_1, R_2	Radius of rollers	
T	Tension force in the destemming timing belt	N
w	Width of pressing plate	mm
w_b	Width of destemming timing belt	mm
α	Destemming belt slack angle	rad
β	Beginning contacting angle	rad
Δ	Stable value	N
θ	The angle between the end contacting point and the driving roller	rad
γ	The ending contacting angle	rad
φ	The inclined angle of the destemming driving roller	rad

Acknowledgment. Huynh, Q.K. was funded by Vingroup Joint Stock Company and supported by the Domestic Master/ Ph.D. Scholarship Programme of Vingroup Innovation Foundation (VINIF), Vingroup Big Data Institute (VINBIGDATA), code VINIF.2019.TS.33 and VINIF.2020.TS.109. The authors also would like to thank Thanh-Binh Bui, Thanh-Hieu Do, Van-Luan Khuu, Dr. Chanh-Nghiem Nguyen and Dr. Quoc-Bao Truong for their enthusiastic participant in this research.

References

1. Vo, T.T.L.: Assessment of agri-product value chains in the mekong delta: problems and solutions. Can Tho University Journal of Science **2**(2016), 100–111 (2016). https://doi.org/10.22144/ctu.jen.2016.001
2. Idrees, S., Hanif, M.A., Ayub, M.A., Hanif, A., Ansari, T.M.: Medicinal plants of south Asia, Chapter 9 – Chili Pepper, pp. 113–124. Elsevier Publisher (2020). https://doi.org/10.1016/B978-0-08-102659-5.00009-4
3. Zhang, D., et al.: A comparative overview on chili pepper (capsicum genus) and sichuan pepper (zanthoxylum genus): From pungent spices to pharma-foods. Trends Food Sci. Technol. (2021, in press). https://doi.org/10.1016/j.tifs.2021.03.004
4. Nguyen, T.T.A., Vo, T.T.L.: Financial efficiency of chili farmers in the Mekong Delta. Can Tho University Journal of Science **48D**(2017), 87–95 (2017). https://doi.org/10.22144/ctu.jvn.2017.633
5. Nagendra, B.K.: System and method of de-stemming produce. The United States Patent and Trademark Office, patent number: US9173431B2 (2014)
6. Ryan, P.H., Dale, E.C., Eduardo, M.G., Anthony, M.H.: Designing a high volume Chile De-stemming machine. In: The ASABE Meeting Presentation, Pennsylvania, USA, 20–23 June, Michigan, USA (2010). https://doi.org/10.13031/2013.30000
7. Ryan, P.H., Dale, E.C., Eduardo, M.G., Anthony, M.H.: Engineering a machine to remove stems from chile peppers – A critical need for the new Mexico chile industry. In: The ASABE Meeting Presentation, Nevada, USA, 21–24 June, Michigan, USA (2009)

8. Huynh, Q.K., Nguyen, V.C., Le, D.K.L., Le, P.H.: A study on chili fruit processing process and new design of chili destemming system. Can Tho University Journal of Science **2A** (2019), 9–16 (2019). https://doi.org/10.22144/ctu.jvn.2019.032
9. Huynh, Q.K., Vo-Nguyen, H.P., Nguyen, V.C.: Mathematical model for identifying destemming force of fresh chilli (Capsicum) fruit. Vietnam Mech Eng J **9**(2020), 24–30 (2020)

A New Approach for Milling Productivity Improvement

Nguyen Hong Linh[1], Trinh Kieu Tuan[2], Nguyen Huu Quang[2], Pham Duc Lam[3], Le Hoang Anh[4], and Tran Thi Phuong Thao[5](✉)

[1] Electric Power University, Hanoi, Vietnam
[2] University of Economics - Technology for Industries, Hanoi, Vietnam
[3] Nguyen Tat Thanh University, Ho Chi Minh City, Vietnam
[4] Vinh Long University of Technology Education, Vinh Long, Vietnam
[5] Thai Nguyen University of Technology, Thai Nguyen, Vietnam
tranphuongthao@tnut.edu.vn

Abstract. In the process of machining mechanical products, high productivity machining and small roughness of the product surface are desirable in all cases. In this study, a new approach to improve milling productivity will be presented on the basis of ensuring the minimum required surface roughness. The experiments have been conducted for 40CrMn steel using TiAlN coated cutting tools. The experimental matrix of 15 experiments has been designed using Box-Behnken method. A regression model showing the relationship between surface roughness and cutting velocity, feed rate and cutting depth has been set up. Sub-regression models representing the relationship between surface roughness and each input parameter have also been created. From these models, the effect of cutting parameters on surface roughness has been determined. The main objective of this study is to determine the value of cutting parameters to improve the cutting productivity while ensuring the minimum surface roughness value.

Keywords: Milling · 40CrMn steel · Surface roughness · Productivity improvement

1 Introduction

Milling is known to be the most productive of all cutting methods. In particular, when milling a plane with a face milling cutter, the productivity is further enhanced because in this case, it is possible to use milling cutter with a large diameter and many blades simultaneously participating in cutting [1]. However, the roughness of the machined surface when milling is another matter. The surface roughness when milling is not comparable with that when using grinding method, or turning method. Therefore, it is necessary to take advantages of productivity of the milling method while ensuring the surface roughness. The surface roughness is mentioned as a parameter to evaluate the machining process because this parameter directly affects the workability and longevity of the product [2, 3]. There have been many published studies on determining the value

D. C. Nguyen et al. (Eds.): ICERA 2021, LNNS 366, pp. 69–78, 2022.
https://doi.org/10.1007/978-3-030-92574-1_7

of parameters of the machining process to simultaneously ensure the minimum surface roughness and the highest machining productivity. These studies have been known as multi-objective optimization of the milling process. Vikor and Waspas methods were employed to determine the value of cutting velocity, feed rate and cutting depth to warrant the minimum surface roughness and the maximum Material Removal Rate (MRR) when milling En-31 steel [4]; Nondominated Sorting Genetic Algorithm II (NGSAII) has been applied to determine the value of cutting velocity, feed rate and cutting depth to confirm the minimum surface roughness and the maximum MRR when milling AF1410 steel [5]; Particle swarm optimization (PSO) algorithm has been applied to determine the value of cutting velocity, feed rate, cutting depth and hardness of workpiece to certify the minimum surface roughness, cutting power and cutting temperature and the maximum MRR when milling AISI H13 steel [6]; Evolutionary algorithm based on decomposition has been used to determine the value of cutting velocity, feed rate and cutting depth to simultaneously ensure the minimum cutting force and surface roughness and the maximum MRR when milling High-strength die steel [7]; Taguchi method and Gray Relational Analysis method (GRA) has been combined to determine the optimal value of the optimal value of cutting velocity, feed rate and cutting depth to confirm the minimum surface roughness and the maximum MRR when milling Al6061/ 3% RHA/ 6% TiC reinforced hybrid composite [8]; Gray fuzzy algorithm and Backpropagation Neural Network-based Genetic Algorithm (BPNN-GA) method have been applied to determine the optimal value of cutting velocity, feed rate, cutting depth and coolant flow to guarantee the minimum surface roughness and the maximum MRR when milling ASSAB XW-42 tool steel [9], etc.

In short, the research for determining the value of parameters of cutting velocity, feed rate and cutting depth to simultaneously ensure the two criteria of the minimum surface roughness and the maximum cutting productivity has been carried out by many studies as mentioned above, and different algorithms have been exploited to find the optimal value of cutting parameters. However, it will be demanding when deployed in production units. The possible difficulties may include the mathematical complexity of algorithms as well as the large time required for running those algorithms on computers. Those reasons may affect the production progress of enterprises, affecting the quality and machining productivity. In this study, a new simple approach will be taken to find the optimal set of cutting parameters to improve the machining productivity while ensuring the small surface roughness value. Through this new approach, the effect of cutting parameters on surface roughness has also been explored.

2 Milling Experiment

HAAS 3-axis CNC milling machine has been used to conduct the experiments (Fig. 1). HAAS operating system has been installed in the machine to fulfill the requirements of the operator. The maximum spindle speed of milling machine is 50,000 rev/min, and the accuracy of movements reaches 0.12 μm.

Fig. 1. HAAS 3-axis CNC milling machine

40CrMns steel samples, a type of medium carbon alloy steel, widely used to manufacture common parts such as shafts, gears, forks, etc., have been used during the experiment. [10]. From the workpiece with 64 mm diameter cylindrical handle, the milling is carried out to create hexagonal billets with the dimension of 45 mm as shown in Fig. 2.

Fig. 2. Experimental sample

The used cutting tool is TiAlN coated chip with a tool nose radius of 0.5 mm, a back edge length of 0.9 mm, a main cutting angle of 900 and an edge with of 7.2 mm. This type of cutting tool has high thermal strength, good cutting performance, high wear resistance and it is well suited for machining alloy steels [11, 12]. The tool body with a diameter of 60 mm has been used to fit 2 symmetrical chips for each experiment.

Each chip is used only once to eliminate the effect of tool wear on surface roughness. The surface roughness has been measured with SJ-201 machine with the set standard length of 2.5 mm.

Cutting velocity, feed rate and cutting depth are selected as parameters for adjustment in each experiment since they can be simply adjusted by the machine operator [13, 14]. A number of studies mentioned in the introduction have also selected these parameters as input parameters for the experimental process. The Box-Behnken design has been used to design the experimental matrix. This type of design is commonly used in the design of experimental matrices in the field of engineering in general and in the mechanical machining in particular [15]. In this design, each variable needs three values. The input variable values at the coding levels have been chosen in accordance with the recommendations of cutting tool manufacturer as shown in Table 1 [16]. Statistical software Minitab 16 has been used to create the experimental matrix of 15 experiments as shown in Table 2. This is an orthogonal matrix, that is, if variables are considered in encrypted form, the sum of products of the rows in any two columns is always zero. The design of orthogonal matrix is to ensure the equal role of each parameter in the matrix. In other words, the number of occurrences of each parameter is the same [15].

Table 1. Value of input parameters at levels

Parameters	Symbol	Unit	Code	Value at levels		
				−1	0	1
Cutting speed	v_c	m/min	x_1	60	100	140
Feed rate	f_z	mm/tooth	x_2	0.12	0.20	0.28
Cutting depth	a_p	mm	x_3	0.20	0.40	0.60

Table 2. Experimental matrix and results

No.	Code			Actual			*Ra (μm)*
	x_1	x_2	x_3	v_c *(m/min)*	f_z *(mm/tooth)*	a_p *(mm)*	
1	1	0	1	140	0.20	0.60	0.590
2	0	0	0	100	0.20	0.40	0.976
3	1	0	−1	140	0.20	0.20	0.569
4	−1	1	0	60	0.28	0.40	1.991
5	−1	0	1	60	0.20	0.60	1.416
6	0	−1	1	100	0.12	0.60	0.521
7	−1	−1	0	60	0.12	0.40	0.755
8	0	0	0	100	0.20	0.40	0.976
9	0	1	1	100	0.28	0.60	1.485
10	0	−1	−1	100	0.12	0.20	0.497
11	1	1	0	140	0.28	0.40	0.895
12	1	−1	0	140	0.12	0.40	0.264
13	0	1	−1	100	0.28	0.20	1.401
14	−1	0	−1	60	0.20	0.20	1.329
15	0	0	0	100	0.20	0.40	0.976

3 Results and Discussion

The order of experiments in Table 2 has been conducted, also, the surface roughness at the experiments has been included in this table. Statistical software Minitab 16 has been used again to analyze the experimental results. A surface roughness model has been set up as shown in the formula (1). This model has a coefficient of determination R^2 of 0.9624 and an adjusted coefficient of determination R^2(adj) of 0.9886. Since both coefficients R^2 and R^2(adj) are very close to 1, it can be confirmed that the model (1) has a very large compatibility (fitness) compared to the experimental data [15].

$$Y = R_a = 0,97595 - 0,39665x_1 + 0,46695x_2 + 0,0272x_3 - 4,6.10^{-17}x_1^2 \\ -2,5.10^{-17}x_2^2 - 2,9.10^{-17}x_3^2 - 0.15105x_1x_2 - 0.0163x_1x_3 + 0.0152x_2x_3 \quad (1)$$

In order to investigate the effect of cutting parameters on surface roughness, various methods have been employed in many published studies such as Pareto chart, Anova analysis method, the chart for reviewing effects of main factors on surface roughness, etc. However, when using those methods, the analysis of experimental results is highly dependent on the understanding of those tools (statistical software) of analyst. In this study, a new method, the survey of individual effect of each input parameter on surface roughness has been applied. It has been proved to be successful in the research of surface grinding and external cylindrical grinding, and has been conducted as follows [17, 18]:

On the basis of Eq. (1), when surveying the effect of one parameter on surface roughness, the other two parameters are fixed.

If two of the three parameters are fixed at −1 and the remaining parameter is changed, the regression equation for the effect of each parameter on surface roughness is presented in the formulas from (2) to (4). The chart showing the effect of each parameter on surface roughness is shown in Fig. 3.

$$Y_{1(-1)} = -4,6.10^{-17}x_1^2 - 0,2293x_1 + 0,4938 \quad (2)$$

$$Y_{2(-1)} = -2,5.10^{-17}x_2^2 + 0,6028x_2 + 1,3291 \quad (3)$$

$$Y_{3(-1)} = -2,9.10^{-17}x_3^2 - 0.0283x_3 + 0.7346 \quad (4)$$

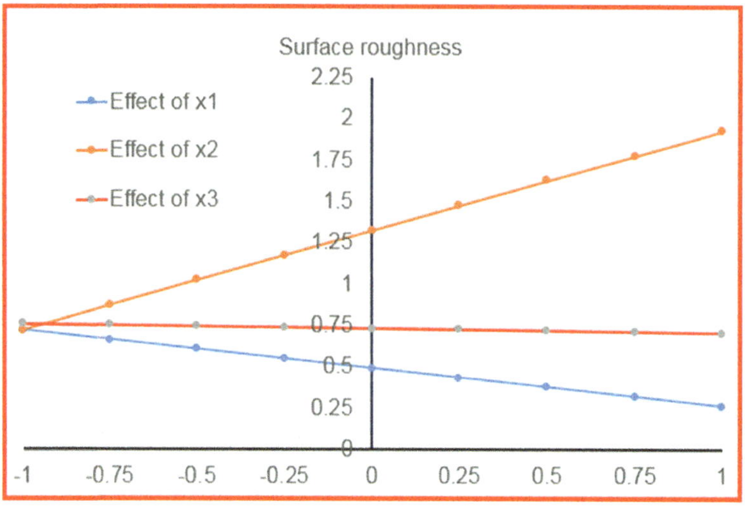

Fig. 3. The curve of the sub-regression model when 2 out of 3 parameters at the coding level −1

If two of the three parameters are fixed at 0 and the remaining parameter is changed, the regression equation for the effect of each parameter on surface roughness is presented in the formulas from (5) to (7). The chart showing the effect of each parameter on surface roughness is shown in Fig. 4.

$$Y_{1(0)} = -4,6.10^{-17}x_1^2 - 0,39665x_1 + 0,97595 \tag{5}$$

$$Y_{2(0)} = -2,5.10^{-17}x_2^2 + 0,46695x_2 + 0,97595 \tag{6}$$

$$Y_{3(0)} = -2,9.10^{-17}x_3^2 + 0.0272x_3 + 0.97595 \tag{7}$$

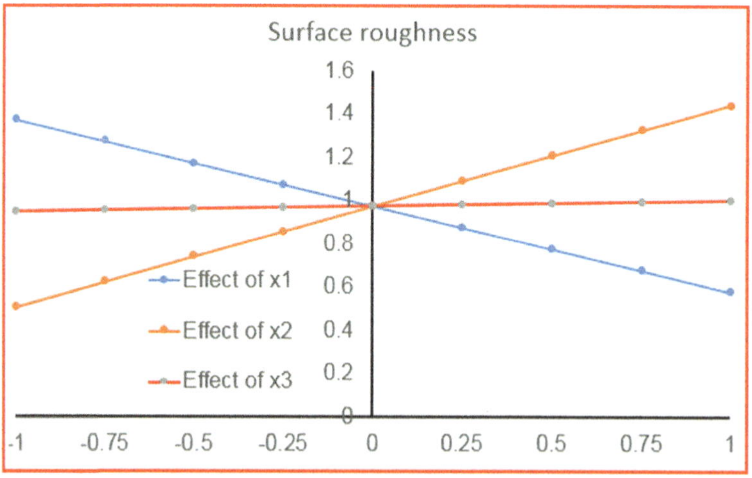

Fig. 4. The curve of the sub-regression model when 2 out of 3 parameters at the coding level 0

If two of the three parameters are fixed at 1 and the remaining parameter is changed, the regression equation for the effect of each parameter on surface roughness is presented in the formulas from (8) to (10). The chart showing the effect of each parameter on surface roughness is shown in Fig. 5.

$$Y_{1(1)} = -4,6.10^{-17}x_1^2 - 0,564x_1 + 1,4853 \tag{8}$$

$$Y_{2(1)} = -2,5.10^{-17}x_2^2 + 0,3311x_2 + 0,5902 \tag{9}$$

$$Y_{3(1)} = -2,9.10^{-17}x_3^2 + 0.0261x_3 + 0.8952 \tag{10}$$

Observing Figs. 3, 4 and 5, it is shown that:

- When the two parameters are fixed at −1 (Fig. 3) and 1 (Fig. 5), the curve of cutting velocity (x1) has the greatest slope. It is shown that in this case, the cutting velocity has the greatest effect on surface roughness.
- When the two parameters are fixed at 0 (Fig. 4), the curve of cutting velocity (x1) and the curve of feed rate (x2) have the same slope. Thus, in this case, cutting velocity and feed rate have the same effect on surface roughness.
- In all three cases, the surface roughness will increase when increasing the feed rate. Conversely, when increasing the cutting velocity, the surface roughness will decrease in all three cases.

A particular finding revealed in all three cases is that the curve of cutting depth (x3) is almost the same as the horizontal axis of the graph, that is, when the cutting depth changes from encoding level −1 to encoding level +1 (corresponding to the change in true value from 0.2 mm to 0.6 mm), the surface roughness is almost unchanged. Thus, it can be concluded that the cutting depth has slight effect on surface roughness.

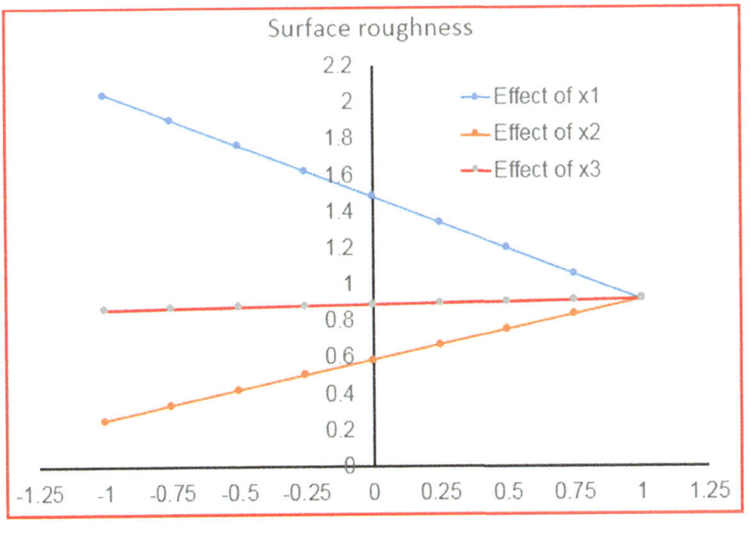

Fig. 5. The curve of the sub-regression model when 2 out of 3 parameters at the coding level 1

4 Determination of Cutting Parameters for Improvement of Productivity

As analyzed above, because the cutting depth has the very little effect on the surface roughness, it is absolutely possible to increase the value of this parameter to improve productivity. However, it is also necessary to carry out the experiments to verify the results. Moreover, from the above analyses, when the cutting velocity value is large and the feed rate value is small, the surface roughness value will be slight. Therefore, the maximum value of cutting velocity and the minimum value of feed rate will be chosen to conduct the experiment. Conducting the milling experiment of 5 steel samples when changing the value of cutting depth, the results are shown in Table 3.

Table 3. Milling results

No.	Code value			Actual value			*Ra (μm)*
	x_1	x_2	x_3	v_c *(m/min)*	f_z *(mm/tooth)*	a_p *(mm)*	
1	1	−1	−1	140	0.2	0.2	0.264
2			−0.5			0.3	0.266
3			0			0.4	0.267
4			0.5			0.5	0.269
5			1			0.6	0.273

The results in Table 3 have shown that when increasing the cutting depth by 3 times (from 0.2 mm to 0.6 mm), the surface roughness change is extremely little, only about 0.09 μm. Thus, in this case, the cutting productivity has increased by three times, while the surface roughness has been almost unchanged. If the milling process is evaluated through surface roughness and cutting productivity, the optimal values of cutting parameters are found to be cutting velocity of 140 m/min, feed rate of 0.2 mm/tooth and cutting depth of 0.6 mm.

5 Conclusion

This study has conducted the experiment of milling 40CrMn steel with TiAlN coated chip on 3-axis CNC milling machine. The method of building a regression function representing the effect of each input parameter on surface roughness has been realized. The determination of the value of cutting parameters to improve cutting productivity while ensuring surface roughness has also been performed. A number of conclusions are drawn as follows:

- The cutting velocity and the feed rate have a great effect on surface roughness. When increasing the cutting velocity, the surface roughness decreases. On the contrary, when increasing the feed rate, the surface roughness grows. Meanwhile, the cutting depth has almost no effect on the surface roughness.

- The cutting productivity can be increased up to 3 times by increasing the cutting depth by 3 times, but the surface roughness is almost unchanged.
- For the minimum value of the surface roughness and the maximum value of cutting power, the values of cutting velocity, feed rate and cutting depth can be selected as 140 m/min, 0.2 mm/tooth and 0.6 mm, respectively.
- A new method to select cutting parameters to improve the productivity on the basis of surface roughness assurance has been successfully realized in this study.

Acknowledgment. This work was supported by Thai Nguyen University of Technology.

References

1. Dich, T.V., Binh, N.T., Dat, N.T., Tiep, N.V., Viet, T.X.: Manufacturing technology. Science and Technics Publishing House, Ha Noi (2003)
2. Shagwira, H., Mbuya, T.O., Mwema, F.M., Herzog, M., Akinlabi, E.T.: Taguchi optimization of surface roughness and material removal rate in CNC milling of polypropylene + 5wt.% quarry dust composites. In: International Conference on Engineering for Sustainable World, vol. 1107, no. 012040, pp. 1–10 (2021). https://doi.org/10.1088/1757-899X/1107/1/012040
3. Kumar, G.V.A., Reddy, D.V.V., Nagaraju, N.: Multi-objective optimization of end milling process parameters in machining of EN 31 Steel: application of AHP embedded with VIKOR and WASPAS Methods. i-Manager's J. Mech. Eng. **8**(4), 39–46 (2018). https://doi.org/10.26634/jme.8.4.14676
4. Venkata Ajay Kumar, G., Reddy, D.V., Naguraju, N.: Multi-objective optimization of end milling process parameters in machining of EN 31 steel: application of AHP embedded with VIKOR and WASPAS methods. i-Manager's J. Mech. Eng. **8**(4), 39–46 (2018)
5. Jin, X., Yan, F., Li, Y., Yang, Z., Li, L.: Multiobjective optimization of milling parameters for ultrahigh-strength steel AF1410 based on the NSGA-II method. Adv Mater Sci Eng **2020**, 1–11 (2020). https://doi.org/10.1155/2020/8796738
6. Vu, N.C., Dang, X.P., Huang, S.C.: Multi-objective optimization of hard milling process of AISI H13 in terms of productivity, quality, and cutting energy under nanofluid minimum quantity lubrication condition. Meas. Control, 1–15 (2020). https://doi.org/10.1177/0020294020919457
7. Tao, F., Weijun, L., Jibin, Z.: Optimization of cutting parameters using multiobjective evolutionary algorithm based on decomposition. J Vibroenginerring **15**(2), 833–844 (2013)
8. Bhasha, A.C., Balamurugan, K.: Multi-objective optimization of high-speed end milling on Al6061/ 3% RHA/ 6% TiC reinforced hybrid composite using Taguchi coupled GRA, In: IEEE International Conference on Computational Intelligence for Smart Power System and Sustainable Energy, pp. 1–6 (2020)
9. Bambang Pramujati, M.: Abdul Wahid, Multi-objective optimization in face milling process with cryogenic cooling using grey fuzzy analysis and BPNN-GA methods. Eng Comput **36**(5), 1542–1565 (2019). https://doi.org/10.1108/EC-06-2018-0251
10. https://equimetals.com/catalogue/chn/gb3077-88/40crmn
11. Klocke, F., Krieg, T.: Coated tools for metal cutting – features and applications. CIRP Ann **48**, 515–525 (1999)

12. Uyen, V.T.N., Son, N.H.: Improving accuracy of surface roughness model while turning 9XC steel using a Titanium Nitride-coated cutting tool with Johnson and Box-Cox transformation. AIMS Mater Sci **8**(1), 1–17 (2021). https://doi.org/10.3934/matersci.2021001
13. Trung, D.D.: Multi-objective optimization of SKD11 steel milling process by reference ideal method. Int J Geol **15**, 1–16 (2021). https://doi.org/10.46300/9105.2021.15.1
14. Trung, D.D.: Influence of cutting parameters on surface roughness during milling AISI 1045 steel. Tribol Ind **42**(4), 658–665 (2020). https://doi.org/10.24874/ti.969.09.20.11
15. Dean, A., Voss, D., Draguljić, D.: Design and Analysis of Experiments, 2nd edn. Springer, New York. https://doi.org/10.1007/b97673 (2007)
16. https://www.hannibalcarbide.com/carbide-tipped-tools/cutters/
17. Trung, D.D., Son, N.H., Nguyen, N.-T.: A Study on productivity improvement for the SKD11 steel grinding by using CBN grinding wheel – a new approach. Univers J Mech Eng **8**(1), 29–33 (2020). https://doi.org/10.13189/ujme.2020.080104
18. Trung, D.D., Son, N.H., Nguyen, N.-T.: Research on improving machining productivity when external cylindrical grinding of 160Cr12Mo steel. Univers J Mech Eng **8**(1), 34–40 (2020). https://doi.org/10.13189/ujme.2020.080105

A Numerical Model for the Composite Sandwich Panel in Vibration by the Homogenization Method

Duong Pham Tuong Minh, Ngo Nhu Khoa, Sy Nguyen-Van, Nguyen Thi Hoa, Ngoc Nguyen-Dinh, Nguyen Quang Hung, Hoang Tien Dat, and Luong Viet Dung(✉)

Thai Nguyen University of Technology, Thai Nguyen, Vietnam
luongvietdung@tnut.edu.vn

Abstract. Sandwich is a material widely used in industries, such as automotive, navy, construction, aviation. It has a high stiffness-to-weight ratio, so it meets the requirements of industrial part weight reduction. In this study, we built a numerical model to analyze the behavior of composite sandwich panels subjected to vibration. However, the sandwich panel was in the form of a shell, so it took a lot of time to perform simulations and build a CAD model. To solve this problem, we proposed a numerical model by homogenization method for this composite sandwich panel. This model was implemented in the finite element software Abaqus and confirmed by comparing the finite element simulation results of the vibration test of the 3D-model and homogenization model.

Keywords: Homogenization · Finite element · Sandwich

1 Introduction

Nowadays, in industry, reducing weight and optimizing the size of parts is an urgent requirement that needs to solve. The composite sandwich panel is a material that has a high stiffness to weight ratio, therefore, it has been used widely and increasingly in industry. With the structure as shown in Fig. 1, the optimal design of a machine part made of the composite sandwich will face many difficulties due to its complex structure. Up to now, many researchers have used different research methods to analyze the mechanical properties of sandwich panels to overcome these difficulties [1–14]. However, there are very few research results that display the difference in CPU computation time between the 3D model and the homogeneous model (or equivalent model) in the study of vibration problems. In the current product design process, there is effective support of computers in creating CAD models and finite element analysis. Therefore, creating a simple CAD model and reducing calculation time are very important. For example, a homogenization model based on the integral method has

D. C. Nguyen et al. (Eds.): ICERA 2021, LNNS 366, pp. 79–88, 2022.
https://doi.org/10.1007/978-3-030-92574-1_8

been developed for corrugated board boxes and the results have shown that the computation time is greatly reduced when using the homogeneous model [15, 16]. In this study, we propose a homogenization model for the honeycomb core to build an alternative homogeneous model for the 3D model of the sandwich panel. Accordingly, the calculation of the mechanical properties for the 3D shell structure of the honeycomb core is carried out in the plane and it is then replaced by an equivalent homogeneous 3D solid core. Vibration test simulation was performed for homogeneous core plate and 3D shell core plate to confirm the accuracy of the proposed homogenous model.

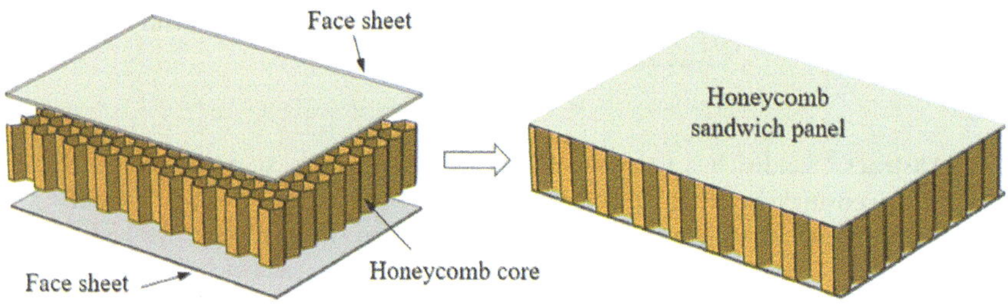

Fig. 1. Honeycomb core sandwich plate model

2 Homogenization Method for the Honeycomb Core of the Sandwich Panel

This method of homogenization is performed based on using a homogeneous fictitious material with equivalent macroscopic properties in place of the heterogeneous real material. We consider a Representative Volume Element (RVE) as presented in Fig. 2. Since the sandwich skin is very stiff compared to the honeycomb core, the deformation of the honeycomb wall will be determined by the skin layers [14], and the effect of tensile or compressive forces for the thin wall will be dominated by their bending effect. The calculation procedure of this homogenization method is based on classical homogeneity theory and recent research results [3, 5, 6, 8, 17].

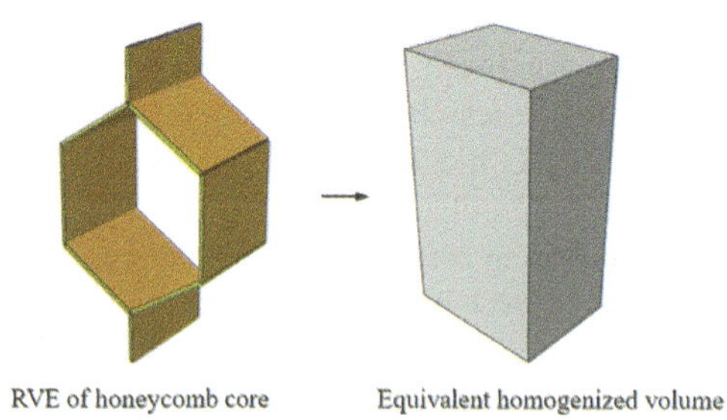

Fig. 2. Honeycomb core RVE

2.1 Determine Young's Modulus E_x

From the REV model, we construct the internal force equation on the five walls EA, AC, CB, CD, and DF. Performing the displacement at the center of the core from h/2 to the end position for points A, C, and D, we have an equivalent structure as presented in Fig. 3. At this point A and B are fixed on the skin; The problem to be determined is what force must be applied at D to get the displacement $u_0 = 1$. We have:

$$\varepsilon_x = \frac{u_0}{2l\cos\theta} \quad ; \quad \varepsilon_y = \frac{v_0}{h + l\sin\theta} = -\nu\varepsilon_x = -\nu\frac{u_0}{2l\cos\theta} \Rightarrow v_0 = -\nu u_0\frac{h + l\sin\theta}{2l\cos\theta} \tag{1}$$

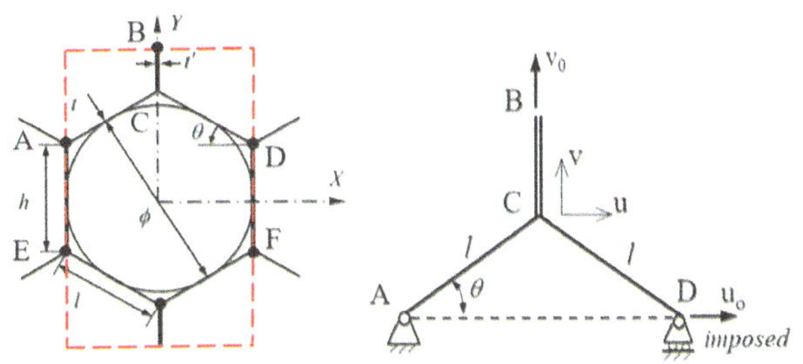

Fig. 3. Determining elastic modulus E_x for a REV of honeycomb core

$$\varepsilon_l = \frac{1}{l}(u\cos\theta + v\sin\theta) \tag{3}$$

$$N_l = \sigma_l.b.t = \frac{Ebt}{l}(u\cos\theta + v\sin\theta) \tag{4}$$

$$N_h = E\frac{v_0 - v}{h}.2.bt \tag{5}$$

Considering the equilibrium of the node at C, we have:

$$N_h = 2N_l\sin\theta \tag{6}$$

From (4), (5), and (6), we have:

$$\frac{2Ebt}{h}(v_0 - v) = 2\sin\theta\frac{Ebt}{l}(u\cos\theta + v\sin\theta) \tag{7}$$

$$\Rightarrow l(v_0 - v) = h\sin\theta(u\cos\theta + v\sin\theta) \tag{8}$$

Another way:

$$\mathrm{u} = \frac{1}{2}\mathrm{u}_0 \quad ; \quad \mathrm{v}_0 = -\nu \mathrm{u}_0 \frac{h + l\sin\theta}{2l\cos\theta} \tag{9}$$

$$\Rightarrow \mathrm{v} = \frac{\mathrm{v}_0 - \frac{h}{4l}\sin^2\theta.\mathrm{u}_0}{1 + \frac{h}{l}\sin^2\theta} \tag{10}$$

We have:

$$P_x = N_l\cos\theta = \frac{Ebt}{l}(\mathrm{u}\cos\theta + \mathrm{v}\sin\theta)\cos\theta \tag{11}$$

$$E_x^* = \frac{\sigma_x^*}{\varepsilon_x^*} = \frac{P_x}{(h + l\sin\theta)b}.\frac{2l\cos\theta}{\mathrm{u}_0} \tag{12}$$

From Eq. (11) and (12), we have:

$$E_x^* = \frac{2E.b.t(\mathrm{u}\cos\theta + \mathrm{v}\sin\theta).\cos^2\theta}{(h + l\sin\theta)b.u} \tag{13}$$

2.2 Determine Young's Modulus $\mathbf{E_y}$

For the case traction, a displacement along the y-direction $\mathrm{v}_0 = 1$ is imposed:

$$\varepsilon_y = \frac{\mathrm{v}_0}{h + lsin\theta} \quad ; \quad \varepsilon_x = \frac{\mathrm{u}_0}{2lcos\theta} = -\nu\varepsilon_y = -\nu\frac{\mathrm{v}_0}{h + lsin\theta} \tag{14}$$

$$\Rightarrow \mathrm{u}_0 = -\nu.\mathrm{v}_0\frac{2l\cos\theta}{h + l\sin\theta} \tag{15}$$

$$\varepsilon_l = \frac{1}{l}(\mathrm{u}\cos\theta + \mathrm{v}\sin\theta) \tag{16}$$

From Eq. (7), we have:

$$\mathrm{v} = \frac{\mathrm{v}_0 - \frac{\mathrm{h}}{2\mathrm{l}}\sin\theta\cos\theta.\mathrm{u}_0}{1 + \frac{\mathrm{h}}{\mathrm{l}}\sin^2\theta} \quad ; \quad \mathrm{u} = \frac{1}{2}\mathrm{u}_0 \tag{17}$$

On the other hand, we have:

$$P_y = N_h = \frac{2Ebt}{h}(\mathrm{v}_0 - \mathrm{v}) \tag{18}$$

$$E_y^* = \frac{\sigma_y^*}{\varepsilon_y^*} = \frac{P_y}{(2l\cos\theta)b}.\frac{h + lsin\theta}{\mathrm{v}_0} \tag{19}$$

From Eq. (18) and (19), we have:

$$E_y^* = \frac{\sigma_y^*}{\varepsilon_y^*} = \frac{E.2tb(\mathrm{v}_0 - \mathrm{v})}{(h.2l\cos\theta)b} \cdot \frac{h + lsin\theta}{\mathrm{v}_0} \tag{20}$$

For sandwich plates with very small core heights, we use the Poisson's ratio like the Poisson's ratio of two skins. However, if the plate has a large core height, we can use Gibson's formula to calculate the Poisson's ratio [18]:

$$v_{yx} = \frac{\left(\frac{h}{l} + sin\theta\right) sin\theta}{\cos^2\theta} \tag{21}$$

$$v_{xy} = \frac{\cos^2\theta}{\left(\frac{h}{l} + sin\theta\right) sin\theta} \tag{22}$$

2.3 Determine Young's Modulus E_z

The elastic modulus E_z is calculated by multiplying the modulus E_s of the honeycomb core by the ratio [13]:

$$E_z^* = E_t \frac{4ht + 4ht}{(2h + 2l\sin\theta)2l\cos\theta} = E_t\left(\frac{t}{l}\right)\frac{h + l}{(h + l\sin\theta)\cos\theta} \tag{23}$$

Poisson's coefficients ν_{zx} and ν_{zy} are assumed to be equal to the Poisson's ratio ν_{12} of the paper layer forming the honeycomb core, that is:

$$v_{zx} = v_{zy} = v_{12-core} \tag{24}$$

The reciprocal relationship allows to determine the remaining 2 Poisson's ratios:

$$v_{xz} = \frac{E_x}{E_z} v_{zx} \tag{25}$$

$$v_{yz} = \frac{E_y}{E_z} v_{zy} \tag{26}$$

2.4 Determine Shear Modulus in the Plane G_{xy}

For in-plane shear, we impose a shear angle γ, so tensile and compression deformation will occur in two inclined walls (see Fig. 4).

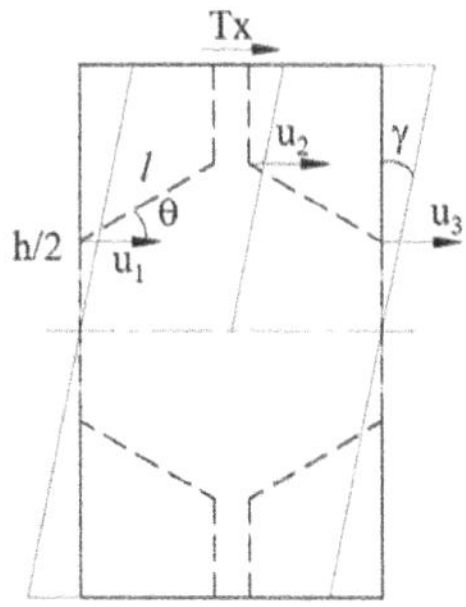

Fig. 4. Calculation model of shear modulus G_{xy} for an RVE

We have:

$$\gamma = \frac{\mathrm{u1}}{\frac{h}{2}} = \frac{\mathrm{u2}}{\frac{h}{2} + l\sin\theta} \tag{27}$$

$$\Rightarrow \varepsilon_s = \frac{(\mathrm{u2{-}u1})cos\theta}{l} = \frac{l\sin\theta.\gamma.cos\theta}{l} \tag{28}$$

On the other hand, we have:

$$\sigma_s = E_s.\varepsilon_s = E_s.\gamma.\sin\theta cos\theta \quad and \quad T_x = \sigma_s.2.t.b.cos\theta \tag{29}$$

For a homogeneous solid core, we have:

$$\tau^*_{xy} = G^*_{xy}\gamma \Rightarrow G^*_{xy} = \frac{\tau^*_{xy}}{\gamma} = \frac{T_x}{2l\cos\theta.b.\gamma} \tag{30}$$

Replace (29)with (30) we have:

$$G^*_{xy} = E_s\left(\frac{t}{l}\right)\sin\theta\cos\theta \tag{31}$$

3 Numerical Validation of the Homogenization Model

To validate the proposed homogenization model, we performed the tests by finite element simulation with the 3D Shell structure and the homogeneous 3D Solid. The two plates used in the simulation had the same length (L = 200 mm) and width (B = 70 mm). The 3D-shell plate using 97918 elements and 98318 nodes of reduced-integration four-node shell elements (S4R). The 3D-solid plate using 1408 elements and 1530 nodes of reduced-integration four-node shell elements (S4R) for the skins, 2816 solid elements C3D8R with 3825 nodes for the solid core. For the shell and core of 3D shell sandwich panels, the mechanical properties are given in Table 1 and

Table 1. Parameters of the layers forming the skins of honeycomb core sandwich plate

E_x (MPa)	E_y (MPa)	ν_{xy}	G_{xy} (MPa)	G_{xz} (MPa)	G_{yz} (MPa)	Thickness (mm)
18000	2000	0.4	8500	10	10	0.2

Table 2. Parameters of the paper forming the honeycomb core of the sandwich plate

E_x (MPa)	E_y (MPa)	ν_{xy}	G_{xy} (MPa)	G_{xz} (MPa)	G_{yz} (MPa)	Thickness (mm)
3292	1594	0.42	788	10	10	0.19

Table 3. Parameters of the paper forming the honeycomb core of the sandwich plate

ϕ (mm)	θ (°)	l = h (mm)	t (mm)	t' (mm)	Height (mm)
8	30	4.62	0.19	0.38	20

Table 2. By using a homogenization method in the previous section, we calculated the core mechanical properties for a homogeneous 3D solid sandwich panel as shown in Table 3.

We first performed simulations to determine the vibration modes for the sandwich panel. To calculate the frequency and vibration mode, the plate is fixed at one end. The three modes obtained by the 3D-shell model and 3D-Solid model are shown in Fig. 5. Simulation with the 3D-shell model consumes 466 s of CPU time and 57 s with the 3D-Solid model. The differences between the three natural frequencies obtained by the two models are less than 6% as indicated in Table 4.

Table 4. Comparison of the natural frequencies calculated by 3D-Shell and 3D-Solid model

	1st mode	2nd mode	3rd mode
3D-Shell	195.34 Hz	833.35 Hz	902.12 Hz
3D-Solid	183.25 Hz	831.13 Hz	895.65 Hz
Difference	6.1%	0.26%	0.71%

Next, we performed shock test simulations for both models. Boundary conditions are applied to the ends of the sandwich panel, an acceleration (A = 5 g) with trapezoidal amplitude, applied vertically to the center of the plate (bottom face of the plate) for a very short time (0.018 s). The deformations, the trapezoidal excitation as well as the responses obtained from two models after the vertical shock are represented in Fig. 6 and Fig. 7. The simulation with the 3D-Shell model is 12.46 times more expensive in CPU time with 10846 s compared to the simulation with our 3D-Solid

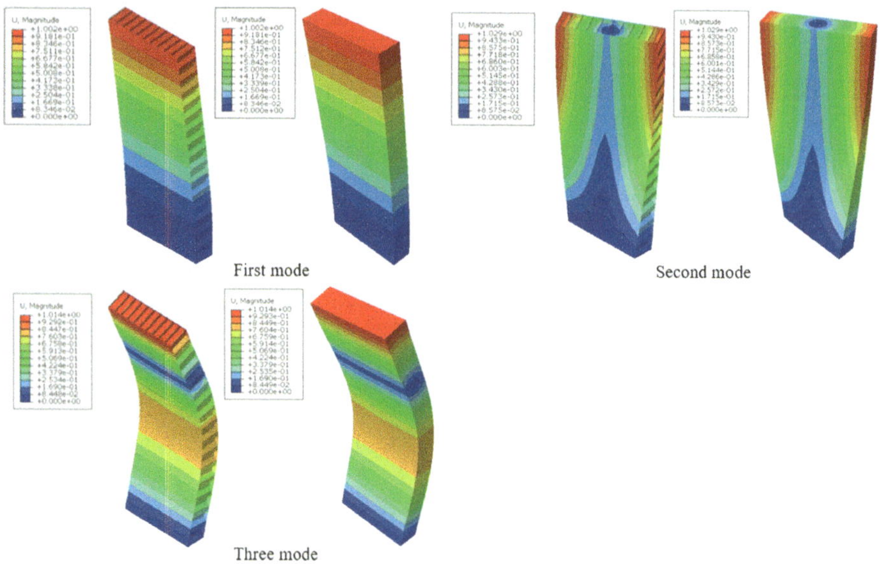

Fig. 5. The three natural frequencies obtained by the 3D-Shell model and 3D-Solid model.

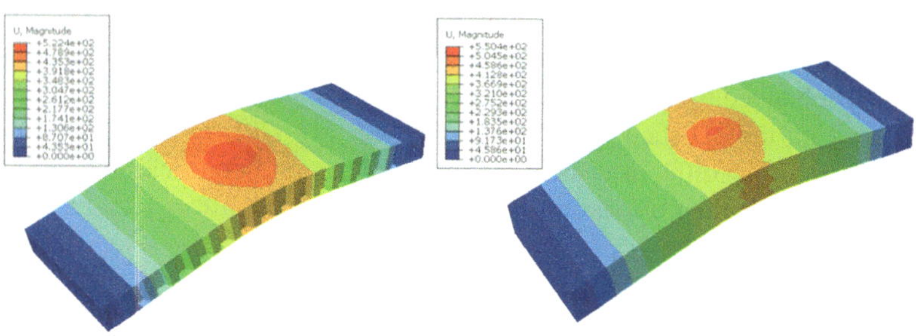

Fig. 6. Deformed plate after vertical shock for the 3D-Shell and 3D- Solid model.

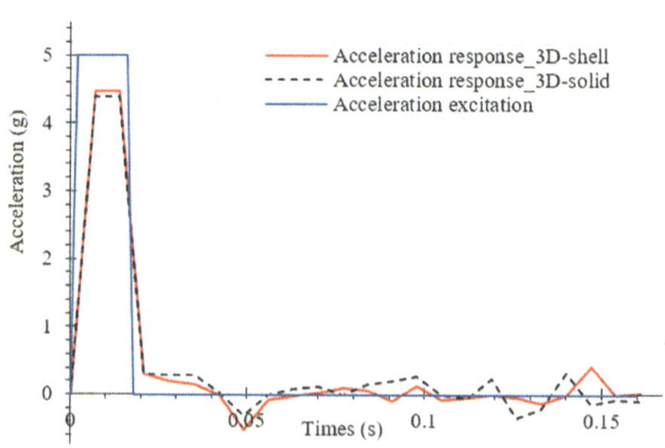

Fig. 7. The excitation acceleration and the response of two models in the shock test.

model with the only 872 s, while their deformation is almost the same. The difference between the amplitudes (for the 3D-Solid model, 4.38 g; for the 3D-Shell model, 4.46 g) of the response is less than 1.79%. From the results obtained when comparing the two models, we can confirm that the homogeneous model for the 3D core can be used to replace the 3D-shell model in finite element analysis.

4 Conclusion

To reduce the calculation and preparation time for the CAD model of the sandwich panel, we proposed the 3D-solid homogeneous model to replace the 3D-shell model. By comparing the numerical simulation results obtained from the simulation of vibration experiments for the two models, it is possible to evaluate the efficiency and accuracy of the proposed homogenous model. Thereby it is possible to confirm that our proposed homogenization method is reliable enough to be applied to honeycomb core sandwich panels. This is an important basis for us to continue to develop homogenization methods for other complex structures.

Acknowledgment. This work was supported by Thai Nguyen University of Technology, Vietnam.

References

1. L.G., Ashby, M.: Cellular solids: structure and properties, Cambridge. Cambridge (199)
2. Carlsson, L.A., Nordstrand, T., Westerlind, B.O.: On the elastic stiffnesses of corrugated core sandwich. J. Sandw. Struct. Mater. **3**(4), 253–267 (2001)
3. Buannic, N., Cartraud, P., Quesnel, T.: Homogenization of corrugated core sandwich panels. Compos. Struct. **59**(3), 299–312 (2003)
4. Cheng, Q.H., Lee, H.P., Lu, C.: A numerical analysis approach for evaluating elastic constants of sandwich structures with various cores. Compos. Struct. **74**(2), 226–236 (2006)
5. Biancolini, M.E.: Evaluation of equivalent stiffness properties of corrugated board. Compos. Struct. **69**, 322–328 (2005)
6. Aboura, Z., Talbi, N., Allaoui, S., Benzeggagh, M.L.: Elastic behavior of corrugated cardboard: experiments and modeling. Compos. Struct. **63**(1), 53–62 (2004). https://doi.org/10.1016/S0263-8223(03)00131-4
7. Zinno, A., Prota, A., Di Maio, E., Bakis, C.E.: Experimental characterization of phenolic-impregnated honeycomb sandwich structures for transportation vehicles. Compos. Struct. **93** (11), 2910–2924 (2011)
8. Tumino, D., Ingrassia, T., Nigrelli, V., Pitarresi, G., Urso Miano, V.: Mechanical behavior of a sandwich with corrugated GRP core: numerical modeling and experimental validation. Frat. ed Integrita Strutt. **30**:317–326 (2014)
9. Ijaz, H., Saleem, W., Zain-ul-Abdein, M., Mabrouki, T., Rubaiee, S., Mahfouz, A.S.B.: Finite element analysis of bend test of sandwich structures using strain energy based homogenization method. Adv. Mater. Sci. Eng. **2017**, 1–10 (2017). https://doi.org/10.1155/2017/8670207

10. Wolff, J., Brysch, M., Hühne, C.: Validity check of an analytical dimensioning approach for potted insert load introductions in honeycomb sandwich panels. Compos. Struct. **202**, 1195–1215 (2018)
11. Farshidi, A., Berggreen, C., Schäuble, R.: Numerical fracture analysis and model validation for disbonded honeycomb core sandwich composites. Compos. Struct. **210**, 231–238 (2019)
12. Garbowski, T., Gajewski, T., Grabski, J.K.: Role of transverse shear modulus in the performance of corrugated materials. Materials **13**(17), 3791 (2020). https://doi.org/10.3390/ma13173791
13. Chen, S., McGregor, O.P.L., Endruweit, A., Harper, L.T., Warrior, N.A.: Simulation of the forming process for curved composite sandwich panels. Int. J. Mater. Form. **13**(6), 967–980 (2019). https://doi.org/10.1007/s12289-019-01520-4
14. Aborehab, A., Kamel, M., Nemnem, A.F., Kassem, M.: Experimental modal testing of a honeycomb sandwich plate. IOP Conf. Ser. Mater. Sci. Eng. **973**, 012034 (2020). https://doi.org/10.1088/1757-899X/973/1/012034
15. Luong, V.D., et al.: Finite element simulation of the strength of corrugated board boxes under impact dynamics. Lecture Notes in Mechanical Engineering (2018)
16. Luong, V.D., Bonnin, A.S., Abbès, F., Nolot, J.B., Erre, D., Abbès, B.: Finite element and experimental investigation on the effect of repetitive shock in corrugated cardboard packaging. J. Appl. Comput. Mech. **7**(2), 820–830 (2021)
17. Oiwa, M., Ogasawara, T., Yoshinaga, H., Oguri, T., Aoki, T.: Numerical analysis of face sheet buckling for a CFRP/Nomex honeycomb sandwich panel subjected to bending loading. Compos. Struct. **270**, 114037 (2021). https://doi.org/10.1016/j.compstruct.2021.114037
18. Gibson, L.J., Ashby, M.F.: Cellular Solids: STRUCTURE and Properties, 2nd edn. Pergamon Press, England (2014)

A Research on the Influences of Forming Parameters on the Error of Width Dimension of Polylactic Acid Products by Fused Deposition Modeling Technology

Nguyen Thanh Nam[1], Nguyen Hoang Hiep[2](✉), and Le Khanh Dien[3]

[1] FME, Industrial University Ho Chi Minh City University (IUH), Ho Chi Minh, Vietnam
[2] Faculty of Mechanical Engineering, Vinh Long University of Technology and Education, Vinh Long, Vietnam
hiepnh@vlute.edu.vn
[3] FME, Saigon Technology University (STU), Ho Chi Minh, Vietnam

Abstract. Nowadays, Fused Deposition Modeling (FDM) technology is being broadly applied in many fields. However, one of the major limitations of the current FDM technology is the accuracy of dimension of the product. This is an important criterion in evaluating the accuracy of surface quality of mechanical products. Therefore, the improvement of the accuracy of dimensions of 3D printed products by FDM technology is indispensable. In the paper, we study the effect of forming parameters on the accuracy of the width dimensions of plastic materials PolyLactic Acid (PLA) parts that were manufactured by FDM technology. The experimental planning method and design of experiment (DOE) are applied to define the optimal forming parameters for enhancing the accuracy of dimensions of PLA products by FDM technology.

Keywords: Forming parameters · 3D printing · FDM technology · PLA · Accuracy of width dimensions

1 Introduction

1.1 Principle of FDM Technology

The FDM technology is based on the principle of melting and sticking the fibers material that are deposited through a thermal nozzle on a series of surfaces of each elevation. First, the product models are drawn by a CAD software in form of CAD file such as JGES files or STL which are accessed by a specialist CAD-CAM software that cut the model to many horizontal layers with different elevation. On each layer, there is a closed orbit that is the intersection of the model and the plane of the layer. The nozzle extrudes the melted material along the orbit from the lowest layer to the adjacent higher one [1]. The thickness of the outlet material can be adjusted from 0.01 "(0.254 mm) to 0.1" (2.54 mm). When a layer of material is completed, the nozzle of the FDM machine moves to the higher adjacent vertical layer with a clearance of the thickness to

D. C. Nguyen et al. (Eds.): ICERA 2021, LNNS 366, pp. 89–99, 2022.
https://doi.org/10.1007/978-3-030-92574-1_9

create the next layer. The extruded material layer will stick to the previous one in time of solidification. The processes are repeated until the product is completely performed. The material of a layer is cooled by different velocity that depends on its dimension and its properties. This event could generate the bending and the modification of the geometrical accuracy of the products.

1.2 Polylactic Acid (PLA) Plastic Material

PLA is a biodegradable thermoplastic derived from renewable sources, such as corn starch, sugar cane, tapioca or even potato starch. This creates the eco-friendliest solution to the environment for 3D printing process. The original nature of PLA is transparent so it can be easily dyed into any color or any shade of color and can glow in the dark [2]. PLA is not as durable and flexible as acrylonitrile butadiene styrene (ABS) plastic, but it is harder and stronger than ABS.

1.3 Researched Parameters

Through the recent research in [3], we can summarize the most important parameters and the range of influence to the error of width dimensions of products that are performed by FDM technology as follows:

- *Layer height:* Layer height is the height of the layer deposited by the nozzle and is a process parameter that has a great influence on the geometrical dimensions. The smaller the thickness of the layer, the bigger the number of layers and the longer of the printing time therefore the better the accuracy on dimensions of the product and inverse. Therefore, we must balance the forming time and the layer thickness before starting the process [4].
- *Raster width:* The raster width is the width of the path of the molten material deposited on each layer. The raster width depends on the properties of the plastic material, and it is about 1.2 ÷ 1.5 times of the nozzle diameter. The smaller the raster width, the better quality of the finished product and the higher the dimensional accuracy of the product [5].
- *Build-Orientation angle:* This is the most notable parameter that are studied by researchers to achieve the best dimension. The optimal build-orientation angle can be adjusted in FDM machine to attain the desired objectives [6].

The output factors are the deviation or the error of the width dimension of the products. The accuracy of the width dimension is depended on the final width dimension of the product to the width dimension of the original CAD model. In industrial FDM technology, the relative errors of the dimensions are ± 0.15% and the tolerance lowest limits is ± 0.2 mm.

Statistics of influential parameters and corresponding values in relation to the error of width dimensions (ΔW) of FDM products are presented in Table 1 and Fig. 1.

Table 1. Statistic of influential parameters and proportional dimensional values

No	Forming parameters	Normal values	The error of width dimensions (ΔW)
01	*Layer height*	0.1 ÷ 0.35 mm	ΔW = 0 ÷ 1%
02	*Raster width*	0.3 ÷ 0.83 mm	ΔW = 0 ÷ 1%
03	*Build-Orientation angle*	$0 \div 30^0$	ΔW = 0 ÷ 1%

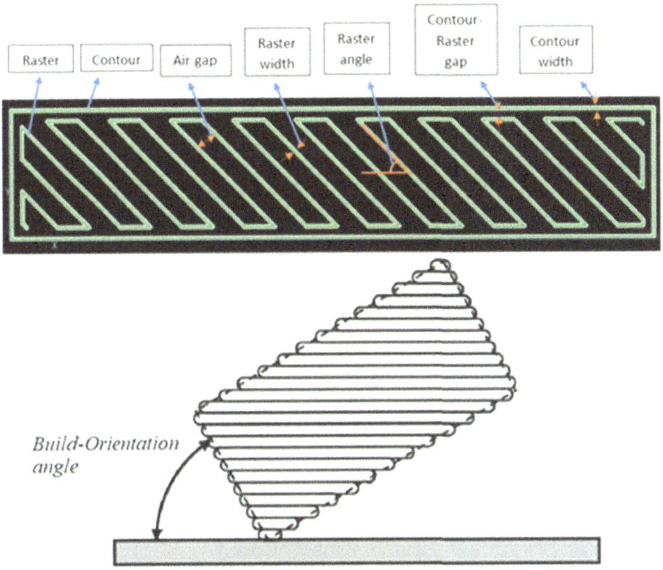

Fig. 1. Illustration of researched parameters [7]

2 Method of Design of Experiments (DOE)

2.1 Dimensional Length of Test Specimen of the Model

To investigate the influence of all 3 above parameters on the error of width dimension of a product that is performed by FDM technology, we choose a rectangular box sample with size of L × W × H: 100 × 60 × 40 mm, with PLA printing material [8] (Fig. 2).

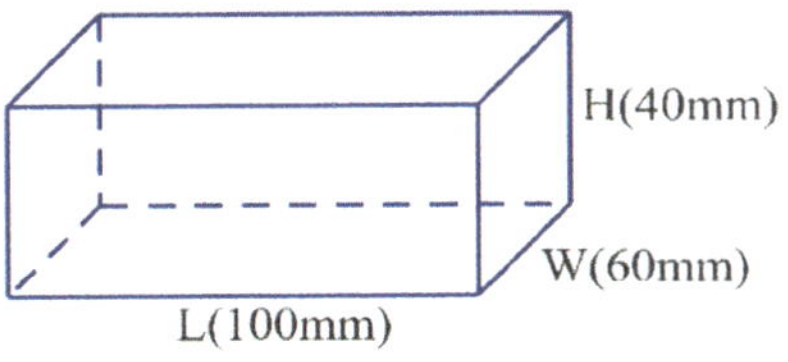

Fig. 2. A rectangular cube test specimen

2.2 Selection of Experimental Planning Parameters

According to the analysis of studies and the publications [9] that are related to the above experiment, we choose the 3 most affected parameters: Layer Height of each printed layer, Raster Width of printed line and the Build-Orientation angle.

With the demand of optimization of the forming parameters, a quadratic experimental design with the polynomial regression equation is selected with the following general form as the following equation:

$$y = b_o + \sum_{i=1}^{k} b_i x_i + \sum_{i=1}^{k} b_{ii} x_i^2 + \sum_{\substack{i<j \\ i,j=1}}^{k} b_{ij} x_i x_j \quad (1)$$

Among the types of quadratic experimental design, the 3-factor Face-Centered Central Composite Design is selected because it has some good properties such as: ability to describe well the properties of research objects in the design space and having contains full factorial design of 3 factors inside. In addition, with the quadratic experimental design, we can optimize to find the suitable technological parameters to gain the best output parameters [10] (Table 2).

Table 2. 3-level values of the input parameters.

No	Parameters	Designation		3 levels values of each factor			Unit
		Natural symbol	Coded	(−1)	(0)	(+1)	
1	*Layer height*	d_h	A	0.1	0.2	0.3	mm
2	*Raster width*	d_w	B	0.3	0.4	0.5	mm
3	*Build-Orientation angle*	α	C	0	15	30	Degree

The following table (Table 3) illustrates the coded DOE matrix of the research.

Table 3. Experimental matrix of coded input factors

No	d_h	d_w	α
1	−1	−1	−1
2	+1	−1	−1
3	−1	+1	−1
4	+1	+1	−1
5	−1	−1	+1
6	+1	−1	+1
7	−1	+1	+1
8	+1	+1	+1
9	−1	0	0
10	+1	0	0
11	0	−1	0
12	0	+1	0
13	0	0	−1
14	0	0	+1
15	0	0	0
16	0	0	0
17	0	0	0

2.3 Manufacturing of Experimental Models

Calculate the number of iterations of the experimental design. Referring to the Student t-distribution table with the statistical significance level of $q = 5\%$ and the degree of freedom $f_y = m - 1 = 29$, we find down $t_b = 2.05$.

The number of iterations of DOE is calculated as:

$$n = \frac{t_b^2 s^2\{y\}(N + d_{tb})}{\beta^2 \bar{y}^2 N} = \frac{2.05^2 \times 148.19 \times (17 + 5.83)}{0.5^2 \times 25.881 \times 17} \approx 4.994 \quad (2)$$

With the selected of number of iterations of 5, total of manufacture models is 17 × 5 = 85. Therefore, we perform all 85 PLA plastic experimental models with the dimensions of L × W × H: 100 × 60 × 40 mm by FDM 3D printer.

3 Results and Discussion

3.1 Analyzing the Result of Experiments

The result of measuring all (17 × 5 = 85) models and the average error of width dimension of each a set of 5 iteration models are displayed in Table 4.

Table 4. Average of error of width dimensions

No	d_h	d_w	α	$\overline{\Delta W}$	$s_{\Delta W}$
1	0.1	0.3	0	0.240	0.012
2	0.3	0.3	0	0.168	0.013
3	0.1	0.5	0	0.154	0.009
4	0.3	0.5	0	0.152	0.015
5	0.1	0.3	30	0.318	0.019
6	0.3	0.3	30	0.240	0.014
7	0.1	0.5	30	0.286	0.011
8	0.3	0.5	30	0.264	0.044
9	0.1	0.4	15	0.286	0.021
10	0.3	0.4	15	0.284	0.023
11	0.2	0.3	15	0.130	0.007
12	0.2	0.5	15	0.102	0.025
13	0.2	0.4	0	0.146	0.037
14	0.2	0.4	30	0.214	0.049
15	0.2	0.4	15	0.162	0.019
16	0.2	0.4	15	0.162	0.022
17	0.2	0.4	15	0.168	0.015

With the application of Minitab® 19 software to analyze the preliminary DOE of the Table 4, we got the result in the Table 5.

Table 5. Result of analyzing of the preliminary DOE from Table 4

Term	Coefficient	SE Coefficient	T-Value	P-Value	Notes
Constant	0.16969	0.00472	35.95	0.000	
d_h	−0.01760	0.00349	−5.05	0.000	
d_w	−0.01380	0.00349	−3.96	0.000	
α	0.04620	0.00349	13.25	0.000	
d_h^2	0.11104	0.00674	16.48	0.000	
d_w^2	−0.05796	0.00674	−8.60	0.000	
α^2	0.00604	0.00674	0.90	0.373	Not significant
$d_h \times d_w$	0.01575	0.00390	4.04	0.000	
$d_h \times \alpha$	−0.00325	0.00390	−0.83	0.407	Not significant
$d_w \times \alpha$	0.01175	0.00390	3.01	0.004	

After eliminating all the Not significant coefficients that have p-Value > 0.05 we reanalyze the DOE with the remained coefficient to receive the value of important significant coefficients in Table 6.

Table 6. The remained valuable coefficient after eliminating by Minitab® 19

Term	Coefficient	SE Coefficient	T-Value	P-Value
Constant	0.17083	0.00453	37.71	0.000
d_h	−0.01760	0.00348	−5.06	0.000
d_w	−0.01380	0.00348	−3.97	0.000
α	0.04620	0.00348	13.29	0.000
d_h^2	0.11309	0.00632	17.90	0.000
d_w^2	−0.05591	0.00632	−8.85	0.000
$d_h \times d_w$	0.01575	0.00389	4.05	0.000
$d_w \times \alpha$	0.01175	0.00389	3.02	0.003

All the Coefficients have the p-Value < 0.05, that means they have significant influence on the regression equation of the output parameters. The results of the application of Analyzing of Variants of (ANOVA) the significant coefficients are displayed in Table 7.

Table 7. The result of ANOVA of the equation of regression

Source	DF	Adj. SS	Adj. MS	F-Value	P-Value
Model	7	0.340964	0.048709	80.59	0.000
Linear	3	0.131732	0.043911	72.65	0.000
d_h	1	0.015488	0.015488	25.63	0.000
d_w	1	0.009522	0.009522	15.76	0.000
α	1	0.106722	0.106722	176.58	0.000
Square	2	0.193787	0.096894	160.32	0.000
d_h^2	1	0.193682	0.193682	320.46	0.000
d_w^2	1	0.047328	0.047328	78.31	0.000
2-Way Interaction	2	0.015445	0.007723	12.78	0.000
$d_h \times d_w$	1	0.009923	0.009923	16.42	0.000
$d_w \times \alpha$	1	0.005523	0.005523	9.14	0.003
Error	77	0.046537	0.000604		
Lack-of-Fit	7	0.007417	0.001060	1.90	0.083
Pure Error	70	0.039120	0.000559		
Total	84	0.387501			

Carrying out a lack-of-fit test of the regression equation and experimental data, we have the p-value of the Lack-Of-Fit source is 0.083 > 0.05 which can conclude that the result regression equation is consistent with the experimental data.

Checking the model summary by R^2 factor, adjusted-R^2 and predicted-R^2, all these values are $\geq 80\%$ that means the model has a very good fitness data. Table 8 displays the model summary by Minitab® 19.

Table 8. The model summary by Minitab® 19

S	R-sq	R-sq(adj)	R-sq(pred)
0.0245841	87.99%	86.90%	85.67%

The regression equation characterizes all the effect of input parameters on the errors of width dimensions is displayed in following equation:

$$\Delta W = -0.0541 - 5.330 d_h + 3.902 d_w - 0.00005\alpha + 11.309 d_h^2 - 5.591 d_w^2 + 1.575 d_h \times d_w + 0.00783 d_w \times \alpha \quad (3)$$

Finally, the construction of the response surface showing the effects of 3 input parameters: Layer height, Raster width and Build-Orientation angle on the errors of width dimension of the specimens:

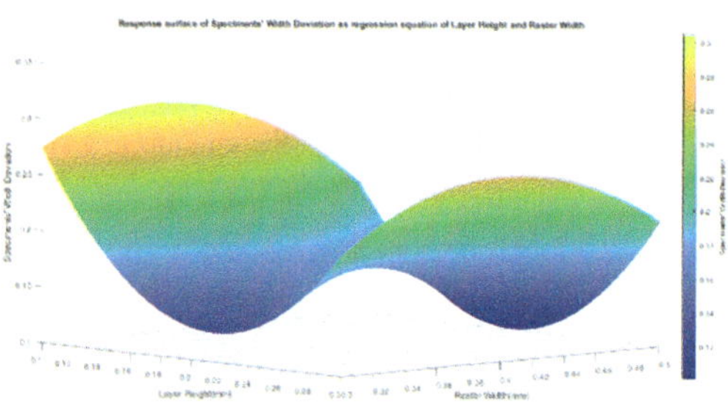

Fig. 3. Diagram of the effects of the layer height and the raster width on the errors of width dimension

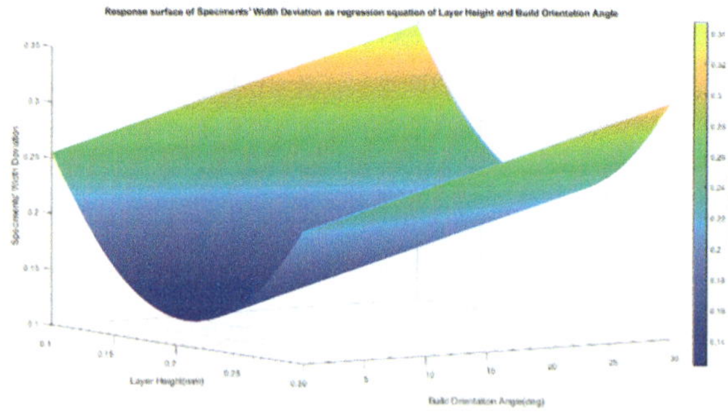

Fig. 4. Diagram of the effects of the layer height and the build-orientation angle on the errors of width dimension

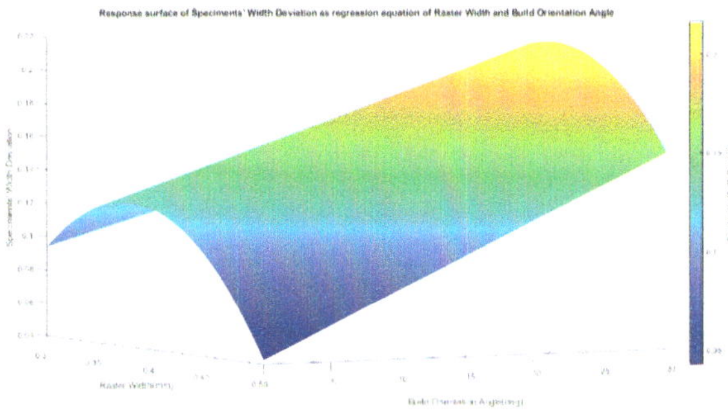

Fig. 5. Diagram of the effects of the raster width and the build-orientation angle on the errors of width dimension

According to Fig. 3 that describes the influences of the layer height and the raster width on the deviation of the width dimension of the specimens, we could conclude that the minimum deviation of error of the absolute width dimension is 0.08 mm when the layer height is 0.21 mm and minimum deviation of error of the absolute width dimension is 0.14 mm when the raster width is 0.42 mm.

According to Fig. 4 that describes the influences of the layer height and the build-orientation angle on the deviation of the width dimension of the specimens, we could conclude that the minimum deviation of error of the absolute width dimension is 0.06 mm when the layer height is 0.20 mm and minimum deviation of error of the absolute width dimension is 0.24 mm when the build-orientation angle of nozzle is about 19°45'.

According to Fig. 5 that describes the influences of the raster width and the build-orientation angle on the deviation of the width dimension of the specimen, we could conclude that the minimum deviation of error of the absolute width dimension is 0.04 mm when the raster width is 0.5 mm and minimum deviation of error of the absolute width dimension is 0.052 mm when the build-orientation angle of nozzle is about 0°.

3.2 Optimization of the Regression Equation to Get the Minimum Error of Width Dimension (Output Parameters)

We use the module "Response Optimizer…" to find the minimum error of width dimension (output parameters) with the boundary condition:

$$\begin{cases} 0.1\,\text{mm} \le d_h \le 0.3\,\text{mm} \\ 0.3\,\text{mm} \le d_w \le 0.5\,\text{mm} \\ 0 \le \alpha \le 30\,(^\circ) \end{cases} \tag{4}$$

Minitab® 19 finds out the minimum value of the errors of width dimension of specimens:

Table 9. Minimum value of the errors of width dimension of specimens

d_h	d_w	α	ΔW Fit
0.201010	0.5	0	0.0431674

From Table 9, we could extract a conclusion that the error of width dimensions reaches the minimum apex at $\begin{cases} d_h = 0.201010\,\text{mm} \\ d_w = 0.5\,\text{mm} \\ \alpha = 0\,(^\circ) \end{cases}$ and the minimum value of $\Delta W = 0.0431674$ mm

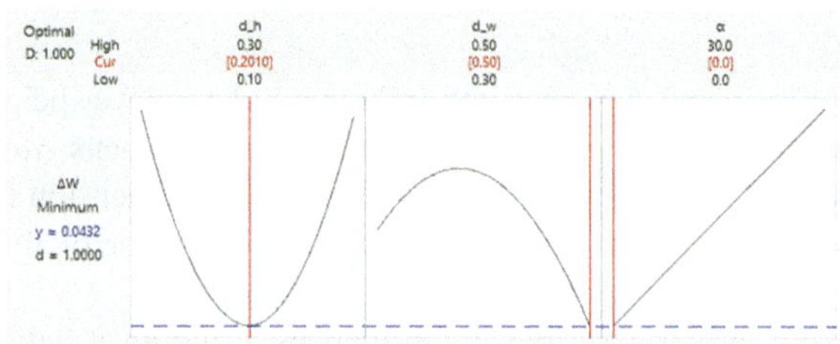

Fig. 6. Diagram of output error of width dimension ΔW via the Layer height (d_h), Raster width (d_w) and Build-orientation angle (α)

In the diagram in Fig. 6, the red values and red vertical lines stand for the critical values of the 3 input variables: the layer height, the raster width and the Build-orientation angle. The minimum absolute deviation of width dimension of the product is 0.0432 mm when the layer width is in the medium value of 0.2 mm and the raster width is in the maximum value of 0.5 mm and the angle of orientation of the nozzle is in the minimum value of 0°.

4 Conclusion

In the conclusion, we devoted our work in the paper to find out a set of optimal input parameters such as the Layer height, the Raster width and the Build-orientation angle of extrusion nozzle by experimental method with the abetment of Minitab® 19 software to enhance the accuracy of the width dimension of the PLA products that are manufactured by FDM technology. The last result shows that in order to get the highest accuracy (the minimum deviation) of the width dimensions of the product we should select the medium value of the Layer height, the maximum of the Raster width and the minimum of the Build-orientation angle of the extrusion nozzle that mean the productivity will be medium, the consummation of the PLA is maximum, and the direction of the extrusion nozzle must be vertical that means it is perpendicular to the layer.

The errors of width dimensions reach the minimum apex at $\begin{cases} d_h = 0.201010\,\text{mm} \\ d_w = 0.5\,\text{mm} \\ \alpha = 0\,(^\circ) \end{cases}$ and the minimum value of $\Delta W = 0.0431674$ mm.

References

1. Wang, C.C., Lin, T.-W., Hu, S.-S.: Optimizing the rapid prototyping process by integrating the Taguchi method with the Gray relational analysis. Rapid Prototyp. J. **13**(5), 304–315 (2007)
2. Sood, A.K., Ohdar, R.K., Mahapatra, S.S.: Improving dimensional accuracy of fused deposition modelling processed part using grey Taguchi method. Mater. Des. **30**, 4243–4252 (2009)
3. Zhang, J., Peng, A.: Process-parameter optimization for fused deposition modeling based on Taguchi method. Adv. Mat. Res. **538**, 444–447 (2012)
4. Nidagundi, V.B., Keshavamurthy, R., Prakash, C.P.S.: Studies on parametric optimization for fused deposition modelling process. Mat. Today Proc. **2**, 1691–1699 (2015)
5. Davim, J.P., Singh, R.: Additive Manufacturing_ Applications and Innovations. Manufacturing Design and Technology Series. 1st edn. CRC Press, Oxfordshire (2019)
6. Nghi, H.H., Tin, N.N., Tuan, N.A., Ha, T.T.T.: Optimizing the process parameters to improve the dimensional accuracy of the FDM product. In: Proceeding of National Conference on Mechanical and Transportation Engineering. VNUHCM Press (2017)
7. Motaparti, K.P., et al.: Effects of build parameters on compression properties for ULTEM 9085 parts by fused deposition modeling. In: 26th Annual International Solid Freeform Fabrication Symposium - an Additive Manufacturing Conference, pp. 964–977. The Universtiy of Texas at Austin, Austin, Texas, USA (2015)
8. Excell, J., Nathan, S.: The Rise of Additive Manufacturing. The Engineer. (2010). https://www.theengineer.co.uk/the-rise-of-additive-manufacturing/. Accessed 20 Jan 2021
9. All3dp.com. https://all3dp.com/2/weekend-project-make-your-own-3d-printed-circuit-board/. Accessed 20 Jan 2021
10. Tam, N.M.: Research on the Effects of Technology Parameters on the Surface Roughness and the Dimensional Accuracy of Products Created by 3D Printer (FDM). Master Thesis, HCMUT (2020)

A Simple Chemical Procedure for Direct Synthesis of NiO on Nickel Foam Electrode Applied in Non-enzymatic Glucose Electrochemical Measurements

Vu Thi Oanh[1], Chu Thi Xuan[1](✉), Le Manh Tu[2,3], and Nguyen Duc Hoa[1]

[1] International Training Institute for Materials Science, Hanoi University of Science and Technology, Hanoi 100000, Vietnam
xuan@itims.edu.vn
[2] Faculty of Materials Science and Engineering, Phenikaa Institute for Advance Study (PIAS), Phenikaa University, Hanoi 100000, Vietnam
[3] A&A Green Phoenix Group, Phenikaa Research and Technology Institute (PRATI), Hanoi 100000, Vietnam

Abstract. It is well-known that the amount of glucose in the blood that exceeds the normal sugar blood range can cause diabetes, which is one of the most dangerous diseases that thread human health. Amongst a lot of methods that have been developed to measure glucose levels, electrochemical glucose sensors have attracted more attention due to their accuracy and efficiency. In our work, a simple, instant, and economical chemical process was conducted to deposit NiO nanostructures on the surface of a nickel foam electrode for non-enzymatic glucose electrochemical sensor. The morphologies and components of synthesised materials were characterised by field emission scanning electron microscopy and energy-dispersive X-ray spectroscopy. Cyclic voltammetry and chronoamperometry were used to measure glucose concentration based on the synthesised materials. The electrode showed a remarkably high sensitivity of 6.8 $mA{\cdot}mM^{-1}{\cdot}cm^{-2}$ and a low limit of detection of 5.7 μM. The results indicated that the fabricated electrode can be a potential candidate for clinical applications.

Keywords: Nickel(II) oxide · Chemical method · Electrochemical sensor · Non-enzymatic glucose sensor

1 Introduction

Glucose detection plays an important role in many kinds of fields such as clinical diagnostic, biomedical, and food industries [1]. Many affordable techniques have been developed to control and monitor the concentration of glucose, namely surface plasmon resonance technique [2], optical technique [3], colorimetric technique [4], and electrochemical technique [5]. Amongst these techniques, the electrochemical method is concerned the most effective one because this method is simple, low-cost, highly

D. C. Nguyen et al. (Eds.): ICERA 2021, LNNS 366, pp. 100–106, 2022.
https://doi.org/10.1007/978-3-030-92574-1_10

sensitivee, and selective even at low concentrations of the sample [5]. The electrochemical analysis involves a simple and quantitative mode of operation and is, therefore, the most widely employed method in glucose sensors. Electrochemical glucose sensors can be used over a broad detection range. Enzymatic and non-enzymatic are two kinds of glucose electrochemical sensors.

Enzymatic glucose sensor is one of the most effective methods to measure glucose concentration because of its sensitivity and selectivity [6]. However, the enzymatic sensor has some disadvantages such as instability, the high cost of enzyme operation, and critical operation. With the advantages such as high sensitivity, long-term stability, and low cost, a non-enzymatic glucose sensor may overcome the disadvantages of the enzymatic glucose sensor [7]. The mechanism of the non-enzymatic glucose sensor is based on the direct oxidation of glucose on the surface of the electrode. Particularly, metal oxide sensors detect glucose via the direct oxidation reaction of glucose with an activated metal oxide contact; the reaction results in an electron transfer to the contact which is recorded by the sensor as a current.

The structure of materials for the working electrode is an important feature in the determination of glucose. Transition metals such as Ni, Cu, Co, and their compound are suitable for anodic catalyst due to their high natural abundance, low cost, and good electrocatalyst. Elemental nickel (Ni) is unstable and oxidizes rapidly in air and solution [8]. Therefore, several Ni-based compounds have been evaluated for their ability to catalyse glucose oxidation [9]. These compounds comprise nickel oxides [10], hydroxides [11], and sulphides [12]. In this report, NiO nanostructures were chosen as functionalized materials for glucose measurement. Here, NiO nanostructures were synthesised directly on commercial nickel foam (NF) electrodes via a facile chemical method. The fabrication of NiO nanohives is easy-operated, fast, and inexpensive. The modified electrode was applied to measure the concentration of glucose with high sensitivity, selectivity, and low limit of detection.

2 Experiment Procedures

Here, two precursors containing $Ni(NO_3)_2.6H_2O$ (Germany) and hexamethylenetramine (HMTA) $C_6H_{12}N_4$ (China) were used for the synthesis of NiO nanostructures. A commercial nickel foam (20 × 5 × 1 mm) after purification with acetone by sonification for 10 min was put into a bottle of 0.1 M $Ni(NO_3)_2.6H_2O$ and 0.1 M HMTA. Then, the bottle was sealed and heated at 90 °C for 10 min. Next, the modified nickel foam electrode was cleaned with deionized water. Finally, NiO/NF electrode was obtained after heating at 300 °C for 2 h.

Electrochemical behaviors of the synthesised materials were measured by cyclic voltammetry and chronoamperometry using PGSTAT302N Autolab workstation (Netherlands) with a three-electrode system, in which the working electrode is the synthesised electrode, the reference electrode is Ag/AgCl electrode (Netherlands), and a platinum wire is the counter electrode. A field emission scanning electron microscopy (FE-SEM) and energy-dispersive X-ray spectroscopy (EDS) were applied to characterise the morphologies and components of the materials.

3 Results and Discussions

3.1 FE-SEM Images and EDS Analysis of Materials

EDS pattern of the synthesised materials is shown in Fig. 1. It is clear that there is the presence of nickel and oxygen in the materials with the atomic percentage of 56.72% and 43.28, respectively. The atomic ratio of approximately 1:1 demonstrates the formation of the NiO phase on the surface of the nickel foam electrode.

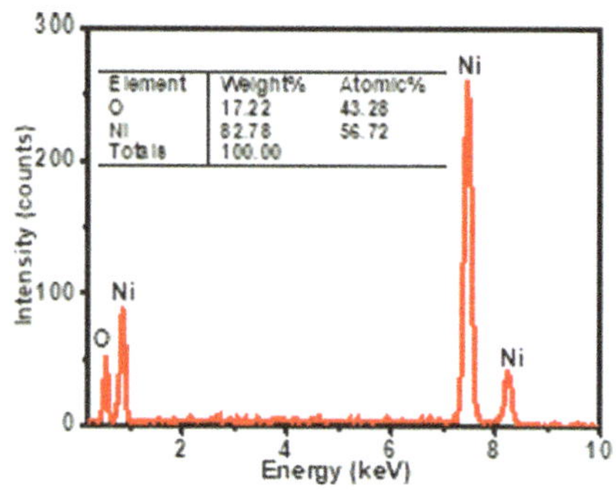

Fig. 1. EDS pattern of the synthesised materials.

FE-SEM images of the synthesised NiO are presented in Fig. 2. The low-magnification FE-SEM image (Fig. 2a) shows that NiO nanohive structures were uniformly formed on the NF surface. When increasing magnification as shown in Fig. 2b, it was seen that NiO nanohives have walls with a thickness of 20–30 nm.

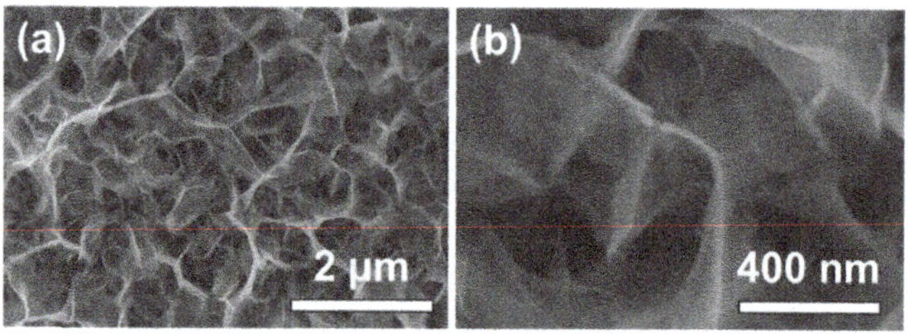

Fig. 2. (a) Low and (b) high magnification FE-SEM images of the synthesised NiO/NF electrode.

3.2 Electrochemical Behaviors and Glucose Measurements

Electrochemical behaviors of the materials were measured in 0.1 M NaOH at room temperature. Figure 3 illustrates the cyclic voltammograms (CVs) of the bare and NiO-modified nickel foam electrodes with a scan rate of 50 $mV{\cdot}s^{-1}$. The CVs of the NiO/NF electrodes show higher peak current signals, which indicates that the electrochemical properties of the electrode were enhanced after deposition of NiO nanohives. Additionally, with no significant shift of CVs plots of three different NiO/NF electrodes synthesised under the same conditions, the synthesised electrodes presented a good reproducibility.

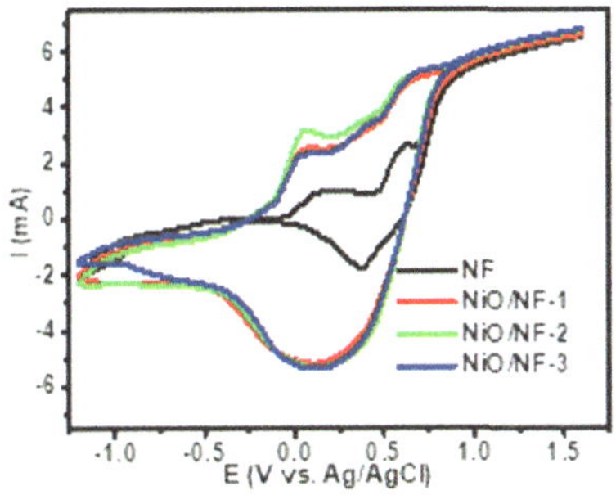

Fig. 3. CVs plots of the NF and three NiO/NF electrodes synthesised under the same conditions.

Glucose measurements. The synthesised electrodes were applied for glucose detection. Figure 4 presents the CVs plots of NiO/NF electrode in 20 mL of 0.1 M NaOH with 0, 3, 4, 5, 6, 7 mM glucose. It is clear that the cathodic peak currents decrease when the glucose concentration rises. The mechanism for the electrocatalytic of NiO toward glucose can be explained by the below equations [13]:

$$NiO + OH^{-} \leftrightarrow NiO(OH) + e^{-} \tag{1}$$

$$2NiO(OH) + glucose \rightarrow 2Ni^{2+} + gluconicacid + H_2O \tag{2}$$

To be more detailed, NiO reduces to NiO(OH), then glucose is oxidised to gluconic acid while NiO(OH) is reduced to Ni^{2+}. The reduction peak current is then reduced when glucose is added to the solution.

To further study the electrocatalytic of NiO/NF for glucose oxidation, chronoamperometry was applied to measure the glucose level. Three voltage values were applied, that are 0.5, 0.6, and 0.7 V. As shown in Fig. 5a, when the glucose rises gradually from 0 to 0.5 mM, the response current at the applied potential of 0.7 V increases linearly. While the potential is at 0.5 or 0.6 V, the response currents rise nonlinearly with the increase of glucose concentration (Fig. 5b). The response current at 0.7 V depends on the glucose concentration as the below function:

$$I = 4.42 \times C_G + 2.39 \tag{3}$$

Where C_G is glucose concentration (mM). The sensitivity of the sensor calculated by dividing the slope to the working electrode area of 0.65 mm^2 is 6.8 $mA \cdot mM^{-1} \cdot cm^{-2}$. Based on the ratio of signal-to-noise of 3 [14], the detection limit (LOD) of the fabricated sensor was calculated at 5.7 μM. These values of the fabricated sensor are compared with some other nickel-based sensors reported before (Table 1). Our sensor shows a high sensitivity as compared to the similar previously reports. All results indicated that the NiO/NF electrode is a capable candidate for clinical glucose detection.

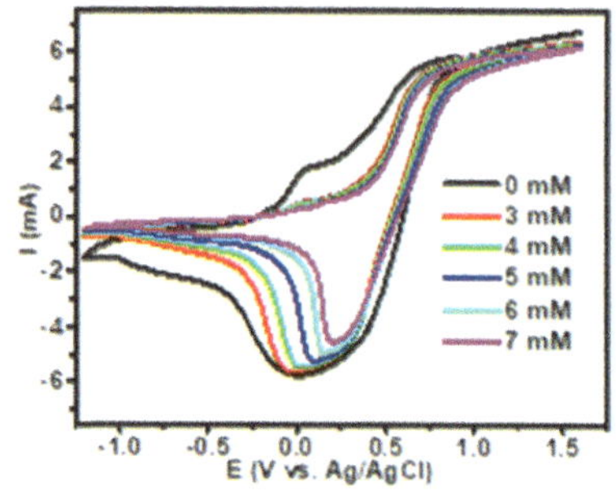

Fig. 4. CVs plots of NiO/NF electrode in 20 mL of 0.1 M NaOH and 0, 3, 4, 5, 6, 7 mM glucose.

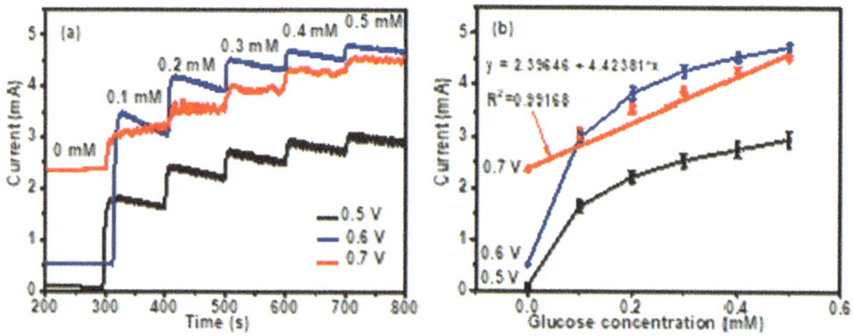

Fig. 5. (a) Amperometric response of NiO/NF toward glucose with a concentration range between 0 and 0.5 mM, and (b) The plot of dependence between response current and glucose concentration at different potentials.

Table 1. To compare the glucose electrochemical sensing of the fabricated sensors with other nickel-based sensors.

Materials	Sensitivity ($mA \cdot mM^{-1} \cdot cm^{-2}$)	LOD (μM)	Reference
NiO nanopetals	0.0039	1	[10]
$Ni(OH)_2$/C	1.3422	0.1	[11]
NiS nanoclusters@NiS nanospere	0.546	0.0083	[12]
NiO nanohives	6.8	5.7	This work

4 Conclusions

In conclusion, NiO nanostructures were successfully synthesised via a simple and low-cost procedure on the surface of nickel foam electrodes without a binder. The fabricated materials have nanohive structures with a thickness of the wall of 20–30 nm. The fabricated electrode was applied for non-enzymatic electrochemical glucose measurements. The sensor shows high sensitivity (6.8 $mA \cdot mM^{-1} \cdot cm^{-2}$), and low limit of detection (5.7 μM) toward glucose. The fabricated sensor is evaluated to be potential for practical applications.

Acknowledgments. This research is funded by the Hanoi University of Science and Technology (HUST) under project number T2020-SAHEP-025.

Vu Thi Oanh was funded by Vingroup Joint Stock Company and supported by the Domestic Master/PhD Scholarship Programme of Vingroup Innovation Foundation (VINIF), Vingroup Big Data Institute (VINBIGDATA), code VINIF.2020.ThS.33.

References

1. Funtanilla, V.D., Candidate, P., Caliendo, T., Hilas, O.: Continuous glucose monitoring: a review of available systems. P T **44**, 550–553 (2019)
2. Srivastava, S.K., Verma, R., Gupta, B.D.: Surface plasmon resonance based fiber optic glucose biosensor. Third Asia Pacific Opt. Sensors Conf. **8351**, 83511Z (2012)
3. Park, E.-Y., Baik, J., Kim, H., Park, S.-M., Kim, C.: Ultrasound-modulated optical glucose sensing using a 1645 nm laser. Sci. Rep. **10**, 13361 (2020)
4. Wang, Q., Dai, T., Sun, P., Wang, X., Wang, G.: A colorimetric and ratiometric glucose sensor based on conformational switch of i-motif DNA. Talanta Open 1:100001 (2020)
5. Teymourian, H., Barfidokht, A., Wang, J.: Electrochemical glucose sensors in diabetes management: An updated review (2010–2020). Chem. Soc. Rev. **49**, 7671–7709 (2020)
6. Lee, H., Hong, Y.J., Baik, S., Hyeon, T., Kim, D.H.: Enzyme-based glucose sensor: from invasive to wearable device. Adv. Healthc. Mater. **7**, 1–14 (2018)
7. Hwang, D.W., Lee, S., Seo, M., Chung, T.D.: Recent advances in electrochemical non-enzymatic glucose sensors – a review. Anal. Chim. Acta **1033**, 1–34 (2018)
8. Li, Q., Wang, L.S., Hu, B.Y., Yang, C., Zhou, L., Zhang, L.: Preparation and characterization of NiO nanoparticles through calcination of malate gel. Mater Lett. **61**, 1615–1618 (2007)
9. Mu, Y., Jia, D., He, Y., Miao, Y., Wu, H.-L.: Nano nickel oxide modified non-enzymatic glucose sensors with enhanced sensitivity through an electrochemical process strategy at high potential. Biosens Bioelectron **26**, 2948–2952 (2011)
10. Mishra, S., Yogi, P., Sagdeo, P.R., Kumar, R.: Mesoporous nickel oxide (NiO) nanopetals for ultrasensitive glucose sensing. Nanoscale Res. Lett. **13**(1), 1–7 (2018). https://doi.org/10.1186/s11671-018-2435-3
11. Chen, L., et al.: Preparation of Ni(OH)2 nanoplatelet/electrospun carbon nanofiber hybrids for highly sensitive nonenzymatic glucose sensors. RSC Adv. **7**, 19345–19352 (2017)
12. Arivazhagan, M., Santhosh, Y.M., Maduraiveeran, G.: Non-enzymatic glucose detection based on nis nanoclusters@nis nanosphere in human serum and urine. Micromachines (2021). https://doi.org/10.3390/mi12040403

13. Baghayeri, M., Amiri, A., Alizadeh, Z., Veisi, H., Hasheminejad, E.: Non-enzymatic voltammetric glucose sensor made of ternary NiO/Fe3O4-SH/para-amino hippuric acid nanocomposite. J. Electroanal. Chem. **810**, 69–77 (2018)
14. Pal, N., Banerjee, S., Bhaumik, A.: A facile route for the syntheses of Ni(OH)2 and NiO nanostructures as potential candidates for non-enzymatic glucose sensor. J. Colloid Interface Sci. **516**, 121–127 (2018)

A Study on Characteristics of Dynamics and Kinematics of the Vehicle Equipped with Dual-Clutch Automatic Transmission

Nguyen Khac Minh[1,2], Dam Huu Vu[2], Mac Duy Hung[2], Vu Van Hai[1], and Nguyen Khac Tuan[2(✉)]

[1] Moscow Automobile and Road Construction State Technical University (MADI), № 64, Leningradsky Prospect, Moscow 125319, Russia
[2] Thai Nguyen University of Technology, № 666, Tich Luong Ward, Thai Nguyen 24131, Vietnam
tuannkcn@tnut.edu.vn

Abstract. In this paper, a simulation model of the vehicle equipped with a 6-speed dual-clutch automatic transmission was built with the help of Simscape driveline/Matlab 2019 tools. The model consists of 4 subsystems: engine, dual-clutch transmission, control block, and vehicle body. Using the model allows determining the influence of structural parameters, working conditions on the vehicle's kinematics and dynamics. Some simulation results of kinematics and dynamics of the vehicle when running on the ECE-15 driving cycle are also analyzed in detail in this paper.

Keywords: Vehicle · Kinetic · Dynamic · Dual clutch · Transmission system · Simulation

1 Introduction

Research advanced technologies to improve performance, to enhance the comfort of passengers, and save energy has become a major trend in the automotive industry. This goal has motivated automobile manufacturers to produce the new transmission system. There have been several types of gearboxes and advanced technology used in the vehicle. Possessing an overall efficiency value of 96.2% manual transmission (MT) is the most efficient transmissions, but manual transmissions do not provide a comfortable working experience. Automatic transmission (AT), equipped with torque converters and planetary gears, is leading the automatic transmission market. But in recent years, it is losing its influence because of the low performance and complex structure of the accompanying technologies.

Automatic manual transmission (AMT) using dry clutches is highly efficient systems, but AMT do not provide comfort during operation due to the existence of torque discontinuities during gear shifting. The continuously variable transmission (CVT) solves the problem of manual gear shifting, allowing the engine to operate in the most energy-efficient way. However, CVT is limited in power and torque transmission and has a very low overall efficiency of 84.6% [1, 2].

D. C. Nguyen et al. (Eds.): ICERA 2021, LNNS 366, pp. 107–116, 2022.
https://doi.org/10.1007/978-3-030-92574-1_11

Compared to the above technologies, the gearbox equipped with dual-clutch transmission (DCT) is a new generation automatic transmission, developed from the parallel shaft gearbox. The odd and even gears are driven on two clutches respectively, and shifting can be achieved by alternating the two clutches, which not only maintains the advantages of a traditional manual transmission but also achieves continuous power transmission and improve energy efficiency and saves energy [3–6]. In addition, the dual-clutch transmission DCT, which is highly flexible in production, can be designed and built based on the existing structure of the manual transmission, and the DCT has become the subject of interest for many automotive manufacturers such as Volkswagen, Getrag, LuK, ZF, Honda, Mitsubishi… The research and development of dual-clutch transmission DCT has also received the attention of scientists around the world.

Author Manish Kulkarni et al. [7] simulated, analyzed, and controlled a dual-clutch automatic transmission DCT. Research has shown that the timing of engagement and disengagement dual-clutch is extremely important to achieve smooth shifting without loss of traction force.

Shift control strategies for the dual-clutch transmission DCT have been clearly defined in a variety of studies such as [8–10]. Author Y. Liu et al. [11] analyzed the dynamics of a dry dual-clutch transmission DCT during vehicle starting and shifting, which is useful for dual-clutch transmission DCT torque control and correction. However, the study only considered a few cases: launch, 1–2 upshift and 5–4 downshift. Goetz et al. in the study [12] presented the detailed model of dual-clutch transmission DCT for the front-wheel drive powertrain, with the synchronizer was being viewed as a simple conical clutch. In the research condition, other influencing factors of the engagement process such as ring unblocking and hub indexing were ignored.

Obviously, modeling and analysis of the vehicle's powertrain to provide the optimal control method is an important factor for the development of the dual-clutch transmission DCT in recent years. However, improving DCT dual-clutch drive control strategies requires a large number of specific experiments, and in each case where the dual-clutch experiments on another vehicle, it is necessary to reset the survey parameters. In this paper, a general model has been built with high flexibility that can be used to survey many different objects by adjusting the input parameter values of the engine, gearbox configuration.

2 Material and Method

Figure 1 shows the schematic of a dual-clutch transmission system with 6-speed dual-clutch transmission (DCT). In this system, the two clutches are connected to two separate sets of gears. The odd gear set (clutches: 1, 3, 5) is connected to one of the clutches and the even set (clutches: 2, 4, 6, R) to the other clutch. It is necessary to preselect the gears to realize the benefits of the DCT system.

To analyze the dynamic and kinetic characteristics of the vehicle equiped with DCT, a computer simulation model of the vehicle system, using the Matlab/Simulink computer software, has been developed. The simulation model includes many sub-models, some of which will simulate the engine, dual-clutch transmission, vehicle body, and transmission controller.

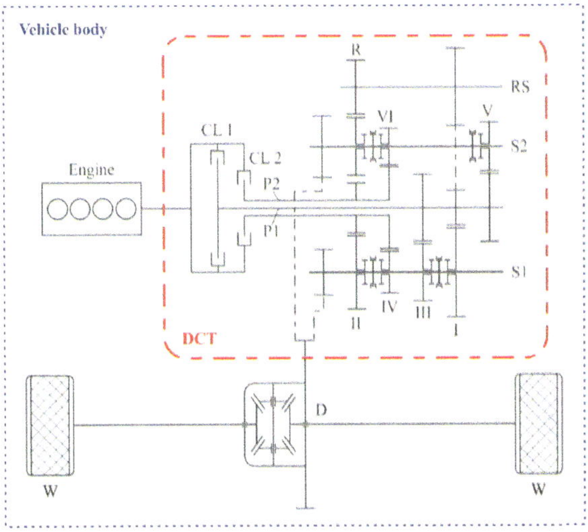

Fig. 1. Schematic of a dual-clutch transmission system [16]

2.1 Engine Model

In this paper, the Generic Engine model [13, 17] was used to model the internal combustion engine (see Fig. 2). By default, the Generic Engine model uses a programmed relationship between torque and speed, modulated by the throttle signal.

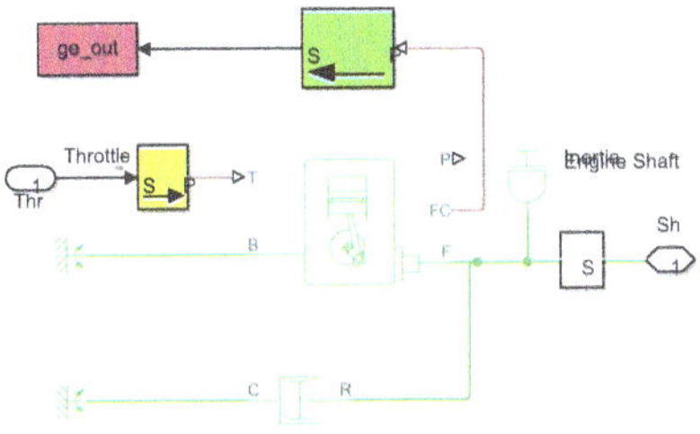

Fig. 2. Engine model

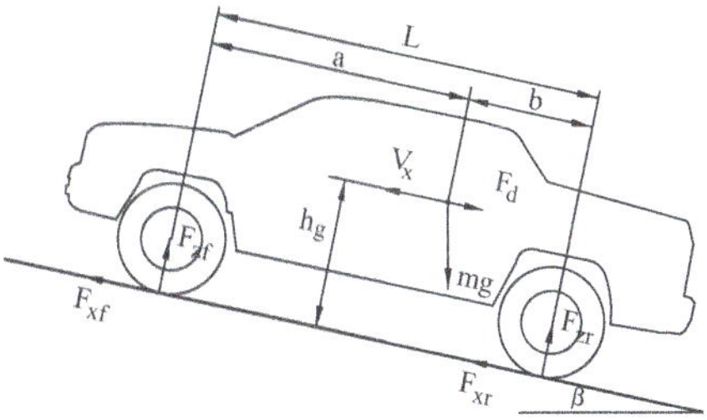

Fig. 3. Free body diagram of the different longitudinal forces acting on the vehicle [17]

The engine model is specified by an engine power demand function g(Ω). The function provides the maximum power available for a given engine speed Ω. The block parameters (maximum power, speed at maximum power, and maximum speed) normalize this function to physical maximum torque and speed values. The normalized throttle input signal T specifies the actual engine power. The power is delivered as a fraction of the maximum power possible in a steady state at a fixed engine speed. It modulates the actual power delivered, P, from the engine: P(Ω, T) = T·g(Ω). The engine torque is τ = P/Ω.

2.2 Vehicle Body and Tire Model

The motion of the vehicle is the result of the forces acting on it [17]. Figure 3 shows a diagram of the forces acting on the vehicle in the general case. The differential equation system describing the dynamics of the vehicle body is written as follows:

$$\begin{aligned} & m\dot{V}_x = F_x - F_d - m.g.sin\beta \\ & F_x = n\left(F_{xf} + F_{xr}\right) \\ & F_d = \tfrac{1}{2} C_d \rho . A(V_x + V_\omega)^2 sgn(V_x + F_\omega) \\ & F_{zf} = \frac{-h\left(F_d + m.g.sin\beta + m.\dot{V}_x\right) + b.m.g.cos\beta}{n(a+b)} \\ & F_{zr} = \frac{h\left(F_d + m.g.sin\beta + m.\dot{V}_x\right) + a.m.g.cos\beta}{n(a+b)} \\ & F_{zf} + F_{zr} = \text{mg}\frac{cos\beta}{n} \end{aligned} \quad (1)$$

Where: g: gravitational acceleration; β: inclining angle; m: mass of the vehicle; H: height of vehicle center of gravity (CG) above the ground; a, b: distance of front and rear axles, respectively, from the normal projection point of vehicle CG onto the common axle plane; V_x: velocity of the vehicle. When $V_x > 0$, the vehicle moves forward. When $V_x < 0$, the vehicle moves backward; V_w: wind speed. When $V_w > 0$, the wind is a headwind. When $V_w < 0$, the wind is a tailwind; n: number of wheels on each axle; F_{xf}, F_{xr}: longitudinal forces on each wheel at the front and rear ground contact points, respectively; F_{zf}, F_{zr}: normal load forces on each wheel at the front and rear ground contact points, respectively; A: effective frontal vehicle cross-sectional area; C_d: Aerodynamic drag coefficient; ρ: Mass density of air; F_d: Aerodynamic drag force.

To model vehicle dynamics and motion, a Simulink model was used (see Fig. 4), in which there are six ports, two input ports: W – Headwind speed, beta – Road incline angle; three output ports: V – longitudinal velocity, NF – Front axle normal force, NR – Rear axle normal force and a conserving port H associated with the horizontal motion of the vehicle body. Tire traction motion is connected to H port.

Since the relationship between drag force F_x and slip is very complex, approximate formulas are often used to describe this relationship. For simulation, we use the Magic formula built by Pajeika [15, 16]. The Magic formula is written as follows:

$$F_x = D\ sin[C\ arctan\{Bx - E(Bx - arctan\ Bx)\}] \quad (2)$$

Where: D, C, B, E are the coefficients built for different types of roads.

To model the tires on the front and rear axles, the tire block (see Fig. 5) is used based on the Magic Formula. This block is taken from the tire & vehicle library in Simscape Driveline. Each tire block includes 5 connections: A, H, N, S, M, in which connection port N is the input port of normal jet signal; connection S is the slip of the tire. The physical signal gate M accepts a four-element vector that corresponds to the Magic Formula Coefficients B, C, D and E.

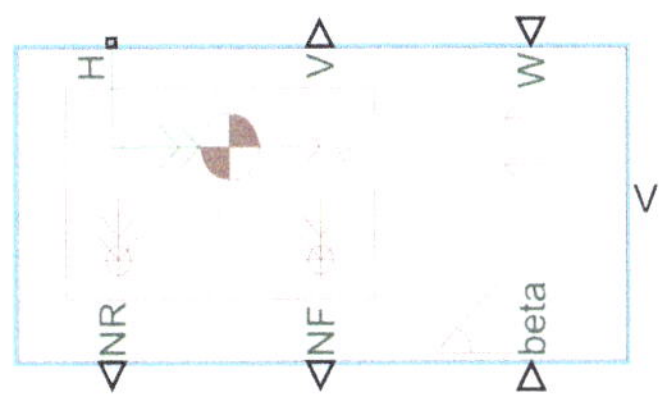

Fig. 4. Longitudinal model of the vehicle dynamics in Simscape drive line

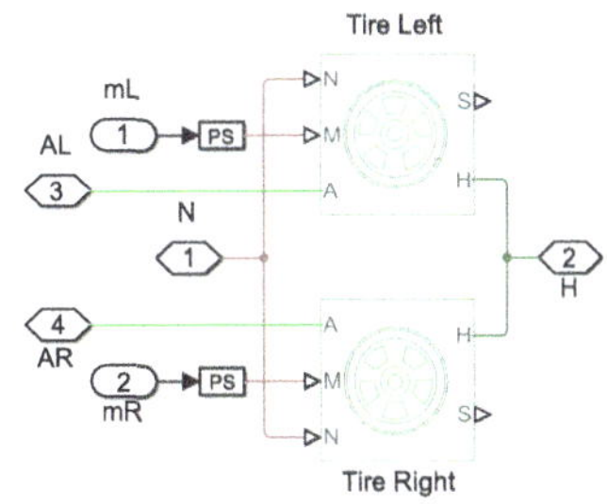

Fig. 5. Tire model in Simscape drive line

2.3 Dual Clutch Transmission System Model

The 6-speed dual-clutch transmission model is built by combining function blocks available in the Simscape driveline library. The model consists of 07 gear transmissions corresponding to 6 forward gear 1, 2, 3, 4, 5, 6, and one reverse gear R. In addition, each pair of gears is linked to a dog clutch controlled by the executive structure. The working status of "Even clutch" and "Odd clutch" depends on the control pressure P_{dem} (see Fig. 6) received from the transmission controller block (see Fig. 7).

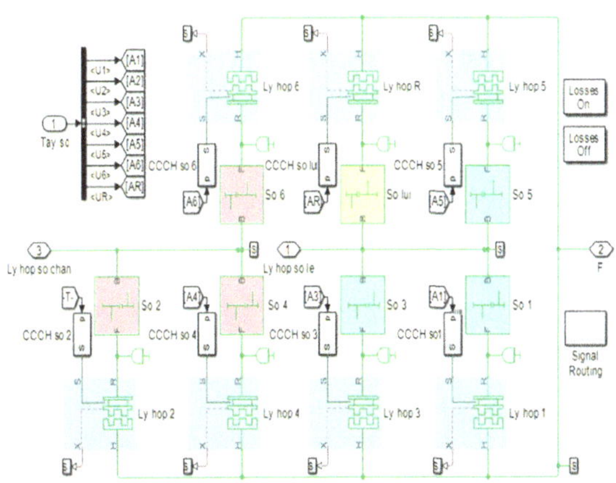

Fig. 6. Model of 6-speed dual clutch transmission

2.4 Transmission Controller Model

The transmission controller model (see Fig. 7) is built based on 4 input signals: clutch slip (C_{slip}), real vehicle speed (V_a); speed demand (V_{cmd}) and engine speed (w_e). The control signals corresponding to the input signal include: gear (G_{dem}), clutch control pressure (P_{dem}), and torque demand from the engine (T_{dem}). In this circumstance, G_{dem} and P_{dem} signals will be sent to the gearbox block to perform appropriate gearshift control. The clutch states and pre-selection of the gears via dog clutches are controlled by a state machine modeled in Stateflow [13].

By linking the model of subsystems built in the above sections, we have a model of the whole vehicle equipped with a dual-clutch drive system as shown in Fig. 7.

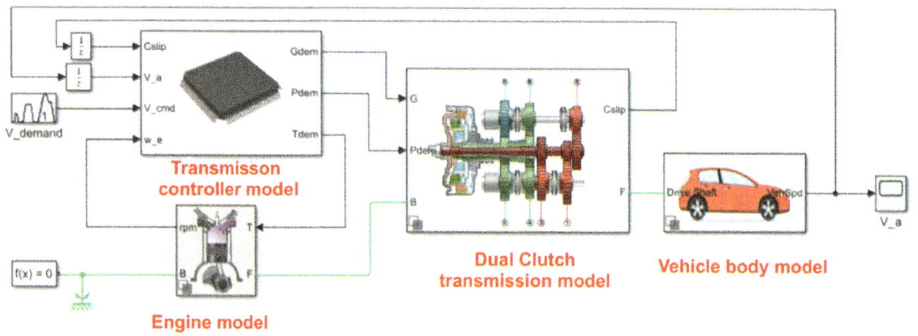

Fig. 7. Complete vehicle model equipped with 6-speeds DCT in Simulink/Simscape driveline

3 Result and Discussion

In this section, some simulation results of kinematic and dynamic parameters of the vehicle equipped with 6-speed dual-clutch transmission system when running on the ECE-15 driving cycle [14] are analyzed. When performing simulation using model in Fig. 7, the vehicle's parameters are selected as follows: a = 1,5 m; b = 1,7 m; h_g = 0,45 m; P_{emax} = 160 kW; n_P = 4600 rpm; n_S = 550 rpm; m_a = 1450 kg; A = 2,6 m^2; C_d = 0,4; f = 0,018; ρ = 1,18 kg/m^3.

It is observed from Fig. 8 that, the ECE cycle can be divided into 3 stages, with each of these stages having phases of pausing (P_i), accelerating (A_i), cruising at a constant speed (C_i) and braking (B_i), total time of the whole cycle is 195 s [14].

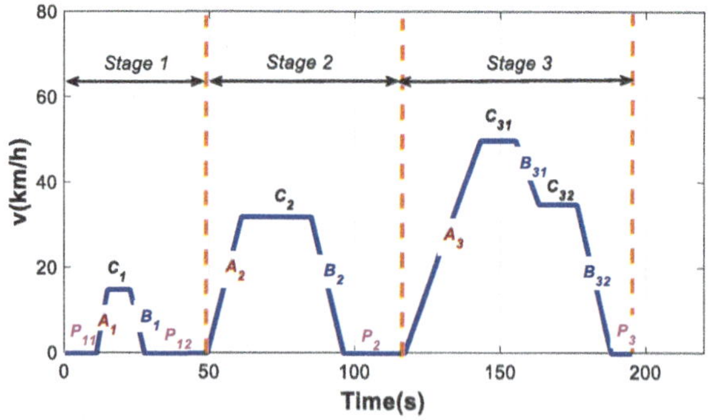

Fig. 8. European standard urban driving cycle (ECE-15)

The time histories of the gear shift and clutch pressures along the ECE-15 are shown in Fig. 9. The half-gear states indicate gear shifts are taking place by engaging and disengaging the dual clutches. During the analysis of the simulation results, it was verified that, in the acceleration scenario A_1 the vehicle accelerates from 0 km/h to 15 km/h in 4 s, there is an upshift 0–1, at 11 s; in the steady-speed phase C_1, the upshift recorded in simulation is 1st–2nd upshift; the B_1 deceleration phase is done in 5 s, there is a 2nd–1st gear downshift at 23 s, due to vehicle speed drop to 0 km/h.

In the second stage, at 49 s the vehicle performs acceleration phase A_2 from 0 to 32 km/h in 12 s, the transmission starts in 1st gear and upshift into 2nd gear, at 11 s, and into 3rd gear, at 13 s. In the cruising phase C_2 at a constant speed of 32 km/h for 24 s, the transmission is kept in 3rd gear. In the deceleration phase B_2, the speed of the vehicle decreases from 32 km/h to zero and there are two downshifts 3rd–2nd gear, at 89 s, and 2nd–1st gear, at 93 s.

In the acceleration scenario of the third stage A_3, we only have upshifts, due to increasing vehicle speed from 0 to 50 km/h. The final upshift recorded in the simulation is the 4th–5th upshift, at 143 s. In the stable driving phase C_{31}, the vehicle's speed is maintained at 50 km/h, the transmission is kept in 5th gear. In the braking phase B_{31}, since the speed is reduced from 50 km/h to 35 km/h, the transmission is forced to do a downshift from 5th to 4th gear and stays in this gear when the speed is kept stable at 35 km/h in 12 s (corresponding to C_{32} phase). In the second deceleration phase of the third stage B_{32}, the vehicle speed is reduced from 35 km/h to 0 km/h in 12 s, we only have downshifts: 4–3 downshift at 177 s; 3–2 downshift at 179 s and 2–1 downshift at 183 s. The simulation results of clutch pressures in Fig. 9 also show that the off-going clutch is released simultaneously as the on-coming clutch is engaged. This gives the uninterrupted torque supply to the driveline during the shifting process.

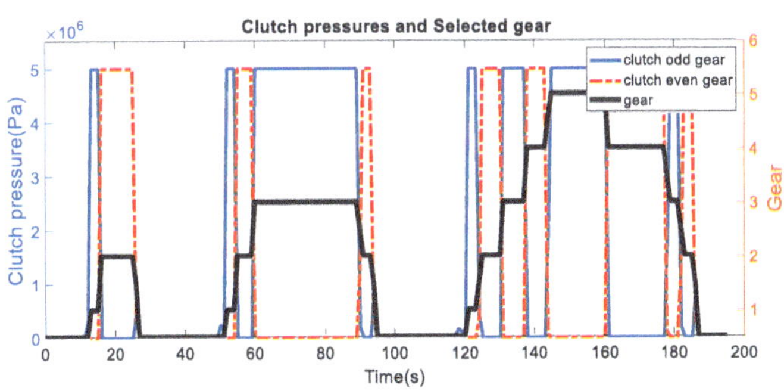

Fig. 9. Clutch pressures and selected gear for the ECE driving cycle

Figure 10 reports the calculation results of the vehicle's acceleration and the required engine torque along the ECE-15. From the simulation results, we can see that in the acceleration phases A_1, A_2, and A_3 (see Fig. 10a), the maximum acceleration value is 1.354 m/s^2 at 12 s in the acceleration phase (A_1) of the 1st-stage, while in the braking phases the maximum value of the acceleration achieved is 2.189 m/s^2 at t = 186 s in the braking phase B_{32} of the 3rd-stage. The larger the difference in the initial and final velocities and the smaller the acceleration time, the greater the value of the vehicle's acceleration.

It is observed from Fig. 10b that, during the acceleration phases, the higher the difference between the initial and final speed values of the vehicle, the larger the value of torque to be provided. The maximum torques demanded from the engine for the acceleration stages A_1, A_2 and A_3 are 52.16 Nm, 73.54 Nm and 84.33 Nm, respectively.

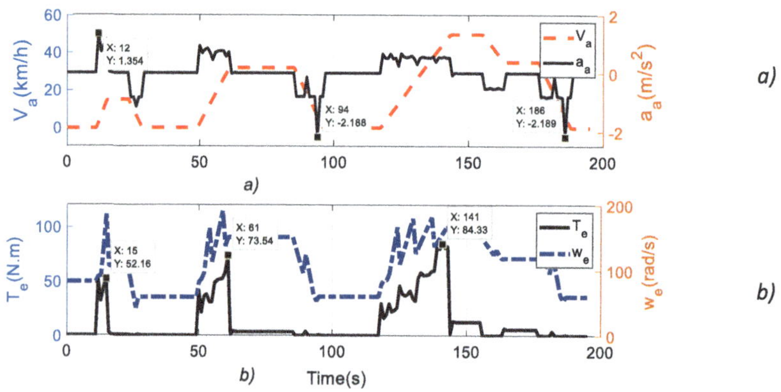

Fig. 10. Vehicle's acceleration, engine torque and engine speed for the ECE-15

Figure 11 shows the longitudinal force (including traction force and brake force) for the ECE-15 driving cycle. It can be seen that: in the acceleration phases, the smaller the vehicle's speed, the greater the traction force at the wheel; The maximum traction forces corresponding to the three phases of acceleration A_1, A_2, and A_3 are 2177 N, 1522 N, and 1047 N respectively. In the braking phases, the larger the difference between the initial and final speeds and the smaller the braking time, the greater the demaned braking force. The maximum demaned braking force P_{bmax} = 1850 N of the phase B_1 corresponding to Δv = 15 km/h and braking time 5 s, the demaned braking force of the phase B_{32} (with Δv = 35 km/h and braking time 12 s) is 3515 N.

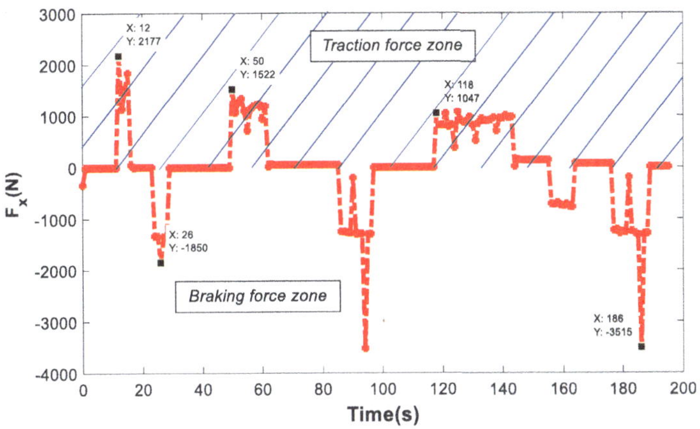

Fig. 11. The longitudinal force–time curve for the ECE-15

4 Conclusions

In this paper, in order to analyze the dynamic and kinetic characteristics of the vehicle, a computer simulation model of the vehicle equipped with a 6-speed dual-clutch transmission, using the Matlab/Simscape driveline computer software, has been developed. Some simulation results of the kinematics and dynamics parameter of the vehicle and gearshift process with ECE-15 driving cycle using the above model are presented. The model is highly flexible, different vehicle variants, with different engine and transmission configurations, can be simulated, only by modifying the parameters. Therefore, this model can be used in the study of the kinematics and dynamics of vehicles with different transmission system configurations at the initial design stage.

Acknowledgment. This research was funded by Thai Nguyen University of Technology.

References

1. Cavina, N., Olivi, D., Corti, E., Poggio, L., Marcigliano, F.: Development of a dual clutch transmission model for real-time applications. IFAC Proc. **45**(30), 440–447 (2012)
2. Kluger, M.A., Long, D.M.: An overview of current automatic, manual and continuously variable transmission efficiencies and their projected future improvements. J. Commer. Veh. **108**(2), 1–6 (1999)
3. Vahidi, A., Druzhinina, M., Stefanopoulou, A., Peng, H.: Simultaneous mass and time varying grade estimation for heavy duty vehicles. In: Proceedings of American Control Conference 6, pp. 4951–4956 (2003)
4. Yoon, Y.-S., Kim, S.J., Kim, K.-S.: Conceptual design of economic hybrid vehicle system using clutchless geared smart transmission. Int. J. Automot. Technol. **14**(5), 779–784 (2013). https://doi.org/10.1007/s12239-013-0086-z

5. Kim, S.J., Song, C., Kim, K.-S., Yoon, Y.-S.: Analysis of the shifting behavior of a novel clutchless geared smart transmission. Int. J. Automot. Technol. **15**(1), 125–134 (2014). https://doi.org/10.1007/s12239-014-0014-x
6. Shin, J.W., Kim, J.O., Choi, J.Y., Oh, S.H.: Design of 2-speed transmission for electric commercial vehicle. Int. J. Automot. Technol. **15**(1), 145–150 (2014). https://doi.org/10.1007/s12239-014-0016-8
7. Kulkarni, M., Shim, T., Zhang, Y.: Modeling and control for launch and shift of dual-clutch transmission vehicles. In: ASME 2005 International Design Engineering Technical Conferences and Computers and Information in Engineering Conference, Long Beach, California, USA, 5b, pp. 597–604 (2005)
8. Goetz, M.: Integrated Powertrain Control for Twin Clutch Transmissions. PhD Thesis, University of Leeds, UK (2005)
9. Kulkarni, M., Shim, T., Zhang, Y.: Shift dynamics and control of dual-clutch transmissions. Mech. Mach. Theory **42**(2), 168–182 (2007)
10. Walker, P.D., Zhang, N., Tamba, R.: Control of gear shifts in dual clutch transmission powertrains. Mech. Syst. Signal Process. **25**(6), 1923–1936 (2011). https://doi.org/10.1016/j.ymssp.2010.08.018
11. Liu, Y., Qin, D., Jiang, H., Zhang, Y.: A systematic model for dynamics and control of dual clutch transmissions. J. Mech. Des. **131**(6), 061012 (2009)
12. Goetz, M., Levesley, M., Crolla, D.: A gearshift controller for twin clutch transmissions, VDI Berichte 2003, pp. 381–400 (2003)
13. https://www.mathworks.com
14. Barlow, T.J., Latham, S., McCrae, I.S., Boulter, P.G.: A Reference Book of Driving Cycles for Use in the Measurement of Road Vehicle Emissions, TRL Limited
15. Pacejka, H.: Tire and Vehicle Dynamics, 3rd ed., Butterworth-Heinemann, Elsevier, 629 p. (2012)
16. Hoan, N.T., Tuan, N.K.: Automotive Transmission System, Vietnam Education and Training Publishing, 231 pp. (in Vietnamese) (2018)
17. Tuan, N.K.: Modeling and simulation of series parallel HEV using Matlab/Simulink. Int. J. Mech. Eng. Technol. **9**(11), 1590–1599 (2018)

A Study on the Effects of Temperature and Heat Retention Time at High Tempering on the Hardness and Strength Limit of the Turbine Shaft Made of 42CrMo Steel

Tran Thi Van Nga[1], Le Hong Ky[2(✉)], Nguyen Ngoc Nhan[2], and Tran Vinh Hung[3,4]

[1] Faculty of Mechanical Engineering, University of Transport and Communications, Hanoi, Vietnam
[2] Vinh Long University of Technology Education, Vinh Long, Vietnam
kylh@vlute.edu.vn
[3] Faculty of Mechanical Engineering and Mechatronics, PHENIKAA University, Hanoi 12116, Vietnam
[4] PHENIKAA Research and Technology Institute (PRATI), A&A Green Phoenix Group JSC, No. 167 Hoang Ngan, Trung Hoa, Cau Giay, Hanoi 11313, Vietnam

Abstract. The turbine shaft is an important part of the turbocharger structure. This part requires quite high hardness, especially when operating it needs high strength to ensure the working conditions with very high speed, pressure and temperature. This paper presents the research results of the effects of temperature and heat retention time at high tempering on the hardness and tensile strength of the turbine shaft in the HX40W model turbocharger structure made of 42CrMo steel.

Keywords: Hardness · Tensile strength · Heat treatment · High tempering · Optimal · Turbine shaft · Turbocharger

1 Introduction

Engines using Turbocharger often have many times higher power than naturally loaded engines with the same cylinder capacity. Turbocharger works with very high temperature, and working speed is usually 100,000 rpm (can be up to 250,000 rpm). The turbine shaft has a very special structure. This is an important part connecting the two turbine blades.

Usually, the shaft part is manufactured in one piece with the flame blade, so in the process of using, the shaft is worn or when there is a problem with the turbine blade, it is required to replace both the shaft and the turbine blade. The manufacture of turbine blades is mainly by casting in metal molds [9, 10]. The heat treatment of turbocharger turbine shafts is still a secret of the companies, so there is almost no scientific

D. C. Nguyen et al. (Eds.): ICERA 2021, LNNS 366, pp. 117–127, 2022.
https://doi.org/10.1007/978-3-030-92574-1_12

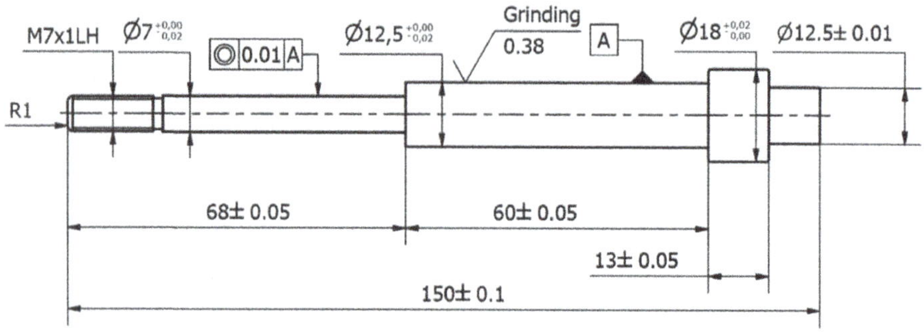

Fig. 1. Turbine shaft

publication. EMAG Corporation [13] introduces the automatic production of turbine shafts (Fig. 2). In the heat treatment part, a robot picks it up from there and transfers it for heat treatment in a MIND-L 1000 from EMAG eldec, which quickly and precisely heats the component to a specific temperature to relieve any possible stresses. However, such an automatic production system is not suitable for the actual conditions of the research facilities. Previously published studies [1–4] mentioned the design and manufacture of turbine blades. [11] the published research results on the effects of temperature and heat retention time at high tempering of steam turbine. High yield strength, tensile strength and percentage elongation are required along with high impact values at the same time. In addition, the material should exhibit good high temperature properties such as creep, elevated temperature strength and thermal stability. The material requires high tensile properties along with good impact strength for proper application. But it is generally observed that these two properties are inversely related to each other, i.e. the efforts to increase the tensile strength of materials results in decrease in impact strength and vice versa. Therefore, one has to judiciously select the heat treatment parameters so as to maintain a good combination of these properties [12].

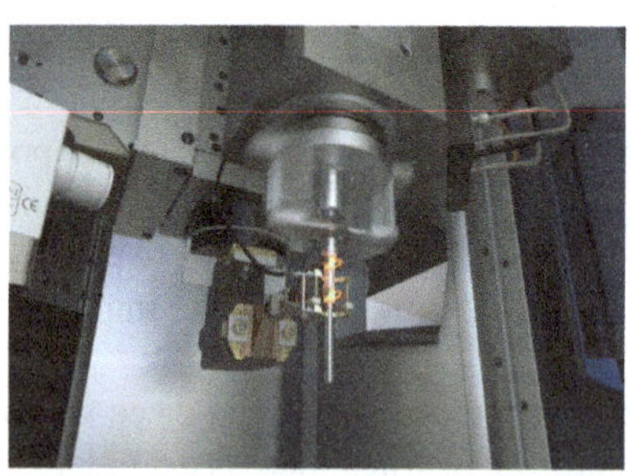

Fig. 2. The heat treatment is performed in a MIND-M 1000 from EMAG eldec

Influence of sub-temperature quenching temperature on the tensile strength, hardness and toughness of 35CrMo steel was studied [14]. The microstructure characteristics of this steel after sub-temperature quenching were analyzed. The results showed in the range of 780–810 °C the strength and hardness of the 35CrMo steel sub-temperature quenched increased with quenching temperature increasing, reaching the highest strength and hardness at 800 °C quenching and then began to decrease, while its elongation decreased with the quenching temperature increasing. Effects of tempering temperature on the microstructures and low temperature impact toughness of 42CrMo4-V steel were investigated [15]. Microstructures of 42CrMo4-V steel after tempering at 570–720 °C for 3 h were experimentally investigated using scanning electron microscopy, electron back-scattered diffraction, transmission electron microscopy and X-ray diffraction.

This paper publishes the research results on the effects of temperature and heat retention time at high tempering on the hardness and tensile strength of the turbine shaft body in the HX40W model turbocharger structure made of 42CrMo steel (Fig. 1). The study also determined the optimal technology parameters when highly tempering the turbine shaft.

2 Experiments

2.1 Sample for Experiment

From the above analysis, in this study, the material is 42CrMo low-alloy steel (Fig. 2a) with the chemical composition given in Table 1 to conduct heat treatment to achieve the microscopic organization of sorbite with the yield strength and low relative elongation mentioned above.

Table 1. Chemical composition of 42CrMo steel for turbine

Steel	Chemical composition						
	C	Si	Mn	Cr	Mo	P	S
42CrMo	0.38–0.45	≤0.40	0.60–0.90	0.90–1.20	0.15–0.30	≤0.035	≤0.035

2.2 Experimental Equipment, Measuring Instruments

With 42CrMo low-alloy steel, to achieve fine sorbite organization, the part must go through high tempering and hardening stages. When hardening, all samples have the same hardening technology mode. This paper focuses on the high tempering section.

The purpose of high tempering is to reduce or lose internal stress, giving the steel high mechanical properties and improving machinability. The workpiece material is hardened 42CrMo steel (Fig. 3a), using a volumetric furnace with LAM – 50 N2 shielding gas (Fig. 3b), heated by a resistor, and controlled by a pulse system to help keep power supply so that the heating is always stable. In addition, the furnace also uses a protective nitrogen gas system to help the parts when the tempering organizations are transformed evenly. Method of single-stage calcination, sparse arrangement in furnace and cooling in SHL QUENCH 100B oil.

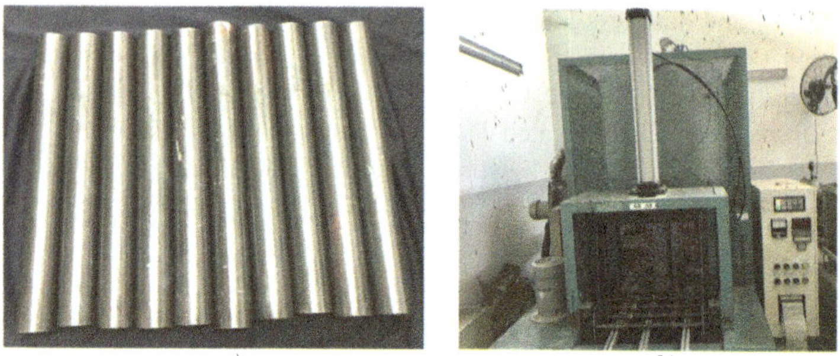

a) *b)*

Fig. 3. Workpieces, equipment and tools of jig for placing workpieces in the experiment

In the study, the Wilson Hardness device was used to determine the hardness of the part in the range of 28–32HRC (Fig. 4). INSTRON's DX300 strength tester as Fig. 5 was used to determine the tensile strength of the specimen $\sigma b = 1044\ N/mm^2$. These measuring devices were also used to measure experimental results.

Fig. 4. Wilson Hardness tester

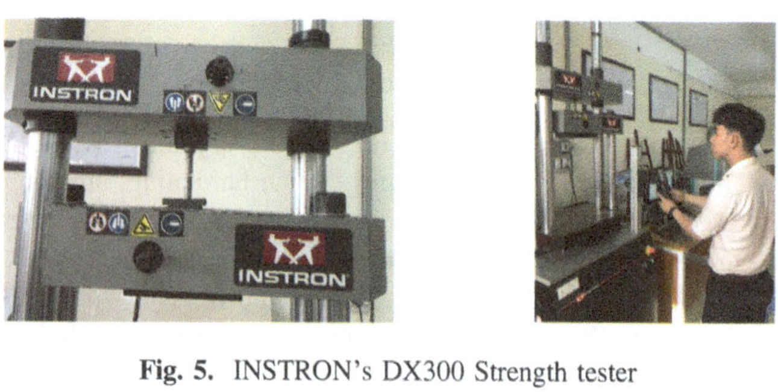

Fig. 5. INSTRON's DX300 Strength tester

2.3 Determining High Tempering Mode

Selecting high tempering temperature of Temp_R: With axial parts, working in continuous rotating conditions at high speed, pressure, strength, as well as the part material with microstructure as sorbite, in the experiment, the temperature range was selected to plan as two edges of high tempering mode: 550 °C (the edge at the high temperature of the low tempering) and 670 °C (the edge at the high temperature of the high tempering) [5, 7].

Selecting tempering time of Time_R: The minimum effective time when high tempering is 120 min, here selected as the lower limit [5, 7]. The mechanical properties of the turbine shaft part to be achieved are the tensile strength and hardness. With 42CrMo steel, the heat retention time is up to 240 min to ensure the Martensite and Austenite organizations to be residual after hardening into the desired Sorbite organization [3].

The purpose of this study is to find the optimal mechanical properties of 42CrMo steel, so in the experiment, the temperature range was selected from 550 – 670 °C. High tempering time is 120–240 min. The input parameters for the experiment are determined as shown in the matrix in Table 2.

Table 2. Input parameters for the experiment

No	Parameters	Unit	Symbol		Design variable value		
			Nature	Encode	(−1)	(0)	(+1)
1	Tempering temperature	°C	Temp_R	A	550	610	670
2	Tempering time	min	Time_R	B	120	180	240

2.4 Experiments

The symmetric mixed quadratic experimental plan of the form B with the polynomial regression equation has the general form as formula (1):

$$y = b_o + \sum_{i=1}^{k} b_i x_i + \sum_{i=1}^{k} b_{ii} x_i^2 \sum_{\substack{i<j \\ i,j=1}}^{k} b_{ij} x_i x_j \ [6] \tag{1}$$

The output parameters to be studied are HRC hardness and σb (σ_b) tensile strength, respectively. The planning matrix is in a natural form like Table 3.

9 experiments were conducted, in which the experiment at the center (Temp_R = 610 °C, time_R = 180 min) with 2 different test specimens to determine the variance of tensile values.

Table 3. High tempering experiment matrix

Experiment	Temp_R	Time_R
1	550	120
2	670	120
3	550	240
4	670	240
5	550	180
6	670	180
7	610	120
8	610	240
9	610	180
10	610	180

The workpiece specimen after high tempering (Fig. 6a) is shaped according to the ASTM E8 standard for tensile test specimens (Fig. 6b). The standard test specimens are measured for hardness on a Wilson Hardness instrument, and for strength on a DX300. The results of the measurements of the specimens after running on Minitab software have the values as shown in Table 4. In which: $\overline{\sigma_b}$ – mean value of tensile strength, s_{σ_b} – variance of tensile value, $\overline{HRC}$ – mean value of hardness, and s_{HRC} – hardness variance.

a) *b)*

Fig. 6. Specimens

Table 4. Results of measuring the hardness and tensile strength of the specimens

No	Temp_R	Time_R	$\overline{\sigma_b}$	s_{σ_b}	$\overline{HRC}$	s_{HRC}
1	550	120	1111	4	35.93	0.2
2	670	120	928	12	23.85	0.5
3	550	240	1093	7	33.97	0.2
4	670	240	888	14	22.5	0.5
5	550	180	1100	4	34.6	0.1
6	670	180	913	13	23.25	0.4
7	610	120	973	7	31.7	0.5
8	610	240	947	13	30.9	0.4
9	610	180	964	13	31.72	0.6
10	610	180	959	13	31.62	0.2

3 Analysis of Experimental Results [6]

3.1 *HRC* Hardness

The synthesis and analysis of the coefficients of the complete regression equation of the hardness are shown in Table 5.

Table 5. The coefficients of the complete regression equation of the hardness

Term	Coef	SE Coef	T-value	P-value
Constant	31.483	0.124	254.81	0.000
Temp_R	−5.817	0.101	−57.66	0.000
Time_R	−0.686	0.101	−6.80	0.000
Temp_R * Temp_R	−2.467	0.160	−15.46	0.000

The regression equation characterizing the influence of temperature and heat retention time at high tempering on the hardness of the specimen has the natural form as follows:

$$\text{HRC} = 162.3 + 0.7390\ \text{Temp\_R} - 0.01144\ \text{time\_}R - 0.000685\ \text{Temp\_R} * \text{Temp\_R} \quad (2)$$

From here, graphs showing the influence of two input parameters, namely tempering temperature and heat retention time, on the HRC hardness of the specimen are constructed as shown in Fig. 6.

Equation (2) and the graphs in Fig. 7 show that when the tempering temperature (Temp_R) increases, the hardness decreases and it decreases a lot when the temperature is higher. As the heat retention time (Time_R) increases, the hardness of the part decreases, but it decreases slightly.

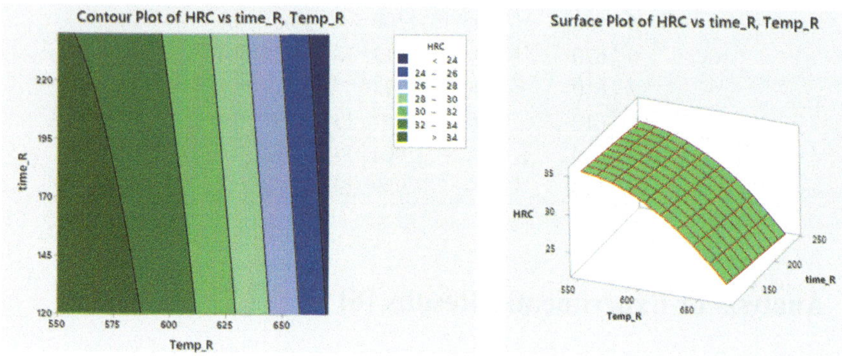

Fig. 7. HRC analysis results

3.2 σ_b Tensile Strength

The synthesis and analysis of the coefficients of the complete regression equation of the tensile strength are shown in Table 6.

Table 6. The coefficients of the complete regression equation of the tensile strength

Term	Coef	SE Coef	T-value	P-value
Constant	960.92	2.96	324.61	0.000
Temp_R	−95.78	2.42	−39.63	0.000
Time_R	−14.00	2.42	−5.79	0.000
Temp_R * Temp_R	44.42	3.82	11.62	0.000

The regression equation characterizing the influence of the input parameters on the strength limit of the specimen has the following natural form:

$$\sigma_b = 6568 - 16.65\, Temp\_R - 0.2333\, time\_R + 0.01234\, Temp\_R * Temp\_R \quad (3)$$

From here, graphs can be constructed showing the influence of two input parameters, namely temperature and tempering time, on the tensile strength of the specimen as shown in Fig. 8.

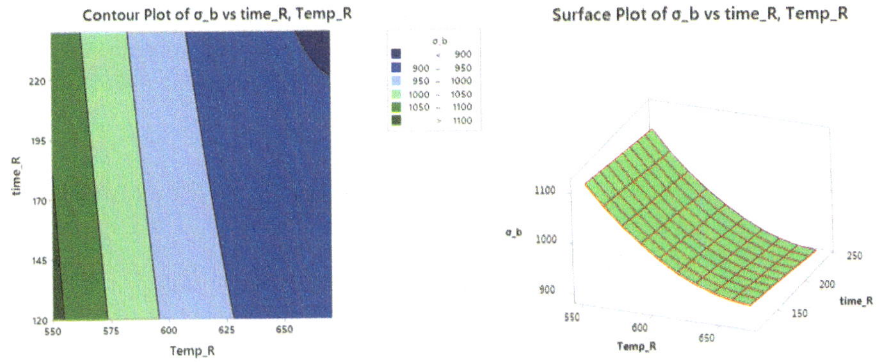

Fig. 8. Strength limit analysis results

Equation (3) and the graphs in Fig. 8 show that as the temperature (*Temp_R*) increases, the strength limit decreases and it tends to decrease a lot as the temperature increases. As for the time factor (*Time_R*), as the time increases, the strength limit of the part decreases, but it decreases very little.

Analysis by Minitab 19 software shows that the technological parameters of high tempering mode affect the mechanical properties of 42CrMo steel, corresponding to the graph experiments showing that as the temperature increases, the strength and hardness limit decreases over time to the high edge of the high tempering limit.

3.3 Optimization of High Tempering Technology Parameters

After having 2 regression equations, two output parameters, namely tempering temperature and heat retention time were optimized. Optimizing these two parameters for a 10% error limit will have the allowable value of the result as shown in Table 7.

Table 7. Allowable value adjustment table of results

Response	Goal	Lower	Target	Upper	Weight	Importance
σ_b	Target	877.0	1044.0	1115.0	1	1
HRC	Target	21.9	31.4	36.1	1	1

Optimum value results:

Solution	Temp_R	Time_R	σ_b Fit	*HRC* Fit	Composite Desirability
1	617.536	120	963.588	31.4000	0.720064

When there is an optimal result, this parameter is taken to perform high tempering to test the actual mechanical properties. Results after tensile test according to ASTM E8 standard:

$$\sigma_b = 1027\,\mathrm{N/mm^2}$$

$$HRC = 30.35$$

Comparing the results with the sample axis:

Sample parts	Parts with optimal parameters
σ_b = 1044 N/mm^2	σ_b = 1027 N/mm^2
HRC = 31.4	HRC = 30.35

From the above comparison results, the hardness and tensile strength of the optimum parts are satisfactory within the allowable planning calculation and experimental data processing.

4 Conclusion

This paper presents the research results on the effects of temperature and heat retention time at high tempering on the hardness and tensile strength of the turbine shaft made of steel. Research results have built regression equation of strength limit and hardness of the turbine shaft parts. The research has also determined the set of technological parameters that have the greatest influence on the hardness and tensile strength of the turbine shaft parts, in which the optimum tempering temperature of *Temp_R* is 620 °C, and the optimal heat retention time when high tempering of *Time_R* is 160 min.

References

1. Ky, L.H., Tam, D.T.: Design, analysis of turbo blade in the engine turbocharger, Vietnam J. Mech. (2020). ISSN: 0866-7056
2. Tam, D.T., Hung, T.V., Ky, L.H.: Assessment of direct working accuracy data of turbine parts from inventor professional software on CNC milling machine, Vietnam J. Mech. (2020). ISSN: 0866-7056
3. Ky, L.H., Thong, N.C.: A study on turbo blade processing in turbocharger structure on CNC HAAS VF2 milling machine, Vietnam J. Mech. (2020). ISSN: 0866-7056
4. Thong, N.C., Hung, T.V., Ky, L.H.: A research on turbine blades in HX40W turbocharger structure on CNC milling machines, Vietnam J. Mech. (2020). ISSN: 0866-7056
5. Ha, N.N.: Building Heat Treatment Mode, HCMUT-VNU-HCM (2018)
6. Loc, N.H.: Planning and Experimental Analysis, HCMUT-VNU-HCM (2011)
7. Cang, N.C.: Handbook of Heat Treatment 1, Hanoi University of Science and Technology
8. Toan, N.C., Tuan, D.D.: Establishment of process and program for pure tensile fatigue testing of metallic materials on LFV 500-HH fatigue testing machine, Transp. Commun. Sci. J. **71**(2) (2020)

9. Takeuchi, Y., Watanabe, T.: Generation of 5-axis control collision-free tool path and post-processing for NC data. CIRP Ann. **41**, 539–542 (1992)
10. Bi, Q.Z., Chen, H., Zou, X., Zhu, L.M., Ding, H.: Five-axis flank milling for design and manufacture of turbocharger compressor impeller. In: Proceedings of ASME Turbo Expo, Turbine Technical Conference and Exposition, pp. 1–11 (2014)
11. Wang, W.Z., Kui, F.-Z.X., Shan, L.Z., Tu, T.: Failure Analysis of the Final Stage Blade in Steam Turbine, School of Mechanical and Power Engineering, East China University of Science and Technology, P.O. Box 402, Shanghai 200237, PR China
12. Ajit Kumar, K., Rajesh, G.: Optimization of Heat Treatment of Steel Developed for Turbine Shafts. BTech Thesis (2012)
13. https://www.emag.com/blog/en/finish-machining-turbocharger-shafts/
14. Chen, R., Sung, W.P.: Effect of sub-temperature quenching temperature on microstructure and property of 35CrMo steel, Adv. Mater. Res. 490–495. https://doi.org/10.4028/www.scientific.net/AMR.490-495.3257
15. Sun, C., Fu, P.-X., Liu, H.-W., Liu, H.-H., Du, N.-Y.: Effect of tempering temperature on the low temperature impact toughness of 42CrMo4-V steel. Metals **8**(4), 232 (2018). https://doi.org/10.3390/met8040232

A Study on the Effects of Tire Vertical Stiffness on Dynamic Load of DVM 2.5 Truck

Nguyen Thanh Tung(✉) and Luong Van Van(✉)

Vinh Long University of Technology Education, 73, Nguyen Hue Street, Vinh Long City, Vietnam
{tungnt, vanlv}@vlute.edu.vn

Abstract. The dynamic load acting on the vehicle has a variable value, it depends on the moving conditions, road quality and tire stiffness, etc. In this paper, the authors study on the effects of the tire vertical stiffness on dynamic load acting on DVM 2.5 truck. The method of structural separation multi-body system and the Newton–Euler equation are used to set the three-dimensional dynamics model. Matlab-simulink software is used to consider the effects of the tire vertical stiffness on dynamic load of DVM 2.5 truck. The results show that, when the tire vertical stiffness is increased, the maximum dynamic load acting on the tires, chassis and road rises and the transmission capacity drops. In order to ensure the dynamics safety and durability, a fully loaded DVM 2.5 truck is run on a class E road according to ISO 8608:2016, the speeds must be less than 50 kph and the tire vertical stiffness must be less than 912 kN/m.

Keywords: Tire vertical stiffness · Dynamic load · Vehicle dynamics safety

1 Introduction

The dynamic load acting on the vehicle is the cause of vibration, loss of dynamics safety, vehicle damage and road destruction. The specifications are used to evaluate the dynamic load acting on vehicle as following formulas:

(i) The maximum dynamic load factor acting on the chassis (η_{max}): This specification is used to evaluate the smooth movement and durability of the vehicle [1–3].

$$\eta_{\max} = \frac{F_{C\max} + F_{K\max}}{F_G};\ \eta_{\max} \leq\ 1.5 \tag{1}$$

(ii) The maximum dynamic load factor acting on the tire (K_{dmax}): This specification is used to evaluate the dynamics safety, smooth movement and durability of the vehicle [1–3].

$$K_{d\max} = \frac{F_{CL\max}}{F_G} + 1;\ K_{d\max} \leq\ 2.5 \tag{2}$$

D. C. Nguyen et al. (Eds.): ICERA 2021, LNNS 366, pp. 128–134, 2022.
https://doi.org/10.1007/978-3-030-92574-1_13

(iii) The minimum dynamic load factor acting on the tire (K_{dmin}): This specification is used to evaluate the vehicle dynamics safety and tire transmission capacity [1–3].

$$K_{d\min} = \frac{F_{CL\min}}{F_G} + 1;\ 0 \leq K_{d\min} \leq 1 \tag{3}$$

$K_{dmin} = 0.5$ within the warning limit; $K_{dmin} = 0$ is the intervention limit.

2 The Three-Dimensional Dynamics Model

The method of separating the structure of the multibody system is used to build a three-dimensional (3D) dynamics model of DVM 2.5 truck is shown in Fig. 1 [4–6].

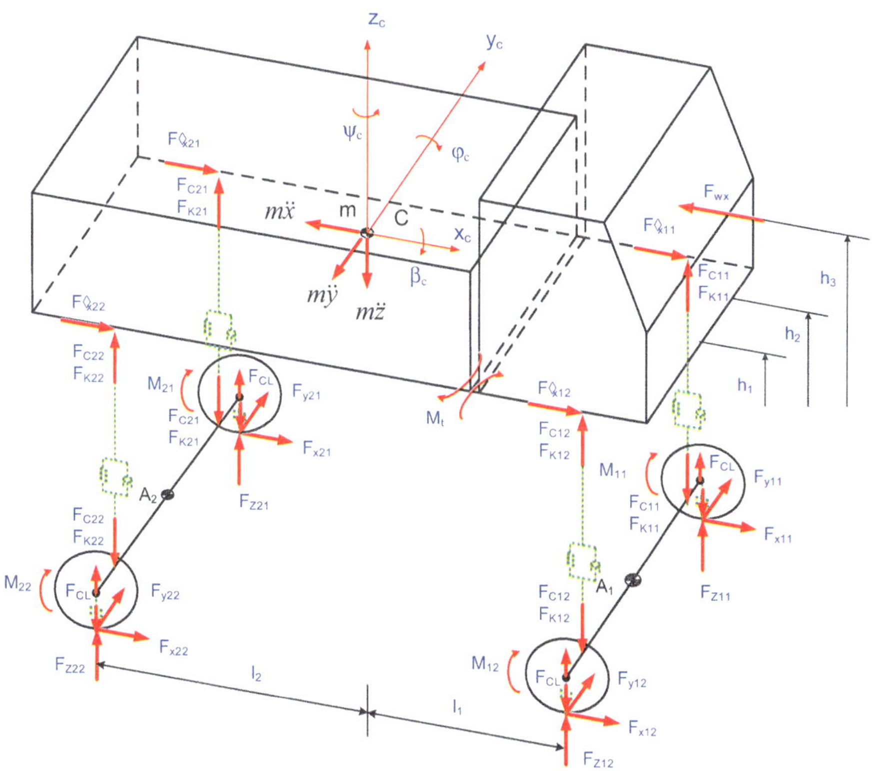

Fig. 1. Dynamics model of the DVM 2.5 truck

The Newton–Euler equations are used to build the system of dynamics equations for DVM 2.5 truck as follows [4–6] (Table 1):

$$m\ddot{x} = F_{x1j}\cos\delta_{1j} - F_{y1j}\sin\delta_{1j} + F_{x2j};\ j = 1\ \textit{is left wheel};\ i = 2\ \textit{is right wheel} \quad (4)$$

$$m\ddot{y} = F_{x1j}\sin\delta_{1j} + F_{y1j}\cos\delta_{1j} + F_{y2j};\ j = 1\ \textit{is left wheel};\ i = 2\ \textit{is right wheel} \quad (5)$$

$$J_z\ddot{\psi} = (F_{x1j}\sin\delta_{1j} + F_{y1j}\cos\delta_{1j})l_1 + (F_{xi2} - F_{xi1})b_i - F_{y2j}l_2 \quad (6)$$

$$m\ddot{z} = F_{Cij} + F_{Kij};\ \ i = 1 \div 2;\ j = 1\ \textit{ is left wheel};\ i = 2\ \textit{ is right wheel} \quad (7)$$

$$J_y\ddot{\varphi} = (F_{C1j} + F_{K1j})l_1 - (F_{C2j} + F_{K2j})l_2 + M_{1j} + M_{2j}; j = 1\ \textit{is lef wheel};\ i = 2\ \textit{is right wheel} \quad (8)$$

$$J_x\ddot{\beta} = (F_{Ci2} + F_{Ki2} - F_{Ci1} - F_{Ki1})w_i;\ i = 1 \div 2 \quad (9)$$

$$m_{A1}\ddot{z}_{A1} = F_{CLij} + F_{KLij} - F_{Cij} - F_{Kij};\ \ i = 1 \div 2;\ j = 1\ \textit{is left wheel};\ i = 2\ \textit{is right wheel} \quad (10)$$

$$m_{A1}\ddot{y}_{A1} = F_{y1j};\ \ j = 1\ \textit{is left wheel};\ \ i = 2\ \textit{is right wheel} \quad (11)$$

$$J_{Ax1}\ddot{\beta}_{A1} = (F_{C11} + F_{K11} - F_{C12} - F_{K12})w_1 + (F_{CL12} + F_{KL12} - F_{CL11} - F_{KL11})b_1 - F_{y11}(r_{11} + \xi_{A11}) - F_{y12}(r_{12} + \xi_{A12}) \quad (12)$$

$$m_{A2}\ddot{z}_{A2} = F_{CL2j} + F_{KL2j} - F_{C2j} - F_{K2j};\ \ j = 1\ \textit{is left wheel};\ \ i = 2\ \textit{is right wheel} \quad (13)$$

$$m_{A2}\ddot{y}_{A2} = F_{y2j};\ \ j = 1\ \textit{is left wheel};\ \ i = 2\ \textit{is right wheel} \quad (14)$$

$$J_{Ax2}\ddot{\beta}_{A2} = (F_{C21} + F_{K21} - F_{C22} - F_{K22})w_2 + (F_{CL22} + F_{KL22} - F_{CL21} - F_{KL21})b_2 - F_{y21}(r_{21} + \xi_{A21}) - F_{y22}(r_{22} + \xi_{A22}) \quad (15)$$

$$J_{Ayij}\ddot{\varphi}_{ij} = M_{Aij} - M_{Bij} - F_{xij}r_{dij};\ i = 1 \div 2;\ j = 1\ \textit{is left wheel};\ \ i = 2\ \textit{is right wheel} \quad (16)$$

Table 1. A list of symbols and abbreviations

Symbols	Units	Explain
i		Axle number of the truck, i = 1 ÷ 2
j		j = 1 is left wheel; j = 2 is right wheel; j = 1 ÷ 2
δ	Degree	Front wheel turn angle
l_i	m	Longitudinal distance of axle i from mass center
h_2	m	Height of mass center from the ground
b	m	Lateral distance of the wheels
φ_{ij}	rad	Rotation angle of the ij wheel
β, φ, ψ	Degree	Rotation angle of the tractor body around the x, y, z axis
m, mA	kg	Mass and un-sprung mass of the truck
J_x, J_y, J_z	kgm^2	Moment of inertia about the x, y, z axis of the sprung mass
J_{Ayij}	kgm^2	Moment of inertia about the y-axis of the ij wheel
F_{Cij}	N	The suspension elastic force of the ij wheel
F_{Kij}	N	The suspension damping force of the ij wheel
F_{CLij}	N	The tire elastic force of the ij wheel
F_{xij}	N	The longitudinal force of the ij wheel
F_{yij}	N	The lateral force of the ij wheel
F_{zij}	N	The vertical force of the ij wheel
F_{Gij}	N	The static weight of the ij wheel
M_{Aij}, M_{Bij}	Nm	Driving torque and braking torque of the ij wheel

3 Survey Results and Discussions

Matlab-Simulink software is used to consider the tire vertical stiffness on dynamic load acting on DVM 2.5 truck. The DVM 2.5 truck is fully loaded and is run at speeds V_0 = [30, 40, 50, 60] kph on a class E road according to ISO 8608:2016 is shown in Fig. 2 [7].

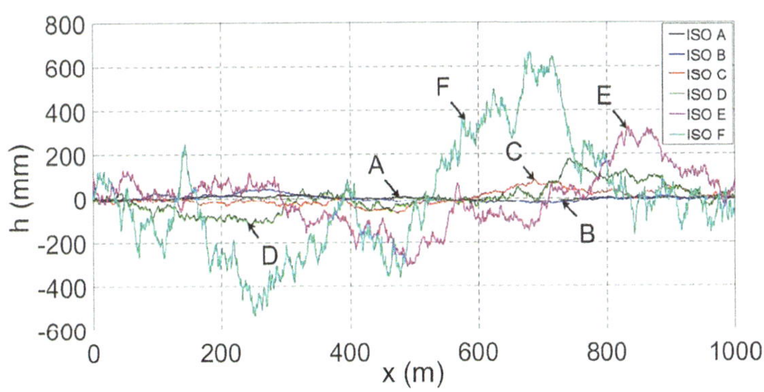

Fig. 2. Road surface according to ISO 8608:2016 standards

The tire vertical stiffness of the DVM 2.5 truck is C_L = 652 kN/m. When surveying, the tire vertical stiffness value is changed as C_{L0} = [0.5, 0.6, 0.7, 0.8, 0.9, 1.0, 1.1, 1.2, 1.3, 1.4, 1.5] C_L, the survey results are as follows:

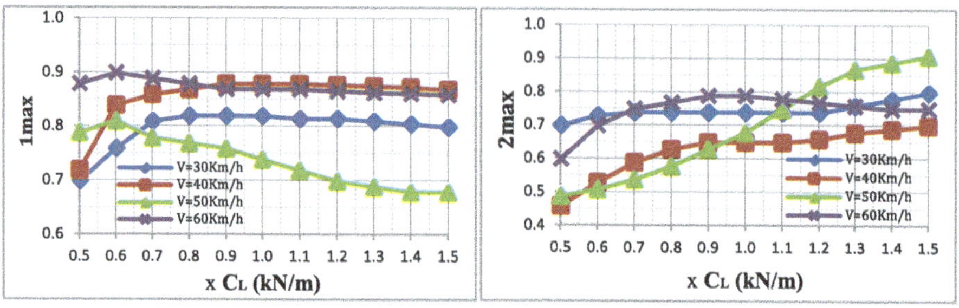

a. Front axle b. Rear axle

Fig. 3. Maximum dynamic load factor acting on the chassis

When the DVM 2.5 truck is run on a class E road at speeds V_0 = [30 ÷ 60] kph, if the tire vertical stiffness is increased from C_{L0} = [0.5 ÷ 1.5] C_L, the maximum dynamic load factor acting on the chassis at the front axle is about η_{1max} = [0.68 ÷ 0.88] shown in Fig. 3a. According to Fig. 3a, the value of the maximum dynamic load factor acting on the chassis at the front axle is $\eta_{1\,max} \leq \eta_{max} = 1.5$, the truck has smooth movement and durability.

When the DVM 2.5 truck is run on a class E road at speeds V_0 = [30 ÷ 60] km/h, if the tire vertical stiffness is increased from C_{L0} = [0.5 ÷ 1.5] C_L, the maximum dynamic load factor acting on the chassis at the rear axle is about η_{2max} = [0.46 ÷ 0.91] shown in Fig. 3b, $\eta_{2\,max} \leq \eta_{max} = 1.5$, the truck has smooth movement and durability.

Figure 4a is a graph showing the maximum dynamic load factor acting on the front tires when the DVM 2.5 truck is run on a class E road at speeds V_0 = [30 ÷ 60] kph. According to Fig. 4a, the tire vertical stiffness is increased from C_{L0} = [0.5 ÷ 1.5] C_L, the maximum

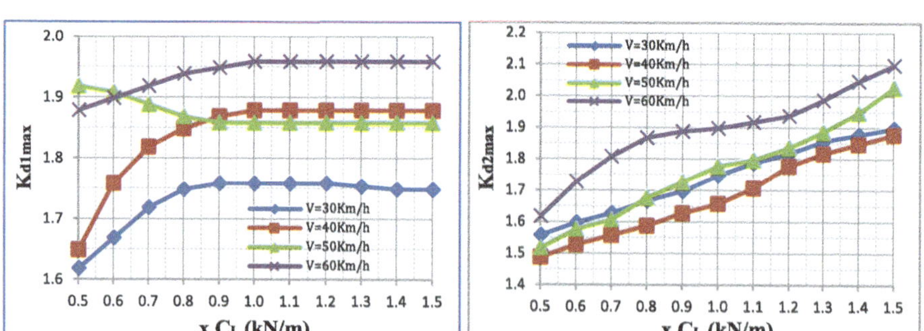

a. Front axle b. Rear axle

Fig. 4. Maximum dynamic load factor acting on the tires

dynamic load factor acting on the front tire is about $K_{d1max} = [1.62 \div 1.96]$, $K_{d1\,max} \leq K_{d\,max} = 2.5$. The front tire of the DVM 2.5 truck has smooth movement and durability, and dynamics safety.

Figure 4b is a graph showing the maximum dynamic load factor acting on the rear tires when the truck is run on a class E road at speeds $V_0 = [30 \div 60]$ kph. When the tire vertical stiffness is increased from $C_{L0} = [0.5 \div 1.5]\ C_L$, the maximum dynamic load factor acting on the rear tires is increased, $K_{d2max} = [1.56 \div 2.1]$, $K_{d2\,max} \leq K_{d\,max} = 2.5$. The rear tires of the DVM 2.5 truck has smooth movement and durability, and dynamics safety.

The minimum dynamic load factor acting on the front tires when the DVM 2.5 truck is run on a class E road at speeds $V_0 = [30 \div 60]$ kph shown in Fig. 5a. When the tire vertical stiffness is increased from $C_{L0} = [0.5 \div 1.5]\ C_L$, the minimum dynamic load factor acting on the front tires drops, $K_{d1min} = [0.36 \div 0]$. When the DVM 2.5 truck is run at $V_0 = [30 \div 50]$ kph, K_{d1min} is within the warning limit, $0 < K_{d1min} \leq 0.5$. The front tires of the DVM 2.5 truck have dynamics safety and tire transmission capacity. When the DVM 2.5 truck is run at a speed of $V_0 = 60$ kph, the value of the minimum dynamic load factor acting on the front tires $K_{d1min} = 0$, K_{d1min} within the intervention limit. The front tires of the DVM 2.5 truck don't have dynamics safety and tire transmission capacity.

The minimum dynamic load factor acting on the rear tires is shown in Fig. 5b. When the tire vertical stiffness is increased from $C_{L0} = [0.5 \div 1.5]\ C_L$, the minimum dynamic load factor acting on the rear tires drops, $K_{d2min} = [0.54 \div 0]$. When $C_{L0} = [0.5 \div 1.4]\ C_L$, K_{d2min} is within the warning limit, $0 < K_{d2min} \leq 0.5$. The rear tire of the DVM 2.5 truck has dynamics safety and tire transmission capacity. When $C_{L0} = 1.5C$, K_{d2min} is within the intervention limit, $K_{d2min} = 0$. The rear tires of the DVM 2.5 truck don't have dynamics safety and tire transmission capacity.

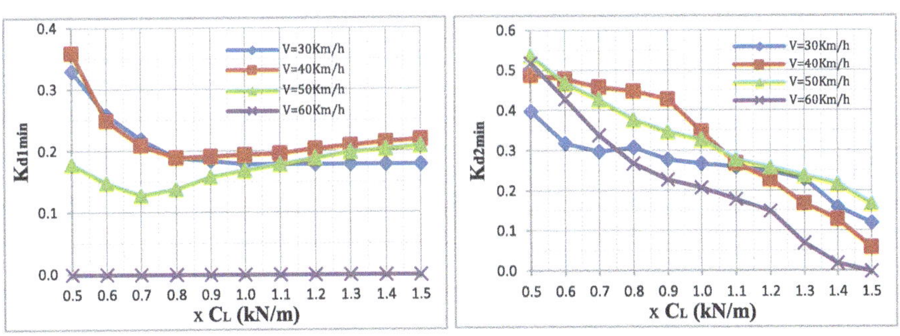

a. Front tire b. Rear tire

Fig. 5. Minimum dynamic load factor acting on the tires

4 Conclusions

When the tire vertical stiffness is increased, the maximum dynamic load value on the tires, chassis and road rises, and the transmission capacity drops. Therefore, the smooth movement and durability, and dynamics safety of the DVM 2.5 truck drop. A fully loaded DVM 2.5 truck is run on a class E road according to ISO 8608:2016, the speeds must be less than 50 kph and the tire vertical stiffness must be less than 912 kN/m (or CL0 $\leq$ 1.4CL), the DVM 2.5 truck is ensured dynamics safety and durability.

References

1. Rajamani, R.: Vehicle Dynamics and Control. Springer, New York (2012)
2. Jazar, R.N.: Vehicle Dynamics Theory and Application. Springer, Switzerland (2017)
3. Huong, V.V.: Vehicle Dynamics. Vietnam Education Publishing House, Vietnam (2014)
4. Nguyen, N.T., Huong, V.V.: The effect of the wheel rotation angle on the braking efficiency of the tractor semi-trailer on the wet roundabout route. Lecture Notes in Networks and Systems, pp. 798–804. Springer, Switzerland (2020)
5. Tung, N.T., Huong, V.V.: Research on the Dynamic Load of the Tractor Semi-Trailer When Braking on the Round Road. Lecture Notes in Mechanical Engineering, pp. 456–461. Springer Verlag, United States (2021)
6. Thanh, T.N.: Research on braking efficiency on the roads with different grip coefficients of the tractor semi-trailer and proposed measures to reduce traffic accidents. Doctoral Thesis, Hanoi University of Science and Technology, Vietnam (2017)
7. ISO 8608:2016: Mechanical Vibration -- Road Surface Profiles. Reporting of Measured Data (2016)

A Three-Phase Transformerless H10 Inverter with Constant Common-Mode Voltage for Photovoltaic Application

Minh-Duc Ngo[1(✉)], Quynh-Van Nong[2], Thuy-Ngan Ngo[3], and Tan-Tai Tran[4]

[1] Thai Nguyen University of Technology, Thai Nguyen, Viet Nam
ngoduc198-tdh@tnut.edu.vn
[2] Thai Nguyen University of Education, Thai Nguyen, Viet Nam
[3] Ha Noi Metropolitan University, Hanoi, Viet Nam
[4] Industrial University of HCM City, Ho Chi Minh, Viet Nam

Abstract. Because of the lack of galvanic isolation, the leakage current causes problems in transformerless three-phase photovoltaic (PV) systems. The conventional three-phase photovoltaic inverter failed to limit the leakage current in a constructive way. In order to solve the leakage current problems in the conventional three-phase photovoltaic inverter, a three-phase transformerless H10 inverter is proposed by combining the conventional four-leg inverter of two switched-boost networks in this paper. The proposed three-phase transformerless H10 inverter has the major features as shoot-through immunity, boost voltage with one-stage conversion. Furthermore, a common-mode voltage of the proposed three-phase transformerless H10 inverter is kept constant during switching cycles. The operating principles for the proposed three-phase transformerless H10 inverter are presented. Finally, the proposed three-phase transformerless H10 inverter topology is tested, and the results validate the effectiveness of the proposed three-phase transformerless H10 inverter topology.

Keywords: Common-mode voltage · Transformerless photovoltaic system · Leakage current

1 Introduction

Recently, transformerless PV inverters [1] have been received more and more attention because of high efficiency and small size. However, due to removing the galvanic isolation between the grid and the dc supply power, the leakage current appears. The common-mode voltage (CMV) with high frequency and amplitude causes the leakage current issue [2]. In a transformerless photovoltaic PV system, the leakage current flows through the parasitic capacitance. As a result, grid current distortions and losses rise. To solve the CMV problems in transformerless photovoltaic (PV) inverters, many topologies have been presented such as H-8 topology in [3] and switched-Capacitor-Voltage-Doubler Based Boost topology in [4]. However, the shoot-through immunity is impossible in these topologies. So, to avoid the shoot-through problem, a dead time between the upper and lower switches is required. A transformerless three-phase four-

D. C. Nguyen et al. (Eds.): ICERA 2021, LNNS 366, pp. 135–141, 2022.
https://doi.org/10.1007/978-3-030-92574-1_14

leg PV inverter with a phase-shifting modulation has been discussed to reduce the CMV in [5]. However, it only operated linearly with a modulation index in the range of 0 to 0.66. A new modulation strategy is introduced in [6] to achieve the constant CMV for Four-Leg PV Inverter. Transformerless Z-Source Four-Leg PV Inverter was presented in [7] with the buck-boost voltage function, constant common-mode voltage and shoot-through immunity. Like in [6], the quasi-switched boost inverter [8] has features such as shoot-through immunity, and single-stage power conversion. The objective of this paper is to introduce a transformerless inverter with shoot-through immunity and low leakage current reduction based on a switched-boost network structure for transformerless PV systems.

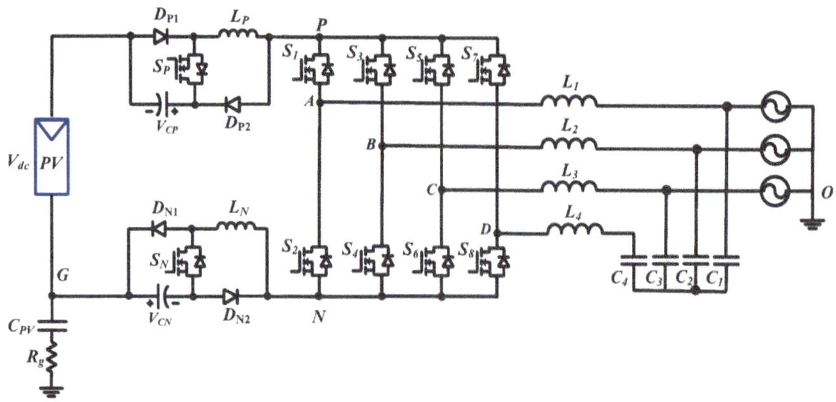

Fig. 1. Proposed three-phase transformerless H10 inverter topology

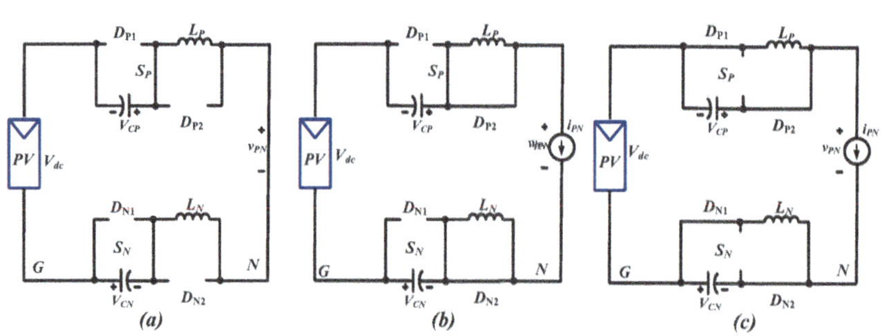

Fig. 2. Equivalent circuits of the proposed three-phase transformerless H10 inverter topology inverter in (a) ST state; (b) NST states 1 and (c) NST state 2.

2 Proposed H10 Inverter Topology and Operation Principles

The proposed H10 inverter is presented in Fig. 1. Two Switched-boost networks are inserted between the DC power (V_{dc}), and the conventional three-phase four-leg inverter. Each Switched-boost network includes one boost inductor, two diodes, one capacitor and an active switch. Unlike the conventional three-phase four-leg inverter, the proposed H10 inverter can work in the shoot-through (ST) state, where both switches of any phase leg are switched on at the same time. The operating states of the proposed H10 inverter are simplified into ST and non-shoot-through (NST) states 1 and NST states 2 as presented in Fig. 2. During ST state as shown in Fig. 2(a), the inverter side is shorted by two power switches of any phase leg. In this state, the time interval is $D.T$, where and T D are the switching period and the duty cycle shoot through of each cycle, respectively. Two switches S_p and S_n are also turned on, while four diodes D_{p1}, D_{p2}, D_{n1}, and D_{n2} are reverse-biased. The two inductors L_p and L_n stores energy while two capacitors C_p and C_n are discharged.

$$\begin{cases} L_p \dfrac{di_{Lp}}{dt} + L_n \dfrac{di_{Ln}}{dt} = V_{dc} + V_{Cn} + V_{Cp} \\ C_p \dfrac{dv_{Cp}}{dt} = C_n \dfrac{dv_{Cn}}{dt} = I_{Lp} = I_{Ln}. \end{cases} \tag{1}$$

During NST states 1 as shown in Fig. 2(b), The time interval is $D_O.T$, where D_O is the duty cycle of two switches S_p and S_n. Two switches S_p and S_n are also turned on. Two diodes D_{p1}, and D_{n1} are reverse-biased while two diodes D_{p2}, and D_{n2} are forward-based.

$$\begin{cases} L_p \dfrac{di_{Lp}}{dt} = L_n \dfrac{di_{Ln}}{dt} = 0 \\ C_p \dfrac{dv_{Cp}}{dt} = C_n \dfrac{dv_{Cn}}{dt} = i_{PN} \\ V_{PN} = V_{dc} + V_{Cn} + V_{Cp}. \end{cases} \tag{2}$$

During non-shoot-through states 2 as shown in Fig. 2(c), two switches S_P and S_N are turned off, while four diodes D_{P1}, D_{P2}, D_{N1}, and D_{N2} are forward-based. The two inductors L_P and L_N discharged while two capacitors C_P and C_N are charged. The time interval is $(1-D_O-D).T$

$$\begin{cases} L_p \dfrac{di_{Lp}}{dt} = -V_{Cp} \; and \; L_n \dfrac{di_{Ln}}{dt} = -V_{Cn} \\ C_p \dfrac{dv_{Cp}}{dt} = I_{Lp} - i_{PN} \; and \; C_n \dfrac{dv_{Cn}}{dt} = I_{Ln} - i_{PN} \\ V_{PN} = V_{dc} + V_{Cp} + V_{Cn}. \end{cases} \tag{3}$$

Assuming inductors L_p, L_n and capacitors C_p, C_n have the same value ($L_p = L_n$ and $C_p = C_n$), the voltage and current of inductors and capacitors are obtained as follows:

$$\begin{cases} V_{Lp} = V_{Ln} = V_L \text{ and } V_{Cp} = V_{Cn} = V_C \\ I_{Lp} = I_{Ln} = I_L \end{cases} \tag{4}$$

The average capacitor voltage, V_C of the proposed H10 inverter is obtained as follows:

$$V_C = V_{Cp} = V_{Cn} = \frac{D}{2(1 - 2D - D_0)} V_{dc}. \tag{5}$$

The boost factor of the proposed H10 inverter can be expressed as:

$$B_p = \frac{V_{PN}}{V_{dc}} = \frac{1 - D - D_0}{1 - 2D - D_0} \tag{6}$$

The CMV (V_{CM}) of the proposed H10 inverter can be defined as follows:

$$V_{CM} = \frac{V_{AG} + V_{BG} + V_{CG} + V_{DG}}{4} \tag{7}$$

The CMV of different switching states during NST states and ST state is indicated in Table 1.

Table 1. Common-mode voltage and switching states

Switching states		V_{CM}
NST states	1111	$(1 + B_p)V_{dc}/2$
	1110, 1101, 1011, 0111	$(2 + B_p)V_{dc}/4$
	1100, 1001, 0011, 0101, 1010, 0110	$V_{dc}/2$
	0001, 0010, 0100, 1000	$(2 - B_p)V_{dc}/4$
	0000	$(1 - B_p)V_{dc}/2$
ST state		$V_{dc}/2$

3 Simulation Results

Simulations are established in PSIM software to confirm the effectiveness of the H10 inverter. The main parameters of the H10 inverter used for simulation are shown in Table 2.

Table 2. Simulation parameters of the H10 inverter

Parameters		Values
Input voltage		300 V
Carrier frequency ($1/T$)		20 kHz
Boost inductor (L_p and L_n)		500 µF
Capacitors (C_p and C_n)		820 µF
Three-phase filter	L_f	1 mH/phase
	C_f	10 µF/phase
Three-phase inductive load (RL)		20 Ω-30 mH/phase

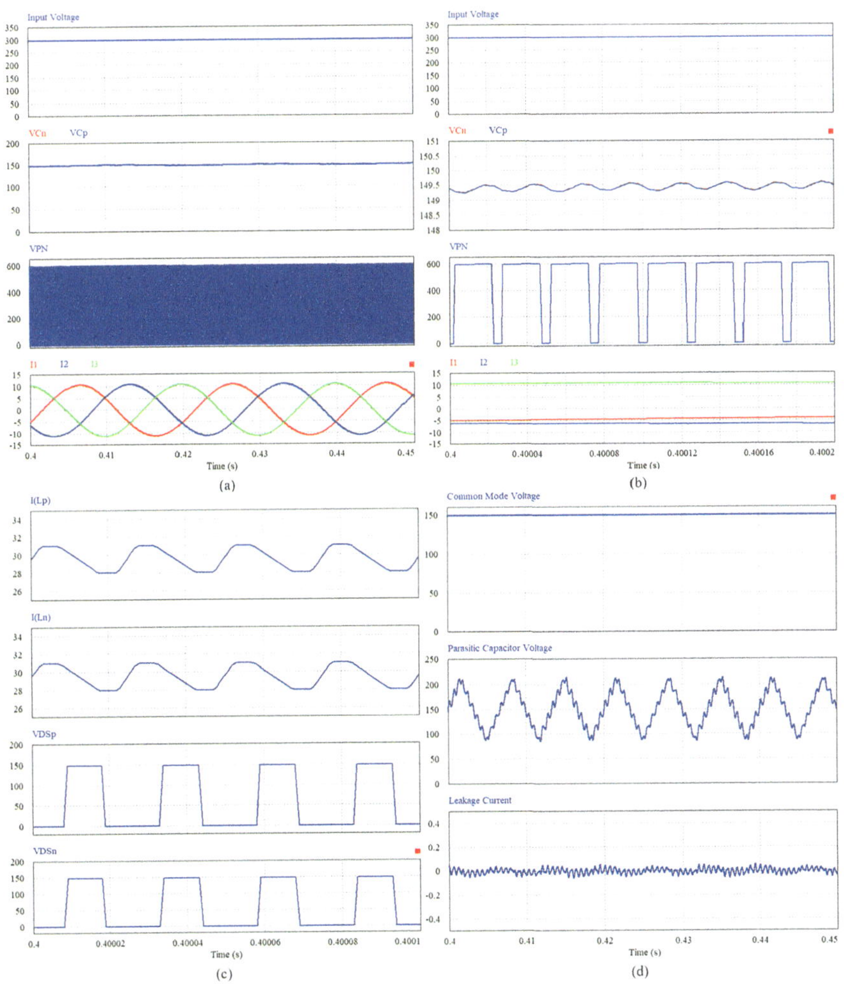

Fig. 3. Simulation results of the H10 inverter. From top to bottom: (a)–(b) input voltage, capacitor C_p and C_n voltages, DC-Link voltage VPN, the phase currents and; (c) Inductor L_p current, inductor L_n current, drain-source voltage of Sp and drain-source voltage of Sn (d) CMV, parasitic capacitor voltage and leakage current.

Figure 3 shows the simulation waveforms of the H10 inverter when $D_0 = 0.4$, $D = 0.2$. As indicated in Fig. 3(a) and Fig. 3(b), the DC-link voltage of the proposed H10 inverter is the square waveform and the peak value of DC-link voltage is 600 V. The peak value of the output phase currents is 10.9 A and their THD is 0.8%. The capacitor C_p and C_n voltages are boosted to 149.5 V as shown in Fig. 3(b). The charging and discharging boost inductors (L_p and L_n) are represented by the inductor currents as indicated in Fig. 3(c). As shown in Fig. 3(d), the CMV is constant without any high-frequency components and its value is 150 V. There are only low-frequency components in the parasitic capacitor voltage as highlighted in Fig. 3(d), therefore, the leakage current is effectively limited. The peak value of leakage current is 55 mA as shown in Fig. 3(d).

4 Conclusions

This paper has presented the three-phase transformerless H10 inverter. The operating principles, steady-state analysis, and verification of the proposed three-phase transformerless H10 inverter are given. The simulation results confirm a good performance of the proposed three-phase transformerless H10 inverter. Because the leakage current is effectively limited with the proposed three-phase transformerless H10 photovoltaic inverter, it is attractive for transformerless three-phase PV systems. The experimental results and detailed theoretical analysis will be shown in the future paper.

Acknowledgement. This research was supported by Research Foundation funded by Thai Nguyen University of Technology.

References

1. Kerekes, T., Teodorescu, R., Liserre, M., Klumpner, C., Summer, M.: Evaluation of three-phase transformerless photovoltaic inverter topologies. IEEE Trans. Power Electron. **24**(9), 2202–2211 (2009)
2. Ngo, M.-D., Nong, Q.-V., Ngo, T.-N., Nguyen, H.-Q., Tran, T.-T., Ahn, S.-J.: A common-ground single-phase boost inverter with suppressed double-frequency ripple for photovoltaic applications. In: Sattler, K.-U., Nguyen, D.C., Vu, N.P., Long, B.T., Puta, H. (eds.) ICERA 2020. LNNS, vol. 178, pp. 7–12. Springer, Cham (2021). https://doi.org/10.1007/978-3-030-64719-3_3
3. Gupta, A.K., Agrawal, H., Agarwal, V.: A novel three-phase transformerless H-8 topology with reduced leakage current for grid-tied solar PV applications. IEEE Trans. Ind. Appl. **55** (2), 1765–1774 (2019)
4. Tran, T.T., Nguyen, M.K., Duong, T.D., Choi, J.H., Lim, Y.C., Zare, F.: A switched-capacitor-voltage-doubler based boost inverter for common-mode voltage reduction. IEEE Access **7**(1), 98618–98629 (2019)
5. Julian, A., Lipo, T., Oriti, G.: Elimination of common-mode voltage in three-phase sinusoidal power converters. IEEE Trans. Power Electron. **14**(5), 982–989 (1999)

6. Guo, X., He, R., Jian, J., Lu, Z., Sun, X., Guerrero, J.M.: Leakage current elimination of four-leg inverter for transformerless three-phase PV systems. IEEE Trans. Power Electron. **31**(3), 1841–1846 (2016)
7. Guo, X., Yang, Y., He, R., Wang, B., Blaabjerg, F.: Transformerless Z-source four-leg PV inverter with leakage current reduction. IEEE Trans. Power Electron. **34**(5), 4343–4352 (2019)
8. Nguyen, M.K., Le, T.V., Park, S.J., Lim, Y.C.: A class of quasi-switched boost inverters. IEEE Trans. Ind. Electron. **62**(3), 1526–1536 (2015)

Adaptive Energy Management in Microgrid Based on New Training Strategy for ANFIS

M. Elsisi[1,2], Minh-Quang Tran[1,3], Vu Thi Lien[4], and Nguyen Thi Thanh Nga[3(✉)]

[1] Industry 4.0 Implementation Center, National Taiwan University of Science and Technology, Taipei 106, Taiwan
minhquang.tran@mail.ntust.edu.tw
[2] Department of Electrical Engineering, Faculty of Engineering (Shoubra), Benha University, 108 Shoubra Street, B.O. Box 11241, Cairo, Egypt
[3] Department of Mechanical Engineering, Thai Nguyen University of Technology, Thai Nguyen, Vietnam
nguyennga@tnut.edu.vn
[4] Falculty of Mechanical Engineering and Mechatronics, PHENIKAA University, Hanoi 12116, Vietnam

Abstract. Managing procedure for charging and discharging battery system plays an essential contributor in improving the performance of energy storage system for example increment of utilizing batteries. This paper aims to develop a new hybrid genetic algorithm-based proportional integral (GA-based PI) controller with an adaptive neuro-fuzzy inference system (ANFIS) for the charging balance of batteries. The dataset is generated by using the GA-based PI controller, then a training strategy is introduced for the ANFIS controller. The proposed approach is evaluated by the GA-based PI controller and the PI controller based on Ziegler Nichols method.

Keywords: ANFIS · Genetic algorithm (GA) · Renewable energy · Introduction

1 Introduction

In recent years, several studies have been conducted for power management control (PMC) in microgrid [1]. The applications of different energy control strategies are concentrated in various systems and smart grids [2–6], water pumping [7–9], and telecommunications [10]. Previous research showed that the PMC is persistently in power balance. Various strategies of PMC have been introduced, such as frequency deviation control in [11] and control with a microcontroller in [12]. These papers use conventional control techniques for PMC, in which the dynamic systems can be controlled by using a mathematical model. In various energy systems, it is difficult to derive an exact linear model for the dynamic system.

With the development of artificial intelligence (AI), up to date, many control techniques have been combining with AI for modeling issues by linguistic, non-distinct control laws derived from skilled knowledge [13–15]. Several papers have additionally

D. C. Nguyen et al. (Eds.): ICERA 2021, LNNS 366, pp. 142–149, 2022.
https://doi.org/10.1007/978-3-030-92574-1_15

been carried out with the use of fuzzy logic (FL) technique and robust control for PMC of hybrid energy networks with storage devices [16–19]. The ANFIS technique needs a suitable training for the design of membership functions and rules to provide an acceptable performance. The GA is used for the training of ANFIS in different applications [20–23]. In [20], a hybrid GA–ANFIS is used for the prediction of wave parameters. A genetic algorithm is utilized to optimize the parameters of ANFIS for brass energy absorbers in [21]. A hybrid genetic algorithm for the training ANFIS with a mutable rule base is introduced in [22]. A self-regulating scheme is developed based on GA-ANFIS for induction motor filter compensation in [23]. The GA proves a satisfactory performance for the training of the ANFIS in the previous papers.

In this work, a new control strategy is developed by the GA-based proportional integral (PI) controller to train the ANFIS controller for battery charging and discharging control. The new controller is applied in an isolated hybrid Wind/Photovoltaic energy scheme. The performance of the suggested control strategy is evaluated with the GA-based PI controller and the traditional PI controller based on the Ziegler Nichols (ZN) technique [24]. The results of proposed approach demonstrate a satisfactory performance for the studied hybrid energy system compared with other techniques.

The rest of the paper is classified as follows. The proposed hybrid GA-PI controller and ANFIS are presented in Sect. 2. In Sect. 3, the simulation results are presented. Finally, the conclusions are provided in Sect. 4.

2 Hybrid GA-PI Controller and ANFIS

In this section, an overview and deployment of ANFIS are described. Generally, ANFIS is a combination of an ANN and an FL inference system. The ANFIS approach has been commonly used for control models owing to great performance with a suitable integration of the ANN and FL. The ANFIS model utilizes a derived dataset based on the inputs and outputs to build up the fuzzy inference system. An ANFIS architecture is illustrated in Fig. 1, in which a multi-layer feed-forward is used. The shape of the membership function is created based on the learning procedure utilizing the back-propagation strategy. More details about the ANFIS architecture were introduced in [25].

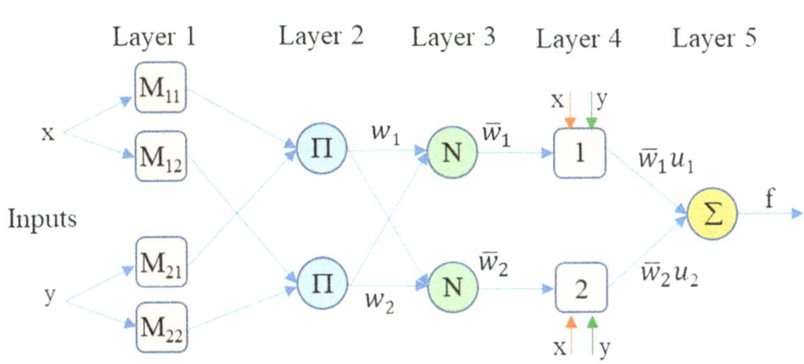

Fig. 1. Descriptions of ANFIS schematic.

The purpose of this research is the deriving of optimal training and testing datasets for the ANFIS technique, in which the data is obtained by using the GA-based PI controller. Assuming that the data is presented in the control (U), the error (e), and the derivative of the error (de/dt) signals. The PI controller parameters are optimized based on GA to provide an acceptable performance as shown in Fig. 2.

The GA is assigned to tune the gains of the PI controller based on minimizing a time-domain performance index which formulated as follows,

$$J = (1 - e^{-\mu})M_p + e^{-\mu}T_s, 0 < \mu < 1 \tag{1}$$

where Mp represents the maximum overshooting, and Ts is the settling time.

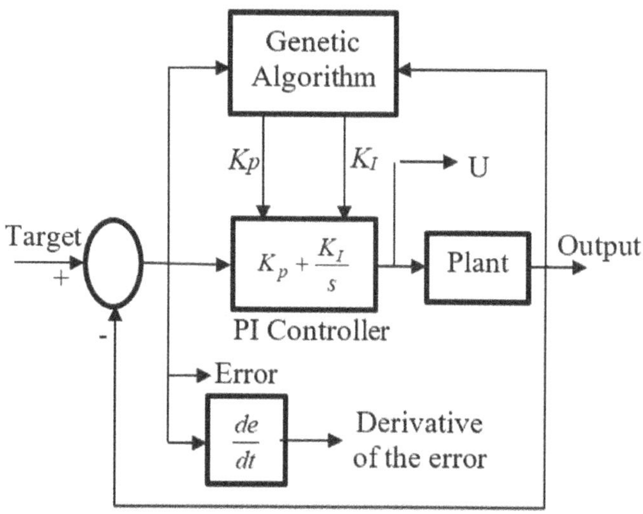

Fig. 2. Description of the GA- based PI control loop.

The performance index J can minimize both Mp and Ts at an acceptable value by setting $\mu = 0.7$. After derived the best gains of the PI controller by GA, the datasets of the GA-based PI controller are utilized to train and test the ANFIS technique. After the training process, the ANFIS controller is used as shown in Fig. 3.

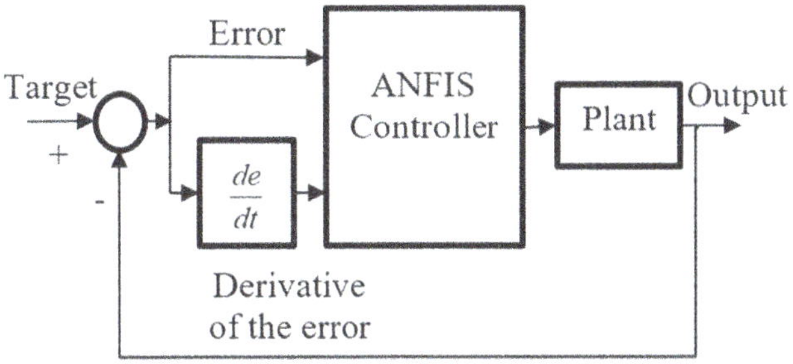

Fig. 3. A schematic diagram of the ANFIS control loop.

3 Results and Discussions

The suggested hybrid renewable energy system in this research consists of a photovoltaic (PV) source, wind source, and batteries [26]. The hybrid system is linked with a DC bus. The PMC which is done by the hybrid GA-based PI and ANFIS technique is linked to the DC to DC converter. The proposed controller is used to maintain the state-of-charge (SOC) of the battery bank at the permissible limits and to expand the lifetime of batteries against the variations in solar irradiance and wind speeds. The synthesis of the under studied hybrid renewable energy scheme is described in Fig. 4.

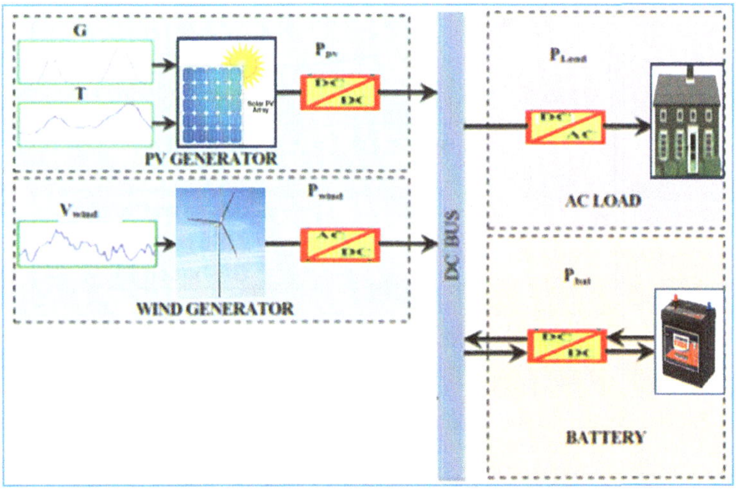

Fig. 4. The schematic of under study energy system.

The control mechanism for battery charge and discharge is done by two ANFIS loops, one loop is specified to maintain the DC bus voltage by adjusting the PWM of the boost converter. While the second loop is utilized to adjust the current of the battery.

The validation of the hybrid control method is carried out by utilizing MATLAB® platform, a 10 kW load is connected to the hybrid energy system. The PI controller gains based on GA and ZN are designed as follows: KP1 = 5.756 & KI1 = 129.404 and KP2 = 69.225 & KI2 = 547.549 for the GA-based PI controller; KP1 = 2 & KI1 = 120 and KP2 = 60 & KI2 = 550 for the ZN-based PI controller. The simulation results are indicated in Figs. 5, 6, and 7. The DC voltage response of the energy system is cleared in Fig. 5. The battery current is shown in Fig. 6. The SOC of the battery is shown in Fig. 7. The response damping characteristics includes; Mp and Ts of the DC voltage and the performance index J values are presented in Fig. 8.

From Fig. 7 and Fig. 8, the simulation results emphasize that the system response with hybrid GA-based PI and ANFIS technique has superior damping performance instead of GA-based PI and ZN-based PI strategies. Furthermore, the proposed controller maintains the SOC of the battery as shown in Fig. 7.

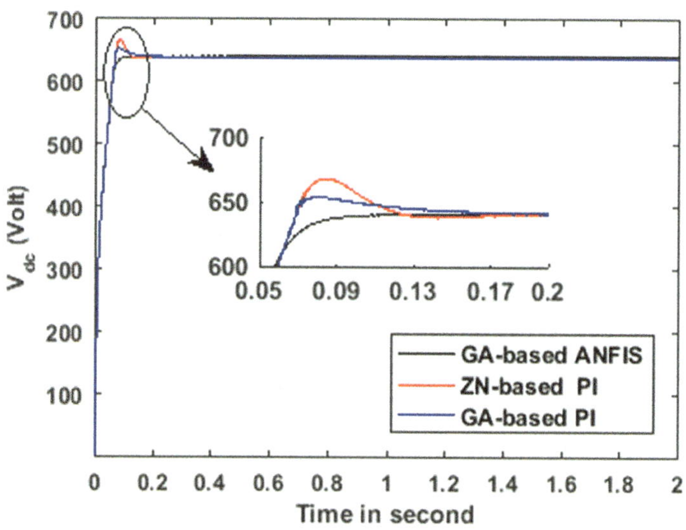

Fig. 5. The system output responses of DC voltage due to the applied controllers.

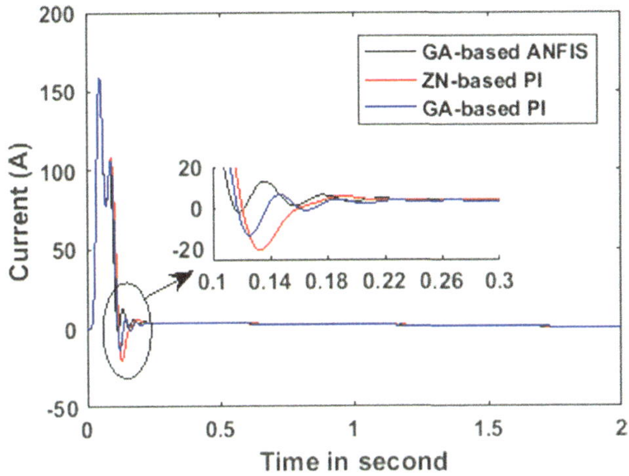

Fig. 6. The system output responses of battery current due to the applied controllers.

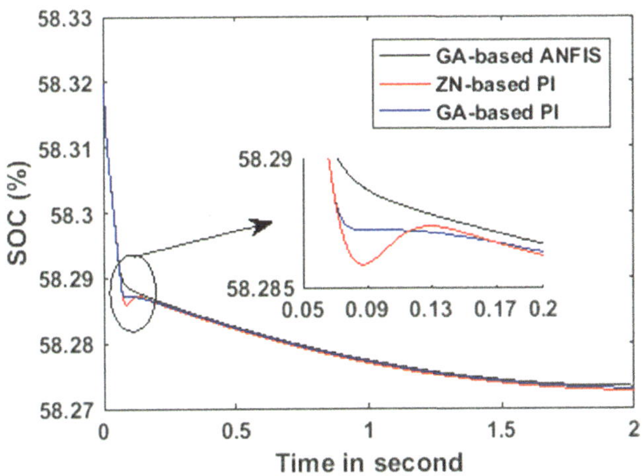

Fig. 7. Change in SOC of battery due to the applied controllers.

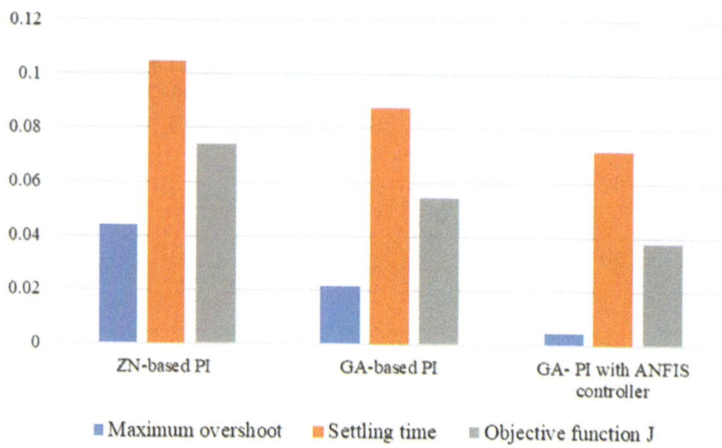

Fig. 8. The performance characteristics of the output DC voltage based on different control strategies.

4 Conclusions

In this paper, a novel hybrid GA-based PI and ANFIS strategy is developed for the energy management of renewable energy systems. The proposed controller is used to enhance the damping characteristics of the renewable energy system which consists of a photovoltaic, wind generator, and batteries. Furthermore, the suggested control strategy maintains the SOC of batteries at the permissible limits to keep the storage batteries work for long time. The test is done using MATLAB/Simulink tools and the results confirm the superiority of the suggested control strategy compared with the GA-based PI controller and the ZN-based PI controller.

Acknowledgment. The authors would like to thank Thai Nguyen University of Technology for supporting this research.

References

1. Nejabatkhah, F., Li, Y.W.: Overview of power management strategies of hybrid AC/DC microgrid. IEEE Trans. Power Electron. **30**(12), 7072–7089 (2015)
2. Miceli, R.: Energy management and smart grids. Energies **6**(4), 2262–2290 (2013)
3. Elsisi, M., Bazmohammadi, N., Guerrero, J.M., Ebrahim, M.A.: Energy management of controllable loads in multi-area power systems with wind power penetration based on new supervisor fuzzy nonlinear sliding mode control. Energy **221**, 119867 (2021)
4. Elsisi, M.: New design of adaptive model predictive control for energy conversion system with wind torque effect. J. Clean. Prod. **240**, 118265 (2019)
5. Kennel, F., Gorges, D., Liu, S.: Energy management for smart grids with electric vehicles based on hierarchical MPC. IEEE Trans. Ind. Inf. **9**(3), 1528–1537 (2013)
6. Elsisi, M., Soliman, M., Aboelela, M.A.S., Mansour, W.: GSA-based design of dual proportional integral load frequency controllers for nonlinear hydrothermal power system. Int. J. Electr. Comput. Energ. Electron. Commun. Eng. **9**(8), 928–934 (2015)

7. Sallem, S., Chaabene, M., Kamoun, M.B.A.: Energy management algorithm for an optimum control of a photovoltaic water pumping system. Appl. Energy **86**(12), 2671–2680 (2009)
8. Palma-Behnke, R., et al.: A microgrid energy management system based on the rolling horizon strategy. IEEE Trans. Smart Grid **4**(2), 996–1006 (2013)
9. Sallem, S., Chaabene, M., Kamoun, M.B.A.: Optimum energy management of a photovoltaic water pumping system. In: Howlett, R.J., Jain, L.C., Lee, S.H. (eds.) Sustainability in Energy and Buildings, pp. 187–197. Springer Berlin Heidelberg, Berlin, Heidelberg (2009). https://doi.org/10.1007/978-3-642-03454-1_20
10. Mekhilef, S., Saidur, R., Safari, A.: A review on solar energy use in industries. Renew. Sustain. Energy Rev. **15**(4), 1777–1790 (2011)
11. Aissou, S., Rekioua, D., Mezzai, N., Rekioua, T., Bacha, S.: Modeling and control of hybrid photovoltaic wind power system with battery storage. Energy Convers. Manage. **89**, 615–625 (2015)
12. Belvedere, B., Bianchi, M., Borghetti, A., Nucci, C.A., Paolone, M., Peretto, A.: A microcontroller-based power management system for standalone microgrids with hybrid power supply. IEEE Trans. Sustainable Energy **3**(3), 422–431 (2012)
13. Elsisi, M., Soliman, M.: Optimal design of robust resilient automatic voltage regulators. ISA Trans. **108**, 257–268 (2021)
14. Colson, C.M., Nehrir, M.H., Pourmousavi, S.A.: Towards real-time microgrid power management using computational intelligence methods. In: Power and Energy Society General Meeting, 2010 IEEE, pp. 1–8. IEEE (2010)
15. Elsisi, M., Abdelfattah, H.: New design of variable structure control based on lightning search algorithm for nuclear reactor power system considering load-following operation. Nucl. Eng. Technol. **52**(3), 544–551 (2020)
16. Schouten, N.J., Salman, M.A., Kheir, N.A.: Energy management strategies for parallel hybrid vehicles using fuzzy logic. Control. Eng. Pract. **11**(2), 171–177 (2003)
17. Elsisi, M.: New design of robust PID controller based on meta-heuristic algorithms for wind energy conversion system. Wind Energy **23**(2), 391–403 (2020)
18. Erdinc, O., Vural, B., Uzunoglu, M.: A wavelet-fuzzy logic based energy management strategy for a fuel cell/battery/ultra-capacitor hybrid vehicular power system. J. Power Sources **194**(1), 369–380 (2009)
19. Elsisi, M.: Optimal design of non-fragile PID controller. Asian J. Control **23**(2), 729–738 (2021)
20. Zanaganeh, M., Mousavi, S.J., Shahidi, A.F.E.: A hybrid genetic algorithm–adaptive network-based fuzzy inference system in prediction of wave parameters. Eng. Appl. Artif. Intell. **22**(8), 1194–1202 (2009)
21. Soleimani, M., Salmalian, K.: Genetic algorithm optimized ANFIS networks for modeling and reduction of energy absorption rate of brass energy absorbers. J Appl Math Is Az Un La **8**, 29–45 (2012)
22. Coy, C.G.: A hybrid-genetic algorithm for training a Sugeno-type fuzzy inference system with a mutable rule base. Doctoral Dissertation, University of Toledo (2010)
23. Othman, A.M., Sharaf, A.M.: A GA-ANFIS self regulating scheme for induction motor filter compensation. Int. J. Energy Eng. **3**(2), 74–96 (2013)
24. Ogata, K., Yang, Y.: Modern Control Engineering, vol. 4. Prentice Hall, India (2002)
25. Elsisi, M., Tran, M.-Q., Mahmoud, K., Lehtonen, M., Darwish, M.M.F.: Robust design of anfis-based blade pitch controller for wind energy conversion systems against wind speed fluctuations. IEEE Access **9**, 37894–37904 (2021)
26. Patel, S., Prajapati, N., Patel, T., Patel, S., Prajapati, N., Patel, T.: Grid connected solar and wind hybrid system. Int. J. **2**, 300–307 (2016)

An Output Observer Integrated Dynamic Surface Control for a Web Handing Section

Ly Tong Thi[1], Trong Dang Van[2], Bach Nguyen Nhu[3], Hung Pham Van[1], Duc Duong Minh[2], Huy Nguyen Danh[2], and Tung Lam Nguyen[2(✉)]

[1] Hanoi University of Industry, Hanoi, Vietnam
[2] Hanoi University of Science and Technology, Hanoi, Vietnam
{duc.duongminh, huy.nguyedanh, lam.nguyentung}@hust.edu.vn
[3] National University of Civil Engineering, Hanoi, Vietnam

Abstract. In this paper, the tension and speed control problem of a rewinding web handling system are considered. Most of the design methods for tension controller of web transport systems require state and output variables feedback. However, when complex systems, as more variables as more feedback signals, need more sensors which makes the system bulky and expensive. The novel of the paper is to design a robust controller for the web transport system based on DSC technique to ensure high accuracy in the tracking and tension controller based on back stepping controller without using tension. The simulation results show the validity and effectiveness of the proposed control.

Keywords: Roll-to-roll system (R2R) · Tension observer · Dynamic surface control (DSC) · Backstepping control · Lyapunov stability

1 Introduction

Nowadays, many practical applications, for example, flexible showcases, shading moving, lighting, and solar cells, using roll-to-roll (R2R) frameworks are commonly winding up in the industry [1–5]. The key point to guarantee the system output quality is to properly handle web moving speed and tension. Based on dynamic modeling of the system depicted in [6] and [7], several papers have been published to enhance the control performance with numerous control schemes, for instance, sliding mode control (SMC) [8] and [9], backstepping technique [10] and [11] to counter effects of unsettling influences for non-linear R2R web transport systems. In particular, there has been a great number of difficulties with sensor mounting, calibrating… during the process of measuring the response of tension force. Therefore, some in-depth research on implementing feedback control is based on the observation set [12] and [13]. In a web rewinding system, information on roll inertia is crucial for control design, a neural network to cope with uncertain roll inertia [14] and [15]. The paper results show good speed and tension tracking performances under unavailability of the inertia. In [16], the proposed controller is designed in the face of the mathematical model of the web

D. C. Nguyen et al. (Eds.): ICERA 2021, LNNS 366, pp. 150–157, 2022.
https://doi.org/10.1007/978-3-030-92574-1_16

transport system that is affected by bounded uncertainties, however, it requires tension feedback signal.

In this paper, by taking the advantages of tension observer, a new control approach which uses tension observer integrated with DSC controller which has not been constructed for an uncertain model of the roll-to-roll system. The contribution of this paper can be summarized as i) the proposal of the control scheme integrated with a tension observer, ii) The control parameters are also analyzed and chosen within a limited range. This paper is organized as follows. The uncertain model of the single-pan roll-to-roll web transport system is presented in Sect. 2. Section 3 is dedicated to developing controllers for web speed and tension regulation. The simulation results are given in Sect. 4. Finally, some concluding remarks are stated in Sect. 5.

2 R2R Web System Modeling

Figure 1 shows a single-span roll-to-roll web control system contains an unwinder and a rewinder. The input torques generated on unwinder and rewinder motors control the web's axial velocity and tension.

Ignoring the influence of the slip, deformation of the material, and the friction between the web and rolls. The dynamics equations can be given as follows:

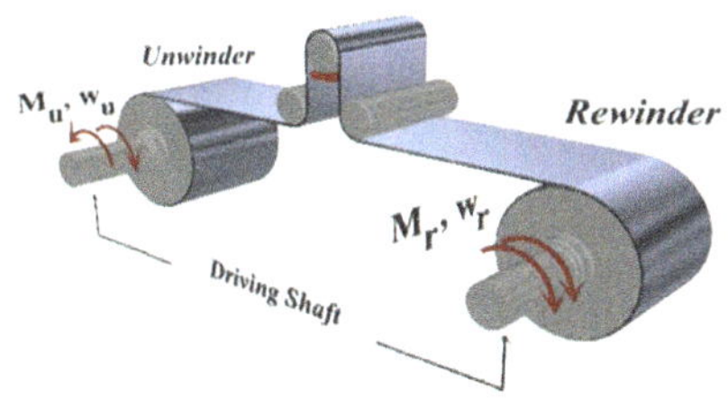

Fig. 1. Single-span R2R web control system

$$\frac{LdT}{dt} = ES(v_r - v_u) + v_u T_{ref} - v_r T; \tag{1}$$

$$\frac{J_u}{R_u}\frac{dv_u}{dt} = R_u T - M_u - \frac{b_{f_u}}{R_u} v_u + \left(aw\rho R_u - \frac{aJ_u}{2\pi R_u^3} \right) v_u^2; \tag{2}$$

$$\frac{J_r}{R_r}\frac{dv_r}{dt} = -R_r T + M_r - \frac{b_{f_r}}{R_r} v_r + \left(\frac{aJ_r}{2\pi R_r^3} - aw\rho R_r \right) v_r^2. \tag{3}$$

Since axial velocity can be calculated as: $v = \omega R$, so the above system of equations are rewritten as follows:

$$\dot{T} = c_1\omega_u + c_2\omega_r T + c_3\omega_r; \tag{4}$$

$$\dot{\omega}_u = c_4 M_u + c_5 T + c_6\omega_u + c_7\omega_u^2; \tag{5}$$

$$\dot{\omega}_r = c_8 M_r + c_9 T + c_{10}\omega_r + c_{11}\omega_r^2. \tag{6}$$

Where: $c_1 = \frac{R_u}{L} T_{\text{ref}} - \frac{ESR_u}{L}$; $c_2 = -\frac{r_r}{L}$; $c_3 = \frac{ESR_r}{L}$; $c_4 = -\frac{1}{J_u}$; $c_5 = \frac{R_u}{J_u}$; $c_6 = -\frac{b_{fu}}{J_u}$; $c_7 = \frac{aw\rho R_u^3}{J_u} - \frac{a}{2\pi R_u}$; $c_8 = \frac{1}{J_r}$; $c_9 = -\frac{R_r}{J_r}$; $c_{10} = -\frac{b_{fr}}{J_r}$; $c_{11} = -\frac{aw\rho R_r^3}{J_r} + \frac{a}{2\pi R_r}$.

The relations among total moment of inertia, operating radius and the thickness of the web are shown in following equations:

$$R_u(t) = R_{u0} - \frac{\theta_u a}{2\pi};\ R_r(t) = R_{r0} + \frac{\theta_r a}{2\pi} \tag{7}$$

$$J_u = J_{u0} + \pi\rho\omega\frac{\left(R_u^4 - R_{u0}^4\right)}{2};\ J_r = J_{r0} + \pi\rho\omega\frac{\left(R_r^4 - R_{r0}^4\right)}{2} \tag{8}$$

Where T: the web tension, J_u: the moment of inertia of the unwind roll and motor, J_r: the moment of inertia of the rewind roll and motor, R_u: the radius of the unwind roll, R_r: the radius of the rewind roll, B_u: the coefficient of vicious friction of the unwind roll, B_r: the coefficient of vicious friction of the unwind roll, E: the elasticity of web, S: the cross sectional area of web, L: the total length of web, a: the thickness of web, ρ: the density of web, w: the width of web.

3 Tension Sensorless Control Design

In this section, we will demonstrate the DSC controller combined with the tension observer for the R2R system. It is noted that, the difficulty in identify system parameters leads to inaccurate coefficients c_i. We assume that system uncertain parameters are described as $c_i = \hat{c}_i + \tilde{c}_i$, rewrite Eqs. (4), (5), and (6) as follows:

$$\dot{T} = \hat{c}_1\omega_u + \hat{c}_2\omega_r T + \hat{c}_3\omega_r + \Delta\xi_T; \tag{9}$$

$$\dot{\omega}_u = \hat{c}_4 M_u + \hat{c}_5 T + \hat{c}_6\omega_u + \hat{c}_7\omega_u^2 + \Delta\xi_u; \tag{10}$$

$$\dot{\omega}_r = \hat{c}_8 M_r + \hat{c}_9 T + \hat{c}_{10}\omega_r + \hat{c}_{11}\omega_r^2 + \Delta\xi_r. \tag{11}$$

Where: $\Delta\xi_u = \tilde{c}_4 M_u + \tilde{c}_5 T + \tilde{c}_6\omega_u + \tilde{c}_7\omega_u^2$; $\Delta\xi_r = \tilde{c}_8 M_r + \tilde{c}_9 T + \tilde{c}_{10}\omega_r + \tilde{c}_{11}\omega_r^2$; and $\Delta\xi_T = \tilde{c}_1\omega_u + \tilde{c}_2\omega_r T + \tilde{c}_3\omega_r$ are lump uncertainties; $\hat{c}_i$ are nominal parameters for the system and are accurately acquired; $\tilde{c}_i$: the error between the calculated model parameters and the actual parameters $(1 \leq i \leq 11)$. The tension observer is determined by:

$$\dot{\hat{T}} = c_1\omega_u + c_2\omega_r\hat{T} + c_3\omega_r \tag{12}$$

At first, a tension controller is considered. Toward this end, defining the error variable between actual tension and desired value as follows:

$$\Delta T_s = T_d - T \tag{13}$$

$$\Delta \hat{T}_s = T - \hat{T} \tag{14}$$

To regulate the error and to limit the phenomenon of "explosion of terms", we suggest a virtual control signal is passed through a low-pass filter defined as:

$$\begin{cases} \tau \dot{F}_{\omega_{ud}} + F_{\omega_{ud}} = \overline{F}_{\omega_{ud}} \\ \overline{F}_{\omega_{ud}}(0) = F_{\omega_{ud}}(0) \end{cases} \tag{15}$$

where τ is the low-pass filter time constant. Proposing the Lyapunov candidate function for error signal of web tension as following:

$$V_T = \frac{1}{2}\Delta \hat{T}_s^2 + \frac{1}{2}(\Delta T_s + \Delta \hat{T}_s)^2 \tag{16}$$

Taking time derivative of (16), it is suggested that the virtual control signal $\overline{F}_{\omega_{ud}}$:

$$\overline{F}_{\omega_{ud}} = \frac{1}{\hat{c}_1}\left(\dot{T}_d - \hat{c}_2\omega_r\hat{T} - \hat{c}_3\omega_r - k_1(\Delta T_s + \Delta \hat{T}_s) - k_\delta(\Delta T_s + \Delta \hat{T}_s)\left(\frac{\delta^2}{2\varepsilon} + k_\Delta\right)\right) \tag{17}$$

where $k_1 > 0$. We have $\dot{V}_T \le 0$ so this satisfies the Lyapunov stability standard. Subsequently, the unwinder angular speed's tracking error is devoted to track virtual control designed in the previous step:

$$\Delta\omega_u = F_{\omega_{ud}} - \omega_u \tag{18}$$

Proposing the Lyapunov candidate function as:

$$V_u = \frac{1}{2}\Delta\omega_u^2 \tag{19}$$

Taking time derivative (25), it is suggested that the control signal is:

$$M_u = \frac{1}{\hat{c}_4}\left[\dot{F}_{\omega_{ud}} - \hat{c}_5\hat{T} - \hat{c}_6\omega_u - \hat{c}_7\omega_u^2 + k_2\Delta\omega_u + k_3 sgn(\Delta\omega_u)\right] \tag{20}$$

From (20), it is clear that $\dot{V}_u \le 0$ with $k_2 > 0$; $k_3 > \delta$, then (19) satisfies the Lyapunov stability standard. The error of the rewinder speed is defined as:

$$\Delta\omega_r = \mathrm{W}_{rd} - \omega_r \tag{21}$$

Similarly, the Lyapunov candidate function is proposed:

$$V_r = \frac{1}{2}\Delta\omega_r^2 \tag{22}$$

To render $\dot{V}_r \leq 0$, we choose the control signal M_r as following:

$$M_r = \frac{1}{\hat{c}_8}\left[\dot{\mathrm{W}}_{rd} - \hat{c}_9\hat{T} - \hat{c}_{10}\omega_r - \hat{c}_{11}\omega_r^2 - k_4\Delta\omega_r - k_5 sgn(\Delta\omega_r)\right] \tag{23}$$

with $k_4 > 0$; $k_5 > \delta$ we prove $\dot{V}_r \leq 0$. Finally, we have defined the error of low-pass filter:

$$\Delta F_f = F_{\omega_{ud}} - \overline{F}_{\omega_{ud}} \tag{24}$$

Taking time derivative of (27), we can write:

$$\Delta\dot{F}_f = -\frac{\Delta F_f}{\tau} + \gamma \tag{25}$$

Proposing the Lyapunov candidate function:

$$V_f = \frac{1}{2}\Delta F_f^2 \tag{26}$$

Low-pass filter time constants τ is chosen such that:

$$\frac{1}{\tau} = 1 + \frac{M^2}{2\varepsilon} + k_0 \tag{27}$$

with the condition $\gamma \leq M$ where $M > 0$ and k_0 is a positive value. Hence, it can be concluded that $\dot{V}_f \leq 0$, thus satisfies the Lyapunov stability standard.

4 Simulation Results

In this section, the performance of the proposed controller is demonstrated through simulations. The parameters of the roll to roll system are utilized in simulation can be listed as following: $R_{u0} = 0.04\,(m)$; $R_{r0} = 0.015\,(m)$; $J_{u0} = J_{r0} = 1\,(kg.m^2)$; $L = 0.3\,(m)$; $\mathrm{w} = 1\,(m)$; $E = 2,5.10^9\,(N/m^2)$; $Bu = Br = 0.00002533\,(kgms/rad)$. Controller parameter selected: $k_1 = 40$; $k_2 = k_3 = 50$; $k_4 = k_\Delta = 20$; $k_0 = k_5 = k_\delta = 10$. At first, simulation results are carried out in Fig. 2, Fig. 3, Fig. 4, and Fig. 5 with 0% parameter variations ($\Delta\xi_T = \Delta\xi_u = \Delta\xi_r = 0$).

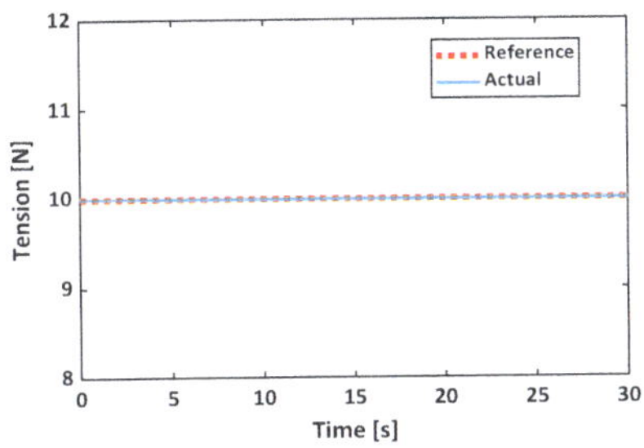

Fig. 2. Web tension response

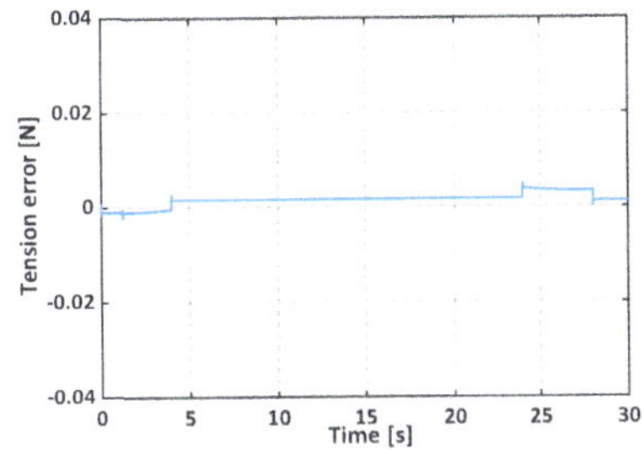

Fig. 3. Tension error

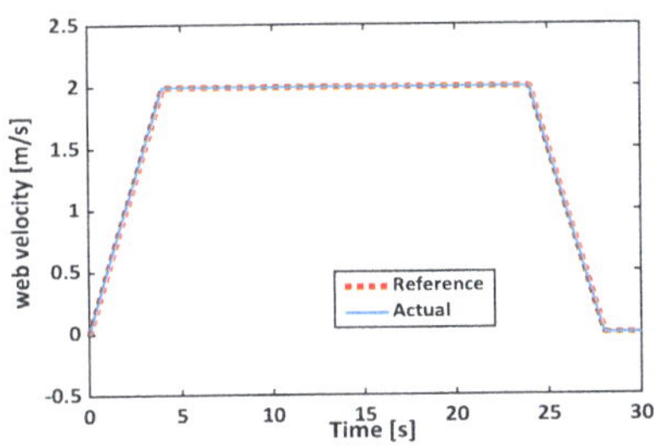

Fig. 4. Velocity response

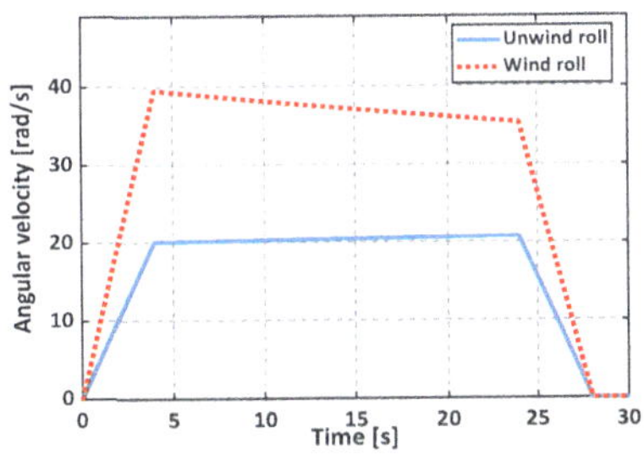

Fig. 5. Angular velocity responses

Figure 2 describes the actual tension when applying the DSC controller and the tension observer against the tension reference Tref = 10 (N). Figure 3 shows the error that is about 0.4% between the actual tension and reference tension. Figure 4 describes the actual velocity following the set trajectories in the shape of a trapezoid. In addition, we also give the trajectory of the angular velocity on the unwind and rewind described in Fig. 5. The simulation results show the correctness of the algorithm. Subsequently, simulation is carried out in Fig. 6 and Fig. 7 with 10% parameter variations ($c_i = \hat{c}_i + 10\%\hat{c}_i$).

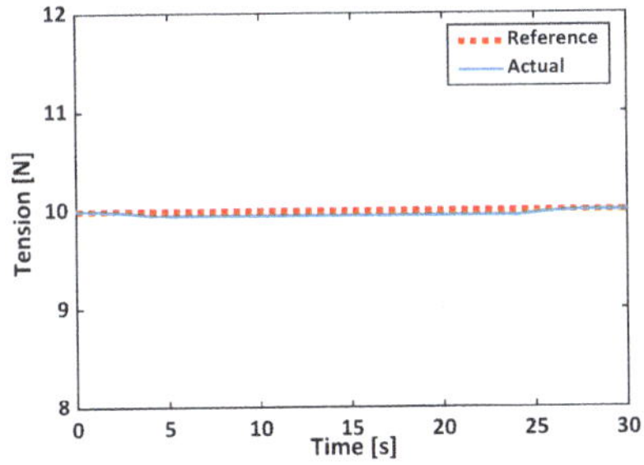

Fig. 6. Tension response

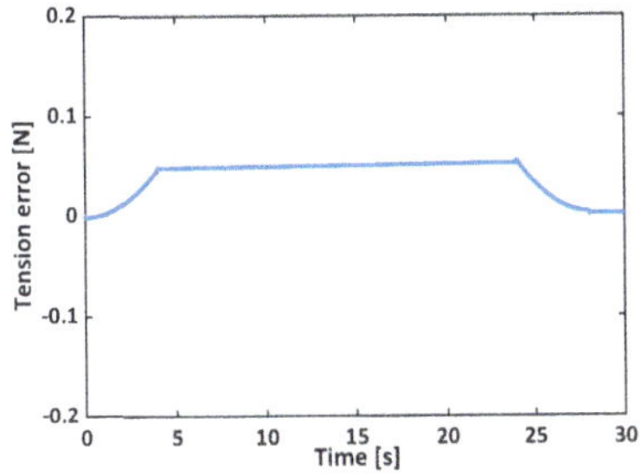

Fig. 7. Tension error

Finally, simulation results are carried out in Fig. 8 and Fig. 9 with 20% parameter uncertainty ($c_i = \hat{c}_i + 20\%\hat{c}_i$).

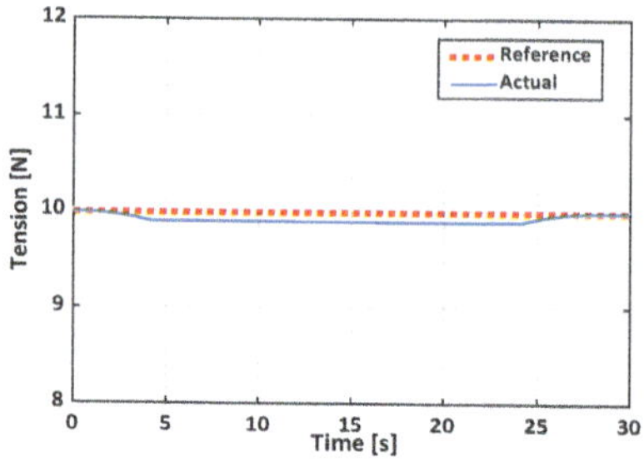

Fig. 8. Tension response

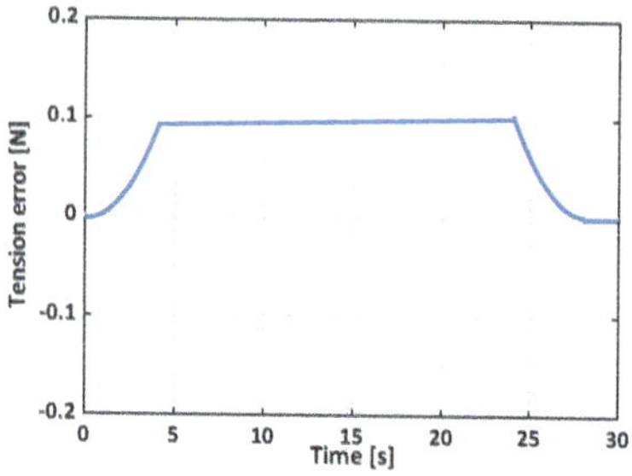

Fig. 9. Tension error

Figure 6, Fig. 7, Fig. 8, and Fig. 9 describe the actual tension and the tension error when applying the DSC controller and the tension observer against the tension reference Tref = 10 (N). In the case with 10% uncertainty of the parameter, the tension error is about 0.6%. In the case with 20% uncertainty of the parameter, the tension error is about 1%. Both cases show high quality of the proposed controller with small error. Based on the presented simulation results, we find that the control structure is able to sustainably adapt to the uncertainties of the object model.

5 Conclusion

This study has successfully built a DSC controller combined with a tension observer to deal with uncertainty in the rewinding system. A tension observer is also integrated to reduce the bulky and cost of the system. The proof of stability for the whole system by Lyapunov theory is also presented in this paper. Simulation results show trajectories of speed and tension. In addition, the simulation results with the parameter uncertainty cases confirmed that the control structure is stable. In the near future, we have two research directions for the future development: i) Experimental verification of the control will be conducted, ii) The research aims at developing a multi-roll system model with uncertain parameters.

Acknowledgement. This research is funded by Hanoi University of Science and Technology (HUST) under project number T2020-PC-022.

References

1. Okada, K., Sakamoto, T.: Adaptive fuzzy control for web tension control system. In: 24th Annual Conference of the IEEE Industrial Electronics Society, Aachen Germany, August 31–September 4 (1998)

2. Chang, K.M., Weng, C.P.: Modeling and control for a coating machine. JSME Int J. Ser. C **44**(3), 656–661 (2001)
3. Kang, C.G., Lee, B.J.: Stability analysis for design parameters of a roll-to-roll printing machine. In: International Conference on Control, Automation and Systems, Seoul, South Korea, October 17–20 (2007)
4. Kang, C.G., Lee, B.J.: MIMO tension modelling and control for roll-to-roll converting machines. IFAC Proc. Vol. **41**(2), 11877–11882 (2008)
5. Sakamoto, T., Fujino, Y.: Modelling and analysis of a web tension control system. In: IEEE International Symposium on Industrial Electronics, Dubrovnik, Croatia, July 10–14 (1995)
6. Choi, K.H., et al.: Web register control algorithm for roll-to-roll system based printed electronics. In: 2010 IEEE Int. Conf. Autom. Sci. Eng. CASE 2010, pp. 867–872 (2010)
7. Choi, K.H., Thanh, T.T., Kim, D.S.: A precise control algorithm for single-span roll-to-roll web system using the back-stepping controller. In: IEEE International Symposium on Industrial Electronics, pp. 1709–1714 (2009)
8. Abjadi, N.R., Soltani, J., Askari, J., Arab Markadeh, G.R.: Nonlinear sliding-mode control of a multi-motor web-winding system without tension sensor. IET Control Theory Appl. **3**(4), 419–427 (2009)
9. Chen, C.L., Chang, K.M., Chang, C.M.: Modeling and control of a web-fed machine. Appl. Math. Model. **28**(10), 863–876 (2004)
10. Choi, K.H., Tran, T.T., Kim, D.S.: Back-stepping controller based web tension control for roll-to-roll web printed electronics system. J. Adv. Mech. Des. Syst. Manuf. **5**(1), 7–21 (2011)
11. Bouchiba, B., Hazzab, A., Glaoui, H., Med-Karim, F., Bousserhane, I.K., Sicard, P.: Backstepping control for multi-machine web winding system. J. Electr. Eng. Technol. **6**(1), 59–66 (2011)
12. Mathur, P.D., Messner, W.C.: Controller development for a prototype high-speed low-tension tape transport. IEEE Trans. Control Syst. Technol. **6**(4), 534–542 (1998)
13. Mokhtari, F., Sicard, P.: Decentralized control design using integrator backstepping for controlling web winding systems. In: 39th Annual Conference of the IEEE Industrial Electronics Society, Vienna, Austria, November 10–13 (2013)
14. Manh, C.N., Van, M.T., Duc, D.N., Tung, L.N., Tien, D.P., Thi, L.T.: Neural network based adaptive control of web transport systems. In: Proceedings of 2019 International Conference on System Science and Engineering, ICSSE 2019, pp. 124–128 (2019)
15. Duc, D.N., Thi, L.T., Tung, L.N.: Imperfection roll arrangement compensation control based on neural network for web handling systems. Eng. Technol. Appl. Sci. Res. **10**(3), 5694–5699 (2020)
16. Thi, L.T., Manh, C.N., Nguyen, T.L.: A control approach to web speed and tension regulation of web transport systems based on dynamic surface control. J. Control Autom. Electr. Syst. **32**(3), 573–581 (2021). https://doi.org/10.1007/s40313-021-00694-0

An Analysis of the Dynamic of the Bus Body Rollover

Nguyen Thanh Quang[1(✉)] and Le H. Quan[2]

[1] Princeton University, Princeton, NJ 08544, USA
nguyenthanhquang@haui.edu.vn
[2] Faculty of Automobile Technology, Hanoi University of Industry, Hanoi, Vietnam

Abstract. The simulation for dynamics of bus body in case of side rollover is applied in this study. Research methodology is digital simulating in Ansys Workbench software. In the results, deformation of bus body, in case of rollover at right angle, has been checked. Level of body deformation is determined by dynamic parameters are deformation value, equivalent stress to compared according to approved stress criterion of body material.

Keywords: Bus body · Rollover · Deformation stress

1 Introduction

Accidents causing bus rollover are the most threatening for passengers and driver because of serious deformation in body structure, which reduces minimum survivable space for passengers. Therefore, bus body durability, which should be enough to absorb crash energy, is important in vehicle design and manufacturing.

According to statistics in USA in 2003, there were over 10.000 car rollover in accidents causing 229.000 deaths and injuries [1]. Simulating and analyzing mechanism and danger of passengers' injury in 300 car falling accidents in Sweden, and investigating injuries of 128 victims show that the highest death rate is due to being throwed out, partially throwed out or completely crashed body [2]. It is important to test crash in vehicle to ensure safety. But it is difficult for manufacturers since testing in a bus costs much more than that in a small car. Finite element method is most commonly used to digitally simulate vehicle's crash process [4, 5]. Generally, in the market, bus chassis manufacturers tend to be independent suppliers (OEM). That is, they become the third party who design and manufacture bus chassis and supply it to assembling factories (part manufacturer, chassis manufacturer and assemblers). Consequently, bus chassis is designed, manufactured and checked according to body standards (in particular) and finished bus standards (in general), that ensures optimization and safety of chassis structure. ECE66 is a standards are used popular [6].

D. C. Nguyen et al. (Eds.): ICERA 2021, LNNS 366, pp. 158–164, 2022.
https://doi.org/10.1007/978-3-030-92574-1_17

2 Simulation of Bus Rollover

2.1 Assumptions of Simulation

Numerical simulation is used to re-describe car's rollover process in order to support the adjustment and optimization in body designing. Following assumptions are used: (1) Complete blocks and chassis that is not impacted by internal complete blocks such as seats, interior cladding, air conditioner and all other interiors are hard-joined; (2) Vehicle rollover at steady speed; (3) Chassis weight is evenly distributed so that the whole body has the same hardness.

Typical forms of bus crash are side crash, rear crash, front crash and side rollover. In motion, the bus wheels roll on the road in a line parallel to bus's longitudinal direction. In case of crash or high speed cornering, the bus turns around. There are 3 forces acting on the bus: tire force (radial force), inertia (centrifugal force) and gravity. Cornering force from tires acts on the ground, below the center block, pushes the bus toward the corner's center. The inertia acts horizontally across the bus's center of gravity at the height position. These 2 forces push the bus outward from the corner. The gravity acts from top to bottom. If tire's force and inertia are stronger than the gravity, the bus begins to rollover. After crash, body deformation can narrow required space for passengers. In case of rollover, body structure should be strong enough to protect passenger space. Body deformation level is determined by the distance from center of gravity to center of front axis (for front crash) or by longitudinal axis (for side perpendicular crash).

In testing process, the body stands in a flat base of 800 mm high and rolls down to a floor underneath by its own gravity with an angular velocity no more than 0.087 rad/s. The invasion of body structure into inner compartment is checked by hand and checking results are decided by the ECE66 standards.

2.2 Simulation Steps

Finite element analysis method and Ansys Workbench software are used for simulation [8].

Step 1. Modeling: The body of 29-seat bus is designed by 3D Solidworks software. All its details are updated to Ansys Workbench software for simulation. Its body consists of 3 mm steel cubic tubes welded into a hard space frame. Body wall consists of low grade thin steel plates 300 M attached to the frame by spot welding. Road surface is hard concrete, having cross-sectional dimension according to specified standards. Ensure 1m extra length of testing road in the front and rear of the vehicle body. A longer road will increase the volume of simulation.

Step 2. By default, the overall coordinate system is installed in the software. There are 2 main connections in the model: internal (inside the vehicle) connection is a hard connection (Bolded) and wheel-surface connection is a frictionless connection.

Body material is made of low carbon steel, an tires are made of rubber. Road surface is hard concrete. The steel specifications are in Table 1.

Table 1. Specifications of steel used in vehicle's body

	Specific gravity (kg/m^3)	Young's modulus (MPa)	Poisson's ratio	Tensile strength (MPa)	Bending strength (MPa)	Cutting strength (MPa)
Low carbon steel (300 M)	7800	182.530	0,3265	1.595	1.958	68.801

C44 (ADMAS) and Triads are gravity components in finite component model (FE). The model is transited in Explicit form with linear components. Components dimension is 5 mm. There are 333.736 components and 367.405 nodes. Using smooth transition method with smooth ratio of 2.272 and transition increment of 1.2, we get results of transition model in Fig. 1.

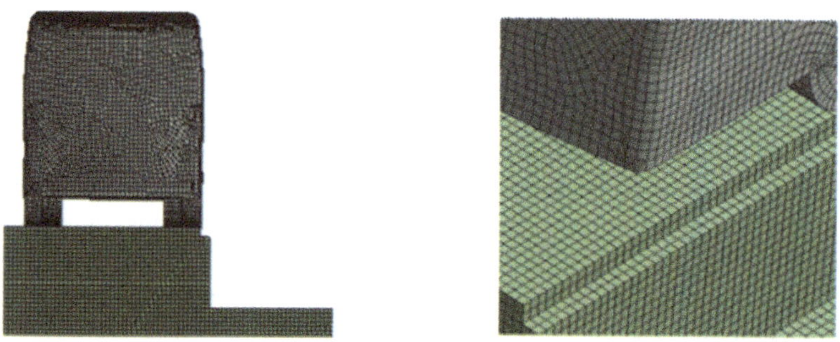

Total transition of vehicle and road surface Form of transition

Fig. 1. Meshing model

Step 3. Setup forces and boundary conditions.

Left horizontal rollover model is set up in the simulation process. To determine gravity height h (the vertical distance from vehicle's center to the position where it contacts with road surface), rollover moment R_M = 0,75 (Nm) acts to the whole body along Edge A and the yellow edge (without label), whose positions in the coordinate system (according to X,Y,Z axes) are −3.0352, 135.12, 182.91 (mm) respectively. Absolutely hard concrete surface is fixed by using Fixed Statement which is limited in all degrees of freedom by all lines in Edge B, Fig. 2.

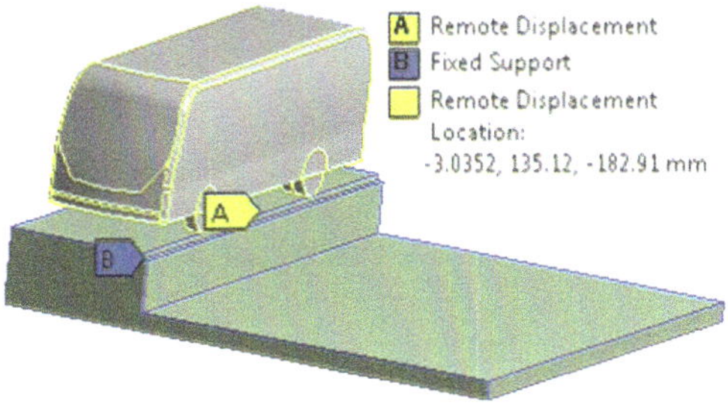

Fig. 2. Setting forces and boundary conditions

The whole simulation process is built according to Explicit Dynamics collision and collision energy analysis (Energy Probe).

3 Analysis of Dynamic Bus Rollover

Figure 3 describes how dynamics (on bus rollover) determines positions of turning points, collision points of body with road surface, and calculates angular acceleration of collision. The rollover process has two phases. In the 1st phase, the bus rollovers from position A to B: the bus rotates around bottom point of rear wheel stopper (O1) until its center of gravity reaches maximum height, which is changing point of the rollover process. In the 2nd phase, the bus rollovers from position B to C: the whole bus falls to the 800 mm level and its left roof edge hits the testing ground causing affected force. This force causes 1st deformation of about 1–1.5 (mm). Then collision force affects the whole bus body, caussing its total deformation.

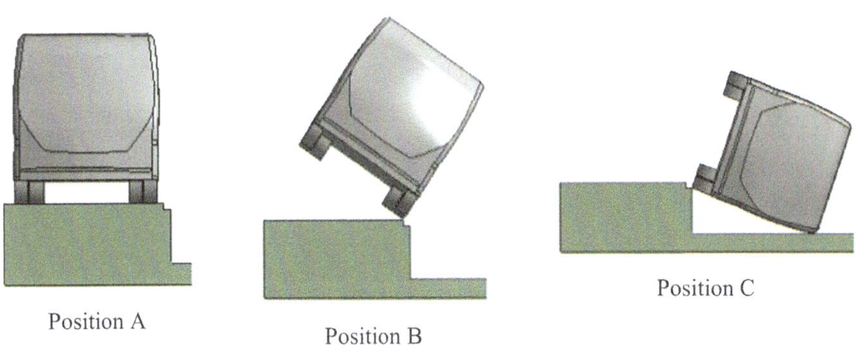

Fig. 3. Dynamical results of bus rollover

Angular velocity ω of the rollover process is determines in formular (1) [7]

$$\omega^2 = \frac{1}{2}\frac{M \times g \times H}{I_{RA}} \quad (1)$$

Where: I_{RA} is inertial moment of bus rollover on X axis, calculated by formular (2)

$$I_{RA} = I_{XX} + M \times D^2 \quad (2)$$

Where: I_{XX} is gravity inertial moment rotating by axis X; H is variation of gravity center height; M is the bus mass, g is gravitational acceleration.

In simulation results, the average rolling speed of rollover is 0.083 rad/s. Kinetic parameters in Table 2 are used in calculation.

Table 2. Kinetic parameters of simulation

Assignment	Low alloy steel, 300 m (low carbon)
Volume	6.5321e + 005 mm^3
Mass	5.1166 kg
Moment of inertia ip1	68019 kg mm^2
Moment of inertia ip2	2.8072e + 005 kg mm^2
Moment of inertia ip3	2.797e + 005 kg mm^2

In results of analysing simulation durability of testing bus body hardness, its material is low carbon steel 300 M in collision, according to ECE-R66 standards shown in Table 3. Maximum stress reaches 3285 MPa (it is much over acceptable value of 320 MPa) and distributes mainly at 2 sides.

Table 3. Values of deformation stress in bus body

	Deformation by X (mm)	Total deformation (mm)	Equivalent deformation (mm/mm)	Equivalent stress (MPa)
Minimum	-6.129	0.0	0.0	0.0
Maximum	3.0702	37.918	3.0362e−002	3285
Average	−1.4572e−003	1.5593	1.0017e−003	162.34

It is caused by instantaneous crash of body top corner with the tesing concrete road, Fig. 4.

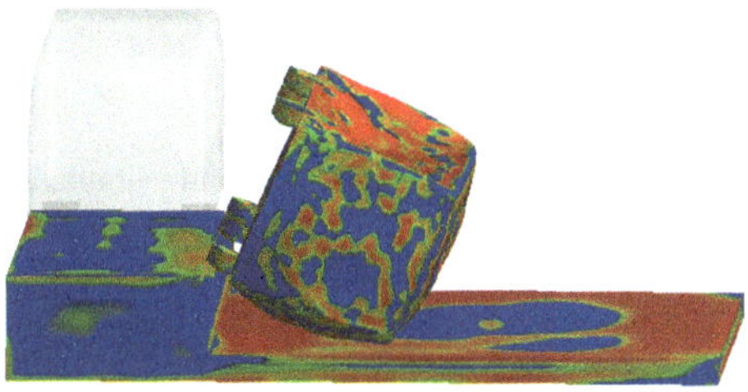

Fig. 4. Deformation stress in bus body

4 Evaluation of Simulation Accuracy

To evaluate the simulation accuracy, we determine the total energy of simulation time, when kinetic energy reduces and changes into internal energy (including stress energy and energy of sliping in component's surface). The energy balance graph shows that different energy distributions in the simulation process and total energy are unchanged. Therefore, the simulation results are correct, Fig. 5.

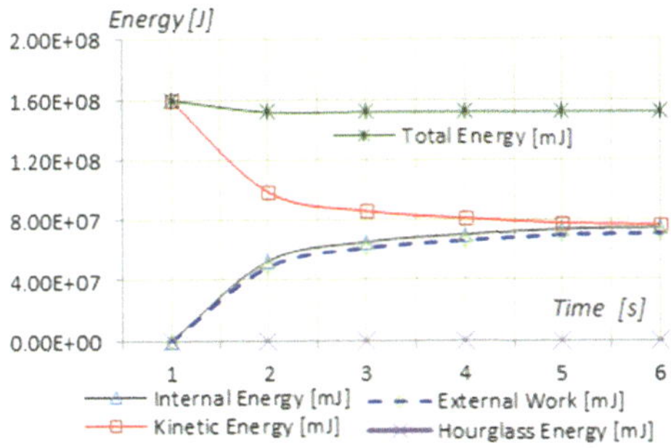

Fig. 5. Energy balance graph of rollover process

5 Conclusion

Using Ansys Workbench software to simulate bus body horizontal rollover (according to AIS 031/ECE-R66) and analyzing bus body durability, we get body deformation stress. Body deformation is determined by transposition values, equivalent deformation, stress, and crash acceleration. Then maximum stress is compared with acceptable stress of body material.

Energy balance method, used in simulation of body rollover, shows that the total energy is unchanged. Therefore, the simulation results are correct and can be used in bus body design.

References

1. Linstromberg, M., Scholpp, G., Scherf, O.: Test and Simulation Tools in a Rollover Protection Development Process. In: Siemens Restraint Systems GmbH, Paper No. 05-0122. Germany (2011)
2. Senthil Kumar, D.: Coach Passenger Injury Risk During Rollover: Influence of The Seat and the Restraint System. Giovanni Belingardi Paolo Martella Lorenzo Peroni Dipartimento di Meccanica Politecnico di Torino C.so Duca degli Abruzzi 24 10129 Torino Italy, Paper Number 05-0439 (2011)
3. Gursel, K.T., Gursesli, S.: Analysis of the superstructure of a designed bus in accordance with regulations ECE R 66. G. U. J. Sci. **23**(1), 71–79 (2010)
4. Satrio Wicaksono, M., Rahman, R.F., Mihradi, S., Nurhadi, I.: Finite element analysis of bus rollover test in accordance with UN ECE R66 standard. J. Eng. Technol. Sci. **49**(6), 799–810 (2017)
5. Sidhu, M.S.: Rollover bus. In: 2012 India HyperWorks Technology Conference, pp. 1–13 (2012)
6. Amendment No 11 (09/2019) To AIS-052 (Rev.1): Code of Practice for Bus Body Design Approval. The automotive research association of India p. B. No. 832, Pune 411 004. https://hmr.araiindia.com/api/AISFiles/AIS_052_Rev1_6d77bc38-f272-438f-be30-589c8e44f1bd.pdf
7. Standard UN-ECE No. 66-02. Strength of Superstructure – Large Passenger Vehicles. ECE – United Nations (2010)
8. Ansys Workbench 2020 R1. Workbench User's Guide. All rights reserved. https://www.ansys.com/academic/free-student-products (2020)

Analysis of Vehicle Tire Stress and Deformation in the Contact Area on Asphalt Concrete

Nguyen Thanh Quang(✉)

Hanoi University of Industry, Hanoi, Vietnam
nguyenthanhquang@haui.edu.vn

Abstract. Tires have an important role in interaction between vehicles and the road because it is the only component of a vehicle contacting the road. Vehicle's loading and running vibration depend on rubber's elasticity and compressed air in tires.

This article analyzes contact area of 4 different tire's patterns including: flat pattern, longitudinal groove pattern, convex lug pattern of off-road vehicles and of trucks. Using Ftire model and simulation in Ansys Worbench software to determine average pressure P_h, frictional stress F_r in contact area, which causes compressive stress and tire pattern's deformation. It then is transmitted to tires and body, causing tire's damage.

Keywords: Tyre · Contact area · Contact stress · Deformation

1 Introduction

Tire is a part of a wheel, together with rim and disk. Its main materials include rubber (for elasticity) and reinforcement fibers in its core (for increasing durability). The main requirements of a tire are: high fatigue strength, high load capacity, long service durability [1]. Tire's main specifications are its speed and loading capacity. Tire's mechanical properties affect vehicle's dynamics because forces are transmitted from tires to wheels, then to suspension system and to the whole body. To determine this effect, an analysis of tire's contact with road surface is required. To do it, an effective method is numerical simulation of a suitable tire's model. Tire models are characterized by their pressure, operating conditions, loads and their nonlinear nature.

FTire (flexible ring tire) model simulates vehicle's wheels, as a solid system interconnected by elastic elements, to analyze tire's vibrations in its contacting with uneven road surface. Ftire is a dynamic and non-linear model, developed by Michael Gisper in 1988. Today, Ftire is popular in automotive industry to analyze vehicle durability, loading capacity and smoothness in motion [2]. Other tire models are used in other researches serving other aims and requirements. In Fiala tire model, lateral and longitudinal forces are used as input parameters for calculating friction coeffitions μ_{max}

D. C. Nguyen et al. (Eds.): ICERA 2021, LNNS 366, pp. 165–170, 2022.
https://doi.org/10.1007/978-3-030-92574-1_18

and μ_{min} in full grip and normal slippage (excluding slippage in breaking and turning conditions) [3]. Magic Formula tire model analyzes general experimental equation in different driving conditions, describes tires behavior in case of lateral forces, torques, longitudinal forces [4]. Swift Model simulates tire dynamics in rotating direction and vehicle motion for analyzing small tires [5]. Hohenheim Model, used for big tires in agriculture, analyzes elasticity and non-linear damping elements representing forces and moments in different direction [6]. Some other models are used to analyze tire status in specific cases. In dynamic analysis, serial models are commonly used, in which horizontal transposition of each tire element is affected by tension of elements [7]. There is a study on traffic accident by estimating physical parameters in results of experiments of tire forces causing tire damages [8]. Analyzing contact time of tires with road surface, in different speed, determines frequency characteristics of the wheel's vertical movement in relation with vehicle's movement stability [9]. Dynamic friction model, which analyzes longitudinal and lateral forces in contact area of tires with road surface, helps to control vehicle's traction at different speeds [10]. A study using finite element analysis (FEM) and Ansys Workbench software to statistically analyze tires (made of natural rubber, with hard structure and linear, isotropic elasticity) shows that tire's technical specifications increase (in vertical load at 5.15 kN), which are lower than that specified in the tire. Increasing specifications are: compress air pressure increases to 515 kg as maximum, equivalent stress (von-miss) increases to 0.9448 Mpa [11]. Tire's vibration, elasticity coefficient, rolling resistance coefficient, heat distribution depend on type of road surface (concrete of asphalt) [12, 13]. The experiment analyzing deformation in tire's surface caused by static and dynamic forces (while running) shows changes in longitudinal direction of contact traces, grip and slip area of tire's lugs [14].

2 Tire Modeling

Tire's surface is a thin, soft belt with friction elements in Ftire model. It moves in the direction of the vehicle's movement and creates contact area between tire and road surface (footprint). Average value of movement force is calculated by a function of deviation and velocity. Movement directions are characterized by stiffness coefficient of each direction: C_{long} is stiffness coefficient of longitudinal direction (direction of vehicle's movement), $C_{bend,\ in\text{-}plane}$ is stiffness coefficient in vertical direction (in vehicle's vertical plane), $C_{bend,\ out\ of\text{-}plane}$ is stiffness coefficient in horizontal direction (in vehicle's horizontal plane), $C_{torsion}$ is movement stiffness coefficient caused by twist in contact area (in vehicle's horizontal plane), Fig. 1.a.

Tire-to-road contact model is showed in Fig. 1.b, where tire is marked A with contact surface S_A and road surface is marked B with contact surface S_B. The contact lies in a plane containing line λ_N. Point P in the tire and point Q in the road create contact gap h which changes by times and by directions causing contact area. The distance h is affected by elastic modulus of tire lug's rubber, number of lugs, height of lugs, tire's cross sectional shape and loading condition.

a) Ftire Model

b) Diagram of contact of two objects

Fig. 1. Model of tire

At time t, h is the distance of tire and road surface (the closest distance between P and Q) and is calculated by Eq. (1).

$$h = \min|P(t) - Q(t)| \tag{1}$$

During contact, stress and deformation on tire surface calculated by Eqs. (2) and (3).

$$\sigma_h = \frac{P_h}{A_h} \tag{2}$$

$$\varepsilon_h = \frac{\Delta L_h}{L_h} \tag{3}$$

Where: P_h is the average pressure to the norman direction, lied in the line connecting P and Q; A_h is the contact area; L_h is the length of contact area.

The average pressure P_h in contact area will cause compressive stress and deformation of tire's lugs, which is then transmitted to tire. The pressure depends on compressed air pressure P_{tire} in the tire and gaps Void between tire's lugs and is calculated by formula (4).

$$P_h = \frac{P_{tire}}{1 - Void} \tag{4}$$

The contact area length L_h, by longitudinal direction, can be approximated by the sliding distance of the wheel according to Eq. (5).

$$L_h = 1 - \frac{v_c}{v_f} \tag{5}$$

Where: v_c is vehicle's velocity; v_f is the wheel velocity.

3 Analyzing Results of Simulating Contact Area

Simulation's parameters include tire's material, disc, friction coeffition. In order to survey tire patterns, static load on tires is considered in specific vehicle. Tire's materials are rubber and butyl (IIR, 30–50% carbon black). Natural rubber and butyl are mixed with carbon black with 30–50% ratio. Road surface material is asphalt. Disc is made of carbon steel. During movement, stress at contact area has 3 directions: σ_X pushes tire forward or backward (it tends to accelerate or decelerate the vehicle); σ_Y pushes tire to the right or to the left (side force); σ_Z pushes tire vertically upwards. Graphs of stress are shown in Fig. 2. Friction stress values of straight groove tire and that of off-road vehicle are similar. Flat tire of racing car causes almost no stress at contact area. These values of heavy truck are biggest (blue line in the graphs) and tend to increase at the end of the survey.

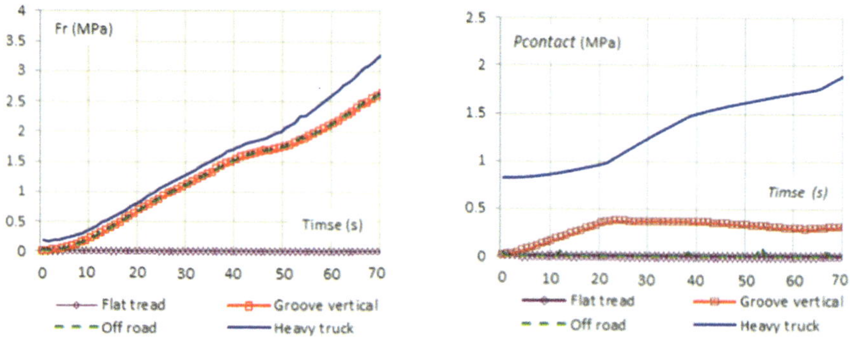

a) Frictional stress F_r (MPa) b) Contact stress P_h (MPa)

Fig. 2. Frictional stress and contact stress at contact area (Color figure online)

Figure 3 demonstrates survey results according to above formulas, stress transposition (deviation) relation and stress deformation relation. Both graphs show that transposition and deformation are proportional to tress at the contact area. In case of increasing load on tires, stress increases and road surface deformation increases to breaking limit. Comparing it with allowable stress in the road design, the determine breaking limit of road surface.

The load is related to stress at contact area. In simulation of 100% rated load with increment of 15%, 30%, 50% and 200% respectively, the find that values of stress (E_s), deviation (d_f) and equivalent deformation (E_e) increase proportionally to the vertical load on tire in Table 1 and Fig. 4.

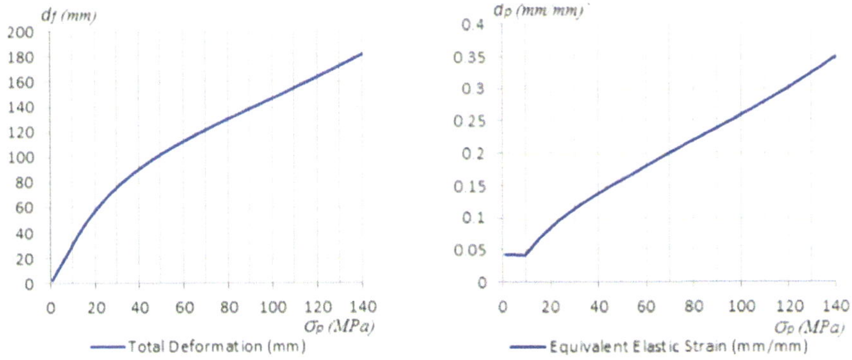

Fig. 3. Survey results of stress and transposition, deformation

Table 1. Values of loads and parameters of contact area

% P_L	Es (MPa)	df (mm)	Ee (mm/mm)
100	136.12	190.64	0.70906
115	152.7	208.74	9.5863
130	160.48	209.15	0.8206
150	181.76	225.28	1.0425
200	262.89	273.72	1.2752

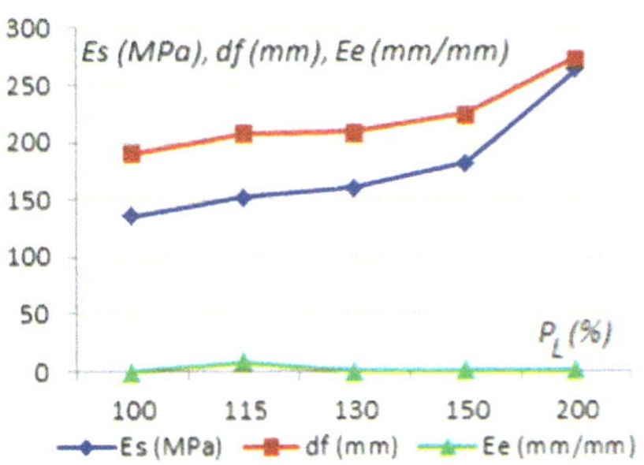

Fig. 4. Graph for Es, df and Ee of contact area

4 Conclusion

Research results show that the interaction of tire and road surface will cause stress to tire surface and road surface. In case of increasing load, stress increases and deformation increases to breaking limit.

Contact area, pressure, stress and deformation are related to tire patterns. For flat tire, contact area is the biggest, but contact pressure and stress are the smallest and evenly distributed. For tire with convex lugs, the contact area smaller, but contact pressure and stress increase and are evenly distributed in lugs' surface. Tire with convex lugs has greater risk for road damage.

Increasing load on tire causes proportional increasing stress and deformation at contact area and greater risk for tire damage. Stress in road surface increases proportionally to that in tires. At the beginning, it increases quickly and reaches maximum value in the 1st half of contact time. Thus, tire surface tends to be damaged in the 1st half of the contact period. Therefore, long time and frequent use are reasons of tire damage.

References

1. Adler, U.: Automotive Handbook, Bosch, SAE Society of Automotive Engineers, 10th edn. (2018). ISBN-10: 0768095670; ISBN-13: 978-0768095678
2. Andrius, R., Henrikas, S.: Magic formula tyre model application for a tyre-ice interaction. In: Proceedings of 10th International Scientific Conference Transbaltica 2017: Transportation Science and Technology, p. 335–341. Elsevier (2017). https://doi.org/10.1016/j.proeng.2017.04.383
3. Wang, X.: Finite Element Method. Tsinghua University Press, Beijing, China (2003)
4. Blundell, M., Harty, D.: Introduction. In: The Multibody Systems Approach to Vehicle Dynamics, pp. 1–22. Elsevier (2004). https://doi.org/10.1016/B978-075065112-7/50003-8
5. Bin, L., Xiaobo, Y., Ankang, J., Yunqing, Z., James, Y.: In-plane flexible ring tire model validation through adams ftire model virtual tests. In: Proceedings of the Asme 2015 International Design Engineering Technical Conferences & Computers And Information In Engineering Conference Idetc/Cie, Paper No: DETC2015-46630, V003T01A032. Boston, Massachusetts. USA, Publisher (2015). https://doi.org/10.1115/DETC2015-46630
6. Gipser, M.: FTire: a physically based application-oriented tyre model for use with detailed MBS and finite-element suspension models. Veh. Syst. Dyn. **43**, 76–91 (2005)
7. Besselink, I.J.M., Schmeitz, A.J.C., Pacejka, H.B.: An improved magic formula/swift tyre model that can handle inflation pressure changes. Veh. Syst. Dyn. **48**(supl), 337–352 (2010). https://doi.org/10.1080/00423111003748088
8. Han, I.: Modelling the tyre forces for a simulation analysis of a vehicle accident reconstruction. Proc. Inst. Mech. Eng. Part D: J. Automob. Eng. **231**(1), 16–26 (2016). https://doi.org/10.1177/0954407016630449
9. Žuraulis, V., Levulytė, L., Sokolovskij, E.: The Impact of road roughness on the duration of contact between a vehicle wheel and road surface. Transport **29**(4), 431–439 (2014). https://doi.org/10.3846/16484142.2014.984330
10. Aguilar, M.J., Alvarez, I.L.: Analysis of tire-road area in control oriented test bed for dynamic friction models. J. Appl. Res. Technol. **13**, 461–471 (2015)
11. Chinedum, O.M., Chika, E.M., Festus, I.A., Ravindranatha Menon, A.R.: static analysis of tyre model developed from natural rubber vulcanizates. In: Proceedings of the World Congress on Engineering, vol. IIWCE, June 29–July 1, 2016. London, U.K. (2016). ISBN: 978-988-14048-0-0 ISSN: 2078-0958 (Print); ISSN: 2078-0966 (Online)
12. Wang, D., Ueckermann, A., Schacht, A., Oeser, M., Steinauer, B., Persson, B.N.J.: Tire–road contact stiffness. Tribol. Lett. **56**(2), 397–402 (2014). https://doi.org/10.1007/s11249-014-0417-x
13. Aguilar-Martínez, J., Alvarez-Icaza, L.: Analysis of tire-road contact area in a control oriented test bed for dynamic friction models. J. Appl. Res. Technol. **13**(4), 461–471 (2015)
14. Roveri, N., Pepe, G., Carcaterra, A.: OPTYRE – a new technology for tire monitoring: evidence of contact patch phenomena. Mech. Syst. Signal Process. **66–67**, 793–810 (2016). https://doi.org/10.1016/j.ymssp.2015.06.019

Analysis Rope Climbing Mechanism

Anh-Tuan Dang[1,3], Dinh-Ngoc Nguyen[2,3(✉)], Nguyen Thi Thanh Nga[1,3], and Nguyen Thi Thu Linh[2,3]

[1] Department of Mechanical Engineering, National Kaohsiung University of Science and Technology Kaohsiung, Kaohsiung, Taiwan

[2] Faculty of International Training, Thai Nguyen University of Technology, Thai Nguyen, Vietnam
ngocnd@tnut.edu.vn

[3] Faculty of Mechanical Engineering, Thai Nguyen University of Technology, Thai Nguyen, Vietnam

Abstract. A kinematic calculating and analyzing method for a robot moving on a horizontal bar using a four-bar mechanism is proposed. By using analytical methods, the kinematic model of the robot is included to examine parameters such as leg tilt, average moving velocity, position, and trajectory of the robot's legs. Using the inversion mechanism method, the basic calculation can determine the movement of the mechanism on the bar during the operation.

Keywords: Four-bar mechanism · Climbing robot · Kinetic analysis · Mechanism inversion

1 Introduction

The four-bar mechanism is one of the first mechanisms designed and analyzed in the history of mechanical engineering. Despite its simple structure, this mechanism is applied in many areas of social life, even in biology and constructing [1]. Variations of the mechanism have been studied and utilized as fundamental models for constructing machine parts (train wheel dampers, brakes, rock crushers, or even robotic arms [2]). From the movement of rods in the links, biomimetic linkages have been such as Jansen, Klann, Chebyshev, Ghassaei walking robots, etc. are developed [3–6] (c.f. Fig. 1).

The links mentioned are specifically employed to move robots in environments not suitable for the use of wheels. The similarity between these machines is the movement of output links corresponds to one rotation of the driver. With the given points acting as landing feet, some of the feet resting on the ground and bearing the weight of the whole machine (group A), some of them are moving forward (group B). For the remaining rotation of the driver, the feet in group A are moving while the weight transfer to group B (now are rested).

D. C. Nguyen et al. (Eds.): ICERA 2021, LNNS 366, pp. 171–179, 2022.
https://doi.org/10.1007/978-3-030-92574-1_19

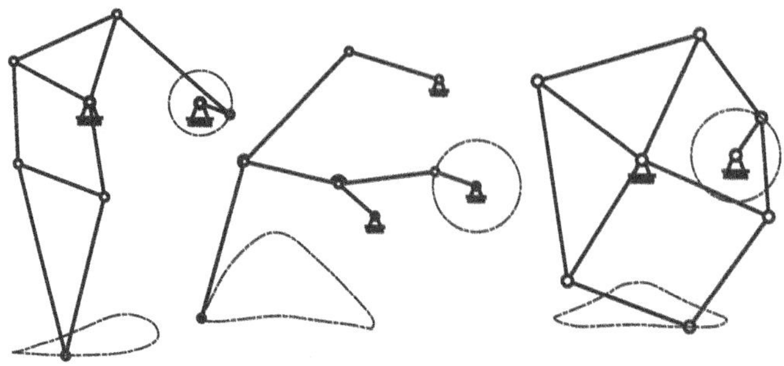

Fig. 1. Moving robot adapted from pin-joint mechanism: Jansen, Klann, Ghassaei linkages.

Designing movement methods for crawling robots is not a new study. There have been a number of studies using different approaches to construct and control the machines' movement [7–9]. Unlike walking robots whose feet set on the ground with the loading compressing the structure, a crawling robot leans its legs on a horizontal bar and tends to stretch the frames in the mechanism (c.f. Fig. 2).

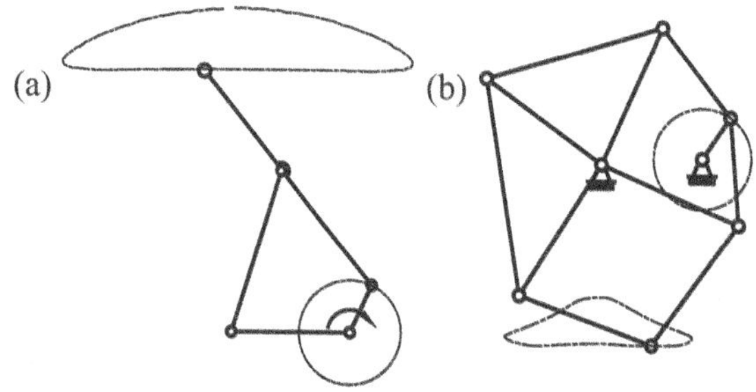

Fig. 2. Compare crawling type (a) and walking type (b) in robot'legs.

Clearly that even without wheels, using the structure from Fig. 2a to move the whole system along a hanging bar is still possible. However, it is not easy to analyze the movement and stability of the machine using movement linkages without simulation.

This study proposes an application of the inversion method in determining the movement of the mechanism. By employing analytical equations, the kinematic parameters of the robot are calculated, includes tilting, velocity, position, and even formation of the robotic legs. A 2D linkage was also constructed in the Working Model environment to verify the accuracy of the proposed model.

2 Kinetic and Dynamic Analysis

According to the locus of the leg from Fig. 2, its movement is separated into two phases: raising phase (a) moving leg from the bar and sliding phase (b) where this leg lean on the bar for the robot to crawl forward. The number of the leg also affects the stability of the mechanism. If the position of the feet is not symmetrical, the robot's center of gravity will be deflected, causing instability and oscillation during the movement.

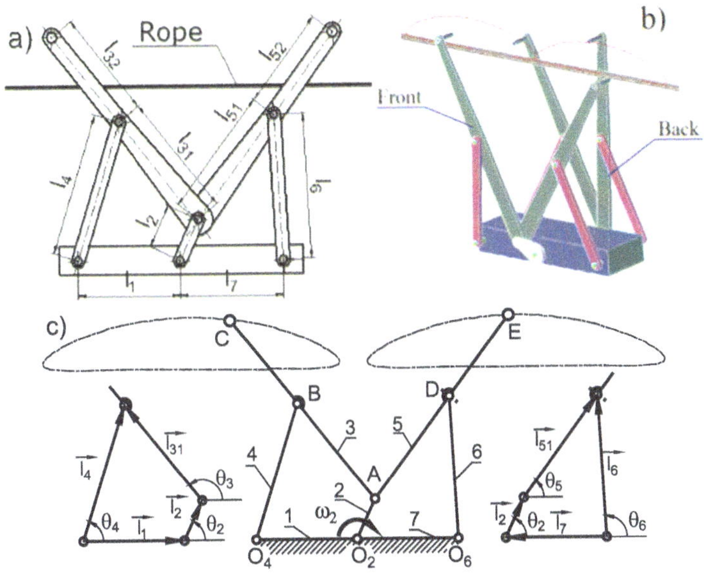

Fig. 3. Mechanism in 2D (a), 3D (b) and diagram of the front linkage (c).

To maintain the stability of the system, a 4-legs model, as shown in Fig. 3, is constructed. This model can ensure the balance of the platform while moving on a bar with the smallest tilting.

The kinetic analysis process will evaluate the position, velocity of joints, and links in the mechanism.

Using close loop equation for two side of legs as in Fig. 1 (with front side loop include links 1, 2, 3, 4 and back side loop are links 2, 5, 6, 7).

For front side loops:

$$\overrightarrow{l_1} + \overrightarrow{l_2} + \overrightarrow{l_{31}} = \overrightarrow{l_4} \tag{1}$$

$$l_1 e^{j\theta_1} + l_2 e^{j\theta_2} + l_{31} e^{j\theta_3} = l_4 e^{j\theta_4} \tag{2}$$

$$\begin{cases} l_4 \sin\theta_4 - l_{31}\sin\theta_3 = l_1\sin\theta_1 + l_2\sin\theta_2 \\ l_4\cos\theta_4 - l_{31}\cos\theta_3 = l_1\cos\theta_1 + l_2\cos\theta_2 \end{cases} \quad (3)$$

For the back side loops:

$$\overrightarrow{l_7} + \overrightarrow{l_2} + \overrightarrow{l_{51}} = \overrightarrow{l_6} \quad (4)$$

$$l_7e^{j\theta_7} + l_2e^{j\theta_2} + l_{51}e^{j\theta_5} = l_6e^{j\theta_6} \quad (5)$$

$$\begin{cases} l_6 \sin\theta_6 - l_{51}\sin\theta_5 = l_7\sin\theta_7 + l_2\sin\theta_2 \\ l_6\cos\theta_6 - l_{51}\cos\theta_5 = l_7\cos\theta_7 + l_2\cos\theta_2 \end{cases} \quad (6)$$

With l_i is the vector of rigid link i in the close loop equations determined by angle θ_i and the given lengths $l_1 = l_7 = 30$; $l_2 = 10$; $l_{31} = l_{51} = l_{32} = l_{52} = l_4 = l_6 = 40$.

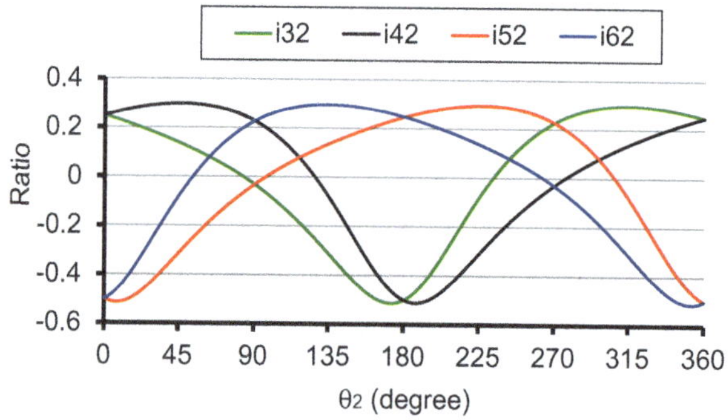

Fig. 4. Ratio of the rotation angle between of links k and link 2.

Solve (3) and (4) to get the corresponding angle θ_3, θ_4 and θ_5, θ_6 according to the movement of link 2 (θ_2). This calculation also can be used to determine the value of angular velocity ω_i in each link and constructing the graph of conversion ratio $i_{k2} = \omega_k/\omega_2$ in Fig. 4.

From the result, positions of point C and E defined by equations:

$$\overrightarrow{R_C} = \overrightarrow{l_2} + \overrightarrow{l_{31}} + \overrightarrow{l_{32}} \quad (7)$$

and

$$\overrightarrow{R_E} = \overrightarrow{l_2} + \overrightarrow{l_{51}} + \overrightarrow{l_{52}} \quad (8)$$

Solve the equations using the results from Fig. 5, the locus of points C and E are acquired:

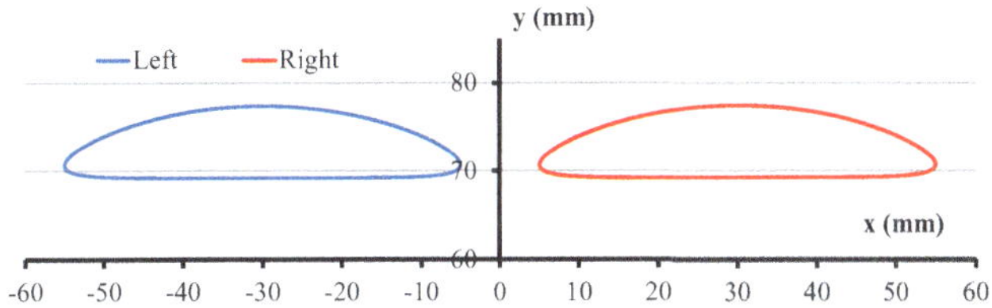

Fig. 5. Trajectories of points C and E.

We also can extract the position graphs of these points in y- direction for a cycle of motor (link 2) for the back side and front side, since the movements of these linkage are 180° difference (Fig. 6).

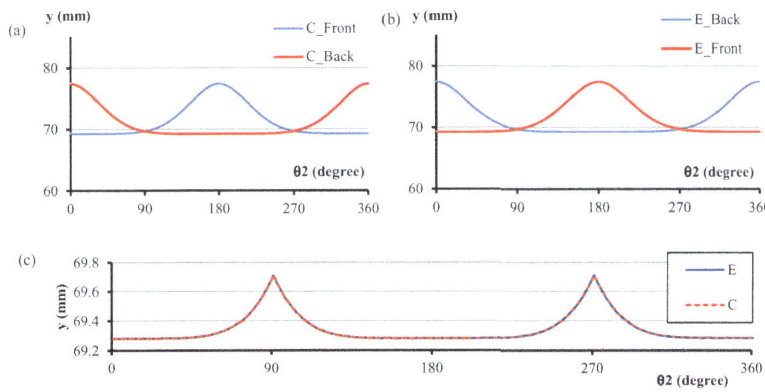

Fig. 6. Horizontal position of four legs (a, b), and the final movement of the structure (c)

It is noticed that during the movement of the mechanism, the period that feet C and E land on bar became fixed points. These point are determined by the y-coordinate of the leg in the back and front sides: $y_C = \min(y_{C\_Front}, y_{C\_Back})$ and $y_E = \min(y_{E\_Front}, y_{E\_Back})$. The result of the selection is the real movement of the legs and shown in Fig. 7.

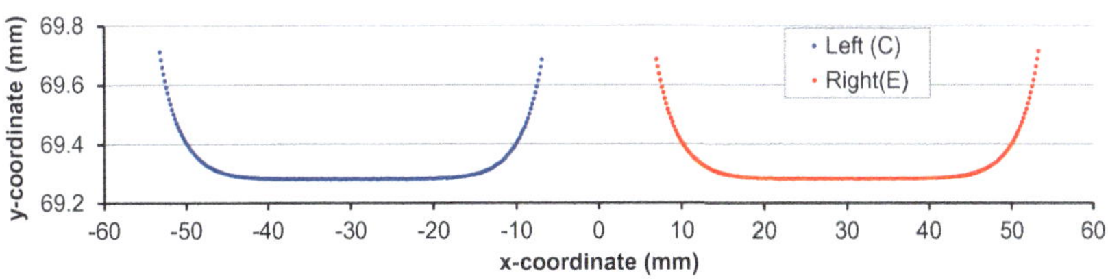

Fig. 7. Position of legs that lay on bar.

From the leg's movement, the inversion method is employed for determining the actual position of the mechanism on the bar. During the period of contact with the bar, the landing point are fixed on the bar while the other links change their positions. By

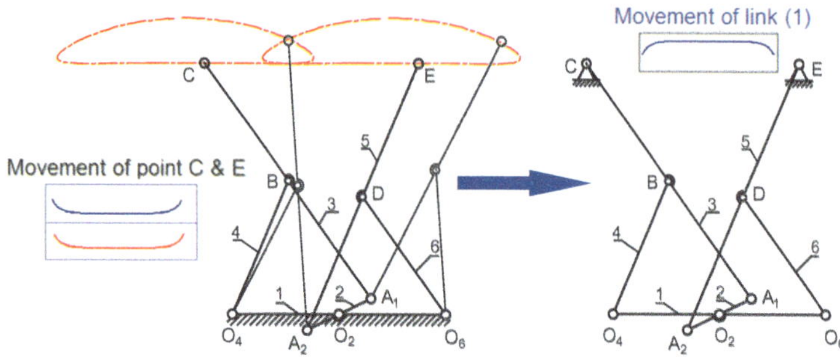

Fig. 8. Result of the mechanism inverse process.

taking the opposite value of the Y-coordinates of the legs in Fig. 7, the movement of the platform can be acquired, as shown in Fig. 8.

It is noticed that the coordinate of two points C and E are placed in the different loop of mechanism, these will exist a small error in their coordinate differences. This value will cause the pitch movement on link 1 and the tilting angle is determined by the equation:

$$\varphi_{tilted} = \arctan\left(\frac{\Delta y}{\Delta x}\right) = \arctan\left(\frac{y_C - y_E}{|x_C - x_E|}\right) \tag{9}$$

Result of the equation is illustrated in Fig. 9:

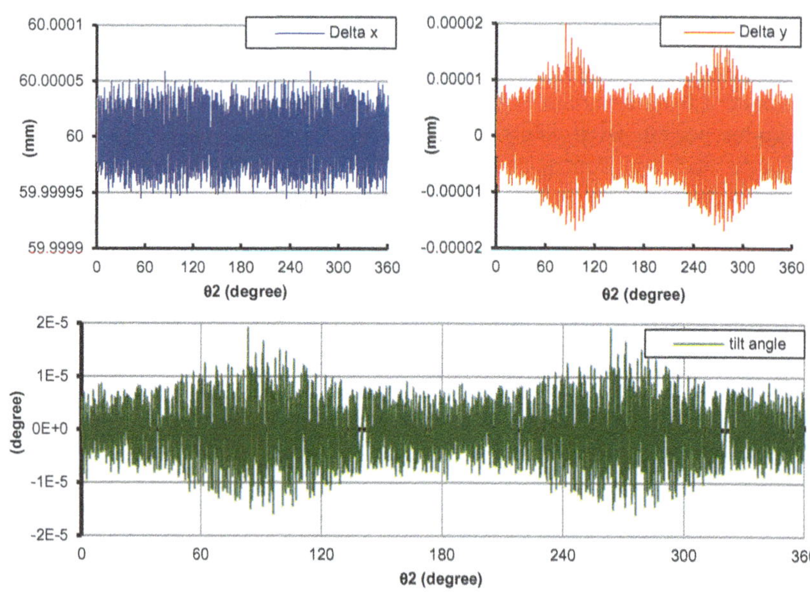

Fig. 9. Difference between position of legs C and E and the tilting angle of link 1.

The results of Fig. 9 show that there is a slight tilt during the movement of the platform (the angle varies from -2×10^{-5} to 2×10^{-5} degrees). These errors appear due to the in-sync of the feet on the left and right sides during their movement.

The mechanism is created in the Working Model environment to verify the accuracy of the calculation method. Figure 10 describe results of the simulation of the new design.

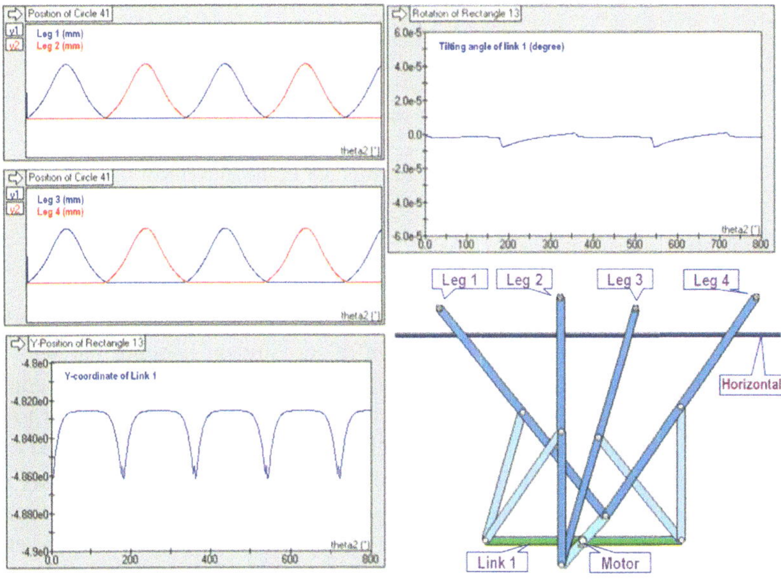

Fig. 10. Using Working Model to simulate the movement of the system.

The results from Fig. 10 show that the proposed method for determining the movement of link 1 is correct: A difference of 180° in the movement of the four legs produces a formation of the same range as the calculation for the platform (which is approximately 0.04 mm). Another notice from the graph in Fig. 10, there are small crooks in the Y-direction movement of link 1 (platform) every 180° rotation of the motor. These crooks can be explained that at these angle, the mechanism changes the landing legs on the b (from legs 1–3 to legs 2–4). Hence, the centroid of the robot changes the loading between the landing legs, causing tilting for the platform and affects the movement of the whole mechanism (Fig. 11).

The matches between results from the simulation model with those from the calculations prove the applicable of inversion method in kinetic analyzing for robotic mechanisms. This method can also be used to calculate the required power of the motor in cases platform is loaded and improve the efficiency of the mechanism in optimizing movement without simulation.

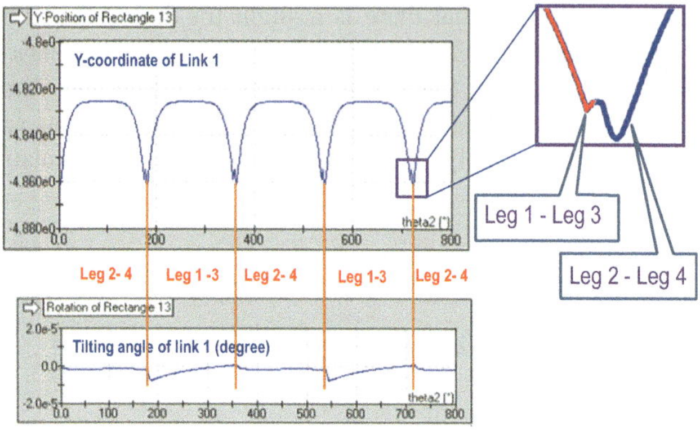

Fig. 11. Detailed explain for the crooked graph in the movement of link 1.

3 Conclusions

The paper proposes a method of calculating and analyzing robot structure formed of a four-bar mechanism. The structure is capable of moving on hanging bar with high stability. Different from solving by graphing method, the calculation using complex digital models helps the calculation process be done accurately and enabled to integrated with program software like Excel, Matlab to quickly give results. Based on the movement of the robot legs, the determination of working conditions (velocity ratio, tilted angled) is simplified and calculated. This method enables us to design for other models of robotic structures and to optimize dimensions of linkage to reduce engine power or raising stability of mechanism.

Acknowledgments. A special thanks to the Thai Nguyen University of Technology for funding this research.

References

1. Muller, M.: A novel classification of planar four-bar linkages and its application to the mechanical analysis of animal systems. Philos. Trans. R. Soc. B **351**, 689–720 (1996)
2. Roy, L., Sen, A., Chetia, R.P., Borah, M.J.: Analysis and synthesis of four-bar mechanism. Int. J. Theor. Appl. Mech. **3**(2), 171–186 (2008)
3. Kazuma, K., Hiroaki, W.: Energy-efficacy comparisons and multi-body dynamics analyses of legged robots with different closed-loop mechanisms. Multi-body Syst. Dyn. **40**(2), 123–153 (2016)
4. Ingram, A.J.: A new type of mechanical walking machine, Ph.D. Thesis. Department of Mechanical and Industrial Engineering Technology, University of Johannesburg, Johannesburg, Department of Physics and Astronomy, South Africa (2006)
5. Ghassaei, A.: The Design and Optimization of a Crank-Based Leg Mechanism, PhD Thesis. Pomona College (2011)
6. Shigley, J.E., Uicker, J.J.: Theory of Machines and Mechanisms. Mc.Graw-Hill (1981)

7. Hewapathirana, N.D., Uddawatta, L., Karunadasa, J.P., Nanayakkara, T.: Analysis on four legged multipurpose rope climbing robot. In: 2009 International Conference on Industrial and Information Systems (ICIIS), pp. 505–510 (2009)
8. Urankar, S., Jain, P., Singh, A., Saxena, A., Dasgupta, B.: Robo-sloth: a rope-climbing robot, pp. 1–10. Department of Mechanical Engineering, Indian Institute of Technology (2003)
9. Koo, Y.C., Elmi, A.B., Wajdi, W.A.F.: Piston mechanism based rope climbing robot. Procedia Eng. **41**, 547–553 (2012)

Application of Graphic Layout Method for Designing Cam Mechanism with Flat-Face Follower

Anh-Tuan Dang[1], Nguyen Thi Hue[2], and Dinh-Ngoc Nguyen[3](✉)

[1] Department of Mechanical Engineering, National Kaohsiung University of Science and Technology Kaohsiung, Kaohsiung, Taiwan
[2] Faculty of Fundamental Science, Thai Nguyen University of Technology, Thai Nguyen, Vietnam
[3] Faculty of International Training, Thai Nguyen University of Technology, Thai Nguyen, Vietnam
ngocnd@tnut.edu.vn

Abstract. This study was conducted to verify the working ability of the cam mechanism mechanisms with flat-face follower designed using envelope theory. The influence of the number of control points and the selection of construction curve for the designing process is verified and simulated using ADAMS software. To evaluate the capacity of the mechanism, kinematic parameters of the follower are also extracted for comparison with the initial designs.

Keywords: Cam mechanism · Envelope theory · Control point · ADAMS

1 Introduction

Cam mechanism is commonly used in industry due to its simple structure and its easy characteristics to design. Two parts of the mechanism consist of the follower and the cam connected by a half-joint, where the follower moving depends on the profile of the cam, which is either manufactured entirely by hand or designed using a machining center. Based on the construction of follower and contacting type between the components, cam mechanisms can be classified as flat-face, roller, or knife-edge followers, as shown in Fig. 1 [1, 2].

Cam mechanism design is the process of constructing the cam's profile based on the motion requirements of the follower while ensuring the kinematic and dynamic properties. A number of documents have been researched and developed about constructing the geometric profiles of cam using the graphic method, approximate methods, or envelope method [3, 4].

D. C. Nguyen et al. (Eds.): ICERA 2021, LNNS 366, pp. 180–188, 2022.
https://doi.org/10.1007/978-3-030-92574-1_20

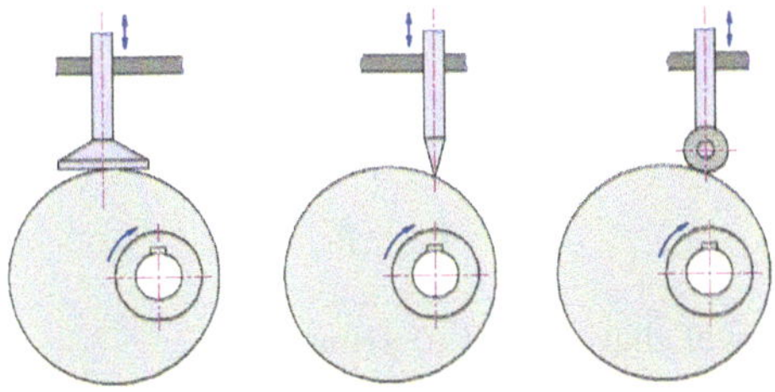

Fig. 1. Three type of follower contact surfaces: flat-face, knife-edge, roller.

Because follower movement is required only for some stages with certain positions while the rest can be varied, choosing motion for followers becomes diverse with the designers. We can choose any type of function (such as cycloid, sine, or harmonic function) for the construction of a single cam product. On that basis, many studies have been conducted and developed on cam profile construction. However, only a few among them are analyzed to evaluate performance of the mechanism after the designing process [5–10].

This paper uses the envelope method [11, 12] to design a cam mechanism with flat-face follower. From the given rule of periodic motion (cycle time and displacement), the angle of follower, the speed of cam, the movement of the follower corresponding to the rotation angle of the cam is calculated, as well as the circumscribed polygon of the cam. Non-Uniform Rational B-Splines (NURB) with different design degrees are then used to construct the final profile of the cam. The designed mechanism then is dynamically simulated in ADAMS to examine the dynamics parameters of the follower as displacement, velocity, and acceleration.

2 Cam Design Process

The problem of determining the displacement graph of the cam mechanism is the basic content of the design process. When the cam performs a counterclockwise rotation from angle φ_i to φ_j, due to the contact between the parts, the position of the follower will vary from Si to Sj (Fig. 2).

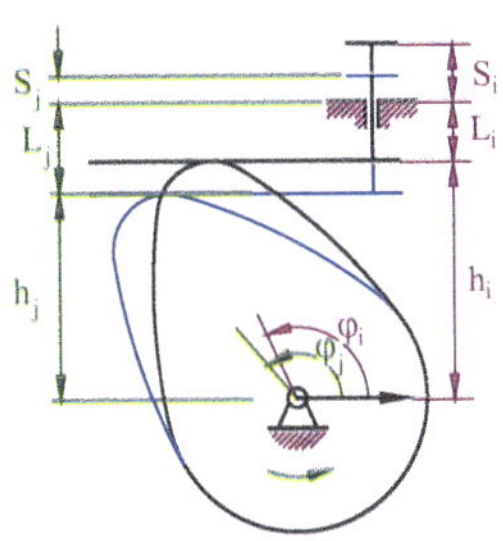

Fig. 2. Determine the displacement of follower when cam rotate from φ_i to φ_j

According to this figure, the relation between cam movement and the envelope construction lines are detected, with:

Distance from the rotation center of cam to ground:

$$h_i + L_i = h_j + L_j \tag{1}$$

The length of follower that contact to ground:

$$L_i + S_i = L_j + S_j \tag{2}$$

Substitute (1) to (2), we have:

$$h_i - h_j = S_i - S_j \tag{3}$$

This means by calculating the value of h (distances between the cam center to the contact position between the cam and the follower), the lifting position of the follower is determined. And reversely, based on the kinematic parameters of the follower, the corresponding contact surface between the components is calculated.

Inversion method is one of the methods to determine the profile of the cam based on the motion of the follower. Cam is considered to be the fixed part and the follower rotates around the cam's center to contact its profile. The cam is then enveloped with the edge of the follower as this follower positions around the cam. Those positions of followers will form a convex polygon enveloping the cam shape (Fig. 3).

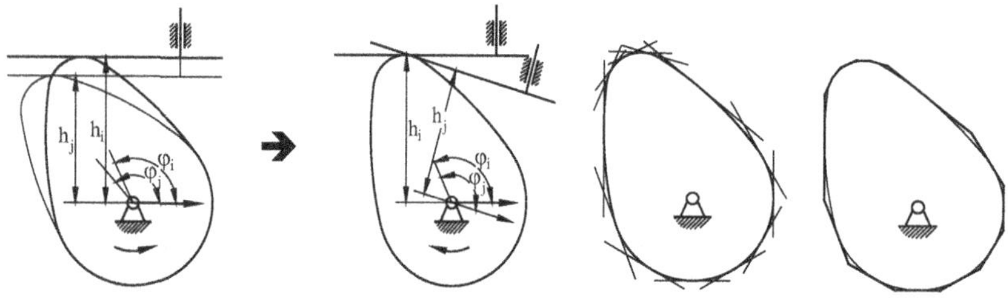

Fig. 3. Determination the displacement of follower using inversion method

Vertices of this polygon can are the inclined points between two adjacent sides, and can be calculated by solving the equation:

$$\begin{cases} x.\cos\varphi_i + y.\sin\varphi_i = h_i \\ x.\cos\varphi_j + y.\sin\varphi_j = h_j \end{cases} \tag{4}$$

The matrix equation of (4):

$$\begin{bmatrix} x \\ y \end{bmatrix} = \begin{bmatrix} \cos\varphi_i & \sin\varphi_i \\ \cos\varphi_j & \sin\varphi_j \end{bmatrix}^{-1} \begin{bmatrix} h_i \\ h_j \end{bmatrix} \tag{5}$$

where (x, y) is the coordinate of the vertices, and h_i, h_j are the position of follower with the corresponding angles φ_i, φ_j.

Various functions can be used to combine the displacement h and the rotation of the follower φ. In this study, by selecting the polynomial function for describing the movement relations, the following equations describing the kinematic properties of the follower are acquired:

$$S_i = h_i + 35(h_{i+1} - h_i)\left(\frac{\theta - \beta_i}{\beta_{i+1} - \beta_i}\right)^4 - 84(h_{i+1} - h_i)\left(\frac{\theta - \beta_i}{\beta_{i+1} - \beta_i}\right)^5 + 70(h_{i+1} - h_i)\left(\frac{\theta - \beta_i}{\beta_{i+1} - \beta_i}\right)^6 - 20(h_{i+1} - h_i)\left(\frac{\theta - \beta_i}{\beta_{i+1} - \beta_i}\right)^7 \tag{6}$$

$$V_i = 140\frac{(h_{i+1} - h_i)}{(\beta_{i+1} - \beta_i)^4}(\theta - \beta_i)^3 - 420\frac{(h_{i+1} - h_i)}{(\beta_{i+1} - \beta_i)^5}(\theta - \beta_i)^4 + 420\frac{(h_{i+1} - h_i)}{(\beta_{i+1} - \beta_i)^6}(\theta - \beta_i)^5 - 140\frac{(h_{i+1} - h_i)}{(\beta_{i+1} - \beta_i)^7}(\theta - \beta_i)^6 \tag{7}$$

$$A_i = 420\frac{(h_{i+1} - h_i)}{(\beta_{i+1} - \beta_i)^4}(\theta - \beta_i)^2 - 1680\frac{(h_{i+1} - h_i)}{(\beta_{i+1} - \beta_i)^5}(\theta - \beta_i)^3 + 2100\frac{(h_{i+1} - h_i)}{(\beta_{i+1} - \beta_i)^6}(\theta - \beta_i)^4 - 840\frac{(h_{i+1} - h_i)}{(\beta_{i+1} - \beta_i)^7}(\theta - \beta_i)^5 \tag{8}$$

where θ is the rotary angle of the cam ($0° \le \theta \le 360°$).

Take the example of a cam structural design problem with the motion of flat-face follower is described as below: the first 90° of cam rotation dwell the follower; the next 120° raise the follower 40 mm from lowest position and the follower continue to dwell for 30° of the rotation then return to lowest position for the last 120° of cam rotation; assuming minimum cam radius r = 100 mm.

The results of the calculation are shown in Fig. 4 with S-V-A graphs of the movement characteristics of the follower.

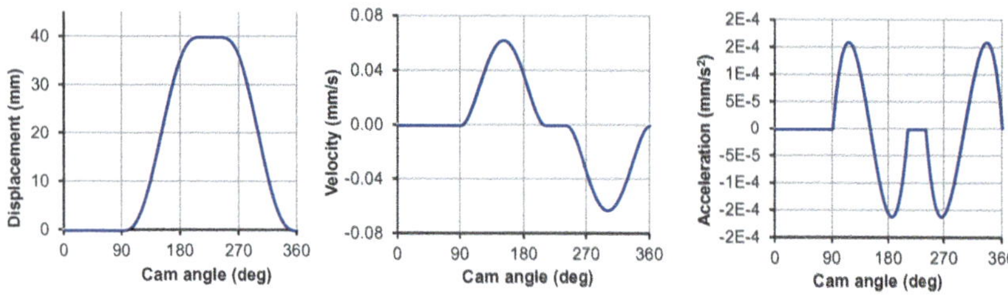

Fig. 4. Kinematic graphs of follower with during a rotation of cam.

By selecting contacting points from the displacement diagram, a polygon enveloping the cam profile is created. For constructing the cam profile base on this envelope polygon, several types of splines were studied [13–16]. Since the popularly used splines (basis spline, Bezier spline, and Non-uniform rational B-spline (NURBS), see Fig. 5) can easy constructed in AutoCAD, we overlook using their function to make it more convenient for the study.

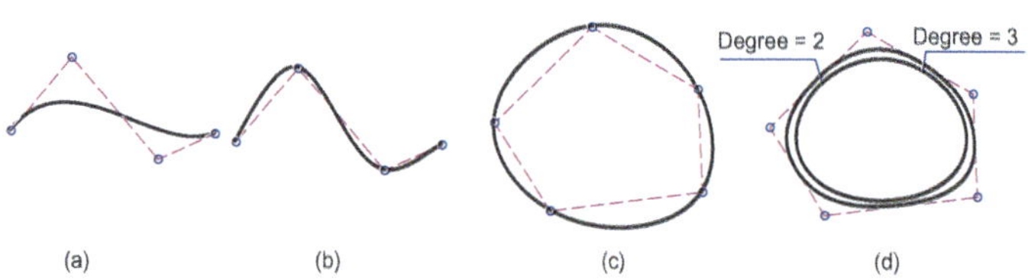

Fig. 5. Basis spline (a), Cubic spline (b) and Spline constructed from control points/control polygon: cubic spline (c) and NURB splines with different degrees (d)

It is noticed from Fig. 5c that the cubic spline is not contacted with the envelope polygon and even overcomes the limit range of this bounding polygon, while the NURB spline with a different degree tends to contact with the edges of this profile.

The selection of the number of vertices for the polygon also decides the fitness of the cam profile with the ideal design: the more the number of vertices is employed, the more quantity of lines or the follower positions can be selected for references and the resulted profile will be more consistent with the designated ideal and more suitable with the designed dynamics conditions.

Determining the vertices of the inscribed polygon is based on the set of parameters (φ_i, h_i) that of the cam angle φ_i and the corresponding position of follower (with the distance $h_i = S_i + r$ measured from the center). Three kinds of polygons with different numbers of vertices are selected as construct parameters of the cam profile, as presented in Fig. 6. Using the interval angle between two adjacent lines $\Delta\varphi$ as the selection parameter, the selection of this value can easily influence the construction of circumscribed polygons. To compare the effect of interval angle $\Delta\varphi$ on the kinematic parameters of the follower, we survey three cases of the cam profile using $\Delta\varphi = 1°$, 5°

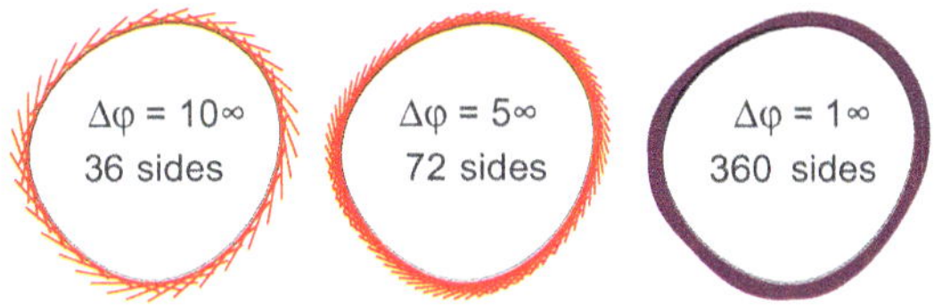

Fig. 6. Circumscribed polygons constructed from inversion method.

and 10°. The spline was then used as NURB spline with 2-degree and 3-degree equations based on the corner of these polygons.

The differences of the constructed cam profiles are compared with the model using $\Delta\varphi = 0.1°$ (nearly fit with the ideal profile). Figure 7 presents the deviation in the radius of these profiles.

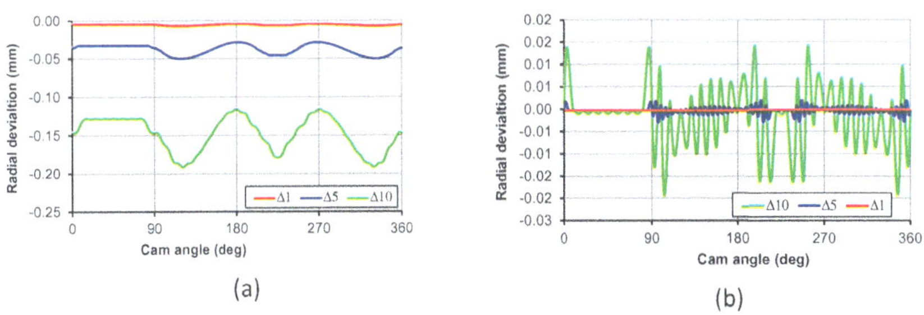

Fig. 7. Deviation of cam radius with different value of $\Delta\varphi$ with 2-degree (a) and 3-degrees (b) of NURB.

The differences in deviation values of the splines can be explained by the principle of the NURB curve using the polygon vertices as the control points of AutoCAD. The smaller the interval angle $\Delta\varphi$, the smaller distance between the cam profile using spline and the envelope polygon, and the smaller errors compared to the ideal kinematic parameter calculated by equations (see Fig. 8).

Simulating the system using ADAMS software with the ideal profile ($\Delta\varphi = 0.1°$), we obtained the corresponding S-V-A kinematic graphs of the follower (see Fig. 9). The result of this structure is similar to the value of the ideal graph from Fig. 4.

Continue the simulation with other types of cam profiles, the corresponding S-V-A kinematic graphs of the follower are obtained, as shown in Fig. 10.

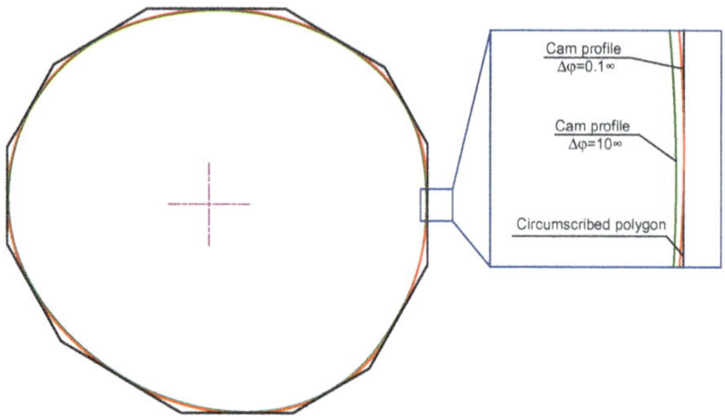

Fig. 8. Influence of $\Delta\varphi$ in constructing the cam profile using 3-degree spline.

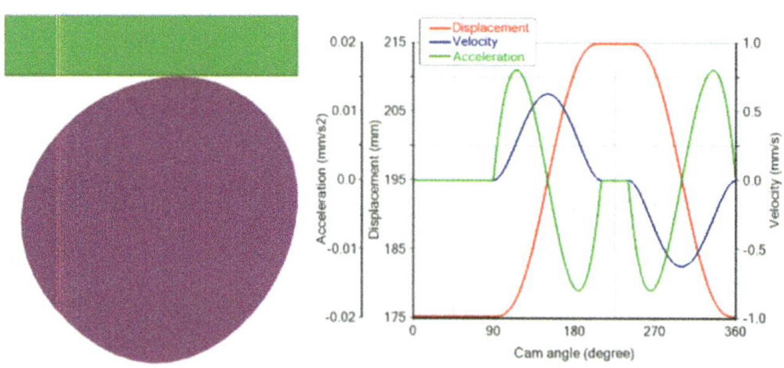

Fig. 9. Cam mechanism and the SVA kinematic graph of follower

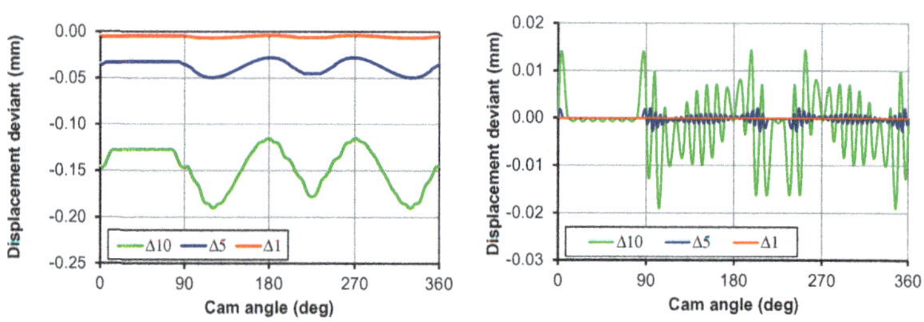

Fig. 10. Deviation of S_V_A graph for 2-degree spline cam profile (left) and 3-degree spline cam profile (right) compare with ideal graph (Fig. 5)

The results from Fig. 11 show that the cam profile using the 3-degree spline can give smaller errors than the 2-degree spline, but the surface is not as smooth and causes fluctuations in the speed and acceleration of the follower. Using 2-degree splines for cam profiles reduce velocity error by 4 times and acceleration by 5 times compared to the result from 3-degree spline).

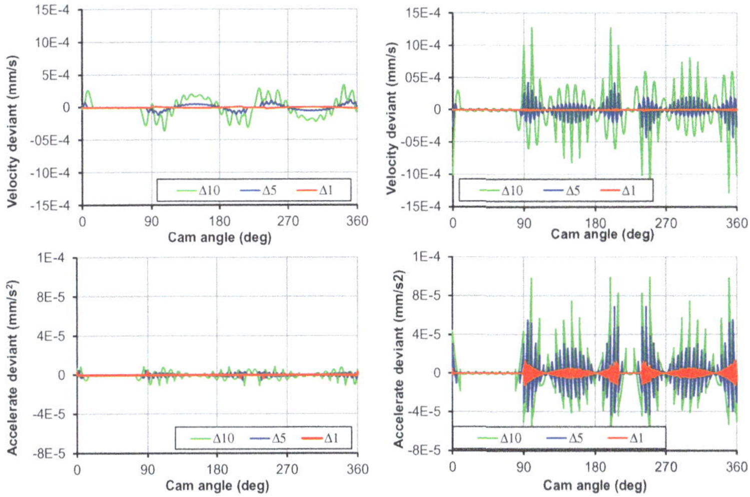

Fig. 11. Deviation of S_V_A graph for 2-degree spline cam profile (left) and 3-degree spline cam profile (right) compare with ideal graph (cont.)

3 Conclusion

The paper proposed a method of constructing the profile for cam structure using flat-face follower. The profile of the cam after the design was drawn in AutoCad in 2D allowing for fabrication by advanced machining methods. The testing process of structural kinematic parameters simulated by ADAMS software gives results quite similar to the model calculated from the polynomial functions. In addition, errors of S-V-A parameters were evaluated to give advice in further design when selecting the interval angle for the mechanism. The proposed envelope method suggests the possibility of automating the process of calculating and designing for cam profile, thereby minimizing the time in the process of fabricating the same structure.

Acknowledgments. The authors wish to thank Thai Nguyen University of Technology for supporting this work.

References

1. Robert, L.N.: Design of Machinery. McGraw-Hill (1992)
2. Harold, A.R.: Cam Design Handbook. McGraw-Hill (2004)
3. Kenneth, J.W., Gary, L.K.: Kinematic, Dynamics and Design of Machinery, 2nd edn. (2003)
4. Shigley, J.E., Uicker, J.J.: Theory of Machines and Mechanisms. McGraw Hill, New York (1995)
5. Zhang, Y., Li, R., Xiao, L., Yang, J.: Study on dynamic simulation of cam mechanism based on Pro/E and ADAMS. In: Proceedings of the 2nd International Conference on Electronic & Mechanical Engineering and Information Technology (2012)

6. Wang, S., Li, A.: Parameterizedly design cam based on excel and solidworks and finite element analysis and simulation by COSMOS. In: 2010 International Conference on E-Product E-Service and E-Entertainment, pp. 1–4 (2010)
7. Ouyang, T., Wang, P., Huang, H.: Cam profile optimization for the delivery system of an offset press. Proc. Inst. Mech. Eng. Part C J. Mech. Eng. Sci. **231**(23), 4287–4297 (2016)
8. Ge, R., Wang, J., Zhang, C.: Model and simulation of globoidal cam mechanisms with oscillating and dual stopping followers. MATEC Web Conf. **108**, 04005 (2017)
9. Quang, T.M., Hung, N.Q.: Computer aided cam design of roller-follower cam mechanism considering kinematic and dynamic requirements. J. Sci. Technol. **139**(09), 125–130 (2015)
10. Zhang, Y., Li, R., Xiao, L., Yang, J.: Study on dynamic simulation of cam mechanism based on Pro E and ADAMS. In: 2nd International Conference on Electronic & Mechanical Engineering and Information Technology, pp. 1229–1232 (2012)
11. Chen, F.Y.: Mechanics and Design of Cam Mechanisms. Pergamon Press, New York (1982)
12. Flores, P., Machado, G.J., Pereira, R.M., Claro, J.P.: Kinematics of the roller motion and cam size optimization of disc cam-follower mechanisms with translating roller followers. In: International Design Engineering Technical Conferences and Computers and Information in Engineering Conference, vol. 49040, pp. 421–432 (2009)
13. Xia, B.-Z., Liu, X.-C., Shang, X., Ren, S.-Y.: Improving cam profile design optimization based on classical splines and dynamic model. J. Cent. South Univ. **24**(8), 1817–1825 (2017)
14. Neamtu, M., Pottmann, H., Schumaker, L.L.: Designing NURBS cam profiles using trigonometric splines. J. Mech. Des. **120**(2), 175–180 (1998)
15. Nguyen, V.T., Kim, D.J.: Flexible cam profile synthesis method using smoothing spline curves. Mech. Mach. Theory **42**(7), 825–838 (2007)
16. Nguyen, T.N., Kurtenbach, S., Hüsing, M., Corves, B.: A general framework for motion design of the follower in cam mechanisms by using non-uniform rational B-spline. Mech. Mach. Theory **137**, 374–385 (2019)

Automatic Navigation Research for Multi-directional Mobile Robots on the Basis of Artificial Intelligence Application, Q-Learning Algorithm

R. V. Hoa[1], T. D. Chuyen[1], N. D. Dien[1], and Dao Huy Du[2](✉)

[1] Faculty of Electrical, University of Economics – Technology for Industries, Hanoi, Vietnam
tdchuyen@uneti.edu.vn
[2] Faculty of Electronics, Thai Nguyen University of Technology, Thai Nguyen, Vietnam
daohuydu@tnut.edu.vn

Abstract. In this article, research on the applications of artificial intelligence in implementing reinforcement learning is a subset of machine learning that deals with learning decisions from rewards given by the environment. Classic reinforcement learning algorithms are usually applied to small sets of states and actions. However, in real applications, the state spaces are of a large scale and this will bring the problems of the generalization and the curse of dimensionality. In this paper, we integrate neural network into reinforcement learning methods to generalize the value of all the states. Simulation results on the Gazebo framework show the feasibility of the proposed method. The robot can complete navigation tasks safely in an unpredicted dynamic environment and becomes a truly intelligent system with strong self-learning and adaptive abilities.

Keywords: Artificial intelligence · Multi-directional mobile robots · Obstacle robots · Reinforcement learning · Autonomous navigation · Q-learning

1 Introduction

Mobile robot's automatic navigation can be divided into three subsystems: information perception, behavioral decision making, and manipulation control. Path planning is the basis of mobile robot navigation and control [1, 2]. The goal of route planning for the multi-directional mobile robot is to find the path from the current location to the target location. The path should be as short as possible, the smoothness of the road must meet the dynamics of the mobile robot and the safe path without collision [3]. Depending on the amount of environmental information known in the road planning process, road planning can be divided into global road planning and local road planning [4]. There are many ways to plan your way. According to specific algorithms and strategies, path planning algorithms can be divided into four categories: pattern matching, artificial lead field, map construction, and artificial intelligence [5, 6]. Each type of route planning

D. C. Nguyen et al. (Eds.): ICERA 2021, LNNS 366, pp. 189–199, 2022.
https://doi.org/10.1007/978-3-030-92574-1_21

algorithm has an optimal application scenario and limitations. Current route planning of mobile robots is mainly based on its surroundings. In addition to the limitations of traditional path planning, robots cannot complete their learning and judgment in complex environments, a bottleneck in the development of research in the field [6]. Therefore, it is especially important to develop a road planning method with low dependence on the environment, which can quickly adapt to the surroundinvixg environment.

Endowing robots with human-like abilities to perform specific skills in a smooth and natural way is one of the important goals of robotics to cover a large range of real-life missions such as delivery, search or rescue missions, security surveillance. Such missions require different levels of autonomy in navigation to react to different dynamic factors such as environmental conditions changes. One of the most common approaches is to train mobile robots via multiple expert demonstrations. These demonstrations provide robots prior experiences from examples, defined as a sequence of state-action pairs that are recorded during the teacher's demonstration of the desired robot behavior. However, in some cases, such as exploring hazardous environments, no prior experience or demonstrations are available to the multi-directional mobile robots.

In this paper, we present a reinforcement learning (RL) based self-learning algorithm in the cases of unknown environment. Experiments are conducted on the autonomous navigation tasks for mobile robots. More specifically, we introduce a neural network structure to generalize and approximate the value of all the states of RL algorithms. The experiments are conductedon the Gazebo simulator by using its open source extension Gym to perform the autonomous navigation tasks [7, 13].

The rest of this paper is organized as follows. Section 2 gives an overview of the related works. Section 3 describes the general methodology and the model free RL based methods. Then, Sect. 4 introduces some experimentation results followed by a discussion of the efficiency of the proposed method. Finally, Sect. 5 concludes our paper and presents our future works.

2 Related Work

Autonomous navigation is a fundamental and critical research area of mobile robotics. There are different reinforcement learning algorithms that have been applied to this problem, such as the fuzzy Q learning presented in [8] or the dueling network architectures in [9].

In addition, a humanoid robot navigation was described in [4, 10] by using a supervised RL approach combined with Gaussian distributed state activation, and the robot successfully performed a backward docking movement used for autonomous recharging. In [11], a neural network reinforcement learning method was studied for visual control of a robot manipulator. A direct mapping from the image space to the actuator command was developed by using two different RL algorithms, Q learning and SARSA, combined with neural networks. A database of representative learning samples was also employed in order to speed up the convergence of the algorithms. This hybrid system could provide a high accuracy of a manipulator positioning in a situation of low resolution images, [15, 16].

3 Neural Network Based Reinforcement Learning Methods

3.1 Model-Free Reinforcement Learning

Reinforcement learning can be formulated as a Markov Decision Process (MDP). Model based RL algorithms can be used if we know the state transition function $T(s, a, s')$. In contrast, we focus more on model free RL that the environment model T is not known in advance, especially in our work of autonomous navigation tasks.

In this paper, we focus on robot control problems in which a mobile robot acts in a stochastic environment by sequentially choosing actions over a sequence of time steps, in order to maximize a cumulative reward. We model the problem as a Markov Decision Process: a state space S, an action space A, a transition dynamics distribution $P(s_{t+1}|s_t, a_t)$ satisfying the Markov property $P(s_{t+1}|s_1, a_1, s_2, a_2, \ldots, s_t, a_t) = P(s_{t+1}|s_t, a_t)$, for any trajectory $s_1, a_1, s_2, a_2, \ldots, s_T, a_T$ in state-action space, and a reward function $r : S \times A \rightarrow R$. A stochastic policy $\pi(s_t, a_t) = P(a_t|s_t)$ is used to select actions and produce a trajectory of states, actions and rewards $s_1, a_1, r_1, s_2, a_2, r_2, \ldots, s_T, a_T, r_T$ over $S \times A \times R$ (Fig. 1).

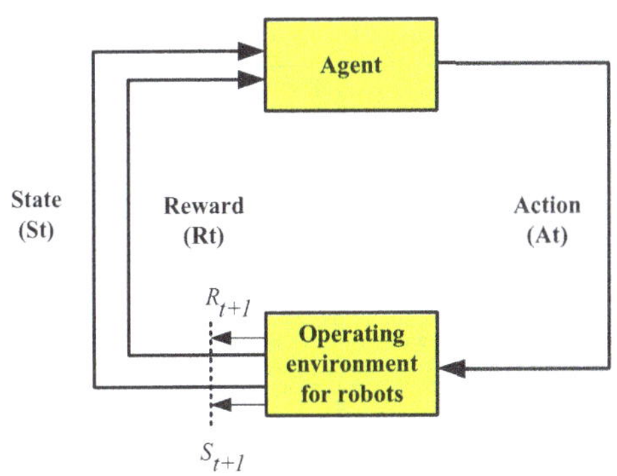

Fig. 1. The schematic diagram of the reinforcement learning model

We study model free RL methods that the robot drives an optimal policy without explicitly learning the model of the environment. More specifically, we focus on Q learning algorithms, one of the most major model free reinforcement learning algorithms.

3.2 The Q – Learning Algorithm

Q – Learning algorithm is an important off-policy model free reinforcement learning algorithm for temporal difference learning. The update of state action values in Q-learning is defined by

$$\Delta(\xi_t, a_t) := \Delta(\xi_t, a_t) + \alpha[r_{t+1} + \gamma \max_a \Delta(\xi_{t+1}, a_t) - \Delta(\xi_t, a_t)] \quad (1)$$

Where α is learning rate, with value between 0 and 1. If the learning rate is set to 0, the Δ values are never updated. If this value set to a value near 1, learning can occur quickly γ is the discount factor, models the importance of future rewards which are worth less than immediate rewards (Fig. 2).

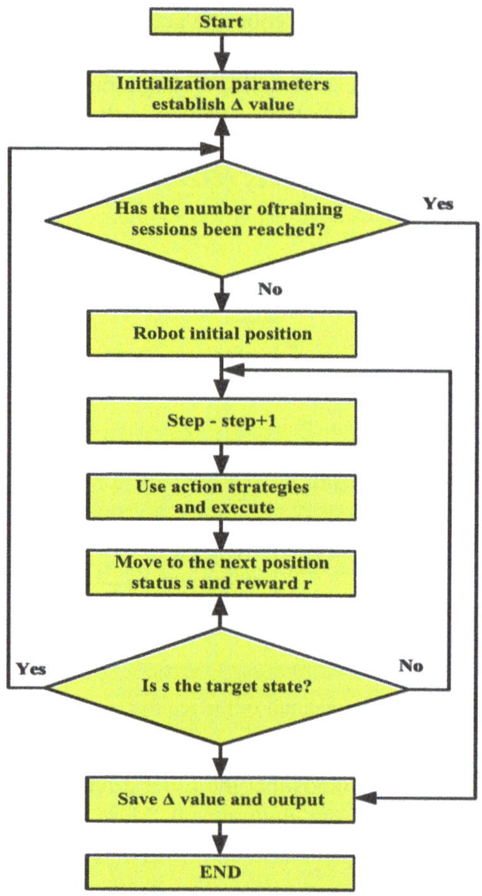

Fig. 2. Schematic diagram of Q-Learning path planning method

3.3 The Neural Network Based Q Learning Algorithm

In this paper, we introduce a new model free reinforcement learning method by integrating a neural network to Q learning. This hybrid method takes advantage of the strengths of both Q learning and neural network. Because the reinforcement learning can be modeled as a MDP, we need first to define the state space S and action space A.

(1) ***State Action Space***

The action space A is defined by six discrete robot actions. Among them, five are basic moving actions: move forward (F), turn left at 30° (L30), turn left at 60° (L60), turn right at 30° (R30), turn right at 60° (R60) and one emergency action: move backward (B). A turn at 90° is not defined because a sharp turn will bring great danger to a real vehicle. In sum,

$$A = \{F, L30, L60, R30, R60, B\} \tag{2}$$

The state space S is composed of a very large but finite number of states, defined by four groups of features, and is expressed in:

$$S = [D_t \quad V_t \quad I_t \quad U_t]^T \tag{3}$$

Where $D_t \in R^{7\times1}$ are sensory degrees of danger, $U_t \in R^{7\times1}$ is a vector to indicate if a wall is detected ($U = 1$) or an object obstacle ($U = 0$) for the seven sensors in D_t. $V_t \in \{1, 2, ..., 9\}$ is the target region and $D_t \in \{0, 1\}$ is the indicator that determines if the robot has detected the target.

(2) ***Reward Function***

The reward function measures the immediate feedback for the action taken at a given state. It evaluates how good or how bad the performed action is. Before giving the reward function, one environment state at each time instant is classified into five state properties: Safe State (SS), Cozy State (CS), Dangerous State (DS), Winning State (WS) and Failure State (FS). The Reward Function can be defined as Table 1.

Table 1. Reward function

State transition	Extra criteria	r
Other states → Winning state		1
Safe state → Cozy state		0
Cozy state → Safe state		0
dangerous state → cozy state		0.5
Cozy state → Dangerous state		0
Dangerous state → Failure state		−1
Dangerous state → Dangerous state (approaching obsacles)	$d_{warn} \leq d^t_{\min} \leq d^{t-1}_{\min} - 2$	−0.3
	$d^t_{\min} = d^{t-1}_{\min} - 1 \leq d_{warn}$	−0.3
	$d^t_{\min} = d^{t-1}_{\min} - 2 \leq d_{warn}$	−0.5
	$d^t_{\min} \leq d^{t-1}_{\min} - 3$	−1
Nonsafe state → Nonsafe state (evading obstacles)	$d_{warn} \leq d^t_{\min}$	0.6
	$d^t_{\min} \geq d_{warn}$	0.3

Component d_{r-o} and d_{r-t} define the distances between the robot and obstacles, and the target, component d_{bou} defines the boundary distance (sensor detection range) of SS and other states, component d_{col} defines the radius of the collision region around the obstacle, component d_{win} defines the radius of the winning region around the target, component d_{bou} defines the cozy distance, component d_{warn} is a warning distance that the robot is approaching too close to an obstacle, component $d_{\min}^{t-1}$ and component $d_{\min}^{t}$ are the minimum distances between the robot and the surrounding obstacles. The total sensor detection range of ϕ is the interval (0; 2π). These distances as in Fig. 3.

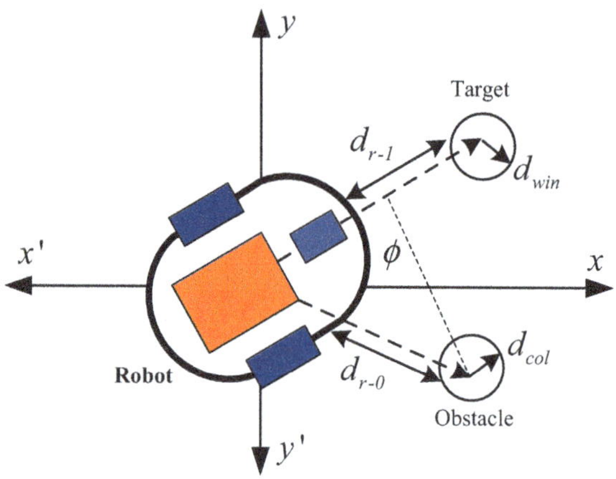

Fig. 3. The mobile robot operation diagram with target and obstacle

(3) ***State Action Value Iteration***

The Δ value function expresses the mapping policy from the perceived state of environment to the executing action. One Δ value $\Delta(\xi_t, a_t)$ corresponds with one specific state and one action in this state. In our method, we propose to predict all state Δ values by using a three layer neural network, as can be seen in the Fig. 4.

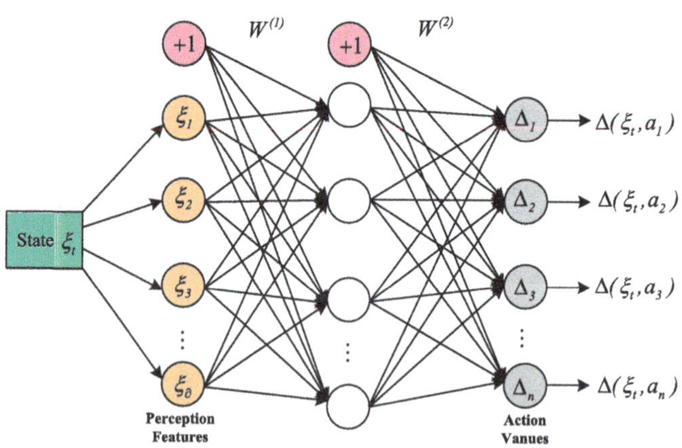

Fig. 4. A three-layer neural network architecture

The action value iteration is realized by updating the neural network by the means of its weights. Different from the classic neural network, the neural network in the reinforcement learning does not has label outputs. Q learning is a process of value iteration and the optimal value after each iteration serves as the target value for neural network training. The update rule is

$$\Delta_{\sigma+1}(\xi_t, a_t) := \Delta_k(\xi_t, a_t) + \alpha[r_t + \gamma \max_a \Delta_\sigma(\mathrm{s}_{t+1}, \mathrm{a}_t) - \Delta_\sigma(\xi_t, \mathrm{a}_t)] \quad (4)$$

Where, component $\Delta_\sigma(\xi_t, a_t)$ is the Δ value in the σth iteration. We employ also the popular stochastic gradient descent (SGD) to train the neural network online. The goal is to minimize the cross entropy cost function.

(4) ***Robot Navigation Using Neural Network based Q-Learning***

After training the robot via the integrated neural network, the resulting policy is still stochastic but near deterministic that used by the robot for future navigation tasks in various environments. The multi directional mobile robot starts its navigation through the environment by finding its current state. If it is a Safe State the robot does not need to follow the policy but changes its orientation towards the target and moves one step forward. It continues moving until entering a Non Safe region where the robot needs to adopt the trained control policy. The robot uses the Neural Network to generate all possible state action Δ values. The robot greedily takes the action that has the biggest Δ value. After that, the robot finds its new state and repeats the process of action selection until the robot reaches its goal or collide an obstacle.

4 Simulation and Experimental Results

4.1 The Research TurtleBot Multi-directional Mobile Robot

Here the authors go to study the model of mobile robot: with the actual hardware architecture of this robot is shown as shown in Fig. 5, in which each hardware module will perform a number of tasks, in the sequence of activities of this mobile robot: such as finding a path, crossing obstacles, etc.

In the tests, the authors perform several tasks in the sequence of the mobile robot's operations, such as the Ubuntupowered Raspberry Pi 3 Model B+. The Raspberry Pi 3 Model B+ embedded computer directly processes information from a range of sensors including the Astra smart camera and the smart sensor then transmits commands to a smart microcontroller. To record images from the environment as well as measure the distance between mobile robots and unknown obstacles, mobile robots are equipped with cameras and smart sensors, in which the smart camera can do 360° laser scanning and ranges within 15 m generate map data to be used for the mapping process. The smart microprocessor control circuit will be the part to receive control signals from Raspberry Pi 3 Model B+ and then send the signal to the MOSFET bridge circuit to operate two motors.

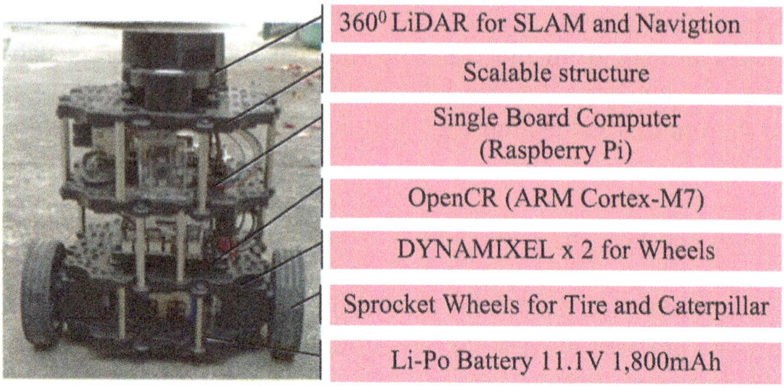

Fig. 5. Pictures of the actual TurtleBot multi-directional robot

4.2 The Simulation Experimental Results

Our objective is to demonstrate the preliminary results obtained by the offline learning phase in Gazebo based simulation [7] and the Open AI Gym extension [12]. We define a navigation mission as the problem of visiting a set of waypoints. The simulation environment in Gazebo is shown in Fig. 6. In this section, the actual mobile robot Turtlebot is tested in a real environment, which is the environment used for real world testing consisting of flowerpots as obstacles. The dynamic obstacles can be randomly generated in the environment. A learning episode of Q learning is acompleted mission. We deployed 6000 learning episodes in simulation.

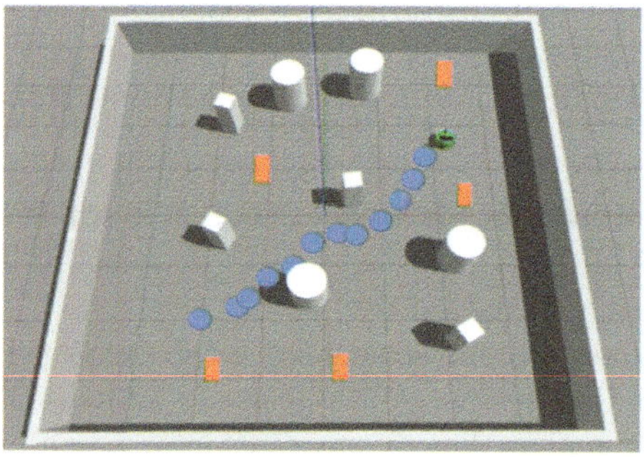

Fig. 6. The constructing visual mapings and robot simulation models in Gazebo

Figure 7 depicts the total reward obtained over 6000 learning episodes. Although there are many variations due to the varying complexity of the environment and the efficiency of the navigation algorithms of existing omnidirectional multi directional mobile robots, the trend line between complex environment and simple environments also indicates the Total reward growth during the learning period is optimal with good results (Figs. 8 and 9).

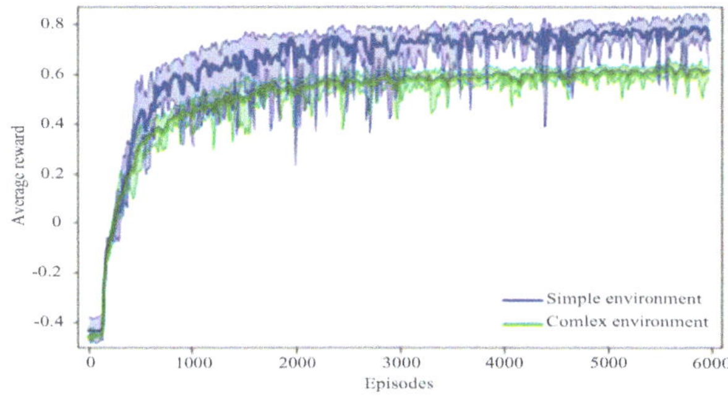

Fig. 7. The total reward of the learning phase of 6000 episodes

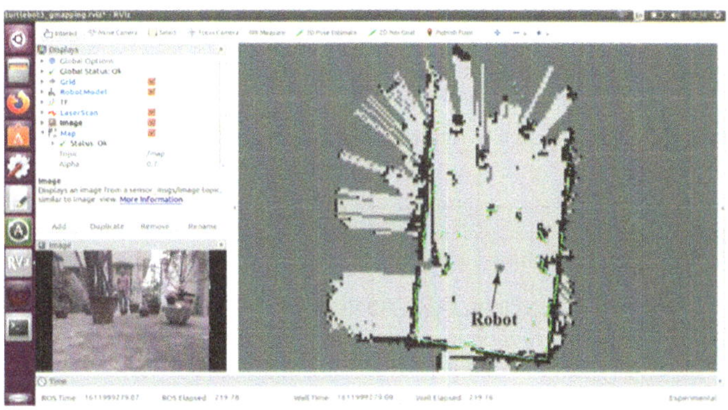

Fig. 8. The Turtlebot mobile robot is in the process of creating concurrent mapping

Fig. 9. Results of navigating the TurtleBot mobile robot on Rviz

5 Conclusions

In this paper, we presented the robot self learning strategy without prior experience under explicit feedback. We explored the mobile robot navigation problem by combining reinforcement learning and neural network. Q-learning is applied to enhance the self learning ability of a mobile robot through trial and error interactions with an unknown environment. We designed new reward expression and introduced the neural network architecture to store and train the large scale Q values, and also to generalize the learning performance to large scale state and action spaces. Experiments are conducted on the autonomous navigation tasks for mobile robots. The simulation results show the stability and feasibility of the hybrid method. This article also performs a successful route planning for mobile robot when interacting with objects that are static and unknown obstacles due to constant update of local path in the environment. Tasks are based on data generated from smart cameras and smart sensors. Robot hardware is also built to facilitate integration of Rviz based peripherals. Furthermore, robot activity can be monitored through visual tools.

Acknowledgments. This research is funded by Thainguyen University of Technology (TNUT).

References

1. Ghosh, S., Panigrahi, P.K., Parhi, D.R.: Analysis of FPA and BA meta-heuristic controllers for optimal path planning of mobile robot in cluttered environment. IET Sci. Meas. Technol. **11**, 817–828 (2017). https://doi.org/10.1049/iet-smt.2016.0273
2. Orozco-Rosas, U., Montiel, O., Sepúlveda, R.: Mobile robot path planning using membrane evolutionary artificial potential field. Appl. Soft Comput. **77**, 236–251 (2019). https://doi.org/10.1016/j.asoc.2019.01.036
3. Han, J., Seo, Y.: Mobile robot path planning with surrounding point set and path improvement. Appl. Soft Comput. **57**, 35–47 (2017). https://doi.org/10.1016/j.asoc.2017.03.035
4. Li, G., Chou, W.: Path planning for mobile robot using self-adaptive learning particle swarm optimization. Sci. China Inf. Sci. **61**(5), 1–18 (2017). https://doi.org/10.1007/s11432-016-9115-2
5. Zhao, Y., Zheng, Z., Liu, Y.: Survey on computationalintelligence-based UAV path planning. Knowl. Based Syst. **158**, 54–64 (2018). https://doi.org/10.1016/j.knosys.2018.05.033
6. Bakdi, A., Hentout, A., Boutami, H., Maoudj, A., Hachour, O., Bouzouia, B.: Optimal path planning and execution for mobile robots using genetic algorithm and adaptive fuzzy-logic control. Robot. Auton. Syst. **89**, 95–109 (2017). https://doi.org/10.1016/j.robot.2016.12.008
7. Zamora, I., Lopez, N.G., Vilches, V.M., Cordero, A.H.: Extending the openai gym for robotics: A toolkit for reinforcement learning using ros and gazebo. arXiv:1608.05742 (2016)
8. Pambudi, A.D., Agustinah, T., Effendi, R.: Reinforcement point and fuzzy input design of fuzzy Q-learning for mobile robot navigation system. In: 2019 International Conference of Artificial Intelligence and Information Technology (ICAIIT) (2019)

9. Ruan, X., Ren, D., Zhu, X., Huang, J.: Mobile Robot Navigation based on Deep Reinforcement Learning. In: 2019 Chinese Control And Decision Conference (CCDC) (2019)
10. Navarro-Guerrero, N., Weber, C., Schroeter, P., Wermter, S.: Real-world reinforcement learning for autonomous humanoid robot docking. Rob. Auton. Syst. **60**(11), 1400–1407 (2012)
11. Miljković, Z., Mitić, M., Lazarević, M., Babić, B.: Neural network reinforcement learning for visual control of robot manipulators. Expert Syst. Appl. **40**, 1721–1736 (2013)
12. Brockman, G., et al.: Openai Gym. arXiv:1606.01540 (2016)
13. Nga, V.T.T., Loc, O.X., Nam, T.H.: Enhanced Learning in Automatic Control with Matlab Simulink. Hanoi Polytechnic Publishing House (2020)
14. Tuan, N.T.: Base Deep learning. The Legrand Orange Book. Version 2 (2020)
15. Quang, H.D., et al.: An approach to design navigation system for omnidirectional mobile robot based on ROS. Int. J. Mech. Eng. Rob. Res. **11**(9), 1502–1508 (2020)
16. Hoa, R.V., Chuyen, T.D., Lam, N.T., Dien, N.D., Son, T.N., Linh, V.T.T.: reinforcement learning based method for autonomous navigation of mobile robots in unknown environments. In: Proceedings of the 2020 International Conference on Advanced Mechatronic Systems. Hanoi, Vietnam, (2020)

Autonomous Active Power-Sharing Control in Photovoltaic Generation-Dominated Microgrids

Trung Thai Tran[1,2] and Minh-Duc Ngo[2(✉)]

[1] RWTH Aachen University, 52074 Aachen, Germany
[2] Thai Nguyen University of Technology, Thai Nguyen City, Vietnam
{tranthaitrung-tdh,ngoduc198-tdh}@tnut.edu.vn

Abstract. This paper proposes a decentralized control strategy that combines Sliding Mode Control (SMC) and Virtual Oscillator Control (VOC) to share the load power among Photovoltaics (PV) generations in islanded microgrids under different solar irradiation levels and load conditions. The SMC based controller of the DC-DC boost converter is used to regulate the DC link voltage. Meanwhile, VOC based DC-AC inverter is used to guarantee the load power-sharing in proportion to the rated powers, the RMS voltage regulation, as well as providing maximum power point tracking (MPPT) function when necessary. The proposed control strategy has the capability of seamlessly switching between different control modes without controller reconfiguration and the support from external devices such as energy storage systems (ESS). The performance of the proposed control strategy is validated in Matlab/Simulink.

Keywords: Photovoltaic generation · Virtual oscillator control · Sliding mode control · Maximum power point tracking · Power sharing

1 Introduction

In the last few decades, the concept of microgrid (MG), that can be operated in both grid-connected and islanded mode, has emerged as controllable interconnected entities in the modern power system, in order to utilize the advantages of integrating renewable energy resources (RE). While operating in islanded mode, a microgrid loses the support of the utility grid for voltage and frequency stability, which now the microgrid controller has to provide, and needs to balance the load demand and power generation on its own. The technical challenges related to stability, reliability, and power quality become more severe in a microgrid with a high penetration level of RE, especially variable RE such as wind, PV generations. In this paper, we focus on technical challenges and solutions for the power-sharing among parallel-connected PVs in low voltage (LV) islanded microgrids. PVs are preferred to operate at an MPP to maximize the produced energy [1]. However, the MPPT algorithm restricts the participation of PVs in regulating the voltage, frequency, and balancing load/generation to ensure the stability of MG.

D. C. Nguyen et al. (Eds.): ICERA 2021, LNNS 366, pp. 200–212, 2022.
https://doi.org/10.1007/978-3-030-92574-1_22

A simple method for the power generation/demand balancing is integrating other dispatchable resources such as energy storage system (ESS) into PV systems [2–4]. However, from a microgrid perspective, installing ESS in many cases is not economically effective.

Another possibility is to implement a proper coordination strategy in the controller of PVs that allow them to flexibly participate in voltage and frequency regulation and track MPP when needed. Different methods have been proposed in the literature combining droop control with maximum power point tracking (MPPT) algorithm to contend with the above issues [5–8]. A multi-mode controller for large capacity centralized PVs is proposed in [5] to achieve full flexibility of dispatching power, regulating voltage, and overcome the system instability. The operation mode selection signals are generated from the higher-level operator. In [6], a hybrid controller consisting of a so-called fast MPPT and slow MPPT for microgrid-integrated PVs is presented in which the value of *dp/dv* is used as a threshold to switch between modes. However, the main drawbacks of the hybrid controller are: 1) it requires proper switching decision to change from one mode to another, 2) the switching function significantly changes the states of the controller causing a significant disturbance in an islanded microgrid. In addition, it is difficult to analyze the stability of such a controller since individual stable modes cannot guarantee the stability of the entire system. A so-called universal controller is presented in [7] in which the MPPT algorithm is embedded into the droop control loop of the two-stage PVs. While the PVs are forced to extract maximum power during normal operation, the controller is seamlessly transformed into droop mode to ensure the system stability during significant load variations without controller reconfiguration. An improved dual droop control scheme is proposed in [8] to control two-stage PVs where the DC-link voltage of each converter is allowed to slightly drop if the PV capacity cannot meet its droop command. By doing this way, the droop power command is reduced which will allow other converters with enough capacity to increase power generation to meet the load demand. In the above methods, a well-known droop control method is used as the primary mechanism for voltage and frequency control, and power-sharing. However, the slow response characteristic of a droop controller causes a relatively long settling time when cooperating with the MPPT algorithm.

This paper proposes a novel decentralized control strategy combining simplified SMC and VOC to control multiple parallel-connected two-stage PVs in islanded AC microgrids. The advantages of VOC over the droop control method are presented in [9, 10]. The proposed method is designed to achieve the following control objectives:

- The PVs shall operate in the load power-sharing mode governed by VOC when they can produce enough power for power-sharing requirements (i.e., proportionally share the load power according to inverter power rating).
- During a low-irradiation situation, the PV controller is automatically switched to MPPT mode. Other PVs that are operated in high irradiation conditions will share the rest of the load in proportion to their rated power.
- The voltage and frequency regulation are guaranteed by VOC in both operating modes.

- The MPPT function and load power-sharing are unified in a single control structure. No switching mechanism and external energy storage systems are needed.

The rest of this paper is organized as follows: in Sect. 2, the general structure of the islanded three-phase AC microgrid with multiple parallel-connected PVs is presented. In Sect. 3, control objectives and related problems will be discussed in detailed. The proposed control strategy is presented in Sect. 4. The validation of the effectiveness of the proposed method in both off-line simulation (Matlab/Simulink) and real-time platform (Opal-RT) is presented in Sect. 4. Section 5 concludes this paper.

2 Architecture of the Tested AC Microgrid

The architecture of a tested AC microgrid is presented in Fig. 1, which consists of a number of PVs connected in parallel to serve loads locally and common loads through physical network. An PV subsystem includes PV array, two-stage power electronic based converter, and an LCL filter to interface with other PVs.

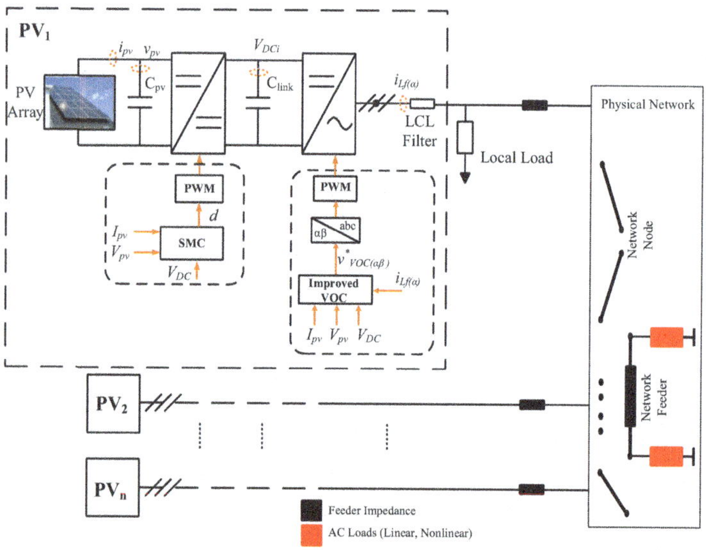

Fig. 1. A general PV based AC microgrid.

The PV control system is composed of two main parts: The SMC is used to regulate the duty cycle of the boost converter to stabilize the DC link voltage. The VOC that controls the DC/AC inverter embeds an MPPT algorithm based on Incremental Conductance (IC) to unify MPP tracking, voltage and frequency regulation, and load power-sharing within the same control structure. This two-stage configuration of PVs provides more degree of freedom for engineers to design a controller, i.e., there are more control variables to simultaneously perform and decouple multiple control objectives (MPPT, reactive power compensation, voltage, and frequency support, etc.).

3 Proposed Autonomous Active Power Sharing Strategy

3.1 Active Power Sharing Problem Formulation

To better understand the possible problem related to the power-sharing control among parallel-connected PVs, Fig. 2 only illustrates the P-V characteristic of a system having two parallel-connected PVs (PV_1 and PV_2) operating in islanded mode to serve a common load at the PCC. Also, the rated power of two PVs is assumed to be equal for simplicity. Without any limitation, the analysis of general microgrids with several PVs connected in parallel can be done correspondingly.

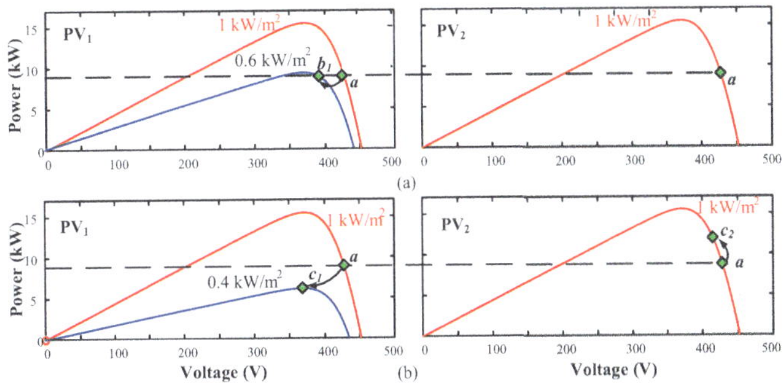

Fig. 2. P-V characteristic of PVs under different irradiation conditions.

In a normal operating condition, i.e., when the generated power from two PVs are enough for load power-sharing requirement (e.g., the 1 kW/m^2 curve in Fig. 2), two PVs operate at the point a on the red curve to share the load equally, governed by VOC (see Fig. 2a). When the solar irradiation of PV_1 reduces to 0.6 kW/m^2, e.g., due to cloud coverage, PV_1 is forced to move its operating point from the red curve to the blue curve. However, the generated power from PV_1 is still enough to share to load power, and hence its operating point moves from a to b_1. Two PVs are still operated in load power-sharing mode and the system stability is maintained.

In a critical condition, i.e., the irradiations of PVs decrease to the point that the power-sharing requirement threshold is met, e.g., the characteristic curve of the PV_1 is dropped to the curve "0.4 kW/m^2" (the green curve in Fig. 2b). In this case, the maximum available power generated by PV_1 (at the point c_1, $P_{c1} < P_a$) is not able to supply the load-sharing requirement. Meanwhile, the operating point of PV_2 moves from point a to point c_2 ($P_{c2} > P_a$), sufficient to compensate for the shortage of supply from PV_1, which is certainly possible because P_{c2} is less than the possible maximum power of PV_2. If the power-sharing ratio is kept the same, the VOC of PV_1 will hence tend to increase its power output by absorbing energy from the DC link capacitor, causing the DC link voltage V_{DC1} to drop. On the other hand, the VOC of PV_2 attempts to decrease its generated power by charging the DC link capacitor, which causes the DC link voltage V_{DC2} to increase. This opposite reaction may result in the collapse of the whole system if the load shedding action is not activated.

3.2 Proposed Autonomous Active Power Sharing Strategy

To solve the problem related to the above-mentioned critical condition, the gains that define the power-sharing ratio in conventional VOC should be changed to regulate the VSC power outputs to follow the changes of irradiation levels and loads. Therefore, this section presents the proposed decentralized active power sharing control strategy that combines VOC with SMC. First, a simplified SMC is designed considering all state variables of boost converter and PV array into the linear manifold to stabilize the DC link voltage by regulating the duty cycle of the boost converter. Then, a necessary modification is applied to the VOC to change the power-sharing ratio in conventional VOC.

Simplified Sliding Mode Control for Boost Converter
Considering the average dynamic model of the boost converter in continuous mode as follows:

$$\frac{dV_{DC}}{dt} = \frac{-V_{DC}}{ZC_{link}} + \frac{1}{C_{link}}(1-u)i_{Lb} \tag{1}$$

$$\frac{di_{Lb}}{dt} = \frac{V_{pv}}{L_b} - (1-u)\frac{V_{DC}}{L_b} \tag{2}$$

$$\frac{dV_{pv}}{dt} = \frac{1}{C_{pv}}i_{pv} - \frac{1}{C_{pv}}i_{Lb} \tag{3}$$

with L_b, R_b being the inductor and its internal resistance, C_{pv} and C_{link} is the capacitor at the input-side and output-side of the boost converter, respectively. The terminal voltage and current of the PV array are denoted as V_{pv} and i_{pv}, respectively. The current i_{DC} is measured at the output of the boost converter. The sliding surface is chosen as a function of the reference tracking error as follows:

$$S_v = K_1 e_{DC} + K_2 e_i + K_3 e_v \tag{4}$$

where: K_1, K_2, K_3 are designed coefficients. $e_{DC} = V_{DC} - V_{DC}^*$, $e_i = i_{Lb} - i_{Lb}^*$, and $e_v = V_{pv} - V_{pv}^*$ are the DC link voltage, inductor current, and PV voltage tracking error, respectively. To avoid multiple required reference control signal, a high-pass filter technique with the transfer function in the Laplace domain $G_{HP}(s) = Ts/(1+Ts)$ with the corner frequency $f_c = 1/T$ is chosen to simplify the sliding surface, as follows:

$$S_v = K_1\left(V_{DC} - V_{DC}^*\right) + K_2\left(i_{HP\_Lb}\right) + K_3\left(V_{HP\_pv}\right) \tag{5}$$

where: i_{HP_Lb}, V_{HP_pv} are the high-pass-filtered measurement of the inductor current and input-side capacitor voltage in place of the reference current i_{Lb}^*, and voltage V_{pv}^*, respectively. The design objective is to obtain a control law for the duty cycle $u^* = u_{re} + u_{sw}$ to control V_{DC} so that it follows the desired constant reference. According to SMC principle, to direct the trajectory of the system state variables toward the desired sliding surface, the following invariance condition must be satisfied:

$$\frac{dS_v}{dt} = 0 \tag{6}$$

Taking the derivative of the sliding surface S_v in (5), then replacing dV_{DC}/dt, di_{Lb}/dt, and dV_{pv}/dt from (1), (2) and (3) to solve (6) we have the reaching term:

$$u_{re} = 1 - \frac{B}{A} \tag{7}$$

where $A = \left(\frac{K_2}{L_b} V_{DC} - \frac{K_1}{C_{link}} i_{Lb}\right)$, and $B = \left(\frac{K_2}{L_b} V_{pv} - \frac{K_1}{ZC_{link}} V_{DC} + \frac{K_3}{C_{pv}} i_{pv} - \frac{K_3}{C_{pv}} i_{Lb}\right)$.

The switching term of the control law is chosen by using a discontinuous *sat* function to reduce the chattering effect, as follows:

$$u_{sw} = \frac{K_4}{A} sat(S_v) \tag{8}$$

In the following, the Lyapunov stability criterion is used to analyze the stability of the proposed controller. The system is considered asymptotically stable if the Lyapunov function W and its time derivative dW/dt satisfy the following conditions:

$$\begin{cases} W \geq 0 \\ \frac{dW}{dt} < 0 \end{cases} \tag{9}$$

Let us choose a Lyapunov function as follows:

$$W = \frac{1}{2} S_v^2 \tag{10}$$

Taking the time derivative of (10), then replacing dS_v/dt from (4) with the control law given in (7) and (8), it yields

$$\frac{dW}{dt} = S_v\left\{A\left(1 - \left(1 - \frac{B}{A} + \frac{K_2}{A} sat(S_v)\right)\right) - B\right\} = -K_4 S_v sat(S_v) \tag{11}$$

It can be seen from (11) that $dW/dt \leq 0 \forall K_4 \geq 0$. Enforcing the Lyapunov criterion explicitly shown in (9), it follows that the proposed SMC controller is asymptotically stable for controlling the DC link voltage if $K_4 > 0$.

Improved Virtual Oscillator Control

Figure 3 shows the general circuit diagram of VOC [11]. It consists of: 1) an RLC circuit to set the system frequency and control the power generated by an inverter, 2) a nonlinear voltage dependent current source to sustain the oscillation.

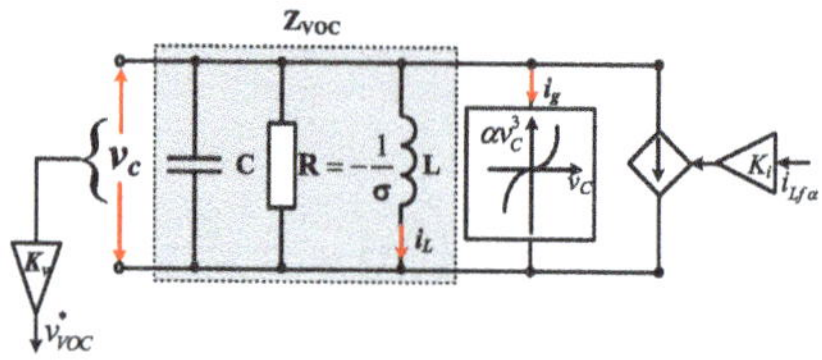

Fig. 3. A general structure of VOC.

VOC takes the alpha component of the filter inductor current $i_{Lf\alpha}$ as an input to generate the reference signal v^*_{VOC} for the PWM block. While a droop based primary controller for inverter includes three loop controllers, VOC can regulate voltage, frequency and power sharing with only a single control loop. Hence, the controller structure is significantly simplified and easy to implement in an actual microcontroller. The equations that govern the dynamics of the VOC are obtained as follows [12]:

$$L\frac{di_L}{dt} = v_c \tag{12}$$

$$C\frac{dv_c}{dt} = -\alpha v_c^3 + \sigma v_c - i_L - K_i i_{Lf} \tag{13}$$

where $\sigma := -1/R$ being a negative conductance, K_v and K_i being the scaling factors to couple VOC input and output to physical electrical signals. v_c and i_L being scaled VOC capacitor voltage and inductor current, respectively.

In the original form of VOC, the current scaling factor K_i is a constant so that the power-sharing ratio between parallel-connected inverters is fixed, regardless of the load conditions. Instead, in this paper, an adaptive MPPT algorithm is adopted in the VOC to regulate the current scaling factor K_i to adjust the power-sharing ratio. This solution allows the PV controller to switch between different modes depending on solar irradiations and load conditions. i.e., based on the ratio dP_{pv}/dV_{pv}, without the need of reconfiguring controller structure and external ESS. The detailed configuration of the proposed method is presented in Fig. 4.

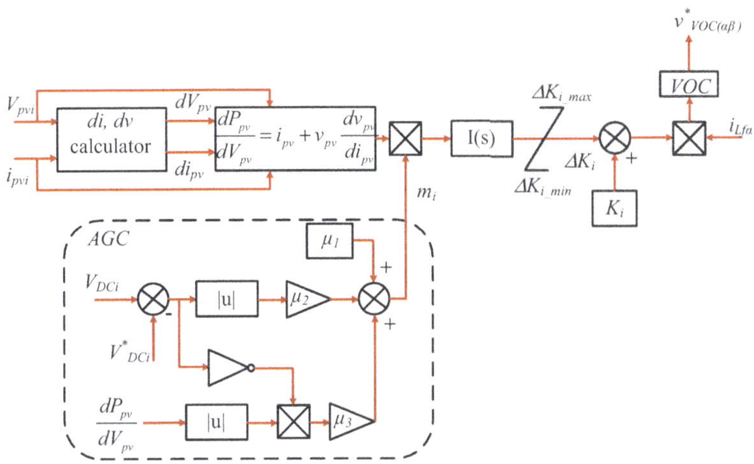

Fig. 4. Detailed configuration of the proposed power control method.

The ratio dP_{pv}/dV_{pv} is used as an input to the integral controller to produce the update of the scaling factor ΔK_i. Thus, the current scaling factor K_i, instead of being constant, now is gradually regulated in the range of $\left[K_i^{(0)} + \Delta K_{i\_min}; K_i^{(0)} + \Delta K_{i\_max}\right]$ where $K_i^{(0)}$ is an initial value of K_i. The lower limit ΔK_{i_min} is *Zero* indicating that PVs are operated at the normal power-sharing condition. The upper limit ΔK_{i_max} is calculated based on the maximum possible power-sharing ratio between PVs. Therefore, in conclusion, the PV controller has two different operating modes, depending on the value of ΔK_i: *Zero* for normal power-sharing mode, *Non-Zero* positive for MPPT.

A simple adaptive gain controller (AGC) is used to improve the performance of the proposed method, and prevent possible interaction with other control loop as follows:

$$m_i = \mu_1 + \mu_2\left|V_{DC} - V_{DC}^*\right| + \mu_3\left(\overline{V_{DC} - V_{DC}^*}\right)\left(\left|\frac{dP_{pv}}{dV_{pv}}\right|\right) \tag{14}$$

where: μ_1 and μ_2 and μ_2 are the designed gains. $\overline{V_{DC} - V_{DC}^*}$ represents the negation logic of the DC link voltage error. Generally, μ_1 is chosen small enough to prevent unexpected oscillation in the steady-state. μ_2 and μ_3 define the dynamics of AGC so that their value can be reasonably larger than μ_1. Considering these statements, the AGC gains is tuned by using the trial-error method to obtain optimal dynamic performance.

4 Simulation Results and Discussion

The effectiveness of the proposed method is validated in MATLAB/Simulink simulation. For ease of the analysis, the tested system consists of three PV systems that has the same rated power operated in parallel to serve the common loads at the PCC. The parameters of PVs and control system is presented in Table 1.

Table 1. Parameters used for tested PVs

Parameters	Values
PV output voltage range (V_{pv})	339.5 – 451.9 V
Voltage at the nominal MPP (V_{MPP})	371 V
Current at the nominal MPP (I_{MPP})	41.8 A
DC-link reference voltage (V_{DC}^{*})	800 V
Rated power of PV inverter (S_{rated})	15 kVA
Rated frequency (f_0)	50 Hz
Rated line-to-line voltage (V_{L-L})	400 V
LCL inverter filter L_f, C_f, L_g	0.95mH, 0.15μF, 0.56mH
SMC (K_1, K_2, K_3, K_4)	0.95, 0.25, 0.93, 4000
AGC (μ_1, μ_2, μ_3)	0.0025, 7, 2.5

4.1 Case 1: Solar Irradiation Change

The objective of this study is to demonstrate the performance of the proposed control strategy under different changes of solar irradiation level. The load comprises a 5-kW resistive load and 15-kW CPL. The load is represented in term of active power to provide more understandable information while analyzing the simulation results. In this case study, the solar irradiation level coming to each PV is changed in different time periods to illustrate: 1) the effectiveness of the proposed control method in changing operating modes, and 2) MPPP performance. The detail of solar irradiation changes is shown in Table 2.

Table 2. Solar Irradiation Changes

Period / **PVs \Unit**	1 (W/m^2)	2 (W/m^2)	3 (W/m^2)	4 (W/m^2)
PV_1	1000	200	200	800
PV_2	1000	1000	200	200
PV_3	1000	1000	1000	1000

The power output and DC link voltage of each PV are presented in Fig. 5. In the period t = *0 – 5* s, all three PVs operate at the maximal standardized irradiation level so that they share the load power in proportion to the rated powers, governed by VOC. Since the inverter filter impedances and the electrical cables connecting PVs with the

PCC are assumed identical, zero power sharing error is observed in this period. In the period t = *5 – 15* s, the solar irradiation coming to PV_1 is reduced, causing its generated power cannot satisfy the load power-sharing requirement. Hence, the controller of the PV_1 is forced to transit to MPPT mode. Meanwhile, PV_2 and PV_3 share the rest of the load equally ($P_2 = P_3 = 8.51$ kW).

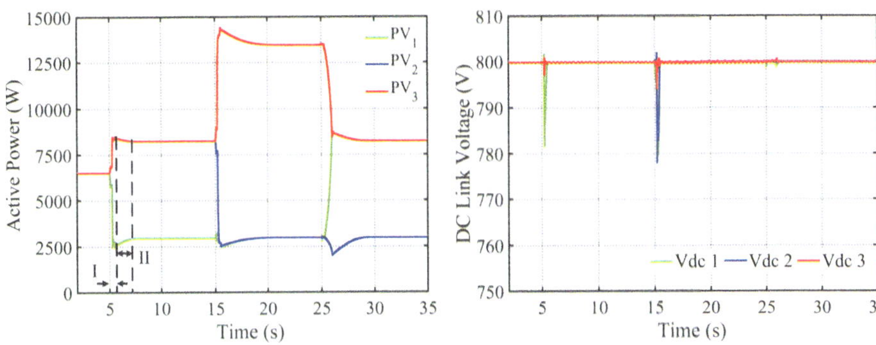

(a) (b)

Fig. 5. System dynamics under different irradiation levels: a) PV power outputs, b) DC link voltage.

Figure 5 shows that the transient period with two distinct phases. Phase I is known as the support phase for DC link voltage control. In this phase, the DC link voltage of PV_1 is quickly brought back to its nominal values by: 1) the effect of SMC, and 2) fast increase of the current scaling factor K_i using integrator with high gain controlled by AGC. In phase II, the system dynamic response jumps to the MPP tracking mode with lower integral gain (compared to the one in phase I) to gradually force a PV array of PV_1 to operate at the MPP. The actual power output of PV_1 in this period is $P_1 =$ 2.98 kW which is equivalent to 98.675% of the expected value (3.02 kW). This steady-state MPP tracking error is deliberately necessary to avoid unexpected variation around the MPP point caused by the calculation of dP/dv in practice.

In the period t = *15 – 25* s, the solar irradiation coming to PV_2 is also reduced from 1000 W/m^2 to 200 W/m^2. Therefore, the PV array of PV_2 is also forced to operate at MPP, with the same power output with PV_1 ($P_2 = 2.98$ kW). Hence, PV_3 must raise up its power output to supply the rest of the load. Like the previous period, a two-phase dynamic response is observed.

In the last period t = *25 – 30* s, the controller of PV_1 is transited back to the power sharing mode since its power output is now enough for the power sharing requirement, to share the load with PV_3 proportionally to their rated powers. Meanwhile, PV_2 remains in MPP tracking mode. It is essential to notice that the controllers of PVs are seamlessly switched between different operating modes with minor transients.

4.2 Case 2: Load Power Changes

In this case study, the objective is to validate the performance of the proposed controller under different load conditions. Three PVs are operated at 500, 600 and 1000 W/m^2 solar irradiation, respectively. In the period t = 0 – 5 s, only the load used in case 1 is connected. Three PVs share the load equally (≈6.67 kW), governed by VOC (see Fig. 6a). At t = 5 s, a second load (4-kW resistive and 6-kW CPL) is connected to the PCC. The controllers of PV1 and PV2, hence, are switched to MPP mode since their power outputs cannot satisfy load sharing requirement ($P_{\text{shared}} = 10\,\text{kW}$). After a short transient period, DC link voltage is maintained at its reference value (see Fig. 6b). The controller of PV1 and PV2 stabilize their power outputs at the MPP ($P_1 = 9.28\,\text{kW}$ and $P_2 = 7.69\,\text{kW}$, equivalent to 99.3% of the expected MPP values). Then PV3 increases its power output to supply the rest of the load. When the second load is disconnected at t = 15 s, the controller of PVs transit back to the power-sharing mode with insignificant transients.

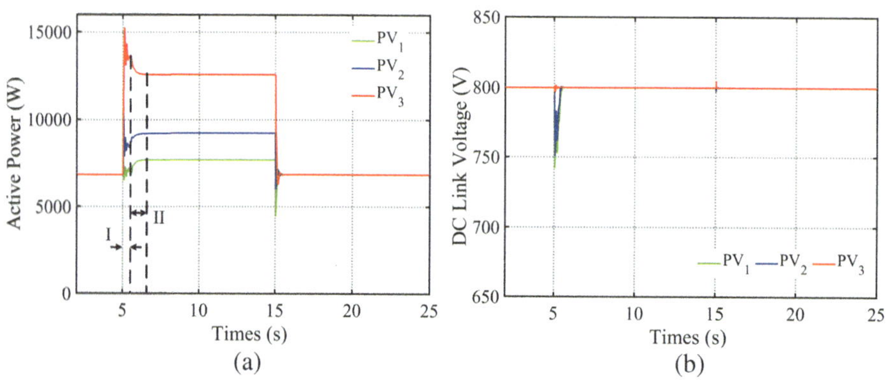

Fig. 6. System dynamics under different load conditions: a) PV power outputs, b) DC link voltage.

4.3 Case 3: Plug-and-Play Function

In this case study, the system dynamics is investigated when a PV is connected to or disconnected from an existing MG. It is assumed that at the time $t = 5$ s the PV_3 is connected to the MG consisting of two PVs supplying a common 20-kW load, and is disconnected at the time $t = 15$ s. Figure 7 shows the active power output transient of each PV in different simulation periods. Notice that despite the addition/removal of PV_3, the controller of PV_1 and PV_2 can seamlessly switch between different control modes to meet the load requirement.

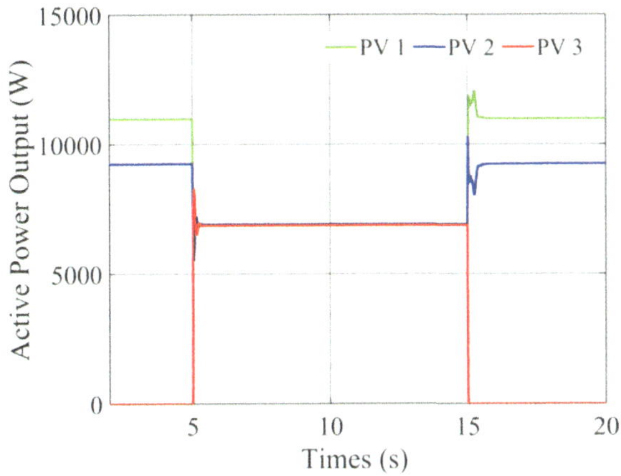

Fig. 7. PV power outputs in case 3.

5 Conclusion

In this paper, a decentralized control strategy is proposed for parallel connected two-stage PVs in islanded MGs, composed of a novel low order SMC and an MPPT based VOC. The proposed control strategy can provide two separated control modes, with seamless transition capability, namely load-power sharing and MPP tracking, under different solar irradiation levels and load conditions, without controller reconfiguration and the need of ESSs. The simulations results show that: a) the proposed control strategy maintains load power-sharing in proportional to their rated powers, governed by VOC when PV array power outputs satisfy load sharing requirements. b) the proposed controller shows superior MPPT performance with the steady-state power tracking error is less than 2%, and c) the effectiveness of the proposed controller is achieved without affecting the intrinsic advantages of conventional VOC.

Acknowledgement. This research was supported by Research Foundation funded by Thai Nguyen University of Technology.

References

1. Bacha, S., Picault, D., Burger, B., Etxeberria-Otadui, I., Martins, J.: Photovoltaics in microgrids: an overview of grid integration and energy management aspects. IEEE Ind. Electron. Mag. **9**(1), 33–46 (2015). https://doi.org/10.1109/MIE.2014.2366499
2. Mahmood, H., Michaelson, D., Jiang, J.: A power management strategy for PV/battery hybrid systems in islanded microgrids. IEEE J. Emerging Sel. Top. Power Electron. **2**(4), 870–882 (2014). https://doi.org/10.1109/JESTPE.2014.2334051
3. deMatos, J.G., eSilva, F.S.F., deSRibeiro, L.A.: Power control in AC isolated microgrids with renewable energy sources and energy storage systems. IEEE Trans. Ind. Electron. **62**(6), 3490–3498 (2015). https://doi.org/10.1109/TIE.2014.2367463

4. Wang, C., Nehrir, M.H.: Power management of a stand-alone wind/photovoltaic/fuel cell energy system. IEEE Trans. Energy Convers. **23**(3), 957–967 (2008). https://doi.org/10.1109/TEC.2007.914200
5. Wandhare, R.G., Agarwal, V.: Advance control scheme and operating modes for large capacity centralised PV-grid systems to overcome penetration issues. In: 2011 37th IEEE Photovoltaic Specialists Conference, pp. 002466–002471 (2011). https://doi.org/10.1109/PVSC.2011.6186445.
6. Elrayyah, A., Sozer, Y., Elbuluk, M.E.: Modeling and control design of microgrid-connected PV-based sources. IEEE J. Emerging Sel. Top. Power Electron. **2**(4), 907–919 (2014). https://doi.org/10.1109/JESTPE.2014.2330797
7. Elrayyah, A., Sozer, Y., Elbuluk, m.: Control of microgrid-connected PV-sources. In: 2012 IEEE Industry Applications Society Annual Meeting, pp. 1–8 (2012). https://doi.org/10.1109/IAS.2012.6374032
8. Liu, H., et al.: An enhanced dual droop control scheme for resilient active power sharing among paralleled two-stage converters. IEEE Trans. Power Electron. **32**(8), 6091–6104 (2017). https://doi.org/10.1109/TPEL.2016.2614324
9. Raisz, D., Thai, T.T., Monti, A.: Power control of virtual oscillator controlled inverters in grid-connected mode. IEEE Trans. Power Electron. **34**(6), 5916–5926 (2019). https://doi.org/10.1109/TPEL.2018.2868996
10. Tran, T., Raisz, D., Monti, A.: Harmonic and unbalanced voltage compensation with VOC-based three-phase four-leg inverters in islanded microgrids. IET Power Electron. **13**(11), 2281–2292 (2020). https://doi.org/10.1049/iet-pel.2020.0174
11. Tran, T.T., Sowa, I., Raisz, D., Monti, A.: An average consensus-based power-sharing among VOC-based distributed generations in multi-bus islanded microgrids. IET Gener. Transm. Distrib. **15**(4), 792–807 (2021). https://doi.org/10.1049/gtd2.12059
12. Johnson, B.B., Sinha, M., Ainsworth, N.G., Dörfler, F., Dhople, S.V.: Synthesizing virtual oscillators to control islanded inverters. IEEE Trans. Power Electron. **31**(8), 6002–6015 (2016). https://doi.org/10.1109/TPEL.2015.2497217

Balancing Energy Consumption for Clustered Wireless Sensor Networks Utilizing Compressed Sensing

Vinh T. Vu[1], Thuong T. K. Nguyen[2], Son Q. Tran[2], and Minh T. Nguyen[2(✉)]

[1] Thai Nguyen University of Information and Communication Technology, Thai Nguyen City 240000, Viet Nam

[2] Thai Nguyen University of Technology, Thai Nguyen City 240000, Viet Nam
nguyentuanminh@tnut.edu.vn

Abstract. Energy saving in Wireless sensor network (WSNs) is always critical since the networks facilitate many applications in different fields. In this paper, we study and analyze unequal clustering algorithms in the literature and propose a new algorithm that combines traditional unequal clustering algorithms and Compressed sensing (CS) techniques for further energy saving and balancing. In the proposed algorithm, sensor nodes are arranged into unequal clusters that the clusters closer to the base-station (BS) have smaller sizes compare to the ones farther from the BS. A greedy tree is proposed to link all cluster-heads (CHs) together as a tree. CS measurements are created at CHs to be sent via the tree to the BS. Based on a certain number of CS measurements achieved, the BS applies CS recovery algorithms to reconstruct all data collected from the network. Energy consumptions for data transmission in the network are analyzed, formulated. Theory and simulation results are provided for verification purposes. The results show promising points to prolong the network lifetime.

Keywords: Wireless sensor networks · Unequal clustering · Compressed sensing · Energy consumption · Greedy tree

1 Introduction

Wireless sensor networks (WSNs) facilitate many applications in different fields. The main functions of the networks focus on sensing data, detecting events [1, 2]. Various applications apply WSNs either to improve the quality of services or to connect other application networks for better performance [3, 4]. However, the networks are still facing energy saving issues while operating. Sensor nodes are often deployed in harsh condition or in the environments that human may not access easily for maintenance [5]. Hence, these networks always require energy-efficient mechanisms to reduce energy consumption to prolong the network lifetime.

Clustering in WSNs is considered as an energy-efficient method to deal with long distance communication links [6–8]. A network is divided into clusters in which each cluster has one cluster-head (CH). The CHs are chosen to collect data from the non-CH sensors to be sent to a base-station (BS) or a data processing center. Hence, non-CH

D. C. Nguyen et al. (Eds.): ICERA 2021, LNNS 366, pp. 213–226, 2022.
https://doi.org/10.1007/978-3-030-92574-1_23

sensor nodes can save energy for not transmitting over long distances to the BS. The CHs are often chosen based on their residual energy among the sensor nodes in the network or randomly to balance the energy consumption [9–12]. The traditional methods show some limitations since the clusters closer to the BS spend less energy than the others, which still causes un-balanced energy consumption in such networks. Then, unequal clustering methods are proposed to relax the issues.

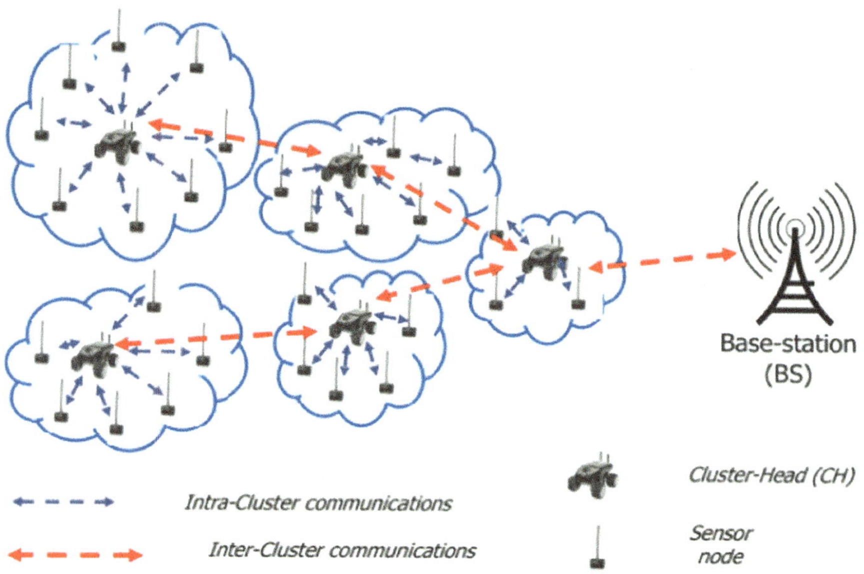

Fig. 1. An Illustration of sensing data is collected through unequal clusters to be sent to the BS for energy consumption balancing purposes.

As shown in Fig. 1, clusters farther from the BS have bigger sizes. The clusters closer to the BS have small sizes since they transmit not only their own data but also relaying data from the farther clusters to the BS. There exist many unequal clustering algorithms that aim to balance energy consumption in WSNs. Thus, several distributed and centralized unequal clustering protocols [13–17] have been developed to solve the problems. The EEUC [13] protocol is proposed to mitigate the hot spot problem of homogeneous WSNs. The network uses unequal clustering and multi-hop routing to balance energy consumption. However, CHs election process does not consider the residual energy of nodes. In protocol RUHEED [14], the CHs are selected based on the highest residual energy. The CH roles are rotated until sensor nodes deplete all energy. In [15], UCR protocol solves the hot spot problem with unequal clustering mechanism. UCR has two parts; one is to create unequal clusters and two is to use a greedy geographic and energy-aware routing protocol for inter-cluster communication. In [16], UCMR protocol is developed to improve the energy consumption in WSNs. The method uses the multi-hop routing with Dijkstra's shortest path algorithm for intra-cluster and inter-cluster transmission. In [17], DEU protocol is proposed based on combining uneven clustering and multi-hop mechanism to prolong the lifetime of WSNs. This method considers both residual energy and communication cost.

Data collection methods in WSNs usually combine with some advanced techniques for further energy saving. Compressed sensing (CS) [18] exploit the high data correlation of sensing data in the networks. The technique offers to reconstruct a vector of sensing data with a small certain number of CS measurements. CS techniques have been used in different data routing methods effectively [19–22]. It seems that almost routing mechanisms can utilize CS to reduce data transmitting in the networks. Random walk [20], tree-based [21], clustering methods [22] integrating with CS techniques reduce significantly energy consumption for data transmission in the networks.

In WSNs with unequal clusters, CS techniques change the way of collecting data in such networks that cause un-balanced energy consumption in the networks. All of the existing unequal clustering methods transmit real data while CS techniques allow WSNs to send only a certain number of CS measurements from the network to the BS. In order to balance the energy consumption for the network utilizing CS, we study further about CS techniques and routing methods and propose new methods to combine them effectively. In the proposal, each cluster only sends a certain number of CS measurements that proportional to the number sensor nodes in the cluster. CHs collect data from their clusters including their own data and create CS measurements. The measurements are forwarded to the BS based on greedy tree as inter-cluster multi-hop routing. We formulate the energy consumption in such networks and show simulation and analysis results.

The rest of this paper is organized as follows. Section 2 formulate the problems including CS techniques applying into WSNs, unequal clustering and greedy tree algorithms. Section 3 provides details about the proposed algorithm. Energy consumption for data transmission in the network with equations are addressed in Sect. 4. Some analysis and simulation results are shown in Sect. 5. Finally, conclusions and future developments are mentioned in Sect. 6.

2 Problem Formulation

2.1 Network Model

A WSN is assumed to have N sensor nodes randomly deployed in a sensing area. The network aims to collect data from all nodes to be sent to a BS or a data processing center. In energy saving terms, the network is divided into unequal clusters where the ones closer to the BS have smaller sizes, as shown in Fig. 1. Each cluster has one CH which is chosen based on a clustering algorithm. The CHs collect data from their own clusters and then they create CS measurements. Finally they forward the CS measurements through the CH closer to the BS based on a routing tree that connects all the CHs. The BS achieves a certain number of CS measurements for data reconstruction.

Our proposed algorithm is based on an unequal clustering algorithm, a greedy tree algorithm and the CS recovery algorithm that are addressed as follows.

2.2 Compressed Sensing Techniques

Compressed sensing (CS) techniques provide novel ideas that allow to recover a huge data based on a small certain number of CS measurements. The number is very small compared to *Shannon-Nyquist* methods. CS techniques require sparse data that include some steps as follows.

A – Sparse Data Presentation
A data vector $X = [x_1 x_2 ... x_N]^T \in RN$, is defined as sparse k in some domains if $\psi = [\psi i, j] \in R^{NxN}$ and $X = \psi\theta$ where θ has k non-zero elements and the rest can be considered as zeros.

B – CS Measurements and Sampling Matrix
CS measurement vector $Y = \Phi X$, where $\Phi = [\varphi_{i,j}] \in R^{MxN}$ includes Gaussian random elements. The vector can be written as $\mathrm{Y} = [\mathrm{y}_1 \mathrm{y}_2 \cdots \mathrm{y}_\mathrm{M}]^\mathrm{T} \in \mathrm{R}^\mathrm{M}$.

C – Data ReconstructionC – Data Reconstruction
With a certain number of CS measurements as $M = O(k \, log \, N/k)$ the whole data could be reconstructed following the CS recovery algorithm as

$$\widehat{\Theta} = argmin||\Theta||_1, \ sbt \ to \ Y = \Phi\Psi\Theta. \tag{1}$$

In real applications, noise could be anywhere in WSN. The CS measurements can be presented as: $Y = \Phi X + e$, trong đó $||e||_2 = \epsilon$. The recovery algorithm is

$$\widehat{\Theta} = argmin||\Theta||_1, \ sbt \ to||Y - \Phi\Psi\Theta||_2 < \in. \tag{2}$$

2.3 Unequal Clustering Algorithm

In our proposed method, we combine some effective unequal clustering algorithms to build an algorithm to be used in our methods. For unequal clusters, we use EEUC protocol [13] to form the network into clusters with different size as shown in Fig. 1. Different to EEUC, the CHs are chosen based on the highest energy residual of sensor nodes following RUHEED protocol [14].

Suppose N_c unequal clusters need to be created from N sensor nodes (s_i, where $i = 1 \ to \ N$). s_i can be chosen as a CH based on its competition range R_i to the BS. R_0 is the maximum competition range which is predefined. It means that the minimum competition range is $(1 - c)R_0$, where c is a constant coefficient in between 0 and 1. The algorithm is run iteratively and finally s_i is chosen as a CH shown in Eq. (3) as follows.

$$R_i = \left(1 - c\frac{d_{max} - d(s_i, BS)}{d_{max} - d_{min}}\right)R_0, \tag{3}$$

where d_{max} and d_{min} are the maximum and the minimum distances between sensor nodes and the BS. $d(s_i, BS)$ is the distance between s_i and the BS.

After each cluster is formed, the cluster role can be rotated among the sensor nodes in the cluster based on the highest residual energy as

$$CH_{prob} = C_{prob} * \frac{E_{residual}}{E_{max}}, \tag{4}$$

where, $E_{residual}$ is the residual energy, E_{max} is the maximum energy of sensor nodes, respectively. C_{prob} is the initial probability which is predefined.

These steps support to form unequal clusters and to choose the most appropriate CH for each cluster in energy saving and balancing manners.

2.4 Greedy Tree Algorithm

In order to build a tree to relay CS measurements from CHs to the BS, a greedy tree is proposed to connect all the CHs to the BS, as known as the root of the tree. At first, all the CHs broadcast their initial number of hops (N_H) to the BS. At first, all the CHs closest the BS have $N_H = 1$. The algorithm runs iteratively and each CH can choose and form its connection to another CH that has the smallest N_H among it neighbors. The greedy algorithm has 8 steps running iteratively that can be written as follows.

Algorithm: *Greedy Multiple-hop Tree formation (GMT)*

1. ***While*** *(the routing paths is changing)*
2. *$N_H(BS) = 0$; $i \in N_c$ CHs*
3. *Nei = set of CH_i's neighbors*
4. ***if*** *distance $[i, j] < R$, where $j \in$ Nei*
5. *CH(i) chooses CH(j) when $N_H(j) = \min\{N_H(Nei)\}$*
6. *Name $N_H(i) = N_H(j) + 1$;*
7. ***end if***
8. ***end while*** *(Until no change of routing paths between CHs)*

---

3 The Main Proposed Algorithm

In this section, we propose the main algorithm that combines other algorithms including the clustering algorithm to form unequal clusters, the greedy tree routing to build a connected network for relaying data, the CS sampling process, and the CS recovery algorithm. The main algorithm is addressed step by step as follows.

Main Algorithm: *Unequal Clustering utilizing Compressed Sensing (UCCS) for WSNs.*

1- Initial Phase

- *N sensor nodes are randomly deployed in a sensing area;*
- *Either sensors or BS can communicate wirelessly and have computational capacity.*

2- Clustering Phase

- *N_c is the number of clusters with unequal sizes is predefined;*
- *EEUC protocol runs iteratively to build N_c clusters based on Eq. (3);*
- *RUHEED protocol runs iteratively to choose N_c CHs based on Eq. (4).*

3- Tree forming Phase

- *GMT algorithm runs to build a tree that connects all N_c CHs to the BS;*
- *The tree is in charge of relaying in multi-hop all data from CHs to the BS.*

4- CS measurement creation and collection Phase

- *All non-CH sensors send their data to the CHs they belong to; N_i denotes as the number of sensors in the ith cluster;*
- *Suppose the BS needs M CS measurements from the network to be able to reconstruct all N sensor readings from the network. M_i denotes the number of CS taking from the ith cluster; ($i = 1$ to N_c);*
- *Each CH multiplies the receive data including its own data to a random Gaussian matrix (Φ) to create M_i CS measurements (*) as*

$$Y_i = \Phi_{N_i \times M_i} X_i, \tag{5}$$

where, X_i denotes data vector including all collected data in the ith cluster;

- *All M CS measurements ($M = \sum_i^{N_c} M_i$) are forwarded through the CH closer to the BS.*

5- Data reconstruction Phase

- *The BS receive enough M CS measurements.*
- *The BS apply Eqs. (1) and (2) to reconstruct all N sensor readings from the network at the BS.*

(*) The number of CS measurements taking from each cluster proportional to the number of sensor nodes in the cluster [2].

4 Energy Consumption Analysis

In this section, we only consider the energy consumption for data transmission that is the functions of distances. Hence, the total energy consumption for the network includes the energy for non-CH sensors send their data to CHs, denoted as $P_{intra-cluster}$, and the energy consumption for relaying CS measurements from CHs to the BS (called

$P_{inter-cluster}$). The total energy consumption, denoted as P_{total} can be calculated as follows [23].

$$P_{total} = (P_{intra-cluster} + P_{inter-cluster}). \tag{6}$$

The following show the details the energy consumption analysis.

4.1 Analysis of $\mathbf{P_{intra-cluster}}$

It is assumed that sensors are uniformly deployed in a square sensing area $L \times L$. The network is divided into N_c clusters. On average, each cluster has N/N_c sensor nodes including 1 CH and $(N/N_c - 1)$ non-CH sensors. On average, we have the total intra-cluster energy consumption can be calculated as

$$P_{intra-cluster} = N_c(N/N_c - 1)E[r^\alpha], \tag{7}$$

where r is a random variable that presents a distance between non-CH sensor node and the CH. α is the path-loss exponent which is chosen as 2 for simplicity. $E[r^\alpha]$ can be calculated based on stochastic problems as

$$E[r^\alpha] = \iint (x^2 + y^2)\rho(x, y)dxdy, \tag{8}$$

$$E[r^\alpha] = \iint r'^2 \rho(r', \theta) r' dr' d\theta, \tag{9}$$

where $\rho(r', \theta)$ present the distribution of the sensor nodes. We can assume that, on average, the cluster has the radius $R = L/\sqrt{\pi N_c}$ and all sensors have uniform distribution, hence, $\rho(r', \theta) = 1/\left(\frac{L^2}{N_c}\right)$. We have

$$E\left[r^2\right] = \frac{1}{L^2/N_c} \int_{\theta=0}^{2\pi} \int_{r'=0}^{R} r'^3 dr' d\theta = \frac{L^2}{2\pi N_c}, \tag{10}$$

and

$$P_{intra-cluster} = \left(\frac{N_c}{N} - 1\right) \frac{L^2}{2\pi}. \tag{11}$$

4.2 Analysis of $\mathbf{P_{\mathit{inter-cluster}}}$

This energy consumption is to forward the total CS measurements from N_c clusters to the BS that can be calculated as follows.

$$P_{inter-cluster} = \sum_{i=1}^{N_c} N_H(i) \times R^{\alpha} \times M_{(i)}, \tag{12}$$

where, $M_{(i)}$ represents the number of CS measurements sampled from the ith cluster following the Lemma 1. $N_H(i)$ is the number of hops away from the ith cluster to the BS. Path-loss exponent α is chosen as 2 for simplicity. R is the communication range of the CHs that can be pre-defined following the GMT algorithm. In fact, the clusters have different sizes and transmit different number of CS measurements to the BS based on the Lemma 1. The number of hops $N_H(i)$ can be calculated on average as follows.

The number of CHs in the sensing area A is assumed to follow Poisson distribution with the mean value as λ

$$\lambda = \frac{N_c}{L \times L} \times A, \tag{13}$$

where $A = 2R(2\theta - cos\theta sin\theta)$ and $\theta = \cos^{-1}\left(\frac{x}{2R}\right)$. x is a random variable that represents the distance between a CH and the BS.

The probability of being able to connect all the CHs to the BS is

$$P(\#\ of\ clusters \geq 1) = 1 - P(\#\ of\ clusters = 0), \tag{14}$$

or

$$P(\#\ of\ clusters \geq 1) = 1 - e^{-\frac{N_c}{L \times L} \times 2R(2\theta - cos\theta sin\theta)}. \tag{15}$$

The probability of being able to make a connection at distance x using n or less hops is denoted as $P_n(x)$. The mean value of the number of hops in a random network is calculated as follows

$$E[n] = Max(N_H) - \sum\nolimits_{n=1}^{Max(N_H)-1} \frac{P(x)}{P_{Max(N_H)}(x)}, \tag{16}$$

where $Max(N_H)$ is the maximum number of hops in the inter-cluster GMT routing.

Based on the calculations from Eqs. (17)–(20), we finally can calculate the $P_{inter-cluster}$ in Eq. (16) as

$$P_{inter-cluster} = \left[Max(N_H) - \sum\nolimits_{n=1}^{Max(N_H)-1} \frac{P(x)}{P_{Max(N_H)}(x)} \right] \times R^2 \times M. \tag{17}$$

5 Simulation Results

In this section, we assume 1000 sensors are randomly deployed in a square sensing area with dimension $L \times L$. The BS is located at the side of the sensing area.

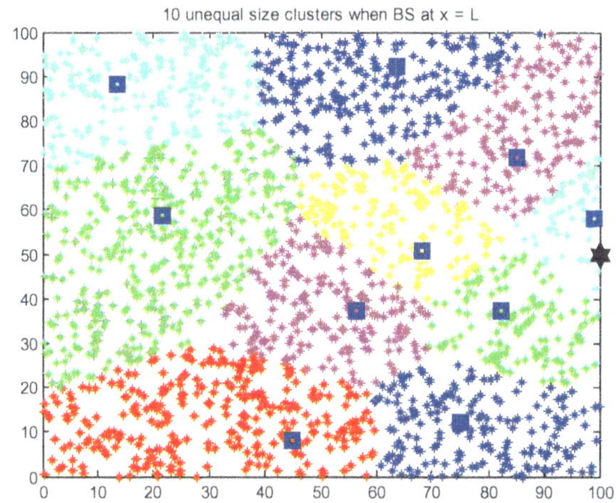

Fig. 2. 10 unequal clusters are created based on the UCCS algorithm.

The UCCS algorithm divides the network into 10 unequal clusters, as shown in Fig. 2. The light blue cluster closest to the BS has the smallest size. The others far from the BS have bigger sizes.

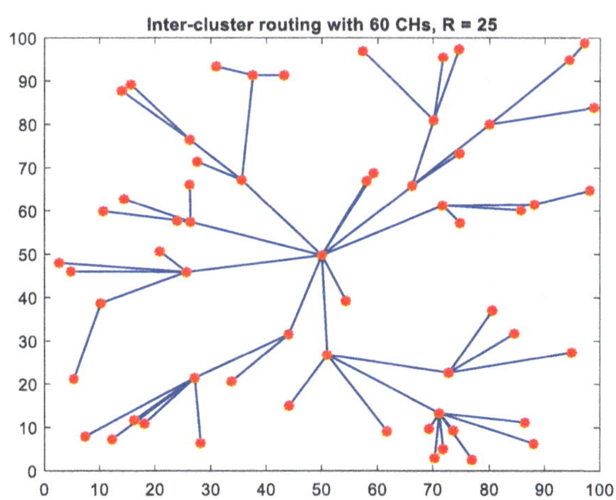

Fig. 3. The greedy multi-hop tree (GMT) is formed to connect all 60 CHs as a tree with the root is the BS at the center of the sensing area.

After the clusters are formed, a tree based on the GMT algorithm should be formed to connect all the CHs to be ready for relaying CS measurements from the CHs to the BS as mentioned in Subsect. 2.4. As shown in Fig. 3, the GMT algorithm creates the tree that connects all 60 CHs to the BS.

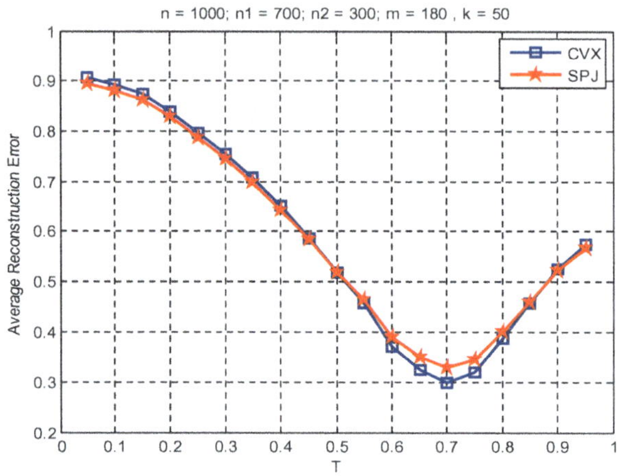

Fig. 4. An Illustration to show a network divided into two unequal clusters. Choosing an appropriate rate as 0.7 can achieve the best performance for the CS recovery.

In Fig. 4, the number of CS measurements are proved to be linearly proportional to the number of sensor nodes in each cluster, as mentioned in the main algorithm.

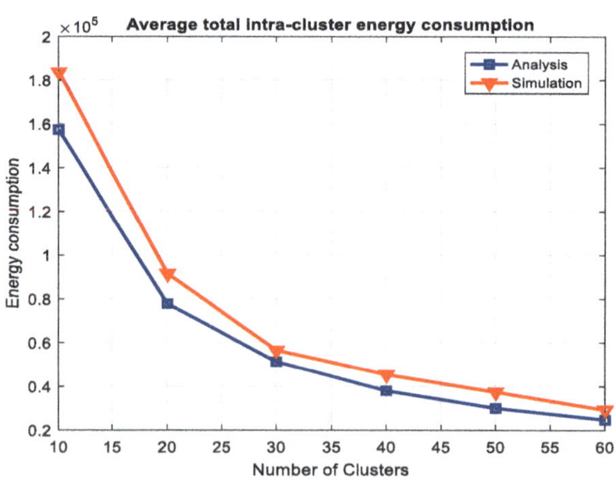

Fig. 5. Total intra-cluster energy consumption from all clusters in the network. The number of clusters change from 10 to 60.

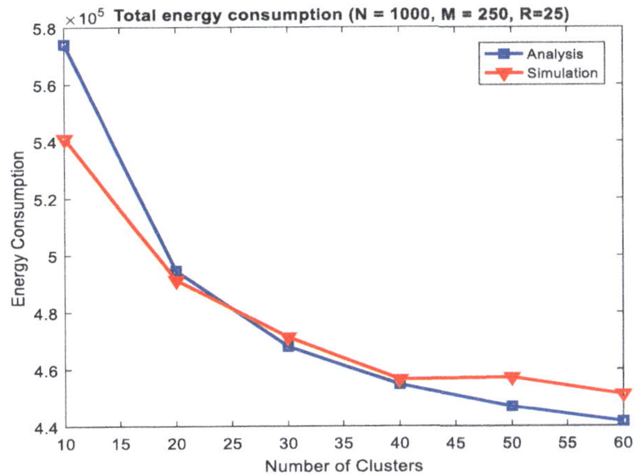

Fig. 6. Total energy consumption for the network.

Figure 5 provides the total energy consumption for non-CH sensors to transmit directly their data to their CHs in all clusters, noted as intra-cluster energy consumption. Both simulation results and analysis results are quite close to each other. We use different number of clusters from 10 to 60. The total intra-cluster energy consumption reduces as we increase the number of clusters to make their sizes smaller. Figure 6 shows the total energy consumption in the network in both analysis and simulation results. We choose the number of CS measurements required to be able to reconstruct data as M = 250). With inter-cluster transmission, we choose the communication range for the CHs as R = 25. The total energy consumption reduces as we increase the number of clusters. The intra-cluster energy consumption reduces. It also means that the number of hops increases that support the network to save more energy with shorter communication distances. However, the trade-off is, this will increase the latency in the network.

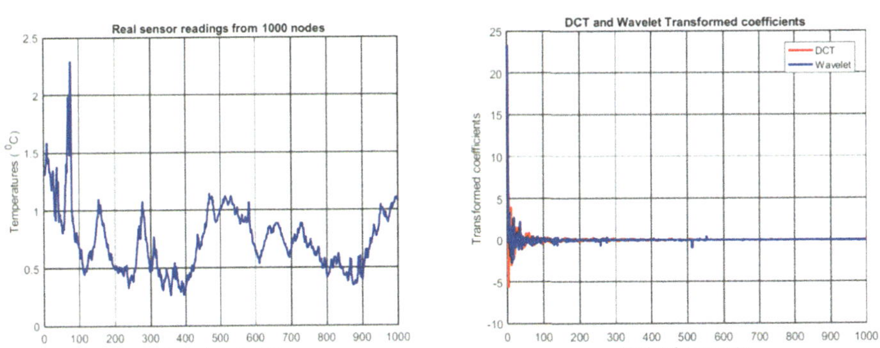

Fig. 7. Real sensor readings from 1000 sensor nodes and transformed coefficients based on DCT and Wavelet domains.

In Fig. 7, real sensor readings from 1000 sensor nodes are presented and then transformed based on both DCT and Wavelet domains. It is obvious that the sensing data is highly correlated and compressible that satisfy the requirements to be able to apply the CS techniques to sample data at the CHs and finally to recover the whole data at the BS based on a certain number of CS measurements.

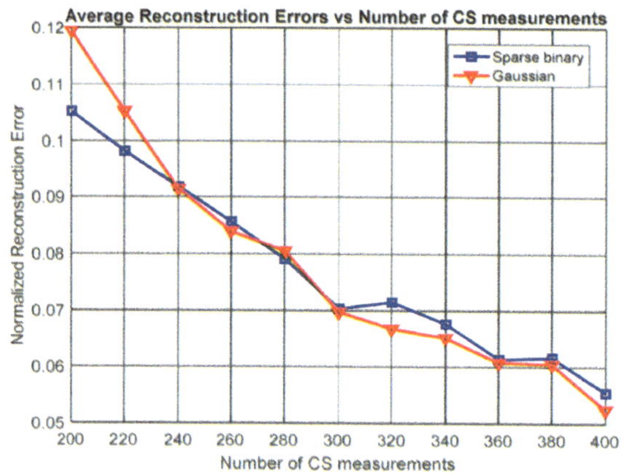

Fig. 8. Average normalized reconstruction errors utilizing both Gaussian and Sparse binary sampling matrices. The number of clusters as $N_C = 15$.

As shown in Fig. 8, the Gaussian matrix works as well as the sparse binary matrix. As the number of CS measurements collected at the BS increases, the reconstruction error reduces. It means that if we collect more CS measurements at each cluster from the network to be sent to the BS, the reconstructed data would be more closely similar to the original data as the reconstruction errors reduce.

6 Conclusion and Future Developments

In this work, we propose an energy-efficient clustering algorithm for data collection in WSNs that either reduces data transmission or balances energy consumption in such networks. We consider existing unequal clustering algorithms and choose two advanced clustering methods to propose a new clustering method. We also propose a greedy multiple hop routing tree that can support CHs to forward data to the BS. We finally combine the cluster-based, tree-based routing methods with CS techniques to reduce data being transmitted to the BS. CS measurements are created at each CHs to be sent to the BS via the routing tree. We prove a Lemma that the number of CS measurements should be linearly proportional to the number of sensor nodes in the networks, in either in mathematic way and in simulation results. We further formulate energy consumption in different scenarios that are compared with simulation results. All the analysis and simulation show that, the methods can either balance energy or save up to 75% energy consumption for the network.

In the future developments, we will exploit more the high correlation of the sensing data in such networks to reduce data transmission in the networks. The equations can be further exploited and optimized. The results show promising points for our future work.

Acknowledgements. The authors would like to thank Thai Nguyen University of Technology, Viet Nam for the support (Project B31-2020).

References

1. Akyildiz, I., Su, W., Sankarasubramaniam, Y., Cayirci, E.: Wireless sensor networks: a survey. Comput. Netw. **38**(4), 393–422 (2002). https://doi.org/10.1016/S1389-1286(01)00302-4
2. Nguyen, M.T., Teague, K.A., Rahnavard, N.: CCS: energy-efficient data collection in clustered wireless sensor networks utilizing block-wise compressive sensing. Comput. Netw. **106**, 171–185 (2016). https://doi.org/10.1016/j.comnet.2016.06.029
3. Dionisis, K., et al.: Applications of wireless sensor networks: an up-to-date survey. Appl. Syst. Innov. **3**(1), 14 (2020)
4. Nguyen, M.T.: An energy-efficient framework for multimedia data routing in internet of things (iots). EAI Endorsed Trans. Ind. Netw. Intell. Syst. **6**(19), 159120 (2019). https://doi.org/10.4108/eai.13-6-2019.159120
5. Nguyen, M.T., Thuong, T.K.N., Teague, K.A.: An energy-efficient combination of sleeping schedule and cognitive radio in wireless sensor networks utilizing compressed sensing. In: International Conference on Engineering Research and Applications. Springer, Cham (2020)
6. Mishra, P., Sharma, S.: A comparative study of existing cluster-based routing protocols in wireless sensor networks. Int. J. Future Gener. Commun. Netw. **9**(7), 43–52 (2016)
7. Nguyen, M.T., Teague, K.A.: Distributed DCT based data compression in clustered wireless sensor networks. In: 2015 11th International Conference on the Design of Reliable Communication Networks (DRCN). IEEE (2015)
8. Kuila, P., Jana, P.K.: Heap and parameter-based load balanced clustering algorithms for wireless sensor networks. Int. J. Commun. Netw. Distrib. Syst. **14**(4), 413–432 (2015)
9. Nguyen, M.T., et al.: Wireless communication technologies and applications for wireless sensor networks: a survey. ICSES Trans. Comput. Netw. Commun. **5**, 1–15 (2019)
10. Nguyen, M.T., Rahnavard, N.: Cluster-based energy-efficient data collection in wireless sensor networks utilizing compressive sensing. In: Military Communications Conference, MILCOM 2013 – 2013 IEEE, pp. 1708–1713 (2013). https://doi.org/10.1109/MILCOM.2013.289
11. Nguyen, M.T., Teague, K.A.: Optimizing number of clusters for energy saving purpose in wireless sensor networks. In: Fujita, H., Nguyen, D.C., Vu, N.P., Banh, T.L., Puta, H.H. (eds.) ICERA 2018. LNNS, vol. 63, pp. 455–461. Springer, Cham (2019). https://doi.org/10.1007/978-3-030-04792-4_59
12. Sert, S.A., Bagci, H., Yazici, A.: MOFCA: multi-objective fuzzy clustering algorithm for wireless sensor networks. Appl. Soft Comput. **30**, 151–165 (2015). https://doi.org/10.1016/j.asoc.2014.11.063
13. Chengfa, L., Mao, Y., Guihai, C., Jie, W.: An energy-efficient unequal clustering mechanism for wireless sensor networks. In: IEEE International Conference on Mobile Adhoc and Sensor Systems Conference, Washington, DC, pp. 598–604 (2005)

14. Nueraili, A., et al.: RUHEED-rotated unequal clustering algorithm for wireless sensor networks. In: 2015 IEEE 29th International Conference on Advanced Information Networking and Applications Workshops. IEEE (2015)
15. Chen, G., Li, C., Ye, M., Wu, J.: An unequal cluster-based routing protocol in wireless sensor networks. Wirel. Netw. **15**, 193–207 (2009)
16. Hari, U., Ramachandran, B., Johnson, C.: An unequally clustered multi-hop routing protocol for wireless sensor networks. In: 2013 International Conference on Advances in Computing Communications and Informatics (ICACCI), Mysore, pp. 1007–1011 (2013)
17. Zhang, C., Liu, F., Wu, N.: A distributed energy-efficient unequal clustering routing protocol for wireless sensor networks. J. Comput. Inf. Syst. **10**, 2369–2376 (2014)
18. Candes, E., Romberg, J., Tao, T.: Robust uncertainty principles: exact signal reconstruction from highly incomplete frequency information. Inf. Theory, IEEE Trans. **52**(2), 489–509 (2006). https://doi.org/10.1109/TIT.2005.862083
19. Li, Y., Liang, Y.: Compressed sensing in multi-hop large-scale wireless sensor networks based on routing topology tomography. IEEE Access **6**, 27637–27650 (2018)
20. Nguyen, M.T., Teague, K.A.: Compressive sensing based random walk routing in wireless sensor networks. Ad Hoc Netw. **54**, 99–110 (2017)
21. Nguyen, M.T., Teague, K.A.: Tree-based energy-efficient data gathering in wireless sensor networks deploying compressive sensing. In: 2014 23rd Wireless and Optical Communication Conference (WOCC). IEEE (2014)
22. Yuan, Y., Liu, W., Wang, T., Deng, Q., Liu, A., Song, H.: Compressive sensing-based clustering joint annular routing data gathering scheme for wireless sensor networks. IEEE Access **7**, 114639–114658 (2019). https://doi.org/10.1109/ACCESS.2019.2935462
23. Nguyen, M.T., Teague, K.A., Rahnavard, N.: Inter-cluster multi-hop routing in wireless sensor networks employing compressive sensing. IEEE Milit. Commun. Conf. **2014**, 1133–1138 (2014). https://doi.org/10.1109/MILCOM.2014.191

Building Intelligent Navigation System for Mobile Robots Based on the Actor – Critic Algorithm

T. D. Chuyen[1], Dao Huy Du[2(✉)], N. D. Dien[1], R. V. Hoa[1], and N. V. Toan[1]

[1] Faculty of Electrical, University of Economics – Technology for Industries, Hanoi, Vietnam
tdchuyen@uneti.edu.vn

[2] Faculty of Electronics, Thai Nguyen University of Technology, Thai Nguyen, Vietnam
daohuydu@tnut.edu.vn

Abstract. This article presents the construction of an intelligent automatic navigation system for mobile robots in a flat environment with defined and unknown obstacles. Programming tools used in the studies are the operating system for mobile robots (Robot Operating System – ROS). From the updated information on maps, operating environment, robot control position and obstacles (Simultaneous Localization and Mapping (SLAM)), we can calculate the motion trajectory of the mobile robot. The navigation system calculates the global and local trajectory for the robot based on the application of Actor-Critic (AC) algorithm. The results of simulation studies in the Gazebo environment and the experimental run on the real Turtlebot mobile robot showed the practical efficiency of automatic navigation for this mobile robot.

Keywords: Intelligent control · Mobile robot · Obstacle robots · Reinforcement learning · Autonomous navigation · Actor-critic algorithm

1 Introduction

The mobile robot's automatic navigation can be divided into three subsystems: information perception, behavioral decision making, and manipulation control. Path planning is the basis of mobile robot navigation and control [1–5]. Reinforcement learning (RL) has shown remarkable promise in recent years, with results on a range of complex tasks such as robotic control and mobile robot (Levine et al., 2016) and playing video games (Mnih et al., 2015) from raw sensory input. The RL algorithms solve these problems by learning a policy that maximizes a reward function that is considered as part of the problem formulation. There is little practical guidance that is provided in the theory of RL about how these rewards should be designed. However, the action environment to control a robot is often represented as a finite state (Markop Decision Process – MDP), and the reinforcement learning algorithms for this context have much to do with the Dynamic planning techniques. The transition probabilities and the gain probabilities in MDP are often random but static during the course of the

D. C. Nguyen et al. (Eds.): ICERA 2021, LNNS 366, pp. 227–238, 2022.
https://doi.org/10.1007/978-3-030-92574-1_24

robot control problem. Unlike supervised learning, in reinforcement learning, there are no correct input/output pairs, and near-optimal actions are not explicitly assessed as true or false. Furthermore, online performance is of interest here, which involves finding a balance between discovery (unmapped territory) and exploitation (existing knowledge). There are two commonly used methods to solve decision problems: searching in strategic space and searching in space of value functions, also known as "strategic iteration" and "value iteration". These two methods are characteristic reinforcement learning algorithms. Besides, in recent studies, scientists have proposed a combined method between the two above methods, which is the actor critic learning method, [5–8].

Previously, we worked on traditional Controllers like PID, Fuzzy PD, PD+I, PI, LQR, and many more classic controllers, [1, 2, 4, 5]. The biggest problem with those methods is that they need to be tuned manually. So, it will cause many errors, the system works inaccurately, making the quality of the controller not high. The working system is not optimal, not good. Optimum values aren't achieved at all for many times. The biggest benefit of reinforcement learning algorithms as controllers is that the model tunes themself reach the optimum values such as: Q-learning algorithm, Deep Q-Network algorithm, DDPG algorithm, SARSA (State Action Reward State Action) algorithm, PG, etc. There are two main reinforcement learning methods, one is value-based and the other is policy-based. Each type has its own advantages and disadvantages. The AC algorithm was born with the aim of merging the two methods, taking the best from each and trying to eliminate all the remaining disadvantages. The AC algorithm is also the basis for approaching modern reinforcement learning algorithms such as A2C (Advantage Actor Critic), A3C (Asynchorous Advantage Actor-Critic), [4, 6, 9, 10].

In this paper, we present a reinforcement learning based self-learning algorithm in the cases of unknown environment. Experiments are conducted on the autonomous navigation tasks for mobile robots. More specifically, we introduce an AC method which is a policy gradient learning algorithm that uses a policy which acts like an actor and a state value function which acts as a critic to generalize and approximate the value of all the states of RL algorithms, [3, 7, 11, 12]. The AC algorithm solves the problem of finding the way and avoiding obstacles (fixed obstacles, obstacles that can be mobile) for mobile robots in the actual working environment and action space.

2 The Control Model for a Mobile Robot

A robot is a complex mechanical system with many masses and many degrees of freedom. Each degree of freedom performs a movement and is controlled by an electric drive system. Furthermore, the robot is a controlled object containing many interrelated motors. To build the control system model for a robot, we consider a mobile robot as shown in Fig. 1. Two coordinate frames can be used to describe the movement of a mobile robot. One is the global coordinate frame (X, Y) defined in the world and the other is the local coordinate frame (Vl and Vr) defined on the mobile robot. The angle between the two coordinate frames is denoted by θ. The robot's motion will be defined for the navigation stack. As the overall coordinate is chosen in Fig. 1, it is clear that the robot's velocity contains three components: the linear velocity along the OX axis and the angular speed along the OY axis.

The path planning task explored in this study relies on a two-wheeled mobile robot and a semi front for arbitrary movement. This mobile robot can control the speed of its two rudders to achieve arbitrary motion trajectories such as linear motion, spin, or circle. Figure 1 shows the robot's posture at adjacent intervals, based on the established kinetic model, this mobile robot can travel on a flat path and can avoid fixed and moving obstacles moving, [2, 5, 7].

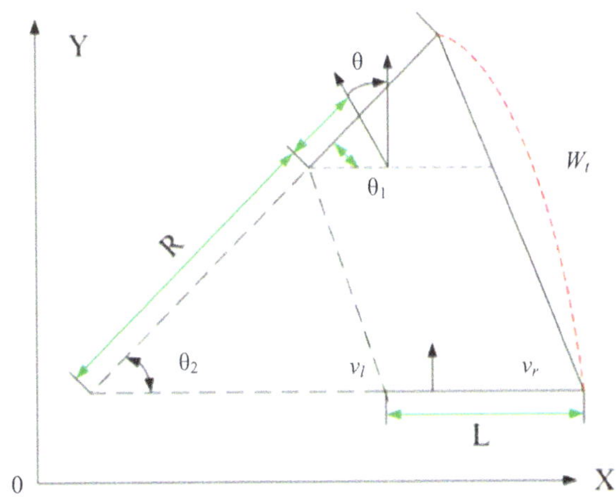

Fig. 1. The path planning motion model of mobile robots

The world coordinate system pose of the mobile robot at time t is set to Wt = [xt, yt, θt]T if the world coordinate pose of the mobile robot at time with (t + Δt) is $W_{t+\Delta t} = [x_{t+\Delta t}, y_{t+\Delta t}]^T$ the distance between the left and right driving wheels is L, the speeds of the left and right driving wheels are vl and vr, and the robot linear speed and angular speed are respectively v and ω, the speed v of the mobile robot in the ideal motion state is:

$$v = \frac{v_l + v_r}{2} \tag{1}$$

The angular velocity of the robot is:

$$\omega = \frac{v_l + v_r}{L} \tag{2}$$

The instantaneous curvature radius R is:

$$R = \frac{v}{\omega} \tag{3}$$

As shown in Fig. 1, $\theta_1 = \theta_2 = \theta$, after Δt, the heading angle of the robot changes as follows:

$$\theta_{t+\Delta t} = \theta_t + \theta \tag{4}$$

The motion from position $W_t = [x_t, y_t, \theta_t]^T$, to $W_{t+\Delta t} = [x_{t+\Delta t}, y_{t+\Delta t}, \theta_{t+\Delta t}]^T$ can be regarded as a circular arc with radius R. If the arc is used to approximate the actual trajectory of the mobile robot, the geometric relationship should be:

$$\begin{bmatrix} x_{t+\Delta t} \\ y_{t+\Delta t} \\ \theta_{t+\Delta t} \end{bmatrix} = \begin{bmatrix} x_t + R(sin(\theta_t + \theta) - sin\theta_t) \\ y_t + R(cos(\theta_t + \theta) - cos\theta_t) \\ \theta_t + \theta \end{bmatrix}, \theta \neq 0 \tag{5}$$

Combining the above equations, the motion equation of the differential mobile robot can be obtained as:

$$\begin{bmatrix} x_{t+\Delta t} \\ y_{t+\Delta t} \\ \theta_{t+\Delta t} \end{bmatrix} = \begin{bmatrix} x_t + \frac{L(v_r + v_l)}{2(v_r - v_l)}(sin(\theta_t + \theta) - sin\theta_t) \\ y_t + \frac{L(v_r + v_l)}{2(v_r - v_l)}(cos(\theta_t + \theta) - cos\theta_t) \\ \theta_t + \theta \end{bmatrix}, \theta \neq 0 \tag{6}$$

The goal is to teach omnidirectional mobile robots to follow a control program based on recognition and data processing, through identification devices such as smart cameras, intelligent sensors and control algorithms so that the robot follows certain trajectories in the space and working environment in accordance with parameters $e_{x,k} \approx 0$, $e_{y,k} \approx 0$, $e_{\theta,k} \approx 0$, in which k is constant. Then the mobile robot will perform awareness in the working environment to make the automatic navigation smoothly (avoiding fixed obstacles, and moving obstacles) perfectly to the destination, [1, 4, 7, 9].

3 Research and Application of AC Algorithm for Mobile Robot

3.1 Model of the Actor Critic Algorithm

In reinforcement learning method, like the Policy Gradient (PG) algorithm, the AC algorithm is an on policy reinforcement learning algorithm that learns online and does not depend on the Model Free model. The AC methods combine some advantages of PG algorithms and Value Function (VF) methods. The AC algorithm also focuses on solving random problems (related to probability distribution).

In this paper, we focus on robot control problems in which a mobile robot acts in a stochastic environment by sequentially choosing actions over a sequence of time steps, in order to maximize a cumulative reward. The main idea of the algorithm is to build a model from two components. The first component is the Actor network. This network takes as input the observed state S, and the corresponding output is the probability distribution of each action in the action space set. This component belongs to the group of policy based methods, playing the role of learning optimal policy. The other component is the critic network, which belongs to the group of value-based methods. This component acts as a function of the value V(s), calculation of an estimate gives an expectation for a long term cumulative reward. From the value function, we can build certain evaluations of the control system, thereby updating the relevant parameters, [4, 8, 12].

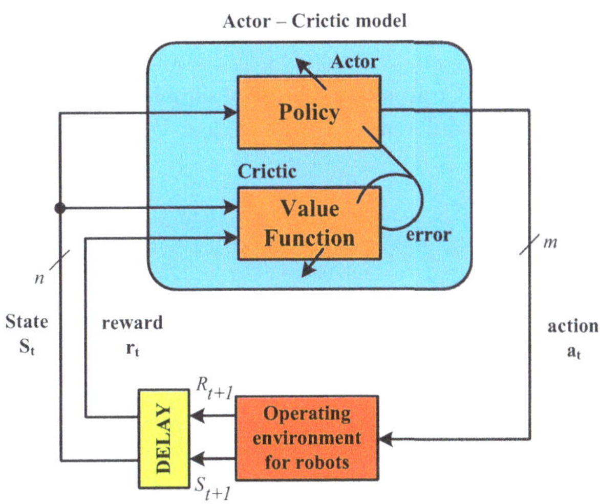

Fig. 2. The schematic diagram of the reinforcement learning model

The sequence of AC algorithm implementation is as follows:

First, randomly initialize model parameter θ_{μ} for network Actor $\mu(s)$ and randomly initialize model parameter θ_V for network critic $V(s)$.

Second, at each training step, the algorithm iterates as follows: Use the existing policy $\mu(S)$ to produce N consecutive samples:

$s_{ts}, a_{ts}, R_{ts+1}, \ldots, s_{ts+N-1}, a_{ts+N-1}, R_{ts+N}, s_{ts+N}$. At the start of training, $ts = 1$. Here for subsequent trainings in the same training set, assign $ts \leftarrow ts + N$. Parameter N is set by "*NumStepsToLookAhead*" argument in "*rlACAgentOptions*" when programming.

Third, compute the profit function at each state in N samples $(G_{ts+1}, \ldots, G_{ts+N})$.

$$G_t = \sum_{k=t}^{ts+N} \gamma^{k-1} R_k + b\gamma^{N-t+1} V(S_{ts+N}|\theta_V)N = P(s_{t+1}|s_t, a_t) \tag{7}$$

Expression (7) is done with:$\forall t = ts+1,\ ts+2, \ldots ts+N$, and γ is discount factor. Fourth, calculate the profit margin function for each of the above positions:

$$\delta_t = G_t - V(s_t),\ \text{when}\ \forall t = ts+1, ts+2, \ldots ts+N \tag{8}$$

Cumulative slope for Actor network over N samples, with μ parameterised actor function:

$$\nabla_{\theta_\mu} J = \sum_{t=1}^{N} \nabla_{\theta_\mu} \ln \mu(s_t).\dot{\delta}_t \tag{9}$$

Cumulative slope for a Critic network over N samples (calculated by minimizing the squared function of the difference between G_t and $V(S_t)$):

$$\nabla_{\theta_V} J = \sum_{t=1}^{N} \nabla_{\theta_V} (G_t - V(s_t))^2 \tag{10}$$

(Here it is necessary to pay attention to the "*EntropyLossWeight*" factor when programming).

Sixth, update the parameters of the networks:

$$\theta_\mu = \theta_\mu + \alpha.\nabla_{\theta_\mu}.J \tag{11}$$

$$\theta_V = \theta_V + \beta.\nabla_{\theta_V}.J \tag{12}$$

In which, α and β are the learning coefficients of the Actor and Critic networks. And finally stop the loop when the algorithm has converged.

Moreover, the critic takes the form of a temporal difference error, it learns and critiques whatever policy currently being followed by the actor. Typically, the critic is a state-value function. After each action selection, it evaluates the new state to determine whether things have gone better or worse than expected. That evaluation is the temporal difference error:

$$\delta_t = r_{t+1} + \gamma.V(s_{t+1}) - V(s_t) \tag{13}$$

Suppose the actor is implemented by a stochastic policy π(a | s, θ) parameterized by vector θ. The critic's error estimate δ_t allows the actor to upgrade its policy according to:

$$\theta_{t+1} = \theta_t + \alpha.\delta_t \frac{d \log \pi(a_t|s_t, \theta_t)}{d\theta_t} \tag{14}$$

Here α is the learning rate for the actor. The critic updates its value function according a common temporal difference procedure given by:

$$V(s_t) = V(s_t) + \beta.\delta_t \tag{15}$$

In robotics, it's often usually to use linear or non-linear feature representations to handle the large scale state action spaces. At that time, the AC algorithm implements action agents for intelligent automatic navigation for mobile robots to perform trajectories to avoid dynamic as well as static obstacles during the movement of the mobile robot. The robot also calculates the shortest trajectory for the robot to move to the destination with the fastest path, [2–4, 9, 10].

3.2 The AC Algorithm Intelligent Navigation for Mobile Robot

In reinforcement learning algorithms, the algorithms are rarely mutually exclusive, most of which will complement and support the other or vice versa. There is always such a relationship between PG algorithm and AC algorithm. Since then, the attention given to these methods has increased greatly, mainly because they have two significant apparent advantages: they require minimal computation to select actions and they can learn an explicitly stochastic policy, [4, 6, 11].

For the AC algorithm (which is an online learning algorithm), it is always updated more often according to the method of instantaneous differential learning (Temporal Differece learning). That is, the system parameter will be updated immediately after a step or a number of iterations (here the number of steps depends on the programming setting, commonly suggested numbers such as N = {32, 64, 128}). This technique will contribute to minimizing the variance for the system. In the process of implementing intelligent navigation for mobile robots, the effectiveness of the learning data also increases, [4]. Next is for the value function V(s). Here comes the role of a critic, directly involved in the calculation of the profit function. More specifically, we can look at the AC algorithm below in the process of implementing smart navigation for mobile robots, [4, 7, 10].The AC algorithm procedure is summarized in the algorithm which is presented in the following Table 1 as follows:

Table 1. The AC algorithm for intelligent navigation perception system for an omnidirectional mobile robot.

Algorithm: The Actor - Critic algorithm

Initialize: Initialize policy $\pi(a \mid s, \theta)$ with initial parameters $\theta = \theta_0$, its

Derivative $d \log \pi(a \mid s, \theta)$, an arbitrarily value function $V^{\pi}(s)$.

Repeat until $\pi(a \mid s, \theta)$ is optimal:

  a) $a_t \leftarrow$ action given by $\pi(a_t \mid s_t, \theta_t)$

  c) Take action a_t, observe next state s_t+1 and reward r_t+1

  Critic evaluation and update:

    b) $\delta_t = r_t+1 + \gamma V(s_t+1) - V(s_t)$

    e) $V(s_t) = V(s_t) + \alpha\delta_t$

  Actor update:

    d) $\theta_{t+1} = \theta_t + \alpha\delta_t \dfrac{d \log \pi(a_t | s_t, \theta_t)}{d\theta_t}$

  Clean out the local buffer

    $s_t \leftarrow s_t+1$

  end

end

At that time, the AC algorithm implements action agents for intelligent automatic navigation for mobile robots to perform trajectories to avoid dynamic as well as static obstacles during the movement of the mobile robot. The robot also calculates the shortest trajectory for the robot to move to the destination with the fastest path, [3, 5]. The specific procedure of AC algorithm for omnidirectional mobile robot navigation is described in details as above.

4 Simulation and Experimental Results

4.1 The Research of TurtleBot Multi-directional Mobile Robot

Here the authors are going to study the model of a mobile robot: the actual hardware architecture of this robot is shown in Fig. 2, in which each hardware module will perform a number of tasks, in the sequence of activities of this mobile robot: such as finding a path, crossing obstacles, etc.

In the tests, the authors perform several tasks in the sequence of the mobile robot's operations, such as the Ubuntu powered Raspberry Pi 3 Model B+. The Raspberry Pi 3 Model B+ embedded computer directly processes information from a range of sensors including the Astra smart camera and the smart sensor then transmits commands to a smart microcontroller (Figs. 3 and 4).

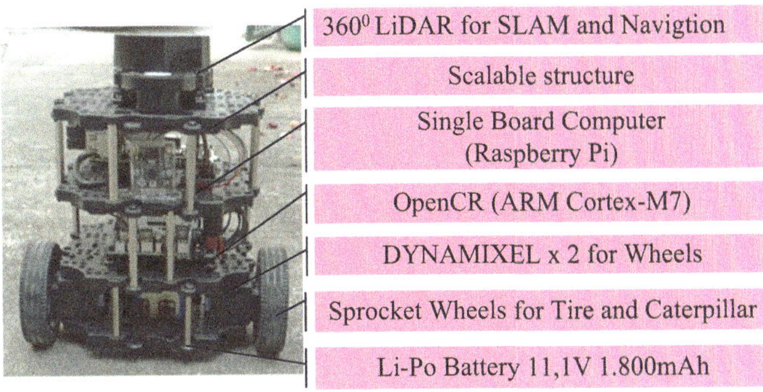

Fig. 3. Pictures of the actual TurtleBot multi-directional robot

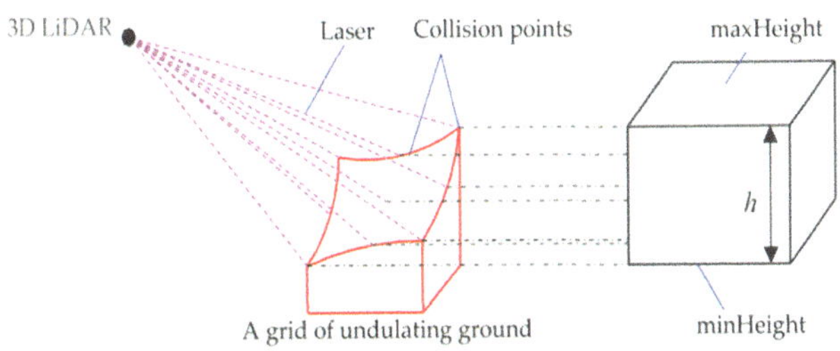

Fig. 4. The structure Diagram of 3D LiDAR terrain detection

To record images from the environment as well as measure the distance between mobile robots and unknown obstacles, mobile robots are equipped with cameras and smart sensors, in which the smart camera can do 360 degrees laser scanning and ranges within 15m to generate map data to be used for the mapping process. The smart microprocessor control circuit will be the part to receive control signals from Raspberry Pi 3 Model B+ and then send the signal to the MOSFET bridge circuit to operate two motors.

4.2 The Simulation Experimental Results

The environment contains a Turtlebot robot with a depth camera and randomly placed obstacles (flower pots) in the plane reality field. This is primarily a visualization tool that can provide live updates of maps generated from the AC reinforcement learning algorithms built above. Furthermore, the robot's trajectory in the map can also be displayed in a reąl world environment. Here, with the process of training, training and identification, the robot knows when to avoid obstacles in the process of moving to the destination safely and perfectly without any collision.

The Fig. 5 shows a map built on Gazebo, the generated map has rectangular obstacles, a circular cylinder, and a mobile robot (green), with a depth camera and other objects. Obstacles are randomly placed in the Gazebo as shown in Fig. 5. The controlled and guided robot automatically moves around the environment to obtain the necessary data that will be used to build the map. The light brown line is the laser scan signal generated from the RPLidar and the robot's current position is updated using geometric measurements.

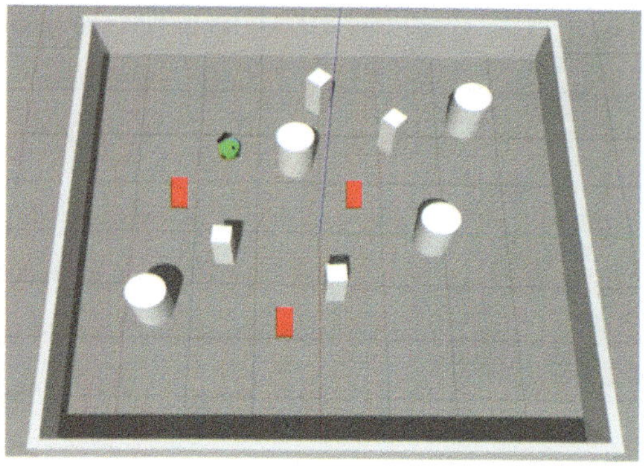

Fig. 5. The build visual maps and robot models in the Gazebo environment

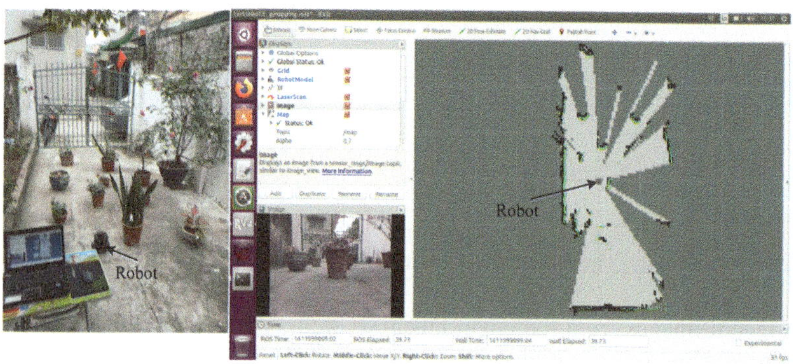

Fig. 6. The interface to execute the AC algorithm and SLAM on ROS with the area around the robot containing obstacle information

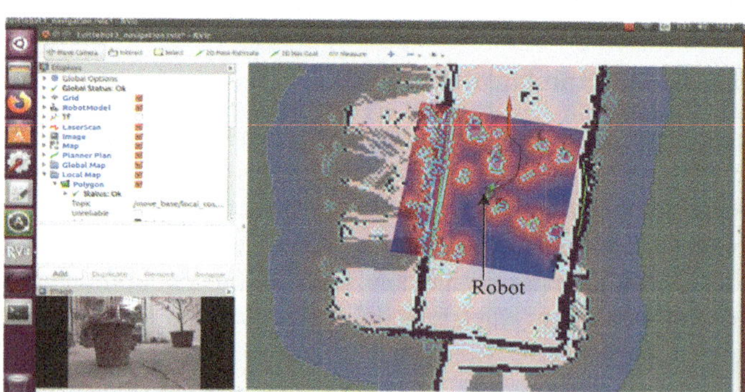

Fig. 7. Results of automatic navigation for the omnidirectional mobile robot Turtlebot in a real map with fixed obstacles in Rviz

Environment for testing mobile robots in a real environment: here the obstacles are flower pots of different sizes that are fixed on the field with a plane suitable for the path of the Turtlebot robot. The results obtained are as follows:

The results show that: Fig. 6 shows the interface for implementing AC and SLAM algorithms on ROS with the area around the robot containing complete obstacle information. In Fig. 7 the results of automatic navigation for the omnidirectional mobile robot Turtlebot in a real map with fixed obstacles in the Rviz environment. Figure 8 shows the results of automatic navigation for robots performing tasks in a real map with unknown obstacles in the Rviz environment.

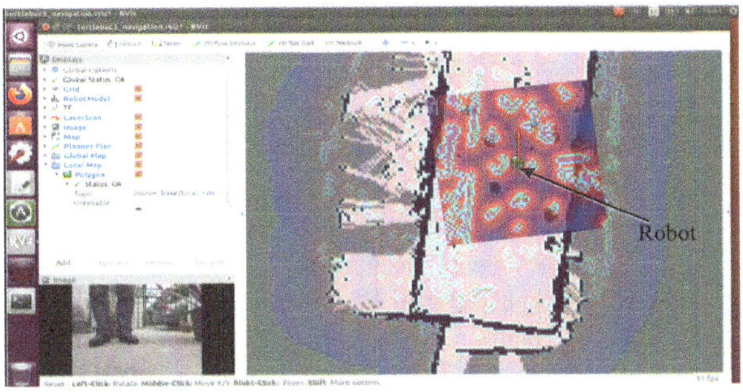

Fig. 8. Result of automatic navigation for omnidirectional mobile robot Turtlebot in real map with unknown obstacle in Rviz

The Actor network takes as input the current state of the mobile robot Turtlebot with a maximum linear velocity of 0.22 m/s and a maximum angular velocity of 2.84 rad/s (162.72 °/s). In the Critic network, the current state value and the action are predicted, using only the fully connected AC algorithm directly to process the input and output states for the robot's control system. Where actions and outputs of the Actor network are concatenated on the basis of a defined algorithm.

5 Conclusions

This article has presented the construction of a test system with an omnidirectional Turtlebot mobile robot. Robotic hardware is also built to facilitate the integration of ROS based peripherals. Furthermore, the robot's activity can be monitored through the visualization tool. The positioning system has calculated the global and local trajectory for the robot based on the application of AC algorithm. Research results show that: experiments in ROS operating system and Rviz environment demonstrate the ability of the robot to navigate automatically to the desired target locations and avoid static and dynamic obstacles in the field. The robot moves safely and effectively in simple and complex environments without collision with any obstacles. These new studies are completely applicable: self-propelled robots, industrial robots, medical robots, especially the control and guidance of mobile robots in industrial factories in Vietnam as well as in the world.

Acknowledgments. This research is funded by Thainguyen University of Technology (TNUT).

References

1. Phuoc, N.D.: The Advanced Control Theory. Science and Technics Publishing House, Viet Nam (2015)
2. Ha, N.T.P.: Modern Control Theory. Ho Chi Minh City National University Press (2016)
3. Tuan, N.T.: Base deep learning. In: The Legrand Orange Book. Version 2, last update, August (2020)
4. Nga, V.T.T., Loc, O.X., Nam, T.H.: Enhanced Learning in Automatic Control with Matlab Simulink. Hanoi Polytechnic Publishing House (2020)
5. Aggarwal, C.C.: Neural Networks and Deep Learning. Springer International Publishing AG, Part of Springer Nature (2018)
6. Ruan, X., Ren, D., Zhu, X., Huang, J.: Mobile robot navigation based on deep reinforcement learning. In: Chinese Control and Decision Conference (2019)
7. Manh, C.N., et al.: Autonomous navigation for omnidirectional robot based on deep reinforcement learning. IJMERR **9**(8), 1134–1139 (2020)
8. Singh, A., Yang, L., Hartikainen, K., Finn, C., Levine, S.: End-to-End Robotic Reinforcement Learning Without Reward Engineering (2019)
9. Ruan, X., Ren, D., Zhu, X., Huang, J.: Mobile robot navigation based on deep reinforcement learning. In: 2019 Chinese Control And Decision Conference (2019)
10. Megalingam, R.K. ROS based autonomous indoor navigation simulation using SLAM algorithm. Int. J. Pure Appl. Math. 199–205 (2018)
11. Fu, J., Singh, A., Larry, D.G., Levine, Y.S.: Variational Inverse Control with Events: A General Framework for Data-Driven Reward Definition. 32nd-NeurIPS 2018, Montréal, Canada (2018)
12. http://wiki.ros.org/ROS/Tutorials, 7-2021.

Calculation and Design of Worm Helical Gearboxes Based on Technique Tools

Trinh Tuan Kieu[1], Nguyen Thi Hong Cam[2], Nguyen Huu Luan[3], Luong Viet Dung[2], Nguyen Anh Tuan[1], Le Hoang Anh[4], Nguyen Manh Cuong[2], and Nguyen Thi Thanh Nga[2(✉)]

[1] University of Economics – Technology for Industries, Hanoi, Vietnam
[2] Thai Nguyen University of Technology, Thai Nguyen, Vietnam
nguyennga@tnut.edu.vn
[3] Nguyen Tat Thanh University, Ho Chi Minh, Vietnam
[4] Vinh Long University of Technology Education, Vĩnh Long, Vietnam

Abstract. In this paper, to get the minimum gearbox volume of a worm helical gearbox, a new method related to the computation of three dimensions of gearboxes is proposed. In the study, the volume of a worm helical gearbox was considered as the objective of the optimization problem. In addition, many main design parameters, comprising the total gearbox ratio, the coefficient of wheel face width of the second stage, the allowable contact stress of the second stage, and the output torque, were discussed to examine their influences on the optimum gear ratios of the gearbox. The technique tools such as the screening experiment, regression, and variance analysis were used to evaluate the effect of the design parameters on the optimum gear ratios. Moreover, a regression model was established for simply computing the optimal gear ratio of the worm unit.

Keywords: Worm helical gearbox · Gear ratio · Design parameters · Systems engineering

1 Introduction

Gearboxes are regularly applied for operating mechanical driven systems due to increased output torque and decreased output revolution of the working shaft. In the design process of gearboxes, the partial gear ratio is the main factor affecting size, mass, and cost of gearboxes. Up to now, many researchers have considered the computation of transmission gear ratios with many different objectives. Cross-section functions have been proposed in [1–8]. In these researches, the gear ratio of gear units has been determined to minimize the cross-section function. Based on calculating the length of gearboxes, the process of gear ratios has been established [9–13]. Mass is one of the most important factors affecting the cost of gearboxes. Therefore, various studies have employed gearbox mass to compute reasonable transmission ratios of gear units [14–18]. In recent years, the height of the gearbox has also been investigated for the design procedure of gearboxes [19, 20].

Regarding methods for calculating transmission ratios of gear units, the practical method presents the optimum transmission ratios via a table [21]. With the graphical

D. C. Nguyen et al. (Eds.): ICERA 2021, LNNS 366, pp. 239–249, 2022.
https://doi.org/10.1007/978-3-030-92574-1_25

method as found in [4, 14, 22–25], partial ratios are described by graphs that display the relation between gear ratios and the total gearbox ratio (see Fig. 1). For the model method, gear ratios are regularly exhibited in mathematical models [3, 5, 26, 27].

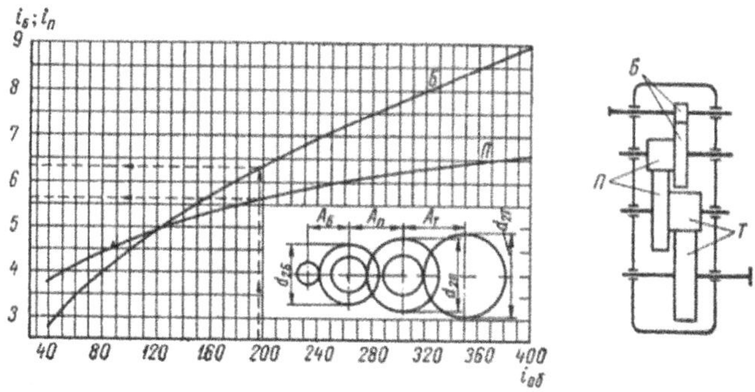

Fig. 1. Graphical method for determining the partial ratios

As discussed in the previous studies, the objective functions have been examined with cross-section, mass, length, and height. In this work, the volume of gearboxes is investigated for a worm helical gearbox since the volume is related to the length, height, and width of the gearbox. For this, the dimensions in the three directions of the gearbox are more reasonably distributed. Additionally, several design parameters were discussed to evaluate their effects on the optimal transmission ratio of the worm unit. Minitab @18 was used to design a screening experiment. Finally, the regression and variance analysis was applied for establishing the explicit equation which can be used to compute the transmission ratio of the worm unit.

2 Optimization Problem

2.1 Determination of the Volume of the Gearbox

The objective function is built based on the volume of the gearbox with dimension parameters, as shown in Fig. 1. V_{gb} is denoted as the volume of the gearbox and is calculated by

$$V_{gb} = H \cdot L \cdot B_1 \quad (1)$$

In here, the dimensions, i.e., H, L, and B_1 can be determined as follows:

$$H = max\left(\frac{d_{w11}}{2} + \frac{d_{w21}}{2}; d_{w22}/2\right) + a_{w1} + \max(d_{w21}/2; d_{w22}/2) + 6.5 \cdot S_G \quad (2)$$

In which, d_{w11}, d_{w21}, d_{w21}, and d_{w22} display the pitch diameters of the pinion and the gear of the worm and the helical gear sets; S_G can be calculated by [16]:

$$S_G = 0.005 \cdot L + 4.5\ [16] \tag{3}$$

In Eq. (2), the center distances of the first stage, a_{w1}, can be expressed as

$$a_{w1} = K_a \cdot \left(K_{HV} \cdot K_{H\beta} \cdot T_{21}/[\sigma_{H1}]^2\right)^{1/3} \tag{4}$$

Where $[\sigma_{H1}]$ is the allowable contact stress of the worm unit and it is calculated in Sect. 2.2. Also, K_a, K_{hV}, and $K_{h\beta}$ are coefficients which can be found in Table 1 (Fig. 2).

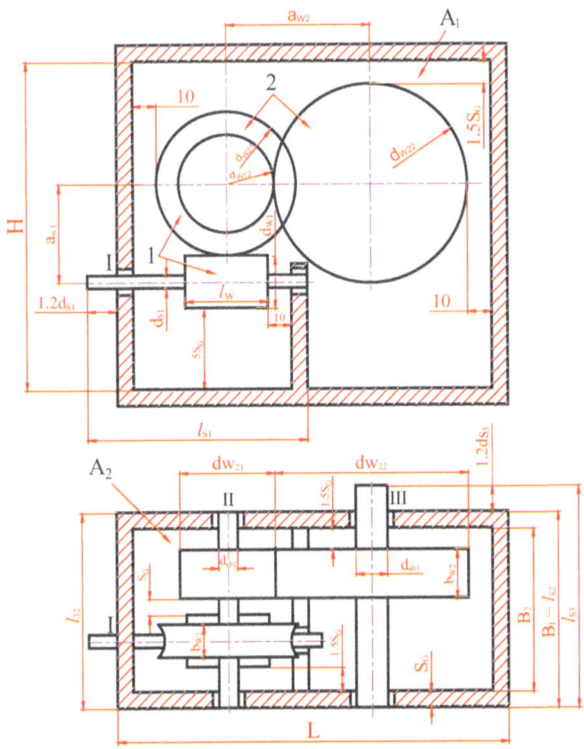

Fig. 2. Worm helical gearbox

Table 1. Value of coefficients

Coefficients	Symbol	Value	References
Material coefficient	K_a	610	[28]
Internal dynamic coefficient	K_{hV}	1–1.3	[28]
Load concentration coefficient	$K_{h\beta}$	1–1.3	[28]

The wheel torque, T_{21}, can be computed as

$$T_{21} = T_{out}/\left(u_2 \cdot \eta_{hg} \cdot \eta_b^2\right) \tag{5}$$

Where, T_{out} is the output torque of the gearbox (Nmm); η_{hg} and η_b are the transmission efficiency of the helical gear unit and a rolling bearing pair. They can be selected in the interval, i.e., $\eta_{hg} = 0.96...0.98$, $\eta_b = 0.99...0.995$ [3]. Taking $\eta_{hg} = 0.97$, $\eta_b = 0.992$, Eq. (5) can be written as

$$T_{21} = 1.0476 \cdot T_{out}/u_2 \tag{6}$$

In Eq. (6), u_2 is the transmission ratio of the helical gear set.
From Fig. 1, the dimensions, L and B_1, can be calculated as follows

$$L = d_{w21}/2 + a_{w2} + d_{w22}/2 + 20 \tag{7}$$

$$B_1 = B_2 + 2 \cdot S_G \tag{8}$$

$$B_2 = b_{w2} + d_{w11} + 4 \cdot S_G \tag{9}$$

Where, a_{w2} is representative for the center distance of the helical gear set. It can be expressed as

$$a_{w2} = k_a \cdot (u_2 + 1) \cdot \sqrt[3]{T_{12} \cdot k_{H\beta}/\left([\sigma_{H2}]^2 \cdot u_2 \cdot X_{ba}\right)} \tag{10}$$

In which, $k_{H\beta}$ represents the contacting load ratio for pitting resistance, $k_{H\beta} = 1.01 \div 1.16$ with the helical gear set [24]; $[\sigma_{H2}]$ is the allowable contact stress of the first stage (MPa); k_a = 43 with steel [24]; X_{ba} is representative for the coefficient of wheel face width of the helical gear set. The pitch diameters $d_{w12}, d_{w22}, d_{w11}$ *and* d_{w21} can be determined by [24]

$$d_{w12} = 2 \cdot a_{w2}/(u_2 + 1) \tag{11}$$

$$d_{w22} = 2 \cdot a_{w2} \cdot u_2/(u_2 + 1) \tag{12}$$

$$d_{w21} = 2 \cdot a_{w1} \cdot z_2/(z_2 + q) \tag{13}$$

$$d_{w11} = d_{w21}/u_1 \tag{14}$$

The number of teeth of worm wheel z_2 can be computed as $z_2 = u_1 \cdot z_1$ with z_1 = 1, 2 or 4 [3]; q is symbolized for the coefficient of the worm pitch diameter which is determined by [24]

$$q = k_q \cdot z_2 \tag{15}$$

with $k_q = 0.25...0.3$.

The allowable contact stress of the worm unit can be computed from the data as found in [24]. Using tin bronze or soft grey iron for the wheel material, the data can be formulated by using the regression technique. The formula is expressed as

$$[\sigma_{H1}] = 5.0515 \cdot v_{sl}^2 - 49.742 \cdot v_{sl} + 189.9 \cdot (R^2 = 0.9906) \tag{16}$$

Where v_{sl} is the slip velocity and is computed as [24]

$$v_{sl} = 0.0088 \cdot \left(P_1 \cdot u \cdot n_1^2\right)^{1/3} \tag{17}$$

In here, P_1 is representative for the power on the first shaft and can be expressed as follows

$$P_1 = T_{11} \cdot n_1 / \left(9.66 * 10^6\right) \tag{18}$$

With using tin bronze for the wheel material, the allowable contact stress of the worm unit can be calculated as

$$[\sigma_{H1}] = K_{HL} \cdot v_{sl} \cdot [\sigma_{H0}] \tag{19}$$

2.2 Determination of the Allowable Contact Stress of the Worm Unit

In Eq. (16), K_{HL} displays the service life ratio and can be computed as [24]

$$K_{HL} = \left(10^7 / N_{HE}\right)^{1/8} \tag{20}$$

The equivalent loading cycle number of the wheel teeth, N_{HE}, can be determined as

$$N_{HE} = 60 \cdot n_2 \cdot t_\Sigma \tag{21}$$

wherein, t_Σ is presentative for the service lifetime (h); n_2 is the rotational speed of the wheel (rpm).

2.3 Objective Function

Based on the above calculation, the objective function depending on the gear ratios can be expressed as follows.

$$\text{Minimize } V_{gb} \tag{22}$$

Subjected to

$$1 \leq u_c \leq 6$$

$$8 \leq u_1 \leq 80 \tag{23}$$

$$1 \leq u_2 \leq 9$$

3 Experiment Design

The optimal transmission ratio of the worm unit u_1 is determined from minimizing the volume of the gearbox as shown in Eq. (22) with the constraints in Eq. (23). To investigate the influence of design parameters on the optimum gear ratio of the worm unit, a screening experiment was set up using Minitab R18 software. The design parameters contain the total gearbox ratio, the coefficient of wheel face width of the second stage, the allowable contact stress of the second stage, and the output torque. Their value range is shown in Table 2. The number of experiments is computed by using 2-level factorial 24 as shown in Table 3.

Table 2. The design parameters and their values

Design parameters	Symbol	Unit	Low value	High value
Total gearbox ratio	u_g	–	50	110
Coefficient of wheel face width of stage 2	X_{ba}	–	0.3	0.4
Allowable contact stress of stage 2	AS_2	MPa	350	420
Output torque	T_{out}	Nm	1000	10000

4 Result Analysis

4.1 The Impact of Design Parameters on the Output Response

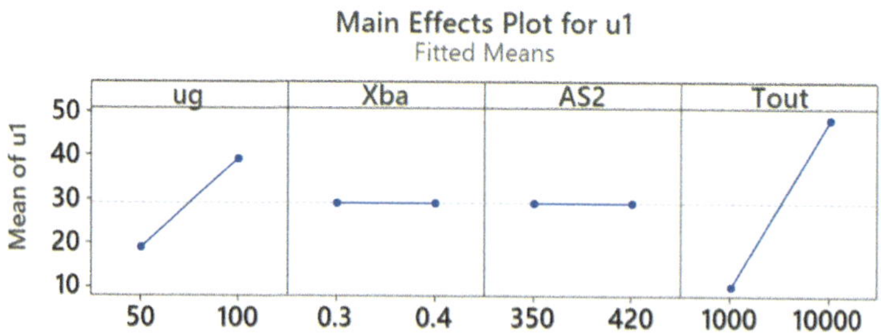

Fig. 3. The relation between the design parameters and the output response

Based on the above analysis of the screening experiment, the effect of the design factors on the optimum transmission ratio of the worm gear u_1 is depicted in Fig. 3. The relation between the design parameter and output response is presented by a straight line. It can be calculated from the figure that the slope angle of the output torque T_{out} is the biggest value, i.e., (48.1928 – 10.1111)/2 = 19.04085. The other factors are sloping gradually, i.e., (39.2437 – 19.06)/2 = 10.09185 for u_g, (29.1519 – 29.1519)/2 = 0 for X_{ba}; (29.1519 – 29.1519)/2 = 0 for AS_2. Therefore, T_{out} has the strongest influence on the optimum transmission ratio of the worm gear. Consequently, u_g also affects the output response enormously. Otherwise, X_{ba} and AS_2 have no effects on the output response.

Figure 4 shows the Pareto chart of the effect with the statistical significance at the 0.5 level. The effect of the design factors on the optimum transmission ratio of the worm gear is also presented by evaluating the magnitude of these factors. It is clearly seen in this figure that the weight of the output torque (factor D) has the greatest value. The following design parameters are the total gearbox ratio (factor A) and the interaction (AD). The other design parameters have no influence on the output response.

Table 3. Experiment design and output response u_1

StdOrder	RunOrder	u_g	X_{ba}	AS_2	T_{out}	u_1
2	1	100	0.3	350	1000	12.22
1	2	60	0.3	350	1000	8.00
8	3	100	0.4	420	1000	12.22
10	4	100	0.3	350	10000	66.27
12	5	100	0.4	350	10000	66.27
11	6	60	0.4	350	10000	30.12
15	7	60	0.4	420	10000	30.12
9	8	60	0.3	350	10000	30.12
7	9	60	0.4	420	1000	8.00
13	10	60	0.3	420	10000	30.12
4	11	100	0.4	350	1000	12.22
14	12	100	0.3	420	10000	66.27
5	13	60	0.3	420	1000	8.00
6	14	100	0.3	420	1000	12.22
16	15	100	0.4	420	10000	66.27
3	16	60	0.4	350	1000	8.00

Besides the effect of the single design parameter on the optimum transmission ratio of the worm unit, Fig. 5 presents the effect of the interactions on the output response. In this figure, there is only the interaction between the output torque, and the total gearbox ratio (Tout*ug) has an influence on the output response. The remaining interactions have an insignificant effect on the output response.

4.2 Regression and Variance Analysis

The regression result is shown in Table 4 with neglecting the insignificant influences on the output response. In this Table, the p-values are symbolized by the star sign. These indicate that these p-values are much smaller than the significant level, i.e., $\alpha = 0.05$. Therefore, they completely have strong effects on the transmission ratio of the worm gear. Additionally, the effects of the design parameters and the interaction are displayed on the second column, and the coded coefficients of these design parameters are indicated in the third column. Therefore, to calculate the optimum transmission ratio of the worm gear, the mathematical model can be established by using the analysis regression technique as the following formula:

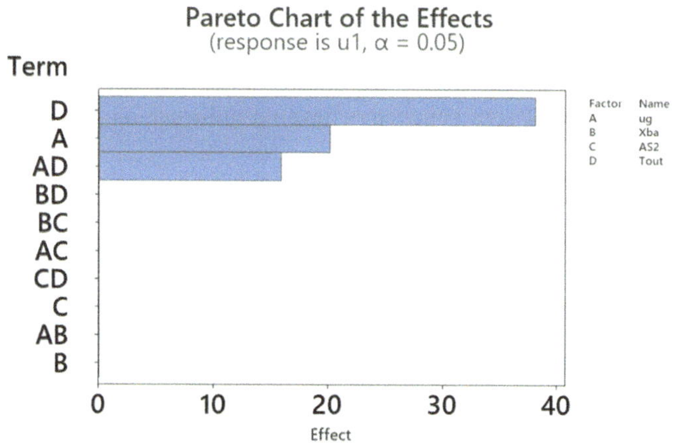

Fig. 4. Weight of effects of the design parameters on the output response

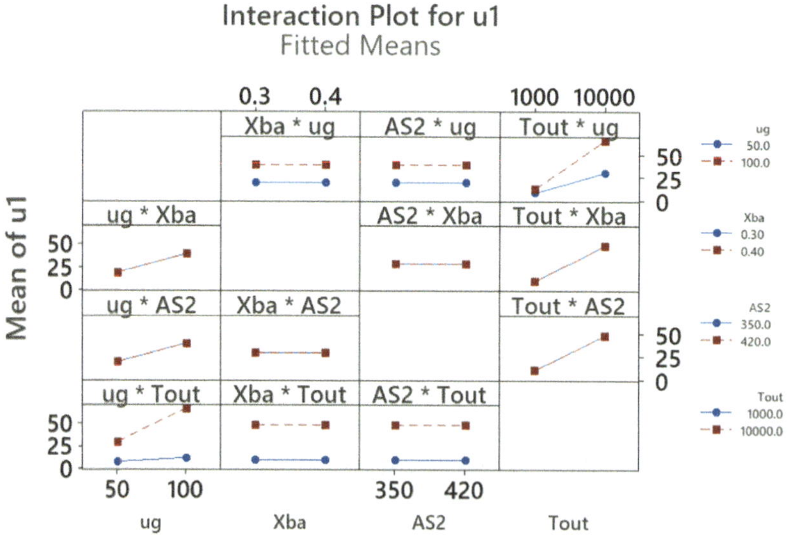

Fig. 5. The influence of interactions on the optimal transmission ratio

$$u_l = 4.867 + 0.01351\ u_g - 0.001089T_{out} + 0.000071u_g * T_{out} \tag{24}$$

To examine the reasonableness of the formula, as shown in Eq. (24), the determination coefficients are 100% (Table 4). The formula in Eq. (24) is thus a good fit for the experimental data. From the formula, this can be easy to determine the optimum transmission ratios of worm gear.

Table 4. The regression model

Coded coefficients						
Term	Effect	Coef	SE coef	T-value	P-value	VIF
Constant		29.15	0.00	*	*	
u_g	20.18	10.09	0.00	*	*	1.00
T_{out}	38.08	19.04	0.00	*	*	1.00
u_g*T_{out}	15.961	7.981	0.000	*	*	1.00
Model summary						
S	R-sq	R-sq(adj)	R-sq(pred)			
0	100.00%	100.00%	100.00%			

Table 5. Variance analysis

Analysis of variance					
Source	DF	Adj SS	Adj MS	F-value	P-value
Model	3	8449.38	2816.46	*	*
Linear	2	7430.34	3715.17	*	*
u_g	1	1629.48	1629.48	*	*
T_{out}	1	5800.86	5800.86	*	*
2-Way Interactions	1	1019.04	1019.04	*	*
u_g * T_{out}	1	1019.04	1019.04	*	*
Error	12	0.00	0.00		
Total	15	8449.38			

Variance analysis was used to more plainly examine the momentous effects of the design parameters and the interactions on the optimum transmission ratio of worm gear. From the observation in Table 5, the p-values with start signs are many tiny values compared to the significant level (i.e., $\alpha = 0.05$). The variance analysis definitely coincides with the regression analysis. Therefore, the formula in Eq. (21) can be used to determine the transmission ratio of the worm unit.

5 Conclusion

In this paper, a new method for calculating the volume of the worm helical gear was proposed. The optimum transmission ratio of the worm unit was obtained by minimizing the objective function and was evaluated via technique tools. From the results of this study, several conclusions are given as follows:

- The influences of the design parameters on the output response demonstrate that the output torque T_{out} has the most important effect on the optimum transmission ratio of the worm unit. The total gearbox ratio u_g and the interaction enormously affect the output response.
- The regression results displayed the effect weight and the coded coefficients of the design parameters on the output response. The regression model was thus established to compute the transmission ratio of the worm unit.
- The variance analysis was also used to examine the regression model carefully. Therefore, this model can be applied to calculate the optimum gear ratio of the worm unit.

Acknowledgement. This work was supported by Thai Nguyen University of Technology.

References

1. Pi, V.N., Tuan, N.K.: Determining optimum partial transmission ratios of mechanical driven systems using a chain drive and a two-step bevel helical gearbox. Int. J. Mech. Eng. Robot. Res. **8**, 708–712 (2019)
2. Vu, N.P., Thao, T.T.P., Hong, T.T., Tuan, N.K., Hung, L.X., Tung, L.A.: Determination of optimum gear ratios of a three stage bevel helical gearbox, Mater. Sci. Eng. 1–8 (2019)
3. Tran, T.H., Cuong, N.V., Ky, L.H., Tuan, N.K., Vu, N.P.: Calculating optimum gear ratios of two step Bevel helical reducer. Int. J. Mech. Eng. Technol. **14**, 3494–3499 (2019)
4. Pi, V.N.: A study on optimal determination of partial transmission ratios of helical gearboxes with second-step double gear-sets. Int. J. Mech. Mechatron. Eng. **2**, 26–29 (2008)
5. Tuan, N.K., et al.: A study on determining optimum gear ratios of mechanical driven systems using two-step helical gearbox with first step double gear sets and chain drive. In: Sattler, K.-U., Nguyen, D.C., Vu, N.P., Tien Long, B., Puta, H. (eds.) ICERA 2019. LNNS, vol. 104, pp. 85–93. Springer, Cham (2020). https://doi.org/10.1007/978-3-030-37497-6_9
6. Vu, N.P., Tuan, N.K., Hung, L.X., Tung, L.A.: A new study on determination of optimum gear ratios of a two-stage helical gearbox. In: Awang, M., Emamian, S.S., Yusof, F. (eds.) Advances in Material Sciences and Engineering. LNME, pp. 107–115. Springer, Singapore (2020). https://doi.org/10.1007/978-981-13-8297-0_14
7. Tu, N.T., et al.: Determining optimal gear ratios of mechanical drive systems using two-stage helical gearbox with second-stage double gear sets and chain drive for minimal system cross section area. In: Sattler, K.-U., Nguyen, D.C., Vu, N.P., Tien Long, B., Puta, H. (eds.) ICERA 2019. LNNS, vol. 104, pp. 238–248. Springer, Cham (2020). https://doi.org/10.1007/978-3-030-37497-6_28
8. Pi, V.N., Tuan, N.K., Hung, L.X., Dung, N.T.Q., Hien, B.T.: Determining optimum gear ratios of a worm—helical gearbox for minimum acreage of the cross section. In: Awang, M., Emamian, S.S., Yusof, F. (eds.) Advances in Material Sciences and Engineering. LNME, pp. 89–96. Springer, Singapore (2020). https://doi.org/10.1007/978-981-13-8297-0_12

9. Cuong, N.V., Ky, L.H., Hong, T.T., Tu, N.T., Vu, N.P.: Splitting total gear ratio of two-stage helical reducer with first-stage double gearsets for minimal reducer length. Int. J. Mech. Prod. Eng. Res. Dev. **9**, 595–608 (2019)
10. Vu, N.P.: Optimal calculation of partial transmission ratios for four-step helical gearboxes with first and third step double gear-sets for minimal gearbox length. Proc. Am. Conf. Appl. Math. 29–32 (2008)
11. Hung, L.X., et al.: Calculation of optimum gear ratios of mechanical driven systems using two-stage helical gearbox with first stage double gear sets and chain drive. In: Sattler, K.-U., Nguyen, D.C., Vu, N.P., Tien Long, B., Puta, Hs. (eds.) ICERA 2019. LNNS, vol. 104, pp. 170–178. Springer, Cham (2020). https://doi.org/10.1007/978-3-030-37497-6_20
12. Tran, T.H., Cuong, N.V., Ky, L.H., Tu, N.T., Pi, V.N.: Calculating optimum gear ratios of mechanical driven systems using worm-helical gearbox and chain drive. Int. J. Appl. Eng. Res. **14**, 3211–3218 (2019)
13. Cam, N.T.H., et al.: Calculating optimum gear ratios of mechanical drive systems using two-stage helical gearbox with second-stage double gear sets and chain drive for minimum gearbox length. Adv. Eng. Res. Appl. 76–84 (2020)
14. Pi, V.N.: Optimal determination of partial transmission ratios for four-step helical gearboxes with first and third step double gear-sets for minimal mass of gears. Appl. Comput. Eng. 53–57 (2008)
15. Hưng, L.X., Pi, V.N., Dự, N.V.: Optimal calculation of partial transmission ratios of four-step helical gearboxes with second and fourth-step double gear-sets for minimal mass of gears. Int. Symp. Mech. Eng. 1–5 (2009)
16. Römhild, I., Linke, H.: Gezielte Auslegung Von Zahnradgetrieben mit minimaler Masse auf der Basis neuer Berechnungsverfahren. Konstruktion, Konstruktion. **44**, 229–236 (1992)
17. Hong, T.T., et al.: Optimization of transmission ratios for two-stage bevel helical gearboxes based on mass function. Int. J. Eng. Res. Technol. **13**, 1692–1699 (2020)
18. Hong, D.T.T.H., et al.: Effects of experimental parameters on the output responses for three-stage bevel helical gearboxes. Int. J. Mech. Prod. Eng. Res. Dev. **10**, 3695–3708 (2020)
19. Vu, N.P.: A new study on the optimal prediction of partial transmission ratios of three-step helical gearboxes with second-step double gear-sets. Appl. Theory Mech. **2**, 229–238 (2007)
20. Thao, T.T.P., et al.: Calculating optimum gear ratios of mechanical driven systems using three step bevel helical gearbox and chain drive for minimum system height. Int. J. Mech. Prod. Eng. Res. Dev. **9**, 211–224 (2019)
21. Milou, G., Dobre, G., Visa, F., Vitila, H.: Optimal design of two step gear units, regarding the main parameters. VDI Berichte **1230**, 227–244 (1996)
22. Kudreavtev, V.N., Gierzaves, I.A.: Design and Calculus of Gearboxes (in Russian). Mashinostroenie Publishing, Sankt Petersburg (1971)
23. Petrovski, A.N., Sapiro, B.A., Saphonova, N.K.: About optimal problem for multi-step gearboxes (in Russian). Vestn. Mashinostr. 13–14 (1987)
24. Chat, T., Uyen, L.V.: Calculation of Mechanical Driven Systems. Education Publisher (1996)
25. Vu, N.P.: A new study on optimal calculation of partial transmission ratios of helical gearboxes with first-step double gear-sets. Lect. Notes Eng. Comput. Sci. 162–165 (2008)
26. Cam, N.T.H., Vu, N.P., Hong, T.T., Ky, L.H., Tung, L.A.: A study on calculation of optimum gear ratios of a two-stage helical gearbox with second stage double gear sets. Int. J. Mech. Prod. Eng. Res. Dev. **9**, 613–620 (2018)
27. Tuan, N.K., et al.: Determining optimal gear ratios of a two-stage helical reducer for getting minimal acreage of cross section. Asia Conf. Mech. Mater. Eng. 1–5 (2018)
28. Grote, K.-H., Antonsson, E.K. (eds.): Springer Handbook of Mechanical Engineering. Springer, Heidelberg (2009). https://doi.org/10.1007/978-3-540-30738-9

Cascade-Loop Control Design for 400 Hz Ground Power Unit

Son Tran Que[1,2], Dong Pham Van[2], Hung Nguyen Dinh Bao[2], Dich Nguyen Quang[2], and Trung Nguyen Kien[3(✉)]

[1] Thai Nguyen University of Technology, Thai Nguyen, Vietnam
[2] Institute for Control Engineering and Automation, Hanoi University of Science and Technology, Hanoi, Vietnam
[3] Department of Industrial Automation, Hanoi University of Science and Technology, Hanoi, Vietnam
trung.nguyenkienl@hust.edu.vn

Abstract. In this paper, a cascade-loop control system in 400-Hz inverter power supply based on unipolar sinusoidal pulse width modulation (SPWM) is proposed for the 400 Hz Ground Power Unit (GPU) in aerospace industry, to improve output voltage quality under the nonlinear load conditions. For the proposed control system, Deadbeat (DB) control, which provides the rapid response for the current especially in the digital system, is utilized in the inner current loop control and the resonant control is performed for the outer voltage loop to reduce the Total harmonic distortion (THD) of output voltage. The computation delay caused by digital processor is taken into consideration and done by using the phase delay compensation technique. The proposed system is tested in testbed systems with a 90-kVA experimental prototype using digital signal processing. The simulation and experiment results show the advantages and effectiveness of the proposed system with relatively small voltage deviation and total harmonic distortion.

Keywords: Cascade-loop control · Unipolar SPWM · Ground power unit

1 Introduction

GPU is the devices supplied to the aircraft externally during maintenance process to avoid the use of airplane's main engines, resulting in reducing the fuel, also removing the acoustic noise and pollution [1].

A lot of power electronic converters are able to apply to the GPU, introduced in many previous literatures. In this research, a three single-phase combined inverter structure is employed to the 400 Hz GPU to supply a stable three phases voltage. To compare to the three-phase three-leg topology in [2], the paralleled interleave [3], and the matrix inverter [4], this structure brings the better performance on unbalanced loads and enhances system adaptability on work conditions. In addition, the replacement of the rectifier circuit in the traditional GPU by a Lithium battery system improves system flexibility and portability. The active neutral-point-clamped topology and its

D. C. Nguyen et al. (Eds.): ICERA 2021, LNNS 366, pp. 250–257, 2022.
https://doi.org/10.1007/978-3-030-92574-1_26

modifications were also introduced in [5, 6]. However, the control complexity has restricted its application in GPU, as contrasted to the proposed structure.

Regarding the control stage, a cascade-loop control system was performed to provide a stable 400 Hz sinusoidal output voltage without regard to the types of loads. For the voltage loop control, many control strategies have been proposed, such as PI control [2], sliding mode control [7], repetitive control [4] etc. Due to the narrow bandwidth, the PI control is impossible in eliminating the steady state error. Sliding mode control is not sensitive to parameter change, and disturbance, however it is complicated to determine the sliding surface. Besides, it cannot provide the good steady state response [7]. Repetitive control can compensate the periodic errors, but the high order of the controller creates challenges for the application in practice. Additionally, most of these strategies do not show the capability of compensating the harmonics distortion, which causes serious impacts to the system performance. In this paper, the resonant control with selective harmonic compensator is exploited in the voltage loop to obtain zero steady-state error tracking, also compensate the harmonic distortion caused by the nonlinear loads. For the inner current loop, it is expected to keep current stable, which contributes to the harmonic distortion compensation at the output voltage, also, develops the system response to load transients. The DB control algorithm, which delivers a very fast response, is selected to get over the constraints of low ratio between the switching and fundamental frequency in the 400 Hz GPU.

The remainder of the paper is organized as follows. Section 2 presents the model of the three-phase 400-Hz GPU. Section 3 focuses on design of the cascade-loop control system. Section 4 and 5 investigate a case study with the simulation and experiment results are analyzed. Finally, conclusive and remarkable points are summarized.

2 Model of GPU

2.1 The Inverter Topology

Figure 1 shows the three single phase combined GPU system, with the DC input voltage is supplied from a Lithium Battery system to improve system flexibility.

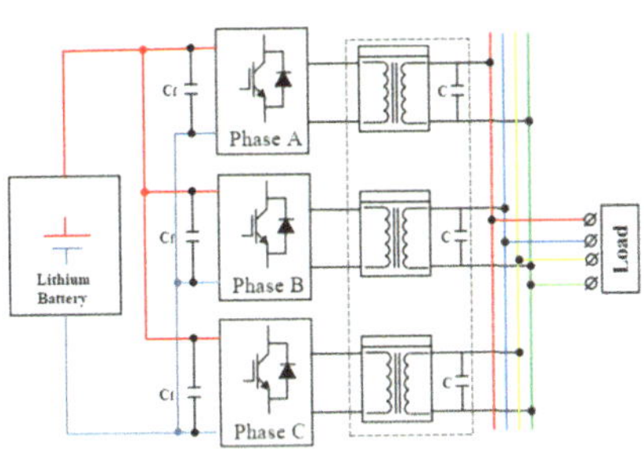

Fig. 1. Three-single phase combined GPU

2.2 GPU Equation

The dynamic equation of the GPU can be expressed as [8]:

$$\begin{cases} C\frac{dv_o}{dt} = i_L - i_O \\ L\frac{di_L}{dt} = -r_L - v_O + v_{inv} \end{cases} \tag{1}$$

In this equation, output voltage v_O and inductor current i_L are considered as state variables, capacitance C, inductance L, inductor resistance r_L, are the GPU parameters, v_{inv} is the output voltage of the inverter, and i_O are load current.

3 Cascade-loop Control Design

3.1 The Overall Control Structure

Figure 2 shows a diagram of the cascade-loop control structure with Feed Forward (FF) technique applied to minimize the coupling components existed in model of GPU [8], in which the Plant model is matched to Eq. (1). Under the assumption that FF is implemented accurately, voltage and current loop control have no effect each other, and can be designed separately.

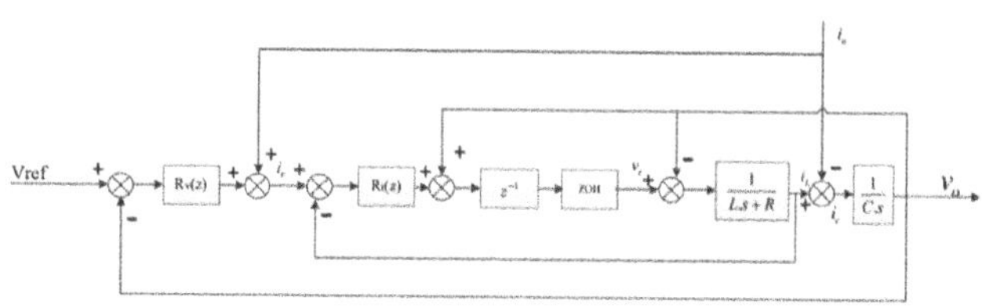

Fig. 2. Cascade-loop control with FF

3.2 The Design of DB Current Loop

The structure of DB current loop is shown in Fig. 3 [8].

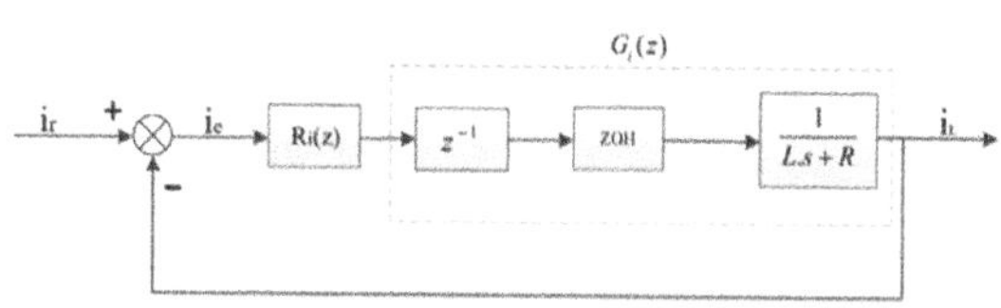

Fig. 3. DB current loop

The total delay time is determined by the sum of sampling time, computing and PWM process and approximates one sampling period. In this design, the output current is supposed to follow the reference current in one sampling period. With the sampling period is T_s, the current loop transfer function, hence, can be derived:

$$H_i(z) = \frac{i_L}{i_r} = z^{-2} \tag{2}$$

The current plant model $G_i(z)$ is determined by [8]:

$$G_i(z) = \frac{1 - e^{\frac{-r_L}{L}T_s}}{r_L} \frac{z^{-1}}{1 - e^{\frac{-r_L}{L}T_s} z^{-1}} \tag{3}$$

Then, the DB control function is:

$$R_i(z) = \frac{r_L}{1 - e^{\frac{-r_L}{L}T_s}} \frac{1 - e^{\frac{-r_L}{L}T_s} z^{-1}}{1 - z^{-2}} \tag{4}$$

3.3 The Design of PR Voltage Loop

Assuming that the FF is done, the voltage loop compensator is reduced to the one shown in Fig. 4, in which, the plant is reduced to the output capacitor impedance, and independent from the current loop and load current.

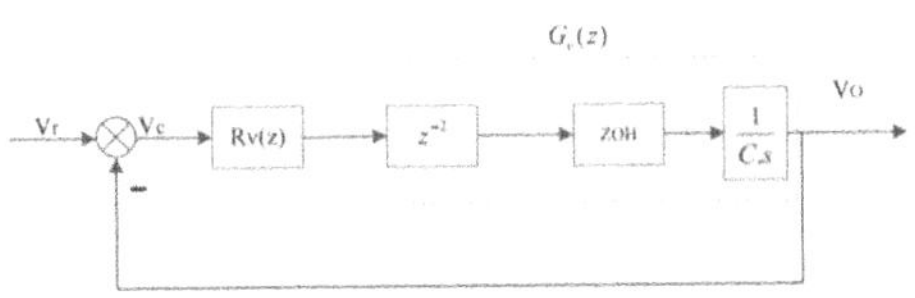

Fig. 4. Resonant voltage loop

The resonant controller transfer function in the continuous domain is as follows [2]:

$$R_v(s) = K_{rn} \frac{s\cos(\theta_n) - \omega_n \sin(\theta_n)}{s^2 + \omega_n^2} \tag{5}$$

The discrete resonant controller can be derived through Zero Order Hold transform:

$$R_v(z) = \frac{(z-1)\sin(\omega_n T_s)\cos(\theta_n) - (z+1)[1 - \cos(\omega_n T_s)]\sin(\theta_n)}{[z^2 - 2\cos(\omega_n T_s)z + 1]\omega_n} \tag{6}$$

Where, K_{rn} is the gain and ω_n is the resonance frequency of the nth compensator. The θ_n is added to the function to implement the phase delay compensation. The output voltage phase difference can be controlled through adjusting θ_n.

4 Case Study

4.1 The Proposed GPU

The system parameters are listed in the Table 1 for the following analysis.

Table 1. System parameters

Parameter	Value
Inductance	200 μH
Capacitance	25 μF
Inductor resistance	0.02 Ω
Switching frequency	18 kHz

4.2 The Simulation Results

The first simulation is designed with the linear load. The results in Fig. 5 shows that the output voltage can track the reference with the phase and amplitude difference are almost zero with only fundamental controller is in active.

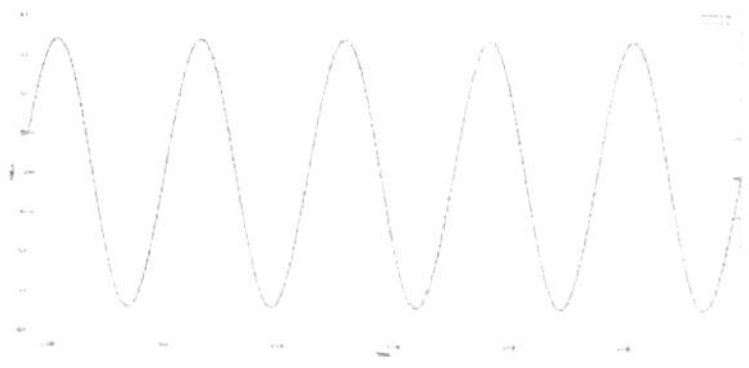

Fig. 5. Resistive load with fundamental controller only

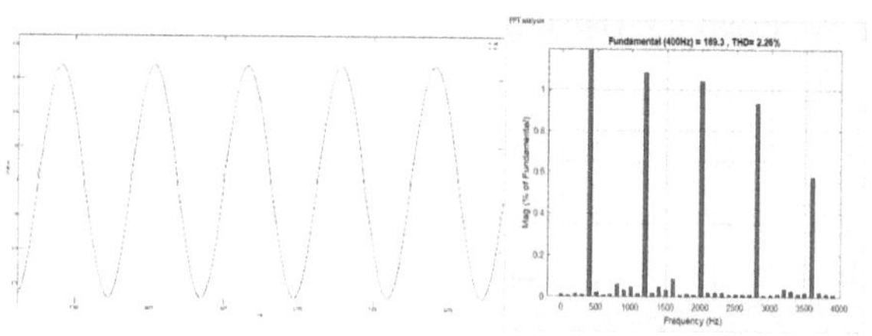

Fig. 6. Output voltage without harmonic compensator under nonlinear load condition.

The dynamic response of the fundamental controller is next examined with the case of nonlinear load included a diode rectifier, a capacitor of 50 μF and a load 1.09 Ω. The output voltage with distortions is depicted in Fig. 6. It can be observed that the THD content is up to 2.26% due to the great increment of the harmonics.

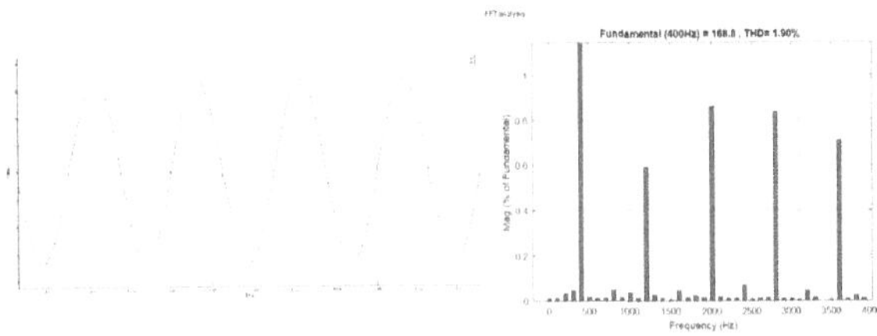

Fig. 7. Output voltage with 3rd harmonic compensator under nonlinear load condition.

The 3rd harmonic compensator is used together with the fundamental controller to compensate the harmonic distortion, with the results shown in Fig. 7. We can see that the 3rd harmonic percentage is decreased to about 0.6%, THD content of the output voltage now is just 1.9%.

The DB control also exhibits an excellent contribution to the system harmonic distortion compensation. The Fig. 8 expresses the output voltage without the DB current controller with more serious distortion and the higher THD content (2.39%) comparing to the results of Fig. 7 when the DB current controller is in active.

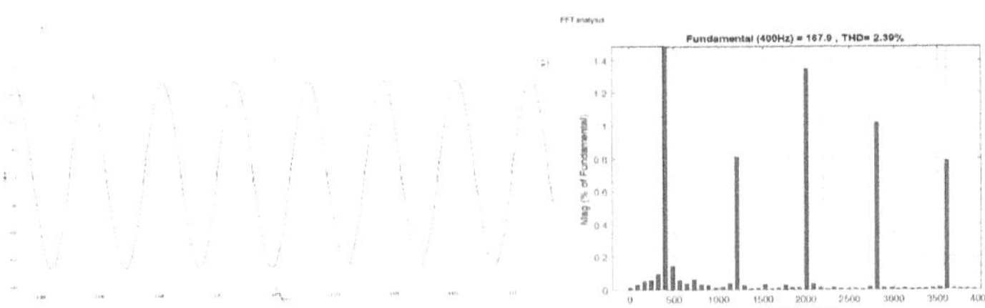

Fig. 8. Output voltage without DB current controller under nonlinear load condition.

In the case of sudden load change, the role of DB is demonstrated through Fig. 9. The system with the DB controller presents a better response, which is remained stable within 0.035 s, smaller than 0.085 s in the case only resonant voltage controller is used.

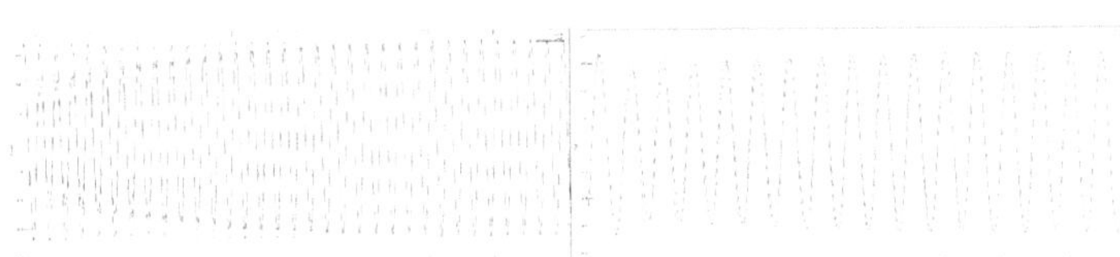

a) b)

Fig. 9. The load change response a) without and b) with the DB controller

4.3 The Experiment Results

A 90 kVA GPU prototype based on Texas Instrument TMS320F28379D microcontroller was built verify the proposed design method (Fig. 10).

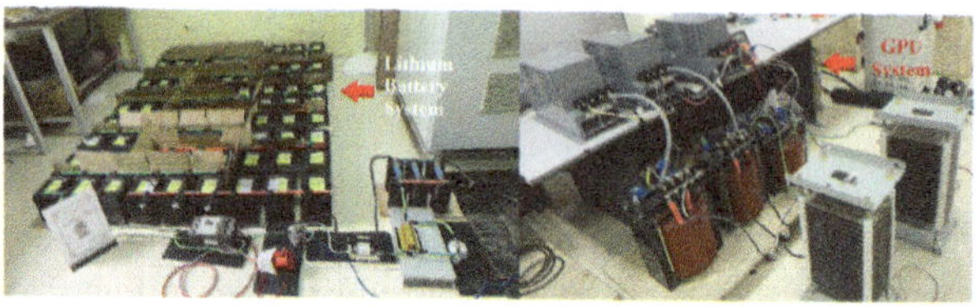

Fig. 10. The GPU experiment system

The experiment is set up to test the system operation with the nonlinear load. A diode rectifier including a capacitor of 50 μF and a load 1.09 Ω is adopted as a nonlinear load. If only the fundamental controller is applied, the output voltage is distorted, THD is up to 3.29% with a third harmonic is 2.25% (Fig. 11). When the 3rd harmonic compensator is added, as in Fig. 12, the 3rd harmonic is depressed to 1.9% and the THD content is significantly reduced to 2.93%.

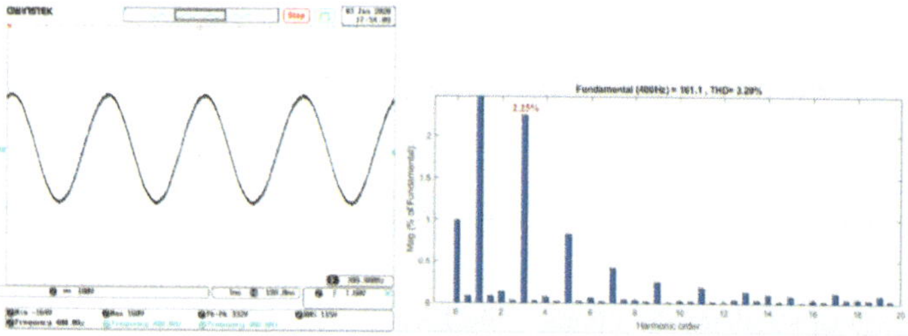

Fig. 11. Output voltage and THD with only fundamental controller

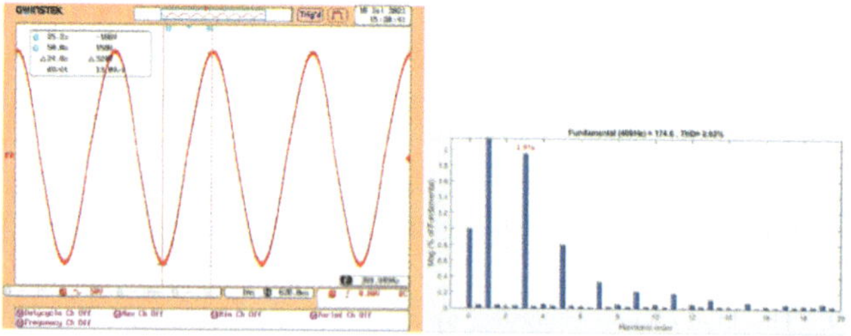

Fig. 12. Output voltage and THD with 3rd compensator is added

5 Conclusion

This paper has proposed a new configuration of GPU in which, the input rectifier is replaced by a lithium battery, three single-phase combined structure is selected due to its advantages particularly with the presence of the unbalanced loads. The control system is designed with cascade resonant-deadbeat control. The simulation and the experiment are performed with different loads demonstrating the efficiency of the resonant voltage loop in regulating the fundamental together with compensating the harmonics. The deadbeat current control also exhibits the capable of controlling the inductor current track the reference as expected.

Acknowledgements. This research is funded by the Hanoi University of Science and Technology (HUST) under project number DTDL.CN-14/18.

References

1. SAE-ARP5015A: Ground equipment - 400 hertz ground power performance requirements. Society of Automotive Engineers, Standard (2011)
2. Li, Z., Li, Y., Wang, P., Zhu, H., Liu, C., Gao, F.: Single-loop digital control of high-power 400-Hz ground power unit for airplanes. IEEE Trans. Ind. Electron. **57**(2) (2010)
3. Mastromauro, R.A., Stasi, S., Gervasio, F., Liserre, M.: A ground power unit based on paralleled interleaved inverters for a more-electric-aircraft. In: 2014 International Symposium on Power Electronics, Electrical Drives, Automation and Motion (2014)
4. Rohouma, W., Zanchetta, P., Wheeler, P.W.: Repetitive control for a four leg matrix converter. In: 5th International Conference on Power Electronics, Machines and Drives (2010)
5. Abarzadeh, M., Kojabadi, H.M., Chang, L.: A modified static ground power unit based on active natural point clamped converter. In: IEEE Energy Conversion Congress and Exposition (ECCE), pp. 3508–3514 (2015)
6. Abarzadeh, M., Kojabadi, H.M.: A static ground power unit based on the improved hybrid active neutral-point-clamped converter. IEEE Trans. Ind. Electron. **63**(12), 7792–7803 (2016)
7. Hou, B., Liu, J., Dong, F., Wang, M., Mu, A.: Sliding mode control strategy of voltage source inverter based on load current sliding mode observer. In: IEEE 8th International Power Electronics and Motion Control Conference (2016)
8. Yao, W., Cui, J., Yao, W.: Single-phase inverter deadbeat control with one- carrier-period lag. Electronics **9**(1), 154 (2020)

Characterization of Particulate Silver Coating for Biomedical Implants

Pham Thi Ly[1(✉)], Nguyen Van Tuan[1], Pham Thi Ha[1], and Le Thu Quy[2]

[1] Institute for Tropical Technology, Vietnam Academy of Science and Technology, Hanoi, Vietnam
ptly@itt.vast.vn
[2] National Key Laboratory for Welding and Surface Treatment Technologies - NARIME, Hanoi, Vietnam

Abstract. In this paper, the particulate silver coating containing nano-silica for biomedical implant was created by the electroplating technique. The characterization of the coating was assessed via its weight, surface morphology images, the concentration of silver ions release in the simulated body fluid (SBF), and the anti-bacterial properties of the silver coating against *Escherichia Coli (E. Coli)* bacteria. The experimental results indicated that the weight and structure of the coating depended on the nano-silica amount adding to the plating solution. The plating solution containing the silica nanoparticles content of 0.1 g/L created the grain silver coating with a fairly uniform dispersion of silica nanoparticles on the silver particle surface. The nano-silica improved the adhesion of the coating. The study results also showed that the silver ion concentration release in SBF after 168 h decreased by 10 times in comparison with the non-nano silica coating. Simultaneously, these coatings still ensured their anti-bacterial ability of up to 99% in 60 min of exposure against *E.Coli* bacteria.

Keywords: Ag/nano-silica · Biomedical implants · Electroplating · Silver coating

1 Introduction

Silver has been widely studied for antimicrobial purposes. The antibacterial properties of silver depend on silver content, particle silver size, and released silver ion concentration into around media. The optimal released silver ion concentration should not be higher than 10 μg/mL to do not cause toxicity to patients [1].

In Vietnam, we created the particulate silver plating on the 316L stainless steel and had the first published paper in 2018 [2]. The silver coating showed good bactericidal ability against *E.Coli* bacteria after 60 min of exposure and the released silver ion amount into the SBF after 168 h can not be toxic to the human body. For stabilization of the released silver ions content in the SBF and enhancing the bactericidal ability of silver coating, in this study, we combined adding the nano-silica in the plating process with heat treatment the coating in argon gas after fabrication.

D. C. Nguyen et al. (Eds.): ICERA 2021, LNNS 366, pp. 258–263, 2022.
https://doi.org/10.1007/978-3-030-92574-1_27

Silver/silica composites have also been mentioned in various forms [3, 4]. A silver-doped silica film showed excellent antimicrobial resistance against both Gram-positive and Gram-negative bacteria [5]. Besides, there are Ag atoms in the SiO_2 matrix and the temperature has an influence on the thermal diffusion of Ag in SiO_2 [6].

2 Material and Method

2.1 Sample Preparation

The silver/nano-silica coating was formed via electroplating technique using plating mode as follows: $AgNO_3$ electrolyte of 0.05M, current density of 0.5 A, and plating time of 10 min [3]. The area of the electrodes was 200 mm^2. Nano-silica with a particle size of 30 ÷ 40 nm was added to the plating solution with different contents in a range of 0.1 ÷ 5 g/L. Dispersing nano-silica by magnetic stirrer VELP Scientific with dispersion speed of about 200 rpm during the plating process. After the plating, the samples were washed with distilled water, dried, then heat-treated in an argon gas environment at 500 °C for 15 min [1]. The nano-silica contents in the plating solution were shown in the sample name. For example, as the nano-silica content in the plating solution of 0.1 g/L, the coating sample was denoted as M0.1.

2.2 Characterization of the Silver Coating

The weight of the coatings was determined by the electronic scale PRECISA XR 205SM-DR with an accuracy of 0.00001 g. The morphology and the chemical element composition on the silver particle surface was analyzed by the scanning electron microscopy SEM-EDX JSM-6010 LV and Field Emission Scanning Electron Microscopy FESEM Hitachi S-4800.

An adhesive tape of 3M (with an acrylic glue layer) was used to determine the adhesion of the silver plating (refer to ASTM D3330). The process of the adhesion test includes steps such as paste adhesive tape covering the entire plating area; peel angle of 90°; repeat the process until the sample weight is constant. The adhesion of the silver plating is directly proportional to the percentage of the residual content of plating after peeling-off tape.

The content of silver and iron ions release in the SBF was determined by atomic absorption spectroscopy (AAS) using SHIMADZU AA6300 equipment. The measurement was performed up to 168 h immersion with each analysis period time of 24 h. The sample was soaked in a beaker with 10 ml fresh SBF maintained at 37 °C during the immersion process. At each time of analysis, the SBF was replaced with 10 mL of a new solution.

The test procedure for the bactericidal ability of silver and silver/nano-silica coating against *E.Coli* ATCC 14169 bacteria was referenced to ISO 221961. The samples were challenged with 6.96×10^8 CFU/ml *E. coli* bacteria in 60 min. After that, the bacterial suspensions were transferred to nutrient Petri plates and incubated at 37 °C for 24 h.

Finally, the living bacteria in each plate were determined by colony counting.

3 Results and Discussion

The Weight of Silver/Nano-Silica Coating

The results in Fig. 1 showed that, as the nano-silica content was in a range of 0.1 ÷ 2.0 g/L, the weight of the silver/nano-silica plating was slightly higher than the silver coating due to the existence of silica nanoparticles in the coating. The coating weight was the largest when the nano-silica content in the plating solution was 0.1 g/L. It means that small nano-silica contents had almost no influence on the plating process. However, when the nano-silica content increased to 3.0 g/L or more, the coating weight tended to decrease. In particular, with the nano-silica concentration of 5 g/L, the coating was not formed because the high silica concentration prevented the silver ions from transferring to the cathode; simultaneously, the nano-silica shielded the cathode surface making the crystal nucleation process taken place more difficult (Table 1).

Table 1. The chemical composition of SBF.

No.	Reagents	Concentration (g/L)
1	NaCl	8.00
2	$NaHCO_3$	0.35
3	KCl	0.40
4	$Na_2HPO_4.2H_2O$	0.48
5	$MgCl_2.6H_2O$	0.10
6	$CaCl_2$	0.18
7	KH_2PO_4	0.06
8	$MgSO_4.7H_2O$	0.10
9	Glucoza	1.00

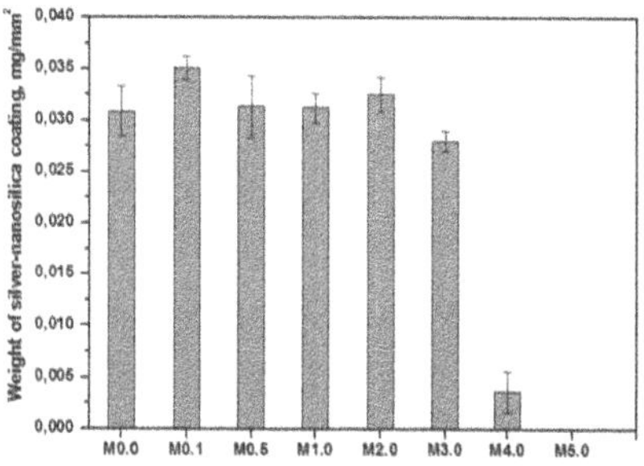

Fig. 1. The weight of silver/nano-silica coating.

Surface Morphology and Chemical Element Composition of Silver Plating

The SEM, FESEM images results (Fig. 2 and 3) confirmed that as the nano-silica content varied from 0.1 to 2 g/L, the silver particle size was bigger than that of the sample without nano-silica because of the existence of the nano-silica on its surface. Besides that, in the plating process, the silica nanoparticles stuck on the cathode surface causing the local polarization phenomena, thus, the crystalline nucleation process did not uniformly perform on the total substrate's surface. As the result, at early crystalline nucleation sites, the silver particles grew up and their size was larger. When the nano-silica content increased to 2.0 or 4.0 g/L, the silver particle size started to decrease due to the nano-silica prevented the growth of silver crystalline nucleation. Even so, the coating was not formed when the added nano-silica content was 5 mg/L.

For the M0.1 sample, nano-silica was evenly distributed on the silver particle surface with its grain size in the range of 30 ÷ 40 nm. Meanwhile, with the nano-silica concentration of 2.0 g/L or more, there was a clustered phenomenon of the nanoparticles on the silver grain surface and the boundary between the silver particles.

The elemental weight percentage of silicon analyzing on the silver particle surface of the M0.1 averaged 0.28%. This result was determined by the EDX.

Because of the largest coating weight and the dispersion uniformly of nano-silica on the silver particles surface, the M0.1 sample is selected for the next investigation with the heat treatment in argon gas at 500 °C for 15 min. The M0.0 silver coating has been shown for comparison purposes.

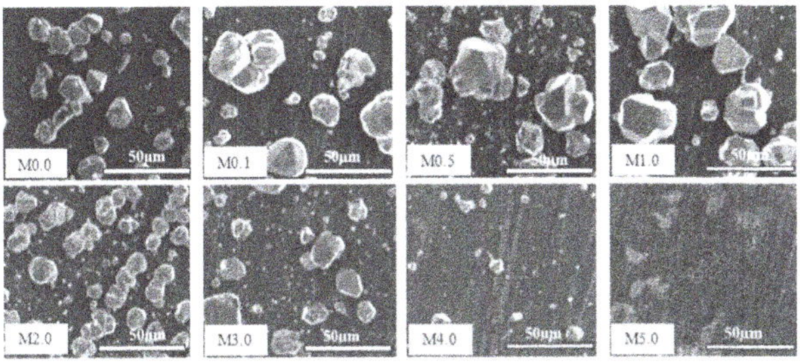

Fig. 2. SEM images (x200) of the silver/nano-silica coating.

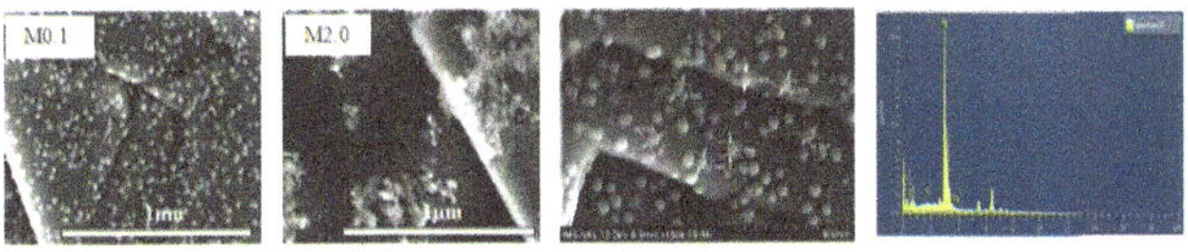

Fig. 3. FESEM images (x20.000) of the particulate silver's surface.

Adhesion

After peeling-off tape, the percentage of the remaining weight of M0.0 and M0.1 samples were 13.86% and 29.19%, respectively. Thus, M0.1 sample had higher adhesion than M0.0 sample. The suitable nano-SiO_2 content made the adhesion between silver particles, silica particles and steel substrate were improved. In addition, after the heat treatment up to 500 °C, Si-O-Ag linkages could be formed [2] that became bridges between silver particles with each other leading to the amount of silver grain separated during the adhesion test reduced, so the adhesion of silver/nano-silica coating increased.

Release of Silver, Iron in the Simulated Body Fluid

The release of silver concentration in the SBF is shown in Fig. 4a. The M0.1 sample had a lower silver ion content in SBF than the M0.0 sample at all times of measurement. After 168 h, the released silver ions amount from the M0.1 sample was 0.017 μg/mL which was about 10 times lower than that of the M0.0 sample. The values were low enough for being not toxic to human cells. The slope of the plots indicated that the silver ion release rate of the M0.1 sample was much slower than the M0.0 sample. The existence of nano-silica on the silver particle surface helped to reduce the silver particle surface area in contact with the SBF leading to a decrease in the amount of released silver ions. In addition, when heat treated up to 500 °C, the Si-O-Ag linkages forming and the diffusion of silver atoms into the silica matrix contributed to control the silver ion release concentration [2].

Regarding the content of iron ions release in the SBF (Fig. 4b), Fe ion contents released from the silver coating samples were higher than that of the steel substrate sample. It might be that, as on the silver coating samples, the Fe/Ag, Ni/Ag, Cr/Ag galvanic couples forming on the surface of 316L steel substrate caused the electro-chemical corrosion phenomena during immersion in the SBF to promote the corrosion of iron, nickel, and chromium. As adding nano-silica, the interface area between silver particles and the substrate decreased due to the intercalation of the nano-silica leading to the inhibition of galvanic corrosion. In addition, the number of galvanic couples could be also reduced because of reducting of the silver crystalline nucleation causing by the shielding of nano-silica on the cathode surface during the plating process. Thus, the Fe content release from galvanic corrosion decreased. However, for this sample, after finishing the plating process, the silica nanoparticles on the cathode are washed off, the 316L steel substrate area in contact with the corrosive solution was larger than the coating without nano-silica. So, the chemical corrosion of the steel substrate was more intense. After 168 h, the concentration of iron ion release of the M0.1 sample was 0.171 μg/mL. It was not toxic for the human body. The allowed limit of the presence of iron ions in the extracellular fluid is 0.84 ÷ 1.12 μg/ml [7].

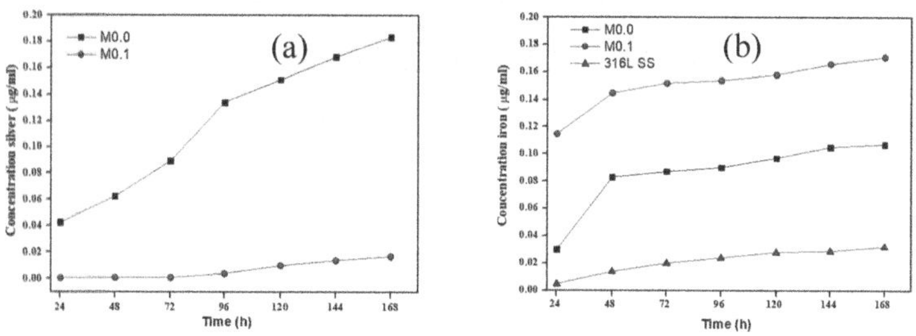

Fig. 4. Silver (a) and iron (b) release concentration in the SBF.

Antibacterial Activity

Table 2. Bactericidal efficacy of samples after 60 min exposure.

Sample	CFU/cm^2	Bactericidal efficacy, %
316L SS	2,58. 104	0
M0.1	2,13. 10^2	99,18
M0.0	2,02. 103	92,17

The bactericidal ability of silver/nano-silica samples was more than 99% in 60 min of exposure against *E.Coli* bacteria. It was higher than that of the silver sample without nano-silica although its silver ion concentration released in the SBF was lower.

The reason might be that the different bactericidal mechanisms, silver particle sizes, and shapes affected the bactericidal ability of the samples [8–10]. The specific bactericidal mechanism in this case has not been studied by the authors (Table 2).

4 Conclusion

In summary, the silver/nano-silica coating has been synthesized by the electroplating method with the silica-nano content in the range of 0.1 ÷ 5 g/L. The results of the coating weight, the surface morphology showed that the most suitable silica-nano amount in the plating solution was 0.1 g/L. For the coating, the adhesion was improved in comparison with silver coating samples without nano-silica. After 168 h, the silver ion concentration released in the SBF was non-toxic to the human body. The bactericidal ability of the sample was more than 99% in 60 min of exposure against *Escherichia Coli* bacteria.

Acknowledgments. This study was financially supported by the Institute for Tropical Technology, Vietnam Academy of Science and Technology.

References

1. Paul, D., Susmita, B., Haluk, B., Amit, B., Lewis, G., Zirkle, M.D.: Antimicrobial particulate silver coatings on stainless steel implants for fracture management. Mater. Sci. Eng. C Mater. Biol. Appl. **32**(5), 1112–1120 (2012)
2. Thi, L.P., et al.: Particulate silver coating on 316L stainless steel for biomedical implants. Vietnam J. Sci. Technol. **56**(3B), 71–78 (2018)
3. Henrika, G., Juha, L., Saima, A., Leena-Sisko, J., Simo-Pekka, H.: Control of the size of silver nanoparticles and release of silver in heat treated SiO_2-Ag composite powders. Materials **11**, 80 (2018)
4. Liang, L., Xiao, L., Yongqi, Y., Ye, S., Miao, Y., Jian, S.: Antibacterial Ag–SiO_2 composite films synthesized by pulsed laser deposition. Mater. Lett. **130**, 79–82 (2014)
5. Baheiraei, N., Moztarzadeh, F., Hedayati, M.: Preparation and antibacterial activity of Ag/SiO_2 thin film on glazed ceramic tiles by sol-gel method. Ceram. Int. **38**, 2921–2925 (2012)
6. Liu, J.D., Liu, Z.W., Chen, Z.Q., Zhang, H.J., Ye, B.J.: Thermal diffusion and epitaxial growth of Ag in Ag/SiO_2/Si probed by XRD, depth-resolved XPS, and slow positron beam. Appl. Surf. Sci. **496**, 143527 (2019)
7. Bischof, H., et al.: Live-cell imaging of physiologically relevant metal ions using genetically encoded FRET-based probes. Cell **8**(5), 492 (2019)
8. Chernousova, S., Epple, M.: Silver as antibacterial agent: ion, nanoparticle, and metal. Angew. Chem. Int. Ed. **52**, 1636–1653 (2013)
9. Prabhu, S., Poulose, E.K.: Silver nanoparticles: mechanism of antimicrobial action, synthesis, medical applications, and toxicity effects. Int. Nano Lett. **2**, 32 (2012)
10. Rai, M., Yakav, A., Gade, A.: Silver nanoparticles as a new generation of antimicrobials. Biotechnol. Adv. **27**, 76–83 (2009)

Combination of the Slider Crank and Cam Mechanisms to Make a Flexible Drive of Automatic Bandsaw Blade Sharpening Machine

Tan Nguyen Dang(✉)

Hanoi University of Mining and Geology, Hanoi, Vietnam
nguyendangtan@humg.edu.vn

Abstract. Wood bandsaw blades differ in thickness, width, length, and tooth configuration. These blades come primarily in the four tooth configurations such as standard, skip, variable and hook. When sharpening different bandsaw blades, it requires changing or adjusting the drive mechanism of blade sharpening machine. The drive mechanism today mainly uses a slider crank with a cam mechanism or combines two cam mechanisms. In order not to have to change the drive mechanism when sharpening different bandsaw blades, this article proposes a flexible drive. When combining movement of slider crank and cam mechanism, kinematics of drive mechanism must be tested to avoid the impact between bandsaw blade and grinding wheel. The kinematics are conventionally determined by using geometry and mathematical method under support of computer software such as Mathcad, Matlab for all working positions in a grinding cycle. This traditional method is not intuitive as well as consumes a lot of time of designer. This article builds the 3D model of slider crank and cam mechanisms, then simulates their movements to determine the kinematics. The simulation results allow to assess the quality of the drive system and calculate the dimensions of drive mechanisms.

Keywords: Bandsaw blade · Sharpening machine · Slide crank mechanism · Cam mechanism · New drive concept

1 Introduction

Bandsaws are manufactured for wood cutting by default, and that includes the type of blade. There are many different blades available for different types of wood cutting needs. The shape of the tooth's cutting edge affects how efficiently the blade can cut through a piece of material while considering such factors as blade life, noise level, smoothness of cuts and chip carrying capacity. The tooth width of a bandsaw blade is referred to as the geometrical thickness of the stellite tip. In the sawmilling practice, a wear-resistant alloy termed stellite is welded directly on top of each swaged tooth prior to sharpening. A sharpened stellite tip forms an axe-shaped tooth, and the thickness of all teeth is made uniform all the way through the blade length using a side dressing

D. C. Nguyen et al. (Eds.): ICERA 2021, LNNS 366, pp. 264–275, 2022.
https://doi.org/10.1007/978-3-030-92574-1_28

machine [1]. The main geometrical parameters of the saw blade are blade width, tooth height, tooth spacing, blade thickness, tooth width (see Fig. 1).

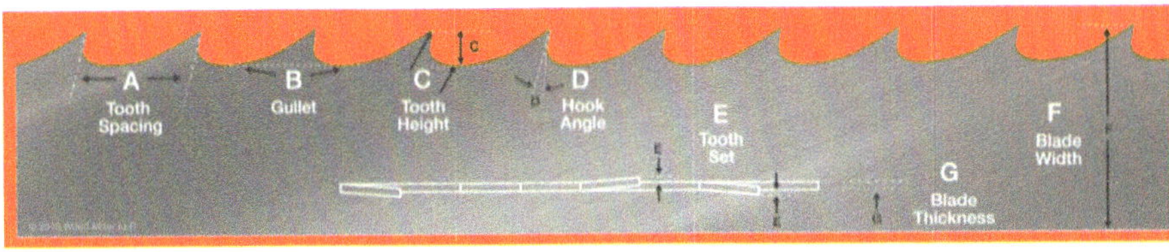

Fig. 1. The geometrical measurements of a bandsaw blade [3]

Tooth spacing is the distance between each tooth from one tip to another. The term "pitch" is also used in reference to tooth spacing as the number of teeth per inch on a bandsaw blade. Shorter spacing is used for resaw purposes, while wider spacing is for higher horsepower sawmill. ***Gullet*** is the area between teeth that captures and removes sawdust while providing strength in the tooth. ***Hook angle*** is the number of degrees that the tooth face leans forward of 90°. The hook angle should be chosen, based on the type of sawing wood [3]. When sharpening the bandsaw blade, tooth spacing, gullet, tooth height, tooth angle must be exactly grinded. The tooth height must be tall enough to allow the gullet to carry out all of the sawdust from the cut. ***Tooth height*** is the distance from the lowest point of the gullet to the tip of the tooth:

- Blades designed for cutting softwoods (balsam, aspen, cottonwood, sycamore, pine, and poplar) have taller teeth
- Blades designed for sawing frozen logs or extreme hardwoods (white oak, hard maple, ash, hickory, and kiln dried) have shorter teeth

Bandsaw blade sharpening machine is designed to sharpen bandsaw blades for wood, bimetallic band saw blades for metals, band saw blades with hard-metal teeth, and special band blades for plastics or food. To sharpen variable pitch band saws, a drive mechanism or grinding program must be reset [2]. The process of sharpening the bandsaw blade sharpening machine requires two basic movements: the movement of the grinding wheel up and down, and the movement of pushing the bandsaw blade to the next tooth [4] (see Fig. 2).

Cam mechanism has the advantage of creating precise movement during grinding, but when changing the pitch of bandsaw blade, it is necessary to change other cam sizes. This means, when sharpening different types of bandsaw blades, different cams are required. In order to flexibly adjust when grinding bandsaw blades with different pitches and tooth height, this paper studies a slider crank and cam mechanism to drive bandsaw blade sharpening machine.

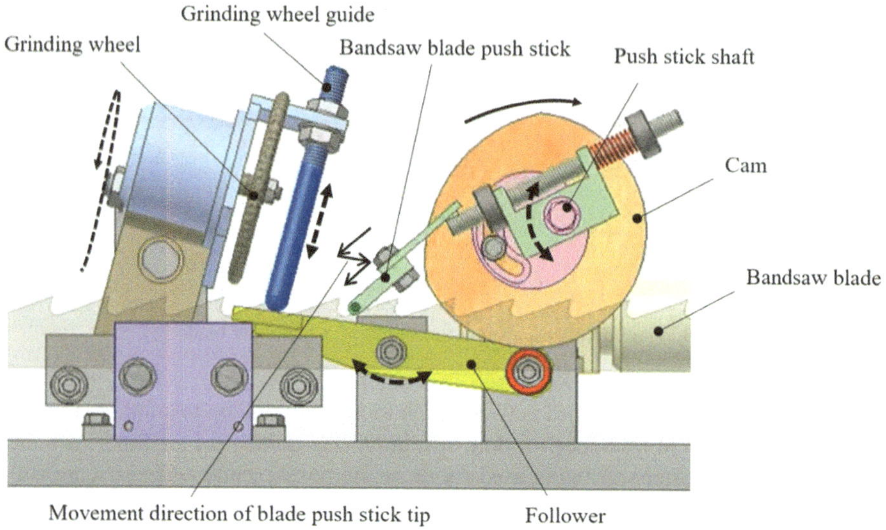

Fig. 2. Cam drive mechanism of bandsaw blade sharpening machine

2 Theoretical Considerations and Study Method

2.1 Requirements of Drive Mechanism

A cam is a rotating or sliding piece in a mechanical linkage used especially in transforming rotary motion into linear motion. It is often a part of a rotating wheel (e.g. an eccentric wheel) or shaft (e.g. a cylinder with an irregular shape) that strikes a lever at one or more points on its circular path. A slider-crank linkage is a four-link mechanism with three revolute joints and one prismatic, or sliding, joint. The rotation of the crank drives the linear movement the slider. The movements *h* of follower and slider according to the rotation angle of the cam and slide crank mechanism φ are shown in Fig. 3.

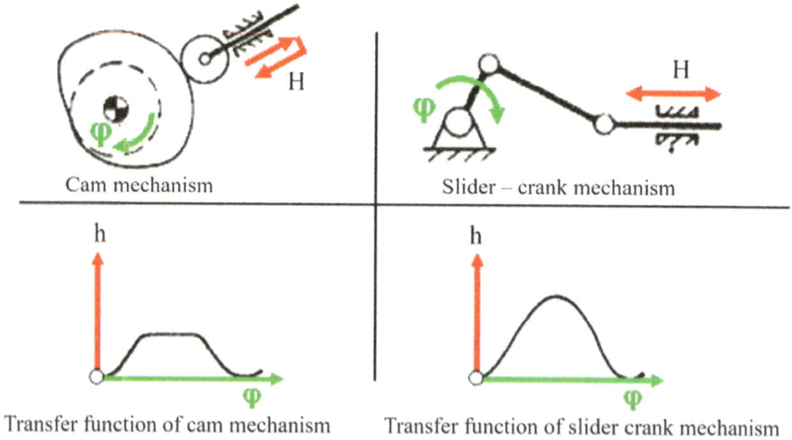

Fig. 3. Working principle of cam mechanism and slider crank [8]

To be able to combine these two mechanisms to drive the bandsaw blade sharpening machine, it is necessary to set their working phase according to the rotation angle φ (see Fig. 4). As proved by this rotation angle, when the grinding wheel moves down to sharpen the bandsaw blade, the bandsaw blade must be held firmly. At this time, the push stick moves backward. After completely grinding a tooth, the push stick moves forward to push next tooth of bandsaw blade to grinding position, the grinding wheel must move up.

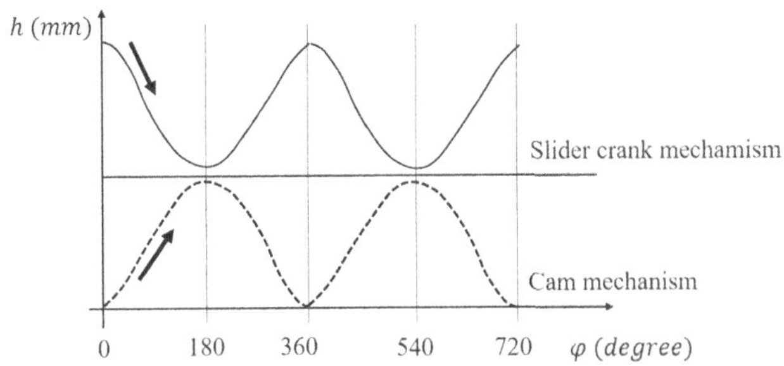

Fig. 4. Motion phase requirement of bandsaw blade sharpening machine

To drive the bandsaw blade sharpening machine, the slider crank and cam mechanism are combined as shown in Fig. 5. The slider of the slider crank mechanism is the cam surface of the cam mechanism.

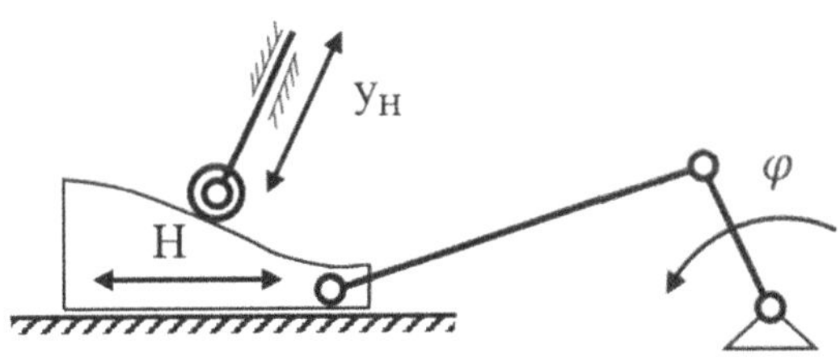

Fig. 5. Combination of slider crank - cam mechanism [7]

Currently, there are two types of cam mechanisms in common use: the swing cam mechanism or the reciprocating cam mechanism. To push the grinding wheel up and down, a cam model with a reciprocating lever is selected. When the crank of the slider crank mechanism rotates, the slider moves with a stroke of H, the slider moves with a stroke of y_H. The motion parameter of the active link (slider) is determined by the variable h. The motion parameter of the passive link (cam follower) is defined as the travel stroke, respectively, and is represented by the function y [7]. The function y is determined as follows:

$$y = y_H f\left(\frac{h}{H}\right) = y_H f(z) \quad (1)$$

Where: y_H the maximum stroke of the cam follower, H the maximum stroke of slider.

For smooth motion of cam follower, there are many ways to select cam profiles such as sine, parabolic or polynomial function. A polynomial has the advantage of smooth contours, simple calculation of constants as well as derivatives [9]. Therefore, this paper selects the cam surface according to the polynomial as follows:

$$f(z) = \sum_{i=0}^{n} A_i z^i \quad (2)$$

Where: A_i are constants and z is the indeterminate.

2.2 Kinematic Parameters of the Slider Crank

To design the slider crank mechanism, it is necessary to determine the position $s = s(\varphi_{21})$, velocity $\dot{s} = v_B$, and acceleration $\ddot{s} = a_B$ of slider. The geometrical parameters of slider crank mechanism are shown in Fig. 6.

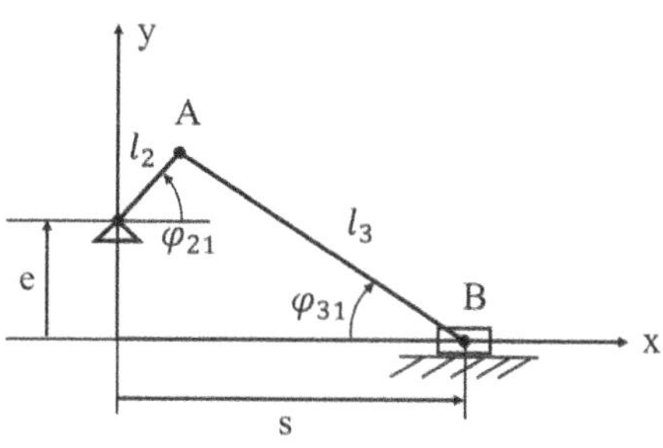

Fig. 6. Geometrical parameters of the slider crank mechanism [11]

Where: the length of the crank l_2, connecting rod l_3, the eccentricity e as well as the rotation angle $\varphi_{12}(t)$.

The ratio between the length of crank and connecting rod is set $\lambda = \frac{l_2}{l_3}$. In this study, the value of eccentricity e is zero. The components of the position, velocity, and acceleration of the slider are determined:

- Position of B point:

$$s = l_2 cos\varphi_{21} + l_3\sqrt{1 - \lambda^2 \sin^2\varphi_{21}} \quad (3)$$

– Velocity of B point:

$$v_B = -l_2 sin\varphi_{21}\left(1 + \frac{\lambda cos\varphi_{21}}{\sqrt{1-\lambda^2\sin^2\varphi_{21}}}\right)\dot{\varphi_{21}} \tag{4}$$

– Acceleration of B point:

$$a_B = -l_2\left[\begin{array}{c} sin\varphi\left(1+\frac{\lambda cos\varphi_{21}}{\sqrt{1-\lambda^2\sin^2\varphi_{21}}}\right)\ddot{\varphi_{21}} + \\ \left(cos\varphi_{21} - \frac{\lambda\left(1-2\cos^2\varphi_{21}-\frac{\lambda^3\sin^2\varphi_{21}\cos^2\varphi_{21}}{1-\lambda^3\sin^2\varphi_{31}}\right)}{\sqrt{1-\lambda^2\sin^2\varphi_{21}}}\right)\dot{\varphi_{21}}^2\end{array}\right] \tag{5}$$

Where: $\varphi_{31} = -\arcsin(\lambda sin\varphi_{21})$

2.3 Determine the Cam Profile

The transfer function of cam mechanism is the law of the movement, which represents the relationship between the moving parameters of cam follower and the cam profile. The mathematical functions that describe the movement of cam follower are called laws of motion. They can be represented as standardized transfer functions, e.g. sine or polynomial (see Fig. 7). In this study, the transfer function $f(z)$ is determined by a polynomial of degree 5.

$$f(z) = A_0 + A_1z + A_2z^2 + A_3z^3 + A_4z^4 + A_5z^5 \tag{6}$$

At the initial position $z = 0$ and $f(0) = 0$, therefore $A_0 = 0$. At the end position $z = 1$ and $f(1) = 1$. To avoid the collision of the cam surface with the cam follower, the second derivative of polynominal at points $z = 0$ and $z = 1$ must be zero: $f''(0) = 0$; $f''(1) = 0$. From there, the constants of polynomial are determined $A_1 = A_2 = 0$.

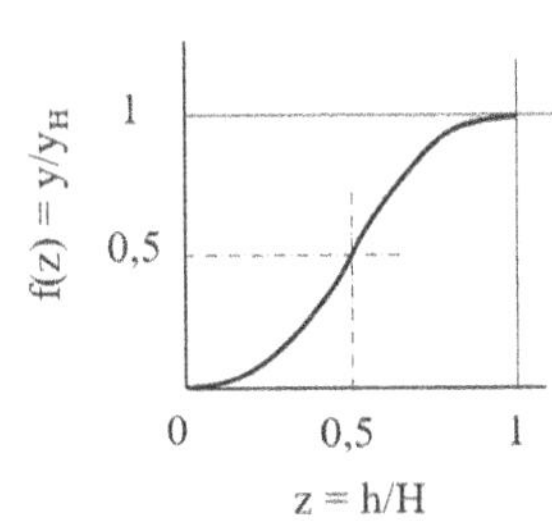

Fig. 7. Normalized symmetrical transfer function [10]

The remaining constants are defined as follows:

$$A_3 = 10 - 6f'(0) - 4f'(1) - \frac{3}{2}f''(0) + \frac{1}{2}f''(1) = 10 \quad (7)$$

$$A_4 = -15 + 8f'(0) + 7f'(1) + \frac{3}{2}f''(0) - f''(1) = -15 \quad (8)$$

$$A_5 = 6 - 3[f'(0) + f'(1)] - \frac{1}{2}[f''(0) - f''(1)] = 6 \quad (9)$$

The polynomial function is rewritten as:

$$f(z) = 10z^3 - 15z^4 + 6z^5 \quad (10)$$

The movement of the slider according to the crank angle is determined

$$h(\varphi_{21}) = l_2 cos\varphi_{21} \quad (11)$$

The movement of cam follower is

$$y(\varphi_{21}) = y_H f\left(\frac{h}{l_2}\right) = y_H\left[1 - \left(10\left(\frac{h}{l_2}\right)^3 - 15\left(\frac{h}{2l_2}\right)^4 + 6\left(\frac{h}{l_2}\right)^5\right)\right], 0 \leq \frac{h}{l_2} \leq 1 \quad (12)$$

Where:

$$h = 0, y = y_H$$

$$h = h_{max} = l_2, y = 0$$

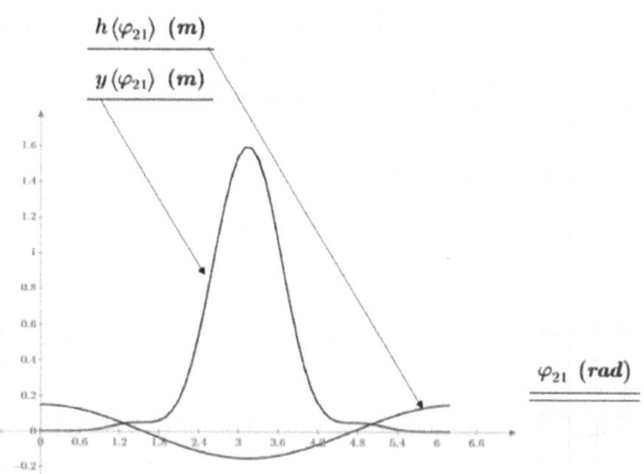

Fig. 8. Displacement of slider and cam follower according to the rotation angle of crank φ_{21}

By the formulas (11) and (12), the graph describing the relationship between the stroke of the slider and the cam follower according to the crank angle is shown in Fig. 8. This motion phase in Fig. 8 is in accordance with the requirements in Fig. 4.

For the slider crank mechanism, when the geometric parameters are available, it is easy to create design drawings. However, with the cam mechanism, the cam profile needs to be determined, it controls not only the stroke of cam follower but also the speed and acceleration of the cam follower. To test the new drive mechanisms, it is necessary to make a new automatic sharpening bandsaw blade machine. However, the manufacturing cost of this new machine is relatively expensive. Therefore, the available automatic sharpening bandsaw blade machine with cam mechanisms was renovated to install and test the new drive mechanisms. The dimensions of the new drive mechanisms must be suitable for the renovated machine. With a given slider length of 0,15 m, a cam follower stroke of 0,05 m, the cam profile by using a polynomial of degree 5 is built and shown in Fig. 9.

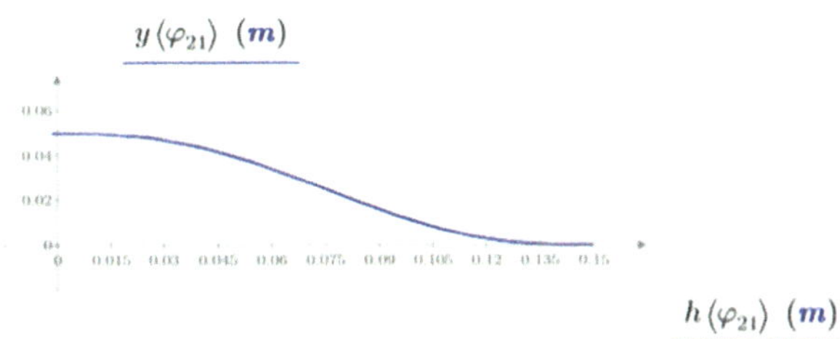

Fig. 9. Profile of translating cam

When designing the cam profile, it is necessary to ensure that there are no impacts between the cam surface and the cam follower. In other words, there are no jumps and kinks on a velocity and an acceleration graph of the cam follower. The velocity $y'(\varphi_{21})$ and acceleration $y''(\varphi_{21})$ graph of the cam follower according to φ_{21} are shown in Fig. 10. According to Fig. 10, the velocity and acceleration curves of the cam follower change softly with the rotation angle of crank φ_{21}.

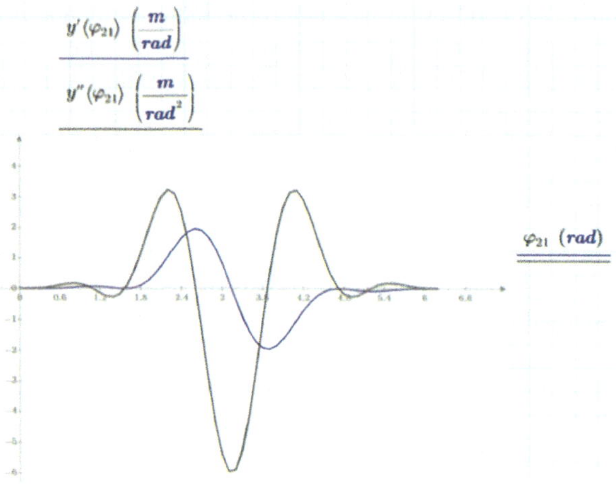

Fig. 10. Velocity and acceleration graph of the cam follower

2.4 Design a Flexible Drive Mechanism

Wood bandsaw blades have a number of tooth per inch namely 1,1; 1,3; 2; 3; 4; 6; 8; 10; 14; 18; and 24 [5, 6]. This means, the pitches are in the range from T = 23,09 mm to T = 1,058 mm. To grind bandsaw blades, the stroke of cam follower and push stick of a bandsaw blade sharpening machine, which is driven by a traditional cam mechanism, are changed by replacing the appropriate cam sizes. In order to push the bandsaw blade with the above pitches, it is necessary to flexibly change the stroke of the push stick. To push a bandsaw blade with a pitch T, the stroke of slider must be greater than the pitch T, so that the push stick can contact the gullet when the slider moves backward. The new driving model for slider crank and cam mechanisms is shown in Fig. 11.

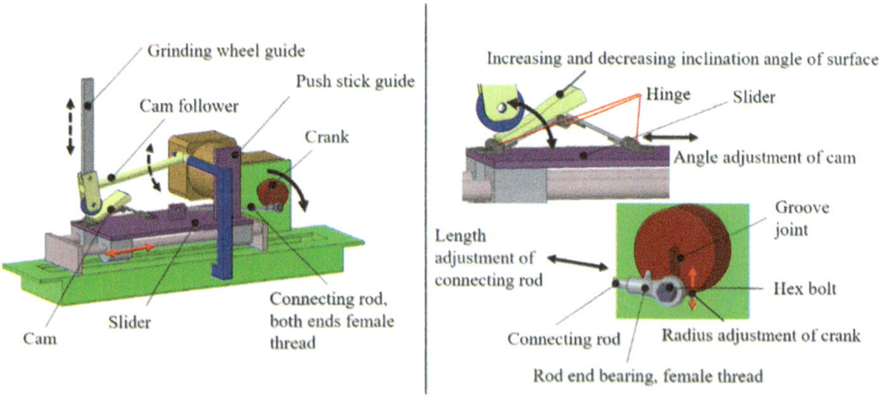

Fig. 11. New drive concept for bandsaw blade sharpening machine

The cam is placed on the small slider-crank with three hinges. By adjusting the position of the hinge fixed on the slider, the angle of translating cam will be altered. When the angle of translating cam is changed, the length of the connecting rod must be adjusted. The connecting rod is made of 2 threaded ends to joint with 2 female threads rod end bearings. By twisting the rod end bearings, the total length of connecting rod is shortened or lengthened. The length of the connecting rod is adjusted in proportion to the pitch T of the bandsaw blade. Although the angle of translating cam alters, the cam profile and roller of cam follower do not change. Therefore, the working quality of the cam mechanism is not affected. The connecting rod is jointed with the crank by a hex bolt and screw nut. A groove is milled on the crank and allows the hex bolt to slide and fix on it. The crank radius is easily adjusted along the groove joint. Therefore, the stroke of the slider and cam follower can be flexibly changed to sharpen different types of bandsaw blades.

By simulation to determine the movement of the slider and cam follower for each pitch of bandsaw blade according to a cycle time, the phase of motion graph of them for each adjustment is established, e.g. in Fig. 12. The adjustments of crank radius, length of connecting rod, angle of cam make displacements of slider and follower different. However, motion phases of the crank and slider do not change.

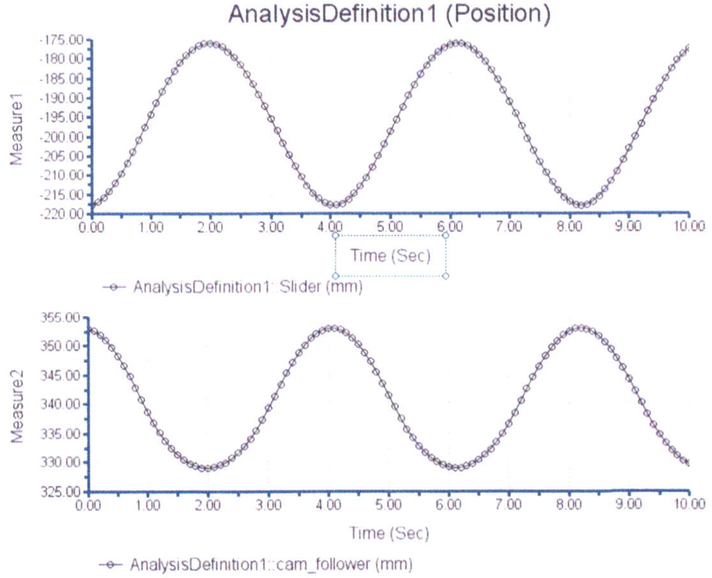

Fig. 12. Motion graph of the slider and cam follower

3 Experiment and Results

Based on the new drive concept for bandsaw blade sharpening machine in Fig. 11, the drive of a bandsaw blade sharpening machine is made (see Fig. 13). In order to be able to change the velocity of push stick and grinding guide, the electric motor is equipped with a frequency converter. Depending on the grinding thickness, the operator will adjust the appropriate motor speed according to each grinding mode.

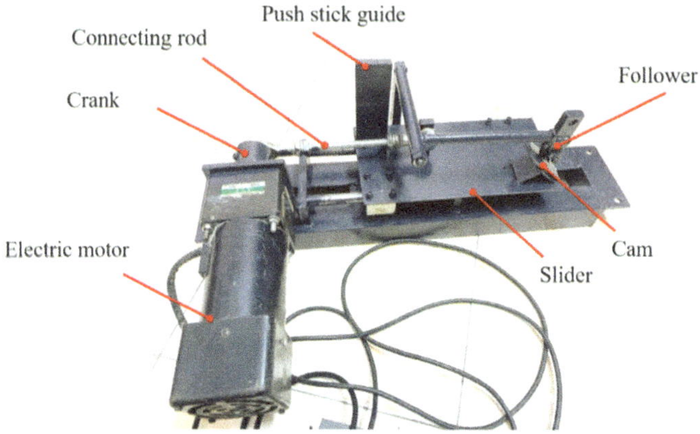

Fig. 13. Drive mechanism of bandsaw blade sharpening machine

Based on the drive mechanism in Fig. 13, the bandsaw blade sharpening machine is renovated and shown in Fig. 14. The bandsaw blade for testing has a value of pitch 12,7 and 8,46 mm. The test results show that the bandsaw blade and grinding wheel move smoothly and precisely. There are no impacts between the bandsaw blade and the grinding wheel. The quality of the bandsaw blade after sharpening is satisfactory.

Fig. 14. Experiment of grinding bandsaw blade with a value of pitch 12,7 mm

4 Conclusions

A bandsaw blade sharpening machine, which normally uses cam mechanisms to drive the push stick and grinding wheel, has the advantage of producing precise motion and trajectories. However, for sharpening different bandsaw blades with different pitches, it is necessary to replace cam sizes. For the purpose of sharpening bandsaw blades with different pitches flexibly without using different cams, this article designs a new drive concept. This new drive concept combines a slider crank mechanism with a cam mechanism. A wood bandsaw blade sharpening machine is built to test the new drive mechanism. This machine grinds two bandsaw blades with a value of pitch 12,7 and 8,46 mm. The machine can adjust easily the stroke of the push stick and displacement of cam follower. The quality of the bandsaw blade after sharpening is good.

References

1. Nor Marzuina, F.K.N., Mohd-Jamil, A.W.: The effective tooth width of bandsaw blade for sawing Malaysian timbers. For. Res. Inst. Malays. (2019)
2. Iseli BNP210: Machine for automatic profile sharpening of band saw blades controlled with 3 CNC axes in wet grinding. https://www.iseli-swiss.com/en/
3. Wood-Minzer: How to choose a sawmill blade profile. https://woodmizer.com/
4. MVM: Band saw sharpening and setting machine. https://www.mvmsrl.it/en/
5. Lenox: Guide to band sawing. https://www.lenoxtools.com/
6. Dakin flathers: Band saw blade teeth per inch (TPI) chart. https://www.dakin-flathers.com/
7. Volmer, J.: Getriebetechnik – Kurvengetriebe. Verlag Technik, Berlin (1989)
8. Berger, M.: Kolloquium Getriebetechnik Tagungsband. Universität Chemnitz Verlag, Chemnitz (2011)
9. Nicholas, J., Hargreaves, J., Hunter, J.: Polynominals, University of Sysney (2006)
10. Dakshin, P., Khurana, S.: Novel CAM Mechanism. Springer, Singapore (2021)
11. Hroncová, D., Binda, M., Kiþák, P.ŠF.: Kinematical analysis of crank slider mechanism using MSC Adams/View. Procedia Eng. **48**, 213–222 (2012)

Cooperative Tracking Framework for Multiple Unmanned Aerial Vehicles (UAVs)

Hai T. Do[1], Hoang T. Hua[1,2], Hoa T. T. Nguyen[1], Minh T. Nguyen[1(✉)], and Hoang T. Tran[3,4]

[1] Thai Nguyen University of Technology, Thai Nguyen City 240000, Vietnam
nguyentuanminh@tnut.edu.vn
[2] University of Rome "La Sapienza", Via Eudossiana 18, 00184 Rome, Italy
[3] Center of Electrical Engineering, Duy Tan University, Da Nang 550000, Vietnam
[4] Faculty of Electrical-Electronic Engineering, Duy Tan University, Da Nang 550000, Vietnam

Abstract. In recent times, multiple Unmanned Aerial Vehicles (UAVs) have been utilized in numerous fields. Applications exploiting UAVs range from civilian to military applications. Multiple UAVs have been used for surveillance, monitoring, penetrating, or event detecting. In these applications, an ability of multiple UAVs to track an interested object plays a crucial role. In this work, a framework of cooperative tracking for a team of multiple UAVs is proposed. The proposed method consists of two main stages. First, an interested object is evaluated by computer vision techniques implemented on UAVs. Each UAV would have its own observation about the object. A consensus algorithm is used to synchronize sensing information of a whole team to decide whether the object is a target or not. In the second stage, a flocking algorithm is used to drive a UAVs team to track the target. Simulations in both cases static and dynamic targets are illustrated tracking performance.

Keywords: Cooperative control · Cooperative tracking · Flocking algorithm · Consensus · Target recognition · Multiple UAVs

1 Introduction

Unmanned aerial vehicles (UAVs) are defined as aircrafts that can remotely operate and do not require human on board. Multiple UAVs systems have been widely deployed in many applications such as traffic monitoring [1, 2], communication bridges [3], smart agriculture and surveilling battlefields [4], etc. In recent times, developments of multiple UAVs systems have extensively progressed. A coordination of multiple UAVs is inspired by formation behaviors in nature. Each UAV member can autonomously react to environments to archive a common mission without the need of a guidance center. Multiple UAVs systems takes advantages of shortening operating time [5], improving energy-efficient [6] and robustness [7].

For example, in the military field, a UAV formation is used as a reconnaissance tool to observe enemy troops and equipment [8]. A network of UAVs can be deployed to

D. C. Nguyen et al. (Eds.): ICERA 2021, LNNS 366, pp. 276–285, 2022.
https://doi.org/10.1007/978-3-030-92574-1_29

help detect intrusions and monitor enemy invasion activities. Besides reconnaissance missions, many UAVs are also used for infiltration missions [9]. UAVs can carry offensive weapons in complex and modern wars. A cooperative missile group increases the chances of a successful attack [10].

In aforementioned applications, an ability to track a common target of an UAV team plays an important role in performance of the whole system. As each UAV member would have different opinions about targeted objects, synchronizing sensing information among the team is crucial to ensure properly operating of the UAV team. In control perspectives, while tracking targets, a UAV team has to fulfill several constraints such as keeping a certain formation, avoiding collisions with other members as well as obstacles or enemies.

In this work, we propose cooperative tracking for multiple UAVs working in a vast area. The UAVs collaborate to each other and are led to a position to track targets. The targets could be static or moveable. The UAVs use consensus algorithms to detect targets. They are led by a flocking control algorithm that support them to avoid obstacles and be able to communicate to their neighbors. We formulate the problems and provide simulation results to verify the problems.

The rest of this paper is organized as follows. Section 2 addresses the literature review including target recognition methods and control methods. Simulation results for the problems are provided in Sect. 3. Finally, conclusion and future developments are address in Sect. 4.

2 Literature Review

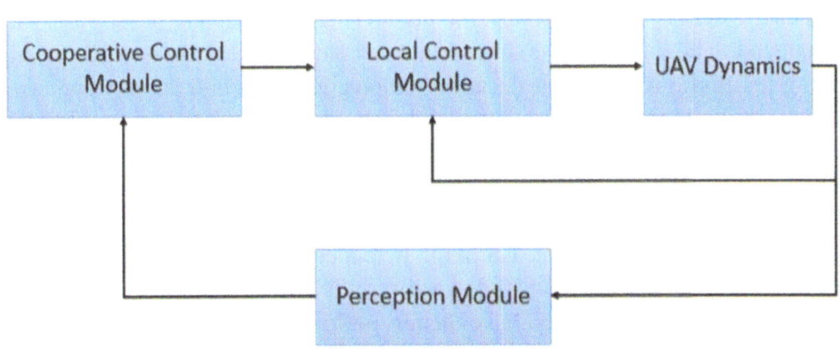

Fig. 1. The general system architecture.

Figure 1 depicts a hierarchical structure of an UAV working in a UAV team. There are three main modules in the structure. The first module is a perception module. UAVs are equipped with visual sensors to obtain information of external environments such as obstacles, or targeted objects. Then, sensing information is fed to cooperative control module, which evaluates sensing data from UAV itself as well as its neighbors to generate trajectories for a UAV. The cooperative controller ensures a UAV team to track an interested object while keeping a desired formation and avoiding collisions. At the lowest stage, a local controller module is used to steer a UAV to follow generated paths. Within the scope of our paper, the local controller is not evaluated.

2.1 Vision-Based Sensing

Detection and Classification

Sensing is an evitable ability in the designing of an autonomous UAVs team. In this paper, each UAV is equipped color cameras. Computer vision techniques are deployed to process visual sensing data. Based on processed data, a cooperative controller performs decision-making for a UAV. This section provides a brief introduction of image processing methods utilized in a sensing module.

Directly using an image for sensing purposes may require complex computation. As on-board computational ability of a UAV is limited, images captured by cameras are need to be processed before utilizing for control purposes. Feature-based methods are effective solutions. In the feature-based method, interesting features from images are detected and extracted. The features are then considered as inputs for object detection or localization. Feature-based methods offer various merits such as robust to image rotation, blur or noise images. The feature-based method consists of two steps. In the first step, feature detection is performed. This step is the lowest level in an image processing algorithm. Features of an image are determined by computing abstractions at every point on the image. Features can be grouped into three main types namely edges, corners, and blobs. The detected features are the most important information of an image. From detected features, we can almost reconstruct an image without the loss of major information. After having important points of an image, we need an algorithm to select data which is needed for specific tasks. Therefore, feature description methods are introduced. These methods extract from detected features data which is required depending on missions. In recent decades, various detectors and descriptors have been extensively studied [11]. After the interest objects are detected, the image classification issue will be utilized to find the desired target. Machine learning (ML) algorithms can be considered effective techniques in image classification tasks. Supervised machine learning (SML) and unsupervised machine learning (USML) are the most popular classification criteria of machine learning. In paper [12, 13] ML models have been successfully used to carry out landslide susceptibility prediction with high accuracy. It observed that the SML models have better performance than the UMSL models. The SML models have strong predictive power for unknown data while USML models have efficient modeling processes, dimensionality reduction, and strong scalability [12]. Furthermore, Deep Learning is a new promising solution. It becomes a particularly useful and robust tool to solve complex problems in many different fields including image classification. In [14] Haq et al. presented a deep learning-based supervised image classification algorithm that applied to very high-resolution remotely sensed images collected from UAVs to assess forest coverage area. The results showed that Deep Learning approach outperform others ML algorithms with overall accuracy of 90%. However, one major disadvantage of deep learning is that it requires better computing facilities compared with traditional ML algorithms.

Graph Theory and Consensus Algorithm for Target Synchronization
Each UAV in a group has distinct observations on the desired target. Under these different observations, objects classification algorithms may result in different predictions [15]. Therefore, a decision-making mechanism is required to synchronize UAVs in a formation. The mechanism should be designed in a distributed form for saving resources and scaling abilities. In order to reach a consensus for UAVs, we have utilized a consensus algorithm based on a graph theory.

The communication topology of a UAV group is described by a graph. A graph $G = (N, E)$ has a set of N nodes (a node presenting a UAV) and a set of E edges. If there exists a communication link between node i and node j, an edge is defined as $E = (i, j)$. It is said that node j receive information from node i. The adjacency matrix is defined as $A = [a_{ij}]$ with $a_{ij} = 1$ if $(j, i) \in E$, otherwise $a_{ij} = 0$. The degree matrix is obtained as $D = diag(d_{in}(i))$. The laplacian matrix of a graph is obtained by $L = D - A$.

The target synchronizaton for UAVs have been tackled by a consensus algorithm to reach a consensus decision. Each UAV exchanges and update its classification result until a consensus among them is reached. The value of prediction results of the jth UAVs at time step t is presented by a n-vector $V_j(t)$. Then, each UAV updates its result as follows:

$$A_j(t+1) = A_j(t) + \sum_{i \in N_i, i \neq j} A_i(t) \tag{1}$$

where N_i is the set of neighboring nodes that can transmit data to node j. The consensus can be reached if the communication topology has a spanning tree. If the consensus value is greater than a threshold, the desired target of a whole formation is detected.

2.2 Flocking Control Theory

An algorithm for a cooperative control module is presented in this section. Flocking algorithm is chosen for controlling a UAV team due to its flexible merits. Different from other formation control algorithms, Each UAV member only needs to keep a certain distance from other members, which provides a high adaptability to external environments.

Consider a group of N UAVs maneuvering with the same altitude. Therefore, the operation of the UAV group is considered in 2D environment. The connection of the UAV team is described by a graph as in Sect. 2.1. At formation control, the dynamics of a UAV can be modeled as a second-integrator model.

$$\begin{cases} \dot{p}_i = v_i \\ \dot{v}_i = u_i \end{cases} ; i \in 1, 2, ..., N \tag{2}$$

where p_i, v_i are positions and velocities of the UAV ith, u_i is control signals.

A UAV has two types of neighbors. The first is actual neighbors, which are UAVs member in a team. The actual neighbors are defined as:

$$N_i^\alpha = \{j \in V : \|p_j - p_i\| \leq R_c, i \neq j\} \tag{3}$$

where $\|p_j - p_i\|$ is the Euclidean norm, R_c is a communication range.

Besides actual neighbors, UAV ith also have virtual neighbors, which are the projection of the UAV ith on obstacles. The virtual neighbors is defined as:

$$N_i^v = \{k \in V_0 : \|p_{ik} - p_i\| \leq r_0, V_0 = 1, 2, ..., K\} \tag{4}$$

where r_0 is an obstacle detection range, V_0 is the set of K obstacles, p_{ik} is 2D position of UAV ith projecting on obstacle k.

In the flocking algorithm [16], control input is the combination of three terms: (1) a gradient-based term, (2) obstacle avoidance, and (3) a navigational term. The first component is to maintain the desired distance between UAVs. It also contains a consensus component to synchronize the velocities of the UAVs group. The second term provides an ability to avoid collisions with obstacles. The third term is used to steer UAVs toward targets.

The gradient-based term is defined as:

$$f_i^f = c_1^\alpha \sum_{j \in N_i^\alpha} \phi_\alpha(\|p_{ik} - p_i\|) n_{ij} + c_2^\alpha \sum_{j \in N_i^\alpha} a_{ij}(v_j - v_i) \tag{5}$$

where c_1^α, c_2^α are positive constant, ϕ_α is the potential function, $n_{ij} = \frac{p_j - p_i}{\sqrt{1 + \varepsilon \|p_j p_i\|^2}}$ is a vector connecting p_i and p_j with a constant $\varepsilon \in (0, 1)$.

The obstacle avoidance term is defined as:

$$f_i^o = c_1^o \sum_{j \in N_i^v} \phi_o(\|p_{ik} - p_i\|) n_{ik} + c_2^o \sum_{j \in N_i^v} b_{ik}(v_{ik} - v_i) \tag{6}$$

where ϕ_o is the repulsive function, n_{ik} is a vector connecting p_i and p_k.

The navigational term is computed as:

$$f_i^n = -c_1(p_i - p_t) - c_2(v_i - v_r) \tag{7}$$

where p_t is the position of a targe, c_1 and c_2 are constants.

3 Simulation Results

3.1 Consensus Algorithm

In this section, the performance of a consensus algorithm has been evaluated. Under flocking control algorithms, UAVs gradually form an α-lattice formation. Each UAV is capable of transmitting or receiving information from UAVs within its communication range. Therefore, it is assumed that there exists a spanning tree in the communication networks of UAVs. A communication of an α-lattice formation with 10 UAVs is depicted in Fig. 2.

Fig. 2. Graph representation.

To test the performance of consensus algorithm, we assigned initial predictions of UAVs as

$$A_1 = 50\%, A_2 = 30\%, A_3 = 90\%, A_4 = 80\%, A_5 = 55\%, A_6 = 88\%,$$
$$A_7 = 99\%, A_8 = 92\%, A_9 = 85\%, A_{10} = 65\%.$$

The result is shown in Fig. 3. UAVs reached an consensus of 74.04% at t = 5 s.

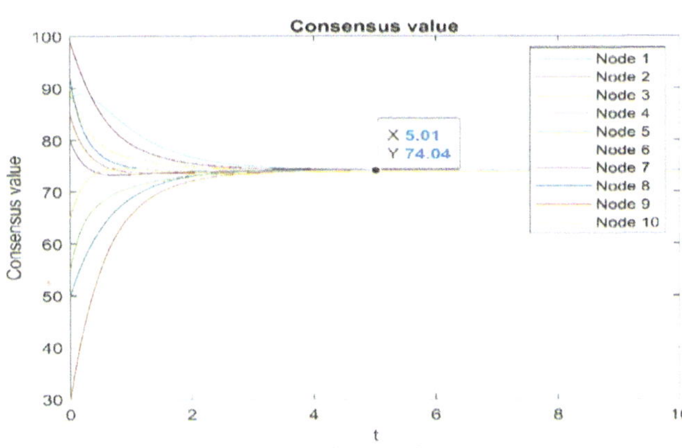

Fig. 3. Ten UAVs with different predicted values reach a consensus.

3.2 Flocking Control Algorithm

In this section, the tracking performance of flocking algorithms is presented. A group of 10 UAVs is distributed with random initial positions. Initial velocities are zero. The communication range is 10. Each UAV can obtain positions of targets or can receive positions of targets by communicating with other members.

In the first case, a UAVs group is required to maneuver to a fixed target. The UAVs are initialized randomly and flock towards a target. The group successfully reaches the target while ensures safe operations with no collisions. The operation is illustrated as in Fig. 4 and Fig. 5.

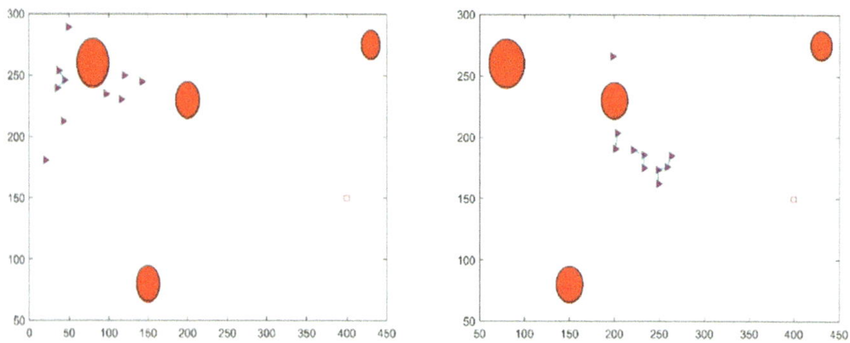

a) Initial positions.

b) A group of UAVs moving to a static target.

Fig. 4. Tracking a static target.

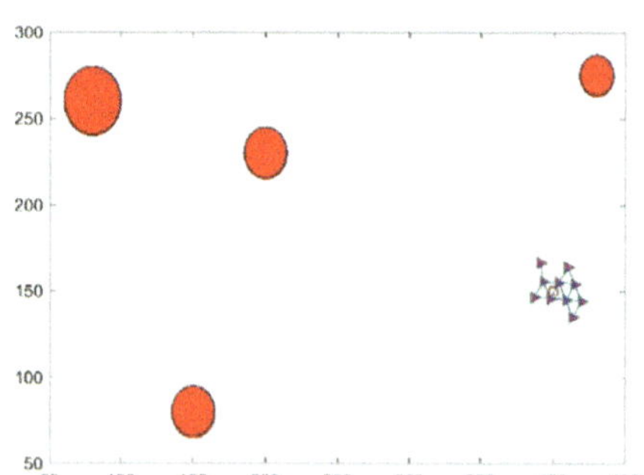

Fig. 5. UAVs gathering at the target.

In the second case, the UAVs group must track a dynamic target. Figure 6 represents the snapshots of UAVs tracking a moving target. It is showed that UAVs track the target and eventually converging to an α-lattice formation around the target. The group is also able to avoid collisions with external obstacles. Simulation results have presented the effectiveness of flocking algorithm in cooperative tracking problems even when connections between UAVs are slightly interrupted.

The above results have been presented assuming that UAVs can get the information of the target by their vision sensing devices. In practice, this information may not be always available due to devices' limitations (e.g., a target is shaded by neighbor UAVs or sensing devices have a small detection range). Normally, UAVs on the outside of a formation will have better observations and distribute the information of the target through UAV networks. However, energy is a critical problem as UAVs keep broadcasting frequently. To solve this problem, Nowzari et al. [17] proposed an event-triggered control scheme for networked systems, in which control signals are updated depending on detections of interesting events. In tracking problems, interesting events

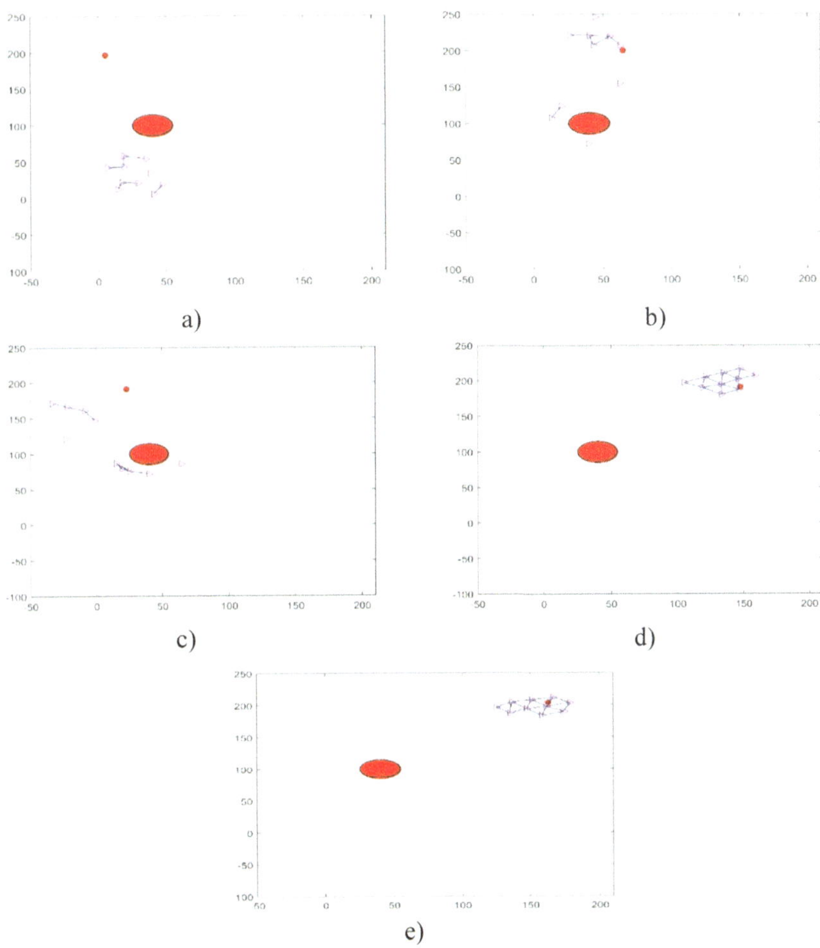

Fig. 6. A few snapshots of initial positions of UAVs (a), avoiding obstacles and while travelling to targets (b, c) and tracking the moving target (d, e).

could be target detection or target relocation. In [18], the authors provided a solution for information sharing among multiple robots networks. Each member shares their information with their neighbors and each member builds an entire map based on shared information. A compressed sensing algorithm was introduced to significantly reduced energy consumption for broadcasting sensing information within networks.

4 Conclusions and Future Developments

In this paper, a cooperative tracking framework for vision-based of multiple UAVs systems is proposed. The visual perception module utilizing image processing techniques is introduced. The design of a cooperative algorithm based on flocking algorithms is addressed. Then, simulations for two scenarios static and dynamic targets demonstrate the effectiveness of the proposed algorithm. The results show the effectiveness of the methods that we proposed. In addition, the whole framework points out the possibility of apply the cooperative methods into practical applications. In future works, the complete system integrating sensing, cooperative control, and local control stage will be explored. Learning techniques based on artificial intelligence also promise significant improvements in the performance of multiple UAV systems. The practical experiments will also be investigated to evaluate the effectiveness of the proposed system in a real-word environment.

Acknowledgements. The authors would like to thank Thai Nguyen University of Technology, Viet Nam for the support.

References

1. Puri, A., Valavanis, K.P., Kontitsis, M.: Statistical profile generation for traffic monitoring using real-time UAV based video data. In: 2007 Mediterranean Conference on Control & Automation, pp. 1–6. IEEE, June 2007
2. Lottes, P., Khanna, R., Pfeifer, J., Siegwart, R., Stachniss, C.: UAV-based crop and weed classification for smart farming. In: 2017 IEEE International Conference on Robotics and Automation (ICRA), pp. 3024–3031. IEEE, May 2017
3. Nguyen, H., Nguyen, M., Truong, L.H., Le, T.: UAV communication networks: benefits, research challenges and opening issues. In: ICSES Transactions on Computer Networks and Communications, pp. 1–5, September 2021, in Press
4. Kim, D., Lee, K., Park, M., Lim, J.: UAV-based localization scheme for battlefield environments. In: MILCOM 2013–2013 IEEE Military Communications Conference, San Diego, CA, USA, 2013, pp. 562–567. https://doi.org/10.1109/MILCOM.2013.102
5. Lin, L., Roscheck, M., Goodrich, M., Morse, B.: Supporting wilderness search and rescue with integrated intelligence: autonomy and information at the right time and the right place. In: Proceedings of the AAAI Conference on Artificial Intelligence, vol. 24, no. 1, July 2010
6. Riehl, J.R., Collins, G.E., Hespanha, J.P.: Cooperative search by UAV teams: a model predictive approach using dynamic graphs. IEEE Trans. Aerosp. Electron. Syst. **47**(4), 2637–2656 (2011)
7. Skorobogatov, G., Barrado, C., Salamí, E.: Multiple UAV systems: a survey. Unmanned Syst. **8**(02), 149–169 (2020)

8. Liu, L., McLernon, D., Ghogho, M., Hu, W., Huang, J.: Ballistic missile detection via micro-Doppler frequency estimation from radar return. Digit. Signal Process. **22**(1), 87–95 (2012)
9. Wang, Y., Dong, S., Ou, L., Liu, L.: Cooperative control of multi-missile systems. IET Control Theory Appl. **9**(3), 441–446 (2015)
10. Lee, J.I., Jeon, I.S., Tahk, M.J.: Guidance law using augmented trajectory-reshaping command for salvo attack of multiple missiles. In: UKACC International Control Conference, pp. 766–771, August 2006
11. Máthé, K., Buşoniu, L.: Vision and control for UAVs: a survey of general methods and of inexpensive platforms for infrastructure inspection. Sensors **15**(7), 14887–14916 (2015)
12. Chang, Z., et al.: Landslide susceptibility prediction based on remote sensing images and GIS: comparisons of supervised and unsupervised machine learning models. Remote Sens. **12**(3), 502 (2020)
13. Huang, F., Yin, K., Zhang, G., Gui, L., Yang, B., Liu, L.: Landslide dis-placement prediction using discrete wavelet transform and extreme learning ma-chine based on chaos theory. Environ. Earth Sci. **75**, 1376 (2016)
14. Nguyen, M.T., Truong, L.H., Le, T.T.H.: Video surveillance processing algorithms utilizing artificial intelligent (AI) for unmanned autonomous vehicles (UAVs). MethodsX **8** (2021). https://doi.org/10.1016/j.mex.2021.101472
15. Harzallah, H., Jurie, F., Schmid, C.: Combining efficient object localization and image classification. In: 2009 IEEE 12th International Conference on Computer Vision, pp. 237–244. IEEE, September 2009
16. Nguyen, M.T., Boveiri, H.R.: Energy-efficient sensing in robotic networks. Measurement **158**, 107708 (2020)
17. Nowzari, C., Garcia, E., Cortés, J.: Event-triggered communication and control of networked systems for multi-agent consensus. Automatica **105**, 1–27 (2019)
18. Nguyen, M.T., La, H.M., Teague, K.A.: Collaborative and compressed mobile sensing for data collection in distributed robotic networks. IEEE Trans. Control Netw. Syst. **5**(4), 1729–1740 (2018). https://doi.org/10.1109/TCNS.2017.2754364

Critical Buckling Load Evaluation of Functionally Graded Material Plate Using Gaussian Process Regression

Huan Thanh Duong[1], Hieu Chi Phan[2(✉)], and Tien-Thinh Le[3,4(✉)]

[1] Faculty of Engineering, Vietnam National University of Agriculture, Trau Quy, Gia Lam, Hanoi 100000, Vietnam

[2] Le Quy Don Technical University, 236 Hoang Quoc Viet, Hanoi 100000, Vietnam
phanchihieu@lqdtu.edu.vn

[3] Faculty of Mechanical Engineering and Mechatronics, PHENIKAA University, Yen Nghia, Ha Dong, Hanoi 12116, Vietnam
thinh.letien@phenikaa-uni.edu.vn

[4] PHENIKAA Research and Technology Institute (PRATI), A&A Green Phoenix Group JSC, No. 167 Hoang Ngan, Trung Hoa, Cau Giay, Hanoi 11313, Vietnam

Abstract. Functionally Graded Materials (FGM) is the advanced material that covers the advantages of both metal and ceramic which contain FGM. Due to the perfect combination, FGM plate is widely developed with the requirement for the practical application to avoid the difficulties from conventional approaches. Consequently, the study establishes the database from analytical model and uses it for developing a machine learning model in the desire of finding an alternative model to predict critical buckling load of the plate. Such a machine learning model is crucial for predicting critical buckling load of the FGM plate without complexity of analytical developments or finite element resources. The Gaussian Process Regression model has been developed based on the database containing 1000 labeled samples created from the analytical model. The Gaussian Process Regression models with and without optimization process are compared and the outstanding of the model with the optimization algorithm is revealed. The proposed model is validated on both train and test set with the R-square values larger than 0.99 and errors are significantly low indicating that the proposed model exhibits the required ability.

Keywords: Functionally graded material · Buckling analysis · Machine learning

1 Introduction

Functionally Graded Materials, FGM, are interesting with the mixture of different proportional of constituent materials, normally made of metal and ceramic, at different locations. This leads to the mechanic properties of FGM changes with the intentional manufacture. The control of proportion of materials is useful to take advantage of each

D. C. Nguyen et al. (Eds.): ICERA 2021, LNNS 366, pp. 286–292, 2022.
https://doi.org/10.1007/978-3-030-92574-1_30

constituent material, the ductile characteristic of metal and the heat resistance of ceramic. Areas with applications of FGMs are expanding such as: nuclear industry, aerospace, medical industry etc.

Along with the demand of FGM for applications, the desire for exact and practical analysis methods for catching the behavior of FGM plate is increasing. Both analytical and numerical methods have been applied for analysis of FGM structures such as in [1–4]. Both of these methods face with the disadvantage of the time consuming algorithm due to the trials and errors process in analytical approach to figure out the proper solution corresponding to the minimum buckling load (i.e. critical load) and the exponential complexity of the problem when the number of elements increases in the finite element analysis [5, 6]. Additionally, applying analytical and finite element analysis for the buckling analysis of FGM plate related to the various advance knowledge and skills that are not always available for engineers.

Consequently, an alternative to avoid these difficulties is establishing a database from conventional models and feeding for the developing model process of machine learning approach [7–9]. The predicting time reduction of machine learning model approach comes from the fact that its models commonly deal with matrix operations rather than solving the integral equations, trials and errors approaches. Thus, it is not surprising to observe the use of machine learning for FGM structures such as [10, 11].

In this study, a machine learning model named as the Gaussian Process Regression, is conducted based on analytical database to provide a practical model predicting critical buckling load for the supported FGM plate. The Gaussian Process Regression model is selected because of its various advantages compared to other machine learning techniques, including nonparametric, robust, and parallelizable properties, etc.

2 Materials and Methods

2.1 Evaluating the Critical Buckling Load of FGM Plate

The basic dimensions of a plate are width, a; length, b and thickness, h, given in Fig. 1. FGM is the mixture of metal and ceramic with the proportional of each molecular material at a given point varying according to its locations. In this study, the proportional is changed through the thickness of the plate.

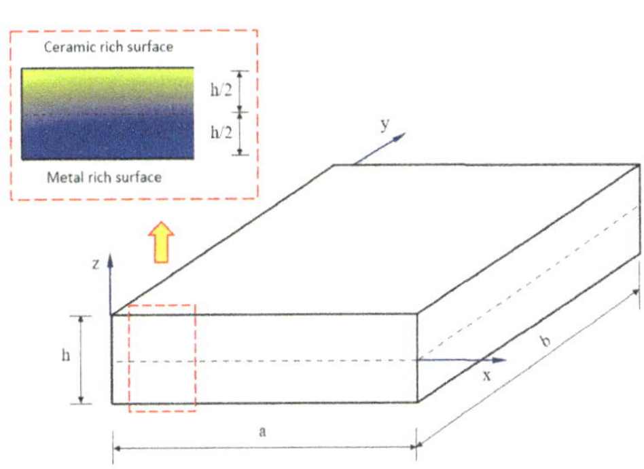

Fig. 1. Geometry and coordinate of FGM plate.

Mechanically, the quantity of Young's modulus, E, is a dependent variable of z, the location of the concerned point along the z-axis. This relationship is explicitly expressed as in Eq. 1 [12]:

$$E(z) = (E_c - E_m).\left(\frac{1}{2} + \frac{z}{h}\right)^p + E_m \quad (1)$$

Where: E_m and E_c are the Young's modulus of metal and ceramic, respectively. p is the volume fraction exponent ($p \geq 0$). ν is the Poisson's ratio (It is considered to be constant in the present study, ν = 0.3).

If the 6 components of displacements u_0, v_0 are the in-plane tangential displacements and w_0 is the transverse displacement of a point (x, y) on the middle surface, ϕ_x, ϕ_y are rotations of the normal to the middle surface about y and x axes respectively. The final equation for the static problem of assessing the critical buckling load of FGM plate is given in Eq. 2 with the multiplication of stiffness matrix [K] and displacement vector u to zero vector.

$$[K] \times \{u\} = \{0\} \quad (2)$$

The critical load, N_{cr}, which is the smallest value of $N^0_{(m,n)}$, the determination of coefficient of matrix [K]:

$$\det|K| = 0 \quad (3)$$

Conventionally, the N_{cr} is obtained by solving Eq. 3. The absolute N_{cr} in this study is replaced by a normalized value as in Eq. 4.

$$\overline{N}_{cr} = N_{cr}a^2/\left(E_m h^3\right) \quad (4)$$

In general, the required inputs of the above problem are k, p, b/h, a/b and E_c/E_m. In the next step, these are randomly generated to create 1000 unlabeled samples with the uniform distributions to obtain a database labeled by the normalized critical buckling, $\overline{N}_{cr}$. Details of the descriptive statistics are given in Table 1.

Table 1. Characteristics of the database.

Parameter	k	p	b/h	a/b	E_c/E_m	$\overline{N}_{cr}$
Role	Input	Input	Input	Input	Input	Output
Min	0.0018	2.0052	20.0246	1.0070	1.0018	0.7153
Mean	2.5019	5.9082	59.7623	3.0348	5.4665	13.6832
Max	4.9902	9.9814	99.7754	4.9950	10.0000	192.5038
StD	1.4282	2.3347	23.2809	1.1600	2.6486	21.0387
CV (%)	*57.1*	*39.5*	*39.0*	*38.2*	*48.5*	*153.8*

2.2 Used Methods

In this study, Gaussian Process Regression technique has been used as a prediction mode because of the following reasons. First, Gaussian Process Regression model was constructed for regression purposes. It is a nonparametric model which is not restrained by a number of data or a form of function. Second, Gaussian Process Regression model is highly robust even in the case of a small size dataset or a large number of input variables. Last but not least, Gaussian Process Regression model can be easily implemented for a numerical purpose. The model is fully described in Table 2 below, including Gaussian Process, the function of the model, and the Kernel function. In this study, the rational quadratic kernel function with default kernel parameters was employed. More details of the model can be referred to Rasmussen [13].

Table 2. Description of Gaussian Process Regression model used in this study.

Characteristics	Equation form	Description of parameters
Gaussian Process G_P	$\begin{cases} \mu(x) = E[f(x)] \\ \psi(x_i, x_j) = E\left[(f(x_i) - \mu(x_i))(f(x_j) - \mu(x_j))^T\right] \end{cases}$	mean $\mu(x)$ and covariance (i.e. kernel) function $\psi(x_i, x_j)$, where x_i, x_j are input vectors of d-dimension and E is mathematical expectation
Gaussian Process Regression model	$f(x) \approx G_P(\mu(x), \psi(x_i, x_j)) = G_P(0, \psi(x_i, x_j))$	$\psi(x_i, x_j)$ Kernel function
Kernel function: rational quadratic	$\psi(x_i, x_j) = \sigma_f^2\left(1 + \frac{r^2}{2\alpha\sigma_l^2}\right)^{-\alpha}$ $r = \sqrt{(x_i - x_j)^T(x_i - x_j)}$	σ_f, σ_l, and α are the signal standard deviation, the characteristic length scale, and the shape parameter, respectively

2.3 Description of the Optimization Problem for Model Parameters

In this section, an optimization procedure is described for obtaining the best hyper-parameters of the Gaussian Process Regression model. Table 3 indicates information of the optimization problem. The signal standard deviation, the characteristic length scale, and the shape parameter (σ_f, σ_l, and α, respectively) are design variables. The optimization was conducted by minimizing fivefold cross validation loss, using mean square error (MSE) as validation [14].

3 Results and Discussion

3.1 Optimization of Model Parameters

Figure 2 presents results of optimization process for Gaussian Process Regression model. It is seen in Fig. 2 that the minimum required number of function evaluations is 12, and the optimizer finds a minimum point (star mark). As a result, the optimal values are 4942.4219, 25.4619, 0.0379, for signal standard deviation σ_f, characteristic length scale σ_l, and shape parameter α, respectively.

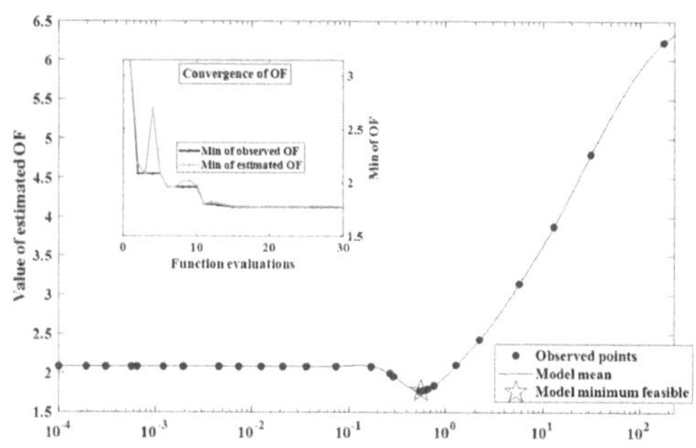

Fig. 2. Results of optimization process: in the small window: min objective versus number of function evaluations; in the large window: estimated objective function versus feasible point.

Table 3. Description of the optimization problem for model parameters.

Parameter	Description
Design variables	σ_f, σ_l, and α
Number of Cross validation folds	5
Validation	Mean square error (MSE)
Objective function (OF)	Loss of 5-fold cross-validation: $C_v(k) = \frac{1}{k}\sum_{i=1}^{k} MSE_i$
Optimization problem	$\underset{x \in \mathbb{R}^3}{\text{minimize}}(C_v(k))$

3.2 Prediction Capability

The main validation results of the developed model are provided in Fig. 3 and Table 4. It can be seen from Fig. 3 that the post-optimized model provides predicted values nearly equal to actual data with the data which near-perfectly follow the diagonal line or the well-known 1:1 line. This implies that the proposed model has the high quality for predicting the interested variable (i.e. $\overline{N}_{cr}$). Also, Fig. 3 provides the improvement of the model with the optimization process when the model without optimization process described in

Sect. 2.3 has nonlinear relationship of actual versus predicted values. Apparently, the optimization process is crucial for the development of the final model.

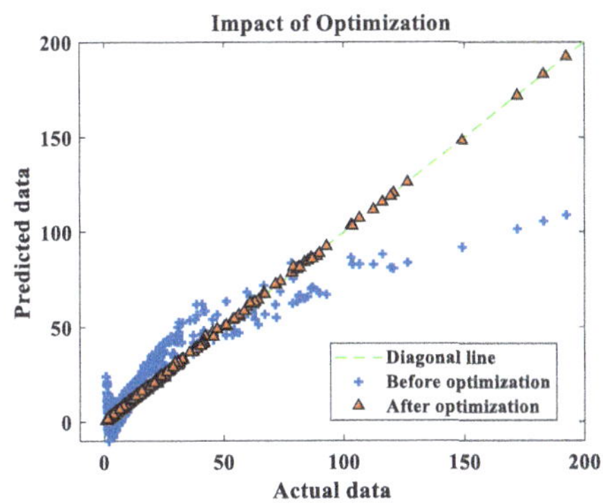

Fig. 3. Predicted values with/without optimization versus actual data.

Table 4. Prediction capability of the optimized model.

Dataset	R^2	RMSE	MAE
Train	0.999738	0.36486	0.2171
Test	0.991002	1.659806	0.908996
All	0.997944	0.958996	0.424669

Further observations on the validating measures of the developed model reinforce the intuition obtained from Fig. 3. The determination coefficient, R^2, of the model on the train, test and all data consistently higher than 0.99 implies that the trend of predicted versus actual values is successfully predicted. The Root Mean Square Errors, RMSE, and Mean Absolute Errors, MAE, are minor on test set at 1.6598 and 0.9090, respectively, compared to the mean value of $\overline{N}_{cr}$ at 13.6832 given in Table 1. This tendency is analogous on the train and test set.

4 Conclusion

The study has developed an efficient and high-quality machine learning models predicting the buckling load of the FGM plate. The Gaussian Process Regression model has proofed its capacity to capture the non-linear relationship of the inputs and output with an efficient optimization included. Further applications can be considered to incorporate this model with advanced technology.

Acknowledgement. This research was funded by ARES-CCD (Académie de Recherche et d'Enseignement supérieur - Commission de la Coopération au Développement) in the framework of the Institutional Support to Vietnam National University of Agriculture (VNUA).

References

1. Thai, H.-T., Choi, D.-H.: An efficient and simple refined theory for buckling analysis of functionally graded plates. Appl. Math. Model. **36**, 1008–1022 (2012). https://doi.org/10.1016/j.apm.2011.07.062
2. Tu, T.M., Quoc, T.H., Long, N.V.: Bending analysis of functionally graded plates using new eight-unknown higher order shear deformation theory. Struct. Eng. Mech. **62**, 311–324 (2017). https://doi.org/10.12989/sem.2017.62.3.311
3. Huan, D.T., Tu, T.M., Quoc, T.H.: Analytical solutions for bending, buckling and vibration analysis of functionally graded cylindrical panel. Vietnam J. Sci. Technol. **55**, 587–587 (2017). https://doi.org/10.15625/2525-2518/55/5/8843
4. Van Long, N., Quoc, T.H., Tu, T.M.: Bending and free vibration analysis of functionally graded plates using new eight-unknown shear deformation theory by finite-element method. Int. J. Adv. Struct. Eng. **8**(4), 391–399 (2016). https://doi.org/10.1007/s40091-016-0140-y
5. Belabed, Z., Ahmed Houari, M.S., Tounsi, A., Mahmoud, S.R., Anwar Bég, O.: An efficient and simple higher order shear and normal deformation theory for functionally graded material (FGM) plates. Compos. B Eng. **60**, 274–283 (2014). https://doi.org/10.1016/j.compositesb.2013.12.057
6. Le, T.-T.: Probabilistic modeling of surface effects in nano-reinforced materials. Comput. Mater. Sci. **186**, 109987 (2021). https://doi.org/10.1016/j.commatsci.2020.109987
7. Phan, H.C., Duong, H.T.: Predicting burst pressure of defected pipeline with principal component analysis and adaptive neuro fuzzy inference system. Int. J. Press. Vessels Pip. **189**, 104274 (2021). https://doi.org/10.1016/j.ijpvp.2020.104274
8. Phan, H.C., Bui, N.D., Pham, T.D., Duong, H.T.: Predicting capacity of defected pipe under bending moment with data-driven model. In: Tien Khiem, N., Van Lien, T., Xuan Hung, N. (eds.) Modern Mechanics and Applications. LNME, pp. 830–840. Springer, Singapore (2022). https://doi.org/10.1007/978-981-16-3239-6_64
9. Phan, H.C., Dhar, A.S.: Predicting pipeline burst pressures with machine learning models. Int. J. Press. Vessels Pip. **191**, 104384 (2021). https://doi.org/10.1016/j.ijpvp.2021.104384
10. Nazari, A., Milani, A.A., Zakeri, M.: Modeling ductile to brittle transition temperature of functionally graded steels by artificial neural networks. Comput. Mater. Sci. **50**, 2028–2037 (2011). https://doi.org/10.1016/j.commatsci.2011.02.003
11. Duong, H.T., Phan, H.C., Tran, T.M., Dhar, A.S.: Assessment of critical buckling load of functionally graded plates using artificial neural network modeling. Neural Comput. Appl. **33**, 16425–16437 (2021). https://doi.org/10.1007/s00521-021-06238-6
12. Reddy, J.N.: Analysis of functionally graded plates. Int. J. Numer. Meth. Eng. **47**, 663–684 (2000). https://doi.org/10.1002/(SICI)1097-0207(20000110/30)47:1/3<663::AID-NME787>3.0.CO;2-8
13. Rasmussen, C.E.: Gaussian processes in machine learning. In: Bousquet, O., von Luxburg, U., Rätsch, G. (eds.) ML -2003. LNCS (LNAI), vol. 3176, pp. 63–71. Springer, Heidelberg (2004). https://doi.org/10.1007/978-3-540-28650-9_4
14. Le, T.-T., Le, M.V.: Development of user-friendly kernel-based Gaussian process regression model for prediction of load-bearing capacity of square concrete-filled steel tubular members. Mater. Struct. **54**(2), 1–24 (2021). https://doi.org/10.1617/s11527-021-01646-5

Demand Response Optimization in Micro-grid Operation with Participation of Renewable Energy Resource and Battery Energy Storage System

V. V. Thang(✉) and N. H. Trung

Faculty of Electrical Engineering, Thai Nguyen University of Technology, Thai Nguyen, Vietnam
{thangvvhtd,nguyenhientrung}@tnut.edu.vn

Abstract. The penetration of renewable energy sources (RS), energy storage, and demand response in energy systems arise challenges for energy management problems. Micro-grid with flexible regulation abilities provides an effective way to solve the problems of RS connected to main grids. This paper proposes a stochastic model in the optimization of demand response program (DRP) of micro-grids which allow simultaneously optimizes both the RS with different technologies and energy storage. The operation cost including costs for purchasing power from the utility grid, the operation cost of RS and energy storage along with emission cost and cost of DRP is minimized in each computed cycle. The uncertainty is modeled by probability density functions, divided into states by the clustering technique and then scenarios are generated by the scenario matrix and reduced in order to reduce the computational burden. The simulation results by the test system show that the proposed model is feasibility and the DRP has significant impacts on efficiency improvement of micro-grids such as reduction of energy cost and emission, and avoidance of connection devices investment.

Keywords: Demand response programs · Optimal scheduling · Micro-grid · Renewable sources and uncertainty

1 Introduction

Nowadays, the micro-grid is one of the important components in smart distribution systems. A micro-grid consists of a low voltage distribution grid, distributed generators with both renewable and conventional sources, and energy storage systems that supply electrical energy to the end-users [1]. Also, micro-grids have a point of common coupling for their connection and disconnection from the utility grid. The micro-grids offer several advantages and benefits such as increased reliability, improved energy efficiency, cost and losses reduction, and emission reduction [2]. However, the appropriate guaranty of power quality, frequency, and voltage for consumers as well as the control, protection, and uncertainty of renewable resources are issues that need to be addressed [3].

D. C. Nguyen et al. (Eds.): ICERA 2021, LNNS 366, pp. 293–301, 2022.
https://doi.org/10.1007/978-3-030-92574-1_31

Also, DRP is a service that involves consumers in the reduction of the electricity demand during peak consumption time based on changes in prices called price-based DRP or financial incentives called incentive-based DRP [3, 4]. Hence, DRP has served to reduce the peak load profile and for deferring investment in generation and distribution systems. Also, DRP can be used as an emergency resource and considered a means of mitigating the uncertainties and intermittencies of renewable resources and improving efficiency of distribution system as well as micro-grids [5].

The important role of the micro-grid in modern power systems has caused several researchers to be carried out in order to analyze different aspects in planning and operation of the micro-grids. In which, an energy management system is proposed to minimize the cost of micro-grids including the RS. The objective function can be the total operation cost [6] or the total annual equipment investment cost, operation cost, and replacement cost [7]. DRPs are considered in an energy resources scheduling method to optimize the hourly operation cost of micro-grids integrating the RS and BESS with various types of DRPs [8] or only a DRP in load reduction and load transfer time [9]. In addition, a multi-objective optimization solution is proposed for the micro-grid energy management problem with hybrid energy sources and battery storage system. Operational cost added emission cost and incentive cost of DRP is optimized for the system. DRPs can consist of load control in the time-based rate DRP and load control in the incentive-based DR programs [10] or price-based DRPs [11]. Moreover, a significant challenge for operating micro-grid due to the uncertainty of RS power, electricity prices, and the energy demand of consumers is also studied and presented in several researches. In [12], a stochastic optimization model is proposed to minimize the daily operation cost of micro-grid considering the random characteristics of only photovoltaic (PV) output and electricity price. Emission and DRP cost is integrated into the objective function and thus the results show the reduction of operation cost in the computed cycle. Similarly, the stochastic short-term scheduling of micro-grids is proposed which considers the price-based DRP and uncertainties of RS and loads. The results illustrate the effectiveness of DRPs in typical micro-grids due to the constructive participation of consumers in changing demand at different time intervals. However, ignoring the effect of emissions on environmental pollution [13] and only using normal probability distribution function (pdf) for different stochastic parameters [14] lead to a reduction in the operational effectiveness of micro-grids. Hence, this paper proposed a stochastic model in the optimization of DRP of micro-grids which allow simultaneously optimizes both the RS with different technologies and battery energy storage system (BESS). The operation cost including costs for purchasing power from the utility grid, the operation cost of RS and BESS along with emission cost and cost of DRP is minimized in each computed cycle.

The rest of the paper consists of Sect. 2 with the micro-grid model, scheduling framework is introduced in Sect. 3, Sect. 4 presents results and discussion, and the conclusive remarks and a few insights for future work is represented in Sect. 5.

2 Micro-grid Modelling

A typical structure of the grid-connected micro-grids as in Fig. 1 is selected in a case study. The load is simultaneously supplied from both RS with PV and wind turbine (WT) resources and the utility grid. Besides, both BESS and incentive-based DRP is considered to enhance the flexibility and effectiveness of micro-grids.

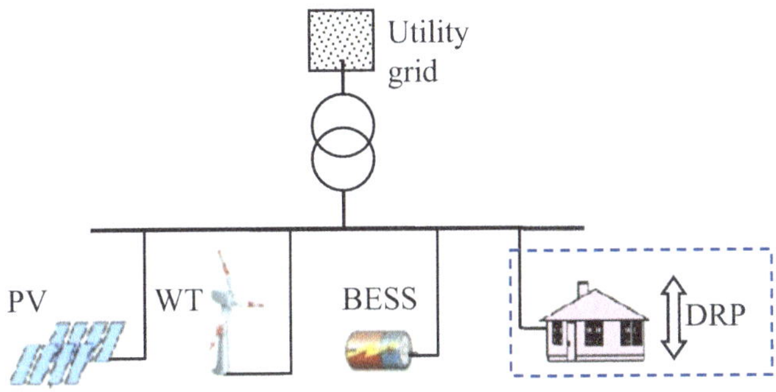

Fig. 1. The structure of micro-grid in case study

2.1 Renewable Energy Resources

In each scenario, the generated power of PV and WT is calculated as expressions (1) and (2), respectively [9, 14]. In which, P_r^{pv}, I_r are rated power and solar irradiance of PV, respectively. I_w, θ_w are operation irradiance and temperature while α_θ, θ_{am} are the power temperature coefficient and standard ambient temperature, respectively. P_w^{wt}, P_r^{wt} are output and rated power together with cut-in speed v_{ci}, rated speed v_{cr}, and cut-off speed v_{co}.

$$P_w^{pv}(I,\theta) = P_r^{pv}.\frac{I_w}{I_r}[1 + \alpha_\theta(\theta_w - \theta_{am})] \tag{1}$$

$$P_w^{wt}(v) = \begin{cases} 0 & v_w \leq v_{ci} \text{ or } v_{co} \leq v_w \\ P_r^{wt}.(v_w - v_{ci})/(v_r - v_{ci}) & v_{ci} \leq v_w \leq v_r \\ P_r^{wt} & v_r \leq v_w \leq v_{co} \end{cases} \tag{2}$$

2.2 BESS

The BESS can charge at hours with low electrical price and load, and then it discharges back to micro-grid at peak hours with high electrical price. The energy balance of BESS in the operation cycle is calculated as Eq. (3) [6, 10]. The charge/discharge of them is decided by two binary variables $I_{t,h}^{be+}$ and $I_{t,h}^{be-}$. The stored energy is limited by lower and upper bounds of state of charge (SoC) to ensure the lifetime of BESS denoted by factors $k_{soc}^{\min}, k_{soc}^{\max}$, respectively. The SoC displays the stored level with the stored energy level at final time $(E_{t,h}^{be})$ and initial time $(E_{t,0}^{be})$.

$$\begin{aligned}
&\sum_{h=1}^{H} \eta_{be}.I_{t,h}^{be-}.P_{t,h}^{be-} = \sum_{h=1}^{H} I_{t,h}^{be+}.P_{t,h}^{be+}; \; I_{t,h}^{be+} + I_{t,h}^{be-} = 1 \\
&E_{t,h}^{be} = E_{t,h-1}^{be} + I_{t,h}^{be-}.P_{t,h}^{be-}.\eta_{be} - I_{t,h}^{be+}.P_{t,h}^{be+} \\
&k_{soc}^{\min}.E_{t}^{be} \le E_{t,h}^{be} \le k_{soc}^{\max}.E_{t}^{be}; \; E_{t,H}^{be} = E_{t,0}^{be}; \; P_{h}^{be+}, P_{h}^{be-} \le P_{\max}^{be}
\end{aligned} \tag{3}$$

2.3 Electrical Price, Load and Demand Response Program

Two price mechanisms widely used in the electrical markets are the time-of-use price (TOU) and the real-time price (RTP). TOU is an economic means to encourage consumers to change the mode of electricity consumption and avoid peak electricity consumption. RTP changes once an hour and depends on the current demand and the economy and safety of the power grid operation [11, 13]. Besides, the load of micro-grids depends on the demand of consumers and often is displayed by the typical day load curve [8, 9].

DRPs can be divided into two types consisting of price-based DRP and incentive-based DRP [13]. In which, the incentive-based DRP through the shifting capability of electrical loads provides benefit of both consumers and operators along with the flexibility of the entire power system. Hence, the incentive-based mechanism should be applied with the incentive subsidy cost for the consumers participating DRPs to creates multiple benefits for stakeholders [3]. The shiftable demand of incentive-based DRP is constrained by Eq. (4) [9, 10]. In which, $k_{\max}^{dr}$ is maximum shiftable load factor. $I_{t,h}^{dr+}, I_{t,h}^{dr-}$ are binary variables deciding the increase or decrease of the power of consumers on DRP. The load curtailment must be shifted to other times and hence total decreased energy must equilibrate total increased energy in the calculated cycle.

$$\begin{aligned}
&0 \le P_{t,h}^{dr+} \le k_{\max}^{dr}.P_{t,h}^{d}.I_{t,h}^{dr+}; \; 0 \le P_{t,h}^{dr-} \le k_{\max}^{dr}.P_{t,h}^{d}.I_{t,h}^{dr-} \\
&\sum_{h=1}^{H} P_{t,h}^{dr+} = \sum_{h=1}^{H} P_{t,h}^{dr-}; \; I_{t,h}^{dr+} + I_{t,h}^{dr-} = 0
\end{aligned} \tag{4}$$

3 Scheduling Framework

3.1 Scenario Generation and Selection

The stochastic parameters of micro-grid are modeled by pdfs and then they are divided into different states with specific values and probabilities in each state by the k-mean clustering technique. In which, the randomness of solar radiation and wind speed is expressed by beta and weibull pdf, respectively [16]. For random variables of loads and electrical prices are modeled by normal pdf [13, 16].

These parameters are integrated into a hybrid model as Eq. (5) [13]. In which, the C_w and $\lambda_w\{C_w\}$ are the matrices that enumerate the possible values and probability of parameters. λ_s^x is the probability of parameter X at each state integrated into scenario w. N_k is the number of parameters X and N_s^x is the number of states of parameter X.

$$M = \{ C_w, \lambda_w\{ C_w\} \} ; \lambda_w\{ C_w\} = \prod_{k=1}^{N_k} \lambda_s^x\{X_s\} ; N_w = \prod_{k=1}^{N_k} N_s^x \tag{5}$$

GAMS/SCENRED program using the fastback forward method is employed to reduce the number of scenarios in order to reduce the computational burden [17]. Then, the system is analyzed under these scenarios as deterministic inputs, and various states are studied by using different scenarios.

3.2 Optimal Model

This section establishes an optimization model for operating grid-connected micro-grids. The objective function is total operation costs in the computed cycle consisting of the purchased energy cost from the utility grid, operation and maintenance cost of RS, incentive cost of DRP, and emission cost as shown in Eq. (6).

$$\begin{aligned} OF = & \sum_{h=1}^{N_h} \sum_{w=1}^{N_w} \lambda_w \left(\pi_{h,w}^{e}.P_{h,w}^{ug} + \sum_{k=1}^{K} C_{om,k}^{rs}.P_{k,h,w}^{rs} \right) + \sum_{h=1}^{N_h} \pi^{dr} (P_h^{dr+} + P_h^{dr-}) \\ & + \sum_{h=1}^{N_h} \sum_{w=1}^{N_w} \pi^{emi}.\lambda_w \left(P_{h,w}^{ug}.\xi_{ug} + \sum_{k=1}^{K} P_{k,h,w}^{rs}.\xi_{rs} \right) \end{aligned} \tag{6}$$

Where, $\pi_{h,w}^{e}$ is electrical prices purchased from the market and and λ_w is probability at scenario w. $P_{h,w}^{ug}$ is the received power from the utility grid while $P_{k,h,w}^{rs}$ is the generated power of RS. $C_{om,k}^{rs}$ is operation and maintenance cost coefficient of RS. P_h^{dr-}, P_h^{dr+} are the decreased and increased power quantities of DRPs at any available time while π^{dr} is the incentive cost of DRPs for the increased or decreased power quantities. π^{emi} is emission taxes. ξ_{rs}, ξ_{ug} are the emission coefficients of RS and traditional sources, respectively [10, 11].

The constraints power balance in each operation scenario of the micro-grid is represented in Eq. (7) including the power of the utility grid, RS, BESS, and DRPs. P_h^d, k_w^d are load profile at hour h and load factor in scenarios w, respectively. $P_{k,h}^{rs}, k_{k,w}^{rs}$ are generated power at hour h and generated power factor in scenarios w of RS, respectively. $P_{\max}^{ug}$ is the maximum power of connected equipment to the utility grid. Besides, constraints of BESS operation and DRP selection are represented in Eqs. (3) and (4), respectively.

$$P^{ug}_{h,w} + \sum_{k=1}^{K} P^{rs}_{k,h,w} = P^{d}_{h,w} - \eta_{\mathrm{be}}.I^{be-}_{\mathrm{h}}.P^{be-}_{h} + I^{be+}_{\mathrm{h}}.P^{be+}_{h} - P^{dr-}_{h} + P^{dr+}_{h}$$
$$P^{d}_{h,w} = k^{d}_{w}.P^{d}_{h};\ P^{rs}_{k,h,w} = k^{rs}_{k,w}.P^{rs}_{k,h};\quad 0 \le P^{ug}_{h,w} \le P^{ug}_{max} \qquad (7)$$

As seen above, the operation framework is a mixed-integer problem thus the GAMS/CPLEX solver is utilized to find out the optimal solution [17].

4 Results and Discussion

Input data of the model includes the mean output power profiles of RS with each technology, and mean load and energy price profiles. The mean output power is displayed by the typical day load curve as in Fig. 2 while the mean load and energy price are represented by the typical day curve in Fig. 3. In each hour, scenarios are generated by the scenario matrix presented in Sect. 3 then 10 scenarios are selected by GAMS/SCENRED.

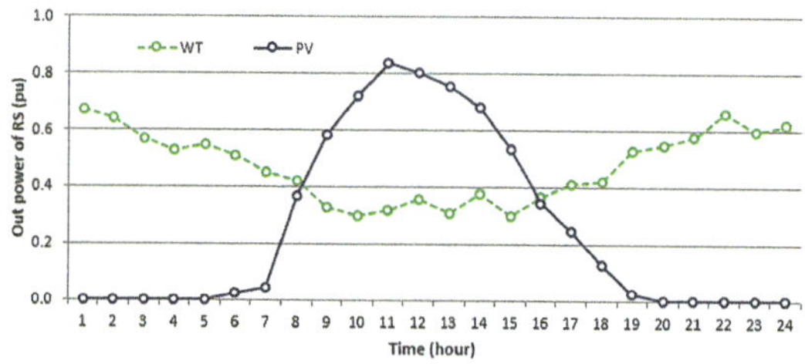

Fig. 2. Mean output power profiles of RS for a typical day

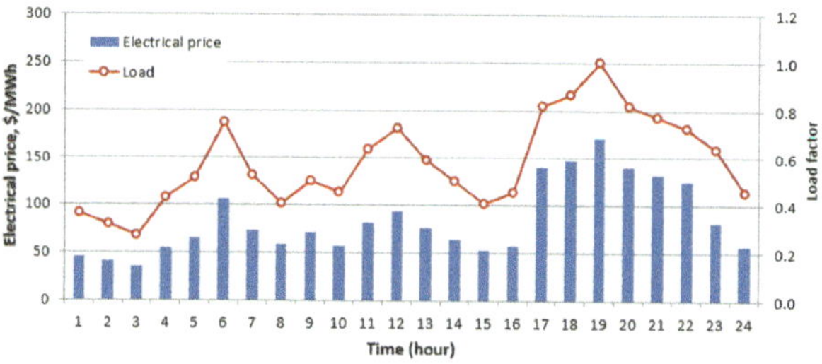

Fig. 3. Mean electrical price and load for a typical day

In addition, the input data of PV, WT, BESS and utility grid is shown in Table 1 [10, 11]. The efficiency of BESS, η_{be}, is about 0.9. The computed cycle is a day corresponding to 24 h. The emission tax of CO_2 is 10\$/ton. It is assumed that 30% of total load demand would participate in DRP with incentive cost for the increased or

Table 1. Input data

No	Equipment	P_r, $P_{max}(kW)$	$E_r(kWh)$	$C_{om}(\$/kW)$	$SoC(\%)$	$\xi(kg/kWh)$
1	PV	300		0.025		0.25
2	WT	200		0.035		0.27
3	BESS	100	500		20–100	
4	Utility grid	500				0.65

decreased power quantities is 12.5\$/MWh [10]. The mean active power of load is assumed about 1000 kW.

4.1 Calculation Result

In this section, two cases with and without DRPs are investigated to analyze the effectiveness of incentive-based DRP on the optimal operation of micro-grids. Optimal operation scheduling of micro-grid in cases is selected and thus economic and technical indicators are also determined as shown in Table 2.

Table 2. Comparison of economic and technical indicators

Economic and technical indicators	Without DRP	With DRP	Comparison
Total operation cost, \$	697.05	663.59	−33.46
Cost of DRPs, \$	0	18.87	18.87
Cost for purchasing electricity, \$	604.02	551.71	−52.31
Operation and maintenance cost of RS, \$	37.62	37.62	0
Emission taxes cost, \$	55.41	55.39	−0.02
Purchased electricity from utility grid, kWh	8021.03	8018.10	−2.97
CO_2 emission, tons	5.55	5.54	−0.01

Comparison between the case with DRP and case without DRP shows that the expected operation cost of micro-grid reduces 33.46\$ corresponding to 4.8% due to the shifted electrical power from peak hours to low hours by the DRPs. The invective cost of DRPs increases 18.87\$ but the total cost for purchasing electricity, operating RS and emission taxes reduce by 52.33\$ lead to the reduction operation cost of the micro-grid. Besides, the participation of consumers in the case with DRPs reduces the peak valley difference of the load and stabilizes the load curve. The charge/discharge power of BEES also reduces, and thus it decreases the purchased electricity amount from the utility grid along with emission. In addition, the total cost of consumers consisting of the reduced energy cost and incentive cost decreases 71.2\$ corresponding with 11.8%.

Optimal operation scheduling of BESS and DRP is selected and the operation power of DRP in typical day is represented in Fig. 4. Consumers should operate the increased power of DPR in low hours as 1st–4th, 10th, 15th, 16th, and 24th h and the decreased power of DPR in peak hours as 6th, 17th–22nd h. The maximum transfer power is 157 kW at the 19th h while the maximum increased demand is 129 kW at the 4th to reduce energy cost and guarantee constraints of DRP. Besides, the power curve received from the utility grid of the 10th scenario which is displayed in Fig. 5 shows the reduction of the peak valley difference. The peak valley difference reduces from 378 kW in the case without DRP to 210 kW in the case with DRP. The maximum power decreases from 688 kW to 562 kW corresponding with 18.31% lead to a reduction in the invested cost of connection equipment to the utility grid.

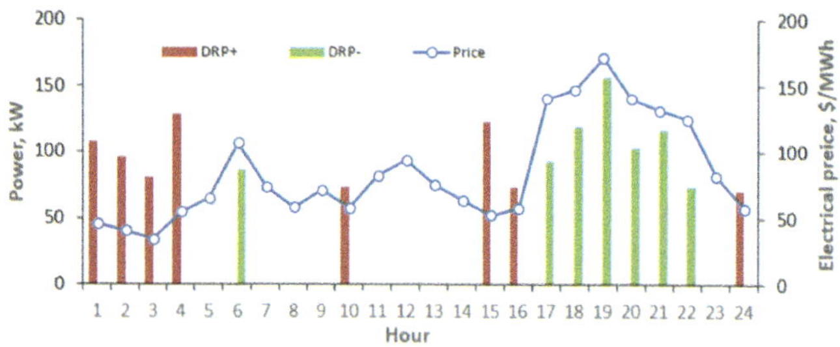

Fig. 4. Power of DRP for a typical day

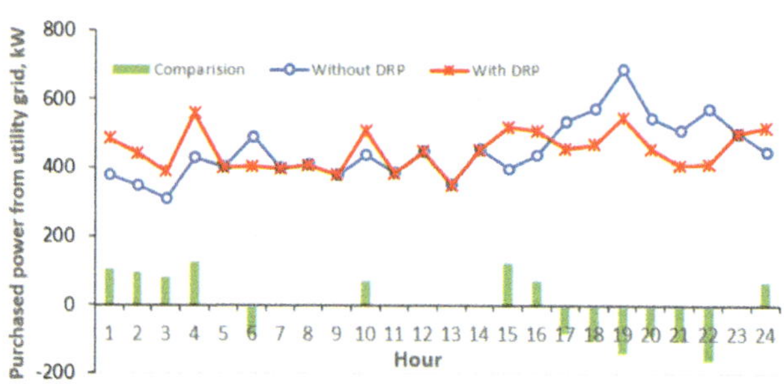

Fig. 5. Load in 10th scenario

5 Conclusion

In this paper, a stochastic model in the optimization of DRP of micro-grids is formulated which allow simultaneously optimizes both the RS with different technologies and BESS. The operation cost including costs for purchasing power from the utility grid, the operation cost of RS and BESS along with emission cost and cost of DRP is minimized in each computed cycle. From the simulation results, it is observed that the proposed model is feasibility and the DRP has a significant impact on the operation cost of micro-grids. After implementation of incentive-based DRP, operation cost, energy cost, and emission reduced by 4.8%, 8.66%, and 0.18%, respectively. In addition, the reduction of peak valley difference and the maximum power received from the utility grid reduce the invested cost of connection equipment of micro-grid.

Acknowledgments. This work is supported by the Thai Nguyen University of Technology (TNUT).

References

1. García Vera, Y.E., Dufo-López, R., Bernal-Agustín, J.L.: Energy management in microgrids with renewable energy sources: a literature review. Appl. Sci. **9**, 1–28 (2019)
2. Shahgholian, G.: A brief review on microgrids: operation, applications, modeling, and control. Int. Trans. Electr. Energy Syst. **31**, 1–28 (2021)
3. Mina-Casaran, J.D., Echeverry, D.F., Lozano, C.A.: Demand response integration in microgrid planning as a strategy for energy transition in power systems. IET Renew. Power Gener. **15**, 889–902 (2021)
4. Imani, M.H, Ghadi, M.J, Ghavidel, S., Li, L.: Demand response modeling in microgrid operation: a review and application for incentive-based and time-based programs. Renew. Sustain. Energy Rev. **94,** 486–499 (2018)
5. Hussain, M., Gao, Y.: A review of demand response in an efficient smart grid environment. Electr. J. **31**, 55–63 (2018)
6. Conteh, P., Tobaru, S., Howlader, H.O.R., Yona, A., Senjyu, T.: Energy management systems for hybrid distributed generation sources in grid connected and stand-alone micro-grids. J. Renew. Sustain. Energy **9**, 1–22 (2017)
7. Yang, L., Huang, K., Kong, X., Luo, Q., Guo, H.: Size optimization of a stand-alone micro-grid integrated with flexible load and the complementary of solar and wind. J. Renew. Sustain. Energy **11**, 1–12 (2019)
8. Zakariazadeh, A., Jadid, S.: Smart microgrid operational planning considering multiple demand response programs. J. Renew. Sustain. Energy **6**, 1–19 (2014)
9. Wang, Y., et al.: Energy management of smart micro-grid with response loads and distributed generation considering demand response. J. Clean. Prod. **197**, 1069–1083 (2018)
10. Murty, V.V.S.N., Kumar, A.: Multi-objective energy management in microgrids with hybrid energy sources and battery energy storage systems. Prot. Control Mod. Power Syst. **5**, 1–20 (2020)
11. Wang, Y., Huang, Y., Wang, Y., Li, F., Zhang, Y., Tian, C.: Operation optimization in a smart micro-grid in the presence of distributed generation and demand response. Sustainability **10**, 1–25 (2018)
12. Zhao, H., Lu, H., Li, B., Wang, X., Zhang, S., Wang, Y.: Stochastic optimization of microgrid participating day-ahead market operation strategy with consideration of energy storage system and demand response. Energies **13**, 1–16 (2020)
13. Nikmehr, M., Najafi-Ravadanegh, S., Khodaei, A.: Probabilistic optimal scheduling of networked microgrids considering time-based demand response programs under uncertainty. Appl. Energy **198**, 267–279 (2017)
14. Farham, H., Alipour, H.: Effects of demand response program and energy storage system on optimal stochastic short-term generation scheduling of grid-connected microgrid. In: Conference on Electrical Power Distribution Networks Conference, pp.80–88 (2017)
15. Rajamand, S.: Effect of demand response program of loads in cost optimization of microgrid considering uncertain parameters in PV/WT, market price and load demand. Energy **194**, 1–16 (2020)
16. Rakipour, D., Barati, H.: Probabilistic optimization in operation of energy hub with participation of renewable energy resources and demand response. Energy **173**, 384–399 (2019)
17. GAMS/SCENRED User Guide (2009). http://www.gams.com/

Determination of Optimum Gear Ratios of Two-Stage Bevel Helical Gearboxes for Getting Minimum Gearbox Volume

Bui Thanh Danh[1], Tran Thi Phuong Thao[2], Tran Ngoc Huy Thinh[3], Nguyen Huu Quang[4], Le Hoang Anh[5], Trinh Kieu Tuan[4], Tran Ngoc Giang[2], and Nguyen Thanh Tu[2](✉)

[1] University of Transport and Communications, Hanoi, Vietnam
[2] Thai Nguyen University of Technology, Thai Nguyen, Vietnam
nguyenthanhtucnvl@tnut.edu.vn
[3] Nguyen Tat Thanh University, Ho Chi Minh City, Vietnam
[4] University of Economics - Technology for Industries, Hanoi, Vietnam
[5] Vinh Long University of Technology Education, Vĩnh Long, Vietnam

Abstract. The gear ratio is a main important factor that affects the size, the mass, the volume, and the cost of a gearbox. Therefore, in calculating and designing gearboxes, the determination of optimum gear ratios is of great interest. This paper introduces a study on determining optimum gear ratios of a two-stage bevel helical gearbox. To find the optimal gear ratios, a problem of optimizing the gearbox for getting the minimum gearbox volume was conducted. In this work, the influence of the input parameters, including the total gearbox ratio, the coefficient of the face width of the bevel and the helical gear sets, the allowable contact stress and the output torque and their interactions on the output, is evaluated. A simulation experiment was designed and conducted with the help of the Minitab R19 software. From the results of the work, regression models to find the optimum gear ratios were proposed.

Keywords: Bevel helical gearbox · Gear ratio · Optimum gearbox design · Volume optimization

1 Introduction

In the optimum design of the reducers, optimizing the partial gear ratios are very important. The reason is that the size, the volume, the mass, and the cost of a gearbox depend largely on the partial gear ratios [1]. Therefore, optimum calculation of partial gear ratio has been mentioned in many researches. For bevel helical gearboxes, so far, there have been several studies on that. V.N. Kudreavtev et al. [1] and Trinh Chat [2] presented a graph method to determine the gear ratios of a two-stage bevel helical gearbox. Besides, Milou et al. [3] presented practical method. In their method, the optimum gear ratios can be found in tabulated form which was build based on practical data. Another method for finding optimum gear ratios is the model method [4–15]. This method is the most useful method for determining optimum gear ratios. In this method, the optimum gear ratios were determined with different objectives, such as the

D. C. Nguyen et al. (Eds.): ICERA 2021, LNNS 366, pp. 302–311, 2022.
https://doi.org/10.1007/978-3-030-92574-1_32

minimum mass of gear [5, 7, 15], the minimum area of gearbox cross section [10, 11, 14], the minimum gearbox length [1], the minimum total mass of the gearbox or the gears [9, 12], or the minimum gearbox cost [16].

Form above analysis, although there have been many studies on the determination of optimum gear ratios for different gearboxes, there is still lack of a work on finding optimum gear ratios for getting minimum gearbox volume for two-stage bevel helical gearbox. That is the reason this study was carried out on this problem. In this study, the effect of the input parameters and and their interactions on the optimum gear ratios was evaluated. In addition, a regression model for calculation of optimum gear ratio u_1 was proposed.

2 Methodology

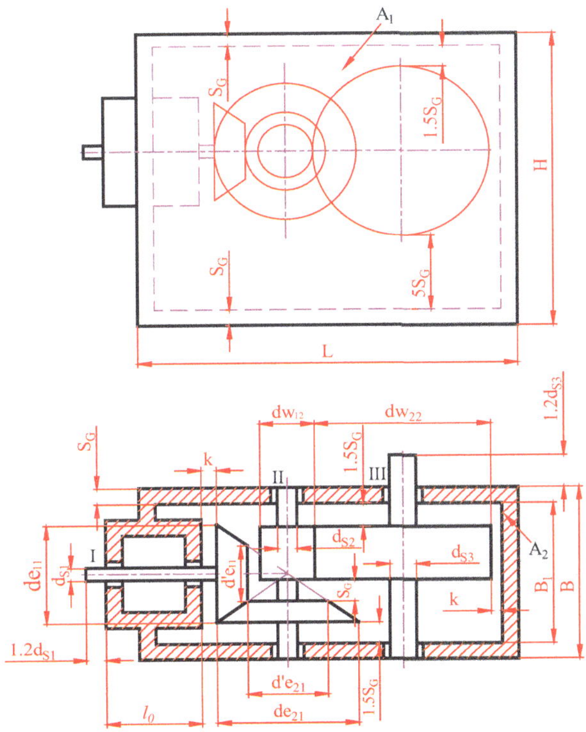

Fig. 1. Schema for determination of gearbox mass

The volume of the gearbox housing, V_{gh} (m^3), can be calculated by the following Equation (Fig. 1):

$$V_{gb} = L \cdot B \cdot H \tag{1}$$

Where, L, H, B and

$$L = 2 \cdot l_0/3 + d_{e21}/2 + d_{w12}/2 + d_{w22} + 2 \cdot k \tag{2}$$

$$H = \max(d_{e21}; d_{w22}) + 8.5 \cdot S_G \tag{3}$$

$$B = b \cdot cos\delta_2 + b_w + 4 \cdot S_G \tag{4}$$

In the above Equations, $k = 8 \div 12$ (mm) [1] and we can choose $k = 10$ (mm); S_G is estimated by:

$$S_G = 0.005 \cdot L + 4.5 \ [13] \tag{5}$$

In Eq. (2), δ_2 is gear pitch cone angle; $tang\delta_2 = u_1 \cdot l_0 = 3 \cdot d_{s1}$ is the diameter of the first shaft which can be determined by [1]:

$$d_{s1} = [T_{11}/(0,2 \cdot [\tau])]^{1/3} \tag{6}$$

In Eq. (2), d_{e21} is the outer pitch diameter of the bevel gear (mm). For a straight bevel gear set, we have [1]:

$$d_{e21} = 2 \cdot u_1 \cdot R_e/\left(1 + u_1^2\right)^{1/2} \tag{7}$$

Where, R_e is the cone distance (mm) wich is determined by [1]:

$$R_e = k_R \cdot \sqrt{(u_1^2 + 1)} \cdot \sqrt[3]{T_{11} \cdot k_{h\beta 1}/[(1 - k_b e) \cdot k_b e \cdot u_1 \cdot [\sigma_H]^2]} \tag{8}$$

In which, k_R is face width coefficient; for a steel straight bevel gear, $k_R = 50$ (MPa) [1]; $k_{H\beta}$ is the contacting load ratio for pitting resistance; for a straight bevel gear set $k_{H\beta} = 1.04 \div 1.18$ [1] and we can chose $k_{H\beta} = 1{\cdot}11$; $k_{be} = b/R_e = 0.25 \div 0.3$ is the face width coefficient and we can chose $k_{be} = 0.27$; T_{11} is the torque on the pinion; d_{e11} and d_{e21} are the outer pitch diameters of the pinion and the gear; For straight bevel gear set, we have:

$$T_{11} = T_{out}/\left(u_t \cdot \eta_{bg} \cdot \eta_{hg} \cdot \eta_{be}^3\right) \tag{9}$$

Wherein, u_t is the total gearbox ratio; η_{bg} is the efficiency of bevel gear set $\eta_{hg} = 0.95 \div 0.97$ [1]; η_{hg} is the efficiency of helical gear set $\eta_{hg} = 0.96 \div 0.98$ [1]; η_{be} is the efficiency of a rolling bearing pair $\eta_{be} = 0.99 \div 9.995$ [1]. Choosing $\eta_{bg} = 0.96$; $\eta_{hg} = 0.97$, $\eta_{be} = 0.992$ we have:

$$T_{11} = 1.1001 \cdot T_{out}/u_t \tag{10}$$

In Eq. (2), d_{w11} and d_{w21} are the pitch diameters of the pinion and the gear of the helical gear set. These diameters can be determined by [1]:

$$d_{w11} = 2 \cdot a_{w1}/(u_1 + 1) \tag{11}$$

$$d_{w21} = 2 \cdot a_{w1} \cdot u_i \cdot /(u_i + 1) \tag{12}$$

Where, a_w is the centre distance of the helical gear set which is found by [1]:

$$a_w = k_a \cdot (u_2 + 1) \cdot \sqrt[3]{T_{12} \cdot k_{H\beta}/([\sigma_{H2}]^2 \cdot u_2 \cdot X_{ba})} \tag{13}$$

With

$$T_{12} = T_{out}/\left(u_2 \cdot \eta_{hg} \cdot \eta_{be}^2\right) \tag{14}$$

With $\eta_{hg} = 0.97$, $\eta_{be} = 0.992$ as mentioned above, we have:

$$T_{12} = 1.0476 \cdot T_{out}/u_2 \tag{15}$$

As the above procedure, the gearbox volume depending on the transmission ratios of the two stages i.e., u_1, and u_2 is considered as the objective function:

$$\text{minimize} \quad V_{gb} \tag{16}$$

This function should be satisfied by the critical constraints:

$$\begin{array}{c} 1 \leq u_1 \leq 9 \\ 1 \leq u_2 \leq 9 \end{array} \tag{17}$$

It is noticed that$u_t = u_1 \cdot u_2$.

In addition, it is known that $u_t = u_1 \cdot u_2$ is the relation between total gearbox ratio and partial ratios. Hence the optimization of u_1 is sufficient, while the optimum ratio of u_2 can be obtained by the expression: $u_2 = u_t/u_1$.

3 Simulation Experiment

To evaluate the influence of the input parameters on the optimum gear ratio u_1, a simulation experiment is designed and performed with 6 input parameters. The input factors and the levels are shown in Table 1. There is 2^6 = 64 number of experimental runs for the experiment. Also, the Minitab 19 software was used to build the experimental plans and analyze the results. The experimental plans and the output response (the optimum gear ratio of the first stage u_1) are described in Table 2.

Table 1. Input parameters and survey level

No.	Parameters	Symbol	Unit	Level	
				Low	High
1	Total gearbox ratio	u_g	–	5	23
2	Coefficient of the face width of bevel gear set	X_{be}	–	0.25	0.3
3	Coefficient of wheel face width of stage 2	X_{ba}	–	0.3	0.4
4	Allowable contact stress of stage 1	AS_1	MPa	360	420
5	Allowable contact stress of stage 2	AS_2	MPa	360	420
6	Output torque	T_{out}	Nm	1000	10000

Table 2. Experimental plans and output response u_1

Std Order	Run Order	CenterPt	Blocks	ug	Xba1	Xba2	AS1	AS2	Tout	u1
1	47	1	1	5	0.25	0.3	360	360	1000	2.11
2	9	1	1	23	0.25	0.3	360	360	1000	5.89
4	5	1	1	23	0.3	0.3	360	360	1000	5.89
5	7	1	1	5	0.25	0.4	360	360	1000	1.93
6	8	1	1	23	0.25	0.4	360	360	1000	5.35
…										
60	44	1	1	23	0.3	0.3	420	420	10000	5.89
61	31	1	1	5	0.25	0.4	420	420	10000	1.93
64	1	1	1	23	0.3	0.4	420	420	10000	5.35

4 Results and Analysis

In Fig. 2, the influence of the main design parameters on the optimum gear ratio u_1 is clearly understood. It was found from this graph that when ug and AS_1 increases, u_1 increases. Conversely, when K_{be}, X_{ba}, and AS_2 increase, u_1 decreases. Besides, T_{out} has a negligible effect on u_1.

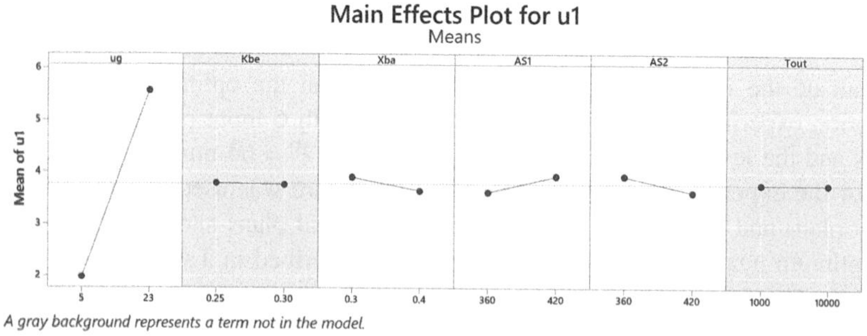

Fig. 2. Main effects plot for the response of u_1

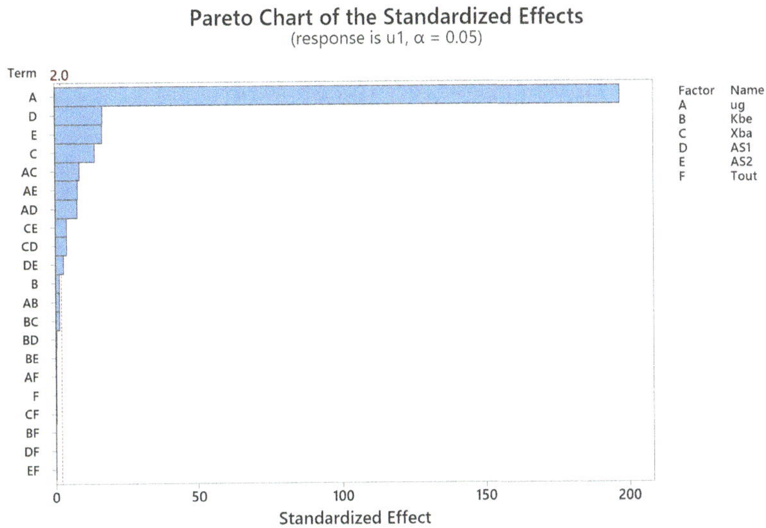

Fig. 3. Pareto chart of the standardized effects on u_1

Figure 3 presents the influence of the input parameters and their interactions on the optimum gear ratio u_1. In the graph, the effect of these parameters is expressed through the length of blue columns. Also, the parameters whose magnitude exceeds the red

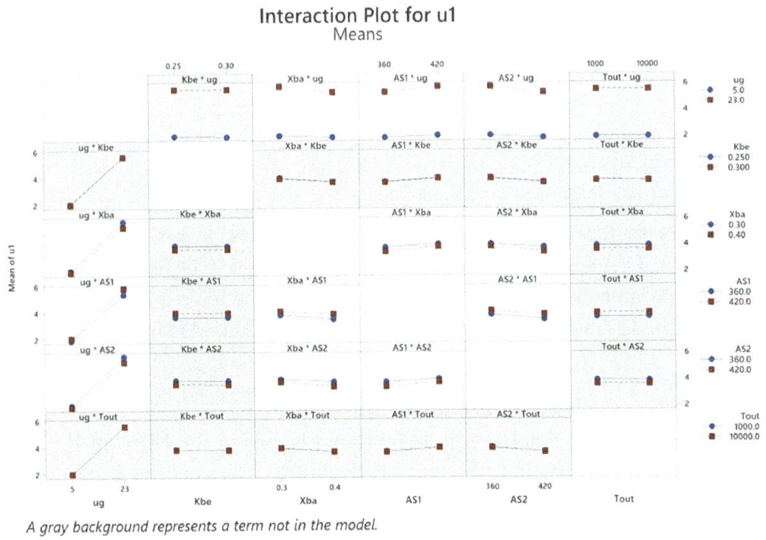

Fig. 4. The interactions between input parameters on the response of u_1

reference line are those that have a significant influence on u_l where the significant level is $\alpha = 0.05$. In this case, the impact of A (u_g) is the largest effect on u_l. Next are parameters C(X_{ba}), D(AS_1), E(AS_2), and interactions AC($u_g * X_{ba}$), AD($u_g * AS_1$), AE ($u_g * AS_2$), CD($X_{ba} * AS_1$), CE($X_{ba} * AS_2$), DE($AS_1 * AS_2$). Besides, the effects of the interactions are clearly shown in Fig. 4.

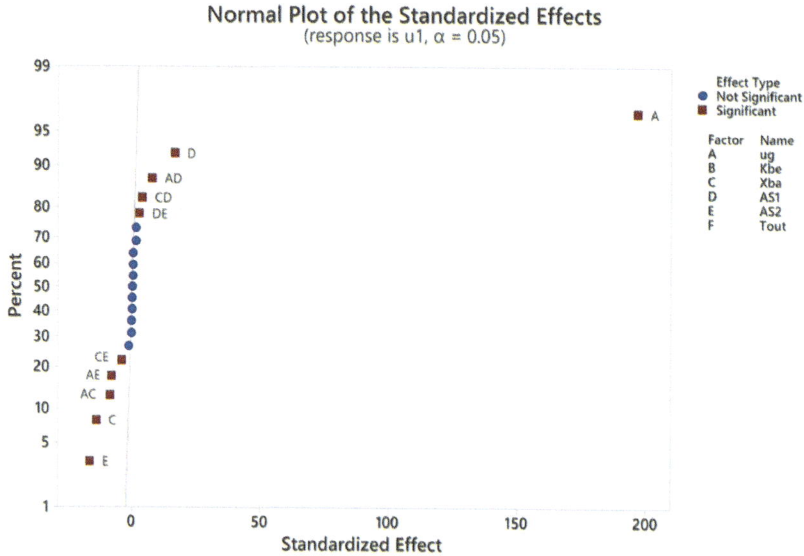

Fig. 5. The evolution of response as a function of input parameters and their interactions u_l

To better determine the trend and influence of the input parameters and interactions, we rely on the normal distribution chart in Fig. 5. From this graph, the parameters marked in red are the ones that have a significant influence on u_l. Also, parameters located further away from the standard curve have a greater influence on u_l. In this case, parameters to the right of the standard curve have a positive influence on u_l including A (u_g), D(AS_1), and interactions AD($u_g * AS_1$), CD($X_{ba} * AS_1$), DE($AS_1 * AS_2$). Besides, parameters to the left of the standard curve have a negative effect on u_l. They are C(X_{ba}), E(AS_2), and interactions AC($u_g * X_{ba}$), AE($u_g * AS_2$), CE($X_{ba} * AS_2$).

Determine the Regression Equation to Calculate u_l

A regression equation for determining u_l is found by using the Minitab R19 software. In the equation, parameters that have no or very little influence on u_l were removed. The estimated coefficients of the input parameters and the interactions between them are described in Table 3.

Table 3. Estimated effects and coefficients for u_1

Term	Effect	Coef	SE Coef	T-Value	P-Value	VIF
Constant		3.77734	0.00870	434.18	0.000	
u_g	3.58531	1.79266	0.00870	206.06	0.000	1.00
X_{ba}	−0.25469	−0.12734	0.00870	−14.64	0.000	1.00
AS_1	0.30344	0.15172	0.00870	17.44	0.000	1.00
AS_2	−0.30156	−0.15078	0.00870	−17.33	0.000	1.00
$u_g * X_{ba}$	−0.15531	−0.07766	0.00870	−8.93	0.000	1.00
$u_g * AS_1$	0.14156	0.07078	0.00870	8.14	0.000	1.00
$u_g * AS_2$	−0.14344	−0.07172	0.00870	−8.24	0.000	1.00
$X_{ba} * AS_1$	0.07156	0.03578	0.00870	4.11	0.000	1.00
$X_{ba} * AS_2$	−0.07344	−0.03672	0.00870	−4.22	0.000	1.00
$AS_1 * AS_2$	0.05219	0.02609	0.00870	3.00	0.004	1.00

Model Summary

S	R-sq	R-sq(adj)	R-sq(pred)
0.0695990	99.88%	99.86%	99.82%

The obtained regression equation can be described as follows:

$$\begin{aligned} u_1 = {} & 5.33 + 0.2609u_g + 0.11X_{ba} - 0.01827AS_1 - 0.00405AS_2 \\ & - 0.1726u_g * X_{ba} + 0.000262u_g * AS_1 - 0.000266u_g * AS_2 \\ & + 0.02385X_{ba} * AS_1 - 0.02448X_{ba} * AS_2 + 0.000029AS_1 * AS_2 \end{aligned} \tag{18}$$

It was noted that all of R-Square and R-Square (adj) (Table 3) are more than 99%. That means the proposed equation is fit very well with the experimental data.

Figure 6 presents the probability distribution of errors (the blue dots) compared to the normal distribution (the solid line). From the graph, most of these errors are distributed very close to the normal distribution. In addition, the errors near zero account for the majority (−0.06 to 0.06). Observing the relationship between the error and the corresponding value of the regression model (versus fit) shows that the points are randomly distributed, showing that u_1 is not significantly affected other than the input parameters which are surveyed.

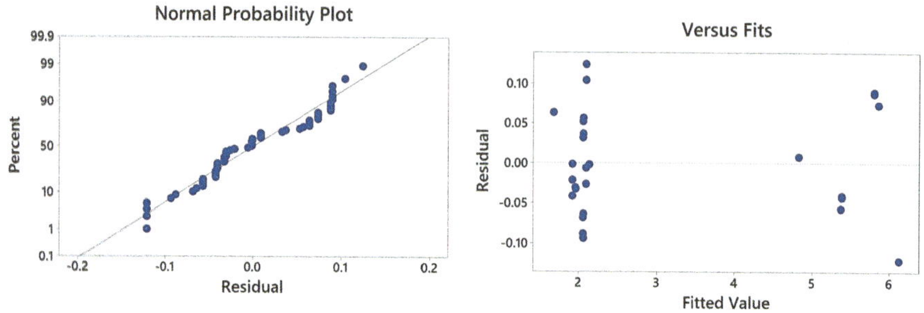

Fig. 6. Graphs estimating errors between experiments and model of u_1

5 Conclusions

In this paper, the results of an optimization study on determining optimum gear ratios of a two-stage cylindrical bevel reducer to get the minimum gearbox volume were presented. In the study, the influence of the input parameters and their interactions on the output response (u_1) was investigated. It was reported that u_g, K_{be}, X_{ba}, AS_1, AS_2 and their interactions are most effected factors on u_1 while the effect of T_{out} can be ignored. In addition, the optimal gear ratio of the reducer is easily determined through the model in the form of a clear function. With the calculation results of R_2 is all greater than 99%, this model has been evaluated as a very good fit to the data. The optimum gear ratios can be determined simply by using proposed explicit models.

Acknowledgment. This work was supported by Thai Nguyen University of Technology.

References

1. Kudreavtev, V.N., Gierzaves, I.A., Glukharev, E.G.: Design and Calculus of Gearboxes. Mashinostroenie Publishing, Sankt Petersburg (1971). (in Russian)
2. Trinh Chat, L.V.U.: Calculation of Mechanical Driven Systems. Education Publisher (1996)
3. Milou, G., Dobre, G., Visa, F., Vitila, H.: Optimal design of two step gear units, regarding the main parameters. VDI Berichte (1230), 227 (1996)
4. Pi, V.N.: A method for optimal calculation of total transmission ratio of two step helical gearboxes. In: Proceedings of the National Conference on Engineering Mechanics, Ha Noi, p. 12 (2001)
5. Hong, T.T., et al.: A study on calculation of optimum gear ratios for three stage bevel helical gearboxes. ARPN J. Eng. Appl. Sci. I(14) (2019)
6. Petrovski, A.N., Sapiro, B.A., Saphonova, N.K.: About optimal problem for multi-step gearboxes. Vestnik Mashinostroenie (10), 13 (1987). (in Russian)
7. Pi, V.N., et al.: Calculating optimum gear ratios of a two-stage helical reducer with first stage double gear sets. In: IOP Conference Series: Materials Science and Engineering (2019). IOP Publishing

8. Pi, V.N., Tuan, N.K.: Optimum determination of partial transmission ratios of three-step helical gearboxes for getting minimum cross section. J. Environ. Sci. Eng. A **5**, 570–573 (2016)
9. Pi, V.N., Binh, N.D., Dac, V.Q., The, P.Q.: Optimal calculation of total transmission ratio of three-step helical gearboxes for minimum mass of gears. J. Sci. Technol. 6 Eng. Univ., 91 (2006). (in Vietnamese)
10. Pi, V.N.: A new method for optimal calculation of total transmission ratio of two step helical gearboxes. In: The Nation Conference on Engineering Mechanics, pp. 179–188 (2001)
11. Pi, V.N.: A new study on optimal calculation of partial transmission ratios of three-step helical reducers for getting minimal cross section dimension. In: The 2nd WSEAS International Conference on Computer Engineering and Applications (CEA 2008), Acapulco, Mexico 25–27 January, pp. 290–293 (2008)
12. Pi, V.N.: Optimal determination of partial transmission ratios of three-step helical gearboxes with first and third-step double gear-sets for minimal mass of gear. In: American Conference on Applied Mathematics (MATH 2008), Harvard, Massachusetts, USA 24–26 March, pp. 385–388 (2008)
13. Romhild, I., Linke, H.: Gezielte Auslegung Von Zahnradgetrieben mit minimaler Masse auf der Basis neuer Berechnungsverfahren. Konstruktion **44**, 229–236 (1992)
14. Pi, V.N.: A study on optimal determination of partial transmission ratios of helical gearboxes with second-step double gear-sets. In: Proceedings of World Academy of Science, International Journal of Mechanical and Mechatronics Engineering, vol. 2, no. 1, pp. 26–29 (2008)
15. Pi, V.N.: Optimal determination of partial transmission ratios for four-step helical gearboxes with first and third step double gear-sets for minimal mass of gears. In: Applied Computing Conference (ACC 2008), Istanbul, Turkey, 27–30 May 2008
16. Vu, N.-P., et al.: The influence of main design parameters on the overall cost of a gearbox. Appl. Sci. **10**(7) (2020)

Determination of Optimum Partial Gear Ratios for Three-Stage Bevel Helical Gearboxes for Cost Function

Nguyen Huu Quang[1], Tran Thi Phuong Thao[2], Tran Quoc Hoang[3], Trinh Kieu Tuan[1], Nguyen Dinh Ngoc[2], Nguyen Thanh Tu[2], Tran Ngoc Giang[2], and Luu Anh Tung[2](✉)

[1] University of Economics - Technology for Industries, Hanoi, Vietnam
[2] Thai Nguyen University of Technology, Thai Nguyen, Vietnam
luuanhtung@tnut.edu.vn
[3] Nguyen Tat Thanh University, Ho Chi Minh City, Vietnam

Abstract. This study aims to present the optimization process of partial gear ratios for three-stage bevel helical gearboxes based on the minimizing cost of the gearbox. Eleven factors considered as the input parameters are selected to investigate their influences on the responses. A simulation experiment is designed and carried out using a computer program. Regression models are proposed to predict the partial gear ratios of u_2 and u_3. In addition to input parameters, their interactions have important effects in which total ratio gearbox ratio (u_g) have the most impact on both u_2 and u_3. Besides, proposed models of two responses are highly consistent with experimental results. The proposed regression equations are based on optimization cost problems.

Keywords: Bevel helical gearbox · Gearbox cost · Gear ratio · Cost optimization

1 Introduction

Gearboxes are important devices in mechanical driven systems. The companies want to increase the selling products it should improve the product quality. Moreover, reducing the costs of gearboxes should be a concerned. It is possible to minimize the gearbox cost if the optimization techniques can be efficiently applied. It is known that the dimension, and the mass of the gearbox is significantly affected by the gearbox ratio. Hence, the cost optimization is satisfied when the gear ratio is optimized.

Optimizing the gear ratio has been profoundly studied by research communities so far [1–14]. There have been three main methods utilized to optimize the gear ratio optimization. These methods are graph method [15, 16], practical method [17], and model method [3, 5, 8, 10, 13, 18–21]. According to these methods, the researchers have investigated the gearboxes with various levels of gear stages such as two-stage gearboxes [2, 4, 5, 8, 10, 12, 14, 20–22], three-stage gearbox [3, 7, 9, 23], and four-stage gearboxes [24, 25]. The gear ratios for the mechanical driven system using a two-step helical reducer optimize by Nguyen K.T et al. [9]. In the mentioned study, the

D. C. Nguyen et al. (Eds.): ICERA 2021, LNNS 366, pp. 312–327, 2022.
https://doi.org/10.1007/978-3-030-92574-1_33

system cross-section area was selected as the objective function of the optimization problem. The regression model was proposed to calculate the optimum ratios. Vu N. P. et al. [17, 19] determined the optimum gear ratios of a three-stage bevel helical gearbox in which the objective functions were chosen as the length of the gearbox. The input parameters, in this case, were the total gearbox ratio, the face width coefficients of the bevel, etc. The influences of input parameters such as on the objective function were selected. A simulation experiment was designed and conducted. A good regression model was proposed to determine the optimum gear ratio. In order to optimize the four-stage gearbox, Vu N. P. et al. [25] performed the optimizing process to find the partial transmission ratios. The objective function aims to find the minimal gear mass. An explicit model was suggested to calculate the optimal ratios actuarially.

It is known that the optimal gear ratios directly impact the cost of the gearbox. Hence, this article is dedicated to studying the cost optimization in terms of calculations of the optimum gear ratio of three-stage helical gearboxes. The optimum gear ratios for second and third stage gear were chosen as the objective functions of the study. Eleven input parameters are purposely selected to probe the effects of each parameter and their interactions on the objective functions. A simulation experiment was planned using a computer program to execute the optimization problem.

2 Optimization Problem

2.1 Cost Analysis of Three-Stage Helical Gearboxes

$$C_{gb} = C_g + C_{gh} + C_s \tag{1}$$

Where, C_g, C_{gh}, and C_s denote to the cost of gears, the cost of gearbox housing and the cost of shafts, respectively.

$$\begin{aligned} C_g &= c_{g.m} \cdot m_g \\ C_{gh} &= c_{gh.m} \cdot m_{gh} \end{aligned} \tag{2}$$

$$\begin{aligned} C_s &= c_{s.m} \cdot m_s \\ m_g &= m_{gh} + m_g + m_s \end{aligned} \tag{3}$$

In which, m_{gh}, m_g, and m_s are orderly the mass of gearbox housing, gears and shafts.

2.2 Gearbox Housing Mass

The mass of gearbox housing m_{gh} can be computed as Eq. (4) in which ρ_{gh} is representative for the weight density of gearbox housing material (kg/m^3), with cast iron material $\rho_{gh} = 7.2$ [29], and V_{gh} is the volume of the gearbox housing (m^3).

$$m_{gh} = \rho_{gh} \cdot V_{gh} \tag{4}$$

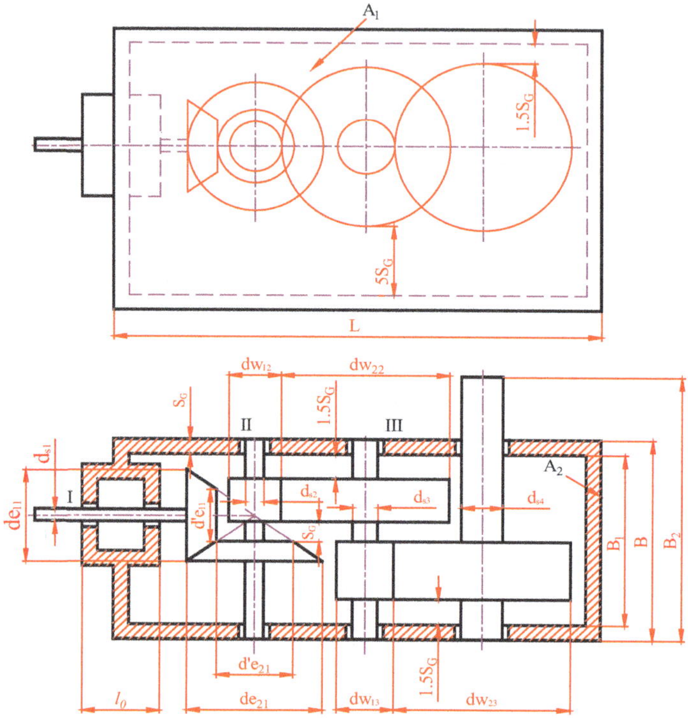

Fig. 1. Schema for determination of gearbox mass

The volume of the gearbox housing can be separated into three parts (c.f Fig. 1) i.e., V_{A1} is the volume of font and back sides of the gearbox; V_{A2} is the volume of the left and right sides, and V_b is the volume of the top and bottom sides. Therefore, V_{gh} can be expressed as

$$V_{gh} = 2 \cdot V_b + 2 \cdot V_{A1} + 2 \cdot V_{A2} \tag{5}$$

In the above equations, $c_{g.m}$, $c_{gh.m}$ and $c_{s.m}$ are the cost per kilogram of the gears, the gearbox housing, and the shafts (USD/kg); m_g, m_{gh}, and m_s are the mass of the gears, the gearbox housing, and the shafts (kg).

From the above equations, to calculate the gearbox cost C_{gb}, it is needed to determined m_g, m_{gh} and m_s.

As seen in Fig. 1, V_b, V_{A1} and V_{A2} can be calculated as:

$$V_b = L \cdot B \cdot S_G \tag{6}$$

$$V_{A1} = L \cdot H \cdot S_G \tag{7}$$

$$V_{A2} = B_1 \cdot H \cdot S_G = (B - 2 \cdot S_G) \cdot H \cdot S_G \tag{8}$$

Where L, H, B_1 and S_G can be determined by:

$$L = (2 \cdot l_0/3 + d_{e21}/2 + d_{w12}/2 + d_{w22}/2 + d_{w13}/2 + d_{w23} + 22.5)/0.975 \text{ [30]} \tag{9}$$

$$H = \max(d_{e21}; d_{w22}; d_{w23}) + 6.5 \cdot S_G \tag{10}$$

$$B = max(b_{w2} + b \cdot cos\delta_2; b_{w2} + b_{w3}) + 6 \cdot S_G \tag{11}$$

$$S_G = 0.005 \cdot L + 4.5 \text{ [30]} \tag{12}$$

Where, $l_0 = 3 \cdot d_{s1}$ [3] with d_{s1} is the diameter of the first shaft which will be determined in the next part; δ_2 is gear pitch cone angle; $tan\delta_2 = u_1$.

2.3 Gear Mass

There are three pairs of gears, i.e., a pair of bevel gears, two pairs of helical gears in the second and third stages as shown in Fig. 1. m_{bg}, m_{hg1}, and m_{hg2} are respectively symbolized for the mass of the three pairs of gears (kg). Thus, the mass of gears in the gearbox can be written as:

$$m_g = m_{bg} + m_{hg2} + m_{hg3} \tag{13}$$

The mass of a pair of bevel gears, m_{bg}, can be expressed as:

$$m_{bg} = m_{bg1} + m_{bg2} \tag{14}$$

In which, m_{bg1} and m_{bg2} can be determined by:

$$m_{bg1} = \rho_g \cdot \{(\pi/3) \cdot [d_{e11}^2 \cdot R_e \cdot cos\delta_1/4 - d_{e11}'^2 \cdot (R_e - b) \cdot cos\delta_1/4]\} \tag{15}$$

$$m_{bg2} = \rho_g \cdot \{(\pi/3) \cdot [d_{e21}^2 \cdot R_e \cdot cos\delta_2/4 - d_{e21}'^2 \cdot (R_e - b) \cdot cos\delta_2/4]\} \tag{16}$$

In Eq. (13), m_{hg2} and m_{hg3} are calculated as:

$$m_{hg2} = \rho_g \cdot (\pi \cdot e_1.d_{w12}^2 \cdot b_{w2}/4 + \pi \cdot e_2.d_{w22}^2 \cdot b_{w2}/4) \tag{17}$$

$$m_{hg3} = \rho_g \cdot (\pi \cdot e_1.d_{w13}^2 \cdot b_{w3}/4 + \pi \cdot e_2.d_{w23}^2 \cdot b_{w3}/4) \tag{18}$$

Where, ρ_g is the weight density of the gear material (kg/m^3); with gear material is steel, $\rho_g = 7.82$ [29]; b is the face with of bevel gears, $b = k_{be}R_e$; d_{e11} and d_{e21} are the outer pitch diameters of the pinion and the gear; with straight bevel gear set, they can be computed as [3]:

$$d_{e11} = 2 \cdot R_e / \left(1 + u_1^2\right)^{1/2} \tag{19}$$

$$d_{e21} = 2 \cdot u_1 \cdot R_e / \left(1 + u_1^2\right)^{1/2} \tag{20}$$

Where, R_e is the cone distance (mm) and it can be expressed as:

$$R_e = k_R \cdot \sqrt{u_1^2 + 1} \cdot \sqrt[3]{T_{11} \cdot k_{h\beta 1} / \left[(1 - k_{be}) \cdot k_{be} \cdot u_1 \cdot [AS_1]^2\right]} \tag{21}$$

For straight bevel gears made by steel materials, k_R is the face width coefficient and $k_R = 50$ [3]; $k_{H\beta 1}$ is representative for the contacting load ratio for pitting resistance and $k_{H\beta 1} = 1.11$; k_{be} is the face width coefficient; T_{11} called the torque on the pinion is calculated by:

$$T_{11} = T_{out} / \left(u_t \cdot \eta_{bg} \cdot \eta_{hg}^2 \cdot \eta_{be}^4\right) \tag{22}$$

Wherein, u_t is the total gearbox ratio; η_{bg}, η_{hg} and η_{be} are efficiencies, accordingly chosen in [3] and their values expressed in Table 1.

Table 1. Efficiency of gearbox components

Type of efficiencies	Symbol	Unit	Value
Efficiency of bevel gear set	η_{bg}	–	0.96
Efficiency of helical gear set	η_{hg}	–	0.97
Efficiency of a rolling bearing pair	η_{be}	–	0.992

In Eqs. (17) and (18), e_1 and e_2 are volume coefficients of the pinion and the gear of the helical gear set; e_1 and e_2 are the volume coefficients of the drive gear and the driven gear, respectively. In practice, these coefficients can be chosen as $e_1 = 1$ and $e_2 = 0.6$; b_{w2} and b_{w3} are the gear widths of the second and the third stages (mm); $b_{w2} = \psi_{ba2} \cdot a_{w2}$ and $b_{w3} = \psi_{ba3} \cdot a_{w3}$ (mm).

The pitch diameters of the second and the third stage can be determined by:

$$d_{w12} = 2 \cdot a_{w2} / (u_2 + 1) \tag{23}$$

$$d_{w22} = 2 \cdot a_{w2} \cdot u_2 \cdot / (u_2 + 1) \tag{24}$$

$$d_{w13} = 2 \cdot a_{w3} / (u_3 + 1) \tag{25}$$

$$d_{w23} = 2 \cdot a_{w3} \cdot u_3 \cdot /(u_3 + 1) \tag{26}$$

Where, a_{w2} and a_{w3} are the center distances of the second and the third which are calculated by [3]:

$$a_{w2} = k_a \cdot (u_2 + 1) \cdot \sqrt[3]{T_{12} \cdot k_{H\beta}/([AS_2]^2 \cdot u_2 \cdot X_{ba2})} \tag{27}$$

$$a_{w3} = k_a \cdot (u_3 + 1) \cdot \sqrt[3]{T_{13} \cdot k_{H\beta}/([AS_3]^2 \cdot u_3 \cdot X_{ba3})} \tag{28}$$

With

$$T_{12} = T_{out}/(u_2 \cdot u_3 \cdot \eta_{hg}^2 \cdot \eta_{be}^3) \tag{29}$$

$$T_{13} = T_{out}/(u_3 \cdot \eta_{hg} \cdot \eta_{be}^2) \tag{30}$$

In which, $k_{H\beta}$ is the contacting load ratio for pitting resistance, $k_{H\beta} = 1.08$ [3]; k_a is the material coefficient, with using steel materials for the helical gears, $k_a = 43$; η_{hg} and η_{be} are shown in Table 1; $[AS_2]$ and $[AS_3]$ are the allowable contact stress of the second and third stages (MPa); X_{ba2} and X_{ba3} are the coefficients of wheel face widths of the second and the third stages.

2.4 Shaft Mass

The mass of all shaft components denoted by m_s can be expressed as

$$m_s = m_{s1} + m_{s2} + m_{s3} + m_{s4} \tag{31}$$

Where, m_{s1}, m_{s2}, m_{s3} and m_{s4} are representative for the mass of the shaft 1, 2, 3, and 4 of the gearboxes, respectively (see Fig. 1). Each shaft mass is normally computed based on multiplying the weight density of shaft material ρ_s and the volume of the shaft. Thus, m_{s1}, m_{s2}, m_{s3} and m_{s4} can be written as

$$m_{s1} = \rho_s \cdot \pi \cdot d_{s1}^2 \cdot l_{s1}/4 \tag{32}$$

$$m_{s2} = \rho_s \cdot \pi \cdot d_{s2}^2 \cdot l_{s2}/4 \tag{33}$$

$$m_{s3} = \rho_s \cdot \pi \cdot d_{s3}^2 \cdot l_{s3}/4 \tag{34}$$

$$m_{s4} = \rho_s \cdot \pi \cdot d_{s4}^2 \cdot l_{s4}/4 \tag{35}$$

In here, l_{s1}, l_{s2}, l_{s3} and l_{s4} are respectively the length of shaft 1, 2, 3 and 4 of the gearbox. As shown in Fig. 1, the length of these shafts can be described by

$$l_{s1} = 1.3 \cdot l_0 + 1.2 \cdot d_{s1} \quad (36)$$

$$l_{s2} = B \quad (37)$$

$$l_{s3} = B \quad (38)$$

$$l_{s4} = B + 1.2 \cdot d_{s4} \quad (39)$$

In which, d_{s1}, d_{s2}, d_{s3} and d_{s4} are preliminary diameters of the shaft 1, 2, 3, and 4 respectively. They can be calculated as

$$d_{s1} = [T_{11}/(0.2 \cdot [\tau])]^{1/3} \quad (40)$$

$$d_{s2} = [T_{12}/(0.2 \cdot [\tau])]^{1/3} \quad (41)$$

$$d_{s3} = [T_{13}/(0.2 \cdot [\tau])]^{1/3} \quad (42)$$

$$d_{s4} = [T_{14}/(0.2 \cdot [\tau])]^{1/3} \quad (43)$$

In Eqs. (40)–(43), $[\tau]$ is the allowable shear stress (MPa) and can be chosen as 17 Mpa [3].

2.5 Optimization Problem

As the above procedure, the gearbox mass depending on the transmission ratios of the three stages i.e., u_1, u_2, and u_3 is considered as the objective function.

$$\text{minimize } C_{gb} \quad (44)$$

Subjected to

$$1 \leq u_1 \leq 6, \quad 1 \leq u_2 \leq 9 \text{ and } \quad 1 \leq u_3 \leq 9 \quad (45)$$

As known that the relation between the partial ratios u_1, u_2, and u_3 can be described as

$$u_g = u_1 \cdot u_2 \cdot u_3 \quad (46)$$

3 Experimental Work

Eleven main design factors (Table 2) are selected as the input parameters orderly assigned by as factors A, B, etc. The influences of input parameters on the objective functions or responses (u_2 and u_3) can be performed using the screening experiment method. All factors will be considered at two values, low and high ones. The factorial experiment used 2-level is adopted by 1/16 fraction. Therefore, $2^{11-4} = 128$ test runs for the experiment are used. This procedure aims to find the optimal gear ratios which can satisfy the objective function of the optimization problem. The plan of experimental work, as well as post analysis, is conducted by using Minitab®19. The output responses are presented in Table 3.

Table 2. Input parameters

Real factor	Minitab®19	Code	Unit	Low	High
Total gearbox ratio	A	ug	–	20	70
Coefficient of wheel face	B	K_{be}	–	0.25	0.3
Coefficient of wheel face width of stage 2	C	X_{ba2}	–	0.3	0.4
Coefficient of wheel face width of stage 3	D	X_{ba3}	–	0.3	0.4
Allowable contact stress of stage 1	E	AS_1	MPa	360	420
Allowable contact stress of stage 2	F	AS_2	MPa	360	420
Allowable contact stress of stage 3	G	AS_3	MPa	360	420
Cost of gearbox housing	H	Cgh	USD/kg	1	5
Cost of gears	I	Cg	USD/kg	2	9
Cost of shafts	J	Cs	USD/kg	1.5	5
Output torque	K	T_{out}	Nmm	1000	10000

Table 3. Experimental plans and output responses

StdOrder	RunOrder	u_g	K_{be}	X_{ba2}	X_{ba3}	AS_1	AS_2	AS_3	C_{gh}	C_g	C_s	T_{out}	u_2	U_3
64	1	70	0.3	0.4	0.4	420	420	360	1	9	5	1000	8.12	5.04
101	2	20	0.25	0.4	0.3	360	420	420	5	2	5	10000	2.15	5.44
46	3	70	0.25	0.4	0.4	360	420	360	5	9	5	1000	8.37	4.89
……														
125	126	20	0.25	0.4	0.4	420	420	420	5	2	1.5	10000	2.9	4.03
92	127	70	0.3	0.3	0.4	420	360	420	1	2	5	10000	5.41	7.57
14	128	70	0.25	0.4	0.4	360	360	360	5	9	1.5	10000	8.5	4.82

4 Results and Discussions

4.1 The Influence of Input Parameters and Their Interactions

Figure 2 presents the variation between the second step (u_2) and third step (u_3) ratios and the input parameters. It can be seen that the total gearbox ratio (u_g) has the strongest impact on u_2, i.e. an increase in u_g leads to increase in u_2 (positive impact), while the remaining factors have little effects on the response of u_2. In case of u_3, the total gearbox ratio (u_g), the cost of gear (C_g), and the cost of shaft (C_s) have the biggest effect on the response of u_3. The first and third mentioned factors have a positive influence, but the second one has a negative influence. In addition to u_g, C_g, C_s, AS_2, AS_3, C_{gh} also have clear influences on the response of u_3.

Another way to evaluate the main effects on the response is considering the Normal plot of the standardized effects. Figure 4 presents the Normal Plot of the standardized effects in which the relationship between response and input parameters, and their interactions are shown. In this case, each factor is denoted using the name in Minitab@19 as presented in Table 2. It was noted that the factor of A is the strongest positive impact on the responses (both u_2 and u_3). The factor of K and J show the largest negative impacts on the responses of u_2 and u_3, respectively. It is noticed that in Fig. 4, the interactions between input parameters are also presented. The interactions of JK and HJ have the biggest positive influences on the responses of u_2 and u_3 respectively, while AG and JK show the domination of negative impacts on the responses of u_2 and u_3, respectively.

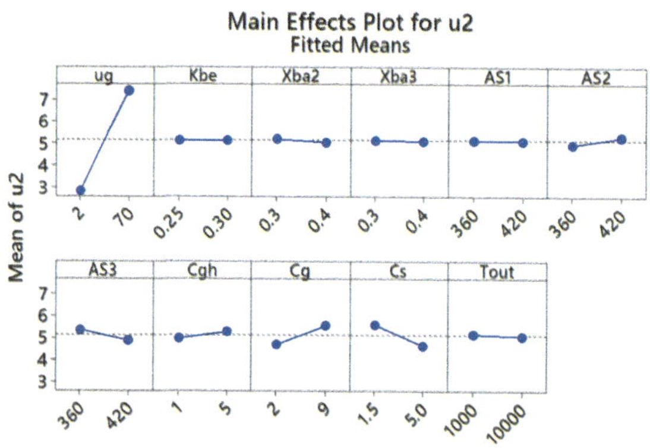

(a)

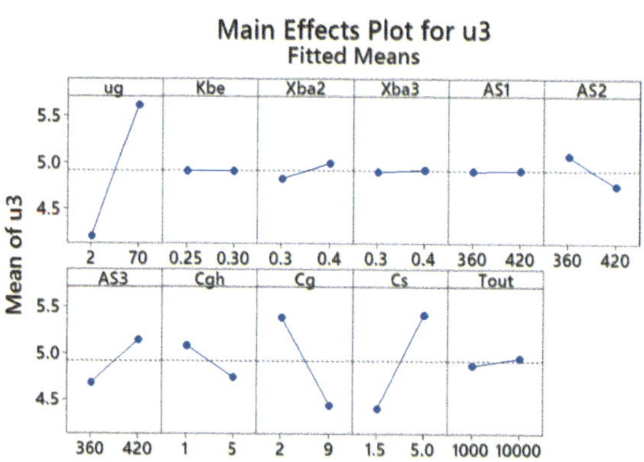

(b)

Fig. 2. Main effects plot for (a) u_2 and (b) u_3.

The Pareto chart of the standardized effects can also be considered as an effective method to show the main impacts of the input parameters on the responses. From the chart (Fig. 4), it can easily see the domination of a factor as previously presented in Fig. 3a and Fig. 3b.

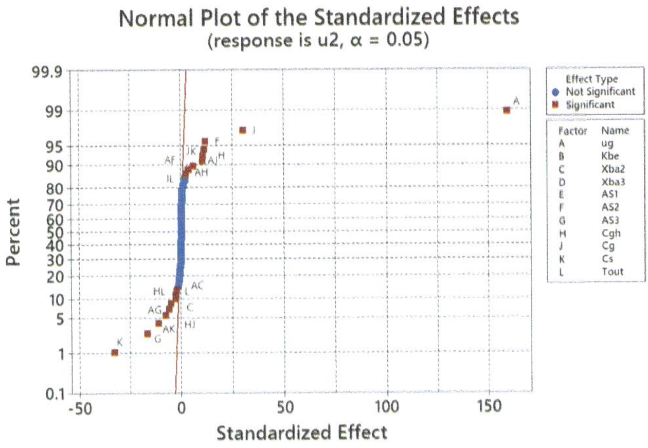

(a)

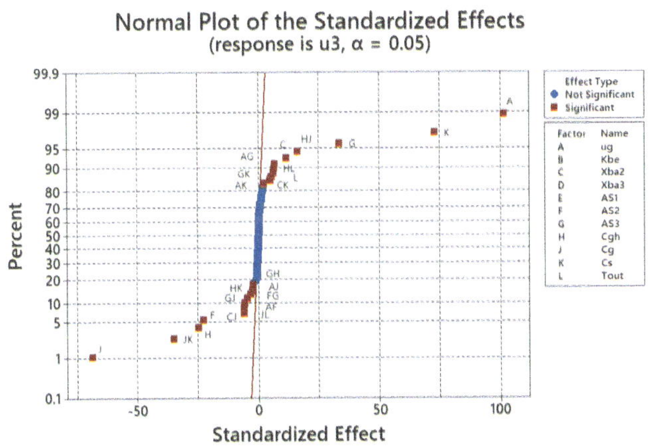

(b)

Fig. 3. Normal distribution charts of parameter effects and interactions for u_2 (a) and u_3 (b)

4.2 Proposed Regression Model of the Response

In order to predict the values of u_2 and u_3, the regression models with two interaction factors is proposed. The equation of 47 is the regression model of u_2, and u_3 respectively. For both, the significant of this regression is $\alpha = 0.05$. The corresponding coefficients of Eqs. (47) and (48) can be listed in Table 4 and Table 5 in which the factors having no influence are eliminated. It is noticed that if the effect of each input parameter, or interactions has P-value higher than significant of α, the response is not strongly impacted by it. Hence, the factors listed in Table 4 and Table 5 are significant to the response u_2 and u_3, because all P-values are inferior to $\alpha = 0.05$. The R-square values for both cases are approximately 0.99. It can be said that the experimental data are crucially consistent with the proposed model.

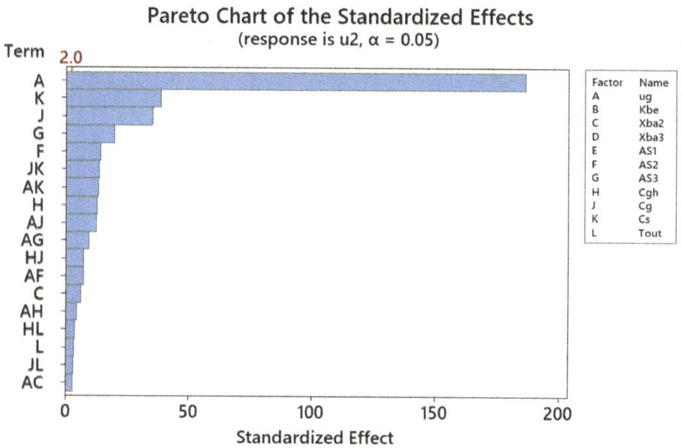

(a)

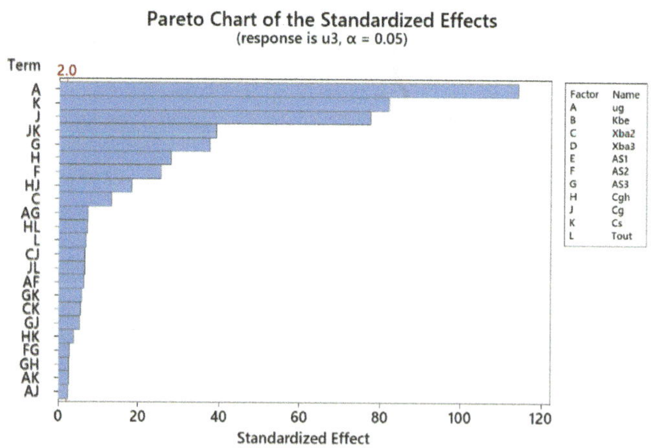

(b)

Fig. 4. Pareto Chart of the Standardized Effects of u_2 (a) and u_3 (b)

Table 4. Estimated Effects and Coefficients for u_2.

Term	Effect	Coef	SE Coef	T-Value	P-Value	VIF
Constant		5.1523	0.0123	419.42	0.000	
u_g	4.5813	2.2906	0.0123	186.47	0.000	1.00
X_{ba2}	-0.1456	-0.0728	0.0123	-5.93	0.000	1.00
AS_2	0.3475	0.1738	0.0123	14.14	0.000	1.00
AS_3	-0.4816	-0.2408	0.0123	-19.60	0.000	1.00
C_{gh}	0.3097	0.1548	0.0123	12.60	0.000	1.00
C_g	0.8697	0.4348	0.0123	35.40	0.000	1.00
C_s	-0.9500	-0.4750	0.0123	-38.67	0.000	1.00
T_{out}	-0.0791	-0.0395	0.0123	-3.22	0.002	1.00
u_g*X_{ba2}	-0.0616	-0.0308	0.0123	-2.51	0.014	1.00
u_g*AS_2	0.1734	0.0867	0.0123	7.06	0.000	1.00
u_g*AS_3	-0.2288	-0.1144	0.0123	-9.31	0.000	1.00
u_g*C_{gh}	0.1025	0.0512	0.0123	4.17	0.000	1.00
u_g*C_g	0.3013	0.1506	0.0123	12.26	0.000	1.00
u_g*C_s	-0.3266	-0.1633	0.0123	-13.29	0.000	1.00
C_{gh}*C_g	-0.1759	-0.0880	0.0123	-7.16	0.000	1.00
C_{gh}*T_{out}	-0.0841	-0.0420	0.0123	-3.42	0.001	1.00
C_g*C_s	0.3288	0.1644	0.0123	13.38	0.000	1.00
C_g*T_{out}	0.0709	0.0355	0.0123	2.89	0.005	1.00

Model Summary

S	R-sq	R-sq(adj)	R-sq(pred)
0.138982	99.72%	99.68%	99.62%

$$\begin{aligned} u_2 =\ & 3.788 + 0.08398\ u_g - 0.804\ X_{ba2} + 0.002731\ AS_2 - 0.003989\ AS_3 \\ & + 0.1451\ C_{gh} + 0.0168\ C_g - 0.3202\ C_s - 0.000007\ T_{out} - 0.01811\ u_g * X_{ba2} \\ & + 0.000085\ u_g * AS_2 - 0.000112\ u_g * AS_3 + 0.000754\ u_g * C_{gh} \\ & + 0.001266\ u_g * C_g - 0.002744\ u_g * C_s - 0.01257\ C_{gh} * C_g - 0.000005\ C_{gh} * T_{out} \\ & + 0.02684\ C_g * C_s + 0.000002\ C_g * T_{out} \end{aligned} \tag{47}$$

$$\begin{aligned} u_3 =\ & -0.80 + 0.01816\ u_g + 1.637\ X_{ba2} + 0.00364\ AS_2 + 0.01392\ AS_3 \\ & - 0.0755\ C_{gh} + 0.1638\ C_g + 0.1121\ C_s + 0.000009\ T_{out} - 0.000039\ u_g * AS_2 \\ & + 0.000045\ u_g * AS_3 - 0.000127\ u_g * C_g + 0.000265\ u_g * C_s - 0.2321\ X_{ba2} * C_g \\ & + 0.3893\ X_{ba2} * C_s - 0.000019\ AS_2 * AS_3 - 0.000268\ AS_3 * C_{gh} \\ & - 0.000310\ AS_3 * C_g + 0.000690\ AS_3 * C_s + 0.016205\ C_{gh} * C_g \\ & - 0.00679\ Cgh * Cs + 0.000005\ C_{gh} * T_{out} - 0.03946\ C_g * C_s - 0.000003\ C_g * T_{out} \end{aligned} \tag{48}$$

Table 5. Estimated Effects and Coefficients for u_3.

Term	Effect	Coef	SE Coef	T-Value	P-Value	VIF
Constant		4.91828	0.00618	796.22	0.000	
u_g	1.40844	0.70422	0.00618	114.01	0.000	1.00
Xba_2	0.16250	0.08125	0.00618	13.15	0.000	1.00
AS_2	-0.31563	-0.15781	0.00618	-25.55	0.000	1.00
AS_3	0.46500	0.23250	0.00618	37.64	0.000	1.00
C_{gh}	-0.34438	-0.17219	0.00618	-27.88	0.000	1.00
C_g	-0.95594	-0.47797	0.00618	-77.38	0.000	1.00
C_s	1.01406	0.50703	0.00618	82.08	0.000	1.00
T_{out}	0.08469	0.04234	0.00618	6.86	0.000	1.00
u_g*AS_2	-0.07875	-0.03938	0.00618	-6.37	0.000	1.00
u_g*AS_3	0.09125	0.04562	0.00618	7.39	0.000	1.00
u_g*C_g	-0.03031	-0.01516	0.00618	-2.45	0.016	1.00
u_g*C_s	0.03156	0.01578	0.00618	2.55	0.012	1.00
Xba_2*C_g	-0.08125	-0.04062	0.00618	-6.58	0.000	1.00
Xba_2*C_s	0.06813	0.03406	0.00618	5.51	0.000	1.00
AS_2*AS_3	-0.03469	-0.01734	0.00618	-2.81	0.006	1.00
AS_3*C_{gh}	-0.03219	-0.01609	0.00618	-2.61	0.011	1.00
AS_3*C_g	-0.06500	-0.03250	0.00618	-5.26	0.000	1.00
AS_3*C_s	0.07250	0.03625	0.00618	5.87	0.000	1.00
$C_{gh}*C_g$	0.22687	0.11344	0.00618	18.36	0.000	1.00
$C_{gh}*C_s$	-0.04750	-0.02375	0.00618	-3.84	0.000	1.00
$C_{gh}*T_{out}$	0.08813	0.04406	0.00618	7.13	0.000	1.00
C_g*C_s	-0.48344	-0.24172	0.00618	-39.13	0.000	1.00
C_g*T_{out}	-0.08094	-0.04047	0.00618	-6.55	0.000	1.00

Model Summary

S	R-sq	R-sq(adj)	R-sq(pred)
0.0698851	99.67%	99.59%	99.49%

Based on the previously analysis, it can be said that the proposed models of u_2 and u_3 can be utilized get the optimum gear ratio of the second and third stage. As a consequence, the optimum gear ration of the first stage can be obtained by $u_1 = u_t/(u_2*u_3)$.

4.3 Validation of Proposed Model

The estimation of errors resulting from the difference between experiments and model of u3 is qualitatively described in Fig. 5. From the Normal Probability Plot, it is observed that the contribution of errors is similar to normal distribution. The versus fits graph reveals that the relation between residual and fitted value of model is random.

Moreover, the Versus Order also exhibits the random relationship between residual and order of data point. The identical tendency is also observed when comparing experiments and proposed model in case of u2 response. The observed phenomena given by the graphs one more time show the reliability of the proposed model which is highly fitted the experiments.

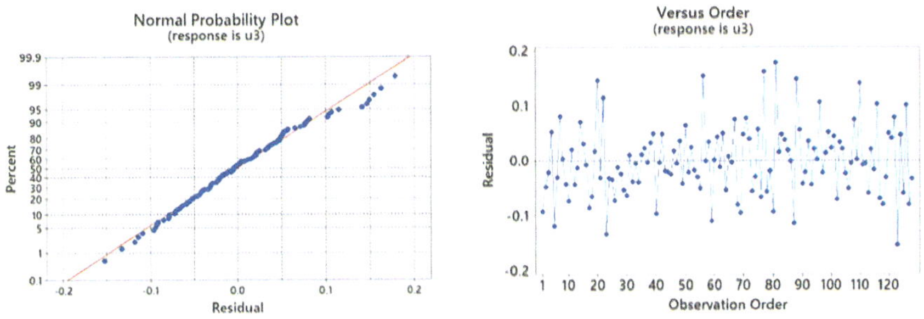

Fig. 5. Graphs estimating errors between experiments and model of u_2.

5 Conclusions

An experimental study to determine the optimum partial gear ratios for Three-stage Helical Gearboxes was conducted. This aim is to be the main principle for obtaining the minimum gearbox cost. In order to perform the experiments, the influences of eleven input parameters and their interactions on the output response of u_2 and u_3 were investigated.

The input parameters include total gearbox ratio, Coefficient of wheel face, Coefficient of wheel face width of stage 2, coefficient of wheel face width of stage 3, allowable contact stress of stage 1, allowable contact stress of stage 2, allowable contact stress of stage 3, output torque, cost of gearbox housing, cost of gears, and cost of shafts. The following conclusion can be made:

- The Total gear ratio has the strongest influence on the both response of u_2 and u_3, while others have little effects on the responses.
- In addition to singular input parameter, their interactions also have the impacts on the u_2, and u_3 responses, but the influential level is different.
- The proposed models of both u_2, and u_3 responses are highly consistent to experimental data. The reliability of the models is validated. It can be said that the proposed models can be applied to optimize the costs of gearbox.

Acknowledgements. This work was supported by Thai Nguyen University of Technology.

References

1. Pi, V.N.: A new method for optimal calculation of total transmission ratio of two step helical gearboxes. In: The Nation Conference on Engineering Mechanics, pp. 179–188 (2001)
2. Pi, V.N., Binh, N.D., Dac, V.Q., The, P.Q.: Optimal calculation of total transmission ratio of three-helical gearboxes for minimum mass of gears. J. Sci. Technol. Techn. Eng. Univ. 170–178 (2006)
3. Romhild, I., Linke, H.: Gezielte Auslegung Von Zahnradgetrieben mit minimaler Masse auf der Basis neuer Berechnungsverfahren. In: Advances in Engineering Research and Application, pp. 249–261 (1992)
4. Tung, L.A., et al.: A study on determination of optimum gear ratios of a two-stage morm gearbox. In: Sattler, K.U., Nguyen, D., Vu, N., Tien Long, B., Puta, H. (eds.) ICERA 2019. LNNS, vol. 104, pp. 76–84. Springer, Cham (2019). https://doi.org/10.1007/978-3-030-37497-6_8
5. Cam, N.T.H., Pi, V.N., Hong, T.T., Tung, L.A.: A study on calculation of optimum gear ratios of a two-stage helical gearbox with second stage double gear sets (2019)
6. Van Cuong, N., Ky, L.H., Hong, T.T., Tu, N.T., Pi, V.N.: Splitting total gear ratio of two-stage helical reducer with first-stage double gear-set for minimal reducer length (2019)
7. Hong, T.T., Cuong, N.V., Ky, L.H., Tu, N.T., Pi, V.N.: Calculating optimum gear ratios of mechanical driven systems using worm-helical gearbox and chain drive (2019)
8. Pi, V.N., Hong, T.T., Thao, T.T.P., Tuan, N.K., Hung, L.X., Tung, L.A.: Calculating optimum gear ratios of a two-stage helical reducer with first stage double gear sets (2019)
9. Pi, V.N., Thao, T.T.P., Hong, T.T., Tuan, N.K., Hung, L.X., Tung, L.A.: Determination of optimum gear ratios of a three stage bevel helical gearbox (2019)
10. Vu, N.P.: A new study on optimal calculation of partial transmission ratios of two-step helical gearboxes (2008)
11. Vu, N.P.: A new study on the optimal prediction of partial transmission ratios of three-step helical gearboxes with second-step double gear-sets (2008)
12. Vu, N.P.: A study on optimal determination of partial transmission ratios of helical gearboxes with second step double gear (2008)
13. Vu, N.P.: Optimal determination of partial transmission ratios of three-step helical gearboxs with second double gears (2008)
14. Pi, V.N., Tuan, N.K.: Determining optimum partial transmission ratios of mechanical driven systems using a chain drive and a two step bevel helical gearbox (2019)
15. Kudreavtev, V.N., Gierzaves, I.A., Glukharev, E.G.: Design and Calculus of Gearboxes, p. 46. Mashinostroenie Publishing (1971)
16. Trinh Chat, L.V.U.: Calculation of Mechanical Driven Systems. Education Publisher (1996)
17. Milou, G., Dobre, G., Visa, F., Vitila, H.: Optimal design of two step gear units, regarding the main parameters. VDI Berichte (1230), 227 (1996)
18. Ky, L.H., et al.: Calculation of optimum gear ratios of mechanical driven systems using worm-helical gearbox and chain drive. In: Sattler, K.U., Nguyen, D., Vu, N., Tien Long, B., Puta, H. (eds.) ICERA 2019. LNNS, vol. 104, pp. 66–75. Springer, Cham (2019). https://doi.org/10.1007/978-3-030-37497-6_7
19. Cam, N.T.H., Pi, V.N., Tuan, N.K., Hung, L.X., Thao, T.T.P.: A study on determination of optimum partial transmission ratios of mechanical driven systems using a chain drive and a three-step helical reducer. In: Fujita, H., Nguyen, D., Vu, N., Banh, T., Puta, H. (eds.) ICERA 2018. LNNS, vol. 63, pp. 91–99. Springer, Cham (2019). https://doi.org/10.1007/978-3-030-04792-4_14

20. Tuan, N.K., et al.: Determining optimal gear ratios of a two-stage helical reducer for getting minimal acreage of cross section, p. 213. In: MATEC Web of Conferences (2018)
21. Tuan, N.K. et al.: A study on determining optimum gear ratios of mechanical driven systems using two-step helical gearbox with first step double gear sets and chain drive. In: Sattler, K. U., Nguyen, D., Vu, N., Tien Long, B., Puta, H. (eds.) ICERA 2019. LNNS, vol. 104, pp. 85–93. Springer, Cham (2020). https://doi.org/10.1007/978-3-030-37497-6_9
22. Hong, T.T., Cuong, N.V., Ky, L.H., Tuan, N.K., Pi, V.N.: Calculating optimum gear ratios of two step bevel helical reducer (2019)
23. Hong, T.T., et al.: A study on determination of optimum gear ratios of a worm helical gearbox (2019)
24. Pi, V.N.: Optimal calculation of partial transmission ratios of four-step helical gearboxes for getting minimal gearbox length. Eng. Technol. Int. J. Mech. Mechatron. Eng. (2008)
25. Pi, V.N.: Optimal determination of partial transmission ratios for four-step helical gearboxes with first and third step double gear-sets for minimal mass of gears. In: Applied Computing Conference, p. 67, no. 3–4 (2008)

Determination of Optimum WEDM Parameters for Minimum Surface Roughness When Cutting SKD11

Nguyen Huu Quang[1], Hoang Xuan Tu[2], Pham Duc Lam[3], Le Hoang Anh[4], Nguyen Anh Tuan[1], Tran Anh Duc[2], Tran Ngoc Giang[2], and Nguyen Thanh Tu[2(✉)]

[1] University of Economics - Technology for Industries, Hanoi, Vietnam
[2] Thai Nguyen University of Technology, Thai Nguyen, Vietnam
nguyenthanhtucnvl@tnut.edu.vn
[3] Nguyen Tat Thanh University, Ho Chi Minh City, Vietnam
[4] Vinh Long University of Technology Education, Vĩnh Long, Vietnam

Abstract. This paper presents a study on determining the optimum input parameter when Wire Electrical Discharge Machining (WEDM) when cutting half round SKD11 tool steel with a radius of 7 mm. In this work, six process parameters, including the cutting voltage, the pulse on time, the pulse off time, the serve voltage, the wire feed, and the cutting speed were investigated. The influences of these parameters on the surface roughness of the workpiece after cutting were analysis. Finally, a set of optimum input parameters was proposed to get the minimum surface finish. It was found from the analysis results that the optimum experimental model was suitable for applying.

Keywords: WEDM · Wire EDM · Surface roughness · SKD11

1 Introduction

In practice, SKD11 steel belongs to difficult-to-machine materials. Therefore, there have been many studies on different methods for machining this material. Many studies on machining this type of material have been done when turning [1, 2], milling [3–5], or grinding [6–8]. These types of machining methods are traditional machining processes.

In addition to traditional machining methods, non-traditional methods are widely used for processing the mentioned materials. In these methods, EDM is a non-conventional machining process that is used to produce difficult-to-machine conductive material. It can machine Ti6Al4V [9, 10], WE43 [11], and also SKD11 [12, 13]. WEDM is a type of EDM in which a thin single-strand metal wire, along with dielectric fluid to cut through conductive materials. WEDM is used very effectively to machine shaped holes, especially narrow slits with curved arcs. Therefore, so far, there have been many studies on this type of machining. In [14], the wire tension control during machining was studied. The influence of process parameters on wire crater sizes was inspected in [15]. Determining optimum input parameters have been learned when

D. C. Nguyen et al. (Eds.): ICERA 2021, LNNS 366, pp. 328–335, 2022.
https://doi.org/10.1007/978-3-030-92574-1_34

machining Inconel 625 [16], hot die steel [17], aluminum HE30 [18], and Ni 55.8 Ti [19]. There are also several studies on WEDM when machining SKD11. The effect of input parameters on the surface roughness was shown in [20] when machining tool steel DC53 or in [21] when processing SKD11. The optimal process parameters were also suggested when machining Al6061 [13], SKD 11 [13, 22], and Ni 55.8 Ti [19]. Optimizing the machining process to achieve the minimum roughness when cutting curved 90CrSi tool steel is mentioned in [23].

This work presents the results of a study for determination of optimum input parameters when cutting half-round SKD11 tool steel with a radius of 7 mm. For the investigation, six input parameters including the cutting voltage, the pulse on time, the pulse off time, the serve voltage, the wire feed, and the cutting speed were taken into account. The effect of these parameters on the surface finish of the workpiece were explored. Also, an experimental model to find the optimum process parameters was recommended.

2 Experimental Setup

Fig. 1. Wire EDM machine used in the experiment

To evaluate the influence of the input parameters on the surface finish, an experiment was design and conducted. In the experiment, six input parameters of WEDM process, including the cutting voltage (VM), the pulse on time (t_{on}), the pulse off time (t_{off}), the servo voltage (SV), the wire feed (WF), and the cutting speed (SPD) were investigated. The factors with their levels are described in Table 1. In addition, the experimental matrix is designed based on the survey levels of the variables and using the Taguchi method with the design L18 ($2^1 \times 3^5$). The minitab R18 software was used to design the experimental matrix. The workpiece material was SKD11 steel and a Wire EDM machine was used in the experiment (Fig. 1). The experimental matrix and the response (the surface roughness) are presented in Table 2.

3 Results and Analysis

Table 1. Input parameters and their levels

No.	Parameter	Code	Unit	Level		
				1	2	3
1	Cutting voltage	VM	V	6	9	12
2	Pulse on time	Ton	μs	6	9	12
3	Pulse off time	Toff	μs	8	12	16
4	Servo voltage	SV	V	24	29	34
5	Wire feed	WF	mm/min	8	10	12
6	Cutting speed	SPD	mm/min	2	4	–

Table 2. Experimental matrix and the surface roughness

Exp. no.	SPD	VM	Ton	Toff	SV	WF	Ra(μm)			Mean	S/N
							Trial 1	Trial 2	Trial 3		
1	2	6	6	8	24	8	1.420	1.427	1.422	1.4230	−3.0641
2	2	9	6	12	29	10	1.352	1.346	1.348	1.3487	−2.5981
3	2	12	6	16	34	12	1.462	1.454	1.458	1.4580	−3.2752
4	2	9	9	8	24	10	1.580	1.583	1.578	1.5803	−3.9750
5	2	12	9	12	29	12	1.595	1.605	1.600	1.6000	−4.0824
6	2	6	9	16	34	8	1.504	1.514	1.518	1.5121	−3.5917
7	2	12	12	8	29	8	1.674	1.670	1.675	1.6730	−4.4699
8	2	6	12	12	34	10	1.655	1.657	1.651	1.6544	−4.3731
9	2	9	12	16	24	12	1.469	1.470	1.464	1.4677	−3.3326
10	4	9	6	8	34	12	1.658	1.668	1.669	1.6651	−4.4286
11	4	12	6	12	24	8	1.573	1.575	1.579	1.5757	−3.9493
12	4	6	6	16	29	10	1.339	1.341	1.332	1.3374	−2.5256
13	4	6	9	8	29	12	1.527	1.520	1.525	1.5240	−3.6597
14	4	9	9	12	34	8	1.693	1.710	1.708	1.7039	−4.6289
15	4	12	9	16	24	10	1.422	1.425	1.429	1.4253	−3.0783
16	4	12	12	8	34	10	1.770	1.774	1.775	1.7730	−4.9742
17	4	6	12	12	24	12	1.628	1.620	1.636	1.6279	−4.2326
18	4	9	12	16	29	8	1.642	1.635	1.638	1.6383	−4.2881

To evaluate the influence of the input parameters on the surface roughness, the Analysis of Variance (ANOVA) method is used with the help of Minitab 18 software and the objective function of the smallest surface texture being the best:

$$S/N = -10log_{10}\left(\frac{1}{n}\sum\nolimits_{i=1}^{n} y_i^2\right) \tag{1}$$

From the results of ANOVA (Table 3), it was noted that T_{on} has the greatest influence on Ra (42.06%); followed by the influence of SPD (22.08%), the remaining parameters have the influence level of T_{off} (16.18%), SV (5.68%), VM (4.15%) and WF (2.21%), respectively.

Table 3. ANOVA for S/N ratios.

Analysis of Variance for SN ratios

Source	DF	Seq SS	Adj SS	Adj MS	F	P	C (%)
SPD	1	2.3964	2.3964	2.3964	17.34	0.006	22.08
VM	2	0.4500	0.4500	0.2250	1.63	0.272	4.15
Ton	2	4.5657	4.5657	2.2829	16.52	0.004	42.06
Toff	2	1.7563	1.7563	0.8782	6.35	0.033	16.18
SV	2	0.6169	0.6169	0.3085	2.23	0.189	5.68
WF	2	0.2400	0.2400	0.1200	0.87	0.466	2.21
Residual Error	6	0.8293	0.8293	0.1382			7.64
Total	17	10.8546					

Model Summary

S	R-Sq	R-Sq(adj)
0.3718	92.36%	78.35%

Table 4 describes the order of influence of the input parameters on S/N. The effect's order is Ton, SPD, Toff, SV, VM, and WF, respectively. With the objective function described in Eq. (1), to achieve the minimum surface texture, the S/N value is maximum for each input parameter. Thus, based on the analysis of the influence of the parameters on the S/N ratio in the chart in Fig. 2, it is possible to determine the optimal set of parameters for the WEDM process. From Fig. 2, it is possible to derive the optimal set of WEDM technology parameters to achieve the smallest surface roughness, as shown in Table 5.

To predict the surface roughness corresponding to the proposed optimum parameters, the analysis tool to predict the results of the Taguchi method from Minitab 18 software was used, and the results are shown in Table 5. Thus, with the optimum set of parameters, the minimum surface roughness achieved is 1.23888 (μm).

Table 4. Order of influence of input parameters on S/N.

Response Table for Signal to Noise Ratios

Smaller is better

Level	SPD	VM	Ton	Toff	SV	WF
1	-3.461	-3.671	-3.265	-4.094	-3.662	-3.987
2	-4.191	-3.763	-3.727	-3.996	-3.731	-3.723
3		-4.043	-4.486	-3.388	-4.085	-3.768
Delta	0.730	0.372	1.222	0.706	0.423	0.264
Rank	2	5	1	3	4	6

Table 5. Prediction of minimum surface roughness.

Prediction

S/N Ratio	Mean	StDev	Ln(StDev)
-2.03955	1.23888	0.0019199	-6.03954

Settings

SPD	VM	Ton	Toff	SV	WF
1	1	1	3	1	2

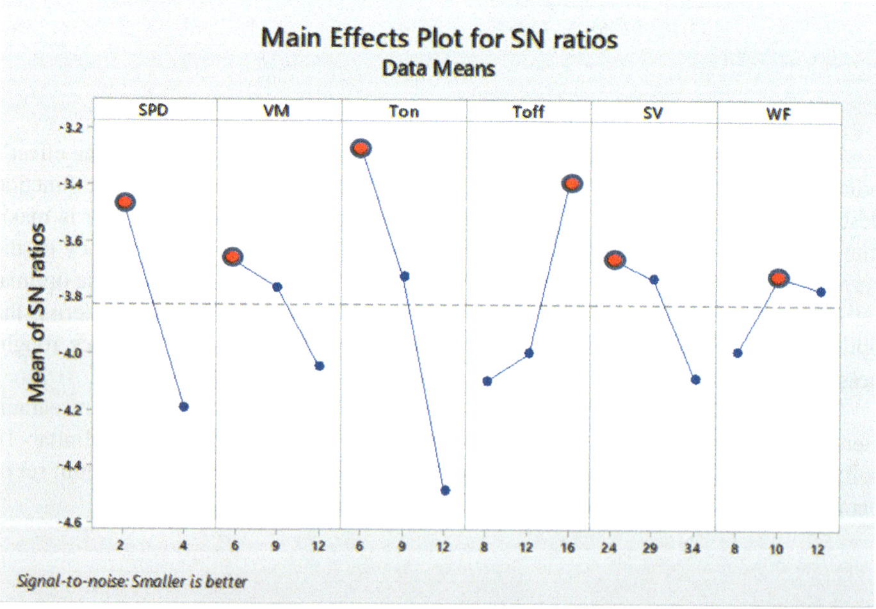

Fig. 2. Effect of WEDM parameters on S/N ratios

4 Evaluation of Experimental Models

The experimental model is evaluated through error distribution charts as shown in Fig. 3. From the chart, it can be seen that in the normal distribution error distribution chart, the errors of the experimental points (the blue points) are close to the normal distribution line (the red solid line). That means the error level is very small. The error frequency chart also shows that the errors are mostly in the range −0.2 to 0.2. The remaining two graphs show the random distribution of experimental errors, which means the built model is largely influenced by the selected input parameters and is not affected by the order of the experiments.

The suitability of the experimental model and the optimum input parameters was checked by the Anderson-Darling method with the use of the Minitab 18 software. From the result (Fig. 4), all of the experimental points (blue dots) are placed within the area limit, with 95% significance level. In addition, the P value is 0.862 - large than the value $\alpha = 0.05$. This indicates that the applied experimental model is suitable.

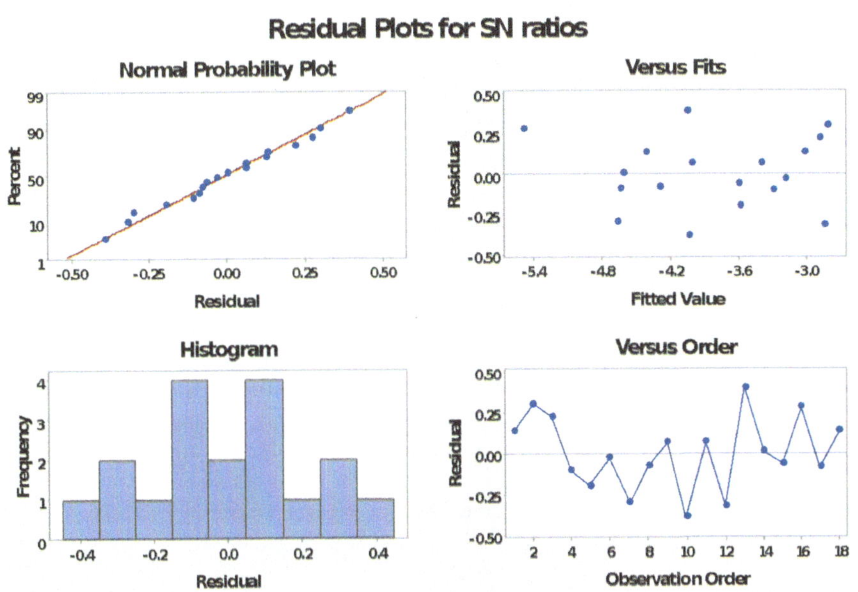

Fig. 3. Probability plot of the fit of the experimental model for Ra

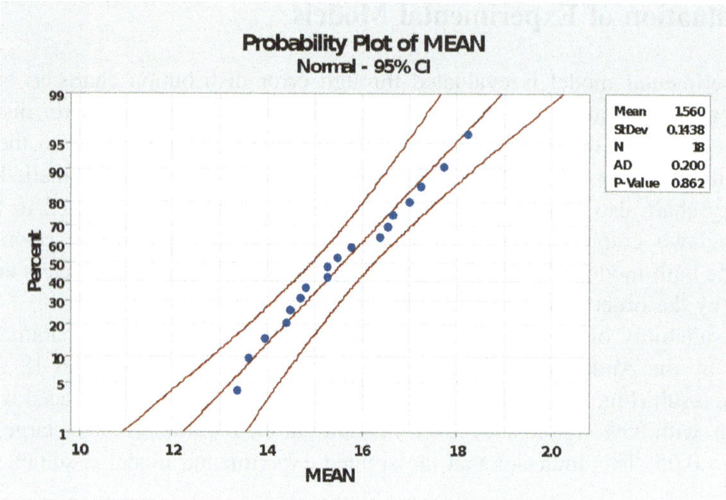

Fig. 4. Residual plots for S/N ratios

5 Conclusion

This study introduces the result of finding the optimum input parameters when WEDM half-round SKD11 tool steel with a radius of 7 mm. In this work, six input parameters, including the cutting voltage, the pulse on time, the pulse off time, the serve voltage, the wire feed, and the cutting speed were investigated. The influences of these parameters on the work-piece surface roughness were inspected. It was found that T_{on} has the greatest impact on the surface roughness (42.06%), followed by the influence of SPD (22.08%), the remaining parameters have the influence level of T_{off}, (16.18%), SV (5.68%), VM (4.15%) and WF (2.21%), respectively. In addition, a regression model to find the optimum input process parameters was proposed. The optimum set of input parameters is: VM = 6; T_{on} = 6; T_{off} = 16; SV = 24; WF = 10; and SPD = 2. Also, it is reported that the proposed experimental model is suitable for applying.

Acknowledgment. This work was supported by Thai Nguyen University of Technology.

References

1. Tzeng, C.-J., Yang, Y.-K.: Determination of optimal parameters for SKD11 CNC turning process. Mater. Manuf. Processes **23**(4), 363–368 (2008)
2. Son, N.H., Thinh, H.X., Nguyen, N.-T., Trung, D.D.: Influence of cutting conditions on the surface roughness when hole turning heat-treated SKD11 steel. In: Sattler, K.-U., Nguyen, D.C., Vu, N.P., Tien Long, B., Puta, H. (eds.) ICERA 2019. LNNS, vol. 104, pp. 426–435. Springer, Cham (2020). https://doi.org/10.1007/978-3-030-37497-6_49
3. Dong, P.Q., Duc, T.M.: Performance evaluation of MQCL hard milling of SKD 11 tool steel using MoS2 nanofluid. Metals **9**(6), 658 (2019)

4. Mac, T.-B., Long, B.T., Nguyen, D.-T.: Optimization of surface roughness and vibration during thermal—assisted milling SKD11 steel using Taguchi method. In: Parinov, I.A., Chang, S.-H., Long, B.T. (eds.) Advanced Materials. SPM, vol. 6, pp. 271–282. Springer, Cham (2020). https://doi.org/10.1007/978-3-030-45120-2_23
5. Gong, F., Zhao, J., Ni, X., Liu, C., Sun, J., Zhang, Q.: Wear and breakage of coated carbide tool in milling of H13 steel and SKD11 hardened steel. SN Appl. Sci. **1**(9), 1–12 (2019). https://doi.org/10.1007/s42452-019-1152-6
6. Tran, T.H., et al.: Improvement of wheel life by optimization of dressing parameters in surface grinding of SKD11 Steel. In: Materials Science Forum. Trans Tech Publ. (2021)
7. Tran, T.H., et al.: Influence of dressing conditions on surface roughness when surface grinding SKD11 steel. In: Materials Science Forum. Trans Tech Publ. (2021)
8. Hong, T.T., et al.: Multi response optimization of dressing conditions for surface grinding SKD11 steel by HaiDuong grinding wheel using grey relational analysis in Taguchi method. In: Sattler, K.-U., Nguyen, D.C., Vu, N.P., Long, B.T., Puta, H. (eds.) ICERA 2020. LNNS, vol. 178, pp. 560–571. Springer, Cham (2021). https://doi.org/10.1007/978-3-030-64719-3_62
9. Fonda, P., et al.: A fundamental study on Ti–6Al–4V's thermal and electrical properties and their relation to EDM productivity. J. Mater. Process. Technol. **202**(1–3), 583–589 (2008)
10. Plaza, S., et al.: Experimental study on micro EDM-drilling of Ti6Al4V using helical electrode. Precis. Eng. **38**(4), 821–827 (2014)
11. Klocke, F., et al.: EDM machining capabilities of magnesium (Mg) alloy WE43 for medical applications. Procedia Eng. **19**, 190–195 (2011)
12. Lee, H.-T., et al.: The study of EDM hole-drilling method for measuring residual stress in SKD11 tool steel. J. Mater. Process. Technol. **149**(1–3), 88–93 (2004)
13. Ming, W., et al.: Investigating the energy distribution of workpiece and optimizing process parameters during the EDM of Al6061, Inconel718, and SKD11. Int. J. Adv. Manuf. Technol. **92**(9–12), 4039–4056 (2017). https://doi.org/10.1007/s00170-017-0488-6
14. Yan, M.-T., Huang, P.-H.: Accuracy improvement of wire-EDM by real-time wire tension control. Int. J. Mach. Tools Manuf. **44**(7–8), 807–814 (2004)
15. Tosun, N., Pihtili, H.: The effect of cutting parameters on wire crater sizes in wire EDM. Int. J. Adv. Manuf. Technol. **21**(10–11), 857–865 (2003)
16. Subrahmanyam, M., Nancharaiah, T.: Optimization of process parameters in wire-cut EDM of Inconel 625 using Taguchi's approach. Mater. Today Proc. **23**, 642–646 (2020)
17. Kuruvila, N.: Parametric influence and optimization of wire EDM of hot die steel. Mach. Sci. Technol. **15**(1), 47–75 (2011)
18. Reddy, V.C., Deepthi, N., Jayakrishna, N.: Multiple response optimization of wire EDM on aluminium HE30 by using grey relational analysis. Mater. Today Proc. **2**(4–5), 2548–2554 (2015)
19. Magabe, R., et al.: Modeling and optimization of Wire-EDM parameters for machining of Ni 55.8 Ti shape memory alloy using hybrid approach of Taguchi and NSGA-II. Int. J. Adv. Manuf. Technol. **102**(5), 1703–1717 (2019)
20. Kanlayasiri, K., Boonmung, S.: An investigation on effects of wire-EDM machining parameters on surface roughness of newly developed DC53 die steel. J. Mater. Process. Technol. **187**, 26–29 (2007)
21. Bobbili, R., Madhu, V., Gogia, A.: Effect of wire-EDM machining parameters on surface roughness and material removal rate of high strength armor steel. Mater. Manuf. Processes **28**(4), 364–368 (2013)
22. Chen, X., et al.: Investigation on MRR and machining gap of micro reciprocated wire-EDM for SKD11. Int. J. Precis. Eng. Manuf. **21**(1), 11–22 (2020)
23. Thi Hong, T., et al.: Effects of process parameters on surface roughness in wire-cut EDM of 9CRSI tool steel. Int. J. Mech. Eng. Technol. **10**(3) (2019)

Determining Optimal Transmission Ratios of Worm Helical Gearbox for Minimum Gearbox Cost

Nguyen Anh Tuan[1], Nguyen Thi Hong Cam[2], Dang Quoc Cuong[3], Le Hoang Anh[4], Nguyen Huu Quang[1], Nguyen Thanh Tu[2], Luong Viet Dung[2], and Tran Ngoc Giang[2(✉)]

[1] University of Economics - Technology for Industries, Hanoi, Vietnam
[2] Thai Nguyen University of Technology, Thai Nguyen, Vietnam
tranngocgiang@tnut.edu.vn
[3] Nguyen Tat Thanh University, Ho Chi Minh City, Vietnam
[4] Vinh Long University of Technology Education, Vĩnh Long, Vietnam

Abstract. The gear ratio is one of the most critical parameters in a gearbox. It is directly related to many other technology parameters. Therefore, the object to be optimized in the gearbox design is often the gear ratio. This paper presents a calculation method to optimize the gear ratio u_2 of a worm helical gearbox according the minimum gearbox's cost objective function. In this study, a simulation experiment was conducted to learn the effect of the main design parameters on the objective function. Also, the influence of the design factors on the optimum gear ratio u_2. Moreover, suitable regression models for the calculation of optimum gear ratios were proposed.

Keywords: Helical worm gearbox · Gear ratio · Optimization · Gearbox design

1 Introduction

In practice, a gearbox is a popular transmission part widely used in the industry. The optimal design of the gearbox has been in research for a long time. However, there remain issues that need to be studied. Most previous studies on optimal gearbox design have been done by determining optimal part ratios [1–16]. Initially, the optimal gear ratios of a gearbox were calculated according to the graph method. The gear ratios were determined from the graph of the relationship between the component ratios and the total gear ratios [1, 2]. Some studies are conducted according to the practical method [9]. Accordingly, the transmission ratio is determined based on actual facts, for example, data provided by manufacturing factories. Finally, the model method is widely utilized [10–13]. In this method, different target functions are set up to calculate the optimum gear ratio, such as mass target, the cross-sectional area target, length target, the total mass target. In addition, there are some other optimization methods such as using optimization algorithms, finite element analysis, which have been used by many researchers [17–19].

D. C. Nguyen et al. (Eds.): ICERA 2021, LNNS 366, pp. 336–349, 2022.
https://doi.org/10.1007/978-3-030-92574-1_35

Based on the results of previous studies, it is obvious that the partial gear ratios have a direct impact on the optimal gearbox design. It is directly related to the size, dimension, mass, and cost of the gearbox. The partial gear ratio optimization has been studied for many helical gearboxes. However, so far, no studies with the objective function of minimum gearbox cost have been done for a helical worm gearbox. In this study, the optimization method to obtain the minimum gearbox cost by optimizing the transmission ratio u_2 is introduced.

2 Methodology

2.1 Cost Analysis

It is known that the cost of a gearbox is typically impacted by the cost components such as the cost of bearing, gear, shaft, and casing. Also, the bearing cost is not taken into account because its calculation is too complicated. Therefore, the cost of a worm-helical gearbox, C_{gb} can be found as:

$$C_{gb} = C_w \cdot m_w + C_{wh} \cdot m_{wh} + C_g \cdot m_g + C_{gh} \cdot m_{gh} + C_s \cdot \tag{1}$$

Where C_w, C_{wh}, C_g,C_{gh}, and C_s are the cost per kilogram (USD/kg) of the worm, worm wheel, gears, gearbox housing, and shafts, respectively. m_w, m_{wh}, m_g, m_s are the mass of the worm, the worm wheel, all gears, and all shafts in the gearbox (kg), respectively.

2.2 The Determination of Gearbox Housing Mass

The mass of the gearbox housing, m_{gh} can be calculated by:

$$m_{gh} = \rho_{gh} \cdot V_{gh} \tag{2}$$

Where, ρ_{gh} is the weight density of gearbox housing material (kg/m^3); as the gearbox housing material is cast iron, $\rho_{gh} = 7.2$ (kg/dm^3); V_{gh} is the volume of the gearbox housing (m^3). The shape of gearbox housing is constituted by different component volumes as shown in Fig. 1.

$$V_{gh} = 2 \cdot V_b + 2 \cdot V_{A1} + 2 \cdot V_{A2} \tag{3}$$

Where, V_b is the volume of bottom housing; V_{A1} and V_{A2} are the volumes of side A_1 and side A_2;

$$V_b = L \cdot B_1 \cdot S_G \tag{4}$$

$$V_{A1} = L \cdot H \cdot S_G \tag{5}$$

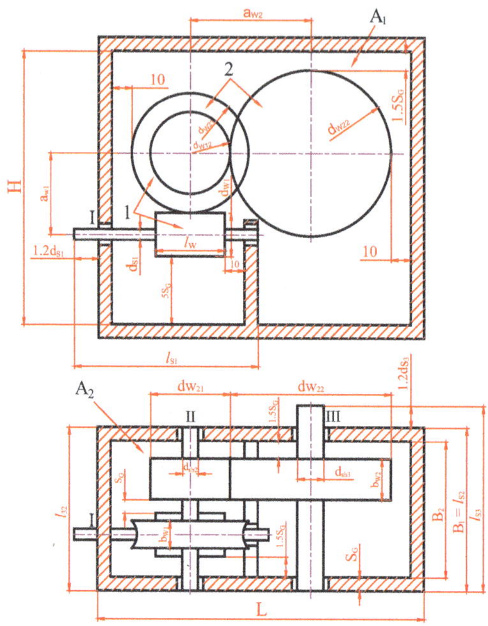

Fig. 1 Calculated schema

$$V_{A2} = B_2 \cdot H \cdot S_G \tag{6}$$

Substituting (4), (5) and (6) into (3):

$$V_{gh} = 2 \cdot L \cdot B_1 \cdot S_G + 2 \cdot L \cdot H.S_G + 2 \cdot B_2 \cdot H \cdot S_G \tag{7}$$

Where, L, H, B_1, B_2 and S_G can be estimated by:

$$L = d_{w21}/2 + a_{w2} + d_{w22}/2 + 20 \tag{8}$$

$$H = max(\frac{d_{w11}}{2} + \frac{d_{w21}}{2}; d_{w22}/2) + a_{w1} + \max(d_{w21}/2; d_{w22}/2) + 6.5 \cdot S_G \tag{9}$$

$$B_2 = b_{w2} + d_{w11} + 4 \cdot S_G \tag{10}$$

$$B_1 = B_2 + 2 \cdot S_G \tag{11}$$

$$S_G = 0.005 \cdot L + 4.5\ [20] \tag{12}$$

In the above equations, d_{w11}, d_{w21}, d_{w21}, and d_{w22} are the pitch diameters of the pinion and the gear of the worm and the helical gear sets. These diameters will be determined in the below sub-sections.

2.3 Worm Mass Calculation

The mass of the worm can be determined by:

$$m_w = \rho_w \cdot \pi \cdot d_{w11}^2 \cdot l_w/4 \tag{13}$$

In which, ρ_w is the weight density of worm material (kg/m^3), with steel worm, $\rho_w = 7.8$ (kg/dm^3) [21]; l_w is the length of worm; In practice, l_w may be approximated by $l_w = 2 \cdot d_{w11}$.

2.4 Worm Wheel Mass Calculation

The mass of the worm wheel is calculated by:

$$m_{wh} = \rho_{wh} \cdot \pi \cdot d_{w12}^2 \cdot b_{w1}/4 \tag{14}$$

In which, ρ_{wh} is the weight density of worm wheel material (kg/m^3), with bronze worm wheel, $\rho_{wh} = 8.88$ (kg/dm^3) [21]; b_{w1} is the width of worm wheel; b_{w1} is determined by $b_{w1} = 0.75 \cdot d_{w11}$ [22].

2.5 Gear Mass Calculations

The mass of gears in the gearbox can be calculated by summing the mass of component gears:

$$m_g = \rho_g \cdot \left(\pi \cdot e_1.d_{w21}^2 \cdot b_{w2}/4 + \pi \cdot e_2.d_{w22}^2 \cdot b_{w2}/4\right) \tag{15}$$

Where ρ_g is the weight density of gear material (kg/m^3), with steel gears, $\rho_g = 7.8$ (kg/dm3) [21]; e_1 and e_2 is the volume coefficients of the drive gear and the driven gear of the first stage, respectively. In practice, e_1 and e_2 can be determined by the values of 1 and 0.6; b_{w2} is the width of the gears calculated by: $b_{w1} = X_{bal} \cdot a_{w1}$ (mm).

2.6 Shaft Mass Calculation

The shaft mass (m_s) can be calculated by the following equation:

$$m_s = m_{s1} + m_{s2} + m_{s3} \tag{16}$$

Where:

$$m_{s1} = \rho_s \cdot \pi \cdot d_{s1}^2 \cdot l_{s1}/4 \tag{17}$$

$$m_{s2} = \rho_s \cdot \pi \cdot d_{s2}^2 \cdot l_{s2}/4 \tag{18}$$

$$m_{s3} = \rho_s \cdot \pi \cdot d_{s3}^2 \cdot l_{s3}/4 \tag{19}$$

In which m_{s1}, m_{s2}, m_{s3} and l_{s1}, l_{s2}, l_{s3} are the mass and length of shaft 1, 2 and 3, respectively; ρ_s is the weight density of the shaft material (kg/m^3).

$$l_{s1} = d_{w21}/2 + l_w/2 + 1.2 \cdot d_{s1} + 20 \tag{20}$$

As noticed above, $l_w = 2 \cdot d_{w11}$. Therefore, we have:

$$l_{s1} = d_{w21}/2 + d_{w11} + 1.2 \cdot d_{s1} + 20 \tag{21}$$

$$l_{s2} = B_1 \tag{22}$$

$$l_{s3} = B_1 + 1.2 \cdot d_{s3} \tag{23}$$

Where:

$$d_{s1} = [T_{11}/(0.2 \cdot [\tau])]^{1/3} \tag{24}$$

$$d_{s2} = [T_{12}/(0.2 \cdot [\tau])]^{1/3} \tag{25}$$

$$d_{s3} = [T_{13}/(0.2 \cdot [\tau])]^{1/3} \tag{26}$$

In which, $[\tau]$ is the allowable shear stress (MPa); $[\tau] = 15 \div 20$ (MPa) and we [22] can choose $[\tau] = 17$ (Mpa).

2.7 Calculating Center Distance and Pitch Diameters of Worm Drive

The worm gear center distance (mm) is calculated by the following equation [22]:

$$a_{w1} = K_a \cdot \left(K_{HV} \cdot K_{H\beta} \cdot T_{21}/[\sigma_{H1}]^2\right)^{1/3} \tag{27}$$

In which, K_a is a coefficient; $K_a = 610$ [22]; K_{hV} is the internal dynamic coefficient; $K_{hV} = 1\ldots1.3$ [22] and we can get $K_{hV} = 1.2$; $K_{h\beta}$ is the load concentration coefficient; $K_{h\beta} = 1\ldots1.3$ [23] and we can get $K_{h\beta} = 1.2$; T_{21} is the wheel torque (N mm) which is determined by:

$$T_{21} = T_{out}/\left(u_2 \cdot \eta_{hg} \cdot \eta_b^2\right) \tag{28}$$

In which T_{out} is the output torque of the system (Nmm); η_{hg} is the transmission efficiency of the helical gear unit ($\eta_{hg} = 0.96\ldots0.98$ [22]); η_b is the transmission efficiency of a rolling bearing pair ($\eta_b = 0.99\ldots0.995$ [22]). Choosing $\eta_{hg} = 0.97$, $\eta_b = 0.992$ and substituting them into (28) gives:

$$T_{21} = 1.0476 \cdot T_{out}/u_2 \quad (29)$$

In Eq. (27), $[\sigma_{H1}]$ is the allowable contact stress of the worm unit (N/mm^2); $[\sigma_{H1}]$ is influenced by the wheel material. If the wheel material is tinless bronze or soft grey iron, $[\sigma_{H1}]$ can be determined by the following regression equation, which was found from the data in [22] (with the determination coefficient $R^2 = 0.9906$):

$$[\sigma_{H1}] = 5.0515 \cdot v_{sl}^2 - 49.742 \cdot v_{sl} + 189.9 \quad (30)$$

Wherein v_{sl} is the slip velocity which is found by [3]:

$$v_{sl} = 0.0088 \cdot \left(P_1 \cdot u \cdot n_1^2\right)^{1/3} \quad (31)$$

Where P_1 is the power on the first shaft; P_1 can be found by:

$$P_1 = T_{11} \cdot n_1/\left(9.66 * 10^6\right) \quad (32)$$

If the wheel material is tin bronze, $[\sigma_{H1}]$ is calculated by [3]:

$$[\sigma_{H1}] = K_{HL} \cdot v_{sl} \cdot [\sigma_{H0}] \quad (33)$$

Where $[\sigma_{H0}]$ is the allowable contact stress when the stress change cycle is 10^7:

$$[\sigma_{H0}] = (0.7\ldots0.9) \cdot \sigma_t \quad (34)$$

In which σ_t is the tensile stress (N/mm^2); the value of σ_t depends on the slip velocity v_{sl}: $\sigma_t = 260$ if $v_{sl} = 5\ldots8$; $\sigma_t = 230$ if $v_{sl} = 8\ldots12$ and $\sigma_t = 285$ if $v_{sl} = 8\ldots25$ [22].

K_{HL} is the service life ratio which is determined as [22]:

$$K_{HL} = \left(10^7/N_{HE}\right)^{1/8} \quad (35)$$

Where N_{HE} is the equivalent loading cycle number for the wheel teeth:

$$N_{HE} = 60 \cdot n_2 \cdot t_\Sigma \quad (36)$$

In which t_Σ is the service lifetime (h); n_2 is the rotational speed of the wheel (rpm). After having a_{w1}, d_{w11} *and* d_{w21} can be calculated by [22]:

$$d_{w21} = 2 \cdot a_{w1} \cdot z_2/(z_2 + q) \quad (37)$$

$$d_{w11} = d_{w21}/u_1 \quad (38)$$

Where, z_2 is the number of worm wheel teeth; $z_2 = u_1 \cdot z_1$ with z_1=1, 2 or 4 [22]; q is the coefficient of the worm pitch diameter which is determined by [22]:

$$q = k_q \cdot z_2 \tag{39}$$

In which, k_q is the coefficient for calculation q; $k_q = 0.25...0.3$ [22].

2.8 Calculating Center Distance and Pitch Diameters of Helical Gear Drive

For the helical gear drive, the center distance a_{w2} is calculated by [22]:

$$a_w = k_a \cdot (u_2 + 1) \cdot \sqrt[3]{T_{12} \cdot k_{H\beta} / \left([\sigma_{H2}]^2 \cdot u_2 \cdot X_{ba} \right)} \tag{40}$$

Where $k_{H\beta}$ is the contacting load ratio for pitting resistance; for the helical gear set $k_{H\beta} = 1.01 \div 1.16$; Therefore, it can be chosen $k_{H\beta} = 1.08$; $[\sigma_{H2}]$ is the allowable contact stress of the first stage (MPa); k_a is the material coefficient; As the gear material is steel, k_a= 43; X_{ba} is the coefficient of wheel face width of the helical gear set; T_{12} is determined by (29).

After obtaining a_{w2}, d_{w12} *and* d_{w22} can be calculated by [22]:

$$d_{w12} = 2 \cdot a_{w2} / (u_2 + 1) \tag{41}$$

$$d_{w22} = 2 \cdot a_{w2} \cdot u_2 / (u_2 + 1) \tag{42}$$

From the above analysis, the optimization problem is defined as:

$$\text{Minimize } C_{gb} \tag{43}$$

With the following constraints

$$\begin{aligned} 1 \le u_c \le 6 \\ 8 \le u_1 \le 80 \\ 1 \le u_2 \le 9 \end{aligned} \tag{44}$$

To solve this optimization problem, it is required to calculate the center distances and the pitch diameters of the worm gear unit and the helical gear unit.

3 Design the Experiment

To evaluate the effect of the design fators on the gear ratio u_2, a simulation experiment was designed. In the experiment, nine main design parameters (Table 1) were investigated. Also, the experimental planning method for the 2^k factorial with k = 9 is used. To reduce the number of experiments while assessing the effects of the input parameters, a separate experimental plan with $2^{9-3} = 128$ experiments is applied. This experiment was designed at resolution VI, so no major factors and two-way interactions are duplicated. To construct the experiment matrix, the Minitab 18 software was applied. The experimental planning matrix is described in Table 2.

Table 1. Input parameters and their levels

No	Parameters	Symbol	Unit	Level	
				Low	High
1	Total gearbox ratio	u_g	–	60	100
2	Coefficient of wheel face width of stage	X_{ba}	–	0.3	0.4
3	Allowable contact stress of stage 2	AS_2	Mpa	350	420
4	Output torque	T_{out}	Nm	3000	50000
5	Cost of gearbox housing	C_{gh}	USD/kg	1.5	5
6	Cost of gears	C_g	USD/kg	3	9
7	Cost of worm	C_w	USD/kg	6	12
8	Cost of wheel worm	C_{wh}	USD/kg	7	15
9	Cost of shafts	C_s	USD/kg	2	5

Table 2. Experimental matrix and optimum gear ratio u_2

Std order	Run order	Center Pt	Blocks	u_g	X_{ba}	AS_2	T_{out}	C_{gh}	C_g	C_w	C_{wh}	C_s	u_2
28	1	1	1	100	0.4	350	50000	5	3	6	7	2	1.82
62	2	1	1	100	0.3	420	50000	5	9	6	7	5	1.81
101	3	1	1	60	0.3	420	3000	1.5	9	12	15	5	1.68
100	4	1	1	100	0.4	350	3000	1.5	9	12	15	5	1.68
14	5	1	1	100	0.3	420	50000	1.5	3	6	15	5	2.86
90	6	1	1	100	0.3	350	50000	5	3	12	15	2	1.68
76	7	1	1	100	0.4	350	50000	1.5	3	12	15	2	1.88
3	8	1	1	60	0.4	350	3000	1.5	3	6	7	5	1.68
...	...	...	...	...	...	...	...	...	...	...	...	...	...
70	127	1	1	100	0.3	420	3000	1.5	3	12	15	2	1.68
126	128	1	1	100	0.3	420	50000	5	9	12	15	2	1.20

4 Results and Discussions

4.1 The Effect of Input Parameters for u_2

The influence of the main design parameters on u_2 is evaluated qualitatively based on the graph in Fig. 2. It was noted from the figure that when X_{ba}, AS_2, T_{out}, C_s, and C_{wh} increase, u_2 increases. Besides, when C_{gh} and C_g increase, u_2 decreases. Finally, u_g and C_w have minimum effect on u_2.

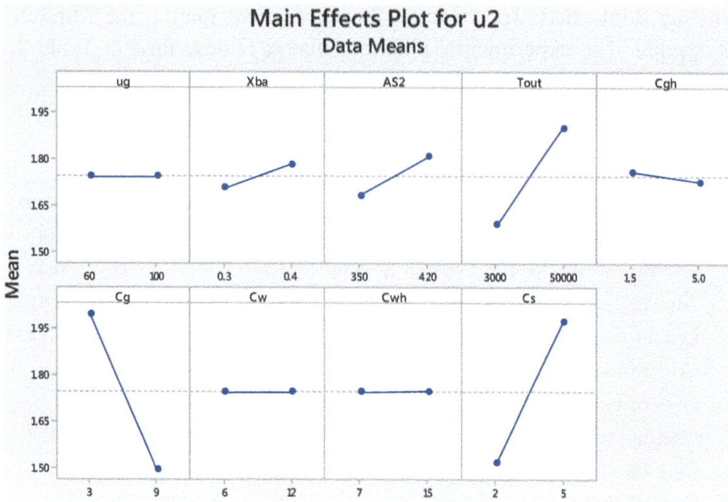

Fig. 2. The influence of main parameters for u_2

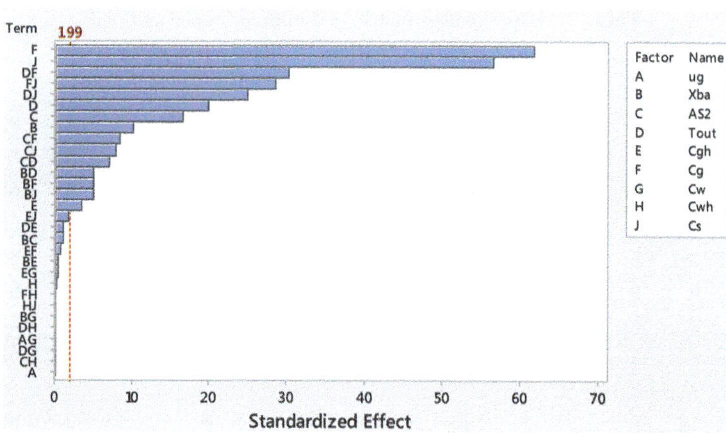

Fig. 3. The Pareto chart normalizes the influence of the input parameters and their interactions for u_2

In the Pareto chart of normalized effects (Fig. 3), the influence of design parameters on u_2 is expressed by the length of the blue columns. The parameters that have magnitude cross the red reference line affect u_2 significantly with significance level $\alpha = 0.05$. Specifically, it can be seen that C_g and C_s have the strongest influence on u_2. Other parameters also have a significant effect on u_2 (including the interactions between them), namely X_{ba}, A_{S2}, T_{out}, C_{gh}, $X_{ba} * T_{out}$, $X_{ba} * C_g$, $X_{ba} * C_s$, $AS_2 * T_{out}$, $AS_2 * Cg$, $AS_2 * Cs$, $T_{out} * C_g$, $T_{out} * C_s$, and $C_g * C_s$. The effects of the interactions are also shown in Fig. 4.

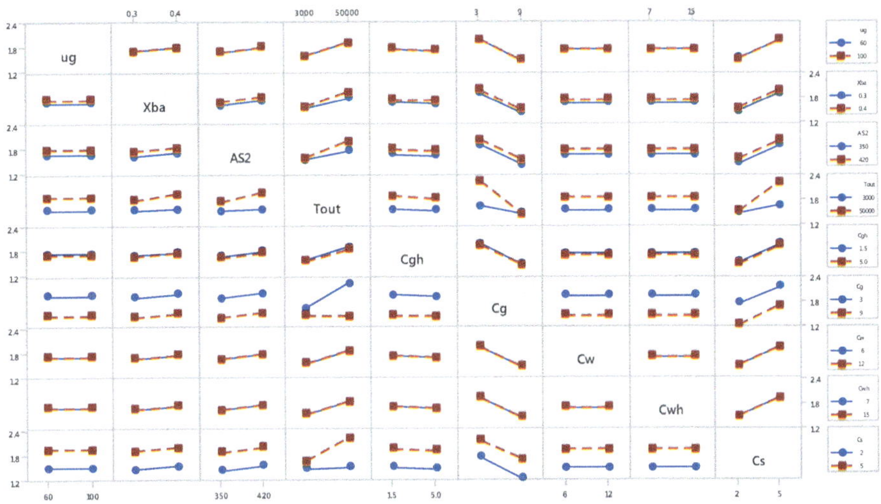

Fig. 4. Interaction plot for u_2

To determine the trend and influence level of the input variables as well as the interactions between them, the normal plot of the standardized effects in Fig. 5 is employed. It is visualized from this chart that the parameters, denoted in red, have a significant effect on u_2. The factors located farther from the standard line have a greater influence. The parameters such as C_s, $C_g * C_s$, $T_{out} * C_s$, T_{out}, AS_2, $AS_2 * Cg$, $AS_2 * T_{out}$, $X_{ba} * T_{out}$, $X_{ba} * C_g$ that are positioned to the right of the standard line have a positive effect on u_2.. On the contrary, those including C_g, $T_{out} * C_g$, $AS_2 * C_s$, $X_{ba} * C_s$, C_{gh}, placed to the left of the standard line, have a negative effect.

4.2 The Regression Equation for u_2

The factors and interactions with slight or no influence on u_2 are excluded. However, due to a large number of variables, the hand calculation is not practicable. So, the regression equation for u_2 is achieved by using the Minitab R18 software. The results obtained are shown in Table 3.

It was reported that all of R-Square and R-Square (adj) (Table 3) are more than 95%. That means the model is well-matched with the data. The regression equation is obtained as follows:

$$\begin{aligned} u_2 = {} & 1.348 + 0.233\,X_{ba} + 0.000515\,AS_2 - 0.000021\,T_{out} - 0.00906\,C_{gh} \\ & - 0.0405\,C_g + 0.0583\,C + 0.000023\,X_{ba} * T_{out} + 0.0057\,X_{ba} * C_g - 0.032\,X_{ba} * C_s \\ & + 0.000000\,AS_2 * T_{out} + 0.000014\,AS_2 * C_g - 0.000040\,AS_2 * C_s \\ & - 0.000002\,T_{out} * C_g + 0.000004\,T_{out} * C_s + 0.00283\,C_g * C_s \end{aligned} \tag{45}$$

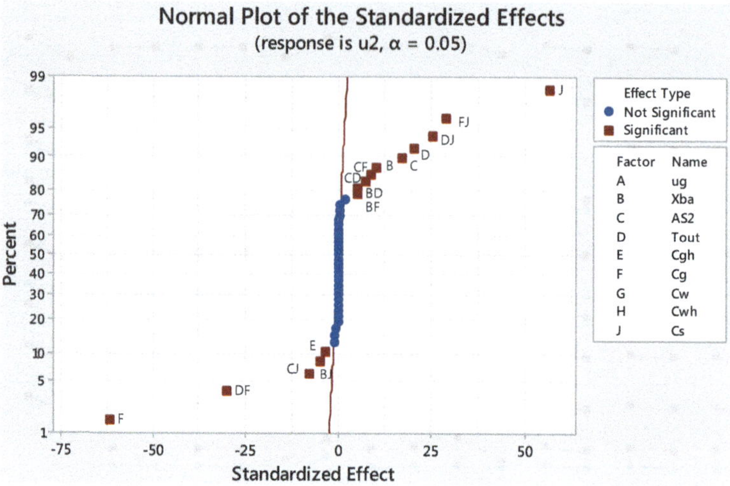

Fig. 5. Normal distribution chart of the influence of parameters and interactions for u_2

Table 3. Coded coefficients for regression equation of u_2.

Coded Coefficients

Term	Effect	Coef	SE Coef	T-Value	P-Value	VIF
Constant		1.74602	0.00820	212.97	0.000	
X_{ba}	0.07516	0.03758	0.00820	4.58	0.000	1.00
AS_2	0.12641	0.06320	0.00820	7.71	0.000	1.00
T_{out}	0.31234	0.15617	0.00820	19.05	0.000	1.00
C_{gh}	-0.03172	-0.01586	0.00820	-1.93	0.056	1.00
C_g	-0.50391	-0.25195	0.00820	-30.73	0.000	1.00
C_s	0.45328	0.22664	0.00820	27.64	0.000	1.00
$X_{ba}*T_{out}$	0.05297	0.02648	0.00820	3.23	0.002	1.00
$X_{ba}*C_g$	0.00172	0.00086	0.00820	0.10	0.917	1.00
$X_{ba}*C_s$	-0.00484	-0.00242	0.00820	-0.30	0.768	1.00
AS_2*T_{out}	0.08359	0.04180	0.00820	5.10	0.000	1.00
AS_2*C_g	0.00297	0.00148	0.00820	0.18	0.857	1.00
AS_2*C_s	-0.00422	-0.00211	0.00820	-0.26	0.797	1.00
$T_{out}*C_g$	-0.32359	-0.16180	0.00820	-19.74	0.000	1.00
$T_{out}*C_s$	0.27297	0.13648	0.00820	16.65	0.000	1.00
C_g*C_s	0.02547	0.01273	0.00820	1.55	0.123	1.00

Model Summary

S	R-sq	R-sq(adj)	R-sq(pred)
0.0927533	96.23%	95.73%	95.08%

4.3 Validation of Proposed Model

The suitability of the model is assessed through a residual evaluation distribution chart to determine the deviation between experimental and u_2 in Fig. 6. The standard probability graph shows that the remainder of the distribution is very close to the normal distribution. Versus Order graph shows points of random distribution, not obeying any rules., which proves that the value of u_2 is not affected by factors other than the survey parameter. Therefore, the relationship between u_2 and survey parameters in (45) has very high reliability.

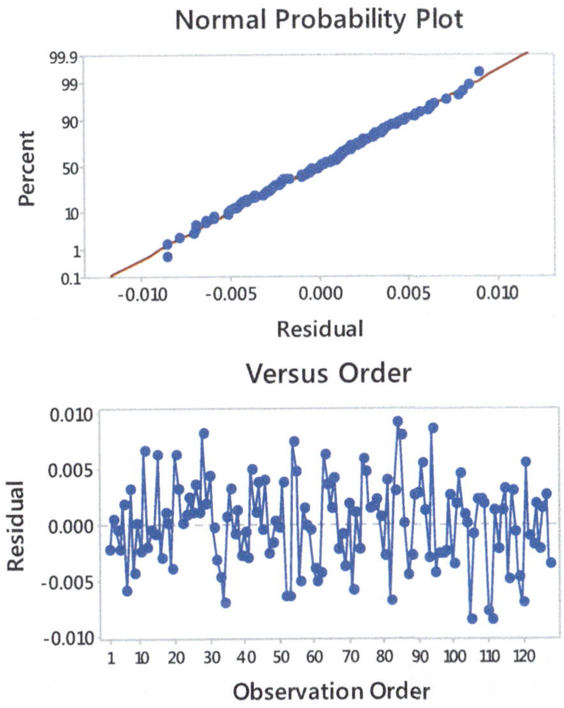

Fig. 6. Normal probability (left) and Versus order plot (right) for u_2

5 Conclusion

In this work, a cost study of a helical worm gearbox reducer is performed by determining the optimum partial gear ratio u_2. To analyze the influence of the main design parameters, a simulation experiment was carried out. The plan of the experiment was created by using the Minitab18 software. Also, nine main process variables were selected to investigate their effect on the optimal gear ratio. The analysis of the results shows that in addition to the impact of the investigating parameters, the interaction between them also has a significant effect on the response u_2. These results coincide with the results of previous studies, which confirmed that the minimum cost of the gearbox can be obtained by optimizing the gear ratio.

Acknowledgment. This work was supported by Thai Nguyen University of Technology.

References

1. Kudreavtev, V.N., Gierzaves, I.A., Glukharev, E.G.: Design and calculus of gearboxes (1971). (in Russian)
2. Chat, T., Uyen, L.V.: Calculation of Mechanical Driven Systems. Education Publisher, Hanoi (1996)
3. Ky, L.H., et al.: Calculation of optimum gear ratios of mechanical driven systems using worm-helical gearbox and chain drive. In: Sattler, K.-U., Nguyen, D.C., Vu, N.P., Tien Long, B., Puta, H. (eds.) ICERA 2019. LNNS, vol. 104, pp. 66–75. Springer, Cham (2020). https://doi.org/10.1007/978-3-030-37497-6_7
4. Hung, L.X., et al.: Calculation of optimum gear ratios of mechanical driven systems using two-stage helical gearbox with first stage double gear sets and chain drive. In: Sattler, K.-U., Nguyen, D.C., Vu, N.P., Tien Long, B., Puta, H. (eds.) ICERA 2019. LNNS, vol. 104, pp. 170–178. Springer, Cham (2020). https://doi.org/10.1007/978-3-030-37497-6_20
5. Cam, N.T.H., Pi, V.N., Tuan, N.K., Hung, L.X., Thao, T.T.P.: A study on determination of optimum partial transmission ratios of mechanical driven systems using a chain drive and a three-step helical reducer. In: Fujita, H., Nguyen, D.C., Vu, N.P., Banh, T.L., Puta, H.H. (eds.) ICERA 2018. LNNS, vol. 63, pp. 91–99. Springer, Cham (2019). https://doi.org/10.1007/978-3-030-04792-4_14
6. Pi, V.N., Thi Hong, T., Thi Phuong Thao, T., Khac Tuan, N., Hung, L.X., Anh Tung, L.: Calculating optimum gear ratios of a two-stage helical reducer with first stage double gear sets. In: IOP Conference Series: Materials Science and Engineering, vol. 542, no. 1 (2019). https://doi.org/10.1088/1757-899X/542/1/012017
7. Tran, T.H., Cuong, N.V., Ky, L.H., Tuan, N.K., Vu, N.P.: Calculating optimum gear ratios of two step bevel helical reducer. Int. J. Appl. Eng. Res. **14**(16), 3494–3499 (2019)
8. Pi, V.N., Tuan, N.K.: Determining optimum partial transmission ratios of mechanical driven systems using a chain drive and a two-step bevel helical gearbox. Int. J. Mech. Eng. Robot. Res. **8**(5), 708–712 (2019). https://doi.org/10.18178/ijmerr.8.5.708-712
9. Milou, G., Dobre, G., Visa, F., Vitila, H.: Optimal Design of two step gear units, regarding the main parameters. VDI Berichte **1230**, 227–244 (1996)
10. The, P.Q., Pi, V.N., Binh, N.D., Dac, V.Q.: Optimal calculation of total transmission ratio of three-step helical gearboxes for minimum mass of gears. J. Sci. Technol. Tech. Eng. Univ. 170–178 (2006)
11. Pi, V.N.: A new method for optimal calculation of total transmission ratio of two step helical gearboxes. In: The National conference on Engineering Mechanics (2001)
12. Pi, V.N.: A new study on optimal calculation of partial transmission ratios of two-step helical gearboxes. Comput. Eng. **2**(1), 162–165 (2008)
13. Pi, V.N.: Optimal determination of partial transmission ratios of three-step helical gearboxs with second double gears (2008)
14. Pi, V.N., Tuan, N.K.: Optimum determination of partial transmission ratios of three-step helical gearboxes for getting minimum cross section dimension. J. Environ. Sci. Eng. A **5** (11) (2016). https://doi.org/10.17265/2162-5298/2016.11.004

15. Tung, L.A., et al.: A Study on determination of optimum gear ratios of a two-stage worm gearbox. In: Sattler, K.-U., Nguyen, D.C., Vu, N.P., Tien Long, B., Puta, H. (eds.) ICERA 2019. LNNS, vol. 104, pp. 76–84. Springer, Cham (2020). https://doi.org/10.1007/978-3-030-37497-6_8
16. Thao, T.T.P., et al.: Determining optimum gear ratios of mechanical driven systems using three stage bevel helical gearbox and chain drive. In: Sattler, K.-U., Nguyen, D.C., Vu, N.P., Tien Long, B., Puta, H. (eds.) ICERA 2019. LNNS, vol. 104, pp. 249–261. Springer, Cham (2020). https://doi.org/10.1007/978-3-030-37497-6_29
17. Zhuang, S.: Gearbox housing topology optimization with respect to gear misalignment gearbox housing topology optimization with respect to gear misalignment. Linköping University, IEI (2012)
18. Xu, X., Chen, Z., Liu, Y., Dong, P., Liu, Y.: Enumerative optimization procedure for the gear train optimization problem of a two-speed dedicated electric transmission. Energies **10** (9) (2017). https://doi.org/10.3390/en10091362
19. Liang, M., Hu, J., Li, S., Chen, Z.: Topology optimization of transmission gearbox under multiple working loads. Adv. Mech. Eng. **10**(11) (2018). https://doi.org/10.1177/1687814018813454
20. Romhild, I., Linke, H.: Gezielte Auslegung Von Zahnradgetrieben mit minimaler Masse auf der Basis neuer Berechnungsverfahren. Konstruktion **44**, 229–236 (1992)
21. American Elements, Specific Gravity. https://www.americanelements.com/specificgravity.html. Accessed 26 Feb 2021
22. Chat, T., Uyen, L.V.: Design and Calculation of Mechanical Transmissions Systems. Educational Republishing House (2007)
23. Antonsson, E.K., Grote, K.-H.: Springer Handbook of Mechanical Engineering, vol. 47, no. 01. Springer, Heidelberg (2009). https://doi.org/10.1007/978-3-540-30738-9

Determining Optimum Gear Ratios for a Four-Stage Helical Gearbox for Getting Minimum Gearbox Cost

Nguyen Huu Quang[1], Nguyen Thanh Tu[2], Nguyen Hong Linh[3], Nguyen Huu Luan[4], Le Hoang Anh[5], Nguyen Anh Tuan[1], Tran Ngoc Giang[2], and Vu Ngoc Pi[2(✉)]

[1] University of Economics - Technology for Industries, Hanoi, Vietnam
[2] Thai Nguyen University of Technology, Thai Nguyen, Vietnam
vungocpi@tnut.edu.vn
[3] Electric Power University, Hanoi, Vietnam
[4] Nguyen Tat Thanh University, Ho Chi Minh City, Vietnam
[5] Vinh Long University of Technology Education, Vĩnh Long, Vietnam

Abstract. This paper presents the research results to determine the optimal partial gear ratios of the four-speed helical gearbox based on the cost optimization problem. To solve that problem, a simulation experiment with 13 input parameters was designed and conducted. The influence of the input parameters on the optimum gear ratios was evaluated. In addition, regression formulas to determine the optimum partial ratios to achieve the smallest gearbox cost have been proposed. These formulas are easy to use because they are all in the form of explicit functions.

Keywords: Helical gearbox · Four-stage gearbox · Gear ratios · Cost optimization

1 Introduction

In practice, there are many drive systems such as mechanical drive, electric drive, hydraulic drive, pneumatic drive, etc. Among them, mechanical drive system is the most commonly used. The reason is that the mechanical transmission has a simple structure, stable working, high reliability, and low cost. A mechanical drive system usually consists of a motor, an operating machine, and a gearbox connecting the motor to the working machine. Some systems may have one or two additional external drives, such as a V-belt or a chain drive. In the elements of the mechanical drive system, the gearbox plays a very important role. It is used to reduce the speed and increase the torque to meet the working requirements of the machine. Therefore, the design and optimal design of the gearbox is a topic that has been and is of interest to many researchers.

So far, there have been quite a few studies on the optimal design of gearboxes. Gearbox dynamics is an important direction covered in [1–3]. Research on the gearbox fault diagnosis is also considered in [4–6]. There have also been many studies interested in noise reduction for gearboxes [7–9]. Research on wind turbine gearbox has been reported in many studies [6, 10, 11].

D. C. Nguyen et al. (Eds.): ICERA 2021, LNNS 366, pp. 350–364, 2022.
https://doi.org/10.1007/978-3-030-92574-1_36

In the optimal design of the gearbox, determining the optimal gear ratios of the gearbox is very important. This is because the gear ratio is a factor that determines the size, weight, and cost of the box. For example, the optimal gear ratio can reduce the gear mass of a gearbox from 242 kg to 178 kg (reduce 26.55%) [12]. Therefore, determining the optimal gear ratio of the gearbox has been chosen as the subject of many studies. In the studies on this issue, determining the optimal gear ratio for the helical gearbox is of the most interest because this is the most commonly used type of gearbox.

With the helical gearbox, the optimal gear ratio has been determined for the 2-stage [13–17], 3-stage [18–20], and 4-stage [21] gearboxes. This problem has also solved with different objective functions such as the minimum gear mass, the smallest gearbox cross-section [14, 17, 19, 20], the minimum gearbox length [15, 16, 21], or the smallest gearbox cost [18, 22], etc… Although there are many studies on calculating the optimal gear ratio for helical gearboxes, so far there is no research for the 4-stage helical gearbox with the cost objective function.

In this paper, a new study on calculating the optimal gear ratio of 4-stage helical gearbox has been reported. In the study, the influence of main design parameters on the optimal gear ratios of all levels was considered. In addition, regression formulas for determining the optimal gear ratios were introduced.

2 Optimization Problem

2.1 Cost Analysis of Three-Stage Helical Gearbox

The cost of a four-stage helical gearbox, C_{gb}, can be determined by:

$$C_{gb} = C_g + C_{gh} + C_s \tag{1}$$

Where, C_g, C_{gh}, and C_s are the cost of gears, the cost of gearbox housing, and the cost of shafts, respectively. These cost components are calculated by:

$$C_g = c_{g.m} \cdot m_g \tag{2}$$

$$C_{gh} = c_{gh.m} \cdot m_{gh} \tag{3}$$

$$C_s = c_{s.m} \cdot m_s \tag{4}$$

In the above Equations, $c_{g.m}$ is the cost per kilogram of gears (USD/kg); m_g is the mass of all gears (kg); $c_{gh.m}$ is the cost per kilogram of gearbox housing (USD/kg); m_{gh} is the mass of the gearbox housing (kg); $c_{s.m}$ is the cost per kilogram of the shaft (USD/kg); m_s is the mass of all shafts in the gearbox (kg).

2.2 The Determination of Gearbox Housing Mass

The mass of gearbox housing (m_{gh}) is calculated by the following equation:

$$m_{gh} = \rho_{gh} \cdot V_{gh} \tag{5}$$

Wherein, ρ_{gh} is the weight density of gearbox housing materials; As the gearbox housing material is cast iron, $\rho_{gh} = 7.2$ (kg/m^3) [23]; V_{gh} is the volume of the gearbox housing (m^3) which can be determined by:

$$V_{gh} = 2 \cdot V_b + 2 \cdot V_{A1} + 2 \cdot V_{A2} \tag{6}$$

Where, V_b, V_{A1}, and V_{A2} are the volumes of the bottom housing, the side A_1, and the side A_2 (kg), respectively. These volumes can be calculated as (see Fig. 1):

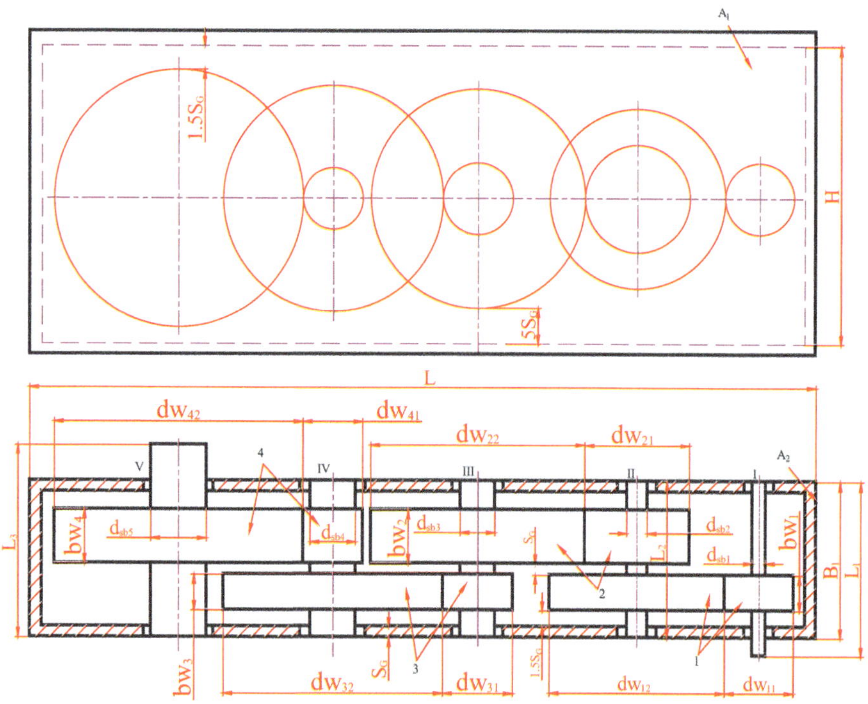

Fig. 1. Schema for determination of gearbox mass.

$$V_b = L \cdot B_1 \cdot 1.5 \cdot S_G \tag{7}$$

$$V_{A1} = L \cdot H \cdot S_G \tag{8}$$

$$V_b = B_2 \cdot H \cdot S_G = (B_1 - 2 \cdot S_G) \cdot H \cdot S_G \tag{9}$$

Substituting (7), (8), and (9) into (6) gets:

$$V_{gh} = 3 \cdot L \cdot B_1 \cdot S_G + 2 \cdot L \cdot H.S_G + 2 \cdot (B_1 - 2 \cdot S_G) \cdot H \cdot S_G \tag{10}$$

In which, L, H, B_1, and SG can be determined by:

$$L = (d_{w11} + d_{w21}/2 + d_{w12}/2 + d_{w22}/2 + d_{w13}/2 + d_{w23}/2 + d_{w14}/2 + d_{w24} + 20 \tag{11}$$

$$H = max(d_{21}, d_{22}, d_{23}, d_{24}) + 6.5 \cdot S_G \tag{12}$$

$$B_1 = \max(b_{w1}, b_{w3}) + \max(b_{w2}, b_{w4}) + 6 \cdot S_G \tag{13}$$

$$S_G = 0.005 \cdot L + 4.5 \ [24] \tag{14}$$

.

2.3 Gear Mass Calculations

For a four-stage helical gearbox, the mass of gears can be found by:

$$m_g = m_{g1} + m_{g2} + m_{g3} + m_{g4} \tag{15}$$

Wherein m_{g1}, m_{g2}, m_{g3}, and m_{g4} are the gear mass of the first, the second, the third, and the fourth stage (kg). The gear mass of stage i ($i = 1 \div 4$) can be determined by the following Equation:

$$m_{gi} = \rho_g \cdot \left(\frac{\pi \cdot e_1.d_{w1i}^2 \cdot b_{wi}}{4} + \frac{\pi \cdot e_2.d_{w2i}^2 \cdot b_{wi}}{4} \right) \tag{16}$$

In which, ρ_g is the weight density of gear material (kg/m^3); with steel gear, $\rho_{gh} = 7.2$ (kg/m^3) [23]; e_1 and e_2 are the volume coefficients of the drive gear and the driven gear, respectively. In practice, we can choose $e_1 = 1$, and $e_2 = 0.6$. b_{wi} is the width of the gears calculated by: $b_{wi} = X_{bai} \cdot a_{wi}$ (mm).

2.4 Shaft Mass Calculation

The mass of the shafts can be determined by:

$$m_s = \sum_{i=1}^{5} m_{si} \tag{17}$$

In which, m_{si} is the mass of shaft i ($i = 1 \div 5$), which is calculated as:

$$m_{si} = \rho_s \cdot \pi \cdot d_{si}^2 \cdot l_{si}/4 \tag{18}$$

Where, m_{si} is the mass of the shaft i of the gearbox (kg); ρ_s is the weight density of shaft material; with shaft material is steel, $\rho_{gh} = 7.2$ (kg/m^3) [23]; l_{si} is the length of shaft i ($i = 1 \div 5$) of the gearbox; l_{si} determined by (cf. Fig. 1):

$$l_{s1} = B_1 + 1.2 \cdot d_{s1} \tag{19}$$

$$l_{s2} = l_{s3} = l_{s4} = B_1 \tag{20}$$

$$l_{s5} = B_1 + 1.2 \cdot d_{s5} \tag{21}$$

Also, d_{si} is the diameter of shaft i ($i = 1 \div 5$) which can be determined by:

$$d_{si} = [T_{1i}/(0.2 \cdot [\tau])]^{1/3} \tag{22}$$

Where, $[\tau]$ is the allowable shear stress; $[\tau]$= 17 Mpa [25].

2.5 Determination of the Centre Distances of the Gear Stages

The center distance of the i stage of the gearbox can be calculated by Eq. (31) [25]:

$$a_{wi} = k_a \cdot (u_i + 1) \cdot \sqrt[3]{T_{1i} \cdot k_{H\beta}/([\sigma_{Hi}]^2 \cdot u_i \cdot X_{ba1})} \tag{23}$$

Where, $k_{H\beta}$ is the contacting load ratio for pitting resistance; In this case, it can be selected by 1.1 [25]; $[\sigma_{Hi}]$ is the allowable contact stress of the i stage ($i = 1 \div 4$) (MPa); k_a is the material coefficient; As the gear material is steel, k_a= 43; X_{ba1} is the wheel face width coefficient of the i stage; T_{1i} is the torque on the drive shaft of the i stage (Nmm) determined by:

$$T_{1i} = \frac{T_r}{\prod_{j=i}^{3} (u_i \cdot \eta_{hg}^{4-i} \cdot \eta_{be}^{5-i})} \tag{24}$$

The pinion and the gear pitch diameters of the i stage ($i = 1 \div 4$) is calculated by [25]:

$$d_{w1i} = 2 \cdot a_{wi}/(u_1 + 1) \tag{25}$$

$$d_{w2i} = 2 \cdot a_{wi} \cdot u_i \cdot /(u_i + 1) \tag{26}$$

2.6 Optimization Problem

From the above analysis, the optimization problem to find optimum gear ratios for getting minimum gearbox cost can be represented as follows:

$$\text{Minime} \quad C_{gb} \tag{27}$$

Satisfy the following constraint:

$$1 \le u_i \le 9 \tag{28}$$

In which i is the stage number ($i = 1 \div 4$).

3 Experimental Setup

To evaluate the influence of the main design parameters on the optimum gear ratios, a simulation experiment was performed. Thirteen main design parameters were selected for this survey (Table 1). The Minitab R19 software was used with the design of screening experiments (1/4 factorial equivalent to $2^{9-1} = 128$ experiments). Table 2 shows the experimental plan and output responses (the gear ratios u_2, u_3, and u_4).

Table 1. Main design parameters and their levels

Factor	Code	Unit	Low	High
Total gearbox ratio	u_t	–	70	150
Wheel face width coefficient of stage 1	X_{ba1}	–	0.3	0.4
Wheel face width coefficient of stage 2	X_{ba2}	–	0.3	0.4
Wheel face width coefficient of stage 3	X_{ba3}	–	0.3	0.4
Wheel face width coefficient of stage 4	X_{ba4}	–	0.3	0.4
Allowable contact stress of stage 1	AS_1	MPa	350	420
Allowable contact stress of stage 2	AS_2	MPa	350	420
Allowable contact stress of stage 3	AS_3	MPa	350	420
Allowable contact stress of stage 4	AS_4	MPa	350	420
Output torque	T_{out}	Nmm	1000	10000
Cost of gearbox housing	C_{gh}	USD/kg	1	5
Cost of gears	C_g	USD/kg	2	9
Cost of shafts	C_s	USD/kg	1.5	5

4 Results and Analysis

4.1 The Influence of Input Parameters

The influences of the main design parameters on optimum gear ratios u_2, u_3, and u_4 are shown qualitatively through the main effects plot (Fig. 2). For the effect of the design factors on u_2, from Fig. 2a, it can be seen that, when the total gearbox ratio u_t changes from 70 (encoding value is –1) to 150 (encoding value is +1), the objective function u_2 changes from 3.93031 to 5.24062. The slope of this graph is (5.24062 – 3.93031)/

2 = 0.655155. Similarly, the slope of u_2 when Xba_1, Xba_2, Xba_3, Xba_4, AS_1, AS_2, AS_3, AS_4, T_{out}, C_{gh}, C_g, and C_s change are: (4.58766 – 4.58328)/2 = 0.00219; (4.59422 – 4.57672)/2 = 0.00875; (4.60297 – 4.56797)/2 = 0.0175; (4.61937 – 4.55156)/2 = 0.033905; (4.63797 – 4.53297)/2 = 0.0525; (4.84359 – 4.32734)/2 = 0.258125; (4.61828 – 4.55266)/2 = 0.03281; (4.69703 – 4.47391)/2 = 0.1156; (4.58766 – 4.58328) = 0.00219; (4.59203 – 4.57891) = 0.00656; (5.09844 – 4.0725) = 0.51297 and (5.0 – 4.17094)/2 = 0.41453. Qualitative comparison shows that the slope of the influence graph of u_t is the largest; next is the graph C_g, C_s, AS_2, AS_4, AS_1, AS_3, X_{ba4}, X_{ba3}, C_{gh}, X_{ba2}, finally T_{out} and X_{ba1}.

Proceed similarly with u_3 and u_4, the influence of the main design parameters on these gear ratios is as follows: With u_3, the greatest influence's factor on u_3 is AS_3; followed by parameters C_g, C_s, AS_2, u_t, AS_4, T_{out}, X_{ba4}, X_{ba3}, C_{gh}, AS_1, X_{ba2} and finally Xba1. With u_4, the most influential parameter is C_s; followed by parameters C_g, AS_4, AS_3, C_{gh}, X_{ba3}, T_{out}, u_t, AS_2, X_{ba4}, AS_1 finally X_{ba1} and X_{ba2}.

Table 2. Experimental plan and the responses

Run order	u_t	X_{ba1}	X_{ba2}	X_{ba3}	X_{ba4}	AS_1	AS2	AS_3	AS_4	T_{out}	C_{gh}	C_g	C_s	u_2	u_3	u_4
1	150	0.4	0.4	0.4	0.3	420	350	420	420	1000	1	2	5	6.47	4.19	5.53
2	70	0.4	0.4	0.3	0.3	350	350	420	420	1000	5	2	1.5	3.95	4.26	3.44
3	150	0.3	0.3	0.3	0.3	350	420	350	350	10000	5	2	5	8.08	4.12	4.51
4	150	0.3	0.4	0.4	0.3	350	420	420	420	10000	1	9	1.5	4.09	3.56	3.56
5	150	0.4	0.4	0.4	0.4	350	420	350	420	1000	1	2	5	6.89	3.84	5.67
6	150	0.3	0.3	0.3	0.4	420	350	420	350	10000	5	2	5	7.38	4.89	4.16
127	70	0.3	0.3	0.3	0.3	350	420	350	350	10000	1	9.	1.5	3.67	3.14	3.30
128	70	0.4	0.4	0.4	0.4	350	420	350	420	1000	5	9	1.5	3.67	3.21	3.65

Figure 3 shows the Normal Plot of the standardized effects for u_2 (Fig. 3a), u_3 (Fig. 3b), and u_4 (Fig. 3c). In Fig. 3a (for u_2), the graph is plotted for the objective function with significance level $\alpha = 0.05$. Brown square points represent the significant factors. These factors include A (variable u_t), E (variable X_{ba4}), F (variable AS_1), G (variable AS_2), H (variable AS_3), J (variable AS_4), M (variable C_g), N (variable C_s) and 2nd order interactions: AD ($u_t \cdot X_{ba3}$), AG ($u_t \cdot AS_2$), AM ($u_t \cdot C_g$), AN ($u_t \cdot C_s$), DL ($X_{ba3} \cdot C_{gh}$), FM ($AS_1 \cdot C_g$)), GM ($AS_2 \cdot C_g$), JM ($AS_4 \cdot C_g$), JN ($AS_4 \cdot C_s$), KM ($T_{out} \cdot C_g$), LN ($C_{gh} \cdot C_s$), MN ($C_g \cdot C_s$). Accordingly, Variable A has the strongest influence, followed by M, N, AM, DL, AN, G, J, AD, AG, F, JM, MN, FM, LN, GM, JN, and E; last H same KM.

For u_3 and u_4, by doing the same as above, with u_3 (Fig. 3b), variable H has the strongest influence, followed by M, N, G, A, MN, J, K, E, D, L, DM, AM, DN, DL, AL, JM, EM, LM, JN, EN, KM, GM, F, AN, AD, GH, FM, HM finally KL. With u_4 (Fig. 3c) variable N has the strongest influence, followed by M, J, H, L, MN, D, LM, K, DM, DN, KL, A, G, KM, JN, JM, EN, E, FG, DL, EM, EL, BF, GK, AM, AL, JL, HN, HM.

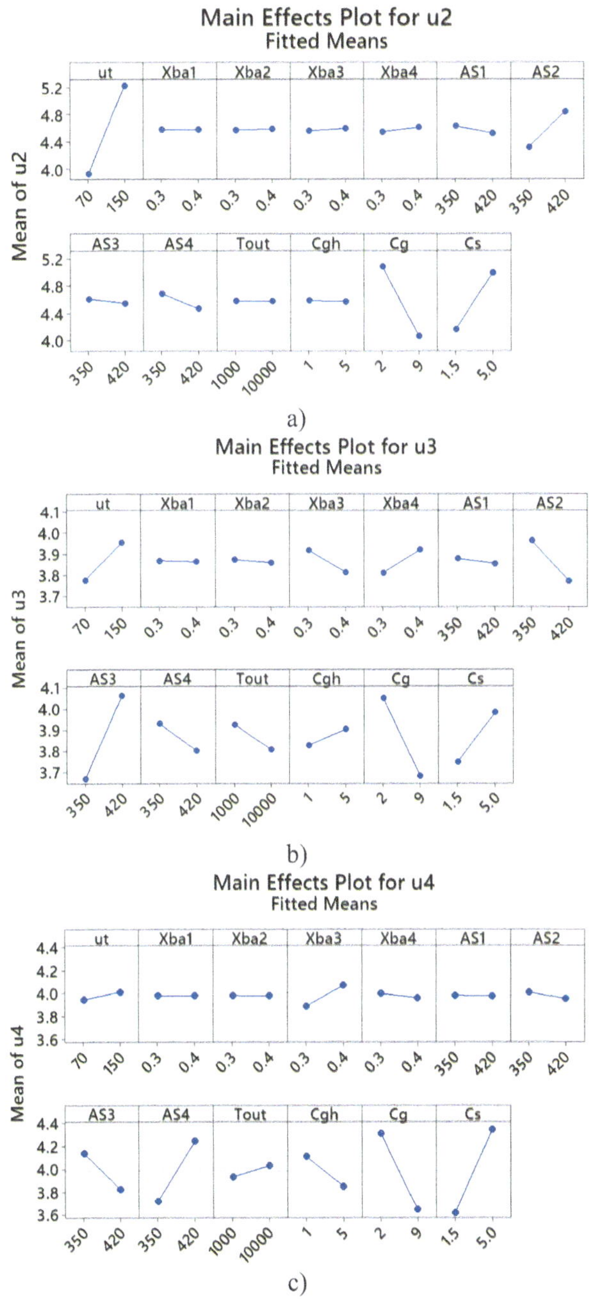

Fig. 2. Main effect plot for a) u_2 b) u_3 c) u_4

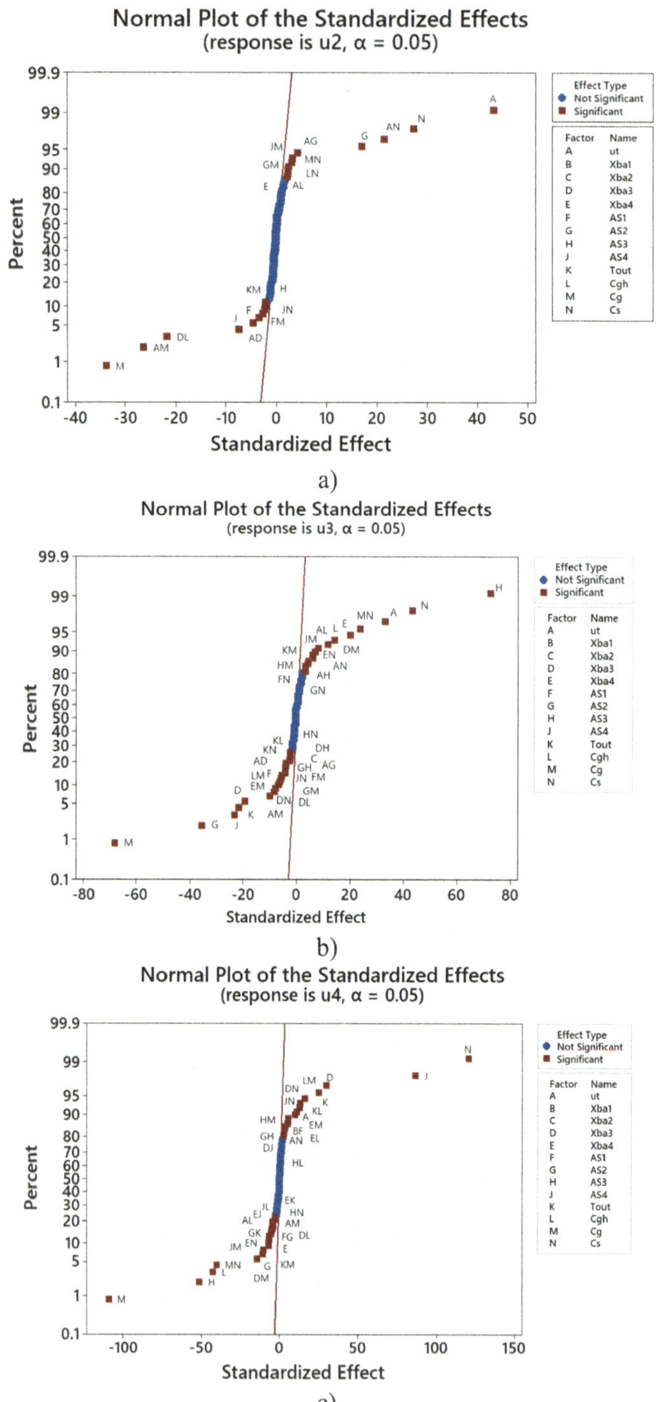

Fig. 3. Normal Plot of the Standardized Effects for a) u_2 b) u_3 c) u_4

4.2 Proposed Regression Models of the Responses

To find models for the determination of the response u_2, u_3, and u_4, a regression process with two interaction factors is performed with the help of the Minitab R19 software. The significance of this regression is $\alpha = 0.05$. The predictable effects and the coefficients for u_2 are displayed in Table 3 in which the factors with no influence on them are removed. It was found that if the impact of each main design parameter or interaction has P-value higher than the significance of α, it does not powerfully affect the response. For instance, the factor of X_{ba1} has p-value of 0.111 superior to $\alpha = 0.05$, which means that X_{ba1} is not significant to the response u_2. The regression equation of the u_2 response can be defined by the following equation:

$$\begin{aligned} u_2 = & -0.933 + 0.01179\ \mathrm{u_t} + 14.077\ \mathrm{X_{ba3}} + 0.678\ \mathrm{X_{ba4}} + 0.000268\ \mathrm{AS_1} \\ & + 0.00082\ \mathrm{AS_2} - 0.000937\ \mathrm{AS_3} - 0.003491\ \mathrm{AS_4} + 0.000011\ \mathrm{T_{out}} \\ & + 1.1145\ \mathrm{C_{gh}} + 0.0161\ \mathrm{C_g} - 0.1277\ \mathrm{C_s} - 0.03500\ \mathrm{u_t} * \mathrm{X_{ba3}} + 0.000045\ \mathrm{u_t} * \mathrm{AS_2} \\ & - 0.002852\ \mathrm{u_t} * \mathrm{C_g} + 0.004641\ \mathrm{u_t} * \mathrm{C_s} - 3.292\ \mathrm{X_{ba3}} * \mathrm{C_{gh}} - 0.000321\ \mathrm{AS_1} * \mathrm{C_g} \\ & + 0.000286\ \mathrm{AS_2} * \mathrm{C_g} + 0.000393\ \mathrm{AS_4} * \mathrm{C_g} - 0.000571\ \mathrm{AS_4} * \mathrm{C_s} \\ & - 0.000002\ \mathrm{T_{out}} * \mathrm{C_g} + 0.01063\ \mathrm{C_{gh}} * \mathrm{C_s} + 0.00768\ \mathrm{C_g} * \mathrm{C_s} \end{aligned} \quad (29)$$

Table 3. Estimated effects and coefficients for u_2

Coded Coefficients

Term	Effect	Coef	SE Coef	T-Value	P-Value	VIF
Constant		4.5855	0.0133	345.75	0.000	
ut	1.3103	0.6552	0.0133	49.40	0.000	1.00
Xba3	0.0350	0.0175	0.0133	1.32	0.190	1.00
Xba4	0.0678	0.0339	0.0133	2.56	0.012	1.00
AS1	-0.1050	-0.0525	0.0133	-3.96	0.000	1.00
AS2	0.5163	0.2581	0.0133	19.46	0.000	1.00
AS3	-0.0656	-0.0328	0.0133	-2.47	0.015	1.00
AS4	-0.2231	-0.1116	0.0133	-8.41	0.000	1.00
Tout	-0.0044	-0.0022	0.0133	-0.16	0.869	1.00
Cgh	-0.0131	-0.0066	0.0133	-0.49	0.622	1.00
Cg	-1.0259	-0.5130	0.0133	-38.68	0.000	1.00
Cs	0.8291	0.4145	0.0133	31.26	0.000	1.00
ut*Xba3	-0.1400	-0.0700	0.0133	-5.28	0.000	1.00
ut*AS2	0.1269	0.0634	0.0133	4.78	0.000	1.00
ut*Cg	-0.7984	-0.3992	0.0133	-30.10	0.000	1.00
ut*Cs	0.6497	0.3248	0.0133	24.49	0.000	1.00
Xba3*Cgh	-0.6584	-0.3292	0.0133	-24.82	0.000	1.00
AS1*Cg	-0.0787	-0.0394	0.0133	-2.97	0.004	1.00
AS2*Cg	0.0700	0.0350	0.0133	2.64	0.010	1.00
AS4*Cg	0.0962	0.0481	0.0133	3.63	0.000	1.00
AS4*Cs	-0.0700	-0.0350	0.0133	-2.64	0.010	1.00
Tout*Cg	-0.0656	-0.0328	0.0133	-2.47	0.015	1.00
Cgh*Cs	0.0744	0.0372	0.0133	2.80	0.006	1.00
Cg*Cs	0.0941	0.0470	0.0133	3.55	0.001	1.00

Model Summary

S	R-sq	R-sq(adj)	R-sq(pred)
0.150046	98.66%	98.36%	97.96%

The last row of the Table 3 shows the regression model evaluation parameters. The decision parameters R^2(symbol R-Sq) and $R - adj^2$ (denoted R-sq(adj)) are at least 97.96%, showing that the model fits the data very well.

Similar to the above, Tables 4 and 5 show the estimated effects and coefficients for u_3 and u_4 after removing the factors with insignificant influence. Also, the regression models for calculating u_3 and u_4 are found by Eqs. (30) and (31):

Table 4. Estimated effects and coefficients for u_3.

Coded Coefficients

Term	Effect	Coef	SE Coef	T-Value	P-Value	VIF
Constant		3.86953	0.00288	1345.76	0.000	
ut	0.17938	0.08969	0.00288	31.19	0.000	1.00
Xba2	-0.01312	-0.00656	0.00288	-2.28	0.025	1.00
Xba3	-0.10500	-0.05250	0.00288	-18.26	0.000	1.00
Xba4	0.10938	0.05469	0.00288	19.02	0.000	1.00
AS1	-0.02406	-0.01203	0.00288	-4.18	0.000	1.00
AS2	-0.19469	-0.09734	0.00288	-33.85	0.000	1.00
AS3	0.39594	0.19797	0.00288	68.85	0.000	1.00
AS4	-0.12688	-0.06344	0.00288	-22.06	0.000	1.00
Tout	-0.11812	-0.05906	0.00288	-20.54	0.000	1.00
Cqh	0.07656	0.03828	0.00288	13.31	0.000	1.00
Cg	-0.37188	-0.18594	0.00288	-64.67	0.000	1.00
Cs	0.23625	0.11812	0.00288	41.08	0.000	1.00
ut*Xba3	-0.02406	-0.01203	0.00288	-4.18	0.000	1.00
ut*AS2	-0.01313	-0.00656	0.00288	-2.28	0.025	1.00
ut*AS3	0.01750	0.00875	0.00288	3.04	0.003	1.00
ut*Cgh	0.04375	0.02187	0.00288	7.61	0.000	1.00
ut*Cq	-0.05469	-0.02734	0.00288	-9.51	0.000	1.00
ut*Cs	0.02406	0.01203	0.00288	4.18	0.000	1.00
Xba3*Cgh	-0.04375	-0.02187	0.00288	-7.61	0.000	1.00
Xba3*Cg	0.06344	0.03172	0.00288	11.03	0.000	1.00
Xba3*Cs	-0.04594	-0.02297	0.00288	-7.99	0.000	1.00
Xba4*Cg	-0.03719	-0.01859	0.00288	-6.47	0.000	1.00
Xba4*Cs	0.03281	0.01641	0.00288	5.71	0.000	1.00
AS1*Cq	-0.02187	-0.01094	0.00288	-3.80	0.000	1.00
AS1*Cs	0.01750	0.00875	0.00288	3.04	0.003	1.00
AS2*AS3	-0.02406	-0.01203	0.00288	-4.18	0.000	1.00
AS2*Cg	-0.03063	-0.01531	0.00288	-5.33	0.000	1.00
AS2*Cs	0.01750	0.00875	0.00288	3.04	0.003	1.00
AS3*Cg	0.02188	0.01094	0.00288	3.80	0.000	1.00
AS3*Cs	-0.01313	-0.00656	0.00288	-2.28	0.025	1.00
AS4*Cq	0.03719	0.01859	0.00288	6.47	0.000	1.00
AS4*Cs	-0.03281	-0.01641	0.00288	-5.71	0.000	1.00
Tout*Cgh	-0.02188	-0.01094	0.00288	-3.80	0.000	1.00
Tout*Cg	0.03281	0.01641	0.00288	5.71	0.000	1.00
Tout*Cs	-0.01531	-0.00766	0.00288	-2.66	0.009	1.00
Cgh*Cg	-0.03500	-0.01750	0.00288	-6.09	0.000	1.00
Cg*Cs	0.12906	0.06453	0.00288	22.44	0.000	1.00

Model Summary

S	R-sq	R-sq(adj)	R-sq(pred)
0.0325310	99.43%	99.20%	98.86%

Table 5. Estimated effects and coefficients for u_4

Coded Coefficients

Term	Effect	Coef	SE Coef	T-Value	P-Value	VIF
Constant		3.98563	0.00299	1334.32	0.000	
ut	0.06582	0.03291	0.00299	11.02	0.000	1.00
Xba1	-0.00224	-0.00112	0.00299	-0.37	0.709	1.00
Xba3	0.18104	0.09052	0.00299	30.30	0.000	1.00
Xba4	-0.03990	-0.01995	0.00299	-6.68	0.000	1.00
AS1	-0.00407	-0.00204	0.00299	-0.68	0.497	1.00
AS2	-0.06385	-0.03193	0.00299	-10.69	0.000	1.00
AS3	-0.31531	-0.15766	0.00299	-52.78	0.000	1.00
AS4	0.52730	0.26365	0.00299	88.26	0.000	1.00
Tout	0.09878	0.04939	0.00299	16.53	0.000	1.00
Cqh	-0.26160	-0.13080	0.00299	-43.79	0.000	1.00
Cg	-0.66691	-0.33345	0.00299	-111.63	0.000	1.00
Cs	0.73431	0.36715	0.00299	122.92	0.000	1.00
ut*Cqh	-0.02706	-0.01353	0.00299	-4.53	0.000	1.00
ut*Cg	-0.02869	-0.01434	0.00299	-4.80	0.000	1.00
ut*Cs	0.01934	0.00967	0.00299	3.24	0.002	1.00
Xba1*AS1	0.03122	0.01561	0.00299	5.23	0.000	1.00
Xba3*AS4	0.01577	0.00789	0.00299	2.64	0.010	1.00
Xba3*Cgh	-0.03344	-0.01672	0.00299	-5.60	0.000	1.00
Xba3*Cg	-0.08731	-0.04366	0.00299	-14.62	0.000	1.00
Xba3*Cs	0.07965	0.03983	0.00299	13.33	0.000	1.00
Xba4*AS4	-0.01718	-0.00859	0.00299	-2.88	0.005	1.00
Xba4*Tout	-0.01651	-0.00826	0.00299	-2.76	0.007	1.00
Xba4*Cgh	0.03300	0.01650	0.00299	5.52	0.000	1.00
Xba4*Cq	0.03308	0.01654	0.00299	5.54	0.000	1.00
Xba4*Cs	-0.04155	-0.02078	0.00299	-6.96	0.000	1.00
AS1*AS2	-0.03889	-0.01944	0.00299	-6.51	0.000	1.00
AS2*AS3	0.01690	0.00845	0.00299	2.83	0.006	1.00
AS2*Tout	-0.03102	-0.01551	0.00299	-5.19	0.000	1.00
AS3*Cgh	0.01518	0.00759	0.00299	2.54	0.013	1.00
AS3*Cg	0.02128	0.01064	0.00299	3.56	0.001	1.00
AS3*Cs	-0.02554	-0.01277	0.00299	-4.28	0.000	1.00
AS4*Cgh	-0.02588	-0.01294	0.00299	-4.33	0.000	1.00
AS4*Cg	-0.04221	-0.02110	0.00299	-7.07	0.000	1.00
AS4*Cs	0.05831	0.02915	0.00299	9.76	0.000	1.00
Tout*Cqh	0.07789	0.03894	0.00299	13.04	0.000	1.00
Tout*Cg	-0.06275	-0.03138	0.00299	-10.50	0.000	1.00
Cgh*Cg	0.15161	0.07580	0.00299	25.38	0.000	1.00
Cq*Cs	-0.24775	-0.12388	0.00299	-41.47	0.000	1.00

Model Summary

S	R-sq	R-sq(adj)	R-sq(pred)
0.0337942	99.80%	99.72%	99.59%

The decision parameters R^2(symbol R-Sq) and $R - adj^2$ (denoted R-sq(adj)) are at least 98.86% for u_3 (Table 4) and at least 99.59% for u_4 (Table 5) showing that these models are very suitable with the data.

$$
\begin{aligned}
u_3 = {} & 1.792 + 0.00344\,\mathrm{ut} - 0.1312\,\mathrm{Xba2} + 0.124\,\mathrm{Xba3} + 1.069\,\mathrm{Xba4} \\
& - 0.000317\,\mathrm{AS1} + 0.001739\,\mathrm{AS2} + 0.008607\,\mathrm{AS3} - 0.001777\,\mathrm{AS4} \\
& - 0.000012\,\mathrm{Tout} + 0.0861\,\mathrm{Cgh} - 0.1007\,\mathrm{Cg} + 0.0566\,\mathrm{Cs} - 0.00602\,\mathrm{ut} * \mathrm{Xba3} \\
& - 0.000005\,\mathrm{ut} * \mathrm{AS2} + 0.000006\,\mathrm{ut} * \mathrm{AS3} + 0.000273\,\mathrm{ut} * \mathrm{Cgh} - 0.000195\,\mathrm{ut} * \mathrm{Cg} \\
& + 0.000172\,\mathrm{ut} * \mathrm{Cs} - 0.2187\,\mathrm{Xba3} * \mathrm{Cgh} + 0.1813\,\mathrm{Xba3} * \mathrm{Cg} - 0.2625\,\mathrm{Xba3} * \mathrm{Cs} \\
& - 0.1062\,\mathrm{Xba4} * \mathrm{Cg} + 0.1875\,\mathrm{Xba4} * \mathrm{Cs} - 0.000089\,\mathrm{AS1} * \mathrm{Cg} + 0.000143\,\mathrm{AS1} * \mathrm{Cs} \\
& - 0.000010\,\mathrm{AS2} * \mathrm{AS3} - 0.000125\,\mathrm{AS2} * \mathrm{Cg} + 0.000143\,\mathrm{AS2} * \mathrm{Cs} \\
& + 0.000089\,\mathrm{AS3} * \mathrm{Cg} - 0.000107\,\mathrm{AS3} * \mathrm{Cs} + 0.000152\,\mathrm{AS4} * \mathrm{Cg} \\
& - 0.000268\,\mathrm{AS4} * \mathrm{Cs} - 0.000001\,\mathrm{Tout} * \mathrm{Cgh} + 0.000001\,\mathrm{Tout} * \mathrm{Cg} \\
& - 0.000001\,\mathrm{Tout} * \mathrm{Cs} - 0.002500\,\mathrm{Cgh} * \mathrm{Cg} + 0.010536\,\mathrm{Cg} * \mathrm{Cs}
\end{aligned}
\tag{30}
$$

$$
\begin{aligned}
u_4 = {} & 1.905 + 0.001445\,\mathrm{ut} - 3.456\,\mathrm{Xba1} + 0.470\,\mathrm{Xba3} + 1.449\,\mathrm{Xba4} \\
& + 0.00293\,\mathrm{AS1} + 0.00309\,\mathrm{AS2} - 0.007285\,\mathrm{AS3} + 0.007629\,\mathrm{AS4} \\
& + 0.000060\,\mathrm{Tout} - 0.1000\,\mathrm{Cgh} + 0.0473\,\mathrm{Cg} + 0.1267\,\mathrm{Cs} - 0.000169\,\mathrm{ut} * \mathrm{Cgh} \\
& - 0.000102\,\mathrm{ut} * \mathrm{Cg} + 0.000138\,\mathrm{ut} * \mathrm{Cs} + 0.00892\,\mathrm{Xba1} * \mathrm{AS1} + 0.00451\,\mathrm{Xba3} * \mathrm{AS4} \\
& - 0.1672\,\mathrm{Xba3} * \mathrm{Cgh} - 0.2495\,\mathrm{Xba3} * \mathrm{Cg} + 0.4551\,\mathrm{Xba3} * \mathrm{Cs} - 0.00491\,\mathrm{Xba4} * \mathrm{AS4} \\
& - 0.000037\,\mathrm{Xba4} * \mathrm{Tout} + 0.1650\,\mathrm{Xba4} * \mathrm{Cgh} + 0.0945\,\mathrm{Xba4} * \mathrm{Cg} \\
& - 0.2374\,\mathrm{Xba4} * \mathrm{Cs} - 0.000016\,\mathrm{AS1} * \mathrm{AS2} + 0.000007\,\mathrm{AS2} * \mathrm{AS3} \\
& - 0.000000\,\mathrm{AS2} * \mathrm{Tout} + 0.000108\,\mathrm{AS3} * \mathrm{Cgh} + 0.000087\,\mathrm{AS3} * \mathrm{Cg} \\
& - 0.000208\,\mathrm{AS3} * \mathrm{Cs} - 0.000185\,\mathrm{AS4} * \mathrm{Cgh} - 0.000172\,\mathrm{AS4} * \mathrm{Cg} \\
& + 0.000476\,\mathrm{AS4} * \mathrm{Cs} + 0.000004\,\mathrm{Tout} * \mathrm{Cgh} - 0.000002\,\mathrm{Tout} * \mathrm{Cg} \\
& + 0.010829\,\mathrm{Cgh} * \mathrm{Cg} - 0.020225\,\mathrm{Cg} * \mathrm{Cs}
\end{aligned}
\tag{31}
$$

Figure 4 shows the probability plots of the validity of proposed models for the response of u_2 (Fig. 4a), u_3 (Fig. 4b), and u_4 (Fig. 4c). These plots were found by using the Anderson–Darling test in Minitab@19. In the graphs, the data set is demonstrative by blue points. The middle line offerings the probability of normal distribution and the other two lines refer to limiting border with significance of 95%. It is reported that the data set for u_2, u_3, and u_4 are sited inside two limiting lines when the P-value of 0.487, 0.553, and 0.974 are much greater than α value of 0.05. This indicates that the converted data fit very well with the normal distribution.

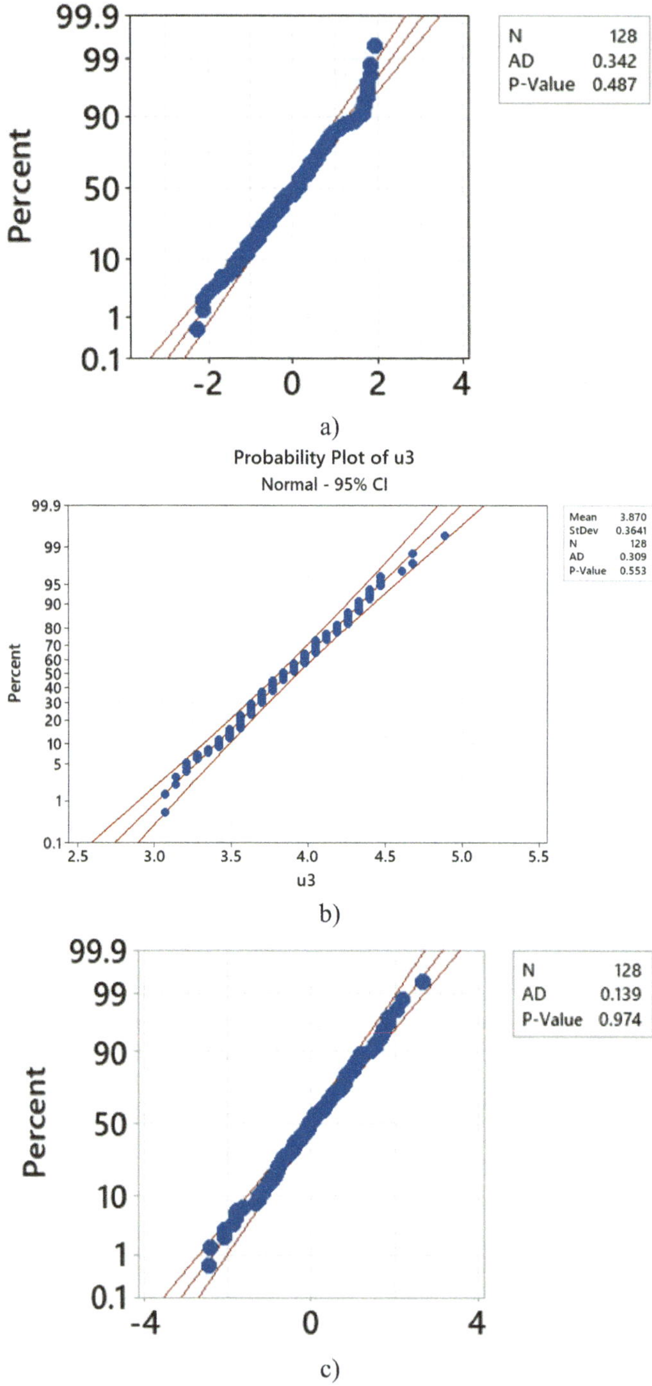

Fig. 4. Probability plots of the validity of regression model for a) u_2 b) u_3 c) u_4

5 Conclusions

This article reported the results of a study on the cost optimization problem to determine the partial gear ratio of a 4-stage helical gearbox. To solve this problem, a simulation experiment was designed and performed. Through the result analysis of the experiment, the influence of main design parameters on the optimal partial gear ratios for a four-stage helical gearbox was evaluated. In addition, regression equations to calculate the optimal gear ratios have been proposed. It was reported that these equations are in good agreement with the experimental data.

Acknowledgment. This work was supported by Thai Nguyen University of Technology.

References

1. Bartelmus, W., et al.: Modelling of gearbox dynamics under time-varying nonstationary load for distributed fault detection and diagnosis. Eur. J. Mech. A/Solids **29**(4), 637–646 (2010)
2. Medina, R., et al.: Vibration signal analysis using symbolic dynamics for gearbox fault diagnosis. Int. J. Adv. Manuf. Technol. **104**(5–8), 2195–2214 (2019). https://doi.org/10.1007/s00170-019-03858-0
3. Chaari, F., Zimroz, R., Bartelmus, W., Fakhfakh, T., Haddar, M.: Model based investigation on a two stages gearbox dynamics under non-stationary operations. In: Fakhfakh, T., Bartelmus, W., Chaari, F., Zimroz, R., Haddar, M. (eds.) Condition Monitoring of Machinery in Non-Stationary Operations, pp. 133–142. Springer, Heidelberg (2012). https://doi.org/10.1007/978-3-642-28768-8_14
4. Fan, X., Zuo, M.J.: Gearbox fault detection using Hilbert and wavelet packet transform. Mech. Syst. Signal Process. **20**(4), 966–982 (2006)
5. Lei, Y., et al.: Planetary gearbox fault diagnosis using an adaptive stochastic resonance method. Mech. Syst. Signal Process. **38**(1), 113–124 (2013)
6. Zhang, Z., Verma, A., Kusiak, A.: Fault analysis and condition monitoring of the wind turbine gearbox. IEEE Trans. Energy Convers. **27**(2), 526–535 (2012)
7. Inoue, K., Yamanaka, M., Kihara, M.: Optimum stiffener layout for the reduction of vibration and noise of gearbox housing. J. Mech. Des. **124**(3), 518–523 (2002)
8. Guan, Y.H., Lim, T.C., Shepard, W.S.: Experimental study on active vibration control of a gearbox system. J. Sound Vib. **282**(3–5), 713–733 (2005)
9. Nayak, U., Aravind, S., Aundhekar, S.: Influence of gear geometry on gearbox noise reduction-an experimental investigation. SAE Int. J. Commer. Veh. **7**(2014-01-9029), 746–752 (2014)
10. Wang, L., et al.: Wind turbine gearbox failure identification with deep neural networks. IEEE Trans. Ind. Inf. **13**(3), 1360–1368 (2016)
11. Zhao, M., Ji, J.: Dynamic analysis of wind turbine gearbox components. Energies **9**(2), 110 (2016)
12. Hong, T.T., et al.: Calculating optimum gear ratios of two step bevel helical reducer. Int. J. Appl. Eng. Res **14**, 3494–3499 (2019)
13. Pi, V.N.: A method for optimal calculation of total transmission ratio of two step helical gearboxes. In: Proceedings of the National Conference on Engineering Mechanics (2001)
14. Tuan, N.K., et al.: Determining optimal gear ratios of a two-stage helical reducer for getting minimal acreage of cross section. In: MATEC Web of Conferences. EDP Sciences (2018)

15. Hung, L.X. et al.: Calculation of optimum gear ratios of mechanical driven systems using two-stage helical gearbox with first stage double gear sets and chain drive. In: Sattler, K.U., Nguyen, D., Vu, N., Tien Long, B., Puta, H. (eds.) ICERA 2019. LNNS, vol. 104, pp. 170–178. Springer, Cham (2020). https://doi.org/10.1007/978-3-030-37497-6_20
16. Cam, N.T.H., et al.: Calculating optimum gear ratios of mechanical drive systems using two-stage helical gearbox with second-stage double gear sets and chain drive for minimum gearbox length. In: Sattler, K.-U., Nguyen, D.C., Vu, N.P., Tien Long, B., Puta, H. (eds.) ICERA 2019. LNNS, vol. 104, pp. 155–163. Springer, Cham (2020). https://doi.org/10.1007/978-3-030-37497-6_18
17. Pi, V.N.: A study on optimal determination of partial transmission ratios of helical gearboxes with second-step double gear-sets. Proc. World Acad. Sci. **27**, 113–116 (2008)
18. Vu, N.-P., et al.: The influence of main design parameters on the overall cost of a gearbox. Appl. Sci. **10**(7), 2365 (2020)
19. Pi, V.N., Tuan, N.K., Hung, L.X., Cam, N.T.H., Thao, T.T.P.: Determining optimum partial transmission ratios of mechanical driven systems using a V-belt drive and a three-stage helical reducer. In: Awang, M., Emamian, S., Yusof, F. (eds.) Advances in Material Sciences and Engineering. LNME, pp. 81–88. Springer, Singapore (2020). https://doi.org/10.1007/978-981-13-8297-0_11
20. Cam, N.T.H., Pi, V.N., Tuan, N.K., Hung, L.X., Thao, T.T.P.: A study on determination of optimum partial transmission ratios of mechanical driven systems using a chain drive and a three-step helical reducer. In: Fujita, H., Nguyen, D., Vu, N., Banh, T., Puta, H. (eds.) ICERA 2018. LNNS, vol. 63, pp, 91–99. Springer, Cham (2019). https://doi.org/10.1007/978-3-030-04792-4_14
21. Pi, V.N.: Optimal calculation of partial transmission ratios for four-step helical gearboxes with first and third-step double gear-sets for minimal gearbox length. In: American Conference on Applied Mathematics (MATH 2008), Harvard, Massachusetts, USA (2008)
22. Pi, V.N., et al.: Cost optimization study of two-step helical gearboxes with first stage double gear sets. Int. J. Mech. Prod. Eng. Res. Dev. (Ijmperd) **10**, 573–584 (2020)
23. Toolbox, T.E.: Density, Specific Weight and Specific Gravity. https://www.engineeringtoolbox.com/density-specific-weight-gravity-d_290.html. Accessed 27 May 2021
24. Römhild, I., Linke, H.: Gezielte Auslegung Von Zahnradgetrieben mit minimaler Masse auf der Basis neuer Berechnungsverfahren. Konstruktion **44**(7–8), 229–236 (1992)
25. Trinh Chat, L.V.U.: Calculation of Mechanical Driven Systems. Education Publisher (2008). (in Vietnamse)

Dynamic Response of Aluminum-Alloy Plates Subjected to Repeated Impacts

Dac Dung Truong(✉) and Nguyen Anh Vu Le

Department of Engineering Mechanics, Nha Trang University, Nha Trang City, Vietnam
{truongdacdung, vulna}@ntu.edu.vn

Abstract. During its service, marine structures may be damaged due to repeated impact loadings arising from various actions including ice/object impact and slamming. Since the aluminum alloy is a light and non-corrosive material, it is used in marine applications such as high-speed vessels. Several impact tests on aluminum alloy plates for investigating the response of structures have been reported in the literature. In this paper, a development of a finite element model was presented to study the response of aluminum alloy plates to repeated impacts. The numerical model was first validated through a comparison with the existing test data. Based on the material properties of the tested models, the strain hardening was defined using the proposed equations. Repeated impact loading was simulated by performing the calculation repeatedly, in which a deformed shape and residual stress/strain from the previous impact were considered in the consecutive impact events. After substantiation of the analysis method, further numerical simulations were carried out by varying several essential parameters such as striker mass and initial impact velocity to examine the impact response of aluminum alloy plates under repeated impacts. The results showed that lateral deformation of plates was accumulated, and impact forces increased when experiencing a repetition of impact loading regardless of the impact scenarios.

Keywords: Repeated impact · Aluminum alloy plate · Permanent deflection · Numerical analysis

1 Introduction

Ships and offshore structures are exposed to various impact loadings induced by slamming, sloshing, green water, and mass impacts during service life. Several studies on the effect of repetition of impact loads on the response of steel structures have been reported [1–11]. The damage extents of steel beams and plates have been extensively explored. Still, they have not yet been thoroughly investigated, especially for aluminum structures, and they are not explicitly considered in relevant rules of classification societies.

Aluminum alloys are vital materials due to their lightweight and non-corrosive properties [12]. The energy absorption capacity per unit density of aluminum is higher than that of conventional steel. Therefore, aluminum alloy panels were widely used for

D. C. Nguyen et al. (Eds.): ICERA 2021, LNNS 366, pp. 365–374, 2022.
https://doi.org/10.1007/978-3-030-92574-1_37

constructing high-speed vessels and offshore structures. To better understand the impact problem and develop reliable design tools, it is essential to thoroughly understand the material and structural dynamic behavior under extreme loadings such as repeated impacts.

Many researchers have only investigated the effect of a single impact on the behavior of aluminum-alloy beams and plates [13–18]. Several researchers have also tried to study the effect of repeated impacts on the response of small aluminum plates [1, 2]. Zhu and Faulkner [1] conducted experimentally and theoretically several repeated impact tests on both mild steel and aluminum plates to examine the damage extents of the plates under repeated loads. Huang et al. [2] performed repeated impact tests to examine the pseudo-shakedown behavior of the clamped aluminum and mild steel plates. They found that the membrane effect due to axial restraint plays a dominant role in the pseudo-shakedown of plates under a repeated impact.

With the recent advances in computational tools and considering the difficulties in conducting experiments, a nonlinear finite element analysis has become the preferable tool for assessing mechanically damaged structures. In particular, it is a useful tool to predict the extent of damage of structural components under impact loading. However, structural designs for ship collision require that numerical prediction of the damaged structures under impact loading be reliably accurate. However, before applying for structural design, the nonlinear dynamic analysis should be validated against experimental data. A full-scale model for impact tests would be valuable to examine the extent of ship damage under realistic test conditions. Still, these tests are costly and thus very rare. One approach is to perform a small-scaled impact test on typical ship structural members to validate the numerical methods for impact analysis. The ship's side plates can be damaged in a collision accident as permanent deformations across areas wider than the frame spacing. Therefore, prior to performing a numerical analysis of the full-scale model of ship structures as the complexity of the finite element model, verifying numerical models of the large deformation of plates would be necessarily considered.

In this study, a nonlinear finite element (FE) analysis is established to investigate the response of aluminum plates to repeated impacts. The developed FE model using the commercial package ABAQUS software is first validated against existing test data. After substantiation of the analysis method, further numerical simulations are then carried out by varying several important parameters such as the mass of striker and initial impact velocity to examine the impact response of aluminum alloy plates under various scenarios of repeated mass impacts.

2 Brief Description of Test Data

A series of tests on aluminum alloy plates under repeated mass impacts has been conducted by Zhu and Faulkner [1] using a runway testing machine at the University of Glasgow, UK. The experimental tests represent repeated impact events. A fully clamped rectangular plate is repeatedly collided by a rigid wedge mass traveling with an initial velocity at the center of the plate, as shown in Fig. 1.

The tip of the wedge was sharp, having a vertical length of $2l_d = 100$ mm. The mass range of the striker could be varied from 15 kg to 60 kg impact speed up to 5 m/s. The plate specimen with 250 mm in length and the width of 150 mm to 250 mm was made from a strain rate insensitive aluminum alloy. The thickness of the plates was 1.65 mm, and its mechanical properties are listed in Table 1. The plates were fixed during impacts. The striker was released from a pre-determined height to run freely along a runway, and the impact-expected position of the striker for the first collision was then configured. After each impact, the central permanent deflection of the plate was measured. Then, the subsequent impact test was performed after setting the same predefined height and position with the previous impact event. Two consecutive impacts were conducted for each model. The impact condition of the two models is summarized in Table 2. Details of the experimental setup can be found in Zhu and Faulkner [1]. Some of the plastic deformation test results of models AL07 and AL10 are employed to validate the FE model established in this study.

Table 1. Mechanical properties of materials [1].

Properties	Units	Values
Density, ρ	kg/m^3	2700
Young's modulus, E	MPa	70000
Poisson ratio, ν	–	0.33
Hardening modulus, E_h	MPa	310
Yield strength, σ_Y	MPa	115

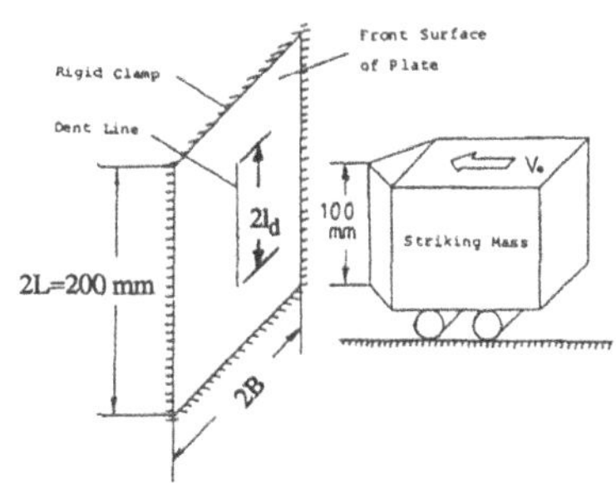

Fig. 1. Schematic of experimental setup [1].

Table 2. Impact test conditions for the models AL07 and AL10 [1].

Model	Mass of striker [kg]	Initial impact velocity [mm/s]		Tip length [mm]
		1st impact	2nd impact	
AL07	12.8	2413	2528	100
AL10	12.8	1826	1802	

3 Finite Element Modeling

3.1 Material Definition

The definition of the specimen plate material is most important. Thus, the material's mechanical properties need to be properly adopted in the simulation to provide a reliable representation of the plate response; the properties available from the relevant reference [19] are used for the current FE models. A simple true stress-strain relation, proposed by Park and Cho [20], is utilized. Note that the validity of the modeling of material behavior was confirmed in Refs. [21–24].

$$\sigma_{tr} = \sigma_Y + \left(\frac{EE_h}{E - E_h}\right)\varepsilon_{pl} \tag{1}$$

where σ_{tr}, ε_{tr} and ε_{pl} ($=\varepsilon_{\pm tr} - \sigma_{tr}/E$) are the true stress, true strain and true plastic strains, respectively.

In this study, since the plastic tested model was only considered, no fracture criterion is defined. Figure 2 shows the true stress-plastic strain curve used in the current analyses implemented in ABAQUS/Explicit [25] as tabulated data. No strain-rate sensitivity is defined in numerical simulations because aluminum is known as a strain-rate insensitive material.

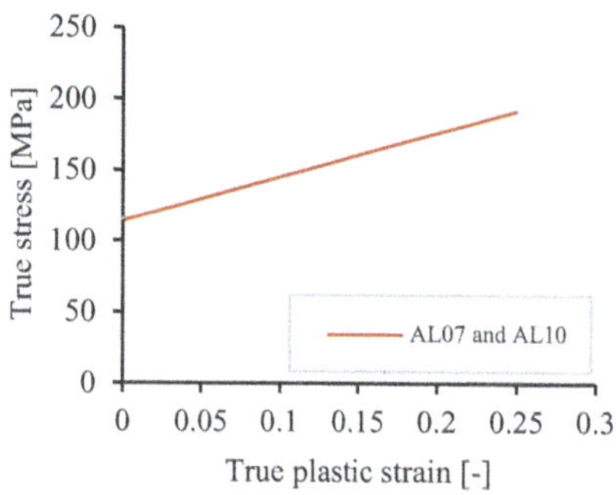

Fig. 2. True stress-plastic strain curve applied for the numerical analysis.

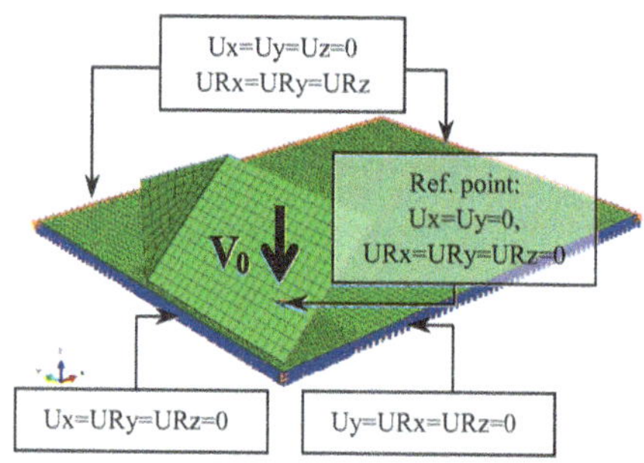

Fig. 3. FE model of the models AL07 and AL10.

3.2 Element and Mesh Size Design

The numerical model designed with two components of a plate and a striker is shown in Fig. 3. The plate is modeled by four-node doubly-curved shell elements with a reduced integration scheme (S4R) from the ABAQUS element library. The striker is modeled as a discrete rigid surface and meshes with the four-node 3-D bilinear rigid quadrilateral element (R3D4). Because of its symmetry conditions, a quarter model is used for all the analyses to save computation time.

A mesh convergence test is carried out to ensure the mesh refinement in the aluminum plate is sufficiently fine enough to capture the stresses and deformations with reasonable accuracy. The contact area of the model is expected to be modeled with fine mesh 1 × 1 mm. The coarse meshes are in the region far away from the contact area. Some median mesh sizes are used in the translation regions between the fine meshing and coarse meshing regions, as shown in Fig. 3.

3.3 Boundary Conditions and Contact Definition

In Zhu and Faulkner's experiments [1], the plate was fixed onto a strong steel frame through the bolts. Hence, the aluminum plate model is constrained in all degrees of freedom in the present FE model at all the plate's edges. Only the vertical translation is free for the striker, in whose direction the initial impact velocity is assigned. The boundary conditions of the numerical model are indicated in Fig. 3.

The contact algorithm, "General contact", is used to simulate the contact behavior between the striker and the plate model during impacts. The "Hard" frictionless contact is defined between striker and model.

The repeated impacts are simulated by performing the calculations repeatedly. The initial velocity is assigned to the striker to simulate the later impacts, and the subsequent impact simulation is started again. In each of the restarted analyses, the deformed shape, residual stress/strain in the current model due to the previous impact are considered.

3.4 Validation of Numerical Analysis

Figure 4 shows the deformation shape after second impacts including von Mises stress distribution on the plates AL10. As seen in the figure, the denting line is formed along the tip of the striker. In addition, the maximum stress can be seen at the ends of denting line due to the corner of the striker tip. Table 2 summarizes the maximum permanent deflection for the two models.

As seen in Fig. 5, the numerical results match well with the existing test data for maximum permanent deflection, especially for the lower kinetic energy model AL10 (i.e., lower value of impact velocity).

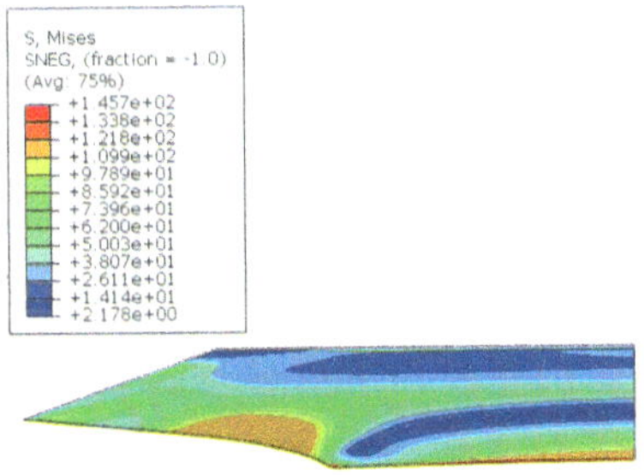

Fig. 4. Deformed shape of the quarter model AL10 after 2nd impact.

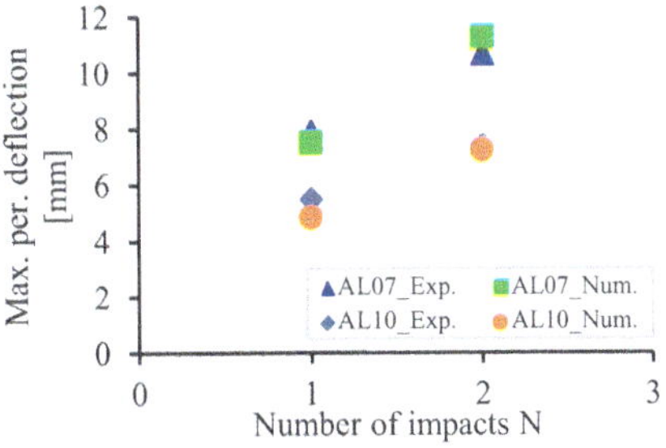

Fig. 5. Comparison of the test results and predictions for maximum permanent deflection.

Table 3 and Fig. 5 show that the maximum permanent deflections are gradually increased with the number of impacts. However, the increments of the permanent deflections can be reduced regardless of the impact velocities as increasing in repetitions of impact loads.

Table 3. A comparison of test and numerical results for maximum permanent deflection.

Model	Experimental [mm]		Numerical [mm]		Bias (Num./Exp.)	
	1st impact	2nd impact	1st impact	2nd impact	1st impact	2nd impact
AL07	7.95	10.71	7.59	11.33	0.95	1.06
AL10	5.55	7.40	4.93	7.29	0.89	0.99

The table and figure show some discrepancies between the tests and numerical predictions for AL07 and AL10 in the second impact. It is believed to be primarily caused by the possible difference in strain hardening model; since no detail in stress-strain curve is available in Ref. [1], a linear relation is assumed based on yield strength, elasticity modulus, and tangent modulus provided in Zhu and Faulkner [19]. The linear true stress-plastic strain curve is then applied with corresponding strains from the statistical standard aluminum group. Also, the bolt connection is not defined, which may lead to a somewhat deviation in predictions. Further experimental and theoretical investigations seem to be required, including the material property related to the strain hardening. However, better predictions on the permanent deflections are observed for lower velocity cases. In general, a good agreement between the test and the numerical results is achieved. It may thus be concluded that the current numerical models can simulate similar impact problems on aluminum alloy plates.

4 Numerical Results and Discussion

4.1 Impact Force

Figure 6 shows the simulated force-deflection curves for two impacts of the model AL10. The first impact yields the lowest impact force and the longer impact duration. As the number of impacts increases, the impact force increases, but the impact duration decreases significantly. In other words, when the number of impacts increases, the material can recover its elasticity.

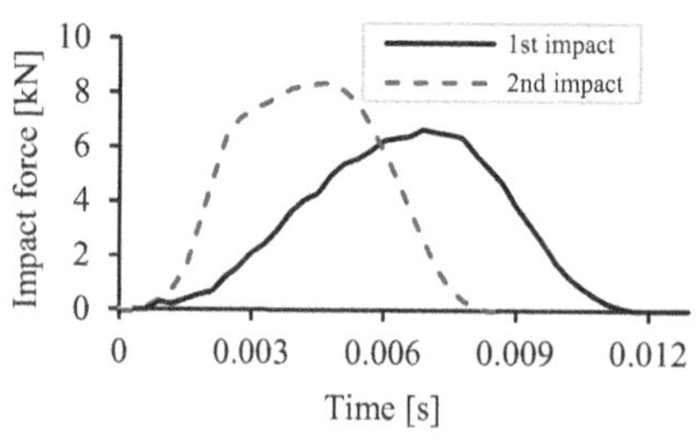

Fig. 6. Time variation of simulated impact forces.

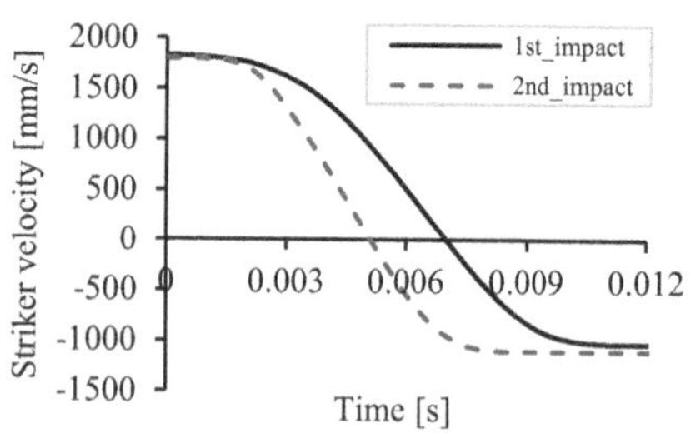

Fig. 7. Time variation of simulated striker impact velocity.

4.2 Velocity of Striker

The velocity-time histories of the striker of the model AL10 during impacts are revealed in Fig. 7. It shows an increase in the rebound velocity when increasing the number of impacts. Accordingly, the increase in the rebound velocity indicates the increase of the accelerations and the impact forces. As seen in the figure, in the second impact, the plate becomes stiffer, indicating a higher slope and rebound velocity than that in the first impact.

4.3 Impact Force-Deflection Curves

It is shown in Fig. 8 that at the beginning of the second impact, the force-deflection curve of the model AL10 rises along the unloading curve of the first impact. When the impact force reaches the maximum impact force of the first impact, the curve will change the slope and rise steadily with the same with the slope of the loading curve of the first impact. With the increase in impact number, the slope of the unloading curve at the second impact is steeper than that in the first impact. These observations are similar to those in Zhu [26].

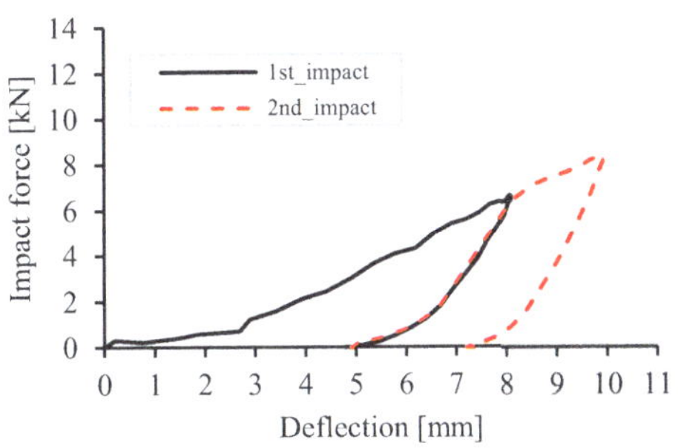

Fig. 8. Force-deflection curves for two consecutive impacts.

5 Parametric Study

The validated numerical model is employed to conduct a relatively extensive parametric study. Effects of some key parameters (striker mass and initial impact velocity) on the response of aluminum alloy plates against repeated impacts are investigated. The numerical analysis results are presented in terms of force-deflection curves. As a basis, all the test conditions of model AL10 showed are adopted in the parametric study.

5.1 Effect of Striker Mass

The impact energy is varied by increasing the mass of the striker. The force-deflection curves with the number of impacts for the different striker masses are shown in Fig. 9. As seen in the figure, the impact force and maximum permanent deflection of the plate increase almost linearly with the increase of impact velocity of the striker for two impacts. The larger the mass of the striker, the larger is the maximum deflection.

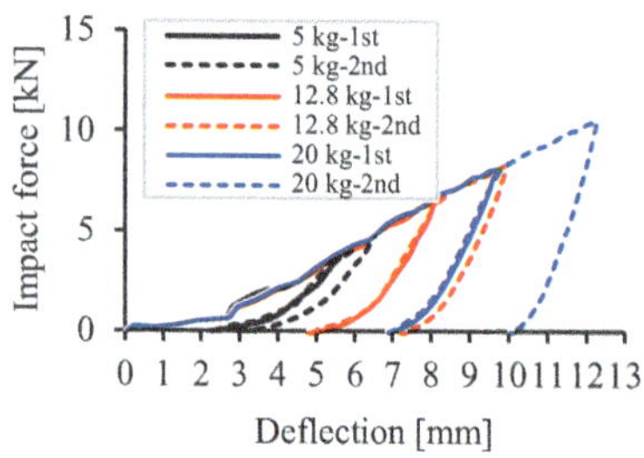

Fig. 9. Force-deflection curves for various striker masses.

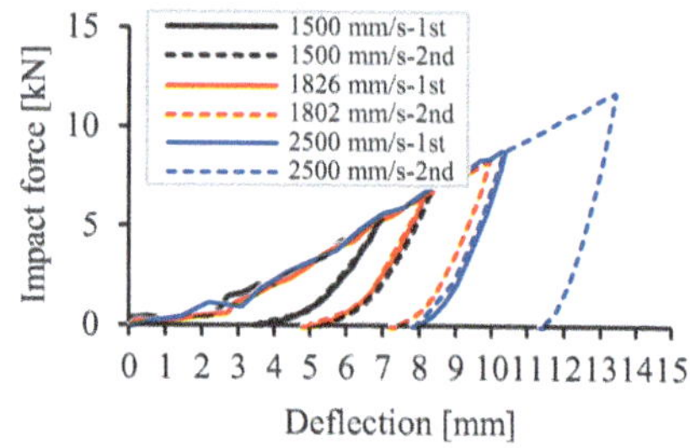

Fig. 10. Force-deflection curves for various initial impact velocities.

5.2 Effect of Striker Velocity

Secondly, the impact energy is varied by increasing the initial impact velocity. The impact energy is proportional to the square of velocity. For higher impact velocity, the energy to be dissipated will result in larger deformations and impact force and longer impact duration for whole impacts. The force-deflection curves for various impact velocities, shown in Fig. 10, indicate the slopes of these curves are the same, implying the plate's stiffness seems unchanged, possibly due to the omission of strain-rate effect. Strain-rate insensitivity is a typical characteristic of aluminum alloy. It is, however, important to note that for other materials, such as mild steel, the strain-rate sensitivity should be included [13].

6 Conclusions

In this paper, a FE model to analyze the dynamic response of aluminum alloy plates subjected to repeated impacts was developed. Impact tests on aluminum plates conducted by other researchers were employed for validating the numerical analysis methodology. The validated model was then used to evaluate the effects of some key parameters, such as the mass of the striker and initial impact velocity, on the repeated impact response of the aluminum plates. Force-deflection curves were used to compare the effects. From the results of this study, the following conclusions may be drawn:

- The FE model developed was successfully used to predict the damage of aluminum plates subjected to repeated impacts with a good agreement compared with tests in terms of permanent deflections. The FE analysis also reveals that the aluminum plate experiences local and global deformations during repeated impact loadings.
- The permanent deflection and its increment are reduced when the number of repeated impacts is increased. The structural stiffness of plates increases with an increase in the transverse deflections owing to membrane force generated during

impacts. The impact force increases, but the impact duration decreases as the impact number increases.

- As the striker mass increased, the damage of structures and collision force increased, regardless of the number of impacts. However, the impact duration can be changed accordingly.
- When the initial impact velocity increases, the deformation of plates increases as expected. At the same time, the stiffness of the structures may be unchanged because of the in-sensitive strain rate of aluminum alloy plates, and the impact duration also decreased with the number of impacts.

The results of this study can be applied to the structural design of ships and offshore structures subjected to repeated impacts from ice floes and/or floating objects as well as cargo floors exposed to luggage objects. However, further experimental and theoretical investigations seem necessary to improve the FE model, especially the material property related to strain rate effects and strain hardening.

Acknowledgments. This work was supported by the Nha Trang University, Vietnam, through Grant-in-aid TR2021-13-01.

References

1. Zhu, L., Faulkner, D.: Damage estimate for plating of ships and platforms under repeated impacts. Mar. Struct. **9**, 697–720 (1996)
2. Huang, Z.Q., Chen, Q.S., Zhang, W.T.: Pseudo-Shakedown in the collision mechanics of ships. Int. J. Impact Eng. **24**, 19–31 (2000)
3. Cho, S.-R., Truong, D.D., Shin, H.K.: Repeated lateral impacts on steel beams at room and sub-zero temperatures. Int. J. Impact Eng. **72**, 75–84 (2014)
4. Truong, D.D., Shin, H.K., Cho, S.-R.: Plastic response of steel grillages subjected to repeated mass impacts. In: Proceedings of the 7th International Conference on Collision and Grounding of Ships and Offshore Structures (ICCGS), Ulsan, South Korea, pp. 173–182 (2016)
5. Truong, D.D., Shin, H.K., Cho, S.-R.: Dynamic response of steel grillages under repeated mass impacts at low temperature. In: Proceedings of the 13th International Symposium on Practical Design of Ships and Other Floating Structures (PRADS), Copenhagen, Denmark (2016)
6. Seo, B.C., Truong, D.D., Cho, S.-R., Kim, D.J., Park, S.K., Shin, H.K.: A study on accumulated damage of steel wedges dead-rise 10° due to slamming loads. Int. J. Naval Arch. Ocean Eng. **10**(4), 520–528 (2018)
7. Truong, D.D., Shin, H.K., Cho, S.-R.: Repeated lateral impacts on steel grillage structures at room and sub-zero temperatures. Int. J. Impact Eng **113**, 40–53 (2018)
8. Truong, D.D., Shin, H.K., Cho, S.-R.: Response of low temperature steel beams subjected to single and repeated lateral impacts. Int. J. Naval Arch. Ocean Eng. **10**(6), 670–682 (2018)
9. Zhu, L., Shi, S., Jones, N.: Dynamic response of stiffened plates under repeated impacts. Int. J. Impact Eng **117**, 113–122 (2018)
10. Zeng, Y., Chen, H., Yu, R., Shen, Z., Yu, Z., Liu, J.: Experimental research on dynamic behavior of circular mild steel plates with surface cracks subjected to repeated impacts in low temperature. Shock Vibr. 3713709 (2020)

11. He, X., Guedes Soares, C.: Experimental study on the dynamic behavior of beams under repeated impacts. Int. J. Impact Eng. **147**, 103724 (2021)
12. Sutherland, L.S., Guedes Soares, C.: Impact behaviour of GRP, aluminium and steel plates. In: Analysis and Design of Marine Structures, pp. 293–300. CRC Press, Boca Raton (2009)
13. Liu, J., Jones, N.: Experimental investigation of clamped beams struck transversely by a mass. Int. J. Impact Eng. **6**(4), 305–335 (1987)
14. Hildrum, H.G., Malo, K.A., Langseth, M., Lademo, O.G.: Numerical modelling of fracture in stiffened aluminium plates subjected to impact loading. Struct. Shock Impact **7**, 133–142 (2002)
15. Fagerholt, F., Grytten, F., Gihleengen, B.E., Langseth, M., Børvik, T.: Continuous out-of-plane deformation measurements of Aa5083-H116 plates subjected to low-velocity impact loading. Int. J. Mech. Sci. **52**, 689–705 (2010)
16. Villavicencio, R., Sutherland, L., Guedes Soares, C.: Numerical simulation of transversely impacted, clamped circular aluminium plates. Ships Offshore Struct. **7**(1), 31–45 (2012)
17. Jones, N., Paik, J.K.: Impact perforation of aluminium alloy plates. Int. J. Impact Eng. **48**, 46–53 (2012)
18. Mohotti, D., Ali, M., Ngo, T., Lu, J.H., Mendis, P., Ruan, D.: Out-of-plane impact resistance of aluminium plates subjected to low velocity impacts. Mater. Des. **50**, 413–426 (2013)
19. Zhu, L., Faulkner, D.: Dynamic inelastic behavior of plates in minor ship collisions. Int. J. Impact Eng. **15**(2), 165–178 (1994)
20. Park, B.W., Cho, S.R.: Simple design formulae for predicting the residual damage of unstiffened and stiffened plates under explosive loadings. Int. J. Impact Eng **32**, 1721–1736 (2006)
21. Truong, D.D., Shin, H.K., Cho, S.-R.: Permanent set evolution of aluminium-alloy plates due to repeated impulsive pressure loadings induced by slamming. J. Mar. Sci. Technol. **23** (3), 580–595 (2017). https://doi.org/10.1007/s00773-017-0494-2
22. Truong, D.D., Jang, B.S., Ju, H.B., Han, S.W., Han, S.: A study on dynamic response of flat stiffened plates to slamming loads considering fluid-structure interaction. In: Proceedings of the 14th International Symposium on Practical Design of Ships and Other Floating Structures (PRADS-2019), Yokohama, Japan, pp. 75–99 (2019)
23. Truong, D.D., Jang, B.S., Ju, H.B., Han, S.W.: Prediction of slamming pressure considering fluid-structure interaction. Part I: numerical simulations. Ships Offshore Struct. (2020). https://doi.org/10.1080/17445302.2020.1816732
24. Truong, D.D., et al.: Benchmark study on slamming response of flat-stiffened plates considering fluid-structure interaction. Marine Struct. **79**, 103040 (2021)
25. ABAQUS User's Manual Version 6.10 (2010)
26. Zhu, L.: Modeling of repeated impacts on ships and offshore platforms. In: International Conference on Safety & Reliability of Ship, Offshore & Subsea Structures, Glasgow, U.K. (2014)

Effects of EDM Parameters on Surface Roughness and Electrode Wear Rate When Processing SKD11 Tool Steel

Nguyen Huu Quang[1], Tran Thanh Hoang[2], Nguyen Huu Luan[3], Le Hoang Anh[4], Nguyen Anh Tuan[1], Bui Thanh Danh[5], Tran Ngoc Giang[2], and Nguyen Manh Cuong[2](✉)

[1] University of Economics - Technology for Industries, Hanoi, Vietnam
[2] Thai Nguyen University of Technology, Thai Nguyen, Vietnam
nmcuong@tnut.edu.vn
[3] Nguyen Tat Thanh University, Ho Chi Minh City, Vietnam
[4] Vinh Long University of Technology Education, Vĩnh Long, Vietnam
[5] University of Transport and Communications, Hanoi, Vietnam

Abstract. This paper introduces a multi-objective optimization study in Electrical Discharge Machining (EDM) of cylindrical shaped parts made of SKD11 steel. In particular, the multi-objective problem consists of two single objectives including the surface roughness (SR) and the electrode wear rate (EWR). In addition, four main parameters of EDM process including the pulse time, the pulse off time, the current, and the servo voltage were selected for this study. The Taguchi method and the gray relation analysis were used to evaluate the effect of the above parameters on the SR and the EWR. Also, an optimal set of input parameters to ensure simultaneous optimization of two objectives, the SR and the EWR, are completely consistent with the experimental data.

Keywords: EDM · Surface roughness · Electrode wear rate · Multi-objective optimization · Cylindrical shaped parts · SKD11

1 Introduction

EDM is one of the most commonly used non-traditional machining methods for machining electrically conductive materials. It is often used to process materials or parts that are difficult to process by traditional methods. Therefore, the study of the EDM process has become an interesting and urgent topic in research.

Up to now, researches on EDM mainly focus on the following issues such as determining EDM process parameters to achieve maximum material removal rate [1, 2], minimum SR [3–8], minimum EWR [9, 10], simultaneously satisfying multiple objectives [11–15], investigating EDM process with different electrode materials such as copper [3–5, 10, 13, 14, 16], brass [13], graphite [13], AlCrNi Coated Aluminum [17], processing with different workpiece materials such as 90CrSi [3, 8–10, 14], SKD11 [16], SS 316 [13], Ti-6Al-4V [11, 17], etc. However, most of the studies to date have concentrated on EDM through-cavities or blind cavities. Recently, there have been some studies on EDM of cylindrical shaped parts [1, 3–5, 7–10, 14, 15]. Nevertheless, these studies have only been done for the workpiece material of 90CrSi.

D. C. Nguyen et al. (Eds.): ICERA 2021, LNNS 366, pp. 375–383, 2022.
https://doi.org/10.1007/978-3-030-92574-1_38

This paper introduces a multi-objective optimization study with two single objectives, SR and EWR, when machining cylindrical shaped parts with the workpiece material of SKD11 steel. In this paper, the impact of key EDM parameters on the SR and the EWR was evaluated. In addition, the optimal set of input EDM parameters to simultaneously obtain the minimum SR and the minimum EWR was found.

2 Experimental Setup

To investigate the influence of input EDM parameters on SR and EWR, four input factors were selected. Table 1 shows these parameters and their levels. The experimental plan was proposed by using the Taguchi method with orthogonal array design L9 (3^4) with the help of Minitab R19 software. Table 2 presents the experimental plan and the measured results of the SR and the EWR. To conduct this experiment, an experimental setup with the use of a sinker EDM machine with model CNC-AG40L, SKD11 steel workpieces and copper electrodes was built (Fig. 1).

Table 1. Input factors and their levels

No.	Parameters	Code	Unit	Level		
				1	2	3
1	Pulse on time	T_{on}	µs	6	16	26
2	Pulse off time	T_{off}	µs	8	18	28
3	Peak current	IP	A	3	7	11
4	Servo voltage	SV	V	3	6	9

Fig. 1. Experimental setup

To solve the multi-objective optimization problem, the Taguchi method and the gray relation analysis (GRA) were applied. In addition, two single objectives, i.e. the minimum SR and the minimum EWR, were employed.

The S/N index of both the SR and the EWR is calculated by the following Equation and described in Table 3:

$$S/N = -10\log_{10}(\frac{1}{n}\sum_{i=1}^{n} y_i^2) \quad (1)$$

Table 2. Experimental plan and measured SR and EWR

Exp. no.	T_{on}	T_{off}	IP	SV	Ra (μm)	EWR (mg/min)
1	6	8	3	3	1.6067	0.353
2	6	18	7	6	2.1993	0.868
3	6	28	11	9	1.8387	0.462
4	16	8	7	9	3.1843	0.467
5	16	18	11	3	2.3107	0.774
6	16	28	3	6	2.0447	0.341
7	26	8	11	6	2.9880	0.329
8	26	18	3	9	2.0310	0.204
9	26	28	7	3	4.1337	0.229

Table 3. S/N values of output parameters

Exp no.	Input parameters				Ra (μm)		EWR (mg/min)	
	T_{on}	T_{off}	IP	SV	Mean	S/N	Mean	S/N
1	6	8	3	3	1.6067	−4.1187	0.35325	9.0383
2	6	18	7	6	2.1993	−6.8675	0.86802	1.2294
3	6	28	11	9	1.8387	−5.3109	0.46209	6.7055
4	16	8	7	9	3.1843	−10.071	0.46729	6.6083
5	16	18	11	3	2.3107	−7.3032	0.77353	2.2304
6	16	28	3	6	2.0447	−6.2235	0.34099	9.3451
7	26	8	11	6	2.988	−9.5106	0.3294	9.6454
8	26	18	3	9	2.031	−6.1677	0.20391	13.8111
9	26	28	7	3	4.1337	−12.363	0.22899	12.8038

For gray relation analysis, the S/N values need to be converted to dimensionless series of quantities. Therefore, the data normalization is required. In this case, the S/N values are normalized by Zi values with $0 \leq Zi \leq 1$. In addition, the values of Zi are determined by the following formula:

$$Z_i = \frac{S/N_i - \min(S/N_i, i = 1, 2, ..., n)}{\max(S/N_i, i = 1, 2, ..., n) - \min(S/N_i, i = 1, 2, ..., n)} \quad (2)$$

Wherein, n is the number of experiments, n = 9.

Determining the Values of the Gray Relation

The gray relation coefficient $y_j(k)$ can be calculated by:

$$y_i(k) = (\Delta_{\min}(k) + \zeta.\Delta_{\max}(k))/(\Delta_i(k) + \zeta.\Delta_{\max}(k)) \tag{3}$$

In which, i = 1, 2,…, n; n is the number of experiments (n = 9); k is the number of output targets (k = 2); Δi(k) = ||Z0(k) – Zi(k)||, is the absolute value of the difference between $Z_0(k)$ (reference value Z0(k) = 1) and$Z_i(k)$) (Z-value of the ith experiment of the kth target); Δmin(k) is the minimum value of i(k); max(k) is the maximum value of i(k); ζ is the discriminant coefficient, determined in the range 0 ≤ ζ ≤, in experimental research ζ = 0.5.

The gray relation degree can be found by the average grey relation value (GRV) of the output objectives:

$$\overline{y_i} = \frac{1}{k}\sum_{j=0}^{k} y_{ij}(k) \tag{4}$$

Where, k is the output target number; y_{ij} is the GRV of the jth output targets in the ith experiment. The calculated results of GRV y_i and mean GRV of the experiments are listed in Table 4.

Table 4. S/N, S/N Z_i, $\mathbf{\Delta_i}$ (k) and grey relation value $\overline{y_i}$

Exp. no	S/N		Z_i		Δ_i (k)		Gray relational value yi		$\overline{y_i}$
	Ra	EWR	$\overline{Ra}$	EWR					
			Reference value		$\overline{Ra}$	EWR	$\overline{Ra}$	EWR	
			1.000	1.000					
1	−4.1187	9.0383	1.000	0.621	0.000	0.379	1.000	0.569	0.784
2	−6.8675	1.2294	0.667	0.000	0.333	1.000	0.600	0.333	0.467
3	−5.3109	6.7055	0.855	0.435	0.145	0.565	0.776	0.470	0.623
4	−10.0708	6.6083	0.278	0.428	0.722	0.572	0.409	0.466	0.438
5	−7.3032	2.2304	0.614	0.080	0.386	0.920	0.564	0.352	0.458
6	−6.2235	9.3451	0.745	0.645	0.255	0.355	0.662	0.585	0.623
7	−9.5106	9.6454	0.346	0.669	0.654	0.331	0.433	0.602	0.517
8	−6.1677	13.8111	0.751	1.000	0.249	0.000	0.668	1.000	0.834
9	−12.3628	12.8038	0.000	0.920	1.000	0.080	0.333	0.862	0.598

3 Results and Analysis

Theoretically, to ensure harmony between the output parameters, the average GRV should be as high as possible. Therefore, the multi-objective problem becomes a single-objective problem with the output parameter being the mean GRV. To solve this problem, the Taguchi method was applied to weigh the effect of the input factors on the mean GRV.

With the objective function being the mean gray relation value as large as possible, the S/N value of $\overline{y_i}$ is calculated according to Eq. 5.

$$S/N = -10\log_{10}(\frac{1}{n}\sum_{i=1}^{n}\frac{1}{y_i^2}) \tag{5}$$

The values of S/N were analyzed by ANOVA method and described in Table 5. From this Table, it is possible to understand the influence of the parameters on $\overline{y_i}$ (Fig. 2). The degree of influence can be described as follows: IP has the greatest influence on $\overline{y_i}$(65.16%), followed by the influence of T_{on} (23.37%), and SV 7.82%. T_{off} has very little influence on (3.66%). The influence order of the input factors on the GRV is presented in Table 6.

Table 5. Analysis of variance for S/N ratios of $\overline{y_i}$

Source	DF	Seq SS	Adj SS	Adj MS	F	P	C (%)
Ton	2	7.5572	7.5572	3.7786	*	*	23.37
Toff	2	1.1833	1.1833	0.5916	*	*	3.66
IP	2	21.0753	21.0753	10.5376	*	*	65.16
SV	2	2.5277	2.5277	1.2639	*	*	7.82
Residual Error	0	*	*	*			
Total	8	32.3435					

Table 6. Influence order of input factors on $\overline{y_i}$

Response Table for Signal to Noise Ratios

Larger is better

Level	Ton	Toff	IP	SV
1	-4.279	-5.005	-2.600	-4.454
2	-6.021	-4.991	-6.083	-5.485
3	-3.924	-4.229	-5.541	-4.286
Delta	2.097	0.776	3.483	1.199
Rank	2	4	1	3

Average gray relation value: 5.594

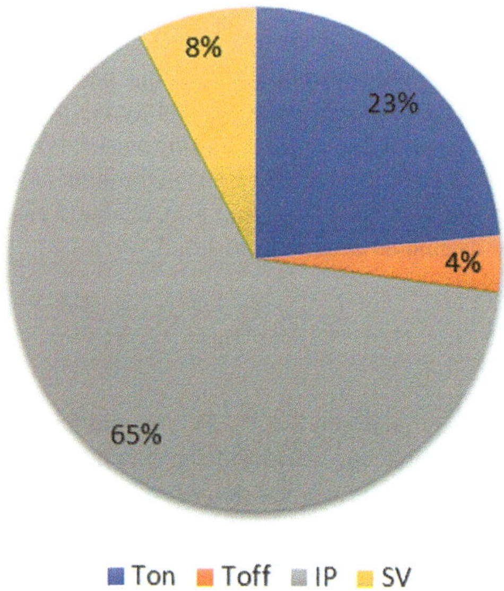

Fig. 2. Influence of input factors on $\overline{y_i}$ (%)

Fig. 3. Main effect plot for S/N ratios of $\overline{y}$

Figure 3 shows the graph of the effect of the input parameters on $\overline{y}$. The influence of each input parameter on yi can be found through this graph. As mentioned above, the input parameters with the highest $\overline{y}$ values are optimal parameters for the multi-objective function. Therefore, from Fig. 3, these optimal values can be simply found (Table 7).

Table 7. Optimal input parameters

Parameters	Code	Unit	Level	Value
Pulse on time	T_{on}	μs	3	26
Pulse off time	T_{off}	μs	3	28
Peak current	IP	A	1	3
Servo voltage	SV	V	3	9

Evaluating Experimental Model

Figure 4 shows the error distribution histograms. These graphs are used to evaluate the experimental model. Some comments can be made based on these graphs: The points representing the experimental error (blue dots) of the standard error distribution plot are close to the normal distribution line (red solid line). That means there is a very small degree of error. In addition, the error frequency graph shows that near-zero errors occur with high frequency. The other two graphs show that the experimental errors are distributed quite randomly. It shows that the built model is largely influenced by the input parameters and is not affected by the experimental order.

To evaluate the fit of the experimental model, the Anderson-Darling method was used. The evaluation results (Fig. 5) show that the data of the experimental points (the blue dots) are all placed within the bounds created by the upper and lower limits. In addition, 95% standard deviation with a P value of 0.295 (greater than $\alpha = 0.05$) shows that the applied empirical model is appropriate.

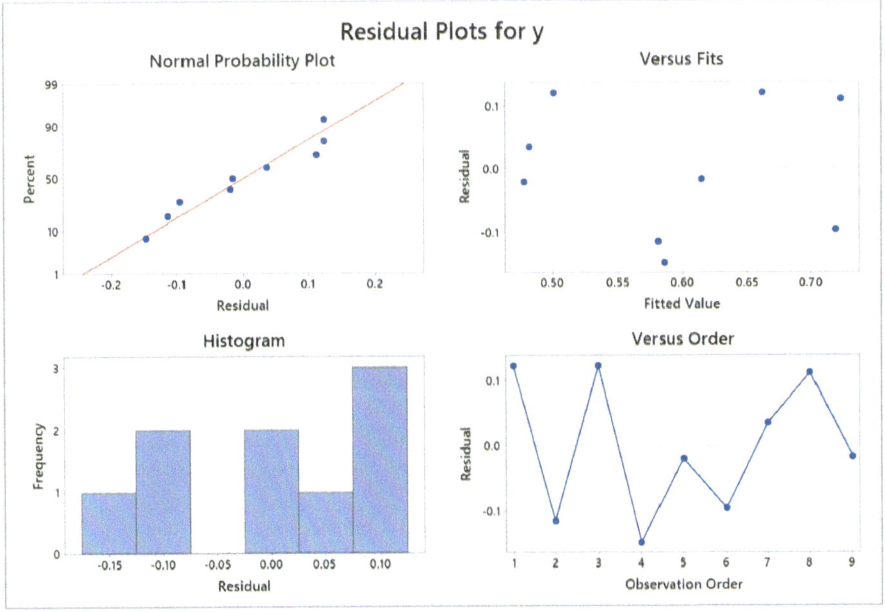

Fig. 4. Residual plots for $\bar{y}$

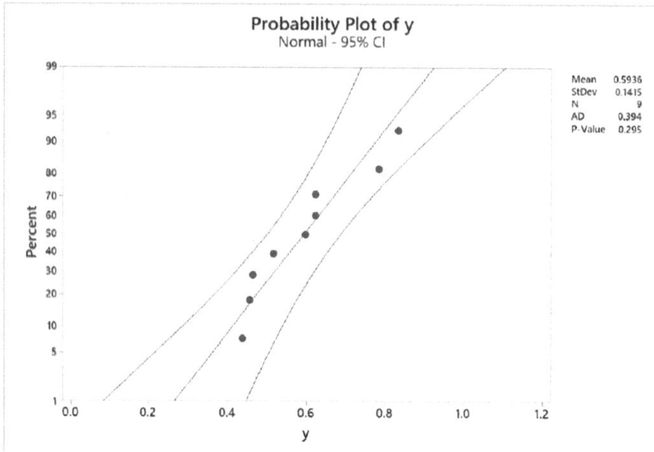

Fig. 5. Probability plot of $\overline{y}$

4 Conclusions

This paper reports the results of a study on multi-objective optimization when processing cylindrical parts machined with SKD11 steel material by EDM method. To accomplish this task, two single objectives including the SR and the EWR were chosen for the optimization problem. In addition, four input parameters including the pulse time, the pulse off time, the current, and the servo voltage were taken into the survey. The effect of these parameters on the SR and the EWR was explored by using the Taguchi method and the gray relation analysis. Furthermore, the optimal EDM factors to warrant two targets simultaneously, the SR, and the EWR, were proposed. It was also noted that the found empirical model is in perfect agreement with the experimental data.

Acknowledgment. This work was supported by Thai Nguyen University of Technology.

References

1. Hong, T.T., et al.: Influence of process parameters on material removal rate in electrical discharge machining cylindrical shaped parts. In: Materials Science Forum. Trans Tech Publ. (2021)
2. Singh, B., Kumar, J., Kumar, S.: Influences of process parameters on MRR improvement in simple and powder-mixed EDM of AA6061/10% SiC composite. Mater. Manuf. Processes **30**(3), 303–312 (2015)
3. Hong, T.T., et al.: Evaluating influences of input parameters on surface roughness in sinking EDM cylindrical shaped parts. In: Materials Science Forum. Trans Tech Publ. (2021)
4. Tran, T.H., et al.: Analysis of effects of machining parameters on surface roughness in electrical discharge machining tablet shape punches using Taguchi method. In: Materials Science Forum (2020)

5. Ky, L.H., et al.: Effect of electrical discharge machining on surface roughness of cylindrical shaped parts. In: Key Engineering Materials. Trans Tech Publ. (2020)
6. Huo, J., et al.: Influence of process factors on surface measures on electrical discharge machined stainless steel using TOPSIS. Mater. Res. Express **6**(8), 086507 (2019)
7. Le Xuan Hung, T.T.H., Pi, V.N.: A study on modelling surface finish in electrical discharge machining tablet shape punches using response surface methodology. J. Environ. Sci. Eng. B **6**, 387–390 (2017)
8. Hoang, T.T., et al.: Modelling surface finish in electrical discharge machining tablet shape punches using response surface methodology. SSRG Int. J. Mech. Eng. **4**(9), 28–30 (2017)
9. Hong, T.T., et al.: Effects of input parameters on electrode wear rate when EDM cylindrical shaped parts. In: Materials Science Forum Trans Tech Publ. (2021)
10. Cuong, N.V., et al.: Influence of process parameters on electrode wear in electrical discharge machining cylindrical shaped parts. In: Key Engineering Materials. Trans Tech Publ. (2020)
11. Kao, J., et al.: Optimization of the EDM parameters on machining Ti–6Al–4V with multiple quality characteristics. Int. J. Adv. Manuf. Technol. **47**(1–4), 395–402 (2010)
12. Marichamy, S., et al.: Optimization of machining parameters of EDM for α-β brass using response surface methodology. Mater. Today Proc. **24**, 1400–1409 (2020)
13. Choudhary, S., Kant, K., Saini, P.: Analysis of MRR and SR with different electrode for SS 316 on die-sinking EDM using Taguchi technique. Glob. J. Res. Eng. (2013)
14. Ky, L.H., et al.: Multi-objective optimization of surface roughness and electrode wear in EDM cylindrical shaped parts. In: Materials Science Forum. Trans Tech Publ. (2021)
15. Hong, T.T., et al.: Multi-objective optimization of process parameters during electrical discharge machining of hardened 90CrSi steel by applying Taguchi technique with Grey relational analysis. In: Sattler, K.-U., Nguyen, D.C., Vu, N.P., Long, B.T., Puta, H. (eds.) ICERA 2020. LNNS, vol. 178, pp. 572–583. Springer, Cham (2021). https://doi.org/10.1007/978-3-030-64719-3_63
16. Muthuramalingam, T., et al.: Experimental investigation of white layer thickness on EDM processed silicon steel using ANFIS approach. Silicon, 1–7 (2019)
17. Phan, N.H., et al.: Tool wear rate analysis of uncoated and AlCrNi coated aluminum electrode in EDM for Ti-6Al-4 V titanium alloy. In: Sattler KU., Nguyen D.C., Vu N.P., Long B.T., Puta H. (eds.) ICERA 2020. LNNS, vol. 178, pp. 832–838. Springer, Cham (2021). https://doi.org/10.1007/978-3-030-64719-3_91

Effect of EDM Parameters on Surface Roughness When Processing SKD11 Tool Steel

Nguyen Hong Linh[1], Tran Ngoc Giang[2], Nguyen Huu Quang[3], Nguyen Huu Luan[4], Le Hoang Anh[5], Nguyen Anh Tuan[3], Nguyen Thanh Tu[2], and Do Thi Tam[2(✉)]

[1] Electric Power University, Hanoi, Vietnam
[2] Thai Nguyen University of Technology, Thai Nguyen, Vietnam
dothitam@tnut.edu.vn
[3] University of Economics - Technology for Industries, Hanoi, Vietnam
[4] Nguyen Tat Thanh University, Ho Chi Minh City, Vietnam
[5] Vinh Long University of Technology Education, Vĩnh Long, Vietnam

Abstract. This paper presents the results of research on the effect of input parameters on the surface roughness (SR) when electrical discharge machining (EDM) of cylindrically shaped parts. In the study, the Taguchi method was used to design the experiment and analyze the results. Besides, the workpiece material is made from SKD11 tool steel, and four input process parameters, including the pulse time, the pulse off time, the current and the surve voltage were investigated. Influence of the input parameters on the SR was evaluated by the Analysis of Variance (ANOVA) method. Furthermore, optimal values of the input parameters have been proposed to achieve the smallest SR.

Keywords: EDM · Electrical discharge machining · Surface roughness · Cylindrical shaped parts

1 Introduction

Non-traditional machining methods were born and developed to meet the increasing demand for processing difficult-to-machine materials. Among the non-traditional machining methods, EDM is the most widely used. EDM process removes conductive material with an electrical pulse. It is regularly used to machine thin-walled parts, stamping and forging dies, small holes, etc. It is, of course, often used for machining difficult-to-machine materials. Consequently, there have been many studies on the process of this type of machining.

So far, studies on the EDM process have been carried out in different directions. Regarding electrodes for EDM, many studies have been carried out on using different electrode materials such as copper [1–4], graphite [4, 5] or Cu-W [6–8]. The wear of electrodes is also a topic of great interest [1]. Research on EDM has also been conducted when processing different materials such as AISI P20 [4], AISI 4340 [8], … In particular, many studies have been focused on determining the optimal input parameters to achieve the maximum material removal rate [2, 3, 9, 10], the minimum surface roughness [2, 3, 9], or the minimum electrode wear rate [3, 9, 10]. Besides, the study on EDM is also carried

D. C. Nguyen et al. (Eds.): ICERA 2021, LNNS 366, pp. 384–391, 2022.
https://doi.org/10.1007/978-3-030-92574-1_39

out by different methods such as the Taguchi method [11–13], gray relational analysis method [11, 14], response surface methodology (RSM) [15, 16], or Topsis [17, 18]. Recently, there have been some efforts to study on processing cylindrically shaped parts [12, 19, 20]. All of these works are conducted for 90CrSi steel.

From the above analysis, it can be seen that so far there have been quite a few studies on the EDM process. However, there is still a lack of research on the machining of SKD11 cylindrically shaped parts. The present work deals with the optimization of EDM process to get the minimum SR when machining the SKD11 cylindrical shaped part. In this work, influence of input factors on the SR was inspected. In addition, optimum input parameters for minimum SR were recommended.

2 Experimental Setup

To evaluate the influence of the input factors on the SR and to find the optimal parameters to achieve the minimum SR, an experiment was carried out. The Taguchi method was used with the support of Minitab 19 software to design and analyze the experimental results. In addition, four input parameters were selected for the survey. Table 1 shows these parameters and their levels. The experimental setup is described in Fig. 1. In the experiment, the workpiece material was SKD11 steel and the electrode material was copper. In addition, the machine used in the experiment was a sinker EDM (model CNC-AG40L). Besides, the experiment was planned with design L9 (3^4) by using the Taguchi method. The experimental plan and the measured SR were listed in Table 2.

Table 1. Input parameters and levels

No	Parameters	Code	Unit	Level		
				1	2	3
1	Pulse on time	T_{on}	µs	6	16	26
2	Pulse off time	T_{off}	µs	8	18	28
3	Peak current	IP	A	3	7	11
4	Servo voltage	SV	V	3	6	9

3 Results and Analysis

To weigh the influence of EDM factors on the SR, the ANOVA analysis tool of Minitab 19 software was used. The objective function is the minimum SR is the best:

$$S/N = -10\log_{10}\left(\frac{1}{n}\sum_{i=1}^{n} y_i^2\right) \quad (1)$$

Table 3 and Fig. 2 show the results of ANOVA. It was noted that IP has the greatest influence on SR (50.43%); followed is the effect of T_{on} (43.13%). The remaining two parameters have the smallest influence on SR with T_{off} of 4.91% and SV of 1.53%.

Table 4 shows the order of influence of the input parameters on S/N with IP, T_{on}, T_{off} and SV, respectively.

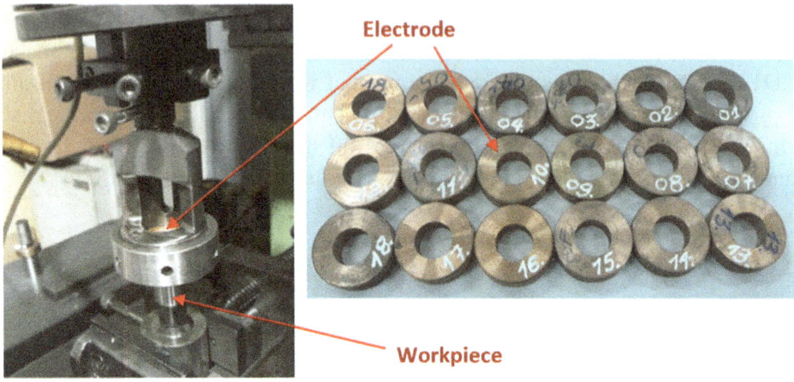

a) b)

Fig. 1. Experimental setup

Table 2. Experimental plan and measured SR

Exp. no.	SPD	VM	Ton	Toff	Ra (μm)			Mean	S/N
					Trial 1	Trial 2	Trial 3		
1	6	8	3	3	1.606	1.593	1.621	1.6067	−4.1187
2	6	18	7	6	2.036	2.153	2.409	2.1993	−6.8675
3	6	28	11	9	1.961	1.892	1.663	1.8387	−5.3109
4	16	8	7	9	2.975	3.349	3.229	3.1843	−10.0708
5	16	18	11	3	2.420	2.047	2.465	2.3107	−7.3032
6	16	28	3	6	2.157	1.908	2.069	2.0447	−6.2235
7	26	8	11	6	3.099	2.927	2.938	2.9880	−9.5106
8	26	18	3	9	2.189	1.975	1.929	2.0310	−6.1677
9	26	28	7	3	3.600	4.383	4.418	4.1337	−12.3628

Table 3. Analysis of variance for S/N ratios

Source	DF	Seq SS	Adj SS	Adj MS	F	P	C (%)
Ton	2	23.4402	23.4402	11.7201	*	*	43.13
Toff	2	2.6671	2.6671	1.3336	*	*	4.91
IP	2	27.4047	27.4047	13.7024	*	*	50.43
SV	2	0.8338	0.8338	0.4169	*	*	1.53
Residual Error	0	*	*	*			
Total	8	54.3458					

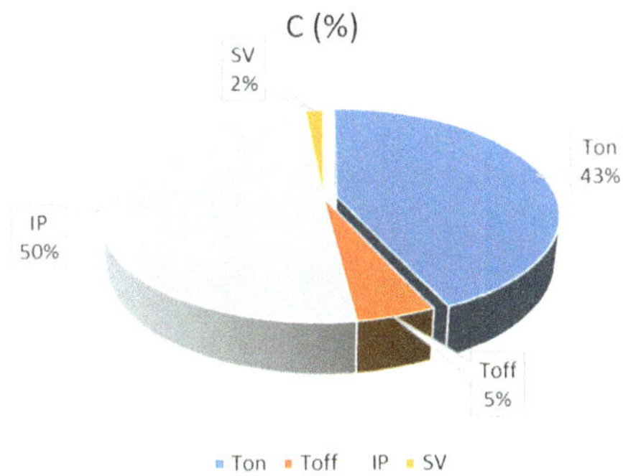

Fig. 2. Influence of input factors on SR (%)

Figure 3 depicts the influence of the input factors on the S/N values, with the goal that the smaller the SR, the larger the S/N value. From the Figure, the SR increases as T_{on} increases from 6 to 26 μs. Besides, the SR decreases as Toff increases from 8 to 18 μs, and it increases again when T_{off} increases from 18 to 28 μs. With IP, the SR increases when IP increases from 3 to 7A. When IP continues to increase from 7 to 11A, the SR decreases but it is still smaller than the value of the SR when IP is 3A. For the servo voltage, when SV increases from 3 to 9 V, the SR decreases. Thus, to achieve the minimum surface roughness, the maximum S/N value for each input parameter. From Fig. 3, a set of optimum input fators for the smallest SR value was found (corresponding with the red points on the Figure) as shown in Table 5. The SR value corresponding to the optimum factors is calculated based on the prediction method by using the Minitab 19 software. The calculated results are shown in Table 6. From the results in Table 6, the minimum SR corresponding to the optimum input factors can reach 0.861667 μm.

Table 4. Response table for signal to noise ratios.

Smaller is better

Level	Ton	Toff	IP	SV
1	-5.432	-7.900	-5.503	-7.928
2	-7.866	-6.779	-9.767	-7.534
3	-9.347	-7.966	-7.375	-7.183
Delta	3.915	1.186	4.264	0.745
Rank	2	3	1	4

Fig. 3. Effect of EDM factors on the S/N ratios

Table 5. Optimum process parameters for minimum SR

Parameter	Code	Unit	Level	Value
Pulse on time	T_{on}	μs	1	6
Pulse off time	T_{off}	μs	2	18
Peak current	IP	A	1	3
Servo voltage	SV	V	3	9

Table 6. Predicted optimum SR

Prediction

S/N Ratio	Mean	StDev	Ln(StDev)
-2.25303	0.861667	0.0264971	-2.87369

Settings

Ton	Toff	IP	SV
1	2	1	3

Evaluation of Experimental Models

The experimental model is evaluated through the standard error distribution charts (Fig. 4). From the graph it can be seen that the errors of the experimental points (corresponding to the blue points) are close to the normal distribution line (red solid line). That shows that the error level is very small. Besides, the Histogram chart shows the frequency of errors. The other two plots show the random distribution of experimental errors. These graphs show that the built model is mainly influenced by the input factors and is not affected by the experimental order.

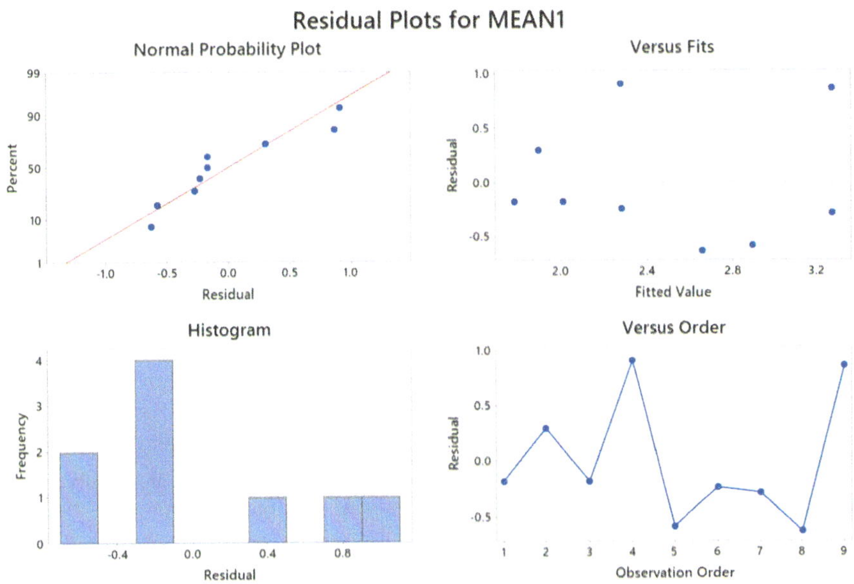

Fig. 4. Residual plots for SR

The fit of the experimental model and the optimal set of parameters was evaluated by the Anderson-Darling method with the use of the Minitab R19 software (see Fig. 5). From Fig. 5, it can be seen that the data corresponding to the experimental points (blue dots) are all within the region bounded by two upper and lower limits with 95% significance level. In addition, a P value of 0.160 greater than a value of $\alpha = 0.05$ indicates that the applied empirical model is in agreement with the above significance level.

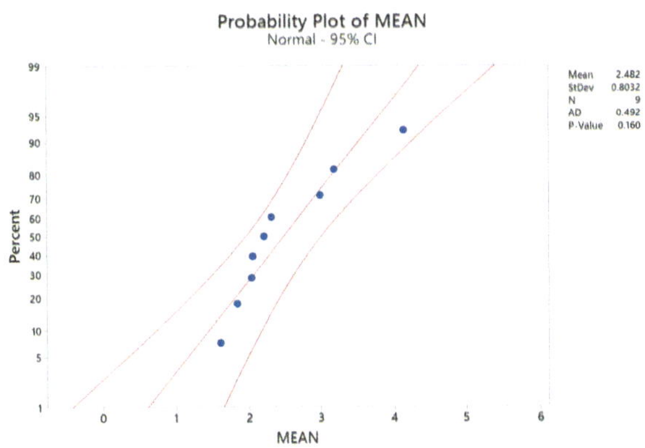

Fig. 5. Probability plot of SR

4 Conclusions

This paper introduces the results of a research on the influence of EDM factors on SR when machining cylindrically shaped parts made from SKD11 steel. To accomplish this problem, an experiment was designed and performed using Taguchi method for experimental design and analysis. In addition, four main input factors including the pulse on time, the pulse off time, the current and the serve voltage were investigated. The effect of the input parameters on the SR were investigated by ANOVA. Moreover, optimal input factors have been suggested to achieve minimum SR.

Acknowledgment. This work was supported by Thai Nguyen University of Technology.

References

1. Khan, A.A.: Electrode wear and material removal rate during EDM of aluminum and mild steel using copper and brass electrodes. Int. J. Adv. Manuf. Technol. **39**(5), 482–487 (2008)
2. Torres, A., Luis, C., Puertas, I.: EDM machinability and surface roughness analysis of TiB2 using copper electrodes. J. Alloy. Compd. **690**, 337–347 (2017)
3. Balasubramanian, P., Senthilvelan, T.: Optimization of machining parameters in EDM process using cast and sintered copper electrodes. Procedia Mater. Sci. **6**, 1292–1302 (2014)
4. Amorim, F.L., Weingaertner, W.L.: The behavior of graphite and copper electrodes on the finish die-sinking electrical discharge machining (EDM) of AISI P20 tool steel. J. Braz. Soc. Mech. Sci. Eng. **29**(4), 366–371 (2007)
5. Tang, Y., et al.: A new technique for the fabrication of graphite EDM electrodes. J. Mater. Process. Technol. **166**(2), 199–204 (2005)
6. Ferreira, J.C.: A study of die helical thread cavity surface finish made by Cu-W electrodes with planetary EDM. Int. J. Adv. Manuf. Technol. **34**(11–12), 1120–1132 (2007)
7. Davim, J., et al.: Performance of cutting tools in machining Cu/W alloys for application in EDM electrodes. Int. J. Refract. Metals Hard Mater. **27**(4), 676–682 (2009)
8. Rizvi, S.A.H., et al.: Modeling of black layer and tool wear in EDM of AISI 4340 using Cu-W electrode. Procedia CIRP **95**, 389–392 (2020)
9. Niamat, M., et al.: Effect of different dielectrics on material removal rate, electrode wear rate and microstructures in EDM. Procedia Cirp **60**, 2–7 (2017)
10. Praveen, L., et al.: Effects of pulse ON and OFF time and electrode types on the material removal rate and tool wear rate of the Ti-6Al-4V Alloy using EDM machining with reverse polarity. In: IOP Conference Series: Materials Science and Engineering. IOP Publishing (2018)
11. Jung, J.H., Kwon, W.T.: Optimization of EDM process for multiple performance characteristics using Taguchi method and Grey relational analysis. J. Mech. Sci. Technol. **24**(5), 1083–1090 (2010)
12. Hong, T.T., et al.: Effects of input parameters on electrode wear rate when EDM cylindrical shaped parts. In: Materials Science Forum. Trans Tech Publ. (2021)
13. Rahang, M., Patowari, P.K.: Parametric optimization for selective surface modification in EDM using Taguchi analysis. Mater. Manuf. Process. **31**(4), 422–431 (2016)
14. Roy, A.K., Kumar, K.: Effect and optimization of various machine process parameters on the surface roughness in EDM for an EN41 material using Grey-Taguchi. Procedia Mater. Sci. **6**, 383–390 (2014)

15. Le Xuan Hung, T.T.H., Pi, V.N.: A study on modelling surface finish in electrical discharge machining tablet shape punches using response surface methodology. J. Environ. Sci. Eng. B **6**, 387–390 (2017)
16. Hoang, T.T., et al.: Modelling surface finish in electrical discharge machining tablet shape punches using response surface methodology. SSRG Int. J. Mech. Eng. **4**(9), 28–30 (2017)
17. Manivannan, R., Kumar, M.P.: Multi-attribute decision-making of cryogenically cooled micro-EDM drilling process parameters using TOPSIS method. Mater. Manuf. Process. **32** (2), 209–215 (2017)
18. Huo, J., et al.: Influence of process factors on surface measures on electrical discharge machined stainless steel using TOPSIS. Mater. Res. Express **6**(8), 086507 (2019)
19. Ky, L.H., et al.: Multi-objective optimization of surface roughness and electrode wear in EDM cylindrical shaped parts. In: Materials Science Forum. Trans Tech Publ. (2021)
20. Hong, T.T., et al.: Evaluating influences of input parameters on surface roughness in sinking EDM cylindrical shaped parts. In: Materials Science Forum. Trans Tech Publ. (2021)

Effects of EDM Parameters on Electrode Wear Rate When Machining SKD11 Steel

Le Hoang Anh[1], Nguyen Manh Cuong[2], Tran Ngoc Huy Thinh[3], Trinh Kieu Tuan[4], Nguyen Anh Tuan[4], Tran Thanh Hoang[2], Tran Ngoc Giang[2], and Nguyen Thanh Tu[2](✉)

[1] Vinh Long University of Technology Education, Vĩnh Long, Vietnam
[2] Thai Nguyen University of Technology, Thai Nguyen, Vietnam
nguyenthanhtucnvl@tnut.edu.vn
[3] Nguyen Tat Thanh University, Ho Chi Minh City, Vietnam
[4] University of Economics - Technology for Industries, Hanoi, Vietnam

Abstract. This paper presents a study on the influence of input parameters on the electrode wear rate (EWR) when electrical discharge machining (EDM) of cylindrical shaped parts made from SKD11 steel. In this study, the Taguchi method was used to design and analyze the results of the experiments. In addition, the input process parameters including the pulse time, the pulse off time, the current, and the serve voltage were investigated. The influence of the input parameters on the EWR was evaluated by applying Analysis of Variance (ANOVA). In addition, optimal values of the input parameters have been proposed to obtain the minimum EWR.

Keywords: EDM · Electrode wear rate · Cylindrical shaped parts · SKD11

1 Introduction

In the mechanical engineering industry, alloy tool steels are widely used. In particular, SKD11 steel is commonly applied to make cold stamping dies, and parts requiring high strength, hardness and good wear resistance due to its high carbon and chromium content, good strength, oxidation resistance, high wear resistance, and good balance between hardness and ductility. Therefore, research on machining SKD11 has been a topic of interest to many researchers.

Up to now, SKD11 steel has been studied with different processing methods. With traditional machining methods, there have been studies on turning [1, 2], milling [3–6], and grinding [7–10]. With non-traditional processing methods, laser [11, 12], Electrical discharge machining (EDM) [13, 14], and Wire EDM [15, 16], etc. were selected to study the processing of SKD11 steel.

Among non-traditional machining methods, EDM is the most commonly used. In EDM, the material is removed from the workpiece by thermal energy. This machining process is highly effective when machining holes and cavities with complex shapes. In EDM processing, the problem of determining the optimal input parameters when processing SKD11 steel has been interested in [17, 18]. The influence of powders mixed into the dielectric solution on the EDM process of SKD11 steel has been evaluated

D. C. Nguyen et al. (Eds.): ICERA 2021, LNNS 366, pp. 392–399, 2022.
https://doi.org/10.1007/978-3-030-92574-1_40

through many studies such as the effect on the material removal rate (MRR) [13, 19], the electrode wear rate (EWR) [13, 20, 21], and the surface roughness (SR) [13, 14]. The influence of EDM process parameters on the machining hole diameter is covered in [22].

From the above analysis, although there have been quite a few studies on machining SKD11 by using EDM method, so far there has been no study on EWR when EDM cylindrical shaped parts with SKD11 steel. Therefore, this paper introduces an optimization study on EWR when EDM the mentioned parts made from SKD11 tool steel. In the paper, the influence of the input factors on the EWR was examined. Also, a set of optimum EDM parameters for minimum EWR was suggested.

2 Experimental Setup

To evaluate the influence of the process factors when processing SKD11 steel, four input parameters were investigated with the levels described in Table 1. The setup of the experiment is presented in Fig. 1. A sinker EDM machine (model CNC-AG40L - Sodick Europe Ltd., UK) was used for the experiment. The electrode material was copper and the workpiece material was SKD11 steel. In addition, the experimental matrix is designed based on the survey levels of the variables using the Taguchi method with design L9 (3^4) and the Minitab R19 software. The experimental matrix and the measured EWR are shown in Table 2.

Table 1. Input factors and their levels

No	Parameters	Code	Unit	Level		
				1	2	3
1	Pulse on time	T_{on}	μs	6	16	26
2	Pulse off time	T_{off}	μs	8	18	28
3	Peak current	IP	A	3	7	11
4	Servo voltage	SV	V	3	6	9

Fig. 1. Experimental setup

Table 2. Experimental matrix and measured EWR

No.	SPD	VM	T_{on}	T_{off}	EWR (mg/min)	Mean	S/N
1	6	8	3	3	0.353	0.35325	9.0383
2	6	18	7	6	0.868	0.86802	1.2294
3	6	28	11	9	0.462	0.46209	6.7055
4	16	8	7	9	0.467	0.46729	6.6083
5	16	18	11	3	0.774	0.77353	2.2304
6	16	28	3	6	0.341	0.34099	9.3451
7	26	8	11	6	0.329	0.32940	9.6454
8	26	18	3	9	0.204	0.20391	13.8111
9	26	28	7	3	0.229	0.22899	12.8038

3 Results and Analysis

To evaluate the influence of the input parameters on the EWR, an analysis of variance (ANOVA) is used with the help of the Minitab 19 software. Also, the objective function is *the smallest EWR is the best*:

$$S/N = -10\log_{10}\left(\frac{1}{n}\sum_{i=1}^{n} y_i^2\right) \quad (1)$$

The results of ANOVA are presented in Table 3 and Fig. 2. It can be learned that T_{on} has the greatest influence on ERW (53.6%), and followed by the effect of TP (24.73%). The other two parameters have a small influence on electrode wear (T_{off} of 16.17% and SV of 5.5%).

Table 3. Analysis of variance for S/N ratios

Source	DF	Seq SS	Adj SS	Adj MS	F	P	C (%)
Ton	2	77.802	77.8019	38.9009	*	*	53.60
Toff	2	23.467	23.4673	11.7337	*	*	16.17
IP	2	35.893	35.8927	17.9463	*	*	24.73
SV	2	7.982	7.9818	3.9909	*	*	5.50
Residual Error	0	*	*	*			*
Total	8	145.144					

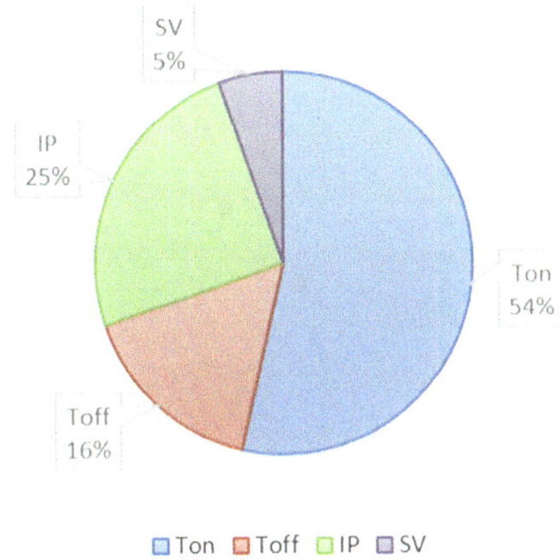

Fig. 2. Influence on EWR (%)

Table 4. The order of influence of input parameters on S/N ratios

Response Table for Signal to Noise Ratios

Smaller is better

Level	Ton	Toff	IP	SV
1	5.658	8.431	10.731	8.024
2	6.061	5.757	6.881	6.740
3	12.087	9.618	6.194	9.042
Delta	6.429	3.861	4.538	2.302
Rank	1	3	2	4

Table 4 displays the influence order of input parameters with a decreasing sequence is T_{on}, IP, T_{off} and SV.

Figure 3 depicts the effect of the input parameters on the S/N ratios. From the figure, it is observed that EWR decreases by a small amount when T_{on} increases from 6 to 16 μs. It decreases by a large amount when T_{on} rises from 16 to 26 μs. Besides, EWR declines when T_{off} grows from 8 to 18 μs and drops again when T_{off} increases from 18 to 28 μs. The EWR decreases when IP increases from 3 to 11A. Also, EWR increases when SV increases from 3 to 6 V. However, when SV continues to increase from 6 to 9 V, EWR decreases. Thus, to achieve the minimum electrode wear, the S/N value is maximum for each input parameter. Therefore, from Fig. 3, a set of optimum parameters for the smallest EWR was found as shown in Table 5.

Fig. 3. Effect of input parameters on the S/N ratios

Table 5. Optimum input parameters for smallest EWR

Parameter	Code	Unit	Level	Value
Pulse on time	T_{on}	μs	3	26
Pulse off time	T_{off}	μs	3	28
Peak current	IP	A	1	3
Servo voltage	SV	V	3	9

Evaluation of Experimental Models

The experimental model is evaluated through error distribution charts as shown in Fig. 4. From Fig. 4, the errors of the experimental points (the blue points) are close to the normal distribution line (the red solid line). That means the error level is very small. The error frequency graph also shows that the errors are mostly in the range −0.2 to 0. Besides, the other two graphs show the random distribution of the mean experimental errors, which means that the built model is largely influenced by the selected input parameters and is not affected by the order of experiments.

The fit of the experimental model and the optimal parameters was checked by Anderson-Darling method with the use of the Minitab 19 software (see Fig. 5). Also, the P value is 0.100 larger than $\alpha = 0.05$. It indicates that the applied experimental model is consistent with the above significance level.

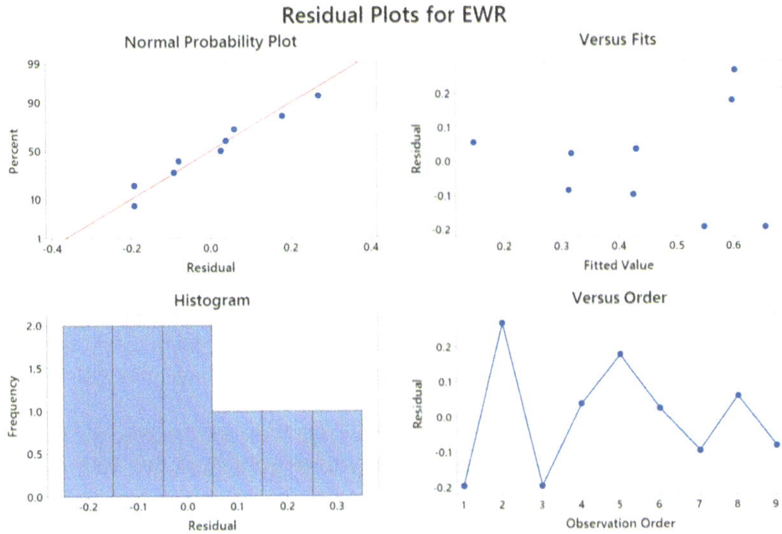

Fig. 4. Residual plots for EWR

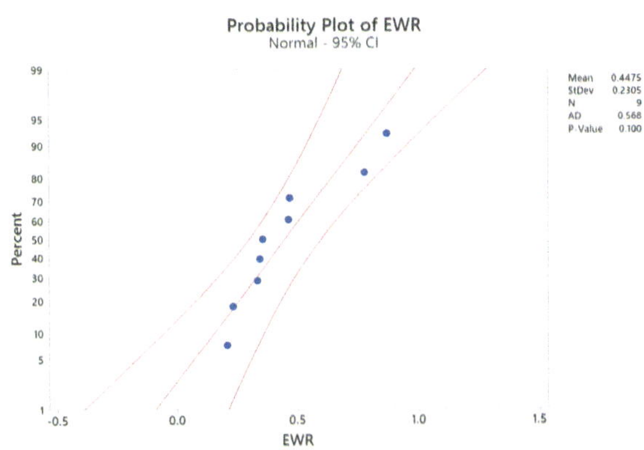

Fig. 5. Probability plot of EWR

4 Conclusions

In this work, an optimization study on EWR rate when EDM cylindrical shaped parts was done. Four input factors including the pulse on time, the pulse off time, the current, and voltage and SKD11 tool steel work material were selected for the investigation. The experiment was planned and analyzed using Taguchi method. The effect of input factors on the EWR were evaluated by ANOVA. In addition, the optimum input parameters were found to get the minimum EWR.

Acknowledgment. This work was supported by Thai Nguyen University of Technology.

References

1. Tzeng, C.-J., et al.: Optimization of turning operations with multiple performance characteristics using the Taguchi method and Grey relational analysis. J. Mater. Process. Technol. **209**(6), 2753–2759 (2009)
2. Lin, W.S.: The study of high speed turning of the hardened mold steel SKD11 by ceramic cutting tools. In: Key Engineering Materials. Trans Tech Publ. (2008)
3. Gong, F., et al.: Fatigue failure of coated carbide tool and its influence on cutting performance in face milling SKD11 hardened steel. Int. J. Refract. Metals Hard Mater. **64**, 27–34 (2017)
4. Li, J., Jing, L., Chen, M.: An FEM study on residual stresses induced by high-speed end-milling of hardened steel SKD11. J. Mater. Process. Technol. **209**(9), 4515–4520 (2009)
5. Dong, P.Q., Duc, T.M.: Performance evaluation of MQCL hard milling of SKD 11 tool steel using MoS2 nanofluid. Metals **9**(6), 658 (2019)
6. Gong, F., Zhao, J., Ni, X., Liu, C., Sun, J., Zhang, Q.: Wear and breakage of coated carbide tool in milling of H13 steel and SKD11 hardened steel. SN Appl. Sci. **1**(9), 1–12 (2019). https://doi.org/10.1007/s42452-019-1152-6
7. Hong, T.T., et al.: Multi response optimization of dressing conditions for surface grinding SKD11 steel by HaiDuong grinding wheel using grey relational analysis in Taguchi method. In: Sattler, K.-U., Nguyen, D.C., Vu, N.P., Long, B.T., Puta, H. (eds.) ICERA 2020. LNNS, vol. 178, pp. 560–571. Springer, Cham (2021). https://doi.org/10.1007/978-3-030-64719-3_62
8. Hong, T.T., et al.: Optimization of dressing parameters in surface grinding SKD11 tool steel by using Taguchi method. In: Sattler, K.-U., Nguyen, D.C., Vu, N.P., Long, B.T., Puta, H. (eds.) ICERA 2020. LNNS, vol. 178, pp. 636–647. Springer, Cham (2021). https://doi.org/10.1007/978-3-030-64719-3_70
9. Tran, T.H., et al.: Influence of dressing conditions on surface roughness when surface grinding SKD11 steel. In: Materials Science Forum. Trans Tech Publ. (2021)
10. Tran, T.H., et al.: Optimizing dressing conditions for minimum flatness tolerance when grinding SKD11 tool steel. In: Materials Science Forum. Trans Tech Publ. (2021)
11. Hirogaki, T., et al.: In-situ heat treatment system for die steels using YAG laser with a machining center. Precis. Eng. **25**(3), 212–217 (2001)
12. Yamada, Z., Watanabe, T., Tamura, T.: Effect of laser pulse shot number on hardness and strength of the welded zone study on micro repair welding by pulsed YAG laser (Report 1) (2002)
13. Mookam, N., Sunasuan, P., Madsa, T., Muangnoy, P., Chuvaree, S.: Effects of graphite and boron carbide powders mixed into dielectric fluid on electrical discharge machining of SKD 11 tool steel. Arab. J. Sci. Eng. **46**(3), 2553–2563 (2021). https://doi.org/10.1007/s13369-020-05156-4
14. Toshimitsu, R., et al.: Improvement in surface characteristics by EDM with chromium powder mixed fluid. Procedia Cirp **42**, 231–235 (2016)
15. Zhang, G., et al.: Modeling and optimization of medium-speed WEDM process parameters for machining SKD11. Mater. Manuf. Process. **28**(10), 1124–1132 (2013)
16. Zhang, G., Zhang, Z., Ming, W., Guo, J., Huang, Y., Shao, X.: The multi-objective optimization of medium-speed WEDM process parameters for machining SKD11 steel by the hybrid method of RSM and NSGA-II. Int. J. Adv. Manuf. Technol. **70**(9–12), 2097–2109 (2013). https://doi.org/10.1007/s00170-013-5427-6

17. Phan, N.H., Van Duc, N., Van Bong, P., Muthuramalingam, T., Vu, N.P., Hung, L.X.: Multi-responses optimization of process parameters in die-sinking EDM process on SKD11 steel using PSI Based Taguchi method. In: Sattler, K.-U., Nguyen, D.C., Vu, N.P., Tien Long, B., Puta, H. (eds.) ICERA 2019. LNNS, vol. 104, pp. 493–499. Springer, Cham (2020). https://doi.org/10.1007/978-3-030-37497-6_57
18. Ming, W., et al.: Green machining: a framework for optimization of cutting parameters to minimize energy consumption and exhaust emissions during electrical discharge machining of Al 6061 and SKD 11. J. Cleaner Prod. **285,** 124889 (2021)
19. Long, B.T., Phan, N.H., Cuong, N., Jatti, V.S.: Optimization of PMEDM process parameter for maximizing material removal rate by Taguchi's method. Int. J. Adv. Manuf. Technol. **87** (5–8), 1929–1939 (2016). https://doi.org/10.1007/s00170-016-8586-4
20. Banh, L.T., Nguyen, P.H., Ngo, C.: Tool wear rate optimization in PMEDM using titanium powder by Taguchi method for die steels. Sci. Technol. Dev. J. **19**(2), 88–97 (2016)
21. Huu, P.-N.: Study of the effects of process parameters on tool wear rate in powder mixed electrical discharge machining by Taguchi method. Sci. Technol. Dev. J. **20**(K7), 55–60 (2017)
22. Chimklin, K., Chungchoo, C.: Design of experiment for analyzing factors affecting to diameters of the hole on SKD11 in the EDM process. In: Proceedings of the 2020 4th International Symposium on Computer Science and Intelligent Control (2020)

Effects of Machining Configurations and Process Parameters on the Machining Damage Generated During Milling CFRP Structures

Nguyen Dinh Ngoc[1(✉)], Pham Van Truong[1], and Nguyen Thi Hue[2]

[1] Faculty of International Training, Thai Nguyen University of Technology, Thái Nguyên, Vietnam
ngocnd@tnut.edu.vn

[2] Faculty of Fundamental Science, Thai Nguyen University of Technology, Thái Nguyên, Vietnam

Abstract. After demolding, in order to get the final form for assembly or other functions, the secondary manufacturing processes are needed to remove the excess materials in the edge of CFRP parts. The machining process of CFRP composites is accompanied by the occurrence of defects induced. Hence, mastering the factors affecting induced defects is crucial. The aims of this study are to conduct the evolution of machining quality, characterized by the ten-point max, R_z when changing the process parameters. Three levels of feed speed and two levels of spindle speed are combined to generate a full experimental design. Both up and down milling is conducted to examine the effects of machining configuration on machining quality. It is found that spindle speed, feed speed and machining length, and machining configuration have impacts on machining quality. However, when milling at longer distance of 3.0 m, the effects of machining configuration is not clear.

Keywords: CFRPs · Machining quality · Milling composite · Surface roughness · Microstructures

1 Introduction

Carbon fiber reinforced plastics (CFRPs), an advanced type of composite materials, have been widely adopted in recent decades due to its outstanding properties such as dimensional stability, high specific strength, and low density. Therefore, CFRPs have been attracted to use in important industries like aerospace, automotive, and shipping. Typically, CFRP parts are fabricated in a near-net-shape process. In order to archive the desired sizes and dimensions as well as other functions some machining operations being milling, drilling, trimming, turning are needed [1, 2]. Because of the mechanical, thermal properties and high abrasiveness of the reinforced components, machining composite materials is hard to be performed. This leads to occur some kinds of machining damage in the wall machined surfaces such as matrix cracking, matrix fiber interface debonding, matrix degradations, fiber pullout and in the edge of machined

D. C. Nguyen et al. (Eds.): ICERA 2021, LNNS 366, pp. 400–406, 2022.
https://doi.org/10.1007/978-3-030-92574-1_41

surfaces such as delamination, burr, or uncut fiber [3–5]. The occurrences of these types of damage can harmfully impact on the integrity of machined surface and on the strengths of CFRP parts [6]. Hence, understanding the factors influencing on the appearance of machining damage is significant to partially or fully control damage occurrence. Machining damage is strongly impacted by some main factors being cutting tool geometries, fiber direction, cutting parameters, machined materials [1, 7–10].

Prakash R et al. [9] investigated the influence of tool geometries on the machining damage in which three tool geometries made of tungsten carbide, namely burr tool with tool shape of trapezoidal (T1), burr tool with tool shape of pyramidal (T2), and four fluted helical end mill (T3) were utilized. The results revealed that tools T1 and T2 generate the minimum degree of delamination due to creating smaller axial cutting force components. However, because of the highest axial cutting force component when compared to those resulting from tools T1 and T2 maximum depth of delamination was seen in the free edge of the composite part machined by tool T3. In addition to tool geometries, process parameters are important factors directly affecting damage occurrences. Several studies in the literature have documented that an increase in feed speed and a decrease in cutting speed typically cause a higher level of machining damage [11]. This is due to the fact that an increase in cutting speed and a decrease in feed speed generate minimum theoretical chip thickness which makes it easy to machine. However, recently in the study of Nguyen-Dinh et al. [12], the results have shown that the combination between high cutting speed and low feed speed generates an extreme level of machining damage. This is due to the fact that machining with a small feed speed will create higher machining time which is combined with the wider contacting area due to high cutting speed. As a result, cutting tools quickly become worn after a short time. Finally, severe machining quality is obtained. Several studies have shown that surface integity is also depended on the machining configuration.

In this study, the influences of process parameters on the machining quality of milling CFRP composite materials will be conducted. Machining quality will be characterized by roughness tester. Moreover, the influence of machining configuration on the appearance of maching damage which directly results on machining quality is also carried out. Finally, analysis of machining quality given by machining configuration is also discussed.

2 Experimental Procedure

The specimens in this study are made of CFRP composites which are laminated by P2352 prepregs according to stacking sequence of [90°/90°/−45°/0°/45°/90°/−45°/90°/45°/90°]s. Before milling, each plate is cut by the dimensions of 300 mm × 150 mm × 5.2 mm in where the cutting direction is along 150 mm length. Information details of CFRPs can be referred to in Table 1. For machining conditions, three levels of feed speed V_f, (500 mm/min, 1000 mm/min, 1500 mm/min) and two levels of spindle speed N, (8000 rpm and 10000 rpm) are combined. A radial depth of cut of

2 mm is steadily adopted herein. In order to examine the effects of machining configuration, down milling and up milling is both conducted. A full factorial design (six conditions) is applied to down milling, but only a level of spindle speed of 8000 rpm, combined to three feed speeds, is conducted with up milling (three cutting conditions). All tests are dry machining. The milling paths will be alternatively changed in both sides of the vise to ensure that down milling is always happened for each machining condition. This means that the odd path and even path follow the first and second arrow directions in Fig. 2 respectively.

After each two paths of cutting tool, machined surface will be unfixed and quantitatively evaluated by measuring surface roughness, R_z. Each value of surface roughness is averagely determined by three measurements from three vicinity positions in machined surface. Six and three new carbide tools with two helix flutes and diameter of 6 mm are chosen for down and up milling respectively (Ref. Fig. 1). For each cutting condition, twenty paths will be taken place. This means that each cutting condition corresponds to 3 m long of machining distance (20 × 15 mm = 3000 mm). Machining test will be conducted in the Mazak vertical center smart 530C (Yamazaki Mazak Corporation, Japan) as presented in Fig. 2. Machining quality is characterized by roughness tester of SJ-210 Mitutoyo (Mitutoyo Corporation, Japan) with cut off length of 0.8 mm and transverse length is 5.0 mm.

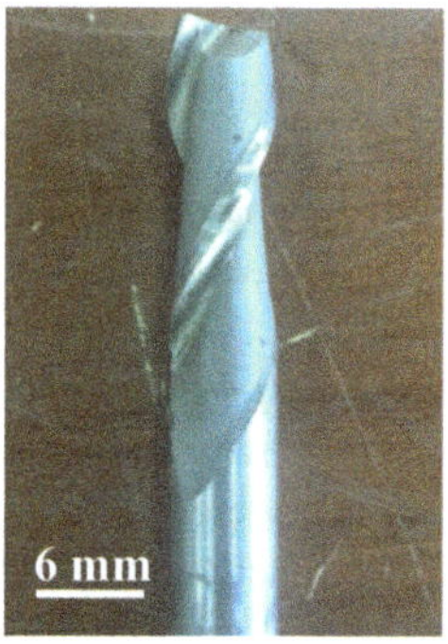

Fig. 1. Carbide milling cutter with two helix cutting edges.

Table 1. Mechanical properties of P2352 prepreg.

Density (g/cm^3)	Longitudinal shear modulus (GPa)	Longitudinal Young's modulus (GPa)	Compressive strength (MPa)	Tensile strength (MPa)	Poisson's ratio
2.63	6.21	162	1552	2844	0.34

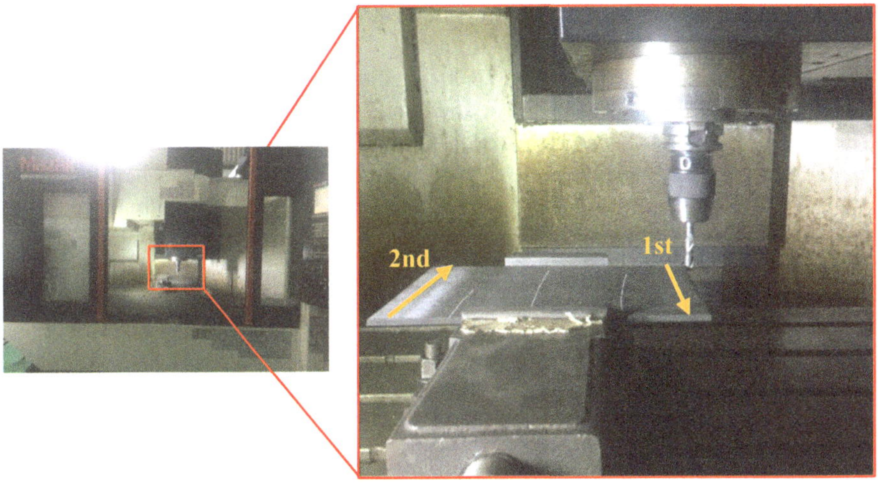

Fig. 2. Experimental setup showing devices used during trimming tests.

3 Results and Discussion

3.1 Influence of Process Parameters on the Machining Quality

In this section, the effects of process parameters on machining quality are presented. Figure 3 shows the evolution of ten-point mean (R_z) as a function of the feed speed, spindle speed, and machining distance. It is noticed that the surface roughness is only measured when machining distance (L_c) reaches 0.15 m corresponding to at the end of the first path, and 3 m corresponding to at the end of the last path or the twentieth path. Regarding the effects of the spindle speed and feed speed after the first cutting tool path or after machining length of 0.15 mm, it is observed that surface roughness increases when the feed speed increases and/or the spindle speed reduces. For instance, R_z increases by 160% and 180% when the feed speed is changed from 500 mm/min to 1500 mm/min for using the spindle speed of 800 rpm and 10000 rpm respectively. For the impacts of the spindle speed, with the similar used the feed speed, the surface quality generated by 10000 rpm is better than that one which is provided by 8000 rpm. However, the influential levels of the spindle speed are lesser than those of the feed speed. Indeed, the maximum percentage reduction is 20% belonging to the feed speed of 500 mm/min. The influences of machining parameters as previously presented have been consisted to those documented by researchers such as [1, 13]. This phenomenon can be interpreted that an increase in the feed speed and/or decrease in the spindle speed typically generate the higher degree of the theoretical chip thickness [11, 14] which is needed higher levels of cutting forces. For this reason, the machined surface is more irregular due to more applying of cutting forces. Hence, cut materials are broken instead of being shared.

At the machining length of 3 m, it is visualized that the surface roughness is more important than those induced at machining length of 0.15 mm irrespective of machining condition. This is because of the effects of the wear phenomenon of cutting tool. Indeed, the authors [15] have shown that cutting edge radius increases after some machining distance, namely rounded cutting edge. This leads the machined surface to rougher. The effects of process parameters on the machining quality are identical to those at machining distance of 0.15 m except for the case of feed speed of 500 mm/min. It is observed that at longer machining distance, the values of recorded surface roughness are 13 μm and 28 μm and they are superior to those at small machining distance. The domination of the surface roughness in case of 500 mm/min feed speed for both spindle speed is due to the severe tool wear as documented by [3, 12].

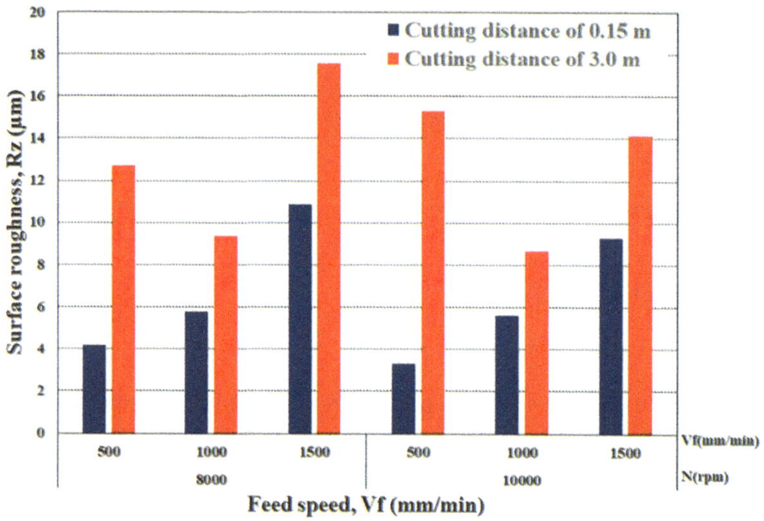

Fig. 3. Evolution of R_z in transverse direction as a function of cutting parameters.

3.2 Influence of Machining Configuration on the Machining Quality

Up and down milling will be tested. While up milling is only conducted for case of 8000 rpm spindle speed, down milling is conducted for full factorial as previous section. The schema of machining path is displayed in Fig. 4. The effects of configuration of milling process are seen in Fig. 5. It is noticed that surface roughness recorded of up and down milling processes are presented at two machining length, i.e. 0.15 m and 3.0 m. At the small machining length, it is observed that surface roughness measured in machined surface of up milling are always smaller than that one given by down milling irrespective of feed speed. These results are consistent with those documented in the literature [11]. This is attributed to the fact that the differences in machining mechanism for each layer with various fiber orientations. Thus, we get different surface integrities. However, at the longer machining length of 3.0 m, it is observed that the effects of machining configuration on surface roughness is unclear. This can be explained that the radius of cutting edge increases at this length of machining or occurrence of the tool wear. The combination between chip formation mechanism and tool wear make for machining quality of down milling is not obvious.

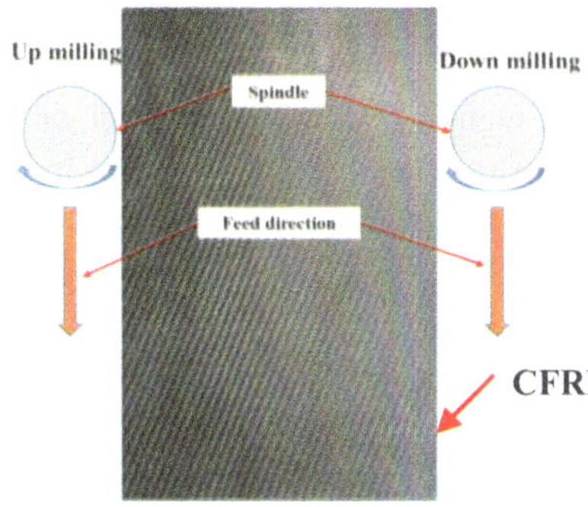

Fig. 4. Schematic machining configuration in both sides of specimens.

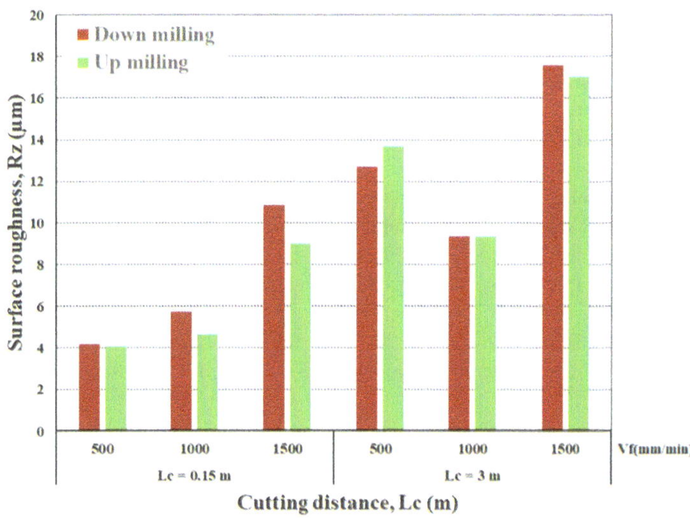

Fig. 5. Influences of machining configuration on machining quality with spindle speed of 8000 rpm with two machining length.

4 Conclusions

The impacts of process parameters and machining configuration on the machining quality when milling CFRP structures are experimentally investigated at two specific values of the machining length in this study. Ten-point max parameter, R_z, is selected to quantify machining quality. Conclusions can be drawn as follows:

- Both feed speed and spindle speed have influences on the machining quality, e.g. an increase in feed speed and/or decrease in spindle speed causes increasing in surface roughness.
- Machining with higher spindle speed and lower feed speed at machining distance of 3.0 m will generate severe tool wear of cutting tools. This makes the machined

surfaces become more irregular. Consequently, the domination of surface roughness at this machining condition (N = 10000 rpm and V_f = 500 mm/min) is observed.
- Machining configuration including up milling and down milling has effects on the machining quality in which surface roughness given by down milling is typically higher than that one obtained by up milling. However, the impact of machining configuration at longer machining distance is not clear.

Acknowledgments. The authors wish to thank Thai Nguyen University of Technology for supporting this work.

References

1. Sheikh-Ahmad, J., Urban, N., Cheraghi, H.: Machining damage in edge trimming of CFRP. Mater. Manuf. Processes **27**(7), 802–808 (2012)
2. Duboust, N., et al.: An optical method for measuring surface roughness of machined carbon fibre-reinforced plastic composites. J. Compos. Mater. **51**(3), 289–302 (2016)
3. Haddad, M., Zitoune, R., Bruno Castanié, F.E.: Machinability and surface quality during high speed trimming of multi directional CFRP. Int. J. Mach. Mach. Mater. **13**, 289–310 (2013)
4. Ngoc, N.D., et al.: The characterization of machined damage of CFRP composite: comparison of 2D and 3D surface roughness performance. In: Advances in Engineering Research and Application, pp. 771–779 (2020)
5. Wang, C., et al.: Occurrence and formation mechanism of surface cavity defects during orthogonal milling of CFRP laminates. Compos. B Eng. **109**, 10–22 (2017)
6. Nguyen-Dinh, N., Bouvet, C., Zitoune, R.: Influence of machining damage generated during trimming of CFRP composite on the compressive strength. J. Compos. Mater. **54**(11), 1413–1430 (2019)
7. Ramulu, M., Wern, C.W., Garbini, J.L.: Effect of fibre direction on surface roughness measurements of machined praphite epoxy composite. Compos. Manuf. **4**, 39–51 (1993)
8. Hintze, W., Hartmann, D., Schütte, C.: Occurrence and propagation of delamination during the machining of carbon fibre reinforced plastics (CFRPs) – an experimental study. Compos. Sci. Technol. **71**(15), 1719–1726 (2011)
9. Prakash, R., et al.: High-speed edge trimming of CFRP and online monitoring of performance of router tools using acoustic emission. Materials **9**(10), 798 2016
10. Palanikumar, K., Karunamoorthy, L., Karthikeyan, R.: Assessment of factors influencing surface roughness on the machining of glass fiber-reinforced polymer composites. Mater. Des. **27**(10), 862–871 (2006)
11. Janardhan, P., Sheikh-Ahmad, J., Cheraghi, H.: Edge trimming of CFRP with diamond interlocking tools (2006)
12. Nguyen-Dinh, N., et al.: Surface integrity while trimming of composite structures: X-ray tomography analysis. Compos. Struct. **210**, 735–746 (2019)
13. Ucar M., Wang, Y.: End-milling machinability of a carbon fiber reinforced laminated composite-M-Ucar.pdf (2003)
14. Sheikh-Ahmad, J., Shahid, A.H.: Effect of edge trimming on failure stress of carbon fiber polymer composites. Int. J. Mach. Mach. Mater. **13,** 331–347 (2013)
15. Nguyen-Dinh, N., et al.: New tool for reduction of harmful particulate dispersion and to improve machining quality when trimming carbon/epoxy composites. Compos. Part A Appl. Sci. Manuf. **131**, 105806 (2020)

Effects of Stearic Acid Coating on the Dispersion of Zeolite 4A in PE Matrix

Thu Ha Pham Thi[1], Thanh Tung Nguyen[1], Thu Trang Pham[1](✉), Trung Duc Ngu-yen[1], and Quoc Toan Tran[2]

[1] Institute of Chemistry, VAST, 18 Hoang Quoc Viet, Cau Giay, Hanoi 122300, Vietnam
[2] Thai Nguyen University, Tan Thinh, Thai Nguyen 24118, Vietnam

Abstract. To evaluate the effects of stearic acid on the dispersibility of zeolite 4A in polyethylene matrix, the master batches containing 20–40 wt% of stearic acid uncoated and coated zeolite 4A and the linear low-density polyethylene (LLDPE) films containing 3–7% stearic acid uncoated and coated zeolite 4A were prepared. The structural change of zeolite 4A which was modified surface with stearic acid was studied using IR spectroscopy. The dispersion of zeolite 4A in LLDPE matrix was evaluated using a scanning electron microscope (master batch samples) and optical microscope (film samples). The change in the melting temperature of LLDPE in the presence of 4A zeolite was evaluated by using differential scanning calorimeter (DSC). The mechanical properties of the film were evaluated according to ASTM D882. Surface-treated zeolite with stearic acid dispersed in the masterbatches and films better than untreated zeolite. Stearic acid did not change the melting temperature of zeolite 4A/PE master batches. The mechanical properties of the films containing stearic acid coated zeolite 4A were higher than those containing unmodified zeolite 4A.

Keywords: Polyethylene · Zeolite 4A · Stearic acid · Dispersion of zeolite 4A · Agglomeration

1 Introduction

Polyethylene is the most commonly used material in the packaging industries because of its many advantages, such as inert, inexpensive, readily available, easy processing, flexible and semi-permeable [1]. To change the thermal, mechanical, permeability of polymer, and reduce production cost, inorganic fillers such as metal oxide, metal powders, carbon black, silica, spherical or fibrous glass [2], talc, calcite [3, 4], kaolin, mica [5] etc. are added.

Zeolites are crystalline aluminosilicate minerals of alkali or alkali earth elements, which have a microporous crystal structure. Zeolites are widely used in the fields, such as catalysis, adsorption, ion-exchange, sensors energy conversion [6]. Recently zeolites are commonly used as fillers in PE films to preserve fresh fruits and vegetables because their pores increase the gas and water vapor permeability for polymer films [7]. However, due to the hydrophilic nature of zeolite, it is necessary to treat chemical surfaces before mixing to increase the compatibility and dispersion of the additive into

D. C. Nguyen et al. (Eds.): ICERA 2021, LNNS 366, pp. 407–415, 2022.
https://doi.org/10.1007/978-3-030-92574-1_42

the PE matrix. To create the hydrophobic surface to increase compatibility of fillers with polymer molecules, fatty acids such as oleic acid, stearic, palm oil [8–10] are often used. Stearic acid is most commonly used because of its advantages, such as low cost and ease of processing. In addition, when modifying the filler surface with stearic acid, it reduces water absorption, prevents the agglomeration of the filler particles, reduces the surface energy to help them disperse easily into the PE matrix.

Hyun K. et al. reported that the zeolite surface modification with stearic acid had improved the flexibility of the low density polyethylene (LDPE), linear low-density polyethylene (LLDPE), high-density polyethylene (HDPE) matrices. Stearic acid modified zeolite dispersed in composites more evenly than unmodified zeolite [11]. Zaharri N. et al. also reported that the mechanical properties of zeolite/PP composite in the presence of stearic acid coupling agent is better than without stearic acid [12]. Therefore, the aim of this study is to evaluate the effects of stearic acid coating on the dispersion of zeolite 4A in PE matrix.

2 Materials and Methods

2.1 Materials

Linear low-density polyethylene (LLDPE) resin was produced by Formosa Plastics Corporation (Yunlin, Taiwan) was used as the matrix material. Its density was 0.924 g/cm^3, and its melt flow index (MFI) was 21 g/10 min at 2.16 kg/190 °C. Zeolite 4A with an average particle size of 6.77 μm was supplied by ChemFine International Co., Ltd (China). Stearic acid is used directly without refining.

2.2 Preparation of Stearic Acid Coated Zeolite 4A

To prepare a 0.1 M stearic acid solution, 17.04 g of stearic acid was placed to a 1 L steel pot, then 400 ml of ethanol and 200 ml of distilled water were added and mixture was stirred constantly at 70 °C to dissolve stearic acid. Thereafter, 100 g of zeolite 4A were added to the stearic acid solution and the mixture was stirred at 70 °C for 5 h. The surface modified zeolite 4A was filtered and dried in a vacuum oven at 60 °C for 6 h, then ground to a fine powder [11].

2.3 Preparation of Zeolite 4A/PE Masterbatches

Master batches of LLDPE and uncoated and stearic acid coated zeolite 4A were prepared on a Brabender Plasticoder device. Zeolite 4A and LLDPE resins were dried at 80 °C for 8 h, then melt mixed at 165 °C for 7 min with a constant screw speed of 45 rpm. The fillers with small particle size such as $CaCO_3$ could added to 60% in a master batch, so with larger zeolite particle size, the ability to evenly disperse into the matrix resin was poorer, we expected to study in the contents of 20–40%. The master batches were then pressed into 1-mm thick sheet on a GoTech press at 170 °C at a pressure of 20 MPa for 5 min.

2.4 Preparation of Zeolite 4A Filled LLDPE Films

LLDPE films with a thickness of 30 μm were made by extrusion blowing by using a XD-35 extruder with a 35 mm screw of L/D 28:1. Because we expected to fabricate films with increased gas and water vapor permeability for application in preserving fresh fruits and vegetables. These films often contained 3–5% inorganic fillers. Therefore, in this paper, the uncoated and stearic acid coated zeolite 4A were added to the film at different contents (3–7%) by using zeolite 4A/PE master batch which loaded 30% of coated zeolite 4A (Table 1).

Table 1. Sample labeling of zeolite 4A filled LLDPE films.

Samples	Components (%)		
	LLDPE	Zeolite	Zeolite/stearic acid
PE	100	–	–
PE-3Ze	97	3	–
PE-5Ze	95	5	–
PE-7Ze	93	7	–
PE-3Ze/St	97	–	3
PE-5Ze/St	95	–	5
PE-7Ze/St	93	–	7

2.5 Characterization

Mechanical Properties. The tensile strength and elongation at break were measured using Instron 5980 Testing Machine (Illinois Tool Works Inc., Norwood, USA). Tensile properties of master barches were conducted according to ASTM D638 at a crosshead speed of 50 mm/min and that of films were conducted according to ASTM D882 at a crosshead speed of 10 mm/min.

Scanning Electron Microscope (SEM). The fracture surface morphology of masterbatches and the surface morphology of films were investigated by using a scanning electron microscopy (SEM) (JEOL 6490, Tokyo, Japan) at 15 kV after the sample surface was sputter-coated with a thin gold layer.

Fourier Transfer Infrared Spectra (FTIR). The chemical structure of uncoated and stearic acid coated zeolite 4A was invested by using a Fourier transform infrared (FTIR) spectrometer (Nicolet Impact model 410, Nicolet, Madison, USA). The equipment was operated with a resolution of 4 cm^{-1} and scanning range from 4000 to 500 cm^{-1}.

Differential Scanning Calorimetry (DSC). The differential scanning calorimetry (DSC) was used to determine the melting temperature of the samples. The DSC was carried out on a NETZSCH DSC 204F1 Phoneix differential scanning calorimetry (Netzsch, Selb, Germany) at a heating rate of 10 °C/min from room temperature to 200 °C in air.

3 Results and Discussion

3.1 Effect of Stearic Acid on the Structure of Zeolite 4A

The FTIR spectra of stearic acid, uncoated and coated zeolite 4A were presented in Fig. 1.

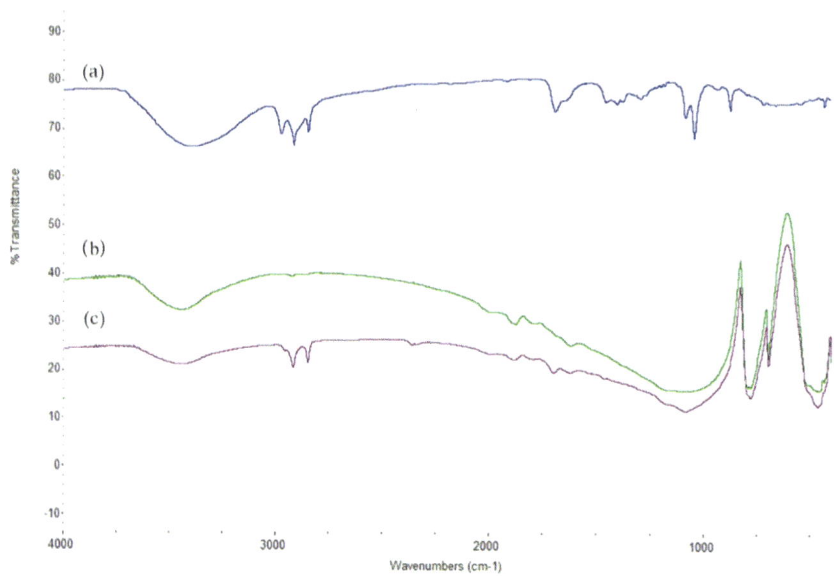

Fig. 1. The FTIR spectra of (a) stearic acid, (b) unmodified zeolite 4A, (c) modified zeolite 4A.

In IR spectrum of stearic acid, the strength peak at 3395.4 cm^{-1} was assigned for –COOH functional group, the strength peak at 1695.7 cm^{-1} was assigned for the stretching of C = O, the peak at 1460.4 cm^{-1} was assigned for the scissoring of -CH_2, the peak at 1297.7 cm^{-1} was assigned for the stretching of C-O [9], the peak at 2916.5 cm^{-1} and 849.0 cm^{-1} was assigned for the asymmetric and symmetric stretching of CH_2, respectively [13].

When comparing the IR spectra of uncoated and coated zeolite 4A, it was discovered that the IR spectra of stearic acid coated zeolite 4A revealed some additional peaks which were near 2920 cm^{-1} and 2850 cm^{-1} corresponding to stretching of CH_2, near 1700 cm^{-1} corresponding to stretching of C = O. This showed that there had been a chemical reaction between stearic acid and zeolite 4A. Using stearic acid to modify wollastonite, Li *et al.* [14] found similar results. A chemical interaction between wollastonite and pimelit acid was also observed by Meng and Duo [15].

3.2 Dispersion of Zeolite 4A Particles in Master Batches

The SEM images taken from the fractured surfaces with liquid nitrogen of uncoated and coated zeolite 4A/PE master batches were shown in Fig. 2.

It could be seen that the population density of the zeolite 4A increased as the zeolite 4A content increased. The results showed that in samples containing uncoated zeolite 4A, the zeolite 4A particles dispersed still relatively uniformly at the zeolite 4A content of 20 and 25%. However, when increasing the uncoated zeolite content to 30–40%, the zeolite 4A particles tended to form agglomeration, not evenly dispersed in the PE

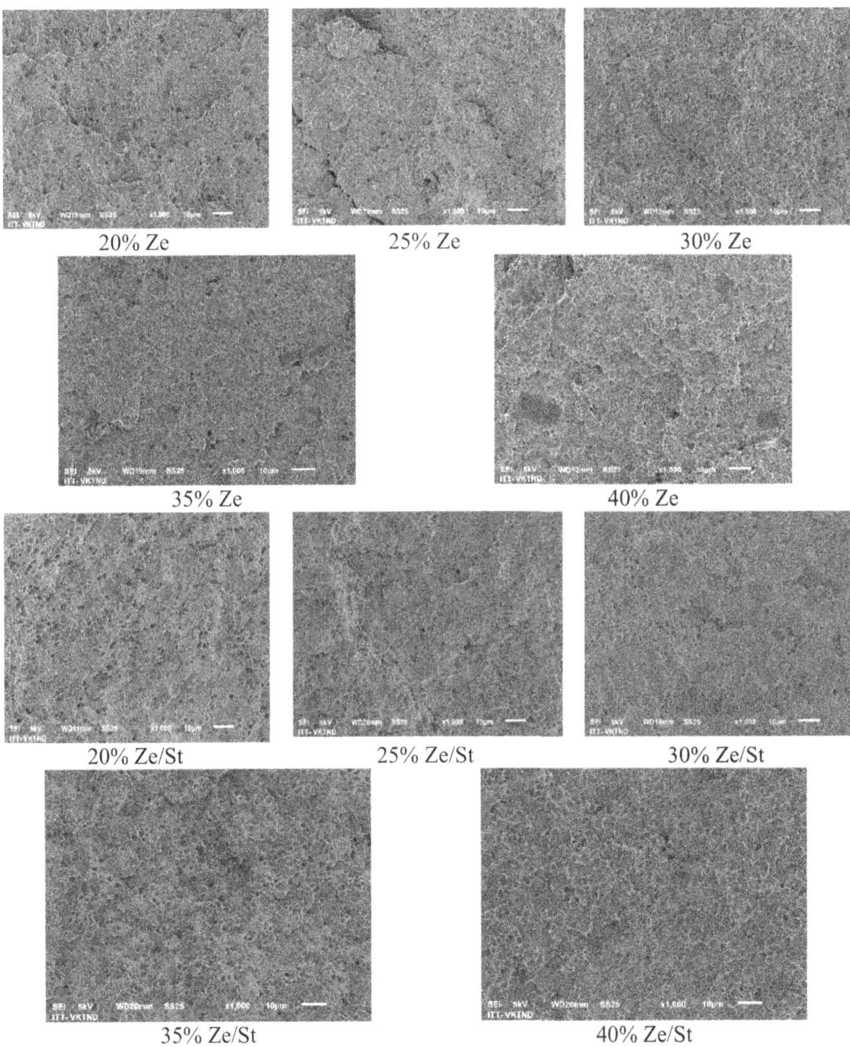

Fig. 2. The morphology of the fracture surface of uncoated zeolite 4A/PE masterbatches and stearic acid coated zeolite 4A/PE masterbatches.

matrix. This was explained that the zeolite particles had high surface energy so the distance between particles was decreased when the density of particles was increased, so there was a possibility of surface interaction of the particles leading to agglomeration.

In samples containing stearic acid coated zeolites, it was found that in the studied concentration range (30–40%) the zeolite particles were still uniformly dispersed on the PE resin. This could be explained because the surface treatment with stearic acid significantly reduced the free energy of the filler particles surface and reduced the interaction between filler particles leading to better dispersion of the filler particles in the PE matrix and reduced agglomeration of particles [9]. Luis F. I. proposed two adsorption modes of fatty acid on the zeolite surface (Fig. 3) [16].

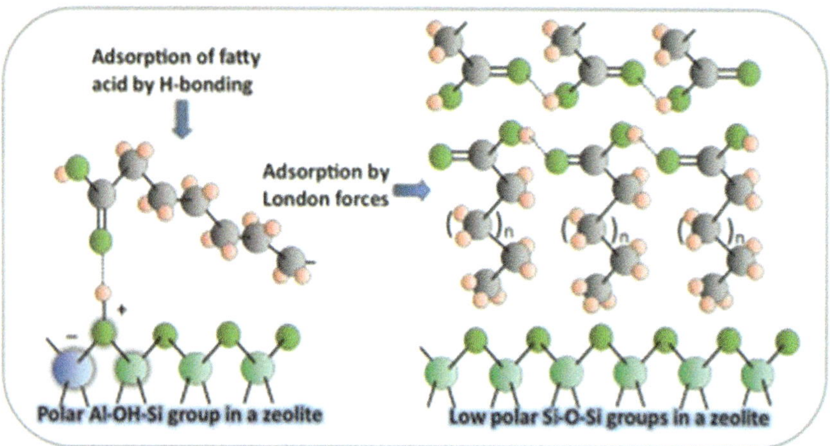

Fig. 3. The adsorption modes of fatty acid on the zeolite surface [16].

3.3 Effect of Stearic Acid on the Thermal Property of Masterbatches

The thermal properties of the masterbatches with different zeolite 4A contents were investigated by Differential Scanning Calorimetry (DSC). The DSC results for zeolite 4A/PE masterbatches were summarized in the Table 2.

Table 2. The melting temperature of the masterbatches.

Zeolite content	Melting temperature of samples (°C)	
	Uncoated zeolite	Coated zeolite
0	127.8	
20%	127.1	128.5
30%	128.6	129.0
40%	127.6	126.3

The results showed that the melting temperature of the master batches containing both uncoated and coated zeolite 4A was not significantly different when compared with pure PE, which indicated that adding zeolite did not change the characteristics of polymeric materials. Daiane et al. also obtained similar results when incorporating silverexchange zeolite Y into polyethylene [17]. Pehlivan et al. also found that incorporating zeolite into polypropylene did not change the melting point of polymer [18].

3.4 Effect of Stearic Acid on the Surface Morphology of the PE Films

Figure 4 showed optical micrographs of the zeolite 4A filled LLDPE films at zeolite 4A concentrations of 3, 5 and 7%.

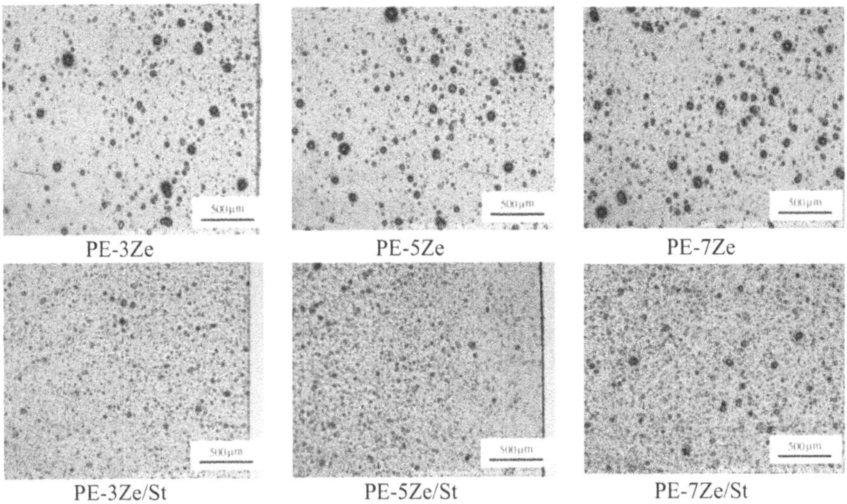

Fig. 4. Optical micrographs of zeolite 4A filled LLDPE films.

The particle-particle interactions led to the agglomeration and heterogeneous distribution of filler particles. In the film samples using unmodified zeolite, the agglomeration of zeolite particles was **observed**. The higher the zeolite content in the PE matrix was, the larger the size of the agglomerated particles was.

Comparing film samples with the same filler content, it could be seen that the stearic acid coated zeolite **particles** dispersed uniformly with smaller particle sizes into the PE matrix. Thus, stearic acid had acted as a binder to increase the adhesion between the zeolite particles and the PE matrix, reducing the agglomeration of zeolite particles. Angel C. D. et al. [9] also found that adding fatty acids as an additive to PP film formulations improved the polymer-filler interactions due to changing surface interactions.

3.5 Effect of Stearic Acid on the Mechanical Properties of the Films

The mechanical properties of film were one of major importance for all applications. Table 3 showed the results of the mechanical properties of pure PE film and zeolite 4A filled polyethylene films.

Table 3. Mechanical properties of the films.

Sample	Tensile strength (MPa)	Elongation at break (%)
PE	26.98 ± 0.3	776.50 ± 4.23
PE-3Ze	22.08 ± 1.76	474.44 ± 23.03
PE-5Ze	21.42 ± 1.91	444.88 ± 52.03
PE-7Ze	22.27 ± 1.08	434.24 ± 12.91
PE-3Ze/St	24.66 ± 1.56	482.63 ± 38.99
PE-5Ze/St	25.95 ± 1.38	486.50 ± 37.46
PE-7Ze/St	24.84 ± 1.28	479.07 ± 21.40

It could be seen that the mechanical properties of pure PE film were higher than uncoated and coated zeolite 4A filled polyethylene films. This was because zeolite 4A particles had limited the mobility of the polymer molecules and formed stress concentration points [12].

The results also showed that both the tensile strength and elongation at break of PE films containing stearic acid coated zeolite 4A were higher than those of PE films containing uncoated 4A zeolite. This was also supported by the results of the optical microscope mentioned earlier. When zeolite 4A was treated with stearic acid, it distributed more uniformly in the PE matrix with a smaller size. In addition, it was observed that when increasing the zeolite 4A content from 3 to 7 wt%, the mechanical properties did not change significantly. This was because the filler particles were still evenly dispersed in the PE matrix at these concentrations. Zhen Huang et al. discovered that when zeolite H-β with contents of 2.5–7.5 wt% was added to PE matrix, the mechanical properties of the PE films changed very little [19].

4 Conclusions

The influence of stearic acid on dispersion, thermal properties, mechanical properties of zeolite 4A/PE master batches and films was investigated. The melting temperature of LLDPE material mixed with modified zeolite 4A did not change much compared to that of LLDPE material mixed with unmodified zeolite 4A. The PE films containing modified zeolite 4A had higher mechanical properties than those containing unmodified zeolite 4A. At the same time, the surface modified zeolite 4A by stearic acid dispersed in the master batches and films better than unmodified zeolite 4A, stearic acid reduced the agglomeration of zeolite particles so that it could be incorporated into the master batch as well as the film with a higher concentration.

1. References

1. Benabid, F.Z., Kharchi, N., Zouai, F., Mourad, A.H.I., Benachour, D.: Impact of co-mixing technique and surface modification of ZnO nanoparticles using stearic acid on their dispersion into HDPE to produce HDPE/ZnO nanocomposites. Polym. Polym. Compos. **27** (7), 389–399 (2019)
2. Bai, S.L., Chen, J.K., Huang, Z.P., Liu, Z.D.: Interface effect on the mechanical behaviour of rigid particle filled polymer. Polym. Int. **50**, 222–228 (2001)
3. Kwon, S., et al.: Tensile property and interfacial dewetting in the calcite filled HDPE, LDPE, and LLDPE composites. Polymer **43**(25), 6901–6909 (2002)
4. Kim, K.J., et al.: Influence of mixing cycle on the degree of mixing of calcite-filled polyethylene upon stretching. J. Appl. Polym. Sci. **87**(2), 311–321 (2002)
5. Verbeek, C.J.R.: Highly filled polyethylene/phlogopite composites. Mater. Lett. **52**(6), 453–457 (2002)
6. Elhassan, A.M.: Study the physical properties of sudan natural and synthesis zeolites modification by PT element. Int. J. Adv. Res. **4**(9), 1692–1711 (2016)
7. Dirim, S.N., Esin, A., Bayindirli, A.: A new protective polyethylene based film containing zeolites for the packaging of fruits and vegetables: films preparation. Turk. J. Eng. Environ. Sci. **27**, 1–9 (2003)
8. El-Hakim, A.A.A., Badran, A.S., Essawy, H.A.: The effect of surface treatment of bentonite on the mechanical properties of polypropylene-bentonite composites. Polym. Plast. Technol. Eng. **43**(2), 555–569 (2004)
9. Angel, C.D., et al.: Mechanical and rheological properties of polypropylene/bentonite composites with stearic acid as an interface modifier. J. Appl. Polym. Sci. **132**(30), 42264–42271 (2015)
10. Othman, N., Ismail, H., Mariatti, M.: Effect of compatibilisers on mechanical and thermal properties of bentonite filled polypropylene composites. Polym. Degrad. Stab. **91**, 1761–1774 (2006)

Experimental Characterization of the Mechanical Properties of the Fused Filament Fabrication (FFF) Printed Polycarbonate Parts

Thao Thi Phuong Phan[1,2(✉)] and Ha Thi Thu Phan[1,3]

[1] Thai Nguyen University of Technology, Thai Nguyen, Vietnam
ttp262@umsystem.edu
[2] Oklahoma State University, Stillwater, OK, USA
[3] University of Missouri, Columbia, MO, USA

Abstract. Fused filament fabrication (FFF) is a well-known technique used to build 3-dimensional objects from thermoplastic materials. In order to make functional parts, the characterization of the printed part has become increasingly desirable. This study aims to study the mechanical properties of printed parts by doing experiments and data analysis through the design of experiments (DOE). The results indicate that those process parameters and their interactions significantly impact the mechanical properties of the printed parts. Further research is needed to identify other factors that could strengthen the effectiveness of printed parts.

Keywords: Mechanical properties · Anisotropic · Design of experiments

1 Introduction

Additive manufacturing (AM), also called 3D printing, has been under development for approximately 30 years and is characterized by introducing freedom in manufacturing. The complex geometries, even entirely moving assemblies, that would be difficult or impossible to achieve with conventional manufacturing methods, can be done in AM. These complex geometries can be more assertive and lighter than their traditionally manufactured counterparts. Crump, S.S. in 1992 patented fused filament fabrication (FFF), and the Stratasys Corporation has produced 3D printers based on FFF for years. FFF is a low-cost process using semi-molten filaments of plastic material on a build platen. FFF products nowadays are used not only for visualization in prototyping but also in the highest standard of performance, such as biomechanics and aerospace, because of their high strength-to-weight (S/W) ratio [1]. In FFF, the feed mechanism is connected to the extrusion nozzles. The extrusion nozzle translates in the XY- plane of the build platform to create the geometry shape in each layer. A new layer is deposited after each successive printed layer. This process is repeated until the final part is complete.

Figure 1 illustrates the process parameter of the FFF printing process is used in this paper. As shown in Fig. 1a, build orientation refers to the orientation angle of the

D. C. Nguyen et al. (Eds.): ICERA 2021, LNNS 366, pp. 416–422, 2022.
https://doi.org/10.1007/978-3-030-92574-1_43

product built on the build platform. Build orientation affects print time, support structure, and surface finish. By doing tensile testing with the same set of process parameters and adjusting only the build orientation, the work in [2] shows that the flat build orientation gives the highest strength, the up-right orientation gives the lowest strength, and the on-edge build orientation gives the intermediate strength in mechanical properties of the printed parts. The raster angle refers to the angle between the raster and the X-axis of the printing platform. The raster angle can be the same or different successive in layers as shown in Fig. 1b. The raster gap refers to the distance between two adjacent rasters in a layer. The raster gap value can be negative (overlapping), zero (in contact), or positive (not in contact) as shown in Fig. 1c. Similar to composite materials where the orientation of the fiber in the matrix governs material properties, the raster gap and the raster angle govern layer properties by defining the density and how the raster is oriented in each layer [3].

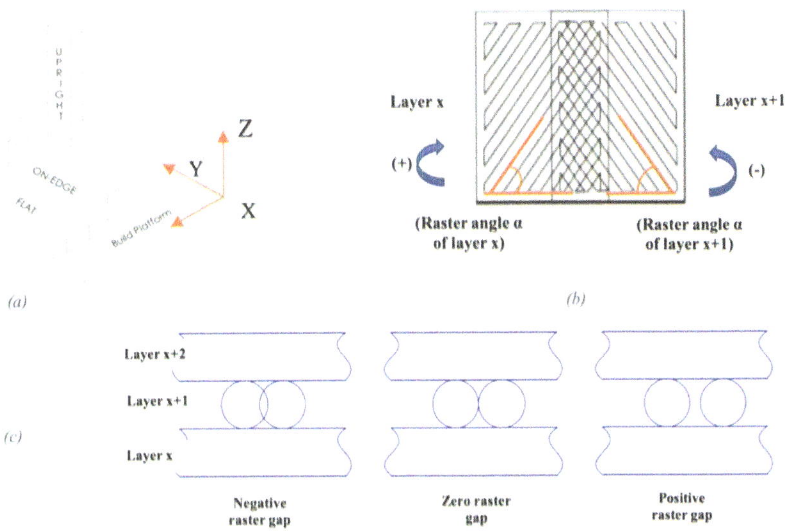

Fig. 1. Build orientation (a), raster angle type (b) and raster gap type (c)

2 Experimental Procedure

The tests were carried out on an ADMET universal testing machine, eXpert2600 with a load cell of 10 kN, axial strain gauge length of 50 mm, and a transverse strain gauge length of 2.54 mm. The fracture surfaces of the specimens were examined with scanning electron microscopy (SEM). All tests were performed at room temperature (23 °C). The testing machine and the SEM are shown in Fig. 2.

The specimens for the tensile tests were designed following the standard ASTM D638 test specimen as shown in Fig. 3a. The specimens were fabricated with a Stratasys Fortus 400 mc with a part dimensional accuracy of ± 0.127 mm. Process parameters were selected as shown in Table 1. Raster gaps were −0.0025 mm, 0.000 mm and +0.0025 mm. The flat (XY), edge (XZ) and up-right (ZX) build orientation was used for all the specimens. The raster angles were selected as [0°, 0°], [+45°, −45°] and [+30°, −60°]. The layer thickness was 0.254 mm. A series of tests followed the orthogonal array L27 [4]. The same general experiment procedure was applied to all the tests in this study. A preload was run once the specimen was loaded into the fixture to removed force from the clamps. After specimen failure, the test was stopped and the process was repeated for another sample. The upright build orientation specimen failure by applying the tensile load perpendicular to the stacked layers pulls these stacked apart as shown in Fig. 3b. Meanwhile, the on-edge and flat build direction specimen failure by applying the tensile load along the path of the deposited filaments pull several individual filaments apart as shown in Fig. 3c. Experimental data were recorded by built-in software at 100 samples/sec. The data set was analyzed to determine the influence of the process parameters on the properties of the printed parts.

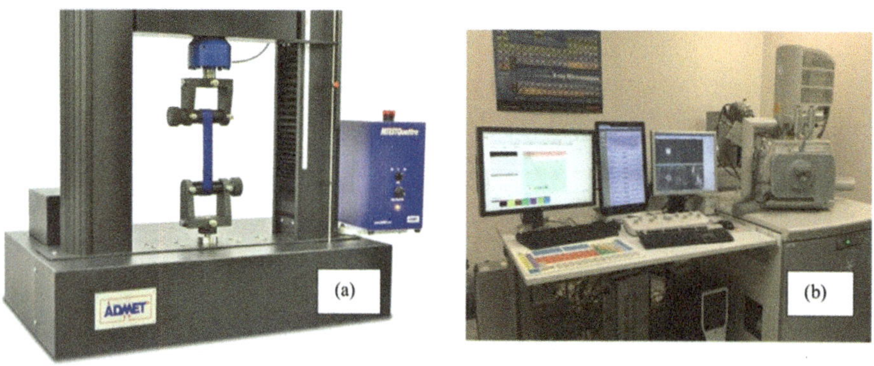

Fig. 2. (a) ADMET universal testing machines (eXpert2600), (b) Quanta 600 – scanning electron microscopy

Table 1. Factors and their alternate level (*: initial parameters)

Factors	Levels		
	1	2	3
(A) Raster Gap, mm	−0.0025	0.000*	0.0025
(B) Raster angle, °	[0°, 0°]	[+45°, −45°]*	[+30°, −60°]
(C) Build orientation	Flat*	Edge	Up-right

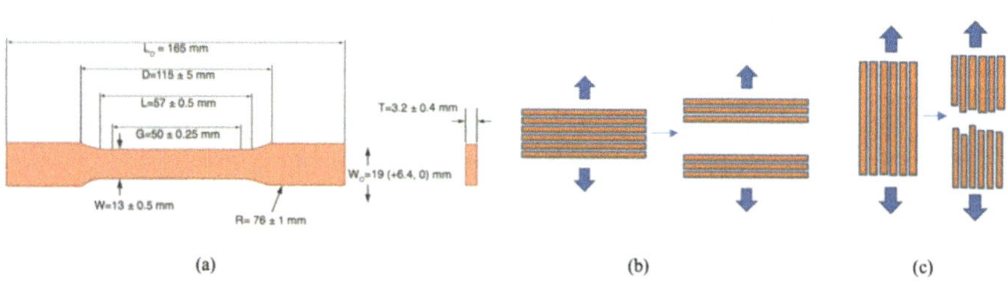

Fig. 3. (a) Specimen design for the tensile test; Tensile load direction is perpendicular to the stacked layers (b) and along the stacked layers (c)

3 Results and Discussion

The tensile strength of the printed parts is presented in Table 2. The number in the parenthesis is the standard deviation. The set of −0.0025 mm raster gap and flat build orientation has the largest tensile strength, the set of 0.000 mm raster gap and edge build orientation has an intermediate tensile strength, the set of + 0.0025 mm raster gap, and up-right orientation has the smallest tensile strength. For all three sets, the set of [+45°, −45°] raster angle has the largest tensile strength value, the set of [+30°, −60°] raster angle has an intermediate tensile strength value, and the set of [0°, 0°] raster angle has the smallest tensile strength value. The experiment data is in good agreement with other studies [5–7].

The SEM images of the failure surface of the tensile specimens of the set [0°, 0°] raster angle, flat build orientation, combined with −0.0025 mm raster gap and 0.0025 mm, are shown in Fig. 4. These images support the idea that the overlapping of the adjacent deposited filaments leads to the higher tensile strength. The negative of the raster gap of the adjacent deposited filaments leads to higher density in the printed parts as shown in Fig. 4a. Meanwhile, the positive raster gap leads to the void between the deposited filaments in the internal structure of the printed parts as shown in Fig. 4b.

Table 2. Tensile strength of printed parts for different set of process parameters

Raster angle, °	[0°, 0°]			[+45°, −45°]			[+30°, −60°]		
Raster gap, mm	−0.0025	0.000	0.0025	−0.0025	0.000	0.0025	−0.0025	0.000	0.0025
Flat build	45.34 (3.66)	40.32 (5.92)	25.12 (4.05)	69.60 (4.54)	51.11 (4.09)	29.45 (4.38)	46.33 (4.82)	41.23 (2.90)	30.25 (4.11)
Edge build	38.05 (2.00)	33.55 (1.45)	22.55 (2.43)	60.37 (2.59)	49.75 (2.89)	20.08 (0.67)	45.69 (2.80)	36.84 (2.96)	27.57 (4.73)
Up-right build	37.52 (2.00)	30.57 (0.80)	18.38 (2.32)	39.26 (0.86)	36.33 (1.84)	16.74 (1.97)	37.88 (4.25)	34.10 (1.10)	25.03 (2.71)

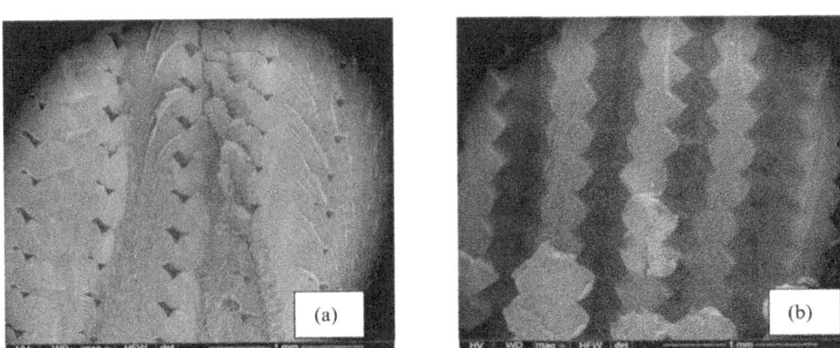

Fig. 4. Failure surface of the tensile specimens of the set [0°, 0°] raster angle, flat build orientation, −0.0025 mm raster gap (a) and 0.0025 mm raster gap (b)

The experiment data are in line with the study of [8–10] about the anisotropic behavior exhibits in different build orientations. For the same raster angle, the tensile strength of the printed part on the flat build is up to 43.6% higher than the up-right build and 16.8% higher than the edge build; the tensile strength of the printed part on the edge build is up to 34.9% higher than the up-right build. The load along the path of the deposited filaments required greater strength to pulls individual filaments apart.

A series of tests are performed through the design of experiments (DOE) to observe the change in process parameters corresponding to the changes in the tensile strength. Applying the Taguchi method helps to reduce the trial of testing. The method is essentially the same as that used in [6, 9, 11, 12] with some alterations in process parameters. The signal-to-noise ratio (S/N) presents the quality characteristics obtained by the Taguchi method in DOE. The S/N ratio categorizes based on the objective of the experiments: lower-the-better, higher-the-better and nominal-the-best. For the mechanical properties of the printed parts with the response of strength, a higher-the-better is used. The optimal condition is also obtained from the S/N ratio. The orthogonal array L27 is used to study the effect of the process parameters on the strength of the printed parts. Moreover, analysis of variance (ANOVA) is used to learn the significant influence of the process parameters and their interactions on the strength of the printed parts.

Figure 5 illustrates the main effects of each factor. The plot with the higher inclination gives the stronger influence. Both factors A (raster gap) and C (build orientation) are very much significant influences while factor B (raster angle) has the most minor influence. Factor B can be set at the fewer level or kept constant. The optimal combination of process parameters is A1B2C1 with the highest number of S/N ratio, −0.0025 mm of the raster gap, [+45°, −45°] of the raster angle and flat direction of the build orientation. The tensile strength at optimal condition increases about 8% of the tensile strength compared to those tested at the initial condition.

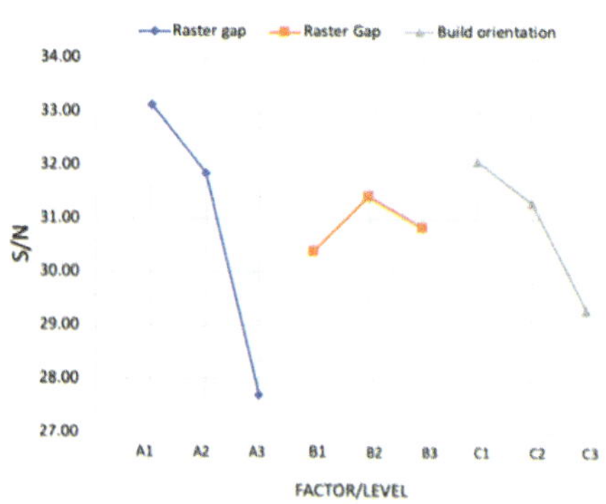

Fig. 5. Main effects plot

The interactions plots illustrate the raster gap versus raster angle (Fig. 6a), the raster gap versus build orientation (Fig. 6b) and the raster angle versus build orientation (Fig. 6c). While non-parallel lines are considered the existence of the interaction, the intersection lines are considered a strong interaction. Therefore, it is evident that a strong interaction exists in AxB (raster gap versus raster angle), and an intermediate interaction exists in BxC (raster angle versus build orientation). The other factors and interactions have intermediate significant influences. They can be applied in customization when there is no need to use an optimal process. The printed parts are still fulfilled functional and economic features.

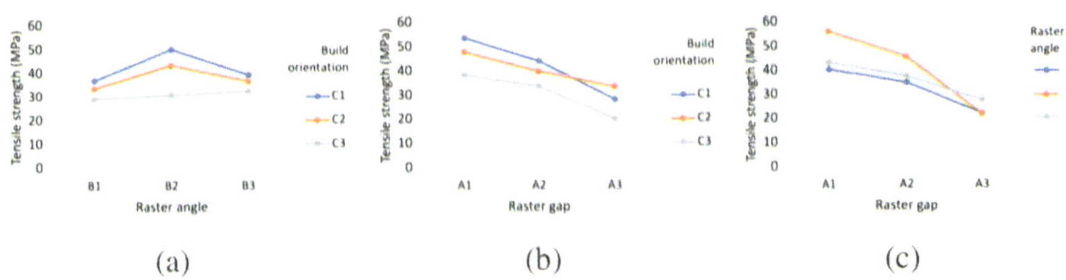

Fig. 6. Interaction plot between raster gap and raster angle (a); Interaction plot between raster gap and build orientation (b); Interaction plot between raster angle and build orientation (c)

4 Conclusion

This work contributes a better understanding of the behavior of printed parts under different process parameters through the design of experiments. Polycarbonate specimens were fabricated with the FFF process and evaluated according to ASTM standard D638 tensile. Three raster gaps (−0.0025 mm, 0,000 mm, 0.0025 mm), three raster angles ([0°, 0°], [+45°, −45°], [+30°, −60°]) and three build orientations (flat, edge, upright) were selected to study the influence of process parameters on the properties of printed parts. The anisotropic behavior exhibits the most at different build orientations. The difference in the tensile strength between the flat versus the up-right build orientation is up to 43.6%. The SEM image of positive raster gaps strengthens the idea of voids between adjacent deposited filaments.

Data analysis methods (Taguchi) are used to study the influence of primary factors and interactions on the properties of printed parts. The factors A (raster gap), B (raster angle) and C (build orientation) are ranked first, second and third, according to the tensile strength of the printed parts. The optimal combination of process parameters is A1B2C1 with the highest number of S/N ratio, -0.0025 mm of the raster gap, [+45°, −45°] of the raster angle and flat direction of the build orientation increasing about 8% of the tensile strength. The interaction of AxC (raster gap versus build orientation) has the most significant influence on the tensile strength at the confidence level of 90%. The other factors and interactions have a significant intermediate effect and can solve a customized problem. Adding thermal factors in further research is promising to solve optimization and prediction of printed parts.

Acknowledgements. The authors would like to acknowledge Dr. El-Gizawy (CEO of ITECH D&M, LLC, USA) and Dr. Hani Salim (Civil Engineering Department, University of Missouri, USA) for facility support.

Declaration of Competing Interest. The authors declare that there are no conflicts of interest regarding the publication of this paper.

References

1. Shashi, G.M., Laskar, M.A.R., Biswas, H., Saha, A.K.: A brief review of additive manufacturing with applications. In: Proceedings of 14th Global Engineering and Technology Conference (2017)
2. Rayegani, F., Onwubolu, G.C.: Fused deposition modelling (FDM) process parameter prediction and optimization using group method for data handling (GMDH) and differential evolution (DE). Int. J. Adv. Manuf. Technol. **73**(1–4), 509–519 (2014). https://doi.org/10.1007/s00170-014-5835-2
3. Phan, T.T.P., Phan, T.Q., El-Gizawy, A.S.: Development of a combined analytical and experimental approach for the determination of the cohesive strength between material extrusion layers using the true area of contact. Addit. Manuf. **30**, 100832 (2019)
4. Phadke, M.S.: Quality Engineering Using Robust Design. Prentice Hall PTR, Hoboken (1995)
5. Liu, X., Zhang, M., Li, S., Si, L., Peng, J., Hu, Y.: Mechanical property parametric appraisal of fused deposition modeling parts based on the gray Taguchi method. Int. J. Adv. Manuf. Technol. **89**(5–8), 2387–2397 (2016). https://doi.org/10.1007/s00170-016-9263-3
6. Deng, X., Zeng, Z., Peng, B., Yan, S., Ke, W.: Mechanical properties optimization of poly-ether-ether-ketone via fused deposition modeling. Materials **11**(2), 216 (2018)
7. Li, H., Wang, T., Sun, J., Yu, Z.: The effect of process parameters in fused deposition modelling on bonding degree and mechanical properties. Rapid Prototyp. J. **24**(1), 80–92 (2018). https://doi.org/10.1108/RPJ-06-2016-0090
8. Torres, J., Cole, M., Owji, A., DeMastry, Z., Gordon, A.P.: An approach for mechanical property optimization of fused deposition modeling with polylactic acid via design of experiments. Rapid Prototyp. J. **22**(2), 387–404 (2016). https://doi.org/10.1108/RPJ-07-2014-0083
9. Chacón, J.M., Caminero, M.A., García-Plaza, E., Núnez, P.J.: Additive manufacturing of PLA structures using fused deposition modelling: effect of process parameters on mechanical properties and their optimal selection. Mater. Des. **124**, 143–157 (2017)
10. Cantrell, J.T., et al.: Experimental characterization of the mechanical properties of 3D-printed ABS and polycarbonate parts. Rapid Prototyp. J. **23**(4), 811–824 (2017). https://doi.org/10.1108/RPJ-03-2016-0042
11. uz Zaman, U.K., Boesch, E., Siadat, A., Rivette, M., Baqai, A.A.: Impact of fused deposition modeling (FDM) process parameters on strength of built parts using Taguchi's design of experiments. Int. J. Adv. Manuf. Technol. **101**(5), 1215–1226 (2019)
12. Heidari-Rarani, M., Ezati, N., Sadeghi, P., Badrossamay, M.R.: Optimization of FDM process parameters for tensile properties of polylactic acid specimens using Taguchi design of experiment method. J. Thermoplast. Compos. Mater. 0892705720964560 (2020)

Experimental Study on Flexural Behavior of Reinforced Concrete Beams Strengthened with CFRP Sheets Under Sustaining Load

Trung Hieu Nguyen(✉), Thi Thu Hien Nguyen, and Duc Viet Le

Hanoi University of Civil Engineering, Hanoi, Vietnam
hieunt@nuce.edu.vn

Abstract. Flexural strengthening of reinforced concrete (RC) beams using carbon fiber reinforced polymer (CFRP) sheets is a solution that shows many advantages in comparison to traditional strengthening solutions. The main goal of this study is to examine the effects of initial load on the effectiveness of strengthening by externally bonded CFRP sheets. Two identical beams were tested. One beam was used as a control specimen while one other beam was strengthened in flexure under sustained load. The value of the initial load was determined based on the cracking of concrete in the tension zone, with a crack width value is 0.3 mm. The obtained results have shown that for RC beams that are already cracked and deformed under sustaining loads, the effectiveness of CFRP strengthening can be demonstrated by the significant increase in the flexural stiffness and the load-carrying capacity of the strengthened RC beams.

Keywords: RC beam · CFRP sheets · Strengthening · Cracking

1 Introduction

The reduction of the actual load-bearing capacity of existing reinforced concrete (RC) beams due to structural deterioration or increasing acting loads on existing buildings due to the change of building functionality addresses the need for structural strengthening. Besides traditional strengthening methods such as widening cross-sections, using post-tensioned reinforcement, and adding steel section, using Fiber-reinforced polymer (FRP) in enhancing the structural performance has been more and more common in recent years. One of the composite materials that has been commonly used for strengthening purposes is Carbon fiber reinforced polymer (CFRP) due to its superior characteristics such as high compressive strength, high elastic modulus, lightweight, and not to be corroded [1–3]. The use of FRP material also shows the convenience of construction process such as quickness, simplicity, and no requirement of much machinery. Figure 1 shown is an example of the FRP strengthening on cracked RC beams.

It is worth noting that the strengthening process is often conducted while structures are still carrying gravity loads. Most of the time reducing the applied load at this time is not available even this technique may improve the strengthening efficiency. For reinforced concrete structural elements subjected to flexural action, it is common that the

D. C. Nguyen et al. (Eds.): ICERA 2021, LNNS 366, pp. 423–430, 2022.
https://doi.org/10.1007/978-3-030-92574-1_44

strengthening process is required when flexural cracks are observed on the tension face, with the crack width ranging from 0.1 mm to 0.3 mm.

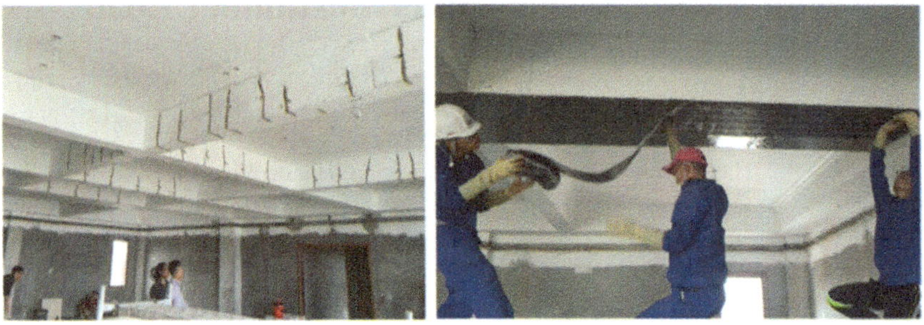

Fig. 1. Strengthening of load-damaged RC beams with CFRP sheets

In the past, there have been several research works on the behavior of virgin beams reinforced externally with CFRP material to increase the load-carrying capacity [3–6]. The obtained results showed that externally bonded CFRP sheets can be effectively used to increase the strength and stiffness of RC beams while maintaining an adequate level of deformability. Besides, it is possible to recognize that the use of CFRP sheets to strengthen load-damaged RC beams, a practical application, has received little attention in experimental research. Some typical studies on the effect of the sustaining load on the performance of RC beams strengthening with CFRP sheets were presented in [7–11]. The reported studies have shown that the initial load is an important factor that affects the ultimate strength of RC beams strengthened with CFRP. Beams strengthened at higher levels of sustaining load have a lower ultimate strength than those of beams strengthened at lower levels of sustaining load.

This paper presents an experimental research on the effectiveness of flexural strengthening of RC beams with CFRP sheets under sustaining load. The findings of this study will be of interest to engineers who are involved in retrofitting and strengthening of RC structures with CFRP materials. The experimental research has been carried out in the Laboratory of Construction Testing and Inspection, Hanoi University of Civil Engineering (HUCE).

2 Experimental Research

2.1 Test Specimens and Materials

In this research, 02 RC beams that have identical rectangular cross-sections, reinforcement, and the strength of concrete were manufactured. All beams tested are 2000 mm long and have the same size b × h = 150 × 200 mm. According to a general calculation to avoid shear failure, two 12 mm-diameter round bars in the tensile area and two 10 mm-diameter round bars in compressive area were used. Shear reinforcement consisted of 6mm-diameter round steel spaced at 100 mm center-center near the bearing support. Among tested beams, one beam, denoted by D-0, was not strengthened to use as a control specimen, and the other beam denoted by D-1. D-1 specimen was subjected to an initial sustaining load which made the maximum crack with of concrete in tensile zone of 0.3 mm. This crack width value is considered as "failure" according to the crack width limit state in TCVN 5574:2018 [12]. The D-1 specimen was strengthened by externally bonded CFRP sheets at the bottom face of the specimen and the strengthening CFRP sheets had dimensions of 1950 × 120 0.111 mm. The reinforcement details of the two-beam specimens were illustrated in Figs. 2 and 3 demonstrates the strengthening solution for D-1 specimen.

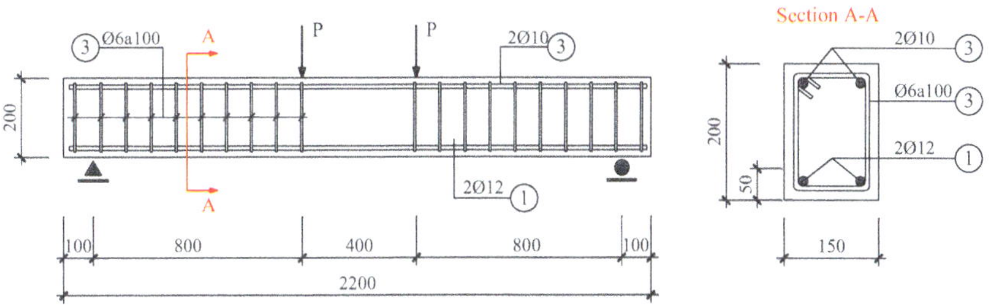

Fig. 2. Detail of tested RC beams

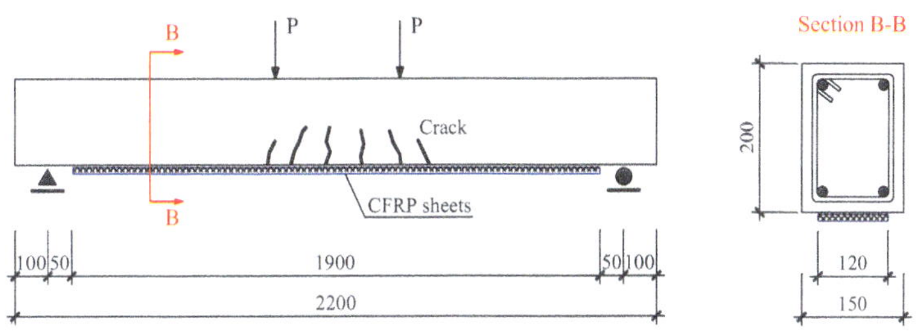

Fig. 3. Strengthening setup

Concrete mix ratio of the test specimens is shown in Table 1. The average cylinder strength at the age of 28 days is 34.7 MPa. Yield strengths of reinforcing bars with diameter ∅10 and ∅12 are 380 MPa and 410 MPa, respectively.

Table 1. Mixture proportions for 1 m^3 of concrete (kg/m^3)

Cement PCB30	Sand	Coarse aggregate	Water	Average cylinder strength R28 (MPa)
380	670	1260	185	34.7

The CFRP sheets used in this study were unidirectional and manufactured by Toray Carbon Co. Ltd (Japan). Table 2 presents the mechanical properties of the CFRP sheets provided by the manufacturer.

Table 2. Mechanical properties for CFRP sheets

Thickness (mm)	Modulus of elasticity (GPa)	Ultimate tensile strength (MPa)	Ultimate tensile elongation (%)
0.111	245.0	3400	1.6

2.2 Generating and Maintaining Initial Load Acting on Strengthened Beam and CFRP Strengthening Process

To generate initial load acting on D-1 beam and constantly maintain this load during the strengthening process, hydraulic jack, load frame, and anchor bolts were used in this experiment. Firstly, 02 concentrated loads had the same value acting on tested beam, spanned a distance of 400 mm and spaced at 800 mm from the support of each side are generated by hydraulic jack, load frame, and load-distributing beam. After creating the required load, 2 anchor bolts were set up to maintain these loads. The crack width of pure bending area of concrete is a fundamental value for determining the value of this load, which had the value of 0.3 mm. To control the behavior of anchor bolts in maintaining load acting on D-1 specimen, Data-logger P3 (manufactured by Vishay – America) was used to follow the modification of tensile strain of two ∅12 longitudinal bars at the middle of the RC beam through 02 strain gauges put on these longitudinal bars before pouring concrete. The results obtained from strain measuring data of the 02 longitudinal bars indicated that the load-maintaining system with anchor bolts has responded to the requirement (Fig. 4).

Fig. 4. Pre-loading on D-1 beam

Figure 5 illustrates the strengthening procedure for D-1 specimen that consists of four main steps, including preparing the bottom face of the beams, apply epoxy to the prepared surface, install the CFRP laminates, and allow the epoxy to cure in 48 h. The cleanness and smoothness of the beam's bottom surface in the preparation process is the key step in this procedure that allow the CFRP sheets to develop their full strength when the beams are loaded.

Fig. 5. The strengthening procedure for D-1 specimen

2.3 Test Setup and Instrumentation

Figure 6 illustrates the typical test setup for the current experimental investigation. These simply supported beams are loaded by a hydraulic jack through a two-loading-point system, creating two equal applied force P. Each loading point is 800 mm away from the beam support. One load-cell is used to measure the applied load P. Three Linear Variable Differential Transducers LVDT-1, LVDT-2, LVDT-3 were used to measure vertical displacements at two supports and the mid-span section of the test beams. At every loading step, all test data including the applied load and vertical deformations are recorded with Data Logger TDS-530.

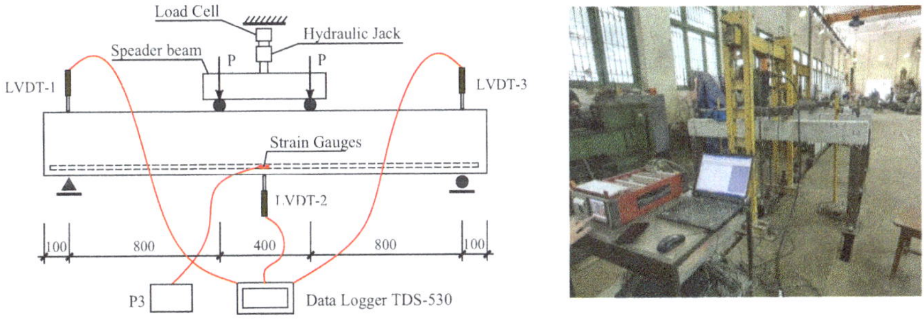

Fig. 6. Test setup

3 Test Results and Discussions

3.1 Load-Deflection Relationships

Figure 7 illustrates load-deflection curves of D-0 and D-1 beam specimens and Table 3 presents the ultimate capacities and deflection characteristics of two tested beams. For the load-deflection curve of the D-0 beam, it can be seen that there are three different behavioral zones. The first zone OA represents the initial stiffness of the uncracked beam D-0. The first visible crack load was determined based on the first change in slope of the load-deflection curve. As can be seen, there is no significant difference in terms of applied load and displacement when the applied load is less than 12 kN. This indicates that the strengthening sheets did not participate in the load-carrying mechanism of the beams when the applied load is small.

The second zone (AB) represents the stiffness of the cracked section. For D-1 beam, the slope of the load-deflection curve was higher than that of the corresponding control beam D-0. As a result, for any given value of load, the deflections of the strengthened beam were lower. This result shows that the contribution of the CFRP layer to the structural stiffness was found to be significant. When the test load on D-0 and D-1 specimens reached about 55 kN, the tensile steel reinforcements were yielded (generally seen as the second change in slope of the load-deflection curve). It can be seen that, for the pre-cracked D-1 beam, the external CFRP layer reduced the stress in the tensile steel reinforcement and hence guaranteed the yield load of the strengthened beam.

The third zone corresponds to damaged tested beams with wide cracks, yielding reinforcement and rupturing the CFRP sheet. As shown in Fig. 7 and Table 3, the ultimate load-carrying capacity of strengthened beam D-1 was higher than that of the control beam D-0. This result again shows the effectiveness of the CFRP sheets for strengthening of RC structures.

Table 3. The ultimate loads of tested specimens

Test specimens	Ultimate loads (kN)	Strengthening effectiveness (%)
D-0	60.5	–
D-1	66.7	9.1%

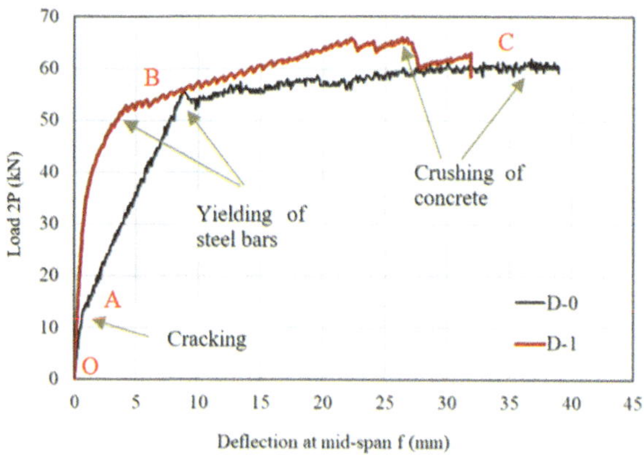

Fig. 7. Load – deflection curves of D-0 and D-1 specimens

3.2 Failure Modes of Test Specimens

Figure 8 presents the failure mode of the strengthened beam D-1. The control beam failed by yielding of steel reinforcement followed by crushing of concrete in the compressive zone at the middle span. The strengthened beam D-1 failed prematurely without warning by rupture of the CFRP sheets after yielding of steel reinforcement.

Fig. 8. Failure mode of D-1 specimen

4 Conclusions

The results of the above-mentioned experimental program have improved the understanding of the behavior of RC beams strengthened with CFRP sheets under sustaining load. Based on the test results of this study, the following conclusions are drawn:

- For RC beams that are already cracked and deformed under sustaining loads, the effectiveness of CFRP strengthening can be demonstrated by the significant increase in the flexural stiffness and the load-carrying capacity of the strengthened RC beams.
- Rupture of the CFRP sheets is the failure mode of the strengthened beam that is observed in the current experimental investigation. This failure mode is a very brittle mode, which is an obvious weakness of this strengthening method.
- Future work is also needed to address the sustainability of the CFRP strengthening method in the Vietnamese climate and weather.

Acknowledgments. The study presented in this paper was financially supported by Hanoi University of Civil Engineering (HUCE), Vietnam through Grant #50-2021/KHXD-TĐ. The financial supports are greatly appreciated.

References

1. ACI 440.2R-08: Guide for the Design and Construction of Externally Bonded FRP Systems for Strengthening Concrete Structures. ACI Committee 440 (2008)
2. FIP-Bulletin No 14: Externally Bonded FRP Reinforcement for RC Structures. Comité Euro-International du Béton (2001)
3. Toutanji, H., Zhao, L., Zhang, Y.: Flexural behavior of reinforced concrete beams externally strengthened with CFRP sheets bonded with an inorganic matrix. Eng. Struct. **28**(4), 557–566 (2006)
4. Meier, U., Kaiser, H.: Strengthening of structures with CFRP laminates. Adv. Compos. Mater. Civil Eng. Struct. 224–232 (1991)

5. Esfahani, M.R., Kianoush, M.R., Tajari, A.R.: Flexural behavior of reinforced concrete beams strengthened by CFRP sheets. Eng. Struct. **29**, 2428–2444 (2007)
6. Bonacci, J.F., Maalej, M.: Externally bonded fiber-reinforced polymer for rehabilitation of corrosion damaged concrete beams. ACI Struct. J. **97**(5), 703–711 (2000)
7. Wang, W.W., Li, G.: Experimental study and analysis of RC beams strengthened with CFRP laminates under sustaining load. Int. J. Solids Struct. **43**, 1372–1387 (2006)
8. Yeong-soo, S., Chadon, L.: Flexural behavior of reinforced concrete beams strengthened with carbon fiber-reinforced polymer laminates at different levels of sustaining load. ACI Struct. J. **100**(2), 231–240 (2003)
9. Wenwei, W., Guo, L.: Experimental study and analysis of RC beams strengthened with CFRP laminates under sustaining load. Int. J. Solids Struct. **43**, 1372–1387 (2005)
10. Shahawy, M., Chaallal, O., Thomas, E.B., Adnan, E.: Flexural strengthening with carbon fiber-reinforced polymer composites of preload full-scale girders. ACI Struct. J. **98**(5), 735–743 (2001)
11. Ceroni, F.: Experimental performances of RC beams strengthened with FRP materials. Constr. Build. Mater. **24**, 1547–1559 (2010)
12. TCVN 5754-2018: Design of Concrete and Reinforced Concrete Structures. Vienamese Standard (2018)

Fusion of Inertial and Magnetic Sensors for Autonomous Vehicle Navigation and Freight in Distinctive Environment

Vo Chi Thanh[1(✉)], Nguyen Ngo Anh Quan[1], Tran Le Thang Dong[1,2], Tran Thuan Hoang[1,2], and Minh T. Nguyen[3]

[1] Center of Electrical Engineering, Duy Tan University, Danang 550000, Vietnam
vochithanhl@dtu.edu.vn

[2] Faculty of Electrical-Electronic Engineering, Duy Tan University, Danang 550000, Vietnam

[3] Thai Nguyen University of Technology, Thai Nguyen 240000, Vietnam

Abstract. The method of navigation is a very important factor that decides the flexibility and accuracy of the AGV in different operating environments. With each method of navigation, we have different techniques for determining the position and angle of the vehicle. In this research, by fusing data from sensors, the angle, direction and position of the AGV then are estimated by the data of the inertial sensor, encoder and calculated by the Kalman filter algorithm. In addition, the minor inaccuracy in measurement of angle, direction and position are corrected when the AGV passes through the magnetic reference points embedded under the floor based on the virtual paths. The simulation results have proved the accuracy of the proposed algorithm.

Keywords: Inertial navigation · Inertial sensor · Magnetic reference point · Automated guided vehicle · AGV

1 Introduction

The study of navigation problems for autonomous guided vehicle (AGV) systems is increasingly being used in the manufacturing and warehousing industries, and is especially important in hazardous environments or in isolated areas. Some of our previous studies have provided some solutions to improve positioning and increase efficiency for motion control [1–3]. In addition, some works [4–8] just stopped using different control methods to control AVG in the environment with many ideal conditions. AGV navigation systems can be divided into two types: fixed path are physical paths such as conductors, paint lines, magnetic tapes, etc., which are fixed on the ground everywhere that the AGV will pass. The advantages of this solution are low cost, ease of implementation and high reliability. However, the disadvantage of this method is less flexibility, poor adaptability to the environment, especially to outdoor environments or where there is a lot of dirt, which increases the maintenance cost of the path; free path is a virtual path that is calculated and determined on a computer based

D. C. Nguyen et al. (Eds.): ICERA 2021, LNNS 366, pp. 431–439, 2022.
https://doi.org/10.1007/978-3-030-92574-1_45

on data of modern navigation sensors such as LiDAR laser sensor, 3D camera, inertial sensor, ultrasonic sensor.... This solution has the advantage of a flexible and highly efficient operating system. However, the disadvantage of these systems is that they are ex-pensive, and it is difficult to improve accuracy.

Recently, solutions using LiDAR laser sensors or cameras are popularly used due to their high accuracy and flexibility, but in some specific environments such as outdoor or dusty, the accuracy of the system will be reduced rapidly. At this point, the inertial sensor is a possible solution as it is suitable for most types of environments. However, due to the nature of the inertial sensor, errors accumulate during operation, the system accuracy will decrease after a period of operation.

In this paper, we will present an AGV navigation solution using an inertial sensor in which the vehicle's direction angle and position are estimated by the data of the inertial sensor, the encoder and calculated through the Kalman Filter algorithm. On this basis, the magnetic sensor is added to correct the accumulated errors, the angular and position measurement errors of the AGV are corrected when passing through the reference points by the magnets embedded in the floor above virtual paths.

2 Kinematic Model of Small Automatic Car

2.1 Dynamic Model Analysis of Autonomous Vehicles

The autonomous vehicle in this research is a moving vehicle with two-wheeled differential, each wheel is attached to an independently controlled electric motor, controlling the vehicle to move and orientation is done by controlling the speed of these two motors.

There are many factors, such as vehicle load, control circuitry, powertrain and body structure, etc., that will have a big impact in the dynamic analysis. In this paper, ignoring some relatively small factors, a relatively simple AGV motion model is presented. The AGV structure is illustrated by the following Fig. 1.

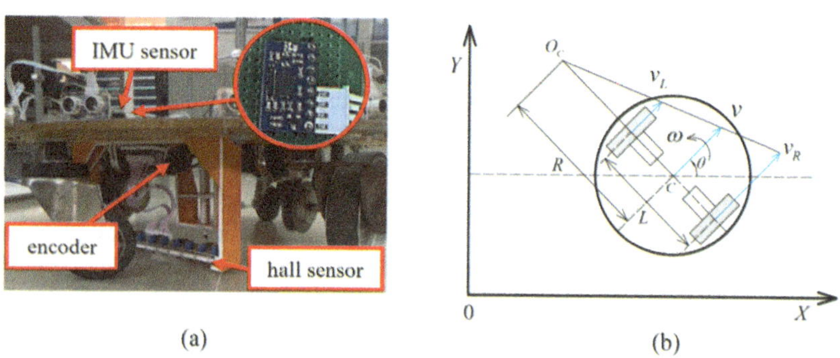

Fig. 1. (a) Action vehicle structure model; (b) Kinetic model of autonomous vehicle

Select the OXY coordinate system associated with the active area plane of the AGV with the origin O being the starting position of the AGV. The center of gravity C of the AGV is the midpoint between the two-wheeled differential drive. AGV will move on an arc with center Oc and radius R. When $v_r = v_l$, the angle of rotation of center OC will coincide with center C of AGV.

Let v, ω the translational speed and angular velocity of AGV. From the kinetic model of AGV in Fig. 1, we have:

$$v = \frac{v_r + v_l}{2} \tag{1}$$

Let x, y be the coordinates of AVG, be the angle between the vertical axis of AVG and the axis OX.

The state of the AGV is represented by the vector $X = [x \quad y \quad \theta]^T$. At time k and time $k - 1$ we have the state of AGV as $X_k = [x_k \quad y_k \quad \theta_k]^T$ and $X_{k-1} = [x_{k-1} \quad y_{k-1} \quad \theta_{k-1}]^T$. Then X_k is calculated in terms of X_{k-1} by the following equation:

$$X_k = \begin{bmatrix} x_k \\ y_k \\ \theta_k \end{bmatrix} = \begin{bmatrix} x_{k-1} \\ y_{k-1} \\ \theta_{k-1} \end{bmatrix} + \begin{bmatrix} cos\theta_{k-1} & 0 \\ sin\theta_{k-1} & 0 \\ 0 & 1 \end{bmatrix} \begin{bmatrix} v_{k-1}T \\ \omega_{k-1}T \end{bmatrix} \tag{2}$$

In there, T is the sampling period of the signal being measured.

2.2 Determine Position and Orientation of Autonomous Vehicles at Reference Points

In order to control AGV, we must track the trajectory of AGV. In this model, the pa-per uses an encoder and an inertial sensor (IMU) to determine the coordinates and direction angle of the AGV. Therefore, trajectory is calculated from the initial input values combined with the feedback values from the sensor, the cumulative error will gradually increase during the movement of the AGV.

So, a magnetic sensor is added to determine the waypoints with known coordinates to recalibrate the trajectory as the AGV passes through these reference points. The magnetic sensor used is a bar consisting of 16 hall sensors placed in a line and separated by a distance of l = 10 mm Thus, the width of the magnetic sensor is 160 mm. The sensor bar will be placed perpendicular to the longitudinal axis of the body at the midpoint of the bar.

When the AGV passes through the magnetic reference points placed on the floor, the hall sensors passing above the reference point will be activated. From the position of the activated hall sensors, the deviation d of the longitudinal axis of the body can be calculated from the reference point as shown in Fig. 2.

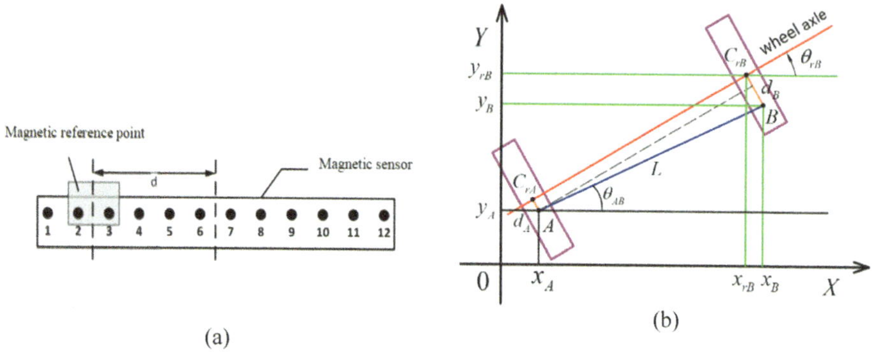

Fig. 2. (a) Magnetic sensor model and magnet reference point; (b) Calculation of position and orientation of AGV at reference point

Assuming two reference points $A(x_A, y_A)$ and $B(x_B, y_B)$ have distance L and known coordinates, the angle of segment AB with respect to axis OX is θ_{AB}. AGV will go from A to B. When the vehicle passes through point A, the magnetic sensor detects the deviation d_A, when the vehicle passes through point B, the sensor will determine the deviation in d_B. From Fig. 2(b), it is possible to calculate the actual position and angle of the vehicle at point B.

Let δ be the angle of deviation between the longitudinal axis of the body and the line segment AB, we have:

$$\delta = arcsin\left[\frac{(d_A - d_B)}{L}\right] \tag{3}$$

Then the actual direction angle of the AGV at point B will be:

$$\theta_{rB} = \theta_{rB} + \delta \tag{4}$$

From there, we can also calculate the coordinates of the actual position of the AGV at point B:

$$\begin{cases} x_{rB} = x_B + d_B sin(\theta_{rB}) \\ y_{rB} = y_B + d_B sin(\theta_{rB}) \end{cases} \tag{5}$$

Thus, at point B, the state of the AGV will be adjusted according to the actual state as follows:

$$X_{rB} = \begin{bmatrix} x_{rB} \\ y_{rB} \\ \theta_{rB} \end{bmatrix} = \begin{bmatrix} x_{rB} = x_B + d_B \sin(\theta_{rB}) \\ y_{rB} = y_B + d_B \sin(\theta_{rB}) \\ \theta_{AB} + arcsin[(d_A - d_B)/L] \end{bmatrix} \tag{6}$$

2.3 Application Filter for Autonomous Vehicles

During operation, the AGV's signal is often affected by measurement disturbances and system noise. To determine the exact value of the state of the AGV, the article applies the Kalman filter, which is widely used in the field of signal processing and control [1, 2].

From (2), we build the system state model system and measure as follows [1, 2]:

$$\begin{aligned} X_k &= AX_{k-1} + Bu_{k-1} + w_{k-1} \\ Z_{k-1} &= HX_{k-1} + v_{k-1} \end{aligned} \tag{7a}$$

In there X_k is the state of AGV at time k, Z_{k-1} is the measurement state at time $k-1$, u_{k-1} is the input control signal, w and v are system noise and measurement noise, A is the state transition matrix, B is the input control matrix, H is the observation transfer matrix, the matrices are specifically defined as follows:

$$A = \begin{bmatrix} 1 & 0 & 0 \\ 0 & 1 & 0 \\ 0 & 0 & 1 \end{bmatrix}, B = \begin{bmatrix} cos\theta_{k-1} & 0 \\ sin\theta_{k-1} & 0 \\ 0 & 1 \end{bmatrix}, u_{k-1} = \begin{bmatrix} v_{k-1}T \\ w_{k-1}T \end{bmatrix}, H = \begin{bmatrix} 1 & 0 & 0 \\ 0 & 1 & 0 \\ 0 & 0 & 1 \end{bmatrix} \tag{7b}$$

The Kalman filter consists of two stages: prediction and correction.

Stage 1: predicts the next state based on the current state and the input control signal value:

$$\begin{aligned} \hat{X}_k^- &= A\hat{X}_{k-1}^- + Bu_{k-1} \\ P_k^- &= AP_{k-1}A^T + Q \end{aligned} \tag{8}$$

Stage 2: recalibrate the estimated value according to the predicted value and the measured signal value:

$$\begin{aligned} K_k &= P_k^- H^T \left(HP_k^- H^T + R\right)^1 \\ \hat{X}_k &= \hat{X}_k^- + K_k\left(Z_k - H\hat{X}_k^-\right) \\ P_k &= (I_n - K_kH)P_k^- \end{aligned} \tag{9}$$

With Q, R are the system noise correlation matrices and the measurement noise is usually determined by taking the variance of the noises.

For the AGV control system in this paper, the Z_k measurement values are the sum of measured values from Encoder, IMU and time reference point during operation.

2.4 Controllers for Autonomous Vehicles

When controlling the AGV to follow the sample trajectory, errors will appear. This error is expressed by the equation [4]:

$$e = \begin{bmatrix} e_1 \\ e_2 \\ e_3 \end{bmatrix} = \begin{bmatrix} cos\theta & sin\theta & 0 \\ -sin\theta & cos\theta & 0 \\ 0 & 0 & 1 \end{bmatrix} \begin{bmatrix} x_r - x \\ y_r - y \\ \theta_r - \theta \end{bmatrix} \tag{10}$$

In there, $X_r = \begin{bmatrix} x_r \\ y_r \\ \theta_r \end{bmatrix}$ is the sample trajectory set up for the operation program of AGV.

From the kinetic model of AGV (2) and the derivative (11) we get:

$$\begin{bmatrix} \dot{e}_1 \\ \dot{e}_2 \\ \dot{e}_3 \end{bmatrix} = \begin{bmatrix} cose_3 & 0 \\ sine_3 & 0 \\ 0 & 1 \end{bmatrix} \begin{bmatrix} u_{r1} \\ u_{r2} \end{bmatrix} + \begin{bmatrix} -1 & e_2 \\ 0 & -e_1 \\ 0 & -1 \end{bmatrix} \begin{bmatrix} u_1 \\ u_2 \end{bmatrix} \tag{11}$$

With $u_r = \begin{bmatrix} u_{r1} \\ u_{r2} \end{bmatrix} = \begin{bmatrix} v_r \\ \omega_r \end{bmatrix}, u = \begin{bmatrix} u_1 \\ u_2 \end{bmatrix} = \begin{bmatrix} v \\ \omega \end{bmatrix}$

The controller for the AGV is built as follows:

$$u = \begin{bmatrix} u_1 \\ u_2 \end{bmatrix} = \begin{bmatrix} v \\ \omega \end{bmatrix} = \begin{bmatrix} u_{r1}cose_3 \\ u_{r2} \end{bmatrix} - \begin{bmatrix} v_1 \\ v_2 \end{bmatrix} \tag{12}$$

In there v_1,v_2 is the adaptive signal of the proposed controller as follows [3]:

$$\begin{bmatrix} v_1 \\ v_2 \end{bmatrix} = \begin{bmatrix} -k & 0 & 0 \\ 0 & -sign(u_{r1}|u_{r1}|k_2) & -k_3 \end{bmatrix} \begin{bmatrix} e_1 \\ e_2 \\ e_3 \end{bmatrix} \tag{13}$$

According to [4] the tracking error e will be 0 when the time $t \to \infty$ follows the Lyapunov stability criterion.

3 Experiment and Discussion

3.1 Simulation Scenario

Use the robot in Fig. 1(a) to experiment. With the distance between 2 wheels $L = 0.47$ m, wheel diameter 0.15 m, maximum translational speed $v_{max} = 1$ m/s. AGV moves along the sample trajectory as show in Fig. 3 from point O(0,0) passing through points A(0,3), B(1.5,3), C(1.5,0), D(4.5,0) and proceeding to destination E(4.5, 3). The AGVR sample orbital target moves as follows: AGVR starts to accelerate with acceleration, then, when it reaches a speed of 0.3 m, the AGV will move steadily straight to point A, when 0.3 m from point A, AGV will decelerate with acceleration of

$-0.15\ m/s^2$ and when it comes to point A, AGV will rotate at an angle of $-90°$, then repeat the acceleration – stabilization – deceleration as before to move towards points B, C, D and E. When moving to the destination E AGV will stop.

When going through the correction points from O, M1, M2, A, M3, B, M4, M5, C, M6, M7, D, M8, M9, the AGV will update the status values of the parameters (including position parameters from Encoder and direction angle from IMU) and the controller based on the above parameters to control the AGV gradually towards the sample trajectory.

3.2 Results and Comments

The results of collecting the trajectories that the robot travels in three cases are shown in Fig. 3 with the true path (true), the path is estimated only by odometry (encoder), estimated path using Kalman filter with IMU sensor (encoder + IMU), and path when adding Hall sensor (Encoder + IMU + Hall).

The results of the actual position and position error of the AGV when following the same trajectory are shown in Fig. 4(a) and Fig. 4(b). Visually, the trajectory when using a combination of encoder sensors, IMU and sensor from Hall is closer to the real path than in the case of using only the Odometry method.

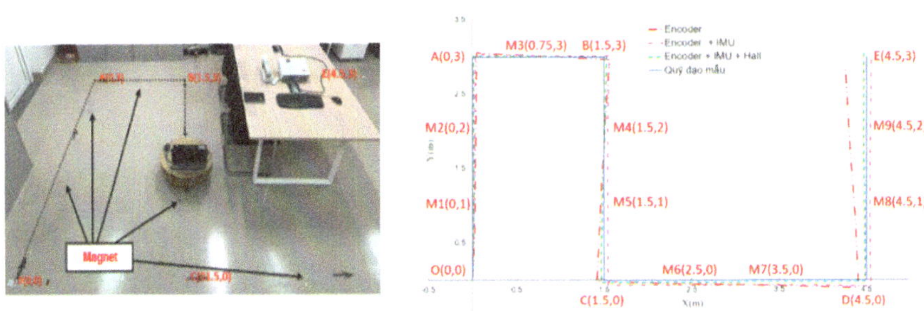

Fig. 3. Experimental images & AGV AGV follows sample trajectory

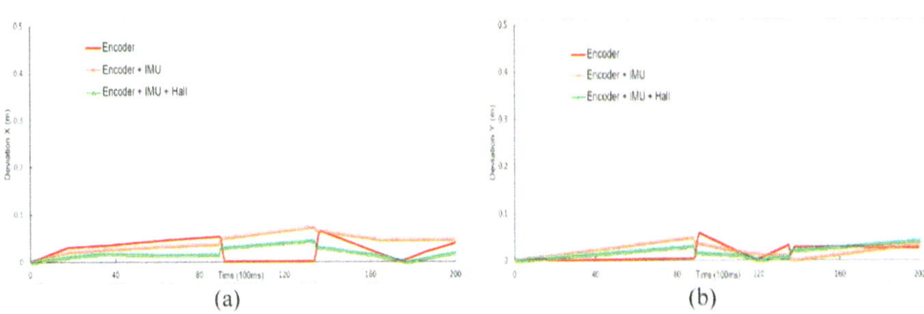

Fig. 4. The deviation between the positions of the AGV is estimated when following the sample trajectory: (a) in the X(m) direction; (b) in the Y(m) direction

Obviously, with the integration of magnetic correction points, it allows AGV to calculate the orbital state in the most realistic way, helping the controller to correct errors to make the AVG move well according to the sample trajectory. Through theory and experiment, it has been shown that: when using the proposed control method for AGV autonomous vehicles using inertial sensors and virtual navigation. The AGV moves according to the sample trajectory and sticks well despite the presence of measurement noise in sensors such as Encoder, IMU. This result has proved the correctness of the proposed algorithm and is a premise for the application of more modern techniques to meet higher requirements in information processing as well as robot control.

4 Conclusions

This paper proposes a method to control AGV using inertial sensors and virtual navigation. The simulation results show that the AGV moves according to the sample trajectory and sticks well despite the presence of measurement noise in sensors such as Encoder, IMU. The results of this paper are the premise for the application of more modern techniques to meet higher requirements in information processing as well as robot control.

Acknowledgements. The authors would like to thank Thai Nguyen University of Technology, Viet Nam for the support.

References

1. Hoang, T.T., Duong, P.M., Van, N.T.T., Viet, D.A., Vinh, T.Q.: Development of an EKF-based localization algorithm using compass sensor and LRF. In: Proceedings of the IEEE 12th International Conference on Control, Automation, Robotics & Vision, pp. 341–346. ICARCV (2012)
2. Hoang, T.T., Duong, P.M., Van, N.T.T., Viet, D.A., Vinh, T.Q.: Multi-sensor perceptual system for mobile robot and sensor fusion-based localization. In: IEEE 1th International Conference on Control, Automation and Information Sciences, pp.259–264. ICCAIS (2012)
3. Hoang, T.T., Hiep, D.T., Duong, P.M., Van, N.T.T., Duong, B.G., Vinh, T.Q.: In: Proceedings of the IEEE 8th Conference on Industrial Electronics and Applications, pp.1308–1313. ICIEA (2013)
4. Klancar, G., Matko, D., Blazic, S.: Mobile robot control on a reference path. In: Proceedings of the 13th Mediterranean Conference on Control and Automation, pp. 1343–1348. Limassol, Cyprus (2005)
5. Ryck, D.M., Mark, V., Debrouwere, F.: Automated guided vehicle systems, state-of-the-art control algorithms and techniques. J. Manuf. Syst. **54**, 152–173 (2020)
6. Moshayedi, A.J., Roy, A.S., Liao, L.: PID tuning method on AGV (automated guided vehicle) Industrial robot. J. Simul. Anal. Novel Technol. Mech. Eng. **12**(4), 53–66 (2020)

7. Hasan, H.S., Abidin, M.S.Z., Mahmud, M.S.A., Muhamad Said, M.F.: Automated guided vehicle routing: Static, dynamic and free range. J. Int. Eng. Adv. Technol. **8**(5C), 1–7 (2019)
8. Kim, S., Jin, H., Seo, M., Har, D.: Optimal path planning of automated guided vehicle using dijkstra algorithm under dynamic conditions. In: 7th International Conference on Robot Intelligence Technology and Applications (RiTA). IEEE (2019)

High-Reliability Three-Phase Switching-Cells Current Source Inverter

Hai Trung Do[1], Duc-Tuan Do[1(✉)], and Tien-The Nguyen[2]

[1] Thai Nguyen University of Technology, Thai Nguyen, Vietnam
doductuan-tdh@tnut.edu.vn

[2] Kyungpook National University, Daegu 41566, Korea

Abstract. The conventional three-phase current source inverter (CSI) suffers from the open-circuit problem. In CSI operation, the overlap-time in the gates signals needed for safe commutations causes the distortion in the output current waveforms. This paper proposes a new three-phase switching-cell current source inverter, called H6-SC-CSI. Based on switching-cells, the inverter reliability is enhanced because open-circuit concerns are eliminated, and the voltage overshoot at turn-of is reduced. In the operation of the proposed CSI, switch overlap-time intervals can be minimized or eliminated. Therefore, the total harmonic distortion (THD) of the output current is reduced. A 1.2-kW prototype of the proposed inverter is designed and tested.

Keywords: Current source inverter · Reliability · Switching-cell · Three-phase

1 Introduction

Compared to voltage source inverters (VSIs), current source inverters (CSIs) have inherent boost functionality [1]. Besides this, the power decoupling element in CSIs is an inductor, which has a longer lifetime compared to electrolytic capacitors used in VSIs [2]. VSIs and CSIs have found their particular applications in solar photovoltaic (PV) systems and power electric drives. However, inverter reliability is one of the major concerns for both conventional CSIs and VSIs. While the conventional VSIs suffer from the shoot-through problem, the major concern for the reliability of the conventional CSIs is the open-circuit problem. In recent years, based on dual-buck power cells, many topologies, such as buck-boost inverter [3] and three-phase inverter [4] have been researched to eliminate the shoot-through problem. Similarly, with the help of the unique operation of switching-cells (SC), the open circuit problem is eliminated [5, 6]. However, most of them are designed for single-phase CSI. In the operation of the conventional three-phase current source inverter, in literature named H6-CSI, the switch overlap-time intervals are added to void the open-circuit problem. However, this overlap-time causes distortion in the output current waveforms [7, 8]. Even with switch overlap-time intervals, the open-circuit is still the major killer of CSIs, especially at some fault conditions.

D. C. Nguyen et al. (Eds.): ICERA 2021, LNNS 366, pp. 440–445, 2022.
https://doi.org/10.1007/978-3-030-92574-1_46

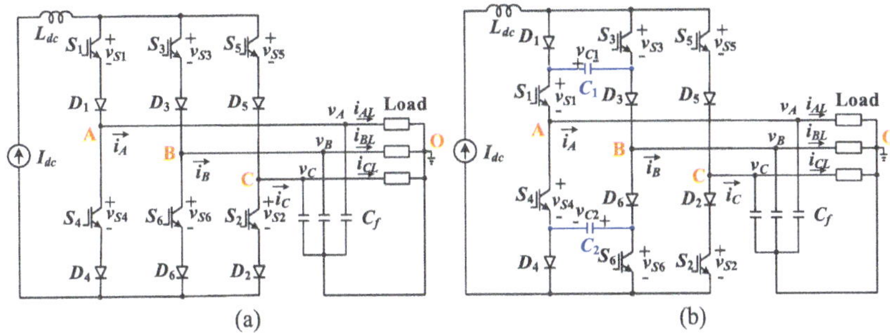

Fig. 1. Schematic of a) conventional three-phase current source inverter (H6-CSI), b) proposed three-phase switching-cell current source inverter (H6-SC-CSI).

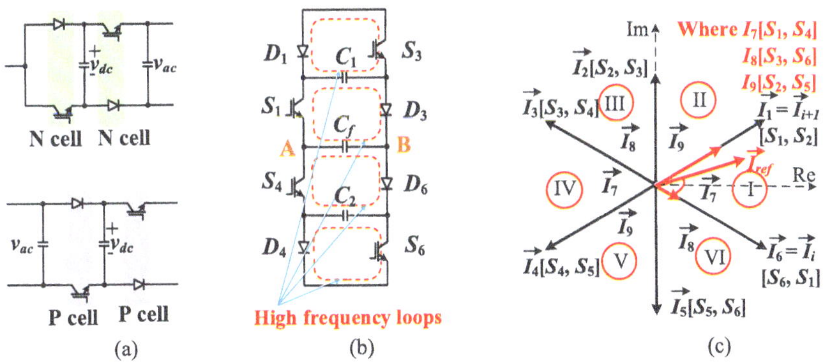

Fig. 2. (a) Series connection of the N-cells and P-cells to form an ac voltage port. (b)The high-frequency loops. (c) Space vector switching diagrams of proposed H6-SC-CSI.

In order to solve the aforementioned problem, a new three-phase switching-cells current source inverter, called H6-SC-CSI, is proposed in this paper. The proposed inverter with switching-cells attains high reliability due to the absence of open-circuit problems. The switch overlap-times used in the proposed H6-SC-CSI can be minimized or eliminated. Therefore, the total harmonic distortion (THD) of the output current is reduced. Besides, the voltage spikes of the semiconductor devices are reduced by minimizing the size of the high-frequency loops.

2 Operation Principle of Proposed Topology

2.1 Proposed H6-SC-CSI

Figure 1(b) shows the proposed H6-SC-CSI. The proposed inverter is based on the basic P- and N-switching cells presented in [5, 6], as shown in Fig. 2(a). Additional capacitors C_1, C_2 combined with diodes (D_1, D_3, D_4, and D_6) provide paths for the inductor current to flow during the switching transition. Therefore, the proposed

H6- SC-CSI avoids the open-circuit problem, and the switch overlap-time intervals can be minimized or eliminated. The space vector of the H6-CSI presented in [7] is used to control the proposed H6-SC-CSI, as shown in Fig. 2(c). However, the switch overlap-time intervals are minimized or eliminated.

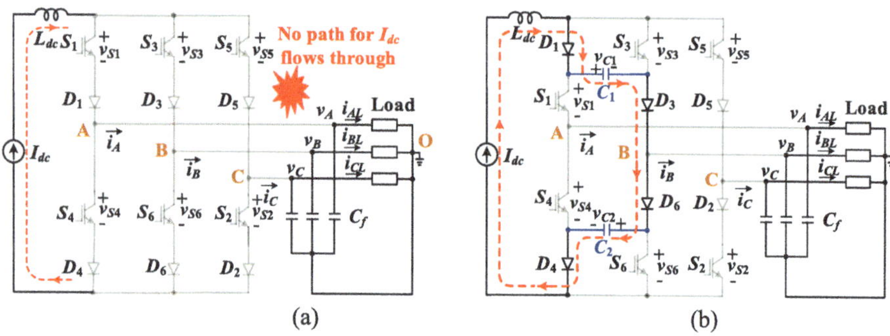

Fig. 3. Simplified equivalent circuit in fault conditions. (a) H6-CSI. (b) H6-SC-CSI.

2.2 Reliability Analysis

Generally, the current source is made by the series connection of a voltage source V_{in} and large inductor. In some fault conditions, if there are some delays, mismatches in control signals of those switches, or all the switching devices are accidentally turned off, the high *di/dt* will appear in the inductor of conventional H6-CSI, generating high-voltage spikes, which may damage power devices, as shown in Fig. 3(a).

However, with the proposed H6-SC-CSI, in the case of some delays, or mismatches between control signals of those switches, additional capacitors are charged. During dead-time intervals (Δt_d), the capacitors C_1 and C_2 are charged, as shown in Fig. 3(b). The increase of the capacitor voltages (ΔV_{C1} and ΔV_{C2}) is given by (1).

$$\Delta V_{C1} = \Delta V_{C2} = \frac{I_{dc}}{C} \Delta t_d \tag{1}$$

Where I_{dc} is the dc-link current, C is the capacitance value.

3 Simulation and Experimental Results

3.1 Simulation Results

In simulations and experiments, a DC voltage source is connected with an inductance of 6 *m*H to represent the current source of 10 A in the DC side, and the output load is resistance (*R*). Table 1 shows the system parameter used for both conventional and proposed CSI. PSIM simulation software was used to verify the operation of CSIs under normal and fault conditions. In the normal operation of H6-CSI, 2 μs overlap time is used. However, for H6-SC-CSI, the overlap-time is minimized to 0.2 μs

because of the elimination of the open-circuit problem. Figure 4(a) shows the output current waveforms of H6-CSI with the THD of 3.6%. As can be seen, the additional overlap-time causes distortion in the output waveforms of H6-CSI. However, there is no significant distortion in the output waveforms of H6-SC-CSI with the THD of 1.8%, as shown in Fig. 4(b). THD of the CSIs is measured by tools of the PSIM simulation software. To verify the operation under fault conditions, 1 μs dead-time, which appears two times during each cycle, is inserted among the control signals. With H6-CSI, when the dead-time appears, the inductor current has no path to flow, and the voltage stress of the switches is increased by huge voltage spikes of about 60 kV, as shown in Fig. 4(c). While the capacitors C_1 and C_2 of H6-SC-CSI are charged by the input current during dead-time intervals, and the voltages of switches are just increased an additional voltage of 100 V, which is the change in capacitor voltages (v_{C1}, v_{C2}), as shown in Figs. 4(d) and (e). It can be seen that H6-CSI suffers from open-circuit problems while proposed H6-SC-CSI can work with some delays or mismatches between control signals of those switches.

Table 1. Testing parameters

Parameters	Value	Parameters	Value
Rated power	1.2 kW	Resistance (R)	30 Ω
Input current (I_{dc})	10 A	Filter capacitance (C_f)	20 μF
Input inductance (L_{dc})	6 mH	Switching frequency	30 kHz
Peak value of line-to-line voltage	270 V	Capacitances ($C_{1,2}$)	0.1 μF

3.2 Experimental Results

A 1.2 kW prototype was implemented with the same simulation parameters. IKW20N60T IGBT and RHRG3060 fast diodes were used in both the conventional and proposed CSI. The film capacitors with the value of 0.1 μF are used for two extra capacitors C_1, C_2. H6-CSI is operated with 2 μs switch overlap-time. In experiments, THD of CSIs is measured with Yokogawa WT1800 power meter. Figure 5(a) shows the output current waveforms of H6-CSI with the THD of 4.8%. With the proposed H6-SC-CSI, the switch overlap-time is minimized. The output current waveform of the proposed H6-SC-CSI with the switch overlap-time of 0.2 μs is shown in Fig. 5(b). The THD of the output current is measured to be equal to 2%. From Figs. 5(a) and (b), it can be seen that the quality of the output current waveform of H6-SC-CSI is improved.

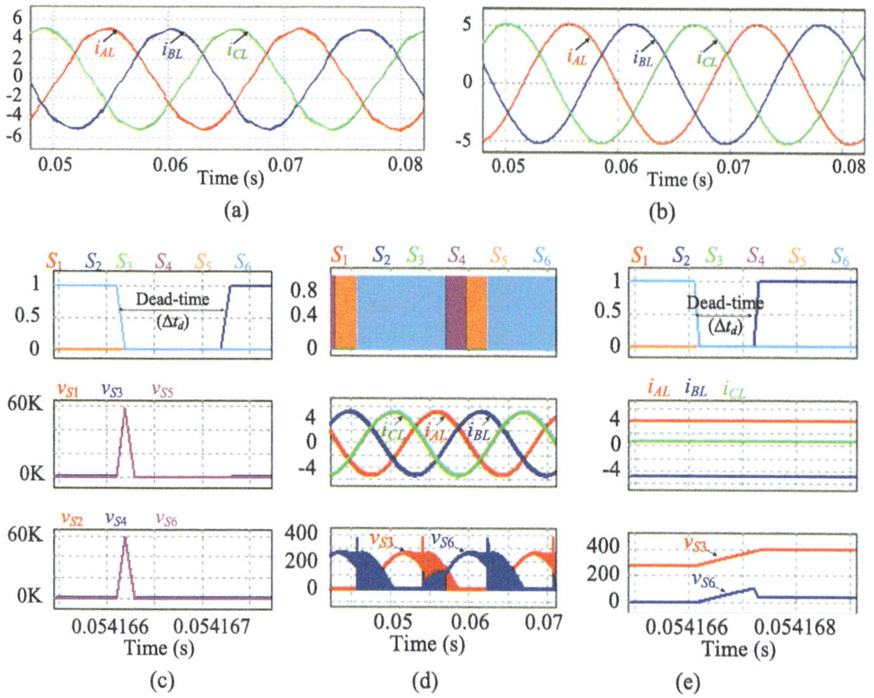

Fig. 4. Simulation results. (a) Waveforms of H6-CSI under normal condition (2 μs overlap-time). b) Waveforms of H6-SC-CSI under normal condition (0.2 μs overlap-time). (c) Waveforms of H6-CSI under fault condition (1 μs dead-time). (d) Waveforms of H6-SC-CSI under fault condition (1 μs dead-time). (e) Magnification of (d) at the dead-time interval.

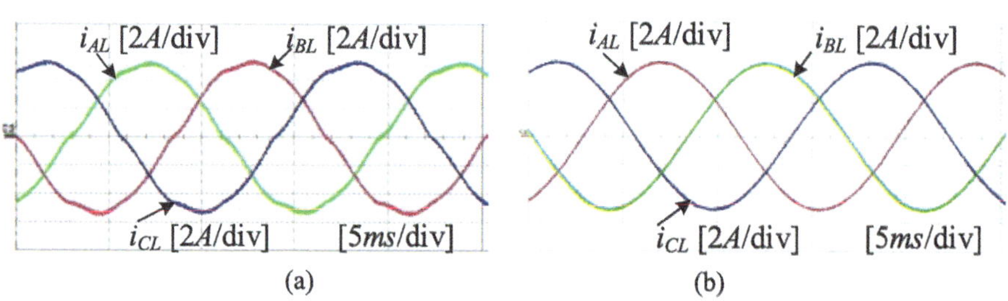

Fig. 5. Experimental results. (a) Waveforms of H6-CSI under normal condition (2 μs overlap-time). b) Waveforms of H6-SC-CSI under normal condition (0.2 μs overlap-time)

4 Conclusion

This paper has presented a new three-phase switching-cell CSI, named H6-SC-CSI, which improves inverter reliability by eliminating the possibility of the open-circuit problem. With the overlap-time between the gate signals is minimized, the quality output waveform of the proposed H6-SC-CSI is also improved. The proposed inverter can be a new alternative candidate for high-power motor drivers and PV system applications.

Acknowledgement. This research was supported by Research Foundation funded by Thai Nguyen University of Technology.

References

1. Komurcugil, H.: Steady-state analysis and passivity-based control of single-phase PWM current-source inverters. IEEE Trans. Ind. Electron. **57**(3), 1026–1030 (2010)
2. Wang, H., Blaabjerg, F.: Reliability of capacitors for DC-link applications in power electronic converters—an overview. IEEE Trans. Ind. Appl. **50**(5), 3569–3578 (2014)
3. Khan, A.A., Cha, H.: Dual-buck-structured high-reliability and high-efficiency single-stage buck-boost inverters. IEEE Trans. Ind. Electron. **65**(4), 3176–3187 (2018)
4. Sun, P., Liu, C., Lai, J., Chen, C., Kees, N.: Three-phase dual-buck inverter with unified pulsewidth modulation. IEEE Trans. Power Electron. **27**(3), 1159–1167 (2012)
5. Tolbert, L.M., Peng, F.Z., Khan, F.H., Li, S.: Switching cells and their implications for power electronic circuits. In: Proceedings of the IEEE IPEMC, pp 773–779 (2009)
6. Do, D.-T., Cha, H.: High-reliability and reduced switch count single-phase dual-output current source inverter using switching-cell structure. IEEE Trans. Power Electron. **36**(6), 6513–6524 (2021)
7. Zmood, D., Holmes, D.: Improved voltage regulation for current source inverters. IEEE Trans. Ind. Appl. **37**(4), 1028–1036 (2001)
8. Morsy, A., Ahmed, S., Enjeti, P., Massoud, A.: An active damping technique for a current source inverter employing a virtual negative inductance. In: Proceedings of the 25th Annual IEEE Applied Power Electronics Conference, pp. 63–67 (2010)

High-Speed Users' Mobility Prediction Scheme Based on Deep Learning for Small Cell and Femtocell Networks

Khoa Dinh Nguyen Dang[1(✉)], Peppino Fazio[1,2], and Miroslav Voznak[1]

[1] VSB – Technical University of Ostrava, 17. listopadu 2172/15, 70800 Ostrava, Czechia
khoa.dinh.nguyen.dang.st@vsb.cz
[2] DSMN, Ca' Foscari University of Venice, via Torino 155, 30172 Mestre (VE), Italy

Abstract. Users' mobility has a huge impact on the performance of cellular networks. Acknowledge users' multiple next locations plays an important role in various aspects which can be mentioned as helping the base stations to pre-calculate and allocate the resource to users faster and more efficiently, shortening the duration of the handover process, reducing significantly the network data congestion, and increasing the overall users' satisfaction. In our article, we focus our attention on multiple users and multi-position ahead prediction for femtocells and small cells, typical of 5G infrastructure. We use Autoregressive Gated Recurrent Units (AR-GRU) to perform the prediction based on acknowledging users' trajectories. We use Simulation of Urban MObility (SUMO) to create our own users' trajectory datasets to train and test the models. In order to prove the effectiveness of the model, we compare its performance with Autoregressive Long Short-Term Memory (AR-LSTM), Deep Learning Neural Network (DNN), Long Short-Term Memory (LSTM), and Gated Recurrent Unit (GRU) models. Then we use the models in two more different datasets from two different simulated regions to prove the ability to work in different contexts.

Keywords: 5G · Mobility prediction · Neural networks · LSTM · GRU

1 Introduction

As shown in the last years, HETerogenous NETworks (HETNETs) are becoming the key solution in 5G technology. In this architecture, as mentioned in [1, 2] and in many more contributions, the femtocell, which is also called Heterogeneous NodeB (HgNB), has been expected to be one of the key factors for the success of the new technology, due to their advantages (easy deployment, low cost, reliability, and flexibility). However, their coverage ranges are short, only about 30–40 m [3, 4]: this leads to the demand for fast data and resource transmission from the cell to the users. Particularly, in the case of fast-moving users (who are driving or traveling by vehicles), these responding speeds should be fast to guarantee the QoS/QoE. On one hand, to achieve

D. C. Nguyen et al. (Eds.): ICERA 2021, LNNS 366, pp. 446–458, 2022.
https://doi.org/10.1007/978-3-030-92574-1_47

this request, engineers can develop and build femtocell and small cell hardware that can afford those requirements. On the other hand, the much easier and cheaper way is helping the Base Stations (BSs) to predict the next location of users. By acknowledging the next several user's locations, they can benefit from it by deciding which is the nearby cells in those locations to pre-calculate users' demand for radio resources and allocate to them at the moment they arrive or preparing the handoff procedure before the users pass by, and so on.

Predict users' next movement based on their mobility history has been redirecting the researcher's attention from the other approaches in the past such as in the paper [5]. According to [6], some approaches that have been studied are State Predictor Method [7], Bayesian network [8] to predict the next users' location. Those researches were not capable of catching the complex patterns in larger contexts, and thus, inefficient to be applied practically in real life.

Another approach is represented by Kalman Filters (KFs) and Extended Kalman Filters (EKFs), which use a series of measurements observed over time. Some contributions based on this approach are described in [9–11]. Also, There has been growing interest in Markovian models, which were introduced in [12] and applied in several wireless mobility pieces of research, which can be mentioned as [13, 14]. Those researches showed that the Markov-based model and Kalman filter work well with the discrete trajectory, but still, they almost do not suit the sampling time trajectory which is much more common in the practical context.

In recent times, the rise of Artificial Neural Network (ANN), one branch of Machine learning marks a huge impact on the trajectory-based mobility prediction problem. In this field, Recurrent neural network (RNN) and two of its variants are GRU and LSTM are well known as the brightest candidates for time-series data problems [15], and since a user trajectory has the space-time relationship among its data, it completely reflects the features of time-series data and the RNNs' based model can be applied. Some contributions to this idea can be mentioned as in [16], where LSTM based model was used to predict pedestrians' motions from video data, and in [17] the authors proposed models using LSTM and GRU to predict the position and speed of vehicles applying on two specific lanes. Their results outperformed the other approaches in higher prediction accuracy, fast deployment, and lower memory consumption.

Given the success of RNNs, plus despite there were still very few contributions focused on HgNB, we have been motivated to apply them to our research. In our work, we proposed an AR-GRU model to predict multiple next locations of multiple users who are moving at high speed for HgNB networks. Two kinds of users that will be considered are users who are in taxis and buses.

2 Problem Formulation and System Model

We are considering users driving taxis and buses into an urban geographical area. The placement of coverage cells will be considered as future work, so now we are focusing on the application of neural networks to users' patterns.

Definition 1: Each user is identified by its unique User Equipment's ID (UE's ID), that is each nth user has the nth ID and the total number of users is N. Indicate $U_n = (u_1, u_2, \ldots u_n)$ is the vector that contains the users' IDs $u_1, u_2, \ldots u_n$, with $U_n = N$. Assign $L_n^{(t)} = \left(X_n^{(t)}, Y_n^{(t)}\right)$ where $X_n^{(t)} = \left(x_n^{(0)}, \ldots x_n^{(t-1)}, x_n^{(t)}\right)$ and $Y_n^{(t)} = \left(y_n^{(0)}, \ldots y_n^{(t-1)}, y_n^{(t)}\right)$ are the vectors containing the x and y coordinates (in meters) of user n at time step t. The trajectory of a user n^{th}, then will be indicated as:

$$T_n = \left(u_n, L_n^{(t)}\right) = \left(u_n, X_n^{(t)}, Y_n^{(t)}\right) \tag{1}$$

Taking the entire trajectory data to train the model consumes a large amount of memory and a lot of calculations. So, the trajectory can be divided into smaller segments. If the lengths of the segments are long enough to emphasize the general pattern of the trajectory, then the model can learn from it to predict the next locations.

Definition 2: Call $[b_1, b_2]$ is a moving window on $[0, t]$. b is assigned as the look-back steps, $0 < b \leq t$ and $b_1 = b_2 - b$. Assign $L_n^{(b)} = \left(X_n^{(b)}, Y_n^{(b)}\right) = \left[\left(x_n^{(b_1)}, \ldots, x_n^{(b_2)}\right), \left(y_n^{(b_1)}, \ldots, y_n^{(b_2)}\right)\right]$ as a set of users' locations from the corresponding segments of their trajectories. The user n^{th}'s segment of trajectory is defined as:

$$T_{n_b} = \left(u_n, L_n^{(b)}\right) \tag{2}$$

Problem 1: Given a segment of b look-back positions of a user, assign p is the probability, s is the number of steps ahead, which illustrates the number of pairs $L_n^{(b_2+s)} = \left(X_n^{(b_2+s)}, Y_n^{(b_2+s)}\right) = \left[\left(x_n^{(b_2+1)}, \ldots, x_n^{(b_2+s)}\right), \left(y_n^{(b_2+1)}, \ldots, y_n^{(b_2+s)}\right)\right]$ in the future time steps. The next s locations of that user can be illustrated as:

$$\widehat{L_n^{(b_2+s)}} = \underbrace{argmax}_{L_n^{(b_2+s)}} p\left(L_n^{(b_2+s)} | T_{n_b}\right) \tag{3}$$

We are provided to use SUMO [18], an open-source package able to simulate and generate users' vehicles trajectories data. Trajectories are simulated and recorded in 6 h to emphasize the real traffic during the working time. The maximum speed for users' vehicles is 120 km/h which is about 32 m/s, and the average speed of the entire traffic flows in the simulated area is 80 km/h, which is approximately 22 m/s. We consider all the trajectory data are collected by GPS every 1 s, and the coordinates for each user are in the meter unit. To serve the comparison in part 3, three datasets (each dataset consists of 3 different sets that are: training set, testing set, and validation set) were created from different cities based on the real maps, which are shown in Fig. 1, and their information is shown in Table 1. Each dataset contains the user's IDs, date-time, and their vehicles' positions.

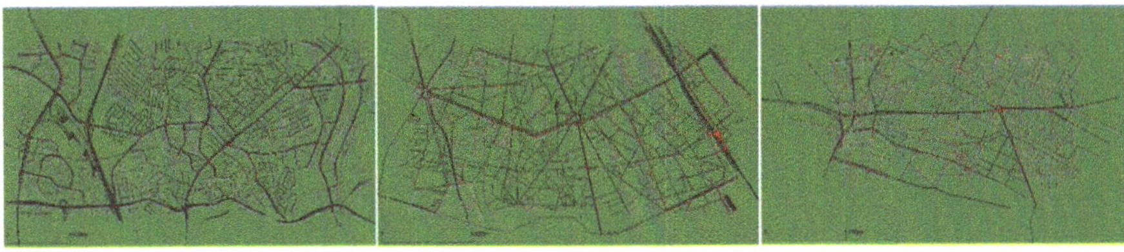

(a) (b) (c)

Fig. 1. A small simulated area in Singapore (a), Paris (b), and Berlin (c).

Table 1. Simulated areas' information.

City	Region's square (km^2)	Total number of vehicles
Singapore	20	60
Berlin	12	40
Paris	10	30

To avoid interference among categorical features (the users' IDs) during the training session, we applied the One-Hot encoding. This encoding represents the categorical variables as binary vectors by mapping each feature with an integer "0" or "1", number "1" indicates the feature's existence and number "0" is for the feature's non–existence. After encoding, the vector U_n is replaced by a binary orthogonal matrix $U_{n(one-hot)}$ that presents the taxis' ID corresponding to U_n as illustrated in (Table 2 and Fig. 2).

Table 2. An illustration of One-Hot encoding applying to our datasets.

ID vector U_n	One-hot encoded ID vector $U_{n\ (one-hot)}$					
	Taxi 1	Taxi 2	...	Bus 1	Bus 2	...
Taxi 1	1	0	...	0	0	...
Taxi 2	0	1	...	0	0	...
...	...	...	...	...	...	...
Bus 1	0	0	...	1	0	...
Bus 2	0	0	...	0	1	...
...	...	...	...	...	...	...

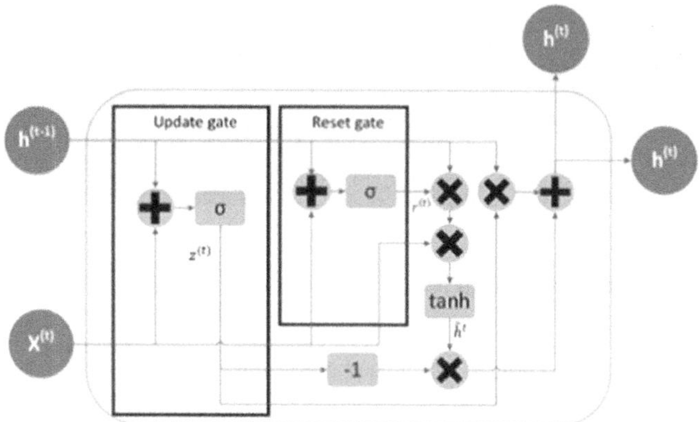

Fig. 2. The structure of a GRU cell.

3 Gated Recurrent Unit Overview and the Proposed Model

GRU is a simple variant of LSTM that comprises only two gates that are: reset gate and update gate, as shown in Fig. 3. GRU cells only have one hidden state transferred between time steps, which can hold both the long-term and short-term dependencies at the same time [19].

We denote $x^{(t)}$ as the input vector, $h^{(t)}$ and $h^{(t-1)}$ as the hidden state at the current state and past state. We indicate two sets of W_z, U_z, B_z and W_r U_r b_r as the weights matrices and biases vector corresponding to each gate and state. Update gate and Reset gate are denoted as $z^{(t)}$ and $r^{(t)}$. The gates' calculations can be presented as [20]:

$$z^{(t)} = \sigma\left(W_z x^{(t)} + U_z h^{(t-1)} + b_z\right) \tag{4}$$

$$r^{(t)} = \sigma\left(W_r x^{(t)} + U_r h^{(t-1)} + b_r\right) \tag{5}$$

Indicate $\tilde{h}^t$ as the candidate output of the current cell, and the set of U_h, W_h, b_h as the Weights matrices and Biases of vector $\tilde{h}^t$. $h^{(t)}$ is calculated in two equations [20]:

$$\tilde{h}^t = \tanh\left[W_h x^{(t)} r^{(t)} \odot \left(U_h h^{(t-1)}\right) + b_h\right] \tag{6}$$

$$h^{(t)} = \left(1 - z^{(t)}\right) \odot h^{(t-1)} + z^{(t)} \odot \tilde{h}^{(t)} \tag{7}$$

Both weights matrices and biases vector in the GRU model will be updated to the new values using through-time backward propagation [20, 21].

At this point, based on the previous calculation, we present our proposed model which is shown in Fig. 4.

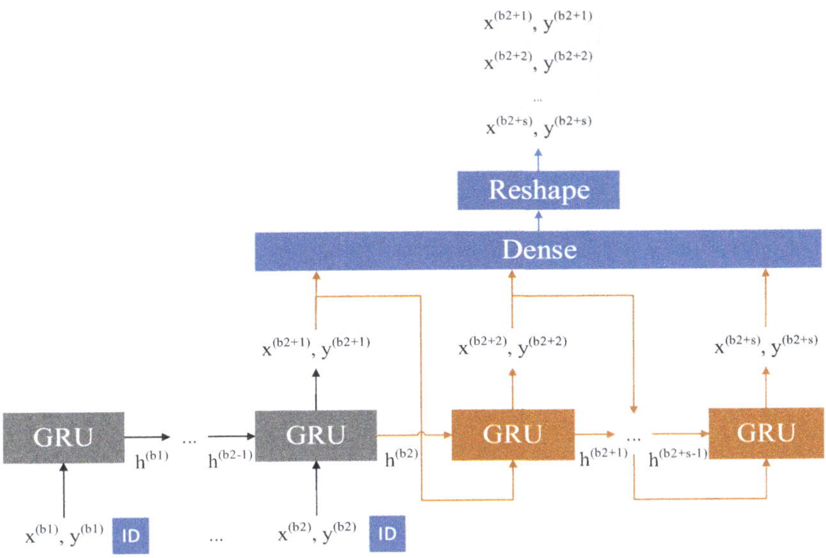

Fig. 3. The structure of the AR-GRU model.

3.1 The Input

The model has two branches. The first branch receives each One-Hot encoded categorical feature corresponding to the user, which are the vectors of n user's IDs, and the second branch takes the coordinates of each vector corresponding to that user. As already mentioned, each subset contains b location samples from a user's full trajectory, so the shape of the input of each branch will be (Batch number, b, n) and (Batch number, b, 2) respectively. The two features are then combined before fitting to the input of the hidden layer.

3.2 The Hidden Layer with the Autoregression Part

We apply the Auto-Regressive (AR) method, which was mentioned in [22]. The benefit of this method is that the output of every prediction at each time step will be fed back to the model's cells and the model will keep outputting the cell state that contains the information they have learned throughout from the past value to the latest future value.

3.3 The Output Layer and Loss Function

The output layer consists of only one dense layer (DL) with two neurons since the aim of the prediction is the pair of location $L_n^{(b_2+s)}$ in the future. By the end of the model's output, all the output vectors corresponding to the time step $b_2+1, b_2+2, \ldots, b_2+s$ were reshaped into a 2 dimensions matrix of $s \times 2$ elements, which is called $\widehat{L_n^{(b_2+s)}}$.

At the output layer, the Mean Squared Error (MSE) is used as the loss function for calculating the error between the actual value and predicted one, after each training epoch. Assuming the dataset contains N samples. The MSE is calculated by:

$$MSE = \frac{1}{N}\sum_{n=1}^{N}\left(L_n^{(b_2+s)} - \widehat{L_n^{(b_2+s)}}\right)^2 \tag{8}$$

This loss value is needed for the gradient that serves the backpropagation process. To speed up the convergence of the loss, we use the Adam optimizer [23], and the learning rate is set to 0.001 to suit the optimizer.

In order to evaluate the goodness of a model, we considered the MSE and R^2 score metrics [24]. The R^2 score varies between 0 and 1, and it indicates how well the model's prediction fits the actual values. It is expressed by:

$$R^2 = 1 - \frac{MSE(model)}{MSE(baseline)} = 1 - \frac{\sum_{n=1}^{N}\left(L_n^{(b_2+s)} - \widehat{L_n^{(b_2+s)}}\right)^2}{\sum_{n=1}^{N}\left(L_n^{(b_2+s)} - \overline{L^{(b_2+s)}}\right)^2} \tag{9}$$

Where $\overline{L^{(b_2+s)}}$ is the mean of all the $L_n^{(b_2+s)}$. The closer the R^2 score to 1 the higher accuracy of the model is in predicting the actual locations in the dataset.

4 The Training Results

In this section, the experiment results performed on Singapore's 60 users testing dataset are presented.

4.1 Different Look-Back Steps

To choose the best number of look-back steps b, we test the model with different values of b, then compare their MSE and R^2 as shown in Fig. 5. The number of steps ahead is set at $s = 2$ for every case of look-backs steps.

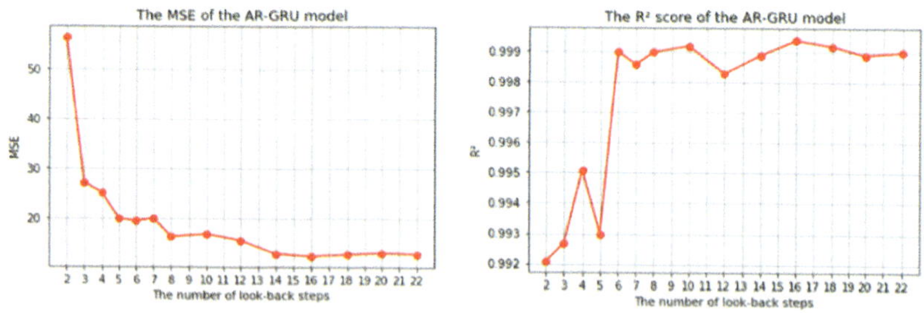

(a) (b)

Fig. 4. The model's MSE (a) and R^2 (b) comparison for different numbers of look-back steps.

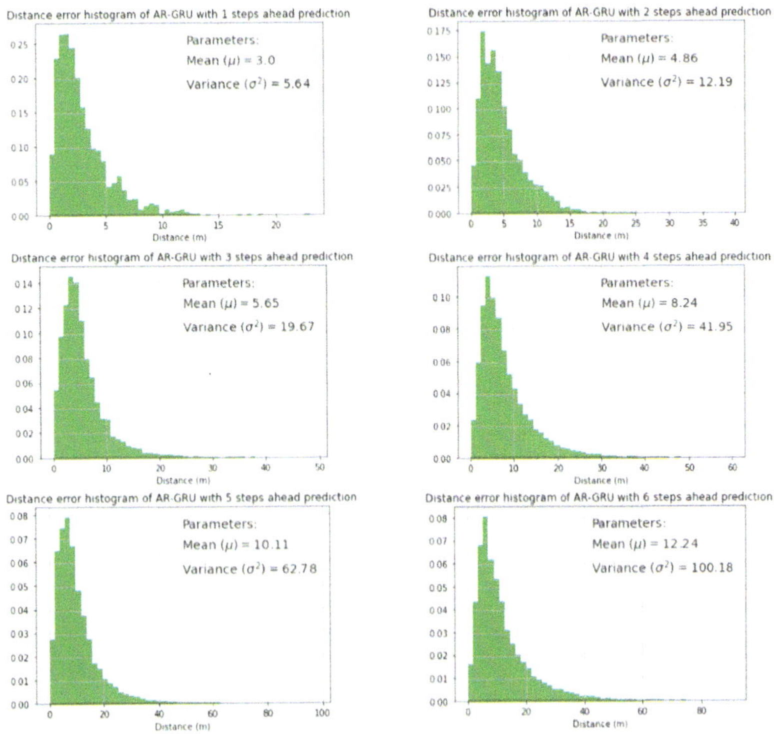

Fig. 5. The model's distance error at the different number of steps ahead.

The MSE is getting lower and the R^2 score is getting higher as the higher value of b is. All the R^2 scores are higher than 0.9, which means the model fits well with the new input data. There is a sharp drop of R^2 scores at step 5, which we believe is caused by the unintended different complexity of the shape of the trajectory segments created when using the moving window methods, and thus increase the $MSE(model)$. When the trajectory segments are long enough for the model to learn and overcome their complexity, the $MSE(model)$ will be lower and we receive the more stable line of R^2 score.

4.2 Different Steps Ahead

Figure 5. shows the distribution of the Euclidean distance between the predicted and the actual positions. As shown in the figure, and as we mentioned above, the average speed of the vehicle in our simulation is more than 80 km/h, which is about 22.22 m/s, and the sampling time of the GPS device is 1 s, by predicting from 1 to 4 near-future locations, the model can predict from a minimum of 22.2 m up to 88.8 m ahead, with the distance errors are almost below 10 m.

To sum up, from those above results, $b = 16$ is the reasonable number of look-back steps for the model to acknowledge the users' trajectory patterns, and 4 positions ahead are the maximum number it can predict for acceptable distance errors.

5 Some Results and Comparisons

Firstly, we compare the efficiency of the model when applied in 3 areas that have different numbers of users and different users' trajectories, using $b = 16$ look-back steps (Table 3).

Table 3. The MSE and R^2 comparison of AR–GRU models for different steps-ahead.

Steps ahead	Singapore area		Berlin area		Paris area	
	MSE	R^2	MSE	R^2	MSE	R^2
1	4.95	0.9999	11.21	0.9999	16.27	0.9996
2	12.43	0.9991	16.70	0.9993	20.57	0.9999
3	20.05	0.9995	37.37	0.9992	21.68	0.9999
4	38.17	0.999	88.35	0.9989	43.49	0.9997

The result shows that the MSE and R^2 scores are not too diverse from each other. So, the model works quite well in the other areas with different trajectories.

Secondly, to verify the goodness of our idea, the AR-GRU model was compared with AR–LSTM, GRU, LSTM, and DNN, whose parameters are shown in Table 4.

Table 4. Some parameters of the models for comparison.

Model	Layers	Neurons per layer	Approximate training time per step ahead (minutes)				Params of hidden layer
			1	2	3	4	
AR-GRU	**1**	**32**	**77**	**146**	**126**	**94**	**9216**
AR-LSTM	1	32	82	83	114	96	12160
GRU	1	16	6	14	20	17	3840
LSTM	1	16	10	19	18	19	5056
DNN	2	32, 62	4	2	4	5	2016

Applying five models on Singapore's testing dataset with $b = 16$ look-back steps, we received the results in Figs. 6 and 7 that respectively show the Cumulative distribution function (CDF) of the distance error and some concrete prediction results. As the figures show, the AR–GRU model outperforms the non–autoregressive model and approximates the MSE and error distance with AR–LSTM model. But, according to Table 4, AR–LSTM requires more time to train and has more params than AR – GRU, therefore, the AR–GRU will be the most reasonable option to implement in the practical context to reduce memory uses and training time consumption.

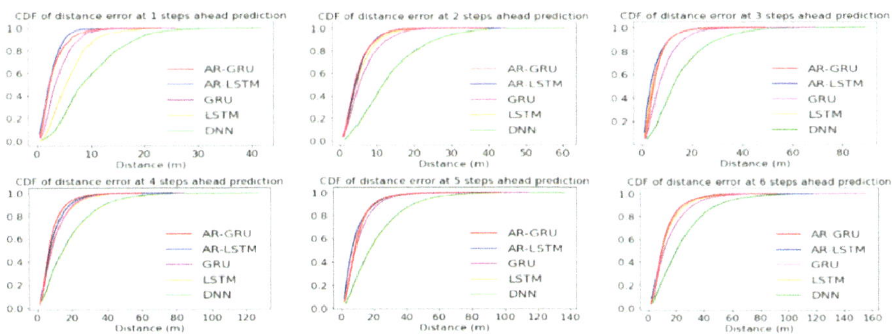

Fig. 6. The CDF comparison between AR-GRU with 4 other models.

6 Conclusion and Future Work

In this paper, we have proposed the AR – GRU model to predict multiple users' trajectories moving at high speed at multiple time steps for HgNB in 5G wireless network.

The results of our experiment on simulated datasets and contexts show the outperform of AR – GRU in various aspects compares to the AR – LSTM and three other sequential models. The model can work well in numerous contexts with different users' trajectories patterns for a reasonable distance error, in the cases from 1 to 4 steps ahead.

This work is mainly focused on how to predict the next multiple users with multiple locations ahead. In the next work, we aim to increase the accuracy of prediction by applying some advanced techniques to the model and resolve the existed drawbacks of using the One-hot encoding method. Besides, the future secondary aim is to make the implementation where these models will be applied in the practical context with the Femtocells and small cells in a heterogeneous network.

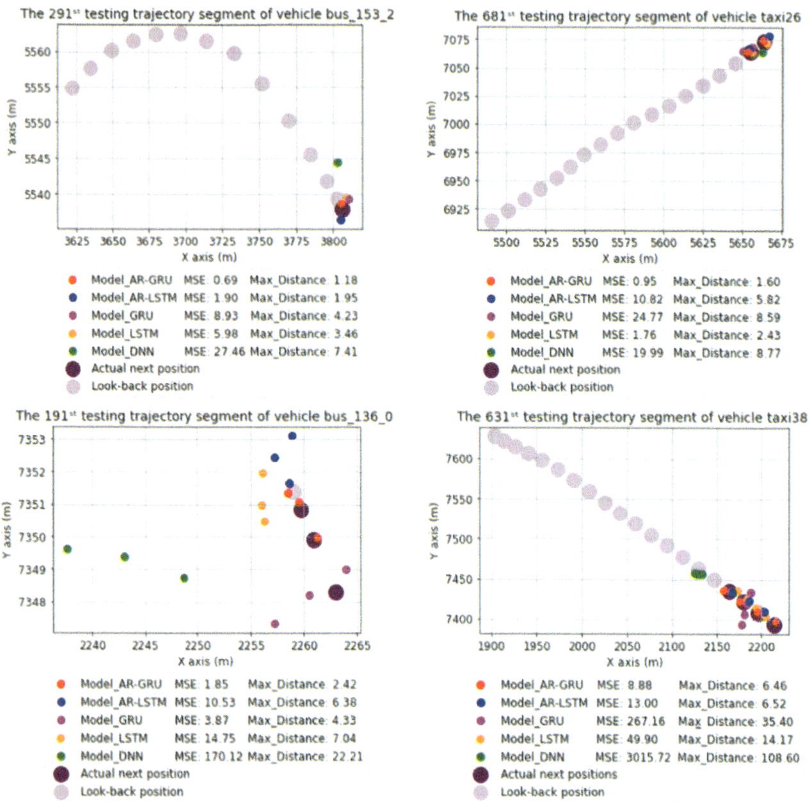

Fig. 7. Some comparisons between AR–GRU and four other models. From upper left to lower right are the examples of 1 step, 2 steps, 3 steps, and 4 steps ahead prediction.

References

1. Mohamed, M.M., El-Badawy, H.M., Abdelhadi, R.H., Ammar, A.A.: Adaptive femtocell accessing control in A 5g heterogeneous network. In: 2020 37th National Radio Science Conference (NRSC), pp. 85–94. IEEE, Cairo (2020)
2. Ghanbarisabagh, M., Vetharatnam, G., Giacoumidis, E., Momeni Malayer, S.: Capacity improvement in 5G networks using femtocell. Wireless Pers. Commun. **105**(3), 1027–1038 (2019)
3. Xin, S., Liang, C., Choi, D., Choi, C.: Power allocation scheme for femto-to-macro downlink interference reduction for smart devices in ambient intelligence. Mob. Inf. Syst. **2016**, 1–10 (2016)
4. Ahmed, A.U., Islam, M.T., Ismail, M.: A review on femtocell and its diverse interference mitigation techniques in heterogeneous network. Wireless Pers. Commun. **78**(1), 85–106 (2014)
5. Santamaria, A.F., Fazio, P., Raimondo, P., Tropea, M., De Rango, F.: A new distributed predictive congestion aware re-routing algorithm for CO_2 emissions reduction. IEEE Trans Veh. Technol. **68**(5), 4419–4433 (2019)

6. Fazio, P., Rango, F.D., Tropea, M.: Prediction and QoS enhancement in new generation cellular networks with mobile hosts: a survey on different protocols and conventional/unconventional approaches. IEEE Commun. Surv. Tutorials **19**(3), 1822–1841 (2017)
7. Amirrudin, N.A., Ariffin, S.H.S., Malik, N.N.N.A., Ghazali, N.E.: User's mobility history-based mobility prediction in LTE femtocells network. In: 2013 IEEE International RF and Microwave Conference (RFM), pp. 105–110. IEEE, Penang (2013)
8. Schreier, M., Willert, V., Adamy, J.: Bayesian, maneuver-based, long-term trajectory prediction and criticality assessment for driver assistance systems. In: 17th International IEEE Conference on Intelligent Transportation Systems (ITSC), pp. 334–341. IEEE, Qingdao (2014)
9. Pathirana, P.N., Savkin, A., Jha, S.: Robust extended Kalman filter applied to location tracking and trajectory prediction for PCS networks. In: Proceedings of the 2004 IEEE International Conference on Control Applications, pp. 63–68. IEEE, Taipei (2004)
10. Mostafavi, S.S., Sorrentino, S., Guldogan, M.B., Fodor, G.: Vehicular positioning using 5G millimeter wave and sensor fusion in highway scenarios. In: ICC 2020 - 2020 IEEE International Conference on Communications (ICC), pp. 1–7. IEEE, Dublin (2020)
11. Zaidi, Z.R., Mark, B.L.: Real-time mobility tracking algorithms for cellular networks based on Kalman filtering. IEEE Trans. Mob. Comput. **4**(2), 195–208 (2005)
12. Hadachi, A., Batrashev, O., Lind, A., Singer, G., Vainikko, E.: Cell phone subscribers mobility prediction using enhanced Markov Chain algorithm. In: 2014 IEEE Intelligent Vehicles Symposium Proceedings, pp. 1049–1054. IEEE, Dearborn (2014)
13. Cheikh, A.B., Ayari, M., Langar, R., Pujolle, G., Saidane, L.A.: Optimized handoff with mobility prediction scheme using HMM for femtocell networks. In: 2015 IEEE International Conference on Communications (ICC), pp. 3448–3453. IEEE, London (2015)
14. Nadembega, A., Hafid, A., Taleb, T.: A destination and mobility path prediction scheme for mobile networks. IEEE Trans. Veh. Technol. **64**(6), 2577–2590 (2015)
15. Wickramasuriya, D.S., Perumalla, C.A., Davaslioglu, K., Gitlin, R.D.: Base station prediction and proactive mobility management in virtual cells using recurrent neural networks. In: 2017 IEEE 18th Wireless and Microwave Technology Conference (WAMICON), pp. 1–6. IEEE, Cocoa Beach (2017)
16. Manh, H., Alaghband, G.J.A.: Scene-LSTM: A Model for Human Trajectory Prediction. arXiv, 1–9 (2018)
17. Jiang, H., Chang, L., Li, Q., Chen, D.: Trajectory prediction of vehicles based on deep learning. In: 2019 4th International Conference on Intelligent Transportation Engineering (ICITE), pp. 190–195. IEEE, Singapore (2019)
18. Lopez, P.A., et al.: Microscopic traffic simulation using SUMO. In: 2018 21st International Conference on Intelligent Transportation Systems (ITSC), pp. 2575–2582 (2018)
19. Sun, S., Chen, J., Sun, J.: Traffic congestion prediction based on GPS trajectory data. Int. J. Distrib. Sens. Netw. **15**(5), 155014771984744 (2019)
20. Li, M.: A Tutorial On Backward Propagation Through Time (BPTT) In The Gated Recurrent Unit (GRU) RNN. Pennsylvania (2016)
21. Aggarwal, M., Murty, M.N.: Deep learning. In: Aggarwal, M., Murty, M.N. (eds.) Machine Learning in Social Networks: Embedding Nodes, Edges, Communities, and Graphs, pp. 35–66. Springer, Singapore (2021)
22. Wang, J., Yan, J., Li, C., Gao, R.X., Zhao, R.: Deep heterogeneous GRU model for predictive analytics in smart manufacturing: application to tool wear prediction. Comput. Ind. **111**, 1–14 (2019)

23. Bianchi, F.M., Maiorino, E., Kampffmeyer, M.C., Rizzi, A., Jenssen, R.: Properties and training in recurrent neural networks. In: Bianchi, F.M., Maiorino, E., Kampffmeyer, M.C., Rizzi, A., Jenssen, R. (eds.) Recurrent Neural Networks for Short-Term Load Forecasting: An Overview and Comparative Analysis, pp. 9–21. Springer International Publishing, Cham (2017)
24. Mandel, J., Mansfield, J.: The statistical analysis of experimental data. Phys. Today **18**(9), 66–68 (1965)

Impact of Input Parameters on Material Removal Speed When PMEDM SKD11 Tool Steel

Le Hoang Anh[1], Hoang Xuan Tu[2], Le Thu Quy[3], Pham Duc Lam[4], Trinh Kieu Tuan[5], Tran Ngoc Giang[2], Nguyen Manh Cuong[2], and Vu Ngoc Pi[2](✉)

[1] Vinh Long University of Technology Education, Vinh Long, Vietnam
[2] Thai Nguyen University of Technology, Thai Nguyen, Vietnam
vungocpi@tnut.edu.vn
[3] National Research Institute of Mechanical Engineering, Hanoi, Vietnam
[4] Nguyen Tat Thanh University, Ho Chi Minh City, Vietnam
[5] University of Economics – Technology for Industries, Hanoi, Vietnam

Abstract. This paper introduces a study on optimizing the powder-mixed electrical discharge machining (PMEDM) for processing cylindrical shaped parts made of SKD11 steel to get the maximum material removal speed. In the study, an experiment was carried out with the use of the Taguchi method for experimental design and analyzing the results. Six input parameters, including the powder concentration, the powder size, the pulse on time, the pulse off time, the current, and the servo voltage, were selected to study the impact of the input factors on the MRS. In addition, optimal values of the input parameters to achieve maximum MRS have been proposed.

Keywords: PMEDM · Electrical discharge machining · Material removal speed · Cylindrical shaped parts · SKD11

1 Introduction

Electrical discharge machining (EDM) is a non-traditional machining method. It is very widely used for processing conductive materials. This method is especially suitable for machining parts with complex shapes and cavities such as dies, cores of dies, molds, etc. EDM method has many advantages, such as it can be machined high hardness materials, low cutting force and noise during machining, etc. However, this method also has some disadvantages such as low material removal rate (MRR); low machined surface quality, high electrode wear rate, etc. The aforementioned disadvantages of the EDM process can be reduced by a solution that has been used by many scientists. It is EDM with a dielectric solution mixed with conductive powder (referred to as PMEDM).

To date, studies have been conducted on various aspects of the PMEDM process. Many studies have been done on the impact of the input process parameters on the MMR [1–6], the SR [4, 5, 7, 8], or the electrode wear rate (EWR) [1, 3, 4, 8], etc.

D. C. Nguyen et al. (Eds.): ICERA 2021, LNNS 366, pp. 459–467, 2022.
https://doi.org/10.1007/978-3-030-92574-1_48

Processing with different powders is also of interest, such as SiC powder [1, 3, 6, 9–12], Al2O3 powder [3, 5], Ti powder, silicon powder [2, 13], or Al powder [6, 14], etc. Different types of electrode material such as copper [1, 3, 4, 9, 12], brass [15], graphite [9, 15, 16], etc. were used for the PMEDM process. In addition, various types of machining materials were also investigated such as 90CrSi [1, 12], SKD11 [6, 8, 14, 17], EN31 [4], H11 [4, 16], AISI D2 die steel [2], Ti-6Al-4V [10, 18], Inconel 718 [3, 11], etc. In an effort to machine cylindrical parts, recently, there have been some studies using the PMEDM method [1, 12]. However, these studies have only been carried out when machining 90CrSi steel.

This paper introduces a study on optimization of the PMEDM process to achieve minimum MRS when machining cylindrical shaped parts with SKD11 steel. In which, the influence of input factors on MRS was learned. Also, optimal process factors to obtain the minimum MRS have been found.

2 Experimental Setup

An experiment to investigate the influence of PMEDM factors on MRS was performed. In this experiment, six input parameters, including the powder concentration (C_p), the powder size (S_p), the pulse on time (T_{on}), the pulse off time (T_{off}), the current (IP), and the servo voltage (SV) are investigated. These parameters and their levels are shown in Table 1. The workpiece material is SKD11 steel, and the electrode material was copper. Besides, SiC powder was used for mixing into the dielectric fluid. Also, a Sodick A30 EDM machine was used for testing. The setup of the experiment is described in Fig. 1.

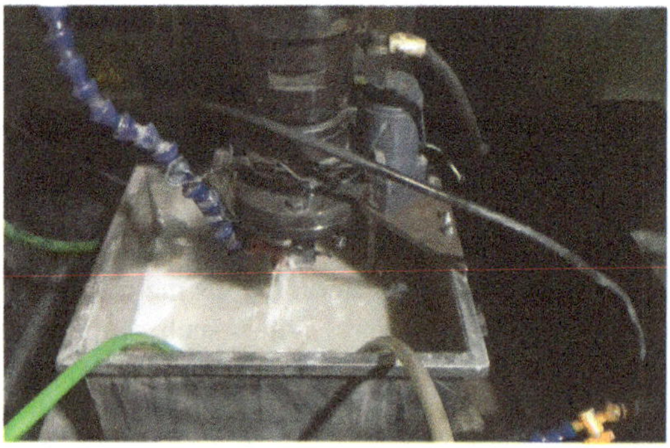

Fig. 1. Experimental setup

Table 1. Input factors and their levels

No	Parameters	Code	Unit	Level					
				1	2	3	4	5	6
1	Powder concentration	C_p	g/l	0	2	3	4	5	6
2	Powder size	S_p	nm	100	500	1000	–	–	–
3	Pulse on time	T_{on}	μs	10	20	30	–	–	–
4	Pulse off time	T_{off}	μs	10	20	30	–	–	–
5	Peak current	IP	A	4	8	12	–	–	–
6	Servo voltage	SV	V	3	5	7	–	–	–

Taguchi method with design L18 (3^4) was used to design the experiment with the help of Minitab R19 software. Table 2 describes the experimental plan and results of the MRS.

Table 2. Experimental plan and the MRS

No	Input parameters						MRS (g/h)	S/N
	C_p	S_p	T_{on}	T_{off}	I_p	SV		
1	0	100	10	10	4	3	9.8773	19.8928
2	0	500	20	20	8	5	16.0979	24.1354
3	0	1000	30	30	12	7	49.6445	33.9174
4	2	100	10	20	8	7	6.7254	16.5544
5	2	500	20	30	12	3	15.8857	24.0201
6	2	1000	30	10	4	5	41.9831	32.4615
7	3	100	20	10	12	5	32.4869	30.2342
8	3	500	30	20	4	7	12.4249	21.8858
9	3	1000	10	30	8	3	11.8259	21.4567
10	4	100	30	30	8	5	28.2703	29.0266
11	4	500	10	10	12	7	8.1986	18.2748
12	4	1000	20	20	4	3	5.2888	14.4672
13	5	100	20	30	4	7	5.3288	14.5327
14	5	500	30	10	8	3	15.1998	23.6368
15	5	1000	10	20	12	5	7.3727	17.3525
16	6	100	30	20	12	3	30.8676	29.7901
17	6	500	10	30	4	5	5.6500	15.0409
18	6	1000	20	10	8	7	5.2931	14.4741

3 Results and Analysis

To weigh the impact of the input factors on the MRS, the ANOVA method in Minitab 19 software was used. Besides, the target function is the maximum MRS is the best:

$$S/N = -10\log_{10}\left(\frac{1}{n}\sum_{i=1}^{n} y_i^2\right) \tag{1}$$

Table 3 shows the results obtained by conducting ANOVA. From this result, it is easy to see that IP has the most significant influence on MRS (49.99%). The parameters that affect the MRS following IP are Cp (17.91%), Sp (10.02%), Toff (9.89%), Ton (5.97%) and SV (4. 83%). Figure 2 shows the influence level (%) of these parameters on MRS. Table 4 shows that the order of influence of parameters on S/N is Ton, Cp, IP, SV, Toff and Sp, respectively.

Table 3. Analysis of variance for S/N ratios.

Analysis of Variance for SN ratios

Source	DF	Seq SS	Adj SS	Adj MS	F	P	C (%)
Cp	5	139.23	139.23	27.847	3.41	0.242	19.16
Sp	2	14.21	14.21	7.103	0.87	0.535	1.96
Ton	2	356.98	356.98	178.488	21.85	0.044	49.13
Toff	2	22.80	22.80	11.399	1.40	0.417	3.14
Ip	2	108.80	108.80	54.402	6.66	0.131	14.97
SV	2	68.27	68.27	34.136	4.18	0.193	9.40
Residual Error	2	16.34	16.34	8.168			2.25
Total	17	726.62					

Model Summary

S	R-Sq	R-Sq(adj)
2.8579	97.75%	80.89%

Figure 3 depicts the influence of the parameters on the S/N value. Since the objective function of MRS, the larger the better, the larger the S/N value, the better. Figure 3 shows the influence of these parameters on the MRS. Specifically as follows: When the concentration of C_p powder risen from 2 to 3 g/l, the MRS decreased slightly; when C_p increased from 3 to 5 g/l, the MRS decreased deeply and it increased slightly at the concentration of 6 g/l. Besides, powder size has no significant influence on MRS. As the powder size increased from 100 to 500 μm, the MRS decreased. However, when increasing the powder size from 500 to 1000 nm, the MRS increased. In addition, T_{on} is the parameter that has the greatest influence on MRS. When T_{on} increases, MRS increases sharply. Unlike the effect of T_{on}, as T_{off} increases from 10 to 20 μs the MRS decreases, but then the MRS increases again as T_{off} continues to increase from 20 to 30 μs. IP has the same effect as T_{on}, but with a lesser degree of influence. When IP increases from level 1 to level 3, the MRS increases. When SV increases from 3 to 5 V SV, MRS increases. However, the MRS decreases with further increase of SV from 5 to 7 V. As stated above, in order to achieve the maximum removal rate, the S/N value must be the maximum for each input parameter. From the chart in Fig. 3, it is easy to determine the reasonable input parameters to achieve the maximum MRS value (Table 5). From Table 5, we see that the optimal pulse mode has $C_p = 0$. That means the material removal rate in the unmixed pulse mode is more significant than when the powder is mixed. The cause may be that the powder disperses the pulse current and reduces the pulse energy.

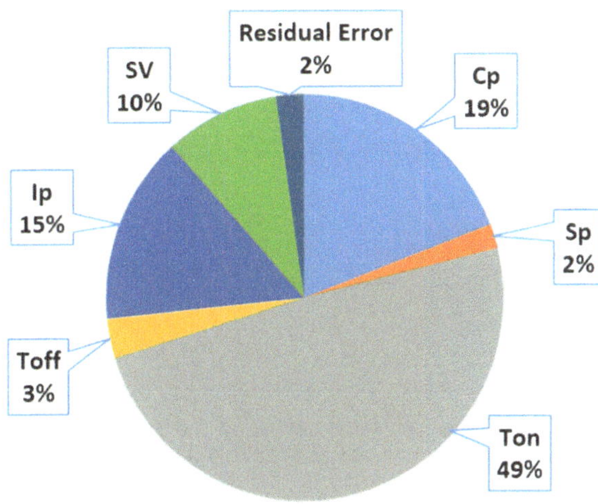

Fig. 2. Influence of the PMEDM factors on the S/N ratio of MRS

Table 4. Order of influence of input parameters on S/N.

Response Table for Signal to Noise Ratios

Larger is better

Level	Cp	Sp	Ton	Toff	Ip	SV
1	25.98	23.34	18.10	23.16	19.71	22.21
2	24.35	21.17	20.31	20.70	21.55	24.71
3	24.53	22.35	28.45	23.00	25.60	19.94
4	20.59					
5	18.51					
6	19.77					
Delta	7.47	2.17	10.36	2.46	5.88	4.77
Rank	2	6	1	5	3	4

The MRS value corresponding to optimum input factors is calculated based on the prediction method using Minitab 19 software. The calculated results are as shown in Table 6. From Table 6, it can be seen that the MRS corresponding to the proposed optimum input parameters is 38.11 (g/h). With a confidence level of 95%, the MRS corresponding to the optimum parameters may appear in the range of 25.56 (g/h) to 50.65 (g/h). These values are much higher than the average MRS of the surveyed experimental plan. Therefore, 95% of MRS was achieved with the optimum input factors higher than 25.56 (g/h).

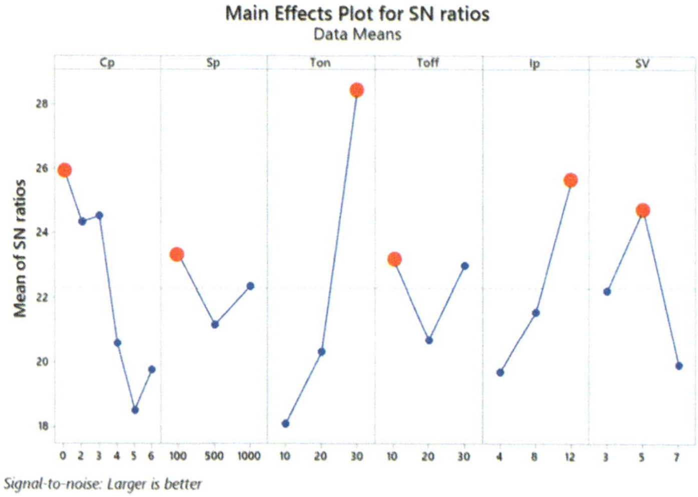

Fig. 3. Effect of input parameters on the S/N ratios

Table 5. Optimum input factors for minimum MRS

Parameter	Code	Unit	Level	Value
Powder concentration	C_p	g/l	1	0
Powder size	S_p	nm	1	100
Pulse on time	T_{on}	µs	3	30
Pulse off time	T_{off}	µs	1	10
Peak current	IP	A	3	12
Servo voltage	SV	V	2	5

Figure 4 shows error distribution histograms to evaluate the experimental model. From this figure, it can be seen that the errors of the experimental points (the blue points) on the normal distribution error distribution chart (Fig. 4a) are close to the normal distribution line (the solid red line). That shows that the model has a small error. Besides, the Histogram chart (Fig. 4c) (for the frequency of errors) shows that the errors appear in the vicinity of 0, accounting for a large proportion. In the remaining two charts (Fig. 4b and d), the experimental errors are randomly distributed. This

Table 6. Predicted optimum MRS

Settings

Variable	Setting
Cp	0
Sp	100
Ton	30
Toff	10
Ip	12
SV	5

Prediction

Type	Fit	SE Fit	CI DF	95% CI	PI DF	95% PI
Conditional	38.11	5.29	6.93	(25.56, 50.65)	4.52	(11.81, 64.41)
Marginal	17.13	7.49	2.65	(-8.60, 42.87)	4.98	(-26.56, 60.83)

means that the experimental model is largely influenced by the selected input parameters and is not affected by order of experiments.

The fit of the experimental model with the optimal input factors was checked by the Anderson-Darling method by using the Minitab 19 software (see Fig. 5). From Fig. 5, it can be seen that the data points corresponding to the experimental points (the blue dots) are all within the bounded area by two upper and lower limits with 95% significance level. In addition, the P-value (0.207) is greater than the $\alpha = 0.05$ value. That means the applied empirical model is consistent with the above significance level.

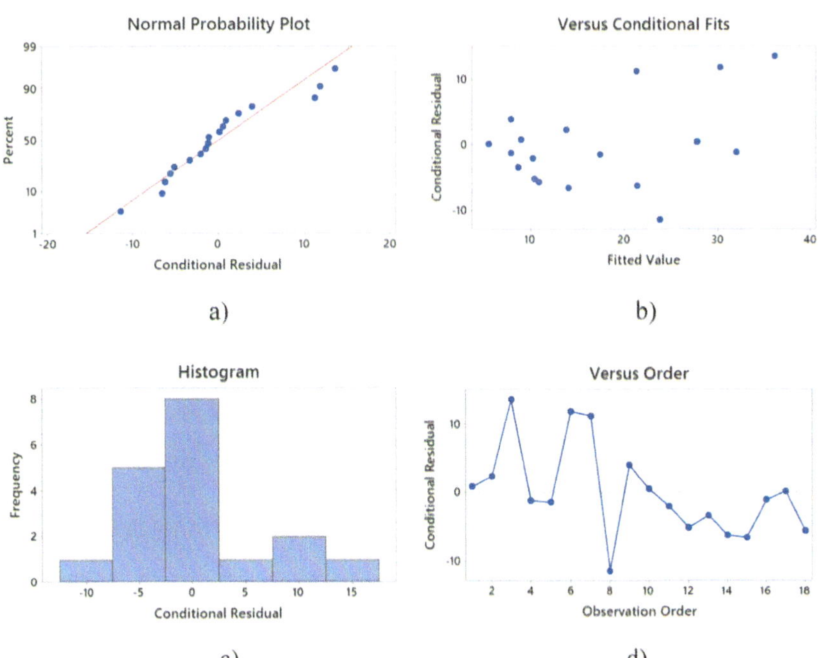

Fig. 4. Conditional residual plots for MRS

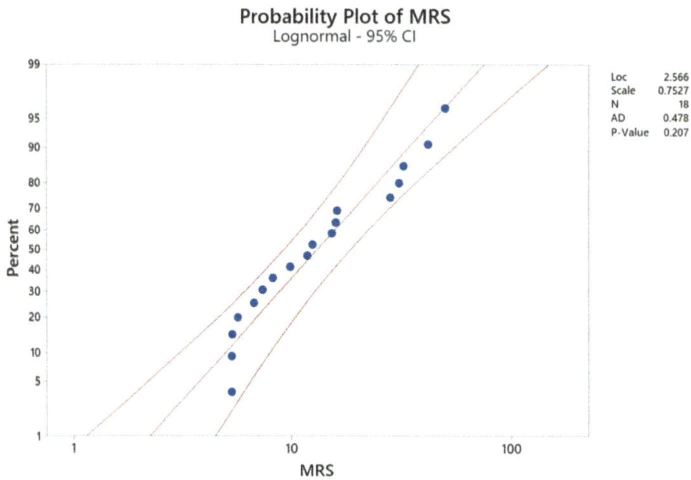

Fig. 5. Probability plot of MRS

4 Conclusions

This paper reports the results of an optimization study on PMEDM when processing cylindrical shaped parts made from SKD11 tool steel to achieve the maximum MRS. In the study, an experiment was designed and performed. The Taguchi method was used to design the experiment and analyze the results. In addition, the influence of six main input factors, including the powder size, the powder concentration, the pulse on time, the pulse off time, the current, and the servo voltage on the MRS was investigated. Furthermore, an optimal set of the input parameters was found to get the maximum MRS.

Acknowledgment. This work was supported by Thai Nguyen University of Technology.

References

1. Hong, T.T., et al.: Multi-objective optimization of PMEDM process of 90CrSi alloy steel for minimum electrode wear rate and maximum material removal rate with silicon carbide powder. In: Materials Science Forum. Trans Tech Publ. (2021)
2. Kansal, H., Singh, S., Kumar, P.: Effect of silicon powder mixed EDM on machining rate of AISI D2 die steel. J. Manuf. Process. **9**(1), 13–22 (2007)
3. Rathi, M.G., Mane, D.V.: Study on effect of powder mixed dielectric in EDM of Inconel 718. Int. J. Sci. Res. Publ. **4**(11), 1–7 (2014)
4. Bhattacharya, A., et al.: Optimal parameter settings for rough and finish machining of die steels in powder-mixed EDM. Int. J. Adv. Manuf. Technol. **61**(5), 537–548 (2012)
5. Kumar, A., et al.: Performance evaluation of Al2O3 nano powder mixed dielectric for electric discharge machining of Inconel 825. Mater. Manuf. Process. **33**(9), 986–995 (2018)

6. Yih-Fong, T., Fu-Chen, C.: Investigation into some surface characteristics of electrical discharge machined SKD-11 using powder-suspension dielectric oil. J. Mater. Process. Technol. **170**(1–2), 385–391 (2005)
7. Talla, G., Gangopadhyay, S., Biswas, C.: Influence of graphite powder mixed EDM on the surface integrity characteristics of Inconel 625. Part. Sci. Technol. **35**(2), 219–226 (2017)
8. Mookam, N., Sunasuan, P., Madsa, T., Muangnoy, P., Chuvaree, S.: Effects of graphite and boron carbide powders mixed into dielectric fluid on electrical discharge machining of SKD 11 tool steel. Arab. J. Sci. Eng. **46**(3), 2553–2563 (2021). https://doi.org/10.1007/s13369-020-05156-4
9. Al-Khazraji, A., Amin, S.A., Ali, S.M.: The effect of SiC powder mixing electrical discharge machining on white layer thickness, heat flux and fatigue life of AISI D2 die steel. Eng. Sci. Technol. Int. J. **19**(3), 1400–1415 (2016)
10. Yaşar, H., Ekmekci, B.: Ti-6Al-4V surfaces in SiC powder mixed electrical discharge machining. In: Advanced Materials Research. Trans Tech Publ. (2014)
11. Sahu, S.K., Jadam, T., Datta, S., Nandi, G.: Effect of using SiC powder-added dielectric media during electro-discharge machining of Inconel 718 superalloys. J. Braz. Soc. Mech. Sci. Eng. **40**(7), 1–19 (2018). https://doi.org/10.1007/s40430-018-1257-7
12. Hong, T.T., et al.: Effect of process parameters on machining time in PMEDM cylindrical shaped parts with silicon carbide powder suspended dielectric. In: Materials Science Forum. Trans Tech Publ. (2021)
13. Luzia, C.A.O., Laurindo, C.A.H., Soares, P.C., Torres, R.D., Mendes, L.A., Amorim, F.L.: Recast layer mechanical properties of tool steel after electrical discharge machining with silicon powder in the dielectric. Int. J. Adv. Manuf. Technol. **103**(1–4), 15–28 (2019). https://doi.org/10.1007/s00170-019-03549-w
14. Sharma, R., Singh, J.: Effect of powder mixed electrical discharge machining (PMEDM) on difficult-to-machine materials – a systematic literature review. J. Manuf. Sci. Prod. **14**(4), 233–255 (2014)
15. Bhui, A.S., et al.: Parametric optimization of ED machining of Ti-6Al-4V in CNTs mixed dielectric medium. Mater. Today Proc. **18**, 1532–1539 (2019)
16. Batish, A., Bhattacharya, A.: Mechanism of material deposition from powder, electrode and dielectric for surface modification of H11 and H13 die steels in EDM process. In: Materials Science Forum. Trans Tech Publ. (2012)
17. Kumar, H.: Development of mirror like surface characteristics using nano powder mixed electric discharge machining (NPMEDM). Int. J. Adv. Manuf. Technol. **76**(1–4), 105–113 (2015)
18. Li, L., et al.: Surface characteristics of Ti-6Al-4V by SiC abrasive-mixed EDM with magnetic stirring. Mater. Manuf. Processes **32**(1), 83–86 (2017)

Impact of Voltage Unbalance and Harmonics on Induction Motor in Operation Mode

Y. Do Nhu[1] and Ngo Xuan Cuong[2](✉)

[1] Hanoi University of Mining and Geology, Ha Noi, Viet Nam
[2] School of Engineering and Technology, Hue University, Hue, Viet Nam
ngoxuancuong@hueuni.edu.vn

Abstract. In industries, induction motors (IM) are widely used because of their simple structure, low cost, and reliable operation. However, the working efficiency of the IM motor is greatly affected by the quality of power supplied from the grid. The current industrial power grid is increasingly unbalanced due to load unbalance and voltage becomes increasingly non-sinusoidal due to the use of more nonlinear loads and power electronic converters. This degrades the power quality leading to a decrease in the efficiency of the IM. This article simultaneously studies the influence of two factors voltage unbalance and harmonics on IM performance efficiencies such as power loss, efficiency, speed, torque, and ripple parameters in IM. In addition, the content of the article also shows the influence of different harmonic components on the working mode of the IM motor. The research results were simulated using Matlab-Simulink software, there are important data for users to have appropriate operating solutions to improve the performance and life of the IM motor in different operating modes.

Keywords: Voltage unbalance · Non-sinusoidal power source · Induction motor

1 Introduction

Induction motors (IM) account for approximately 75% of industrial energy consumption worldwide. Therefore, to reduce electricity consumption globally, it is essential to improve the efficiency of IM [1]. Today's industrial power grids are increasingly unbalanced due to unbalanced consumption loads and the increase of nonlinear loads causing harmonics. These are the two main factors that degrade the power quality of the IM power supply. Under unbalanced voltage conditions, the equivalent circuit of the IM can be split into forward and reverse order equivalent circuits, causing the stator currents in forward and reverse order to increase loss [2, 3].

In the study the impact of unbalanced voltage supply on the performance of 3-phase induction motor. The results show that voltage unbalances lead to reduce motor efficiency and reduce power factor. Voltage unbalances in the voltage source can lead to problems such as excessive losses, overvoltages, mechanical oscillations, and interference with electrical equipment [4]. Unbalanced voltage has a marked effect on the losses in the IM which greatly affects the winding temperature rise and the slip

D. C. Nguyen et al. (Eds.): ICERA 2021, LNNS 366, pp. 468–478, 2022.
https://doi.org/10.1007/978-3-030-92574-1_49

coefficient of the IM. The effect of voltage asymmetry not only causes windings to heat up much faster than the effects of other power quality standards but also affects power, torque, and ripple on IM [5–7].

Harmonics are generated when the voltage is distorted, and the current waveform is non-sine caused by nonlinear loads or power electronic converters. When the voltage is distorted by the waveforms of the current and flux density in the IM. Therefore, the working efficiency and loss of the IM will be more complicated under the condition of unbalanced voltage and non-sinusoidal current [2]. The consequences of harmonics are becoming more and more important, operating the motor on a harmonic system leads to a reduction in the efficiency and life of the motor due to an increase in iron and copper losses at harmonic frequencies. Also, voltage and current harmonics can affect torque. Furthermore, harmonics in the rotating machine can cause vibration and noise [8–10].

The inverter's PWM supply causes an additional loss, which is manifested as a temperature rise in the IM. The common losses are slightly reduced as higher switching frequencies in the PWM supply are adopted. The temperature rise due to additional losses is significant compared with machines operated from sinusoidal sources [11–14].

Chauhan and Singh analyzed the influence of voltage unbalance and harmonic with 3rd and 5th components on 3-Phase IM in the state of Undervoltage and Overvoltage, but did not consider total harmonic distortion (THD) and other higher-order components [15]. This does not fully reflect the IM operation in its actual state.

The above-analyzed studies rarely mention simultaneously voltage unbalance and harmonics, or the research assumptions are incomplete. However, these two factors often coexist on the power system, the analysis of just one factor above cannot fully assess the influence of power quality on the working efficiency of the IM. The content of this article analyzes the simultaneous influence of both voltage unbalance and harmonic on the IM, the influence of each harmonic order component on the performance of the IM. Research results are important data for users to have appropriate operating solutions to improve the performance and life of the IM in different operating modes.

2 Theoretical Basis

2.1 Building a IM Model Under Asymmetric Voltage

Formulate a model to replace the positive and negative-sequence components on a phase as shown in Fig. 1 [2, 4].

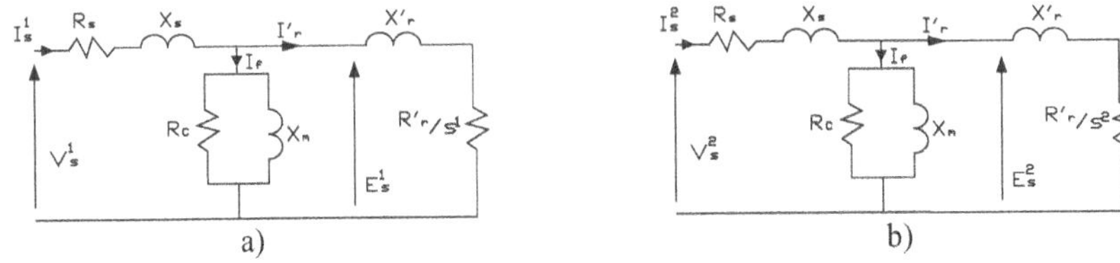

Fig. 1. Substitution diagram of positive-sequence (a) and negative-sequence component (b).

When the three-phase voltage is asymmetrical, the slip coefficient consisting of 2 mentioned components is defined by:

$$s^1 = \frac{\omega_s - \omega_r}{\omega_s} = 1 - \frac{\omega_r}{\omega_s} = 1 - K_u; s^2 = \frac{-\omega_s - \omega_r}{-\omega_s} = 1 + \frac{\omega_r}{\omega_s} = 1 + K_u \quad (1)$$

where: s^1 - positive-sequence slip coefficient; s^2 - Negative-sequence sliding coefficient (the default index "1" corresponds to positive-sequence, index "2" corresponds to negative-sequence).

Let:

$$K_u = \frac{\omega_r}{\omega_s} \quad (2)$$

The voltage unbalance factor is defined as the ratio of the negative sequence voltage component (V_n) to the positive sequence voltage component (V_p). This definition is consistent with IEC standard 61000-4-30:

$$K_V(\%) = \frac{V_n}{V_p} 100 \quad (3)$$

2.2 Effect of Asymmetric Phenomena

a. Influence on the Electromagnetic Torque of the Three-Phase IM

When the three-phase voltages are asymmetry, the total electromagnetic power of the motor is determined:

$$P_{dt} = P^1_{dt} + P^2_{dt}; P^1_{dt} = \frac{3R_r\left(V^1_s\right)^2}{s^1\left[\left(\frac{R_r}{s^1}\right)^2 + (X_r)^2\right]}; P^2_{dt} = \frac{3R_r\left(V^2_s\right)^2}{s^2\left[\left(\frac{R_r}{s^2}\right)^2 + (X_r)^2\right]} \quad (4)$$

where: V_s^1 and V_s^2- positive-sequence and negative-sequence voltage on stator side.

When the three-phase voltage is not symmetrical, the electromagnetic torque consists of two above components and is determined by the formula:

$$T = T^1 + T^2 = \frac{3R_r\left(V^1_s\right)^2}{s^1\left[\left(\frac{R_r}{s^1}\right)^2 + (X_r)^2\right].\omega_s} - \frac{3R_r\left(V^2_s\right)^2}{s^2\left[\left(\frac{R_r}{s^2}\right)^2 + (X_r)^2\right].\omega_s} \quad (5)$$

When the three-phase voltage is asymmetrical the starting torque of the motor determined by equation:

$$T_{KD} = \frac{3R_r\left(V^1_s\right)^2}{[(R_r)^2 + (X_r)^2].\omega_s} - \frac{3R_r\left(V^1_s\right)^2}{[(R_r)^2 + (X_r)^2].\omega_s} \quad (6)$$

When the three-phase voltage is asymmetric, the torque of the motor is reduced by a quantity equal to the negative-sequence torque.

b. Effect on Current and Loss in Motor

The losses on the rotor are determined by the formula:

$$\Delta P_{Cu2} = 3R_r I_r^2 = sP_{dt} = \frac{3R_r V_s^2}{\left[\left(\frac{R_r}{s}\right)^2 + (X_r)^2\right]} \quad (7)$$

When the three-phase voltage is not symmetrical, the loss consists of two components:

$$\Delta P_{Cu2} = \Delta P^1_{Cu2} + \Delta P^2_{Cu2} = \frac{3R_r (V_s^1)^2}{\left[\left(\frac{R_r}{s^1}\right)^2 + (X_r)^2\right]} + \frac{3R_r (V_s^2)^2}{\left[\left(\frac{R_r}{s^2}\right)^2 + (X_r)^2\right]} \quad (8)$$

The losses on the stator are similarly calculated according to the formula:

$$\Delta P_{Cu1} = \Delta P^1_{Cu1} + \Delta P^2_{Cu1} = 3R_s (I_s^1)^2 + 3R_s (I_s^2)^2 \quad (9)$$

When the three-phase voltage is symmetrical, the loss on the motor also increases by an amount equal to the negative-sequence component.

2.3 Effect of Harmonic

The distorted voltage waveform has enormous harmonics components, which distort the waveforms of the currents and flux densities in the induction motor. Thus, efficiency will be compromised and the loss characteristics will be more complex under unbalanced and distorted voltage conditions [2]. Total harmonic distortion (THD) in the current can be calculated to that given as follows:

$$THD_i = \frac{\sqrt{\sum_{n=2}^{\infty} I_n^2}}{I_1} = \frac{\sqrt{I_2^2 + I_3^2 + I_4^2 + I_5^2 + I_6^2 + \ldots}}{I_1} \quad (10)$$

Copper losses on rotor and stator are determined according to as follows:

$$P_{Cu2} = R_r'(I_{ra}^2 + I_{rb}^2 + I_{rc}^2); \quad P_{Cu1} = R_s(I_{sa}^2 + I_{sb}^2 + I_{sc}^2) \quad (11)$$

Under the influence of harmonics, the motor losses in the above formulas increase, leading to a decrease in motor efficiency.

2.4 Performance Evaluation Model

The physical model is considered to be linear, the inductance is assumed to be constant, so saturation of the magnetic circuit is not considered here [14].

Synchronous speed and rated torque:

$$N_s = \frac{60 . f}{p}; T_n = \frac{30 . P_n}{\pi . N_n} \quad (12)$$

Calculate the power factor (PF), mechanical power and mechanical losses are calculated according to the following formula:

$$PF = \frac{P_{in}}{\sqrt{P_{in}^2 + Q_{in}^2}}; P_{\_M} = T_e . \omega_m; P_{\_loss\_m} = F . \omega_m^2 \quad (13)$$

where P_{in} - the active input power, in W; Q_{in} - the reactive input power, in Var; T_e - the electromagnetic torque, in Nm; ω_m - the angular velocity of the rotor, in rad/s; F - friction factor, in N.m.s.

Copper loss on the rotor and stator are determined by the formula:

$$P_{Cu1} = R_s(I_{sa}^2 + I_{sb}^2 + I_{sc}^2); P_{Cu2} = R_r'(I_{ra}^2 + I_{rb}^2 + I_{rc}^2) \quad (14)$$

where, $I_{sa}, I_{sb}, I_{sc}; I_{ra}, I_{rb}, I_{rc}$ - RMS value ofstator current and rotor current in phase a, b, c; R_s - stator resistance; R'_r - rotor resistance.

Finally, when ignoring components of iron loss, auxiliary loss, the IM efficiency is determined by the following formula:

$$\eta = \frac{P_{\_M}}{P_{\_M} + P_{cu1} + P_{cu2} + P_{\_loss\_m}} . 100 \quad (15)$$

3 Building a Research Model

The motor parameter to be studied is a 250 kW motor, whose nominal data and other characteristic dimensions are indicated in Table 1. This is a 3-phase squirrel cage rotor asynchronous motor, class F insulation, connected in the shape of Y.

Table 1. Motor parameter.

Parameter	Value	Parameter	Value
Motor name	A4-355L-4	Stator full Load current (I_s)	29 A
Motor type	Asynchronous motor	Moment of inertia of the rotor (J)	6.25 kg. m^2
Rotor type	Squirrel cage	Starting/rated current (I_s/I_n)	6.0
Rated power (P_n)	250 kW	Starting/rated torque (T_s/T_n)	1.2
Rated line-to-line voltage (V_n)	6 kV	Maximum pull out/rated torque (T_{br}/T_n)	2.2
Frequency (f_n)	50 Hz	Power factor *(PF)*	0.88
Rated speed (N_n)	1480 rpm	Rated efficiency (η_n)	94.2%
Pole pairs (p)	2		

The studied IM model is built on given in Fig. 2. The paper uses a 3-phase asynchronous motor model in the Matlab-Simulink library version R2014a, the simulation parameters of IM are based on the mathematical model with an inparameter Table 1.

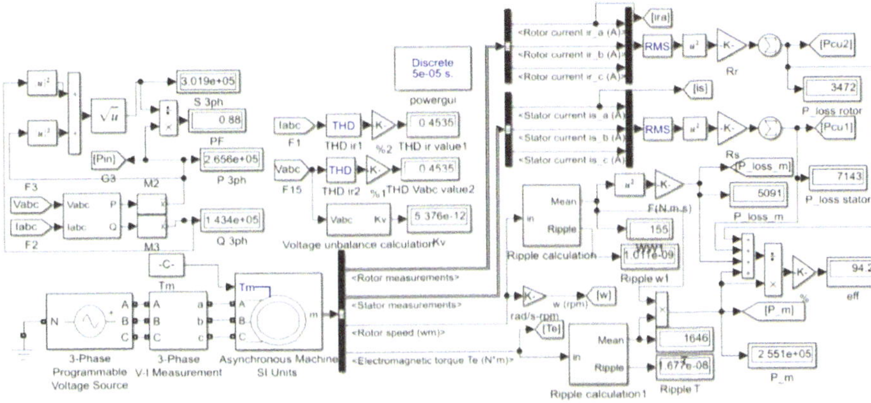

Fig. 2. Research simulation model.

Investigate motor working in normal mode with symmetrical, sinusoidal three-phase power supply and motor's load-carrying factor K = 100%. The survey results on the model are shown in Table 2 and the graph in Fig. 3.

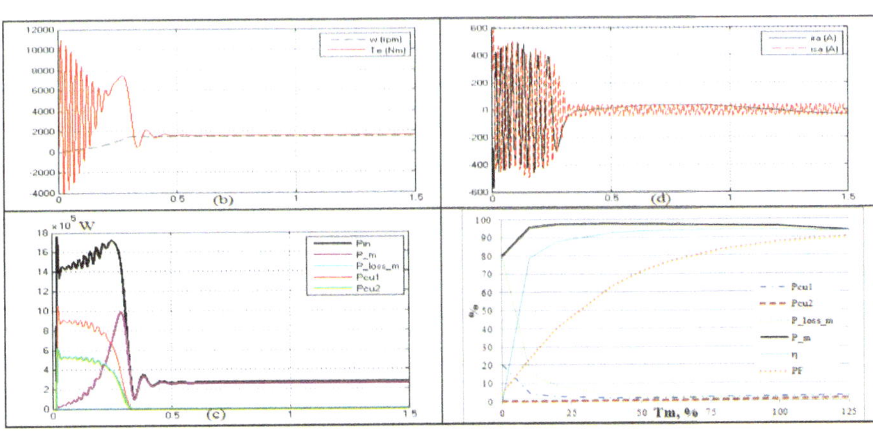

Fig. 3. IM parameters on the model. (a) Source waveform; (b) Motor speed and torque; (c) Power characteristic; (d) Stator and rotor currents.

Investigate the working parameters on the motor research model in the rated load mode, the working voltage of the motor is symmetrical, the sinusoidal results are given in Table 2.

Table 2. Typical parameters of asynchronous motor in symmetry and sinusoidal voltages

Motor Load, %	Parameter						
	THDu	P_{cu1}, W	P_{cu2}, W	N, rpm	T_e, Nm	PF	η, %
100	0	7143	3472	0	0	0.88	94.20

From the results shown in Table 2, it can be seen that the parameters of the motor after running the simulation model have values consistent with Table 1. Thus, the model is built reliably to make a research model evaluate different cases in the working process.

4 Research Results and Discussion

According to the standard NEMA [16], It is not recommended that induction motors operate under more than 5% unbalanced supply, in fact, this is rarely violated. THD of inverter supply is generally less than 15%. Consider inverter supply conditions (ignoring high-order harmonics) [2]. In this study, it is assumed that the power supply for the IM has an unbalanced voltage of up to 5% and voltage THD reaches 15% for each harmonic component of 5th, 7th, and 11th order.

The results are shown in Fig. 4, giving the results of surveying transient characteristics of the motor, torque, and loss corresponding to the case of voltage unbalance coefficient Kv = 3% and harmonic of 5th order with THD = 5%.

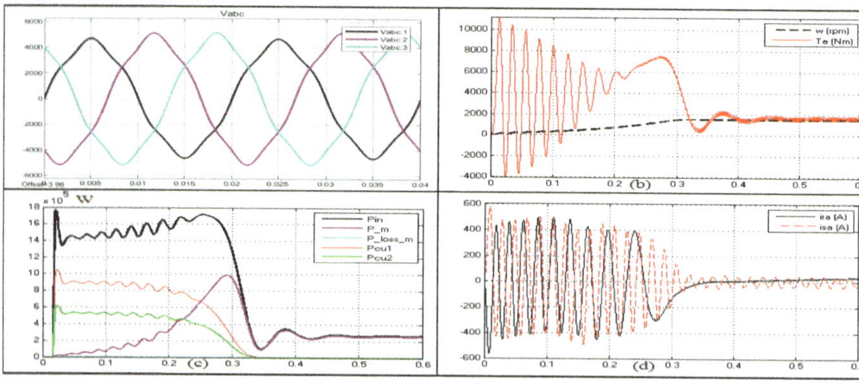

Fig. 4. Transient characteristics under mode Kv = 3% and THD = 5%. (a) Source waveform; (b) Motor speed and torque; (c) Power characteristic; (d) Stator and rotor currents.

From the results shown in Fig. 4, it can be seen that when the motor power supply is voltage-unbalance with a non-sinusoidal waveform, the speed, torque of the motor is vibrated, the loss of the motor increases, leading to a decrease in the efficiency of the motor compared to the normal working mode.

Using the simulation model built in Fig. 2, conduct a survey of the working mode of the IM in different cases, corresponding to the case of the corresponding voltage-unbalance Kv% {0 ÷ 5%} and the influence of harmonic components of harmonics of the 5th, 7th and 11th order (corresponding to the symbols on the figure are H5, H7 and H11) with harmonic amplitude THD has values in the intervals {0%, 5%, 10%, 15%}. The results of the power loss of the motor and the motor efficiency are shown in Fig. 5 and the speed and torque characteristics of the motor are shown in Fig. 6.

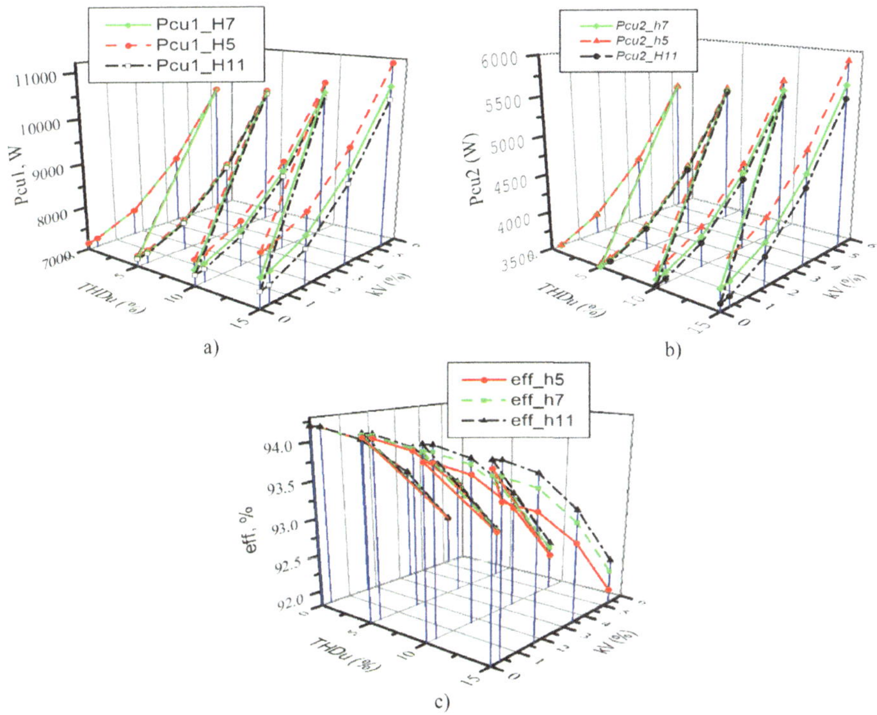

Fig. 5. Motor copper loss. (a) Stator loss; (b) Rotor loss; (c) IM efficiency characteristics.

From Fig. 5 it can be seen that:

- With the three harmonic components studied, under the same conditions, The 5th harmonic component (H5) is the largest loss gain, followed by the 7th harmonic component (H7), and finally the 11th harmonic component (H11).
- When the motor power supply voltage at the same time has voltage unbalance coefficient Kv = 5% and harmonic with THD = 15% according to NEMA standards compared to the sinusoidal symmetry power supply: Stator copper loss in

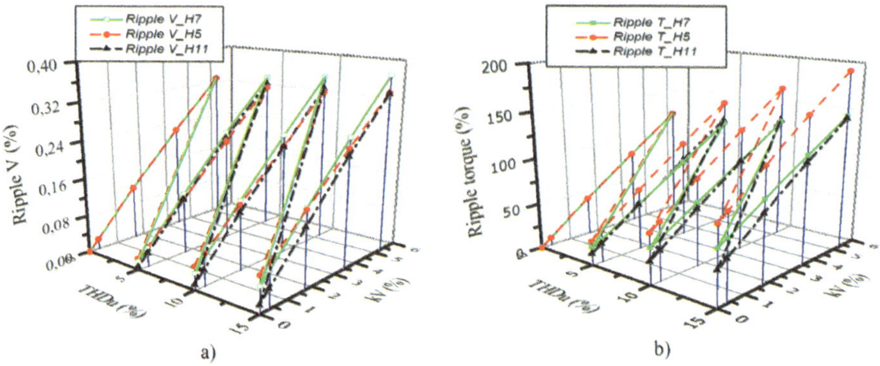

Fig. 6. IM speed (a) and torque (b) ripple.

motor increased to 1.56 times; Rotor copper loss in motor increased to 1.7 times and IM Efficiency (η_n) reduced by 2.21%.

- When motor power supply voltage at the same time has voltage unbalance coefficient Kv = 5% and harmonic with THD = 0% compared to the sinusoidal symmetry power supply: Stator copper loss in motor increased to 1.42 times; Rotor copper loss in motor increased to 1.53 times and IM Efficiency (η_n) reduced by 1.68%.
- When the motor power supply voltage at the same time has voltage unbalance coefficient Kv = 0% and harmonic with THD = 15% compared to the sinusoidal symmetry power supply: Stator copper loss in motor increased to 1.14 times; Rotor copper loss in motor increased to 1.17 times and IM Efficiency (η_n) reduced by 0.2%.

From the results in Fig. 6, it can be seen that:

- With the three harmonic components studied, under the same conditions, The 5th harmonic component (H5) is the increase in torque ripple, followed by the 7th harmonic component (H7), and finally the 11th harmonic component (H11).
- When the motor power supply voltage at the same time has voltage unbalance coefficient Kv = 5% and harmonic with THD = 15% according to NEMA standards compared to the sinusoidal symmetry power supply: Torque ripple gain 194.5% and Speed ripple gain 0.34%.
- When the motor power supply voltage at the same time has voltage unbalance coefficient Kv = 5% and harmonic with THD = 0% compared to the sinusoidal symmetry power supply: Torque ripple gain 122.1% and Speed ripple gain 0.33%.
- When the motor power supply voltage at the same time has voltage unbalance coefficient Kv = 0% and harmonic with THD = 15% compared to the sinusoidal symmetry power supply: Torque ripple gain 81.8% and Speed ripple gain 0.07%.

5 Conclusion

The article presented the influence of voltage unbalance and harmonics on the working mode of the IM. The results show that, when the power supply to the motor is unbalanced voltage and harmonics, the speed and torque of the IM decrease and cause a strong ripple, copper losses on stator and rotor increase dramatically, leading to high total losses in the motor, resulting in reduced IM efficiency.

Under the same operating conditions as the unbalanced voltage power supply and harmonic with three harmonic components, losses and torque ripple increase the most by the 5th harmonic component, followed by the 7th harmonic component, then by the 11th harmonic component.

The worst case is when the motor is operating under the conditions of parameters allowed by NEMA standards with Kv = 5% and total THD = 15%, the total copper loss in the motor increases nearly 1.7 times, the torque ripple of the motor increases to nearly 194.5% causing vibration when the IM is working.

The above analysis shows that study the effects of voltage unbalance and harmonics on the working mode of the IM is necessary to determine the loss, efficiency of IM, thereby recommending users to have a suitable operating solution to improve the life of the mine drainage pump motor.

References

1. Pablo, D., et al.: Voltage unbalance and harmonic distortion effects on inductionmotor power, torque and vibrations. Electr. Power Syst. Res. **140**, 866–873 (2016)
2. Zhang, D., et al.: Effect of voltage unbalance and distortion on the loss characteristics of three-phase cageinduction motor. IET Electr. Power Appl. **12**(2), 264–270 (2018)
3. Anwari, M., Hiendro, A.: New unbalance factor for estimating performance of a three-phase induction motor with under- and overvoltage unbalance. IEEE Trans. Energy Convers. **25** (3), 619–625 (2010)
4. Mirabbasi, D., et al.: Effect of unbalanced voltage on operation of induction motors and its detection. In: International Conference on Electrical and Electronics Engineering-ELECO pp.189–192. Publisher IEEE (2009)
5. Gnacinski, P.: Effect of unbalanced voltage on windings temperature, operational life and load carrying capacity of induction machine. Energy Conv. Manag. **49**(4), 761–770 (2008)
6. Wang, Y.J.: Analysis of effects of three-phase voltage unbalance on induction motors with emphasis on the angle of the complex voltage unbalance factor. IEEE Trans. Energy Conv. **16**(3), 270–275 (2001)
7. Dekhandji, F.Z., Refoufi, L., Bentarzi, H.: Quantitative assessment of three phase supply voltage unbalance effects on induction motors. Int. J. Syst. Assur. Eng. Manag. **8**(1), 393–406 (2015). https://doi.org/10.1007/s13198-015-0401-3
8. Rata, M., Rata, G.: Study solution of induction motor dynamic braking. In: Proceedings of the 2016 International Conference on Development and Application Systems (DAS), pp. 33–37, Suceava, Romania (2016)
9. Raţă, G., et al.: Induction motor speed estimator using rotor slot harmonics. Adv. Electr. Comput. Eng. **9**(1), 70–73 (2009)

10. Beleiu, H.G., et al.: Harmonics consequences on drive systems with induction motor. Appl. Sci. **10**, 2–15 (2020)
11. Ding, X., Mi, C.C.: Impact of inverter on losses and thermal characteristics of induction motors. Int. J. Power Electron. **3**(6), 641–651 (2011)
12. Al-Badri, M., Pillay, P., Angers, P.: A novel in situ efficiency estimation algorithm for three-phase induction motors operating with distorted unbalanced voltages. IEEE Trans. Ind. Appl. **53**(6), 5338–5347 (2017)
13. El-Kharashi, E., et al.: The impact of the unbalance in both the voltage and the frequency on the performance of single and cascaded induction motors. Energy **181**, 561–575 (2019)
14. Ngo, X.C., Do, N.Y., Tran, Q.H.: The influence of voltage quality on the performance of EKG excavators in open pit mines. Inżynieria Mineralna **1**(2), 139–145 (2020)
15. Chauhan, S., Singh, S.B.: Effects of voltage unbalance and harmonics on 3-phase induction motor during the condition of undervoltage and overvoltage. In: 2019 6th International Conference on Signal Processing and Integrated Networks, pp. 1141–1146 (2019)
16. Motor and Generators: Effects of unbalanced voltages on the performance of polyphase induction motors, NEMA Standards. Annual Pulp and Paper Industry Technical Conference. IEEE (1999)

Impacts of Dressing Conditions on Wheel Lifetime When External Grinding of SKD11 Steel

Hoang Anh Le[1], Dinh Ngoc Nguyen[2], Ngoc Huy Thinh Tran[3], Huu Quang Nguyen[4], Kieu Tuan Trinh[4], Hong Linh Nguyen[5], Thanh Tu Nguyen[2], and Xuan Tu Hoang[2(✉)]

[1] Vinh Long University of Technology Education, Vinh Long, Vietnam
[2] Thai Nguyen University of Technology, Thai Nguyen, Vietnam
hxt_dhktcn@tnut.edu.vn
[3] Nguyen Tat Thanh University, Ho Chi Minh City, Vietnam
[4] University of Economics - Technology for Industries, Ha Noi, Vietnam
[5] Electric Power University, Ha Noi, Vietnam

Abstract. The influence of dressing parameters on the tool life of grinding wheel when grinding of internal cylindrical grinding of hardened SKD11 steel was studied in this article. Also, the coarse dressing depth, the number of coarse dressing, the fine dressing depth, the number of fine dressing, the non-feeding dressing, and the dressing feed speed can be considered as input parameters. Besides, the Taguchi method and the Analysis of Variance (ANOVA) method were applied to find the optimum set of dressing parameters which can improve the tool life of the grinding wheel. The results reveal that the tool life of grinding wheel will maximize at 16 min when the set of dressing parameters is optimum.

Keywords: Grinding dressing · External grinding · Surface roughness · Multi-optimization · SKD11 steel

1 Introduction

In order to get the final dimension, grinding operations are always needed [1–11]. This means that the shape and dimensions of grinding wheel significantly influences on the tolerance of both dimension and position of machined parts. For this reason, dressing process is adopted to eliminate the removed chips which are stick on the surface of grinding wheel [12]. Dressing operation is significantly impacted by feed rate, rotational directional [13–19]. Although experimental and simulation studies have been conducted by research community, the studies dealing with optimizing dressing parameters has been still inefficient [1, 2, 20]. Taguchi method is one of the procedures applied for optimize the influences of dressing parameters on various objective functions, i.e. maximum tool life and material removal rate, minimum surface roughness. Taguchi method combined with Grey Relational Analysis (GRA) is used also to determine the optimal set of dressing parameters [15]. Besides, a regression model supporting for predicting optimum parameters is proposed.

D. C. Nguyen et al. (Eds.): ICERA 2021, LNNS 366, pp. 479–487, 2022.
https://doi.org/10.1007/978-3-030-92574-1_50

In this study, the coarse dressing depth, the number of coarse dressing, the fine dressing depth, the number of fine dressing, the non-feeding dressing, and the dressing feed speed are purposely selected as input parameters to find their optimum values leading to minimum surface roughness. Optimal set of dressing parameters will be found by using Taguchi method. Additionally, experimental results are used to find the ANOVA method which can interpret the significance of each parameter or its interactions on the responses.

2 Experimental Setup

Three main stages of optimum process can be schematically shown in Fig. 1. They are input parameters, induced parameters, and outcome parameters. Cantext Aquatex 3810 with a concentration of 3% and a flow of 10 l/min is adopted to cool the grinding area. Besides, the feed rate (S_d) of 1,8 m/min, the depth per cut of 0.005/single stroke, the speed of grinding wheel (V_d) of 29.3 m/s, and the total depth of 0.05 mm were applied for the input values. All tests are conducted similarly cutting condition as previously mentioned. The tool used for dressing grinding wheel and devices for experiments are seen in Fig. 2 and Table 1 respectively. Dimensions of workpiece made of SKD11 steel: ∅25 × 200. The chemical composition of SKD11 steel can be referred in Table 2. The external grinding process is shown in Fig. 3.

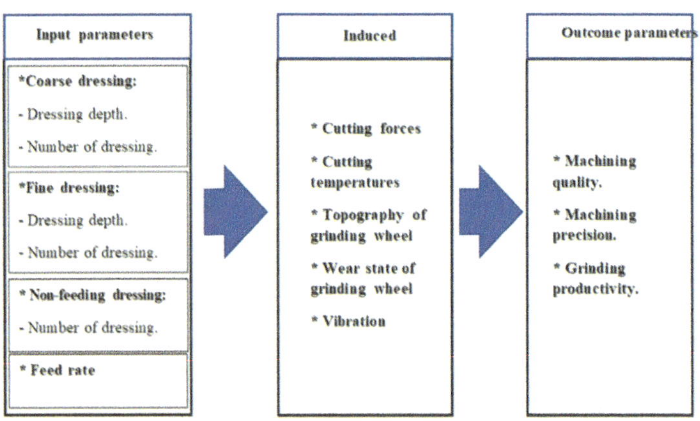

Fig. 1. Process of optimization of dressing parameters.

Table 1. Experimental devices for grinding and dressing process

Machine and equipment	Specification
Grinding machine (Japan origin)	CONDO-Hi-45 HTS
Grinding wheel (Vietnam origin)	Ct80MV1-G 400 × 40 × 203 35 m/s
Dressing tool (Russian origin)	3908-0088C type 2
Coarseness stylus (Japan origin)	Mitutoyo 178-923-2A, SJ-201

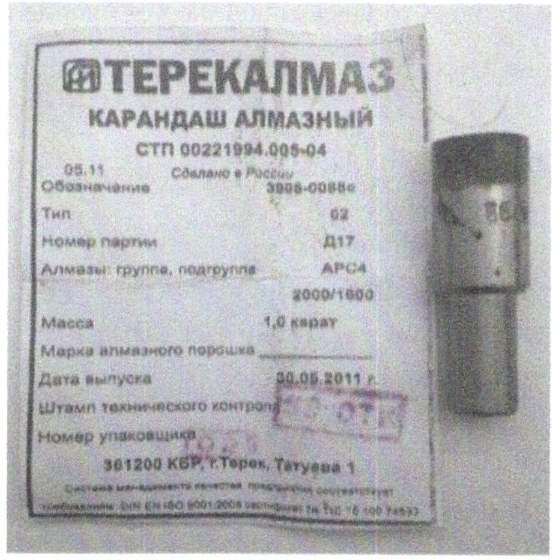

Fig. 2. Diamond grinding tool.

Table 2. Chemical composition of SKD11 steel.

Element	C	Cr	Mo	V	Si	Mn	S	P
Percentage (%)	0.9–1.5	12	1.0	0.3–5	0.2–5	0.4–5	≤0.025	≤0.025

Fig. 3. External grinding experimentation setup.

The levels of the input parameters for investigating are presented in Table 3. It is observed that the coarse dressing depth and the feed speed are considered of two levels, while the remains are conducted by four levels. The orthogonal array of experiments will be selected by L_{16} in which 14° of freedom is 14, and random errors are measured by two remains. All mentioned are performed by using Minitab@19. The possible tests involving values of each input parameters are displayed in Table 4.

Table 3. Input parameters and their examining levels

TT	Input parameters	Denoting	Level of investigating			
			1	2	3	4
1	Coarse dressing depth (mm)	a_r	0.03	0.04		
2	Number of coarse dressing	n_r	1	2	3	4
3	Fine dressing depth (mm)	a_f	0.005	0.01	0.015	0.02
4	Number of Fine dressing	n_f	0	1	2	3
5	Non-feeding dressing	n_0	0	1	2	3
6	Feed speed (m/min)	S_d	1	1.5		

Table 4. Distributions of examining levels of input parameters

Exp. no	a_f	n_f	n_0	n_r	a_r	S_d
1	0.005	0	0	1	0.03	1.0
2	0.005	1	1	2	0.03	1.5
3	0.005	2	2	3	0.04	1.0
4	0.005	3	3	4	0.04	1.5
5	0.010	0	1	3	0.04	1.5
6	0.010	1	0	4	0.04	1.0
7	0.010	2	3	1	0.03	1.5
8	0.010	3	2	2	0.03	1.0
9	0.015	0	2	4	0.03	1.5
10	0.015	1	3	3	0.03	1.0
11	0.015	2	0	2	0.04	1.5
12	0.015	3	1	1	0.04	1.0
13	0.020	0	3	2	0.04	1.0
14	0.020	1	2	1	0.04	1.5
15	0.020	2	1	4	0.03	1.0
16	0.020	3	0	3	0.03	1.5

3 Results and Discussions

Experiments are necessary to carry out before optimizing dressing process. Table 5 shows the plans of tests according to Taguchi method and the mean values of grinding wheel tool life. Sixteen tests with the detail's information are listed. The tool life of grinding wheel shown in the table is averagely calculated by three measurements. The experimental results are created based on the conditions, e.g. the larger is better for surface roughness (Table 5) and measuring three times. Moreover, the signal-noise ratios are also determined and exhibited in tables.

The minimum number of potential tests can be determined thanks to the orthogonal array. An S/N ratio is opted for conducting the performance characteristics and calculating the contribution of each parameter of the process. The signal-to-noise ratio is examined to be larger is better as described as follows:

$$\frac{S}{N} = -10.log10[(\sum_{1}^{n} 1/y_i^2)/n]$$

Where: y_i is the observed data; n is the number of experiments that are repeated.

Table 5 shows the experimental results and S/N rations where each representative value of tool life is measured three times. The input factors involve the coarse dressing depth (a_r), the coarse dressing times (n_r), the non-feeding dressing (n_o), the fine dressing times (n_f), the fine dressing depth (a_f), and the dressing feed rate (S_d). S/N ratio is signal to noise one receiving by using the Minitab R19 software. It is seen that the tool life of grinding wheel varies from 8.22 min to 16.48 min. The ANOVA method was adopted to check the significance of singular input parameters on the objective function. It is observed that the non-feeding dressing (n_0) exhibits the most impact on the tool life of grinding wheel, 39.42%. The second factor having strong effect is the number of the fine dressing, n_f, with 25.07%. On the contrary, the coarse dressing depth, a_r, has little impact on the response, 0.79%. The effects of input parameters can be also graphically shown in Fig. 4. These results also reveal that the proposed model can fit well experiments when R-square value reaches 98.16% (Table 6).

Table 5. The obtained experimental results for tool life of grinding wheel.

Exp. no	a_f	n_f	n_0	n_r	a_r	S_d	Wheel life (minute)			Means	S/N
							Trial 1	Trial 2	Trial 3		
1	0.005	0	0	1	0.03	1.0	9.47	8.59	9.80	9.29	19.3153
2	0.005	1	1	2	0.03	1.5	8.87	8.97	9.10	8.98	19.0630
3	0.005	2	2	3	0.04	1.0	16.93	15.50	17.02	16.48	24.3172
4	0.005	3	3	4	0.04	1.5	8.80	8.21	8.54	8.52	18.5949
5	0.010	0	1	3	0.04	1.5	9.88	10.20	9.50	9.86	19.8675
6	0.010	1	0	4	0.04	1.0	9.73	9.80	9.12	9.55	19.5873
7	0.010	2	3	1	0.03	1.5	9.13	9.20	8.80	9.04	19.1227
8	0.010	3	2	2	0.03	1.0	9.08	9.35	9.68	9.37	19.4270
9	0.015	0	2	4	0.03	1.5	12.60	13.01	12.83	12.81	22.1510
10	0.015	1	3	3	0.03	1.0	10.60	10.80	11.05	10.82	20.6781
11	0.015	2	0	2	0.04	1.5	9.37	9.86	9.63	9.62	19.6568
12	0.015	3	1	1	0.04	1.0	9.27	9.32	8.98	9.19	19.2618
13	0.020	0	3	2	0.04	1.0	9.03	8.82	9.15	9.00	19.0829
14	0.020	1	2	1	0.04	1.5	10.17	10.29	10.50	10.32	20.2704
15	0.020	2	1	4	0.03	1.0	10.87	11.35	11.42	11.21	20.9874
16	0.020	3	0	3	0.03	1.5	8.03	8.13	8.50	8.22	18.2910

Table 6. Analysis of variance for means

Source	DF	Seq SS	Adj SS	Adj MS	F	P	C (%)
a_f	3	5.3937	5.3937	1.7979	1.57	0.517	8.68
n_f	3	15.5729	15.5729	5.1910	4.53	0.330	25.07
n_0	3	24.4915	24.4915	8.1638	7.12	0.267	39.42
n_r	3	11.4784	11.4784	3.8261	3.34	0.378	18.48
a_r	1	0.4892	0.4892	0.4892	0.43	0.632	0.79
S_d	1	3.5511	3.5511	3.5511	3.10	0.329	5.72
Residual Error	1	1.1461	1.1461	1.1461			1.84
Total	15	62.1229					

Model Summary

S	R-Sq	R-Sq(adj)
1.0706	98.16%	72.33%

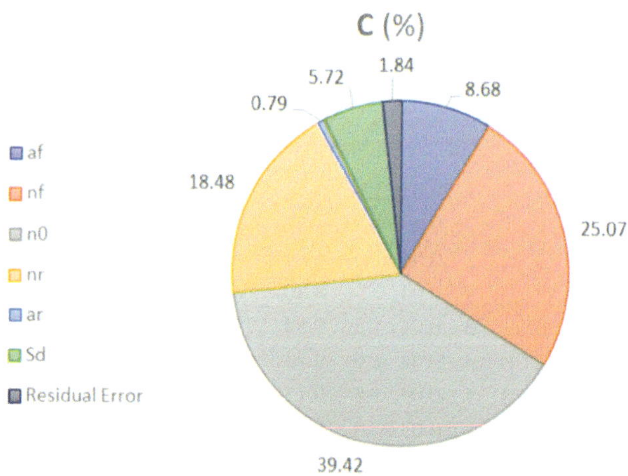

Fig. 4. The influential percent of input parameters on the responses

The main effects plot as shown in Figs. 5 and 6 can be used to describe the main impacts of input parameters on the tool life of grinding wheel and signal to noise ratio (S/N) respectively. The evolution of dressing parameters with their levels in Fig. 5 shows that a_f, n_f, n_0, n_r have significant impacts on the response. The effects of signal-to-noise ratio on the surface roughness are separately presented in Fig. 6. The stable tendency of a_r and S_d parameters are clearly seen (Table 7).

Table 7. Response table for signal to noise ratios

Level	a_f	n_f	n_0	n_r	a_r	S_d
1	10.816	10.240	9.169	9.459	9.968	10.614
2	9.457	9.916	9.810	9.242	10.318	9.672
3	10.609	11.590	12.247	11.346		
4	9.688	8.824	9.345	10.523		
Delta	1.359	2.766	3.078	2.103	0.350	0.942
Rank	4	2	1	3	6	5

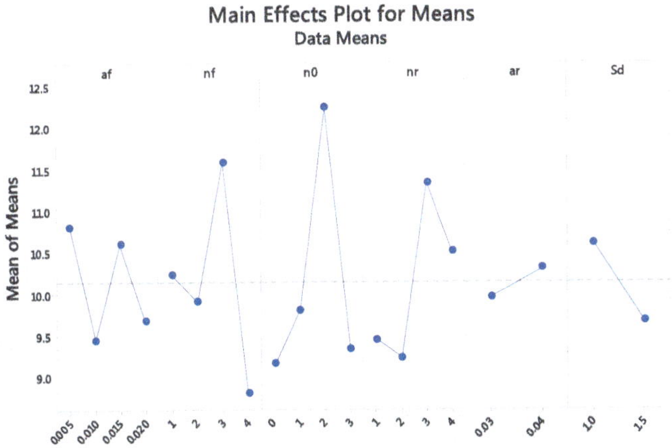

Fig. 5. The influence of input parameters on the tool life.

Thc experimental results reveal that the tool life of grinding wheel will maximize when the set of dressing parameters is optimum. The optimum set consists of the fine dressing depth at level of 3, the number of fine dressings at level of 3, the non-feeding dressing at level of 3, the number of coarse dressing at level of 3, the coarse dressing depth at level of 2, and the feed speed at level of 1. Finally, it is found that the optimum value of grinding wheel tool life is 16 min, corresponding to previously present dressing parameters.

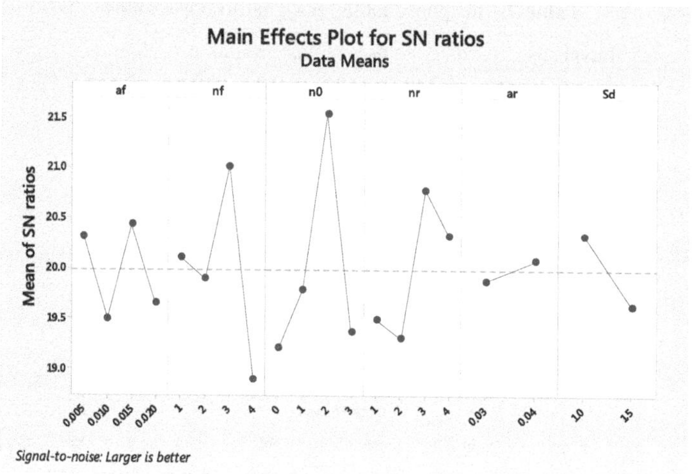

Fig. 6. The influence of input parameters on the signal to noise ratio.

4 Conclusions

The current study finds the optimal set of dressing parameters in order to maximize tool life of grinding wheel when conducting external grinding process of hardened SKD 11 steel. Taguchi method and the ANOVA method are utilized to determine the orthogonal array for experimental plans. The following conclusions can be made:

- It is observed that the non-feeding dressing (n_0) exhibits the most impact on the tool life of grinding wheel, 39.42%. The second factor having strong effect is the number of fine dressings, n_f, with 25.07%. On the contrary, the coarse dressing depth, a_r, has little impact on the response, 0.79%.
- The tool life of the grinding wheel gets the maximum value when using the Fine dressing depth at level of 3, the number of fine dressings at level of 3, the non-feeding dressing at level of 3, the number of coarse dressing at level of 3, the coarse dressing depth at level of 2, and the feed speed at level of 1.
- The optimum value of the tool life of grinding wheel is 16 min when using the optimum dressing parameters.

Acknowledgment. This work was supported by Thai Nguyen University of Technology.

References

1. Tu, H.X., et al.: Influence of dressing parameters on surface roughness of workpiece for grinding hardnened 9XC tool steel (2019)
2. Hung, L.X., Pi, V.N., Tung, L.A., Tu, H.X., Jun, G., Long, B.: Determination of optimal exchanged grinding wheel diameter when internally grinding alloy tool steel 9CrSi (2018)

3. Tu, H.X., et al.: Calculation of optimum exchanged grinding wheel diameter when external grinding tool steel 9CrSi. Int. J. Mech. Eng. Robot. Res. **8**, 59–64 (2018)
4. Tran, T.H., et al.: Analysis of effects of machining parameters on surface roughness in electrical discharge machining tablet shape punches using Taguchi method. Mater. Sci. Forum **977**, 12–17 (2020)
5. Tung, L.A., et al.: A study on optimization of surface roughness in surface grinding 9CrSi tool steel by using Taguchi method. In: Advances in Engineering Research and Application, pp. 100–108 (2019)
6. Tu, H.X., Pi, V.N., Jun, G.: A study on determination of optimum parameters for lubrication in external cylindrical grinding base on Taguchi method. Key Eng. Mater. **796**, 97–102 (2019)
7. Tran, T.H., et al.: A study on cost optimization of external cylindrical grinding. Mater. Sci. Forum **977**, 18–26 (2020)
8. Hung, L.X., et al.: A study on coolant parameters in internal grinding of 9CrSi steel. Mater. Sci. Forum **950**, 24–31 (2019)
9. Tran, T.H., et al.: A study on calculation of optimum exchanged grinding wheel diameter when surface grinding stainless steel. Mater. Sci. Forum **977**, 3–11 (2020)
10. Hung, L.X., et al.: A study on calculation of optimum exchanged grinding wheel diameter when internal grinding. Mater. Today: Proc. **18**, 2840–2847 (2019)
11. Vu Ngoc, P., et al.: Experimental determination of optimum exchanged diameter in surface grinding process. J. Environ. Sci. Eng. A **6**(2), 85–89 (2017)
12. Palmer, J., et al.: An experimental study of the effects of dressing parameters on the topography of grinding wheels during roller dressing. J. Manufact. Process. **31**, 348–355 (2018)
13. Prusak, Z., Webster, J.A., Marinescu, I.D.: Influence of dressing parameters on grinding performance of CBN/Seeded Gel hybrid wheels in cylindrical grinding. Int. J. Prod. Res. **35** (10), 2899–2916 (1997)
14. Puerto, P., et al.: Evolution of surface roughness in grinding and its relationship with the dressing parameters and the radial wear. Procedia Eng. **63**, 174–182 (2013)
15. Tung, L.A., Pi, V.N., Lien, V.T., Hong, T.T., Hung, L.X., Long, B.T.: Optimization of dressing parameters of grinding wheel 9CrSi tool steel using the Taguchi method with grey relational analysis (2019)
16. Novák, M., Naprstkova, N., Kasuga, H.: Influence of dressing parameters on surface roughness of workpiece for grinding hardnened 9XC tool steel (2016)
17. Tran, T.H., et al.: Improvement of wheel life by optimization of dressing parameters in surface roughness grinding of SKD11 steel (2019)
18. Dadaso, D., Mohite, S.M.J.: An investigation of effect of dressing parameters for minimum surface roughness using CNC cylindrical grinding machine (2016)
19. Novák, M., Naprstkova, N., Kasuga, H.: Influence of grinding wheel dressing on the roughness of final surface and cutting force during GGG60 grinding. Key Eng. Mater. **686**, 218–223 (2016)
20. Vu, N.P., Long, B.T., Vu, H.K., Nguyen, H.Q.: Profit optimization of external cylindrical grinding (2014)
21. Hoang, X.T., et al.: Influence of dressing parameters on surface roughness of workpiece for grinding hardened 9XC tool steel (2019)
22. Pi, V.N., Lu, A.T., Hung, L.X., Long, B.T.: Cost optimization of surface grinding process. J. Environ. Sci. Eng. **5**, 606–611 (2016)

Influence of Abrasive Density on the Material Removal Mechanisms in Diamond Wire Sawing of Silicon Crystal

Van-Nhat Le[1(✉)] and Van-Dung Tran[2]

[1] Faculty of International Training, Thai Nguyen University of Technology, Thai Nguyen, Vietnam
lenhat@tnut.edu.vn
[2] National Research Institute of Mechanical Engineering, Ha Noi, Vietnam

Abstract. In recent years, silicon wafer based solar cells have dominated the market due to their long lifetime and higher photovoltaic efficiency in power generation. Wafers are mainly manufactured by slicing Si ingots using fixed diamond wire sawing process (DWS). This paper presents an investigation of the influences of diamond abrasive density on the material removal mechanisms in DWS based on the model of wire profile model. Results show that the abrasive density directly affects the distribution of cutting depth in different cutting zones. While brittle material removal is still the major way of material removal at the bottom of the wire heading to the ingot, material removal at the side shifts to pure ductile cutting mechanism happens when the abrasive density increases. The developed model could be useful in determining combinations of sawing parameters with given abrasive density for obtaining better DWS process with high efficiency and good wafer surface quality.

Keywords: Wire sawing · Material removal mechanism · Abrasive machining

1 Introduction

In recent years, solar power generation has been an increasing interest globally. Silicon (Si) wafer based solar cells have dominated the market due to their long lifetime and higher photovoltaic efficiency in power generation. Wafers are mainly manufactured by slicing Si ingots using fixed diamond wire sawing process (DWS). Compared to the loose abrasive slurry which was popular in the past, the DWS has multiple advantages including smaller kerf loss, lower wafer cost and better environmental friendliness. Recently, there have been two trends in making wafer cost reduction including (1) increasing the number of wafers obtained from a given Si ingot by reducing the wafer thickness and (2) the use of thin wires to reduce the kerf loss which is the material removed between wafers [1]. Together, these two trends are expected to be used for slicing thin wafers with thin wires in the future. Since the wire sawing is a mechanical process for cutting hard and brittle material, the understanding of material removal mechanisms and its influencing factors are basic problems in studying the process. Several studies focusing on the use of different wires have been conducted.

D. C. Nguyen et al. (Eds.): ICERA 2021, LNNS 366, pp. 488–496, 2022.
https://doi.org/10.1007/978-3-030-92574-1_51

Suzuki et al. conducted experiments and reported that material removal in ductile mode is enhanced by using finer diamond abrasive [2]. The study also shows that when smaller abrasives are used, the machining mode transits from brittle to ductile. The influence of core wire diameters on the mechanical strength of wafer has been studied in other report [3]. The result shows that the mechanical strength of thin wafers sliced differed greatly when using different wire sizes. Nevertheless, these studies have not reported the influence of abrasive density in DWS. The diamond abrasive density varies in wide range of value as shown in Fig. 1 and directly affects cutting mechanisms and sliced wafer properties.

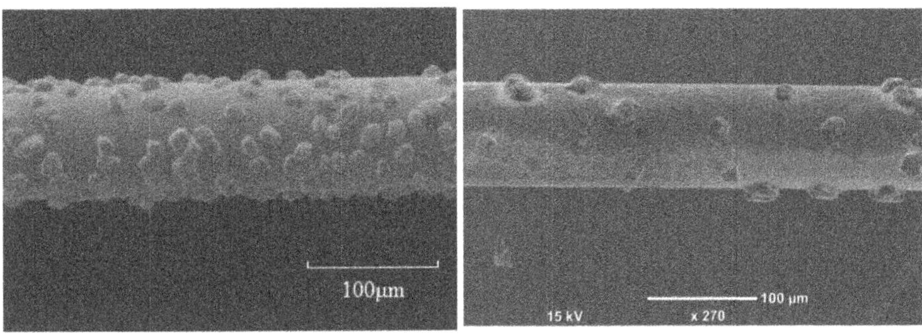

Fig. 1. SEM images of the diamond wires with different abrasive density [4, 5]

This paper presents an investigation of the influence of diamond abrasive density on the material removal mechanisms in DWS based on the wire profile model. Results show that the abrasive density directly affects the distribution of cutting depth in different cutting zones. While brittle material removal is still the major way of material removal at the bottom of the wire heading to the ingot during wire sawing, material removal at the side shifts to pure ductile cutting regime takes place when increasing the abrasive density. The model developed in this study could be useful in determining combinations of sawing parameters with given abrasive density for obtaining better DWS process with high efficiency and better wafer surface quality.

2 Material Removal Mechanisms of Abrasives in DWS

Material removal mechanism during DWS can be regarded as a scribing process of a single abrasive. Studies on ductile cutting mechanism of crystalline silicon have been reported [6–8]. The critical cutting depth, d_c, at which the cutting transits from ductile to brittle mode can be predicted according to previous work [6, 9]:

$$d_c = 0.15\left(\frac{E}{H}\right)\left(\frac{K_c}{H}\right)^2 \tag{1}$$

where K_c is the fracture toughness, H is the hardness, and E is the elastic modulus of the specimen.

The cutting depth of each abrasive during DWS can be used to identify cutting mechanism and volume of material removed by that abrasive. At the cutting depths lower than critical cutting depth, the material removed area is estimated by projected area of abrasive tip following sawing direction as shown in the diagram in Fig. 2.

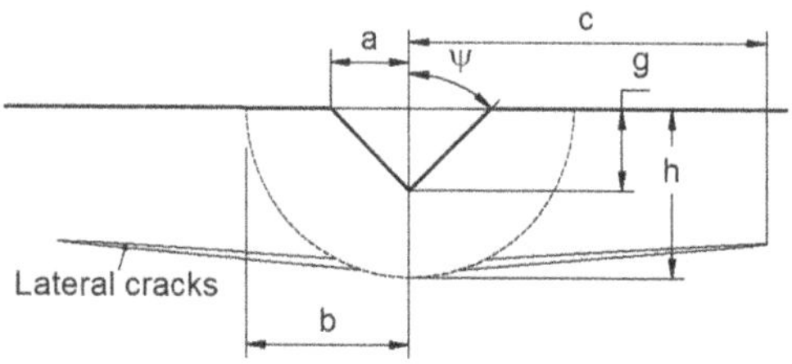

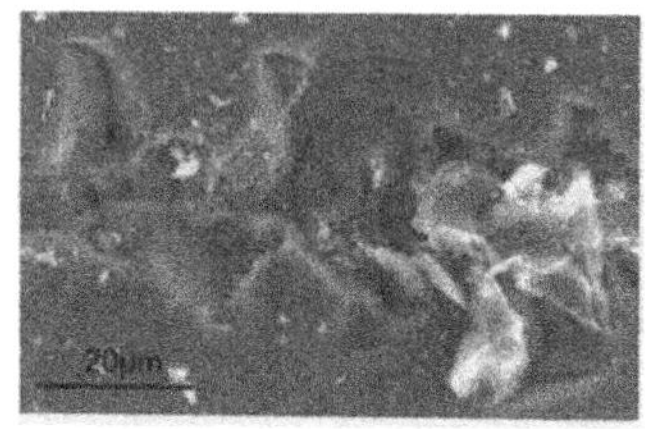

Fig. 2. Indentation model of single pyramidal indenter [10] (left) and brittle cutting mode by single diamond scribing at 0.772 μm cutting depth [6] (right).

In brittle mode, taken at higher cutting depth, the cracks propagated during loading helps to remove more material. It is assumed that the portion of material above the lateral cracks will be removed after unloading. The thickness of the removed material, h, can be identified with the depth of the plastic zone [10, 11].

$$h \approx b \approx \left(\left(\frac{E}{H}\right)^{\frac{1}{2}} / H^{\frac{1}{2}}\right)(\cot\psi)^{\frac{1}{3}} P^{\frac{1}{2}} \tag{2}$$

and

$$c = \frac{\zeta_L^{\frac{1}{2}}}{M^{\frac{1}{4}}}(\cot\psi)^{\frac{5}{12}} \frac{E^{\frac{1}{2}}}{K_c^{\frac{1}{2}} H^{\frac{1}{2}}} P^{\frac{5}{8}} \tag{3}$$

Where $\zeta_L = 25 \times 10^{-3}$, $M = 3/4$, E is the material Young's modulus, H is the hardness, ψ is the half tip angle, P is the normal load on the abrasive and b is the depth of plastic zone.

The length of lateral crack, c, induced by each abrasive is assumed to be about the size of plastic zone, b, according to experimental work and stress analysis conduct by Wu et al. [12]. The relationship between the plastic zone and impact area can be estimated by experimental observations [6].

$$\frac{b}{a} \approx \left(\frac{E}{H}\right)^m (\cot\psi)^{\frac{1}{3}} \text{ in which } m = 1/2 \tag{4}$$

3 DWS Model for the Investigation of Abrasive Density to Material Removal Mechanisms

During steady DWS process, an ingot section is modeled as illustrated in Fig. 3. In the figure, v_x represents wire speed and f_z represents ingot feeding speed. Coordinate system is set with the origin O fixed at the wire center. During a time interval, Δt, a length of the wire L crosses the ingot section forming a cutting profile represented by array $T_i(\theta_i, r_i^t)$. The cutting profile directly interacts with the ingot section profile represented by array $W_i(\theta_i, r_i^w)$ at each time interval as illustrated in the right diagram of Fig. 3

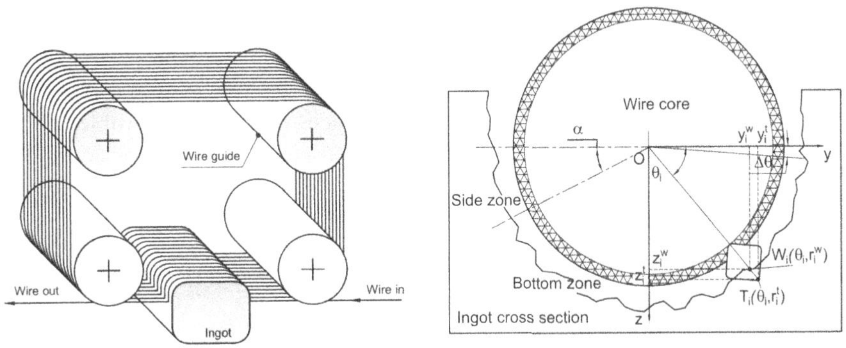

Fig. 3. Schematic diagram of DWS (left) and the model of ingot section (right).

The cutting area of DWS can be divided into two zones: side and bottom cutting. While abrasives at the side zone directly contribute to the generation of sliced wafer surfaces, the abrasives at bottom zone primarily contribute to the material removal. The angle defining side zone, α, is 20° according to study of Li, X., et al., [13]. The influences of abrasive density to the material removal mechanism in DWS can be investigated by this cutting profile model and its interaction with the ingot section profile. The cutting profile is divided by N points equivalent to $(N-1)$ segments. The polar element angle is defined as: $\Delta\theta = 180^0/(N-1)$. In the array $T_i(\theta_i, r_i^t)$, r_i^t is the radical distance from the ith point on the cutting profile to the coordinate origin O. The polar angle specifying the ith position on the cutting profile is calculated as: $\theta_i = (i-1)\Delta\theta$. The random position of abrasive is defined by the possibility of existence of an abrasive, which is:

$$s = \frac{K}{N} = \frac{\eta\pi d v_x \Delta t}{2N} \tag{5}$$

where K is the total number of abrasives at the wire bottom half crossing the ingot section during a time interval, estimated by: $K = 0.5\pi d v_x \Delta t\eta$, where η is abrasive density and d is the wire diameter.

The abrasive is assumed to have conical shape with half tip angleφ. The half tip angle is randomly selected from 55° to 80° according to previous report [14]. The size of diamond abrasives, r_g, can be identified according to probability density function as:

$$p(r_g) = \frac{1}{\sqrt{2\pi}\sigma} e^{\frac{-(\mu + \log(r_g))^2}{2\sigma^2}} \tag{6}$$

where μ and σ are abrasive mean size and standard deviation of abrasive size range, respectively.

If there is an abrasive at point i, the element r_i^t of the cutting profile array $T_i(\theta_i, r_i^t)$ is defined as:$r_i^t = 0.5d + 2r_g$. The cutting profile at the neighboring points, j^{th}, is also updated as: $r_j^t \approx r_i^t - r_i^t\theta_{ij}\cot\psi = r_i^t(1 - \Delta\theta|i-j|\cot\psi)$.

After the cutting profile is formed, the ingot section profile $W_i(\theta_i, r_i^w)$ is then generated by interaction between cutting profile and existing ingot section profile. If there exists an abrasive at the point i, the cutting depth, g_{ci}, of that abrasive is expressed as $g_{ci} = (r_i^t - r_i^w)$ in which r_i^t and r_i^w are radical distances of the cutting profile and ingot section profile with the same polar angle θ_i. The ingot section profile is updated based on the amount of material removed by active abrasives as follows (illustrated in Fig. 4):

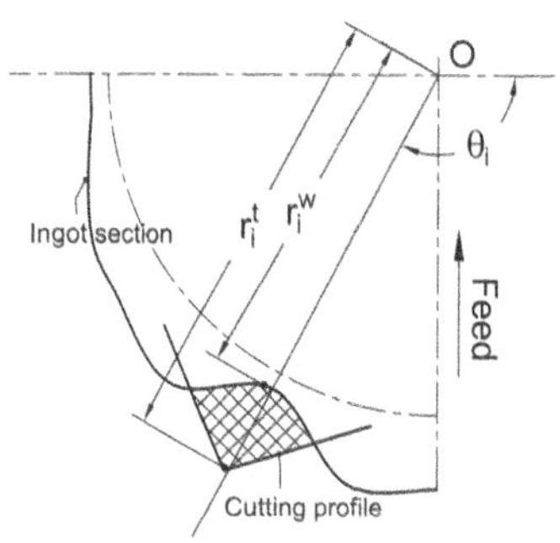

Fig. 4. Interactions between cutting profile and ingot section profile.

- If $g_{ci} \leq 0$, the abrasive does not penetrate into the ingot meaning that does not occur.
- If the abrasive is in contact with the ingot and $g_{ci} \leq g_c$, material removal is in ductile mechanism. Then, $r_i^w = \max\{r_i^w, r_i^t\} = r_i^t$.
- If $g_{ci} > g_c$, it means that the abrasive is cutting mechanism, the element r_i^w is updated as: $r_i^w = r_i^t + \frac{b}{a}g_i$ where indicates the ratio between the thickness of material above the lateral cracks and the cutting depth given in Eq. (4).

After each time interval, the ingot feed speed is simulated by transforming the ingot section profile toward the wire center O by the value $\Delta z = f_z\Delta t$. The ingot section profile after transformation is then updated based on Cartesian coordinate system as: $W_i(y_i^w, z_i^w, t + \Delta t) = W_i(y_i^w, z_i^w - \Delta z, t)$. The polar angle specifying the position of section profile after transforming is also changed to new polar angle accordingly.

4 Simulation Procedure and Input Parameters

In order to investigate the influence of abrasive density to the material removal mechanism of DWS, simulation runs with two values of abrasive density referred to previous report that are 1200 (pcs/mm^2) [2] and 650 (pcs/mm^2) [5]. Process parameters, material properties and wire parameters are given in Table 1. The programs of number simulation were compiled in Matlab program in which the time interval Δt, polar element angle, and slicing length are set at 0.001s, 0.1°, and 500 μm, respectively.

Table 1. Input parameters for the simulations of DWS

Input parameters	Value	Material properties of Si	Value
Wire core diameter (μm)	100	Hardness (Gpa)	10
Abrasive size (μm)	6–12	Young's Modulus (Gpa)	127
Wire velocity (m/min)	1000	Fracture toughness (MPa.m$^{1/2}$)	0.7
Feeding velocity (mm/min)	0.7	Critical cutting depth for brittle-ductile transition (μm)	0.5

5 Results and Discussions

In abrasive machining, the number of active abrasives and their cutting depth are not measurable. This is the reason why simulations are good tool to study effects of wire and process parameters to the cutting depth and material removal mechanisms during DWS. Figure 5 shows the distribution of active abrasives with respect to the cutting depth. At higher abrasive density, the number of abrasives cutting in ductile mechanism

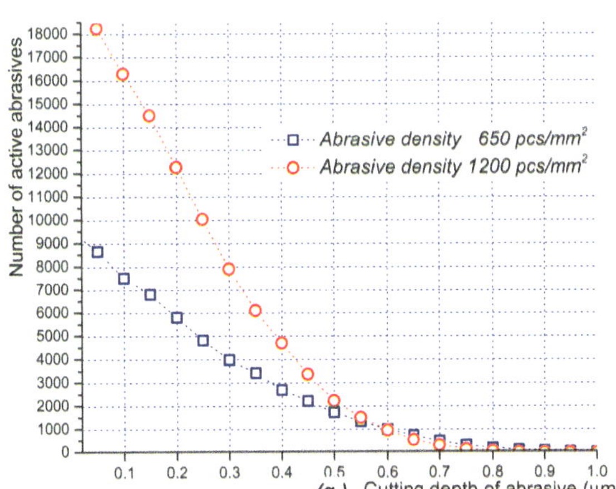

Fig. 5. Distribution of cutting depth with different values of abrasive density.

is much higher than that at lower density. However, numbers of brittle cutting abrasives with depth of cut higher than 0.5 μm are not much distinct between two cases meaning that brittle cutting mechanism always exists and contribute to the material removal of DWS.

Figure 6 shows the comparison of active abrasives with respect to cutting depth at side and bottom cutting zone. It can be observed that the number of brittle cutting abrasives at bottom zone is much higher than that at the side zone. The plot indicates that it is possible to select the combination of abrasive density and other parameters for the optimal DWS in which pure ductile cutting takes place at side zone for better surface quality while keeping the material removal rate high at the bottom zone.

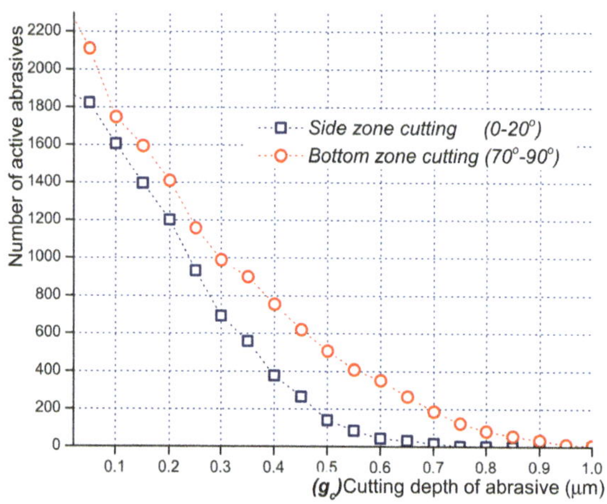

Fig. 6. Active abrasives at side and bottom cutting zone (Abrasive density: 650 pcs/mm^2).

The influence of the abrasive density to the cutting depth and material removal mechanisms can be illustrated in Fig. 7. The number of brittle cutting abrasives is much reduced in the case of using higher abrasive density. Therefore, the trend of using small wire with and higher abrasive density will be beneficial in DWS industry.

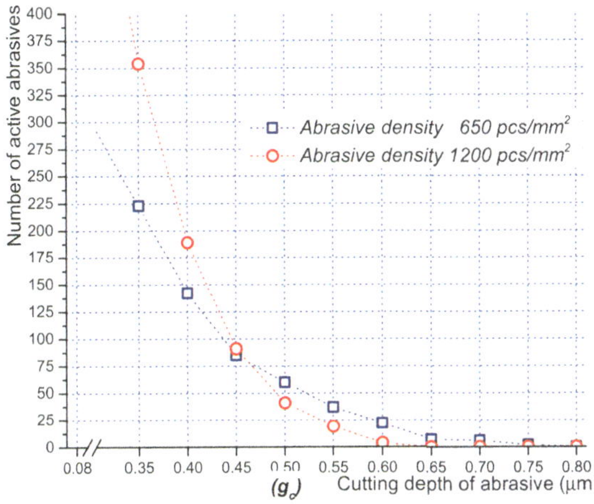

Fig. 7. Cutting depth distribution at side cutting zone under different abrasive density.

6 Conclusions

This study presents an investigation of the influence of abrasive density on the material removal mechanisms in DWS based on the model wire and ingot section model. Results show that the abrasive density directly affects the distribution of cutting depth in different cutting zones. While brittle material removal is still the major way of material removal at the bottom of the wire heading to the ingot during wire sawing, material removal at the side shifts to pure ductile cutting mechanism occurs with the increase of the abrasive density. The developed model could be used in determining combinations of sawing parameters with given abrasive density for obtaining optimal DWS process with high efficiency and better wafer sliced surface quality.

Acknowledgements. The work described in this paper was supported by Thai Nguyen University of Technology.

References

1. International Technology Roadmap for Photovoltaics (ITRPV): 10th ed. (2019). https://itrpv.vdma.org/ (2018)
2. Suzuki, T., Nishino, Y., Yan, J.: Mechanisms of material removal and subsurface damage in fixed-abrasive diamond wire slicing of single-crystalline silicon. Precis. Eng. **50**, 32–43 (2017)
3. Sekhar, H., et al.: The impact of subsurface damage on the fracture strength of diamond-wire-sawn monocrystalline silicon wafers. Jap. J. Appl. Phys. **57**, 08RB08 (2018)
4. Wang, L., et al.: Analytical prediction of subsurface microcrack damage depth in diamond wire sawing silicon crystal. Mater. Sci. Semicond. Process. **112**, 105015 (2020)

5. Sekhar, H., et al.: Mechanical strength problem of thin silicon wafers (120 and 140 μm) cut with thinner diamond wires (Si kerf 120 → 100 μm) for photovoltaic use. Mater. Sci. Semicond. Process. **119**, 105209 (2020)
6. Wu, H., Melkote, S.N.: Study of ductile-to-brittle transition in single grit diamond scribing of silicon: application to wire sawing of silicon wafers. J. Eng. Mater. Technol. **134**(4), 1–8 (2008)
7. Arif, M., et al.: A predictive model of the critical undeformed chip thickness for ductile–brittle transition in nano-machining of brittle materials. Int. J. Mach. Tools Manuf. **64**, 114–122 (2013)
8. Wang, J.-J.J., Liao, Y.-Y.: Critical depth of cut and specific cutting energy of a microscribing process for hard and brittle materials. J. Eng. Mater. Technol. **130**(1), 011002 (2007)
9. Bifano, T.G., Dow, T.A., Scattergood, R.O.: Ductile-regime grinding. a new technology for machining brittle materials. J. Eng. Ind. **113**(2), 184–189 (1991)
10. Marshall, D.B., Lawn, B.R., Evans, A.G.: Elastic/plastic indentation damage in ceramics: the lateral crack system. J. Am. Ceram. Soc. **65**(11), 561–566 (1982)
11. Evans, A.G., Gulden, M.E., Rosenblatt, M.: Impact damage in brittle materials in the elastic-plastic response regime. **361**(1706), 343–365 (1978)
12. Liang, Z., et al.: Experimental study on brittle – ductile transition in elliptical ultrasonic assisted grinding (EUAG) of monocrystal sapphire using single diamond abrasive grain. Int. J. Mach. Tools Manuf. **71**, 41–51 (2013)
13. Li, X., et al.: The effect of cut depth and distribution for abrasives on wafer surface morphology in diamond wire sawing of PV polycrystalline silicon. Mater. Sci. Semicond. Process. **91**, 316–326 (2019)
14. Mahmoud, T.A., Tamaki, J., Yan, J.: Three-dimensional shape modeling of diamond abrasive grains measured by scanning laser microscope. Key Eng. Mater. 238–239 (2003)

Influence of Cutting Parameters on Surface Roughness When Surface Grinding C250 with Segmented

Do Duc Trung[1], Nguyen Anh Tuan[2], Pham Duc Lam[3], Le Dang Ha[1,2,3,4], and Le Thi Phuong Thao[4](✉)

[1] Ha Noi University of Industry, Ho Chi Minh City, Vietnam
[2] University of Economics – Technology for Industries, Hanoi, Vietnam
[3] Nguyen Tat Thanh University, Ho Chi Minh City, Vietnam
[4] Thai Nguyen University of Technology, Thai Nguyen, Vietnam
lethao828@tnut.edu.vn

Abstract. C250 steel is a Nickel-based alloy steel with really high hardness, strength and ductility. An excellent feature of this steel is its ability to limit surface cracks during machining. In this article, a study on the influence of cutting parameters on surface roughness when grinding this steel is presented. Segmented grinding wheels with 18 grooves were used and the experimental process was carried out with a total of 15 experiments in the form of a Box-Behnken design matrix. At each experiment the parameters were selected as the input parameters including the velocity of workpiece, the feed rate and the cutting depth. The surface roughness has been selected as the output parameter of the experimental process. The analysis of experimental results has determined the influence of the input parameters on the surface roughness. A regression model showing the relationship between the input parameters and the surface roughness was also developed. Genetic Algorithm (GA) was chosen as the instrument to solve the optimization problem. The results of the optimization problem have determined the values of the cutting parameters to ensure the minimum surface roughness. The experiments to verify the optimal results of the cutting parameters were also conducted. The results show that the optimal value has been achieved with a very high reliability degree, in which the deviation between the experimental and predicted values is only 5.79%. Some further directions when studying the grinding technology with segmented grinding wheel have also been proposed in this article.

Keywords: C250 steel · Segmented grinding wheel · Surface roughness · Optimization · Genetic algorithm

1 Introduction

C250 material is a material with high hardness and toughness, suitable for making parts of aircraft and space [1, 2]. This material is considered difficult to process because during the machining process a large amount of cutting heat is generated, which makes the cutting tool shelflife less durable as well as greatly affects the quality of the

D. C. Nguyen et al. (Eds.): ICERA 2021, LNNS 366, pp. 497–506, 2022.
https://doi.org/10.1007/978-3-030-92574-1_52

machined surface [3–5]. Therefore, the solutions to reduce the cutting heat when machining this steel are necessary to be studied.

To own the finished products of the mechanical industry, the machining methods commonly used include milling, turning, drilling, broaching, grinding, etc. in which, grinding is often chosen as the final machining method for the surfaces with high requirements for accuracy, small surface roughness [6]. So as to reduce the negative impact of cutting heat on the quality of the machined surface, when grinding C250 steel, CBN grinding wheels are often used [7]. However, it can be seen that the cost of CBN grinding wheels is many times higher than that of ordinary grinding wheels (for example, aluminum oxide grinding wheel). In Vietnam, with the same size of grinding wheel, the cost of CBN grinding wheel is about 8 to 10 times higher than that of aluminum oxide grinding wheel. In addition, for CBN grinding wheels, the thickness of the grains layer is very small compared to the thickness of the inner aluminum disc (the thickness of the grains layer is only a few millimeters), which makes the cost of the grinding wheel very high. Thus, the problem here is to find a solution to reduce the cost of the grinding wheels but still ensure the quality of the machined surface.

In the grinding technology, some solutions are suggested to reduce the cutting heat, including using reasonable lubricant parameters [8], rational applying of the cutting parameters and the wheel dressing parameters [6], etc. A number of other solutions to reduce the cutting heat when grinding have also been carried out such as improvement of chip breaker conditions, heat dissipation and supply of coolant into the cutting zone by reducing the formation of an encirclement of solid air around the wheel, which is done by covering the grinding wheel with a rubber seal [9]; The grinding wheel before being put into use has been impregnated in Mo2S solution or Sol–Gel Alumina and a Synthetic Organosilicon Polymer – Based Impregnate solution, which reduces the grains wear, the friction between the grinding wheel and the machined surface, while reducing rust on the surface of the workpiece [10, 11].

In addition to some solutions to reduce the cutting heat when grinding with ordinary grinding wheels as described above, the use of special textured wheels to reduce the cutting heat when grinding has also been carried out by a number of studies. Among them, two types of grinding wheel with discontinuous working surface have been examined by several studies including the slotted grinding wheel and the segmented grinding wheel. Using a slotted grinding wheel can reduce the cutting heat by 40–80% compared to the ordinary grinding wheels [12]. Using a segmented grinding wheel results in a reduction in both surface roughness and cutting force compared to using an ordinary grinding wheel [13–15]. This presents a very good prospect that the grinding wheels with discontinuous surfaces are well suited for grinding hard-to-machine materials such as C250 steel.

In case of comparing the slotted grinding wheel and the segmented grinding wheel, it can be seen that the use of the segmented grinding wheel has more advantages than the slotted grinding wheel. Firstly, with the same thickness of the grinding wheel, when using a segmented grinding wheel, it will allow to carry out with a larger horizontal feed rate compared to when using a slotted grinding wheel. It can be said that using a segmented grinding wheel will allow a maximum feed rate equal to the wheel thickness to be used. However, with a slotted grinding wheel, the feed rate will be less than half of the wheel thickness, as when the feed rate exceeds this limit, there will be a part that is not grinded on the surface of the workpiece; Secondly, when making the grinding wheels by forming press and then sintering, only one die mold is needed for a segmented grinding wheel, whereas for a slotted grinding wheel, the number of die molds will depend on the number of slots as per the circumference of the wheel, and at least 2 (if the number of slots is 1, the number of die mold is equal to 2). However, up to this point of time, there have been no studies published on the use of segmented aluminum oxide grinding wheels to grind C250 steel. So as to reduce the cost of grinding wheel in the production process, in this study, the author will conduct an experiment to grind C250 steel with segmented Aluminum Oxide grinding wheel. The parameter selected to evaluate the grinding process is the surface roughness. The optimum values of the cutting parameters have also been determined to ensure a minimum surface roughness.

2 Grinding Experiment

2.1 Experimental Materials

C250 steel was selected as the experimental material in this study. Some basic parameters of this steel are shown in Table 1. The experimental steel samples have the length, width and height of 45 mm, 45 mm and 12 mm, respectively.

Table 1. Some parameters of C250 steel.

Ni	Mo	Co	Ti	Al	Density	Coefficient of thermal expansion	Melting point	Specific heat capacity	Thermal conductivity
%					g/cm^3 at 20 °C	1/K at 20 °C	°C	J/(kg*K) at 20 °C	W/(m*K) at 20 °C
18	5	8.5	0.4	0.1	8.2	1.3 × 10–5	1425–1505	440	20–25

2.2 Experimental System

The Taiwan-made APSG-820/2A surface grinder was used to carry out the experiments. The Aluminum Oxide grinding wheel employed in this study has the symbol as WA46J7V1A. The grinding wheel has an outer diameter of 180 mm, a thickness of 13 mm and a hole diameter of 31.75 mm. On the grinding wheel, there are 18 grooves, each of which is a semicircle with a diameter of 12 mm. The number of grooves equal to 18 was determined to be the best value when evaluating the grinding process of DIN 1.2379 steel through the criteria including the surface roughness, the system vibration in three directions X, Y, Z and the material removal rate (MRR) [16]. When evaluating the grinding process of SAE420 steel through the surface roughness and the system's vibration components, the number of grooves equal to 18 was also determined to be the best value for the slotted grinding wheels [17]. The cutting heat also has the minimum value when the number of grooves on the wheel is 18 in the case of grinding SKD11 steel [18].

The surface roughness is measured by Machine SJ-201, a product of MITUTOYO company. The standard length of the measurement has been set up on the machine is 0.8 mm, and the measuring direction is made in the direction perpendicular to the cutting speed and made at least 3 times on each experimental sample. The surface roughness value at each experiment is calculated as the average value of consecutive measurements.

2.3 Experimental Design

The experiments were designed according to the Box-behnken-type experimental design matrix with three input parameters as cutting parameters including the velocity of the workpiece, the feed rate and the cutting depth. According to this matrix design, each input parameter will have three degrees of values corresponding to three encoding degrees, including −1, 0 and 1. The values of the parameters at the degrees when grinding with a segmented grinding wheel have been selected within their range according to the studies [16, 17]. The values of the input parameters at the encoding degrees are shown in Table 2, whereas the experimental matrix is shown in Table 3.

Table 2. The input parameters.

Parameters	Symbol	Unit	Value at levels		
			−1	0	1
Workpiece velocity	v_w	m/min	5	10	15
Feed rate	f	mm/stroke	2	4	6
Depth of cut	a_p	mm	0.01	0.015	0.02

3 Grinding Experiment

The experiments were conducted in the order displayed in Table 3. The experiments were carried out when the grinding wheel's velocity was 26 m/s, the wheel dressing depth was 0.01 mm, the wheel dressing feed rate was 100 m/min, the coolant used was 10% emulsion with a flow rate of 4.6 little/min. These values have been chosen depending on the technological capabilities of the machine as well as the studies published when applying to the segmented grinding wheel [16, 17]. The surface roughness values at each experiment have also been included in Table 3. These values are all in roughness grades from 8 to 9, consistent with the ability to achieve the surface roughness when machining by general grinding methods [6, 19]. In other words, the segmented aluminum oxide grinding wheel is perfectly suitable for grinding C250 steel. Using the segmented Aluminum Oxide grinding wheel will reduce the costs more than using CBN grinding wheel.

Figure 1 exhibits a Pareto chart, which shows the influence of the input parameters on the surface roughness. From this chart, it can be seen that the velocity of the workpiece is the parameter with the greatest influence on the surface roughness, followed by the influence of the feed rate, while the cutting depth has a negligible influence on the surface roughness. These results are also consistent with using the segmented grinding wheels when grinding DIN 1.2379 steel [16] and grinding SAE420 steel [17]. This problem can be explained as follows: when the velocity of the workpiece is in change, it will change the time when the heat source transmits to the machined surface, thereby resulting in change of the plastic deformation of the surface metal layer, and in that way changing the surface roughness. For the feed rate, when this parameter changes, it will change the number of granule scratches into the workpiece surface, which also makes the surface roughness change.

From the experimental results in Table 3, a regression model of the surface roughness has also been developed as shown in Eq. (1). This model has a coefficient of determination R-Sq = 0.9984 and an adjusted coefficient of determination R-Sq (adj) = 0.9955. The calculation of these coefficients as well as their significance can be found in many documents on experimental planning [20, 21]. The values of these coefficients are all very close to 1, which proves that the regression model has a very high compatibility with the experimental data. In other words, this regression function has a high accuracy when it is used to predict the surface roughness for the values of the input parameters.

Table 3. The experimental matrix and results.

No.	Code value			Actual value			*Ra* (μm)
	v_w	f	a_p	v_w (m/min)	f (mm/stroke)	a_p (mm)	
1	0	0	0	10	4	0.015	0.413
2	−1	0	−1	5	4	0.01	0.321
3	0	1	1	10	6	0.02	0.492
4	1	1	1	15	6	0.015	0.582
5	−1	−1	0	5	2	0.015	0.255
6	0	0	0	10	4	0.015	0.424
7	1	0	1	15	4	0.02	0.504
8	1	−1	0	15	2	0.015	0.426
9	1	0	−1	15	4	0.01	0.503
10	0	−1	1	10	2	0.02	0.335
11	−1	1	0	5	6	0.015	0.400
12	0	0	0	10	4	0.015	0.426
13	−1	0	0	5	4	0.02	0.323
14	0	1	−1	10	6	0.01	0.490
15	0	−1	−1	10	2	0.01	0.355

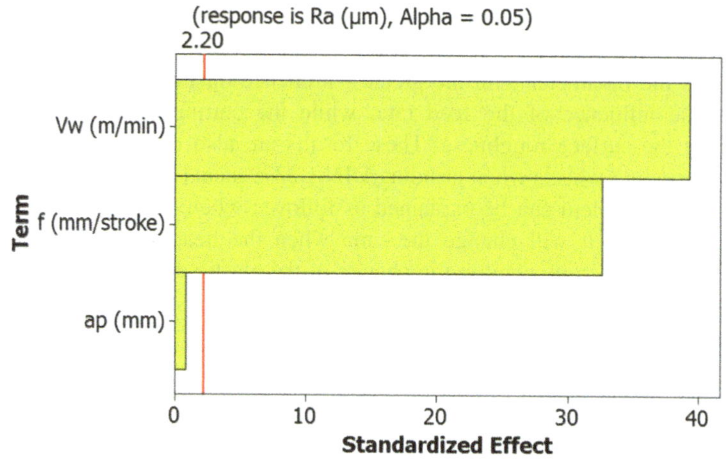

Fig. 1. Parato chart of the influence of the parameters on the surface roughness.

$$R_a = 0.09388 + 0.02115 \cdot v_w + 0.02606 \cdot f + 1.12500 \cdot a_p - 0.00021 \cdot v_w^2 \\ - - 120 \cdot a_p^2 + 0.00028 \cdot v_w \cdot f - -0.01 \cdot v_w \cdot a_p + 0.55 \cdot f \cdot a_p \quad (1)$$

This regression model was used to perform the determination of the optimal value of the cutting parameters. To solve the optimization problem to determine the value of the cutting parameters so as to ensure the minimum surface roughness, it is required to define the restricting domain of the parameters. Specifically, the value of the cutting parameters must be within their investigated range. Then the optimization problem is written as in Eq. (2).

$$\begin{cases} R_a = f(v_w, f, a_p) \rightarrow min \\ R_a > 0 \\ 5(m/min) \le v_w \le 15(m/min) \\ 2(mm/stroke) \le f \le 6(mm/stroke) \\ 0.01(mm) \le a_p \le 0.02(mm) \end{cases} \quad (2)$$

Genetic Algorithm has been used to solve the optimization problem. The parameters of the algorithm include the number of populations of 100, with the probability of hybridization of 0.25, and the probability of mutation of 0.05, which are the values chosen according to the studies [22–24]. The algorithm has been written in Visual Basic code in Excel. In Fig. 2, the graph of the adaptation function against the surface roughness objective function is presented. The optimal values of the cutting parameters as well as the values of the output parameters are shown in Table 4.

The optimal values set of the input parameters in Table 4 was used to repeat the experiments to verify the optimal results. Through five experiments that are carried out, the value of the output parameters is also shown in Table 4. From the results of comparing the value of the surface roughness when doing experiment with and predicting by genetic algorithm in this table, it shows that the experimental and predicted values are very close. The average deviation between the predicted results and the experimental results is only 5.79%, which shows that the optimal value of the input parameters has been found to have a very high reliability degree. Thus, the value of the cutting parameters in Table 4 is the set of values to be determined.

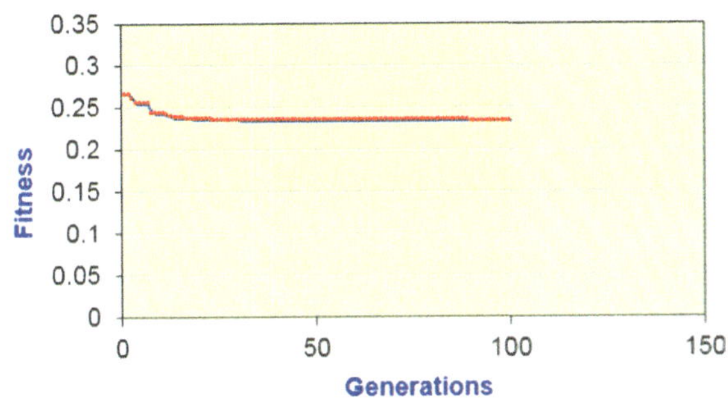

Fig. 2. The graph of the adaptation of R_a

Table 4. Surface roughness when doing experiment based on the optimal values of the cutting parameters

No.	Optimization value of cutting parameters			Ra (μm)		Deviation (%)
	v_w *(m/min)*	*f (mm/stroke)*	a_p *(mm)*	Experimented	Predicted	
1	5.000	2.000	0.020	0.251	0.234	6.77
2				0.248		5.65
3				0.248		5.65
4				0.246		4.88
5				0.249		6.02
Mean				0.248		5.79

4 Conclusions

In this study, the C250 steel grinding experiment was conducted with the segmented grinding wheel, in which the grinding wheel material is aluminum oxide, the number of grooves on the wheel is 18. At each experiment, the values of the velocity of the workpiece, the feed rate and the cutting depth were changed. Surface roughness has been selected as the output parameter of the experimental process. Genetic algorithm has been applied to solve the optimization problem. Some conclusions are drawn as follows:

When the segmented grinding wheels were initially used to grind C250 steel, the results showed that the surface roughness when grinding in this case is also suitable for most general grinding methods. In terms of the surface roughness, the segmented grinding wheels are perfectly suitable for grinding this steel. Using the segmented aluminum oxide grinding wheel will save much cost compared to using CBN grinding wheel.

Among the three cutting parameters investigated, the velocity of the workpiece is the parameter that has the greatest influence on the surface roughness, followed by the influence of the feed rate. The cutting depth has no significant influence on the surface roughness. The influence of three mode parameters on the surface roughness when grinding C250 steel is similar to that when grinding DIN 1.2379 steel and SAE420 steel.

A regression model has been developed, showing the relationship between the input parameters and the surface roughness. This model has a very high compatibility with the experimental data.

For the surface roughness to have the minimum value, the velocity of the workpiece, the feed rate and the cutting depth are 5 m/min, 2 mm/stroke and 0.02 mm, respectively.

This study and the studies published [16, 17] only consider the number of grooves on the grinding wheel, but do not pay attention to the size of the grooves as well as the grinding wheel material, the number of grinding wheel dressing parameters and the lubricating cooling technology parameters, etc. Therefore, in the further studies, these parameters will need to be considered. In addition to the surface roughness investigated as in this article, the SEM images to analyze the surface morphology and the parameters of cutting force, grains wear, etc. should also be performed to evaluate the grinding process more generally when using the segmented grinding wheel.

Acknowledgment. This work was supported by Thai Nguyen University of Technology.

References

1. Guo, Z., Sha, W., Li, D.: Quantification of phase transformation kinetics of 18 wt.% Ni C250 maraging steel, Materials Science and Engineering: A, vol. Investigation of Temperature and Energy partition during Maraging Steel Micro-grinding, (1–2), 10–20 (2004). https://doi.org/10.1016/j.msea.2004.01.040
2. Ding, Z., Li, B., Fergani, O., Shao, Y., Liang, S.Y.: Investigation of temperature and energy partition during maraging steel micro-grinding. Procedia CIRP **56**, 284–288 (2016). https://doi.org/10.1016/j.procir.2016.10.084
3. Varghese, V., Jagmalpuria, A., Badiger, P.V., Ramesh, M.R.: Optimisation of machining parameters for end milling of maraging steel MDN 250 using TiAlSiN and TiSiN coated WC-Co inserts. AIP Conf. Proc. **2204**(040031), 1–14 (2020). https://doi.org/10.1063/1.5141604
4. Guo, Y.B., Sahni, J.: A comparative study of hard turned and cylindrically ground white layers. Int. J. Mach. Tools Manuf **44**, 135–145 (2004). https://doi.org/10.1016/j.ijmachtools.2003.10.009

5. Ding, Z., Li, B., Zou, P., Liang, S.Y.: Material phase transformation during grinding. Adv Mat Res **1052**, 503–508 (2014). https://doi.org/10.4028/www.scientific.net/AMR.1052.503
6. Malkin, S., Guo, C.: Grinding Technology: Theory and Applications of Machining with Abrasives. Industrial Press, New York (2008)
7. Ding, Z., Li, B., Liang, S.Y.: Phase transformation and residual stress of Maraging C250 steel during grinding. Mater. Lett. **154**, 37–39 (2015). https://doi.org/10.1016/j.matlet.2015.04.040
8. Dung, H.T., Trung, D.D., Thien, N.V., Ky, L.H., Sonpheth, K.: Influence of Lubricant Parameters on Surface Roughness of Workpiece When Grinding SKD11 Steel. In: Sattler, K.-U., Nguyen, D.C., Vu, N.P., Tien Long, B., Puta, H. (eds.) ICERA 2019. LNNS, vol. 104, pp. 436–447. Springer, Cham (2020). https://doi.org/10.1007/978-3-030-37497-6_50
9. Guha, S., Das, P.P., Chakraborty, S.: Improvement in the performance with less stiff air layer formation around the rubber tube-pasted grinding wheel. J Mech. Eng. Sci. **233**(15), 5175–5189 (2019). https://doi.org/10.1177/0954406219844534
10. Kapłonek, W., Nadolny, K., Sutowska, M., Mia, M., Pimenov, D.Y., Gupta, M.K.: Experimental studies on MoS2-treated grinding wheel active surface condition after high-efficiency internal cylindrical grinding process of INCONEL ® alloy 718. Micromachines. **10**(255) 2019. https://doi.org/10.3390/mi10040255
11. Kapłonek, W., et al.: Internal cylindrical grinding process of INCONEL® alloy 600 Using grinding wheels with sol–gel alumina and a synthetic organosilicon polymer-based impregnate. Micromachines **11**(2), 115 (2020). https://doi.org/10.3390/mi11020115
12. Lee, K.W., Wong, P.K., Zhang, J.H.: Study on the grinding of advanced ceramics with slotted diamond wheels. J. Mater. Process. Technol. **100**(1–3), 230–235 (2000). https://doi.org/10.1016/S0924-0136(00)00403-9
13. Jin, D.X., Meng, Z.: Research for discontinuous grinding wheel with multi- porous grooves. Key Eng. Mater. **259**, 117–121 (2004). https://doi.org/10.4028/www.scientific.net/KEM.259-260.117
14. Fan, X., Miller, M.H.: Force analysis for grinding with segmental wheels. Mach. Sci. Technol. **10**(4), 435–455 (2006). https://doi.org/10.1080/10910340600996142
15. Handigund, P.B., Miller, M.H.: Abrasive Wear and Forces in Grinding of Silicon Carbide. Michigan Technological University, Houghton, MI (2001)
16. Trung, D.D., Thien, N.V., Nguyen, N.-T.: Application of TOPSIS method in multi-objective optimization of the grinding process using segmented grinding wheel. Tribol. Ind. **43**(1), 12–22 (2021). https://doi.org/10.24874/ti.998.11.20.12
17. Trung, D.D., Nguyen, N.–T., Tien, D.H., Dang, H.L.: A research on multi-objective optimization of the grinding process using segmented grinding wheel by Taguchi-Dear method, EUREKA: Phys. Eng. (1) 67–77, (2021). https://doi.org/10.21303/2461-4262.2021.001612
18. Phuong, N.T., Giang, N.T.P., Dong, N.T.: A research on the affect of technologycal parameters on cutting temperature when machining use segmented grinding wheel. International Journal of Electronics Communication and Computer Engineering **8**(3), 208–212 (2017)
19. Marinescu, L.D., itchiner, M., Uhlmann, E., Rowe, W.B.: Handbook of Machining with Grinding Wheels. CRC Press (2006)
20. Dean, A., Voss, D. (eds.): Design and Analysis of Experiments. Springer-Verlag, New York (1999). https://doi.org/10.1007/b97673
21. A. Dean, D. Voss, D. Draguljić, *Design and Analysis of Experiments - Second Edition*, Springer, 2007

22. Khoi, P.B., Trung, D.D., Cuong, N., Man, N.D.: Research on optimization of plunge centerless grinding process using genetic algorithm and response surface method. Int. J. Sci. Eng. Technol. **4**(3), 207–211 (2015). https://doi.org/10.17950/ijset/v4s3/319
23. Saravanan, R., Sachithanandam, M.: Genetic algorithm (GA) for multivariable surface grinding process optimisation using a multi – objective function model. Int. J. Adv. Manuf. Technol. **17**, 330–338 (2001). https://doi.org/10.1007/s001700170167
24. Trung, D.D., Cuong, N., Man, N.D., Tuan, N.Q., Khoi, P.B.: Multi-objective optimization of centerless grinding process for 20X– carbon infiltration steel by genetic and generalized reduced gradient algorithms. Am. J. Eng. Res. **6**(1), 91–96 (2017)

The Influence of Drivetrain Layout on Lateral Vehicle Dynamic

Nguyen Khac Minh[1,2], Do Tien Dung[2], and Nguyen Khac Tuan[2(✉)]

[1] Moscow Automobile and Road Construction State Technical University (MADI), № 64, Leningradsky Prospect, Moscow 125319, Russia
[2] Thai Nguyen University of Technology, № 666 Tich Luong Ward, Thai Nguyen 24131, Vietnam
tuannkcn@tnut.edu.vn

Abstract. In this paper, a mathematical model describing the vehicle dynamics is built based on a nonlinear tire model and vehicle dynamic model of longitudinal and lateral motion. The influence of two drivetrain layouts FWD and RWD on vehicle lateral dynamics are considered. The analysis is done in MATLAB/SIMULINK environment. The results confirm that the inertia and lateral forces generated at the tires when the vehicle is cornering depend on many different parameters, in which these forces have a proportional relationship with the vehicle velocity and steering angle. The greater the vehicle velocity and/or steering angle, the greater the inertia and lateral forces generated. Besides that, when cornering, the lateral forces of the tires and the inertia forces in the longitudinal and transversal directions of the FWD vehicle are greater than those of RWD vehicle.

Keywords: Lateral dynamics · Drivetrain layout · Simulation · Lateral force · Inertia force · Non-linear tire model

1 Introduction

In recent years, improving vehicle safety and performance are issues of particular concern. The lateral dynamics directly affect the safety of vehicles, so this research direction has attracted the attention of scientists [1–8].

The drivetrain plays an important role in the overall performance of the vehicle. Especially, it has a significant influence on the dynamic characteristics, fuel consumption, and stability of the vehicle [9, 10]. Currently, there are many types of vehicle drivetrain layout, however, the two most commonly used options are front-wheel drive (FWD) and rear-wheel drive (RWD) [10].

Analysis of published literature related to vehicle powertrains shows that these works mainly focus on studying dynamic loads generated in drivetrains [10–12], or survey acceleration characteristics, fuel consumption for vehicles equipped with different drivetrains [13–15].

The literature also shows that there are very few studies related to the influence of drivetrain layout on vehicle dynamics and stability… The objective of this paper is to

D. C. Nguyen et al. (Eds.): ICERA 2021, LNNS 366, pp. 507–516, 2022.
https://doi.org/10.1007/978-3-030-92574-1_53

study the influence of the drivetrain on the vehicle's lateral dynamics, especially focusing on the side tire force and inertia force generated when the vehicle is cornering.

1.1 Tire Model

When the vehicle is in motion, most of the forces interact with the vehicle through the tire at the contact area between the tire and the road surface. Understanding the nature of this interaction is fundamental to controlling vehicle processes such as turning, braking, or accelerating. Figure 1 shows the dynamics of the tire [16].

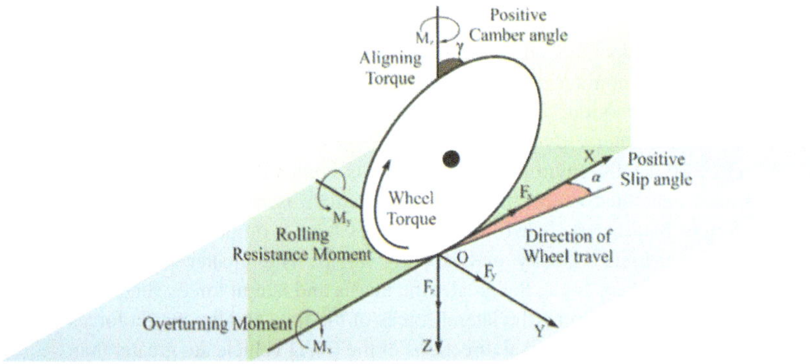

Fig. 1. The dynamics of the tire

The origin of the coordinate system is the center of contact between the tire and the road surface. The x-axis is positive in the forward direction, the y-axis is positive to the right, and the z-axis is positive down. F_x – tractive force is the component in the x-direction of the resultant force exerted on the tire by the road; F_y – lateral force is the component in the y-direction; F_z – normal force is the component in the z-direction; α – slip angle is the angle formed between the direction of the velocity of the center of the tire and the plane x-z, and other parameters have been described in Fig. 1.

To describe the nonlinear tire model, several empirical formulas can be used, typically the Magic Formula. In essence, the characterization of the tire is an approximation problem of an unknown analytic function with specific constraints. On the basis of experimentally building a nonlinear tire model, the lateral force is a function of the variables α and F_z expressed in the following formula [16–19]:

$$F_y(\alpha, F_z) = \mu F_z \left[1 - \exp\left(-\frac{C_\alpha \alpha}{\mu F_z}\right)\right] \tag{1}$$

In which:

F_y – lateral force; F_z – normal force; α – slip angle; μ – represents the ratio between the normalized tire friction and the lateral force; C_α – cornering stiffness.

2 Vehicle Dynamics Model

2.1 Model Description

The mathematical model of the vehicle is a combination of the nonlinear tire model and vehicle dynamics model with consideration of the longitudinal and lateral motion. Usually, there are many degrees of freedom involved in vehicle dynamics, and the choice of how many variables to analyze will depend on the purpose of the study. Figure 2 depicts the Single track model used in this study.

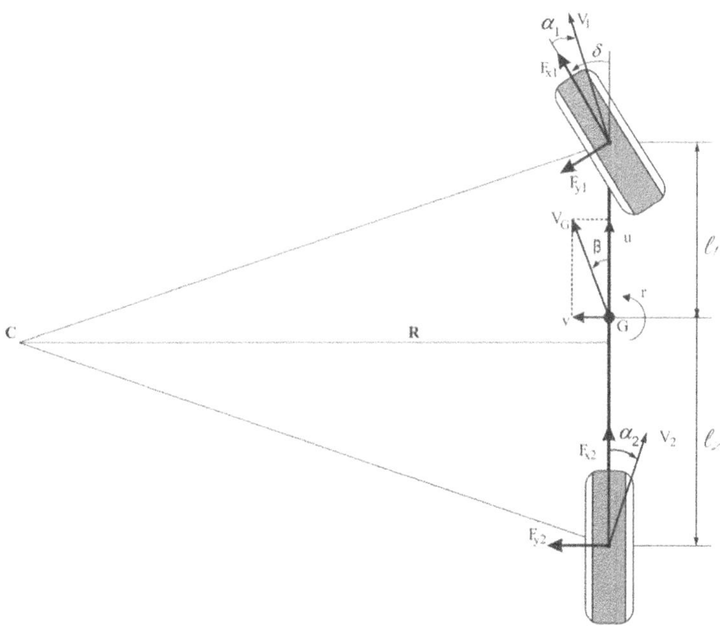

Fig. 2. The Single track model

In Fig. 2 the vehicle is simulated as a mass rigid body having mass m and center of gravity G; the center gravity of the vehicle located distances l_1 from the front axle and l_2 from the rear axle; C – instantaneous rotation center; u – forward velocity; v – lateral velocity; r – yaw velocity; F_{x1} – rolling resistance at front tire; F_{x2} – rolling resistance at rear tire; F_{y1} – lateral force at front tire; F_{y2} – lateral force at rear tire; α_1 – slip angles of the front tire; α_2 – slip angles of the rear tire; δ – steer angle front wheels; β – vehicle slip angle.

The survey was conducted with the following assumptions: the vehicle is considered to be a rigid body; the vehicle is symmetric through the z-x plane; constant vehicle parameters; only consider motion in the x-y plane; good quality road surface; no delay in lateral force generation; vehicles without suspension; vehicles with stiff steering, such that the wheel rotation angle is equivalent to the steering angle. The equilibrium equation is written on the basis of Newton's Law of a rigid body moving in a plane:

$$\begin{aligned} \sum X &= m.a_x \\ \sum Y &= m.a_y \\ \sum Z &= J.\dot{r} \end{aligned} \tag{2}$$

Where: ΣX, ΣY and ΣZ are the sum of longitudinal, transversal and rotational forces respectively. a_x – longitudinal acceleration; a_y – lateral acceleration; mass m and moment of inertia J. With the assumption that the vehicle is symmetric about the x-z plane, and that the torques delivered to the two wheels of the same axis are equal, expression (2) is developed:

$$\begin{aligned} m(\dot{u} - vr) &= F_x1 - F_y1\delta + F_x2 - F_{xa} \\ m(\dot{v} + ur) &= F_x1\delta + F_y1 + F_y2 \\ J.\dot{r} &= (F_x1\delta + F_y1)l_1 - F_y2l_2 \end{aligned} \tag{3}$$

We can consider that the longitudinal velocity is always bigger than the lateral one cause by the yawning rate [15, 19]. Mathematically, we have:

$$u = |v + rl_1|$$

$$u = |v - rl_2|$$

and so, we can approximately change the arc with its tangent and have relations:

$$\delta - \alpha_1 = \frac{v + rl_1}{u}$$

$$-\alpha_2 = \frac{v - rl_2}{u}$$

Finally, we have:

$$\begin{aligned} \alpha_1 &= \delta - \frac{v + rl_1}{u} \\ \alpha_2 &= -\frac{v - rl_2}{u} \end{aligned} \tag{4}$$

And referring to the tyre model, following equation described Non-Linear Tyre Model with Relaxation Length is adopted [15]:

$$\frac{d}{u}\dot{F}_{yi} + F_{yi} = \mu F_{zi}\left\{1 - \exp\left[-\frac{C_{\alpha i}\alpha_i}{\mu F_{zi}}\right]\right\} \tag{5}$$

in which: d – relaxation length; $C_{\alpha 1}$ – cornering stiffness front tyre; $C_{\alpha 2}$ – cornering stiffness rear tyre.

Where the subscript is i = *1,2*

From the above analysis, the vehicle lateral dynamics model of the rear-wheel drive vehicle (RWD) combined with Non-Linear Tyre Model with Relaxation Length is shown by the following equations:

$$\begin{aligned} m(\dot{v}+ur) &= F_{y1}+F_{y2} \\ J.\dot{r} &= F_{y1}l_1 - F_{y2}l_2 \\ \tfrac{d}{u}\dot{F}_{y1}+F_{y1} &= \mu F_{z1}\left\{1-\exp\left[-\frac{C_{\alpha 1}}{\mu F_{z1}}\left(\delta-\frac{v+rl_1}{u}\right)\right]\right\} \\ \tfrac{d}{u}\dot{F}_{y2}+F_{y2} &= \mu F_{z2}\left\{1-\exp\left[\frac{C_{\alpha 2}}{\mu F_{z2}}\left(\frac{v-rl_2}{u}\right)\right]\right\} \end{aligned} \tag{6}$$

In Eq. (6) we consider the longitudinal velocity constant and rear traction. In this case, the equations are uncoupled and the first equation is an algebraic one. However, the mathematical problem is uncoupled only when the steer and traction have been at the different axis, but in this study, it happens only in the rear traction case. Also, note that F_{x2} does not appear in equilibrium equations even if it is the traction force applied by the tyre during the vehicle motion.

For the front-wheel drive(FWD) vehicle, the mathematical model described the vehicle lateral dynamics as follows:

$$\begin{aligned} m(\dot{v}+ur) &= F_{y1}+F_{y2}+F_{x1}\delta \\ J.\dot{r} &= F_{y1}l_1 - F_{y2}l_2 + F_{x1}l_1\delta \\ \tfrac{d}{u}\dot{F}_{y1}+F_{y1} &= \mu F_{z1}\left\{1-\exp\left[-\frac{C_{\alpha 1}}{\mu F_{z1}}\left(\delta-\frac{v+rl_1}{u}\right)\right]\right\} \\ \tfrac{d}{u}\dot{F}_{y2}+F_{y2} &= \mu F_{z2}\left\{1-\exp\left[\frac{C_{\alpha 2}}{\mu F_{z2}}\left(\frac{v-rl_2}{u}\right)\right]\right\} \end{aligned} \tag{7}$$

In which the tractive force F_{x1} at the front wheel has been described by the following relationship:

$$F_x 1 = m(\dot{u}-vr)+F_{y1}\delta+F_{xa} \tag{8}$$

In (6), F_{xa} is the aerodynamics force applied by the air resistance, can be calculated by the following equation:

$$F_{ax} = \frac{1}{2}C_x.\rho.u^2.A \tag{9}$$

where

C_x: drag coefficient

ρ: air density

A: projected area of a vehicle from the front.

Knowing v,u, and r, the acceleration in the longitudinal and lateral direction can be found by:

$$\begin{aligned} a_y &= (\dot{u}-vr) \\ a_x &= (\dot{v}+ur) \end{aligned} \tag{10}$$

Therefore, inertia forces in X and Y directions can be determined as follows:

$$\begin{aligned} F_{ay} &= m(\dot{v} + ur) \\ F_{ax} &= m(\dot{u} - vr) \end{aligned} \tag{11}$$

2.2 Simulation Setup

A simulation model based on the mathematical model of FWD and RWD vehicles (Eqs. (6) and (7)) is designed using Matlab – Simulink. The main parameters of the model are provided in Table 1:

Table 1. Parameters and values

Parameters	Units	Symbol	Values
Mass of vehicle	kg	m	1362
Longitudinal velocity x-direction	m/s	u	20
Moment of inertia	kg.m^2	J	1623.8
Distance from center of gravity to front axle	m	l_1	0.9552
Distance from center of gravity to rear axle	m	l_2	1,5128
Relaxation length	m	d	0.25
Coefficient of road adhesion	–	μ	0.9
Cornering stiffness front tyre	Ns/rad	$C_{\alpha 1}$	84085
Cornering stiffness rear tyre	Ns/rad	$C_{\alpha 2}$	87342
Steer angle front wheels	rad	δ	0.085

To study the influence of the drivetrain layout on the vehicle lateral dynamics, the author first performed a computational simulation, calculating the lateral force at the wheels and the force of inertia in the X and Y directions. The input parameters when calculating are selected as shown in Table 1. In the next step, the authors perform simulations and calculate these force values corresponding to the vehicle speed changing from 10 m/s to 30 m/s and three steering angles $\delta_1 = 0.1$; $\delta_2 = 0.2$ and $\delta_3 = 0.3$ rad/s, respectively.

3 Simulation Results and Discussion

Figure 3 shows the calculation results of the lateral force of the front tires F_{y1} and the rear tires F_{y2}, Fig. 4 is the result of calculating the inertia force F_{ax}, F_{ay} in the X and Y directions respectively to the input parameters as shown in Table 1.

The obtained simulation results show that the variations of forces with time are the same for both FWD and RWD. In the early stage before reaching the maximum value, the inertia forces in X and Y directions and lateral forces of the tires tend to increase rapidly, after reaching the maximum value, these forces decrease and quickly reach a steady state. However, for FWD vehicles, the corresponding forces are larger than those of RWD vehicle. Specifically, for the input data as shown in Table 1, the maximum lateral force for the FWD and RWD are $F_{y1f\_max} = 2649N$ and $F_{y1r\_max} = 2610N$ respectively, i.e. the largest lateral force of the front tires for the FWD vehicle is about 1.5% larger than the RWD vehicle. Similarly, the value of the lateral force of the rear tires, the inertia forces in the X and Y direction of the FWD vehicle are 4.97%, 13.7%, and 5.17% larger than those of the RWD vehicle, respectively.

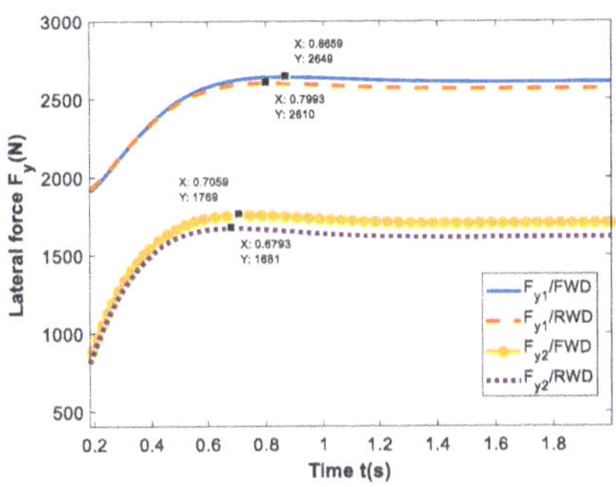

Fig. 3. Lateral force-time curve

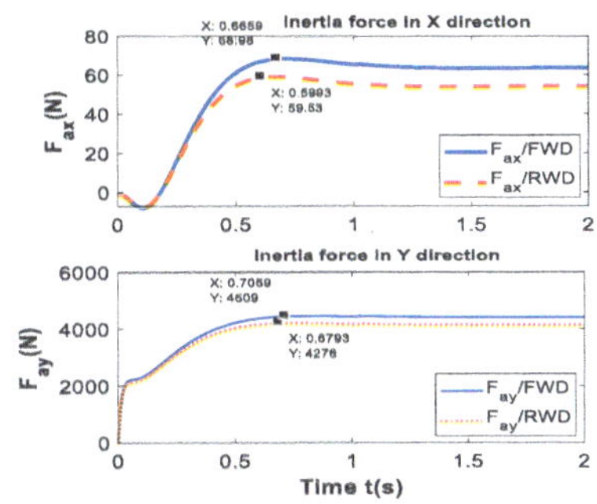

Fig. 4. Inertia force-time curve

Figure 5 shows the relationship between the maximum inertia force in the X and Y directions when the vehicle cornering at different speeds (10–30 m/s). When calculating, the steering angle is taken $\delta 1 = 0.2$ rad, other parameters are kept the same as in Table 1. Analysis of the results in Fig. 5 shows that when vehicle cornering, the maximum value of the inertia forces in the X and Y directions are proportional to the velocity of the vehicle. The greater the vehicle velocity, the greater the inertia forces generated and vice versa. However, the value of inertia force in the X direction is much smaller than that in the Y direction. In addition, the maximum value of the inertia force in both the X and Y directions of the FWD vehicle is larger than those of the RWD vehicle.

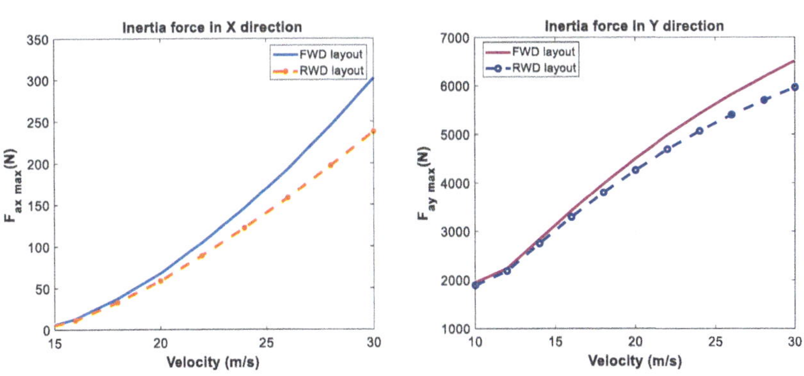

Fig. 5. Maximum inertia force in X and Y directions vs vehicle velocity

Figures 6 and 7 show the results of calculating the relationship between the horizontal force of the tire and the time corresponding to different speeds and steering angles of the vehicle. From these figures, we can see that the variation of lateral forces follows the same law for both FWD and RWD vehicles. At small steering angles, the lateral forces of the front and rear tires of the FWD vehicle are greater than those of the RWD vehicle. However, at large steering angles, the lateral forces of the tires are not significantly different between the two drivetrain layouts. Furthermore, the value of the lateral force of the front and rear tires is directly related to the vehicle velocity and steering angle. The greater the speed or steering angle the greater the lateral force of the tires generated and vice versa.

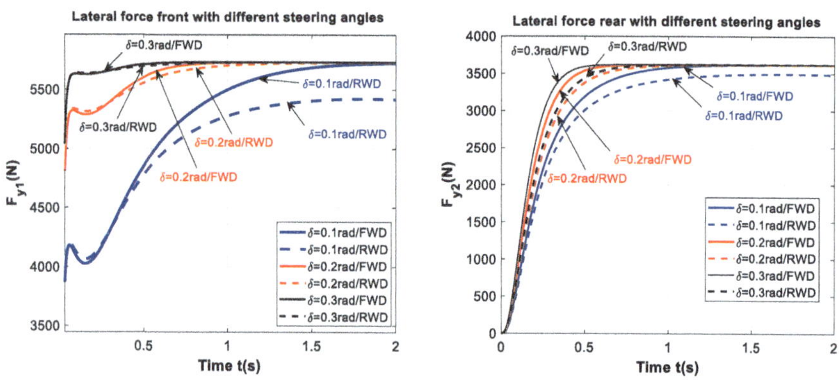

Fig. 6. Lateral force-time curve with different steering angles

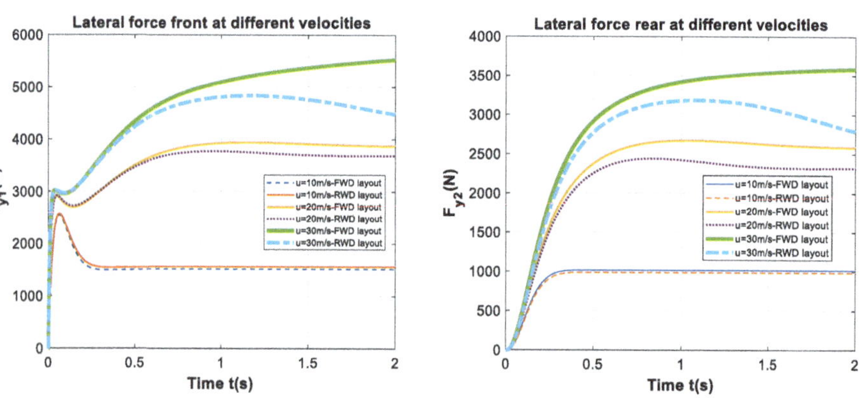

Fig. 7. Lateral forces – time curve with different vehicle velocities

4 Conclusions

In this paper, a mathematical model describing the vehicle dynamics is built based on a nonlinear tire model and vehicle dynamic model of longitudinal and lateral motion. The influence of two drivetrain layouts FWD and RWD on vehicle lateral dynamics are considered. The analysis is done in MATLAB/SIMULINK environment. The main following conclusions are drawn from the results:

1. The inertia and lateral forces generated at the tires when the vehicle is cornering depend on many different parameters, in which these forces have a proportional relationship with the vehicle velocity and steering angle. The greater the vehicle velocity and/or steering angle, the greater the inertia and lateral forces generated;
2. When cornering, the lateral forces of the tires and the inertia forces in the longitudinal and lateral directions of the FWD vehicle are greater than those of the RWD vehicle;
3. For both FWD and RWD vehicles, the lateral forces of the front tires are always greater than those of the rear tires.

Acknowledgment. This research was funded by Thai Nguyen University of Technology.

References

1. Sierra, C., Tseng, E., Jain, A., Peng, H.: Cornering stiffness estimation based on vehicle lateral dynamics. Veh Sys Dyn **44**(1), 24–38 (2006)
2. Li, L., Wang, F.Y., Zhou, Q.: Integrated longitudinal and lateral tire/road friction modeling and monitoring for vehicle motion control. IEEE Tran. Intell. Transp. Syst. **7**(1), 1–19 (2006). https://doi.org/10.1109/TITS.2005.858624
3. Beji, L., Bestaoui, Y.: Motion generation and adaptive control method of automated guided vehicles in road following. IEEE Trans. Intel. Transp. Syst. **6**(1), 113–123 (2005). https://doi.org/10.1109/TITS.2004.833758
4. Liaw, D.-C., Chung, W.-C.: A feedback linearization design for the control of vehicle's lateral dynamics. Nonlinear Dyn. **52**(4), 313–329 (2008). https://doi.org/10.1007/s11071-007-9280-8
5. Kiencke, U., Daiß, A.: Observation of lateral vehicle dynamics. Control Eng. Pract. **5**(8), 1145–1150 (1997). https://doi.org/10.1016/S0967-0661(97)00108-1
6. Ackermann, J., Guldner, J., Sienel, W., Steinhauser, R., Utkin, V.I.: Linear and nonlinear controller design for robust automatic steering. IEEE Trans. Control Syst. Technol. **3**(1), 132–143 (1995). https://doi.org/10.1109/87.370719
7. Akar, M., Kalkkuhl, J.C.: Lateral dynamics emulation via a four-wheel steering vehicle. Veh. Syst. Dyn. **46**(9), 803–829 (2008). https://doi.org/10.1080/00423110701632925
8. Wielitzkam M., Dagen, M., Ortmaier, T.: State estimation of vehicle's lateral dynamics using Unscented Kalman Filter. The IEEE Conference on Decision and Control, Los Angeles, USA, 5015–5020 (2014)
9. Cornel Stan.: Alternative Propulsion for Automobiles, Springer (2016)
10. Hoan, N.T., Tuan, N.K.: Automotive Transmission System. Vietnam Educational Publishing House (2018)

11. Tsitovich, I.S., Algin, V.B.: Vehicle Dynamics. Minsk: Nauka and Tekhnika, (Monograph in Russian) (1981)
12. Tuan, N.K., Quynh, L.V.: Modeling and simulation of vehicle verticalvibration from powertrain and road excitation. In: International Symposium on Technology for Sustainability, Bangkok, Thailand, 512–515 (2012)
13. Doumiat, M. et al.: A method to estimate the lateral tire force and the sideslip angle of a vehicle: Experimental validation. In: IEEE 2010 American Control Conference, USA, 6936–6942 (2010)
14. Rajamani, R.: Vehicle Dynamic and Control. Springer (2006)
15. Tarasik, V.P. (2006) The Theory of Automobile Movement SPb: BHV-Petersburg
16. David. A. (ed.) Crolla.: Automotive Engineering – Powertrain, Chassis System and Vehicle Body. 1st ed. United States of America (2009)
17. Guiggiani, M.:The Science of Vehicle Dynamic, Springer (2017)
18. Genta, G.: Motor Vehicle Dynamics. World Scientific Publishing, Singapore (1997)
19. Wong, J.Y.: Theory of Ground Vehicles, 4th edn. John Wiley and Sons Inc, New York (2008)

Influence of EDM Factors on Surface Roughness and Material Removal Speed When Machining SKD11 Steel

Trinh Kieu Tuan[1], Do Thi Tam[2], Tran Ngoc Huy Thinh[3], Le Hoang Anh[4], Nguyen Huu Quang[1], Bui Thanh Danh[5], Tran Ngoc Giang[2], and Nguyen Manh Cuong[2(✉)]

[1] University of Economics - Technology for Industries, Hanoi, Vietnam
[2] Thai Nguyen University of Technology, Thai Nguyen, Vietnam
nmcuong@tnut.edu.vn
[3] Nguyen Tat Thanh University, Ho Chi Minh City, Vietnam
[4] Vinh Long University of Technology Education, Vin Long, Vietnam
[5] University of Transport and Communications, Hanoi, Vietnam

Abstract. In this paper, the results of a multi-objective optimization problem of electric discharge machining (EDM) of cylindrically shaped parts are reported. Two objectives, surface roughness (SR) and material removal speed (MRS), were selected for the multi-objective problem. In addition, SKD11 steel and 4 main EDM parameters, including the pulse on time, the pulse off time, the servo current, and the servo voltage, were investigated. The influence of these parameters on the SR and the MRS was analyzed using Taguchi and Gray Relation Analysis methods. Moreover, the optimal input factors to ensure simultaneous two targets, the SR and the MRS, have been shown. The proposed model has been also evaluated with the conclusion that it is completely consistent with the experimental data.

Keywords: EDM · Surface roughness · Material removal speed · Multi-objective optimization · Cylindrical shaped parts

1 Introduction

In fact, EDM research is always an exciting topic for scientists. This is because it is the most commonly used non-traditional machining method. Until now, EDM studies have focused heavily on the optimization of machining processes. EDM process optimization problem has been investigated with single-objective and multi-objective problems. There are many single-objective problems that have been investigated, such as determining the optimal pulse mode to achieve the minimum SR [1–4], the largest material removal rate (MRR) [5, 6], the smallest electrode wear rate (EWR) [6–8], etc. The multi-objective problem is also studied by many researchers. These problems have been established to determine the optimal pulse mode to satisfy multiple objectives simultaneously, such as SR and MRR [9, 10], Ra and EWR [11], the SR and fractal dimensions [12], MRR and electrode wear, and SR, MMR and EWR [13–16], etc.

D. C. Nguyen et al. (Eds.): ICERA 2021, LNNS 366, pp. 517–526, 2022.
https://doi.org/10.1007/978-3-030-92574-1_54

EDM process optimization problems have been solved by various methods such as the Taguchi method [4, 7, 8], response surface methodology (RSM) [2, 3, 5, 6], Taguchi method and Gray Relational Analysis [9, 11, 16], TOPIS method [12, 13], or desirability-based multi-objective particle swarm optimization (DMPSO) [14, 15], etc.

The aforementioned studies all focused on machining concave parts. Recently, there have been a number of studies on processing cylindrical shaped parts [3, 4, 11, 17]. However, these studies were carried out with 90CrSi steel.

This study presents the results of a multi-objective optimization study with two single objectives, SR and MRS, when machining cylindrical shaped parts with workpiece material is SKD11 steel. The influence of the main EDM parameters on the objective function was analyzed. In addition, the optimal input factors in achieving the minimum SR and maximum MRS simultaneously has been proposed.

2 Experimental Setup

To evaluate the influence of the input parameters on SR and MRS, four EDM process parameters were selected with survey levels described in Table 1. The experimental matrix was designed based on survey levels of variables by using the Taguchi method with L9 (3^4) orthogonal array design with the use of Minitab R19 software. The experimental matrix, the measurement results of the SR and the MRS are described in Table 2. The experimental setup is described as shown in Fig. 1. In which, the workpiece is made of SKD11 steel, and the electrode is made of copper. In addition, a sinker EDM machine with model CNC-AG40L was used for the experiment.

Table 1. Input parameters and levels

No	Parameters	Code	Unit	Level		
				1	2	3
1	Pulse on time	T_{on}	μs	6	16	26
2	Pulse off time	T_{off}	μs	8	18	28
3	Peak current	IP	A	3	7	11
4	Servo voltage	SV	V	3	6	9

To optimize multi-objectives, the gray relation analysis (GRA) method combined with Taguchi was used. Also, two single objectives are carried out, which is the minimum SR and the maximum MRS.

The S/N index of Ra and MRS is calculated according to the following formulas:

With the minimum SR:

$$S/N = -10log_{10}\left(\frac{1}{n}\sum\nolimits_{i=1}^{n} y_i^2\right) \tag{1}$$

With the maximum MRS:

$$S/N = -10log_{10}\left(\frac{1}{n}\sum\nolimits_{i=1}^{n}\frac{1}{y_i^2}\right) \quad (2)$$

Table.2 Experimental plan and measured SR and MRS

Exp. No.	T_{on}	T_{off}	IP	SV	R_a (μm)	MRS (g/h)
1	6	8	3	3	1.6067	0.7693
2	6	18	7	6	2.1993	1.1348
3	6	28	11	9	1.8387	0.9634
4	16	8	7	9	3.1843	2.6826
5	16	18	11	3	2.3107	2.8785
6	16	28	3	6	2.0447	0.6974
7	26	8	11	6	2.9880	0.7622
8	26	18	3	9	2.0310	0.5853
9	26	28	7	3	4.1337	1.6701

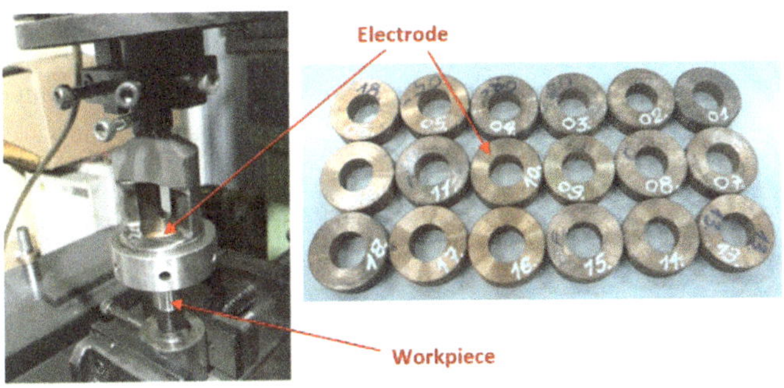

Fig. 1. Experimental setup

The calculated S/N ratios of the two output targets are described in Table 3.

To be able to analyze gray relationships based on S/N values, it is required to convert to comparison series with dimensionless quantities. Therefore, it is necessary to normalize the data. Here, the S/N index calculation results are normalized in the range (0 ÷ 1). The S/N index is normalized by the Zi value with 0 ≤ Zi ≤ 1. Zi is determined according to the following formula to avoid the effect of using different dimensions and to reduce variability:

$$Z_i = \frac{S/N_i - min(S/N_i, i = 1, 2, \ldots, n)}{max(S/N_i, i = 1, 2, \ldots, n) - min(S/N_i, i = 1, 2, \ldots, n)} \quad (3)$$

Where n is the number of experiments, n = 9.

Determine the value of gray relation:

The gray relation coefficient $y_j(k)$ is determined by the formula:

$$y_i(k) = \frac{\Delta_{min}(k) + \xi \cdot \Delta_{max}(k)}{\Delta_i(k) + \xi \cdot \Delta_{max}(k)} \quad (4)$$

With i = 1, 2, …, n.

In which:

- n: is the number of experiments (n = 9); k is the number of output targets (k = 2);
- $\Delta_i(k)$ is the absolute deviation of the reference sequence which is determined by the formula:
- $\Delta_i(k) = \|Z_0(k) - Z_i(k)\|$, is the absolute value of the difference between $Z_0(k)$ (reference value Z0(k) = 1) and $Z_i(k)$) (Z-value of the ith experiment of the kth target).
- $\Delta_{min}(k)$ is the minimum value of i(k)
- max(k) is the maximum value of i(k)
- ζ is the discriminant coefficient, determined in the range $0 \leq \zeta \leq$, in experimental research $\zeta = 0.5$.

Table 3. S/N, S/N Zi, Δ_i (k) and GRV $\overline{y_i}$

Exp. No	S/N		Zi		Δ_i (k)		Gray relational value yi		$\overline{y_i}$
	Ra	MRS	$\overline{Ra}$	MRS	$\overline{Ra}$	MRS	$\overline{Ra}$	MRS	
			Reference	Value					
			1.000	1.000					
1	−4.1187	−2.2784	1.000	0.172	0.000	0.828	1.000	0.376	0.688
2	−6.8675	1.09861	0.667	0.416	0.333	0.584	0.600	0.461	0.531
3	−5.3109	−0.3238	0.855	0.313	0.145	0.687	0.776	0.421	0.598
4	−10.0708	8.57106	0.278	0.956	0.722	0.044	0.409	0.919	0.664
5	−7.3032	9.18324	0.614	1.000	0.386	0.000	0.564	1.000	0.782
6	−6.2235	−3.1299	0.745	0.110	0.255	0.890	0.662	0.360	0.511
7	−9.5106	−2.359	0.346	0.166	0.654	0.834	0.433	0.375	0.404
8	−6.1677	−4.6522	0.751	0.000	0.249	1.000	0.668	0.333	0.501
9	−12.3628	4.45508	0.000	0.658	1.000	0.342	0.333	0.594	0.464

The degree of gray relationship is determined through the average GRV (GRV) of the output objectives:

$$\overline{y_i} = \frac{1}{k}\sum\nolimits_{j=0}^{k} y_{ij}(k) \tag{5}$$

In which:

- k is the output target number.
- y_{ij} is the GRV of the jth output targets in the ith experiment.

The calculated results of GRV y_i and mean GRV of the experiments are shown in Table 3.

3 Results and Analysis

To ensure the harmony between the output factors, the higher the average GRV, the better. Thus, the multi-objective problem becomes a single-objective problem with the output being the mean GRV. The Taguchi method was used to evaluate the influence of EDM parameters on the mean GRV.

With the goal that the mean GRV is as large as possible, the S/N value of is calculated according to formula 2.

The calculated results of S/N value were analyzed by ANOVA method and shown in Table 4.

From the analysis results of the influence of the EDM parameters on the mean GRV (Table 5), it is noted that, T_{on} has the largest influence on $\overline{y_i}$ (57.61%) and followed by the influence of SV (34.14%) and T_{off}. (7.27%). Besides, IP has very little effect on $\overline{y_i}$ (0.98%). This notice is also described in Fig. 2.

Table 4. Analysis of variance for S/N ratios of $\overline{y_i}$

Analysis of Variance for SN ratios

Source	DF	Seq SS	Adj SS	Adj MS	F	P	C (%)
Ton	2	15.3337	15.3337	7.66687	*	*	57.61
Toff	2	1.9349	1.9349	0.96746	*	*	7.27
IP	2	0.2605	0.2605	0.13027	*	*	0.98
SV	2	9.0868	9.0868	4.54339	*	*	34.14
Residual Error	0	*	*	*			
Total	8	26.6160					

The order of influence of the parameters on the GRV obtained by ANOVA analysis is described in Table 5.

The chart of the influence of the parameters on $\overline{y}$ is shown in Fig. 3. From this graph, the trend of influence of each input parameter on y_i can be seen. As analyzed above, the set of input parameters with the highest $\overline{y}$ value is the optimal parameter for the multi-objective function. From Fig. 3, these optimal values can be easily determined (given in Table 6).

With the obtained optimal parameters, it is possible to calculate the output results by using the Taguchi method with the support of Minitab 19 software. From the predicted results, it is found that, with the optimal EDM parameters, the output SR is Ra≈2.31067 (μm) and the MRS≈2.87847 (g/h) (Table 7).

Table 5. Order of influence of the parameters on $\overline{y_i}$

Larger is better

Level	Ton	Toff	IP	SV
1	-4.404	-4.892	-5.028	-4.018
2	-3.841	-4.546	-5.241	-6.401
3	-6.848	-5.656	-4.825	-4.675
Delta	3.007	1.110	0.417	2.383
Rank	1	3	4	2

Mean gray relation value: 0.571

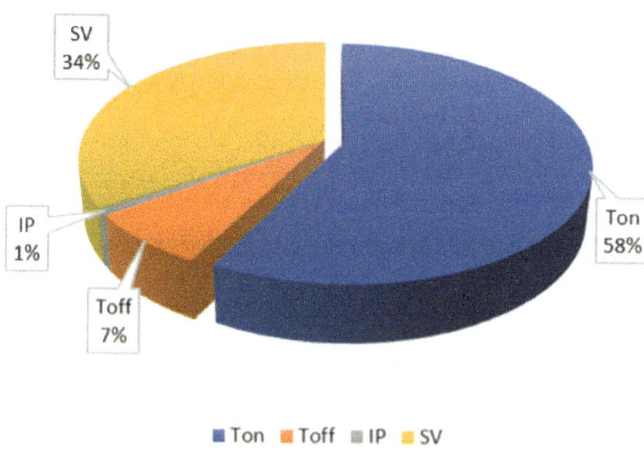

Fig. 2. Influence of input factors on $\overline{y_i}$ (%)

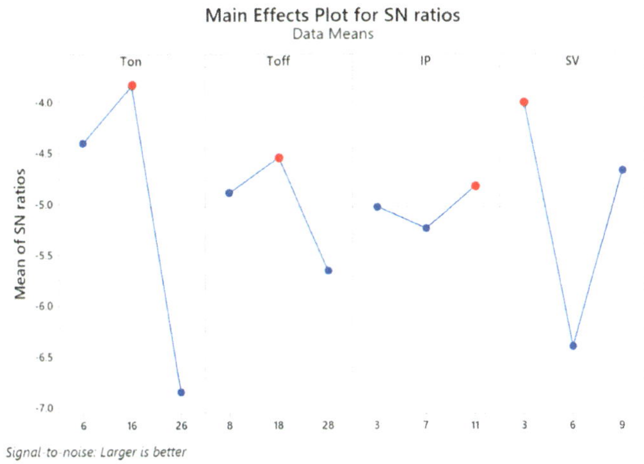

Fig. 3. Main effect plot for S/N ratios of $\bar{y}$

Table 6. Optimum input parameters

Parameter	Code	Unit	Level	Optimum value
Pulse on time	Ton	μs	2	16
Pulse off time	Toff	μs	2	18
Peak current	IP	A	3	3
Servo voltage	SV	V	1	3

Table 7. Predicted output values

Prediction

Response	S/N Ratio	Mean	StDev	Ln(StDev)
Ra	-7.3032	2.31067	0.229448	-1.472
MRS	9.18324	2.87847	-	-

Settings

Ton	Toff	IP	SV
2	2	3	1

Evaluate of experimental model:

To evaluate the proposed model, error distribution charts have been created in Fig. 4. It can be seen from the graphs that in the standard error distribution chart, the errors of the experimental points (blue dots) lie close to the normal distribution line (red solid line). It shows that the error level is very small. The Histogram shows the frequency of errors.

This graph shows that errors near 0 occur with high frequency. The other two plots show the random distribution of experimental errors. It means that the built model is largely influenced by the selected input parameters and is not affected by order of experiments.

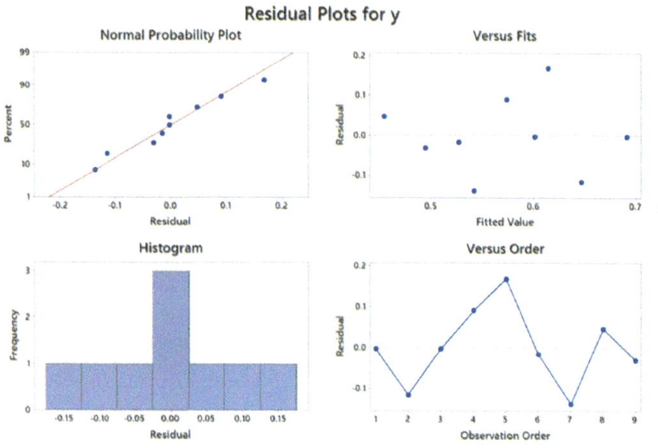

Fig. 4. Residual plots for $\bar{y}$

The appropriateness of the experimental model is also verified by the Anderson-Darling method (Fig. 5). The Figure shows that the data corresponding to the experimental points (blue dots) are all within the region bounded by 2 upper and lower limits. In addition, the standard deviation 95% with P-value of 0.757 – greater than the value $\alpha = 0.05$. This indicates that the applied empirical model is suitable.

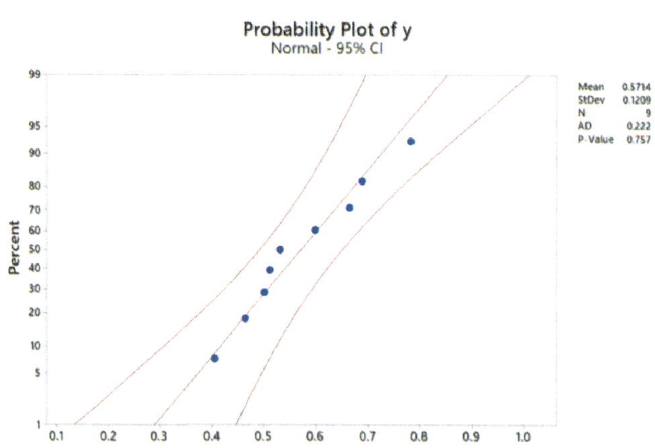

Fig. 5. Probability plot of $\bar{y}$

4 Conclusions

This paper presents the results of a multi-objective optimization study of the EDM processing cylindrical shaped parts. In this study, the multi-objective problem consists of two single objectives, namely SR and MRS. Besides, SKD11 steel and four input parameters including the pulse on time, the pulse off time, the current, and the servo voltage were selected for this investigation. The influence of these parameters on the SR and the MRS was evaluated by applying Taguchi method and gray relation analysis. In addition, optimal input parameters to ensure simultaneous two targets, the SR and the MRS, were found. The proposed empirical model has also been evaluated and it is in perfect agreement with the experimental data.

Acknowledgment. This work was supported by Thai Nguyen University of Technology.

References

1. Rahman, M.M., Khan, A.R., Noor, M.M., Kadirgama, K., Bakar, R.A.: Optimization of machining parameters on surface roughness in EDM of Ti-6Al-4V using response surface method. Adv. Mater. Res. **213**, 402–408 (2011)
2. Hoang, T.T., et al.: Modelling surface finish in electrical discharge machining tablet shape punches using response surface methodology. SSRG Int. J. Mech. Eng **4**(9), 28–30 (2017)
3. Hung, L.X., Hoang, T.T., Pi, V.N.: A study on modelling surface finish in electrical discharge machining tablet shape punches using response surface methodology. J. Environ. Sci. Eng. B **6**(7), 387–390 (2017)
4. Hong, T.T., et al.: Evaluating influences of input parameters on surface roughness in sinking EDM cylindrical shaped parts. Mater. Sci. Forum **1018**, 85–90 (2021)
5. Pradhan, M.K., Biswas, C.K.: Modelling of machining parameters for MRR in EDM using response surface methodology (2008)
6. Iqbal, A., Khan, A.A.: Modeling and analysis of MRR, EWR and surface roughness in EDM milling through response surface methodology. Am. J. Eng. Appl. Sci. **3**(4), 611–619 (2010)
7. Van Cuong, N., et al.: Influence of process parameters on electrode wear in electrical discharge machining cylindrical shaped parts. Key Eng. Mater. **853**, 24–28 (2020)
8. Phan, N., et al.: Tool wear rate analysis of uncoated and AlCrNi coated aluminum electrode in EDM for Ti-6Al-4 V titanium alloy. In: Sattler, K.-U., Nguyen, D.C., Ngoc Pi, V., Long, B.T., Puta, H. (eds.) Advances in Engineering Research and Application: Proceedings of the International Conference on Engineering Research and Applications, ICERA 2020, pp. 832–838. Springer International Publishing, Cham (2021). https://doi.org/10.1007/978-3-030-64719-3_91
9. Tharian, B.K., et al.: Multi-objective parametric optimization in EDM using grey relational analysis. Mater. Today: Proc. **16**, 1013–1019 (2019)
10. Kumar, S., Dhanabalan, S., Narayanan, C.S.: Application of ANFIS and GRA for multi-objective optimization of optimal wire-EDM parameters while machining Ti–6Al–4V alloy. SN Appl. Sci. **1**(4), 1–12 (2019)
11. Ky, L.H., et al.: Multi-objective optimization of surface roughness and electrode wear in EDM cylindrical shaped parts. Mater. Sci. Forum **1018**, 59–64 (2021)
12. Prabhu, S., Vinayagam, B.K.: Multiresponse optimization of EDM process with nanofluids using TOPSIS method and genetic algorithm. Archiv. Mech. Eng. **63**(1), 45–71 (2016)

13. Pattnaik, S., et al.: Multi objective optimization of EDM process parameters using fuzzy TOPSIS method. In: 2015 International Conference on Innovations in Information, Embedded and Communication Systems (ICIIECS). IEEE (2015)
14. Majumder, A., et al.: An approach to optimize the EDM process parameters using desirability-based multi-objective PSO. Prod. Manuf. Res. **2**(1), 228–240 (2014)
15. Majumder, A.: Process parameter optimization during EDM of AISI 316 LN stainless steel by using fuzzy based multi-objective PSO. J. Mech. Sci. Technol. **27**(7), 2143–2151 (2013)
16. Radhika, N., et al.: Multi-objective optimization of EDM parameters using grey relation analysis. J. Eng. Sci. Technol. **10**(1), 1–11 (2015)
17. Hong, T., et al.: Effects of input parameters on electrode wear rate when EDM cylindrical shaped parts. Mater. Sci. Forum **1018**, 79–83 (2021)

Influence of Input Parameters on Sieving Efficiency for Dewatering Vibration Screen Systems

Thi Thanh Nga Nguyen[1(✉)], Anh Tung Luu[1], and Duy Thien Mai[2]

[1] Faculty of Mechanical Engineering, Thai Nguyen University of Technology, Thai Nguyen, Vietnam
nguyennga@tnut.edu.vn
[2] Z129 Factory, Ministry of Defence, Hanoi, Vietnam

Abstract. In dewatering vibration screen, increasing the sieving efficiency plays an important role for mining processes. This paper proposes the influence of the input parameters on the sieving efficiency for dewatering vibration screen systems. The input parameters include the frequency of cyclone motor, the frequency of vibrating motor, and the inclination angle of vibrating screen. The experiment design was conducted based on the Taguchi method. By using ANOVA analysis, the effect of the input parameters on the sieving efficiency was evaluated. Moreover, the optimal input parameters were found in order to maximize the sieving efficiency. The analysis results were examined with the experiment results. The difference between analysis and experiment is 1.18%. For this, the analysis result is significant for predicting the sieving efficiency of the dewatering vibration screen system.

Keywords: Dewatering vibration screen systems · Ore waste · Taguchi method · Sieving efficiency

1 Introduction

Dewatering vibration screen systems are widely used in industrial fields such as mining, metallurgy, construction and agriculture [1]. In mining, the process of separating particles of different sizes plays an important role because of its effect on sieving efficiency [2]. A number of studies on vibrating screen systems for separating different size of particles is displayed in [3–5]. The kinematic characteristics of vibrating screen and moisture of particles in coal mining have been analyzed by automatic dynamic simulation analysis using the ADMAS software [6]. Regarding the design of vibrating screen systems, Wang and Tong [7] uses 3D-DEM to simulate the effect of screening efficiency and screen length of a linear vibrating screen. Zhao et al. [8] has been also used 3D-EDM for simulating the screening process on a circularly vibrating screen. Zhanfu Li et al. [4] proposed a model of vibrating screens in order to simulate the effect of working parameters on screening efficiency.

The mining of minerals not only has exploited dry materials, but also the solid-liquid separation has been investigated. Doerffer and Heinrich [9] presented a range of

D. C. Nguyen et al. (Eds.): ICERA 2021, LNNS 366, pp. 527–535, 2022.
https://doi.org/10.1007/978-3-030-92574-1_55

processing equipment for its dewatering capabilities and operating parameters. Keller and Stahl [10] proposed the effect of vibration frequency and acceleration of solid-liquid with different moistures on sieving efficiency when exploiting coal and limestone. Sung and Turian [11] investigated the filtration and dewatering of narrow-side coal particles for fine coal particles.

In order to evaluate the capacity of vibration screen systems, Monhanty et al. [12, 13] proposed a comparative examination of fine particle size separation technologies. This work shows that the size separation performance using Pansep screen brings significantly better than performance of the traditional technologies. Sieving is a process in which certain types of materials with different size are separated by one-deck or multi-deck screens. In theory, particles larger than the mesh aperture remain on the screen surface and leave the screen surface as they pass through the end of the screen, and these particles are called overflow particles. However, other smaller particles pass through the sieve by the mesh holes that are called fine particles. Zhang et al. [14] investigated the relation between sieving efficiency and working parameters of vibration screen systems such as vibrating screen such as mesh aperture size, screen length, inclination angle, vibration amplitude, and vibration frequency. Hydrocyclones are used in dewatering vibration screening systems by application of the centrifugal force in the prevailing swirling flow field. Performance efficiency of dewatering cyclone has been investigated based on governing maximum solid recovery and minimum water recovery [15].

This paper deals with the evaluation of input parameters on the sieving efficiency for dewatering vibration screen systems. To do this, the experiment data was designed by using the Taguchi method. The experiment data was analyzed based on the ANOVA analysis. The analysis results were verified by experiments with the optimal input parameters.

2 Experimental Design

A set of experiments was conducted on the dewatering vibration screen system as shown in Fig. 1a with model of the system, 150ZJ-I-C42. Ore waste contains solid concentration in ore waste (17%–20%) and P_2O_5 (3%–5%). The experiment is carried out with the three input parameters including the frequency of cyclone motor, the frequency of vibrating motor and the inclination angle of the vibrating screen as seen Fig. 1b, and their values are shown in Table 1.

Table 1. Input parameters and their value range

Input parameters	Symbol	Unit	Level 1	Level 2	Level 3
Frequency of cyclone motor	f_c	Hz	30	35	40
Frequency of vibrating motor	f_v	Hz	38	42	46
Inclination angle of vibrating screen	Alpha	Degree	2	6	10

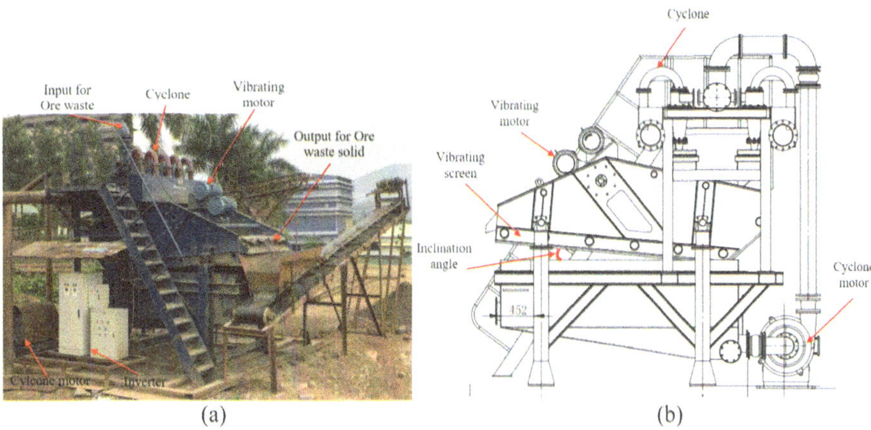

Fig. 1. (a) Experiment of dewatering vibration screen system; (b) Schema of dewatering vibration screen system

The experiment has been conducted on the basis of the Taguchi method in order to design and analysis data of the experiments. For this, the influence of the input parameters on the sieving efficiency for dewatering of ore waste was investigated. The experiment data was designed by using the Taguchi method with L9 orthogonal array as shown in Table 2.

Table 2. Experiment data

Test run no.	f_c	f_v	Alpha	MRR (Ton/h)	
				Mean	S/N
1	30	38	2	35	30.88
2	30	42	6	34.7	30.81
3	30	46	10	34.2	30.68
4	35	38	6	34.3	30.71
5	35	42	10	30.5	29.69
6	35	46	2	35.2	30.93
7	40	38	10	30.6	29.71
8	40	42	2	32.9	30.34
9	40	46	6	34.8	30.83

3 Result and Discussion

3.1 Influence of Input Parameters on Sieving Efficiency

The results analyzed by ANOVA are shown in Table 3 and Table 4. It is clearly seen that the inclination angle of the vibrating screen has strongest influence on the sieving efficiency (54.57%). The other parameters, which have significant influence on the sieving efficiency, are the frequency of vibrating motor (24.15%) and the frequency of cyclone motor (20.27%). Figure 2 shows the relation between the input parameters and the sieving efficiency. When the value of the frequency of cyclone motor increases, the sieving efficiency decreases. With the frequency of vibrating motor, values from level 1 to level 2 make the decreasing values of the sieving efficiency; otherwise, values of sieving efficiency increase when values of the frequency of vibrating motor change from level 2 to level 3. The effect of the inclination angle of the vibrating screen on the sieving efficiency is inverse to the influence of the frequency of vibrating motor.

Table 3. Influence of input parameters on the sieving efficiency at different levels

Level	f_c	f_v	Alpha
1	**34.63**	33.30	34.37
2	33.33	32.70	**34.60**
3	32.77	**34.73**	31.77
Delta	1.87	2.03	2.83
Rank	3	2	1

Table 4. ANOVA analysis for sieving efficiency

Source	DF	Seq SS	Adj SS	Adj MS	F	P	C%
f_c	2	5.4956	5.4956	2.7478	24.01	0.040	20.27
f_v	2	6.5489	6.5489	3.2744	28.61	0.034	24.15
Alpha	2	14.8422	14.8422	7.4211	64.84	0.015	54.74
Residual error	2	0.2289	0.2289	0.1144			0.84
Total	8	27.1156					100.00

Model summary

S	R-Sq	R-Sq(adj)
0.3383	99.16%	96.62%

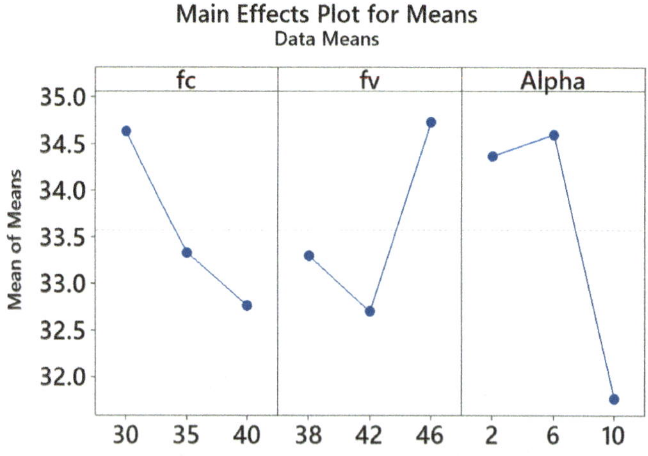

Fig. 2. Influence of the input parameters on the sieving efficiency

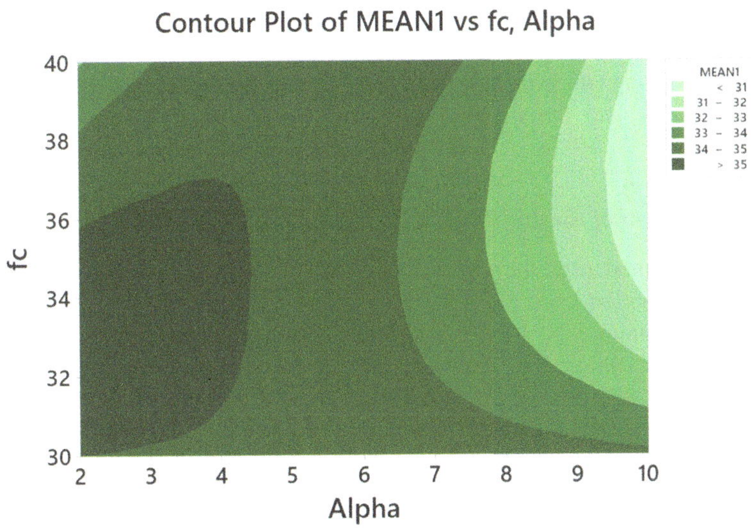

Fig. 3. Influence of the frequency of cyclone motor and the inclination angle of vibrating screen on the sieving efficiency

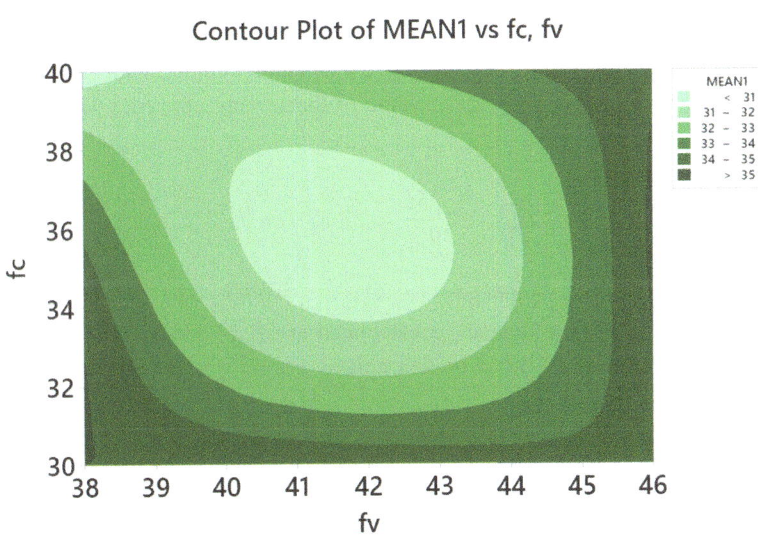

Fig. 4. Influence of the frequency of cyclone motor and the frequency of vibrating motor on the sieving efficiency

Figure 3 presents the effect of the frequency of cyclone motor and the inclination angle of vibrating screen on the sieving efficiency. As seen this figure, the sieving efficiency obtains the maximum value at the dark green region. At this region, the optimal values of the frequency of cyclone motor are in the interval, i.e., [30, 36] (Hz);

the optimal inclination angle of vibrating screen are determined in the interval, i.e., [2, 4]. Figure 4 shows the frequency of cyclone motor and the frequency of the vibrating screen on the sieving efficiency. It can be seen that the optimal values of the frequency of cyclone motor and the inclination angle of vibrating screen are in the dark green zone.

3.2 Optimization of Sieving Efficiency

The Taguchi method uses the term "signal-to-noise" (S/N) to measure the quality characteristic deviation from the desired value. The optimal factors can be obtained based on the maximum S/N ratio. As observed from Table 3 and Fig. 2 that the optimal values of the input parameters are obtained as $f_c = 30$ Hz at level 1, $f_v = 46$ Hz at level 3 and $Alpha = 6^0$ at level 2.

The mean value of the optimal sieving efficiency is determined from the optimal parameters at their level. Thus, the formula for computing the mean values of the sieving efficiency (*SE*) is

$$\overline{SE_{op}} = \overline{f_{c1}} + \overline{f_{v3}} + \overline{Alpha_2} - 2.\overline{T_{gg}}, \tag{1}$$

where $\overline{f}_{c1}$ is mean value of the sieving efficiency at the frequency of cyclone motor with level 1, i.e., $\overline{f}_{c1} = 34.63\, Ton/h$; $\overline{f}_{v3}$ is mean value of the sieving efficiency at the frequency of vibrating motor with level 3, i.e., $\overline{f}_{v3} = 34.73\, Ton/h$; $\overline{Alpha_2}$ is mean value of the sieving efficiency at the inclination angle with level 2, i.e., $\overline{Alpha_2} = 34.6\, Ton/h$; $\overline{T}_{gg}$ is the mean value of the sieving efficiency for all test runs and it can be computed as

$$\overline{T_{gg}} = \frac{\sum_1^9 SE}{9} = 33.58\ Ton/h \tag{2}$$

Substituting all of the parameters, i.e., $\overline{f}_{c1}, \overline{f}_{v3}, \overline{Alpha_2}$, and $\overline{T}_{gg}$ into Eq. (1), the mean value of the sieving efficiency is calculated as $\overline{SE_{op}} = 36.57\, Ton/h$.

A confidence interval (CI) can be determined as

$$CI = \pm\sqrt{F_{\alpha}(1, f_e).V_e.\left(\frac{1}{N_e} + \frac{1}{R}\right)}, \tag{3}$$

where $f_e = 2$ is the degree of freedom of error; $V_e = 0.1144$ is the mean of error; $R = 3$ is number of trials in each experiment; $F_{\propto} = 8.5263$ is a coefficient for the confidence level $\alpha = 90\%$; N_e is calculated as

$$N_e = \frac{Number\ of\ test\ runs}{1 + Sum\ of\ degrees\ of\ freedom} = \frac{27}{1+2+2+2} = 3.857 \quad (4)$$

Substituting all parameters, i.e., f_e, V_e, $F_{\propto}$, R, and N_e into Eq. (3), the confidence interval is obtained as $CI = \pm 0.76$.

With $\alpha = 90\%$, the predicted optimal sieving efficiency with the optimal input parameters at their levels $\overline{f}_{c1}/\overline{f}_{v3}/\overline{Alpha}_2$ can be determined as follows

$$(36.57 - 0.76)\ Ton/h \leq \overline{SE}_{op} \leq (36.57 + 0.76)\ Ton/h \quad (5)$$

An experiment is implemented with the optimal parameters. The comparison of analysis and experiment is shown in Table 5. As seen that the difference between the analysis and experiment is 1.18%. For this, the analysis result is significant for predicting the sieving efficiency of the dewatering vibration screen system.

Table 5. Comparison between analysis and experiment results

Output factor	Optimal input parameters		
	Analysis	Experiment	Error (%)
	$f_{c1}/f_{v3}/Alpha_2$	$f_{c1}/f_{v3}/Alpha_2$	
SE [*Ton/h*]	36.57	37	1.18

3.3 Evaluation of the Data

In addition to evaluate the data of the experiment, Fig. 5 depicts the Jonhson-Darling Transformation for the sieving efficiency. As seen in the probability of original data, the data distributes further conference line since the P-value is smaller than 0.05. For this, the transformation of the probability must be applied as shown in the transform data of this figure. From this result, it can be said that the data is good fit with the standard line.

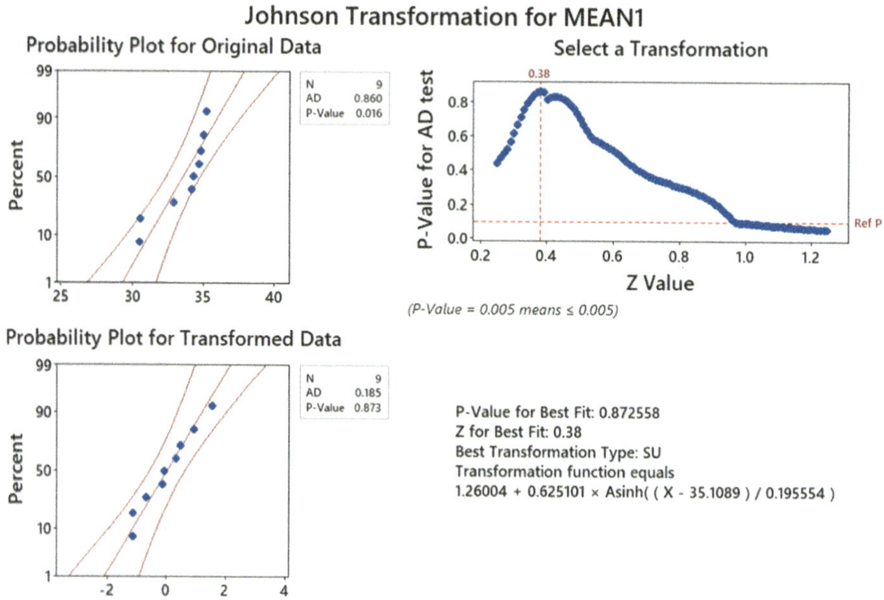

Fig. 5. Probability plot for transformed data

4 Conclusion

In this paper, the proposed design experiment using Taguchi method with L9 orthogonal array was carried out to evaluated the influence of the input parameters on the sieving efficiency and find the optimal parameters. The conclusion can be drawn as follows

- All input parameters have significant influences on the sieving efficiency. It was found that the inclination angle of the vibrating screen has strongest influence on the sieving efficiency;
- The optimal input parameters were obtained by using ANOVA analysis.
- An experiment was implemented to verify the analysis result. Due to this the calculation method can be used to predict the sieving efficiency;
- The data was also evaluated by the probability plot for transformed data.

Acknowledgements. The authors thank Thai Nguyen University of Technology (TNUT) for supporting this work.

References

1. Li, Z., Tong, X., Zhou, B., Ge, X., Ling, J.: Design and efficiency research of a new composite vibrating screen. Shock Vib. **2018** (2018). https://doi.org/10.1155/2018/1293273
2. Moncada, M., Rodríguez, C.G.: Dynamic modeling of a vibrating screen considering the ore inertia and force of the ore over the screen calculated with discrete element method. Shock Vib. **2018** (2018). https://doi.org/10.1155/2018/1714738
3. Standish, N., Bharadwaj, A.K., Hariri-Akbari, G.: A study of the effect of operating variables on the efficiency of a vibrating screen. Powder Technol. **48**, 161–172 (1986). https://doi.org/10.1016/0032-5910(86)80075-4
4. Li, Z., Tong, X., Zhou, B., Wang, X.: Modeling and parameter optimization for the design of vibrating screens. Miner. Eng. **83**, 149–155 (2015). https://doi.org/10.1016/j.mineng.2015.07.009
5. Design and structure optimization of a reconfigurable vibrating screen for the mining and mineral processing industries (2016)
6. Jiang, H., et al.: Kinematics characteristics of the vibrating screen with rigid-flexible screen rod and the behavior of moist coal particles during the dry deep screening process. Powder Technol. **319**, 92–101 (2017). https://doi.org/10.1016/j.powtec.2017.06.036
7. Wang, G., Tong, X.: Screening efficiency and screen length of a linear vibrating screen using DEM 3D simulation. Min. Sci. Technol. **21**, 451–455 (2011). https://doi.org/10.1016/j.mstc.2011.05.026
8. Zhao, L., Zhao, Y., Liu, C., Li, J., Dong, H.: Simulation of the screening process on a circularly vibrating screen using 3D-DEM. Min. Sci. Technol. **21**, 677–680 (2011). https://doi.org/10.1016/j.mstc.2011.03.010
9. Doerffer, M., Heinrich, R.: Efficient dewatering solutions on vibrating screens. In: Tailings Mine Waste 2008 - Proceedings of 12th International Conference, pp. 101–111 (2009). https://doi.org/10.1201/9780203882306.ch10
10. Keller, K., Stahl, W.: Vibration screens for dewatering—theory and practice. Min. Metall. Explor. **14**(1), 27–34 (1997)
11. Sung, D.J., Turian, R.M.: Chemically enhanced filtration and dewatering of narrow-sized coal particles. Sep. Technol. **4**, 130–143 (1994). https://doi.org/10.1016/0956-9618(94)80016-2
12. Mohanty, M.K., Palit, A., Dube, B.: A comparative evaluation of new fine particle size separation technologies. Miner. Eng. **15**, 727–736 (2002). https://doi.org/10.1016/S0892-6875(02)00169-3
13. Mohanty, M.K.: Fine coal screening performance enhancement using the Pansep screen. Int. J. Miner. Process. **69**, 205–220 (2003). https://doi.org/10.1016/S0301-7516(02)00124-2
14. Zhang, B., Gong, J., Yuan, W., Fu, J., Huang, Y.: Intelligent prediction of sieving efficiency in vibrating screens. Shock Vib. **2016** (2016). https://doi.org/10.1155/2016/9175417
15. Dubey, R.K., Climent, E., Banerjee, C., Majumder, A.K.: Performance monitoring of a hydrocyclone based on underflow discharge angle. Int. J. Miner. Process. **154**, 41–52 (2016). https://doi.org/10.1016/j.minpro.2016.07.002

Influence of PMEDM Factors on Surface Roughness When Processing SKD11 Steel

Huu Quang Nguyen[1], Manh Cuong Nguyen[2], Thu Quy Le[3], Thanh Tu Nguyen[2], Quoc Hoang Tran[4], Thanh Hoang Nguyen[5], Ngoc Pi Vu[2], and Xuan Tu Hoang[2(✉)]

[1] University of Economics - Technology for Industries, Ha Noi, Vietnam
[2] Thai Nguyen University of Technology, Thai Nguyen, Vietnam
hxt_dhktcn@tnut.edu.vn
[3] National Research Institute of Mechanical Engineering, Ha Noi, Vietnam
[4] Nguyen Tat Thanh University, Ho Chi Minh City, Vietnam
[5] Bac Kan Colleges, Bac Kan, Vietnam

Abstract. In this paper, the results of a study on the influence of process parameters on surface roughness (SR) when powder-mixed electrical discharge machining (PMEDM) of cylindrical shaped parts of SKD11 steel are presented. In this work, an experiment was performed with the use of the Taguchi method for experimental design and result's analysis. In the experiment, six input parameters, including the powder concentration, the powder size, the pulse on time, the pulse off time, the current, and the servo voltage, were investigated. The effect of the input parameters on the SR was evaluated by analysis of variance (ANOVA). In particular, optimal values of the input parameters to achieve the minimum SR were found.

Keywords: PMEDM · Electrical discharge machining · Surface roughness · Cylindrical shaped parts

1 Introduction

Electrical discharge machining (EDM) is one of the most commonly non-traditional machining methods for machining electrically conductive materials. It is very popularly used for machining cavity-shaped parts, complex shapes, such as dies, cores of stamping dies, molds as well as for machining hard materials, etc. This processing method has many advantages, such as it can process metals and alloys after heat treatment; low cutting force and noise, etc… On the other hand, EDM also has some disadvantages, including low material removal rate (MRR); the electrode wears out quite quickly, and the machined surface quality is not high.

To overcome the disadvantages of EDM process, a highly effective solution is EDM with the use of a dielectric solution mixed with conductive powder (or PMEDM). Therefore, this processing method has been studied by many scientists.

So far, there have been quite a few studies on PMEDM. Studies have evaluated the influence of PMEDM process parameters on SR [1–4], MMR [1, 4–7], electrode wear (EWR) [1, 4, 5, 7], etc… Studies on PMEDM have also been conducted when

D. C. Nguyen et al. (Eds.): ICERA 2021, LNNS 366, pp. 536–544, 2022.
https://doi.org/10.1007/978-3-030-92574-1_56

machining various types of workpiece materials such as 90CrSi [3, 4, 6, 8, 9], AISI D2 die steel [10, 11], Inconel 718 [12, 13], Ti-6Al-4V [14, 15], etc. Different electrode materials have been investigated during PMEDM such as copper [16–18], brass [19, 20], Cu-W [1, 16], or graphite [18, 21], etc. The use of different powders for this process has also been investigated such as machining with SiC [2, 10, 12, 14, 17], Ti [22], Al [2, 13], Al_2O_3 [16, 23, 24], silicon powder [1, 25], etc. The powder size is also the object of many studies such as tungsten powder with the size of 10–15 μm [22], chromium powder with size of 45–55 μm [5], Al2O3 with size of 0.51 μm [1], 45–50 μm [23, 24], etc. Recently, there have been several studies on PMEDM of cylindrical shaped components [3, 6–9]. However, these studies were carried out with the workpiece material of 90CrSi steel.

This paper presents a study on optimization of PMEDM process to achieve the minimum SR when machining cylindrical parts made of SKD11 steel. The impact of the input factors on the SR was evaluated. In addition, optimal input parameters to achieve the minimum SR have been proposed.

2 Experimental Setup

To evaluate the influence of the PMEDM factors on the SR, six input parameters, including the powder concentration (C_p), particle size (S_p), time pulse generation time (T_{on}), pulse stop time (T_{off}), pulse current strength (IP) and discharge gap voltage (SV), were included in the survey. These parameters and their levels were described in the Table 1. The experimental setup is presented in Fig. 1. In the setup, the workpiece material was SKD11 steel, and the electrode material was copper. Also, a sinker EDM (Sodick A30) was used.

Table 1. Input factors and their levels

No.	Parameters	Code	Unit	Level					
				1	2	3	4	5	6
1	Powder concentration	C_p	g/l	0	2	3	4	5	6
2	Powder size	S_p	nm	100	500	1000	–	–	–
3	Pulse on time	T_{on}	μs	10	20	30	–	–	–
4	Pulse off time	T_{off}	μs	10	20	30	–	–	–
5	Peak current	IP	A	4	8	12	–	–	–
6	Servo voltage	SV	V	3	5	7	–	–	–

The experimental matrix is designed based on the survey levels of the variables with the use of the Taguchi method and the design L18 (3^4). Also, the Minitab R19 software was used to design the experimental matrix. The experimental matrix and the SR's results are described in Table 2.

Table 2. Experimental matrix and output SR

No.	Input parameters						Ra (μm)			Mean	S/N
	C_p	S_p	T_{on}	T_{off}	IP	SV	Trial 1	Trial 2	Trial 3		
1	0	100	10	10	4	3	2.285	2.130	1.405	1.94000	−5.9226
2	0	500	20	20	8	5	2.315	2.390	2.749	2.48467	−7.9305
3	0	1000	30	30	12	7	3.321	2.802	3.459	3.19400	−10.1206
4	2	100	10	20	8	7	2.047	1.972	2.010	2.00967	−6.0635
5	2	500	20	30	12	3	1.634	2.698	2.279	2.20367	−7.0309
6	2	1000	30	10	4	5	1.180	1.247	1.290	1.23900	−1.8672
7	3	100	20	10	12	5	2.437	2.345	2.146	2.30933	−7.2817
8	3	500	30	20	4	7	1.120	1.181	1.076	1.12567	−1.0345
9	3	1000	10	30	8	3	2.202	2.134	2.230	2.18867	−6.8051
10	4	100	30	30	8	5	2.736	2.034	2.397	2.38900	−7.6264
11	4	500	10	10	12	7	2.004	1.776	1.751	1.84367	−5.3302
12	4	1000	20	20	4	3	0.935	1.023	1.135	1.03100	−0.2925
13	5	100	20	30	4	7	2.193	1.649	1.709	1.85033	−5.4196
14	5	500	30	10	8	3	2.064	2.267	2.675	2.33533	−7.4181
15	5	1000	10	20	12	5	1.121	1.379	1.132	1.21067	−1.7023
16	6	100	30	20	12	3	3.234	2.420	2.848	2.83400	−9.1074
17	6	500	10	30	4	5	1.148	1.142	1.417	1.23567	−1.8846
18	6	1000	20	10	8	7	2.613	2.288	2.815	2.57200	−8.2363

3 Results and Analysis

To evaluate the influence of the input parameters on the SR, the ANOVA method is used with the support of Minitab 19 software. In addition, the objective function is the minimum surface texture is the best:

$$S/N = -10\log_{10}\left(\frac{1}{n}\sum_{i=1}^{n} y_i^2\right) \tag{1}$$

Fig. 1. Experimental setup

The results of ANOVA described in Table 3. It was found from the Table that IP has the greatest influence on the SR (49.99%); followed by the influence of C_p (17.91%), S_p (10.02%), T_{off} (9.89%), T_{on} (5.97%) and SV (4.83%). The percentage influence of these parameters is also clearly depicted in Fig. 2.

The order of influence of input parameters is described in detail in Table 4. From the Table, the order of influence of parameters on S/N is IP, C_p, T_{off}, S_p, T_{on} and SV.

Table 3. Analysis of variance for S/N ratios

Analysis of Variance for SN ratios

Source	DF	Seq SS	Adj SS	Adj MS	F	P	C (%)
Cp	5	27.087	27.087	5.417	5.15	0.171	17,91
Sp	2	15.152	15.152	7.576	7.20	0.122	10,02
Ton	2	9.030	9.030	4.515	4.29	0.189	5,97
Toff	2	14.959	14.959	7.479	7.10	0.123	9,89
Ip	2	75.591	75.591	37.796	35.90	0.027	49,99
SV	2	7.298	7.298	3.649	3.47	0.224	4,83
Residual Error	2	2.106	2.106	1.053			1,39
Total	17	151.222					

Model Summary

S	R-Sq	R-Sq(adj)
1.0261	98.61%	88.16%

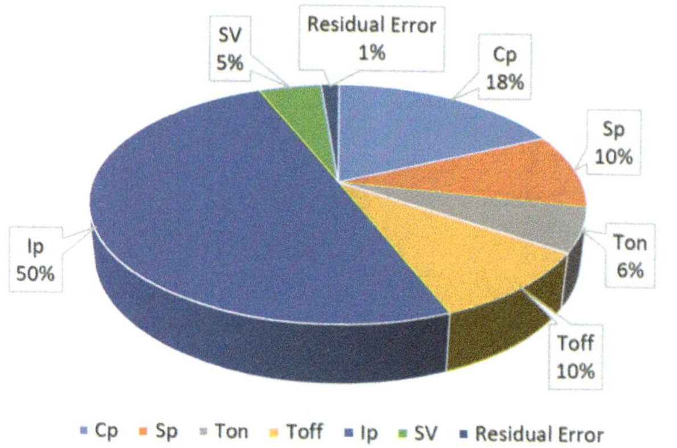

Fig. 2. Influence of the input parameters on the S/N ratio of Ra

Table 4. Order of influence of input parameters on S/N

Smaller is better

Level	Cp	Sp	Ton	Toff	Ip	SV
1	-7.991	-6.904	-4.618	-6.009	-2.737	-6.096
2	-4.987	-5.105	-6.032	-4.355	-7.347	-4.715
3	-5.040	-4.837	-6.196	-6.481	-6.762	-6.034
4	-4.416					
5	-4.847					
6	-6.409					
Delta	3.575	2.066	1.578	2.126	4.610	1.381
Rank	2	4	5	3	1	6

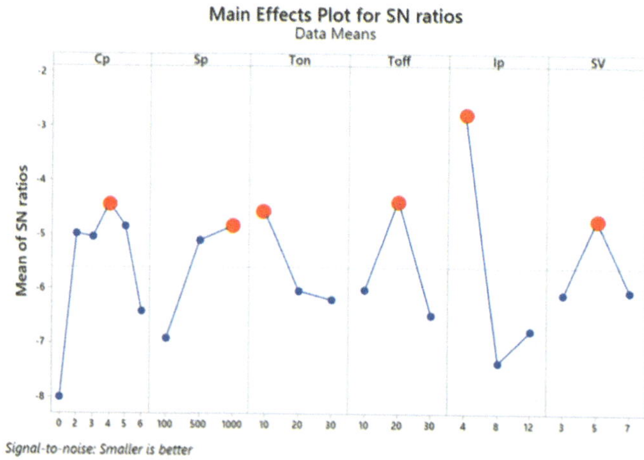

Fig. 3. Effect of input parameters on the S/N ratios

Figure 3 depicts the influence of the parameters on the S/N value. Since the target surface roughness is as small as possible, the larger the S/N value. From Fig. 3, it can be clearly seen that the influence of these parameters is as follows: For powder concentration C_p: adding powder additive to the pulse process significantly reduces surface roughness. SR is highest when $C_p = 0$ and it will decrease when the additive powder is added. When C_p increases from 2 to 4%, Ra decreases. However, if we continue to increase C_p to 5 and 6%, Ra will increase again. The powder size also had a significant effect on SR. As the powder size increased, Ra decreased, but when the powder size increased from 500 to 1000 nm, Ra decreased insignificantly. When T_{on} increases, SR increases. Besides, SR increased as T_{off} increased from 10 to 20 µs but Ra decreased as T_{off} continued to increase from 20 to 30 µs. Minimum SR when IP is low 4 (A). Ra arise when IP move up from 4 to 8 A. If the IP continues to increase, Ra will decrease slightly. Besides, SR decreases when SV increases from 3 to 5 V, but SR increases again when SV increases from 5 to 7 V. As stated above, in order to achieve the minimum SR, the maximum S/N value corresponding to each input parameter. Therefore, from Fig. 3, it is possible to determine the optimal set of PMEDM parameters to minimize Ra (Table 5). Using the Minitab 19 software, the SR value corresponding to the optimum input parameters was found (Table 6). From Table 6, the SR corresponding to the found optimal input parameters is 0.593 (µm). Also, with a confidence level of 95%, surface roughness results can be obtained in the range of 0.004 to 1.181 µm. Therefore, 95% of the SR is achieved with the optimal input factors less than 1.181 (µm).

Table 5. Optimum input factors for minimum SR

Parameter	Code	Unit	Level	Value
Powder concentration	C_p	g/l	4	4
Powder size	S_p	nm	3	1000
Pulse on time	T_{on}	µs	1	10
Pulse off time	T_{off}	µs	2	20
Peak current	IP	A	1	4
Servo voltage	SV	V	2	5

Table 6. Predicted optimum SR

Type	Fit	SE Fit	CI DF	95% CI	PI DF	95% PI
Conditional	0.593	0.181	2.90	(0.004, 1.181)	2.33	(-0.447, 1.632)

Evaluation of Experimental Model:

The experimental model is evaluated through error distribution charts, as shown in Fig. 4. From the chart, it can be seen that the experimental errors (the blue points) lying close to the normal distribution line (the red solid line). That means the deviation level is very small. Besides, the frequency of errors (the Histogram chart) shows that the errors appear in the range −0.1 to 0.1, accounting for a large proportion. Also, the remaining two graphs show the random distribution of experimental errors. That means that the proposed model is largely influenced by the selected input parameters and is not affected by the order of the experiments.

The suitability of the experimental model and the optimal input parameters was checked by the Anderson-Darling method using the Minitab 19 software (Fig. 5). It can be found from the chart that all the experimental points (the blue dot) are in the region bounded by 2 upper and lower limits with 95% significance, and P value of 0.283 is larger than $\alpha = 0.05$. This indicates that the experimental model is consistent with the above significance level.

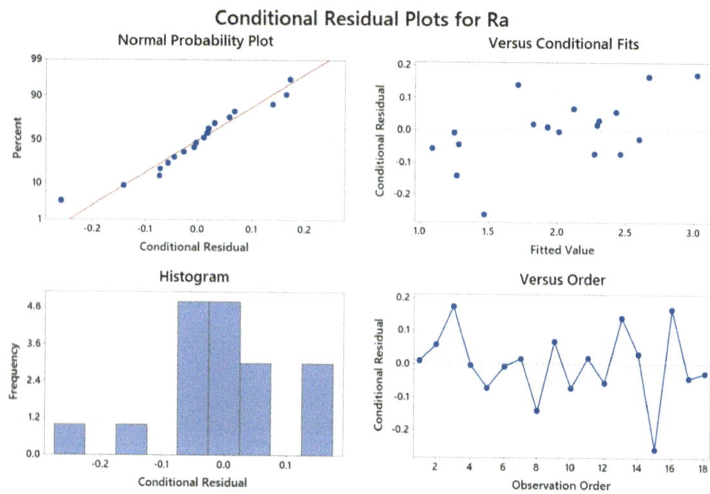

Fig. 4. Residual plots for SR

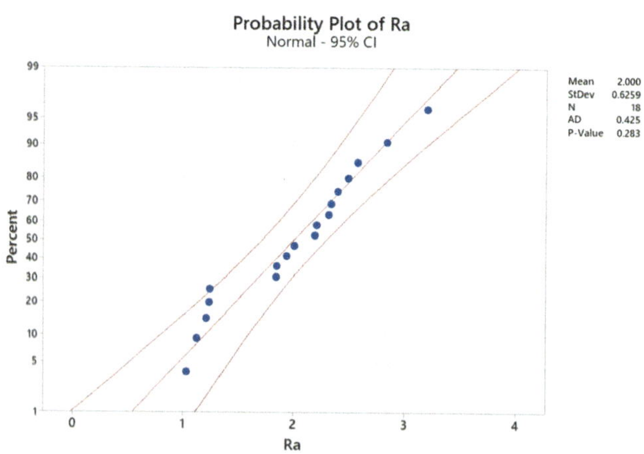

Fig. 5. Probability plot of SR

4 Conclusions

In this paper, the results of a study on the effect of PMEDM's parameters on the SR when processing cylindrical shaped parts made from SKD11 tool steel. To solve this problem, an experiment was designed and implemented using the Taguchi method for test design and result analysis. In addition, six key input factors, including the powder size, the powder concentration, the pulse on time, the pulse off time, the current, and the servo voltage, were investigated. It was reported that IP is the greatest effected factor on the SR (49.99%); followed by Cp (17.91%), Sp (10.02%), Toff (9.89%), Ton (5.97%) and SV (4.83%). The influence of the input parameters on the SR was evaluated using the ANOVA method. In addition, optimal input factors have been proposed to achieve the minimum SR.

Acknowledgment. This work was supported by Thai Nguyen University of Technology.

References

1. Kansal, H., Singh, S., Kumar, P.: Performance parameters optimization (multi-characteristics) of powder mixed electric discharge machining (PMEDM) through Taguchi's method and utility concept (2006)
2. Ganachari, V., Kavade, M., Mohite, S.: Effect of mixture of Al and SiC powder on surface roughness in PMEDM using Taguchi method with GRA optimization. Int. J. Adv. Eng. Res. Stu. **2**(4), 04–07 (2013)
3. Hong, T.T., et al.: A study on influence of input parameters on surface roughness in PMEDM cylindrical shaped parts. Mater. Sci. Forum **1018**, 65–70 (2021)
4. Hong, T.T., et al.: Multi-objective optimization of PMEDM input factors for processing cylindrical shaped parts. Mater. Sci. Forum **1018**, 71–77 (2021)
5. Ojha, K., Garg, R., Singh, K.: Experimental investigation and modeling of PMEDM process with chromium powder suspended dielectric. Int. J. Appl. Sci. Eng. **9**(2), 65–81 (2011)
6. Nguyen, M.C., et al.: Influence of input factors on material removal rate in PMEDM cylindrical shaped parts with silicon carbide powder suspended dielectric. Key Eng. Mater. **861**, 129–135 (2020)
7. Hong, T.T., et al.: Multi-objective optimization of PMEDM process of 90CrSi alloy steel for minimum electrode wear rate and maximum material removal rate with silicon carbide powder. Mater. Sci. Forum **1018**, 51–58 (2021)
8. Tran, T.-H., et al.: Electrical discharge machining with SiC powder-mixed dielectric: an effective application in the machining process of hardened 90CrSi steel. Machines **8**(3), 36 (2020)
9. Hong, T.T., et al.: Effect of process parameters on machining time in PMEDM cylindrical shaped parts with silicon carbide powder suspended dielectric. Mater. Sci. Forum **1018**, 97–102 (2021)
10. Al-Khazraji, A., Amin, S.A., Ali, S.M.: The effect of SiC powder mixing electrical discharge machining on white layer thickness, heat flux and fatigue life of AISI D2 die steel. Eng. Sci. Technol. Int. J. **19**(3), 1400–1415 (2016)
11. Hosni, N.A.J., Lajis, M.A.: The influence of Span-20 surfactant and micro-/nano-Chromium (Cr) powder mixed electrical discharge machining (PMEDM) on the surface characteristics of AISI D2 hardened steel. IOP Conf. Ser. Mater. Sci. Eng. **342**, 012095 (2018)

12. Sahu, S.K., Jadam, T., Datta, S., Nandi, G.: Effect of using SiC powder-added dielectric media during electro-discharge machining of Inconel 718 superalloys. J. Braz. Soc. Mech. Sci. Eng. **40**(7), 1–19 (2018)
13. Patel, S., Thesiya, D., Rajurkar, A.: Aluminium powder mixed rotary electric discharge machining (PMEDM) on Inconel 718. Aust. J. Mech. Eng. **16**(1), 21–30 (2018)
14. Yaşar, H., Ekmekci, B.: Ti-6Al-4V surfaces in SiC powder mixed electrical discharge machining. Adv. Mater. Res. **856**, 226–230 (2013)
15. Shabgard, M., Khosrozadeh, B.: Investigation of carbon nanotube added dielectric on the surface characteristics and machining performance of Ti-6Al-4V alloy in EDM process. J. Manuf. Process. **25**, 212–219 (2017)
16. Assarzadeh, S., Ghoreishi, M.: Modeling and optimizing powder mixed electrical discharge machining (PMEDM) process using neural networks. In: Proceedings of the 2nd Tehran International Congress on Manufacturing Engineering (TICME 2007) Iran University of Science and Technology (2007)
17. Hu, F., et al.: Surface properties of SiCp/Al composite by powder-mixed EDM. Procedia CIRP **6**, 101–106 (2013)
18. Selvarajan, L., et al.: An experimental investigations and optimization of performance measures in EDM using copper and graphite electrodes. Mater. Today Proc. **46**, 9341–9347 (2020)
19. Ahmad, S., et al.: Surface roughness and surface topography of Inconel 718 in powder mixed dielectric electrical discharge machining (PMEDM). Int. J. Integr. Eng. **10**(5), 181–186 (2018)
20. Surekha, B., et al.: Response surface modelling and application of fuzzy grey relational analysis to optimise the multi response characteristics of EN-19 machined using powder mixed EDM. Aust. J. Mech. Eng. **19**(1), 19–29 (2021)
21. Long, B.T., et al.: Machining properties evaluation of copper and graphite electrodes in PMEDM of SKD61 steel in rough machining. Int. J. Eng. Adv. Technol. ISSN, 2249–8958 (2015)
22. Singh, B., Kumar, J., Kumar, S.: Experimental investigation on surface characteristics in powder-mixed electrodischarge machining of AA6061/10% SiC composite. Mater. Manuf. Processes **29**(3), 287–297 (2014)
23. Assarzadeh, S., Ghoreishi, M.: A dual response surface-desirability approach to process modeling and optimization of Al 2 O 3 powder-mixed electrical discharge machining (PMEDM) parameters. Int. J. Adv. Manuf. Technol. **64**(9–12), 1459–1477 (2013)
24. Assarzadeh, S., Ghoreishi, M., Shariyyat, M.: Response surface methodology approach to process modeling and optimization of powder mixed electrical discharge machining (PMEDM). In: Proceedings of the 16th International Symposium on Electromachining (ISEM-XVI) (2010)
25. Luzia, C.A.O., Laurindo, C.A.H., Soares, P.C., Torres, R.D., Mendes, L.A., Amorim, F.L.: Recast layer mechanical properties of tool steel after electrical discharge machining with silicon powder in the dielectric. Int. J. Adv. Manuf. Technol. **103**(1–4), 15–28 (2019)

Investigating Effects of Distance Air-Gaps on Iron-Core Shunt Reactors

Tu Pham Minh, Hung Bui Duc(✉), Thinh Tran Van, Dung Dang Chi, and Vuong Dang Quoc

School of Electrical Engineering, Hanoi University of Science and Technology, Hanoi, Viet Nam
hung.buiduc@hust.edu.vn

Abstract. The shunt reactor plays an important role for high and super-high voltages in transmission and distribution systems. It is used to absorb excess reactive powers generated by capacitive powers on the lines for no-load or low-load cases. In addition, it is also an important component to balance reactive powers in the system, to avoid overvoltage at the end of the lines and to maintain voltage stability at the specified level as well. In order to reduce the magnetic flux density and avoid saturation of the magnetic circuit, the reluctance value has to be increased by air gaps along the iron core. This leads to decreasing the influence of fringing fluxes around the air gaps. Non-magnetic materials are often used at the air gaps to separate between iron core blocks. The fringing flux departs and re-enters the lamination perpendicularly, thus the core blocks of a large oil-immersed shunt reactor are generally made from radially laminated silicon steels. In this paper, the correction of the distance between the air gaps is proposed to reduce the fringing flux and the value of the winding inductance of shunt reactors. All parameters of the shunt reactors are calculated via an analytic method based on the theory of magnetic circuit model, then a finite element method is used to compute and simulate the winding inductance for changing the distance between the air gaps.

Keywords: Shunt reactors · Air gaps · The distance between the air gaps · Inductances · Analytic method · Finite element method

1 Introduction

As we have known, the shunt reactor is an important component used widely to improve stability and efficiency in power transmission systems. High and super-high voltages in power transmission systems usually have relatively large line lengths. When no loads or low loads, the parasitic capacitance will appear in transmission lines. Especially, for the long line, its value is very large, which will increase the voltage along the line, causing overvoltage at the end of the line.

In order to maintain voltage stability at the specified level, it is necessary to use shunt reactors to absorb the excess reactive power generated by the line capacitance, balancing the reactive power in electrical systems [1, 2]. In addition, so as to reduce the magnetic flux density and avoid saturation of the magnetic circuit, the reluctance needs

D. C. Nguyen et al. (Eds.): ICERA 2021, LNNS 366, pp. 545–557, 2022.
https://doi.org/10.1007/978-3-030-92574-1_57

to be increased by adding air gaps along the iron core of shunt reactors. However, these are also reasons to occur fringing fluxes around the air gaps, this leads to an increase in the winding inductance of the shunt reactors [3, 4]. The volume of the air gaps depends on the main parameters of the shunt reactor. i.e., reactive power, magnetic flux density, frequency of power system. The area and length of the air gap can be determined via this volume. the volume of the air gaps is usually a constant. Hence, in order to reduce the air gap length (l_g), the core area (A_g) will be increased. This is corresponding to an increase in the ratio of the area to the length of the air gap (A_g/l_g), which leads to an increase in the area of iron core and yoke. In another way, to reduce the influence of the fringing flux and the winding inductance, the air gap is divided into several smaller air gaps distributed along the iron core. For the same number of the air gaps, the fringing flux and winding inductance (which is one of the important parameters of shunt reactors) are reduced via changing distance between the air gaps. Relating to shunt reactors, many authors have recently studied the compensation reactive power in the power system [1, 2, 5], but these researches mainly focus on the analysis of magnetic fields [7, 8], modeling and simulation of Shunt Reactors [6, 14, 15] or the interaction between shunt reactors and the power system in different transient conditions [9, 10], or the vibration analysis [11, 12].

In order to overcome this drawback, this paper proposed an analytical method based on the theory of magnetic circuit model to determine the main parameters of the shunt reactors and then a finite element method (FEM) is introduced to compute the winding inductance for changing the distance between the air gaps, thereby giving the curving relationship between the winding inductance and the distance between the air gaps.

2 Analytic Method

The shunt reactors can be built in single-phase or three-phase units. All dimensions of the single-phase shunt reactor are shown in Fig. 1, where W_w, H_w and D_c are respectively the width, the length of winding, and the diameter of iron core reactors. The iron yoke is rectangular with a height H_y and a depth D_y being equal to the diameter of the iron core D_c.

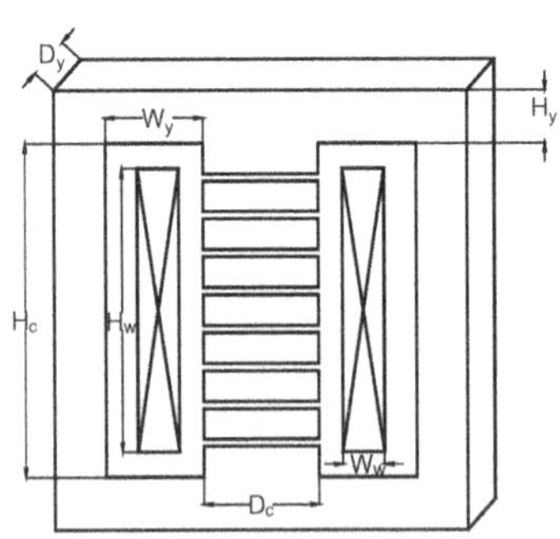

Fig. 1. The structure of the shunt reactor.

The volume of air gaps can be defined via the magnetic circuit model. It should be noted that the volume of the air gaps depends on the main parameters of the shunt reactor, i.e., reactive power, magnetic flux density, frequency of power system, energy storage in the winding space and the air gaps.

The reluctance of iron core Rc is very small in comparison with the reluctance of the air gaps Rg. It can be thus neglected in the equivalent magnetic circuit. From the relationship between the magnetomotive force (MMF), the magnetic flux and the reluctance of the magnetic circuit, the current relationship with the gap parameters in the core can be determined. The MMF and the electric current can be derived as below

$$F = I.N = \Phi.R_{g,} \tag{1a}$$

$$I = \left(\frac{1}{\sqrt{2}}\right)\frac{B_m.l_g}{\mu_0.N} \tag{1b}$$

Where $\phi, R_g, B_m, l_g, N \text{and} \mu_0$ are respectively the magnetic flux, the air gap reluctance, maximum flux density, air gap length, the number of turns in the winding and the vacuum permeability.

The winding resistance is normally very small in comparison with the inductance of the winding, it can be thus neglected when defining parameters of the electric circuits. The electromagnetic force (EMF) is defined as

$$E = \left(\frac{2\pi}{\sqrt{2}}\right).f.N.B_m.A_g \tag{2}$$

Based on the Eqs. (1b) and (2), the volume (V_g) is determined via the air gap length (l_g), and the air gap area (A_g), which is

$$V_g = A_g.l_g = \frac{Q}{\frac{\pi}{\mu_0}.f.B_m^2}, \tag{3}$$

Where ϕ_m, A_g, f, Q are respectively the maximum magnetic flux, the area of the air gap, frequency of power system and reactive power.

If the air gap area is equal to the iron core area, the diameter of the iron core is written as

$$D_c = \sqrt{\frac{4.A_g}{\pi}}. \tag{4}$$

From the area and the diameter of core in the Eq. (4), the iron yoke area can be defined as.

$$D_y = D_c; \tag{5}$$

$$H_y = \frac{A_g}{2.D_y} v \tag{6}$$

The fringing flux departs and re-enters the lamination perpendicularly, so the core blocks of a large oil-immersed shunt reactor are generally made from radially laminated silicon steels. The winding inductance can be calculated via the equivalent electric circuit or the equivalent magnetic circuit, i.e.

$$L = \frac{U^2}{\omega.Q} \text{or} L = N^2.\mu_0.\left(\frac{A_g}{l_g}\right) = N^2.P_g \tag{7}$$

From Eq. (7), the number of turns can be defined as

$$N = \sqrt{\frac{L}{\mu_0.(\frac{A_g}{l_g})}}. \tag{8}$$

The cross-sectional area of the winding A_w can be defined through the current density and the number of turns as below:

$$A_w = W_w.H_w = \frac{N.I}{k_u.J}, \tag{9}$$

Where k_u is the filled factor of the windings in the magnetic window, and J is the electric current density.

As analyzed above, a critical phenomenon appearing around the air gaps is called the fringing flux. The flux lines get distorted as it crosses the distributed air gaps in the axial direction shown in Fig. 2. The fringing fluxes spread out of the core blocks and return to the neighboring core blocks, increasing the air gaps permeance. Based on the total length of the air gap, it needs to be split into several gaps (g) to reduce the effects of the fringing fluxes, thereby reducing the fringing inductance of the shunt reactor. When the reactive power, magnetic flux density and frequency of power systems are considered as a constant, the volume of the air gaps is also a constant. Thus, in order to reduce the air gap length lg, the area (Ag) will be increased. This is corresponding to an increase of the ratio of the area to the length of air gap Ag/lg, which leads the used iron to be increased and reducing the number of turns.

In each case of the number of air gaps, when changing the distance between the air gaps from the minimum value (Hg_Min) to the maximum value (Hg_Max) make a change in the fringing flux lines, thus changing the value of the inductance.

Figure 2 points out the magnetic flux lines in case of 5-air gaps when the distance between the air gaps is changed.

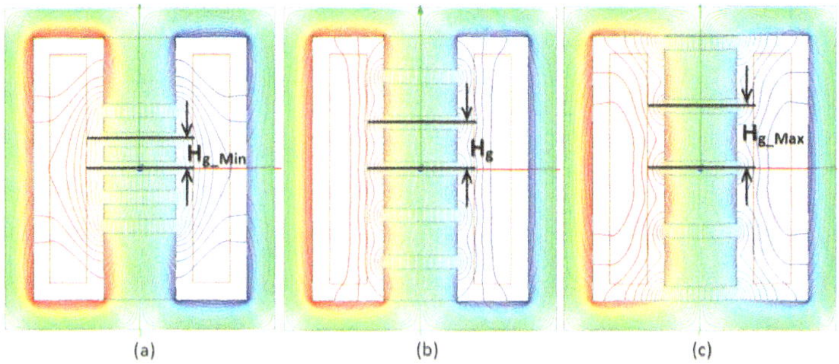

Fig. 2. The changes in the fringing flux with different H_g.

The minimum distance between the air gaps is where two air gaps touching together, this means that all of the air gaps become only one air gap. Thus, the minimum distance between the air gaps H_{g_min} can be chosen equal to twice the length of each air gap. The maximum distance between the air gaps H_{g_max} when the top and bottom core block are close to the top and bottom iron yoke as shown in Fig. 2(*c*). The parameters H_{g_min} and H_{g_max} are defined as

$$H_{g\_Min} = \frac{2l_g}{g}, \tag{10a}$$

$$H_{g\_Max} = \frac{g(H_w + 2.b_{wc}) - l_g}{g(g-1)}. \tag{10b}$$

When equally distributing the air gaps along the iron core, the distance between the air gaps can be calculated as

$$H_g = \frac{g(H_w + 2.b_{wc}) + l_g}{g(g+1)}. \tag{11}$$

The permeance of the fringing flux is defined with the iron core, i.e. [13]

$$P_f = \mu_0 . \frac{D_c}{\sqrt{\pi}} \left(1 + ln \frac{\pi h_c}{2l_g}\right). \tag{12}$$

Where h_c is the height of a core block.

The permeance of the air gaps taken the fringing flux into account is expressed as

$$P_{gf} = P_g + P_f. \tag{13}$$

The winding inductance is finally written via the term

$$L = N^2.(P_c + P_{gf}). \quad (14)$$

Where P_c is the permeance of the core.

The calculated results of five cases of the modeled shunt reactor are given in Table 1.

Table 1. Parameters of five cases of the modeled shunt reactor by an analytic method

Parameters	Symbols	Values				
		Case-1 Ag/lg = 0.5	Case-2 Ag/lg = 0.75	Case-3 Ag/lg = 1.0	Case-4 Ag/lg = 1.25	Case-5 Ag/lg = 1.5
Reactive power	Q (MVAr)	35				
Voltage	U (kV)	$500/\sqrt{3}$				
Current	I (A)	121.24				
Inductace	L (H)	7.5788				
The number of the air gaps	g	Varies from 2 to 20 air gaps along the iron core				
The total length of the air gaps	l_g (mm)	546	446	386	345	315
Diameter iron core	D_c (mm)	590	653	701	741	776
Height of iron core	H_c (mm)	2082	1907	1793	1710	1646
Turn number	N (turns)	2644	2262	2018	1846	1704
Height of winding	H_w (mm)	1812	1637	1523	1440	1376
Width of winding	W_w (mm)	302	273	254	240	229

3 Modeling of the Shunt Reactors via a FEM

A FEM is a widely acknowledged numerical method employed in calculations of stationary and transient, linear and nonlinear, isotropic and anisotropic electromagnetic problems. Maxwell® is an interactive software package of ANSYS software that uses the FEM to analyze, simulate and solve two and three-dimensional electromagnetic field problems through different forms of Maxwell's equations:

$$\text{Faraday's law} \quad \nabla \times \mathrm{E} = -\frac{\partial \mathrm{B}}{\partial \mathrm{t}}, \quad (15)$$

$$\text{Gauss' magnetism law} \quad \nabla \bullet \mathrm{B} = 0 \quad (16)$$

$$\text{Ampere's law } \nabla \times H = J + \frac{\partial D}{\partial t}, \tag{17}$$

Where B is the magnetic flux density, H is the magnetic field, E is the electric field, J is the electric current and D is the electric flux density. The density of the energy (w_m) and the inductance (L) of the winding can be computed as

$$w_m = \frac{1}{2}\vec{H}.\vec{B} \tag{18}$$

$$W_m = \iiint w_m dv \tag{19}$$

$$L = \frac{2.W_m}{I^2} \tag{20}$$

In this part, the modeling and simulation for all cases of the modeled shunt reactor with parameters given in Table 1 are performed by a FEM. The accuracy and validity of the FEM are validated by recent papers [14, 15].

Due to the anisotropy of magnetic properties, a set of 3D laminated coordinate systems was defined based on silicon steel lamination rules, in which the anisotropy of magnetic properties for laminated silicon steel in the rolling direction (RD), the transverse direction (TD) and the lamination direction (LD) were considered [12]. Firstly, the model of all parts (yoke, core blocks, winding) in 3D coordinate systems is created with core blocks being generally cylindrical as Fig. 3(a). Then the lamination direction of the yoke and all core blocks are chosen in 3D laminated coordinate systems. The core blocks of a large oil-immersed shunt reactor are generally made from radially laminated silicon steels as Fig. 3(c) [15]. The mapping between the laminated coordinate systems and the spatial rectangular coordinate systems is depicted in Table 2.

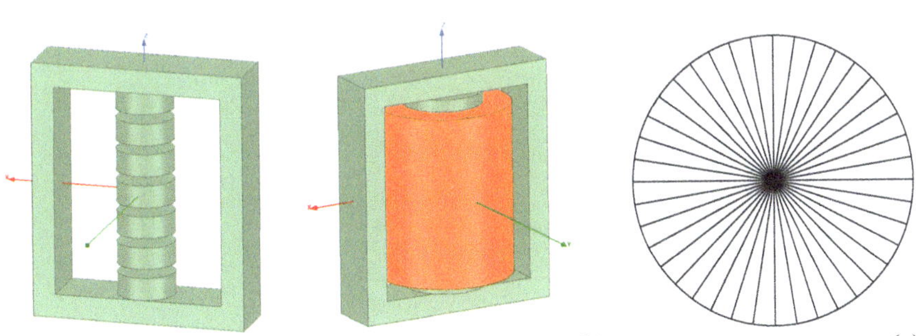

(a) (b) (c)

Fig. 3. 3D-model of the single-phase shunt reactor and radially laminated silicon steels.

Table 2. The mapping between the two coordinate systems.

Part	RD	TD	LD
Core blocks	z-axis	Perpendicular z-axis	Radial as a fan-shaped
Top and bottom yokes	x-axis	z-axis	y-axis
Side yokes	z-axis	x-axis	y-axis

In each case of the number of the air gaps from 2 to 20 air gaps, the winding inductance value is computed with changing the distance between the air gaps, from the minimum value H_{g_Min} to the maximum value H_{g_Max}. In order to determine the relationship between the winding inductance and the distance between the air gaps, it needs to create variables for modeling objects. Table 3 presents an example that has 6 air gaps, corresponding to 7 iron core blocks. All core blocks are cylindrical, in which the center position is marked from C1 to C7.

The total inductance corresponds to different air gaps in case 1 is pointed out in Fig. 4. In each case of the number of air gaps, the inductance value is computed for a distance of the air gaps varied from the minimum value H_{g_Min} to the maximum value H_{g_Max}.

As presented in Fig. 4, when (H_g) increases from H_{g_Min}, then the total inductance is decreased. This means that the fringing effect around the air gaps is also reduced. However, after reaching the minimum inductance, if H_g continues to increase up to the value H_{g_max}, the inductance turned to increase again. By looking at all the curves in Fig. 4, there is an optimum value for the distance between the air gaps to reach the minimum inductance, denoted H_{g_opt}. The values of H_{g_opt} to get the minimum inductance for each case of the number of air gaps are illustrated in Fig. 5.

Table 3. The value of the center position on rectangular coordinate systems

Center position	Value (x, y, z)
C1	0mm, 0mm, (Hw+2*bwc)/2
C2	0mm, 0mm, ((g-1)/2*lc+(g-2)/2*lg/g)
C3	0mm, 0mm, ((g-3)/2*lc+(g-4)/2*lg/g)
C4	0mm, 0mm, ((g-5)/2*lc+(g-6)/2*lg/g)
C5	0mm, 0mm, -((g-5)/2*lc+(g-4)/2*lg/g)
C6	0mm, 0mm, -((g-3)/2*lc+(g-2)/2*lg/g)
C7	0mm, 0mm, -(Hw+2*bwc)/2

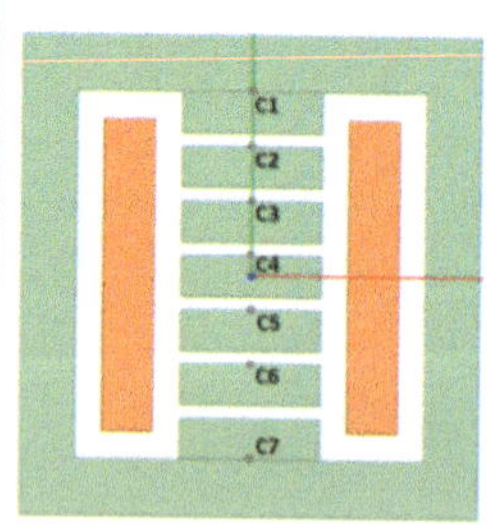

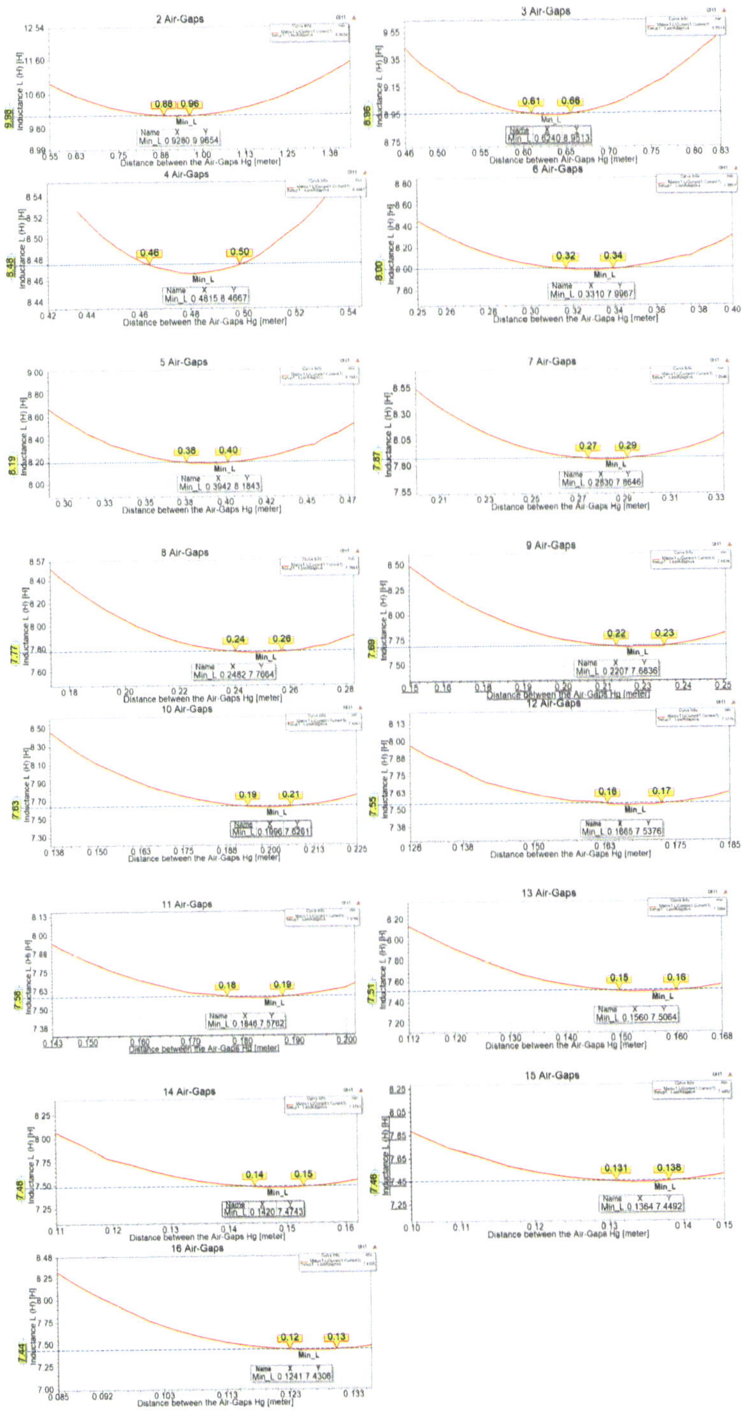

Fig. 4. The inductance versus the distance between the air-gaps H_g for $g = 2$ to 16 air-gaps.

Figure 5 presents the comparison of results Hg between two methods (FEM and analytic method). The values of Hg_Analytical calculated by an analytic method are checked to be close to the values Hg_opt obtained by FEM. The terms Hg-lo and Hg-up are the lower and upper limits of Hg, so that the error on the values of inductance is lower than 0.1% in comparison with the minimum inductance. As results shown in Fig. 5, the values of Hg calculated by the analytic method are in the range from Hg_lo to Hg_up for the values of the number of air gaps being bigger than 8.

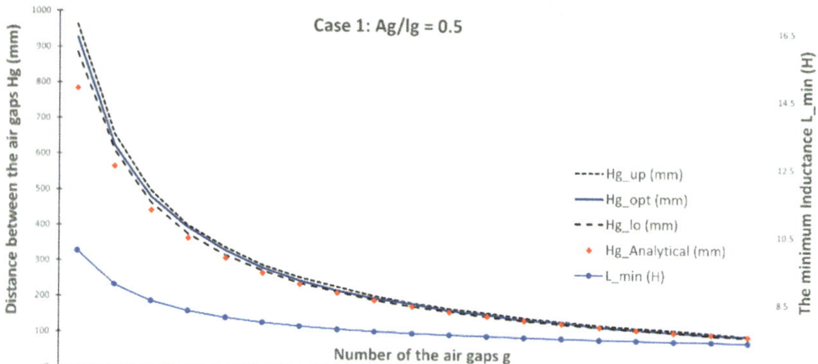

	2	3	4	5	6	7	8	9	10	11	12	13	14	15	16	17	18	19	20
Hg_up (mm)	963	655	498	401	339	291	257	231	206	187	172	160	147	138	129	122	117	109	103
Hg_opt (mm)	928	624	481.5	394.2	331	283	248.2	220.7	199.6	184.6	166.5	156	142	136.4	124.1	118.1	111.3	105.7	101.3
Hg_lo (mm)	885	610	464	376	317	275	241	217	194	177	163	150	140	131	123	116	110	103	99
Hg_Analytical (mm)	785	566	443.5	365.2	310	270	239.3	214.7	194.6	177.6	163.5	152	141	132.4	124.1	117.1	111.3	105.7	100.3
L_min (H)	9.965	8.951	8.467	8.184	7.997	7.865	7.766	7.684	7.626	7.576	7.538	7.506	7.474	7.449	7.431	7.413	7.395	7.39	7.371

Fig. 5. The changes of the distance between the air gaps versus the number of the air gaps for case 1.

Figure 5 also includes the values of the inductance according to variations of the number of the air gaps, where the increasing number of the air gaps, the decreasing the values of the inductance. Because of decreasing the length of each air gap, the fringing reluctance around the air gap is increased, therefore the fringing flux and fringing inductance decreased. As a result, when splitting into 11 air gaps, the rated winding inductance value is obtained, the value of H_{g_opt} to get the minimum inductance is 184.6 (mm). The distance between the air gaps H_g can be chosen as a value in the range from 177 to 187 (mm) or 177.6 (mm), which is calculated by analytical Eq. (11).

Similar to case 1, the simulation results for different designs of the other cases according to variations of H_g are shown in Fig. 6.

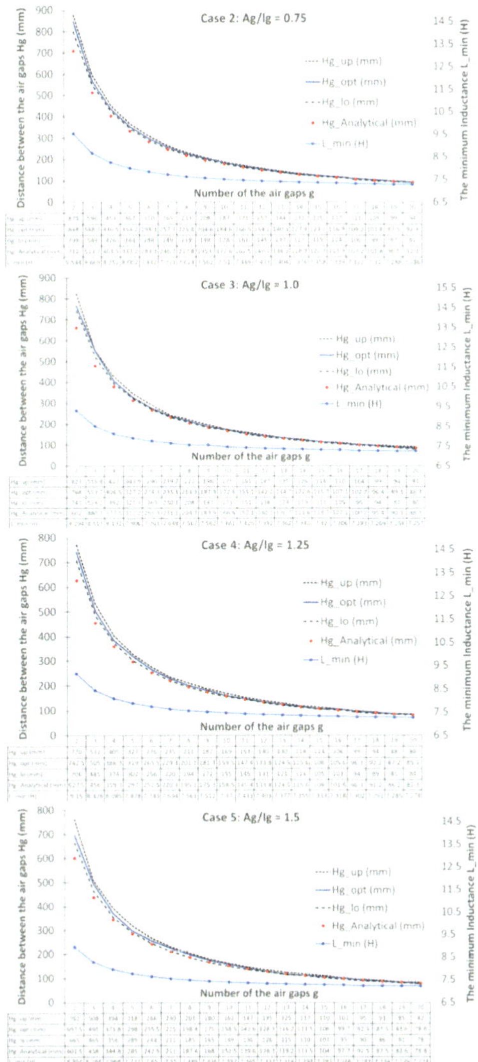

Fig. 6. The changes in the distance between the air gaps versus the number of the air gaps, for different cases.

In each case, the desired value of the winding inductance is achieved for a certain number of the air gaps. In case 1, the ratio of area to the total length of the air gaps $A_g/l_g = 0.5$, the total length of the air gaps is the longest of all cases $l_g = 546$ (mm), then it needs to be split into 11 air gaps. However, the A_g/l_g ratio is increased up to 1.5, as in case 5, the total length of the air gaps is the shortest $l_g = 315$ (mm). It should be noted that the required inductance is obtained for splitting into 7 air gaps.

The results of this study have shown that how to choose the distance between the air gaps on the iron core to obtain the minimum inductance for each case when designing shunt reactors.

4 Conclusions

In this study, the main parameters corresponding to five cases of the ratio Ag/lg of the shunt reactor have been successfully presented. In addition to the analytic method, a FEM by using Ansys Maxwell software has been applied to compute and simulate the winding inductance for changing the distance between the air gaps *Hg*, thereby giving the relationship between the winding inductance and the distance between the air gaps. When increasing *Hg* from minimum to maximum value causes changes in the fringing flux, thus also changing in the value of the winding inductance, by that way determining the optimal distance H_{g_opt} to get the minimum value of the inductance. This study has helped researchers and manufacturers to determine the distance between the air gaps of the shunt reactors in practice.

Acknowledgment. This research is funded by Hanoi University of Science and Technology (HUST) under grant number T2021-PC-006. The authors also gratefully acknowledge Quy Nhon University which created favorable conditions for the authors to use the copyright-supported Ansys software program to compute and simulate the practical problem in this research. This software is the package belonging to ANSYS Electronics Desktop V19.R1.

References

1. Divya Swarna Sri, A.: Depiction and compensation of Ferranti effect in transmission line. Int. J. Res. Appl. Sci. Eng. Technol. (IJRASET) **6**(III) (2018). ISSN: 2321–9653
2. Nashawati, E., Fischer, N., Le, B., Taylor, D.: Impacts of shunt reactors on transmission line protection. Originally presented at the 38th Annual Western Protective Relay Conference, October 2011
3. Van den Bossche, A., Valchev, V., Filchev, T.: Improved approximation for fringing permeances in gapped inductors. In: Conference Record of the 2002 IEEE Industry Applications Conference. 37th IAS Annual Meeting (Cat. No.02CH37344), vol. 2, pp. 932–938 (2002). https://doi.org/10.1109/IAS.2002.1042670
4. Yang, Z., Suryanarayana, H., Wang, F.: An improved design method for gapped inductors considering fringing effect. In: 2019 IEEE Applied Power Electronics Conference and Exposition (APEC), pp. 1250–1256 (2019). https://doi.org/10.1109/APEC.2019.8721811
5. Zhang, Y.G., et al.: Shunt reactors compensation research of UHV accessed to Jiangxi power grid. In: Advanced Materials Research, vol. 1070–1072, Trans Tech Publications, Ltd., pp. 1029–1034, December 2014. https://doi.org/10.4028/www.scientific.net/amr.1070-1072.1029
6. Minh, T.P., et al.: Finite element modeling of shunt reactors used in high voltage power systems. Eng. Technol. Appl. Sci. Res. **11**, 7411–7416 (2021)
7. Ishihara, Y., Morino, T., Todaka, T.: Analysis of magnetic fields of a delta type shunt reactor. IEEE Trans. Mag. **25**(4), 2846–2848 (1989). https://doi.org/10.1109/20.34302

8. Ma, C., Bai, B., An, Z., Dong, Y.: Study on main magnetic field of ultra-high voltage magnetically controlled saturated reactor. In: 2014 17th International Conference on Electrical Machines and Systems (ICEMS), pp. 3615–3619 (2014). https://doi.org/10.1109/ICEMS.2014.7014118
9. Lee, C.-Y., Chen, C.-J., Chen, C.-R., Hsu, Y.-F.: Comparison of Transient phenomena when switching shunt reactors on the line's two terminals and station Busbar. Presented at POWERCON 2004, Singapore, 21–24 November 2004 (2004)
10. Hutahaean, R.P., Sianipar, G.: Transient analysis of 150 kV shunt reactor at Bulukumba substation. In: Proceedings of the 2011 International Conference on Electrical Engineering and Informatics, pp. 1–6 (2011). https://doi.org/10.1109/ICEEI.2011.6021818
11. Gao, Y., Muramatsu, K., Fujiwara, K., Ishihara, Y.: Vibration analysis of a reactor driven by an inverter power supply considering electromagnetism and magnetostriction. Trans. Mag. **45**(10). 4789–4792 (2009)
12. Lü, F., Guo, J., Niu, L., Geng, J., Pan, Y.: A New 3D method for reactor core vibration based on silicon steel lamination rules and application in UHV shunt reactors. Math. Prob. Eng. 1–11 (2019). https://doi.org/10.1155/2019/7290536
13. Balakrishnan, A., Joines, W.T., Wilson, T.G.: Air-gap reluctance and inductance calculations for magnetic circuits using a Schwarz-Christoffel transformation. IEEE Trans. Power Electron. **12**(4), 654–663 (1997)
14. Arabul, A.Y., Kurt, E., Senol, I., Arabul, F.K.: An ınvestigation on flux density of three phase distributed Air-Gap 3–5 legged shunt reactor. In: IRES-27th ICIET, Amsterdam, Netherlands, 25 December 2015 (2015)
15. Dawood, K.: Modeling of distribution transformer for analysis of core losses of different core materials using FEM. In: 2019 8th International Conference on Modeling Simulation and Applied Optimization (ICMSAO) (2019)
16. Shunt reactors for medium and highvoltage networks. https://new.siemens.com/

Knot Vector Optimization of NURBS for High Speed Cam Mechanisms Based on Dynamic Characteristics

Thi Thanh Nga Nguyen[1(✉)], Van Sy Nguyen[1], and Minh Quang Tran[2]

[1] Faculty of Mechanical Engineering, Thai Nguyen University of Technology, Thai Nguyen, Vietnam
nguyennga@tnut.edu.vn

[2] Industry 4.0 Implementation Center, Center of Cyber-Physical System Innovation, National Taiwan University of Science and Technology, Taipei 10607, Taiwan

Abstract. In high speed cam mechanism, dynamic characteristics are issues that are always taken care since these affect the vibration of cam systems. This paper deals with a knot vector optimization of NURBS for high speed cam in order to minimize contact force between a cam and follower. The NURBS curve is used for displacement motion of the follower. The cam motions are improved based on the optimization of the contact force. The influence of the knot vector on the motion curves (displacement, velocity, acceleration and jerk) was investigated. The results demonstrate that the kinematics and dynamics of high speed cam mechanisms were improved.

Keywords: Cam mechanism · Knot vector · NURBS · Contact force

1 Introduction

Cam mechanisms are widely applied in many machine systems such as combustion engines, automatic machines, and cutting machines. In the cam design, the displacement motion is one of the main parameters that decide kinematic and dynamics of cam systems. Designers chose a displacement functions depending on the output motions because selecting the displacement function not only affects the velocity, acceleration, and jerk but also the working requirements. Standard functions including harmonic, cycloidal, trapezoidal, and polynomial have been described for the displacement motion [1–4]. Many researches proposed spline functions for cam design [5–11].

Regarding dynamics of cam mechanisms, Gatti and Mundo [7] studied the control of follower vibrations based on the elasticity of the mechanical components. Hu et al. [12] also presented a dynamic model of a cam mechanism with elastic roller follower. The dynamics of a globoidal cam systems was proposed [13]. This research illustrated the effectiveness of the torque compensation of the high speed cam systems. Kaplan [14] proposed a mathematical modeling and simulation of high speed cam mechanisms in order to minimize residual vibration. The dynamic performance of cam systems

D. C. Nguyen et al. (Eds.): ICERA 2021, LNNS 366, pp. 558–564, 2022.
https://doi.org/10.1007/978-3-030-92574-1_58

using rational B-splines and nonparametric rational B-spline for synthesizing the cam follower motion has been investigated in order to improve the efficiency of the cam systems [15–17].

This paper deals with an optimization of the knot vector of NURBS for cam mechanisms based on the dynamic characteristics. The NURBS curve is used to describe the follower motion. The contact force is considered as an objective function to optimize the NURBS. The organization of this paper is as follows. Section 2 presents the NURBS curve for cam displacement. The dynamics model of cam systems for computing the contact force between cam and the follower is proposed in Sect. 3. Then Sect. 4 shows the results and discussions. Finally, conclusion is drawn in Sect. 5.

2 Description of Cam Motion

The cam motion is described by the follower motion. The follower motion, denoted by y, can be expressed as the following NURBS curve in Eq. (1)

$$y(\theta) = \sum_{i=1}^{n} R_{i,p}(\theta)P_i, \theta \in [a, b], i = 1, 2, \ldots, n, \tag{1}$$

Where θ is the angle of camshaft; P_i are control points; p is degree of the NURBS curve; $R_{i,p}(\theta)$ are basis functions of NURBS and determined as

$$R_{i,p}(\theta) = \frac{N_{i,p}(\theta)w_i}{\sum_{j=1}^{n} N_{j,p}(\theta)w_j}. \tag{2}$$

Here, w_i are weights; $N_{i,p}(\theta)$ are calculated as.

$$N_{i,0}(\theta) = \begin{cases} 1 & nếu \quad \theta_i \le \theta \le \theta_{i+1} \\ 0 & Cho\ các\ trường\ hợp\ khác \end{cases}$$
$$N_{i,p}(\theta) = \frac{\theta - \theta_i}{\theta_{i+p} - \theta_i} N_{i,p-1}(\theta) + \frac{\theta_{i+p+1} - \theta}{\theta_{i+p+1} - \theta_{i+1}} N_{i+1,p-1}(\theta), \tag{3}$$

where θ_i are elements in the knot vector. The knot vector, denoted by $\boldsymbol{V}$, can be expressed as

$$V = \{\underbrace{a, \ldots, a,}_{(p+1)} \theta_{p+2}, \ldots, \theta_{m-p-1}, \underbrace{b, \ldots, b}_{(p+1)}\} \tag{4}$$

The velocity and acceleration of cam motion using NURBS can be expressed as

$$y'(\theta) = \frac{dy(\theta)}{d\theta} = \sum_{i=1}^{n} R^1_{i,p}(\theta)P_i, \quad \theta \in [a,b]. \tag{5}$$

$$y''(\theta) = \frac{d^2y(\theta)}{d\theta^2} = \sum_{i=1}^{n} R^2_{i,p}(\theta)P_i, \quad \theta \in [a,b]. \tag{6}$$

$$y'''(\theta) = \frac{d^3y(\theta)}{d\theta^3} = \sum_{i=1}^{n} R^3_{i,p}(\theta)P_i, \quad \theta \in [a,b]. \tag{7}$$

In Eqs. (5) and (6), $R^1_{i,p}(\theta)$ and $R^2_{i,p}(\theta)$ are presentative for the first and the second derivatives of the basis functions of NURBS, respectively.

3 Optimization Statement

Wear of cam surface has been occurred since the large contact force between cam and the follower. Therefore, the contact force should be minimized. To compute the contact force, the dynamic model of single degree of freedom (DOF) of cam systems is considered (see Fig. 1) [18]. Applying the Newton's second law, the motion equation can be expressed as.

$$m\ddot{y}_{out} + (c + c_f)\dot{y}_{out} + (k + k_f)y_{out} = ky + c\dot{y} \tag{8}$$

Where m is equivalent mass of the follower system; c and k are respectively presented for the damping coefficient and equivalent stiffness coefficient of the follower; c_f and k_f are representative for the damping coefficient and equivalent stiffness coefficient of the spring, respectively; y_{out} is out motion and y is cam motion as shown in Eq. (1).

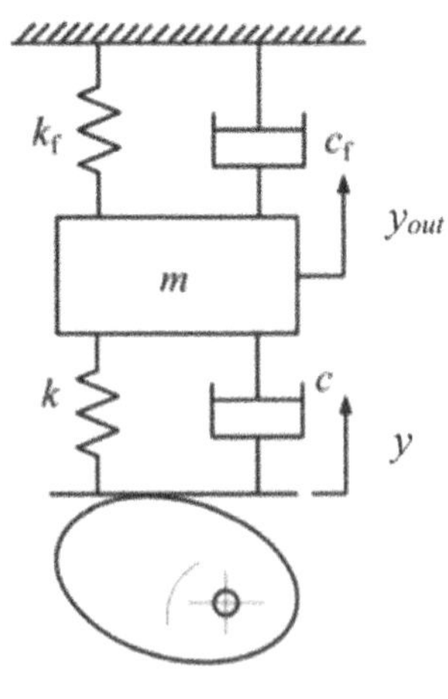

Fig. 1. Dynamic model of a single DOF of cam systems

From Eq. (8), the contact force, denoted by F, can be calculated as follows

$$\begin{aligned} F &= k(y - y_{out}) + c(\dot{y} - \dot{y}_{out}) + F_0 \\ &= m\ddot{y}_{out} + c_f\dot{y}_{out} + k_f(y_{out} + y_0). \end{aligned} \tag{9}$$

In Eq. (9), F_0 is initial contact force, i.e., $F_0 = k_f y_0$.
To minimize the contact force, the objective function can be expressed as.

$$\text{Min}F(X) = m\ddot{y}_{out} + c_f\dot{y}_{out} + k_f(y_{out} + y_0). \tag{10}$$

where X is design variables described for the knot vector of NURBS. As shown in Eq. (4), the $(p + 1)$ first knots are equal to a, and the $(p + 1)$ end knots are equal to b,. X can be written as

$$X = \left[\theta_{p+2}, \theta_{p+3}, \ldots, \theta_{m-p-1}\right]^T. \tag{11}$$

Subjected to

$$\begin{aligned} a \le \theta_i \le b \\ \theta_{i-1} \le \theta_i \le \theta_{i+1} \end{aligned} \tag{12}$$

4 Result and Discussion

This section considers the cam systems with three segments i.e., rise, dwell, and fall. The original cam mechanism used the seventh polynomial function for motion curves. The optimal knots as shown in Eq. (12) are obtained by minimizing the objective function in Eq. (10). For this example, the design variables can be written as $X = [\theta_7, \theta_8]^T$. The input parameters are used in this example as shown in Table 1.

Table 1. Input parameters and their value range

Input parameters	Symbol	Unit	Value
Equivalent mass	m	Kg	1
Equivalent stiffness	k	N/mm	3500
Spring stiffness	k_f	N/mm	40
Equivalent damping	c	Ns/m	70
Spring damping	c_f	Ns/m	8
Cam speed	n	r/min	1000

Figure 2 shows the influence of the knots on the displacement curve. It can be seen from Fig. 2 that the knot vector has large effect on the kinematic characteristics of cam systems. With the small variation of the knots, the displacement curves have enormous changes. For this, the velocity, acceleration and jerk curves also have tremendous variations so that the dynamic characteristics have been affected by these changes.

Figure 3 presents the kinematic characteristics i.e., the displacement, velocity, acceleration and jerk curves with the optimal design parameters, $\boldsymbol{X}_{op} = [0.3201, 0.6746]^T$ for NURBS. From the observation of this figure that the kinematics of the follower motion using NURBS are resulted much better than that using

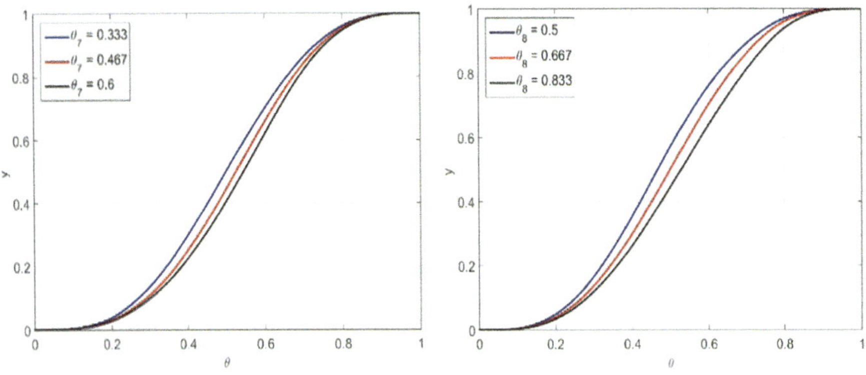

Fig. 2. Influence of the knot on the normalized displacement motion

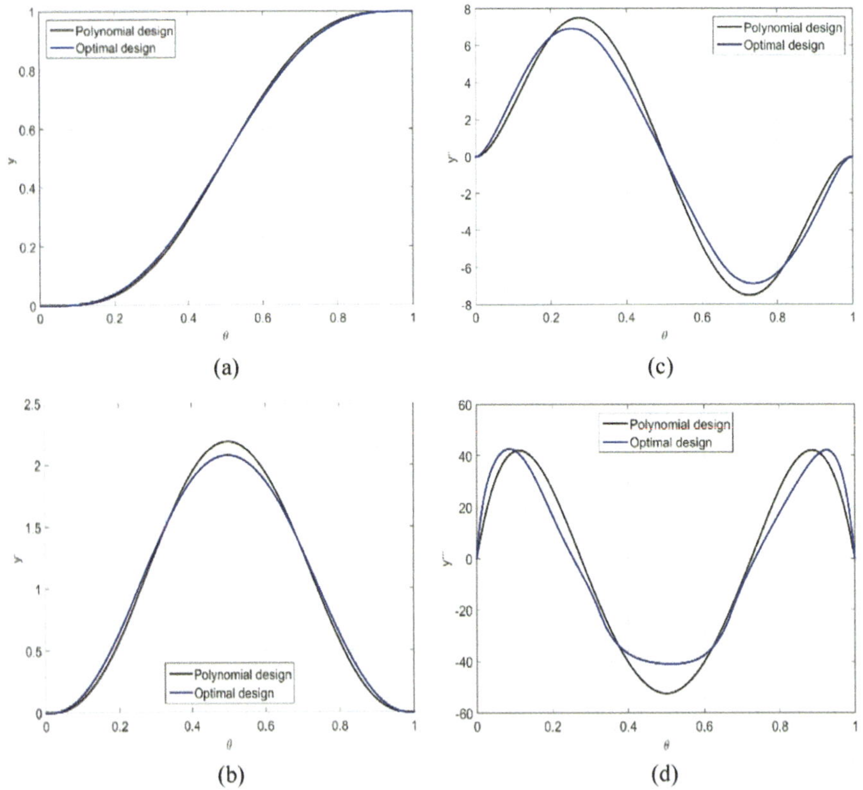

Fig. 3. Normalized kinematic characteristics of the rise motion of cam mechanisms: (a) Displacement; (b) Velocity; (c) Acceleration; (d) Jerk

polynomial. The maximum values of velocity acceleration, and jerk are $y'_p = 2.1875$; $y''_p = 7.5107; y'''_p = 52.5$ for original curves and $y'_{op} = 2.0745$; $y''_{op} = 6.9162$; $y'''_{op} = 42.5521$ for optimal NURBS.

The comparison of normalized contact force between the original motion and optimal motion is depicted in Fig. 4. It is clearly seen that the maximum value of the contact force is reduced by using the NUBRS curve for the follower motion compared with the polynomial curve.

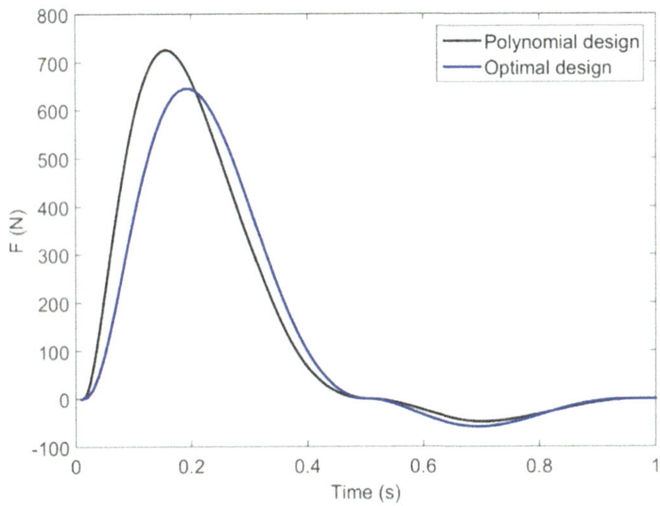

Fig. 4. Normalized contact force of the rise motion

5 Conclusion

In this paper, the optimization of the knot vector using NURBS for follower motion is proposed. The NURBS curve is generally presented for cam curves i.e., the displacement, velocity, acceleration and jerk functions. The contact force is considered for the objective function. In order to minimizing the contact force, the knot vector is obtained. The example of the cam system demonstrates that the kinematic characteristics using NURBS of the cam system is improved compared to the original cam curve. The maximum value of the contact force is also reduced. For this, in high speed cam systems, this is greatly significant to reduce wear of cam surface.

Acknowledgements. The authors thank Thai Nguyen University of Technology (TNUT) for supporting this work.

References

1. Norton, R.L.: Design of Machinery. McGraw-Hill (2003)
2. Uicker, J.J., Pennock, G.R., Shigley, J.E.: Theory of Machines and Mechanisms, 5th edn. United State of America by Oxford University Press, New York (2017)
3. Chen, H., Nguyen, T.T.N., et al.: Application of a Cam workbench for education in mechanical engineering. New Adv. Mech. Mech. Transm. Robot. Mech. Mach. Sci. 177–186 (2017). https://doi.org/10.1007/978-3-319-45450-4
4. Nguyen, T.T.N., Duong, T.X., Nguyen, V.S.: Design general Cam profiles based on finite element method. Appl. Sci. **11**, 6052 (2021). https://doi.org/10.3390/app11136052
5. Al-Samarai, R.A., Al-Douri, Y., Haftirman, Ahmad, K.R.: Tribological properties of WS2 nanoparticles lubricants on aluminum-silicon alloy and carbon steels, Walailak J. Sci. Technol. **10**, 277–287 (2013). https://doi.org/10.2004/wjst.v10i3.253
6. Mandal, M., Naskar, T.K.: Introduction of control points in splines for synthesis of optimized cam motion program. Mech. Mach. Theory **44**, 255–271 (2009). https://doi.org/10.1016/j.mechmachtheory.2008.01.005
7. Gatti, G., Mundo, D.: On the direct control of follower vibrations in Cam-follower mechanisms. Mech. Mach. Theory **45**, 23–35 (2010)
8. Nguyen, T.T.N., Kurtenbach, S., Hüsing, M., Corves, B.: Improving the kinematics of motion curves for Cam mechanisms using NURBS. In: Dede, M.I.C., İtik, M., Lovasz, E.-C., Kiper, G. (eds.) IFToMM 2017. MMS, vol. 52, pp. 79–88. Springer, Cham (2018). https://doi.org/10.1007/978-3-319-60702-3_9
9. Nguyen, T.T.N., Kurtenbach, S., Hüsing, M., Corves, B.: Evaluating the knot vector to synthesize the Cam motion using NURBS. In: Zeghloul, S., Romdhane, L., Laribi, M.A. (eds.) Computational Kinematics. MMS, vol. 50, pp. 209–216. Springer, Cham (2018). https://doi.org/10.1007/978-3-319-60867-9_24
10. Nguyen, T.T.N., Kurtenbach, S., Hüsing, M., Corves, B.: A general framework for motion design of the follower in Cam mechanisms by using non-uniform rational B-spline. Mech. Mach. Theory **137**, 374–385 (2019)
11. Nguyen, T.T.N., Nguyen, V.S., Nguyen, T.B.N., Vu, T.L.: An evaluation of B-spline for synthesis of Cam motion with a large number of output conditions, pp. 173–180 (2020)
12. Hu, B., Zhou, C., Ma, L.: Dynamics model of a cam mechanism with an elastic oscillating roller follower. In: 2015 IFToMM World Congress Proceedings, IFToMM 2015, pp. 1–5 (2015). https://doi.org/10.6567/IFToMM.14TH.WC.FA.019
13. Kuang, J.H., Hsu, C.M., Hu, C.C.: Dynamic behavior of globoidal cam systems with torque compensation mechanisms. Mech. Mach. Theory **45**, 1201–1214 (2010). https://doi.org/10.1016/j.mechmachtheory.2010.02.006
14. Kaplan, H.: Mathematical modeling and simulation of high-speed Cam mechanisms to minimize residual vibrations. Proc. Inst. Mech. Eng. Part C J. Mech. Eng. Sci. **228**, 2402–2415 (2014). https://doi.org/10.1177/0954406213519436
15. Sateesh, N.: CAM-follower systems using nurbs, pp. 15–21 (2014)
16. Jamkhande, A.K., Tikar, S.S., Ramdasi, S.S., Marathe, N.V.: Design of high speed engine's Cam profile using B-spline functions for controlled dynamics. SAE Technical Paper (2012)
17. Jiang, J.K., Iwai, Y.R.: Improving the B-spline method of dynamically-compensated cam design by minimizing or restricting vibrations in high-speed Cam-follower systems. J. Mech. Des. Trans. ASME **131**, 0410031–0410038 (2009)
18. Xia, B.-Z., Liu, X.-C., Shang, X., Ren, S.-Y.: Improving cam profile design optimization based on classical splines and dynamic model. J. Cent. South Univ. **24**(8), 1817–1825 (2017). https://doi.org/10.1007/s11771-017-3590-x

Microscopic Strain of Random Discontinuous Fiber Composites Subject to Various Macroscopic Strain Conditions

Ngo Nhu Khoa[1], Hoang Thi Hai Yen[2], Van-Truong Nguyen[3], Nguyen Thi Hoa[4], and Tien-Dat Hoang[5](✉)

[1] Department of Mechanical Engineering, Thai Nguyen University of Technology, Thai Nguyen, Vietnam
[2] Faculty of Political Theory, Thai Nguyen University of Technology, Thai Nguyen, Vietnam
[3] Faculty of Fundamental Sciences, Thai Nguyen University of Technology, Thai Nguyen, Vietnam
[4] Department of Automotive Engineering, Thai Nguyen University of Technology, Thai Nguyen, Vietnam
[5] Faculty of International Training, Thai Nguyen University of Technology, Thai Nguyen, Vietnam
hoangdat@tnut.edu.vn

Abstract. With the wide applications of composite materials, virtual simulation tools become a big matter of concern in the new development process to predict the behaviors of materials before fabricating. The mechanical properties, behaviors of composites are strongly dependent on their microstructures. To link the gap between microstructure to macroscopic properties, the homogenization technique is used. Conversely, the localization technique is utilized to investigate the behavior of constituent materials at microstructure under macroscopic loading conditions. In this paper, the asymptotic homogenization (AH) method is applied to investigate the behavior of fibrous composite materials under different macroscopic strain conditions which are very complicated to carry out by experiment works. For demonstration, a short fiber reinforced plastic model with random fiber orientation, random fiber arrangement, and random fiber length is analyzed. The influences of strain conditions on this complex microstructure on the microscopic strain distribution are observed and visualized on any arbitrary cross-section. The results show that the high microscopic strain arises more in the case of coupled shear strain conditions through histograms, whereas the largest microscopic strain appears in the case of coupled transverse strain and shear strain conditions. This analysis is worthy of the discontinuous fiber composite material design, initial damage prediction, and damage propagation analysis for further steps.

Keywords: Microscopic strains · Fibrous composite materials · Homogenization method

D. C. Nguyen et al. (Eds.): ICERA 2021, LNNS 366, pp. 565–573, 2022.
https://doi.org/10.1007/978-3-030-92574-1_59

Acronyms and Symbols

σ	Stress
C	Stiffness tensor
ε	Strain
x	Macroscopic scale
y	Microscopic scale
v	Poisson ratio
u	Displacement
χ	Characteristic displacement
δu	Virtual displacement
E	Macroscopic strain
RVE	Representative volume element
SVE	Stochastic volume element
n	Level of standard deviation
i, j, k, l, p, q	Index (from 1 to 3)
T	Traction
Γ	Boundary
Λ	Scale ratio
Y	Domain of RVE
$\|Y\|$	Volume of RVE
Mean	Mean value
S.D	Standard deviation
AH	Asymptotic homogenization

1 Introduction

Composite materials have many applications in a broad range of engineering fields and industries such as aerospace, automobile, civil engineering and so on. However, modelling and predicting the properties, microscopic strain of these materials remain a challenge due to their heterogeneities in microstructures. Homogenization techniques are well-known tools to efficiently derive those homogenized composite properties analytically or numerically from the constituent properties and the microstructures of heterogeneous materials. Various micromechanical models have been developed to predict the macroscopic properties and simulate the behavior of composite materials. In the analytical homogenization branch, Eshelby proposed a method to predict the effective moduli of composite considering the ellipsoidal inclusion in 1957 [1]. Later, Mori and Tanaka [2] developed a technique based on Eshelby's inclusion method to estimate the average internal stress in a matrix containing inclusion with eigenstrains. However, these analytical homogenization methods can only consider uniform strain or stress inside microstructures.

Among numerical homogenization methods with the help of finite element method as a numerical tool, AH method is very popular and is used to model, predict the material properties, and compute the behavior of complex composite materials accurately. Terada and Kikuchi have employed the homogenization method to characterize the response of heterogeneous solids undergoing inelastic deformations in three-dimensional elasto-plastic problems and presented a rigorous computational tool of using digital image-based modeling with asymptotic homogenization method to study complex micromechanical characteristics of composite materials [3]. Fish et al. [4] have developed a nonlocal damage theory for brittle composite materials using homogenization method and localization techniques based on the double scale asymptotic expansion of damage. In homogenization methods with the help of finite element method, representative volume element (RVE) is used to analyze complex composite materials. In these works, the RVE models of the materials were only investigated after the materials were fabricated. In designing and manufacturing new materials, there is a need to predict the characteristics and behaviors of the materials before fabrications.

In this paper, microscopic strain distributions of discontinuous fibrous composite materials due to various macroscopic strain conditions are predicted. The simulation is performed by stochastic homogenization method. In the numerical example, a random short fiber composite model is used to demonstrate the proposed prediction. Axial, shear, and coupled macroscopic strains are applied on the model. The obtained results can capture not only microscopic strain concentrations but also influences of macroscopic strain conditions on the microscopic strain distributions. This investigate is worth of further steps to predict stochastic microscopic damage behavior of short fiber reinforced composite materials due to complex loading conditions.

2 Homogenization and Localization in Linear Analysis

This section presents in general the AH technique in elasticity domain. In AH approach, composite structures are assumed to be composed of many periodic unit cells which is so call RVE. The behavior of a whole structure can be obtained from the analysis of the RVE. In hierarchical analysis, thee scales including microscale, mesoscale and macroscale are linked together. The homogenization technique provides the properties or response of a structure at higher scale given the properties or response of the structure's constituent materials at lower scale. Conversely, localization techniques provides the response of the constituents given the response of the structure. During multiscale analysis, a particular stage in the analysis procedure can function on both levels simultaneously. In this paper, only two-scaled problems between microscale and macroscale are considered.

The constitutive equation of elasticity is addressed by

$$\sigma_{ij} = C_{ijkl}\varepsilon_{kl} \tag{1}$$

where C_{ijkl} is the stiffness tensor. σ_{ij} and ε_{kl} are the stress and strain of the RVE.

The displacements are approximated by an asymptotic expansion of a scale ratio λ which is the ration between macroscopic scale x and microscopic scale y [5].

$$u_i(x) = u_i^{(0)}(x,y) + \lambda u_i^{(1)}(x,y) + \lambda^2 u_i^{(2)}(x,y) + \lambda^3 u_i^{(3)}(x,y) + \cdots \quad (2)$$

where u_i^0 is a macroscopic displacement, and $u_i^{(1)}, u_i^{(2)}, u_i^{(3)}, \ldots$ are perturbed displacements due to the microscopic heterogeneity of microstructure.

Suppose that the structure is subjected to a traction T on the boundary Γ. The body force, thermal force are neglected. By taking the limit of scale ratio λ to be 0, the homogenization is achieved. In this analysis, for reducing computation cost, only first order perturbed displacement is considered. Using the separating variable method, the perturbed displacement can be expressed as

$$u_i^{(1)}(x,y) = -\chi_i^{pq} \frac{\partial u_p^{(0)}}{\partial x_q} \quad (3)$$

where χ_i^{pq} is a characteristic displacement mode which is very important to bridge the gap between microscopic scale to macroscopic scale. Characteristic displacement is a periodic function at microscale.

The microscopic equation can be derived as follows [6]

$$\int_Y \left(C_{ijkl} - C_{ijpq} \frac{\partial \chi_p^{kl}}{\partial y_q} \right) \frac{\partial \delta u_i^{(1)}}{\partial y_j} dY = 0 \text{ for } \forall \delta u_i^{(1)} \quad (4)$$

where $\delta u_i^{(1)}$ is virtual displacement, Y and $|Y|$ are the domain and volume of RVE, respectively.

The microscopic strain in RVE is expressed as [7]

$$\varepsilon_{ij}(y) = E_{ij} - \frac{\partial \chi_i^{kl}}{\partial y_j} E_{kl} \quad (5)$$

where E_{ij} and E_{kl} are the uniform macroscopic strains.

Finite element method is applied to solve the characteristic displacement in Eq. 4. Following that, the homogenized matrix and microscopic strain of RVE can be computed. If taking into account the randomness or uncertainty of material model, the displacement can be expressed by an asymptotic expansion concerning random variables in the same way as in Ref. [7]. It results in the variance of microscopic strain as in Eq. (6). Where, ε_{mean} and $\varepsilon_{S.D.}$ are mean and standard deviation of microscopic strain, respectively. n is the level of standard deviation.

$$\varepsilon = \varepsilon_{mean} + n \times \varepsilon_{S.D.} \quad (6)$$

3 Numerical Simulations

For investigating the behaviors of discontinuous fibrous composites due to various macroscopic strain conditions, a short fiber reinforced plastic model is analyzed. The mechanical properties of constituents are listed in Table 1. In this simple example, the fiber and resin matrix are considered as isotropic materials. The uncertainty is considered in the matrix phase as same as that in Ref. [8].

Table 1. Elastic module of glass fiber and epoxy resin [9]

Material	E (GPa)	ν
Silenka E-Glass 1200tex	72	0.26
MY750/HY917/ DY063 epoxy	3.79	0.37

The fiber orientation pattern and arrangement have dominant influences of injection molded short fiber composites. The composite is stiffer in the direction of the longitudinal orientation but much weaker in the transverse direction. For large samples like injection molded plates, the composite molding has a typical skin–core–skin structure with two skin layers. The fibers are highly oriented in the flow direction and a core layer whereas fibers are mainly aligned transversely to the flow direction. The proposed method can be used to simulate the behaviors of both skin and core layers. However, for reducing the computational time, in this study, the illustrated model was used to investigate the skin layer.

In the example, a fiber length distribution is selected randomly from the measure data [8]. The model has 16 short glass fibers with 3 fiber lengths of 0.435 mm, 0.543 mm, and 0.645 mm. The size of the SVE is 224 mm × 224 mm × 672 mm suitable to model the skin layer and core layer as well. The volume fraction of fiber is 4.5%., the behaviors of short composite material under different macroscopic strains have not been yet investigated in the author' work [8]. In this example, those behaviors will be investigated comprehensively. The model is subject to various macroscopic strain conditions including five study cases E_{22}, E_{23}, E_{23} coupled E_{31}, and E_{22} coupled E_{23}. The fiber orientation and fiber arrangement are randomly in the RVE. Where $E_{22} = E_{23} = E_{31} = 0.5\%$. The model is constrained by periodic boundary condition. The mesh model has 330,869 tetrahedral 10 nodes-elements and 447,727 nodes. The formulation is solved by an in-house FORTRAN program.

As the computed results, microscopic strain of all elements were computed. For observing the strain concentration and microscopic strain distributions in an arbitrary cross-sections can be visualized. However, only two typical cross-sections were selected in this paper (Fig. 1).

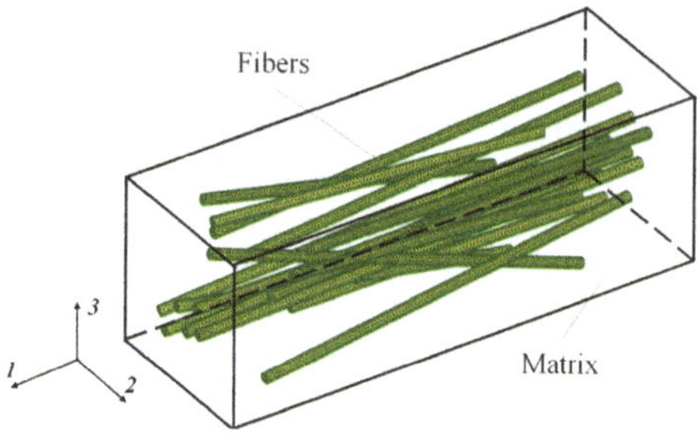

Fig. 1. Random short fiber composite RVE

As the obtained results, Fig. 2 shows the microscopic strain distributions on the middle cross-section 2–3, one of arbitrary cross-sections, which reveals the deformations of fibers and strain concentrations near the fibers in the five study cases. The behaviors of the model are much more complex than those of unidirectional composite. The strain concentrations appear in the locations where fibers are close each other. The damage can initiate from these locations. The cross-sections 1–2 and 1–3 in Fig. 2 provide the evident strain concentrations along with the fibers. The high strains arise near the ends of fibers in case of E_{22}, and E_{23} coupled E_{31}. However, in case of strain condition E_{22} coupled E_{23} the high strain arises more near the middle of fibers because of the transverse strain E_{22}. It can be pointed out that the material model is weaker in the transverse direction. The high strains appear more in case of coupled strain conditions. Especially, the largest strain can be found in case of transverse strain E_{22} and shear strain E_{23}. Additionally, the high strain areas are not only affected by strain loading conditions but also fiber arrangements. Moreover, this approach can be used to predict the stochastic damage propagations considering uncertainties at microstructures where these propagations are impossible to capture by experiments.

The statistical histograms of microscopic strains due to different strain loading conditions are shown in Fig. 3. Surprisingly, the estimated number of high strain elements in the case of E_{23} coupled E_{31} is larger than that in the case of E_{22} coupled E_{23}. It may result in more catastrophic damage behavior due to coupled shear strain conditions. Obviously, the model is safer due to only single shear strain E_{23}. It is worth noting that a normal distribution can accurately represent the microscopic strain distribution in the case of E_{23}, whereas other non-normal distributions should be used for better representation of the histogram shapes in case of coupled strain conditions. This prediction is crucial to assume the suitable strain distribution of damage criteria during reliability analysis. From Fig. 3, it is obvious to know that the effective strains in the matrix phase is around 10 times larger than those in the fiber phase. These obtained results can be used to figure out the damage behavior of fiber and matrix phases (Fig. 4).

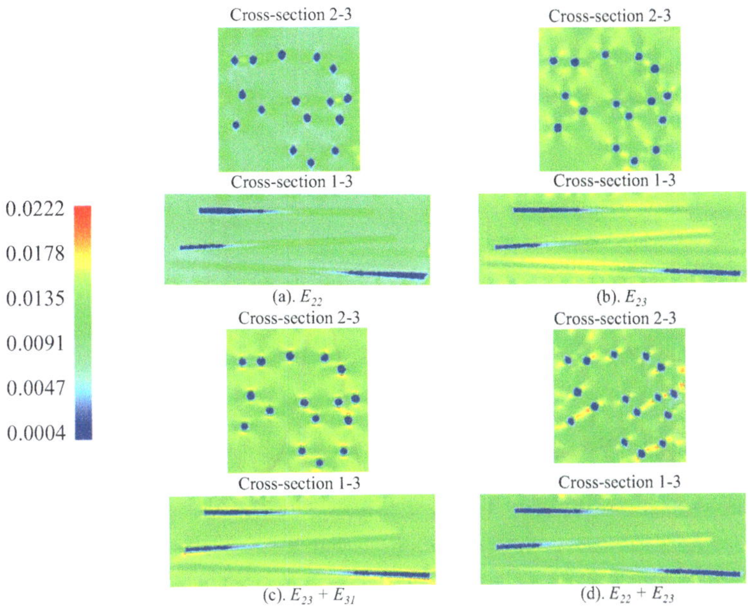

Fig. 2. Effective microscopic strain distribution on middle cross-sections 2–3 and 1–3

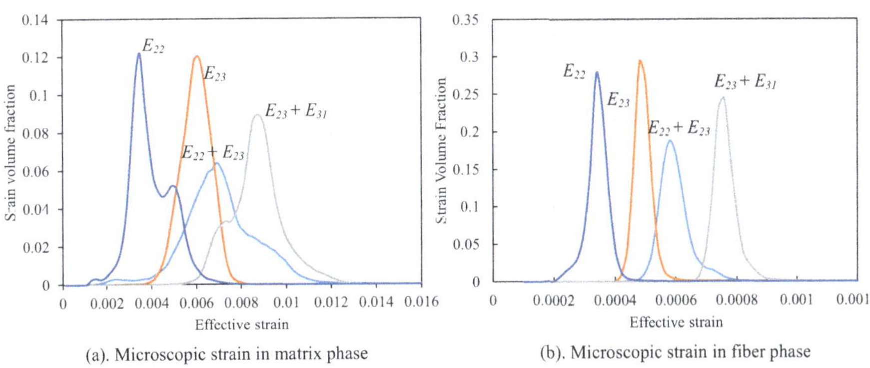

Fig. 3. Histograms of Mises microscopic strains subject to different macroscopic strains

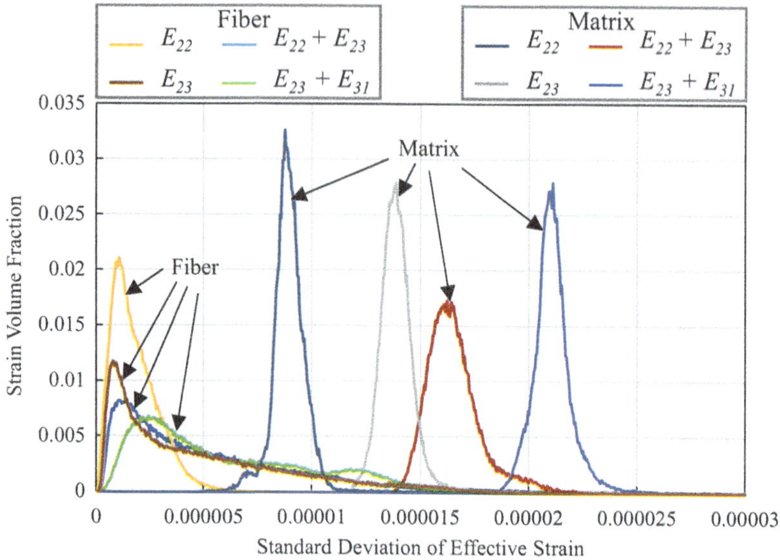

Fig. 4. Standard deviation of Mises microscopic strains subject to uncertainty in matrix phase

To improve the reliability of the prediction, the standard deviation of microscopic strains should be figured out. Although, in this example, only physical parameters of the matrix is considered as a random distribution with 3% of standard deviation, the variances of microscopic strains of both fiber and matrix phases were still captured. Those variances are very challenged to capture by experiments to explain the variation of obtained results. This simulation scheme can be used to point out the complex behaviors of composite materials since it can figured out the influence of randomness or uncertainty in microstructures. The values of the standard deviations are quite small because the randomness in mechanical properties is considered only for matrix with small variability. However, they should not be ignored to reduce risk during prediction or design new materials.

4 Conclusions

In this paper, the stochastic homogenization method was applied to investigate the behaviors of random short fiber reinforced composite material due to various strain conditions. Two-scaled problem of RVE model with periodic boundary condition was used to link the gap between microscale and macroscale. Therefore, the microscopic strains can be obtained by solving the microscopic equation considering macroscopic conditions. Four study cases were carried out to figure out the influences of various strain conditions on the microscopic strain distributions. The results show the high strains appear near the ends of fibers in case of E_{22}, and E_{23} coupled E_{31}, whereas the high strains appear near the middle of fibers in case of strain condition E_{22} coupled E_{23}. It means the material model is weaker in the transverse direction and becomes worse under coupled strain conditions. Additionally, the high strain areas, where damages can happen first, also depend on fiber arrangement and randomness of mechanical properties of constituent materials. Furthermore, this approach can be used to analyze the behavior of nanocomposites based polymers. However, the model of interface or adhesive layer between fibers and matrix phases should be formulated since the effect of interfaces is significant in the case of nanocomposites.

Acknowledgement. This work was supported by Thai Nguyen University of Technology for a scientific project.

References

1. Eshelby, J.D.: The determination of the elastic field of an ellipsoidal inclusion, and related problems. Proc. R. Soc. Lond. A **241**, 376–396 (1957)
2. Mori, T., Tanaka, K.: Average stress in matrix and average elastic energy of materials with misfitting inclusions. Acta Metall. **21**, 571–574 (1973). https://doi.org/10.1016/0001-6160(73)90064-3
3. Terada, K., Miura, T., Kikuchi, N.: Digital image-based modeling applied to the homogenization analysis of composite materials. Comput. Mech. **20**, 331–346 (1997). https://doi.org/10.1007/s004660050255
4. Fish, J., Yu, Q., Shek, K.: Computational damage mechanics for composite materials based on mathematical homogenization. Int. J. Numer. Methods Eng. **45**, 1657–1679 (1999)
5. Takano, N., Ohnishi, Y., Zako, M., Nishiyabu, K.: The formulation of homogenization method applied to large deformation problem for composite materials. Int. J. Solids Struct. **37** (44), 6517–6535 (2000)
6. Takano, N., Ohnishi, Y., Zako, M., Nishiyabu, K.: The formulation of homogenization method applied to large deformation problem for composite materials. Int. J. Solids Struct. **37**, 6517–6535 (2000)
7. Hoang, T.-D., Takano, N.: First-order perturbation-based stochastic homogenization method applied to microscopic damage prediction for composite materials. Acta Mech. **230**(3), 1061–1076 (2018). https://doi.org/10.1007/s00707-018-2337-6
8. Hoang, T.-D., Abe, Y., Nakamura, S., Miyoshi, A., Takano, N.: Stochastic nonlinear multiscale computational scheme for short fiber reinforced composites to study the influence of microstructural variability on damage propagation. SN Appl. Sci. **2**(2), 1–16 (2020). https://doi.org/10.1007/s42452-020-1961-7
9. Soden, P.D., Hinton, M.J., Kaddour, A.S.: Lamina properties, lay-up configurations and loading conditions for a range of fibre reinforced composite laminates. Fail Criteria Fibre-Reinf.-Polym. Compos **58**, 30–51 (2004)

Model Predictive Control for Vehicle Active Suspension Systems

Lai Thi Thanh Hoa[1(✉)], K. L. Lai[1], and Tran Ngoc Son[2]

[1] Thai Nguyen University of Technology, Thái Nguyên, Vietnam
laithithanhhoa@tnut.edu.vn
[2] University of Economics - Technology for Industries, Hanoi, Vietnam

Abstract. Model Predictive Control is an advanced control strategy that is suitable for multi-target optimal control problems with multiple inputs and multiple outputs that has been applied to many fields. This paper presents an optimal control design method based on the model for the active suspension system, including building the control structure, define the input and output states of the system and the signals constraints. Simulation examples of the control systems with specific parameters are also presented in the paper. The simulation results show the effectiveness of the proposed method.

Keywords: Model predictive control · Vehicle active suspension · Optimal control structure · MPC

1 Introduction

As is well known, the vehicle suspension plays an important role in vehicle performance, which is a hot topic that has long attracted the research interest of scientists. Quality requirements for vehicle suspension systems include ride comfort, traction and suspension deflection. However, these requirements are contradictory so a compromise must be reached between them in order to the system to achieve the best possible quality. To achieve this goal, a significant number of research has been carried out [3, 6]. Among the proposed solutions, active suspension is a popular approach to improve suspension quality and has great potential to meet user requirements [2, 4, 8]. The purpose of active suspension control is to improve ride comfort, while keeping the suspension stroke and tire deflection at an acceptable level. In recent years, much attention has been paid to active damping control. Some important results based on active suspension control approaches using different control techniques have been proposed, such as fuzzy logic control and neural networks [1], adaptive control [5], Distributed Backstepping [7], H_∞ Loop-Shaping robust control [9], etc.

Model Predictive Control (MPC), as the name suggests, is a feedback control technique that based on a model. It's also relies on a real-time optimizer. Other names for MPC are receding horizon control and quadratic programming control. We can also call MPC a new name which is Optimization and Prediction Control (OPC). It is a made-up name by the authors to help the reader remember that MPC also carries an optimization element. The MPC's cost function is made of four elements:

D. C. Nguyen et al. (Eds.): ICERA 2021, LNNS 366, pp. 574–580, 2022.
https://doi.org/10.1007/978-3-030-92574-1_60

$$J(zk) = J_y(z_k) + J_u(z_k) + J_{\Delta u}(z_k) + J_\varepsilon(z_k) \quad (1)$$

Where

z_k is the sequence of manipulated variables from sample k to k + p − 1
J_y is the cost function for output reference tracking (or tracking error)
J_u is the cost function for manipulated variable tracking (or deviation from nominal manipulated variable).
$J_{\Delta u}$ is the cost function for change in manipulated variables
J_ε is the cost function for constraint violations

Currently MPC has become an advanced control strategy and has been widely applied in many different fields, such as petrochemical filtration technology, food processing technology, space technology, paper technology, etc.

The application of MPC to active suspension will be an effective solution to solve the problem of multi-goal optimization, managing the balance between conflicting factors in the active suspension include ride comfort, traction and suspension deflection.

The rest of this article is organized as follows. The problem of mathematical description is presented in Sect. 2, the controller design is presented in Sect. 3. Section 4 provides an example of a design and finally some conclusion comments are made in Sect. 5.

2 Mathematical Model of Suspension

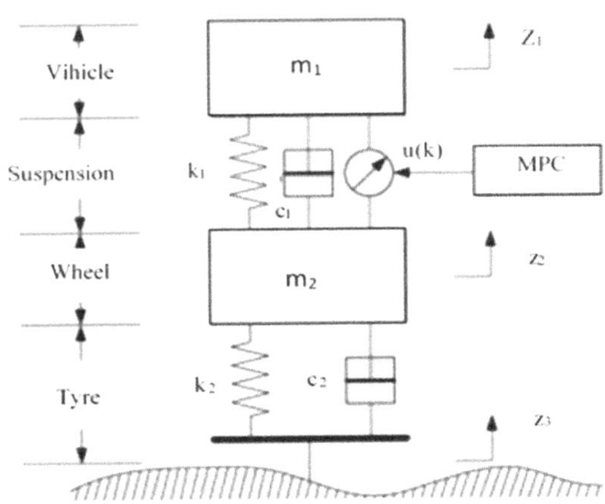

Fig. 1. Quarter-car model with an active suspension

Consider the model shown in Fig. 1. In this figure, m_1 is the sprung mass, representing the car chassis; m_2 is the unsprung mass, representing the mass of the wheel cluster; c_1 and k_1 are damping and stiffness of the suspension system, respectively; k_2 and c_2 stand for compressibility and damping capacity of the pneumatic tires respectively; z_1 and z_2

are the displacements of the sprung and unsprung masses, respectively; z_3 is the transposition input of the path; u is the active input of the suspension system.

The ideal dynamic equations of the sprung and unsprung masses are given by

$$\begin{cases} m_1\ddot{z}_1(t) + b_1[\dot{z}_1(t) - \dot{z}_2(t)] + k_1[z_1(t) - z_2(t)] = u(t) \\ m_2\ddot{z}_2(t) - b_1[\dot{z}_1(t) - \dot{z}_2(t)] - k_1[z_1(t) - z_2(t)] + k_2[z_2(t) - z_3(t)] + b_2[\dot{z}_2(t) - \dot{z}_3(t)] = -u(t) \end{cases} \tag{2}$$

When designing control laws for active suspension, we need to consider the following factors:

Ensure ride comfort: It has been accepted that the ride comfort is closely related to body acceleration. Therefore, when we design the controller, one of our main goals is to reduce body acceleration:

$$\ddot{z}_1 = 0 \tag{3}$$

- Ensure mechanical strength: due to mechanical structure, the suspension stroke should not exceed the maximum allowed, that is

$$|z_1(t) - z_2(t)| \le z_{max} \tag{4}$$

- In addition, in order to make sure the car safety, we should ensure the firm uninterrupted contact of wheels to road and the dynamic tire load should be small, that is

$$k_1(z_2(t) - z_3(t)) < (m_1 + m_2)g \Rightarrow \frac{k_1(z_2(t) - z_3(t))}{(m_1 + m_2)g} < 1 \tag{5}$$

Another hard constraint imposed on active suspensions is from the limited power of the hydraulic actuator

$$|u(t)| \le u_{max} \tag{6}$$

3 Build Controllers

In this section, we will design a predictive controller to generate a signal u(t) acting on the vehicle suspension so that the expressions (2), (3), (4) are satisfied with the constraint (5)

First, we need to represent expression (1) in the state space in a form suitable for the design of the MPC controller. Set the state variables as follows

$$x_1(t) = (z_1 - z_2)$$
$$x_2(t) = (z_2 - z_3)$$
$$x_3(t) = \underline{z}_1$$
$$x_4(t) = \underline{z}_2$$
$$x(t) = [x_1(t) \quad x_2(t) \quad x_3(t) \quad x_4(t)]^T$$

where $x_1(t)$ represents the suspension deflection, $x_2(t)$ is the tire deflection, $x_3(t)$ is the sprung mass speed; $x_4(t)$ is the unsprung mass speed. Noise input is defined

$$w(t) = \dot{z}_3(t)$$

The kinetic Eqs. (1) can be written

$$\dot{x}(t) = A\begin{bmatrix} x_1(t) \\ x_2(t) \\ x_3(t) \\ x_4(t) \end{bmatrix} + B\begin{bmatrix} u(t) \\ w(t) \end{bmatrix}$$

where

$$A = \begin{bmatrix} 0 & 0 & 1 & -1 \\ 0 & 0 & 0 & 1 \\ -\frac{k_1}{m_1} & 0 & -\frac{b_1}{m_1} & \frac{b_1}{m_1} \\ \frac{k_1}{m_2} & -\frac{k_2}{m_2} & \frac{b_1}{m_2} & -\frac{b_1+b_2}{m_2} \end{bmatrix}; B = \begin{bmatrix} 0 & 0 \\ 0 & -1 \\ 1/m_1 & 0 \\ -1/m_2 & b_2/m_2 \end{bmatrix}$$

$$C = \begin{bmatrix} 0 & 0 & 1 & 0 \\ 1 & 0 & 0 & 0 \\ 0 & \frac{k_1}{(m_1+m_2)g} & 0 & 0 \end{bmatrix}; D = \begin{bmatrix} 0 & 0 \\ 0 & 0 \\ 0 & 0 \end{bmatrix}$$

From the above equations, we build a predictive controller according to the model consisting of 2 inputs, 3 outputs including:

- Input 1: u(t) is the measurable input, unit N
- Input 2: undulating road surface r(t) is random noise, unmeasurable, unit (m)
- Output 1: body acceleration (m/s^2); measurable quantity
- Output 2: Suspension stroke $(z_1 - z_2)$ (m) unmeasurable
- Output 3: traction (kg m/s^2) $\frac{k_1(z_2(t)-z_3(t))}{(m_1+m_2)g} < 1$

Constraint: $|u(t)| \le u_{\max}$.

4 Simulation Results

To compare the performance of the proposed method with Hun's H∞ control results [9], we set the parameters of the active suspension system to be the same as [9], shown in Table 1.

Table 1. Parameters of 1/4 suspension system

m_1	m_2	k_1	k_2	b_1	b_2
973 kg	114 kg	92720 N/m	101115 N/m	1095 Ns/m	14.6 Ns/m

- Quality requirements $\ddot{z}_1 \to 0$; $|z_1(t) - z_2(t)| \le z_{max} = 0.08$ m; $\frac{k_1(z_2(t)-z_3(t))}{(m_1+m_2)g} < 1$
- Constraint conditions $|u(t)| \le u_{max} = 1500$ N; $\underline{x}_1 < 0.01$ m/s^2; $x_2 < 0.08$ m

Simulation diagram as Fig. 2

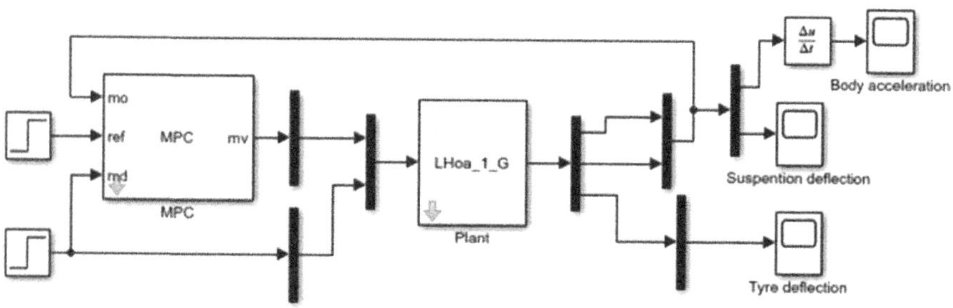

Fig. 2. Schematic diagram of the vehicle's suspension control system

The simulation results are shown in Fig. 3, Fig. 4 and Fig. 5. Table 2 presents a comparison of the results of the proposed paper with those published in [9] and [10]. From the simulation results, we can see those parameters such as body acceleration, suspension deformation and traction are within the allowable limits. Although the parameters of the proposed method have more fluctuations, the amplitude of fluctuations is much smaller than the results published by [9] and [10].

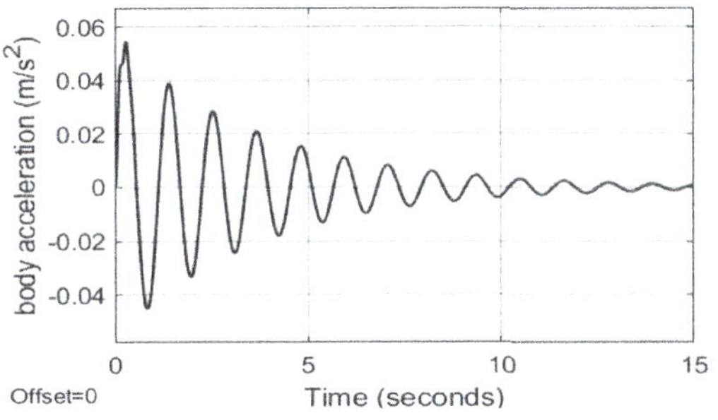

Fig. 3. Response of body acceleration

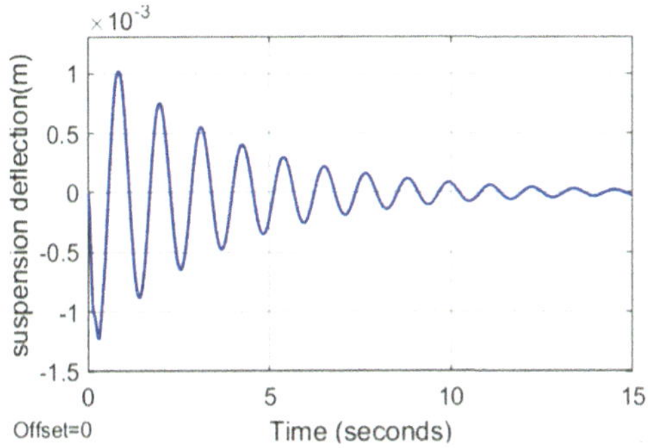

Fig. 4. Suspension deflection (z_1 - z_2)

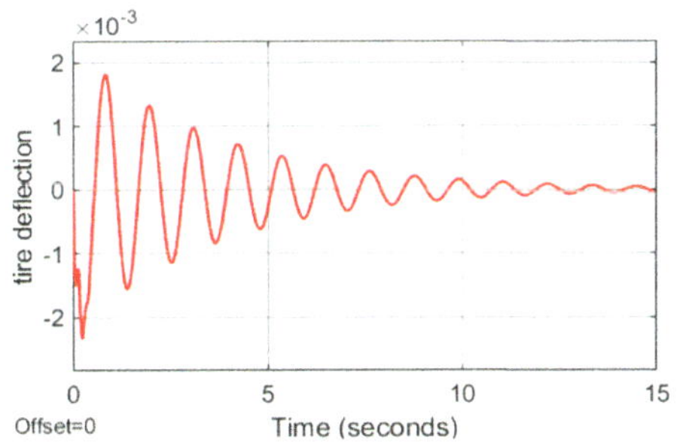

Fig. 5. Type deflection

Table 2. Comparing the results of this study with those already published in [9] and [10]

Parameters	Results in this paper	Results in [9]	Results in [10]
Max of of body acceleration (m/s^2)	0.057	2	4
$(z_1 - z_2)_{max}$ (m)	10–3	$4 * 10^{-2}$	10^{-2}
Max of deflection (m)	$1.7 * 10^{-3}$	0.15	–
Transitional time (second)	10	2	2.5

5 Conclusion

In this paper, a model predictive controller with multiple inputs and outputs is first applied to an active damping system. In this way, it is easy to compromise the performance characteristics of the active suspension to achieve an acceptable set of performance parameters while create comfort and safety riding. The simulation results with specific data have shown the effectiveness and feasibility of the proposed solution. However, this paper only mentions applying MPC to 1/4 vehicles. In the next step, we will study the application of MPC for the whole vehicle.

References

1. Al-Holou, N., Lahdhiri, T., Joo, D.S., Weaver, J., Al-Abbas, F.: Sliding mode neural network inference fuzzy logic control for active suspension systems. IEEE Trans. Fuzzy Syst. **10**(2), 234–246 (2002)
2. Chen, H., Liu, Z.Y., Sun, P.Y.: Application of constrained control to active suspension system on half-car models. J. Dyn. Syst. Meas. Control **127**, 345–354 (2005)
3. Du, H., Lam, J., Sze, K.Y.: Non-fragile output feedback vehicle suspension control using genetic algorithm. Eng. Appl. Artif. Intell. **16**(7–8), 667–680 (2003)
4. Du, H., Lam, J., Sze, K.Y.: Design of non-fragile controller for active vehicle suspensions. J. Vibr. Control **11**, 225–243 (2005)
5. Fialhm, I., Balas, G.J.: Road adaptive active suspension design using linear parameter-varying gain-scheduling. IEEE Trans. Control Syst. Technol. **10**(1), 43–54 (2002)
6. Hrovat, D.: Survey of advanced suspension developments and related optimal control applications. Automatica **33**(10), 1781–1817 (1997)
7. Sezgin, A., Hacioglu, Y., Yagiz, N.: Control of vehicle active suspensions with actuator delay via distributed backstepping. Int. J. Sci. Eng. Invest. **6**(69), 140–231 (2017)
8. Williams, R.A.: Automotive active suspensions. Proc. Inst. Mech. Eng. J. Part D: Automob. Eng. **211**, 415–444 (1997)
9. Gao, H., Sun, W., Shi, P.: Robust sampled-data control for vehicle active suspension system. IEEE Trans. Control Syst. Technol. **18**(1), 238–245 (2010)
10. Nekoui, M.A., Hadavi, P.: Optimal control of an active suspension system. In: 14th International Power Electronics and Motion Control Conference, EPE-PEMC 2010, pp. 60–63 (2010)

Modeling and Simulation the Heat Changing Process in Automotive Dry-Clutch

Nguyen Khac Minh[1,3](✉), Karpukhin Kirill Evgenievich[1,2], Mac Duy Hung[3], and Vu Van Hai[3](✉)

[1] Moscow Automobile and Road Construction State Technical University (MADI), № 64, Leningradsky Prospect, Moscow 125319, Russia
[2] The Central Research and Development Automobile and Engine Institute (NAMI), № 2, Avtomotornaya Street, Moscow 125438, Russia
[3] Thai Nguyen University of Technology, № 666 Tich Luong Ward, Thai Nguyen 24131, Vietnam
vuvanhai@tnut.edu.vn

Abstract. During clutch engagement, a large amount of heat is generated. This high temperature adversely affects the working efficiency of the clutch: increased wear, thermal deformation, surface cracking, even damage to friction surfaces. Therefore, to improve the working life and efficiency of the clutch assembly, the study of heat generation and heat distribution is a necessary task. In this study, the author built a diagram describing the heat generated and its propagation in the clutch. Conducting surveys according to typical exploitation scenarios of automotive dry-clutch, the authors estimated the heat generated at the main components: flywheel, friction lining, and pressure plate. The results show that the relative sliding velocity and the engagement time have a great influence on the temperature generated in the clutch. Among them, the friction lining witnessed the highest temperature variation, and the temperature rise occurred faster when repeating the closed-opened clutch process many times.

Keywords: Friction clutch · Friction material · Heat transfer · Flywheel · Friction lining · Pressure plate

1 Introduction

The clutch assembly is an important element used in a mechanical system to link the shafts and transmit torque between them. A well-functioning clutch is a key factor in driving comfort: smooth start, moving frequently, and shifting gears (vehicles equipped with manual transmission). During engagement times, the clutch torque is generated by friction between the working surfaces and then transmitted to the input shaft of the automotive transmission by the friction disc. High temperatures occur in the exposed surface of the clutch (friction disc, flywheel, and pressure plate) due to the relative motion between these parts during sliding. This high temperature is the main cause of some disadvantages such as increased wear, thermal deformation of the friction surface, generation of surface cracks. In some cases, these defects can cause the working

D. C. Nguyen et al. (Eds.): ICERA 2021, LNNS 366, pp. 581–590, 2022.
https://doi.org/10.1007/978-3-030-92574-1_61

surface to fail before its intended life. Therefore, the task of studying the problem of heat generation and heat transfer in the clutch is very necessary.

While the majority of mechanical assemblies in automobiles have high requirements for lubricating conditions and reducing friction at working surfaces, the material used in the clutch assembly is characterized by a rather high coefficient of friction. However, repeated engagement in the clutch assembly can cause the temperature of the clutch discs to rise significantly, resulting in rapid wear of the friction surfaces. In most cases, wear of mechanical components was identified as the underlying cause of the deterioration in extraction quality and reduced equipment performance [1–4].

The authors Kimming and Agner [5] showed that the temperature increase above the threshold of 350 °C–400 °C will reduce the bonding quality between the friction surfaces and can lead to permanent damage to the clutch system.

M. Azeem Ashraf et al. [6] studied the phenomenon of friction plate wear, the results indicated that the wear process depends on many factors and parameters (geometric parameters of the contact surface, normal force, relative velocity, hardness of material, influence of temperature).

Author Anosh Ali and colleagues [7] built a mathematical model and used the finite element method to analyze the overheating of the clutch disc with different materials and compare the temperature distribution across the clutch system. Roland Biczó et al. in the work [8] analyzed the properties of composites used as friction plates in dry clutches of automobiles. The results of this study provide an effective reference and new guidance for material determination methods for similarly complex materials.

The authors E. Mouffak and M. Bouchetara in the study [9] selected three geometric models of clutch discs designed in 3D using Solid Works and then converted them into FEM models. The authors analyzed the influence of the parameters: slip time, angular velocity, lining material, and pressure force on the thermal characteristics of the clutch, to arrive at the relationship between the selected disc parameters and generated temperature.

The author Van der Heijden et al. [10] showed the need for a system to control the engagement of the dry clutch in automobiles, the authors have compared and evaluated many different control methods. In the study [11], the authors Oday I. Abdullah and Josef Schlattmann built a mathematical model of the dry clutch (single disc clutch) to find the temperature field during the sliding time between the working surfaces. Research conditions are assumed to have stability in pressure and wear.

El-Sherbiny et al. [12] studied the temperature distribution in friction discs, flywheels, and pressure plates. The thermoelastic finite difference method was used to investigate the influence of contact area contact on the rising temperature. The temperature is estimated around 100 °C–120 °C for one engagement and about 280 °C–300 °C for multiple engagements including slip times up to 2 s.

In this study, a mathematical model describing the heat-changing process of the automotive dry-friction clutch is built based on the law of conservation of energy. The temperature variation of main components in the clutch assembly flywheel, friction lining, and pressure plate under typical working conditions are investigated.

2 Thermal Model of Automotive Clutch

2.1 Powertrain Modeling

In general, passenger car powertrains consist of the following basic elements: internal combustion engine; clutch device; gearbox and differential; drive shafts. The operating principle diagram of the basic automotive transmission system is shown in Fig. 1.

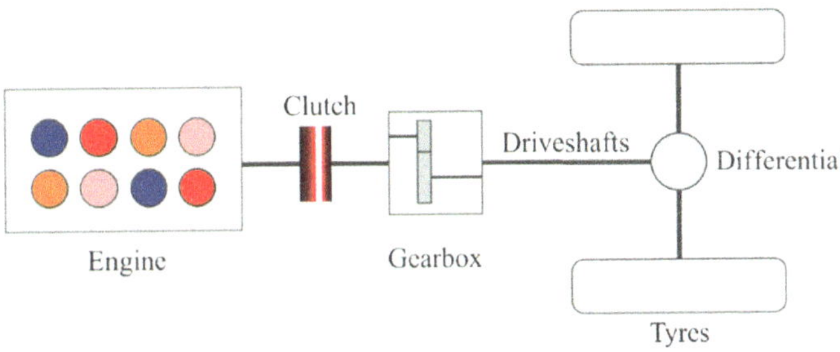

Fig. 1. Schematic overview of the powertrain.

Single-disc friction clutches have advantages: simple structure, efficiency, and high reliability, so they are widely used in the automotive powertrain. The clutch system, shown in Fig. 2, consists of housing, pressure plates, friction plates, a clutch disc with torsion dampers, and a release mechanism.

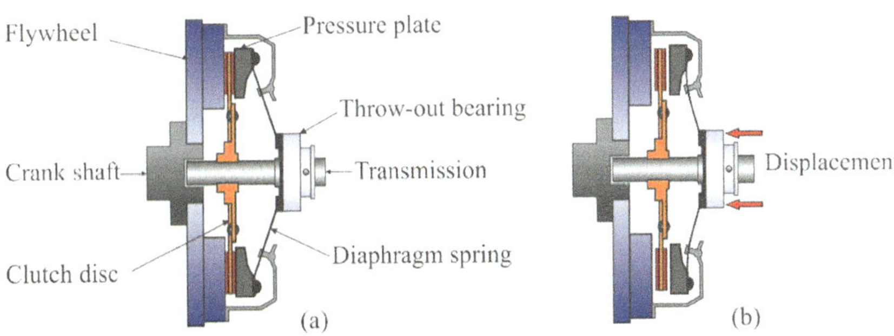

Fig. 2. Schematic overview of a dry clutch. (a) Engaged (b) Disengaged.

The clutch is normally closed, as the diaphragm spring is pre-tensioned when assembled. The clutch disc is angularly fixed to the transmission input shaft and can therefore rotate at a different speed than the engine and flywheel. When the clamp load becomes greater than zero a friction force will arise if there is a speed difference. This force will result in torque around the crankshaft and input shaft working to reduce the speed difference. With a larger normal force, a larger torque will be transmitted and if the speed difference gets reduced to zero the clutch will lock up, i.e. static friction will take place. When locked up, the clutch acts as a solid unit and transmits the engine torque as long as it does not exceed the stiction torque.

2.2 Mathematical Model

The heat generated in the friction surface is calculated based on the law of conservation of energy. Two cases of heat transfer occurring in the clutch should be considered:

- The heat transfers between solid surfaces in direct contact;
- The convective heat transfers through the surrounding air.

The heat transfer is illustrated in the model shown in Fig. 3.

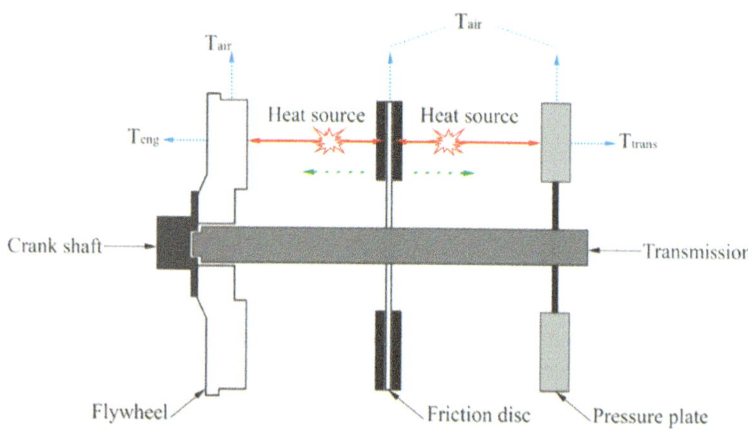

Fig. 3. The thermal model of the main elements in the clutch.

The friction clutch considered in this study has three main components: flywheel *(fw)*, friction lining *(fl)*, and pressure plate *(pp)*. During the working process, there are two sources of heat generated by the phenomenon of friction: i) The contact between the flywheel and the friction lining; ii) The contact between friction lining and pressure plate.

The thermal model of the clutch used is based on the law of conservation of energy. Therefore, any change in the temperature of each component is the result of an increase or decrease in the total amount of heat generated [13]:

$$m_i c_i \frac{\partial T}{\partial t} = \sum \dot{q} \tag{1}$$

The total amount of heat generated during friction between the clutch working surfaces is calculated by the formula:

$$\dot{q} = Z\frac{2\mu F_n \omega (r_o^3 - r_i^2)}{3(r_o^2 - r_i^2)} \quad (2)$$

where:

Z - the number of frictional interfaces; μ - the interfacial coefficient of friction; r_o, r_i - the inner and outer radius of the clutch disc; F_n - the normal contact load; ω - the relative angular velocity.

The heat balance equations for the main components of friction clutch are presented:

**Friction disc*

$$(mc)_{fd}\frac{\partial T}{\partial t} = \dot{q}_{fw/fd} + \dot{q}_{pp/fd} - \dot{q}_{fd-fw} - \dot{q}_{fd-pp} - \dot{q}_{fd,i} - \dot{q}_{fd,o} \quad (3)$$

**Flywheel*

$$(mc)_{fw}\frac{\partial T}{\partial t} = \dot{q}_{tot-fw} + \dot{q}_{fd-fw} - \dot{q}_{fw,i} - \dot{q}_{fw,o} - \dot{q}_{fw,e} \quad (4)$$

**Pressure plate*

$$(mc)_{pp}\frac{\partial T}{\partial t} = \dot{q}_{tot-pp} + \dot{q}_{fd-pp} - \dot{q}_{pp,i} - \dot{q}_{pp,o} - \dot{q}_{pp,t} \quad (5)$$

In which:

$\dot{q}_{fw/fd}$- the heat of the friction disc in the contact between the friction disc and the flywheel
$\dot{q}_{pp/fd}$ - the heat of the friction disc in the contact between the friction disc and the pressure plate
$\dot{q}_{tot-pp}$ - the heat of the pressure plate in the contact between the pressure plate and friction disc
$\dot{q}_{fd-fw}$ - the heat from the friction disc conducted to the flywheel
$\dot{q}_{fd-pp}$ - the heat from the friction disc conducted to the pressure plate
$\dot{q}_{fd,i}$; $\dot{q}_{fd,o}$ - the heat convected from the inner and outer radius of the friction disc
$\dot{q}_{tot-fw}$ - the heat of the flywheel in the contact between the flywheel and friction disc
$\dot{q}_{fw,i}$; $\dot{q}_{fw,o}$ - the heat convected from the inner and outer radius of the flywheel
$\dot{q}_{fw,e}$ - the heat dissipated from the engine side of the flywheel
$\dot{q}_{fd-pp}$ - the heat from the friction disc conducted to the pressure plate
$\dot{q}_{pp,i}$; $\dot{q}_{pp,o}$ - the heat convected from the inner and outer radius of the pressure plate
$\dot{q}_{pp,t}$ - The heat dissipated from the transmission side of the pressure plate

3 Simulations and Discussions

3.1 Determine the Parameters of the Model

Based on the mathematical model describing the thermal process in the automotive dry-clutch assembly built in Sect. 2. The authors have simulated and calculated the temperature changes of the main components of the clutch with the parameters shown in Table 1. The calculation aims to determine the thermal process and the influence of some working parameters of the clutch on the temperature variation of the flywheel, friction disc, and pressure plate under some specific working conditions.

Table 1. Model parameters and material properties.

Parameters	Units	Values
Thickness of pressure plate	m	0.055
Thickness of flywheel	m	0.05
Thickness of friction lining	m	0.01
Outer radius of flywheel	m	0.150
Inner radius of flywheel	m	0.013
Outer radius of pressure plate	m	0.145
Inner radius of pressure plate	m	0.1
Outer radius of friction material	m	0.145
Inner radius of friction material	m	0.1
Number of friction surfaces	–	2
Thermal Conductivity of pressure plate and flywheel	W/mK	55
Specific heat of pressure plate and flywheel	J/kg·K	360
Density of pressure plate and flywheel	kg/m3	7200

During the simulation, consider the change of clutch slip velocity ω_s depending on slip time t_s and the angular velocity of flywheel ω_{ofw} according to the linear relationship as in the formula:

$$\omega_s = \omega_{ofw}\left(1 - \frac{t}{t_s}\right) \tag{6}$$

Figure 4 (a) shows the change of clutch slip velocity ω_s corresponding to different slip time values $t_{s1} = 0.3$ s; $t_{s\,2} = 0.5$ s; $t_{s3} = 0.7$ s and the constant angular velocity of the flywheel $\omega_{fw} = 700$ rpm. Meanwhile, Fig. 4 (b) illustrates the variation of clutch slip velocity ω_s corresponding to the constant slip time $t_s = 0.5$ s and the angular velocity of the flywheel varies respectively with the values $\omega_{fw1} = 700$ rpm; $\omega_{fw2} = 1000$ rpm; và $\omega_{fw3} = 1500$ rpm.

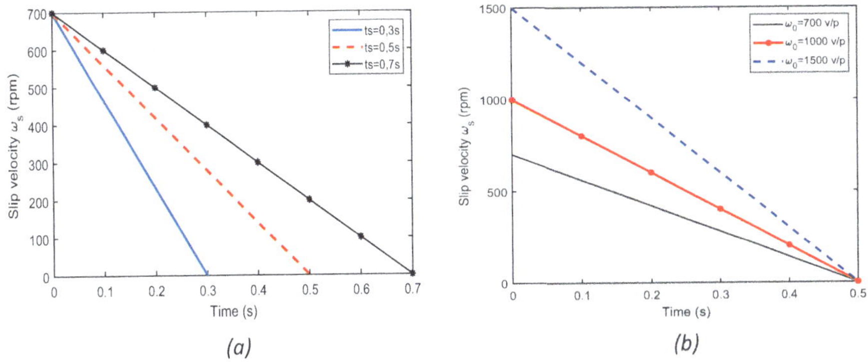

Fig. 4. Variation of clutch slip speed. (a) the case of different sliding time values, (b) the case of different values the initial angular velocity of the flywheel

3.2 Result Analysis

The authors carried out the research on the changing temperature of the flywheel, friction lining, and pressure plate in the typical case of the clutch opening and closing continuously when the driver performs the gear shift.

Figure 5 illustrates the changing friction lining temperature when examining an engagement time and full engagement clutch, corresponding to four different scenarios. The estimated results show that for the engagement time $t_s = 0.8$ s the temperature of the friction lining increases about $\Delta T = 9$ K, from the initial temperature $T_o = 298$ K (25 °C) to $T_1 = 307$K. After that, increasing the engagement time $t_s = 1$s the temperature of the friction lining increased rapidly about $\Delta T = 19$ K, which is twice as large as in the case of the previous investigation. In addition, Fig. 5 shows that during the full engagement, the temperature of the friction lining tends to decrease. The longer the engagement time, the lower the temperature of the friction lining. It is explained that during the full engagement, the heat of the friction lining is transferred to other surrounding assemblies (see Fig. 6), causing the temperature of the friction disc to decrease. Comparing simulation results with experimental results [16] showed a very small difference between the results.

Figure 6 describes the variations in temperature of the main components in the clutch assembly when repeating the closed-opened clutch process many times with the engagement time $t_s = 1$ s and the full engagement time $t_e = 5$ s (the disengagement period is assumed to be instantaneous). Analysis the results on the graph shows that, in the sliding stage, the temperature of the components: flywheel, friction lining, and pressure plate all increase, whereas in the full engagement clutch phase, their temperature tends to decrease. The obtained results also show that the temperature of the main parts in the clutch assembly increases rapidly as the sliding time increases and the value of the friction lining temperature is the highest compared to other parts; after six cycles of the closed-opened process with engagement time $t_s = 1$ s, time of full engagement $t_e = 5$ s, from the initial temperature of clutch assembly $T_o = 298$ K, the temperature of the surface flywheel, plate pressure, the friction lining increases and reaches the values T_{fw} = 304 K, T_{pp} = 320 K, and T_{fl} = 352 K respectively.

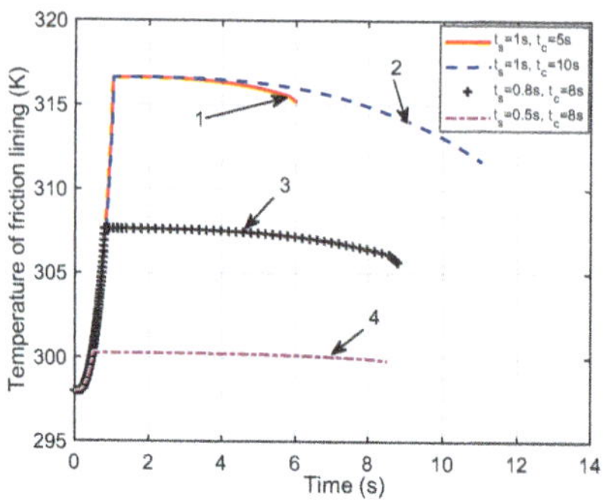

Fig. 5. The temperature rise of friction lining

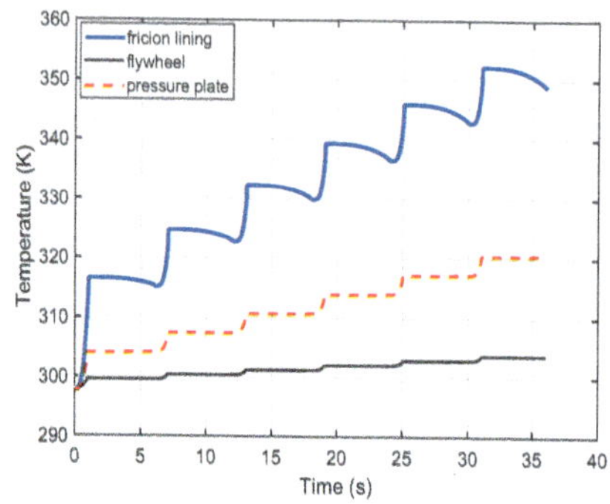

Fig. 6. The temperature rise of main components in the clutch

Figure 7 indicates the result of calculating the temperature change of the friction lining in case the vehicle starts and changes the gear ratio with manual transmission from i_{t1} to i_{t4} respectively; when performing the simulation, the engagement time values are t_{s1} = 1 s, $t_{s2} = t_{s3}$ = 0.8 s, t_{s4} = 0.5 s, and the full engagement time values are t_{e1} = 5 s, t_{e2} = 5 s, t_{e3} = 10 s, t_{e4} = 15 s respectively. The Figure shows that the temperature of the friction lining increases fastest in the first stage (t_{s1}) when the vehicle starts at gear ratio i_{t1}, the temperature increases from 297 K to 317 K; in the remaining engagement times, illustrate the process of changing the gear ratio of the manual transmission when the vehicle is moving (respectively t_{s2}, t_{s3}, and t_{s4}), which process sees an increase in the friction lining temperature. However, the temperature rise is slower than in the previous period. As a result, in the starting stage of the vehicle (t_{s1}), the engine speed must reach a value large enough to generate the starting torque, at the clutch the slip speed has a very large value, resulting in the largest amount of heat generated during this period. In the next process, change the gear ratio while the vehicle is in motion. At this time, the sliding velocity is much smaller than the starting stage, so the temperature generated is also smaller than the first stage.

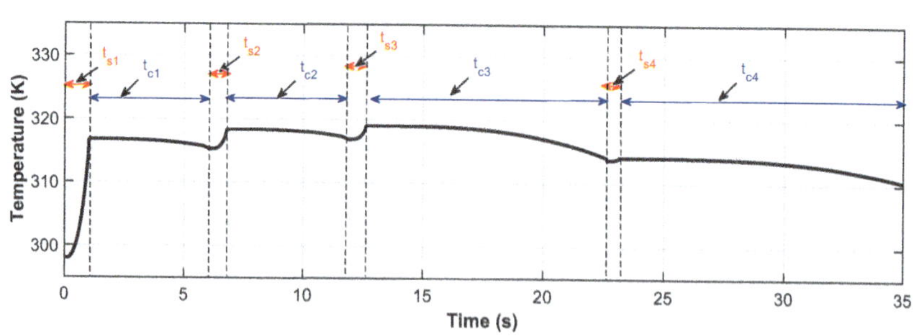

Fig. 7. Temperature variation of friction lining in case of gear ratio change with manual transmission

During the full engagement times t_{e1}, t_{e2}, t_{e3}, t_{e4} with gear ratios i_{t1}, i_{t2}, i_{t3}, i_{t4} respectively, the temperature of the friction lining witnessed a slight decrease. As the full engagement time increases, the friction plate temperature decreases, in this simulation, the temperature drops from 1 K to 4 K. The temperature of the clutch assembly increases or decreases depending on the opened or closed clutch process, in this study with the total survey time t_{tot} = 35 s, the highest temperature of the friction lining reached value T_{max} = 319 K, at the end of the survey, the temperature of the friction lining increased ΔT_{fl} = 12 K, from 298 K to 310 K.

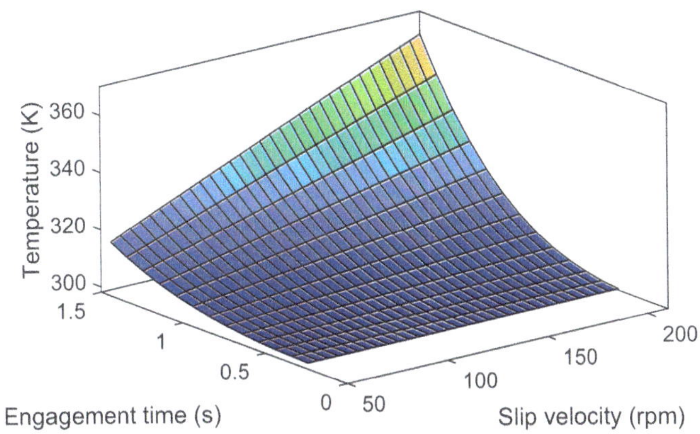

Fig. 8. Effect of the engagement time and slip speed to the highest temperature of friction disc

Figure 8 shows the simulation results of the relationship of the engagement time and slip speed of the clutch to the highest temperature of friction lining. Simulation conditions are defined as follows: slip time from 0.3 s to 1.5 s; slip speed varies from 50 rad/s to 200 rad/s, clutch assembly initial temperature 298 K. Research results show that, as the slip time of thc clutch increases, the highest temperature value of the friction lining rises; The greater the clutch slip speed, the higher the maximum temperature value of the friction lining. In case of the slip speed ω_f = 200 rad/s and slip time t_s = 1.5 s, the highest temperature of the friction lining is T_{max} = 366 K, an increase of ΔT = 68 K compared to the initial temperature.

4 Conclusion

In this study, a mathematical model describing the heat-changing process of the automotive dry-friction clutch is built based on the law of conservation of energy. The temperature variation of main components in the clutch assembly flywheel, friction lining, and pressure plate under typical working conditions are investigated.

The following remarks would be useful for further studies:

i) During the working process, the temperature of the flywheel, the friction lining, the pressure plate increases because of the relative sliding of the friction surfaces in the clutch, the largest temperature rise corresponds to the first slip period when the clutch initial temperature is minimum.

ii) The temperature of friction lining shows the highest value compared to other parts. When investigating the average clutch slip time $t_s = 0.4$ s in a closed and opened clutch process, the temperature of the friction lining increases about $\Delta T = 14$ K compared to the initial temperature;

iii) In the engagement time, the greater the slip time, the higher temperature of the friction lining; The greater the relative sliding speed, the higher the highest temperature of the friction lining.

Acknowledgment. This research was funded by Thai Nguyen University of Technology.

References

1. Blau, P.J.: Fifty years of research on the wear of metals. Tribol. Int. **30**(5), 321–331 (1997)
2. Zhao, E.-H., Ma, B., Ly, H.-Y.: Study on the high temperature friction and wear behaviors of Cu-based friction pairs in wet clutches by pin-on-disc tests. Adv. Mater. Sci. Eng. **2017**, 1–8 (2017)
3. Quinn, T.F.J.: The role of wear in the failure of common tribosystems. Wear **100**(1–3), 399–436 (1984)
4. Akhtar, M.M.J., Abdullah, O.I., Schlattmann, J.: Transient thermoelastic analysis of a dry clutch system. Mach. Des. **5**(4), 141–150 (2013)
5. Kimming, K., Agner, I.: Double clutch – wet or dry, this is the question. In: LuK Symposium, Germany (2006)
6. Azeem Ashraf, M., Sobhi-Najafabadi, B., Göl, Ö., Sugumar, D.: Numerical simulation of sliding wear for a polymer-polymer sliding contact in an automotive application. Int. J. Adv. Manuf. Technol. **41**(11–12), 1118–1129 (2009)
7. Anosh, A., et al.: Dry friction clutch disc of an automobile under transient thermal load: a comparison of friction lining materials. In: MATEC Web of Conferences, vol. 124 (2017)
8. Biczó, R., Kalácska, G.: Micromechanical model and thermal properties of dry-friction hybrid polymer composite clutch facings. Int. J. Eng. Manag. Sci. (IJEMS) **4**(1), 335–340 (2019)
9. Mouffak, E., Bouchetara, M.: Transient thermal behavior of automotive dry clutch discs by using Ansys software. Mechanika **22**(6), 562–570 (2016)
10. Van Der Heijden, A.C., Serrarens, A.F.A., Camlibel, M.K., Nijmeijer, H.: Hybrid optimal control of dry clutch engagement. Int. J. Control **80**(11), 1717–1728 (2007)
11. Abdullah, O.I., Schlattmann, J.: Thermal behavior of friction clutch disc based on uniform pressure and uniform wear assumptions. Friction **4**(3), 228–237 (2016). https://doi.org/10.1007/s40544-016-0120-z
12. El-Sherbiny, M., Newcomb, T.P.: Temperature distributions in automotive dry clutches. Proc. Inst. Mech. Eng. **190**, 359–365 (1976)
13. Mills, A.F.: Basic Heat and Mass Transfer, 2nd edn. Pearson College, England (2013)
14. Centea, D., Rahnejat, H., Menday, M.T.: Non-linear multi-body dynamic analysis for the study of clutch torsional vibrations (judder). Appl. Math. Model. **25**(3), 177–192 (2021)
15. Yang, L.-K., Heyan, L., Ahmadian, M., Ma, B.: Analysis of the influence of engine torque excitation on clutch judder. J. Vibr. Control **23**(4), 645–655 (2017)
17. Velardocchia, M., Amisano, F., Flora, R.: A linear thermal model for an automotive clutch. J. Passeng. Cars **109**(6), 1219–1225 (2000)

Modeling of Five-Axis Ball-End Milling with Tool Orientation Effects

Minh-Quang Tran[1,2], Vu Thi Lien[3], Meng-Kun Liu[4], and Thi Thanh Nga Nguyen[2(✉)]

[1] Industry 4.0 Implementation Center, Center of Cyber-Physical System Innovation, National Taiwan University of Science and Technology, Tapei 10607, Taiwan

[2] Department of Mechanical Engineering, Thai Nguyen University of Technology, Thai Nguyen, Vietnam
nguyennga@tnut.edu.vn

[3] Falculty of Mechanical Engineering and Mechantronics, PHENIKAA University, Hanoi 12116, Vietnam

[4] Department of Mechanical Engineering, National Taiwan University of Science and Technology, Tapei 10607, Taiwan

Abstract. An accurate and fast estimation of cutting forces is necessary for improving the performance of five-axis milling. However, the engagement between cutter and workpiece is very complex and irregular, it represents the main challenge for the prediction of cutting forces. This research aims to present a cutting force model of five-axis ball end milling, in which the effects of lead and tilt angles are investigated. The simulation results show that the resultant cutting force and total engagement area between cutter and workpiece are increased rapidly when the tilt and lead angles are close to zero. The cutting forces under different tool orientations are analyzed. The results of the proposed approach are successfully evaluated by the existing model.

Keywords: Five-axis ball-end milling · Cutting force model · Tool orientation

1 Introduction

In recent decades, five-axis ball end milling has been commonly utilized to manufacture sculptured surfaces and free-form surfaces which have complex shapes in many fields for example aerospace, automotive, and biomedical industries. This is because the tool accessibility of the five-axis ball milling has been improved. Choosing appropriate cutting conditions is a key contributor to increasing machining productivity and precision of products. Modeling of chip thicknesses and cutting forces during the cutting, therefore, are required to optimize the performance of machining. However, modeling of dynamic cutting force in five-axis milling is very complicated owing to tool orientation effects on the model of cutting force [1]. The tool orientations and part geometries lead to variations of engagement between tool and workpiece during the cutting, they are the most challenging as cutting force simulation.

D. C. Nguyen et al. (Eds.): ICERA 2021, LNNS 366, pp. 591–600, 2022.
https://doi.org/10.1007/978-3-030-92574-1_62

It is different from three-axis milling machine, the tool axis in five-axis machine is able to change its orientation. The tool orientation that is presented by the tilt and lead angles directly effects on machining surface topography, tool wear, and machining efficiency [2, 3]. Several approaches to modeling five-axis ball end milling have been investigated. The engagement between the end mill cutter and workpiece is proportional to the cutting force, which is represented by a function of instantaneous chip thickness [4–6]. A cutting force model was developed for ball-end milling using the cutting area and the specific cutting force was proposed by [7]. Layegh et al. introduced an approach to simulate the cutting forces based on a mechanistic force model [8]. An analytical modeling of cut geometry of five-axis machining was used to estimate the cutting force [9]. Modeling of five-axis milling forces developed from geometric modeling of three-axis ball end milling, in which the milling force is modeled using orthogonal to oblique transformation [6]. The tool orientation effects on the cutting stability of variable pitch ball-end milling were presented using a developed comprehensive dynamic model [10].

In this research, modeling of cutting force of five-axis ball end milling including the effects of lead and tilt angles is presented. The kinematic and force models are developed from three-axis operations. Several simulations of chip thickness and cutting force under variations of the tool orientation have been performed. The results of the proposed approach are successfully evaluated by the existing method.

2 Geometry Modeling of Ball End Mill

In order to model the ball end mill, the cutting edges are separated into a group of j circular sectors by discretizing along the end mill cutter axis as seen in Fig. 1. Each cutting edge point $M_{j,i}$ is determined by the Cartesian tool coordinate frame (X_t, Y_t, Z_t) located in the center of the ball-portion as shown in Eq. (1).

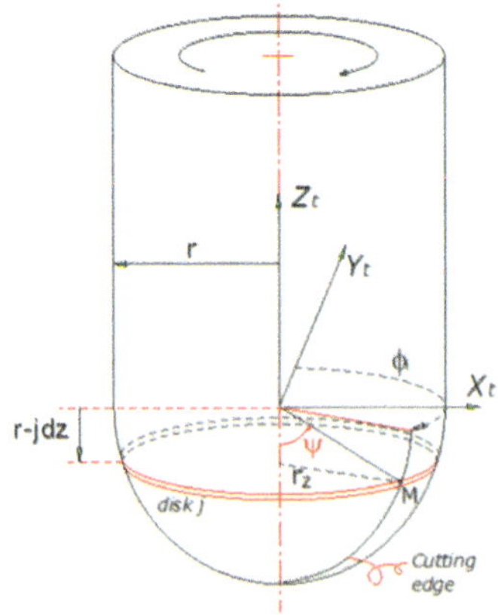

Fig. 1. Geometry of ball end mill.

$$\vec{M}_{j,i} = \begin{bmatrix} r\sin(\psi_j)\cos(\phi_{j,i}) \\ r\cos(\psi_j)\sin(\phi_{j,i}) \\ -r\cos(\psi_j) \end{bmatrix} \tag{1}$$

$$\psi_j = \cos^{-1}(1 - \frac{jdz}{r}) \tag{2}$$

The tool immersion angle is determined as follows,

$$\phi_{j,i} = \Phi + \frac{(i-1)2\pi}{N_t} + \delta_j \tag{3}$$

The lag angle is caused by the helix angle, γ, on the cutting edge, it is calculated as:

$$\delta_j = [1 - \cos(\psi_j)]\frac{r}{z}\gamma \tag{4}$$

The length of the infinitesimal curved cutting edge, *db*, is expressed by Fig. 2.

$$db = r\varphi_j = r(\psi_j - \varphi'_j) \tag{5}$$

$$\varphi'_j = \cos^{-1}(1 - \frac{dz(j-1)}{r}) \tag{6}$$

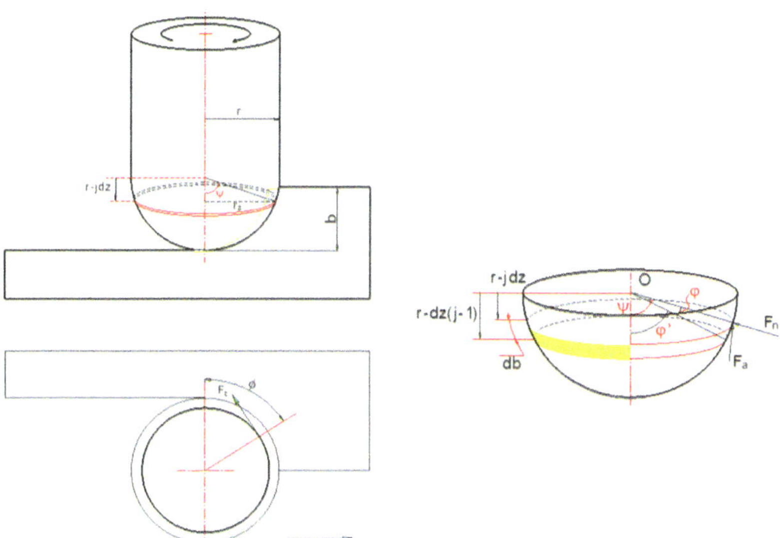

Fig. 2. Geometry of ball end mill for chip width calculation.

3 Ball End Mill Geometry in Five-Axis Milling

The inclination of the tool in five-axis milling is defined by lead and tilt angle, as illustrated in Fig. 3. In this model, the feed coordinate frame (X_f, Y_f, Z_f) is utilized to determine the lead and tilt angle, X_f, is along to the feed direction, Y_f is perpendicular to the feed direction, and Z_f is normal to the machining surface. The tilt angle is positive when the tool is tilted to the right and facing feed direction in case of the down milling. Likewise, the lead angle becomes a positive value when the tool is leaning forward to the feed direction. The transformation matrix that transforms the tool coordinate frame to workpiece coordinate frame is represented as follows,

$$^{W}T_t = Rot(x,\beta)Rot(y,\alpha) = \begin{bmatrix} 1 & 0 & 0 \\ 0 & \cos(\beta) & -\sin(\beta) \\ 0 & \sin(\beta) & \cos(\beta) \end{bmatrix} \begin{bmatrix} \cos(\alpha) & 0 & \sin(\alpha) \\ 0 & 1 & 0 \\ -\sin(\alpha) & 0 & \cos(\alpha) \end{bmatrix} \quad (7)$$

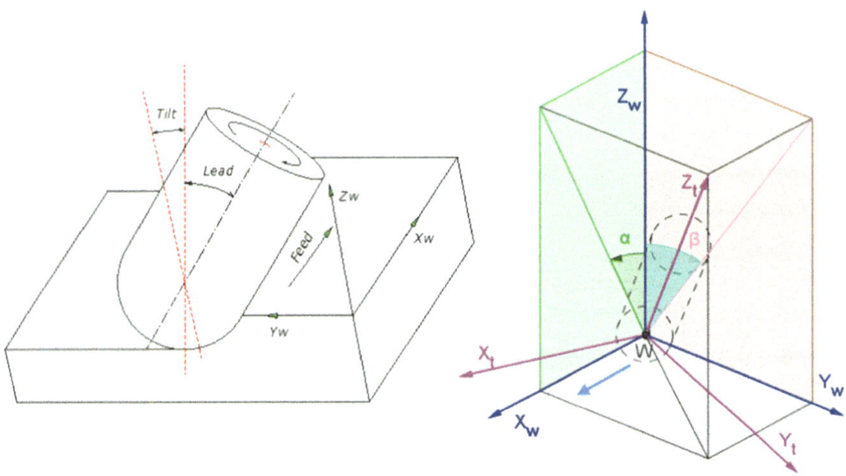

Fig. 3. Tool inclination angle and coordinate systems of ball end mill.

$$^{W}T_t = \begin{bmatrix} \cos(\alpha) & 0 & \sin(\alpha) \\ \sin(\beta)\sin(\alpha) & \cos(\beta) & -\sin(\beta)\cos(\alpha) \\ -\cos(\beta)\sin(\alpha) & \sin(\beta) & \cos(\beta)\cos(\alpha) \end{bmatrix} \quad (8)$$

In which α and β present the lead and tilt angle, they are clearly defined in Fig. 3. During the cutting process, the lead and tilt angle are calculated and updated based on specific tool location which generated by the NC tool path code [11].

4 Engagement Boundaries

The engagement between the cutter and workpiece could be identified following the start and exit angles of the cut. The start and exit angles in the ball end mill depend on up or down milling.

For up Milling:

$$\begin{cases} \theta_s = \theta_1 \\ \theta_e = \cos^{-1}(1 - \frac{a}{r}) - \theta_1 \end{cases} \tag{9}$$

For down milling:

$$\begin{cases} \theta_s = \pi - \cos^{-1}(1 - \frac{a}{r}) + \theta_1 \\ \theta_e = \pi - \theta_1 \end{cases} \tag{10}$$

where $\theta_1 = \frac{(r-b)\tan(\alpha)}{\sqrt{r^2-(r-b)^2}}$; a is radial depth of cut, and b is axial depth of cut. The engagement limit is determined following the equations:

$$\psi_{\min} = \begin{bmatrix} \alpha, a \geq 0 \\ 0, a < 0 \end{bmatrix} \tag{11}$$

$$\psi_{\max} = \alpha + \cos^{-1}(1 - \frac{b}{r}). \tag{12}$$

5 Modeling of Cutting Force

Figure 4 illustrates a procedure to evaluate the effects of the tool orientation on the cutting force model in five-axis ball-end milling. The inclination of the tool is represented by the tilt and the lead angles which are generated from the tool location data defined by the NC tool path as shown in Fig. 5. The engagement between the end mill and workpiece, cutter parameters and the cutting conditions are required to modeling the chip thickness.

The chip thickness of the cutting can be formulated as follows,

$$h(\psi_j, \phi_i) = f_t \sin(\psi_j)\sin(\phi_i)\cos(\lambda) \pm f_t \sin(\psi_j)\sin(\lambda) \tag{13}$$

where λ is the feed inclination angle.

The different cutting forces in radial, axial, and tangential direction are defined as,

$$\begin{aligned} dF_r &= K_r h(\psi_j, \phi_i)db + K_{re}db \\ dF_a &= K_a h(\psi_j, \phi_i)db + K_{ae}db \\ dF_t &= K_t h(\psi_j, \phi_i)db + K_{te}db \end{aligned} \tag{14}$$

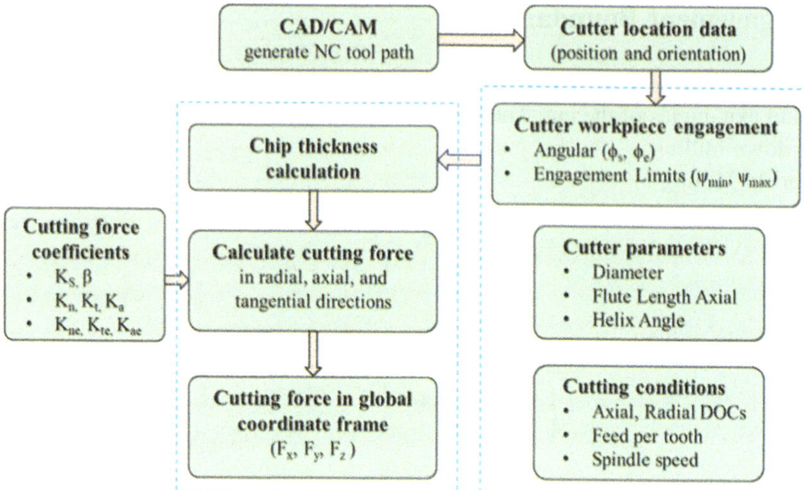

Fig. 4. Flow chart of force model in five-axis ball-end milling.

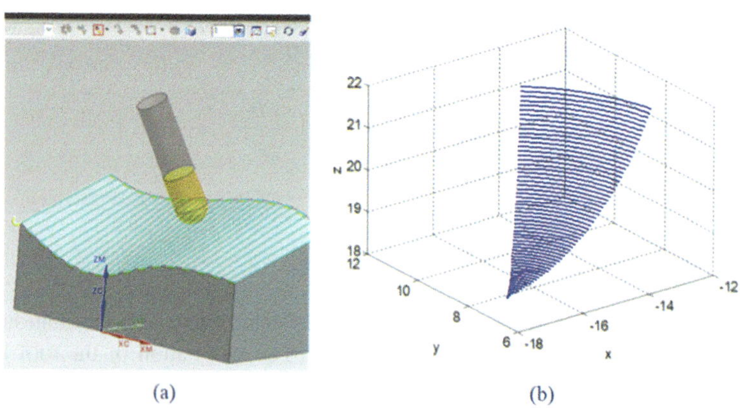

Fig. 5. (a) Tool location generation and (b) discrete cutter workpiece contact.

The differential cutting force respects tool coordinate frame is represented as follows,

$$\begin{bmatrix} dF_X \\ dF_Y \\ dF_Z \end{bmatrix}^t = T_t \begin{bmatrix} dF_t \\ dF_n \\ dF_a \end{bmatrix} = Rot(z, \phi_i)Rot(x, -[90 + \psi_j]) \begin{bmatrix} dF_t \\ dF_n \\ dF_a \end{bmatrix} \tag{15}$$

The matrix T_t transforms differential cutting force in tangential, radial and axial directions to tool coordinate frame (X_t, Y_t, Z_t).

$$T_t = \begin{bmatrix} \cos(\phi_i) & \sin(\phi_i)\sin(\psi_j) & -\sin(\phi_i)\cos(\psi_j) \\ \sin(\phi_i) & -\cos(\phi_i)\sin(\psi_j) & \cos(\phi_i)\cos(\psi_j) \\ 0 & -\cos(\psi_j) & -\sin(\psi_j) \end{bmatrix} \tag{16}$$

The differential cutting force respects workpiece coordinate frame is represented in Eq. (17).

$$\begin{bmatrix} dF_X \\ dF_Y \\ dF_Z \end{bmatrix}^w = T_w \begin{bmatrix} dF_X \\ dF_Y \\ dF_Z \end{bmatrix}^t = T_w T_t \begin{bmatrix} dF_t \\ dF_n \\ dF_a \end{bmatrix} \tag{17}$$

6 Results and Discussions

Several simulations have been performed. The cutting tool parameters and cutting conditions shown in Table 1 are used for slot milling. The workpiece material is an aluminum alloy. The cutting force coefficients obtained from experiments are described in Table 2.

Table 1. Parameter of ball end mill and cutting conditions

Tool diameter	Flutes	Helix angle	Axial depth of cut	Radial depth of cut	Feed per tooth	Rotational speed
(mm)	-	(°)	(mm)	(mm)	(mm)	(rpm)
12	2	30	3	12	0.1	12000

Table 2. Defined force coefficients

Ks	β	K_n	K_t	K_a	K_{ne}	K_{te}	K_{ac}
(N/mm^2)	(°)	(N/mm^2)	(N/mm^2)	(N/mm^2)	(N/mm)	(N/mm)	(N/mm)
813	65.9	331.8	742.2	331.8	12	23	12

In order to eliminate the tool tip contact, the lead angle should be positive. The simulation results of cutting force prediction for two cases are presented in Fig. 6 and Fig. 7. The simulation results obtained from the proposed model show that the cutting force in y direction increase rapidly compared to in x and z directions as the tilt angle changes from −10° to −15°. A comparison of the cutting force results between the proposed model and Ozturk's model [12]. Figure 7 shows that the simulated cutting forces in three directions are very similar and well fit to the results from Ozturk's model. Furthermore, the effects of tool inclination on engagement area and cutting force are illustrated in Fig. 8 and Fig. 9 respectively. It indicates that the resultant cutting force and total engagement area between cutter and workpiece are increased rapidly as tilt and lead angles close to zero.

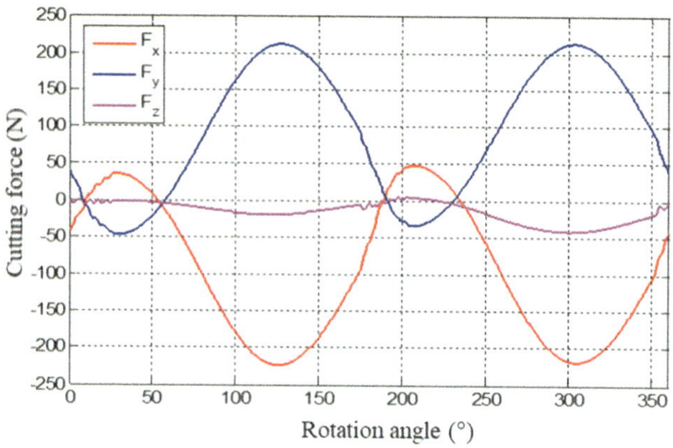

Fig. 6. Cutting force for lead angle of 15°, tilt angle of − 15°, surface inclination angle of 10°.

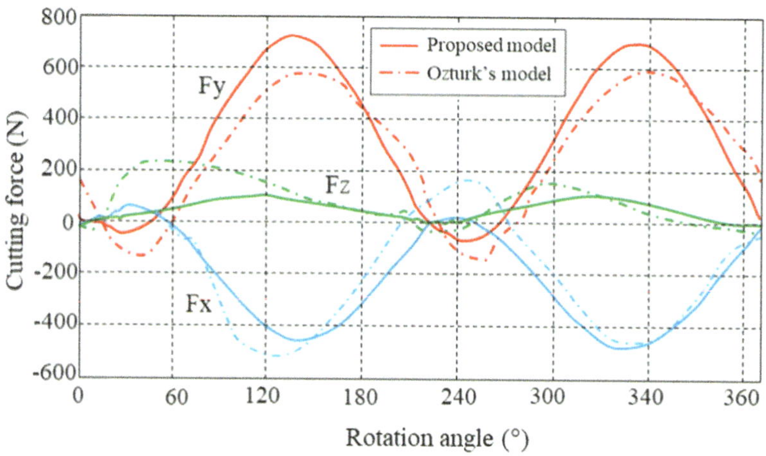

Fig. 7. Cutting force for lead angle of 15°, tilt angle of −10°.

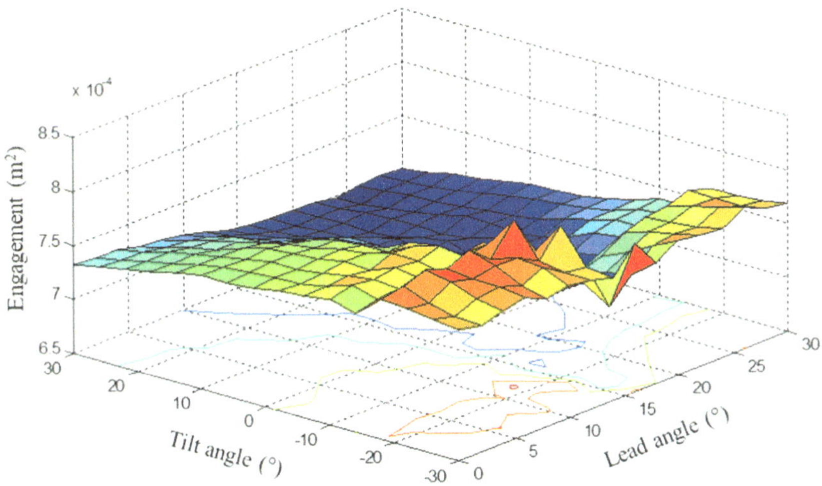

Fig. 8. The engagement area for variation of tool inclination angles.

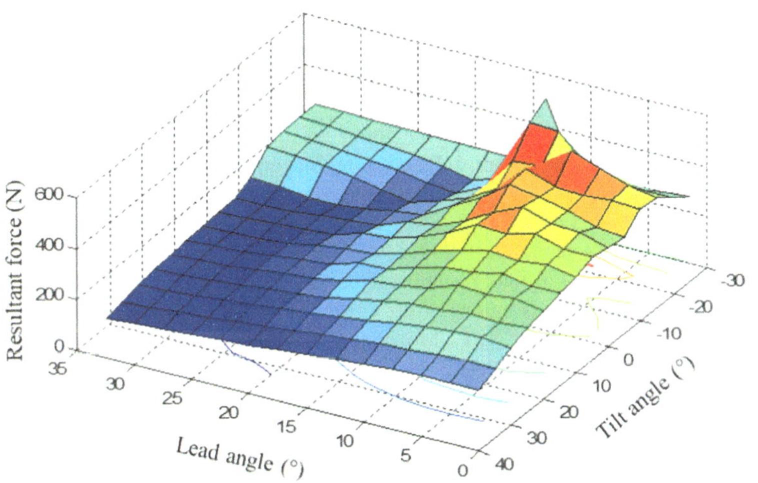

Fig. 9. The cutting force for variation of tool inclination angles.

7 Conclusions and Future Works

In this paper, kinematics of ball end milling in five-axis was presented. The model used to predict cutting force including effects of tool orientation was proposed. The simulation results carried out by a developed program show relationship between lead, tilt angle and cutting force. Thus, optimal cutting conditions can be determined to enhance the performance of the five-axis ball end milling. In our future work, the effects of tool inclination angle on tool wear, surface topography, and stability of process will be investigated. Appropriate tool orientation then can be yielded in order to increase the effectiveness of the five-axis ball end milling process.

Acknowledgment. This work was financially supported by the Thai Nguyen University of Technology (TNUT).

References

1. Ozturk, E., Tunc, L.T., Budak, E.: Investigation of lead and tilt angle effects in 5-axis ball-end milling processes. Int. J. Mach. Tools Manuf. **49**(14), 1053–1062 (2009)
2. Han, S.G., Zhao, J.: Effect of tool inclination angle on surface quality in 5-axis ball-end milling. Adv. Mater. Res. **97–101**, 2080–2084 (2010)
3. Tan, L., Yao, C., Ren, J., Zhang, D.: Effect of cutter path orientations on cutting forces, tool wear, and surface integrity when ball end milling TC17. Int. J. Adv. Manuf. Technol. **88**(9–12), 2589–2602 (2016). https://doi.org/10.1007/s00170-016-8948-y
4. Liu, M.-K., Tran, M.-Q., Chung, C., Qui, Y.-W.: Hybrid model- and signal-based chatter detection in the milling process. J. Mech. Sci. Technol. **34**(1), 1–10 (2020). https://doi.org/10.1007/s12206-019-1201-5
5. Tran, M.-Q., Liu, M.-K., Tran, Q.-V.: Analysis of milling chatter vibration based on force signal in time domain. In: Sattler, K.-U., Nguyen, D.C., Vu, N.P., Long, B.T., Puta, H. (eds.) ICERA 2020. LNNS, vol. 178, pp. 192–199. Springer, Cham (2021). https://doi.org/10.1007/978-3-030-64719-3_22
6. Chung, C., Tran, M.-Q., Liu, M.-K.: Estimation of process damping coefficient using dynamic cutting force model. Int. J. Precis. Eng. Manuf. **21**(4), 623–632 (2020). https://doi.org/10.1007/s12541-019-00297-5
7. Chiang, S.-T., Tsai, C.-M., Lee, A.-C.: Analysis of cutting forces in ball-end milling. J. Mater. Process. Technol. **47**(3), 231–249 (1995)
8. Layegh, E., Erdim, H., Lazoglu, I.: Offline force control and feedrate scheduling for complex free form surfaces in 5-axis milling. Proc. CIRP **1**, 96–101 (2012)
9. Budak, E., Ozturk, E., Tunc, L.T.: Modeling and simulation of 5-axis milling processes. CIRP Ann. **58**(1), 347–350 (2009)
10. Zhan, D., Jiang, S., Niu, J., Sun, Y.: Dynamics modeling and stability analysis of five-axis ball-end milling system with variable pitch tools. Int. J. Mech. Sci. **182**, 105774 (2020)
11. Boz, Y., Erdim, H., Lazoglu, I.: Modeling cutting forces for 5-axis machining of sculptured surfaces (2010)
12. Ozturk, E., Budak, E.: Modeling of 5-axis milling processes. Mach. Sci. Technol. **11**(3), 287–311 (2007)

Multi-objective Optimization for Minimum Surface Roughness and Maximum Wheel Life When External Cylindrical Grinding SKD11 Steel

Anh Tuan Nguyen[1], Thi Thanh Nga Nguyen[2], Quoc Hoang Tran[3], Huu Quang Nguyen[1], Hoang Anh Le[4], Hong Linh Nguyen[5], Ngoc Giang Tran[2], and Xuan Tu Hoang[2](✉)

[1] University of Economics - Technology for Industries, Hanoi, Vietnam
[2] Thai Nguyen University of Technology, Thai Nguyen, Vietnam
hxt_dhktcn@tnut.edu.vn
[3] Nguyen Tat Thanh University, Ho Chi Minh City, Vietnam
[4] Vinh Long University of Technology Education, Vĩnh Long, Vietnam
[5] Electric Power University, Hanoi, Vietnam

Abstract. Increasing the quality of workpieces and tool life is essential in grinding processes. This paper proposes a grey relation method to optimize surface roughness and wheel life in external cylindrical grinding SKD11 steel. The parameters comprising the rough dressing depth of cut, the number of rough dressings, the finish dressing depth of cut, the number of finish dressing, the dressing number without the depth of cut, and the longitudinal feed rate were evaluated by the grey relation coefficient at different levels. The optimum parameters were found based on the highest value of the grey relation coefficient in each factor at different levels. The results are reliable due to the combination with examining the influences of these parameters on the output responses, i.e., surface roughness and wheel life. Applying the optimal values of these parameters in external cylindrical grinding SKD11 steel the surface roughness, and the wheel life respectively obtain 0.143 μm and 14.5 min.

Keywords: External cylindrical grinding · Grinding parameters · Surface roughness · Wheel life

1 Introduction

Grinding is one of the methods for machining the final surface of workpieces because this method can bring the quality of the surface finish. Abrasive grains are used as a tool during machining processes. They are divided by grain sizes, bonding materials, wheel diameters, wheel thickness. During the grinding process, abrasive grains remove very small particles from the surface of the workpiece to obtain a better-quality surface finish. Many types of researches focus on optimization of grinding processes. Studies on the determination of optimum exchanged grinding wheel diameter play a vital role since this numerously affects the cost of grinding processes [1–6]. In these researches,

D. C. Nguyen et al. (Eds.): ICERA 2021, LNNS 366, pp. 601–609, 2022.
https://doi.org/10.1007/978-3-030-92574-1_63

the optimum diameter of grinding wheels was suggested to replace another grinding wheel. For this, the grinding cost was greatly reduced. For cost optimizations, several researchers have proposed the effects of grinding parameters on the grinding cost [7–11]. These studies concentrated on finding out the optimal grinding parameters in order to minimize the cost functions. The cost functions are established by the cost of a machining product.

Besides, the effects of parameters on the machining quality in the grinding process have been investigated. Saravanakumar et al. [12] proposed the evaluation of grinding parameters for surface grinding. In this work, several parameters such as the speed of the grinding wheel, the table speed, and the depth of cut were examined in to improve the surface roughness when grinding carbon steel AISI 1042. Jeevanantham and Sivaram [13] presented the influences of machining parameters on the material removal rate and surface roughness for grinding EN8 and EN31 steel. Dressing parameters have also been important factors that affect the quality of the workpiece. Daneshi et el. [14] and [15] have displayed the impact of dressing parameters on the grinding forces, the quality of workpiece surface, and the grinding wheel wear. Many factors, including the initial grinding wheel diameter, the grinding wheel width, the total depth of dressing cut, the Rockwell hardness of workpiece, the wheel life, and the radial grinding wheel wear per dress and the cost components were evaluated and showed the influences of these parameters on the quality of the workpiece [16]. The surface roughness was also mentioned in [17, 18]. In this study, several parameters such as the coolant concentration, the cross feed, the table speed, the coolant flow, the depth of cut were, etc., considered to minimize surface roughness for machining 9CrSi steel.

This study deal with using the grey relation for minimizing roughness and maximizing wheel life in external cylindrical grinding. The experiment was implemented by SKD11 steel. Six parameters containing the rough dressing depth of cut, the number of rough dressings, the finish dressing depth of cut, the number of finish dressing, the dressing number without the depth of cut, and the longitudinal feed rate were discussed based on the evaluation of the grey relation coefficient. From this, the optimal parameters were found in order to improve the quality of the product (minimum roughness and maximum wheel life).

2 Experimental Setup

The steel SKD11 with dimensions Ø25x200 was used for experiments. The chemical composition of SKD11 includes C: 0.9–1.5%, Cr: 12%, Mo: 1.0%, V: 0.35%, Si: 0.25%, Mn: 0.45%, S: $\leq$ 0.025%, P: $\leq$ 0.025%. The experiment system is shown in Table 1.

In the experiment, cutting parameters for the external cylindrical grinding of the workpiece SKD11 steel are shown as follows:

Longitudinal feed rate: 1.8 (m/min).
Depth of cut: 0.005 mm/ stroke.
Total depth of cut: 0.05 mm.
Grinding wheel speed: 29.3 (m/s).

Table 1. Experiment system

Experimental equipment	Symbol	Origin
External cylindrical grinding machine	CONDO-Hi-45 HTS	Japan
Grinding wheel	Ct80MV1-G 400x40x203 35m/s	Vietnam
Dresser	3908-0088C type 2	Russia
Surface roughness measurement	Mitutoyo 178-923-2A, SJ-201	Japan

Table 2. Parameters and their values at different levels

Parameters	Symbol	Unit	Levels			
			1	2	3	4
Rough dressing depth of cut	a_r	mm	0.03	0.04	–	–
Number of rough dressings	n_r	–	1	2	3	4
Finish dressing depth of cut	a_f	mm	0.005	0.01	0.015	0.02
Number of finish dressing	n_f	–	0	1	2	3
Dressing number without depth of cut	n_0	–	0	1	2	3
Longitudinal feed rate	S_d	m/min	1	1.5	–	–

Table 3. Experiment matrix

Exp. No	a_f	n_f	n_0	n_r	a_r	S_d
1	0.005	0	0	1	0.03	1.0
2	0.005	1	1	2	0.03	1.5
3	0.005	2	2	3	0.04	1.0
4	0.005	3	3	4	0.04	1.5
5	0.010	0	1	3	0.04	1.5
6	0.010	1	0	4	0.04	1.0
7	0.010	2	3	1	0.03	1.5
8	0.010	3	2	2	0.03	1.0
9	0.015	0	2	4	0.03	1.5
10	0.015	1	3	3	0.03	1.0
11	0.015	2	0	2	0.04	1.5
12	0.015	3	1	1	0.04	1.0
13	0.020	0	3	2	0.04	1.0
14	0.020	1	2	1	0.04	1.5
15	0.020	2	1	4	0.03	1.0
16	0.020	3	0	3	0.03	1.5

The experimental design was established by six parameters with four different levels, as shown in Table 2. The parameters include the rough dressing depth of cut, the number of rough dressings, the finished dressing depth of cut, the number of finish dressing, the dressing number without depth of cut, and the longitudinal feed rate. Based on the Taguchi method to design the experiment for the six parameters, L16 was used to create the experimental matrix, as shown in Table 3.

To enhance the accuracy of the experiment, each test was performed 3 times. Table 4 presents the experimental plans and the response (the surface roughness).

Table 4. Results of the wheel life and S/N with three trials

Exp. No	a_f	n_f	n_0	n_r	a_r	S_d	Wheel life, T (min)			Means	S/N
							Trial 1	Trial 2	Trial 3		
1	0.005	0	0	1	0.03	1.0	9.47	9.29	19.3153	0.488	0.493
2	0.005	1	1	2	0.03	1.5	8.87	8.98	19.0630	0.405	0.433
3	0.005	2	2	3	0.04	1.0	16.93	16.48	24.3172	0.229	0.242
4	0.005	3	3	4	0.04	1.5	8.80	8.52	18.5949	0.342	0.353
5	0.010	0	1	3	0.04	1.5	9.88	9.86	19.8675	0.266	0.245
6	0.010	1	0	4	0.04	1.0	9.73	9.55	19.5873	0.334	0.327
7	0.010	2	3	1	0.03	1.5	9.13	9.04	19.1227	0.211	0.195
8	0.010	3	2	2	0.03	1.0	9.08	9.37	19.4270	0.301	0.292
9	0.015	0	2	4	0.03	1.5	12.60	12.81	22.1510	0.319	0.307
10	0.015	1	3	3	0.03	1.0	10.60	10.82	20.6781	0.243	0.218
11	0.015	2	0	2	0.04	1.5	9.37	9.62	19.6568	0.338	0.312
12	0.015	3	1	1	0.04	1.0	9.27	9.19	19.2618	0.303	0.289
13	0.020	0	3	2	0.04	1.0	9.03	9.00	19.0829	0.313	0.327
14	0.020	1	2	1	0.04	1.5	10.17	10.32	20.2704	0.353	0.368
15	0.020	2	1	4	0.03	1.0	10.87	11.21	20.9874	0.235	0.223
16	0.020	3	0	3	0.03	1.5	8.03	8.22	18.2910	0.317	0.324

Table 5. Results of the surface roughness and S/N with three trials

Exp. No	a_f	n_f	n_0	n_r	a_r	S_d	Surface roughness, Ra (μm)			Means	S/N
							Trial 1	Trial 2	Trial 3		
1	0.005	0	0	1	0.03	1.0	0.488	0.493	0.507	0.496	6.0892
2	0.005	1	1	2	0.03	1.5	0.405	0.433	0.424	0.421	7.5179
3	0.005	2	2	3	0.04	1.0	0.229	0.242	0.220	0.230	12.7462
4	0.005	3	3	4	0.04	1.5	0.342	0.353	0.326	0.340	9.3573
5	0.010	0	1	3	0.04	1.5	0.266	0.245	0.251	0.254	11.8981
6	0.010	1	0	4	0.04	1.0	0.334	0.327	0.306	0.322	9.8280
7	0.010	2	3	1	0.03	1.5	0.211	0.195	0.227	0.211	13.4977
8	0.010	3	2	2	0.03	1.0	0.301	0.292	0.285	0.293	10.6704
9	0.015	0	2	4	0.03	1.5	0.319	0.307	0.336	0.321	9.8729
10	0.015	1	3	3	0.03	1.0	0.243	0.218	0.229	0.230	12.7569
11	0.015	2	0	2	0.04	1.5	0.338	0.312	0.324	0.325	9.7666
12	0.015	3	1	1	0.04	1.0	0.303	0.289	0.284	0.292	10.6891
13	0.020	0	3	2	0.04	1.0	0.313	0.327	0.337	0.326	9.7406
14	0.020	1	2	1	0.04	1.5	0.353	0.368	0.338	0.353	9.0393
15	0.020	2	1	4	0.03	1.0	0.235	0.223	0.227	0.228	12.8265
16	0.020	3	0	3	0.03	1.5	0.317	0.324	0.335	0.325	9.7512

3 Results and Discussions

To find the optimal dressing parameters for multi-objectives, the grey relation was applied to analyze the effect of the parameters on the wheel life and surface roughness. Signal-to-noise (S/N) in the Taguchi method is used to examine the output responses. S/N in each experiment is shown in Table 4 and 5. With a larger S/N ratio, the results will be less effect of the noise. Therefore, the factor Z_{ij} ($0 < Z_{ij} < 1$) is characterized for the standardization of S/N ratio to avoid the effects of using different units and reducing variability. Z_{ij} can be computed as

$$Z_i = \frac{SN_i - \min(SN_i, = 1, 2, ..n)}{\max(SN_i, j = 1, 2, ..n) - \min(SN_i, = 1, 2, ..n)} \tag{1}$$

In here, n is the number of experiments (n = 16).

The grey relational coefficient is determined by

$$y_i(k) = \frac{\Delta_{min} + \xi.\Delta_{max}}{\Delta_i(k) + \xi.\Delta_{max}}, \text{ for } i = 1, 2, \ldots, n \tag{2}$$

Where k is the number of output parameters; $\Delta_j(k)$ is the absolute value, $\Delta_j(k) = \|Z_0(k) - Z_j(k)\|$ in which Z0(k) and Zj(k) are the reference and the specific comparison sequences, respectively; Δ_{min} and Δ_{max} are representative for the smallest and largest values of Δi(k); ζ presents the distinguishing coefficient, $0 \leq \zeta \leq 1$.

From Eq. (3), the mean of the grey relational coefficient can be simply computed as

$$\overline{y_i} = \frac{1}{k}\sum_{j=0}^{k} y_{ij}(k) \tag{3}$$

Table 6 shows the results of the calculated parameters. It is a fact that a higher value of the grey relational coefficient demonstrates better product quality. For the observation in Table 6, the highest grey relational coefficient value is dropped on experiment No. 3 with $\bar{y}_i = 0.916$. However, this value has not yet been an optimal level since the mean value of the grey relational coefficient has to be determined for each parameter at different levels.

Table 6. Results of calculated variables

Exp. No	S/N		Zi		$\Delta_i(k)$		y_i		$\overline{y_i}$
	Ra	T	Ra	T					
			Value of reference		Ra	T	Ra	T	
			1.000	1.000					
1	6.0892	19.3153	0.000	0.170	1.000	0.830	0.333	0.376	0.355
2	7.5179	19.063	0.193	0.128	0.807	0.872	0.383	0.364	0.373
3	12.7462	24.3172	0.899	1.000	0.101	0.000	0.831	1.000	0.916
4	9.3573	18.5949	0.441	0.050	0.559	0.950	0.472	0.345	0.409
5	11.8981	19.8675	0.784	0.262	0.216	0.738	0.698	0.404	0.551
6	9.828	19.5873	0.505	0.215	0.495	0.785	0.502	0.389	0.446
7	13.4977	19.1227	1.000	0.138	0.000	0.862	1.000	0.367	0.684
8	10.6704	19.427	0.618	0.189	0.382	0.811	0.567	0.381	0.474
9	9.8729	22.151	0.511	0.641	0.489	0.359	0.505	0.582	0.544
10	12.7569	20.6781	0.900	0.396	0.100	0.604	0.833	0.453	0.643
11	9.7666	19.6568	0.496	0.227	0.504	0.773	0.498	0.393	0.445
12	10.6891	19.2618	0.621	0.161	0.379	0.839	0.569	0.373	0.471
13	9.7406	19.0829	0.493	0.131	0.507	0.869	0.496	0.365	0.431
14	9.0393	20.2704	0.398	0.328	0.602	0.672	0.454	0.427	0.440
15	12.8265	20.9874	0.909	0.447	0.091	0.553	0.847	0.475	0.661
16	9.7512	18.291	0.494	0.000	0.506	1.000	0.497	0.333	0.415

Using ANOVA analysis, the effects of the parameters on the output responses are shown in Table 7. It is noted that the order influences on the output factors are descendingly dropped on the number of finish dressing n_f (43.39%), the number of rough dressings n_r (26.53%), the dressing number without the depth of cut n_0 (20.89%), the longitudinal feed rate S_d(5.56%), the finish dressing depth of cut a_f (1.83%), and the rough dressing depth of cut a_r (0.03%).

Table 7. Influence of parameters on the output response

Analysis of Variance for Means

Source	DF	Seq SS	Adj SS	Adj MS	F	P	C (%)
af	3	0.005877	0.005877	0.001959	0.35	0.812	1.83
nf	3	0.139565	0.139565	0.046522	8.21	0.250	43.39
n0	3	0.067201	0.067201	0.022400	3.95	0.350	20.89
nr	3	0.085343	0.085343	0.028448	5.02	0.314	26.53
ar	1	0.000099	0.000099	0.000099	0.02	0.916	0.03
Sd	1	0.017885	0.017885	0.017885	3.16	0.326	5.56
Residual Error	1	0.005668	0.005668	0.005668			1.76
Total	15	0.321638					100.00

Model Summary

S	R-Sq	R-Sq(adj)
0.0753	98.24%	73.57%

To get the optimal values of the parameters, the grey relational coefficient was analyzed by ANOVA at different levels as shown in Table 8. It can be observed from Table 8 that the largest value of the grey relational coefficient for each factor is determined at different levels, i.e., a_f at level 2, n_f at level 3, n_0 at level 3, n_r at level 3, a_r at level 1 and S_d at level 1. The average of the grey relational coefficient is 0.516. Furthermore, the optimal parameters are also depicted in Fig. 1. The optimal values of the parameters are shown in Table 9, i.e., $a_f = 0.01$(mm), $n_f = 3$, $n_0 = 2$, $n_r = 3$, $a_r = 0.03$(mm) and $S_d = 1$(m/min). By the experiment, the surface roughness and the wheel life respectively obtain 0.143 μm and 14.5 min.

Table 8. Response table for means

Level	af	nf	n0	nr	ar	Sd
1	0.5131	0.4700	0.4153	0.4874	**0.5186**	**0.5495**
2	**0.5386**	0.4757	0.5141	0.4310	0.5136	0.4827
3	0.5258	**0.6764**	**0.5934**	**0.6313**		
4	0.4868	0.4423	0.5415	0.5147		
Delta	0.0518	0.2341	0.1782	0.2003	0.0050	0.0669
Rank	5	1	3	2	6	4
$\overline{y_i} = 0.516$						

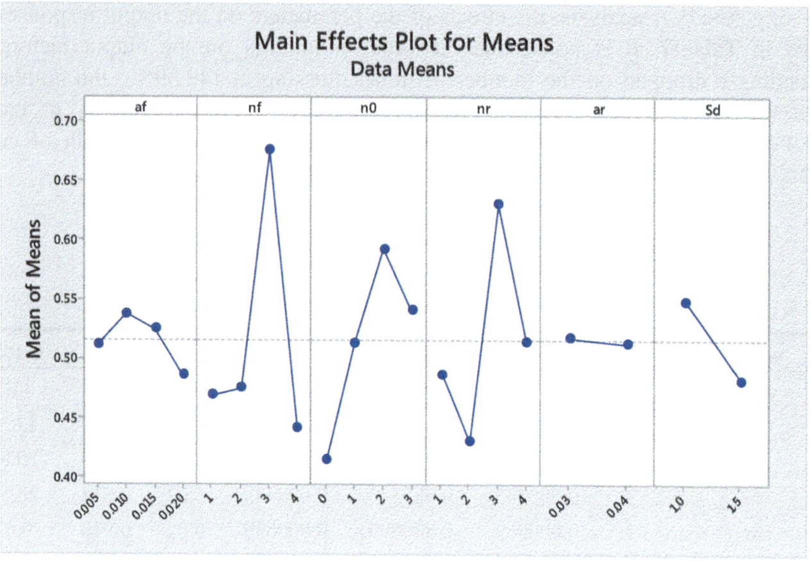

Fig. 1. Influences of parameters on the output responses

Table 9. Optimal values of parameters

Predicted values			
S/N Ratio	**Mean**	**StDev**	**Ln(StDev)**
15.2043	0.143229	0.0092459	-4.72806

Settings					
a_f	n_f	n_o	n_r	a_r	S_d
0.01	3	2	3	0.03	1

4 Conclusion

This paper presents the results of a multi-objective optimization study for minimum surface roughness and maximum wheel life when external cylindrical grinding SKD11 steel. By using the Taguchi method to optimize the multi-output responses, the conclusions have been obtained as follows:

The influence of the output responses is descendingly arranged by the number of finish dressing, the number of rough dressings, the dressing number without the depth of cut, the longitudinal feed rate, the finished dressing depth of cut, and the rough dressing depth of cut;

The grey relation is used to find out the optimal values of the parameters. The optimal values are $a_f = 0.01$(mm), $n_f = 3$, $n_0 = 2$, $n_r = 3$, $a_r = 0.03$(mm) and $S_d = 1$(m/min).

The values of the surface roughness and wheel life are 0.143 μm and 14.5 min at these optimal values of the parameters.

Acknowledgment. This work was supported by Thai Nguyen University of Technology.

References

1. Tu, H.X., Jun, G., Hung, L.X., Tung, L.A., Pi, V.N.: Calculation of optimum exchanged grinding wheel diameter when external grinding tool steel 9CrSi. Int. J. Mech. Eng. Robot **8**, 59–64 (2019)
2. Tran, T.H., et al.: Optimization of replaced grinding wheel diameter for minimum grinding cost in internal grinding. Appl. Sci. **9**, 1363 (2019)
3. Vu, N.P., Luu, A.T., Le, X.H., Nguyen, V.N.: Experimental determination of optimum exchanged diameter in surface grinding process. J. Environ. Sci. Eng. A. **6**, 85–89 (2017)
4. Hung, L.X., Pi, V.N., Tung, L.A., Tu, H.X., Jun, G., Long, B.: Determination of optimal exchanged grinding wheel diameter when internally grinding alloy tool steel 9CrSi. IOP Conf. Ser. Mater. Sci. Eng. **417**, 012026 (2018)
5. Tran, T.H., et al.: A study on calculation of optimum exchanged grinding wheel diameter when surface grinding stainless. Mater. Sci. Forum. **977**, 3–11 (2020)
6. Pi, V.N., Tung, L.A., Hong, T.T., Nga, N.T.T., Hung, L.X., Long, B.T.: An optimization of exchanged grinding wheel diameter when surface grinding alloy tool steel 9CrSi. Mater. Today Proc. **18**, 2225–2233 (2019)
7. Vu, N.P., Luu, A.T., Le, X.H., Banh, T.L.: Cost optimization of surface grinding process. J. Environ. Sci. Eng. A. **5**, 606–611 (2016)
8. Vu, N.P., Le, X.H., Luu, A.T., Banh, T.L.: Cost optimization of internal grinding. J. Mater. Sci. Eng. B. **6**, 291–296 (2016)
9. Vu, N.P., Phan, Q.T., Vu, H.K., Nguyen, N.H.: Cost optimization of external cylindrical grinding. Appl. Mech. Mater. **312**, 982–989 (2013)
10. Tran, T.H., et al.: A study on cost optimization of external cylindrical grinding. Mater. Sci. Forum. **977**, 18–26 (2020)
11. Tran, T.H., et al.: A study on cost optimization of external cylindrical grinding, Mater. Sci. Forum. 977 MSF, 18–26 (2020)
12. Saravanakumar, A., Dhanabal, S., Jayanand, E., Logeshwaran, P.: Analysis of process parameters in surface grinding process. Mater. Today Proc. **5**, 8131–8137 (2018)
13. Jeevanantham, S. Robinson Smart, D.S., Sivaram, N.M., Nallusamy, S., Manikanda Prabu, N.: Effect of machining parameters on MRR and surface roughness in internal grinding using EN8, EN31 steel, Int. J. Appl. Eng. Res. **12**, 2963–2968 (2017)
14. Daneshi, A., Jandaghi, N., Tawakoli, T.: Effect of dressing on internal cylindrical grinding. Proc. CIRP **14**, 37–41 (2014)
15. Xuan Tu, H., et al.: Influence of dressing parameters on surface roughness of workpiece for grinding hardened 9XC tool steel. IOP Conf. Ser. Mater. Sci. Eng. **542** (2019)
16. Hoang, T.D., Tran, T.H., Van Cuong, N., Le, H.K., Thi, N., Nga, T.: An optimization study on surface grinding stainless steel **7**, 6621–6625 (2018)
17. Luu, A.T., Vu, N.P., Do, T.T.H., Le, X.H., Tien Long, B.: A study on optimization of surface roughness in surface grinding 9CrSi tool steel by using Taguchi method. Adv. Eng. Res. Appl. 100–108 (2019)
18. Ky, L.H., et al.: Optimization of grinding parameters for surface roughness when grinding tablet by CBN wheel on CNC milling machine. Int. J. Mech. Eng. Technol. 960–967 (2019)

Multi-objective Optimization of AA7075 Aluminum Alloy Drilling Process

Nguyen Lam Khanh and Nguyen Van Cuong(✉)

University of Transport and Communications, Hanoi, Vietnam
nguyencuong@utc.edu.vn

Abstract. AA7075 aluminum alloy has the same hardness as steel but its weight is only about one-third that of steel. Thanks to this outstanding advantage, this material is widely applied in many different fields, especially in the field of automobile and aviation. Therefore, it is very necessary to study the multi-objective optimization when machining this material. In this article, a study on multi-objective optimization when drilling this material is presented. The experiments were conducted in accordance with a matrix designed by the Taguchi method. Spindle speed and feed rate were two parameters of which optimal values should be determined. Reference Ideal Method (*RIM*) was used to solve the multi-objective problem. The purpose of this study is to determine the values of the above two cutting parameters to simultaneously ensure the maximum value of Material Removal Rate (*MRR*), the minimum value of Diameter Error (*DE*) and the very minimum value of Roundness Error (*RE*). In addition, the effect of cutting parameters on *DE* and *RE* was also discovered. Finally, the directions for further study were also mentioned in this paper.

Keywords: Drill · AA7075 aluminum alloy · Material removal rate · Diameter error · Roundness error · Optimization · Taguchi · RIM

1 Introduction

Drilling is the most common machining method for making holes in solid materials. This method is often used as a preparation machining step for the next ones such as piercing and boring [1, 4]. In many cases, the hole surface after drilling can also be used immediately to participate in the assembly process. Many criteria area are used to evaluate the drilling process such as surface roughness, cutting force, tool wear, thrust force, torque, entry burr height, exit burr height, etc. Many studies have been carried out to determine the optimal value of parameters of the machining process to ensure one or several criteria of the drilling process when machining each specific material.

Vedant et al. [5] used the Taguchi method to design the experimental matrix when drilling HE30 (6082) aluminum alloy. The purpose of this study is to determine the optimal value of spindle speed, feed rate and drill point angle, for the purpose of ensuring the minimum value of surface roughness. Kumar et al. [6] also used the Taguchi method to design the experimental matrix when drilling AA5052 aluminum alloy. They selected spindle speed, feed rate and drill bit diameter as input parameters for each experiment. They found the optimal value of input parameters to ensure the

D. C. Nguyen et al. (Eds.): ICERA 2021, LNNS 366, pp. 610–622, 2022.
https://doi.org/10.1007/978-3-030-92574-1_64

minimum surface roughness. Haleel [7] also used the Taguchi method to design the experimental matrix when drilling AI 5050 aluminum alloy. Spindle speed, feed rate and drill bit diameter were selected as input parameters for each experiment. He determined the optimal value of input parameters to ensure the minimum value of surface roughness. Osman et al. [8] designed an experimental matrix to drill AISI D2 tool steel by the Taguchi method with input parameters including spindle speed, feed rate and drill bit material. They found optimal set of input parameters to ensure the minimum DE. Pereria et al. [9] determined two optimal sets of spindle speed, feed rate and pecking cycle when drilling aluminum aerospace alloys on an industrial robot. While an optimal set of these parameters was intended to ensure the minimum surface roughness, the other was to ensure the minimum exit burr height. The experimental design and optimization method used in this study was the Taguchi method. Alagarsamy et al. [10] also used Taguchi method to design the experimental matrix when drilling 7075 aluminum alloy. Spindle speed, feed rate and cutting depth were selected as input parameters for each experiment. They determined the two optimal sets of input parameters, of which a set was intended to ensure the minimum surface roughness and the other was to ensure the maximum MRR. Navanth et al. [11] also designed an experimental matrix in the form of Taguchi for the experiment of dry drilling of AL2014 aluminum alloy. Spindle speed, feed rate and drill point angle were selected as input parameters. They determined the two optimal sets of input parameters, of which a set was intended to ensure the minimum surface roughness and the other was to ensure the minimum DE. Singh et al. [12], when experimenting with drilling of Al6082 aluminum alloy, also designed the experimental matrix by the Taguchi method with input parameters including cooling oil, feed rate and drill bit diameter. They determined the two optimal sets of input parameters, of which a set was intended to ensure the minimum surface roughness and the other was to ensure the minimum burr height. Kanagaraju et al. [13], when experimenting with drilling of 6061-T6 aluminum alloy, also designed the experimental matrix by the Taguchi method in which spindle speed and feed rate were selected as input parameters. They determined the two optimal sets of input parameters, of which a set was intended to ensure the minimum thrust force and the other was intended to ensure the minimum torque. Sreenivasulu et al. [14], when experimenting with drilling of 6061 aluminum alloy, also designed the experimental matrix by the Taguchi method in which spindle speed, feed rate, drill bit diameter, drill point angle and drill clearance angle were selected as five input parameters. They determined the two optimal sets of five parameters mentioned above to ensure either the minimum thrust force or the minimum torque. Abhishek et al. [15] combined Taguchi and Principal Component Analysis (PCA) methods to determine the optimal value of drill bit diameter, cutting speed and feed rate when drilling Metal Matrix Composite (MMC) aluminum silicon carbide. The purpose of this study was to simultaneously ensure the four criteria of the minimum thrust force, the minimum torque, the minimum error circularity and the minimum error circularity out. Golshan et al. [16] determined the optimal value of drill bit diameter, spindle speed and feed rate to simultaneously ensure the minimum surface roughness and the minimum DE. The experimental material used in this study was 7075 aluminum alloy. The combination of Taguchi and Non-dominated Sorting Genetic Algorithm (NSGA) was used to solve the multi-objective optimization problem. Raj et al. [17], when experimenting with drilling of AA6101 T6 aluminum alloy, designed the experimental matrix by the Taguchi method with input parameters of spindle speed, feed rate and drill bit diameter. GRA method was also used to determine the optimal value of input parameters to simultaneously ensure the minimum surface roughness and the minimum ovality. Venkatesulu et al. [18], when experimenting with drilling of Al6063/B4C composites, also designed the experimental matrix by the Taguchi method. The five parameters of reinforcement percentages, spindle speed, feed rate, drill bit material and coolant type were selected as input parameters. They have combined AHP method and Topsis method to determine the optimal value of input parameters to simultaneously ensure the minimum cutting power, the minimum thrust force, the minimum torque, the minimum cutting temperature, the minimum surface roughness and the very minimum drill bit vibration. Soepangkat et al. [19] used the combination of Back Propagation Neural Network (BPNN) and Genetic Algorithm (GA) methods for multi-objective optimization of the drilling process of stainless steel-GFRP stacks. They determined the optimal value of spindle speed, feed rate and drill point angle to simultaneously ensure that all four parameters of thrust force, torque, hole surface roundness, and delamination were kept to a minimum value. Solage et al. [20] determined the material type of workpiece, spindle speed, feed rate and material type of cutting tool to simultaneously ensure the minimum thrust force, the minimum torque and the minimum surface roughness. In this study, they designed the experimental matrix by the Taguchi method. The optimal method used was GRA. Ay et al. [21] also combined Taguchi and GRA methods for multi-objective optimization when drilling Hybrid aluminum matrix composites. Spindle speed, feed rate and graphite content in the workpiece material were selected as three input parameters for each experiment. They determined the optimal value of three input parameters to simultaneously ensure the minimum surface roughness and the minimum cutting force. Kannan et al. [22] experimented with drilling of stainless steel, in which each experiment changed the two parameters, namely spindle speed and feed rate. The experimental matrix was built with the Taguchi method. The GA algorithm was applied to solve the multi-objective optimization problem. Finally, they determined the value of two cutting parameters to simultaneously ensure the three criteria of the minimum surface roughness, the minimum entry burr height and the minimum exit burr height.

Through an overall study of a number of above studies, it shows that there are many parameters to evaluate the drilling process in general and the aluminum alloy drilling in particular. It is extremely difficult to consider all those parameters within a single study. MRR, DE and RE are the parameters greatly determining the efficiency of drilling process, and are often considered when studying the drilling process [23–25]. If these three parameters are often considered in the drilling process, it is because MRR is the parameter directly reflecting machining productivity, while DE and RE are the two factors greatly affecting the accuracy of the joints as well as affecting the machining stages after drilling (if necessary) [26–28]. However, so far, there have been no published studies considering all of these three parameters.

Also, through the study of a number of studies in this section, it shows that Taguchi method has been very successfully applied in building the experimental matrix. The use of the entire Taguchi method for single-objective optimization of the drilling process has been carried out in a number of studies. The combination of Taguchi with some other methods (GA, GRA, AHP, etc.) has also succeeded in multi-objective optimization of the drilling process in some studies. RIM is also a method for multi-objective optimization, which has been successful when combined with Taguchi method for multi-objective optimization in some cases, such as in the selection of military aircraft for Spanish military [30], in multi-objective optimization of the milling process [29, 31, 32]. However, so far, there have been no published studies applying the combination of Taguchi and RIM methods for multi-objective optimization of the drilling process.

From the above analysis, in this study, we built an experimental matrix by the Taguchi method to experiment with drilling of AA7075 aluminum alloy, and RIM method was applied for multi-objective optimization of the drilling process. The purpose of this study is to determine the values of some parameters of the drilling process to ensure the maximum MRR, the minimum DE and the minimum RE.

2 RIM Method

RIM is a method to solve the multi-objective optimization problem. This method is based on the concept of "ideal solution", which is carried out in the following steps [33]:

Step 1: Normalization process.

In this stage, the reference ideal interval will be determined in accordance with the formula (1).

$$d_{min}(x, [C, D]) = \min(|x - C|, |x - D|) \quad (1)$$

In which:

x is the value of a criterion at a given option.

[C, D] is the reference ideal interval.

The next stage of normalization process is to determine the normalized value in accordance with the following formula:

$$f(x, [A, B], [C, D]) = \begin{bmatrix} 1 & if\ x \in [C, D] \\ 1 - \frac{d_{min}(x,[C,D])}{A-C} & if\ x \in [A, C]\ and\ A \neq C \\ 1 - \frac{d_{min}(x,[C,D])}{D-B} & if\ x \in [D, B]\ and\ D \neq B \end{bmatrix} \quad (2)$$

Where [A, B] is the range of values from the smallest to the largest of a certain criterion.

Step 2. Normalize the valuation matrix X with the reference ideal

$$Y = \begin{bmatrix} f(x_{11}, t_1, s_1) & \ldots & f(x_{1n}, t_n, s_n) \\ f(x_{21}, t_1, s_1) & \ldots & f(x_{2n}, t_n, s_n) \\ \ldots & \ldots & \ldots \\ f(x_{m1}, t_1, s_1) & \ldots & f(x_{mn}, t_n, s_{1n}) \end{bmatrix} \quad (3)$$

Where the function f has been calculated in accordance with the formula (2), n is the number of criteria, m is the number of options.

Step 3: Determine the weight for each criterion, where i is the number of criteria.

$$\sum_{i=1}^{n} w_i = 1 \ and \ 0 < w_i < 1 \tag{4}$$

Step 4. Calculate the weighted normalized matrix Y'.

$$Y' = Y.W = \begin{bmatrix} y_{11}.w_1 & \ldots & y_{1n}.w_n \\ y_{21}.w_1 & \ldots & y_{2n}.w_n \\ \ldots & \ldots & \ldots \\ y_{m1}.w_1 & \ldots & y_{mn}.w_n \end{bmatrix} \tag{5}$$

Step 5. Calculate the variation to the normalized reference ideal for each alternative.

$$I_i^+ = \sqrt{\sum_{j=1}^{m} \left(y'_{ij} - w_j\right)^2} \tag{6}$$

$$I_i^- = \sqrt{\sum_{j=1}^{n} \left(y'_{ij}\right)^2} \tag{7}$$

In which:
i = 1, 2, … m (number of options)
j = 1, 2, … n (number of criteria)
Step 6. Calculate the relative index

$$R_i = \frac{I_i^-}{I_i^+ + I_i^-} \tag{8}$$

In which: 0 < Ri < 1, i = 1, 2, …. m

Step 7. Rank the options in accordance with the value of Ri. The option with the largest Ri is the best one.

3 Experimental Study

The experiments were conducted on Ikeda vertical milling machine (Japan). The number of spindle speed grades and the levels of feed rate were finite. The adjustment of the values of these parameters was through the lever system. The experimental billet was AA7075 aluminum alloy, whose chemical composition is shown in Table 1. This is a material with high hardness equivalent to steel, but it is much lighter in weight. This material is widely used in the automotive and aviation industries [34]. Figure 1 shows the sample after drilling. The cutting tool used in the experiments was a hard alloy drill bit from WIDIN (Korea), with its diameter of 10 mm (Fig. 2). This is the type of drill bit commonly used to drill aluminum alloy [35]. Each experiment used a new drill bit to eliminate the effect of tool wear on output parameters.

Table 1. Chemical composition of AA7075 aluminum alloy

Components	Si	Mg	Mn	Cu	Fe	Cr	Zn	Ti	Al
%	0.62	1.1	0.12	0.32	0.68	0.33	0.22	0.14	95.52

Fig. 1. Experimental sample

The two selected parameters were variables changing during the experiment, inclusive of spindle speed (nW) and feed rate (f). The Taguchi method was used to design the experimental matrix with a total of 9 experiments, in which each cutting parameter had three levels of values as shown in Table 2. The values of cutting parameters in this table were selected in accordance with the ability of experimental machine to adjust those values as well as in accordance with the recommendations of tool manufacturer [35], and in accordance with a number of published studies on AA7075 aluminum drilling [36–38].

Table 2. Experimental matrix

No.	Code value		Real value	
	n_w	f	n_w (rpm)	f (rev/min)
1	1	1	790	0.12
2	1	2	790	0.14
3	1	3	790	0.16
4	2	1	1020	0.12
5	2	2	1020	0.14
6	2	3	1020	0.16
7	3	1	1280	0.12
8	3	2	1280	0.14
9	3	3	1280	0.16

MRR was calculated in accordance with the formula (9). Where D is the drill bit diameter, D = 10 mm. DE and RE were measured using a Mitutoyo Crysta - Plus M544 3D coordinate measuring machine. In which, DE was calculated as the difference of hole diameter size minus the drill bit size.

$$MRR = \frac{1}{4}\pi \cdot D^2 \cdot n_w \cdot f \tag{9}$$

4 Results and Discussion

The experimental results are shown in Table 3. In Fig. 2 and Fig. 3, the charts of the effect of cutting parameter on DE and RE are shown. The interaction effects of cutting parameters on DE and RE are shown in Figs. 4 and 5, respectively.

Table 3. Experimental results

No.	n_w (rpm)	f (rev/min)	MRR (mm^3/min)	DE (mm)	RE (mm)
1	790	0.12	7445.575	0.023	0.032
2	790	0.14	8686.504	0.028	0.042
3	790	0.16	9927.433	0.029	0.035
4	1020	0.12	9613.274	0.012	0.018
5	1020	0.14	11215.486	0.017	0.031
6	1020	0.16	12817.698	0.025	0.028
7	1280	0.12	12063.716	0.031	0.053
8	1280	0.14	14074.335	0.026	0.049
9	1280	0.16	16084.954	0.034	0.071

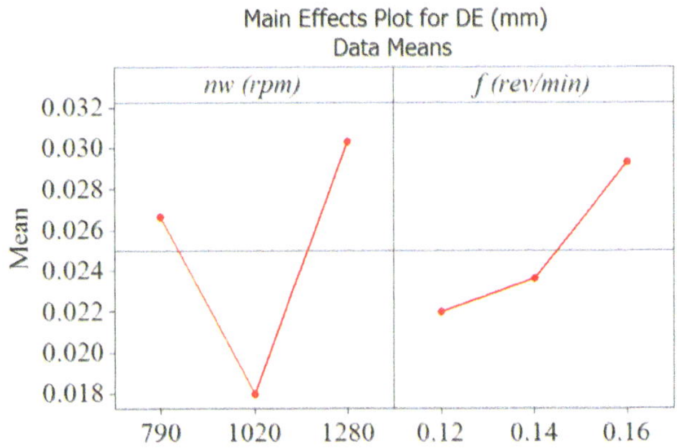

Fig. 2. Main effects for DE

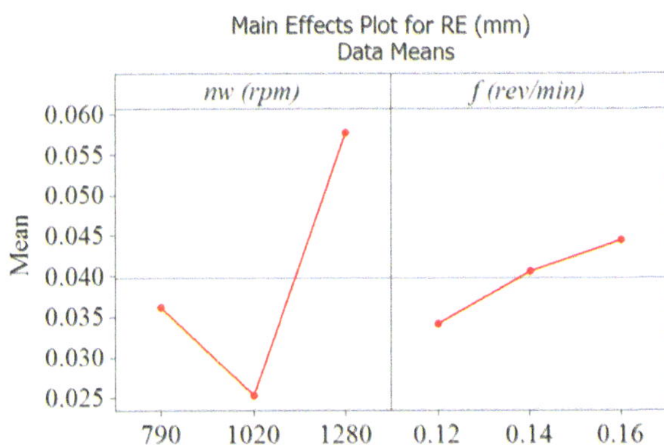

Fig. 3. Main effects for RE

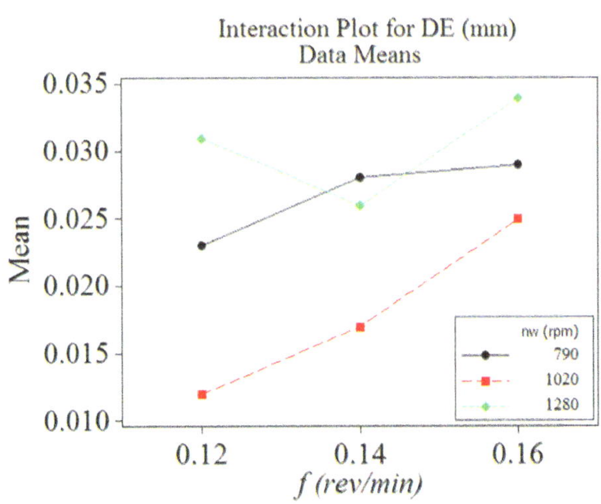

Fig. 4. Interaction plot for DE

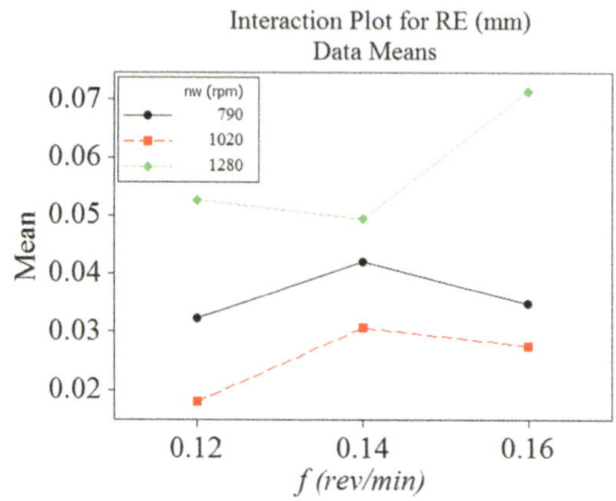

Fig. 5. Interaction plot for RE

Observing the above figures, it shows that:

- Feed rate has a greater effect on DE than spindle speed.
- Spindle speed has a greater effect on RE than feed rate.
- When the spindle speed is equal to 790 rpm and 1,020 rpm, if the feed rate increases, it makes DE increase. When the spindle speed is equal to 1,280 rpm, if the feed rate increases from 0.12 mm/rev to 0.14 mm/rev, it makes DE increase. However, if the feed rate continues to increase, it makes DE increase.
- When the spindle speed is equal to 790 rpm and 1,020 rpm, if the feed rate increases from 0.12 mm/rev to 0.14 mm/rev, it makes RE increase. However, if the feed rate continues to increase, it makes RE decrease. On the contrary, when the spindle speed is equal to 1,280 rpm, if the feed rate increases from 0.12 mm/rev to 0.14 mm/rev, it makes RE increase. However, if the feed rate continues to increase, it makes RE increase.

For the MRR, it is obvious that the increase of spindle speed and feed amount makes the MRR increase.

The above analysis shows that the extent and effect law of cutting parameters on MRR, DE and RE is quite complex and sometimes contradictory. Therefore, it is excessively difficult to determine the value of cutting parameters to simultaneously ensure the maximum MRR, the minimum DE and the minimum RE if only the graphs in Figs. 2, 3, 4 and 5 are observed. The experiments in Table 3 show that: The MRR has the largest value in experiment #9. However, also in this experiment, DE and RE are also the largest. DE and RE both have the smallest value in experiment #4, but the MRR in this experiment is only moderate. Therefore, if only the data in Table 3 is observed, it is impossible to determine which experiment simultaneously ensures the maximum MRR, the minimum DE and the minimum RE. In fact, it is also not possible to determine any experiments in which the concepts of "largest" and "smallest" can be achieved perfectly. Determining which values achieve the desired results can only be made on the sound mathematical basis. The RIM method will be applied to solve this problem.

5 Multi-objective Optimization Using RIM Method

From the experimental data in Table 3, it shows that the MRR has the minimum and maximum values of 7,445.575 m^3/min and 16,084.954 m^3/min; the minimum and maximum values of DE are 0.012 mm and 0.034 mm, respectively; the minimum and maximum values of RE are 0.0180 mm and 0.0714 mm, respectively. Thus, we have:

[A, B] = [7445.575 16084.954 0.012 0.034 0.0180 0.0714].

And:

[C, D] = [16084.954 16084.954 0.012 0.012 0.0180 0.0180].

Applying the formula (1) to determine the reference ideal intervals of MRR, DE and RE. Applying the formula (2) to determine the normalize values of MRR, DE and RE. Applying the formulas (3) to (7), we can calculate the values of parameters of Ii+ and Ii−, in which the weights of MRR, DE and RE share the same value, i.e. $w_1 = w_2 = w_3 = 1/3$. All these values are shown in Table 4.

Table 4. Parameters in RIM

No.	d_{min} (MRR, [C, D])	d_{min} (DE, [C, D])	d_{min} (RE, [C, D])	f(MRR)	f(DE)	f(RE)	I_i^+	I_i^-
1	7445.575	0.023	0.0322	2.00000	0.50000	0.73408	1.72225	2.18835
2	8686.504	0.028	0.042	1.85636	0.27273	0.55056	1.53964	1.95540
3	9927.433	0.029	0.0348	1.71273	0.22727	0.68539	1.42756	1.85872
4	9613.274	0.012	0.018	1.74909	1.00000	1.00000	1.70096	2.24929
5	11215.486	0.017	0.0306	1.56364	0.77273	0.76404	1.37558	1.90416
6	12817.698	0.025	0.0275	1.37818	0.40909	0.82210	1.15600	1.65608
7	12063.716	0.031	0.0527	1.46545	0.13636	0.35019	1.14925	1.51287
8	14074.335	0.026	0.0494	1.23273	0.36364	0.41199	0.90333	1.34966
9	16084.954	0.034	0.0714	1.00000	0.00000	0.00000	0.81650	1.00000

Applying the formula (8), we can calculate Ri. After that, use the value of Ri to rank the options, the results are shown in Table 5.

Table 5. Ranking of options

No.	n_w (rpm)	f (rev/min)	MRR (mm^3/min)	DE (mm)	RE (mm)	R_i	Ranking
1	790	0.12	7445.575	0.023	0.032	0.55959	7
2	790	0.14	8686.504	0.028	0.042	0.55948	8
3	790	0.16	9927.433	0.029	0.035	0.56560	6
4	1020	0.12	9613.274	0.012	0.018	0.56941	4
5	1020	0.14	11215.486	0.017	0.031	0.58058	3
6	1020	0.16	12817.698	0.025	0.028	0.58892	2
7	1280	0.12	12063.716	0.031	0.053	0.56830	5
8	1280	0.14	14074.335	0.026	0.049	0.59905	1
9	1280	0.16	16084.954	0.034	0.071	0.55051	9

From the data in Table 5, it shows that experiment #8 is the best option, while experiment #9 is the worst option. Thus, it can be said that for the purpose of determining the values of input parameters so that the MRR is "maximum", the DE is "minimum and the RE is "minimum", then experiment #8 is the best option. Since then, we come to the conclusion that in the experimental system conditions of this study, in order to simultaneously ensure the maximum MRR, the minimum DE and the minimum DE, the spindle speed is equal to 1,280 rpm and the feed rate is equal to 0.16 mm/rev.

6 Conclusion

In this study, the experiment of AA7075 aluminum alloy drilling was conducted with a hard alloy cutting tool. At each experiment, the spindle speed and the feed rate were changed. MRR, DE and RE were determined in each experiment. A number of conclusions are drawn as follows:

- The feed rate has a greater effect on DE than the spindle speed. Meanwhile, the spindle speed has a greater effect on RE than the feed rate.
- The RIM method was first applied to solve the multi-objective optimization problem in this study. The results determine that, in order to simultaneously ensure the maximum MRR, the minimum DE and the minimum RE, the spindle speed is equal to 1280 rpm, and the feed rate is equal to 0.16 mm/rev.
- Determining the optimal value of parameters of coolant (coolant type, concentration, pressure, flow) to simultaneously ensure the criteria when drilling AA7075 aluminum alloy is the work that will be done by us in the next study.

References

1. Sreenivasulu, R., Rao, C., Ravindra, K.: Effect of thrust and torque exerted during drilling to optimize exit burr height and thickness by choosing variable drill bit geometry: a simplified theoretical model approach. Int. J. Data Netw. Sci. **4**, 43–56 (2020)
2. Singh, G.R., Sharma, V.S.: Sustainable drilling of Aluminium 6061–T6 alloy by using nanofluids and Ranque-Hilsch vortex tube assisted by MQL: an optimisation approach. Int. J. Mach. Mach. Mater. **20**(3), 252–273 (2018)
3. Devitte, C., Souza, G.S.C., Souza, A.J., Tita, V.: Optimization for drilling process of metalcomposite aeronautical structures. Sci. Eng. Compos. Mater. **28**, 264–275 (2021)
4. Das, R., Barik, T.: An experimental study on the burr formation in drilling of aluminum channels of rectangular section, 5th International & 26th All India Manufacturing Technology. Des. Res. Conf. **831**, 1–5 (2014)
5. Vedant, P.U., Dhiraj, D.P., Nakeed, P.R., Sagar, A.S., Sakharkar, N., Mohan, V.M.: Experimental optimization of machinability parameters in drilling of He30 Aluminium alloy Kingpin Bush **5**(7), 25–31 (2017)
6. Kumar, M.D.S., Prasad, J., Krishna, D.A., Narayana, M.V., Anusha, M., Saravanakumar, A.: Parametric optimization of aluminium alloy milling using Taguchi method for surface roughness. Int. J. Sci. Res. Rev. **7**(3), 516–522 (2018)

7. Haleel, A.J.: Optimization drilling parameters of aluminum alloy based on Taguchi method. Al-Khwarizmi Eng. J. **14**(2), 14–21 (2018)
8. Osman, M.H.B., Rahman, M.H.B.A., Ahmad, M.N.B., Wahid, M.K.B., Maidin, N.A.B.: Optimization of drilling parameters on diameter accuracy in dry drilling process of AISI D2 tool steel **12**(20), 9644–9652 (2017)
9. Pereira, B., Griffiths, C.A., Birch, B., Rees, A.: Optimization of an autonomous robotic drilling system for the machining of aluminum aerospace alloys. Res. Square 1–19 (2021)
10. Alagarsamy, S.V., Sagayaraj, S.A.V., Raveendran, P.: Optimization of drilling process parameters on surface roughness & material removal rate by using Taguchi method. Int. J. Eng. Res. Gener. Sci. **4**(2), 290–298 (2016)
11. Navanth, A., Sharma, T.K.: A study of Taguchi method based optimization of drilling parameter in dry drilling of AL 2014 alloy at low speeds. Int. J. Eng. Sci. Emerg. Technol. **6** (1), 65–75 (2013)
12. Singh, J., Garg, K., Singh, P.: Optimization of performance parameters in drilling process for minimizing the burr formation. Asian J. Eng. Applied Technol. **7**(S2), 127–131 (2018)
13. Kanagaraju, T., Samuel John Peter, J., Rajan Samuel, D., Prince Prakash, J.: Optimization of drilling parameters for thrust force and torque in friction drilling process. Middle-East J. Sci. Res. **24**(4), 1577–1582 (2016)
14. Sreenivasulu, R., Rao, C.S.: Effect of drilling parameters on Thrust force and Torque during Drilling of Aluminium 6061 Alloy – based on Taguchi design of experiments. J. Mech. Eng. **46**, 41–48 (2016)
15. Abhishek, K., Sonkar, V., Datta, S., Mahapatra, S.S.: Optimization in drilling of MMC composites: a case research. J. Basic Appl. Eng. Res. **1**(2), 79–83 (2014)
16. Golshan, A., Ghodsiyeh, D., Gohari, S., Ayob, A., Baharudin, B.T.H.T.: Optimization of machining parameters during drilling of 7075 aluminium alloy. Appl. Mech. Mater. **248**, 20–25 (2013)
17. Raj, R., Rathore, P.: Multi objective parameter optimisation in drilling of aluminium alloy towards minimum ovality and surface roughness. Int. J. Sci. Res. Dev. **7**(4), 1090–1092 (2019)
18. Venkatesulu, M., Kotaiah, K.R.: Optimization of process parameters in drilling of Al6063/B4C composites using AHP-Topsis method. J. Eng. Appl. Sci. **13**(24), 10461–10467 (2018)
19. Soepangkat, B.O.P., Norcahyo, R., Effendi, M.K., Sateria, A., Pramujati, B.: Application of BPNN-GA to optimize drilling process of stainless SteelGFRP stacks. Int. J. Mech. Mech. Eng. **19**(4), 30–39 (2019)
20. Solage, R.D., Shinde, S.M.: Experimental study and optimization of process parameters in drilling of B4c and Mg reinforced hybrid AMMCs manufactured by powder metallurgy. Int. J. Eng. Res. Technol. **7**(3), 196–205 (2018)
21. Ay, M., Altunpak, Y., Hartomacıoğlu, S.: The grey-based Taguchi method: optimisation of drilling of hybrid aluminum matrix composites. Spec. Issue 6th Int. Congr. Exhib. **131**(3), 551–554 (2017)
22. Kannan, T.D.B., Kumar, B.S., Kannan, G.R., Umar, M., Khan, M.C.: Application of genetic algorithm technique for machining parameters optimization in drilling of stainless steel. Mech. Mech. Eng. **23**, 271–276 (2019)
23. Motorcu, A.R., Ekici, E.: The optimization of drilling parameters for hole diameter dimensional accuracy and minimum roundness error in the drilling of Al/10B4C/3Gr hybrid composites with the cryogenically treated HSS drills. In: 8th International Advanced Technologies Symposium, pp. 624–631 (2017)
24. Prakash, C., et al.: Investigating the efficacy of adhesive tape for drilling carbon fibre reinforced polymers. Materials **14**(1699), 1–17 (2021)

25. Pallapothu, H., Kumar, U.A., Laxminarayana, P.: Micro drilling of glass fibre reinforced polymer composites. Mater. Today: Proc. 1–5 (2020). (Article, in press)
26. Aamir, M., Tu, S., Giasin, K., Tolouei-Rad, M.: Multi-hole simultaneous drilling of aluminium alloy: a preliminary study and evaluation against one-shot drilling process. J. Mater. Res. Technol. **9**(3), 3394–4006 (2020)
27. Sterle, L., Krajnik, P., Pu Savec, F.: The effects of liquid-CO2 cooling, MQL and cutting parameters on drilling performance. CIRP Ann. – Manuf. Technol. (14), 1–4 (2021)
28. Venkatesan, K., Devendiran, S., Bhupatiraju, S.C.S.R., Kolluru, S., Kumar, C.P.: Experimental investigation and optimization of micro-drilling parameters on Inconel 800 superalloy. Mater. Manuf. Processes 1–15 (2020)
29. Mehmet Alper Sofuoğlu: A new hybrid decision making model to optimize machining operations. Online J. Sci. Technol. **8**(2), 5–8 (2018)
30. Sánchez-Lozano, J.M., Rodríguez, O.N.: Application of Fuzzy Reference Ideal Method (FRIM) to the military advanced training aircraft selection. Appl. Soft Comput. J. **88** (106061), 1–12 (2020)
31. Gürgen, S., Çakır, F.H., Sofuoğlu, M.A., Orak, S., Kuşhan, M.C., Li, H.: Multi-criteria decision-making analysis of different non-traditional machining operations of Ti6Al4V. Soft Comput. **23**, 5259–5272 (2019)
32. Trung, D.D.: Multi-objective optimization of SKD11 steel milling process by Reference Ideal Method. Int. J. Geol. **15**, 1–18 (2021)
33. Cables Perez, E., Lamata, M.T., Verdegay, J.L.: RIM-reference ideal method in multicriteria decision making. Inf. Sci. **337–338**(10), 1–10 (2016)
34. https://en.wikipedia.org/wiki/7075_aluminium_alloy
35. http://hoahaoco.com/en/san-pham/dao-phay-widin-korea/
36. Yaşar, N., Boy, M., Günay, M.: The effect of drilling parameters for surface roughness in drilling of AA7075 alloy. MATEC Web Conf. **112**(0101), 1–8 (2017)
37. Yaşar, N., Boy, M., Günay, M.: The effect of drilling parameters for surface roughness in drilling of AA7075 alloy. MATEC Web Conf. **112**(01018), 1–8 (2017)
38. Yıldız, T., Sur, G.: Investigation of drilling properties of AA7075/Al2O3 functionally graded materials using gray relational analysis. Proc. Inst. Mech. Eng. Part B: J. Engineering Manuf. 1–14 (2021, first published online)

Multi-objective Optimization of PMEDM Process for Maximum Material Removal Speed and Minimum Electrode Wear Rate When Machining SKD11 Steel

Anh Tuan Nguyen[1], Thi Tam Do[2], Thu Quy Le[3], Ngoc Huy Thinh Tran[4], Hoang Anh Le[5], Ngoc Giang Tran[2], Ngoc Pi Vu[2], and Xuan Tu Hoang[2](✉)

[1] University of Economics – Technology for Industries, Ha Noi, Vietnam
[2] Thai Nguyen University of Technology, Thai Nguyen, Vietnam
hxt_dhktcn@tnut.edu.vn
[3] National Research Institute of Mechanical Engineering, Ha Noi, Vietnam
[4] Nguyen Tat Thanh University, Ho Chi Minh City, Vietnam
[5] Vinh Long University of Technology Education, Vinh Long, Vietnam

Abstract. This paper presents the results of a multi-objective optimization study when powder-mixed electrical discharge machining (PMEDM) of cylindrically shaped parts made of SKD11 steel to achieve the maximum removal speed (MRS), and minimum electrode wear rate (EWR). To carry out this study, an experiment was conducted with six input parameters, including the powder concentration, the powder size, the pulse on time, the pulse off time, the current, and the servo voltage. Besides, the Taguchi method combined with the gray relation analysis (GRA) method was used to design experiment, and analyze experimental results. The impact of the input parameters on the MRS and the EWR was analyzed. In addition, optimal input parameters to get the maximum MRS, and the minimum EWR simultaneously were found.

Keywords: PMEDM · EDM · Multi-objective optimization · Cylindrically shaped parts · SKD11

1 Introduction

Electrical discharge machining (EDM) is a machining method where the workpiece material removal is accomplished by sparks generated continuously between the electrode and the workpiece, which are separated by a dielectric. EDM is very suitable for processing complex parts or recessed cavities, such as stamping dies, forgings or castings, etc. Besides, this type of processing also has several disadvantages, such as the quality of the machined surface is not high, the electrodes usually wear out quickly, etc. In order to limit these disadvantages, a metal powder was mixed into the dielectric liquid, and so the EDM process became the PMEDM process.

Up to now, research on PMEDM has always been a topic of interest to researchers. Many studies on the PMEDM process have been carried out. Studies on this type of

D. C. Nguyen et al. (Eds.): ICERA 2021, LNNS 366, pp. 623–633, 2022.
https://doi.org/10.1007/978-3-030-92574-1_65

machining have been conducted to improve the machined surface quality [1–4], increase the material removal rate [1, 4–10], the machining time [11], or reduce the electrode wear rate [7, 8, 12]. Many studies have been done on PMEDM with mixing different metal powders into the dielectric liquid such as SiC [3–5, 9–11, 13], Al [1, 14], Al2O3 [15], W [8, 12], Cr [2], or Ti [5, 16, 17]. Besides, PMEDM with different machining materials has received much attention such as SKD11 [16, 17], AISI D2 die steel [5], 90CrSi [9–11], Ti-6Al-4V [13, 18], EN8 steel [2], Inconel 718 [14, 15], etc. Electrode materials are also the subject of many studies, such as copper [3, 4, 8], graphite [5, 19], and brass [20]. Recently, there has been some research on machining cylindrically shaped parts [3, 4, 9–11].

This work deals with a study to simultaneously optimize two single-objective functions, maximum MRS and minimum EWR, when processing cylindrically shaped parts of SKD11 steel using the PMEDM process. In the study, an experiment was carried out with the use of the Taguchi method combined with the gray relation analysis (GRA) method to design the experiment and analyze the experimental results. The influence of the input process parameters on the above objective functions was investigated. Furthermore, optimal input parameters to achieve the mentioned two objectives simultaneously have been proposed.

2 Experimental Setup

In this study, an experiment was carried out to solve the multi-objective optimization problem with two objectives performed simultaneously, including the maximum MRS and the minimum EWR, an experiment was performed. Six input process parameters were selected to evaluate their effects on MRS and EWR. These factors and their levels for the experiment are given in Table 1. Figure 1 shows the experimental setup. In this experiment, a Sodick A30 EDM machine, SKD11 steel samples, copper electrodes, and SiC powder for mixing in the dielectric fluid were used.

Table 1. Input factors and their levels

No	Parameters	Code	Unit	Level					
				1	2	3	4	5	6
1	Powder concentration	C_p	g/l	0	2	3	4	5	6
2	Powder size	S_p	nm	100	500	1000	–	–	–
3	Pulse on time	T_{on}	µs	10	20	30	–	–	–
4	Pulse off time	T_{off}	µs	10	20	30	–	–	–
5	Peak current	IP	A	4	8	12	–	–	–
6	Servo voltage	SV	V	3	5	7	–	–	–

Fig. 1. Experimental setup

The Taguchi method with the design L18 (3^4) with the support of the Minitab R19 software was used to create the experimental plan. Table 2 presents the plan of the experiment and the average values of the measured results of the MRS and the EWR.

Table 2. Experimental plan and the mean of MRS and EWR

Exp. No.	Input factors						Mean of MRS (g/h)	Mean of EWR (mg/min)
	C_p	S_p	T_{on}	T_{off}	IP	SV		
1	0	100	10	10	4	3	9.8773	13.5333
2	0	500	20	20	8	5	16.0979	99.8345
3	0	1000	30	30	12	7	49.6445	29.6923
4	2	100	10	20	8	7	6.7254	35.9337
5	2	500	20	30	12	3	15.8857	85.4975
6	2	1000	30	10	4	5	41.9831	16.9221
7	3	100	20	10	12	5	32.4869	55.6941
8	3	500	30	20	4	7	12.4249	11.3585
9	3	1000	10	30	8	3	11.8259	10.2848
10	4	100	30	30	8	5	28.2703	24.1109
11	4	500	10	10	12	7	8.1986	10.4982
12	4	1000	20	20	4	3	5.2888	18.9691
13	5	100	20	30	4	7	5.3288	15.2504
14	5	500	30	10	8	3	15.1998	64.6147
15	5	1000	10	20	12	5	7.3727	24.7060
16	6	100	30	20	12	3	30.8676	67.6757
17	6	500	10	30	4	5	5.6500	10.7152
18	6	1000	20	10	8	7	5.2931	32.8711

In this study, it is necessary to conduct multi-objective optimization or simultaneous optimization of two objectives, which are the maximum material removal rate and the minimum electrode wear. To do that, the Taguchi method combined with the GRA method was used.

The S/N indexes of MRS and EWR can be calculated by:

For maximum MRS:

$$S/N = -10\log_{10}(\frac{1}{n}\sum_{i=1}^{n}\frac{1}{y_i^2}) \tag{1}$$

For minimum EWR:

$$S/N = -10\log_{10}(\frac{1}{n}\sum_{i=1}^{n}y_i^2) \tag{2}$$

Table 3 shows the calculated S/N indexes of the two objectives.

Table 3. The calculated S/N value of MRS and EWR

No.	Input factors						MRS (g/h)		EWR (mg/min)	
	C_p	S_p	T_{on}	T_{off}	IP	SV	Mean	S/N	Mean	S/N
1	0	100	10	10	4	3	9.8773	19.8928	13.5333	−22.6281
2	0	500	20	20	8	5	16.0979	24.1354	99.8345	-39.9856
3	0	1000	30	30	12	7	49.6445	33.9174	29.6923	-29.4529
4	2	100	10	20	8	7	6.7254	16.5544	35.9337	-31.1100
5	2	500	20	30	12	3	15.8857	24.0201	85.4975	-38.6391
6	2	1000	30	10	4	5	41.9831	32.4615	16.9221	-24.5691
7	3	100	20	10	12	5	32.4869	30.2342	55.6941	-34.9162
8	3	500	30	20	4	7	12.4249	21.8858	11.3585	-21.1064
9	3	1000	10	30	8	3	11.8259	21.4567	10.2848	-20.2439
10	4	100	30	30	8	5	28.2703	29.0266	24.1109	-27.6443
11	4	500	10	10	12	7	8.1986	18.2748	10.4982	-20.4223
12	4	1000	20	20	4	3	5.2888	14.4672	18.9691	-25.5609
13	5	100	20	30	4	7	5.3288	14.5327	15.2504	-23.6656
14	5	500	30	10	8	3	15.1998	23.6368	64.6147	-36.2066
15	5	1000	10	20	12	5	7.3727	17.3525	24.7060	-27.8560
16	6	100	30	20	12	3	30.8676	29.7901	67.6757	-36.6087
17	6	500	10	30	4	5	5.6500	15.0409	10.7152	-20.6000
18	6	1000	20	10	8	7	5.2931	14.4741	32.8711	-30.3363

To implement the GRA, the S/N values need to be transformed to dimensionless quantities or need to normalize the data. In this case, the S/N indexes are normalized in the range (0 ÷ 1) and by Z_i ($0 \leq Z_i \leq 1$) which is calculated by the following Equation:

$$Z_i = \frac{S/N_i - \min(S/N_i, i = 1, 2, ..., n)}{\max(S/N_i, i = 1, 2, ..., n) - \min(S/N_i, i = 1, 2, ..., n)} \tag{3}$$

In which, n is the number of experiments; n = 18.

The gray relation (GR) coefficient $y_j(k)$ is determined by:

$$y_i(k) = \frac{\Delta_{\min}(k) + \zeta.\Delta_{\max}(k)}{\Delta_i(k) + \zeta.\Delta_{\max}(k)} \tag{4}$$

In which, i = 1, 2,…, n; n = 18 is the experimental number; k = 2 is the single objective number; Δi(k) = ||Z0(k) – Zi(k)||, is the absolute value of the difference between $Z_0(k)$ (reference value Z0(k) = 1) and $Z_i(k)$. (Z-value of the ith experiment of the kth target); Δmin(k) is the minimum value of i(k); max(k) is the maximum value of i (k); ζ is the discriminant coefficient; $0 \leq \zeta \leq 1$; For a experimental study, we can chose ζ = 0.5.

The GR degree can be determined through the average GR value of the output objectives:

$$\overline{y_i} = \frac{1}{k}\sum_{j=0}^{k} y_{ij}(k) \tag{5}$$

Where, k is the single objective number; y_{ij} is the GR value of the jth output objectives in the ith experiment.

The results of the calculated GR value y_i and the mean GR value of the experiments are listed in Table 4.

Table 4. S/N, S/N Z_i, Δ_i (k) and GR value $\overline{y}_i$

No	S/N		Zi		Δ_i (k)		GR value yi		$\overline{y}_i$
	Ra	EWR	$\overline{Ra}$	EWR					
			Reference value		MRS	EWR	MRS	EWR	
			1.000	1.000					
1	−5.9226	−22.6281	0.4271	0.8792	0.721	0.121	0.409	0.805	0.636
2	−7.9305	−39.9856	0.2228	0.0000	0.503	1.000	0.499	0.333	0.362
3	−10.1206	−29.4529	0.0000	0.5335	0.000	0.466	1.000	0.517	0.425
4	−6.0635	−31.1100	0.4128	0.4496	0.893	0.550	0.359	0.476	0.468
5	−7.0309	−38.6391	0.3144	0.0682	0.509	0.932	0.496	0.349	0.385
6	−1.8672	−24.5691	0.8398	0.7809	0.075	0.219	0.870	0.695	0.726
7	−7.2817	−34.9162	0.2889	0.2568	0.189	0.743	0.725	0.402	0.408
8	−1.0345	−21.1064	0.9245	0.9563	0.619	0.044	0.447	0.920	0.894
9	−6.8051	−20.2439	0.3373	1.0000	0.641	0.000	0.438	1.000	0.715
10	−7.6264	−27.6443	0.2538	0.6251	0.251	0.375	0.665	0.572	0.486
11	−5.3302	−20.4223	0.4874	0.9910	0.804	0.009	0.383	0.982	0.738
12	−0.2925	−25.5609	1.0000	0.7307	1.000	0.269	0.333	0.650	0.825
13	−5.4196	−23.6656	0.4783	0.8267	0.997	0.173	0.334	0.743	0.616
14	−7.4181	−36.2066	0.2750	0.1914	0.529	0.809	0.486	0.382	0.395
15	−1.7023	−27.8560	0.8566	0.6144	0.852	0.386	0.370	0.565	0.671
16	−9.1074	−36.6087	0.1031	0.1711	0.212	0.829	0.702	0.376	0.367
17	−1.8846	−20.6000	0.8380	0.9820	0.971	0.018	0.340	0.965	0.860
18	−8.2363	−30.3363	0.1917	0.4888	1.000	0.511	0.333	0.494	0.438

3 Results and Analysis

To ensure harmony between output targets, the average GR value should be as high as possible. In this case, the multi-objective optimization problem becomes a single-objective optimization problem with the average GR as the output. To solve this problem, the Taguchi method is used to evaluate the influence of the input PMEDM parameters on the average GR value. The S/N value of was calculated according to Eq. 2 to get the mean GR as large as possible. The analysis of variance (ANOVA) for S/N ratios are given in Table 5.

Table 5. Analysis of variance for S/N ratios of $\bar{y}_i$.

Source	DF	Seq SS	Adj SS	Adj MS	F	P	C (%)
Cp	5	14.002	14.002	2.8004	1.27	0.495	21.91
Sp	2	2.247	2.247	1.1234	0.51	0.662	3.52
Ton	2	19.317	19.317	9.6586	4.38	0.186	30.23
Toff	2	9.726	9.726	4.8629	2.21	0.312	15.22
Ip	2	12.359	12.359	6.1797	2.81	0.263	19.34
SV	2	1.836	1.836	0.9181	0.42	0.706	2.87
Residual Error	2	4.406	4.406	2.2028			6.90
Total	17	63.893					

From Table 5, it can be seen that the influence of the parameters on the average GR value is as follows: T_{on} has the greatest influence (30.23%); followed by the influence of C_p (21.91%), IP (19.34%), and T_{off} (15.22%). The other two parameters with negligible influence are S_p (3.52%), and SV (2.87%). This result is also shown in Fig. 2. This result is also described in Fig. 2.

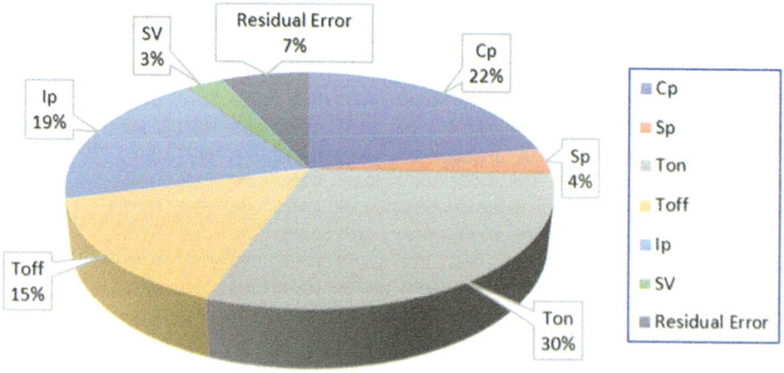

Fig. 2. Effect of input factors on $\bar{y}_i$ (%)

The influence of the input parameters on the GR value through ANOVA analysis is evaluated in order as shown in Table 6. Table 6 shows the order of input parameters ranked according to their influence on y as T_{on}, Cp, IP, T_{off}, Sp, and SV, respectively.

Table 6. Influence order of input parameters on S/N.

Larger is better

Level	Cp	Sp	Ton	Toff	Ip	SV
1	-4.783	-5.303	-4.734	-4.943	-4.17	-5.579
2	-5.732	-5.448	-6.549	-6.108	-6.192	-4.869
3	-3.717	-4.637	-4.105	-4.337	-5.026	-4.940
4	-4.551					
5	-6.416					
6	-5.577					
Delta	2.699	0.811	2.444	1.771	2.022	0.710
Rank	1	5	2	4	3	6

Average gray relation value: 0.567

The graph of the influence of the parameters on y is shown in Fig. 3. Theoretically, the value of each input parameter for which the value of the S/N ratio of $\bar{y}$ is the largest will be the optimal value of that parameter. Therefore, from Fig. 3, it is easy to determine the optimal values of the input parameters for the multi-objective function (see Table 7).

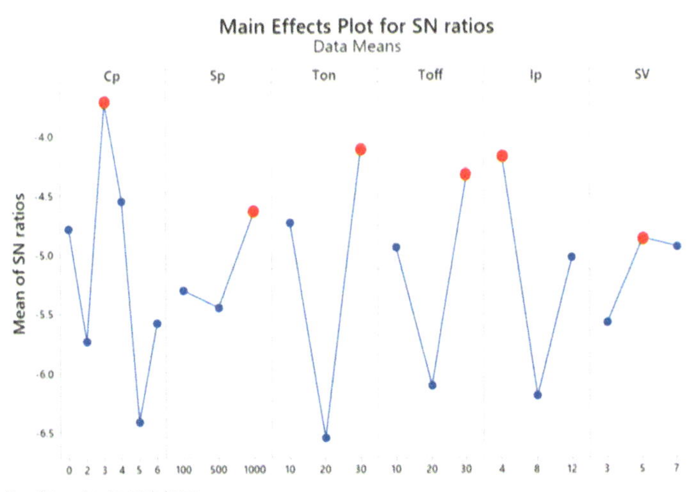

Fig. 3. Influence of input factors on the S/N ratios of $\bar{y}$

Figure 4 depicts error distribution histograms. These graphs allow evaluating the appropriateness of the experimental model. From Fig. 4, the following observations can be drawn: On the normal error distribution chart (Fig. 4a), the points representing the errors of the experiments (the blue points) are close to the line. normally distributed (the solid red line). It shows that the error level is very small. The graph of error occurrence frequency or the Histogram (Fig. 4c) shows that errors near 0 (−0.05 to 0.05) appear with high frequency. The remaining two graphs (Fig. 4b and d) show that the experimental errors are distributed quite randomly. It shows that the built model is highly influenced by the input parameters and is not affected by the order of experiments.

Table 7. Optimum input factors

Parameter	Code	Unit	Level	Value
Powder concentration	C_p	g/l	3	3
Powder size	S_p	nm	3	1000
Pulse on time	T_{on}	µs	3	30
Pulse off time	T_{off}	µs	3	30
Peak current	IP	A	1	4
Servo voltage	SV	V	2	5

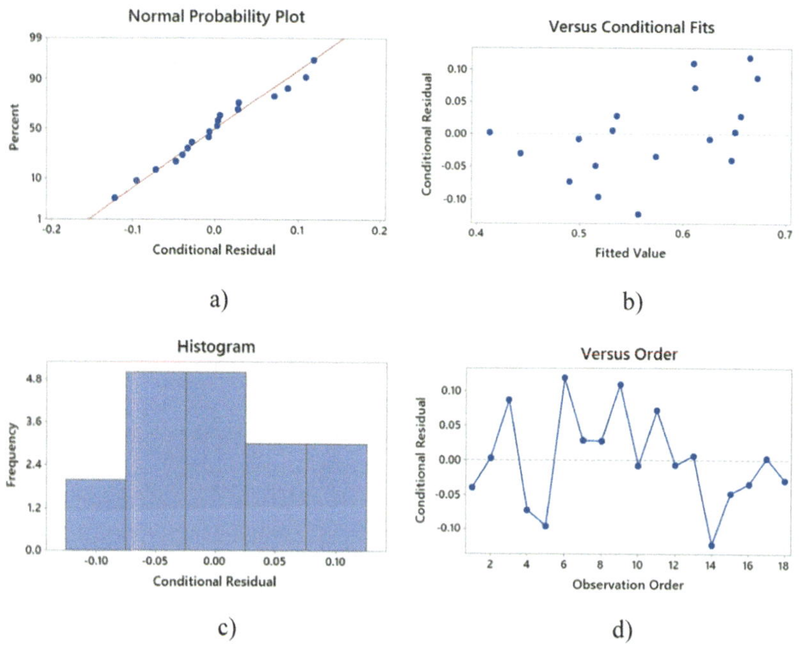

Fig. 4. Residual plots for $\bar{y}$

Figure 5 depicts the fit of the experimental model. This graph was constructed by using the Anderson-Darling method. From the graph, it can be seen that the experimental points (the blue dots) are all within the limit region defined by two upper and lower limit lines with a standard deviation of 95%. In addition, the P-value of 0.236 is much larger than the value of $\alpha = 0.05$, indicating that the experimental model is suitable.

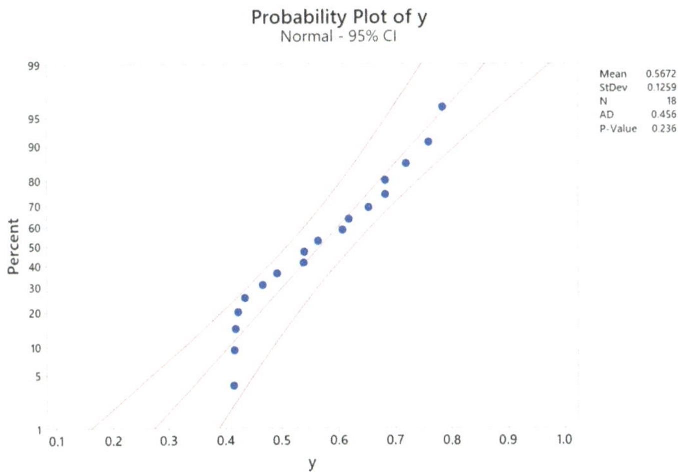

Fig. 5. Probability plot for $\overline{y}$

4 Conclusions

In this paper, the results of a multi-objective optimization problem when PMEDM cylindrically shaped parts made from SKD11 tool steel have been presented. This multi-objective problem consists of two single-objective functions, including the maximum MRS and the minimum EWR. To solve this problem, an experiment was conducted. The Taguchi method combined with the GRA method was used to design the experiment and analyze the experimental results. From the results of the experiment, the effect of six input process parameters including the powder size, the powder concentration, the pulse time, the pulse off time, the current, and the servo voltage on the multi-target function was evaluated. It was reported that the order of influence of the input parameters on the multi-objective function is T_{on} (30.23%), C_p (21.91%), IP (19.34%), T (15.22%), Sp (3.52%), and SV (2.87%). In addition, optimal input parameters to simultaneously optimize two single objective functions (the MRS and the EWR) have been proposed. The optimal parameters are $C_p = 3$ (g/l), Sp = 1000 (nm), $T_{on} = 30$ (μs), $T_{off} = 30$ (μs), IP = 4 (A), and SV = 5 (V).

Acknowledgment. This work was supported by Thai Nguyen University of Technology.

References

1. Jabbaripour, B., et al.: Investigating surface roughness, material removal rate and corrosion resistance in PMEDM of γ-TiAl intermetallic. J. Manuf. Process. **15**(1), 56–68 (2013)
2. Ojha, K., Garg, R., Singh, K.: Effect of chromium powder suspended dielectric on surface roughness in PMEDM process. Tribol. Mater. Surf. Interfaces **5**(4), 165–171 (2011)
3. Tran, T.-H., et al.: Electrical discharge machining with SiC powder-mixed dielectric: an effective application in the machining process of hardened 90CrSi steel. Machines **8**(3), 36 (2020)
4. Hong, T.T., et al.: Multi-response Optimization in PMSEDM process using Taguchi-Grey method. In: Sattler, K.-U., Nguyen, D.C., Vu, N.P., Long, B.T., Puta, H. (eds.) ICERA 2020. LNNS, vol. 178, pp. 594–606. Springer, Cham (2021). https://doi.org/10.1007/978-3-030-64719-3_65
5. Al-Khazraji, A., Amin, S.A., Ali, S.M.: The effect of SiC powder mixing electrical discharge machining on white layer thickness, heat flux and fatigue life of AISI D2 die steel. Eng. Sci. Technol. Int. J. **19**(3), 1400–1415 (2016)
6. Kansal, H., Singh, S., Kumar, P.: Technology and research developments in powder mixed electric discharge machining (PMEDM). J. Mater. Process. Technol. **184**(1–3), 32–41 (2007)
7. Kung, K.-Y., Horng, J.-T., Chiang, K.-T.: Material removal rate and electrode wear ratio study on the powder mixed electrical discharge machining of cobalt-bonded tungsten carbide. Int. J. Adv. Manuf. Technol. **40**(1–2), 95–104 (2009)
8. Kumar, N., Ahsan, R.: Study of PMEDM efficiency on EN-31 steel using tungsten powder in dielectric fluid. Int. J. Emerg. Technol. Eng. Res. **5**(5), 17–24 (2017)
9. Nguyen, M.C., et al.: Influence of input factors on material removal rate in PMEDM cylindrical shaped parts with silicon carbide powder suspended dielectric. In: Key Engineering Materials. Trans Tech Publ. (2020)
10. Hong, T.T., et al.: Optimization of PMEDM parameters for improving MMR in machining 90CrSi steel – a Taguchi approach. In: Sattler, K.-U., Nguyen, D.C., Vu, N.P., Long, B.T., Puta, H. (eds.) Advances in Engineering Research and Application: Proceedings of the International Conference on Engineering Research and Applications, ICERA 2020, pp. 648–657. Springer International Publishing, Cham (2021). https://doi.org/10.1007/978-3-030-64719-3_71
11. Hong, T.T., et al.: Effect of process parameters on machining time in PMEDM cylindrical shaped parts with silicon carbide powder suspended dielectric. In: Materials Science Forum. Trans Tech Publ. (2021)
12. Singh, B., Kumar, J., Kumar, S.: Investigation of the tool wear rate in tungsten powder-mixed electric discharge machining of AA6061/10% SiCp composite. Mater. Manuf. Process. **31**(4), 456–466 (2016)
13. Yaşar, H., Ekmekci, B.: Ti-6Al-4V surfaces in SiC powder mixed electrical discharge machining. In: Advanced Materials Research. Trans Tech Publ. (2014)
14. Patel, S., Thesiya, D., Rajurkar, A.: Effect of aluminium powder concentration on powder mixed electric discharge machining (PMEDM) of inconel-718. J. Mach. Form. Technol. **8** (1/2), 29 (2016)
15. Patel, S., Thesiya, D., Rajurkar, A.: Aluminium powder mixed rotary electric discharge machining (PMEDM) on Inconel 718. Aust. J. Mech. Eng. **16**(1), 21–30 (2018)
16. Long, B.T., Phan, N.H., Cuong, N., Jatti, V.S.: Optimization of PMEDM process parameter for maximizing material removal rate by Taguchi's method. Int. J. Adv. Manuf. Technol. **87** (5–8), 1929–1939 (2016). https://doi.org/10.1007/s00170-016-8586-4

17. Huu, P.-N.: Study of the effects of process parameters on tool wear rate in powder mixed electrical discharge machining by Taguchi method. Sci. Technol. Dev. J. **20**(K7), 55–60 (2017)
18. Rouniyar, A.K., Shandilya, P.: Multi-objective optimization using Taguchi and grey relational analysis on machining of Ti-6Al-4V alloy by powder mixed EDM process. Mater. Today Proc. **5**(11), 23779–23788 (2018)
19. Ibrahim, A.N., Amin, S.A., Ali, S.M.: Effect of SiC powder mixing (PMEDM) on surface residual stresses using copper and graphite electrodes (2015)
20. Bhattacharya, A., Batish, A., Kumar, N.: Surface characterization and material migration during surface modification of die steels with silicon, graphite and tungsten powder in EDM process. J. Mech. Sci. Technol. **27**(1), 133–140 (2013)

Multi-objective Optimization of PMEDM Process for Minimum Surface Roughness and Maximum Material Removal Speed When Processing SKD11 Steel

Huu Quang Nguyen[1], Manh Cuong Nguyen[2], Thu Quy Le[3], Ngoc Huy Thinh Tran[4], Hoang Anh Le[5], Ngoc Giang Tran[2], Ngoc Pi Vu[2], and Xuan Tu Hoang[2](✉)

[1] University of Economics – Technology for Industries, Ha Noi, Vietnam
[2] Thai Nguyen University of Technology, Thai Nguyen, Vietnam
hxt_dhktcn@tnut.edu.vn
[3] National Research Institute of Mechanical Engineering, Ha Noi, Vietnam
[4] Nguyen Tat Thanh University, Ho Chi Minh City, Vietnam
[5] Vinh Long University of Technology Education, Vinh Long, Vietnam

Abstract. This paper introduces a study on multi-objective optimization of powder-mixed discharge machining (PMEDM) for machining cylindrical parts made of SKD11 steel to achieve the minimum surface roughness (SR) simultaneously and maximum material removal speed (MRS). In this study, an experiment was conducted with the design and analysis of the results according to the Taguchi method. Six input parameters including powder concentration, powder size, pulse generation time, pulse off time, servo current and voltage were selected to study the influence of input factors on SR and MRS. In particular, the optimal values of the input PMEDM parameters to achieve the minimum SR and maximum MRS simultaneously have been proposed.

Keywords: PMEDM · Electrical discharge machining · Material removal speed · Cylindrical shaped parts · SKD11

1 Introduction

Non-traditional machining is a machining method where the cutting tool does not come into direct contact with the work piece during processing. This type of machining includes laser machining, electrical discharge machining (EDM), electrochemical machining, electrochemical grinding, etc. Among these methods, EDM is the most commonly used method for processing conductive parts. The reason is that this method has many advantages such as low cutting force and noise when machining; easily process parts with very complex surfaces; capable machine parts with high hardness etc… However, EDM also has some limitations such as low machined surface quality; the material removal rate (MRR) is not high; fast wear electrodes etc. In order to help the EDM process overcome these limitations, a solution has been proposed. The solution is that in the EDM process, metal powders are mixed into dielectric fluid when

D. C. Nguyen et al. (Eds.): ICERA 2021, LNNS 366, pp. 634–644, 2022.
https://doi.org/10.1007/978-3-030-92574-1_66

processing EDM. This process is called powder-mixed electrical discharge machining (PMEDM).

To date, there are quite a few studies on the PMEDM process. This process has been investigated with different electrode materials such as copper [1–4], graphite [5, 6], brass [7], etc. The workpiece material is also an important object of the research. Many studies on PMEDM has been conducted for different workpiece materials, such as Ti-6Al-4V [8], Inconel 718 [9, 10], Inconel 625 [11, 12], SKD11 [13, 14], 90CrSi [1, 2, 4, 15, 16], H11 die steel [3], etc… The PMEDM process was investigated with various types of metal powders including SiC [1, 2, 4, 11, 15], Al2O3 [10, 17], Ti [13, 14], etc… In particular, many studies have focused on optimizing the PMEDM process. There have been several optimization studies with single-objective [1, 4, 11, 14, 18], and multi-objective [2, 19–21] functions. The single objective functions are usually conducted as minimum SR [4, 9, 11, 18], maximum MRR [10, 13, 15], minimum EWR [10, 14], minimum machining time [1], etc… Multi-objective studies can consist of two [2, 16, 22] or three single-objective [19, 21] functions. The optimization problems on PMEDM process are solved by many methods such as the Taguchi method [10, 13, 18], the Taguchi method in combination with the gray relation analysis (GRA) [2, 16, 22, 23], Taguchi and Topsis [24, 25], Response surface methodology (RSM) [19], Topsis and GRA [3], etc. Recently, there have been several studies on the PMEDM process when processing cylindrical parts [1, 2, 4, 15, 16, 18]. However, these studies were only learned about the process of machining 90CrSi steel.

In this paper, a study on multi-objective optimization when PMEDM SKD11 steel is introduced. The multi-objective problem consists of two single objectives, the smallest SR and the largest MRS. The influence of the input factors on the RS and the MRS was explored. In addition, optimal input parameters to achieve the above two objectives simultaneously have been proposed.

2 Experimental Setup

To solve the problem of simultaneous optimization of two single objectives including the minimum SR and the maximum MRS objectives, an experiment was conducted. Figure 1 describes the setup of the experiment which consists of a Sodick A30 EDM, copper electrodes, SKD11 steel samples, and SiC powder for mixing into the dieletric liquid. Also, six input parameters were selected for this experiment to evaluate their effects on the SR and the MRS. Table 1 presents these parameters and their levels.

Table 1. Input factors and their levels

No	Parameters	Code	Unit	Level					
				1	2	3	4	5	6
1	Powder concentration	Cp	g/l	0	2	3	4	5	6
2	Powder size	Sp	nm	100	500	1000	–	–	–
3	Pulse on time	Ton	μs	10	20	30	–	–	–
4	Pulse off time	Toff	μs	10	20	30	–	–	–
5	Peak current	IP	A	4	8	12	–	–	–
6	Servo voltagez	SV	V	3	5	7	–	–	–

Fig. 1. Experimental setup

The experiment was built according to the design L_{18} ($3^{\wedge 4}$) with the use of the Taguchi method of Minitab R19 software. The experimental plan and the obtained test results (the SR and the MRS) are given in Table 2.

Table 2. Experimental plan and the response (SR and MRS)

No.	Input factors						SR (μm)		MRS (g/h)	
	Cp	S_p	T_{on}	T_{off}	IP	SV	Mean	S/N	Mean	S/N
1	0	100	10	10	4	3	1.94000	−5.9226	9.8773	19.8928
2	0	500	20	20	8	5	2.48467	−7.9305	16.0979	24.1354
3	0	1000	30	30	12	7	3.19400	−10.1206	49.6445	33.9174
4	2	100	10	20	8	7	2.00967	−6.0635	6.7254	16.5544
5	2	500	20	30	12	3	2.20367	−7.0309	15.8857	24.0201
6	2	1000	30	10	4	5	1.23900	−1.8672	41.9831	32.4615
7	3	100	20	10	12	5	2.30933	−7.2817	32.4869	30.2342
8	3	500	30	20	4	7	1.12567	−1.0345	12.4249	21.8858
9	3	1000	10	30	8	3	2.18867	−6.8051	11.8259	21.4567
10	4	100	30	30	8	5	2.38900	−7.6264	28.2703	29.0266
11	4	500	10	10	12	7	1.84367	−5.3302	8.1986	18.2748
12	4	1000	20	20	4	3	1.03100	−0.2925	5.2888	14.4672
13	5	100	20	30	4	7	1.85033	−5.4196	5.3288	14.5327
14	5	500	30	10	8	3	2.33533	−7.4181	15.1998	23.6368
15	5	1000	10	20	12	5	1.21067	−1.7023	7.3727	17.3525
16	6	100	30	20	12	3	2.83400	−9.1074	30.8676	29.7901
17	6	500	10	30	4	5	1.23567	−1.8846	5.6500	15.0409
18	6	1000	20	10	8	7	2.57200	−8.2363	5.2931	14.4741

To optimize multi-objectives, in this study, GRA combined with the Taguchi method is used. Also, the two objectives are the minimum SR and the maximum MRS.

The S/N indexes of the SR and the MRS are calculated according to the following formulas:

For the minimum SR:

$$S/N = -10\log_{10}(\frac{1}{n}\sum_{i=1}^{n} y_i^2) \quad (1)$$

For the maximum MRS:

$$S/N = -10\log_{10}(\frac{1}{n}\sum_{i=1}^{n} \frac{1}{y_i^2}) \quad (2)$$

The calculated S/N indexes of the two targets SR and MRS are given in Table 2.

To be able to analyze gray relationships, the S/N values need to be converted to comparison series with dimensionless quantities or need to normalize the data. In this case, the S/N index calculation results are normalized in the range (0 ÷ 1). The S/N index is normalized by the value of Zi with 0 $\leq$ Zi $\leq$ 1Zi is determined according to the following formula to avoid the effect of using different dimensions and to reduce the variability:

$$Z_i = \frac{S/N_i - \min(S/N_i, i = 1, 2, ..., n)}{\max(S/N_i, i = 1, 2, ..., n) - \min(S/N_i, i = 1, 2, ..., n)} \quad (3)$$

Where, n is the number of experiments; In this case, n = 18.

The gray relation coefficient yj(k) is determined by the formula:

$$y_i(k) = \frac{\Delta_{\min}(k) + \zeta.\Delta_{\max}(k)}{\Delta_i(k) + \zeta.\Delta_{\max}(k)} \quad (4)$$

In (4), i = 1, 2,…, n; n is the number of experiments (n = 18); k is the number of output targets (k = 2); Δi(k) = ||Z0(k) − Zi(k)||, is the absolute value of the difference between $Z_0(k)$ (reference value Z0(k) = 1) and $Z_i(k)$ (Z-value of the ith experiment of the kth target); Δmin(k) is the minimum value of i(k); max(k) is the maximum value of i(k); ζ is the discriminant coefficient, determined in the range 0 $\leq \zeta \leq$, in experimental research ζ = 0.5.

The degree of gray relation can be calculated through the average grey relational (GR) value of the output objectives:

$$\overline{y_i} = \frac{1}{k}\sum_{j=0}^{k} y_{ij}(k) \quad (5)$$

Where, k is the output target number; y_{ij} is the GR value of the ith output targets in the ith experiment.

The calculated results of GR value y_i and the mean GR value of the experiments are listed in Table 3.

Table 3. S/N, S/N Zi, Δ_i (k) and GR value $\bar{y}_i$

No	S/N		Zi		Δ_i (k)		GR value yi		$\bar{y}_i$
	Ra	MRS	$\overline{Ra}$ (Reference value 1.000)	MRS (Reference value 1.000)	$\overline{Ra}$	MRS	$\overline{Ra}$	MRS	
1	−5.9226	19.8928	0.4271	0.2789	0.573	0.721	0.466	0.409	0.438
2	−7.9305	24.1354	0.2228	0.4971	0.777	0.503	0.391	0.499	0.445
3	−10.1206	33.9174	0.0000	1.0000	1.000	0.000	0.333	1.000	0.667
4	−6.0635	16.5544	0.4128	0.1073	0.587	0.893	0.460	0.359	0.409
5	−7.0309	24.0201	0.3144	0.4911	0.686	0.509	0.422	0.496	0.459
6	−1.8672	32.4615	0.8398	0.9251	0.160	0.075	0.757	0.870	0.814
7	−7.2817	30.2342	0.2889	0.8106	0.711	0.189	0.413	0.725	0.569
8	−1.0345	21.8858	0.9245	0.3814	0.075	0.619	0.869	0.447	0.658
9	−6.8051	21.4567	0.3373	0.3594	0.663	0.641	0.430	0.438	0.434
10	−7.6264	29.0266	0.2538	0.7485	0.746	0.251	0.401	0.665	0.533
11	−5.3302	18.2748	0.4874	0.1958	0.513	0.804	0.494	0.383	0.439
12	−0.2925	14.4672	1.0000	0.0000	0.000	1.000	1.000	0.333	0.667
13	−5.4196	14.5327	0.4783	0.0034	0.522	0.997	0.489	0.334	0.412
14	−7.4181	23.6368	0.2750	0.4714	0.725	0.529	0.408	0.486	0.447
15	−1.7023	17.3525	0.8566	0.1483	0.143	0.852	0.777	0.370	0.573
16	−9.1074	29.7901	0.1031	0.7878	0.897	0.212	0.358	0.702	0.530
17	−1.8846	15.0409	0.8380	0.0295	0.162	0.971	0.755	0.340	0.548
18	−8.2363	14.4741	0.1917	0.0004	0.808	1.000	0.382	0.333	0.358

3 Results and Analysis

Theoretically, to ensure the harmony between output parameters, the higher the average gray relation value, the better. Thus, the objective function of the multi-objective problem becomes a single-objective problem with the output being the mean gray relation value. The Taguchi method was used to evaluate the influence of the PMEDM parameters on the mean GR value.

With the goal that the mean GR value is as large as possible, the S/N value of is calculated according to formula (2). Calculation results of S/N values analyzed by ANOVA method are shown in Table 4.

Table 4. Analysis of variance for S/N ratios of $\overline{y_i}$

Analysis of Variance for SN ratios

Source	DF	Seq SS	Adj SS	Adj MS	F	P	C (%)
Cp	5	5.027	5.027	1.0053	0.81	0.632	8.33
Sp	2	7.044	7.044	3.5218	2.85	0.260	11.68
Ton	2	16.585	16.585	8.2925	6.71	0.130	27.49
Toff	2	1.996	1.996	0.9980	0.81	0.553	3.31
Ip	2	18.558	18.558	9.2790	7.50	0.118	30.77
SV	2	8.638	8.638	4.3190	3.49	0.223	14.32
Residual Error	2	2.473	2.473	1.2367			4.10
Total	17	60.321					

Model Summary

S	R-Sq	R-Sq(adj)
1.1121	95.90%	65.15%

From the result in Table 4, it was noted that IP has the greatest influence on $\overline{y_i}$ $\overline{\overline{y}}$ (30.77%); followed by the influence of Ton (27.49%) and SV (14.32%), Sp (11.68%), Cp (8.33%) and Toff having the least influence (3.31%). This result is also shown in Fig. 2.

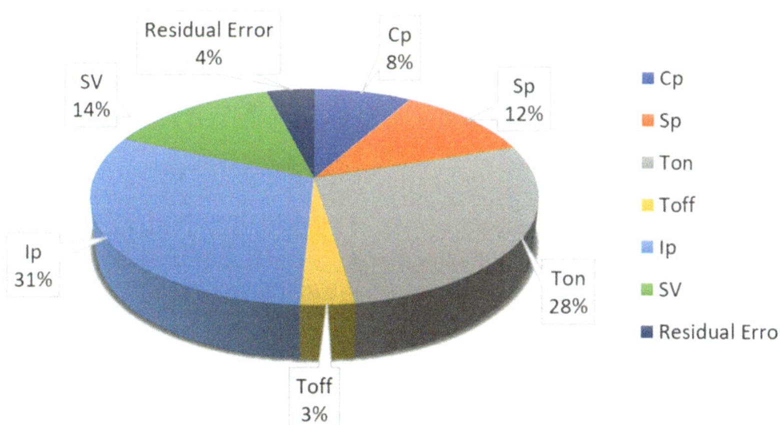

Fig. 2. Influence of input factors on $\overline{y_i}$ (%)

The effect order of the input factors on the GR value obtained by ANOVA analysis is shown in Table 5.

As analyzed above, the parameter set with the highest value of y for each input parameter is the set of optimal parameters. From Fig. 3, it can be found the influence trend of each input parameter on y and it is easy to determine the optimal set of input parameters for the multi-objective function (as shown in Table 6).

From the obtained optimal parameters, using the Minitab 19 software, it is possible to calculate the results of the objective functions (Table 7). From the predictive results, it is known that when processing with the optimal parameters, the output results are the SR of approximately 1.083 μm, and the MRS of approximately 29.21 g/h.

Table 5. Effect order of input factors on S/N

Larger is better

Level	Cp	Sp	Ton	Toff	Ip	SV
1	-5.910	-6.418	-6.566	-6.157	-4.844	-6.203
2	-5.439	-6.137	-6.476	-5.382	-7.236	-4.874
3	-5.260	-4.973	-4.486	-5.989	-5.449	-6.451
4	-5.381					
5	-6.509					
6	-6.557					
Delta	1.297	1.445	2.080	0.775	2.392	1.578
Rank	5	4	2	6	1	3
Average gray relation value: 0.522						

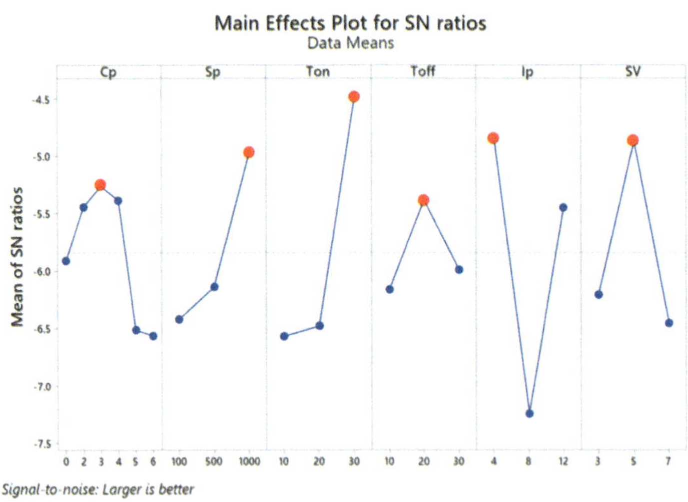

Fig. 3. Effect of input factors on the S/N ratios of $\overline{y_i}$

Table 6. Optimum input factors

Parameter	Code	Unit	Level	Value
Powder concentration	C_p	g/l	3	3
Powder size	S_p	nm	3	1000
Pulse on time	T_{on}	μs	3	30
Pulse off time	T_{off}	μs	2	20
Peak current	IP	A	1	4
Servo voltage	SV	V	2	5

Figure 4 is used to evaluate the experimental model. Figure 4a shows the error of experimental points (the blue points) on the normal distribution error distribution graph close to the normal distribution line (the red solid line). This result shows that the model has a small error. Figure 4c (Histogram) shows the error frequency. The graph shows that the errors appearing in the vicinity of 0 account for a large proportion. Figure 4b and d show randomly distributed experimental errors. In other words, the experimental model is largely influenced by the selected input parameters and it is not affected by the order of experiments.

Table 7. Predicted optimum R_a and MRS

Settings

Variable	Setting
Cp	3
Sp	1000
Ton	30
Toff	20
Ip	4
SV	5

Prediction

Ra	Fit	SE Fit	CI DF	95% CI	PI DF	95% PI
Conditional	1.083	0.181	2.900	(0.495, 1.672)	2.33	(0.044, 2.123)
Marginal	2	0.382	4.780	(1.003, 2.997)	6.07	(0.013, 3.986)
MRS	**Fit**	**SE Fit**	**CI DF**	**95% CI**	**PI DF**	**95% PI**
Conditional	29.21	5.290	6.930	(16.67, 41.75)	4.52	(2.91, 55.51)
Marginal	17.13	7.490	2.650	(-8.60, 42.87)	4.98	(-26.56, 60.83)

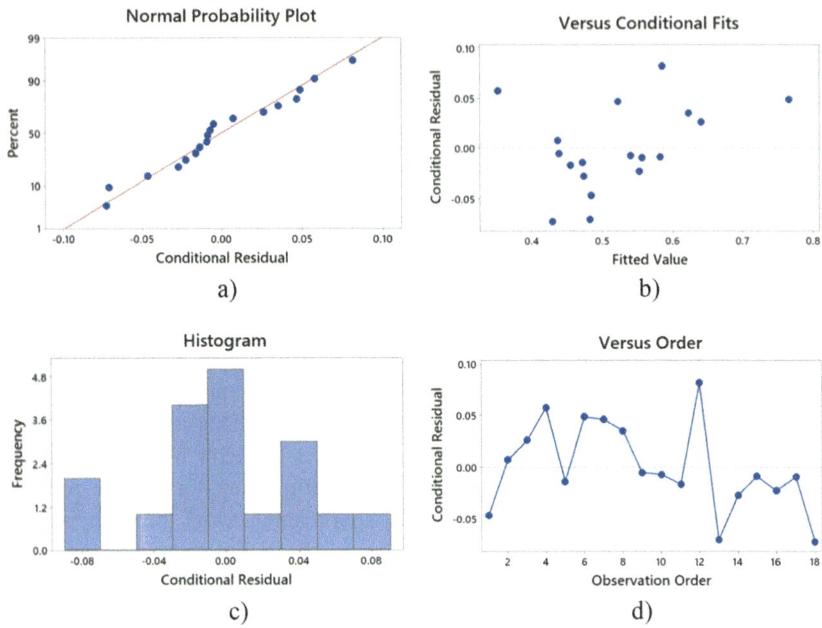

Fig. 4. Conditional residual plots for $\overline{y_i}$

The fit of the experimental model was evaluated by the Anderson-Darling method (Fig. 5). The results show that the experimental results (blue dots) are within the

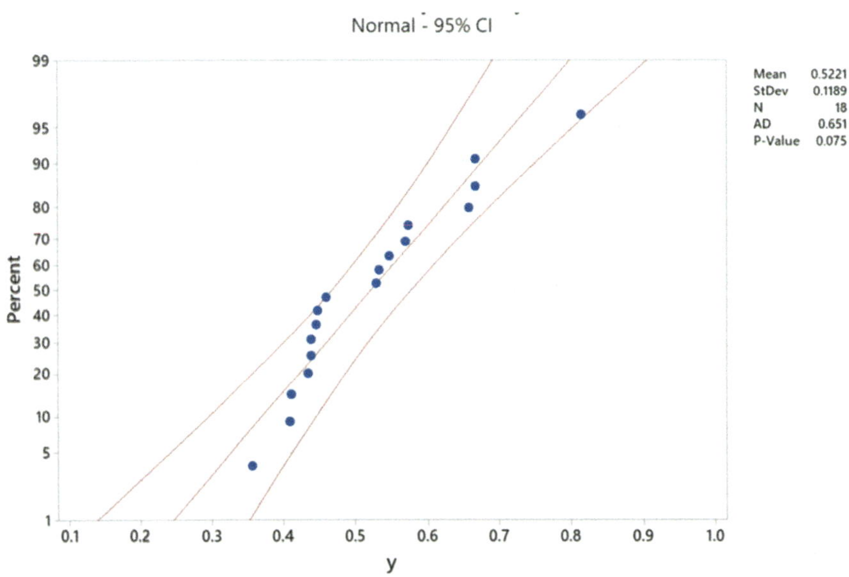

Fig. 5. Probability plot for $\overline{y_i}$

bounded area by two upper and lower limits with a standard deviation of 95%. In addition, the P-value of 0.075 is greater than the value of $\alpha = 0.05$, showing that the applied experimental model is suitable.

4 Conclusions

This paper reports the results of a multi-objective optimization study of the PMEDM process when machining cylindrical parts made from SKD11 tool steel. Two single-objective functions, the smallest SR and the largest MRS, were chosen for the multi-objective optimization problem. To solve this problem, an experiment was designed and performed using the Taguchi method to design the experiment and analyze the results. The influence of six input PMEDM factors including the powder size, the powder concentration, the pulse on time, the pulse off time, the current, and the servo voltage on the multi-objective function was evaluated. The order of influence of input parameters on y is IP (30.77%), T_{on} (27.49%), SV (14.32%), S_p (11.68%), C_p (8.33%), and T_{off} (3.31%). In addition, optimal input parameters have been proposed to obtain the minimum SR and maximum MRS simultaneously. These parameters are $C_p = 2$ (g/l), $S_p = 1000$ (nm), $T_{on} = 30$ (μs), $T_{off} = 20$ (μs), IP = 4 (A), and SV = 5 (V).

Acknowledgment. This work was supported by Thai Nguyen University of Technology.

References

1. Hong, T.T., et al.: Effect of process parameters on machining time in PMEDM cylindrical shaped parts with silicon carbide powder suspended dielectric. In: Materials Science Forum. Trans Tech Publ.
2. Hong, T.T., et al.: Multi-objective optimization of PMEDM process of 90CrSi alloy steel for minimum electrode wear rate and maximum material removal rate with silicon carbide powder. In: Materials Science Forum. Trans Tech Publ. (2021)
3. Tripathy, S., Tripathy, D.: Multi-attribute optimization of machining process parameters in powder mixed electro-discharge machining using TOPSIS and grey relational analysis. Eng. Sci. Technol. Int. J. **19**(1), 62–70 (2016)
4. Tran, T.-H., et al.: Electrical discharge machining with SiC powder-mixed dielectric: an effective application in the machining process of hardened 90CrSi steel. Machines **8**(3), 36 (2020)
5. Long, B.T., et al.: Machining properties evaluation of copper and graphite electrodes in PMEDM of SKD61 steel in rough machining. Int. J. Eng. Adv. Technol. (IJEAT), pp. 2249–8958 (2015)
6. Al-Khazraji, A., Amin, S.A., Ali, S.M.: The effect of SiC powder mixing electrical discharge machining on white layer thickness, heat flux and fatigue life of AISI D2 die steel. Eng. Sci. Technol. Int. J. **19**(3), 1400–1415 (2016)
7. Bhattacharya, A., Batish, A., Kumar, N.: Surface characterization and material migration during surface modification of die steels with silicon, graphite and tungsten powder in EDM process. J. Mech. Sci. Technol. **27**(1), 133–140 (2013)
8. Yunpeng, Z., Guangbiao, S., Anzhou, Z.: Effect of mixed powder ultrasonic vibration on surface structure and mechanical properties of Ti-6Al-4V in electro-discharge machining. Rare Met. Mater. Eng. **43**(1), 189–193 (2014)

9. Ahmad, S., et al.: Surface roughness and surface topography of Inconel 718 in powder mixed dielectric electrical discharge machining (PMEDM). Int. J. Integr. Eng. **10**(5) (2018)
10. Patel, S., Thesiya, D., Rajurkar, A.: Aluminium powder mixed rotary electric discharge machining (PMEDM) on Inconel 718. Aust. J. Mech. Eng. **16**(1), 21–30 (2018)
11. Talla, G., Gangopadhyay, S., Biswas, C.: Effect of powder-suspended dielectric on the EDM characteristics of Inconel 625. J. Mater. Eng. Perform. **25**(2), 704–717 (2016)
12. Talla, G., Gangopadhyay, S., Biswas, C.: Influence of graphite powder mixed EDM on the surface integrity characteristics of Inconel 625. Part. Sci. Technol. **35**(2), 219–226 (2017)
13. Long, B.T., Phan, N.H., Cuong, N., Jatti, V.S.: Optimization of PMEDM process parameter for maximizing material removal rate by Taguchi's method. Int. J. Adv. Manuf. Technol. **87** (5–8), 1929–1939 (2016). https://doi.org/10.1007/s00170-016-8586-4
14. Banh, L.T., Nguyen, P.H., Ngo, C.: Tool wear rate optimization in PMEDM using titanium powder by Taguchi method for die steels. Sci. Technol. Dev. J. **19**(2), 88–97 (2016)
15. Nguyen, M.C., et al.: Influence of input factors on material removal rate in PMEDM cylindrical shaped parts with silicon carbide powder suspended dielectric. In: Key Engineering Materials. Trans Tech Publ. (2020)
16. Hong, T.T., et al.: Multi-response optimization in PMSEDM process using Taguchi-Grey method. In: Sattler, K.-U., Nguyen, D.C., Vu, N.P., Long, B.T., Puta, H. (eds.) ICERA 2020. LNNS, vol. 178, pp. 594–606. Springer, Cham (2021). https://doi.org/10.1007/978-3-030-64719-3_65
17. Taherkhania, A., et al.: Investigation of surface quality in cost of goods manufactured (COGM) method of μ-Al2O3 powder-mixed-EDM process on machining of Ti-6Al-4V (2021)
18. Hong, T.T., et al.: A study on influence of input parameters on surface roughness in PMEDM cylindrical shaped parts. In: Materials Science Forum. Trans Tech Publ. (2021)
19. Kazi, F., Waghmare, C., Sohani, M.: Optimization and comparative analysis of silicon and chromium powder-mixed EDM process by TOPSIS technique. Eng. Appl. Sci. Res. **48**(2), 190–199 (2021)
20. Dharmendra, B., Kodali, S.P., Rao, B.N.: A simple and reliable Taguchi approach for multi-objective optimization to identify optimal process parameters in nano-powder-mixed electrical discharge machining of INCONEL800 with copper electrode. Heliyon **5**(8), e02326 (2019)
21. Tripathy, S., Tripathy, D.: Surface Characterization and Multi-response optimization of EDM process parameters using powder mixed dielectric. Mater. Today Proc. **4**(2), 2058–2067 (2017)
22. Rouniyar, A.K., Shandilya, P.: Multi-objective optimization using Taguchi and grey relational analysis on machining of Ti-6Al-4V alloy by powder mixed EDM process. Mater. Today Proc. **5**(11), 23779–23788 (2018)
23. Rajavel, R., et al.: Investigation on machinability of Al 2024& 7.5% Si3N4 metal matrix composite with PMEDM using Taguchi based GRA. Mater. Today Proc. **46**, 9449–9453 (2020)
24. Kumar, D., Payal, S., Beri, N.: Parameters optimized specifying for EDM with PM/ conventional electrodes on M2 tool steel using Taguchi and TOPSIS. Ind. J. Sci. Technol. **10** (22), 1–8 (2017)
25. Nguyen, H.-P., Pham, V.-D., Ngo, N.-V.: Application of TOPSIS to Taguchi method for multi-characteristic optimization of electrical discharge machining with titanium powder mixed into dielectric fluid. Int. J. Adv. Manuf. Technol. **98**(5–8), 1179–1198 (2018). https://doi.org/10.1007/s00170-018-2321-2

Multiple Criteria Decision Making When Turning by Taguchi–Vikor Method

Nguyen Lam Khanh and Nguyen Van Cuong(✉)

University of Transport and Communications, Hanoi, Vietnam
nguyencuong@utc.edu.vn

Abstract. In this study, a multiple criteria decision making study is presented when turning S55C steel. The experimental process was conducted in the order of the Taguchi matrix with a total of 16 experiments. At each experiment, the coolant type, cutting velocity, feed rate and depth of cut were changed. The cutting tool used was a TiCN coated insert. Surface roughness, Material Removal Rate (*MRR*) and Tool Wear (*TWR*) were selected as the criteria for evaluating the turning process. The Vikor method was applied to determine the type of coolant and the value of cutting parameters to simultaneously ensure three criteria: minimum surface roughness, maximum MRR and minimum TWR. The orientation for further research has also been recommended in this article.

Keywords: Turning · Surface roughness · MRR · TWR · Taguchi · Vikor

1 Introduction

With machining methods in general and machining by turning in particular, the surface roughness parameters, MRR and TWR are always concerned. Surface roughness affects the corrosion resistance, wear resistance, fatigue strength and accuracy of the joint. MRR is a parameter to evaluate the performance of the machining process, and TWR not only directly affects the tool life but also affects the machining accuracy, especially the accuracy of the workpiece's diameter [1–3]. Therefore, many studies have been conducted to determine the value of the parameters of the machining process to ensure the minimum surface roughness, maximum MRR and minimum TWR. Maheswara Rao et al. [4] conducted an experiment of AA7075 aluminum alloy turning with a tungsten carbide insert. In this study, they designed a matrix according to the Taguchi method, the selected input parameters include spindle speed, feed rate and depth of cut. The analysis of the Sign-to-Noise (S/N) ratio has determined that in order to obtain minimum surface roughness, the spindle speed, feed rate and depth of cut should be 1000 rev/min, 0.2 mm/rev and 0.5 mm, respectively. Gajanana et al. [5] conducted an experiment of OF15-5PH stainless steel turning with a coated carbide tool. They also designed an experimental matrix according to the Taguchi method with input parameters including spindle speed, feed rate and depth of cut. They also applied the method of analyzing the S/N ratio to determine the value of input parameters to ensure the minimum surface roughness. As a result, they found the optimal values of spindle speed, feed rate and depth of cut to be 750 rev/min, 0.15 mm/rev and 0.9 mm, respectively. Kajal et al. [6] used CNMG 120408 GT cutting tool to turn EN354 steel.

D. C. Nguyen et al. (Eds.): ICERA 2021, LNNS 366, pp. 645–655, 2022.
https://doi.org/10.1007/978-3-030-92574-1_67

In this study, they also built an experimental matrix according to the Taguchi method with input parameters including cutting velocity, feed rate and depth of cut. By analyzing the S/N ratio they determined that in order for obtaining minimum surface roughness, the cutting velocity was 225 m/min, the feed rate was 0.05 mm/rev and the depth of cut was 1.2 mm. Kumar et al. [7], when conducting an experiment of EN19 steel turning, also designed an experimental matrix according to the Taguchi method with input parameters including spindle speed, feed rate and depth of cut. The cutting tool material used was TNMG carbide. By conducting an analysis of the S/N ratio they also determined that in order for obtaining minimum surface roughness, the spindle speed was 1500 rev/min, the feed rate was 0.1 mm/rev and the depth of cut was 0.4 mm; MRR obtained the maximum value when the spindle speed was 1500 rev/min, the feed rate was 0.15 mm/rev and the depth of cut was 0.4 mm. Vijaykumar et al. [8] designed a matrix according to the Taguchi method to conduct an experiment of AISI 52100 steel turning with a cutting tool as CCMT-32.52MT-TT8020. The spindle speed, feed rate, and depth of cut were also selected as input parameters for the experimental process. As a result of the analysis of the S/N ratio, they determined that in order for obtaining the smallest surface roughness, the spindle speed was 450 rev/min, the feed rate was 0.05 mm/rev and the depth of cut was 0.4 mm; The optimal values of spindle speed, feed rate and depth of cut to ensure the maximum MRR was 1120 rev/min, 0.12 mm/rev and 0.4 mm. Akkus [9] used a diamond insert of DNMG150608 type to turn AISI 1040 steel. He also built an experimental matrix according to the Taguchi method with input parameters including cutting velocity, feed rate, and depth of cut. As a result of analyzing the S/N ratio, he found the value of input parameters to ensure the minimum surface roughness: the cutting speed was 250 m/min, the feed rate was 0.15 mm/rev and the depth of cut was 3.5 mm.

Some of the above studies have shown that the Taguchi method has been successfully applied in many studies to determine the optimal value of the machining process parameters in some cases. However, if only the Taguchi method is used to design the experiments and then the S/N ratio analysis method is used to analyze the experimental results, only a certain criterion of the turning process can be achieved (e.g. minimum surface roughness or maximum MRR, etc.). In order to overcome this limitation, the Taguchi method has been used in combination with a number of other methods to simultaneously solve multiple objectives of the machining process. Pawar et al. [10] designed an experimental matrix according to the Taguchi method when turning high-speed steel (M2). Cutting velocity, feed rate, depth of cut and nose radius were selected as input parameters for each experiment. The cutting tool used was TiN coated carbide inserts. The ANOVA analysis method has been applied and it has been determined that in order to simultaneously ensure the minimum surface roughness and the maximum MRR, the cutting velocity was 350 m/min, the feed rate was 0.15 mm/rev, and the depth of cut was 0.75 mm and the nose radius was 1.2 mm. Satynarayana et al. [11], when turning EN18 steel carried out in two cases: dry turning and Minimum Quantity Lubrication (MQL), the cutting tool was coated tool insert. They also designed an experimental matrix according to the Taguchi method with three input parameters: cutting velocity, feed rate and depth of cut. The ANOVA method was also applied to determine the value of cutting parameters to simultaneously ensure the minimum surface roughness and the maximum MRR. They have determined that in both dry turning and MQL cases the optimal values of cutting velocity, feed rate and depth of cut were 100 m/min, 0.05 mm/rev and 0.4 mm, respectively. Ambigai et al. [12] designed a matrix according to the Taguchi method when turning Ti6Al4V titanium alloy using a TiAlN coated insert. At each experiment, three parameters were also changed, namely cutting velocity, feed rate and depth of cut. They also applied the Gray Relation Analysis (GRA) method and determined that in order to simultaneously ensure the minimum surface roughness and the maximum MRR, the cutting velocity was 35 m/min, the feed rate was 0.15 mm/rev and the depth of cut was 1.5 mm. Surya et al. [13] designed a matrix according to the Taguchi method to turn 7075 Aluminum using a cutting tool as High-Speed Steel (HSS). The ANOVA and GRA analysis methods were also applied to determine the optimal value of spindle speed, feed rate and depth of cut. The result determined that in order to simultaneously ensure the minimum surface roughness, maximum MRR and minimum TWR, the values of spindle speed, feed rate and depth of cut were 325 rev/min, 0.05 mm/rev and 0.4 mm, respectively. Dureja et al. [14] used the TiSiN-TiAlN coated cutting tool to turn AISI D3 steel. They also designed an experimental matrix according to the Taguchi method with input parameters including cutting velocity, feed rate and depth of cut. The Response Surface Method (RSM) was applied to determine the value of cutting parameters to ensure simultaneously the minimum surface roughness and TWR. As a result, the required values of parameters were the cutting velocity of 130 m/min, the feed rate of 0.13 mm/rev and the depth of cut of 0.21 mm. Gupta et al. [15] used the TiAlN coated cutting tool to turn unidirectional glass fiber reinforced plastics (UD-GFRP). The Taguchi method was applied to design an experimental matrix with the input parameters of tool nose radius, tool rake angle, feed rate, cutting velocity, cutting environment and depth of cut. The Principal Component Analysis (PCA) method was used to solve the multi-objective optimization problem. The results showed that in order to ensure the minimum surface roughness and maximum MRR at the same time, the tool nose radius was 0.8 mm, the tool rake angle was 00, the feed rate was 0.2 mm/rev, and the cutting velocity was 159.66 m/min, the cutting environment was cooled, and the depth of cut was 1.4 mm.

Vikor is known as a multiple criteria decision-making method that has been applied in many fields such as: for soil management assessment [16], in EDM machining [17, 18], in milling [19]... However, up to now, according to the authors of this article, the studies applying the Vikor method in multiple criteria decision-making for the turning method are still quite limited. From these analyses, this study will conduct a multiple criteria decision-making study on the turning process by a combination of Taguchi method and the Vikor method. The machining parameters considered in this study include coolant type, cutting velovity, feed rate and depth of cut. Three parameters selected as criteria for evaluating the turning process include surface roughness, MRR and TWR.

2 Vikor Method

The Vikor method is a prioritization method. The content of this method is as follows [20]:

It is assumed that there are i options A = { Ai| i = 1,2,…,m} and j criteria C = { Cj| j = 1,2,…,n}; the evaluation value of option i compared to criterion j is F = { fij| i = 1,2,…,m; j = 1,2,…,n}; weight of criteria W = { wj| i = 1,2,…,m}. Then the information about Vikor decision matrix consists of 4 components I(A,C,F,W) as shown in Table 1. Calculation steps for classification are as follows:

Table 1. Vikor decision matrix

	C_1	C_2	...	C_j	...	C_n
A_1	f_{11}	f_{12}	…	f_{1j}	…	f_{1n}
A_2	f_{21}	f_{22}	…	f_{2j}	…	f_{2n}
…	…	…	…	…	…	…
A_n	f_{m1}	f_{m2}	…	f_{mj}	…	f_{mn}
Max (f_j)	f^*_1	f^*_2	…	f^*_j	…	f^*_n
Min (f_j)	f^-_1	f^-_2	…	f^-_j	…	f^-_n
W	w_1	w_2	…	w_j	…	W_n

Step 1: determine the best value f^*_j and the worst value f^-_j of all criteria C_j *(with* $j = 1,2,...n$*).*

- If criterion j is a positive criterion then: f * j = max fij and f − j = min fij.
  - If criterion j is a negative criterion (bad criterion), then: f * j = min fij and f − j = max fij.

Step 2: normalize the matrix and calculate Si, Ri
Set:

$$\text{rij} = (|\text{f} * \text{j} - \text{fij}|)/(|\text{f} * \text{j} - \text{f} - \text{j}) \tag{1}$$

The Si (utility) and Ri (separate regret) values are calculated as follows:

$$S_i = \sum_{j=1}^{n} w_j\left(\left|f_j^* - f_{ij}\right|\right)/\left(\left|f_j^* - f_j^-\right|\right) = \sum_{j=1}^{n} w_j r_{ij} \tag{2}$$

$$R_i = max\left[w_j\left(\left|f_j^* - f_{ij}\right|\right)/\left(\left|f_j^* - f_j^-\right|\right)\right] = max\left[w_j r_{ij}\right] \tag{3}$$

Step 3: Calculate Q_i

$$Q_i = v(S_i - S^*)/(S^- - S^*) + (1 - v)(R_i - R^*)/(R^- - R^*) \tag{4}$$

With:

$$0 \leq v \leq 1$$

Of which: v is the weight of the positive group. With three criteria to evaluate and assuming that the weight of each criterion is the same, so $v = 1/3$ [20].

$1 - v$ is the weight of the negative group.

$$S^* = \min S_i \tag{5}$$

$$S^- = \max S_i \tag{6}$$

$$R^* = \min R_i \tag{7}$$

$$R^- = \max R_i \tag{8}$$

Step 4: Rank options A_i on the basis of the values S_i, R_i, Q_i in order from small to large. The option A_i with the largest Q_i is the best one.

3 Experimental Process

S55C steel was used during the experiment. Such steel contains medium carbon content. Such steel is often used to make parts such as plastic molds, brick molds, parts in the field of boilers, construction mechanics. The results of the spectroscopic analysis show the chemical composition of elements of steel as shown in Table 2. The workpiece has a diameter of 45 mm, a length of 130 mm. During the turning, the workpiece is clamped in a three-jaw chuck, the turning process is performed only on the 80 mm length of the workpiece. Each TiCN-coated insert was used for one experiment, with a nose radius of 0.8 mm. Such insert is well suited for turning steels with medium carbon content, such as S55C material [21].

Table 2. Chemical composition of S55C steel

Element	C	Si	Mn	P	S	Cu	Ni	Cr
%	0.54	0.26	0.75	0.026	0.032	0.026	0.018	0.012

Dainichi F250 CNC lathe was used to perform the experiments (Fig. 1). Three types of cooling lubricants were used, including Machine Coolants 128 FL, Tectylcol 1240 and Caltex Aquatex 3180. These are all commonly used oils in machining in general and in CNC turning in particular. These oils have all been mixed with water according to the concentration recommended by the manufacturer. Specifically, Machine Coolants 128 FL is mixed with a concentration of 4%, Tectylcol 1240 is mixed with a concentration of 6% and Caltex Aquatex 3180 is mixed with a concentration of 3.5%.

The Taguchi method was used to design an orthogonal matrix with 16 experiments (L16). Four variables changed in each experiment including coolant type (C), cutting velocity (*V*), feed rate (*F*) and depth of cut (*D*). Each variable has been selected four values corresponding to four coding levels 1, 2, 3 and 4. The experimental matrix is presented in Table 3.

Fig. 1. Experimental machine

Surface roughness was measured with the SJ201 machine of Mitutoyo – Japan (Fig. 2), with the standard length of the measurement of 0.8 mm. The VHX-6000 digital microscope was used to measure tool wear (Fig. 3). This instrument performs the measurement in compliance with ISO 4287, JIS-G0551, ISO-643, ASTM-E1382, ISO-16232, ISO25178 standards, the instrument's image magnification is up to 5000 times [22].

Table 3. Orthogonal experimental matrix L16

Trial	Code variable				Real variable			
	C	V	F	D	C (–)	V (m/min)	F (mm/rev)	D (mm)
1	1	1	1	1	Machine Coolants 128 FL	90	0.09	0.2
2	1	2	2	2	Machine Coolants 128 FL	120	0.12	0.3
3	1	3	3	3	Machine Coolants 128 FL	150	0.15	0.4
4	1	4	4	4	Machine Coolants 128 FL	180	0.18	0.5
5	2	1	2	3	Tectylcol 1240	90	0.12	0.4
6	2	2	1	4	Tectylcol 1240	120	0.09	0.5
7	2	3	4	1	Tectylcol 1240	150	0.18	0.2
8	2	4	3	2	Tectylcol 1240	180	0.15	0.3
9	3	1	3	4	Tectylcol 1290	90	0.15	0.5
10	3	2	4	3	Tectylcol 1290	120	0.18	0.4
11	3	3	1	2	Tectylcol 1290	150	0.09	0.3
12	3	4	2	1	Tectylcol 1290	180	0.12	0.2
13	4	1	4	2	Caltex Aquatex 3180	90	0.18	0.3
14	4	2	3	1	Caltex Aquatex 3180	120	0.15	0.2
15	4	3	2	4	Caltex Aquatex 3180	150	0.12	0.5
16	4	4	1	3	Caltex Aquatex 3180	180	0.09	0.4

MRR is calculated according to the formula $MRR = (N.\pi.\ D_w.\ F.D)/60$. Of which: N is the number of revolutions of the workpiece per minute, D_w is the diameter of the workpiece, F is the feed rate and D is the depth of cut.

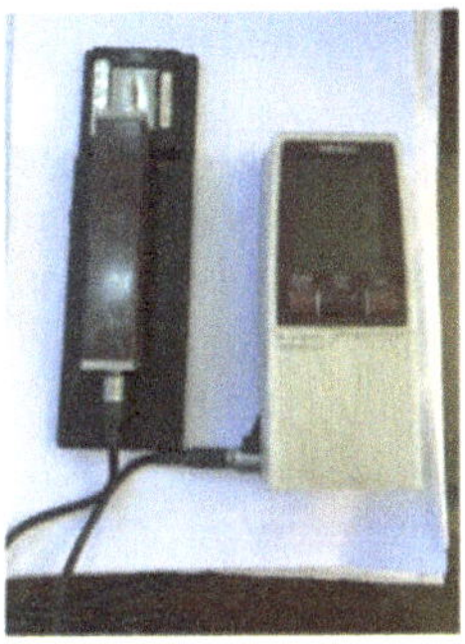

Fig. 2. Surface roughness tester SJ201

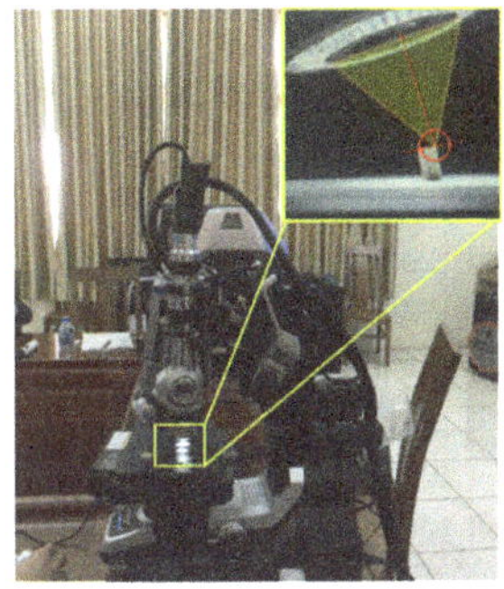

Fig. 3. Digital Microscope: Controller VHX-6000

4 Experimental Results and Discussion

The experimental results are presented in Table 4. Observation of the data in this table shows that the surface roughness has the smallest value in experiment 14 (Ra = *0.238 μm*). However, in this experiment, the tool wear is also the largest (TWR = *4.80 μm*). *MRR* has the largest value in experiment 4 ($MRR = 270\ mm^3/min$). However, in this experiment, the tool wear was also the largest (TWR = *4.80 μm*). *TWR* has the smallest value in experiment 15 (TWR = *1.63 μm*). But the surface roughness in this experiment was larger than the same in experiment 14, and the *MRR* in this experiment was smaller than the same in experiment 4.

Thus, it can be stated that it is not possible to choose an experiment that gives absolute assurance that such experiment has the smallest surface roughness, maximum *MRR* and minimum *TWR*. We can only choose an experiment where the surface roughness is considered the "smallest", the MRR is considered the "largest" and the *TWR* is considered the "smallest". And of course, this work cannot be done by just observing the data in Table 4, but it can only be done when using a certain mathematical method to make a multiple criteria decision. This work will be solved when applying the Vikor method, which will be presented in the next part of this article.

Table 4. Experimental results

Trial	C (–)	V (m/min)	F (mm/rev)	D (mm)	Ra (μm)	MRR (mm^3/min)	TWR (μm)
1	Machine Coolants 128 FL	90	0.09	0.2	0.828	27.00	2.98
2	Machine Coolants 128 FL	120	0.12	0.3	0.394	72.00	2.29
3	Machine Coolants 128 FL	150	0.15	0.4	0.283	150.00	1.64
4	Machine Coolants 128 FL	180	0.18	0.5	0.783	270.00	3.48
5	Tectylcol 1240	90	0.12	0.4	0.428	72.00	2.34
6	Tectylcol 1240	120	0.09	0.5	0.404	90.00	3.22
7	Tectylcol 1240	150	0.18	0.2	0.555	90.00	1.38
8	Tectylcol 1240	180	0.15	0.3	0.514	135.00	3.49
9	Tectylcol 1290	90	0.15	0.5	0.564	112.50	2.12
10	Tectylcol 1290	120	0.18	0.4	0.602	144.00	2.99
11	Tectylcol 1290	150	0.09	0.3	0.365	67.50	3.25
12	Tectylcol 1290	180	0.12	0.2	0.515	72.00	1.59
13	Caltex Aquatex 3180	90	0.18	0.3	0.561	81.00	4.80
14	Caltex Aquatex 3180	120	0.15	0.2	0.238	60.00	4.80
15	Caltex Aquatex 3180	150	0.12	0.5	0.239	150.00	1.63
16	Caltex Aquatex 3180	180	0.09	0.4	0.827	108.00	3.27

5 Multiple Criteria Decision Making

To facilitate multiple criteria decision making, we set the surface roughness as criterion C_1, *MRR* as criterion C_2, and *TWR* as criterion C_3. Then we have a decision matrix as shown in Table 5.

Table 5. Vikor decision matrix for the turning process

Trial	C_1	C_2	C_3
1	0.828	27.00	2.98
2	0.394	72.00	2.29
3	0.283	150.00	1.64
4	0.783	270.00	3.48
5	0.428	72.00	2.34
6	0.404	90.00	3.22
7	0.555	90.00	1.38
8	0.514	135.00	3.49
9	0.564	112.50	2.12
10	0.602	144.00	2.99
11	0.365	67.50	3.25
12	0.515	72.00	1.59
13	0.561	81.00	4.80
14	0.238	60.00	4.80
15	0.239	150.00	1.63
16	0.827	108.00	3.27
Max (f_j)	0.828	270	4.8
Min (f_j)	0.238	27	1.38

The results of calculating the quantities r_{ij} (with $i = 1 \div 16; j = 1 \div 3$) according to formula (1) are presented in Table 6.

Table 6. Values of r_{ij}

Trial	r_1	r_2	r_3
A_1	1401.9864	1.0000	0.5322
A_2	1402.7220	0.8148	0.7339
A_3	1402.9102	0.4938	0.9240
A_4	1402.0627	0.0000	0.3860
A_5	1402.6644	0.8148	0.7193
A_6	1402.7051	0.7407	0.4620
A_7	1402.4492	0.7407	1.0000
A_8	1402.5186	0.5556	0.3830
A_9	1402.4339	0.6481	0.7836
A_{10}	1402.3695	0.5185	0.5292
A_{11}	1402.7712	0.8333	0.4532
A_{12}	1402.5169	0.8148	0.9386
A_{13}	1402.4390	0.7778	0.0000
A_{14}	1402.9864	0.8642	0.0000
A_{15}	1402.9847	0.4938	0.9240
A_{16}	1401.9881	0.6667	0.4474

The results of calculating quantities S_i, R_i and Q_i according to the formulas (2) to (8) are presented in Table 7.

Table 7. Values of Si, R_i, Q_i and ranking

Trial	S_i	R_i	Q_i	Ranking
A_1	467.8395	467.3288	0.1826	14
A_2	468.0903	467.574	0.8013	4
A_3	468.1093	467.6367	0.9365	2
A_4	467.4829	467.3542	0.0509	16
A_5	468.0662	467.5548	0.7506	6
A_6	467.9693	467.5684	0.7281	7
A_7	468.0633	467.4831	0.6056	9
A_8	467.8191	467.5062	0.5269	11
A_9	467.9552	467.478	0.5401	10
A_{10}	467.8057	467.4565	0.4207	13
A_{11}	468.0192	467.5904	0.7977	5
A_{12}	468.0901	467.5056	0.6645	8
A_{13}	467.7389	467.4797	0.4328	12
A_{14}	467.9502	467.6621	0.9059	3
A_{15}	468.1342	467.6616	0.9990	1
A_{16}	467.7007	467.3294	0.1126	15

From the data in Table 7, it can be seen that: option A_{15} is the best one, and option A4 is the worst one. In option A_{15}, *TWR* has the smallest value among the total of 16 options ($TWR_{15} = 1.63\ \mu m$). The surface roughness in option A_{15} is only a very small amount (0.001 μm) larger than the same in option A_{14}. *MRR* at option A_{15} is only smaller than *MRR* at option A_4. Thus, it can be summarized that *TWR* at option A_{15} is the best one, surface roughness at option A_{15} is ranked 2nd, *MRR* at option A_{15} is ranked 2nd (along with option A_3). Therefore, it can be affirmed that A_{15} is the best option. In order to achieve this best solution, during the machining process, it is required to choose a cooling oil which is Caltex Aquatex 3180, the cutting velocity is 150 m/min, the feed rate is 0.12 mm/rev and the depth of cut is 0.5 mm.

6 Conclusion

The turning process of S55C steel with TiCN cutting piece was performed in this study. The experimental process was designed according to the Taguchi matrix with 16 experiments. In each experiment, the coolant type, cutting velocity, feed rate and depth of cut were changed. At each experiment, surface roughness and TWR were measured and MRR was calculated. The Vikor method has been applied to make a multiple criteria decision. The results have determined that in order to simultaneously ensure the multiple criteria including minimum surface roughness, minimum TWR and maximum MRR, it is required to choose a cooling oil of Caltex Aquatex 3180 in the machining process. The parameters of cutting velocity, feed rate and depth of cut are 150 m/min, 0.12 mm/rev and 0.5 mm, respectively. Determining the parameters of the cutting tool, the flow and the pressure of the cooling oil to ensure the multiple criteria of turning this steel is the work that will be done by the author of this article in the coming time.

References

1. Prajapati, N.K., Patel, S.M.: A review on optimization of process parameters for surface roughness and MRR for S.S.316 on CNC machine. Int. J. Adv. Eng. Res. Stud. **2**(1), 100–103 (2012)
2. Aslantas, K., et al.: Investigations on surface roughness and tool wear characteristics in micro-turning of Ti-6Al-4V alloy. Materials **13**(2998), 1–20 (2020)
3. Mia, M., et al.: Taguchi S/N based optimization of machining parameters for surface roughness, tool wear and material removal rate in hard turning under MQL cutting condition. Measurement **8**(4), 1–36 (2018)
4. Maheswara Rao, C., Venkatasubbaiah, K.: Optimization of surface roughness in CNC turning using Taguchi method and ANOVA. Int. J. Adv. Sci. Technol. **93**, 1–14 (2016). https://doi.org/10.14257/ijast.2016.93.01
5. Gajanana, S., Chakradhar, D., Sai Alok Reddy, N.: Optimization of process parameters in hot machining of 15–5 PH stainless steel using Taguchi method. Int. J. Eng. Sci. Invent. Res. Dev. **3**(3), 236–244 (2016)
6. Kajal, S., Yadav, S.: Optimization of CNC turning parameters for surface roughness on EN 354 steel using Taguchi method. J. Mater. Sci. Mech. Eng. **2**(10), 54–57 (2015)

7. Kumar, R., Bala Kumar, G.: Optimization of machining parameters in CNC turning on surface finish and MRR for EN19 steel using Taguchi Method. Int. J. Innov. Res. Sci. Eng. Technol. **6**(15), 136–146 (2017)
8. Vijaykumar, H.K., Siddiq, A., Sinan, M.: Optimization of turning parameters using Taguchi technique for MRR and surface roughness of hardened AISI 52100 steel. J. Eng. Res. Appl. **5**(5), 39–44 (2014)
9. Akkus, H.: Optimising the effect of cutting parameters on the average surface roughness in turning process with the Taguchi Method. Mater. Technol. **52**(6), 781–785 (2018)
10. Pawar, K., Palhade, R.D.: Multi-objective optimization of CNC turning process parameters for high speed steel (M2) using Taguchi and ANOVA method. Int. J. Hybrid Inf. Technol. **8** (4), 67–80 (2015)
11. Satynarayana, K., Swathi, A., Neeraja, K., Samaikhya, M., Rajkiran, K.: Taguchi optimisation in machining EN18 using coated tool under dry and MQL condition. E3S Web Conf. **184**, 01013 (2020). https://doi.org/10.1051/e3sconf/202018401013
12. Ambigai, R., Prabhu, S., Gandhi, R.: Taguchi based optimization of coated and uncoated tool insert for turning Ti6Al4V using grey relation analyst. IOP Conf. Ser. Mater. Sci. Eng. **912**(032048), 1–12 (2020)
13. Surya, M.S., Vepa, K.S., Karanam, M.: Optimization of machining parameters using ANOVA and grey relational analysis while turning Aluminium 7075. Int. J. Recent Technol. Eng. **8**(2), 5682–5686 (2019)
14. Dureja, S., Sing, R., Bhatt, M.S.: Optimizing flank wear and surface roughness during hard turning of AISI D3 steel by Taguchi and RSM methods. Prod. Manuf. Res. J. **2**(1), 767–783 (2014)
15. Gupta, M., Kumar, S.: Investigation of surface roughness and MRR for turning of UD-GFRP using PCA and Taguchi method, Eng. Sci. Technol. Int. J., 1–12 (2014)
16. Dinh, L.C.: The integrated model of GIS and AHPVIKOR for evaluating sustainable land-use management. Sci. Technol. Dev. **19**(4), 97–104 (2016)
17. Gangil, M., Pradhan, M.K.: Optimization the machining parameters by using VIKOR method during EDM process of Titanium alloy. Mater. Today Proc. **5**, 7486–7495 (2018)
18. Bhuyan, R.K., Routara, B.C.: Optimization the machining parameters by using VIKOR and entropy weight method during EDM process of Al–18% SiCp metal matrix composite. Decis. Sci. Lett. **5**, 269–282 (2016)
19. Kumar, V.A., Reddy, D.V.V., Nagaraju, N.: Multi-objective optimization of end milling process parameters in machining of EN 31 steel: application of AHP embedded with VIKOR and WASPAS methods. J. Mech. Eng. **8**(4), 39–46 (2018)
20. Opricovic, S., Tzeng, G.H.: Compromise solution by MCDM method: a comparative analysis of VIKOR and TOPSIS. Eur. J. Oper. Res. **156**, 445–455 (2004)
21. Talib, R.J., Zaharah, A.M., Selamat, M.A., Mahaidin, A.A., Fazir, M.F.: Friction and wear characteristics of WC and TiCN-coated insert in turning carbon steel workpiece. Procedia Eng. **68**, 716–722 (2013)
22. https://www.keyence.com/products/microscope/digital-microscope/vhx-6000/models/vhx-6000/

Multiple Objective Optimization of the 1655 Steel Milling Process

Nguyen Lam Khanh and Nguyen Van Cuong(✉)

University of Transport and Communications, Hanoi, Vietnam
nguyencuong@utc.edu.vn

Abstract. 1655 steel is a medium carbon steel, used very commonly to manufacture parts such as shafts, gears, or rollers. Therefore, the study on multiple objective optimization when machining this type of steel will contribute to improve the efficiency of the production process. In this study, the milling of this steel was performed by the Titanium Nitride-coated cutting tool. The Taguchi method was applied to design an experimental matrix. In each experiment, parameters including nose radius, spindle speed, feed rate and depth of cut were changed. The surface roughness (*SR*) and flank wear (*VB*) were measured and material removal rate (*MRR*) was calculated. The Topsis method was also applied to solve the multiple objective optimization problem. The results determined the value of the nose radius and the cutting parameters to simultaneously ensure the criteria as the minimum SR, the minimum VB and the maximum MRR.

Keywords: Milling · 1655 steel · Surface roughness · Flank wear · Material removal rate · Taguchi · Topsis

1 Introduction

SR, *VB* and *MRR* are the parameters that are often particularly interested in machining methods in general and in milling method in particular [1–4]. These three parameters are particularly paid attention to because *SR* directly affects the workability and life of the product, *VB* directly affecting the life of the tool also has a great influence on the quality of the product, and *MRR* is a parameter reflecting the productivity of the machining process. Therefore, machining processes often aims to ensure small *SR*, small *VB* and large *MRR*. Consequently, many studies have been carried out to find solutions to ensure one or at the same time the above objectives in each specific condition and by different methods. This work is known as milling optimization.

Thakre [5] used the carbide inserts cutting tool to test the milling of AISI 1040 steel. He used the Taguchi method to determine the optimal values of cutting velocity, the feed rate, the depth of cut and the flow of the coolant to ensure the minimum SR. Gajanana et al. [6] also applied the Taguchi method to determine the value of the two optimal value sets of cutting velocity, the feed rate and the depth of cut when milling 15-5PH stainless steel with the insert SNMG 120408 cutting tool. While tne optimal value set of the cutting parameters ensures the minimum *SR*, the other set ensures the maximum *MRR*. Vansh et al. [7] also determined two optimal values sets of the

D. C. Nguyen et al. (Eds.): ICERA 2021, LNNS 366, pp. 656–668, 2022.
https://doi.org/10.1007/978-3-030-92574-1_68

parameters, including the cutting velocity, the feed rate and the depth of cut when milling Titanium Grade 12 Alloy with CBN tool. One set ensures the minimum *SR* and the other ensures the maximum *MRR*. The determination of the optimal value of the cutting parameters in this study was also performed by the Taguchi method. Taguchi method was applied by Jurina et al. [8] to determine the optimal values of the cutting speed, the depth of cut, the feed per tooth, the feed speed and the radial depth of cut to ensure that when milling DIN X38CrMoV53 steel with the coated cemented carbide cutting tool, the minimum *VB* would be obtained. Ranganath et al. [9] also used the Taguchi method to determine two optimal value sets of the cutting velocity, the feed rate and the depth of cut when milling 6061 aluminum alloy with cemented carbide cutting tool. Of the two sets of optimal parameters they found, one was to ensure the minimum *SR* and the other was to ensure the maximum *MRR*. Kumar et al. [10] used an uncoatedcarbide insert to mill ADC12 Piston Alloy. In this study, they also used the Taguchi method to determine three optimal value sets of the cutting velocity, the feed rate and the depth of cut. It showed that one set was to ensure the minimum *SR*, whereasn another one was to to ensure the maximum *MRR* and the remaining one was to ensure the minimum *VB*.

In addition to using the only Taguchi method to perform single-objective optimization as in some of the above studies, the Taguchi method was also combined with a number of other methods to perform multiple objective optimization of the milling process in some studies. Norcahyo et al. [11] studied to determine the optimal value of coolant flow, the cutting velocity, the feed rate and the depth of cut to ensure simultaneously the criteria including minimum *SR*, minimum *VB* and maximum *MRR* when milling ASSAB XW-42 tool steel with the end mill solid carbide cutting tool. In this study, the Taguchi method was emplyed in combination with gray relational analysis (GRA). Lusi et al. [12] also combined the Taguchi method and GRA for multiple objective optimization when milling ASSAB XW-42 tool steel with the solid carbide end mill cutting tool. The purpose of this study was to determine the coolant type, the cutting velocity, the feed rate and the depth of cut in order to ensure the minimum *SR* and maximum *MRR* at the same time. Ali et al. [13] also used the Taguchi and *GRA* methods to determine the optimal value of the cutting velocity, the feed rate, the depth of cut and the tool path strategy to simultaneously ensure the minimum *SR*, maximum *MRR* and minimum cutting time when milling Al2024-T4 alloy with the Torus-end mill (HSS) cutting tool. The combination of Taguchi and GRA methods for multiple objective optimization of milling SUH03 steel with uncoated carbide inserts was also performed by Yazhini et al. [14]. In the study, they determined the tool nose radius and the values of the cutting velocity, the feed rate and the depth of cut to simultaneously ensure the minimum *SR*, minimum *VB* and maximum *MRR*. Trung [15] combined the Taguchi method and the Reference Ideal Method (RIM) method to determine the tool material type and the optimal values of the cutting velocity, the feed rate and the depth of cut to simultaneously ensure the minimum *SR*, maximum *MRR* and minimum cutting force when milling SKD11 steel. Ulfiyah et al. [16] combined the Taguchi and Weighted Principal Component Analysis (WPCA) methods for multiple objective optimization when milling 6061 aluminum alloy with the high spedd steel (HSS) cutting tool. They determined the type of milling operation and the values of the cutting velocity, the feed rate and the depth of cut to simultaneously ensure the minimum *SR*, minimum *VB* and maximum *MRR*.

The analysis of some of the above studies shows that the Taguchi method has been successfully used in single-objective optimization of the milling process. Combining the Taguchi method with a certain mathematical method has also been very successful in optimizing the multiple objectives of the milling process.

The Topsis method is a very commonly used in multiple objective optimization. This method was successful when being combined with the Taguchi method in multiple objective optimization of some cases such as multiple objective optimization of bearing grinding process [17], multiple objective optimization of AISI 52100 steel turning process [18], multiple objective optimization of M42 steel turning process [19], multiple objective optimization of EDM machining process of Al6061 material [20], multiple objective optimization of surface grinding process of DIN 1.2379 steel by segmented grinding wheel [21], etc. However, until now, according to the authors of this article, there have been no published studies on the combination of the Taguchi and Topsis methods to optimize multi-objective of the milling process.

1655 steel is marked according to Sweden's SS standard (equivalent to steel types: C55 – EU standard, C55 – standard DIN, S55C – JIS standard, 55 – standard GB, 12060 – standard CSN). This is a medium carbon steel with good machinability, commonly used to make parts such as shafts, gears, rollers, etc. Among those parts, the milling method occupies a relatively large workload. Therefore, multiple objective optimization research when milling this steel will bring great efficiency to the production process. However, according to the authors of this article, the studies on multiple objective optimization when milling this steel is still very limited.

In order to overcome the limitations as analyzed above, in this study, multiple objective optimization of the 1655 steel milling process will be conducted. The combination of Taguchi and Topsis methods will be performed to determine the optimal value of the parameters of the machining process to simultaneously ensure the criteria including the minimum *SR*, minimum *VB* and maximum *MRR*.

2 Topsis Method

The Topsis method is a commonly used to solve multiple objective optimization problems. This method is aimed to find out the best solution (the most ideal solution) from many given options. The steps performed in the Topsis method are described as follows [21–24]:

Step 1: Sort the options according to a matrix.

$$X = \begin{bmatrix} x_{11} & \ldots & x_{1j} & x_{1n} \\ x_{21} & \ldots & x_{2j} & x_{2n} \\ \ldots & \ldots & \ldots & \ldots \\ x_{i1} & \ldots & x_{ij} & x_{in} \\ \ldots & \ldots & \ldots & \ldots \\ x_{m1} & \ldots & x_{mj} & x_{mn} \end{bmatrix} \tag{1}$$

Of which: x_{ij}, is the value of criterion j in option *i; n* is the number of indicators; *m* is the number of options.

Step 2: Determine the conversion values according to formula (2).

$$x'_{ij} = \frac{x_{ij}}{\sqrt{\sum_{i=1}^{n} x_{ij}^2}} \quad (2)$$

Step 3: Assign weight w_j to the criteria.
Step 4: Calculate the normalized matrix according to formula (3).

$$Y = w_j.x'_{ij} \quad (3)$$

Step 5: Determine the best solution and the worst solution for the criteria according to (4) and (5)

$$A^+ = \left\{y_1^+, y_2^+, \ldots, y_j^+, \ldots, y_n^+\right\} \quad (4)$$

$$A^- = \left\{y_1^-, y_2^-, \ldots, y_j^-, \ldots, y_n^-\right\} \quad (5)$$

Of which: y_j^+ and y_j^- are the best and worst values of the jth criterion, respectively.
Step 6: Determine the values S_i^+ and S_i^- according to (6) and (7).

$$S_i^+ = \sqrt{\sum_{j=1}^{n} \left(y_{ij} - y_j^+\right)^2} \quad i = 1, 2, \ldots, m \quad (6)$$

$$S_i^- = \sqrt{\sum_{j=1}^{n} \left(y_{ij} - y_j^-\right)^2} \quad i = 1, 2, \ldots, m \quad (7)$$

Step 7: Determine the values C_i^* according to (8).

$$C_i^* = \frac{S_i^-}{S_i^+ + S_i^-} \quad i = 1, 2, \ldots, m; \quad 0 \leq C_i^* \leq 1 \quad (8)$$

Step 8: Rank the options according to the principle that the one with the largest C_i^* is the best one.

3 Experimental Process

Samples of 1655 steel were used during the experiment. The steel sample has a hardness of 46 HRC, with the length of 60 mm, the width of 40 mm and the height of 40 mm. 3-axis CNC milling machine of HAAS brand was used to perform the experiments. The Titanium Nitride-coated cutting tool was used with several different values for the nose radius. This type of cutting tool is very commonly used for machining on CNC machines [25–27]. The nose radius, the spindle speed, the feed rate and the depth of cut were chosen to be variable in each experiment. Each parameter was selected with 4 levels of values as shown in Table 1. The values of the selected nose radius are those available in the market. The values of the selected cutting parameters are based on the recommendations of the cutting tool manufacturer [28].

Table 1. Input parameters

Parameter	Sign	Unit	Code	Value at levels			
				1	2	3	4
Nose radius	r	mm	x_1	0.3	0.5	0.8	1.2
Spidle speed	n_w	rpm	x_2	540	720	900	1080
Feed rate	f_z	mm/tooth	x_3	0.09	0.12	0.15	0.18
Depth of cut	a_p	mm	x_4	0.5	0.75	1	1.25

The Taguchi method was applied to design an experimental matrix. The advantage of such method was to build an experimental matrix with many input parameters, each input parameter had many value levels but the total number of experiments was not much. In addition, being able to choose input parameters as qualitative or as discrete ones was also an unique advantage of this method [29–31]. Of the four input parameters, changing the cutting tool (changing the nose radius) took more time than changing the rest. Therefore, the nose radius was chosen as the first parameter to build an experimental matrix. With 4 input parameters, each parameter had 4 values, the experimental matrix consisted of 16 experiments as shown in Table 2.

Table 2. L16 Orthogonal matrix

Trial	Code variable				Real variable			
	x_1	x_2	x_3	x_4	r (mm)	n_w (rpm)	f_z (mm/tooth)	a_p (mm)
1	1	1	1	1	0.3	540	0.09	0.5
2	1	2	2	2	0.3	720	0.12	0.75
3	1	3	3	3	0.3	900	0.15	1
4	1	4	4	4	0.3	1080	0.18	1.25
5	2	1	3	3	0.5	540	0.12	1
6	2	2	1	4	0.5	720	0.09	1.25
7	2	3	4	1	0.5	900	0.18	0.5
8	2	4	3	2	0.5	1080	0.15	0.75
9	3	1	3	4	0.8	540	0.15	1.25
10	3	2	4	3	0.8	720	0.18	1
11	3	3	1	2	0.8	900	0.09	0.75
12	3	4	2	1	0.8	1080	0.12	0.5
13	4	1	4	2	1.2	540	0.18	0.75
14	4	2	3	1	1.2	720	0.15	0.5
15	4	3	2	4	1.2	900	0.12	1.25
16	4	4	1	3	1.2	1080	0.09	1

At each experiment, 02 new chips were installed on the tool body with a diameter of 14 mm. The reason that each cutting piece was used only once was to eliminate the influence of tool wear on the output parameters. The *SR* was measured with SJ201 (Mitutoyo – Japan), with the pre-set standard length of 0.8 mm. *SR* at each experiment was the average value of at least 3 consecutive measurements. The VHX600 digital microscope was used to measure the *VB* on the back of tool (Fig. 1). This instrument performed the measurement in compliance with ISO 4287, JIS-G0551, ISO-643, ASTM-E1382, ISO-16232, ISO25178 standards, the instrument's magnification was up to 5000 times [32]. *MRR* was the volume of material removed in one minute, calculated by Eq. (9).

$$MRR = V_f \cdot a_p \cdot b_w (mm^3/min) \quad (9)$$

Of which:

V_f is the feed rate in minutes. With the number of chips mounted on the tool body equal to 2, then $V_f = 2.n_w.f_z$;

b_w is the milling width. In this case, a symmetrical milling was performed, which means that the milling width is equal to the diameter of the tool, i.e. $b_w = 14\ mm$.

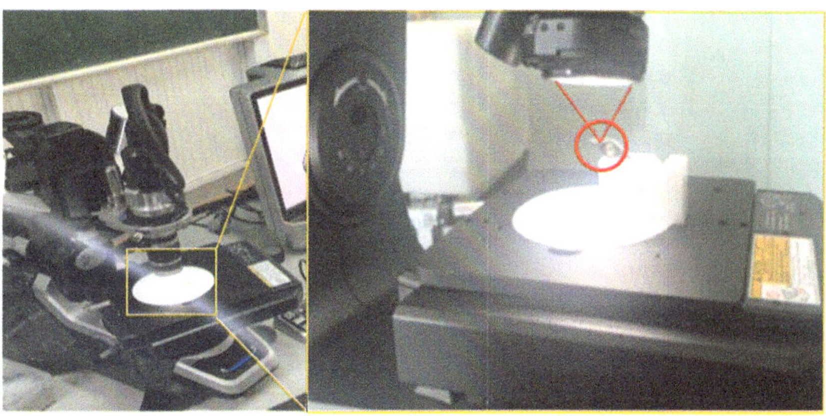

Fig. 1. Digital microscopeVHX600

4 Experimental Results and Discussion

The experimental results are presented in Table 3. In Figs. 2, 3 and 4, the graph shows the influence of input parameters on output parameters. Observing these figures shows that the influence of input parameters on output parameters is different. Some examples can be given to clarify this statement as follows: (1) The nose radius has a great influence on *VB* but not much on *SR*. This parameter is not present in formula (9), so it does not affect the *MRR*; (2) *MRR* is greatly influenced by spindle speed, but spindle speed has little effect on *SR* and *VB*. (3) The feed rate also has a great influence on the

MRR but not much on the *SR* and *VB*. (4) *VB* and *MRR* are greatly influenced by the depth of cut, but the depth of cut has little effect on *SR*. Thus, it can be seen that if only observing the graphs in Figs. 2, 3 and 4, there will be no base for choosing the values of input parameters to simultaneously ensure the minimum *SR*, minimum *VB* and minimum *MRR*.

Table 3. Experimental results

Trial	*r (mm)*	n_w *(rpm)*	f_z *(mm/tooth)*	a_p *(mm)*	*SR (μm)*	*VB (μm)*	*MRR (mm*^{3}*/min)*
1	0.3	540	0.09	0.5	0.994	25.688	680.40
2	0.3	720	0.12	0.75	0.473	19.740	1814.40
3	0.3	900	0.15	1	0.341	14.137	3780.00
4	0.3	1080	0.18	1.25	0.941	29.998	6804.00
5	0.5	540	0.12	1	0.514	20.171	1814.40
6	0.5	720	0.09	1.25	0.485	27.756	2268.00
7	0.5	900	0.18	0.5	0.666	11.896	2268.00
8	0.5	1080	0.15	0.75	0.617	30.084	3402.00
9	0.8	540	0.15	1.25	0.677	18.274	2835.00
10	0.8	720	0.18	1	0.722	25.774	3628.80
11	0.8	900	0.09	0.75	0.438	28.015	1701.00
12	0.8	1080	0.12	0.5	0.618	13.706	1814.40
13	1.2	540	0.18	0.75	0.673	41.376	2041.20
14	1.2	720	0.15	0.5	0.286	41.372	1512.00
15	1.2	900	0.12	1.25	0.287	14.051	3780.00
16	1.2	1080	0.09	1	0.992	28.187	2721.60

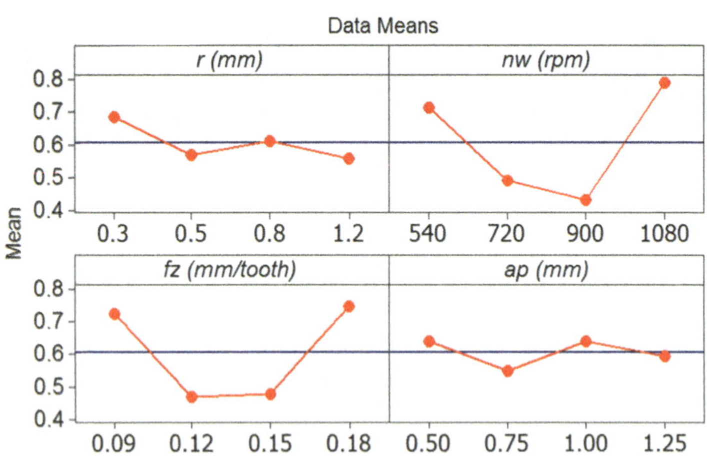

Fig. 2. Main effect plot for *SR*

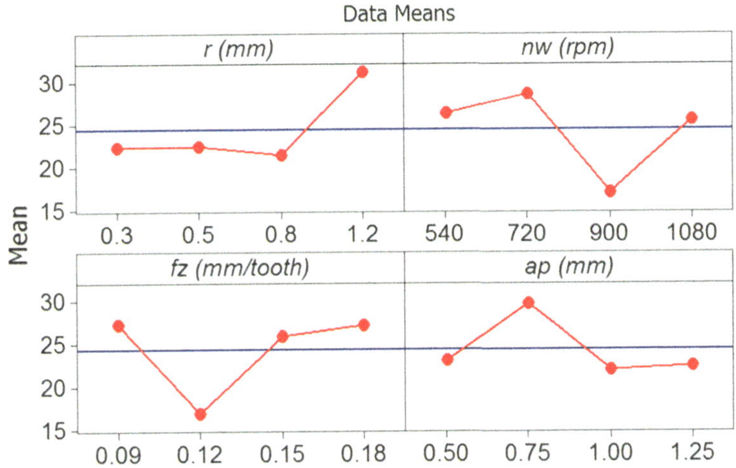

Fig. 3. Main effect plot for *VB*

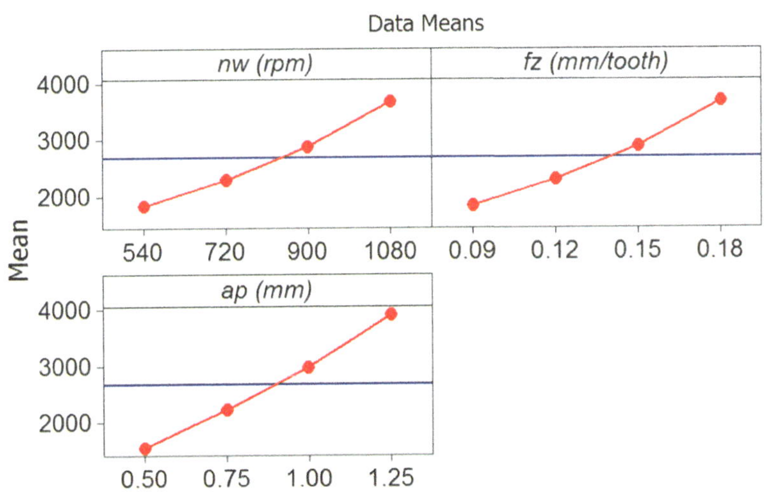

Fig. 4. Main effect plot for *MRR*

On the other hand, when observing the data in Table 3, it shows that: In experiment 14, *SR* has the smallest value, but also in this experiment, *VB* is very large (*VB* in experiment 14 is only smaller than *VB* in experiment 13), *MRR* in this experiment was only average. In experiment 7, *VB* has the smallest value, but *SR* in this experiment is quite large, the *MRR* in this experiment is also just average. *MRR* has the largest value in experiment 4, but also in this experiment, *SR* is very large (*SR* in experiment 4 is only smaller than *SR* in experiments 1 and 16). Therefore, it can be said that if only observing the data in Table 3, it is not possible to determine which experiments that simultaneously ensure the minimum *SR*, minimum *VB* and maximum *MRR*. And it can also be asserted that there is no experiment that perfectly ensures all three of these criteria. In fact, we can only determine one experiment where *SR* is considered "minimum", *VB* is considered "minimum" and *MRR* is considered "maximum", and of course this also work must be based on mathematical scientific arguments. The Topsis method will be applied to solve this work and is presented in the next section of this article.

5 Multiple Objective Optimization

For the specific case in this study, the matrix of options of the expenditures to be optimized is written as follows:

$$X = \begin{bmatrix} SR_1 & VB_1 & MRR_1 \\ SR_2 & VB_2 & MRR_2 \\ \ldots & \ldots & \ldots \\ SR_{15} & VB_{15} & MRR_{15} \\ SR_{16} & VB_{16} & MRR_{16} \end{bmatrix}$$

Applying formula (2) to determine the conversion values for the criteria as shown in Table 4.

Table 4. Conversion value in Topsis

No.	x_{SR}	x_{VB}	x_{MRR}
1	0.3843	0.2476	0.0565
2	0.1828	0.1903	0.1507
3	0.1318	0.1363	0.3139
4	0.3638	0.2892	0.5650
5	0.1987	0.1945	0.1507
6	0.1875	0.2676	0.1883
7	0.2575	0.1147	0.1883
8	0.2385	0.2900	0.2825
9	0.2617	0.1762	0.2354
10	0.2791	0.2485	0.3013
11	0.1693	0.2701	0.1412
12	0.2389	0.1321	0.1507
13	0.2602	0.3989	0.1695
14	0.1106	0.3988	0.1256
15	0.1109	0.1355	0.3139
16	0.3835	0.2717	0.2260

Applying formula (3) to determine the normalized matrix as shown in Table 5. The weight of the 3 selected output parameters has equal values, ie $w_j = 1/3;\ j = 1 \div 3$.

From the data in Table 5, the best and worst solutions for the options have been determined according to formulas (4) and (5), presented in Table 6.

Table 5. Normalized Matrix

No.	Y_{SR}	Y_{VB}	Y_{MRR}
1	0.1281	0.0825	0.0188
2	0.0609	0.0634	0.0502
3	0.0439	0.0454	0.1046
4	0.1213	0.0964	0.1883
5	0.0662	0.0648	0.0502
6	0.0625	0.0892	0.0628
7	0.0858	0.0382	0.0628
8	0.0795	0.0967	0.0942
9	0.0872	0.0587	0.0785
10	0.0930	0.0828	0.1004
11	0.0564	0.0900	0.0471
12	0.0796	0.0440	0.0502
13	0.0867	0.1330	0.0565
14	0.0369	0.1329	0.0419
15	0.0370	0.0452	0.1046
16	0.1278	0.0906	0.0753

Table 6. Best and worst solutions

	SR	*VB*	*MRR*
A+	0.0369	0.0382	0.1883
A−	0.1281	0.1330	0.0188

The values S_i^+, S_i^- and C_i^* are calculated according to the respective formulas (6), (7) and (8), presented in Table 7. In this table, the options have also been ranked.

Therefore, according to the data in Table 7, experiment 15 is the best one, while experiment 1 is the worst. Looking back at the experimental results in Table 3, iexperiment 1, *SR* had the largest value, *MRR* had the smallest value, and *VB* had a fairly large value, so the obvious statement that experiment 1 was the worst. In experiment 15, *SR* was very small (*SR* in experiment 15 was only 0.01 μm larger than *SR* in experiment 14), *VB* was also very small (*VB* in experiment 15 was only larger than *VB* in experiments 7 and 12), *MRR* stood at number 2 (*MRR* in experiment 15was only smaller than *MRR* in experiment 4). Although the concepts of "minimum" and "maximum" cannot be obtained for the parameters in experiment 15. However, as analyzed in Sect. 4, it is not possible to find an experiment that perfectly achieves requirements of the criteria. Therefore, on the basis of a comprehensive review of the criteria, it can be affirmed that experiment 15 is the best one. Thus, the optimal values of the parameters are as follows: the nose radius is 1.2 mm, the spindle speed is 720 rpm, the feed rate is 0.15 mm/tooth and the depth of cut is 0.5 mm.

Table 7. Several values in Topsis

No.	Si^{+}	Si^{-}	C_i	*Ranking*
1	0.1975	0.0505	0.2035	16
2	0.1424	0.1017	0.4165	9
3	0.0843	0.1487	0.6383	2
4	0.1025	0.1736	0.6288	3
5	0.1436	0.0973	0.4038	10
6	0.1379	0.0903	0.3958	11
7	0.1347	0.1127	0.4555	6
8	0.1187	0.0968	0.4490	7
9	0.1225	0.1037	0.4583	5
10	0.1134	0.1020	0.4736	4
11	0.1517	0.0882	0.3677	12
12	0.1447	0.1061	0.4230	8
13	0.1698	0.0560	0.2479	15
14	0.1744	0.0941	0.3505	13
15	0.0840	0.1529	0.6456	1
16	0.1542	0.0707	0.3143	14

6 Conclusion

The experimental process of milling 1655 steel with Titanium Nitride-coated cutting tool was performed in this study. The combination of Taguchi and Topsis methods was performed to solve the multiple objective optimization problem. Some conclusions are drawn as follows:

- In order to simultaneously ensure the criteria including the minimum *SR*, minimum *VB* and maximum *MRR*, the optimal values of the nose radius, the spindle speed, the feed rate and the depth of cut are 1.2 mm, 720 rpm, 0.15 mm/tooth and 0.5 mm, respectively.
- The combination of Taguchi and Topsis methods to solve multiple objective optimization problems has not only been successful in this study and in published studies [17–21] but it also leads to a new promising prospect when this combination is applied to solve multiple objective optimization problems in other machining processes.

References

1. Klocke, F., Arntz, K., Cabral, G.F., Stolorz, M., Busch, M.: Characterization of tool wear in high-speed milling of hardened powder metallurgical steels. Adv. Tribol. **2021**, 1–14 (2021)
2. Nurhaniza, M., Ariffin, M.K.A.M., Mustapha, F., Baharudin, B.T.H.T.: Analyzing the effect of machining parameters setting to the surface roughness during end milling of CFRP-aluminium composite laminates. Int. J. Manuf. Eng. **2016**, 1–10 (2016)
3. Li, W., Guo, Y.B., Barkey, M.E., Jordon, J.B.: Effect Tool wear during end milling on the surface integrity and fatigue life of inconel 718. Procedia CIRP **14**, 546–551 (2014)
4. Ersvik, E., Khalid, R.: Milling in hardened steel – a study of tool wear in conventionaland dynamic milling. Uppsala Universitet (2015)
5. Ali, R.A., Mia, M., Khan, A.M., Chen, W., Gupta, M.K.: Catalin Iulian Pruncu, multi-response optimization of face milling performance considering tool path strategies in machining of Al-2024. Materials **12**(1013), 1–19 (2019)
6. Gajanana, S., Chakradhar, D., Sai Alok Reddy, N.: Optimization of process parameters in hot machining of 15–5 PH stainless steel using Taguchi method. Int. J. Eng. Sci. Invent. Res. Dev. **3**(3), 236–244 (2016)
7. Vansh, D.P., Patel, D.: Effect of cutting parameters on the surface roughness and MRR of titanium alloys using VMC. Int. J. Innovative Res. Sci. Technol. **4**(6), 1–10 (2017)
8. Jurina, F., Vopat, T., Kuruc, M., Simna, V.: The tool wear observation of milling tools in high feed machining of hardened steels. In: Katalinic, B. (ed.) Proceedings of the 30th International DAAAM Symposium 2019, pp. 0753–0757. DAAAM International Vienna (2019). https://doi.org/10.2507/30th.daaam.proceedings.103
9. Ranganath, M.S., Mishra, R.: Optimization of surface roughness and material removal rate on conventional dry turning of aluminium (6061). Int. J. Adv. Res. Innovation **1**, 62–72 (2014)
10. Kumar, H.D., Ilangovan, S., Radhika, N.: Optimization of cutting parameters for MRR, tool wear and surface roughness characteristics in machining ADC12 piston alloy using DOE. Tribol. Ind. **42**(1), 32–40 (2020)
11. Norcahyo, R., Soepangkat, B.O.P.: Optimization of multi response in end milling process of ASSAB XW-42 tool steel with liquid nitrogen cooling using taguchi-grey relational analysis. In: AIP Conference Proceedings 1855(020011), pp. 1–9 (2017)
12. Lusi, N., Pamuji, D.R., Fiveriati, A., Afandi, A., Prayogo, G.S.: Application of Taguchi and grey relational analysis for parametric optimization of end milling process of ASSAB-XW 42. Adv. Eng. Res. **198**, 514–517 (2020)
13. Ali, R.A., Mia, M., Khan, A.M., Chen, W., Gupta, M.K., Pruncu, C.I.: Multi-response optimization of face milling performance considering tool path strategies in machining of Al-2024. Materials **12**(7), 1–19 (2019)
14. Yazhini, B., Rajeswari, S., Sivasakthivel, D.: Optimization of machining parameters of valve steel SUH03 (X45CrSiMo10-2) using gray based Taguchi Method. Appl. Mech. Mater. **813–814**, 376–381 (2015)
15. Trung, D.D.: Multi-objective optimization of SKD11 steel milling process by reference ideal method. Int. J. Geol. **15**, 1–16 (2021)
16. Ulfiyah, L., Pramujati, B., Soepangkat, B.O.P.: Optimization of tool wear, surface roughness and material removal rate in the milling process of Al 6061 using Taguchi and weighted principal component analysis (WPCA). Appl. Mech. Mater. **493**, 535–540 (2014)
17. Wisnuadi, A.R., Damayanti, R.W., Pujiyanto, E.: Multi response optimization of internal grinding process parameters for outer ring using taguchi method and PCR-TOPSIS. AIP Conf. Proc. **1931**(1), 1–6 (2018)

18. Umamaheswarrao, P., Raju, D.R., Suman, K.N.S., Sankar, B.R.: Topsis based optimization of process parameters while hard turning of AISI 52100 steel. Acta Mech. Malaysia **2**(2), 28–31 (2019)
19. Sagar, P.B., Pantanwane, P.D., Rajiv, B.: Optimisation of hard turning of M42 steel using PCR-Topsis method. In: 5th International & 26th All India Manufacturing Technology, Design and Research Conference, vol. 325, pp. 1–6. AIMTDR (2014)
20. Sankar, S.S., Reddy, M.V.K.: Experimental investigation and optimization of EDM process parameters on Al6061 by using TOPSIS and comparison with genetic algorithm. Int. J. Eng. Adv. Technol. **8**(2S2), 76–83 (2019)
21. Trung, D.D., Thien, N.V., Nguyen, N.-T.: Application of TOPSIS method in multi-objective optimization of the grinding process using segmented grinding wheel. Tribol. Ind. **43**(1), 12–22 (2021)
22. Temucin, T., Tozan, H., Vayvay, O., Harnicarova, M., Valicek, J.: A fuzzy based decision model for nantraditional machining process selection. Int. J. Adv. Manuf. Technol. **70**, 2275–2282 (2014)
23. Chen, S.-J., Hwang, C.-L.: Fuzzy Multiple Attribute Decision Making. Springer Berlin Heidelberg, Berlin, Heidelberg (1992)
24. Hwang, C.L., Yoon, K.: Multiple Attribute Decision Making Methods and Applications. Springer-verlag, Berlin (1981)
25. Klocke, F., Krieg, T.: Coated tools for metal cutting – features and applications. CIRP Ann. **48**, 515–525 (1999)
26. Prengel, H.G., Pfouts, W.R., Santhanam, A.T.: State of the art in hard coatings for carbide cutting tools. Surf. Coat. Technol. **102**, 183–190 (1998)
27. Uyen, V.T.N., Son, N.H.: Improving accuracy of surface roughness model while turning 9XC steel using a Titanium Nitride-coated cutting tool with Johnson and Box-Cox transformation. AIMS Mater. Sci. **8**(1), 1–17 (2021)
28. https://ruko.de/en/blog/tin-the-titanium-nitrite-coating
29. Phadke, M.S.: Quality Engineering Using Robust Design. Printice Hall (1989)
30. Karna, S.K., Singh, R.V., Sahai, R.: Application of Taguchi method in Indian industry. Int. J. Emerging Technol. Adv. Eng. **2**(11), 387–391 (2012)
31. Karna, S.K., Sahai, R.: An overview on Taguchi method. Int. J. Math. Eng. Manage. Sci. **1**, 11–18 (2012)
32. https://www.keyence.com/products/microscope/digital-microscope/vhx-6000/models/vhx-6000/

Nanogold as an Additive in Iron Oxide Electrode

Bui Thi Hang[1(✉)] and Doan Ha Thang[2]

[1] International Training Institute for Materials Science, Hanoi University of Science and Technology, 1 Dai Co Viet, Hanoi, Vietnam
hang.buithi@hust.edu.vn, hang@itims.edu.vn
[2] Office of Vietnam Space Committee, Ministry of Science and Technology, 113 Tran Duy Hung, Hanoi, Vietnam

Abstract. In the present work, we have investigated the influence of gold nanoparticles (nanogold) on redox reactions of iron/air battery anode. The Fe_2O_3 nanoparticles and gold nanoparticles were used as iron sources and additives, respectively. Acetylene black carbon (AB) was introduced as an additive to improve the conductivity of the iron oxide electrode. A blend of Fe_2O_3 and gold nanoparticles (Fe_2O_3/Au) was prepared by mixing Fe_2O_3 nanoparticles in a solution of pure gold nanoparticles. The grain size and morphology of gold nanoparticles were observed by Transmission Electron Spectroscopy (TEM) whereas the particle size, morphology of Fe_2O_3 and Fe_2O_3/Au mixture were examined by Scanning Electron Spectroscopy (SEM). The electrochemical characteristics of Fe_2O_3, Fe_2O_3/Au, and Fe_2O_3/Au/AB electrodes in alkaline solution were investigated using cyclic voltammetry (CV) measurement. Results suggest that Fe_2O_3/Au mixture has a uniform distribution of gold nanoparticles on an iron surface and thereby improves the cyclability of Fe_2O_3/Au electrode. The positive effect of AB on the electrochemical properties of Fe_2O_3/Au/AB composite electrodes was revealed via improved cyclability of Fe_2O_3/Au/AB electrode as evidenced by higher redox reaction rate and larger redox current.

Keywords: Gold nanoparticles · Nanogold · Pure nanogold solution · Fe_2O_3 nanoparticles · Fe-air battery anode

1 Introduction

Nanomaterials have received considerable attention from researchers because they have special properties compared to bulk materials. Current studies and applications of nanomaterials have focused on biomedicine, energy, and the environment [1–4]. At the nanoscale, the metal nanoparticles are very active. The properties that metals do not manifest at microscale such as bactericidal capability, the catalyst for many reactions occurring at room temperatures or low temperatures (minus degree) have been depicted at the nanoscale. When reducing the dimension from a bulk material to a nanomaterial, the metals exhibit changes in properties such as optical properties, electrical properties, magnetic properties, etc. This makes them attractive for use in a wide range of industrial and commercial applications.

D. C. Nguyen et al. (Eds.): ICERA 2021, LNNS 366, pp. 669–674, 2022.
https://doi.org/10.1007/978-3-030-92574-1_69

At the nanoscale, gold metal has special properties and they are applied in medicine and some other fields [5–8]. Extending nanoparticle applications in a metal /air battery, iron oxide nanoparticles (nm-Fe_2O_3) used as iron/air battery anode exhibited better cyclability than Fe_2O_3 microparticles [9].

In this study, we fabricated a mixture of Fe_2O_3/Au material as an electrode material for use in metal/air batteries. Nanocarbon is used as an additive to improve the conductivity and electrochemical behavior of iron electrodes.

2 Experiments

Fe_2O_3 nanoparticles (Wako Pure Chemical Co.) and acetylene black (AB, Denki Kagaku Co. Ltd.) were used as iron sources and carbon additives, respectively. Pure gold nanoparticles solution was also used as the electrode additive to improve the conductivity of electrode and iron redox reactions rate.

The mixture of iron oxide and gold nanoparticles (Fe_2O_3/Au) was prepared by adding 1.6 g Fe_2O_3 and 10 ml deionized water in various nanogold solutions viz., 5 ml, 10 ml, and 20 ml. The mixture was stirred at room temperature for 2 h, then dried at 40 °C for 24 h to obtain Fe_2O_3/Au powder. XRD and SEM analyses were performed on the resulting Fe_2O_3/Au powder. TEM and UV-VIS analyses were carried out on pure gold nanoparticles. To investigate the electrochemical properties of the synthesized Fe_2O_3/Au, electrode sheets were prepared by mixing 90 wt% Fe_2O_3/Au with 10 wt% polytetrafluoroethylene (PTFE) binder followed by mechanical grinding and then rolling. The effect of AB carbon on the electrochemical properties of Fe_2O_3/Au was also studied using Fe_2O_3/Au/AB electrode. The Fe_2O_3/Au/AB electrode sheet was also fabricated in the same way with 45 wt% of Fe_2O_3/Au, 45 wt% AB and 10 wt% PTFE. Fe_2O_3/Au and Fe_2O_3/Au/AB electrodes were punched into the form of a pellet of 1 cm diameter. The electrode was then pressed onto a titanium mesh at a pressure of about 150 kg/cm^2.

Cyclic voltammetry (CV) was carried out using three-electrode glass cells with Fe_2O_3/Au or Fe_2O_3/Au/AB as the working electrode, Pt mesh as the counter electrode, and Hg/HgO as the reference electrode. The electrolyte was 8 mol dm^{-3} KOH aqueous solution. CV measurements were taken at a scan rate of 5.0 mV s^{-1} and within a range of −1.3 V to −0.1 V. In all electrochemical measurements, we used fresh electrodes without pre-cycling.

3 Results and Discussion

To observe the shape and grain size of the commercial Fe_2O_3, SEM measurement was done and the result is shown in Fig. 1. Fe_2O_3 particles are relatively large, smooth, and not uniform. Some Fe_2O_3 granules are of round shape while a majority of particles are smooth, with dimensions ranging from several hundred nanometers to micrometers. The difference in size and shape of Fe_2O_3 particles will affect the electrochemical properties of the Fe_2O_3 electrode.

TEM analysis of pure gold solution has been observed to study the morphology and grain size, as shown in Fig. 2. The gold grains are non-uniform with a size of few nanometers. The gold grains tend to aggregate together to form larger particles of few tens of nanometers. With a much smaller size than Fe_2O_3, the gold nanoparticles may well be distributed on the surface of Fe_2O_3 particles, and such distribution is expected to be a bridge for better electronic transfer between iron oxide elements and consequently support for electrochemical reactions of iron.

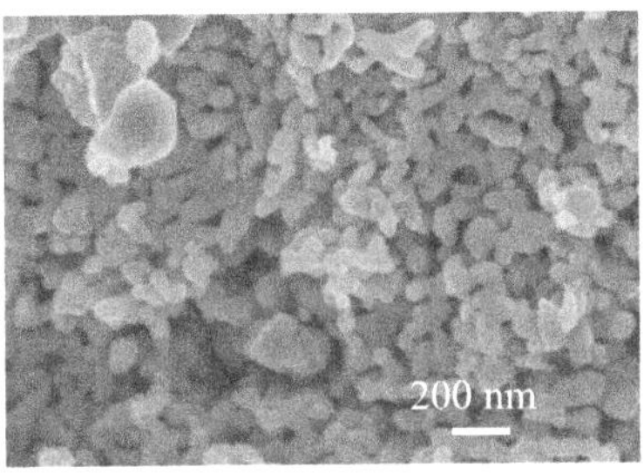

Fig. 1. SEM image of Fe_2O_3 powder

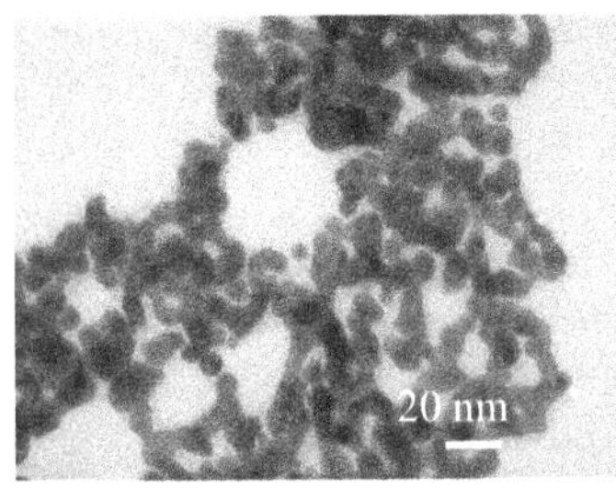

Fig. 2. TEM image of nanogold

UV-VIS absorption spectrum of gold nanoparticles in the deionized water is shown in Fig. 3. The sharp peak occurs at about 550 nm together with a broad hump at longer wavelengths indicating that this solution contains pure gold nanoparticles and they tend to aggregate in this medium. This is in agreement with the observed TEM result.

SEM measurements were carried out on the mixer of iron oxide, nanogold solution, Fe_2O_3/Au material and the results are depicted in Fig. 4. The existence of small particles on bigger particle surfaces demonstrates that nanogold particles are distributed on Fe_2O_3 surfaces. Such distribution of Au will enhance the conductivity of Fe_2O_3 electrode and thus support the redox reaction of Fe_2O_3. A composite anode of Fe_2O_3 active material with a nano-sized conductor AB and Au is expected to solve the problems of iron anodes of Fe-air batteries.

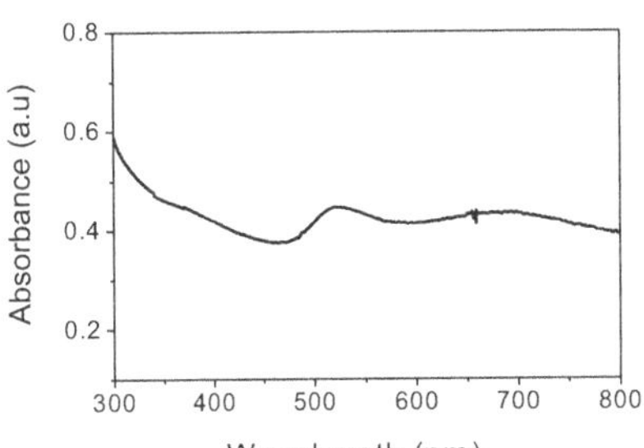

Fig. 3. UV-VIS absorption spectra for pure gold nanoparticles solution

Fig. 4. SEM image of Fe_2O_3/Au powder

To investigate the electrochemical characterization of the Fe_2O_3 electrode, the cyclic voltammetry (CV) was performed in 8M KOH electrolyte solution and the results are shown in Fig. 5. The redox peaks are small, with only one oxidation peak like hump observed while the reduction peak is observable close to the hydrogen evolution peak. This means that the redox reaction rate of iron oxide is very slow and it can be ascribed due to a large internal resistance of semiconductor iron oxides.

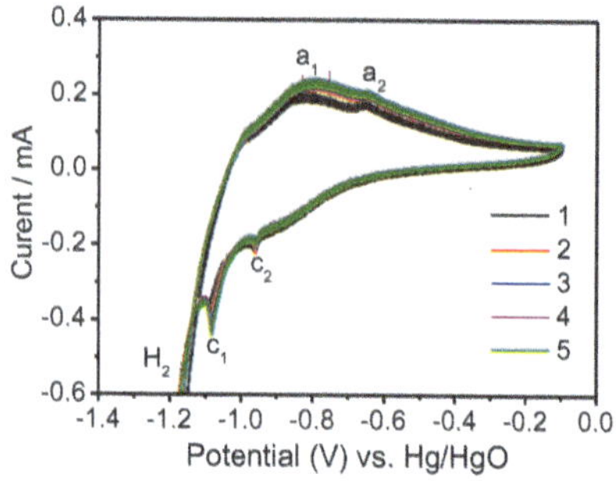

Fig. 5. Voltammogams of Fe_2O_3 electrode

Using the conductive material as an additive is mandatory to improve the conductivity of the iron oxide electrode. Figure 6 presents the CV profiles of the iron oxide electrode using Au as an additive. We observe in 5 ml nanogold solution sample (Fig. 6a), CV profile is similar to that of the iron oxide without an nanogold additive (Fig. 5) whereas samples with 10 and 20 ml nanogold solutions (Fig. 6b, c) show well-defined CV curves evidenced by the redox couple peaks (a_1/c_1, a_2/c_2) during forward and backward scans, respectively. These results demonstrate the positive effects of nanogold particles in improving the redox reaction rate of the iron electrode. When adding 5 ml nanogold solution to Fe_2O_3 material, the amount of nanogold particles is small so its effect on electrochemical behavior is negligible (Fig. 6a). When nanogold solution is increased to 10 ml (Fig. 6b), one couple peak a_1/c_1 appears while two couple peaks a_1/c_1, a_2/c_2 occur with 20 ml sample nanogold solution (Fig. 6c). The a_1/c_1 corresponds to Fe/Fe (II) redox couple while a_2/c_2 corresponds Fe(II)/Fe(III) redox couple. Thus, the sample using 20 ml nanogold solution provides better cyclability than samples with 5 and 10 ml.

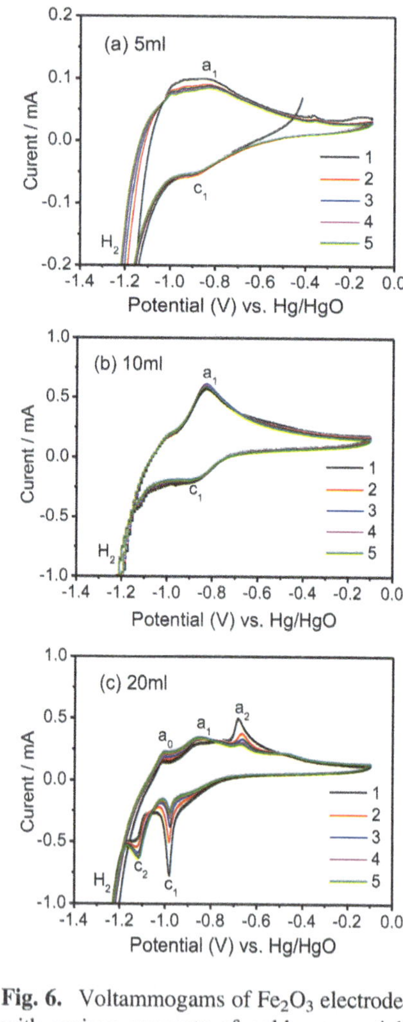

Fig. 6. Voltammogams of Fe_2O_3 electrodes with various amounts of gold nanoparticle solutions: (a) 5 ml, (b) 10 ml and (c) 20 ml.

To increase the conductivity of Fe_2O_3 electrode, acetylene black carbon (AB) was used as an additive to mix with Fe_2O_3/Au at a ratio of Fe_2O_3/Au:AB:PTFE = 45:45:10 wt.%. The CV voltammograms are depicted in Fig. 7. With the presence of AB in the electrode, all the samples 5 (Fig. 7a), 10 (Fig. 7b) and 20 ml (Fig. 7c) exhibit larger and higher redox peaks. However, when the nanogold amount is increased from 5 to 20 ml, the oxidation peaks tend to move toward more positive potentials while reduction peaks shift toward more negative potentials causing an increase in over-potential. In addition, the deposition peak of iron (c_1) is overlapped by the hydrogen evolution peak. This is the cause for low charge/discharge efficiency and high self-discharge rate of Fe_2O_3/Au/AB electrodes.

To evaluate the effect of gold nanoparticle additives on electrochemical properties of Fe_2O_3/Au/AB electrode, the discharge capacity of Fe_2O_3/Au/AB electrode is calculated from CV data in Fig. 7 and presented in Fig. 8. The discharge capacity contains the capacity of oxidation reactions of Fe/Fe(II) and Fe(II)/Fe(III). The discharge capacities of all samples with various amounts of gold nanoparticle solutions are larger than that of pure Fe_2O_3 electrode, increasing with repeated cycling and then gradually decreasing. This suggests that the presence of nanogold additives in the electrode supported the dissolution of iron to Fe (II) and Fe(II) to Fe(III). However, this effect depends on the nanogold amount. For example, discharge capacities of Fe_2O_3/Au/AB electrodes with 5, 10 ml nanogold solutions were larger than those with 20 ml nanogold solutions. The gradual decrease of discharge capacity with repeated cycling is attributed to the formation of $Fe(OH)_2$ passive layer during the oxidation of the iron electrode. Comparison of discharge capacity results of all the samples in Fig. 8 reveals that the best performance is delivered with 10 ml nanogold.

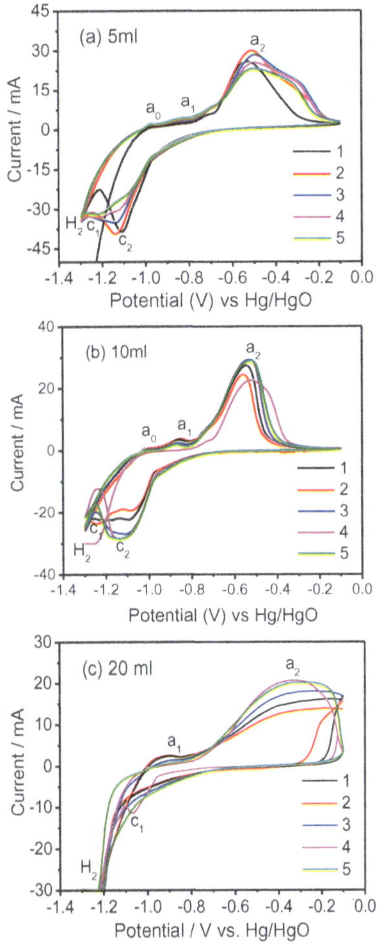

Fig. 7. Voltammogams of Fe_2O_3/Au/AB electrodes withvarious amounts of gold nanoparticle solutions: (a) 5 ml, (b) 10 ml and (c) 20 ml

Thus, in this experimental condition, 10 ml nanogold solution seems to be the most suitable for Fe_2O_3/Au/AB electrode. However, the discharge capacity still gradually decreases with further cycling and shall be reported in our next communication.

4 Conclusions

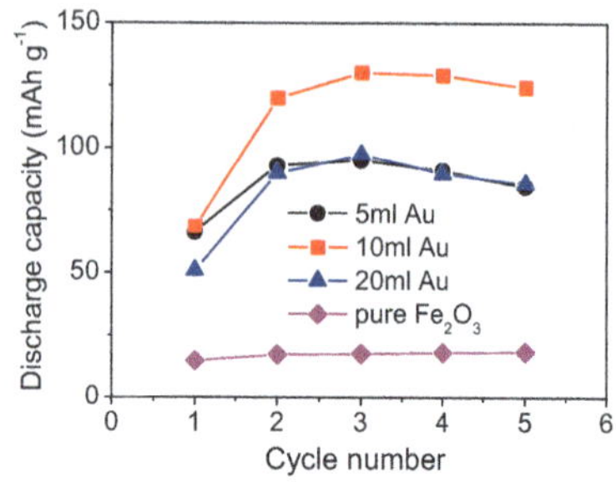

Fig. 8. Discharge capacity of Fe_2O_3/Au/AB electrodes with various amounts of gold nanoparticle solutions

The Fe_2O_3/Au and Fe_2O_3/Au/AB composite electrodes were prepared using gold nanoparticles and AB as electrode additives. The effect of nanogold and AB which have a grain size of few tens and hundred nanometers respectively were investigated by CV measurements. Both the gold nanoparticles and AB show positive effects on the electrochemical properties of Fe_2O_3 electrodes as evidenced by improved redox reaction rate and cyclability of iron oxide electrodes. In Fe_2O_3/Au mixture, gold nanoparticles were distributed on the iron oxide surface and thus the cyclability of the Fe_2O_3/Au electrode has been improved. For Fe_2O_3/Au/AB composite electrode, AB with high conductivity, large surface area supports the redox reaction of iron and consequently enhances high redox currents. The effect of gold nanoparticles depends on the amount of nanogold and 10 ml nanogold solution exhibits the highest performance.

Acknowledgment. This research is funded by the Hanoi University of Science and Technology (HUST) under project number T2020-SAHEP-018.

References

1. Koumoulos, E.P., et al.: Metrology and nano-mechanical tests for nano-manufacturing and nano-bio interface: challenges & future perspectives. Mater. Des. **137**, 446–462 (2018)
2. Sadeghnejad, M., Shafabakhsh, G.: Use of Nano SiO_2 and Nano TiO_2 to improve the mechanical behaviour of stone mastic asphalt mixtures. Constr. Build. Mater. **157**, 965–974 (2017)
3. Zhu, C.-S., Fang, X.-Q., Liu, J.-X.: Surface energy effect on buckling behavior of the functionally graded nano-shell covered with piezoelectric nano-layers under torque. Int. J. Mech. Sci. **133**, 662–673 (2017)
4. Ge, X., et al.: Nanomaterial-enhanced paper-based biosensors. TrAC Trends Anal. Chem. **58**, 31–39 (2014)
5. Ghosh, P., Han, G., De, M., Kim, C., Rotello, V.M.: Gold nanoparticles in delivery applications. Adv. Drug Deliv. Rev. **60**(11), 1307–1315 (2008)
6. Haume, K., et al.: Gold nanoparticles for cancer radiotherapy: a review. Cancer Nanotechnol. **7**(1), 1–20 (2016). https://doi.org/10.1186/s12645-016-0021-x
7. Hu, C., Niestroj, M., Yuan, D., Chang, S., Cheng, J.: Treating cancer stem cells and cancer metastasis using glucose-coated gold nanoparticles. Int. J. Nanomed. **10**, 2065–2077 (2015)
8. Dimitriou, N.M., et al.: Gold nanoparticles, radiations and the immune system: current insights into the physical mechanisms and the biological interactions of this new alliance towards cancer therapy. Pharmacol. Ther. **178**, 1–17 (2017)
9. Hang, B.T., Anh, T.T., Thang, D.H.: Effect of Fe_2O_3 morphology on the electrochemical properties of Fe_2O_3/C composite electrode as Fe-Air battery anode. J. Nanosci. Nanotechnol. **16**, 7999–8006 (2016)

Nonlinear Control of Axial Gap Magnetic Bearing Motors: A Disturbance Observer-Based Method

Tung Ngo Manh[1,2(✉)], Duc Thinh Le[2], Phuong Nguyen Huy[2], Dang Pham Quang[2], Dich Nguyen Quang[2], and Tung Lam Nguyen[2(✉)]

[1] Hanoi University of Industry, Hanoi, Vietnam
tung_nm@haui.edu.vn
[2] Hanoi University of Science and Technology, Hanoi, Vietnam
lam.nguyentung@hust.edu.vn

Abstract. The paper proposes a control strategy using integral backstepping control for an axial flux permanent motor. Filters are integrated into each control design step to eliminate tracking errors and undesirable effects. Moreover, the controller needs to reduce the effect of load and disturbances, and the high-gain observer is performed to estimate the uncertainty components of the control system. As a result, the speed trajectory can be more effectively held at high speed and medium speed. Finally, the simulation results show that the proposed control system can obtain the prescribed output tracking transient performance and robustness for the system uncertainties.

Keywords: Axial gap magnetic bearing · Dynamic surface · High-gain observer · Backstepping control

1 Introduction

In recent years, motors with integrated magnetic bearings have been increasingly studied and applied due to their advantages compared to traditional ball bearings. The magnetic bearing motor operates with noncontact rotor levitation to run at a very high-speed range with no friction, no lubrication, and contamination-free. Compared with conventional motors, the structure of magnetic bearing motors is larger in size and heavier in weight, so they are not widely applied in industry. Therefore, a form of axial flux permanent motor, as shown in Fig. 1, is studied and overcome these limitations, which is the combination of a disc motor and axial magnetic bearing. It is simple in structure and control than conventional magnetic bearing-motor since the hardware component can be reduced [1]. The motor rotor is arranged with permanent magnets, and it is placed between two stators. When operating, the motor air gap between the stator and the rotor is constantly varying, affecting the motor parameters and increasing the nonlinearity of the system. To deal with this problem, the control process needs to separate the coupling between the generating torque and the force acting on the axial

D. C. Nguyen et al. (Eds.): ICERA 2021, LNNS 366, pp. 675–684, 2022.
https://doi.org/10.1007/978-3-030-92574-1_70

position. Hence, based on Flux Oriented Control (FOC), the coupling problem can be solved via controlling flux forming and torque generating current [2–4].

Sliding mode control approach has been extensively studied owing to its robustness against model uncertainties, parameter perturbations and external disturbances [5, 6]. The dominant drawback of sliding mode control is the chattering problem which restricts its application. Robust velocity control of PMSM using integral sliding mode control was proposed in [10] to reduce the chattering phenomenon. Nevertheless, designing a specific sliding function is a difficult task. Backstepping control is introduced as a recursive and systematic design methodology for nonlinear systems, particularly for the systems with matched uncertainties [7–9]. To improve backstepping control, an integral action is integrated into the design. The added integrator improves the system's robustness against modeling uncertainties and external disturbance, thus improving steady-state control accuracy. However, the use of backstepping is problematic because of the "explosion of terms," while sliding mode control cannot generally be used for a system with mismatched uncertainties [10].

In order to avoid the drawback of both backstepping and sliding mode controls, a nonlinear control technique called dynamic surface control (DSC) has been developed [11, 12]. The DSC method is composed of multi sliding surfaces and a series of first-order low-pass filters to avoid the mathematical difficulties of an "explosion of terms" [13, 14]. Moreover, the existence of DSC gains and filter time constants for semi-global stability was theoretically proved, and it was extended to adaptive dynamic surface control to consider a specific class of parametric uncertainty [15]. Several applications of DCS in electrical drive control systems can be found in [10, 16].

During system operation, there are always disturbances and uncertainty components. Therefore, the paper proposes a high-gain disturbance observer (HGDOB) to compensate for these disturbance components for the system. The feasibility of the proposed methods is demonstrated through simulation of the control system. Simulation results show that both position response and speed response follow the set trajectory. The system is also stable to the coupling effects of the two control loops.

2 Mathematical Model

When presenting the mathematical model of the motor, it is important to note that the permanent magnet arrangement on the rotor affects the inductance of the stator winding. The paper considers the salient-pole motor with stator inductance as a function of the rotor angle position and air gap between the stator and the rotor. This inductance is inversely proportional to the air gap and it is presented by the following approximation formula [1]:

$$L_{sd} = 3L'_{sd0}/2g + L_{sl}; L_{sq} = 3L'_{sq0}/2g + L_{sl} \tag{1}$$

In which, L'_{sd0} and L'_{sq0} are the d-axis and q-axis inductance per gap unit; L_{sl} denotes leakage inductance. Clearance between stator and rotor g is given:

$$g = g_0 \pm z \tag{2}$$

where g_0 and z is clearance at equilibrium position and displacement from the equilibrium position. The voltage vector equation of a stator on the $\alpha\beta$ coordinate system is described as follows:

$$U_s = i_s R_s + L_s di_s/dt + E_s \tag{3}$$

where i_s denote the stator current vector, U_s is the stator voltage vector. The electromotive force induction vector is shown as follows:

$$E_s = d\lambda_m/dt \tag{4}$$

where λ_m is the magnitude of the flux linkage between the rotor and the stator, ω_e is the rotor speed, θ_e is the position of the rotor flux vector. Equation (3) is converted to dq coordinate system as follows:

$$\begin{cases} u_{sd} = R_s i_{sd} + L_{sd} di_{sd}/dt - \omega_e L_{sq} i_{sq} \\ u_{sq} = R_s i_{sq} + L_{sq} di_{sq}/dt + \omega_e L_{sd} i_{sd} + \omega_e \lambda_m - E_{s\beta}/L_s \\ \lambda_{sd} = L_{sd} i_{sd} + \lambda_m \\ \lambda_{sq} = L_{sq} i_{sq} \end{cases} \tag{5}$$

with λ_{sd}, λ_{sq} are the stator flux on dq coordinate. The current distribution principles of each stator are determined as [2, 17]:

$$\begin{cases} i_{sq1} = i_{sq2} = i_{sq} \\ i_{sd1} = i_{sd0} - i_{sd} \\ i_{sd2} = i_{sd0} + i_{sd} \end{cases} \tag{6}$$

where i_{sd1} và i_{sd2} are components of the axial current on the two stators that generate the attraction force F_1 và F_2. i_{sd0} is the offset current, and it has a minimal value or approximate zero. If displacement z is zero or very small compared to the air gap at the equilibrium point g_0, the equations for calculating the total torque and total axial force are given as follows [2, 4]:

$$F = 4K_{Fd} i_f i_{sd}; T = 2K_T i_{sq} \tag{7}$$

where K_{Fd} and K_T are the axial suction coefficient and the axial suction coefficient. i_f is regarded as the direct current generating a permanent magnet magnetic field. In fact, when the rotor rotates and moves axially, there is always the presence of load torque and applied axial load force. At the same time, the motor itself also has an uncertain component in the value of the inductance. Therefore, if the disturbances and uncertainty components can be estimated, the system control quality can be improved. The dynamic equations of the rotating motion and the axial motion of the rotor can be expressed as follows:

$$\begin{cases} m\ddot{z} = F - F_L + \Delta g_2 \\ \dot{\omega}_e = T - T_L + \Delta g_3 \end{cases} \tag{8}$$

From Eqs. (7) and (8), the dynamic equations are rewritten as follows:

$$\begin{cases} \ddot{z} = \underbrace{\frac{4K_{Fd}I_f}{m}}_{g_2} i_{sd} \underbrace{- \frac{F_L}{m} + \frac{\Delta g_2}{m}}_{d_2} \end{cases} \tag{9}$$

$$\begin{cases} \dot{\omega}_e = \underbrace{\frac{2K_T}{J}}_{g_3} i_q \underbrace{- \frac{T_L}{J} + \frac{\Delta g_3}{J}}_{d_3} \end{cases} \tag{10}$$

where m is the mass of the rotor, J is an inertial moment. d_2, d_3 are disturbances of the motor caused by load force, load torque, and uncertainties Δg_2 and Δg_3. These uncertainties can be defined as:

$$\Delta g_2 = 3\Delta L'_{sd_0} \Delta i_f i_{sd} / g_0^2; \Delta g_3 = \Delta J/J * d\omega_e/dt + 3p_c \Delta L'_{sd_0} \Delta i_f i_{sq} / g_0 J \tag{11}$$

where $\Delta J = J_a - J, \Delta L'_{sd0} = L'_{sd0,a} - L'_{sd0}, \Delta i_f = i_{f,a} - i_f$, in which J, L'_{sd0} and i_f represent the nominal parameter values, $J_a, L'_{sd0,a}$ and $i_{f,a}$ represent actual parameter values.

3 High-Gain Disturbance Observer Design

When operating an axial gap PMSM control system, disturbances as load force, load torque, and uncertain parameters affect the quality and robustness of the controller. Thus the disturbance observer is proposed to tackle the above problems. Define the state variables $z = x_1, \dot{z} = x_2, \omega_e = x_3$, Eqs. (9) and (10) are rewritten as:

$$\begin{cases} \dot{x}_1 = x_2 \\ \dot{x}_2 = g_2 i_{sd} + d_2 \\ \dot{x}_3 = g_3 i_{sq} + d_3 \end{cases} \tag{12}$$

Let $\hat{d}_2$ and $\hat{d}_3$ are the estimated disturbances. The derivative of $\hat{d}_2$ and $\hat{d}_3$ according to the high-gain observer as in [18] are written as follows:

$$\dot{\hat{d}}_2 = (\dot{x}_2 - g_2 i_{sd} - \hat{d}_2)/\varepsilon_2; \dot{\hat{d}}_3 = (\dot{x}_3 - g_3 i_{sq} - \hat{d}_3)/\varepsilon_3 \tag{13}$$

where $1/\varepsilon_2$ and $1/\varepsilon_3$ are the observer gains. Define auxiliary state variables ξ_2 and ξ_3:

$$\xi_2 = \hat{d}_2 - x_2/\varepsilon_2; \xi_3 = \hat{d}_3 - x_3/\varepsilon_3 \tag{14}$$

Equation (13) is rewritten as follows:

$$\begin{aligned} \dot{\xi}_2 &= -(\xi_2 + x_2/\varepsilon_2)/\varepsilon_2 - (g_2 i_{sd})/\varepsilon_2 \\ \dot{\xi}_3 &= -(\xi_3 + x_3/\varepsilon_3)/\varepsilon_3 - (g_3 i_{sq})/\varepsilon_3 \end{aligned} \tag{15}$$

The following disturbance estimation error dynamics is obtained as:

$$\dot{\tilde{d}}_i = -\tilde{d}_i/\varepsilon_i + \dot{d}_i, i = [2, 3] \tag{16}$$

So $\left|\tilde{d}_i\right| \leq e^{-\left(\frac{1}{\varepsilon_i}\right)t}\left|\tilde{d}_i(0)\right| + \varepsilon_i \rho_i(t)$ and the upper bound of $\left|\tilde{d}_i(\infty)\right|$ becomes smaller as ε_i gets smaller.

4 Integral Backstepping Controller Design

4.1 Axial Displacement Control

The radial displacement of the rotor is kept stable by the radial magnetic bearing. Thus the axial deviation is independent of the radial displacement and is calculated as follows:

$$\begin{cases} \dot{x}_1 = x_2 \\ \dot{x}_2 = g_2 i_{sd} + d_2 \end{cases} \tag{17}$$

Step 1: Define the boundary layer errors as: $y_1 = x_1^* - x_{1d}$. In the first step, the axial displacement is forced to track its reference. The axial displacement tracking error e_1 defined as $e_1 = x_1^* - x_1$. With x_{1d} is a new state variable through the first-order filter:

$$x_1^* = \lambda_1 \dot{x}_{1d} + x_{1d} \tag{18}$$

with $\lambda_1 > 0$. A position error signal is defined as:

$$e_1 = x_{1d} - x_1 + k_{i1}\int_0^t (x_{1d} - x_1)dt \tag{19}$$

where k_{i1} is an integral action constant. The derivation of Eq. (19) to time yields:

$$\dot{e}_1 = \dot{x}_{1d} - \dot{x}_1 + k_{i1}(x_{1d} - x_1) \tag{20}$$

Lyapunov candidate function V_1 is defined as:

$$V_1 = (e_1^2 + y_1^2)/2 \tag{21}$$

Then the time derivative of V_1 is:

$$\dot{V}_1 = e_1\dot{e}_1 + y_1\dot{y}_1 = e_1(\dot{x}_{1d} - x_2 + k_{i1}(x_{1d} - x_1)) + y_1(\dot{x}_1^* - y_1/\lambda_1)$$

The control law is chosen as:

$$x_2^* = \dot{x}_{1d} + k_{i1}(x_{1d} - x_1) + k_1 e_1 \tag{22}$$

where k_1 and k_{i1} are positive constants and must be satisfied $k_1 > k_{i1}$.

Step 2: Define the error:$y_2 = x_2^* - x_{2d}$. The same as Step 1, x_{2d} also is a new state variable through the first-order filter as:

$$x_2^* = \lambda_2\dot{x}_{2d} + x_{2d} \tag{23}$$

where the constant $\lambda_2 > 0$. The error signal is defined:

$$e_2 = x_{2d} - x_2 + k_{i2}\int_0^t (x_{2d} - x_2)dt \tag{24}$$

where k_{i2} is an integral action constant, the derivative of Eq. (24)

$$\dot{e}_2 = \dot{x}_{2d} - \dot{x}_2 + k_{i2}(x_{2d} - x_2) \tag{25}$$

with $k_2 > k_{i2} > 0$. The Lyapunov candidate function is given as:

$$V_2 = V_1 + e_2^2/2 + \tilde{d}_2^2/2 + y_2^2/2 \tag{26}$$

To ensure its negativity and the stability of the displacement loop, the i_{sd}^* virtual control law is chosen as:

$$i_{sd}^* = \left[\dot{x}_{2d} + k_{i2}(x_{2d} - x_2) + k_2 e_2 - \hat{d}_2\right]/g_2 \tag{27}$$

4.2 Speed Control

Define the boundary layer error as $y_3 = x_3^* - x_{3d}$, where x_3^* is the desired rotor speed. A new state variable x_{3d} is obtained through the filter, the first-order filter is defined as follows:

$$x_3^* = \lambda_3\dot{x}_{3d} + x_{3d} \tag{28}$$

with constant $\lambda_3 > 0$. The speed error signal is defined as follows:

$$e_3 = x_{3d} - x_3 + k_{i3} \int_0^t (x_{3d} - x_3) dt \tag{29}$$

where k_{i3} is an integral action constant. The derivative of Eq. (29) gives:

$$\dot{e}_3 = \dot{x}_{3d} - \dot{x}_3 + k_{i3}(x_{3d} - x_3) = -k_3 e_3 \tag{30}$$

with $k_3 > k_{i3} > 0$. The Lyapunov candidate function is chosen to prove the stability of the speed loop as:

$$V_3 = V_2 + (e_3^2 + \tilde{d}_3^2)/2 \tag{31}$$

To ensure its negativity and then the stability of the current loop, the control law i_{sq}^* is chosen as:

$$i_{sq}^* = (\dot{x}_{3d} + k_{i3}(x_{3d} - x_3) + k_3 e_3 - \hat{d}_3)/g_3 \tag{32}$$

The control law i_{sq}^* ensures that the current x_3 tracks the virtual input x_{3d}. This virtual input is computed to assure speed tracking.

5 Simulation and Result

The system structure parameters are given: phase resistance is 2.6 Ω; the rotor mass is 0.28 kg; the moment of inertia is 10.6 × 10−6 kgm^2; the amplitude of the flux generated by the permanent magnet is λ_m = 0.022 Wb; inductance value L_{s0}' = 11 × 10−6 Hm; leakage inductance L_{sl} = 5 × 10−3 H; Air gap between stator and rotor is 1.2 mm.

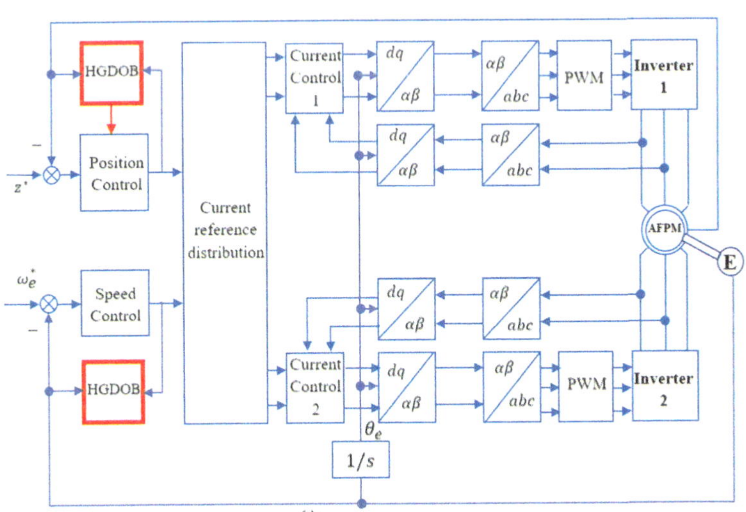

Fig. 1. Block diagram of vector control

To evaluate the quality of the observer, the simulation process assumes that the load torque $T_L = 1\ N$ is put into the system at time 1 s, the uncertainty error is 10%. Assume that initial position $z = 0.5$ mm, the uncertainty parameter varies in the range 10%, example equilibrium position $g'_0 = g_0(1 + 10\%.rand())$, the load torque and *disturbance* of torque are described in Fig. 4 with $T_L = 0{,}1\ N$, $\Delta L = 0.01\ sin(20t)$, the load axial force and *disturbance* of axial force are described in Fig. 5, where $F_L = 5\ N$, $\Delta F_L = 0.5\ sin(20t)$. Assuming that at times 6 s and 14 s, appears the axial load acting on the system and at times 2 s and 10 s, the system is subjected to a load torque.

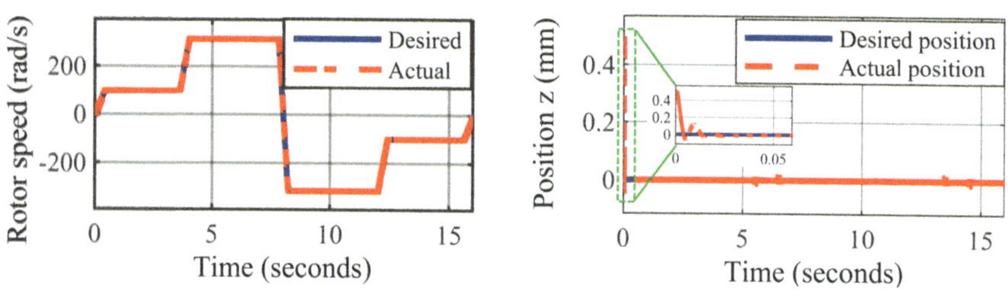

(a) (b)

Fig. 2. The speed rotor and axial position response

Figure 2(a) shows the response of the motor speed when the setpoint is from 100 *rad/s* to 300 *rad/s*. The output speed value after a time interval of 0.025 s follows the set trajectory with a small overshoot. Even when reversing the motor rotation, the actual speed closely follows the set value. At a time like 2 s, 6 s, 10 s and 14 s, the disturbance and load appear, but these components are not affected much. From Fig. 2 (b), we can see the axial position response to the initial deflection after a very short time interval of 0.005 s and returning to the equilibrium position. When axial load disturbance is applied, the position oscillation is minimal and quickly stabilizes in the equilibrium position. Thus, the axial position and speed controllers both show the ability to adjust according to the set trajectory and eliminate the influence of disturbances.

The axial force and load torque disturbance responses are shown in Fig. 3(a, b). The high-gain observer performs the estimation of disturbance, and the uncertainty components are sinusoidal functions. The results show that the estimated value closely follows the real disturbance value, the estimation error is not significant. These values are then sent to the controller to compensate for disturbance to improve system accuracy.

Figure 4 shows the response of the i_{sd} and i_{sq} reference current values. Figure 4(a) shows that when a load is applied, the axial position controller calculates an additional amount of current i_{sd} to counteract the interference. Similar to Fig. 4(b), when a load torque occurs, the speed controller also calculates an amount of current i_{sq} to compensate for the load torque. At the same time, the graph also shows that the coupling effect between the two loops of axial position and speed has been significantly limited.

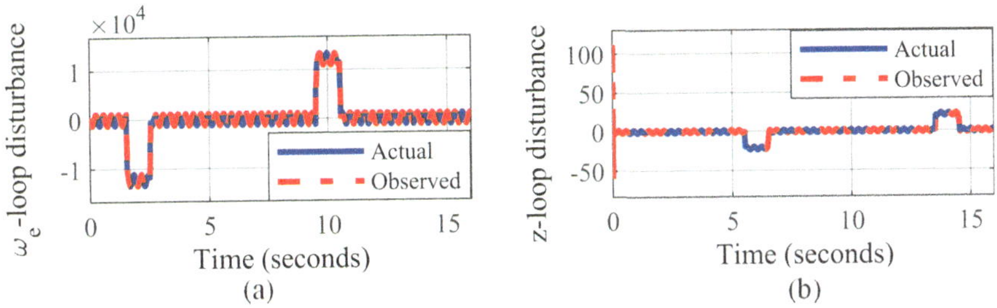

Fig. 3. The observer disturbance response of the position and the speed loop

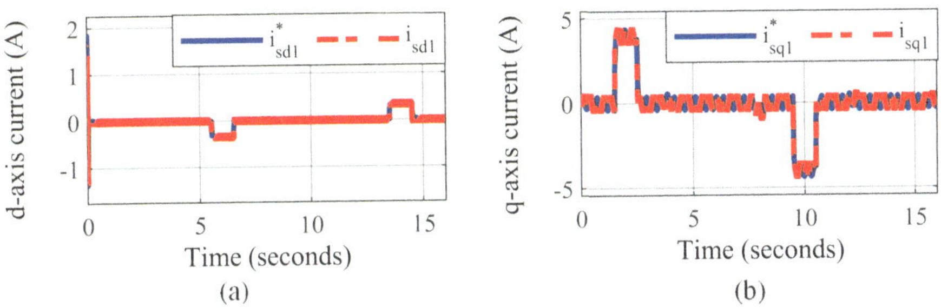

Fig. 4. Responses of *dq* currents

6 Conclusion

This paper investigates a combined controller based on a robust backstepping control method and a high-gain observer for estimated disturbance. An improved backstepping method included adding integral actions and using the first-order filter at each step of the algorithm, bring a result to ensure the rotor speed and the axial position to track the trajectories references efficiently. The lump disturbance including the uncertainties parameter, load torque, and axial load force, is considered in the backstepping controller. To improve the robustness, a nonlinear high-gain disturbance observer method is estimated for the lump disturbance in the system, which is used for the feed-forward compensation control. Simulation and experiment results prove that the designed control system has excellent transient speed response and strong robustness.

Acknowledgement. This research is funded by Hanoi University of Science and Technology (HUST) under project number T2021-PC-001.

References

1. Quang, D.N., Satoshi, U.: Salient pole permanent magnet axial-gap self-bearing motor. Magn. Bear. Theory Appl. 61–83 (2010). https://doi.org/10.5772/intechopen.83966

2. Nguyen, Q., Ueno, S.: Salient pole permanent magnet (2009). https://doi.org/10.5772/intechopen.83966
3. Nguyen, Q.D., Ueno, S.: Analysis and control of nonsalient permanent magnet axial gap self-bearing motor. IEEE Trans. Ind. Electron. **58**(7), 2644–2652 (2011). https://doi.org/10.1109/TIE.2010.2076309
4. Nguyen, Q.D., Ueno, V.: Axial position and speed vector control of the inset permanent magnet axial gap type self bearing motor. IEEE/ASME Int. Conf. Adv. Intell. Mechatron. AIM, 130–135 (2009). https://doi.org/10.1109/AIM.2009.5230025
5. Wang, H., Li, S., Zhao, Z.: Design and implementation of chattering free sliding mode control method for PMSM speed regulation system. In: Proceedings of the IEEE Int. Conf. Ind. Technol., vol. 2015-June, no. June, pp. 2069–2074 (2015). https://doi.org/10.1109/ICIT.2015.7125401
6. Feng, L., Deng, M., Xu, S., Huang, D.: Speed regulation for PMSM drives based on a novel sliding mode controller. IEEE Access **8**, 63577–63584 (2020). https://doi.org/10.1109/ACCESS.2020.2983898
7. Karabacak, M., Eskikurt, H.I.: Design, modelling and simulation of a new nonlinear and full adaptive backstepping speed tracking controller for uncertain PMSM. Appl. Math. Model. **36**(11), 5199–5213 (2012). https://doi.org/10.1016/j.apm.2011.12.048
8. Xue, W., Zhang, M., Wang, R.: Synchronization of fractional-order PMSM chaotic model based on backstepping control method. In: Proceedings of the 2017 9th Int. Conf. Model. Identif. Control. ICMIC 2017, vol. 2018-March, no. Icmic, pp. 647–651 (2018). https://doi.org/10.1109/ICMIC.2017.8321535
9. Yu, Y.J., Fang, J.C., Xiang, B., Wang, C.E.: Adaptive backstepping tracking control for rotor shaft tilting of active magnetically suspended momentum wheel. ISA Trans. **53**(6), 1892–1900 (2014). https://doi.org/10.1016/j.isatra.2014.07.002
10. Swaroop, D., Hedrick, J.K., Yip, P.P., Gerdes, J.C.: Dynamic surface control for a class of nonlinear systems. IEEE Trans. Automat. Contr. **45**(10), 1893–1899 (2000). https://doi.org/10.1109/TAC.2000.880994
11. Wang, C., Lin, Y.: Robust adaptive dynamic surface control for a class of MIMO nonlinear systems with unknown non-symmetric dead-zone. Asian J. Control **16**(2), 478–488 (2014). https://doi.org/10.1002/asjc.708
12. He, R., Han, Q.: Dynamics and stability of permanent-magnet synchronous motor. Math. Probl. Eng. **2017**, 1–8 (2017). https://doi.org/10.1155/2017/4923987
13. Lan, Y.-H., Zhou, L.: Backstepping control with disturbance observer for permanent magnet synchronous motor. J. Control Sci. Eng. **2018**, 1–8 (2018). https://doi.org/10.1155/2018/4938389
14. Xia, P., Deng, Y., Wang, Z., Li, H.: Speed adaptive sliding mode control with an extended state observer for permanent magnet synchronous motor. Math. Probl. Eng. **2018**, 1–13 (2018). https://doi.org/10.1155/2018/6405923
15. Hedrick, B.S.J.K.: Dynamic surface control of uncertain nonlinear systems
16. Li, G., Xu, W., Zhao, J., Wang, S., Li, B.: Precise robust adaptive dynamic surface control of permanent magnet synchronous motor based on extended state observer. IET Sci. Meas. Technol. **11**(5), 590–599 (2017). https://doi.org/10.1049/iet-smt.2016.0252
17. Nguyen, Q.D., Ueno, S.: Sensorless speed control of inset type axial gap self-bearing motor using extended EMF. In: 2010 Int. Power Electron. Conf. – ECCE Asia, pp. 2260–2264. IPEC (2010). https://doi.org/10.1109/IPEC.2010.5542012
18. Won, D., Kim, W., Shin, D., Chung, C.C.: High-gain disturbance observer-based backstepping control with output tracking error constraint for electro-hydraulic systems. IEEE Trans. Control Syst. Technol. **23**(2), 787–795 (2015). https://doi.org/10.1109/TCST.2014.2325895

NURBS Curve Trajectory Tracking Control for Differential-Drive Mobile Robot by a Linear State Feedback Controller

Nguyen Hong Thai[1(✉)] and Trinh Thi Khanh Ly[2]

[1] Department of Mechanical Design and Robotics, School of Mechanical Engineering, Hanoi University of Science and Technology, Hanoi, Vietnam
thai.nguyenhong@hust.edu.vn

[2] Faculty of Automation Technology, Electric Power University, Hanoi, Vietnam

Abstract. This paper presents a method to design a linear state feedback controller with time-varying parameters for the differential-drive mobile robot to track a NURBS curve trajectory. First, the linear state feedback controller is designed based on a kinematic error model of the robot. Then, determine the time-varying parameters of the controller according to the varying angular velocity for the desired trajectory. A robot model has been experimental designed and manufactured for experimental verification of the proposed controller. The simulation and experimental results show that the kinematic controller developed by this study has high-performance trajectory tracking with minor tracking errors. Therefore, it shows can apply that the results of this research in many different fields.

Keywords: Feedback controller · Mobile robot · Trajectory tracking · NURBS curve · Kinematic error model

1 Introduction

In recent years, wheeled mobile robots (WMR) play an increasingly crucial role in modern industries [1]. Meanwhile, the differential drives mobile robot (DDMR) is a popular choice due to its simplicity of construction and ease of control [2, 3]. Improving accuracy, including position and orientation, is always a challenge, so it is a topic that has been and is still being interested in many researchers. Research in this area includes: Hossein [4] developed optimization-based nonlinear control laws using the prediction of DDMR responses to kinematic and dynamic trajectory tracking control; Laurens et al. [5] proposes a linear-parameter varying (LPV) controller design approach for trajectory tracking of DDMR based on a kinematic model; Haoyu et al. [6] proposes a control strategy integrating the non-linear extended state observer and the non-singular terminal sliding mode control for the trajectory tracking of DDMR subject to bounded disturbances; Javier et al. [7] proposed an output dynamic feedback control for trajectory tracking of WMR to overcome the lack of attitude and Cartesian velocity measurements. Also, there are some other similar studies [8–14] on trajectory

D. C. Nguyen et al. (Eds.): ICERA 2021, LNNS 366, pp. 685–696, 2022.
https://doi.org/10.1007/978-3-030-92574-1_71

tracking of WMR with simple straight and curved trajectories. Regarding the robot's trajectory, one problem is that it is necessary to smooth the bends to avoid sudden changes in the robot's speed. Research in this area currently has the following methods: Dubins [16] using arcs to fillet the folded segments; use B-spline to path planning [20]; Using the NURBS curve to trajectory design control of robotic manipulator drilling blasting holes in tunnel construction [15]; Or use the NURBS curve to design trajectories for AGVs in production workshops [19] etc.

This study states the trajectory tracking problem and constructs the linear state feedback controllers with time-varying parameters for DDMR according to the NURBS curve trajectory. The linearized kinematic model is used to develop a tracking linear controller with a pole placement approach. Verified the results under simulations and experiments of differential drive mobile robot platforms.

2 Kinematic Error Model of Differential Drive Mobile Robot

In the fixed coordinate system $\vartheta_f\{O_f x_f y_f\}$ associated with the working environment, consider the motion of a mobile robot that follows a given desired trajectoryξ_d. In practice, the robot's position and posture during move often deviate from the desired trajectory, as depicted in Fig. 1. This error is described as the deviation between a virtual reference robot and the actual robot. It is expressed as follows: e_x that gives the error in the direction of driving, e_y that gives the error in the perpendicular direction, e_φ that gives the error in the orientation. Thus, error vector e defined as:

$$\mathbf{e}(t) = \begin{bmatrix} e_x & e_y & e_\phi \end{bmatrix}^T = \mathbf{q}_d(t) - \mathbf{q}(t) \tag{1}$$

Wherein: $\mathbf{q}_d(t) = \begin{bmatrix} x_d(t) & y_d(t) & \phi_d(t) \end{bmatrix}^T$ is the virtual reference posture of the robot determined in the coordinate system $\vartheta_f\{O_f x_f y_f\}$ and $\mathbf{q}(t) = \begin{bmatrix} x(t) & y(t) & \phi(t) \end{bmatrix}^T$ is the actual posture of the robot during real movement.

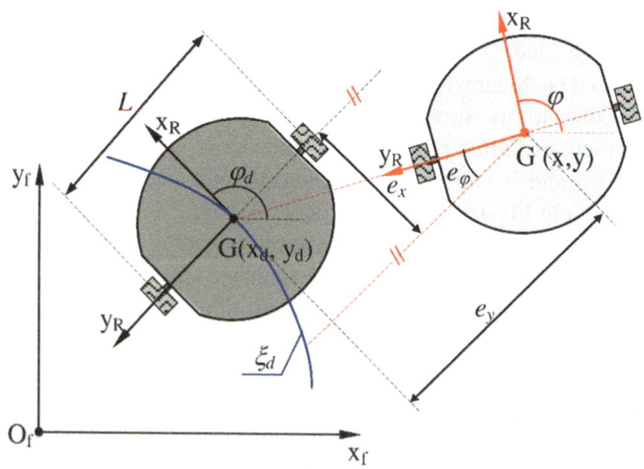

Fig. 1 Schematic of determining the position error of a mobile robot in local coordinates.

Set $\vartheta_R\{Gx_Ry_R\}$ is the coordinate system mounted on the robot, transform $\mathbf{e}(t)$ from ϑ_f to ϑ_R of the robot:

$$\mathbf{e}_R(t) = \mathbf{Q}(\phi(t))^T\,\mathbf{e}(t) \tag{2}$$

Where: $\mathbf{Q}(\phi(t)) = \begin{bmatrix} \cos\phi(t) & -\sin\phi(t) & 0 \\ \sin\phi(t) & \cos\phi(t) & 0 \\ 0 & 0 & 1 \end{bmatrix}$

Take the derivative Eq. (2) according to t:

$$\dot{\mathbf{e}}_R(t) = \mathbf{Q}(\phi(t))^T\,(\dot{\mathbf{q}}_d(t) - \dot{\mathbf{q}}(t)) + \dot{\mathbf{Q}}(\phi(t))^T\,\mathbf{Q}(\phi(t))\,\mathbf{e}_R(t) \tag{3}$$

Where: $\dot{\mathbf{q}}(t) = \begin{bmatrix} \dot{x}(t) \\ \dot{y}(t) \\ \dot{\phi}(t) \end{bmatrix} = \begin{bmatrix} V\cos\phi(t) \\ V\sin\phi(t) \\ \Omega(t) \end{bmatrix} = \begin{bmatrix} \cos\phi(t) & 0 \\ \sin\phi(t) & 0 \\ 0 & 1 \end{bmatrix} \begin{bmatrix} V(t) \\ \Omega(t) \end{bmatrix}$ and $\dot{\mathbf{q}}_d(t) = \begin{bmatrix} \dot{x}(t) \\ \dot{y}(t) \\ \dot{\phi}(t) \end{bmatrix} = \begin{bmatrix} V_d\cos\phi_d(t) \\ V_d\sin\phi_d(t) \\ \Omega_d(t) \end{bmatrix} = \begin{bmatrix} \cos\phi_d(t) & 0 \\ \sin\phi_d(t) & 0 \\ 0 & 1 \end{bmatrix} \begin{bmatrix} V_d(t) \\ \Omega_d(t) \end{bmatrix}$ With $V_G(t)$ and $\Omega(t)$ are the velocities and angular velocity of the robot during real movement, while $V_{Gd}\,(t)$ and $\Omega_d(t)$ are the robot's desired velocity and angular velocity.

From Eq. (3), the posture error model can be written as follows:

$$\dot{\mathbf{e}}_R(t) = \begin{bmatrix} \cos e_\phi(t) & 0 \\ \sin e_\phi(t) & 0 \\ 0 & 1 \end{bmatrix} \begin{bmatrix} V_d(t) \\ \Omega_d(t) \end{bmatrix} + \begin{bmatrix} -1 & e_{yR} \\ 0 & -e_{xR} \\ 0 & -1 \end{bmatrix} \begin{bmatrix} V(t) \\ \Omega(t) \end{bmatrix} \tag{4}$$

Equation (4) is the kinematic error model of the differential drive mobile robot, which is nonlinear with the input signal $\mathbf{u} = [\,V(t) \quad \Omega(t)\,]^T$.

3 Design of Moving Trajectory of Robots by the NURBS Curve

Trajectory design for mobile robots has so far included many different methods, such as [16] using arcs to soften intersections; Manas et al. [17] use of Dubins path of optimum turning radius to plan the trajectory between two waypoints; in [18, 19] used the NURBS curve to generate trajectory planning for AGVs in production plants; in [20] use B-spline to path planning and trajectory planning simultaneously for omnidirectional mobile robots. Therefore, in this study, we use the NURBS curve to establish the desired trajectory planning ξ_d of the mobile robot in the coordinate system $\vartheta_f\{O_fx_fy_f\}$, thereby determining $\mathbf{q}_d(g) = [\,x_d(g) \quad y_d(g) \quad \phi_d(g)\,]^T$. Thus, according to [18] the non-uniform B – Spline rational curve (NURBS) is defined by:

$$\mathbf{P}(g) = \sum_{i=1}^{n+1} \mathbf{B}_i R_{i,k}(g) \tag{5}$$

Where $\mathbf{B}_i$ is the vertices of the interpolated polygons in 2D space, and $R_{i,k}(g)$ is the basis function of the B-Spline rational curve and is given by:

$$R_{i,k}(g) = \frac{h_i N_{i,k}(g)}{\sum_{i=1}^{n+1} h_i N_{i,k}(g)} \tag{6}$$

With $h_i \geq 0$ for all values of i, $N_{i,k}(g)$ given by the recursive formula:

$$\begin{cases} N_{i,1}(g) & = \begin{bmatrix} 1 & x_i \leq g \leq x_{i+1} \\ 0 & \end{bmatrix} \\ N_{i,k}(g) & = \frac{(g-x_i)N_{i,k-1}(g)}{x_{i+k-1}-x_i} + \frac{(x_{i+k}-t)N_{i+1,k-1}(g)}{x_{i+k}-x_{i+1}} \end{cases} \tag{7}$$

x_i is the value of the node vector and satisfies the condition $x_i \leq x_{i+1}$. The tangent to the curve is determined by:

$$\tau = \dot{\mathrm{P}}(g) = \sum_{i=1}^{n+1} \mathrm{B}_i \left[\frac{h_i \dot{N}_{i,k}(g)}{\sum_{i=1}^{n+1} h_i N_{i,k}(g)} - \frac{h_i N_{i,k}(g) \sum_{i=1}^{n+1} h_i \dot{N}_{i,k}(g)}{\left(\sum_{i=1}^{n+1} h_i N_{i,k}(g)\right)^2} \right] \tag{8}$$

Thus, at each coordinate point $G_d(x_d(g), y_d(g))$ on the interpolate NURBS to generate desired trajectory planning ξ_d can we determine entirely the robot posture angle $\phi_d = \angle(\vec{\tau}(t), x_f)$.

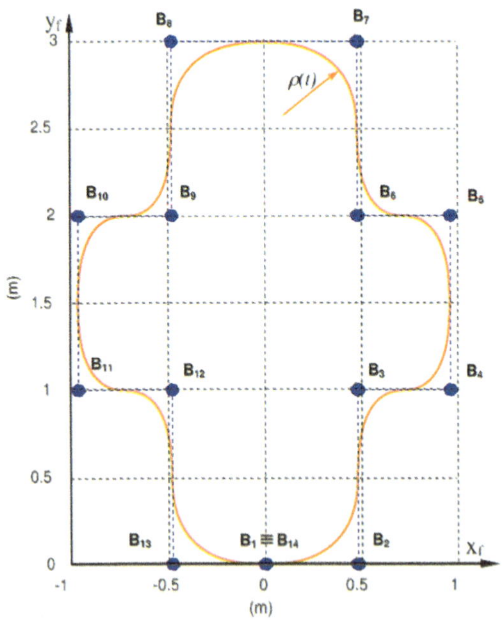

Fig. 2 The NURBS curve establishes desired trajectory planning.

Figure 2 is the desired trajectory planning for mobile robots interpolated by NURBS curves through the coordinate grid nodes given in Table 1.

Table 1. Control point grid of the NURBS trajectory.

B_1	B_2	B_3	B_4	B_5	B_6	B_7
$\begin{bmatrix}0\\0\end{bmatrix}$	$\begin{bmatrix}0.48\\0\end{bmatrix}$	$\begin{bmatrix}0.48\\1\end{bmatrix}$	$\begin{bmatrix}0.96\\1\end{bmatrix}$	$\begin{bmatrix}0.96\\2\end{bmatrix}$	$\begin{bmatrix}0.48\\2\end{bmatrix}$	$\begin{bmatrix}0.48\\3\end{bmatrix}$
B_8	B_9	B_{10}	B_{11}	B_{12}	B_{13}	B_{14}
$\begin{bmatrix}-0.48\\3\end{bmatrix}$	$\begin{bmatrix}-0.48\\2\end{bmatrix}$	$\begin{bmatrix}-0.96\\2\end{bmatrix}$	$\begin{bmatrix}-0.96\\1\end{bmatrix}$	$\begin{bmatrix}-0.48\\1\end{bmatrix}$	$\begin{bmatrix}-0.48\\0\end{bmatrix}$	$\begin{bmatrix}0\\0\end{bmatrix}$

4 Development of a Linear State Feedback Controller

Apply a first-order Taylor expansion that approximates Eq. (4) to a linear equation at the neighborhood of the reference point (xd, yd) around the zero-error are obtained:

$$\begin{bmatrix}\dot{e}_x\\\dot{e}_y\\\dot{e}_\phi\end{bmatrix}=\begin{bmatrix}0&\Omega_d&0\\-\Omega_d&0&V_d\\0&0&0\end{bmatrix}\begin{bmatrix}e_x\\e_y\\e_\phi\end{bmatrix}+\begin{bmatrix}-1&0\\0&0\\0&-1\end{bmatrix}\begin{bmatrix}V\\\Omega\end{bmatrix}\tag{9}$$

Thus, the controller designed to have three states $x(t)$, $y(t)$, $\varphi(t)$ and two inputs $V(t)$, $\Omega\,(t)$. In the linear state space, the feedback control law has the form [21]:

$$\begin{bmatrix}V\\\Omega\end{bmatrix}=\begin{bmatrix}-k_1&0&0\\0&-sign(V_d)k_2&-k_3\end{bmatrix}\begin{bmatrix}e_x\\e_y\\e_\phi\end{bmatrix}\tag{10}$$

Where: k_1, k_2 and k_3 are the coefficients of the controller.

In this controller is a combination of feedforward and linear feedback control based on pole placement approach, so a new input is given by:

$$\begin{cases}V(t)=V_d-u_1\\\Omega(t)=\Omega_d-u_2\end{cases}\tag{11}$$

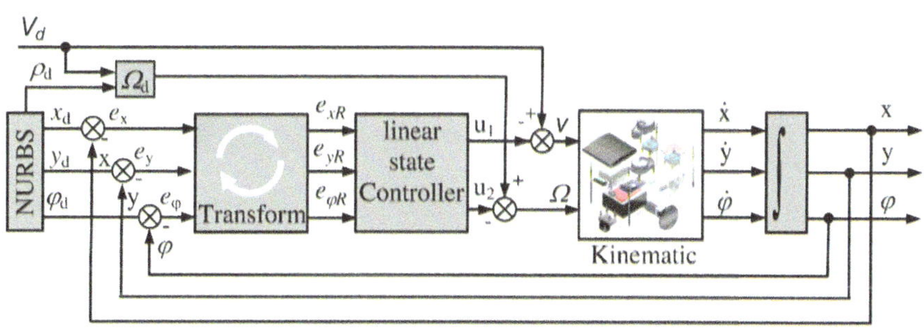

Fig. 3 The linear state feedback controller NURBS curve trajectory tracking.

From Eqs. (9) to (11) and the database established in Sect. 3, we have a block diagram of a linear state feedback controller tracking the NURBS curve trajectory as described in Fig. 3.

In the controller shown in Fig. 3, Ω_d (t) determined from the setting velocity V_d and $\rho(t)$ according to the following expression:

$$\Omega_d(t) = \frac{V_d}{\rho(t)} \tag{12}$$

Wherein: $\rho(t)$ is the radius of the desired path trajectory determined from the NURBS curve; V_d is the desired velocity.

According to [21] the coefficients k_1, k_2 and k_3 are determined by:

$$\begin{cases} k_1(t) = k_3(t) = 2\zeta\sqrt{\Omega_d^2(t) + gV_d^2(t)} \\ k_2(t) = gV_d(t) \end{cases} \tag{13}$$

With: $g > 0$, while $0 < \zeta < 1$.

5 Simulation and Experimental Results

Figure 4 is the robot platform designed and manufactured by this research conducted at the Department of Design of Machines and Robots, Hanoi University of Science and Technology. In which: (1) Shell material: is 5 mm thick mica to ensure lightness and rigidity; (2) Drive: Two differential drive wheels connected to integrated DC servo motor with integrated planetary gear reducer with gear ratio 1:14 and encoder 500CPR (2 channels A-B, 500 pulses); (3) Hardware: The central processor is an embedded computer Raspberry Pi 4 model b, control module Arduino ATmega 2560 R3; (4) Communication: connect between host and robot by module Wi-Fi WHR-G301N; (5) Robot control software is developed on ROS operating system.

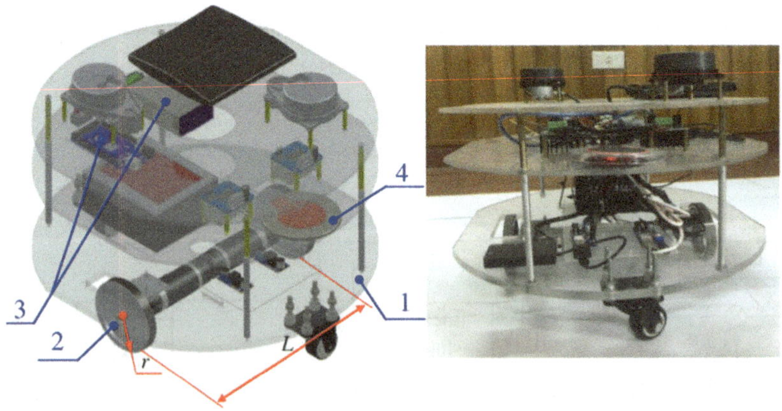

(a) (b)

Fig. 4 Structure of differential-drive mobile Robot with (a) Design on Solidworks software and (b) Robot prototype.

From the database of the path trajectory in Fig. 2 set up in Sect. 3 and the linear state controller set up in Sect. 4, along with the setting parameters:

(1) Dimensional of the robot with (i) Distance of two wheels with control $L = 295$ mm and (ii) Wheel radius $r = 47.5$ mm (see Fig. 4a).

(2) Controller settings: Set velocity $V_d = 0.8$ m/s at point G of the robot. From Vd, the desired trajectory data is determined in Sect. 3 combined with Eqs. (12) and (13), after surveying the minimum average error, the controller parameter is selected as follows: $\zeta = 0.9$, $g = 300$, $k_2 = 240$ and $k_1(t)$, $k_3(t)$ are time-varying parameters.

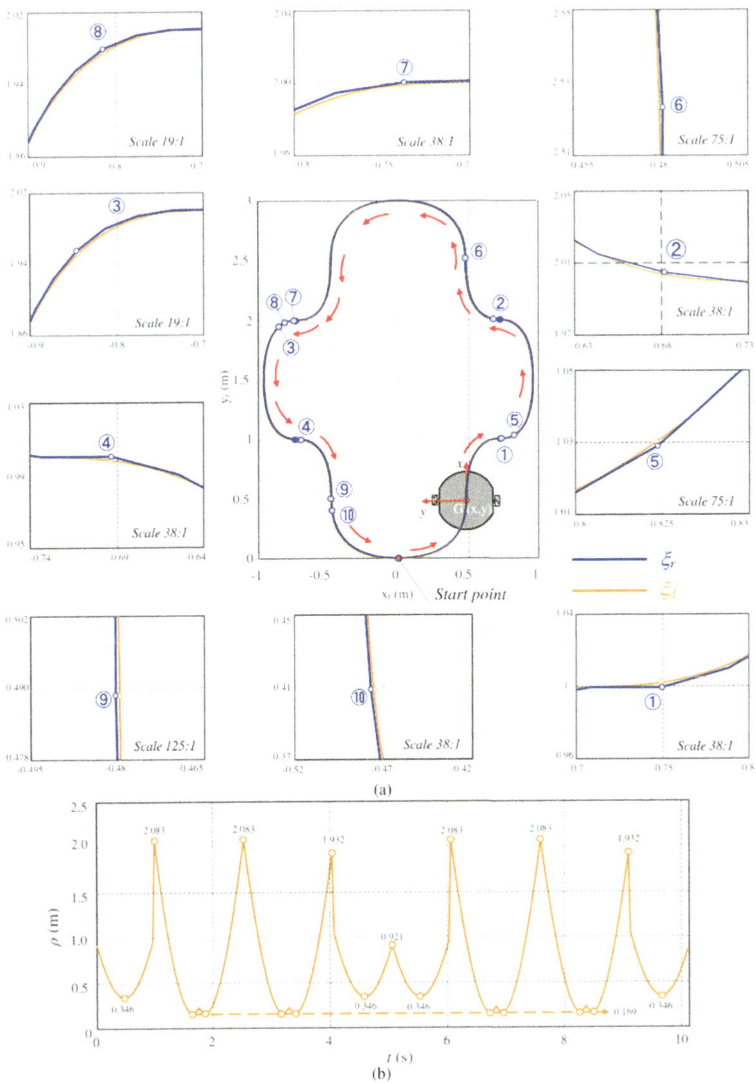

Fig. 5 The movement trajectory of the robot is simulated on Matlab's Simulink with (a) Actual and desired robot path trajectories and (b) radius of curvature of desired path trajectory.

With the settings as shown in Fig. 5, the simulation results of the robot's actual trajectory, while Fig. 6a is the moving speed of the robot, including the velocity V at G and the angular velocity Ω of the robot when the radius of curvature $\rho_d(t)$ of the desired trajectory ξ_d varies from 0.169 m to 2.83 m (see Fig. 5b).

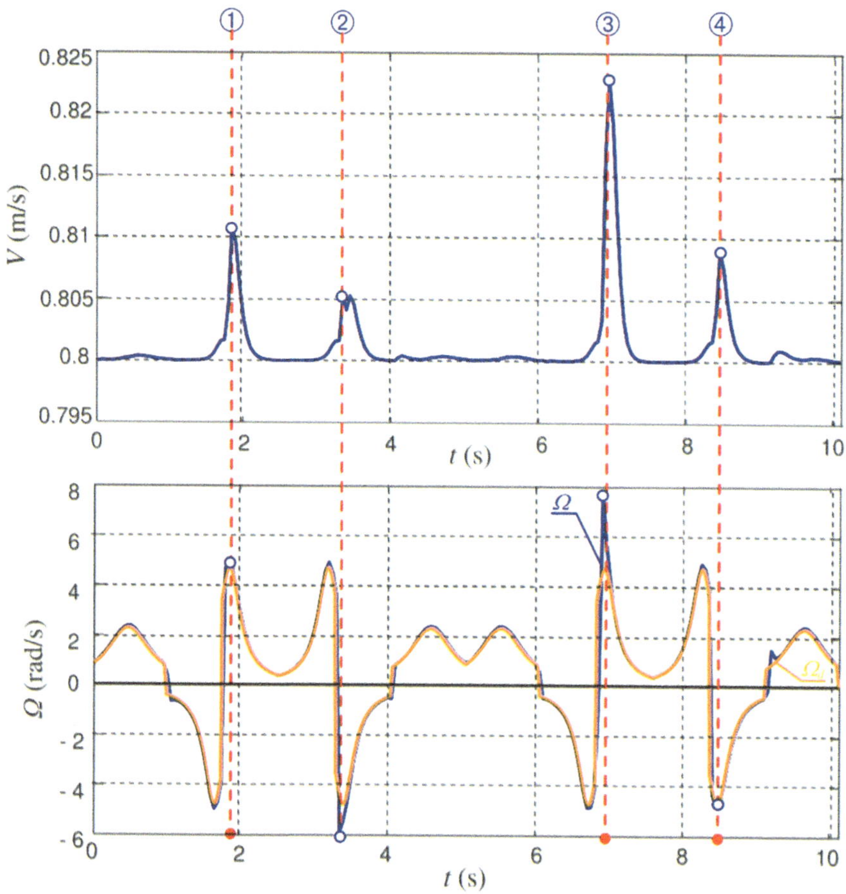

Fig. 6 The moving speed of the robot with (a) The moving speed of the robot and (b) The actual and desired angular velocity when the robot moves the trajectory tracking NUBRS.

From Figs. 5 and 6, since the desired angular velocity Ω_d determined from V_d and $\rho_d(t)$ according to Eq. (12), the angular velocity error of the robot at each time $\Delta\Omega = \Omega - \Omega_d$ varies from 0 rad/s to 3.28 rad/s (see Fig. 6b). The reason at positions ⬟, ⬢, ⬢, ⬢ is that the robot positions change direction from clockwise to counterclockwise to follow the desired trajectory ξ_d as depicted in Fig. 6a. This leads to the graph of tracking errors of the robot's trajectory and posture, as shown in Fig. 7.

From Fig. 5a and Fig. 7, it is easy to see that, when the robot moves: (1) From the starting point to ⬟, there will be a change of direction from clockwise to counterclockwise and, in turn, follow tracking from the inside to out of the desired trajectory with a position error is 2.8 mm and posture error of approximately 1°; (2) From position ⬟ to position ⬟ on the trajectory, the robot starts to track from the outside to the inside of the desired trajectory (reversing the direction to the inward starting from position ●) with the position error decreasing from 2.8mm to 1.2 mm and the error of direction is about 4.6°; (3) From position ⬟ to position ⬢, the robot reverses the tracking direction from the outside to the inside of the desired trajectory and vice versa at position ⬢, the error in the x-direction is 0.5 mm, at position + is the reverse direction with an angular error of 6.3° produces an error in the y-direction of 4 mm; (4) From position ⬢ to position ⬢, the robot follows tracking from the outside to the inside of the desired trajectory with the maximum error of 1.68 mm in the x-direction, 3 mm in the y-direction, and 7.7° in the reverse posture; (5) From ⬢ to starting point, the robot follows tracking inside to the outside of the desired trajectory through points + and +. Point + reverses direction with an angular error of 1.3°, making point + an error in the x-direction is 1 mm.

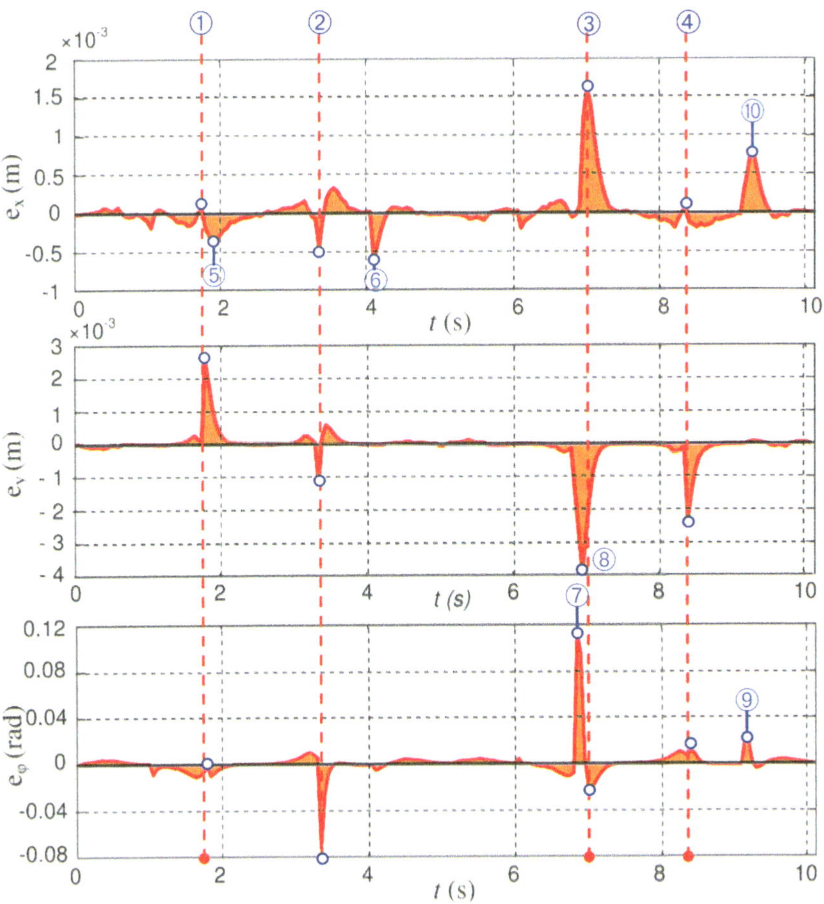

Fig. 7 Error in position and posture of the robot during movement.

Angular Velocity of Wheels 1 and 2

From Fig. 1, in the coordinate system $\vartheta_R\{Gx_Ry_R\}$, the angular velocity of the two wheels is determined by:

$$\begin{cases} \omega_1(t) = \frac{1}{r}\left(V(t) - \frac{\Omega(t)L}{2}\right) \\ \omega_2(t) = \frac{1}{r}\left(V(t) + \frac{\Omega(t)L}{2}\right) \end{cases} \tag{14}$$

Thus, the graph of Fig. 8 is the speed control parameter of the two wheels for the robot to tracking follow the desired NURBS trajectory.

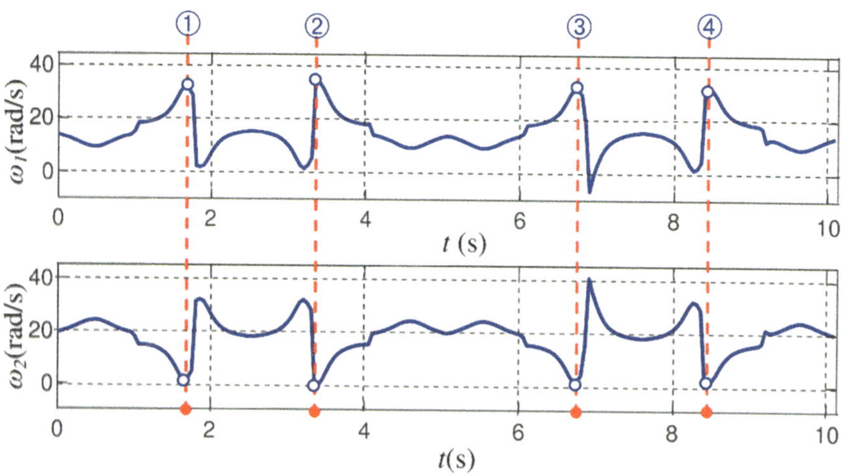

Fig. 8 Angular velocities of wheel 1 and wheel 2 as the robot moves along the NURBS trajectory.

6 Conclusion

From the above calculation, simulation and experimental results. In this paper, a differential-drive mobile robot has experimentally designed and manufactured with the following results:

i) A linear state feedback kinematics controller has developed for the differential-drive mobile robot can produce high-performance tracking with minor tracking errors.
ii) Design the desired path of the robot by the NUBRS interpolation method to ensure that the path is a smooth curve to avoid sudden speed changes of the robot.
iii) A robot platform is proposed to be designed with defined hardware to be able to experimental verification of the controller proposed by this study.

Thereby, it can apply the research results presented in this paper in many different fields: in hospitals, in residential settings and in an industrial environment etc.

Acknowledgments. This research was funded by the Ministry of Industry and Trade in a ministerial level scientific and technological research project, conducted in 2020, code: DTKHCN.076/20.

References

1. Ahmadi, S.M., Taghadosi, M.B., AmirReza, H.M.: A state augmented adaptive backstepping control of wheeled mobile robots. Trans. Inst. Measur. Control **43**(2), 434–45 (2021)
2. Leena, N., Saju, K.K.: Modelling and trajectory tracking of wheeled mobile robots. Procedia Technol. **24**, 538–545 (2016)
3. Alves, T.G., Lages, W.F., Henrique, R.V.B.: Parametric identification and controller design for a differential drives mobile robot. Int. Fed. Autom. Control **51**(15), 437–442 (2018)
4. Mirzaeinejad, H.: Optimization-based nonlinear control laws with increased robustness for trajectory tracking of non-holonomic wheeled mobile robots. Transp. Res. Part C Emerging Technol. **101**, 1–17 (2019)
5. Jacobs, L., Preter, A.D., Anthonis, J., Swevers, J., Pipeleers, G.: Trajectory tracking of AGVs by linear parameter-varying control: a case study. Int. Fed. Autom. Control **51**, 43–48 (2018)
6. Wang, H., Zuo, Z., Wang, Y., Yang, H.: Composite control for trajectory tracking of wheeled mobile robots with NLESO and NTSMC. IET Control Theory **15**(12), 1–9 (2021)
7. Pliego-Jiménez, J., Martínez-Clark, R., Cruz-Hernández, C., Arellano-Delgado, A.: Trajectory tracking of wheeled mobile robots using only Cartesian position measurements. Automatica **133**, 109756 (2021)
8. Zhang, H., Wu, D., Yao, T.: Research on AGV trajectory tracking control based on double closed-loop and PID control. J. Phys. Conf. Ser. **1074**, 012136 (2018)
9. Chen, W., Zhang, T.: An indoor mobile robot navigation technique using odometry and electronic compass. Int. J. Adv. Rob. Syst. **14**(3), 1–15 (2017)
10. Ngo, H.Q.T., Nguyen, T.P., Nguyen, H.: Research and develop of AGV platform for the logistics warehouse environment. In: Arai, K., Bhatia, R., Kapoor, S. (eds.) FTC 2018. AISC, vol. 881, pp. 455–465. Springer, Cham (2019). https://doi.org/10.1007/978-3-030-02683-7_32
11. García-Sánchez, J.R., et al.: Tracking Control for Mobile Robots Considering the Dynamics of All Their Subsystems: Experimental Implementation. Hindawi Complexity, vol. 2017, pp. 1–18 (2017)
12. Kim, T., Kim, W., Choi, S., Kim, H.J.: Path Tracking for a skid-steer vehicle using model predictive control with on-line sparse gaussian process. Int. Fed. Autom. Control **50**(1), 5755–5760 (2017)
13. Rayguru, M.M., Elara, M.R., Ramalingam, B., Muthugala, M.A.V.J., Samarakoon, S.M.B. P.: A path tracking strategy for car like robots with sensor unpredictability and measurement errors. Sensors **20**(11), 1–17 (2020)
14. Sun, Z., Xie, H., Zheng, J., Man, Z., He, D.: Path-following control of Mecanum-wheels omnidirectional mobile robots using nonsingular terminal sliding mode. Mech. Syst. Signal Process. **147**, 1–14 (2021)
15. Nguyen, T.H., Nguyen, T.Q.: A kinematic control algorithm for blasthole drilling robotic arm in tunneling. Sci. Technol. Dev. J. **20**(K5), 13–22 (2017)

16. Cook, G.: Mobile Robots Navigation Control and Remote Sensing. John Wiley & Sons Inc, Hoboken, NJ, USA (2011)
17. Manas, C., Leena, V., Rangan, B.: Towards optimal computation of energy optimal trajectory for mobile robots. Adv. Control Optim. Dyn. Syst. Kanpur India **2014**, 83–87 (2014)
18. Ly, T.T.K., Thai, N.H., Dzung, L.Q., Thanh, N.T.: Determination of kinematic control parameters of omnidirectional AGV robot with mecanum wheels track the reference trajectory and velocity. In: Sattler, K.-U., Nguyen, D.C., Ngoc Pi, V., Long, B.T., Puta, H. (eds.) Advances in Engineering Research and Application: Proceedings of the International Conference on Engineering Research and Applications, ICERA 2020, pp. 319–328. Springer International Publishing, Cham (2021). https://doi.org/10.1007/978-3-030-64719-3_36
19. Thai, N.H., Ly, T.K.T., Dzung, L.Q.: Roadmap, routing and obstacle avoidance of AGV robot in the static environment of the flexible manufacturing system with matrix devices layout. Sci. Technol. Dev. J. **24**(3), 2091–2099 (2021)
20. Sun, Y., Zhang, C., Liu, C.: Collision-free and dynamically feasible trajectory planning for omnidirectional mobile robots using a novel B-spline based rapidly exploring random tree. Int. J. Adv. Rob. Syst. **18**(3), 1–16 (2021)
21. Tzafestas, S.G.: Introduction to Mobile Robot Control. Elsevier inc. (2017)

Optimization for Replaced Diameter of Aluminum Oxide Nozzle in Abrasive Blasting Systems

Le Hoang Anh[1], Nguyen Thi Quoc Dung[2], Tran Ngoc Huy Thinh[3], Nguyen Huu Quang[4], Nguyen Anh Tuan[4], Nguyen Dinh Ngoc[2], Tran Ngoc Giang[2], and Vu Ngoc Pi[2](✉)

[1] Vinh Long University of Technology Education, Vinh Long, Vietnam
[2] Thai Nguyen University of Technology, Thai Nguyen, Vietnam
vungocpi@tnut.edu.vn
[3] Nguyen Tat Thanh University, Ho Chi Minh City, Vietnam
[4] University of Economics - Technology for Industries, Hanoi, Vietnam

Abstract. Blasting system has been widely used in industry so far. The optimization to reduce the blasting cost is necessary. The current study presents on a method to optimize the replaced nozzle diameter made of aluminum oxide of blasting system. The solved issue is based on finding the objective function of minimizing the blasting cost. The Design of Experiment (DOE) method is applied using Minitab19 to investigate the influence of seven input parameters on the response, the optimum replaced nozzle diameter. According to the obtained results, it is known that the optimum replaced nozzle diameter is crucially/highest impacted by the initial nozzle diameter d_{N0}. Furthermore, the interactions between input parameters also have important effects in which the initial nozzle diameter is the key factor when combined with others and has significant impacts on the response. In addition, the effectively suggested regression model is wholly suitable to the experimental data. With this result, it is simple to calculate the optimum replaced nozzle diameter.

Keywords: Abrasive blasting · Sand blasting · Replaced nozzle diameter · Aluminum oxide nozzle

1 Introduction

Cleaning surfaces in the industry has been an important step in the industry to serve various processes such as painting, engraving, welding, coating, etc. There have been different techniques to satisfy the requirements of cleaned surfaces so far. These are the laser surface cleaning [1], the high-speed air jet [2], the blasting system [3, 4], etc. Among the mentioned technologies, blasting system has been broadly utilized. In this system, the abrasive will be sprayed to the cleaning surfaces with high air pressure [5]. For a given modern blasting system, it can be used with different kinds of abrasive, i.e. garnet, steel shot, copper or sand. The documented results in [6] show that the demand for using abrasives over the world rises from $8.6 billion to $11.4 billion between 2014 and 2019. Hence, the calculations to reduce the cost, as well as enhance the

D. C. Nguyen et al. (Eds.): ICERA 2021, LNNS 366, pp. 697–707, 2022.
https://doi.org/10.1007/978-3-030-92574-1_72

productivity or profit rate for the blasting system, is a crucial task and subjected to the research community. There have been many studies dealings with optimizing the blasting system to reduce the cost, such as the nozzle diameter, cleaning time, process parameters, etc. Vu et al. [7] optimized the cost system to find the optimum nozzle diameter. In the study, the influence of process parameters such as nozzle diameter, air pressure, efficiency, cost components, and the nozzle wear on the cleaning cost was investigated. The results showed that cleaning with the optimum nozzle diameter, the cleaning cost and the cleaning time can economize. Moreover, the study also proposed that instead of making the nozzle body by full wear-resistant materials, a thin layer of 0.2 mm thickness of these materials covering the inside wall of the nozzle hole is enough. Vu [8] proposed a regression model to optimize the exchanged nozzle diameter of the sandblasting system. The proposed model is useful due to flexible in applying for various nozzle materials. Additionally, the results give the fact that the profit rate and the cleaning time can be importantly reduced when applying the prosed model to calculate the optimum nozzle diameter. Yue et al. [9] constructed a model which is based on the experimental data when an automatic abrasive blasting system is used. The result of the model has a great improvement on the blasting quality, productivity, and reducing labor cost. McPhee et al. [10] also suggested a mathematical model validated by experiments. It was shown that the efficiency and economics of the blasting system could increase by more than 50%. In regard to the steel shot blasting system, Vu et al. [4] optimized the cost of the blasting systems. The effect of the input parameters, including the air pressure, the nozzle diameter, the cost components, and the nozzle wear, on the optimum nozzle diameter were investigated. Besides, the regression model gives a new method to determine the replaced nozzle diameter and by this way the cleaning cost strongly reduced. Another study dealing with the steeling shot blasting system was conducted by Vu et al. [11]. The mentioned study presents a profit optimization of the steel shot blasting system. The regression model is also proposed in this study, and there is the fact that using the model, the profit and blasting time can hugely decrease.

Based on the above analyses, it is realized that the above studies have not been conducted by using the Design of Experiment technique, which has been adopted in many studies such as [12–15]. Moreover, it is noticed that the nozzle can be made of variety of materials such as aluminum oxide, tungsten carbide, silicon carbide composite, and boron carbide [16]. Among these materials, aluminum oxide offers good service life and a lower price than others. On the other hand, the study of blasting systems with these nozzle materials has few publications in the literature. In this study, seven blasting process parameters, including the initial nozzle diameter, the nozzle wear rate per hour, the time for changing a nozzle, the compressor power, the machine cost per hour, the nozzle cost per piece, and the cost of sand have been selected to investigate their influences on the optimum nozzle diameter. A computer program will be constructed to solve the requirements of the optimized issue. Additionally, the DOE technique is accepted to support the study. Also, a mathematical model will be proposed to find the optimum replaced nozzle diameter, which made of aluminum oxide.

2 Methodology

Typically, the cost of each process in sand blasting system is assigned to the cost for cleaning one square meter,C_{cl} (USD/m^2), and can be determined as:

$$C_{cl} = \left(C_{m,h} + C_{a,h} + C_{n,h}\right)/v_{cl} \quad (1)$$

In this case, $C_{m,h}$, $C_{a,h}$ and $C_{n,h}$ are correspondingly denoted to the cost per hour (USD/h) of the cleaning system, abrasive and nozzle; v_{cl} is the rate of cleaning (m^2/h).

The abrasive cost is calculated as:

$$C_{a,h} = 3600 \cdot \dot{m}_a \cdot C_{a,m} \quad (2)$$

Where $C_{a,m}$ is the abrasive cost per kilogram (USD/kg); $\dot{m}_a$ is the abrasive mass flow rate (kg/h) which can be identified by the following equation [17]:

$$\dot{m}_a = 15.14 \cdot d_n^{-0.0214} \cdot P^{1.0359} \quad (3)$$

And the nozzle cost per hour can be determined by:

$$C_{n,h} = C_{n,p}/L_n \quad (4)$$

$C_{n,p}$ and L_n in this Equation is the nozzle cost per piece (USD/piece) and the nozzle lifetime (h) respectively.

For aluminum oxide nozzle, the nozzle lifetime can last from 10 to 30 (h) [18]. Also, a nozzle will be replaced when it wears 1/16-inch (or 1.59 mm) beyond its original size [19]. Therefore, the wear rate WR can be established by:

$$WR = 1.59/L\_n = 1.59/(10 \div 30) \approx 0.053 \div 0.159(mm/h) \quad (5)$$

The data in [18] show that the rate of cleaning v_{cl} can be established by the following regression model, which is confirmed by the determination coefficient $R^2$of 0.9984:

$$v_{cl} = 6.6101 \cdot 10^{-6} \cdot d_n^{2.0121} \cdot p^{1.5988} \quad (6)$$

Where, p is the air pressure (kPa); From the data in [18], the following regression equation (with the determination coefficient $R^2 = 0.9774$) was found to determine the air pressure:

$$p = 3854.7 \cdot d_n^{-2.5783} \cdot P^{1.1801} \quad (7)$$

Where, P is the power (kW); d_n is nozzle diameter (mm).

The variation of the air pressure and the cleaning rate with the nozzle diameter is presented in Fig. 1. It is seen that the nozzle diameter has an important effect on the air pressure, i.e. the air pressure decreases with an increase in the nozzle diameter corresponding to the nozzle lifetime growths. On the other hand, the cleaning rate depends on nozzle diameter and air pressure (see Eq. (3)). When the nozzle diameter increases, its grows significantly (c.f. Fig. 2).

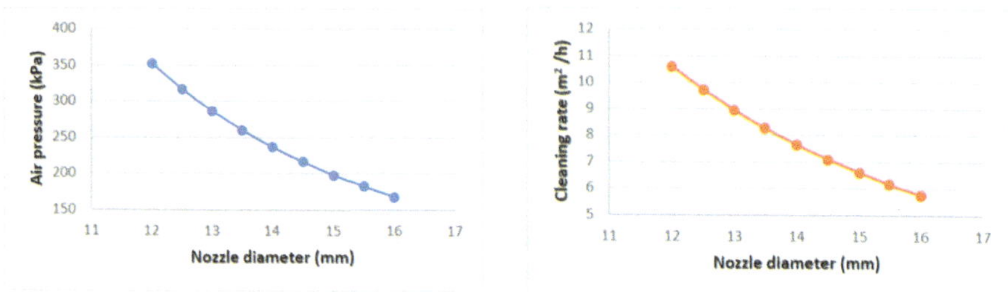

a) b)

Fig. 1. Effect of nozzle diameter on a) air pressure and b) cleaning rate

Figure 2 presents the relation between the cleaning cost and the nozzle diameter corresponding to $C_{m,h}$ = 10 (USD/h); $C_{a,h}$ = 0.1 (USD/kg); $C_{n,p}$ = 200 (USD/piece); WR = 0.002 (mm/h); the power P = 30 (kW). It is observed that the cleaning cost strongly depends on the replaced nozzle diameter. Moreover, it exists the optimum value of the replaced nozzle diameter d_{op} where the cleaning cost minimizes.

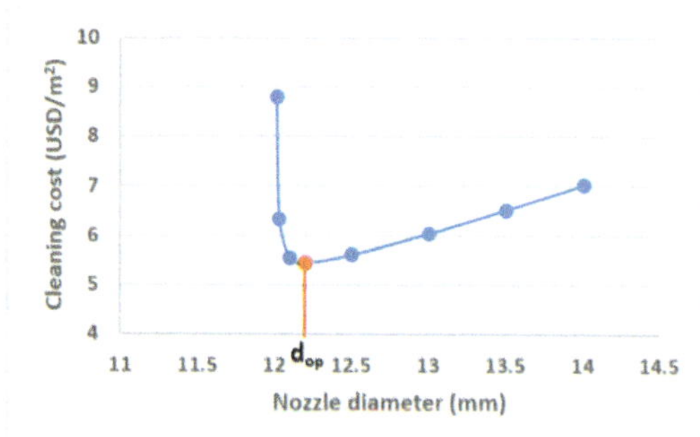

Fig. 2. Cleaning cost versus nozzle diameter

It is interesting to know that when replacing the optimum nozzle diameter (d_{op} = 12.2 (mm) into the system, the cleaning cost reduces by 22.56% compared to using replaced nozzle diameter calculated by the traditional method, 13.9 mm.

Based on the previously mentioned analyses of the literature, in this study, the optimization problem will be conducted by establishing a computer program. Seven main process parameters are purposely chosen and investigated to find the optimum replaced nozzle diameter. These investigated parameters are the air pressure, the abrasive mass flow rate, the initial nozzle diameter, the time for changing a nozzle, the machine cost including the labor cost and the overhead cost, the sand cost, the nozzle cost per piece, the electric power cost, and the average of the nozzle wear rate. The optimum issue can be found in terms of solving the objective functions with the constraints as:

$$\min C_l = f(d_n) \tag{8}$$

With the following constraints:

$$\begin{gathered} 5 \leq d_{n0} \leq 12. \\ 0.001 \leq WR \leq 0.002 \\ 4 \leq P_c \leq 80 \\ 1 \leq t_{cn} \leq 15 \\ 5 \leq C_{m,h} \leq 100 \\ 50 \leq C_{n,p} \leq 300 \\ 0.4 \leq C_s \leq 4.5 \end{gathered} \tag{9}$$

3 Design the Experiment

A plan of simulation experiment is set up in which full experimental design, including all potential tests, is conducted. The evolution of the response, the optimum replacednozzle diameter made of aluminum oxide dop, will be investigated under the influences of seven input parameters. A screening experiment that can provide mathematical models is used in this situation. As a result, $2^7 = 128$ tests are performed. The input parameters are correspondingly assigned a capitalized letter when inputting the data into the Minitab R19 software. The details of the given data conducted at two levels of values can be shown in Table 1. The results of all tests are shortly presented in Table 2.

Table 1. Input parameters and survey level

Real factor	Symbol in Minitab19	Name	Unit	Low	High
Initial nozzle diameter	A	d_{NO}	mm	5	12.5
Nozzle wear rate per hour	B	WR	10^{-3} mm/h	0.053	1.59
Time for changing a nozzle	C	t_{cn}	min	1	15
Compressor power	D	P_c	kW	4	80
Machine cost per hour	E	$C_{m,h}$	USD/h	5	10
Nozzle cost per piece	F	$C_{n,p}$	USD/piece	1	50
Cost of sand	K	C_s	USD/kg	0.4	4.5

Table 2. Experimental plans and output response

Run order	CenterPt	Blocks	d_{NO}	WR	t_{cn}	P_c	$C_{m,h}$	$C_{n,p}$	C_s	d_{op}
1	1	1	5	0.053	15	80	100	1	0.4	5.18
2	1	1	12.5	0.053	15	4	100	50	4.5	12.84
3	1	1	5	0.053	15	4	5	1	4.5	5.18
4	1	1	5	0.159	1	4	5	50	0.4	5.74
….										
127	1	1	5	0.159	15	80	100	1	4.5	5.3
128	1	1	12.5	0.053	15	4	5	1	4.5	12.78

4 Results and Discussions

4.1 The Influence of Main Parameters and Their Interactions

In this part, the influence of the main parameters on the optimum replaced diameters dop, selected as the output response, will be detailed. Figure 3 exhibits the evolution of the initial optimum diameters dop as a function of the input parameters. We can see that the response is strongest impacted by the initial nozzle diameter dN0, and this is the positive impact. It means that when the dN0 value changes from 5 to 12.5 the response increases. In addition, the response is also positively influenced by t_{cn} (C), $C_{n,p}$ (F), and WR (B). Regarding the negative influence, it is observed that P_c (D), C_s (G), $C_{m,h}$ (E) have an orderly negative impact on the response. It is visualized that exception for the case of dN0 (A) it is difficult to clearly realize both positive and negative influence of input parameters on the response. For this reason, it should be adopted by other techniques to support the mentioned issue.

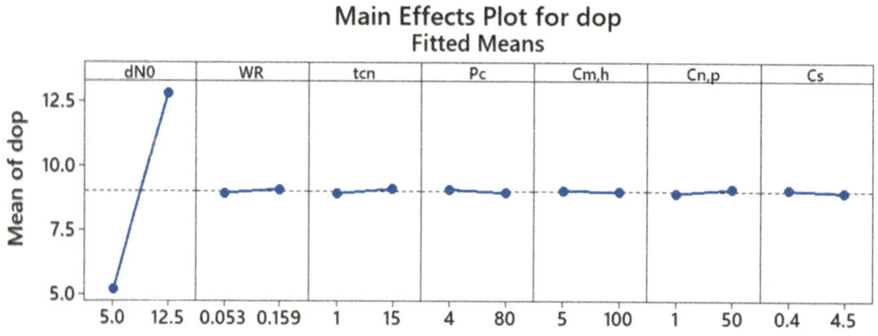

Fig. 3. Main effects plot for optimum replaced nozzle diameter

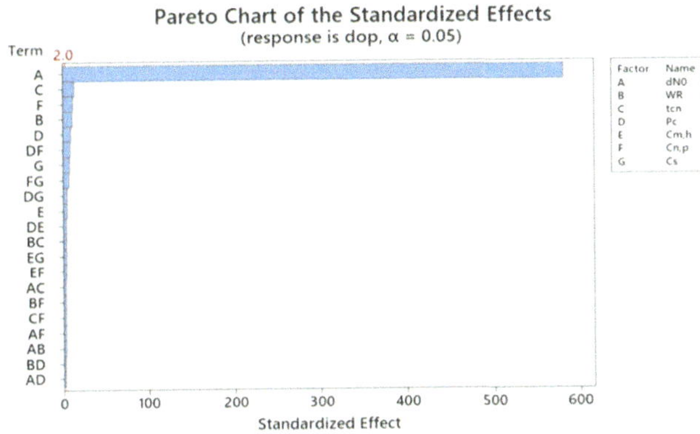

Fig. 4. Pareto Chart of the Standardized Effects

Among the possible techniques, Pareto Chart and Normal Plot of the Standardized Effects can meet the requirements (c.f. Figs. 4 and 5). Figure 4 shows that the factors having a strong impact are far arranged for the standardized line. Each parameter is considered to have statistical significance at the level of 0.05. It is easy to see the influencing domination of the initial nozzle diameter d_{N0} when compared to the others. The impacts of the input parameters on the response can be listed in order as follows: Initial nozzle diameter (d_{N0}), Time for changing a nozzle (t_{cn}), Nozzle cost per piece ($C_{n,p}$), Nozzle wear rate per hour (WR), Compressor power (P_c), Cost of sand (C_s), Machine cost per hour ($C_{m,h}$).

The detail of the response evolution can also be graphically expressed in Fig. 5 using the Normal Plot. In this case, the positive and negative impacts are separately arranged. Moreover, the blue points present for the factors which not have a significant impact on the response. For the positive impacts put above like the factors of A (d_{N0}), C (t_{cn}), F ($C_{n,p}$), B (WR), DG (P_c and C_s), DE (P_c and $C_{m,h}$), BC (WR and t_{cn}), EG ($C_{m,h}$ and C_s), AC (d_{N0} and t_{cn}), BF (WR and $C_{n,p}$), AF (d_{N0} and $C_{n,p}$), while the negative ones are placed under such as the factors of D (P_c), DF (P_c and $C_{n,p}$), G (C_s), FG (Cn,p and C_s), E ($C_{m,h}$), EF ($C_{m,h}$ and $C_{n,p}$), CF (t_{cn} and $C_{n,p}$), BD (WR and P_c). It is noted that the interactions between the single input parameters on the response are also shown in Fig. 5.

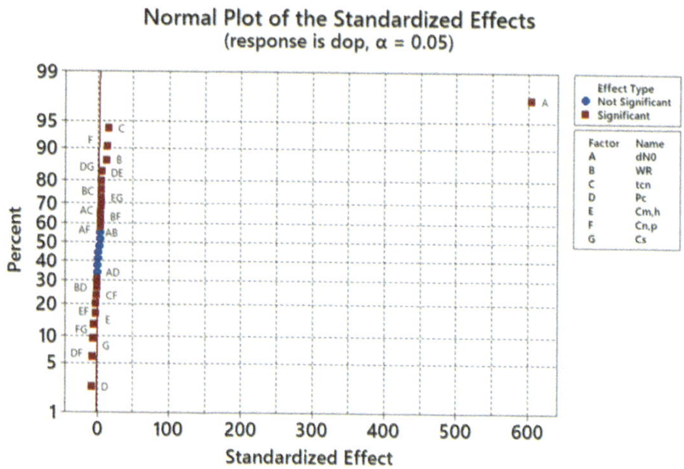

Fig. 5. Normal Plot for d_{op}

4.2 The Proposed Regression Model of the Response

In order to better understand the relation between the input parameters and the response, as well as better prediction, using the regression model is essential. In this case, the significance of the model can be selected with the level of $\alpha = 0.05$, and two interaction factors are included using Minitab19. In this model, the factors having a strong effect on the response are only taken into account when their p-values are smaller than the significance ($\alpha = 0.05$). All of the coefficients of the model are presented in Table 3. In addition, it is realized that the experimental data and the suggested regression model (Eq. 10) is greatly persistent when the minimum value of R-square is approximately 100% (see Table 3).

$$\begin{aligned} d_{op} &= 0.1037 + 1.00058d_{N0} - 0.002WR + 0.00181t_{cn} - 0.000133Pc - 0.001255C_{m,h} \\ &+ 0.00616C_{n,p} - 0.03126C_s + 0.0892d_{N0}.WR + 0.000801d_{N0}.t_{cn} - 0.000104d_{N0}.P_c \\ &+ 0.000202d_{N0}.C_{n,p} + 0.0646WR.t_{cn} - 0.00756WR.P_c + 0.01558WR.C_{n,p} \\ &- 0.000109t_{cn}.C_{n,p} + 0.000014P_c.C_{m,h} - 0.000058P_c.C_{n,p} + 0.000396P_c.C_S \\ &- 0.000020C_{m,h}.C_{n,p} + 0.000238C_{m,h}.C_S - 0.000857C_{n,p}.C_S \end{aligned} \tag{10}$$

Table 3. Estimated effects and coefficients for d_{op}

Coded Coefficients

Term	Effect	Coef	SE Coef	T-Value	P-Value	VIF
Constant		9.02977	0.00656	1375.98	0.000	
dN0	7.62922	3.81461	0.00656	581.28	0.000	1.00
WR	0.14578	0.07289	0.00656	11.11	0.000	1.00
tcn	0.18047	0.09023	0.00656	13.75	0.000	1.00
Pc	-0.12172	-0.06086	0.00656	-9.27	0.000	1.00
Cm,h	-0.05391	-0.02695	0.00656	-4.11	0.000	1.00
Cn,p	0.15359	0.07680	0.00656	11.70	0.000	1.00
Cs	-0.09828	-0.04914	0.00656	-7.49	0.000	1.00
dN0*WR	0.03547	0.01773	0.00656	2.70	0.008	1.00
dN0*tcn	0.04203	0.02102	0.00656	3.20	0.002	1.00
dN0*Pc	-0.02953	-0.01477	0.00656	-2.25	0.027	1.00
dN0*Cn,p	0.03703	0.01852	0.00656	2.82	0.006	1.00
WR*tcn	0.04797	0.02398	0.00656	3.65	0.000	1.00
WR*Pc	-0.03047	-0.01523	0.00656	-2.32	0.022	1.00
WR*Cn,p	0.04047	0.02023	0.00656	3.08	0.003	1.00
tcn*Cn,p	-0.03734	-0.01867	0.00656	-2.85	0.005	1.00
Pc*Cm,h	0.05172	0.02586	0.00656	3.94	0.000	1.00
Pc*Cn,p	-0.10828	-0.05414	0.00656	-8.25	0.000	1.00
Pc*Cs	0.06172	0.03086	0.00656	4.70	0.000	1.00
Cm,h*Cn,p	-0.04547	-0.02273	0.00656	-3.46	0.001	1.00
Cm,h*Cs	0.04641	0.02320	0.00656	3.54	0.001	1.00
Cn,p*Cs	-0.08609	-0.04305	0.00656	-6.56	0.000	1.00

Model Summary

S	R-sq	R-sq(adj)	R-sq(pred)
0.0742455	99.97%	99.96%	99.95%

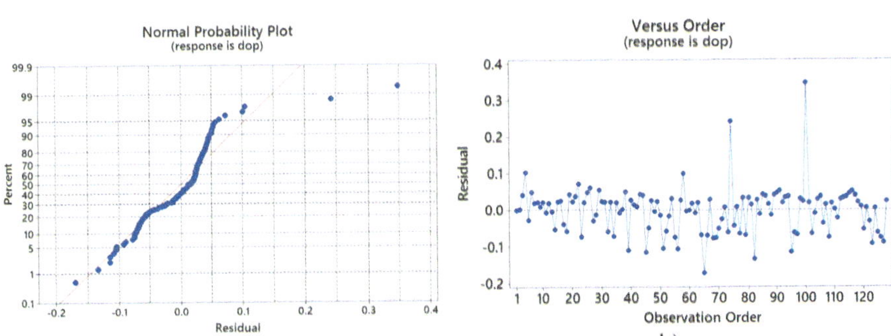

a) b)

Fig. 6. Normal Probability and Versus order plot for d_{op}

4.3 Validation of the Proposed Model

The proposed model expects to fit the experimental data. For this reason, the validation of the model is necessary. The Normal Probability plot and Observation Order presented in Fig. 6 can meet the validating requirements. Based on the results given by the figure, we can see that the data in Fig. 6.a reaches the normal distribution when almost data placed near the red line. Regarding the Versus Order (c.f. Fig. 6.b), it can be visualized that the data arbitrarily scatter on the graph. That means the obtained data are independent on any controlling exception for the input parameters. This phenomenon provides important reliability between input data and the response.

5 Conclusion

In the current study, the cost optimization of the blasting system was conducted. The objective function is purposely selected to find the optimum replaced nozzle diameter made aluminum oxide. The optimization problem was solved by using a computer program, while the experiment was design based on the software of the Minitab R19 software. The impacts of the input data on the optimum replaced nozzle diameter (d_{op}) of the blasting system have been performed. From the obtained results, the following conclusion should be made:

- The optimum replaced nozzle diameter (d_{op}) of the blasting system was highest affected by the initial nozzle diameter (d_{N0}). Nevertheless, the influences of six other input parameters on the response are not clear.
- In addition to the initial nozzle diameter, the interactions of some input parameters have great significance on the optimum repalced nozzle diameter.
- The factors of Normal plots and Versus observation exhibit the high insistence between the prediction model and experimental data which are arbitrarily scattered.

The suitableness between the data and the regression model is reliable. This can support for the research going further. The regression model can provide a simple way to calculate for the optimum replaced nozzle diameters.

Acknowledgment. This work was supported by Thai Nguyen University of Technology.

References

1. Song, W.D., Hong, M., Zhang, L., Lu, Y.F., Chong, T.C.: Laser cleaning technology and its application. In: Proceedings Second International Symposium on Laser Precision Microfabrication, vol. 4426, (2002). https://doi.org/10.1117/12.45683
2. Gotoh, K., et al.: Enhancement of particle removal performance of high-speed air jet by setting obstacle in jet flow. Part. Sci. Technol. **33**(5), 567–571 (2015)
3. Pi, V.N., Duc, T.M.: A study on multi-objective optimization of abrasive blasting systems. Adv. Mater. Res. **126–128**, 29–34 (2010)
4. Pi, V.N., Chinh, P.C.: A study on cost optimization of steel shot blasting systems. Appl. Mech. Mater. **52–54**, 568–572 (2011)

5. Hansel, D.: Abrasive Blasting System. Clemco Industries Corp, Burlingame, Calif (2000)
6. G.A.M.R: https://www.esherpamarketreports.com/reports/%EF%BB%BFglobal-abrasives-market (2020). Accessed 2020
7. Pi, V.N., Hoogstrate, A.M.: Cost optimization of abrasive blasting systems: a new and effective way for using blasting nozzles. Key Eng. Mater. **329**, 323–328 (2007)
8. Pi, V.N.: Profit optimization of abrasive blasting systems. Key Eng. Mater. **389–390**, 392–397 (2008)
9. Yue, L., Tee, K.F., Liu, B., Feng, T.: Automatic abrasive blasting system for large diameter pipe bends. Int. J. Manuf. Technol. Manage. **33**(3/4), 268–282 (2019)
10. William, S., et al.: High productivity vacuum blasting system. AR–26-98FT40367-04, United States (2000)
11. Pi, V.N., Thai, P.H., Hung, T.Q.: Profit optimization of steel shot blasting processes. Appl. Mech. Mater. **312**, 976–981 (2013)
12. Nguyen, T.H.C., Pi, V.N., Hong, T.T., Ky, L.H., Tung, L.A.: A study on calculation of optimum gear ratios of a two-stage helical gearbox with second stage double gear sets. Int. J. Mech. Prod. Eng. Res. Dev. **9**(2), 613–620 (2019)
13. Pi, V.N., Thao, T.T.P., et. al.: Determination of optimum gear ratios of a three stage bevel helical gearbox. Conf. Ser. Mater. Sci. Eng. **542**(1) (2019). https://doi.org/10.1088/1757-899X/542/1/012007
14. Tung, L.A., et al.: A study on determination of optimum gear ratios of a two-stage worm gearbox. In: Sattler, K.-U., Nguyen, D.C., Ngoc Pi, V., Long, B.T., Puta, H. (eds.) ICERA 2019. LNNS, vol. 104, pp. 76–84. Springer, Cham (2020). https://doi.org/10.1007/978-3-030-37497-6_8
15. Vu, N.-P., et al.: The influence of main design parameters on the overall cost of a gearbox. Appl. Sci. **10**(7), 2365 (2020)
16. Momber, A.: Blast Cleaning Technology. Springer, Berlin, Heidelberg (2008). https://doi.org/10.1007/978-3-540-73645-5
17. Vu, N.-P., et al.: Optimizing replaced nozzle diameter of abrasive blasting systems using experiment technique design. Appl. Sci. **10**(11), 3920 (2020)
18. Norton Sand Blasting Equipment, N.S.G.a.I.: http://www.nortonsandblasting.com/nsbnozzles2.html%20. Accessed 2020
19. https://clemcoindustries.com/products/accessories/boron-carbide-lined-metal-jacketed-bsd-series. Accessed 2020

Optimization of Gear Ratios for Three-Stage Bevel Helical Gearboxes Based on Gearbox Volume Function

Trinh Kieu Tuan[1], Bui Thanh Danh[2], Nguyen Huu Luan[3], Nguyen Anh Tuan[1], Le Hoang Anh[4], Nguyen Thanh Tu[5], Tran Ngoc Giang[5], and Luu Anh Tung[5](✉)

[1] University of Economics - Technology for Industries, Hanoi, Vietnam
[2] University of Transport and Communications, Hanoi, Vietnam
[3] Nguyen Tat Thanh University, Ho Chi Minh, Vietnam
[4] Vinh Long University of Technology Education, Vinh Long, Vietnam
[5] Thai Nguyen University of Technology, Thai Nguyen, Vietnam
luuanhtung@tnut.edu.vn

Abstract. This paper introduces the results of an optimization study on gear ratios for three-stage helical gearboxes based on the objective function of the minimum volume gearbox. To solve this problem, a simulation experiment is designed and performed using a computer program. In this work, eight main design parameters were selected to investigate their influence on the optimum gear ratios. In addition, regression models to calculate the optimal gear ratios u_1 and u_2 have been proposed. The study also confirms that these proposed formulas are well suited to the experimental data.

Keywords: Bevel helical gearbox · Gearbox design · Gear ratio · Gearbox volume

1 Introduction

A mechanical transmission system consists of several main components such as an electric motor, a gearbox and/or a V-belt drive or a chain drive and couplings. Among these elements, the gearbox is the main component used to reduce the speed of the motor, increase the torque to meet the working requirements of the working machine. Due to its important role, research on the optimal design of the gearbox has been interested in many researchers.

In gearbox's optimal design, calculating the optimal gear ratio plays a very important role because it greatly affects the size, mass, volume, and cost of a gearbox. That is why there has been a lot of research to date to determine the optimal gear ratios. Determination of the optimal gear ratios has been carried out for different types of gearboxes such as helical gearboxes [1–8], bevel gearboxes [9–15], worm gearboxes [16–19], etc… with two stages [1, 2, 5–7, 20], three stages [3, 4, 9–11, 21], and four stages [22]. The optimal gear ratio is also found with different objective functions such as minimum gearbox length [23–27], minimum gearbox cross-section [2, 3, 9, 11, 18–

D. C. Nguyen et al. (Eds.): ICERA 2021, LNNS 366, pp. 708–720, 2022.
https://doi.org/10.1007/978-3-030-92574-1_73

21], minimum gear mass [15, 27], or reasonable gearbox structure [17]. Studies on the calculation of the optimal gear ratios have also been carried out with the mechanical transmission system with a V-belt drive [3, 7, 13, 28, 29], or a chain drive [1, 9, 12, 20, 23, 24].

In this paper, a new study on determining the optimal gear ratios of a four-stage helical gearbox in order to have the smallest gearbox volume. The influence of the input factors on the optimal gear ratios was analyzed. In addition, regression formulas to calculate the optimal gear ratios have been proposed.

2 Optimization Problem

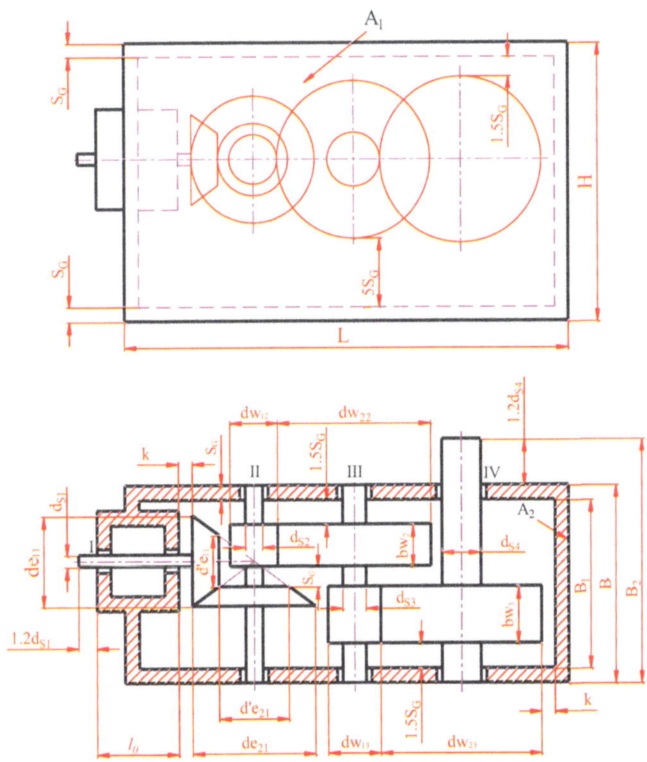

Fig. 1. Schema for determination of gearbox volume

The volume of a three-stage bevel heliacal gearbox V_{gb} (m^3) is determined by (Fig. 1):

$$V_{gb} = L \cdot B \cdot H \quad (1)$$

In which, L, H, and B are calculated by the following Equations:

$$L = 2 \cdot l_0/3 + d_{e21}/2 + d_{w12}/2 + d_{w22}/2 + d_{w13}/2 + d_{w23} + 2 \cdot k \quad (2)$$

$$H = \max(d_{e21}; d_{w22}; d_{w23}) + 8.5 \cdot S_G \quad (3)$$

$$B = max(b_{w2} + b \cdot cos\delta_2; b_{w2} + b_{w3}) + 6 \cdot S_G \quad (4)$$

In the above Equations, k = 8 ÷ 12 (mm) [30] and it can be choosen as $k = 10$ (mm); S_G is estimated by [31]:

$$S_G = 0.005 \cdot L + 4.5 \quad (5)$$

In Eq. (2), δ_2 is gear pitch cone angle; $tang\delta_2 = u_1$; $l_0 = 3.d_{s1}$ with d_{s1} is the diameter of the first shaft which is calculated by [30]:

$$d_{s1} = [T_{11}/(0,2 \cdot [\tau])]^{1/3} \quad (6)$$

Where, T_{11} called the torque on the pinion which is calculated as:

$$T_{11} = T_{out}/(u_t \cdot \eta_{bg} \cdot \eta_{hg}^2 \cdot \eta_{be}^4) \quad (7)$$

In which, u_t is the total gearbox ratio; η_{bg}, η_{hg} and η_{be} are the efficiency of bevel gear, helical gear set and a pair of rolling bearing, respectively. Approximate can get $\eta_{bg} = 0.96$; $\eta_{hg} = 0.97$ and $\eta_{be} = 0.992$ [3].

In Eqs. (2), (3) and (4), b is the face with of bevel gears, $b = k_{be}R_e$; d_{e11} and d_{e21} are the outer pitch diameters of the pinion and the gear; with straight bevel gear set, they can be computed as [3]:

$$d_{e11} = 2 \cdot R_e/\left(1 + u_1^2\right)^{1/2} \quad (8)$$

$$d_{e21} = 2 \cdot u_1 \cdot R_e/\left(1 + u_1^2\right)^{1/2} \quad (9)$$

Where, R_e is the cone distance (mm) and it can be expressed as:

$$R_e = k_R \cdot \sqrt{u_1^2 + 1} \cdot \sqrt[3]{T_{11} \cdot k_{h\beta1}/\left[(1 - k_{be}) \cdot k_{be} \cdot u_1 \cdot [AS_1]^2\right]} \quad (10)$$

For straight bevel gears made by steel materials, k_R is the face width coefficient and $k_R = 50$ [3]; $k_{H\beta1}$ is representative for the contacting load ratio for pitting resistance and $k_{H\beta1} = 1.11$; k_{be} is the face width coefficient.

In Eq. (4), b_{w2} and b_{w3} are the gear widths of the second and the third stages (mm); $b_{w2} = \psi_{ba2} \cdot a_{w2}$ and $b_{w3} = \psi_{ba3} \cdot a_{w3}$ (mm).

Where, a_{w2} and a_{w3} are the center distances of the second and the third stages which are calculated by [3]:

$$a_{w2} = k_a \cdot (u_2 + 1) \cdot \sqrt[3]{T_{12} \cdot k_{H\beta}/([AS_2]^2 \cdot u_2 \cdot X_{ba2})} \quad (11)$$

$$a_{w3} = k_a \cdot (u_3 + 1) \cdot \sqrt[3]{T_{13} \cdot k_{H\beta}/([AS_3]^2 \cdot u_3 \cdot X_{ba3})} \quad (12)$$

With

$$T_{12} = T_{out}/(u_2 \cdot u_3 \cdot \eta_{hg}^2 \cdot \eta_{be}^3) \quad (13)$$

$$T_{13} = T_{out}/(u_3 \cdot \eta_{hg} \cdot \eta_{be}^2) \quad (14)$$

In the above Equations, $k_{H\beta}$ is the contacting load ratio for pitting resistance, $k_{H\beta} = 1.08$ [3]; k_a is the material coefficient, with using steel materials for the helical gears, k_a= 43; $[AS_2]$ and $[AS_3]$ are the allowable contact stress of the second and third stages (MPa); X_{ba2} and X_{ba3} are the coefficients of wheel face widths of the second and the third stages.

After having a_{w2} and a_{w3}, the pitch diameters of the second and the third stage can be determined by:

$$d_{w12} = 2 \cdot a_{w2}/(u_2 + 1) \quad (15)$$

$$d_{w22} = 2 \cdot a_{w2} \cdot u_2 \cdot /(u_2 + 1) \quad (16)$$

$$d_{w13} = 2 \cdot a_{w3}/(u_3 + 1) \quad (17)$$

$$d_{w23} = 2 \cdot a_{w3} \cdot u_3 \cdot /(u_3 + 1) \quad (18)$$

As the above procedure, the gearbox volume depending on the transmission ratios of the three stages i.e., u_1, u_2, and u_3 is considered as the objective function

$$minimize V_{gb} \quad (19)$$

subjected to

$$1 \leq u_1 \leq 6, 1 \leq u_2 \leq 9 \text{ and } 1 \leq u_3 \leq 9 \quad (20)$$

3 Simulation Experiment

To evaluate the impact of the input parameters on the optimal gear ratios, a simulation experiment was designed and carried out. In which, 8 main design parameters were taken into the survey process (Table 1). Besides, a design of 2^k with k = 8 and 1/2 factorial was applied with the support of Minitab R19 software. That means a number of 128 experiments (2^{8-1}) was run. Also, the test plan and the optimal gear ratio values of u_1 and u_2 are shown in Table 2.

4 Results and Analysis

To evaluate the impact of the main design factors on the gear ratiosu_1, and u_2, the graphs in Fig. 2 were used. From these graphs, it was found that, with u_1 (Fig. 2a), whenu_g, Kbe, AS_1, T_{out} increases, u_1 increases. On the other hand, if X_{ba3}, AS_2, and AS_3 increase, then u_1 decrease. Besides, X_{ba2} has negligible effect on u_1. Similarly, for the effect of input parameters on u_2 (Fig. 2b), when u_g, X_{ba2}, AS_2 increases, u_2 increases. In addition, with the increase of Kbe, X_{ba3}, AS_1, AS_3, and T_{out}, u_2 decreases.

Table 1. Main design paramters and their levels

No.	Parameters	Symbol	Unit	Level	
				Low	High
1	Total gearbox ratio	u_g	–	20	70
2	Coefficient of the face width of bevel gear set	K_{be}	–	0.25	0.3
3	Coefficient of wheel face width of stage 2	X_{ba2}	–	0.3	0.37
4	Coefficient of wheel face width of stage 3	X_{ba3}	–	0.34	0.4
5	Allowable contact stress of stage 1	AS_1	MPa	360	420
6	Allowable contact stress of stage 2	AS_2	MPa	360	420
7	Allowable contact stress of stage 3	AS_3	MPa	360	420
8	Output torque	T_{out}	Nm	1000	10000

Table 2. Experimental plan and the optimum ratios

Run order	u_g	K_{be}	X_{ba2}	X_{ba3}	AS_1	AS_2	AS_3	T_{out}	u_1	u_2
1	70	0.25	0.37	0.34	420	360	420	1000	5.97	4.71
2	70	0.25	0.30	0.34	360	360	360	10000	5.90	4.92
3	20	0.25	0.37	0.34	360	360	360	10000	2.82	3.73
4	70	0.30	0.37	0.34	360	420	420	10000	5.90	5.27
5	70	0.30	0.37	0.40	420	360	360	10000	5.90	4.99
6	20	0.30	0.37	0.34	420	360	360	10000	3.10	3.52
...										
127	70	0.25	0.30	0.34	420	360	420	10000	5.90	4.43
128	20	0.30	0.37	0.34	360	420	420	1000	2.68	3.87

Figure 3 shows a normalized Pareto plot of the influence of input parameters and their interactions on u_1 and u_2. From Fig. 3, it is possible to understand the influence of the input parameters and their interactions on u_1 (Fig. 3a) and u_2 (Fig. 3b). The impact of the input factors is shown in the length of the blue column. In this case, the parameters whose magnitude exceeds the red reference line are those that have a significant influence on u_1, u_2 with significance level $\alpha = 0.05$. Specifically, the influence of A(u_g) on u_1 is the largest. Other parameters including their interactions have a significant effect on u_1. It can be described as follows: B (K_{be}), D (X_{ba2}), E (AS_1), G (AS_3), H (T_{out}) and the interactions AB (u_g * K_{be}), AE (u_g * AS_1), AG (u_g * AS_3), AH (u_g * T_{out}), EF (AS_1 * AS_2), FG (AS_2 * AS_3). For the gear ratio u_2, the effect of A(u_g) on u_2 is also the largest. In addition, other parameters that affect u_2 include: B (K_{be}), C (X_{ba2}), D (X_{ba3}), E (AS_1), F (AS_2), G (AS_3), H (Tout) and image. The effects of interactions on u_2 are: AB (u_g * K_{be}), AC (u_g * X_{ba2}), AD (u_g * X_{ba3}), AE (u_g * AS1), AF (u_g * AS_2), AG (u_g * AS_3), AH (u_g * T_{out}), EF (AS_1 * AS_2).

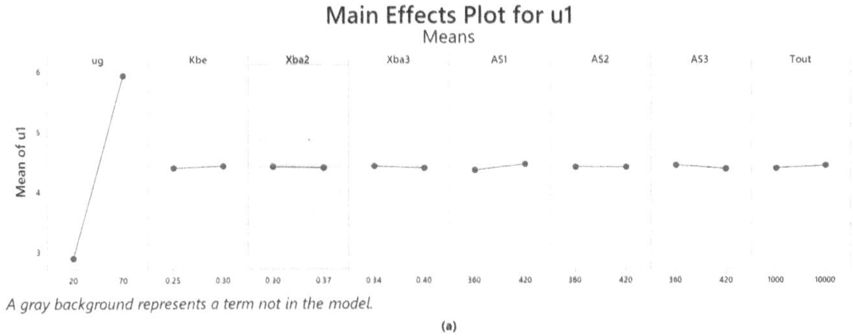

a)

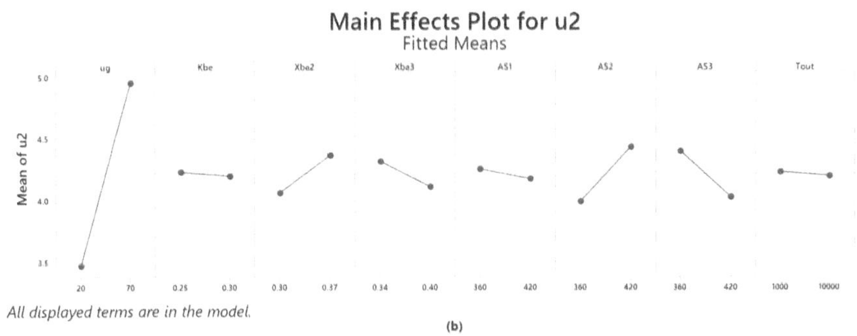

b)

Fig. 2. Main effect plot for a) u_1 b) u_2

To better determine the trend and influence of the input parameters and interactions on the optimum gear ratios, the normal plot of standardized effects is used (Fig. 4). From Fig. 4a, the parameters marked in red are the ones that have a significant influence on u_1. Parameters located further away from the standard line are the parameters that have a greater effect on u_1. The parameters to the right of the calibration curve are those that have a positive influence on u_1 including: A (u_g), B (K_{be}), E (AS_1), H (T_{out}) and AG interactions (u_g * AS_3), EF (AS_1 * AS_2). The parameters to the left of the standard curve are those that have a negative influence on u_1, including: D (X_{ba3}), G (AS_3) and interactions AB (u_g * K_{be}), AE (u_g * AS_1), AH (u_g * T_{out}), FG (AS_2 * AS_3).

For the gear ratio u_2, from Fig. 4b it was found that parameters A (u_g), C (X_{ba2}), F (AS_2) and interactions AB (u_g * K_{be}), AC (u_g * X_{ba2}), AE (u_g * AS_1), AF (u_g * AS_2), AH (u_g * T_{out}) have a positive effect on u_2. Besides, parameters B (K_{be}), D (X_{ba3}), E (AS_1), G (AS_3)), H (T_{out}) and the interactions AD (u_g * X_{ba3}), AG (u_g * AS_3), EF (AS_1 * AS_2) have an inverse effect on u_2.

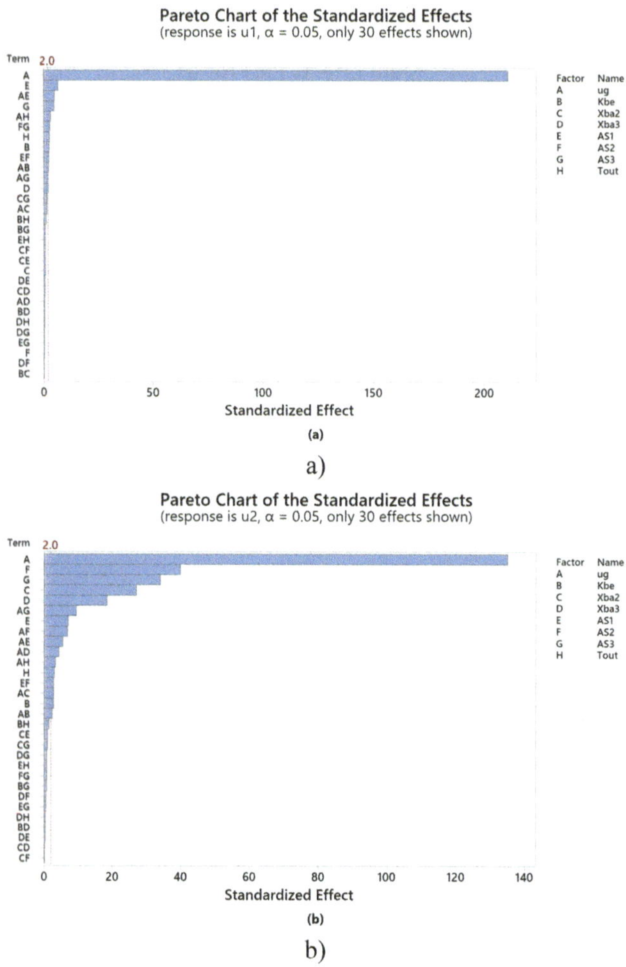

Fig. 3. Pareto charts for the standardized effects for a) u_1 b) u_2

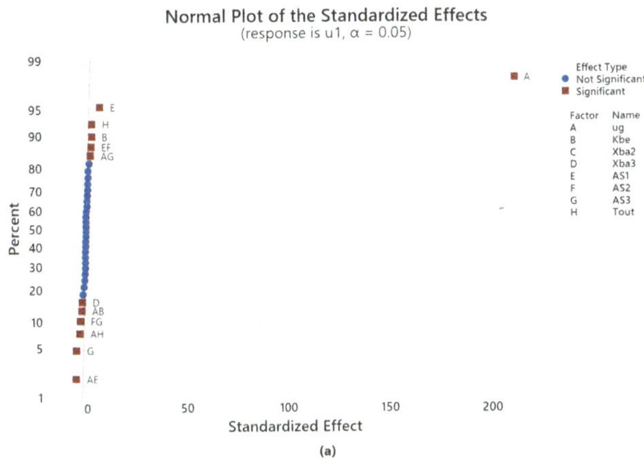

a)

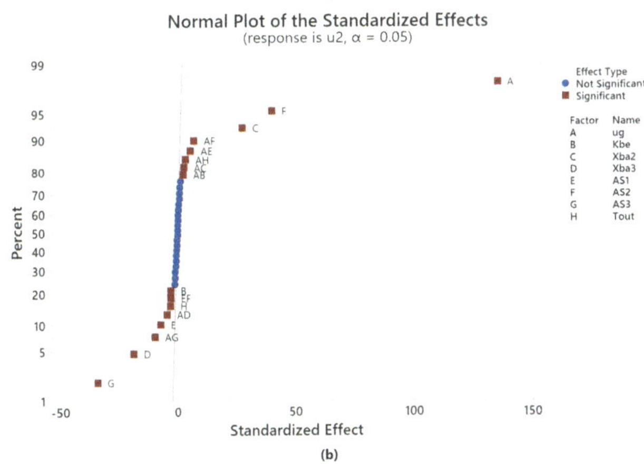

b)

Fig. 4. Normal Plot of the Standardized Effects for a) u_1 b) u_2

Regression Models for Optimum Gear Ratios

To determine the regression equations for u_1 and u_2, Minitab 19 software was used to remove parameters that had no or little influence on them (with P-values smaller than 0.05). The estimated coefficients of the input parameters and the interactions are given on Table 3 for u_1 and Table 4 for u_2. The regression equations to find u_1 and u_2 are described by the following equations:

$$u_1 \quad \begin{aligned} & -0.14 + 0.08052\ \mathrm{u_g} + 2.030\ \mathrm{K_{be}} - 0.529\ \mathrm{X_{ba3}} - 0.00390\ \mathrm{AS_1} \\ & + 0.00180\ \mathrm{AS_2} + 0.00754\ \mathrm{AS_3} + 0.000015\ \mathrm{T_{out}} - 0.0271\ \mathrm{u_g} * \mathrm{K_{be}} \\ & - 0.000050\ \mathrm{u_g} * \mathrm{AS1} + 0.000021\ \mathrm{u_g} * \mathrm{AS_3} - 0.000000\ \mathrm{u_g} * \mathrm{T_{out}} \\ & + 0.000020\ \mathrm{AS_1} * \mathrm{AS_2} - 0.000025\ \mathrm{AS_2} * \mathrm{AS_3} \end{aligned} \tag{21}$$

$$u_2 \quad \begin{aligned} &0.760 + 0.01994\ \mathrm{u_g} - 1.619\ \mathrm{K_{be}} + 3.481\ \mathrm{X_{ba2}} - 1.932\ \mathrm{X_{ba3}} \\ &+ 0.00367\ \mathrm{AS_1} + 0.01185\ \mathrm{AS_2} - 0.003070\ \mathrm{AS_3} - 0.000011\ \mathrm{T_{out}} \\ &+ 0.02188\ \mathrm{u_g} * \mathrm{K_{be}} + 0.01813\ \mathrm{u_g} * \mathrm{X_{ba2}} - 0.03281\ \mathrm{u_g} * \mathrm{X_{ba3}} \\ &+ 0.000042\ \mathrm{u_g} * \mathrm{AS1} + 0.000052\ \mathrm{u_g} * \mathrm{AS_2} - 0.000071\ \mathrm{u_g} * \mathrm{AS_3} \\ &+ 0.000000\ \mathrm{u_g} * \mathrm{T_{out}} - 0.000018\ \mathrm{AS_1} * \mathrm{AS_2} \end{aligned} \tag{22}$$

Table 3. Estimated effects and coefficients for u_1

Coded Coefficients

Term	Effect	Coef	SE Coef	T-Value	P-Value	VIF
Constant		4.40648	0.00687	641.77	0.000	
ug	3.02422	1.51211	0.00687	220.23	0.000	1.00
Kbe	0.04047	0.02023	0.00687	2.95	0.004	1.00
Xba3	-0.03172	-0.01586	0.00687	-2.31	0.023	1.00
AS1	0.09953	0.04977	0.00687	7.25	0.000	1.00
AS2	-0.00547	-0.00273	0.00687	-0.40	0.691	1.00
AS3	-0.07328	-0.03664	0.00687	-5.34	0.000	1.00
Tout	0.04266	0.02133	0.00687	3.11	0.002	1.00
ug*Kbe	-0.03391	-0.01695	0.00687	-2.47	0.015	1.00
ug*AS1	-0.07547	-0.03773	0.00687	-5.50	0.000	1.00
ug*AS3	0.03172	0.01586	0.00687	2.31	0.023	1.00
ug*Tout	-0.04922	-0.02461	0.00687	-3.58	0.000	1.00
AS1*AS2	0.03609	0.01805	0.00687	2.63	0.010	1.00
AS2*AS3	-0.04484	-0.02242	0.00687	-3.27	0.001	1.00

Model Summary

S	R-sq	R-sq(adj)	R-sq(pred)
0.0776810	99.77%	99.74%	99.71%

Table 4. Estimated effects and coefficients for u_2

Coded Coefficients

Term	Effect	Coef	SE Coef	T-Value	P-Value	VIF
Constant		4.21398	0.00523	805.77	0.000	
ug	1.48203	0.74102	0.00523	141.69	0.000	1.00
Kbe	-0.03172	-0.01586	0.00523	-3.03	0.003	1.00
Xba2	0.30078	0.15039	0.00523	28.76	0.000	1.00
Xba3	-0.20453	-0.10227	0.00523	-19.55	0.000	1.00
AS1	-0.07984	-0.03992	0.00523	-7.63	0.000	1.00
AS2	0.43859	0.21930	0.00523	41.93	0.000	1.00
AS3	-0.37516	-0.18758	0.00523	-35.87	0.000	1.00
Tout	-0.03391	-0.01695	0.00523	-3.24	0.002	1.00
ug*Kbe	0.02734	0.01367	0.00523	2.61	0.010	1.00
ug*Xba2	0.03172	0.01586	0.00523	3.03	0.003	1.00
ug*Xba3	-0.04922	-0.02461	0.00523	-4.71	0.000	1.00
ug*AS1	0.06234	0.03117	0.00523	5.96	0.000	1.00
ug*AS2	0.07766	0.03883	0.00523	7.42	0.000	1.00
ug*AS3	-0.10609	-0.05305	0.00523	-10.14	0.000	1.00
ug*Tout	0.03828	0.01914	0.00523	3.66	0.000	1.00
AS1*AS2	-0.03172	-0.01586	0.00523	-3.03	0.003	1.00

Model Summary

S	R-sq	R-sq(adj)	R-sq(pred)
0.0591679	99.55%	99.49%	99.40%

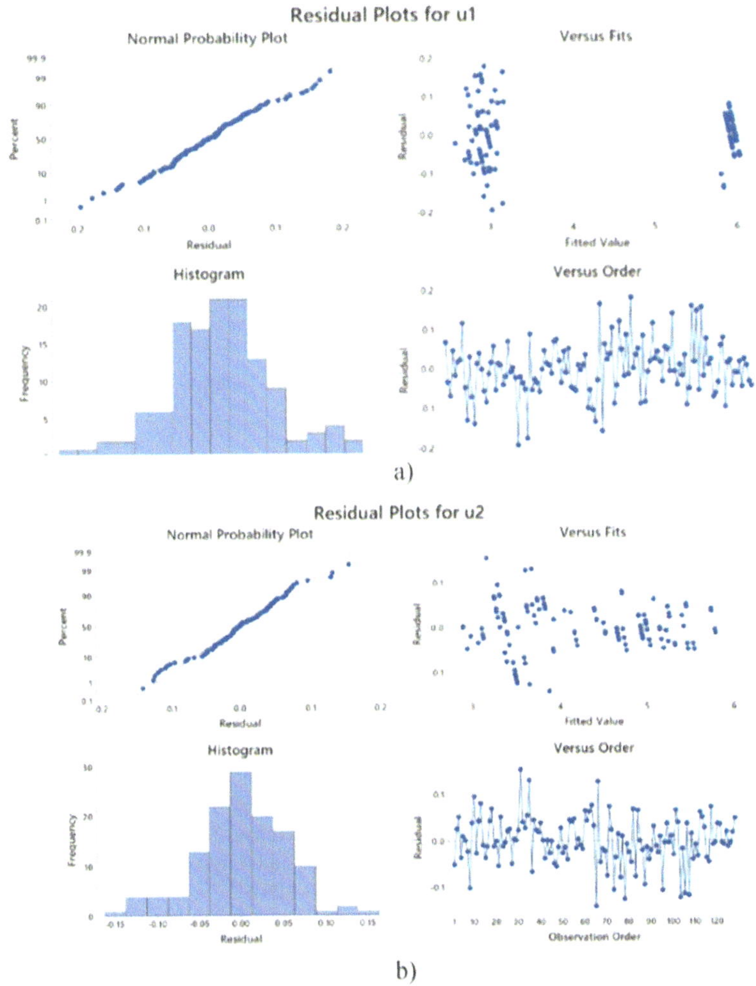

Fig. 5. Normal probability plots of the validity of a) u_1 b) u_2

It was noted that R^2(symbol R-Sq) and $R - adj^2$ (denoted $R_{\text{-sq(adj)}}$) are at least 99.71% for u_1 (Table 3) and 99.4% for u_2 (Table 4). That means the proposed models fit the data fabulously.

The residual plot for u_1 and u_2 are described in Fig. 5. From Figs. 5a, b it is found that the errors of both u_1 and u_2 (shown by blue dots) compared with the normal distribution (solid line) show that they are even distribution very close to the normal distribution. On the graph of the frequency of errors, the errors near 0 account for the majority (−0.1 to 0.1 for both u_1, u_2). Besides, the graph of the relation between the error and the corresponding values of the regression model (versus fit) shows that the points are randomly distributed. That proves that u_1 and u_2 are not significantly affected other than the parameters input to the survey.

5 Conclusions

The paper describes an optimization study on determining optimum gear ratios for a three-stage helical gearbox. The objective function has been minimized to obtain the optimal gear ratios. Screening experiments were also performed with eight input parameters that had a significant influence on the optimal transmission ratio. The experimental results show a significant influence of the input parameters and their interactions on the optimal gear ratios. It can be seen that the biggest influence on the gear ratios belongs to the total gear ratio ut. In addition, regression formulas to calculate optimal gear ratios u_1 and u_2 have been proposed. These formulas are evaluated to be in good agreement with the experimental data.

Acknowledgements. This work was supported by Thai Nguyen University of Technology.

References

1. Pi, V.N., Tuan, N.K., Hung, L.X.: A new study on calculation of optimum partial transmission ratios of mechanical driven systems using a chain drive and a two-stage helical reducer. In: Awang, M., Emamian, S.S., Yusof, F. (eds.) Advances in Material Sciences and Engineering. LNME, pp. 97–105. Springer, Singapore (2020). https://doi.org/10.1007/978-981-13-8297-0_13
2. Pi, V.N., Tuan, N.K., Hung, L.X., Tung, L.A.: A new study on determination of optimum gear ratios of a two-stage helical gearbox. In: Awang, M., Emamian, S.S., Yusof, F. (eds.) Advances in Material Sciences and Engineering. LNME, pp. 107–115. Springer, Singapore (2020). https://doi.org/10.1007/978-981-13-8297-0_14
3. Pi, V.N., et al.: Determining optimum partial transmission ratios of mechanical driven systems using a V-belt drive and a three-stage helical reducer. In: Awang, M., Emamian, S.S., Yusof, F. (eds.) Advances in Material Sciences and Engineering. LNME, Springer, Singapore (2020). https://doi.org/10.1007/978-981-13-8297-0
4. Cam, N.T.H., Pi, V.N., Tuan, N.K., Hung, L.X., Thao, T.T.P.: A study on determination of optimum partial transmission ratios of mechanical driven systems using a chain drive and a three-step helical reducer. In: Fujita, H., Nguyen, D.C., Vu, N.P., Banh, T.L., Puta, H.H. (eds.) ICERA 2018. LNNS, vol. 63, pp. 91–99. Springer, Cham (2019). https://doi.org/10.1007/978-3-030-04792-4_14
5. Pi, V.N., et al.: Calculating optimum gear ratios of a two-stage helical reducer with first stage double gear sets. IOP Conf. Ser. Mater. Sci. Eng. **542**, 012017 (2019)
6. Pi, V.N., et al.: Determination of optimum gear ratios of a two-stage helical gearbox with second stage double gear sets. IOP Conf. Ser. Mater. Sci. Eng. **635**(1), 012012 (2019)
7. Pi, V.N., Hong, T.T., Hung, L.X.: Determining optimal partial transmission ratios of mechanical driven systems using a V-belt drive and a two-stage helical reducer. IOP Conf. Ser. Mater. Sci. Eng. **635**(1), 012011 (2019)
8. Tuan, N.K., et al.: Determining optimal gear ratios of a two-stage helical reducer for getting minimal acreage of cross section. MATEC Web Conf. **213**, 01008 (2018)

9. Thao, T.T.P., et al.: Determining optimum gear ratios of mechanical driven systems using three stage bevel helical gearbox and chain drive. In: Sattler, K.-U., Nguyen, D.C., Ngoc Pi, V., Long, B.T., Puta, H. (eds.) Advances in Engineering Research and Application: Proceedings of the International Conference on Engineering Research and Applications, ICERA 2019, pp. 249–261. Springer International Publishing, Cham (2020). https://doi.org/10.1007/978-3-030-37497-6_29
10. Hong, T.T., et al.: Calculating optimum gear ratios of mechanical driven systems using three step bevel helical gearbox and chain drive for minimum system height. Int. J. Mech. Prod. Eng. Res. Dev. **9**(5), 211–224 (2019)
11. Pi, V.N., Thao, T.T.P., Hong, T.T., Tuan, N.K., Hung, L.X., Tung, L.A.: Determination of optimum gear ratios of a three stage bevel helical gearbox. IOP Conf. Ser. Mater. Sci. Eng. **542**(1), 012007 (2019)
12. Pi, V.N., Tuan, N.K.: Determining optimum partial transmission ratios of mechanical driven systems using a chain drive and a two-step bevel helical gearbox. Int. J. Mech. Eng. Rob. Res **8**, 708–712 (2019)
13. Tuan, N.K., Thao, T.T.P., Cam, N.T.H., Hung, L.X., Pi, V.N.: Optimum calculation of partial transmission ratios of mechanical driven systems using a V-belt and a three-step bevel helical gearbox. In: Fujita, H., Nguyen, D.C., Vu, N.P., Banh, T.L., Puta, H.H. (eds.) ICERA 2018. LNNS, vol. 63, pp. 469–476. Springer, Cham (2019). https://doi.org/10.1007/978-3-030-04792-4_61
14. Tuan, N.K., et al.: Culation of optimum gear Rati stage Bevel Helical Gearbo. Technology **9**, 11 (2018)
15. Hong, T.T., et al.: Optimization of transmission ratios for two-stage Bevel Helical Gearboxes based on mass function
16. Tung, L.A., et al.: A study on determination of optimum gear ratios of a two-stage worm gearbox. In: Sattler, K.-U., Nguyen, D.C., Vu, N.P., Tien Long, B., Puta, H. (eds.) ICERA 2019. LNNS, vol. 104, pp. 76–84. Springer, Cham (2020). https://doi.org/10.1007/978-3-030-37497-6_8
17. Cuong, N.M., et al.: Calculation of optimum gear ratios of two-step worm gearbox. In: Sattler, K.-U., Nguyen, D.C., Vu, N.P., Tien Long, B., Puta, H. (eds.) ICERA 2019. LNNS, vol. 104, pp. 179–188. Springer, Cham (2020). https://doi.org/10.1007/978-3-030-37497-6_21
18. Pi, V.N., Tuan, N.K., Hung, L.X., Dung, N.T.Q., Hien, B.T.: Determining optimum gear ratios of a worm—helical gearbox for minimum acreage of the cross section. In: Awang, M., Emamian, S.S., Yusof, F. (eds.) Advances in Material Sciences and Engineering. LNME, pp. 89–96. Springer, Singapore (2020). https://doi.org/10.1007/978-981-13-8297-0_12
19. Pi, V.N., et al. A study on determination of optimum gear ratios of a worm-helical gearbox. in IOP Conference Series: Materials Science and Engineering. 2019. IOP Publishing
20. Tuan, N.K., et al.: A study on determining optimum gear ratios of mechanical driven systems using two-step helical gearbox with first step double gear sets and chain drive. In: Sattler, K.-U., Nguyen, D.C., Ngoc Pi, V., Long, B.T., Puta, H. (eds.) Advances in Engineering Research and Application: Proceedings of the International Conference on Engineering Research and Applications, ICERA 2019, pp. 85–93. Springer International Publishing, Cham (2020). https://doi.org/10.1007/978-3-030-37497-6_9
21. Pi, V.N., et al.: A study on calculating optimum gear ratios of a three-stage helical gearbox
22. Pi, V.N.: Optimal calculation of partial transmission ratios for four-step helical gearboxes with first and third-step double gear-sets for minimal gearbox length. In: American Conference on Applied Mathematics (MATH 2008), Harvard, Massachusetts, USA (2008)

23. Cam, N.T.H., et al.: Calculating optimum gear ratios of mechanical drive systems using two-stage helical gearbox with second-stage double gear sets and chain drive for minimum gearbox length. In: Sattler, K.-U., Nguyen, D.C., Vu, N.P., Tien Long, B., Puta, H. (eds.) ICERA 2019. LNNS, vol. 104, pp. 155–163. Springer, Cham (2020). https://doi.org/10.1007/978-3-030-37497-6_18
24. Hung, L.X., et al.: Calculation of optimum gear ratios of mechanical driven systems using two-stage helical gearbox with first stage double gear sets and chain drive. In: Sattler, K.-U., Nguyen, D.C., Ngoc Pi, V., Long, B.T., Puta, H. (eds.) Advances in Engineering Research and Application: Proceedings of the International Conference on Engineering Research and Applications, ICERA 2019, pp. 170–178. Springer International Publishing, Cham (2020). https://doi.org/10.1007/978-3-030-37497-6_20
25. Van Cuong, N., et al.: Splitting total gear ratio of two-stage helical reducer with first-stage double gearsets for minimal reducer length. Int. J. **9**(6), 595–608 (2019)
26. Cam, N.T.H., Pi, V.N., Hong, T.T.: A study on calculation of optimum gear ratios of a two-stage helical gearbox with second stage double gear sets. **500**, 1 (2019)
27. Pi, V.N.: Optimum determination of partial transmission ratios of three-step helical gearboxes. WSEAS Trans. Appl. Theor. Mech. **3**, 42–43 (2008)
28. Cam, N.T.H., Pi, V.N., Tuan, N.K., Hung, L.X., Thao, T.T.P.: Determining optimal partial transmission ratios of mechanical driven systems using a V-belt drive and a helical reducer with second-step double gear-sets. In: Fujita, H., Nguyen, D.C., Ngoc Pi, V., Banh, T.L., Puta, H.H. (eds.) Advances in Engineering Research and Application: Proceedings of the International Conference, ICERA 2018, pp. 261–269. Springer International Publishing, Cham (2019). https://doi.org/10.1007/978-3-030-04792-4_35
29. Tuan, N.K., Pi, V.N.: Optimum calculation of partial transmission ratios of mechanically driven systems using a V-belt and a helical gearbox with first-step double gear sets. Int. J.. Mech. Eng. Robot. Rese **8**(2), 323–326 (2019)
30. Trinh Chat, L.V.U.: Calculation of Mechanical Driven Systems. Education Publisher (1996)
31. Römhild, I., Linke, H.: Gezielte Auslegung Von Zahnradgetrieben mit minimaler Masse auf der Basis neuer Berechnungsverfahren. Konstruktion **44**(7–8), 229–236 (1992)

Optimization of Grinding Technology Parameters to Surface Roughness and Material Extracting Productivity of Turbine Shaft Parts

Nguyen Ngọc Nhan, Nguyen Chi Thong, Nguyen Hoang Thien, and Le Hong Ky(✉)

Vinh Long University of Technology Education, Vinh Long, Vietnam
kylh@vlute.edu.vn

Abstract. The turbine shaft is an important part of the turbocharger structure. This part re-quires not only high mechanical properties, precision but also high surface smoothness to ensure the working conditions with very high speed and tempera-ture. This paper presents the research results of optimization of grinding technol-ogy parameters to surface roughness and material extracting productivity of tur-bine shaft parts made of 42CrMo steel of the turbocharger structure.

Keywords: Grinding · Surface roughness · Extracting productivity · Turbine shaft · Turbo-charger

1 Introduction

The turbine shaft of the turbocharger (Fig. 1) has a very special structure. This is an important part connecting the two turbine blades. Usually, the shaft part is manufactured in one piece with the compressor blade, so in the process of using, the shaft is worn or when there is a problem with the turbine blade, it is required to replace both the shaft and the turbine blade. Previously published studies [9–12] mentioned the design and manufacture of turbine blades, the article [8] published research on the influence of heat treatment on mechanical properties of steel as part of turbine shaft body. Research [14] mentioned the fabrication of turbine shafts by different methods such as free forging, open die forging. Studies [15, 16] fabricating turbine shafts by turning method with optimal cutting technology parameters. This paper presents the research results of optimization on some parameters of the machining process when grinding turbine shaft parts made of 42CrMo steel to ensure the minimum surface roughness and maximum material removal rate (*MRR*). MRR plays a significant role in evaluating the product performance i.e. product with low cost and high quality. MRR and surface roughness are important for describing the quality of the product. MRR directly affects the cost of the product and time of manufacturing and is the primary concern in manufacturing industries even with utilizing the highly automated machines like CNC machine [13].

D. C. Nguyen et al. (Eds.): ICERA 2021, LNNS 366, pp. 721–730, 2022.
https://doi.org/10.1007/978-3-030-92574-1_74

2 Experiments

2.1 Sample for Experiment

The sample for experiment is the workpiece of the turbine shaft that has been heat treated and performed the turning process. After hard turning, the workpiece has dimensions shown in Fig. 1.

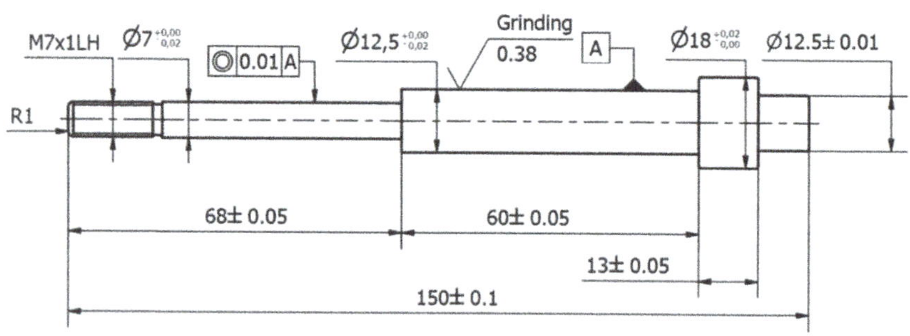

Fig. 1. Turbine shaft

2.2 Testing Machine and Grindstone

The study uses a TKG-30 hydraulic external circular grinder, Fig. 2.a. The grinding machine has high precision and can grind all kinds of parts by the surface of the circumference or the face-grindstone. With the use of jigs, it is possible to grind many types of parts such as round cylinders, planes, step grooves according to the shape of the grindstone.

Fig. 2. Grinding machine and workpiece

The grindstone used in this study is manufactured by Hai Duong Grindstone Factory, with the symbol Cn80-G-V1-160-18-75 × 32 m/s. In which: Cn-grinding grain type is aluminum oxide, 80-grain size of stone, G-binder is ceramic, V1-type of cylindrical stone, 160-outside diameter of stone, 18-thickness of stone, 75 – hole diameter, 32 m/s-maximum allowable cutting speed.

2.3 Measuring Device

The Mitutoyo SJ-210 roughness meter was used to measure the surface roughness of the samples after grinding. At each experimental sample, measurements were made at least 3 times, the measurement was made in the direction perpendicular to the cut velocity vector. The surface roughness value of each sample was calculated as the average value of consecutive measurements.

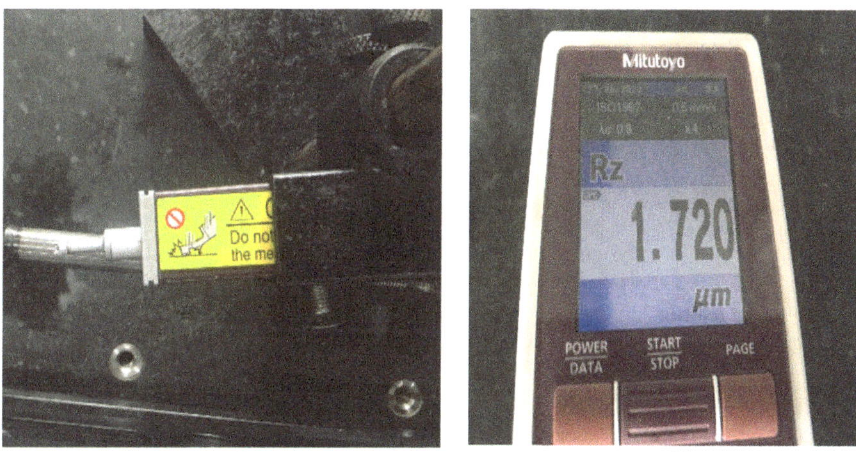

a) *b)*

Fig. 3. SJ-210 roughness meter

MRR is calculated as the amount of material removal in one minute. This quantity is calculated by subtracting the volume of the workpiece before grinding from the volume of the workpiece after grinding and then dividing it by the grinding time. In which the grinding time is calculated by the grinding length divided by the displacement speed of the grindstone mound (feed rate).

2.4 Experiment Design

Taguchi's experimental matrix design method was applied to design the experimental matrix. Three parameters of cutting mode were selected as input parameters including cutting speed (speed of grindstone), part speed and the feed rate. In order to be able to adjust the cutting speed as required, the grinder was connected to a frequency converter (frequency transformer). Each cutting mode parameter was adjusted with three levels of values as shown in Table 1. The experimental matrix designed as an orthogonal Taguchi matrix with 9 experiments (L9) is presented in Table 2.

Table 1. Values of input parameters at levels

Parameters	Symbol	Unit	Values at levels		
			1	2	*3*
Cutting speed	V_d	rpm	2550	2700	2850
Part speed	V_{ct}	rpm	460	520	580
Feed rate	S	rpm	0.05	0.075	0.09

Table 2. L9 Orthogonal matrix

No.	Encryption values			Real values		
	V_d	V_{ct}	S	V_d *(m/rev)*	V_{ct} *(rpm)*	S *(mm/rev)*
1	−1	−1	−1	2550	460	0.050
2	−1	0	0	2550	520	0.075
3	−1	1	1	2550	560	0.090
4	0	−1	0	2700	460	0.075
5	0	0	1	2700	520	0.090
6	0	1	−1	2700	560	0.050
7	1	−1	1	2800	460	0.090
8	1	0	−1	2800	520	0.050
9	1	1	0	2800	560	0.075

2.5 Experimental Conditions

The experiments were carried out under the following conditions: depth of cut 0.015 mm; repairing stone with stone repair tools with 1 diamond grain, stone repair depth 0.01 mm, stone repair feed rate 80 mm/min; using N-600 industrial oil (made in Vietnam) with a concentration of 12%, flow rate of 7.8 L/min. These values have been chosen according to the grindstone manufacturer's recommendations.

2.6 Results and Discussion

The experiment was carried out in the order of the experiments in Table 2, measuring the surface roughness with the SJ-210 device as shown in Fig. 3, calculating the MRR for each experiment, presented in Table 3. The Figs. 4 and 5 show the Pareto chart of the influence of the input parameters on the output parameters, with the significance level chosen to be 0.05.

Table 3. Experiment results

No.	V_d *(m/min)*	V_{ct} *(rpm)*	*S (mm/rev)*	*Ra (μm)*	*MRR* (mm^3/min)
1	2550	460	0.050	0.48	75.104
2	2550	520	0.075	0.72	112.656
3	2550	560	0.090	1.12	135.187
4	2700	460	0.075	0.32	119.282
5	2700	520	0.090	0.88	143.139
6	2700	560	0.050	0.54	79.522
7	2800	460	0.090	0.74	148.440
8	2800	520	0.050	0.52	82.467
9	2800	560	0.075	0.54	123.700

Figure 4 shows a line showing the influence of feed rate on *Ra* beyond the red limit line in the Pareto chart (the line parallel to the vertical axis). It means that this parameter has a significant influence on *Ra*. The workpiece speed and cutting speed have little effect on *Ra*.

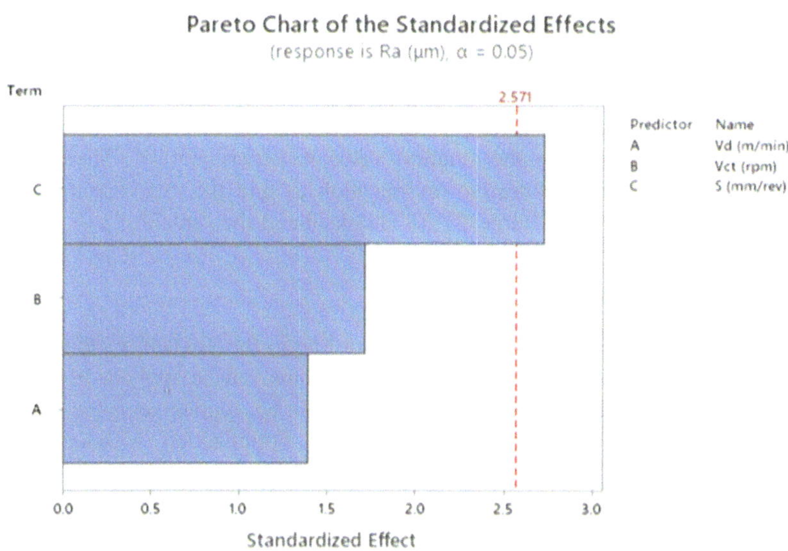

Fig. 4. Pareto chart of the influence of the parameters on *Ra*.

Figure 5 shows that feed rate and cutting speed have a significant effect on *MRR*, where the influence of feed rate on *MRR* is larger than that of cutting speed. The workpiece speed has little significant effect on the *MRR*.

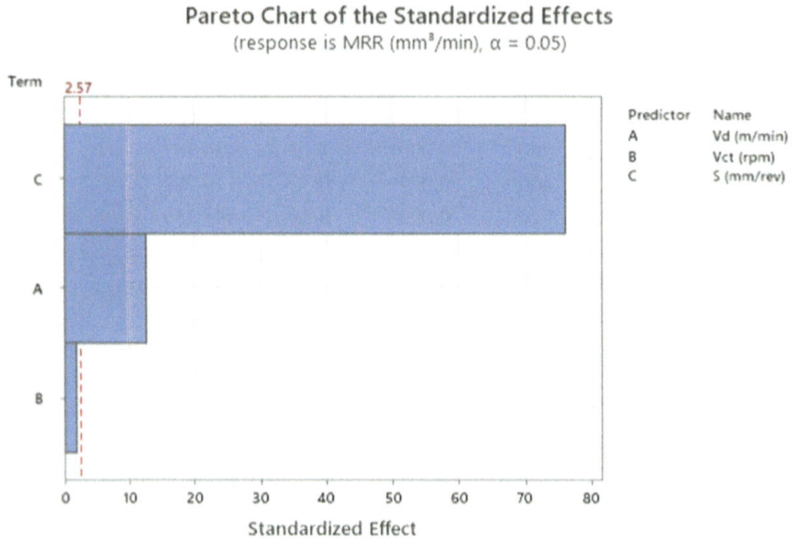

Fig. 5. Pareto chart of the influence of the parameters on *MRR*

Thus, it can be seen that the influence of the input parameters on the output parameters is not the same. Therefore, it is difficult to adjust the value of the input parameters to ensure the purposes for the output parameters (minimum *Ra*, maximum *MRR*). On the other hand, the data in Table 3 show that: *MRR* has the largest value in experiment #7, while Ra has the smallest value in experiment #4. Therefore, it is necessary to perform multi-objective optimization to find the experiment where the *MRR* is considered to be the "largest", *Ra* is considered the "smallest".

2.7 Multi-objective Optimization

When the experimental matrix is designed according to the Taguchi method, to solve the multi-objective problem, many different algorithms can be applied. In this study, the PSI (preference selection index) algorithm will be applied to solve the multi-objective problem. PSI performs multi-objective optimization without assigning weights to the criteria. This method is performed according to the following steps [7]:

Step 1: Define the goal

Step 2: Establish a decision matrix based on available information

Step 3: Normalize the attributes

$$N_{ij} = \frac{x_{ij}}{x_j^{max}} \quad \text{for the beneficial type} \tag{1}$$

$$N_{ij} = \frac{x_j^{min}}{x_{ij}} \quad \text{for disadvantaged type} \tag{2}$$

Where i is the ordinal number of rows in the matrix (I = 1 ÷ n), j is the ordinal number of columns in the matrix (j = 1 ÷ m), x_{ij} is the value of the norm in row i and column j.

Step 4: Calculate the mean values of the normalized data

$$N = \frac{1}{n}\sum_{i=1}^{n} N_{ij} \tag{3}$$

Step 5: Determine the priority value from the mean

$$\varphi_j = \sum_{i=1}^{n} \left[N_{ij} - N\right]^2 \tag{4}$$

Step 6: Determine the deviation in the priority value

$$\emptyset_j = \left[1 - \varphi_j\right] \tag{5}$$

Step 7: Determine the overall priority value for the norms

$$w_j = \frac{\emptyset_j}{\sum_{j=1}^{m} \emptyset_j} \tag{6}$$

Step 8: Calculate the PSI priority selection index of each alternative

$$\theta_j = \sum_{j=1}^{m} x_{ij}.w_j \tag{7}$$

Step 9: Rank the alternatives, whichever with θi the large is the best one

Multi-objective Optimization of Grinding Process by PSI Method

Step 1: Define the goal. The objective of this study was to determine the best experiment in Table 3. At that experiment *Ra* was considered "smallest", *MRR* was considered "largest".

Step 2: Establish a decision matrix based on available information. From the data in Table 3, the last two columns in the table form a matrix of 9 rows and 2 columns. This is the matrix where the "best" row of the 9 rows needs to be determined.

Step 3: The normalized values of the attributes are calculated according to the formulas (1), (2), presented in Table 4.

Table 4. Normalized values of attributes

No.	*Ra* (μm)	*MRR* (mm^3/min)	N_{Ra}	N_{MRR}
1	0.48	75.104	0.6667	0.5060
2	0.72	112.656	0.4444	0.7589
3	1.12	135.187	0.2857	0.9107
4	0.32	119.282	1.0000	0.8036
5	0.88	143.139	0.3636	0.9643
6	0.54	79.522	0.5926	0.5357
7	0.74	148.440	0.4324	1.0000
8	0.52	82.467	0.6154	0.5556
9	0.54	123.700	0.5926	0.8333
Mean			0.5548	0.7631

Step 4: The mean of the normalized value is calculated according to formula (3). This value has also been included in Table 4.

Step 5: Determine priority value from the mean according to the formula (4): $\varphi_{Ra} = 0.3533$; $\varphi_{MRR} = 0.2859$

Table 5. θ value in PSI and ratings

No.	*Ra* (μm)	*MRR* (mm^3/min)	θ	Rankings
1	0.48	75.104	25.8577	9
2	0.72	112.656	38.7865	6
3	1.12	135.187	46.6356	3
4	0.32	119.282	40.9092	5
5	0.88	143.139	49.2691	2
6	0.54	79.522	27.3902	7
7	0.74	148.440	51.0318	1
8	0.52	82.467	28.3902	8
9	0.54	123.700	42.4990	4

Step 6: Determine the deviation in the priority value according to the formula (5): $\varnothing_{Ra} = 0.6467$; $\varnothing_{MRR} = 0.7148$

Step 7: Determine the overall priority value for the norms according to the formula (6): $w_{Ra} = 0.4752$; $w_{MRR} = 0.5248$.

Step 8: Calculate the priority selection index θ of each alternative according to formula (7). The data are presented in Table 5.

Step 9: Rank the alternatives in Table 5.

From the ranking results in Table 5, it shows that experiment #7 is the best experiment among 9 experiments. Besides, experiment #1 is the worst experiment. In experiment #7 it is clear that *MRR* has the largest value among 9 experiments (*MRR* = 148,440 mm3/min); *Ra* is 0.74 μm, smaller than the value of *Ra* in experiments #3 and #5. Although *Ra* in experiment #7 is not their smallest value among 9 experiments, for the purpose of multi-objective optimization, it can be affirmed that experiment #7 is the "best" experiment. Thus, in the investigated values of the parameters of the cutting mode, when the cutting speed is 2800 rpm (corresponding to 23.46 m/s), the speed of the workpiece is 460 rpm (corresponding to 18.06 m/s). min) and a feed rate of 0.09 mm/rev would be the optimal set of values to ensure "minimum" surface roughness and "maximum" *MRR*.

3 Conclusion

This study conducted round grinding experiments on Turbine shafts, made of 42CrMo steel. Three parameters were selected as input parameters of the experimental process including cutting speed, workpiece speed, and feed rate. Ra and MRR were selected as the output parameters of the experimental process. Analysis of experimental results has determined the influence of input parameters on output parameters. The PSI method has been applied to solve the multi-objective problem. Some conclusions are drawn as follows:

- The feed rate is a parameter that has a significant influence on *Ra*. Meanwhile, cutting speed and workpiece speed have little influence on *Ra*.
- The feed rate has the greatest influence on the *MRR*, followed by the degree of influence of the cutting speed, while the workpiece speed has a negligible effect on the MRR.
- To ensure both norms including: *Ra* "smallest" and *MRR* "maximum", the values of cutting speed, workpiece speed and feed rate are 23.46 m/s, 18.06 m/min and 0.09 respectively mm/rev.

References

1. Manoj, M., Jinu, G.R., Muthuramalingam, T.: Multi response optimization of AWJM process parameters on machining TiB2 particles reinforced Al7075 composite using Taguchi-DEAR methodology. SILICON **10**(5), 2287–2293 (2018). https://doi.org/10.1007/s12633-018-9763-x
2. Temuçin, T., Tozan, H., Vayvay, Ö., Harničárová, M., Valíček, J.: A fuzzy based decision model for nontraditional machining process selection. Int. J. Adv. Manuf. Technol. **70**(9–12), 2275–2282 (2013). https://doi.org/10.1007/s00170-013-5474-z
3. Chen, S.-J., Hwang, C.-L.: Fuzzy Multiple Attribute Decision Making. Springer, Berlin, Heidelberg (1992). https://doi.org/10.1007/978-3-642-46768-4
4. Hwang, C.L., Yoon, K.: Multiple Attribute Decision Making Methods and Applications. Springer, Berlin, Heidelberg (1981). https://doi.org/10.1007/978-3-642-48318-9
5. Opricovic, S., Tzeng, G.H.: Compromise solution by MCDM method: a comparative analysis of VIKOR and TOPSIS. Eur. J. Oper. Res. **156**, 445–455 (2004)
6. Cables Perez, E., Lamata, M.T., Verdegay, J.L.: RIM-Reference ideal method in multicriteria decision making. Inf. Sci. **337–338**(10), 1–10 (2016). https://doi.org/10.1016/j.ins.2015.12.011

7. Maniya, K., Bhatt, M.G.: A selection of material using a novel type decision-making method Preference selection index method. Mater. Des. **31**, 1785–1789 (2010). https://doi.org/10.1016/j.matdes.2009.11.020
8. Ky, L.H., Nhan, N.N.: Research on the effects of high Ram mode on mechanical properties of 42CrMo steel for the Turbine shaft. Vietnam J. Mech. ISSN0866-7056 (2021)
9. Ky, L.H., Tam, D.T.: Design analysis of turbine blade in the turbocharger. Vietnam J. Mech. ISSN0866-7056 (2020)
10. Tam, D.T., Hung, T.V., Ky, L.H.: Evaluation on the accuracy of direct machining of turbine blade parts from inventor professional software on CNC milling machines. Vietnam J. Mech. ISSN0866-7056 (2020)
11. Ky, L.H., Thong. N.C.: Research on Turbo wing processing in turbocharger structure on CNC HAAS VF2 milling machine. Vietnam J. Mech. ISSN0866–7056 (2020)
12. Thong, N.C., Hung, T.V., Ky, L.H.: Research on processing turbine blades in HX40W turbocharger structure on CNC milling machine. Vietnam J. Mech. ISSN0866-7056 (2020)
13. Chaturvedi, P.K., Narang, H.K., Sahu, A.K.: Optimization of MRR and surface roughness of AlMg3 (AA5754) Alloy in CNC lathe machine by using Taguchi method. Appl. Mech. Mater. **877**, 110–117 (2018). https://doi.org/10.4028/www.scientific.net/AMM.877.110
14. Lukaszek-Sołek, A., Chyła, P., Bednarek, S.: Forging of gas turbine shaft – numerical modelling of technological process, Metal 2015 Brno Czech Republic EU (2015)
15. Xu, X., Yu, Z.: Failure analysis of a locomotive turbocharger main-shaft. Eng. Fail. Anal. **16**, 495–502 (2009)
16. Daniello, R.: Design and Manufacturing of a Miniature Turbojet Engine. Worcester Polytechnic Institute (2018)

Optimization of PMEDM Parameters for Minimizing Electrode Wear Rate When Processing SKD11 Steel

Anh Tuan Nguyen[1], Thi Tam Do[2], Thu Quy Le[3], Quoc Cuong Dang[4], Kieu Tuan Trinh[1], Thanh Tu Nguyen[2], Ngoc Pi Vu[2], and Xuan Tu Hoang[2](✉)

[1] University of Economics - Technology for Industries, Ha Noi, Vietnam
[2] Thai Nguyen University of Technology, Thai Nguyen, Vietnam
hxt_dhktcn@tnut.edu.vn
[3] National Research Institute of Mechanical Engineering, Ha Noi, Vietnam
[4] Nguyen Tat Thanh University, Ho Chi Minh, Vietnam

Abstract. A study on optimization of powder-mixed electrical discharge machining (PMEDM) when machining cylindrical parts made of SKD11 steel to minimize the electrode wear rate has been introduced in this paper. In the study, an experiment was conducted in which the Taguchi method was used to design the experiment and analyze the results. Besides, six input factors containing the powder concentration, the powder size, the pulse on time, the pulse off time, the current, and the servo voltage were taken to study their effect on the tool wear rate. In addition, an optimal set of input factors to limit the tool wear rate has been proposed.

Keywords: PMEDM · Electrical discharge machining · Electrode wear rate · Cylindrically shaped parts · SKD11

1 Introduction

In practice, electrical discharge machining (EDM) has been widely used to process conductive parts. This method is very suitable for processing stamping dies, cores of dies, molds, etc. Although EDM has many advantages such as being able to process parts with complex shapes; the cutting force and noise during machining are small; it can machine materials with high hardness etc..., this method also has some disadvantages such as low quality of machined surfaces; the material removal rate (MRR) is not high; the electrodes wear out quickly, etc... To limit the disadvantages of the EDM process, a solution has been used for the EDM process in which the dielectric solution is mixed with conductive powder (referred to as PMEDM).

Up to now, quite a few studies on the PMEDM process have been conducted. The optimization problem to determine the optimal input parameters of the PMEDM process has been solved for single-objective [1–6], and multi-objective [7–9] problems. The single objective functions are usually the maximum MRR [2, 4, 10–12], the machining time [1], the minimum surface roughness (SR) [3–6, 10], and the minimum

D. C. Nguyen et al. (Eds.): ICERA 2021, LNNS 366, pp. 731–738, 2022.
https://doi.org/10.1007/978-3-030-92574-1_75

electrode wear rate (EWR) [10–12]. The multi-objective problems will typically combine two or three single-objective functions such as the SR and the MRR [8, 9, 13], the EWR and the MRR [7, 14], or the SR, the MRR, and the EWR [15, 16]. Various powders such as SiC [1, 4, 7], Si [17], Al [3, 17], Al_2O_3 [6, 10], etc. were used for the study on PMEDM. The use of different electrode materials is also of great interest, such as copper [1, 8, 16, 18], graphite [8, 18, 19], or brass [12] electrodes. Many workpiece materials have also been investigated such as 90CrSi [1, 7], Inconel 625 [17], Inconel 718 [3, 10, 12], SKD11 [2, 8], Ti-6Al-4V [5, 20] etc... Recently there have been several works processing on cylindrically shaped parts with 90CrSi steel [1, 7]. However, so far, there is no PMEDM with the workpiece material of SKD11 steel.

This paper presents the results of a study on optimizing the PMEDM process to achieve minimum EWR when machining cylindrical parts with SKD11 steel. The influence of the input PMEDM parameters on the EWR was evaluated. In addition, optimal input factors to minimize EWR have been proposed.

2 Experimental Setup

An experiment to investigate the effect of PMEDM factors on the EWR was conducted. The experimental setup (Fig. 1) includes an EDM Sodick A30, copper electrodes, and SKD11 steel samples. Also, SiC powder was used to mix into the dielectric fluid. Six input parameters have been carefully selected to investigate their influence on the EWR. These parameters, including the powder concentration (C_p), the powder size (S_p), the pulse on time (T_{on}), the pulse off time (T_{off}), the current (IP), and the servo voltage (SV). Table 1 shows the input parameters and their levels.

Table 1. Input parameters and their levels

No	Parameters	Code	Unit	Level					
				1	2	3	4	5	6
1	Powder concentration	Cp	g/l	0	2	3	4	5	6
2	Powder size	Sp	nm	100	500	1000	–	–	–
3	Pulse on time	Ton	μs	10	20	30	–	–	–
4	Pulse off time	Toff	μs	10	20	30	–	–	–
5	Peak current	IP	A	4	8	12	–	–	–
6	Servo voltage	SV	V	3	5	7	–	–	–

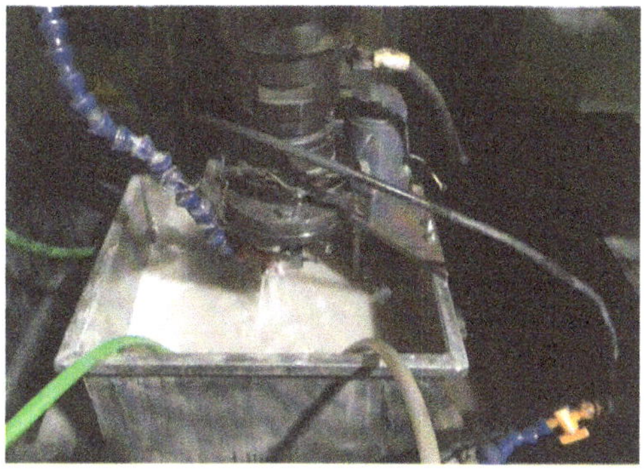

Fig. 1. Experimental setup

Design L18 (3^4) was selected for the experimental design using the Taguchi method in Minitab R19 software. The test plan and results of the EWR are presented in Table 2.

Table 2. Experimental plan and the EWR

No	Input parameters						EWR (mg/min)	S/N
	C_p	S_p	T_{on}	T_{off}	I_p	SV		
1	0	100	10	10	4	3	13.5333	−22.6281
2	0	500	20	20	8	5	99.8345	−39.9856
3	0	1000	30	30	12	7	29.6923	−29.4529
4	2	100	10	20	8	7	35.9337	−31.1100
5	2	500	20	30	12	3	85.4975	−38.6391
6	2	1000	30	10	4	5	16.9221	−24.5691
7	3	100	20	10	12	5	55.6941	−34.9162
8	3	500	30	20	4	7	11.3585	−21.1064
9	3	1000	10	30	8	3	10.2848	−20.2439
10	4	100	30	30	8	5	24.1109	−27.6443
11	4	500	10	10	12	7	10.4982	−20.4223
12	4	1000	20	20	4	3	18.9691	−25.5609
13	5	100	20	30	4	7	15.2504	−23.6656
14	5	500	30	10	8	3	64.6147	−36.2066
15	5	1000	10	20	12	5	24.7060	−27.8560
16	6	100	30	20	12	3	67.6757	−36.6087
17	6	500	10	30	4	5	10.7152	−20.6000
18	6	1000	20	10	8	7	32.8711	−30.3363

3 Results and Analysis

The ANOVA method in the Minitab R19 software was used to investigate the influence of PMEDM parameters on EWR. In addition, it is best to use the objective function as the minimum EWR:

$$S/N = -10\log_{10}\left(\frac{1}{n}\sum_{i=1}^{n} y_i^2\right) \tag{1}$$

From the results of ANOVA (Table 3), it was noted that IP has the most significant influence on EWR (35.23%); followed by T_{on} (29.08%), C_p (15.91%), SV (7.18%), T_{off} (5.47%), and S_p (5.24%). The influence of these factors is also depicted in the chart in Fig. 2. The influence of the input parameters on the S/N ratios is sorted by decreasing influence as T_{on}, IP, C_p, SV, T_{off}, and S_p (Table 4).

Table 3. Analysis of variance for S/N ratios

Analysis of Variance for SN ratios

Source	DF	Seq SS	Adj SS	Adj MS	F	P	C (%)
Cp	5	118.63	118.63	23.726	3.38	0.244	15.91
Sp	2	39.06	39.06	19.532	2.78	0.264	5.24
Ton	2	216.79	216.79	108.397	15.44	0.061	29.08
Toff	2	40.78	40.78	20.392	2.90	0.256	5.47
Ip	2	262.70	262.70	131.350	18.71	0.051	35.23
SV	2	53.56	53.56	26.782	3.81	0.208	7.18
Residual Error	2	14.04	14.04	7.022			1.88
Total	17	745.58					

Model Summary

S	R-Sq	R-Sq(adj)
2.6499	98.12%	83.99%

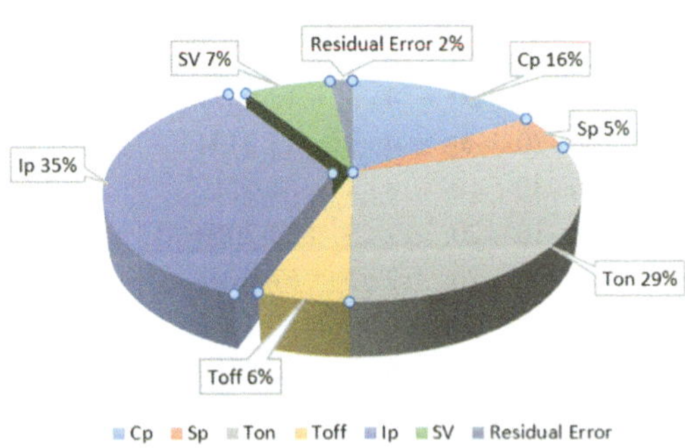

Fig. 2. Effect of input parameters on the S/N ratio of EWR

Table 4. Order of influence of input parameters on S/N

Smaller is better

Level	Cp	Sp	Ton	Toff	Ip	SV
1	-30.69	-29.43	-23.81	-28.18	-23.02	-29.98
2	-31.44	-29.49	-32.18	-30.37	-30.92	-29.26
3	-25.42	-26.34	-29.26	-26.71	-31.32	-26.02
4	-24.54					
5	-29.24					
6	-29.18					
Delta	6.90	3.16	8.37	3.66	8.29	3.97
Rank	3	6	1	5	2	4

The influence of the PMEDM parameters on the S/N ratio is shown in Fig. 3. With the objective of the EWR being as small as possible, the larger the S/N value, the better. From the figure, the following observations can be drawn: When C_p increases, the EWR will be significantly reduced. When C_p increased from 2 to 3 and then 4 g/l, EWR decreased sharply. However, if C_p continues to increase to 5 and 6 g/l, the EWR will decrease sharply. The increase in S_p powder size from 100 to 500 μm causes a slight increase in EWR. However, when the powder size was increased from 500 to 1000 nm, the EWR decreased. When T_{on} increases from 10 to 20 μs, EWR increases sharply. If T_{on} continues to increase to 30 μs, the EWR decreases. Besides, it is easy to see that EWR increases as T_{off} increases from 10 to 20 μs, but it decreases as T_{off} continues to increase from 20 to 30 μs. The influence of IP and SV on S/N is described as follows: When IP increases from 4 to 12 A, EWR increases sharply. When SV increases from 3 to 7 V, EWR increases. As mentioned above, the optimal input parameters for minimum EWR will correspond to the maximum S/N values. Therefore, from Fig. 3, it is easy to determine the optimal PMEDM factors for the minimum EWR (see Table 5).

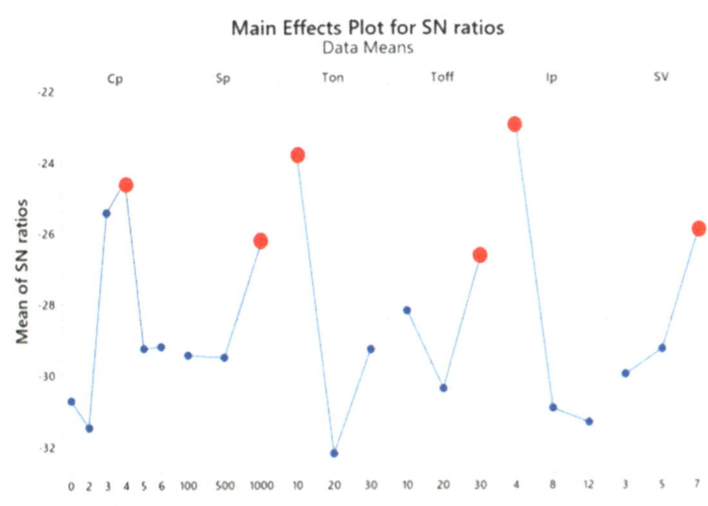

Fig. 3. Effect of input parameters on the S/N ratios

Table 5. Optimum input parameters for minimum EWR

Parameter	Code	Unit	Level	Value
Powder concentration	C_p	g/l	4	4
Powder size	S_p	nm	3	1000
Pulse on time	T_{on}	µs	1	10
Pulse off time	T_{off}	µs	3	30
Peak current	IP	A	1	4
Servo voltage	SV	V	3	7

To evaluate the experimental model, the error distribution histograms were established (Fig. 4). From Fig. 4, the following observations can be made: In the chart of the normal distribution of error (Fig. 4.a), the errors of the experimental points (the blue points) are located close to the distribution line standard (solid red line). So it can be said that the error level is small. The Histogram chart (Fig. 4.c) is a graph that shows the frequency of errors occurring. This chart shows the errors that occur in the interval around 0 occupy a large proportion. From the other two graphs (Figs. 4.b and 4.d), it can be seen that the experimental errors are the random distribution. It also means that the model that has been built is largely influenced by the selected input parameters and is not affected by order of experiments.

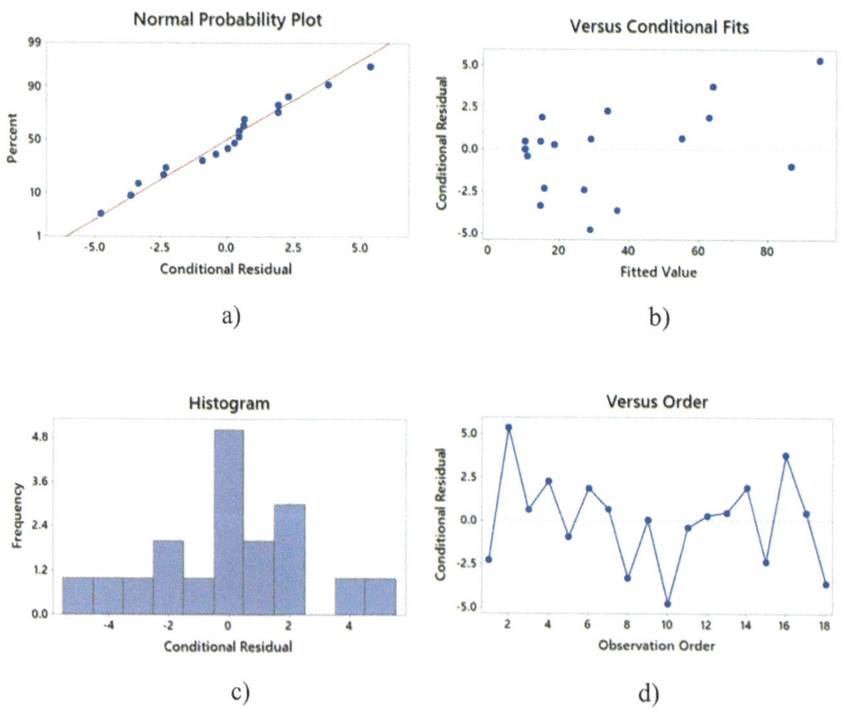

Fig. 4. Conditional residual plots for EWR

The fit of the experimental model with the optimal PMEDM parameters is checked by the Anderson-Darling method with the use of Minitab 19 software. The test results are shown in Fig. 5. From this figure, the experimental data (the blue dots) are in the region bounded by two upper and lower limits with a 95% significance level. In addition, the P-value = 0.272 is greater than the $\alpha = 0.05$ value. That shows that the applied experimental model is consistent with the above significance level.

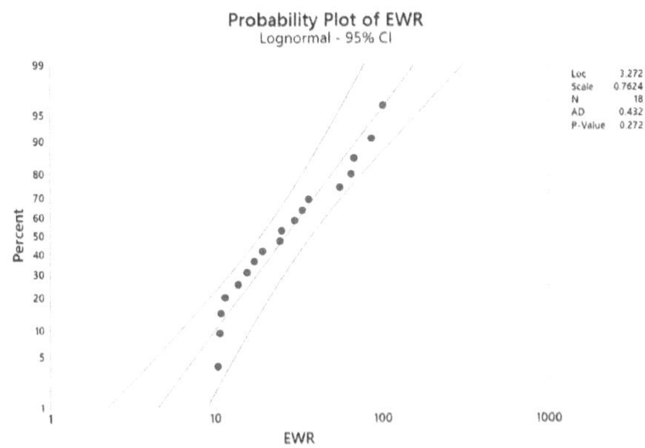

Fig. 5. Probability plot of EWR

4 Conclusions

The present work deals with an optimization study on PMEDM when machining cylindrically shaped parts made from SKD11 steel to get the minimum EWR. In this study, an experiment was conducted. Also, the Taguchi method was used for experimental design and result's analysis. Moreover, six main input parameters, including the powder size, the powder concentration, the pulse on time, the pulse off time, the current, and the servo voltage on the MRS were taken into the experiment for evaluation of their effect on the EWR. Finally, optimum input PMEDM factors were recommended for getting the minimum EWR.

Acknowledgment. This work was supported by Thai Nguyen University of Technology.

References

1. Hong, T.T., et al.: Effect of process parameters on machining time in PMEDM cylindrical shaped parts with silicon carbide powder suspended dielectric. Mater. Sci. Forum (2021)
2. Long, B.T., Phan, N.H., Cuong, N., Jatti, V.S.: Optimization of PMEDM process parameter for maximizing material removal rate by Taguchi's method. Int. J. Adv. Manuf. Technol. **87** (5–8), 1929–1939 (2016)

3. Ahmad, S., et al.: Surface roughness and surface topography of Inconel 718 in powder mixed dielectric electrical discharge machining (PMEDM). Int. J. Integr. Eng. **10**(5) (2018)
4. Das, A., Tharra, B., Siva Rao Sammeta, V.V.N., Barma, J.D.: Attribute of SiC powder additive mixed EDM on machining performance and surface integrity aspects of inconel 625. In: Pandey, K.M., Misra, R.D., Patowari, P.K., Dixit, U.S. (eds.) Recent Advances in Mechanical Engineering. LNME, pp. 869–879. Springer, Singapore (2021). https://doi.org/10.1007/978-981-15-7711-6_86
5. Haque, R., et al.: Improvement of surface quality of Ti-6al-4v alloy by powder mixed electrical discharge machining using copper powder. In: Mechanical Engineering. Facta Universitatis (2021)
6. MangapathiRao, K., Vinaykumar, D., Chandra Shekar, K., Kumar, R.R.: Investigation and analysis of EDM process – a new approach with Al2O3 nano powder mixed in sunflower oil. IOP Conf. Ser. Mater. Sci. Eng. **1057**(1), 012059 (2021)
7. Hong, T., et al.: Multi-objective optimization of PMEDM process of 90CrSi alloy steel for minimum electrode wear rate and maximum material removal rate with silicon carbide powder. Mater. Sci. Forum **1018**, 51–58 (2021)
8. Huu-Phan, N., et al.: Multi-criteria decision making using preferential selection index in titanium based die-sinking PMEDM. J. Korean Soc. Precis. Eng. **36**(9), 793–802 (2019)
9. Padhee, S., et al.: Multi-objective parametric optimization of powder mixed electro-discharge machining using response surface methodology and non-dominated sorting genetic algorithm. Sadhana **37**(2), 223–240 (2012)
10. Patel, S., Thesiya, D., Rajurkar, A.: Aluminium powder mixed rotary electric discharge machining (PMEDM) on Inconel 718. Aust. J. Mech. Eng. **16**(1), 21–30 (2018)
11. Kung, K.-Y., Horng, J.-T., Chiang, K.-T.: Material removal rate and electrode wear ratio study on the powder mixed electrical discharge machining of cobalt-bonded tungsten carbide. Int. J. Adv. Manuf. Technol. **40**(1–2), 95–104 (2009)
12. Boopathi, R., Sundaram, S.: Influence of process parameters for electrical discharge machine using nano particle and brass electrode. Int. J. Res. Advent Technol. **2**(11), 45–49 (2014)
13. Rouniyar, A.K., Shandilya, P.: Multi-Objective optimization using Taguchi and grey relational analysis on machining of Ti-6Al-4V alloy by powder mixed EDM process. Mater. Today Proc. **5**(11), 23779–23788 (2018)
14. Le, Q.D., et al.: Effect of low-frequency vibrations on MRR, EWR and Ra in powder-mixed electrical discharge machining. Int. J. Modern Physics B **34**(22n24), 2040153 (2020)
15. Bhaumik, M., Maity, K.: Multi-objective optimization of PMEDM using response surface methodology coupled with fuzzy based desirability function approach. Decis. Sci. Lett. **6**(4), 387–394 (2017)
16. Kumar, S., et al.: Multi-objective parametric optimization using the Grey-Taguchi approach for powder mixed electric discharge machining of grade-II titanium alloy with cryogenically treated electrodes (2017)
17. Talla, G., Gangopadhyay, S., Biswas, C.: Effect of powder-suspended dielectric on the EDM characteristics of Inconel 625. J. Mater. Eng. Perform. **25**(2), 704–717 (2016)
18. Al-Khazraji, A., Amin, S.A., Ali, S.M.: The effect of SiC powder mixing electrical discharge machining on white layer thickness, heat flux and fatigue life of AISI D2 die steel. Eng. Sci. Technol. Int. J. **19**(3), 1400–1415 (2016)
19. Long, B.T., et al.: enhanced material removal rate and surface quality of H13 steel in electrical discharge machining with graphite electrode in rough machining. Int. J. Sci. Eng. Technol. **4**(2), 104–109 (2015)
20. Yaşar, H., Ekmekci, B.: Ti-6Al-4V surfaces in SiC powder mixed electrical discharge machining. Adv. Mater. Res. (2014)

Monitoring and Evaluating the Fermentation Level of Black Tea Using the Random Forest Model

Du Huy Dao[1(✉)], Nhung Cam Tang[1], and Binh Thanh Pham[2]

[1] Faculty of Electronics, Thai Nguyen University of Technology, Thai Nguyen, Vietnam
daohuydu@tnut.edu.vn
[2] Northern Mountainous Agriculture and Foretry Science Institute, Thai Nguyen, Vietnam

Abstract. In the production of black tea, the fermentation stage plays a key role Abstract: In the production of black tea, the fermentation stage plays a key role in determining the quality of the finished tea. The color will often change during each stage of the fermentation process. In order to assess the degree of fermentation of black tea, tea producers must observe the change of color directly with their eyes, until the color turns red and the fermentation process has reached. To be able to accurately determine the color of standard fermented tea by manual method is not high, resulting in substandard tea quality. In this paper, the author proposes a model to evaluate the fermentation level of black tea based on the computer vision system combined with the Random Forest (RF) model on the HSV chart. The database for analysis and evaluation was collected by the author directly from Song Lo black tea factory, Tuyen Quang province. Besides, the author also compares this algorithm with algorithms such as SVM (Support Vector Machine) and MLP (Multi Layer Perceptron) to see that the classification efficiency of the RF algorithm is the highest.

Keywords: Black tea · Fermentation · Random forest · HSV

1 Introduction

Black tea is a popular beverage in the world, in addition to being processed into many different attractive drinks [1], black tea also brings a lot of benefits to human health such as preventing cancer, degeneration. nerve, prevent diabetes, support treatment of cardiovascular disease, enhance metabolism [2, 3]. Black tea production accounts for nearly 80% of total tea production worldwide, in which there are 4 largest black tea producing countries in the world [4]: China, India, Kenya, Sri Lanka. VietNam is the 7th largest producer of black tea in the world.

In the production of black tea includes five stages. The first stage is started with the plucking of green tea, where two leaves and a bud is the standard. The next step is withering, where tea leaves are spread on a withering bed for them to lose moisture. There is then the cut, tear, and curl step, here tea leaves are cut and torn to open them up for oxidation. The fermentation stage is where tea reacts with oxygen to produce

D. C. Nguyen et al. (Eds.): ICERA 2021, LNNS 366, pp. 739–753, 2022.
https://doi.org/10.1007/978-3-030-92574-1_76

compounds that are responsible for the quality of tea. Heat is passed through tea in the drying stage to remove moisture, and the last is sorting where tea is put into various categories based on their quality. So fermentation is an extremely important stage, it determines the characteristics of tea both in color, flavor and shape of tea [5]. The characteristic pigments for black tea are mainly formed during fermentation. The color of black tea is a combination of many different color compounds including the following main types: Theaflavins (TFs), Thearubigins (TRs), Anthocyanidin [6, 7].

Moreover, black tea has a significant change in color during fermentation from turquoise to greenish yellow, yellowish brown and finally dark brown. The color of tea depends on the absolute content of Theaflavin (yellow) and Thearubigin (red) and depends on the content of these two color groups. The red color of black tea is due to the transformation of Dicatechin and Theaflavin. If we want the tea to have a dark color, we must prolong the fermentation, but the enhancement of the color of the mixed water will reduce the flavor of the tea due to the aroma forming only reaching the maximum limit at the third fermentation time. Therefore, to increase the fermentation time to obtain a dark water color, the protection of fragrance can be said to be impossible. The color of the tea only increased for a certain time when fermenting then the color decreased. Color is an important property and image features recognition in computer vision systems. Color information in tea leaves can be accurately identified and described using computer vision system to collect images and extract color characteristics. This color change process can be observed and distinguished by the human eye but it is very difficult to determine the threshold of color. The determination of the human color threshold is only qualitative. However, in large-scale black tea production, the degree of fermentation of black tea and sensory quality is currently predicted through experience through color variation. Therefore, the fermentation process is sometimes not accurate because the fermentation process has not reached or exceeded and the color of the tea is not uniform and the tea flavor is not predicted correctly. Specifically, colors relate to the level of fermentation as follows:

- Red pink: fermented enough
- Brown: excessive fermentation
- Green speckled color: fermentation is not enough, in this case it is necessary to find out the cause to overcome it, possibly due to previous stages such as wilting and crumbling.

Based on the color characteristics during fermentation can be used as a basis to optimize the factors affecting fermentation such as temperature, humidity, fermentation time. Currently, there has been no research project to control the factors affecting colorbased fermentation, but only separate studies for each stage. Surajit Borah [8, 9] has set up a detection system, which can be used to detect the degree of fermentation of broken black tea. In this system, images of tea leaves are collected intermittently during fermentation. Histogram HIS and RGB are extracted, the image with the highest sensory level will be considered as the standard image in the group, then the distance (DPV value) between the measured image and the target image. The standard is determined using the Manhattan distance algorithm; If the sample has a DPV value of 3 RGB channels lower than 0.3, it is considered that the fermentation is not achieved. Xueyan Wu [10] based on the optimal Entropy calculation algorithm for Wavelet resolution and KRecent Neighbor algorithm to develop tea learning system based on machine learning and machine vision for the purpose of classifying tea each other automatically and accurately. Chunwang Dong's article demonstrated the image color characteristics and the nonlinear method that can be used to evaluate changes in quality indicators (eg. Sensory quality and pigment).during fermentation of black tea [11]. Research project to build the optimal automation system for processing and preserving agricultural products of Nguyen Xuan Quynh has put the optimal control technology into black tea production line to optimize automation. The process of processing and preserving tea in both OTD and CTC processing methods [12].

2 Proposed Technical

With the method applied in this paper, the author will build a system to identify and classify the degree of black tea fermentation based on three time points of 60 min, 90 min and 120 min. The identification model will give the degree of fermentation of black tea based on the comparison of images collected from the dark and compared with a set of standardized samples and assessed the level of fermentation by tea experts working at factory.

2.1 Building an Image Acquisition System from the Black Tea Brewing Line

Black tea fermentation line at Song Lo tea factory – Tuyen Quang is illustrated as shown in Fig. 1

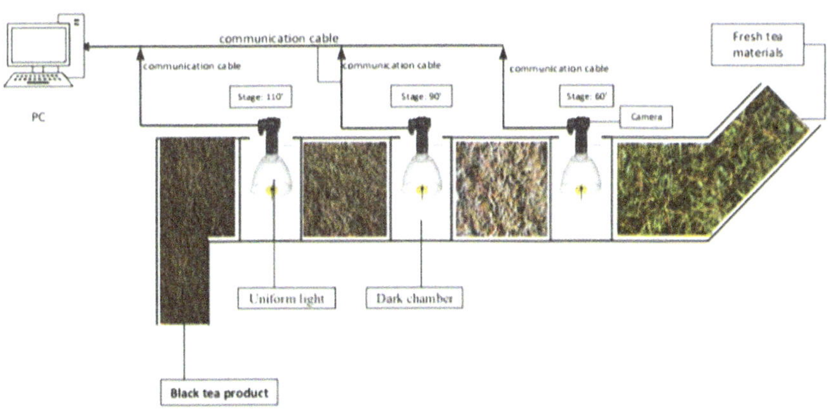

Fig. 1. Images data collection system in black tea fermentation line

Through experimental research at the time points: 60 min, 90 min and 120 min, the image of fermented tea changed the most. The system uses information from images collected automatically in the darkroom at the three locations mentioned above. Figure 2 illustrates the dark chamber designed and installed at the black tea production line

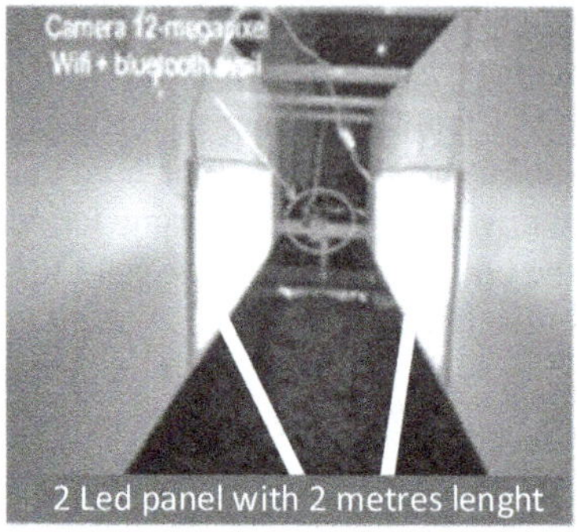

Fig. 2. The dark chamber is designed and installed in the black tea production line

The black tea image device acquisition part plays an important role in the construction of fermentation evaluation model. In particular, the black tea line is placed in the environment is a factory with low light intensity, limited network communication, high humidity (because of high humidity tea fermentation). Therefore, to collect highly accurate data, it is necessary to satisfy factors such as: light uniform, not affected by interference from the outside environment, area size large enough sampling.

Collected Images Database

The database has collected standard samples at the input time points, that is when tea reaches the level of about 0–30% at 03 stages, about 30–60% at 03 stages, about 60–90% at 03 stages; greater than 90% at 03 periods; compared with the fermentation standard.

Stage	Level 1 (The level of about 0% ÷ 30%)	Level 2 (The level of about 31% ÷ <60%)	Level 3 (The level of about 61% ÷ <90%)	Level 4 (The level of about >90%)
60 min	100 images	170 images	150 images	120 images
90 min	125 images	157 images	150 images	120 images
120 min	90 images	115 images	130 images	90 images

Figures 3, 4, 5 and 6 are some standard images of tea fermenting at various degrees at the 60-min period provided by the brewery's experts.

Fig. 3. Image of fermented tea reaching 0%–30%

Fig. 4. Image of fermented tea reaching 31%–60%

Fig. 5. Image of fermented tea reaching 61%–90%

Fig. 6. Image of standard fermented tea

2.2 Establishment the Identity Model

The Basis of Feature Extraction Design

In the manual tea quality assessment method, experts rely on the color of the tea leaves to change. However, in the visual system computer, we need to pay attention to image processing techniques to enhance the performance system results. For example, the same color may differ in the image if there is a difference changes in lighting direction, photographic range, and light source intensity In the paper, the authors use analytical methods the colors of tea fermenting in stages by taking the value of the average color of the image, Figs. 7 and 9 illustrates the degree of fermentation of tea input, after 30 min, 60 min Fig. 7.

Fig. 7. Photo of tea and typical color of the input stage

Fig. 8. Photo of tea and typical color after 30 min

Fig. 9. Photo of tea and typical color of 60 min stage

Through illustrated in Figs. 7, 8 and 9, we can see the sharp turn through the stages doimau. Therefore, in the following sections, the characteristics of tea images through the fermentation stages will be analyzed and extracted on the basis of color.

Color Histogram Comparison Method

Although established to create an environment of constant illumination in the process of collecting images, however, still a number of dissimilar elements may appear in practice. This problem can be overcome by choosing to use the appropriate color model. A color histogram represents the distribution of colors in an image. For digital images, the color histogram represents the number of pixels that have a value within a fixed range, spanning the color space of the image. A color histogram is a vector where each dimension of the vector stores the number of pixels of a certain color in the image. The axes of the chart are evenly divided with fixed intervals called "bin". Swain and Ballard [13] stated in their study that the histogram is invariant for rotation and translation of different viewing axes, and changes only slightly with different viewing angles. Therefore, the color histogram comparison technique was applied to compare the color in the tea image taken at the present time with the sample image set during the same incubation period. The measure of similarity between two images was introduced by Swain and Ballard in color histogram intersection method. According to this method, given 2 color histograms H(I) and H(Q) of 2 images I and Q respectively, each color histogram consists of n bin, the similarity based on the color histogram intersection is calculated as follows:

$$S(H(I), H(Q)) = \frac{\sum_{j=1}^{n} min\{h_j(I), h_j(Q)\}}{N_Q \times M_Q} \tag{1}$$

Where:

- $\boldsymbol{h_j(I)}$**:** Number of pixels with value j in the image I
- $\boldsymbol{N_Q \times M_Q}$**:** the standard image size
- n: the number of bin in color chart
- S: similarity

The above formula can be expressed by another method, measuring the difference between two color histograms by the Manhattan distance (L1 norm) is calculated as follows:

$$D(H(I),(h(Q)) = \frac{1}{N_I \times M_I} \times \sum_{j=1}^{n} \left|h_j(I) - h_j(Q)\right| \quad (2)$$

Where:

- $h_j(I)$: Number of pixels of image I in bin j
- n: The number of bin in color chart
- $N_I \times M_I$: image size I
- D: dissimilarity

With a given threshold T, two images are said to have similar colors if the dissimilarity D < T.

Fig. 10 illustrates the experimentt for calculating the dissimilarity between 2 images in the same fermentation time period:

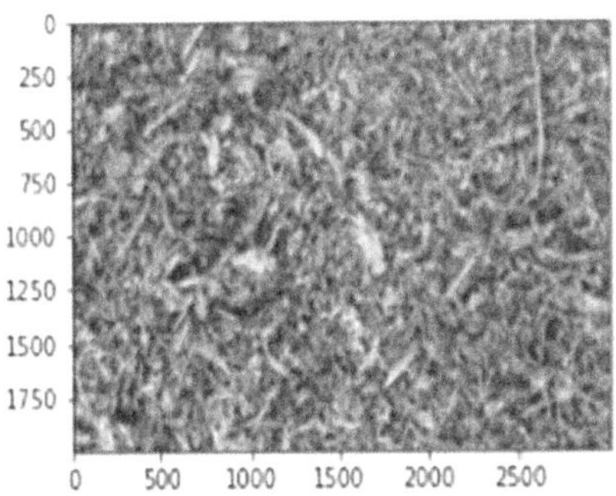

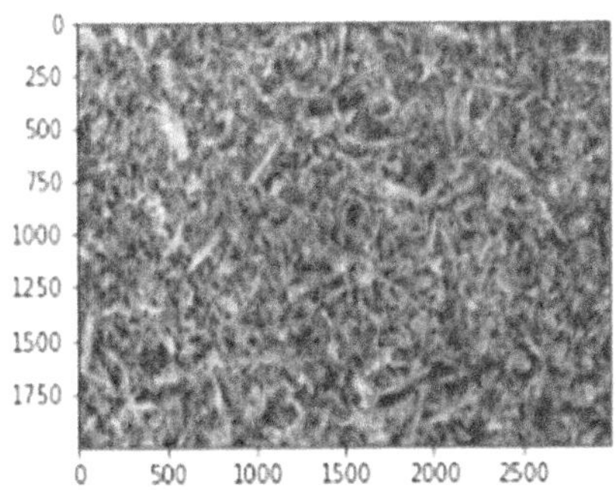

Fig. 10. Two images of tea fermenting in the same at 60 min stage

The distance values D calculated by formula (3.2) in each color space are respectively:

In the RGB color channel:

- Red channel: D = 0.12
- Green channel: D = 0.11
- Blue channel: D = 0.10

In the HSV color channel:

- Hue channel: D = 0.09
- Saturation channel: D = 0.06
- Value channel: D = 0.12

In the HSV color channel:

- Value channel D = 1.41
- a channel: D = 1.98
- b channel D = 1.54

In terms of color histogram variation, the LAB channel is quite large, so it is not suitable to evaluate the similarity. Evaluation of the color histogram on the 2 color channels RGB and HSV is illustrated in Figs. 11 and 12.

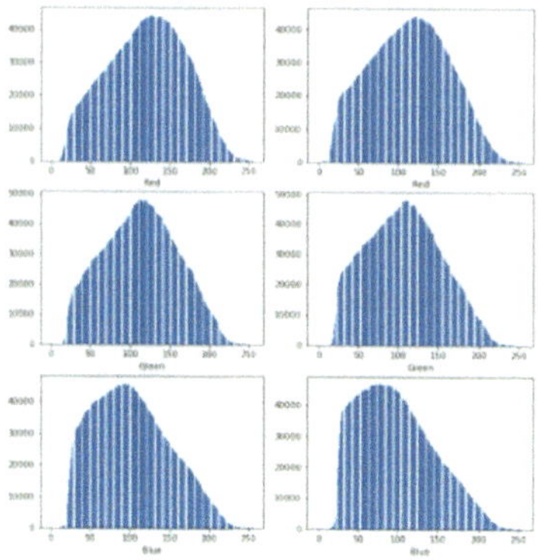

Fig. 11. RGB color channel histogram of 2 images in the same fermentation stage.

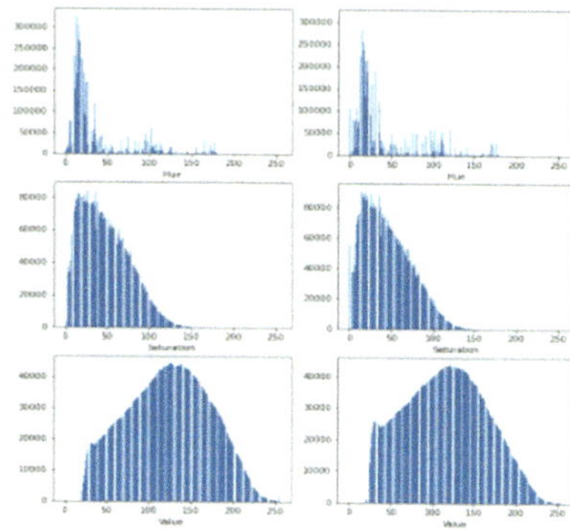

Fig. 12. HSV color channel histogram of 2 images in the same fermentation stage

Realizing that there is a large variation between the color chart in 2 photos reviewed by channel RGB color by color chart RGB affected by the lighting, and information on each color channel R, G, B are including light intensity – variable factor in each tea photo. - The two color histograms in Hue and Saturation color channels between the two images have the same shape because the light intensity color channel has been separated into a separate color channel, so it is less affected by noise. The Value color channel represents highly variable luminance. From these comments, it can be concluded that the image features are obtained from two Hue channels, and that the Saturation channel in the HSV color histogram is effective in assessing the degree of black tea fermentation.

Dissimilarity Thresholding

To determine the maximum threshold of the disagreement between the two images in the same period of fermentation standards in mold 60 min, the author implement any similarities calculate cross between pictures. The results obtained in Table 1 by Hue channel and Table 1 by Saturation channel (Tables 2, 3 and 4):

Table 1. Dissimilarity of fermentation images reaching Hue channel

	0.jpg	1.jpg	2.jpg	3.jpg	4.jpg	5.jpg
0.jpg	0	**0.26**	**0.105**	**0.183**	**0.186**	**0.256**
1.jpg	0.26	0	**0.218**	**0.139**	**0.149**	**0.145**
2.jpg	0.105	0.218	0	**0.122**	**0.119**	**0.196**
3.jpg	0.183	0.139	0.122	0	**0.045**	**0.142**
4.jpg	0.186	0.149	0.119	0.045	0	**0.117**
5.jpg	0.256	0.145	0.196	0.142	0.117	0

Table 2. Dissimilarity of fermentation images with Saturation channel

	0.jpg	1.jpg	2.jpg	3.jpg	4.jpg	5.jpg
0.jpg	0	**0.058**	**0.059**	**0.079**	**0.102**	**0.151**
1.jpg	0.058	0	**0.086**	**0.062**	**0.072**	**0.102**
2.jpg	0.059	0.086	0	**0.052**	**0.086**	**0.167**
3.jpg	0.079	0.062	0.052	0	**0.035**	**0.120**
4.jpg	0.102	0.072	0.086	0.035	0	**0.090**
5.jpg	0.151	0.102	0.167	0.120	0.090	0

The maximum value between dissimilarity values in the same standard set is called the dissimilarity threshold:

$$D_{th} = \max_{1,Q=1\to k}\left\{\frac{1}{N_I \times M_I} \times \sum_{j=1}^{n}\left|h_j(I) - h_j(Q)\right|\right\} \tag{3}$$

The purpose of thresholding is the maximum value of dissimilarity in the database is to select the two images with the most dissimilarity colors in the same period of fermentation. If the images are arranged in a certain color order, the selected images will have colors located at the two extremes of the database. Therefore, those images will be considered as standard images for the qualified black tea fermentation database for color comparison with images from the tea database at other stages. Specifically, an It image will be considered qualified if the following conditions are satisfied and vice versa:

$$D(I_{max}, I_t) < D_{th} \text{ và } D(I_{min}, I_t) < D_{th} \tag{4}$$

Feature Extraction

Based on the analysis results in the color histogram comparison method, we see two components of color and saturation. The purpose of analyzing the color of tea images is to compare the color of images of unfermented tea. standard with the color of the standard fermented tea image. Two standard images (selected after considering Dth) of the standard fermented tea image database as the basis for additional color features; are located at the two extremes of the standard database, so those two images are called Imax, Imin of the database. Then, the dissimilarity is calculated for both the color space and saturation of each image in the entire database for these two standard images. These values are used as tea image features. In addition, the correlation between the 2 color histograms of the 2 channels Hue and Saturation colors are also selected as image features:

The following features are selected for tea image as color description to represent the image according to color content:

The following features are selected for tea image as color description to represent the image according to color content:

- Average value of H color channel
- Maximum dissimilarity of H color channel
- Minimum dissimilarity of H channel
- Average value of S color channel
- Maximum dissimilarity of S color channel
- Minimum dissimilarity of H color channel
- The correlation between the H and S color channel histograms

From the above characteristics, the author will proceed to build machine learning models to classify the degree of fermentation of black tea according to 4 groups as follows:

- Tea at the input stage: 0–30%
- Tea at the stage: 30–60%
- Tea at 60–90% stage
- High quality fermented tea >90% From the above characteristics, the author will proceed to build machine learning models to

Classify the degree of fermentation of black tea according to 4 groups as follows:

- Tea at the input stage: 0–30%
- Tea at the stage: 30–60%
- Tea at 60–90% stage
- High quality fermented tea >90%

Construction Machine Learning Models to Classify the Degree of Fermentation From the features extracted from the methods mentioned in section Feature extraction, visualize the database consisting of 7-dimensional feature vectors divided into 4 groups before being included in the classification model.

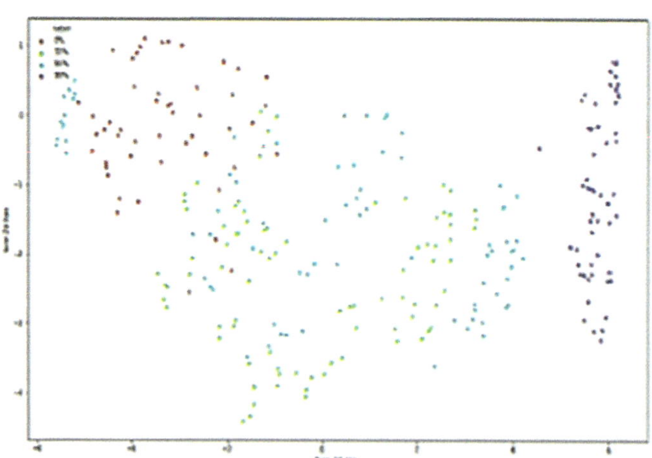

Fig. 13. The result data classification degree of fermented black tea

From Fig. 13, it can be seen that the input and output layers are now clearly separated - Layers at the stage of 30% and 60% are being partially overlapped by noise due to errors in image acquisition and image quality assessment from labelers. In this paper, we use 03 machine learning models to classify different groups of black tea fermentation levels including: • Random Forest Model • Support Vector Machine Model • Multi Layer Perceptron Model The classification results are shown in Sect. 3 test results.

3 Result of Experiment

Rating Measure

There are many metrics to measure the effectiveness of a machine learning model. Depending on different problems, we use different metrics. The simplest and most commonly used way is accuracy. This evaluation simply calculates the ratio between the number of correctly predicted points and the total number of points in the test dataset. In this problem, it is often defined that the more important data class that needs to be correctly identified is the Positive class (P-positive), the other class is called Negative (N-negative). We define True Positive (TP), False Positive (FP), True Negative (TN), False Negative (FN) as follows (Fig. 14):

- True Positive occurs when the predicted fermentation level is correct compared to the actual assessment.
- False Positive occurs when the predicted result is to reach the correct level of fermentation but in fact at that level the tea has over-fermented or under-fermented.
- True Negative occurs when the predicted result is not satisfactory and the actual assessment of fermented tea is also unsatisfactory.
- False Negative is the result that occurs when the fermentation level has been reached according to the expert's assessment but the system recognizes it as unsuccessful.

Table 3. Types of errors in prediction

	Prediction is possitive	Prediction is negative
True is possitive	True possitive (TP)	False negative (FN)
True is negative	False possitive (FP)	True negative (TN)

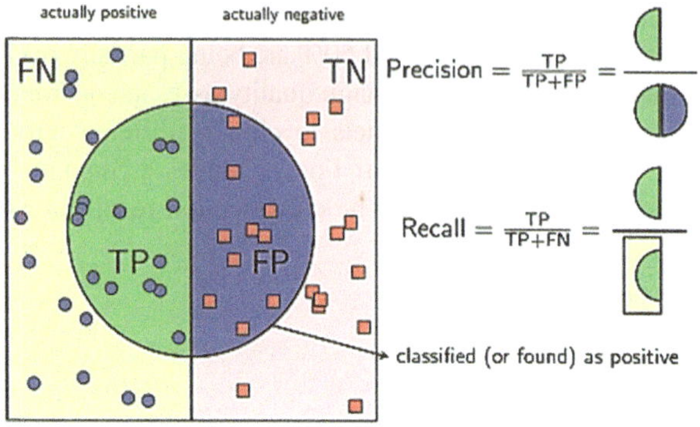

Fig. 14. Precision and recall definition and relationship

Precision is defined as the ratio of the number of true positives among the points classified as positive (TP + FP) or the number of accuracy of the points found.

Recall is defined as the ratio of true positives among those that are actually positives (TP + FN) or the percentage that do not miss positives.

Precison and Recall are two fractions with the same numerator but different denominators. Both Precision and Recall are non-negative numbers less than or equal to one.

$$Precision = \frac{\mathrm{TP}}{\mathrm{TP}+\mathrm{FP}} \tag{5}$$

$$Recall = \frac{\mathrm{TP}}{\mathrm{TP}+\mathrm{FN}} \tag{6}$$

High precision means that the accuracy of the points found is high.

High Recall means high True Positive Rate, which means that the rate of missing really positive points is low. Also, F1-score, is the harmonic function of precision and recall (assuming that these two quantities are non-zero):

$$\frac{2}{F_1} = \frac{1}{\text{precision}} + \frac{1}{\text{recall}}$$

$$F_1 = 2\frac{1}{\frac{1}{\text{precision}} + \frac{1}{\text{recall}}} = 2\frac{\text{precision} * \text{recall}}{\text{precision} * \text{recall}} \tag{7}$$

F1-score has a value in the half range (0,1). The higher the F1, the better the model.

When both recall and precision are equal to 1 (best possible), F1 = 1. Thus, a classifier with precision = recall = 0.5 is better than another classifier with precision = 0.3, recall = 0.8 in this measure.

Experiment Results

Classifier machine learning models Experiment:

- Random Forest (RF)
- Support Vector Machine (SVM)
- Multi Layer Perceptron (MLP)

Table 4. Test results on models

Model	Precision	Recall	F1- Score	Accuracy
SVM	0.95	0.92	0.92	0.93
RF	0.94	0.93	0.94	0.94
MLP	0.89	0.87	0.87	0.87

The confusion matrix Table Figs. 15, 16 and 17 illustrate in detail the classification results between the predicted results and the actual results of the expert assessment (ground-truth) (Fig. 18).

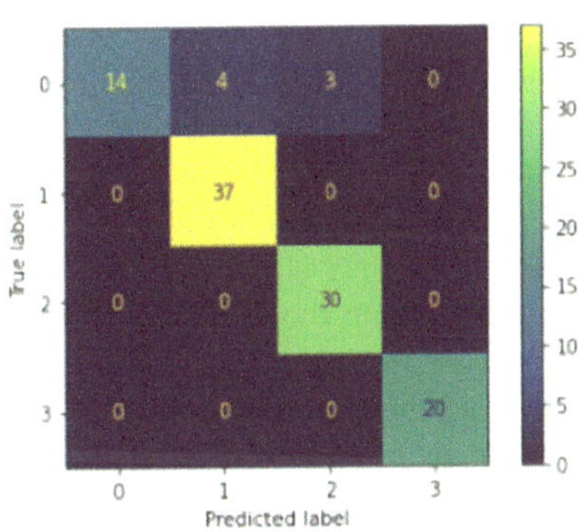

Fig. 15. Confusion Matrix of classes division results of SVM model

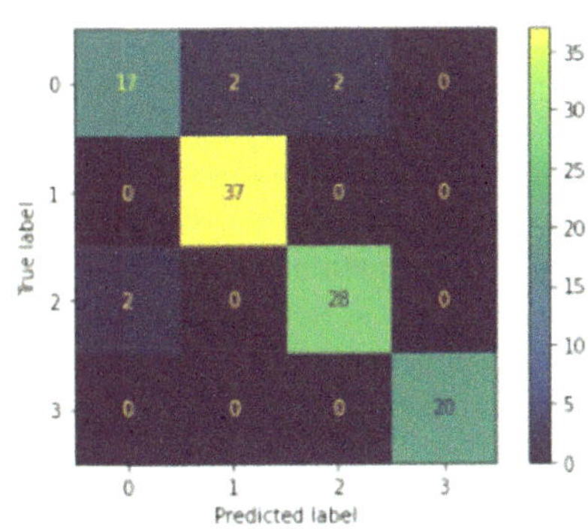

Fig. 16. Confusion Matrix of classes division results of RF model

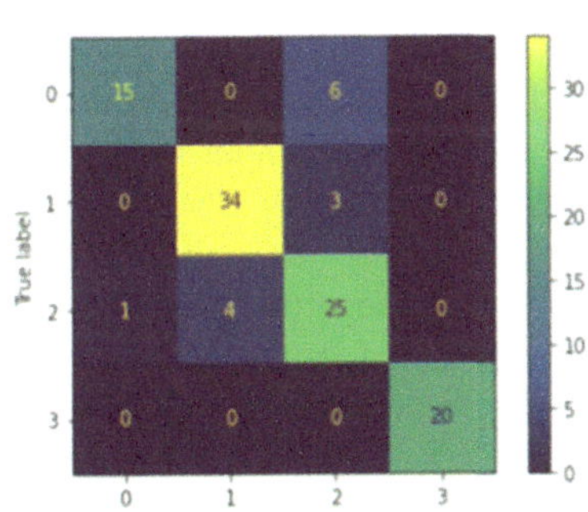

Fig. 17. Confusion Matrix of classes division results of MLP model

Software Interface

Fig. 18. Software interface

On the interface screen, there are 3 windows to evaluate 3 stages of tea brewing after 60 min, 90 min, 120 min. Each stage will run an independent model to be able to evaluate the quality of tea in each stage. Thereby, the system can assess the quality of tea that has met the standard according to the criteria of time. If the quality is not satisfactory or has excessive fermentation, the fan system, the rear humidification nozzle will increase or decrease. Assuming that at the standard position in the 60-min period, the evaluation result is that the fermentation value is only about 50–80%, that is, the tea color is not reached, then it is necessary to speed up the fan, impregnated nozzle to increase the amount of gas. Touch the tea block more than make the color change faster. Similarly, if at the standard position 60 min, the evaluation result is only about 20–50%, the corresponding humidity level must be increased. In [14], we use the color model of CIE lab and use 03 algorithms: SVM, RF and MLP. However, in this paper, the results about the degree of distinction between the algorithms are clearer.

4 Conclusions and Future Work

Applying automation and artificial intelligence into the production process is the trend following industry 4.0, which is the goal of today's businesses.

In this article, the author has applied computer vision to the process of evaluating the quality of black tea fermentation, bringing great benefits such as saving labor costs, increasing productivity and quality. The research results on the obtained data set as stated in the project show that the recognition ability is feasible for the classification machine learning models. The results of the actual test of the article prove the accuracy of the model and the factors used for evaluation analysis.

Acknowledgments. This research is funded by Thai Nguyen University of Technology (TNUT).

References

1. Sigley, G.: Tea and China's rise: Tea, nationalism and culture in the 21st century. Int. Commun. Chin. Cult. **2**, 319–341. Detection and Prevention using Android Application; April 2015, IJISET – International Journal of Innovative Science, Engineering & Technology. **2**(4) (2015)
2. Yiannakopoulou, E.: Green Tea Catechins: Proposed Mechanisms of Action in Breast Cancer Focusing on the Interplay Between Survival and Apoptosis. Anti-Cancer Agents Med. Chem. **14**, 290–295 (2014). [CrossRef]
3. Miura, K., Hughes, M.C.B., Arovah, N.I., Van Der Pols, J.C., Green, A.C.: Black Tea Consumption and Risk of Skin Cancer: An 11-Year Prospective Study. Nutr. Cancer. **67**, 1049–1055 (2015). [CrossRef] [PubMed]
4. General statistics office of VietNam: The statistical analysis reports and forecasts in 2019
5. Xu, M., Wang, J., Gu, S.: Rapid identification of tea quality by E-nose and computer vision combining with a synergetic data fusion strategy. J. Food Eng. **241**, 10–17 (2019). [CrossRef]
6. Jayabalan, R., Malbaša, R.V., Lonˇcar, E.S., Vitas, J.S., Sathishkumar, M.: A review on Kombucha tea-microbiology, composition, fermentation, beneficial effects, toxicity, and tea fungus. Compr. Rev. Food Sci. Food Saf.**13**, 538–550 (2014). [CrossRef]
7. Jolvis Pou, K.: Fermentation: The Key Step in the Processing of Black Tea. J. Biosyst. Eng.41, 85–92 (2016). [CrossRef]
8. Tiep, P.T., Giang, V.H., Hang, T.T.: Curriculum to Prevent Pests and Diseases of Tea, Ministry of Agriculture and Rural Development (2012)
9. Moallem, P., Razmjooy, N.: Optimal threshold computing in automatic image thresholding using adaptive particle swarm optimization. J. Appl. Res. Technol. **10**(5), 703–712 (2012)
10. Choras, R.S.: Image feature extraction techniques and their applications for CBIR and biometrics systems, Int. J. Biol. Biomed. Eng. **1**(1) (2007)
11. Dong, C., Liang, G., Hu, B., Yuan, H., Jiang, Y., Zhu, H., Qi, J.: Prediction of Congou Black Tea Fermentation Quality Indices from Color Features Using Non-Linear Regression Methods. Scientific reports, Published July (2018)
12. Quynh, N.X.: Research Project to build the optimal automation system for processing and preserving agricultural products technology. Ministry Science and Technology (2015)
13. Ballard, D.H., Swain, M.J.: Color indexing. Int. J. Comput. Vis. (1991)
14. Binh, P.T., Du, D.H., Nhung, T.C.: Control and optimize black tea fermentation using computer vision and optimal control algorithm. In: Proceedings of the International Conference on Engineering Research and Applications, ICERA (2019)

Optimization of Transmission Ratios for Two-Stage Bevel Helical Gearboxes Based on Cost Function

Nguyen Hong Linh[1], Tran Thi Phuong Thao[2], Dang Quoc Cuong[3], Trinh Kieu Tuan[4], Le Hoang Anh[5], Nguyen Anh Tuan[4], Nguyen Thanh Tu[2], and Tran Ngoc Giang[2(✉)]

[1] Electric Power University, Hanoi, Vietnam
[2] Thai Nguyen University of Technology, Thai Nguyen, Vietnam
trannngocgiang@tnut.edu.vn
[3] Nguyen Tat Thanh University, Ho Chi Minh City, Vietnam
[4] University of Economics - Technology for Industries, Hanoi, Vietnam
[5] Vinh Long University of Technology Education, Vinh Long, Vietnam

Abstract. This study aims to present the results of an optimization study to find optimum partial gear ratios for two-stage bevel helical gearboxes based on minimizing gearbox cost. In the study, ten main design factors considered as the input parameters are selected to investigate their influences on the responses. Also, a simulation experiment is designed and carried out using computer program. It is found that in addition to the input parameters, their interactions also have important effects in which total ratio gearbox ratio (u_g) have the most impact on the u_2. Besides, a regression model was proposed to predict the second gear ratios u_2. The proposed model is highly consistent to the experimental results. From this model, the other partial gear ratios of the other components are easily determined.

Keywords: Cost optimization · Bevel helical gearbox · Gear ratio · Gearbox cost · Systems engineering

1 Introduction

In gearbox design, determination of optimum gear ratios is very important. This is because they are the most important factors affecting the mass, the dimensions, the size, and the cost of a gearbox. That is the reason why up to now many studies have been carried out on finding optimum gear ratios of different gearboxes. Studies have been conducted on gearboxes with many stages such as two-stage gearboxes [1–5], three-stage gearboxes [6–9], four-stage gearboxes [10] etc. Also, different methods have been applied to optimize the gear ratios, such as graph method [11], practical method [12], and model method [13, 14]. In the model method, models for finding optimum gear ratios were suggested for different objectives. For example, a model for calculating the minimum height of two state bevel helical gearboxes was proposed in [15]. For three stage bevel helical gearboxes, the minimum length of the gearbox was chosen as the objective of the optimization problem [16]. The objectives of the optimization studies

D. C. Nguyen et al. (Eds.): ICERA 2021, LNNS 366, pp. 754–766, 2022.
https://doi.org/10.1007/978-3-030-92574-1_77

can be the minimum area of gearbox cross section [2, 5, 18], the minimum total mass of gearbox [19], or the gearbox cost [20].

It is aware that the cost of a gear box is directly impacted by its gear ratios. Hence, this article dedicates to study a cost optimization study in terms of calculating the optimum gear ratios of a bevel helical gearbox. In the study, ten input parameters are purposely selected to explore their dependences on the cost function. Also, a simulation experiment is performed using computer programming to execute the optimization problem. Finally, a regression equation to calculate the optimum gear ratio u_2 was presented.

2 Methodology

2.1 Cost Analysis

In practice, cost components such as the bearing cost, the gear cost, the shaft cost, and the gearbox housing cost have been known to impact the cost of a gearbox. If the cost of the bearing is ignored because its determination is very complicated, the cost of a two-stage bevel helical gearbox, C_{gb}, can be found as:

$$C_{gb} = C_{bg} + C_{hg} + C_{gh} + C_s \quad (1)$$

Where, C_{bg}, C_{hg}, C_{gh} and C_s are the bevel gear cost, the helical gear cost, the gearbox housing cost, and the shaft cost, respectively. These cost elements are calculated by:

$$C_{bg} = c_{bg.m} \cdot m_{bg} \quad (2)$$

$$C_{hg} = c_{hg.m} \cdot m_{hg} \quad (3)$$

$$C_{gh} = c_{gh.m} \cdot m_{gh} \quad (4)$$

$$C_s = c_{s.m} \cdot m_s \quad (5)$$

Where, $c_{bg.m}$ and $c_{hg,m}$ are the cost per kilogram of bevel and helical gears (USD/kg); m_{bg} and m_{hg} are the mass of all bevel and helical gears in the gearbox (kg). $C_{gh,m}$, $C_{s,m}$ (USD/kg) are the cost per kilogram of gearbox housing and shaft (USD/kg); m_{gh}, m_s (kg) denote the mass of gearbox housing, and all shafts, respectively. Replacing Eqs. (2) to (5) into (1) gets:

$$C_{gb} = c_{bg.m} \cdot m_{bg} + c_{hg} \cdot m_{hg} + c_{gh.m} \cdot m_{gh} + c_{s.m} \cdot m_s \quad (6)$$

2.2 Gear Mass Calculations

The mass of bevel gears and helical gears (m_{bg} and m_{hg}) can be determined by:

$$m_{bg} = m_{bg1} + m_{bg2} \tag{7}$$

$$m_{hg} = m_{hg1} + m_{hg2} \tag{8}$$

With

$$m_{bg1} = \rho_g \cdot \left\{ (\pi/3) \cdot \left[d_{e11}^2 \cdot R_e \cdot cos\delta_1/4 - d_{e11}'^2 \cdot (R_e - b) \cdot cos\delta_1/4 \right] \right\} \tag{9}$$

$$m_{bg2} = \rho_g \cdot \left\{ \left(\frac{\pi}{3}\right) \cdot \left[d_{e21}^2 \cdot R_e \cdot cos\delta_2/4 - d_{e21}'^2 \cdot (R_e - b) \cdot cos\delta_2/4 \right] \right\} \tag{10}$$

$$m_{hg} = \rho_g \cdot \left(\pi \cdot e_1 \cdot d_{w11}^2 \cdot \frac{b_w}{4} + \pi \cdot e_2 \cdot d_{w21}^2 \cdot \frac{b_w}{4} \right) \tag{11}$$

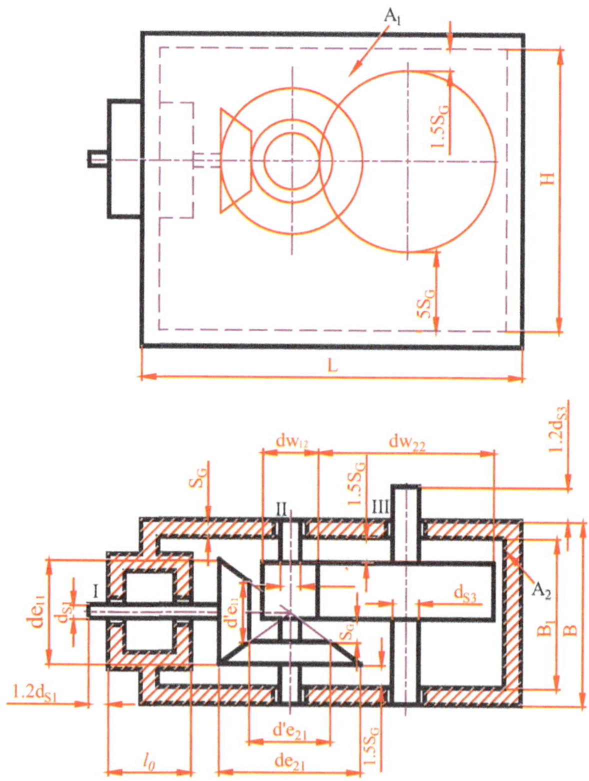

Fig. 1. Schema for determination of gearbox mass

Where, m_{bg1} and m_{bg2} are the mass of the pinion and the bevel gear (kg); m_{hg1} and m_{hg2} are the mass of the pinion and the helical gear (kg); ρ_g is the weight density of gear material (kg/m^3); with the gear material as steel, $\rho_g = 7.82$ [21]; e_1 and e_2 are the volume coefficients of the pinion and the gear of the helical gear set; $e_1 = 1$ and $e_2 = 0.6$ [23]; b_w is the gear width of the helical gear set (mm); $b_w = \psi_{ba} \cdot a_w$ (mm); The external cone distance R_e is determined based on the pitting resistance of the straight bevel gear set (mm) [23]:

$$R_e = k_R \cdot \sqrt{u_1^2 + 1} \cdot \sqrt[3]{T_{11} \cdot \frac{k_{h\beta 1}}{\left[(1 - k_{be}) \cdot k_{be} \cdot u_1 \cdot [\sigma_H]^2\right]}} \tag{12}$$

In which, k_R is the face width coefficient; for a steel straight bevel gear, $k_R = 50$ (MPa) [3]; $k_{H\beta}$ is the contacting load ratio for pitting resistance; for a straight bevel gear set $k_{H\beta} = 1.04 \div 1.18$ [23] and we can choose $k_{H\beta} = 1.11$; $k_{be} = b/R_e = 0.25 \div 0.3$ is the face width coefficient and it is possible to take $k_{be} = 0.27$; T_{11} is the torque on the pinion; d_{e11} and d_{e21} are the outer pitch diameters of the pinion and the gear; For straight bevel gear set, we have [23]:

$$T_{11} = T_{out} / \left(u_t \cdot \eta_{bg} \cdot \eta_{hg} \cdot \eta_{be}^3\right) \tag{13}$$

$$d_{e11} = 2 \cdot R_e / \left(1 + u_1^2\right)^{1/2} \tag{14}$$

$$d_{e21} = 2 \cdot u_1 \cdot R_e / \left(1 + u_1^2\right)^{1/2} \tag{15}$$

Wherein u_t is the total gearbox ratio; η_{bg} is the efficiency of bevel gear set $\eta_{hg} = 0.95 \div 0.97$ η_{hg} is the efficiency of helical gear set $\eta_{hg} = 0.96 \div 0.98$ [23]; η_{be} is the efficiency of a rolling bearing pair $\eta_{be} = 0.99 \div 9.995$ [23]. Choosing $\eta_{bg} = 0.96$; $\eta_{hg} = 0.97$, $\eta_{be} = 0.992$ we have:

$$T_{11} = 1.1001 \cdot T_{out} / u_t \tag{16}$$

$$d_{w11} = 2 \cdot a_{w1} / (u_1 + 1) \tag{17}$$

$$d_{w21} = 2 \cdot a_{w1} \cdot u_i / (u_i + 1) \tag{18}$$

Where, a_w is the centre distance of the helical gear set which is found by [23]:

$$a_w = k_a \cdot (u_2 + 1) \cdot \sqrt[3]{T_{12} \cdot k_{H\beta} / \left([\sigma_{H2}]^2 \cdot u_2 \cdot X_{ba}\right)} \tag{19}$$

With

$$T_{12} = T_{out} / \left(u_2 \cdot \eta_{hg} \cdot \eta_{be}^2\right) \tag{20}$$

With $\eta_{hg} = 0.97$, $\eta_{be} = 0.992$ as mentioned above, we have:

$$T_{12} = 1.0476 \cdot T_{out}/u_2 \tag{21}$$

Where,

- $k_{H\beta}$ is the contacting load ratio for pitting resistance; for the helical gear set $k_{H\beta} = 1.01 \div 1.16$ [16]; Therefore, it is possible to take $k_{H\beta} = 1.08$;
- $[\sigma_{H2}]$ is the allowable contact stress of the first stage (MPa);
- k_a is the material coefficient; As the gear material is steel, k_a = 43 [23];
- X_{ba} is the coefficient of wheel face width of the helical gear set;

2.3 Determination of Gearbox Housing Mass

The mass of the gearbox housing m_{gh} can be identified as:

$$m_{gh} = \rho_{gh} \cdot V_{gh} \tag{22}$$

Where ρ_{gh} is the weight density of gearbox housing material (kg/m^3); when gearbox housing material is cast iron, $\rho_{gh} = 7.2$ [21]; V_{gh} is the volume of the gearbox housing (m^3). Figure 1 shows the different component volumes of the of gearbox housing which is calculated as:

$$V_{gh} = 2 \cdot V_b + 2 \cdot V_{A1} + 2 \cdot V_{A2} \tag{23}$$

Where V_b is the volume of bottom housing; V_{A1} and V_{A2} are the volumes of side A_1 and side A_2;

$$V_b = L \cdot B \cdot S_G \tag{24}$$

$$V_{A1} = L \cdot H \cdot S_G \tag{25}$$

$$V_{A2} = B_1 \cdot H \cdot S_G = (B - 2 \cdot S_G) \cdot H \cdot S_G \tag{26}$$

Substituting (23), (24) and (25) into (22):

$$V_{gh} = 2 \cdot L \cdot B \cdot S_G + 2 \cdot L \cdot H \cdot S_G + 2 \cdot (B - 2 \cdot S_G) \cdot H \cdot S_G \tag{27}$$

Where L, H, B_1 and S_G can be estimated by:

$$L = (2 \cdot l_0/3 + d_{e21}/2 + d_{w12}/2 + d_{w22} + 22.5)/0.975 \quad [22] \tag{28}$$

$$H = \max(d_{e21}; d_{w22}) + 6.5 \cdot S_G \tag{29}$$

$$B = b \cdot cos\delta_2 + b_w + 4 \cdot S_G \tag{30}$$

$$S_G = 0.005 \cdot L + 4.5 \quad [22] \tag{31}$$

In this case, $l_0 = 3 \cdot d_{s1}$ [23] with d_{s1} as the diameter of the first shaft which will be determined in the next part; δ_2 is the gear pitch cone angle; $tang\delta_2 = u_1$.

2.4 Shaft Mass Calculation

According to the same protocol as previous section, the shaft mass (m_s) can be obtained by summing the mass of shaft components:

$$m_s = m_{s1} + m_{s2} + m_{s3} \tag{32}$$

Where,

$$m_{s1} = \rho_s \cdot \pi \cdot d_{s1}^2 \cdot l_{s1}/4 \tag{33}$$

$$m_{s2} = \rho_s \cdot \pi \cdot d_{s2}^2 \cdot l_{s2}/4 \tag{34}$$

$$m_{s3} = \rho_s \cdot \pi \cdot d_{s3}^2 \cdot l_{s3}/4 \tag{35}$$

With m_{s1}, m_{s2} and m_{s3}; l_{s1}, l_{s2} and l_{s3} as the mass and length of shaft 1, 2 and 3 of the gearbox, respectively; ρ_s is the weight density of shaft material (kg/m^3).

$$l_{s1} = 1.3 \cdot l_0 + \left(d_{e1} - d_{e1}^{'}\right)/(2 \cdot tang\delta_1) + 1.2 \cdot d_{s1} \tag{36}$$

$$l_{s2} = B \tag{37}$$

$$l_{s3} = B + 1.2 \cdot d_{s3} \tag{38}$$

Where:

$$d_{s1} = [T_{11}/(0,2 \cdot [\tau])]^{1/3} \tag{39}$$

$$d_{s2} = [T_{12}/(0,2 \cdot [\tau])]^{1/3} \tag{40}$$

$$d_{s3} = [T_{13}/(0,2 \cdot [\tau])]^{1/3} \tag{41}$$

In which, $[\tau]$ is the allowable shear stress (MPa); $[\tau] = 15 \div 20$ (MPa) [23] and it can be chosen as $[\tau] = 17$ (Mpa).

2.5 Optimization Problem

The optimum problem can be described as following objective function, C_{gb}.

$$\text{Minimize } C_{gb} \tag{42}$$

This function should be satisfied by the critical constraints:

$$\begin{aligned} 1 \leq u_1 \leq 9 \\ 1 \leq u_2 \leq 9 \end{aligned} \tag{43}$$

It is noticed that $u_t = u_1 \cdot u_2$ is the relation between the total gearbox ratio and partial ratios. Therefore, it is only necessary to optimize the second ratio u_2 while the optimum ratio of u_1 can be obtained by the expression: $u_1 = u_t/u_2$.

3 Experimental Work

To evaluate the effects of the input parameters on the optimum gear ratio u_2 of the gearbox, a simulation experiment is conducted with ten input parameters shown in Table 1. In order to reduce the number of experiments while assessing the influence of input parameters on the output response, the design is arranged with (2^{k-p}) or $2^{10-4} = 128$ number of experiments. Also, to perform the experiment, a computer program was constructed based on Eqs. (42) and (43). In addition, the Minitab R18 software is used for design the experiment and analyze the results. The experimental plan and the output response (u_2) are expressed in Table 2.

Table 1. Input parameters

No.	Parameters	Symbol	Unit	Level	
				Low	High
1	Total gearbox ratio	u_g	–	7	35
2	Face width coefficient of bevel gear set	K_{be}	–	0.25	0.3
3	Coefficient of wheel face width of stage 2	X_{ba}	–	0.3	0.4
4	Allowable contact stress of stage 1	AS_1	MPa	350	420
5	Allowable contact stress of stage 2	AS_2	MPa	350	420
6	Output torque	T_{out}	Nm	1000	10000
7	Cost of gearbox housing	C_{gh}	USD/kg	1	5
8	Cost of bevel gears	C_{bg}	USD/kg	7	12
9	Cost of helical gears	C_{hg}	USD/kg	3	9
10	Cost of shafts	C_s	USD/kg	1.5	5

Table 2. Experimental plans and output response u_2.

Std order	Run order	U_g	K_{be}	X_{ba}	AS_1	AS_2	T_{out}	C_{gh}	C_{bg}	C_{hg}	C_s	u_2
1	1	7	0.25	0.3	350	350	1000	1	12	9	5	3.6269
2	97	35	0.25	0.3	350	350	1000	1	7	9	1.5	5.8333
3	125	7	0.3	0.3	350	350	1000	1	7	3	5	4.5455
4	45	35	0.3	0.3	350	350	1000	1	12	3	1.5	6.1511
5	95	7	0.25	0.4	350	350	1000	1	7	3	1.5	3.8462
6	50	35	0.25	0.4	350	350	1000	1	12	3	5	6.2057
...												
127	19	7	0.3	0.4	420	420	10000	5	7	9	1.5	3.139
128	104	35	0.3	0.4	420	420	10000	5	12	9	5	5.8333

4 Results and Discussion

4.1 The Influence of Input Parameters on the Optimum Transmission Ratio

Figure 2 illustrates an assessment of the influence of main parameters on the second gear ratio u_2. It can be seen that the relation between the optimal transmission ratio and each factor is described by the straight lines. It can be recognized from Fig. 2 that the parameter of total gearbox ratio *ug* has the strongest influence on the optimum transmission, while the remaining factors have smaller effects on the output respond of u_2 compared to the total gearbox ratio. In addition, when the parameter of total gearbox ratio (u_g), the Coefficient of the face width of bevel gear set (K_{be}), the allowable contact stress of stage 2 (AS_2), the bevel gear cost (C_{bg}), the output torque (T_{out}) and the Cost of shafts (C_s) increase, u_2 rises. Also, the allowable contact stress of stage 1 (AS_1), the Cost of helical gears (C_{hg}) and Cost of gearbox housing (C_{gh}) grow, u_2 decreases. The Coefficient of wheel face width of stage 2 (X_{ba}) has very little effect on u_2.

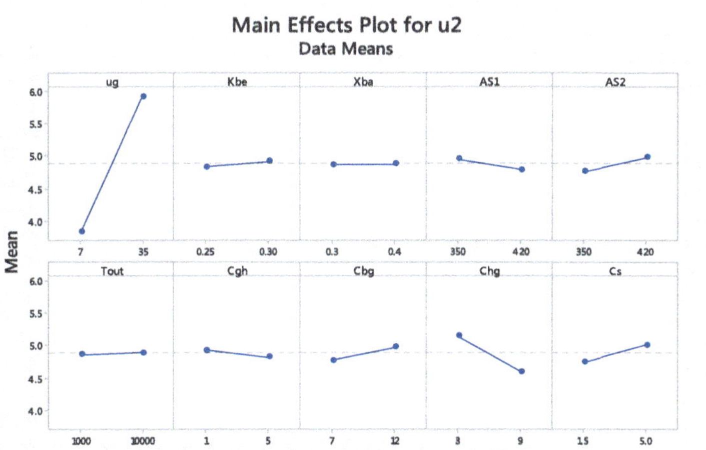

Fig. 2. Main effects plot for the response of u_2.

The Pareto chart in Fig. 3 shows the influence of input parameters and their interactions on u_2. The influence of the parameters is displayed in the blue column. Here, the parameters whose magnitude exceeds the red reference line are those that have a significant influence on u_2 with the significant level $\alpha = 0.05$. Specifically, it is noted that the influence of u_g on u_2 is the largest. Other parameters with a significant influence on u_2 include interactions, namely: B (K_{be}), D (AS_1), E (AS_2), F (T_{out}), G (C_{gh}), H (C_{bg}), J (C_{hg}), K (C_s) and interactions AB (u_g * K_{be}), AD (u_g * AS_1), AE (u_g * AS_2), AG (u_g * C_{gh}), AH (u_g * C_{bg}), AJ (u_g * C_{hg}), AK (u_g * C_s), BJ (K_{be} * C_{hg}), CF (X_{ba} * T_{out}), DH (AS_1 * C_{bg}), DJ (AS_1 * C_{hg}), EJ (AS_2 * C_{hg}), FG (T_{out} * C_{gh}), GJ (C_{gh} * C_s), HJ (C_{bg} * C_{hg}), JK (C_{hg} * C_s). The effect of the interactions is also clearly presented in Fig. 4.

To determine the trend and influence of the input parameters and interactions, the normal distribution chart in Fig. 5 is analyzed. From this graph, the parameters marked in red are the parameters that have a significant influence on u_2, meanwhile the parameters that are further away from the standard curve have a greater influence on u_2. The parameters located on the right of the calibration curve have a positive influence on u_2. They include A (u_g), B (K_{be}), E (AS_2), F (T_{out}), H (C_{bg}), K (C_s), AD (u_g * AS_1), AJ (u_g * C_{hg}), AG (u_g * C_{gh}), DJ (AS_1 * C_{hg}), FG (T_{out} * C_{gh}), GJ (C_{gh} * C_s). The parameters to the left of the calibration curve have a negative influence on u_2, including: D (AS_1), G (C_{gh}), J (C_{hg}), AB (u_g * K_{be}), AE (u_g * AS_2), AH (u_g * C_{bg}), AK (u_g * C_s), BJ (K_{be} * C_{hg}), CF (X_{ba} * T_{out}), DH (AS_1 * C_{bg}), EJ (AS_2 * C_{hg}), HJ (C_{bg} * C_{hg}), JK (Ch_g * C_s).

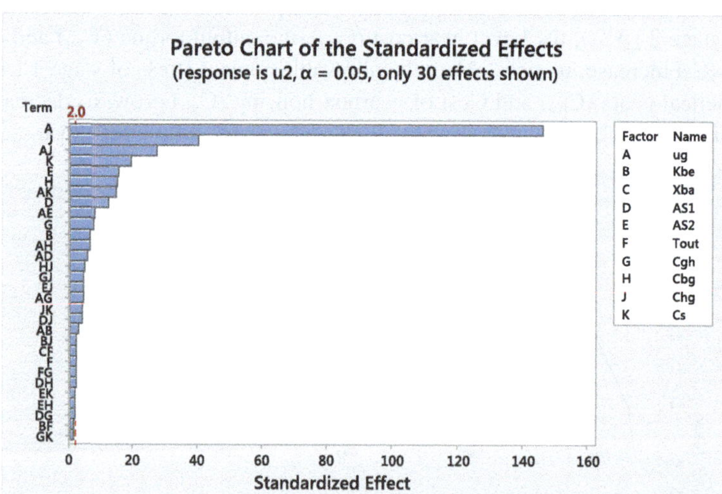

Fig. 3. Pareto chart of the standardized effects

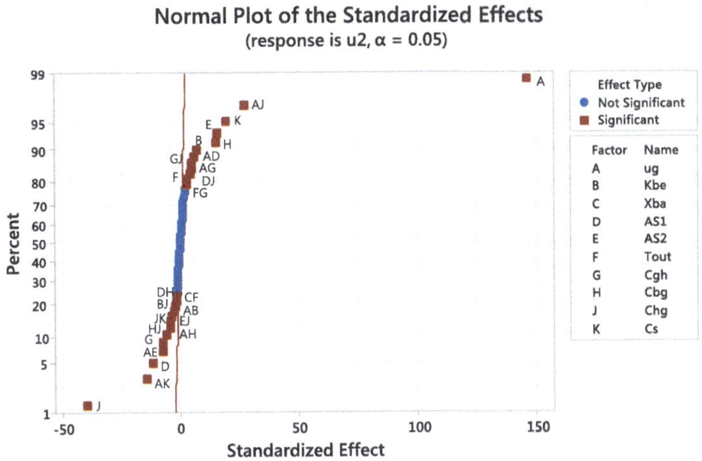

Fig. 4. Normal plot of the standardized effects

4.2 Determining the Regression Model of the Response u2

After removing the parameters have no or very little influence on u_2. Because the number of variables is very large, so the calculation by hand is not feasible, we determine the regression equation for u_2 by using Minitab 18 software. The estimated coefficients of the input parameters and also the interactions between them are described in Table 3. From Table 3, all factors listed in Table 3 are significant to the response u_2. This is because all of P-values are smaller than the significant level ($\alpha = 0.05$). By this way, the regression equation of the response u_2 is proposed as Eq. (44):

$$\begin{aligned} u_2 &= 1.454 + 0.0942\ u_g + 4.295\ K_{be} + 0.487\ X_{ba} - 0.005723\ AS_1 \\ &+ 0.007258\ AS_2 + 0.000021\ T_{out} - 0.0914\ C_{gh} + 0.06114\ C_{bg} \\ &- 0.0795\ C_{hg} + 0.1933\ Cs - 0.0467\ u_g * K_{be} \\ &+ 0.000064\ u_g * AS_1 - 0.000094\ u_g * AS_2 + 0.000951\ u_g * C_{gh} \\ &- 0.000941\ u_g * C_{bg} + 0.003905\ u_g * C_{hg} \\ &- 0.003726\ u_g * C_s - 0.254\ K_{be} * C_{hg} - 0.000065\ X_{ba} * T_{out} \\ &+ 0.000327\ AS_1 * C_{hg} - 0.000373\ AS_2 * C_{hg} + 0.000002\ T_{out} * C_{gh} \\ &+ 0.00588\ C_{gh} * C_{hg} - 0.00695\ C_{hg} * C_s \end{aligned} \tag{44}$$

Table 3. Estimated effects and coefficients for u_2.

Term	Effect	Coef	SE Coef	T-Value	P-Value	VIF
Constant		4.89075	0.00879	556.23	0.000	
u_g	2.09679	1.04839	0.00879	119.23	0.000	1.00
K_{be}	0.08947	0.04473	0.00879	5.09	0.000	1.00
X_{ba}	0.01303	0.00652	0.00879	0.74	0.460	1.00
AS_1	-0.16929	-0.08465	0.00879	-9.63	0.000	1.00
AS_2	0.21272	0.10636	0.00879	12.10	0.000	1.00
T_{out}	0.02963	0.01481	0.00879	1.68	0.095	1.00
C_{gh}	-0.10852	-0.05426	0.00879	-6.17	0.000	1.00
C_{bg}	0.20691	0.10345	0.00879	11.77	0.000	1.00
C_{hg}	-0.53965	-0.26982	0.00879	-30.69	0.000	1.00
C_s	0.25691	0.12846	0.00879	14.61	0.000	1.00
u_g*K_{be}	-0.03270	-0.01635	0.00879	-1.86	0.066	1.00
u_g*AS_1	0.06260	0.03130	0.00879	3.56	0.001	1.00
u_g*AS_2	-0.09249	-0.04625	0.00879	-5.26	0.000	1.00
u_g*C_{gh}	0.05325	0.02663	0.00879	3.03	0.003	1.00
ug*C_{bg}	-0.06585	-0.03293	0.00879	-3.74	0.000	1.00
u_g*C_{hg}	0.32803	0.16401	0.00879	18.65	0.000	1.00
u_g*C_s	-0.18259	-0.09130	0.00879	-10.38	0.000	1.00
K_{be}*C_{hg}	-0.03813	-0.01906	0.00879	-2.17	0.032	1.00
X_{ba}*T_{out}	-0.02920	-0.01460	0.00879	-1.66	0.100	1.00
AS_1*C_{hg}	0.06871	0.03435	0.00879	3.91	0.000	1.00
AS_2*C_{hg}	-0.07831	-0.03915	0.00879	-4.45	0.000	1.00
T_{out}*C_{gh}	0.02945	0.01472	0.00879	1.67	0.097	1.00
C_{gh}*C_{hg}	0.07055	0.03527	0.00879	4.01	0.000	1.00
C_{hg}*C_s	-0.07296	-0.03648	0.00879	-4.15	0.000	1.00

Model Summary

S	R-sq	R-sq(adj)	R-sq(pred)
0.0994786	99.38%	99.23%	99.04%

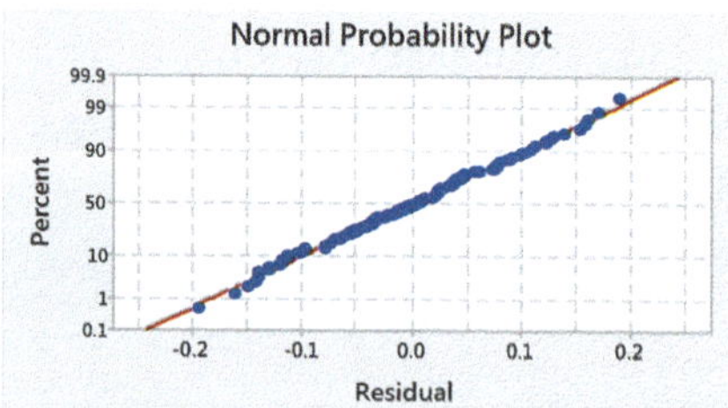

a)

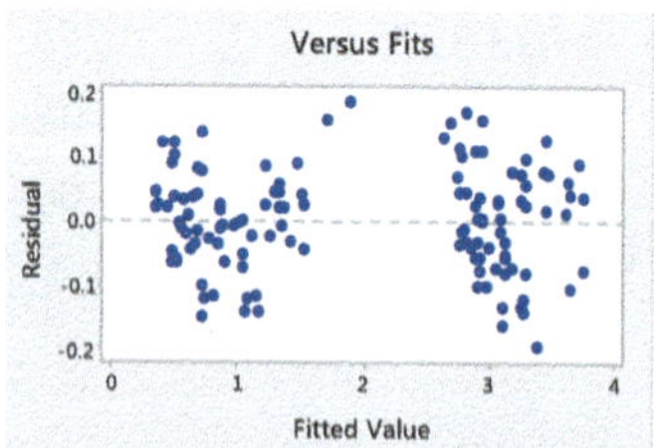

b)

Fig. 5. The normal probability plot and versus order plot for u_2.

From the normal probability plot (Fig. 5a), the residuals (the blue points) are distributed very close to the standard line (the straight red line). That means the errors are very small. This is also clearly shown in the versus fit plot (Fig. 5b), the residuals near point 0 account for the majority. In addition, it can be found from the plot (Fig. 5b), the points distribute randomly. That means u_2 is not significantly affected by other parameters except the input parameters. Due to these results in the charts above, combined with the calculation results of the coefficients R^2, all of determination coefficients R-Square are all greater than 99% (Table 3). It can be said that the experimental data are crucially consistent with the proposed model.

5 Conclusions

In this paper, a study on determination of the optimum gear ratios of a two-stage bevel helical gearbox to get the minimum gearbox cost was presented. In the study, a simulation experiment with ten main design parameters was performed to evaluate their influence with the output which is the second gear ratio u_2. Also, a regression model for calculation of the optimum gear ratio u_2 for getting the minimum gearbox cost was suggested. Using these explicit models, the optimum gear ratios can be determined simply. This model has been rated with a coefficient R^2 greater than 99%, which proves that the model fits the data very well.

Acknowledgements. This work was supported by Thai Nguyen University of Technology.

References

1. Pi, V.N.: A new study on optimal calculation of partial transmission ratios of two-step helical gearboxes. In: CEA 2008, Acapulco, Mexico, 25–27 January 2008, pp. 162–165 (2008)
2. Pi, V.N.: A study on optimal determination of partial transmission ratios of helical gearboxes with second-step double gear-sets. Proc. World Acad. Sci. Int. J. Mech. Mechatron. Eng. **2** (1), 26–29 (2008)
3. Tung, L.A., et al.: A study on determination of optimum gear ratios of a two-stage worm gearbox. In: Sattler, K.-U., Nguyen, D.C., Vu, N.P., Tien Long, B., Puta, H. (eds.) ICERA 2019. LNNS, vol. 104, pp. 76–84. Springer, Cham (2020). https://doi.org/10.1007/978-3-030-37497-6_8
4. Tuan, N.K., et al.: Determining optimal gear ratios of a two-stage helical reducer for getting minimal acreage of cross section. In: 2018 MATEC Web of Conferences. EDP Sciences (2018)
5. Hong, T.T., et al.: Calculating optimum gear ratios of two step bevel helical reducer. Int. J. Appl. Eng. Res. **14**(16), 3494–3499 (2019)
6. Pi, V.N.: A new study on the optimal prediction of partial transmission ratios of three-step helical gearboxes with second-step double gear-sets. WSEAS Trans. Appl. Theor. Mech. **2** (11), 156–163 (2007)
7. Pi, V.N., Binh, N.D., Dac, V.Q., The, P.Q.: A new and effective method for optimal splitting of total transmission ratio of three step bevel-helical gearboxes. In: The 6th Vietnam Conference on Automation, Hanoi (2005)

8. Pi, V.N., et al.: Determination of optimum gear ratios of a three stage bevel helical gearbox. In: IOP Conference Series: Materials Science and Engineering. IOP Publishing (2019)
9. Hong, T.T., et al.: A study on calculation of optimum gear ratios for three stage bevel helical gearboxes. ARPN J. Eng. Appl. Sci. **14**(14), 2592–2598 (2019)
10. Pi, V.N.: Optimal determination of partial transmission ratios for four-step helical gearboxes with first and third step double gear-sets for minimal mass of gears. In: Applied Computing Conference, ACC 2008, Istanbul, Turkey, 27–30 May 2008 (2008)
11. Kudreavtev, V.N., Gierzaves, I.A., Glukharev, E.G.: Design and Calculus of Gearboxes. Mashinostroenie Publishing (1971)
12. Milou, G., Dobre, G., Visa, F., Vitila, H.: Optimal design of two step gear units, regarding the main parameters. VDI Berichte 1996, vol. 1230, p. 227 (1996)
13. Pi, V.N., Binh, N.D., Dac, V.Q., The, P.Q.: Optimal calculation of total transmission ratio of three-step helical gearboxes for minimum mass of gears. J. Sci. Technol. 6 Eng. Univ. **55**, 91–93 (2006)
14. Pi, V.N.: A new study on optimal calculation of partial transmission ratio of three-step helical reducer. In: The 3rd IASME/WSEAS International Conference on Continuum Mechanics, Cambridge, UK, 23–25 February (2008)
15. Pi, V.N.: A new and effective method for optimal calculation of total transmission ratio of two step bevel - helical gearboxes. In: International Colloquium on Mechanics of Solids, Fluids, Structures & Interaction, Nha Trang, Vietnam, pp. 716–719 (2000)
16. Pi, V.N.: A study on optimal calculation of partial transmission ratios of three-step bevel helical gearboxes. In: International Workshop on Advanced Computing and Applications, Ho Chi Minh City, Vietnam, pp 277–286 (2008)
17. Pi, V.N., Cam, N.T.H., Tuan, N.K.: Optimum calculation of partial transmission ratios of mechanical driven systems using a V-belt and two-step bevel helical gearbox. J. Environ. Sci. Eng. A **5**, 566–570 (2016)
18. Pi, V.N.: A new method for optimal calculation of total transmission ratio of two step helical gearboxes. In: 2001 The Nation Conference on Engineering Mechanics, pp. 179–188 (2001)
19. Vu, N.P.: Optimal determination of partial transmission ratios of three-step helical gearboxs with second double gears (2008)
20. Vu, N.-P., et al.: The influence of main design parameters on the overall cost of a gearbox. Appl. Sci. **10**(7), 2365 (2020)
21. Toolbox, T.E.: Density, specific weight and specific gravity. https://www.engineeringtoolbox.com/density-specific-weight-gravity-d_290.html. Accessed 15 Nov 2019
22. Römhild, I., Linke, H.: Gezielte Auslegung Von Zahnradgetrieben mit minimaler Masse auf der Basis neuer Berechnungsverfahren. Konstruktion **44**(7–8), 229–236 (1992)
23. Trinh Chat, L.V.U.: Calculation of Mechanical Driven Systems. Education Publisher (1996)

Optimization on WEDM for Maximum Material Removal Rate When Cutting Half Round of SKD11 Steel

Trinh Kieu Tuan[1], Hoang Xuan Tu[2], Dang Quoc Cuong[3], Le Hoang Anh[4], Nguyen Anh Tuan[1], Tran Anh Duc[2], Nguyen Thanh Tu[2], and Tran Ngoc Giang[2(✉)]

[1] University of Economics - Technology for Industries, Ho Chi Minh City, Vietnam
[2] Thai Nguyen University of Technology, Thai Nguyen, Vietnam
tranngocgiang@tnut.edu.vn
[3] Nguyen Tat Thanh University, Ho Chi Minh City, Vietnam
[4] Vinh Long University of Technology Education, Vinh Long, Vietnam

Abstract. This work deals with the determination of optimum Wire Electrical Discharge Machining (WEDM) process parameters when machining half-round SKD11 steel with a radius of 7 mm for getting the maximum Material Removal Rate (MRR) . In this work, six input parameters, counting the pulse on time, the pulse off time, the cutting voltage, the serve voltage, the cutting speed, and the wire feed were examined. The impact of these parameters on the material removal rate of the machining process was investigated. Moreover, optimum process parameters to get the maximum MRR were found. It was also noted that the optimum parameters were appropriate for use.

Keywords: WEDM · Wire EDM · Material removal rate · SKD11 steel

1 Introduction

In practice, among tool steels, SKD11 is used quite commonly. This is because it is steel with very good tensile strength and high wear resistance. In addition, this steel has a balance between hardness and ductility. Therefore, up to now, there have been quite a few studies on processing this steel. These studies are usually about traditional types of machining, such as turning [1, 2], milling [3, 4], or grinding [5–7]. Besides, non-traditional machining methods such as laser, EDM, WEDM, etc. are also used to process SKD11 steel. Studies using laser processing to process this type of steel can include [8, 9]. Among the non-traditional types of machining, EDM machining is used the most for SKD11 machining. The authors in [10] studied using EDM to drill holes when measuring residual stress on materials with high hardness and strength, including SKD11. In [11] investigated the difference in energy distribution during EDM processing of Al6061, Inconel718, and SKD11. The influence of parameters in the EDM process on the hole diameters when machining SKD11 was investigated in [12].

Wire EDM is a type of EDM that is widely used to cut cylindrical holes of any shape, narrow slits with arbitrary arcs. Therefore, WEDM has attracted the attention of

D. C. Nguyen et al. (Eds.): ICERA 2021, LNNS 366, pp. 767–774, 2022.
https://doi.org/10.1007/978-3-030-92574-1_78

many researchers. In [13], the results of the multi-objective optimization problem were given to achieve the largest MRR when processing SKD11. In [14], gray relational analysis was used to determine the optimal WEDM parameters when processing SKD11 tool steel. Optimal process parameters when machining SKD 11 are also proposed in [11, 13]. Besides, the influence of the input parameters on the surface finished was reported in [15] when processing SKD11.

In this paper, a study for finding the optimum WEDM parameters when cutting half round SKD11 tool steel with a radius of 7 mm for maximum MRR was introduced. Also, main process parameters containing the cutting voltage, the pulse on time, the pulse off time, the serve voltage, the wire feed, and the cutting speed were investigated. The influence of these parameters on the MMR were inspected, and the optimum WEDM process parameters was proposed.

2 Experimental Setup

Fig. 1. WEDM machine used in the experiment

Table 1. Levels of input parameters

No	Parameter	Code	Unit	Level		
				1	2	3
1	Cutting voltage	VM	V	6	9	12
2	Pulse on time	T_{on}	μs	6	9	12
3	Pulse off time	T_{off}	μs	8	12	16
4	Servo voltage	SV	V	24	29	34
5	Wire feed	WF	mm/min	8	10	12
6	Cutting speed	SPD	mm/min	2	4	–

To investigate the impact of WEDM parameters on the MRR, an experiment was carried out. Figure 1 shows the WEDM machine used in the experiment. The workpiece material is SKD11 steel. Also, six main input parameters including the cutting voltage (VM), the pulse on time (Ton), the pulse off time (Toff), the servo voltage (SV), the wire feed (WF), and the cutting speed (SPD) were investigated with the levels as present in Table 1. Besides, the experimental plan with the design L18 ($2^1 \cdot 3^5$) is designed by using the Taguchi method and the Minitab R18 software. Table 2 shows the experimental plan and the MRR values which were calculated from the measurements of the material removal volume and the machining time.

3 Results and Analysis

Experimental results were analyzed by the Analysis of Variance (ANOVA) method on the Minitab 18 software. Since the objective function is MRR, it is best to have the largest MRR:

$$S/N = -10log_{10}\left(\frac{1}{n}\sum_{i=1}^{n}\frac{1}{y_i^2}\right) \quad (1)$$

From the results of ANOVA analysis in Table 3 and Fig. 1, it is found that T_{on} has the greatest influence on MRR (59.01%); followed by the influence of VM (21.15%). The effects of the remaining parameters are T_{off} (5.9%), SPD (4.47%), SV (4.46%) and WF (3.65%), respectively.

Table 2. Experimental plan and the MRR

Exp. No	SPD	VM	Ton	Toff	SV	WF	MRR (mm^3/min.)			Mean	S/N
							Trial 1	Trial 2	Trial 3		
1	2	6	6	8	24	8	0.123	0.128	0.118	0.1230	−18.2163
2	2	9	6	12	29	10	0.087	0.072	0.075	0.0780	−22.2427
3	2	12	6	16	34	12	0.139	0.140	0.142	0.1403	−17.0578
4	2	9	9	8	24	10	0.127	0.120	0.129	0.1253	−18.0513
5	2	12	9	12	29	12	0.119	0.112	0.105	0.1120	−19.0497
6	2	6	9	16	34	8	0.156	0.157	0.142	0.1517	−16.4099
7	2	12	12	8	29	8	0.193	0.189	0.195	0.1923	−14.3211
8	2	6	12	12	34	10	0.185	0.180	0.188	0.1843	−14.6921
9	2	9	12	16	24	12	0.158	0.152	0.152	0.1540	−16.2539
10	4	9	6	8	34	12	0.156	0.155	0.149	0.1533	−16.2927
11	4	12	6	12	24	8	0.137	0.130	0.141	0.1360	−17.3440
12	4	6	6	16	29	10	0.093	0.087	0.099	0.0930	−20.6666
13	4	6	9	8	29	12	0.145	0.139	0.148	0.1440	−16.8417
14	4	9	9	12	34	8	0.164	0.167	0.159	0.1633	−15.7439
15	4	12	9	16	24	10	0.137	0.145	0.139	0.1403	−17.0643
16	4	12	12	8	34	10	0.207	0.218	0.212	0.2123	−13.4655
17	4	6	12	12	24	12	0.173	0.173	0.179	0.1750	−15.1426
18	4	9	12	16	29	8	0.165	0.172	0.170	0.1690	−15.4463

Table 4 shows the order of influence of parameters on S/N is Ton, SPD, Toff, SV, VM, and WF, respectively. The objective function described in Eq. 1. It is used to achieve the maximum MMR corresponding to the S/N value of each input parameter. The analysis of the influence of the parameters on the S/N ratio was presented in the chart in Fig. 2. From the chart in Fig. 2, it is possible to derive the optimal set of WEDM parameters to achieve the maximum MMR. The specific values of these optimum parameters are shown in Table 5.

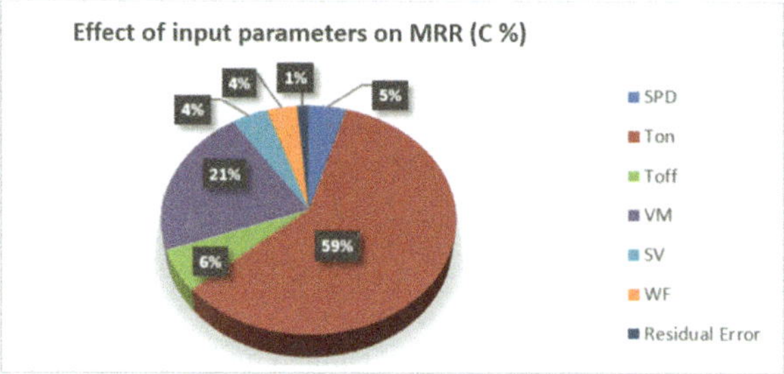

Fig. 2. Order of influence of input parameters on MRR

Table 3. Analysis of Variance of input parameters on MRR

Analysis of Variance for Means

Source	DF	Seq SS	Adj SS	Adj MS	F	P	C (%)
SPD	1	0.000873	0.000873	0.000873	19.76	0.004	4.47
Ton	2	0.011525	0.011525	0.005762	130.45	0.000	59.01
Toff	2	0.001152	0.001152	0.000576	13.04	0.007	5.90
VM	2	0.004131	0.004131	0.002066	46.76	0.000	21.15
SV	2	0.000871	0.000871	0.000435	9.85	0.013	4.46
WF	2	0.000713	0.000713	0.000356	8.07	0.020	3.65
Residual Error	6	0.000265	0.000265	0.000044			1.36
Total	17	0.019529					

Model Summary

S	R-Sq	R-Sq(adj)
0.0066	98.64%	96.15%

The results of MRR corresponding to the set of optimal parameters obtained are given in Table 6. This result is obtained by predicting the results of Taguchi planning by using Minitab 18 software. Thus, with the optimal set of parameters, the MMR is 0.23722 (mm^3/min) (Fig. 3).

Table 4. Order of influence of input parameters on S/N

Response Table for Signal to Noise Ratios

Larger is better

Level	SPD	Ton	Toff	VM	SV	WF
1	-17.37	-18.64	-16.20	-17.01	-16.25	-16.99
2	-16.45	-17.19	-17.37	-18.09	-17.70	-17.34
3		-14.89	-17.15	-15.61	-16.77	-16.38
Delta	0.92	3.75	1.17	2.48	1.45	0.95
Rank	6	1	4	2	3	5

Fig. 3. Effect of input parameters on S/N ratio

Table 5. Optimum WEDM parameters

Parameters	Code	Unit	Level	Value
Cutting voltage	VM	V	3	12
Pulse on time	Ton	µs	3	12
Pulse off time	Toff	µs	1	8
Servo voltage	SV	V	1	24
Wire feed	WF	mm/min	3	12
Cutting speed	SPD	mm/min	2	4

Table 6. Predicted MMR

Prediction

S/N Ratio	Mean	StDev	Ln(StDev)
-11.2428	0.237222	0.0029358	-5.75802

Settings

SPD	Ton	Toff	VM	SV	WF
2	3	1	3	1	3

Evaluation of Experimental Models: The experimental model is evaluated through error distribution charts as shown in Fig. 4. From the chart, it can be seen that the errors of the experimental points (the blue points) on the histogram very close to the normal distribution line (red solid line). It shows that the deviation is very small. Also, the error frequency chart also shows that the errors are mostly in the range −0.4 to 0.4. The remaining two graphs describe the random distribution of the errors of the experimental points. From these graphs, it found that the built model is largely influenced by the selected input parameters and is not affected by order of the experiments. The suitability of the experimental model and the optimal set of parameters is checked by Anderson-Darling method by using Minitab 18 software (Fig. 5). From Fig. 5, all points correspond to the data (blue dots) are in the region bounded by 2 upper and lower limits with 95% significance. In addition, the P value of 0.977 is larger than the value of $\alpha = 0.05$. This indicates that the applied empirical model is suitable.

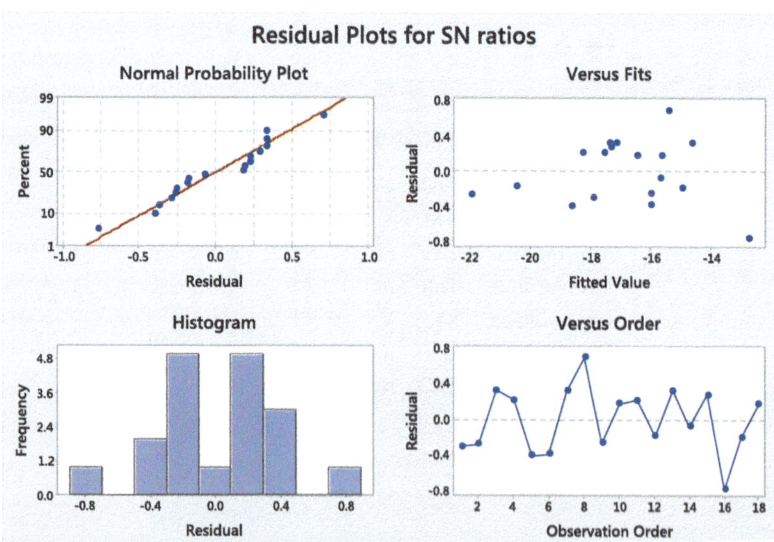

Fig. 4. Residual plots for S/N ratios

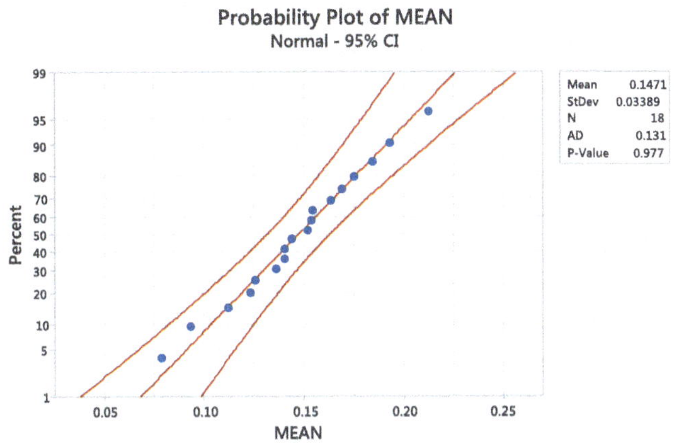

Fig. 5. Probability plot of the fit of experimental model for MRR

4 Conclusions

This work presents a study on determination of optimum WEDM parameters when processing half round SKD11 steel with a radius of 7 mm. To do that, six input process parameters counting the cutting voltage, the pulse on time, the pulse off time, the serve voltage, the wire feed and the cutting speed were considered. The effects of these parameters on the MRR were examined. It was reported that T_{on} has the greatest influence on MRR (59.01%), followed by the influence of VM (21.15%). The effects of the remaining parameters are T_{off} (5.9%), SPD (4.47%), SV (4.46%) and WF (3.65%), respectively. Moreover, the following set of optimum WEDM parameters was proposed: VM = 12; T_{on} = 12; T_{off} = 8; SV = 24; WF = 12; and SPD = 4. Also, it is noted that the proposed model is appropriate for applying in practice.

Acknowledgements. This work was supported by Thai Nguyen University of Technology.

References

1. Tzeng, C.-J., et al.: Optimization of turning operations with multiple performance characteristics using the Taguchi method and Grey relational analysis. J. Mater. Process. Technol. **209**(6), 2753–2759 (2009)
2. Dogra, M., et al.: Tool wear, chip formation and workpiece surface issues in CBN hard turning: a review. Int. J. Precis. Eng. Manuf. **11**(2), 341–358 (2010)
3. Duc, T.M., Dong, P.Q., Van Thanh, D.: Evaluation of MQCL technique using MoS 2 nanofluids during hard milling process of SKD 11 tool steel. Int. J. Mech. Eng. Appl. **7**(4), 91 (2019)

4. Amin, A., Saad, M.H.B.M., Arif, M.D.: Modeling & Optimization of Surface Roughness & Vibration Amplitude in Heat Assisted End Milling of SKD 11 Tool Steel using Ball Nose Tool. Trans Tech Publ, In Advanced Materials Research (2012)
5. Tran, T.H., et al.: Improvement of Wheel Life by Optimization of Dressing Parameters in Surface Grinding of SKD11 Steel. Trans Tech Publ, In Materials Science Forum (2021)
6. Tran, T.H., et al.: Influence of Dressing Conditions on Surface Roughness when Surface Grinding SKD11 Steel. Trans Tech Publ, In Materials Science Forum (2021)
7. Hong, T.T., et al.: Multi Response Optimization of Dressing Conditions for Surface Grinding SKD11 Steel by HaiDuong Grinding Wheel Using Grey Relational Analysis in Taguchi Method. In: Sattler, K.-U., Nguyen, D.C., Vu, N.P., Long, B.T., Puta, H. (eds.) ICERA 2020. LNNS, vol. 178, pp. 560–571. Springer, Cham (2021). https://doi.org/10.1007/978-3-030-64719-3_62
8. Hirogaki, T., et al.: In-situ heat treatment system for die steels using YAG laser with a machining center. Precis. Eng. **25**(3), 212–217 (2001)
9. Kitamuraa, Y., et al.: Nanostructured tool steel fabricated by combination of laser melting and friction stir processing. In: Proceedings of the 1st International Joint Symposium on Joining and Welding: Osaka, Japan 2013, Woodhead Publishing (2013)
10. Lee, H.-T., et al.: The study of EDM hole-drilling method for measuring residual stress in SKD11 tool steel. J. Mater. Process. Technol. **149**(1–3), 88–93 (2004)
11. Ming, W., et al.: Investigating the energy distribution of workpiece and optimizing process parameters during the EDM of Al6061, Inconel718, and SKD11. Int. J. Adv. Manuf. Technol. **92**(9–12), 4039–4056 (2017). https://doi.org/10.1007/s00170-017-0488-6
12. Chimklin, K., Chungchoo, C.: Design of experiment for analyzing factors affecting to diameters of the hole on SKD11 in the EDM process. In: Proceedings of the 2020 4th International Symposium on Computer Science and Intelligent Control (2020)
13. Chen, X., et al.: Investigation on MRR and machining gap of micro reciprocated wire-EDM for SKD11. Int. J. Precis. Eng. Manuf. **21**(1), 11–22 (2020)
14. Huang, J., Liao, Y.: Optimization of machining parameters of wire-EDM based on grey relational and statistical analyses. Int. J. Prod. Res. **41**(8), 1707–1720 (2003)
15. Bobbili, R., Madhu, V., Gogia, A.: Effect of wire-EDM machining parameters on surface roughness and material removal rate of high strength armor steel. Mater. Manuf. Processes **28**(4), 364–368 (2013)

Performance Evaluation of the Combined Differential Evolution and Jaya Algorithm for Structural Optimization Under Transient Excitations and 26 Mathematical Benchmark Functions

Sy Nguyen-Van(✉), Thi Thanh Nga Nguyen, Luong Viet Dung, Duong Pham Tuong Minh, Nguyen Quang Hung, Nguyen Van Trang, and Nguyen Thi Hoa

Thai Nguyen University of Technology, Thai Nguyen, Vietnam
vansy@tnut.edu.vn

Abstract. Recently, the combined differential evolution and the Jaya algorithm (CDJ) was proposed for shape and size optimization of truss structures under multiple frequencies. However, its extended applications for such problems under transient excitations haven't been investigated yet so far. Also, its performance still requires more insightful verifications and this study aims to perform CDJ on 26 famous mathematical benchmark functions. The obtained results of the above-mentioned problems reveal that CDJ gives the best performances compared with those of other algorithms available in the literature.

Keywords: The combined differential evolution and Jaya algorithm (CDJ) · 26 mathematical benchmark functions · Shape and size optimization · Truss · Transient excitations

1 Introduction

In structural optimization, especially for truss structures, most of the researches focus on minimizing the truss weight under static loads or frequency constraints, and the works that consider constraints related to transient excitations are very limited [1]. However, in nature, most of the structural and mechanical systems are under dynamic loads. Hence, the optimization of such systems under transient loads or dynamic loads is very promising to be extended for application in practice.

Over the past decades, a variety of optimization algorithms have been proposed for such structural optimization such as Differential Evolution [2], Particle Swarm Optimization [3], Genetic Algorithm [4], Jaya algorithm. Recently, the combined differential evolution and Jaya algorithm (CDJ) [5] has been proposed for the shape and size optimization of truss structures. However, the application of CDJ in truss optimization under excitation loads hasn't been investigated yet so far. In addition, for more insightful verification of CDJ, it was also tested on 26 mathematical benchmarks functions which are the most used functions for verification of any new metaheuristic algorithms.

D. C. Nguyen et al. (Eds.): ICERA 2021, LNNS 366, pp. 775–785, 2022.
https://doi.org/10.1007/978-3-030-92574-1_79

2 The Objective Function

Regarding the structural optimization of truss structures under excitation loads, cross-sectional areas and nodal coordinates are design variables. The weight of the truss structure is the objective function. The strength, displacement, and buckling are considered as constraints. Hence, the objective function can be given as follows:

$$\text{Minimize} : Mass(\mathbf{A}, \mathbf{X}) = \sum\nolimits_{i=1}^{t} L_i(\mathbf{X})\rho_i\mathbf{A}_i \tag{1}$$

The strength constraint is given as follows [6]:

$$\frac{P_r}{\phi P_n} - 1 \leq 0 \tag{2}$$

where P_r and P_n denote the required and nominal axial (tensile or compressive) strengths for the ith truss member, respectively; ϕ stands for the resistance factor (0.85 and 0.9 for compressive and tensile members, respectively). Concretely, the nominal axial strength is expressed as follows:

$$P_n = \begin{cases} F_y A_g, \text{for tensile members} \\ F_{cr} A_g, \text{for compressive members} \end{cases} \tag{3}$$

Where F_y and A_g denote the yield stress and the gross cross-sectional area of the member, respectively; F_{cr} stands for the critical stress related to flexural buckling which is expressed as follows:

$$P_n = \begin{cases} F_y A_g, & if \quad \lambda_c = (Kl/r\pi)\sqrt{F_y/E} \leq 1.5 \\ F_{cr} A_g, & if \quad \lambda_c = (Kl/r\pi)\sqrt{F_y/E} > 1.5 \end{cases} \tag{4}$$

where l, K, r, and E stand for the laterally unbraced length of the member, effective length factor, governing radius of gyration about the buckling axis, and Young's elastic modulus, respectively. Also, the displacement constraint is considered as follows:

$$\frac{d_{m,n}}{\left[d_{m,n}^{a}\right]} - 1 \leq 0 \tag{5}$$

Where $d_{m,n}$ and $\left[d_{m,n}^{a}\right]$ stand for the calculated and associated allowable displacements at node m^{th} along with n(x, y, z) direction, respectively. The last constraint is the slenderness ratio for truss members and is provided as follows:

$$\begin{cases} (Kl)/(300r) - 1 \leq 0, \text{ for tensile members} \\ (Kl)/(200r) - 1 \leq 0, \text{ for compressive members} \end{cases} \tag{6}$$

It shows that Eq. (1) presents a constrained optimization problem, and thus to a constrained optimization into a corresponding unconstrained one, the penalty function approach is used in this study [2, 7]. As a result, Eq. (1) can be rewritten as follows:

$$\text{Minimize} : f_{objective}(\mathbf{A}, \mathbf{X}) = (1 + \alpha_1 \upsilon)^{\alpha_2} Mass(\mathbf{A}, \mathbf{X}), \tag{7}$$

$$\text{With} : \upsilon = \sum\nolimits_{z=1}^{m} \max[0, g_z(\mathbf{A}, \mathbf{X})], \tag{8}$$

where υ is the sum of violated constraints; $g_z(\mathbf{A}; \mathbf{X})$ and m are the zth constraint and the number of constraints, respectively; the parameters α_1 and α_2 are dependent on the exploration and exploitation rates of the search space, respectively. In this research, α_1 is set to be 1, and α_2 is equal to 1.5 in the first population and constantly increased to 6.

3 Optimization Algorithm

The CDJ was successfully proposed for shape and size optimization of the truss structures under multiple constraints [5]. Herein, in this study, the CDJ is adopted as an optimizer. Concretely, CDJ can be expressed fully as follows:

- *Initialization*
  Firstly, NP individuals are randomly generated in a given search space. Each ith individual ($i = 1, 2, \ldots, NP$) is a vector of D design variables and given as follows:

$$X_{i,j}^{g=0} = L_j + rand_{i,j}[0, 1](H_j - L_j), \quad j = 1, 2, \ldots, D, \tag{9}$$

  where L, H are the lower and upper boundaries, respectively; NP is the number of individuals in the population, D is the number of design variables, and rand[0; 1] is a random value within the range of 0 and 1; g is the gth population. The ith individual vector is given as follows.

$$X_i^g = \left\{X_{i,1}^g, X_{i,2}^g, \ldots, X_{i,j}^g, \ldots, X_{i,D}^g\right\}, \tag{10}$$

- *Mutation*
  To enhance the diversity of the population, a vector $\mathbf{V}_i^g$ is generated by using hybrid operators of DE and Jaya for achieving good balancing of the global and local searches. Concretely, the global searching ability is enhanced by using, *DE/rand/1,* and Jaya. These operators are expressed as follows:

$$\text{DE/rand/1} : \mathbf{V}_i^g = \mathbf{X}_{R_1}^g + F\left(\mathbf{X}_{R_2}^g - \mathbf{X}_{R_3}^g\right), \tag{11}$$

$$\text{Jaya operator} : \mathbf{V}_i^g = \mathbf{X}_{R_1}^g + rand[0, 1]\left(\mathbf{X}_{best}^g - \left|\mathbf{X}_{R_2}^g\right|\right) - rand[0, 1]\left(\mathbf{X}_{worst}^g - \left|\mathbf{X}_{R_3}^g\right|\right) \tag{12}$$

To switch between *DE/rand/1* and Jaya, a value $\sigma = \text{rand}[0; 1]$ is used. Moreover, if $\sigma > 0.3$, its new location is updated by using Eq. (10), otherwise, it is generated by Eq. (11).

In the local search, DE/best/1 operator is used. DE/best/1 operator is presented as follows:

$$\text{best}/1 : \mathbf{V}_i^g = \mathbf{X}_{best}^g + F\left(\mathbf{X}_{R_1}^g - \mathbf{X}_{R_2}^g\right), \tag{13}$$

It should be noted that R_1; R_2; R_3; R_4 and R_5 are random differential numbers chosen from [*1,2,3,…,NP*]; F is a random number between 0 and 1, and $\mathbf{X}_{best}^g$ is the best individual vector in the current population.

To transform from the global search to the local search, an adaptive parameter δ is used and is defined as follows:

$$\delta = |f_m/f_b - 1|, \tag{14}$$

Where f_m and f_b are the mean and best objective functions in the previous population, respectively. The values of δ are decreased during optimization. If δ is bigger than a constant value of 10^{-4}, called "*epsilon*", the mutation operators of *DE/rand/1* and Jaya operator are performed, otherwise, the *DE/best/1* operator is utilized.

- *Crossover*

To get good optimal solutions, the crossover stage is used. The ith trial vector $\mathbf{U}_i^g = \left\{\mathbf{U}_{i,1}^g, \mathbf{U}_{i,2}^g, \ldots, \mathbf{U}_{i,j}^g, \ldots, \mathbf{U}_{i,D}^g\right\}$ is generated by mixing the target vector $\mathbf{X}_i^g$ and the mutant vector $\mathbf{V}_i^g$ as follows:

$$U_{i,j}^g = \begin{cases} V_{i,j}^g, & \text{if } j = Rand \text{ or } rand_{i,j}[0,1] \leq Cr, \\ X_{i,j}^g, & \text{otherwise}, \end{cases} \tag{15}$$

where Rand is a random number within the range of [1; NP], and the crossover value Cr is randomly chosen in [0.7, 1].

- *Selection*

For the selection of new individuals for next-generation, an elitist selection approach [8] is applied to choose the best individuals for the next iteration.

Table 1. Data for shape and size optimization of truss problems with multiple frequency constraints

Example	Young's modulus E $(N = m^2)$	Material density $\rho(kg = m^3)$	Bounds of cross-sectional areas, $A(m^2)$	Yeild stress Fy (MPa)	Nodal coordinate bounds (m)
37 bar-truss	2.1×10^{11}	7860	$1 \times 10^{-4} \leq A \leq 2.0 \times 10^{-4}$	240	[0, 0.06]

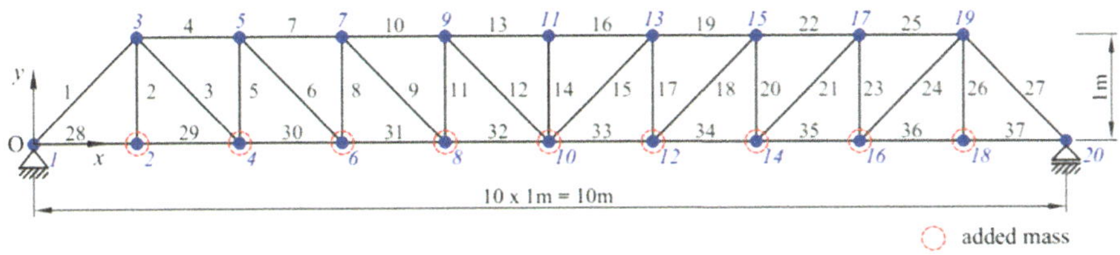

Fig. 1. The 37-bar truss.

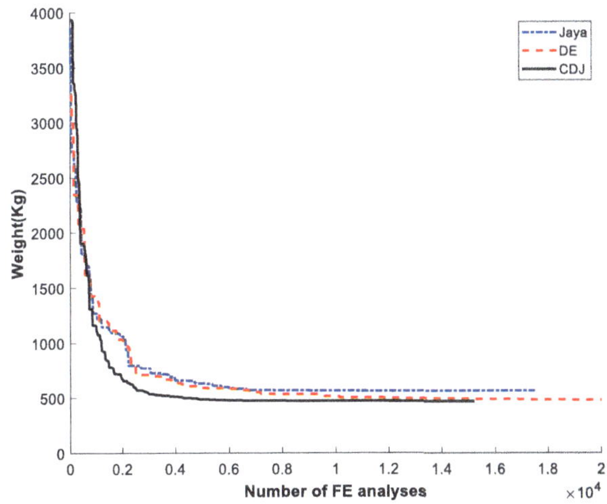

Fig. 2. The weight convergence of the 37-bar truss by using the DE, Jaya and CDJ.

4 Numerical Examples

4.1 Shape and Size Optimization

In this section, the optimization problem is a 37-bar truss as shown in Fig. 1.

Herein, at all free nodes in the lower chord, a lumped mass of 10 kg is added. Design variables are all nodal coordinates in the higher chord and truss member sizes. Strength, displacement, and buckling conditions are constraints. The structure is under excitation load $(P(t))$. Concretely, the excitation load of $P(t) = P_0.cos(2f\pi t)$ with an amplitude of 2.0 kN and a frequency (f) of 1.5 Hz is applied at node 10 along the $y-$ direction. The data in terms of material properties, lower and upper bounds of design variables, and displacement constraints are shown in Table 1. The results of CDJ, DE, and Jaya are reported and compared. All algorithms have the population size NP of 20. In the optimization procedure of such algorithms, they are stopped whether the value of δ is less than or equal to 10^{-6} or the number of FEAs reaches the maximum number of 20000. 10 differential runs of each algorithm are performed and data of the number of FEAs, average weight, worst weight, and standard deviation of the weight are also reported. It should be noted that to obtain the strength, displacement, and buckling constraints, the Newmark-Beta method [9] is used, where the damping matrix is given as follows:

$$\mathbf{C} = \delta_M \mathbf{M} + \delta_K \mathbf{K} \tag{16}$$

Where δ_M and δ_K are the mass and stiffness proportional damping factors, respectively and are both equal to 0.05.

Table 2. Optimal outcomes of the 37-bar truss optimization under transient excitations.

Design variable	This current study		
Y_j(m);A_i(cm^2)	DE	JAYA	CDJ
Y_3,Y_{19}	1.00192	1.05957	1.00004
Y_5,Y_{17}	1.21257	1.03682	1.24107
Y_7,Y_{15}	1.58522	1.05474	1.53097
Y_9,Y_{13}	1.77927	1.12600	1.78558
Y_{11}	1.91020	1.63562	1.86149
A_1,A_{27}	6.59525	6.28319	6.28391
A_2,A_{26}	3.80673	3.14246	3.14251
A_3,A_{24}	6.49386	6.28775	6.28347
A_4,A_{25}	4.60608	5.76641	4.30487
A_5,A_{23}	3.80793	4.35027	3.14193
A_6,A_{21}	6.32021	18.86576	6.28367
A_7,A_{22}	4.84921	11.24564	5.14671
A_8,A_{20}	4.14181	7.12107	3.31935
A_9,A_{18}	6.71813	19.50701	6.85805
A_{10},A_{19}	6.91575	10.56145	5.82321
A_{11},A_{17}	3.76857	6.23933	3.73696
A_{12},A_{15}	7.96760	17.31180	7.82434
A_{13},A_{16}	6.03004	6.27464	6.71705
A_{14}	4.25487	6.58204	3.96246
Best weight (kg)	477.501	565.218	**470.119**
Number of FE analyses	20000	17500	**15220**
Worst weight (kg)	491.509	911.126	475.929
Average weight (kg)	480.677	744.496	472.279
Standard deviation	4.324	105.052	1.829

Table 2 shows optimal solutions of Jaya, DE, and CDJ. It shows that the optimal weight gotten by the CDJ is better than those of others (477.501 kg for Jaya, 565.218 for DE, and 470.119 for CDJ). For the convergence speed, the CDJ is the best one with only 14440 FEAs, while the DE and the Jaya require 20000 FEAs and 17500 FEAs, respectively. In addition, the convergence speeds are shown in Fig. 2. It reveals that the CDJ has the fastest convergence speed.

Figure 3 and Fig. 4 illustrate the slenderness constraints and displacement and strength obtained by CDJ, respectively. It shows that there are no violated constraints. Figure 5 shows the change of fundamental frequency during the whole optimization procedure. It reveals that the fundamental frequency aims to bypass the frequency of the excitation load. Hence, it entails that frequency constraints are involved in the truss optimization under transient excitations.

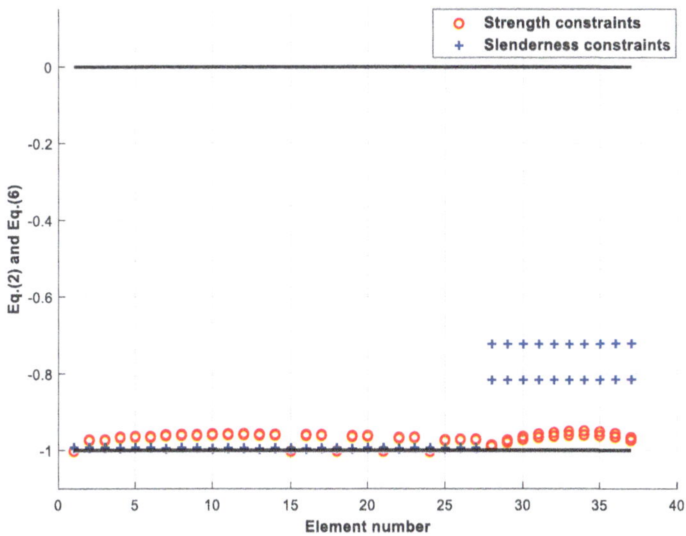

Fig. 3. Stress and slenderness constraint results of the 37-bar truss at the optimal design computed by the CDJ

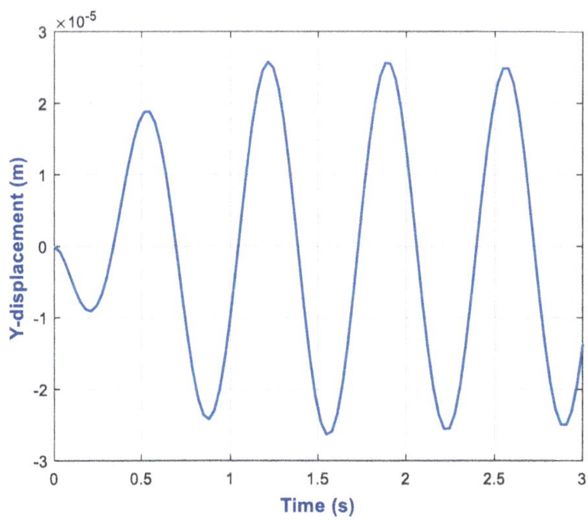

Fig. 4. Time-history displacement at node 10 in y-direction of the 37-bar truss under cosine loading (frequency = 1.5 Hz)

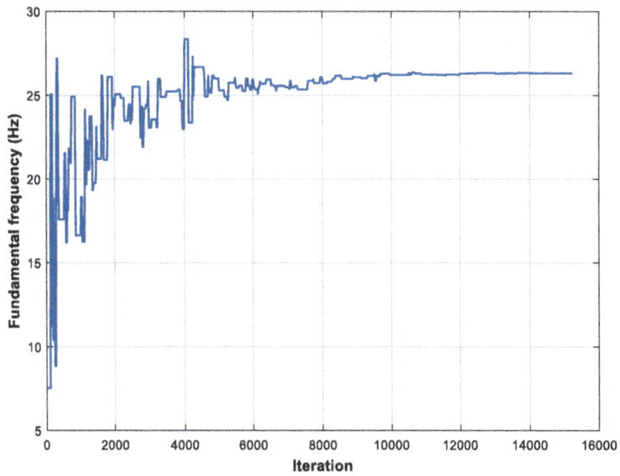

Fig. 5. Variation of fundamental frequency (Hz) during the optimization under the excitation load of F(t) = 2000cos(3πt) on the 37-bar truss.

4.2 26 Benchmark Functions

This section evaluates the performances of CDJ for 26 benchmark mathematical functions. The CDJ performance is compared with those of Jaya [10], DE [11], BA [12], FPA [12] and GWO [13]. For the detailed formation of 26 benchmark functions, the readers can find them in [14].

In such 26 functions, the number of variables ranges from 2 to 30, and functions are also branched into four groups which are unimodal-nonseparable (UN), unimodal-separable (US), multimodal- nonseparable (MN) and multimodal-separable (MS). The maximum number of iterations and number of the population are 10,000 and 50, respectively. In addition, if any value is smaller than 1×10^{-12}, it is set to 0.

There are 30 differential runs for all algorithms. In Table 3, the results reveal that CDJ is the best one because CDJ find 23/26 global optimum values, while the Jaya, DE, BA, FPA and GWO find 11, 16, 10, 11 and 18 global optimum values, respectively.

Table 3. Comparison of optimal results for 26 benchmark functions

No.	Functions	Min		Jaya (2016)	DE (1997)	BA (2010)	FPA (2012)	GWO (2016)	CDJ
			Best	0	0	0	0	5.17E-11	0
1	Beale	0	Mean	0.0254	0	0.0508	0.0000	0.0000	0
			StdDev	0.1391	0	0.1933	0.0000	0.0000	0
			Best	-1	-1	-1	-1	-1	-1
2	Easom	-1	Mean	-1	-1	-3.33E-02	-1	-1	-1
			StdDev	0	0	1.83E-01	0	1.05E-09	0
			Best	0	0	0	0	0	0
3	Matyas	0	Mean	0	0	0	0	0	0
			StdDev	0	0	0	0	0	0
			Best	0	0	0.88281	0	0	0
4	Bohachev-sky1	0	Mean	0	0	2.14814	0	0	0
			StdDev	0	0	0.73791	0	0	0
			Best	0	0	0	0	1.77E-09	0
5	Booth	0	Mean	0.3382	0.0169	0.2367	0	0	0.0845
			StdDev	0.2432	0.0926	0.2574	0	0	0.1923
			Best	-1.800	-1.8013	-1.8013	-1.8013	-1.8013	-1.8013
6	Micha-lewicz2	-1.8013	Mean	-1.698	-1.8013	-1.6215	-1.8013	-1.8013	-1.8013
			StdDev	0.1002	0.0000	0.3336	0.0000	0.0000	0.0000
			Best	0	0	0.0437	0	0	0
7	Schaffer	0	Mean	0	0.0000	0.4184	0	0	0
			StdDev	0	0.0000	0.1432	0	0	0
			Best	-1.0316	-1.03163	-1.03163	-1.03163	-1.03163	-1.03163
8	Six Hump Camel Back	-1.0316	Mean	-1.0316	-1.03163	-0.97722	-1.03163	-1.03163	-1.03163
			StdDev	0	0	0.20707	0	4.61E-11	0
			Best	0	0	0	0	0	0
9	Boachev-sky2	0	Mean	0	0	0.09956	0	0	0
			StdDev	0	0	0.13229	0	0	0
			Best	0	0	0.22626001	0	0	0
10	Boachev-sky3	0	Mean	0	0	1.40797	0	0	0
			StdDev	0	0	0.82645	0	0	0
			Best	-186.73	-186.73	-186.73	-186.73	-186.73	-186.73
11	Shubert	-186.73	Mean	-186.73	-186.73	-168.02	-186.73	-186.73	-186.73
			StdDev	0.00	0.00	37.17	3E-04	2E-03	0
			Best	7E-04	0	1E-08	0	1.17E-08	0
12	Colville	0	Mean	6E-02	0	4E-01	0	0.3405634	0
			StdDev	0.096449589	0	1.5222605	0	0.6303513	0
			Best	-3.9987	-4.6877	-4.4958932	-4.6877	-4.6877	-4.6877
13	Micha-lewicz5	-4.6877	Mean	-2.8082	-4.6849	-3.3145	-4.6877	-4.4330	-4.6785
			StdDev	0.4166	0.0106	0.6659	0.0000	0.3482	0.0295
			Best	0	0	0	0	0	0
14	Zakharov	0	Mean	1.19E-12	0	0	0	0	0
			StdDev	1.14E-12	0	0	0	0	0
			Best	-4.7126	-9.6602	-8.35604	-9.43771	-9.58073	-9.6602
15	Micha-lewicz10	-9.6602	Mean	-4.0029	-9.6421	-6.0459	-8.9719	-8.2217	-9.6266
			StdDev	0.3976	0.0260	1.1159	0.2758	0.7511	0.0442
			Best	1.1E-07	2.3E-12	1.3E+04	9.6E-09	5.9E-08	0
16	Step	0	Mean	1.3E-05	7.1E-11	4.3E+04	2.4E-07	3.2E-01	0
			StdDev	4.0E-05	7.6E-11	6.6E+03	3.9E-07	2.5E-01	0
			Best	1.5E-07	5.8E-12	3.0E+04	1.2E-08	0	0
17	Sphere	0	Mean	2.5E-05	1.3E-10	4.2E+04	1.5E-07	0.0E+00	0
			StdDev	6.1E-05	2.4E-10	4.9E+03	1.8E-07	0.0E+00	0
			Best	3.5E-08	0	0	6.5E-10	0	0
18	SumSquares	0	Mean	1.7E-06	6.46E-12	0	2.55E-08	0.00E+00	0
			StdDev	2.9E-06	6.99E-12	0	4.18E-08	0.00E+00	0
			Best	4.8E-03	2.8E-03	3.3E-02	3.1E-03	7.1E-06	9.1E-04
19	Quartic	0	Mean	1.0E-02	2.0E-02	8.4E-02	1.1E-02	3.8E-05	2.1E-03
			StdDev	3.5E-03	7.0E-03	3.3E-02	4.3E-03	2.2E-05	5.6E-04
			Best	9.3E-05	1.9E-06	9.3E+01	2.3E-04	0	0
20	Schwefel 2.22	0	Mean	7.3E-03	7.3E-06	1.2E+02	6.3E-04	0.0E+00	2.47E-06
			StdDev	1.2E-02	8.0E-06	1.1E+01	3.0E-04	0.0E+00	9.45E-06
			Best	5.2E-06	2.0E-11	3.5E+05	1.4E-07	0	0
21	Schwefel 1.2	0	Mean	4.6E-05	1.2E-09	5.8E+05	1.4E-06	0.0E+00	0
			StdDev	8.1E-05	3.0E-09	7.9E+04	1.6E-06	0.0E+00	0
			Best	5.5E+00	6.2E+00	1.2E+06	9.5E+00	2.4E+01	0
22	Rosenbrock	0	Mean	4.9E+01	1.1E+01	5.1E+06	2.1E+01	2.6E+01	2.7E-01
			StdDev	3.1E+01	1.1E+01	2.8E+06	3.4E+00	8.7E-01	1.0E+00
			Best	6.7E-01	6.7E-01	6.7E-01	6.7E-01	6.7E-01	6.7E-01
23	Dixon Price	0	Mean	7.9E-01	6.7E-01	7.0E-01	6.7E-01	6.7E-01	6.7E-01
			StdDev	6.4E-01	6.6E-09	7.4E-02	1.3E-06	1.9E-09	9.94E-08
			Best	1.7E+01	1.0E+01	1.4E+02	2.3E+01	0	5.0E+00
24	Rastrigin	0	Mean	3.2E+01	5.1E+01	2.0E+02	5.0E+01	0.0E+00	5.2E+01
			StdDev	7.8E+00	3.3E+01	3.0E+01	9.8E+00	0.0E+00	4.5E+01
			Best	5.6E-07	4.3E-12	4.6E+02	6.9E-08	0	0
25	Griewank	0	Mean	5.6E-03	9.9E-04	5.6E+02	3.1E-03	0.0E+00	0.00214
			StdDev	8.0E-03	3.0E-03	5.5E+01	4.6E-03	0.0E+00	0.00442
			Best	1.5E-04	5.0E-07	2.0E+01	1.1E-05	0	0
26	Ackley	0	Mean	1.1E-01	2.9E-06	2.0E+01	1.0E-01	0.0E+00	0
			StdDev	3.3E-01	2.6E-06	1.7E-03	3.9E-01	0.0E+00	0
Count of algorithm found the global minimum				11/26	16/26	10/26	11/26	18/26	23/26

5 Conclusion

This study successfully verifies the performances of CDJ on shape and size optimization of the truss under transient loads and 26 mathematical functions. It reveals that CDJ gives the best performances compared to the classical DE and Jaya. In the truss optimization under transient loads, it entails that frequency constraints are involved in the truss optimization under transient excitations. Resultantly, the CDJ is propitious for application in other engineering optimization such as path optimization, optimal synthesis of mechanisms, shells, plates, and so on.

Acknowledgment. This research was supported by Thai Nguyen University of Technology.

References

1. Azad, S.K., Bybordiani, M., Azad, S.K., Jawad, F.K.J.: Simultaneous size and geometry optimization of steel trusses under dynamic excitations. Struct. Multidiscip. Optim. **58**(6), 2545–2563 (2018). https://doi.org/10.1007/s00158-018-2039-7
2. Lieu, Q.X., Do, D.T.T., Lee, J.: An adaptive hybrid evolutionary firefly algorithm for shape and size optimization of truss structures with frequency constraints. Comput. Struct. **195**, 99–112 (2018)
3. Millan-Paramo, C., Abdalla Filho, J.E.: Size and shape optimization of truss structures with natural frequency constraints using modified simulated annealing algorithm. Arab. J. Sci. Eng. **45**(5), 3511–3525 (2019). https://doi.org/10.1007/s13369-019-04138-5
4. Cazacu, R., Grama, L.: Steel truss optimization using genetic algorithms and FEA. Procedia Technol. **12**, 339–346 (2014). https://doi.org/10.1016/j.protcy.2013.12.496
5. Nguyen-Van, S., Nguyen, T.T.N., Nguyen-Dinh, N., Lieu, Q.X.: Truss optimization under frequency constraints by using a combined differential evolution and Jaya algorithm. In: Sattler, K.-U., Nguyen, D.C., Vu, N.P., Long, B.T., Puta, H. (eds.) ICERA 2020. LNNS, vol. 178, pp. 861–873. Springer, Cham (2021). https://doi.org/10.1007/978-3-030-64719-3_95
6. (AISC), A.I. ofSteel C.: Manual of Steel Construction: Load and Resistance Factor Design. Chicago (1994)
7. Kaveh, A., Zolghadr, A.: Democratic PSO for truss layout and size optimization with frequency constraints. Comput. Struct. **130**, 10–21 (2014)
8. Padhye, N., Bhardawaj, P., Deb, K.: Improving differential evolution through a unified approach. J. Glob. Optim. **55**, 771–799 (2013)
9. Reddy, J.N.: Introduction to the finite element method. McGraw-Hill Education (2019)
10. Rao, R.V.: Jaya: A simple and new optimization algorithm for solving constrained and unconstrained optimization problems. Int. J. Ind. Eng. Comput. **7**, 19–34 (2016). https://doi.org/10.5267/j.ijiec.2015.8.004
11. Storn, R., Price, K.: Differential evolution–a simple and efficient heuristic for global optimization over continuous spaces. J. Glob. Optim. **11**, 341–359 (1997)
12. Yang, X.S.: Nature-Inspired Optimization Algorithms. Elsevier. https://doi.org/10.1016/C2013-0-01368-0

13. Mirjalili, S., Lewis, A.: The whale optimization algorithm. Adv. Eng. Softw. **95**, 51–67 (2016). https://doi.org/10.1016/j.advengsoft.2016.01.008
14. Nguyen-van, S., Nguyen, K.T., Luong, V.H., Lee, S., Lieu, Q.X.: A novel hybrid differential evolution and symbiotic organisms search algorithm for size and shape optimization of truss structures under multiple frequency constraints. Expert Syst. Appl. **184**:11553 (2021). https://doi.org/10.1016/j.eswa.2021.115534

Prediction Model for Tensile Modulus of Carbon Nanotube–Polymer Composites

Tien-Thinh Le[1,2(✉)] and Minh Vuong Le[3]

[1] Faculty of Mechanical Engineering and Mechatronics, PHENIKAA University, Yen Nghia, Ha Dong, Hanoi 12116, Viet Nam
thinh.letien@phenikaa-uni.edu.vn
[2] A&A Green Phoenix Group JSC, PHENIKAA Research and Technology Institute (PRATI), No. 167 Hoang Ngan, Trung Hoa, Hanoi 11313, Cau Giay, Viet Nam
[3] Laboratoire Modélisation et Simulation Multi Echelle, Université Paris-Est, MSME UMR 8208 CNRS, 5 bd Descartes, 77454 Marne-la-Vallée, France

Abstract. This work develops a Neural Network (NN) model for the prediction of the tensile modulus of carbon nanotube (CN)/polymer nanocomposites, based on experimental database. A data set composed of 282 configurations is collected from available resources. Considered input variables of the dataset are such as mechanical properties of separated phases, density of polymer matrix, processing method, geometry of CN, modification method at the CN surface, etc. while the problem output is the tensile modulus of nanocomposite. Parametric studies have been performed in finding optimum architecture of the proposed NN model.

Keywords: Neural Network · Carbon nanotubes · Nanocomposites · Polymer · Tensile modulus

1 Introduction

With the constant development of technologies and economic, and from an environmental point of view, more and more materials need to be developed. However, finding and characterizing novel materials are conventionally done by experiments [1]. Historically, the average timeframe to fully comprehend the behavior of materials using laboratory experiments is 10 to 20 years. These types of experiments, especially for composite materials are costly in terms of resources and equipment. In addition, they can only study a limited number and/or mixture proportion when considering a novel material.

In order to deal quickly and efficiently with such large datasets, a computational revolution in computational materials science has come involving Machine Learning (ML) methods. Application of machine learning algorithms in material science has been proven successful with several published papers in the literature. Nouira et al. [2] have employed Generative Adversarial Networks for discovering new crystallographic structures. Such applications on crystal architecture have also been investigated by Ryan et al. [3] using Deep Learning and by Oliynyk et al. [4] using Support Vector

D. C. Nguyen et al. (Eds.): ICERA 2021, LNNS 366, pp. 786–792, 2022.
https://doi.org/10.1007/978-3-030-92574-1_80

Machine. In terms of material properties, Zheng et al. [5] have employed Convolutional Neural Networks for predicting material properties from the periodic table. Kim et al. [6] have studied dielectric properties of materials based on Random Forest regression. The overall thermal properties of composite materials have been predicted by Wei et al. [7] using Gaussian Process regression.

To evaluate the feasibility of using ML in mechanical and materials science, in this work, we develop and optimize the architecture of a Neural Network model to predict the tensile modulus of nanocomposites (polymer based reinforced by carbon nanotubes). The model was developed and trained based on experimental data, which were gathered from available literature.

2 Materials and Methods

2.1 Neural Network Model

A Neural Network (NN) model has been chosen to predict the tensile modulus of nanocomposites because of its various advantages compared to other machine learning models. NNs are machine learning models that are inspired by the function of biological neural networks. NN models are designed for regression and classification problems. The main principle of NN is to learn through detection of patterns and relationships within the input data [8]. In general, NN models are composed of 3 main layers: (i) input layer which consists of the input data; (ii) hidden layers whose functions are training and testing the model using input data from input layer; (iii) output layer which exports the output response [9].

One of the advantages of NN models is the capability to train with any type of input data [10], which can help us to save a lot of computational time, especially when input data is collected from several sources. Another advantage of NN is that the information is stored the entire network instead of a database, which helps speed up the training and testing processes. In addition, NN is very suite for parallel computing, which can significantly reduce the computational time.

2.2 Database

As mentioned above, the database used in this study has been gathered from available resources in the literature. Indeed, it is composed of 31 types of polymers, combining with 29 CN incorporating methods, 30 CN surface modifications and 5 mechanical testing methods. Six parameters of polymers and carbon nanotubes have been used as variables for the ML model, such as: (a) polymer matrix type; (b) mechanical properties of polymer matrix (its tensile modulus, tensile strength, yield strength); (c) physical's characteristics of CN (its density, length, diameter); (d) mechanical properties of CN (its tensile modulus); (e) incorporation factors (CN weight fraction, processing method, CN surface modification); (f) tensile testing method. The 282 data used in this work were randomly divided into two sub-datasets (using a uniform distribution), 70% were served for training the AI models and 30% were served as the validating part [11]. The influence of random splitting was also investigated using a Monte Carlo scheme and statistical analysis such as estimator of convergence. The training dataset was next scaled into the range of [−1; 1], as it is common in machine-learning for minimizing the bias between variables. Scaling parameters of the training dataset were also employed for scaling the testing dataset as well, in order to prevent pre-correlation. Unknown values in the database have also been processed when training AI models.

3 Results and Discussion

3.1 Optimization of NN's Architecture by Means of a Parametric Study

In a NN architecture, hidden layers, defined as the layers between input and output layers, can be set up in many different manners. The analysis of the architecture of hidden layers is crucial as the performance of the predictor is directly affected by the number layers and the neurons in each hidden layer. In this study, taking R^2, RMSE, MAE and the Slope Angle as the criteria for comparison, the influence of the number of neurons in hidden layer as well as the number of hidden layers on the predicted results is analyzed. Summary of this parametric study is presented in Table 1. Concerning the problem, the number of neurons in input layer is fixed as 17, for output layer is 1, the activation function for hidden layer is chosen as the sigmoid function, activation function for output layer is the linear function, the training ANN algorithm is Levenberg-Marquardt and the cost function is the mean square error function. The number of hidden layers is varied from 1 to 2, whereas the number of neurons in each hidden layer is varied from 1 to 30 neurons.

Table 1. ANN's characteristics.

Type	Parameter	Value and description
Fixed	Number of neurons in input layer	17
	Number of neurons in output layer	1
	Activation function for hidden layer	Sigmoid
	Activation function for output layer	Linear
	Training algorithm	Levenberg-Marquardt
	Cost function	Mean square error
Parametric study	Number of hidden layers	N_{layer}, varying from 1 to 2
	Number of neurons in hidden layers	N_{neuron}, varying from 1 to 30
Information on weight parameter matrices	Size of weight matrix of hidden layer	$N_{neuron} \times N_{layer}$
	Size of bias vector of hidden layer	$1 \times N_{neuron}$
	Size of weight matrix of output layer	$1 \times N_{neuron}$
	Size of bias vector of output layer	1×1

The average of quality assessment criteria for the training and testing datasets are plotted in Fig. 1. It can be seen that with a number of neuron inferior to 5, the quality of the NN predictor is poor. The optimal choice in function of time-consuming and prediction quality is 15 neurons in the hidden layer, confirmed by all the quality assessment criteria (R^2, RMSE, MAE and Slope angle) (Fig. 2) ,

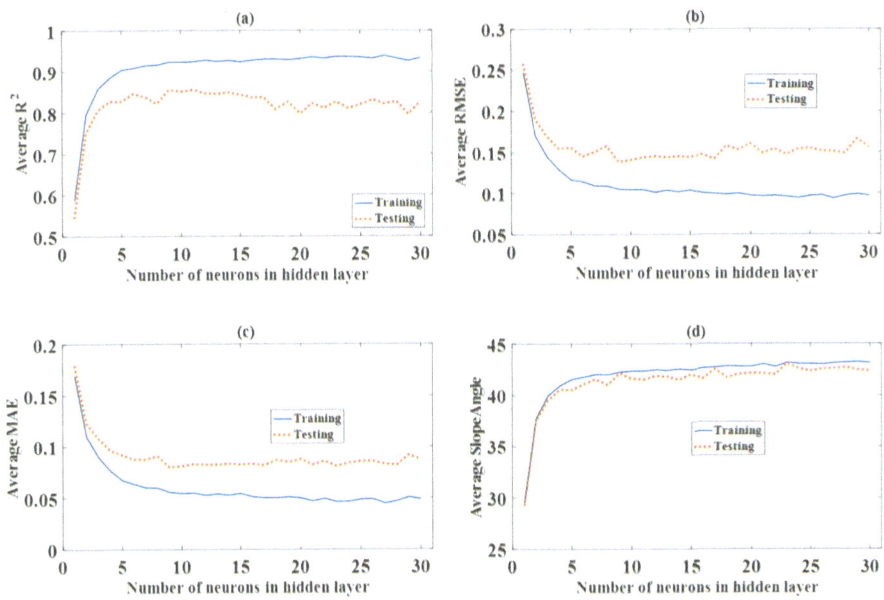

Fig. 1. Average of quality assessment criteria in function of number of neurons in hidden layer (case of one hidden layer architecture): (a) R^2, (b) RMSE, (c) MAE and (d) SlopeAngle.

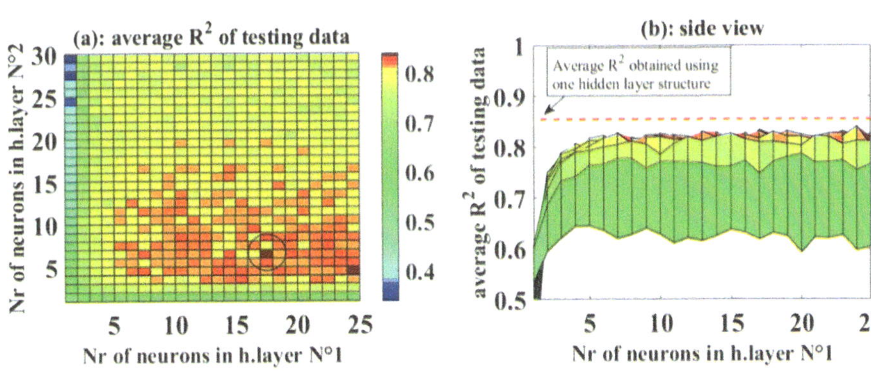

Fig. 2. Parametric study for the two hidden layer architecture: (a) average R^2 of testing data in function of number of neurons in each layer, (b) comparison between one and two hidden layer architectures.

3.2 Prediction Capability

In this section, results of prediction for a typical configuration are presented, using one hidden layer with 15 neurons. It is worth noticed that an analysis on the cost function in training the optimal NN model is performed, previously to the principal results presented herein. The optimal epoch is deduced as 6 to prevent overfitting problem from the NN algorithm. The performance of the NN in terms of linear fitting is displayed in Fig. 3. It can be seen that the model exhibits very good performance for both training and testing datasets. The correlation coefficient in both cases is satisfied (R^2 = 0.908 for the training part and R^2 = 0.911 for the testing part). Details of other performances of the NN model is presented in Table 2:

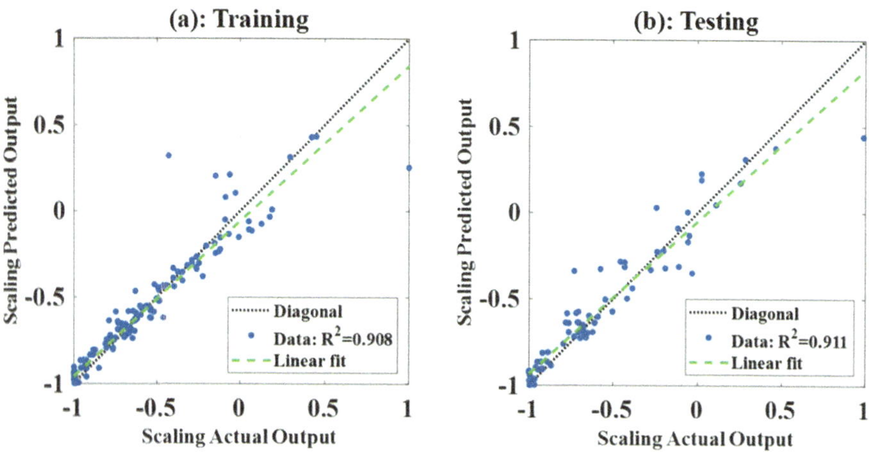

Fig. 3. Regression graphs for (a) training and (b) testing data

Table 2. Summary of prediction capability.

Dataset	R^2	RMSE	MAE	Error Mean	Error Std	Slope	Slope Angle
Training	0.908	0.111	0.063	−0.008	0.111	0.900	41.981
Testing	0.911	0.126	0.085	−0.020	0.125	0.883	41.441

The comparison between the measured and predicted tensile modulus with respect to each sample is presented in Fig. 4. In this example, the tensile modulus values are scaled in a [−1, 1] range and the R^2 value is 0.910. In summary, the NN with 15 neurons and one hidden layer have a strong capability in predicting the tensile modulus of nanocomposite.

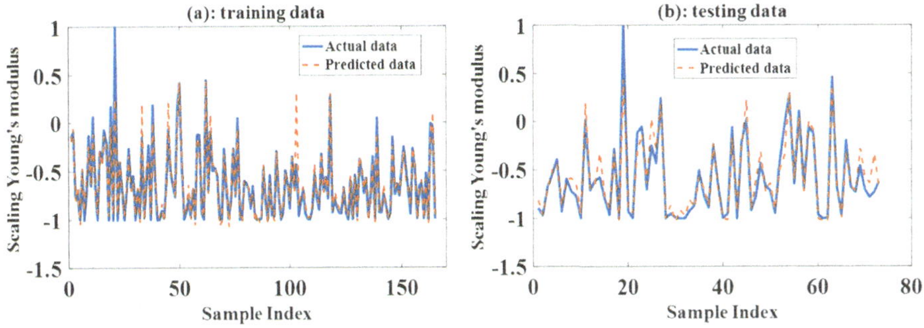

Fig. 4. Graphs of scaling tensile modulus in function of sample index for (a) training and (b) testing data.

4 Conclusion

In this study, a neural network model has been employed to predict the tensile modulus of CN/polymer nanocomposites, based on experimental database collected from available literature. The obtained results are promising; with the correlation coefficients for the training and testing datasets are 0.908 and 0.911, respectively. Two other quality assessments (RMSE and MAE) have also been used for the performance of the model, which is quite reasonable. A good correlation between measured and predicted data has also been obtained by the proposed method.

Acknowledgement. This research is funded by PHENIKAA University under grant number 2–05.2020.04.

References

1. Chen, L.: The materials genome initiative and advanced materials. Engineering. **1**, 169 (2015). https://doi.org/10.15302/J-ENG-2015056
2. Nouira, A., Sokolovska, N., Crivello, J.-C.: CrystalGAN: Learning to Discover Crystallographic Structures with Generative Adversarial Networks. arXiv:1810.11203 [cs, stat]. (2019)
3. Ryan, K., Lengyel, J., Shatruk, M.: Crystal structure prediction via deep learning. J. Am. Chem. Soc. **140**, 10158–10168 (2018). https://doi.org/10.1021/jacs.8b03913
4. Oliynyk, A.O., Adutwum, L.A., Harynuk, J.J., Mar, A.: Classifying crystal structures of binary compounds ab through cluster resolution feature selection and support vector machine analysis. Chem. Mater. **28**, 6672–6681 (2016). https://doi.org/10.1021/acs.chemmater.6b02905
5. Zheng, X., Zheng, P., Zhang, R.-Z.: Machine learning material properties from the periodic table using convolutional neural networks. Chem. Sci. **9**, 8426–8432 (2018). https://doi.org/10.1039/C8SC02648C
6. Kim, C., Pilania, G., Ramprasad, R.: From organized high-throughput data to phenomenological theory using machine learning: the example of dielectric breakdown. Chem. Mater. **28**, 1304–1311 (2016). https://doi.org/10.1021/acs.chemmater.5b04109

7. Wei, H., Zhao, S., Rong, Q., Bao, H.: Predicting the effective thermal conductivities of composite materials and porous media by machine learning methods. Int. J. Heat Mass Transf. **127**, 908–916 (2018). https://doi.org/10.1016/j.ijheatmasstransfer.2018.08.082
8. Ho, N.X., Le, T.-T., Le, M.V.: Development of artificial intelligence based model for the prediction of Young's modulus of polymer/carbon-nanotubes composites. Mech. Adv. Mater. Struct. 0, 1–14 (2021). https://doi.org/10.1080/15376494.2021.1969709
9. Le, T.-T.: Practical Hybrid Machine Learning Approach for Estimation of Ultimate Load of Elliptical Concrete-Filled Steel Tubular Columns under Axial Loading. Adv. Civ. Eng. e8832522 (2020). https://doi.org/10.1155/2020/8832522
10. Thanh Duong, H., Chi Phan, H., Le, T.-T., Duc Bui, N.: Optimization design of rectangular concrete-filled steel tube short columns with balancing composite motion optimization and data-driven model. Structures. **28**, 757–765 (2020). https://doi.org/10.1016/j.istruc.2020.09.013
11. Le, T.-T., Le, M.V.: Development of user-friendly kernel-based Gaussian process regression model for prediction of load-bearing capacity of square concrete-filled steel tubular members. Mater. Struct. **54**(2), 1–24 (2021). https://doi.org/10.1617/s11527-021-01646-5

Removal of Phosphate from Water Using Sonochemically Synthesized ZnO Nanoparticles

Pham Huong Quynh[1], Dang Van Thanh[2], Tran Quoc Toan[3], Nguyen Nhat Huy[4,5], and Nguyen Thi Thuy[5,6](✉)

[1] Hanoi University of Industry, 298 Cau Dien Street, Bac Tu Liem District, Hanoi, Vietnam

[2] TNU-University of Education, 20 Luong Ngoc Quyen Street, Thai Nguyen, Vietnam

[3] Department of Chemistry, Thai Nguyen University of Education, Thai Nguyen, Vietnam

[4] Faculty of Environment and Natural Resources, Ho Chi Minh City University of Technology (HCMUT), Ho Chi Minh, Vietnam

[5] Vietnam National University Ho Chi Minh City, Linh Trung Ward, Thu Duc City, Ho Chi Minh, Vietnam

ntthuy@hcmiu.edu.vn

[6] Department of Environmental Engineering, International University, Thu Duc City, Ho Chi Minh, Vietnam

Abstract. In this study, we aimed to synthesize ZnO nanomaterial and applied it for phosphate removal in aqueous solutions. ZnO material was prepared by a sonochemical route which is a simple and efficient method for large-scaled production of ZnO in nanosize. The surface characteristics and material structure of ZnO were determined via using electron scanning microscopy and X-ray diffraction methods. Batch experiments were conducted for evaluating the adsorption capacity of ZnO material. The effects of solution pH, adsorption time, adsorbent amount, and initial phosphate concentration on the adsorption capacity were investigated to determine the suitable condition for phosphate removal. Results showed a high phosphate removal ability at a wide pH range of 3–7 after 90 min with an adsorbent dosage of 0.6 g/L. The adsorption of phosphate followed Langmuir isotherm with a maximum adsorption capacity of 100 mg/g. The removal of phosphate in wastewater of Lam Thao Fertilizers and Chemicals JSC (Phu Tho Province) was also conducted with a high removal efficiency of 88.75%.

Keywords: Nano ZnO · Phosphate · Adsorption · Wastewater

1 Introduction

Phosphate ($(PO_4)^{3-}$) is a nutrient that often presents in natural water sources receiving from domestic and industrial wastewater. At low concentrations, phosphate does not affect human health. When combined with nitrogen at high concentrations, phosphate

D. C. Nguyen et al. (Eds.): ICERA 2021, LNNS 366, pp. 793–798, 2022.
https://doi.org/10.1007/978-3-030-92574-1_81

will cause eutrophication that is a phenomenon accelerating the process of algae growth and after the algae dies, green water phenomenon, unpleasant odors and releasing toxins can occur, which leads to fish death and water pollution [1]. For these reasons, the removal of phosphate is necessary to reduce pollution in water. In recent years, many studies have suggested methods to remove phosphate from water such as precipitation, adsorption, biological treatment, and ion exchange. In the above methods, the method of adsorption of phosphate with nanomaterials is a remarkable topic.

ZnO is an environmentally friendly material. In addition, ZnO is one of the most researched materials for photocatalyst applications, gas sensors and solar cells [2]. Recently, phosphate adsorption research using ZnO or Zn–Al layered double hydroxides has been tested with very promising results [3, 4]. However, ZnO nanomaterials in these studies were created with a long time and high temperature to increase the adsorption efficiency of materials. Therefore, finding a method of manufacturing ZnO nanomaterials with a short time and at a low temperature is necessary for industrial application, especially in water and wastewater treatment.

In the previous study, we synthesized ZnO nanomaterials and applied them to treat H_2S gas [5]. The results showed that the nanomaterials had a large surface area of 20 m^2/g with relatively homogenous particle size of 20–70 nm and were able to fully adsorb H_2S in 30 min. In this study, we continue to develop the application of the above nanomaterials for the adsorption of phosphate in water by the batch adsorption method. The ability to remove other ions of ZnO nanomaterials has also been tested.

2 Materials and Methods

All chemicals have PA purity. ZnO nanomaterials were prepared based on the method published in a previous study [5]. 180 mL of 0.3 M NaOH was slowly added at a rate of 2 mL/min to 180 mL of 0.15 M $ZnSO_4$. The mixture was contained in a beaker and then placed in an ultrasonic device (40 kHz, 400 W) and stirred at speed of 5 rpm. After 60 min, the mixture was filtered and washed with DI water until the washed pH values remain constant. The material was dried at 100 °C for 24 h, denoted as UZ. The characteristics of the material were determined using SEM (FESEM Hitachi SU8000) and XRD (D2 PHASER).

The experiments were conducted by the batch adsorption method at room temperature ($\sim$25 °C) with a shaking speed of 200 rpm. pH value was adjusted via using 0.1 M HNO_3 and 0.1 M NaOH solutions. Investigating parameters included pH, adsorption time, and initial phosphate concentration. After adsorption, the samples were centrifuged at a speed of 4000 rpm for 10 min. Phosphate concentration was determined by the ascorbic acid method using UH5300 UV-Visible Double-Beam Spectrophotometer at the wavelength λ = 880 nm. The experiments were repeated 3 times and the average values were reported. Samples of real wastewater were taken at the Lam Thao Fertilizers and Chemicals Corporation according to TCVN 6663 - 1: 2011.

Adsorption efficiency was calculated based on formula (1) [5]:

$$H\% = \frac{(C_0 - C_e)}{C_0} \times 100\% \tag{1}$$

Where C_o is the inlet concentration of pollutant (mg/L) and C_e is the concentration of the pollutant at adsorption equilibrium (mg/L). H is adsorption efficiency (%).

3 Results and Discussions

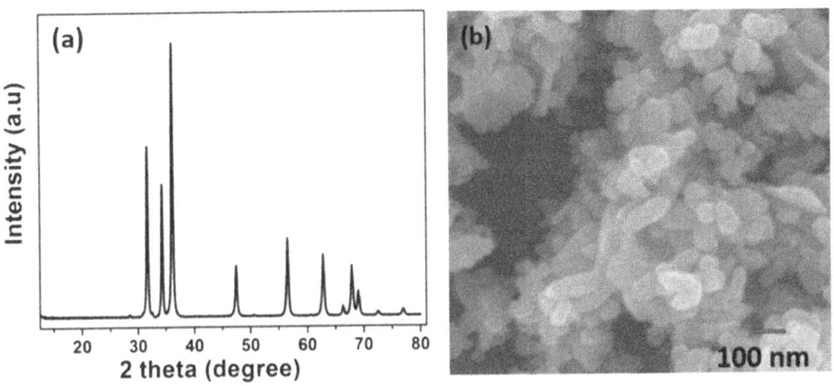

Fig. 1. (a) XRD diagram and (b) SEM image of the UZ nanomaterial

Figure 1a is the XRD result of the ZnO nanomaterial. The wurtzite hexagonal structure of the material was identified with characteristic peaks at 2θ of 31.77, 34.40, 36.22, 47.61, 56.58, 62.85, 66.41, 67.93, 69.08, 72.54, and 76.85°, corresponding to the planes of (100), (002), (101), (102), (110), (103), (200), (112), (201), (004), and (202) [5, 6]. The SEM result in Fig. 1b shows the granular shape of the ZnO material with particle size in the range of 50–100 nm.

The point of zero charge for ZnO was measured to be 7.04 (Fig. 2a). Figure 2b shows the effect of pH on the phosphate adsorption of ZnO. At a pH value in the range of 1–2, the adsorption capacity was very low at 30%. At pH value in the range of 3–7, the adsorption capacity of the material was very high (98–99%). When the pH value increased from 7 to 8, adsorption efficiency at pH 8 reduced by 30% as compared to the adsorption efficiency at pH 7. The reason is that at a pH value lower than the point of zero charge, the surface of ZnO may be protonated according to the following process.

$$\equiv \mathrm{Zn\text{ - }OH} + \mathrm{H}^+ \rightleftarrows \equiv \mathrm{Zn\text{ - }OH_2^+} \tag{2}$$

Where $\equiv$Zn-OH_2^+ represents a group of positively charged surfaces that facilitate the ion exchange. At a low pH, phosphate exists in solution in the form of $H_2PO_4^-$. Therefore, the electrostatic suction force is created between the adsorbent surface and the adsorbed substance. Under high pH value of the solution, $H_2PO_4^-$ is gradually replaced by HPO_4^{2-} and the number of positive charge points is reduced. For that reason, the saturation time of the material is reduced. In addition, when the pH value is high, the surface of the material also occurs the reduction of ≡Zn-OH.

$$\equiv \text{Zn - OH} - \text{H}^+ \rightleftarrows \equiv \text{Zn - O}^- + 2\text{H}^+ \tag{3}$$

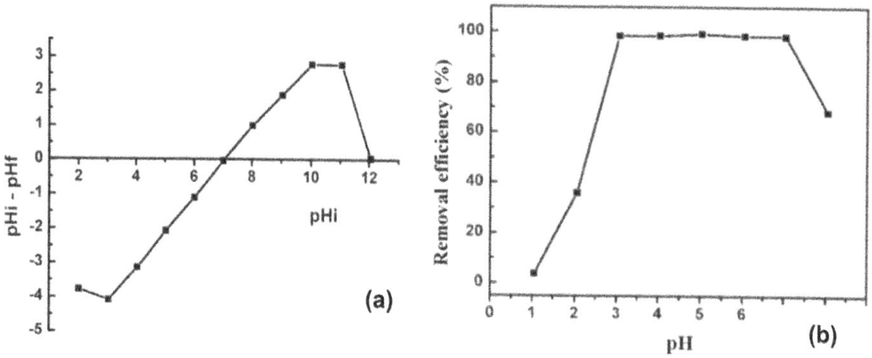

Fig. 2. (a) The pH_{pzc} of UZ and (b) Effect of pH on the phosphate adsorption by ZnO

At this time, the surface of the material is negatively charged so that an electrostatic repulsion interaction occurs among negatively charged ions in the solution. In addition, the high concentration of OH^- ion competes with phosphate ions ($H_2PO_4^-$, HPO_4^{2-}, PO_4^{3-}) for the adsorption center of the adsorbent [4, 7].

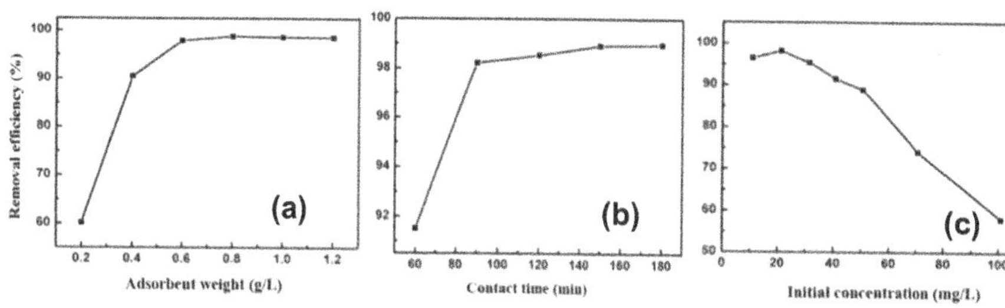

Fig. 3. (a) Effect of ZnO amount, (b) Contact time, and(c) Initial phosphate concentration on the phosphate adsorption by ZnO

As seen in Fig. 3a, the adsorption efficiency increased with the amount of adsorbent. When the amount of material increased from 0.2 to 0.6 g/L, the adsorption efficiency increased 37.81%, due to the increase in surface area available for adsorption. However, the amount of phosphate adsorbed on a unit of adsorbent (i.e., adsorption capacity) decreased. Therefore, 0.6 g/L was chosen as a suitable amount of ZnO for adsorption of phosphate.

Adsorption efficiency reached 91.52% at 60 min, as observed in Fig. 3b. From 90 to 180 min, the adsorption efficiency did not change significantly. This may be due to the fact that as the adsorption process gradually reaches equilibrium, the adsorbent surface is covered by ion PO_4^{3-}. As a result, the adsorption capacity is not significantly increased. Therefore, 90 min was chosen as the equilibrium time for further tests. Figure 3c presents the effect of initial phosphate concentration on the adsorption. It is obvious that the phosphate adsorption efficiency decreased with the increase of initial phosphate concentration. It indicates that 0.6 mg/L of ZnO was not sufficient for high initial concentration of phosphate. Depending on the initial phosphate concentration as well as the requirement of treatment to be achieved, testing the treatment with higher ZnO concentrations is recommended.

The experimental data was then applied for the Langmuir isotherm model and the result is illustrated in Fig. 4. According to Fig. 4b, the maximum adsorption capacity (q_{max}) of UZ is calculated as 100 mg/g with a high correlation coefficient R^2 (0.997).

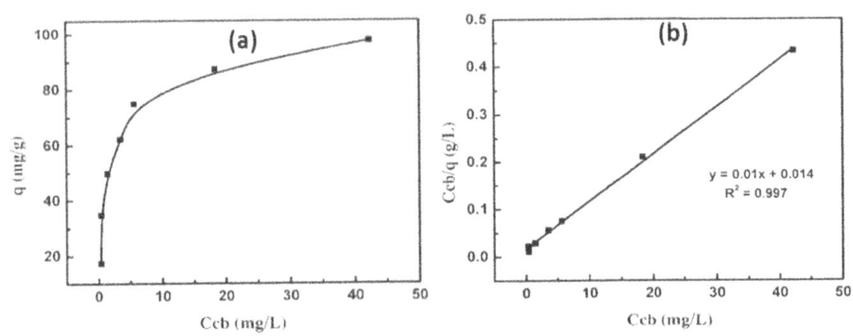

Fig. 4. The graphs of (a) Langmuir isotherm and (b) the dependence of C_{cb} and C_{cb}/q (C_{cb}: phosphate concentration at equilibrium)

Table 1. UZ material application for Lam Thao Fertilizers and Chemicals Corporation's wastewater

No.	Parameter	Unit	Inlet	Outlet	Removal efficiency (%)	QCVN 40:2011/BTNMT (column B)*
1	pH	–	2.3	5.14	–	5.5–9
2	Color	Pt/Co	<5	–	–	150
3	COD	mg/L	264.8	50	81.10	150
4	BOD_5 (20 °C)	mg/L	84.0	–	–	50
5	TSS	mg/L	8280	–	–	100
6	T-N	mg/L	19	–		40
7	NH_4^+	mg/L	10.8	0.17	98.43	10
8	T-P	mg/L	50	–	–	6
9	PO_4^{3-}	mg/L	44.9	5.05	88.75	–

*National technical regulation on industrial wastewater (with $K_q = 1$ and $K_f = 1$)

The experiment was then conducted for real wastewater treatment with an initial phosphate concentration of 44.9 mg/L at conditions of pH 5, ZnO amount of 0.6 g/L, and adsorption time of 90 min. Results showed that adsorption efficiency reached 88.75% and the outlet concentration of phosphate after adsorption was 5.05 mg/L (Table 1). The adsorption efficiency of real wastewater adsorption was lower than that of synthetic wastewater samples (9.47%). This could be explained by the fact that wastewater has many factors that interfere with adsorption processes such as organic compounds and other ions. Besides, the removal efficiencies for COD and NH_4^+ were 81.1 and 98.4%, respectively.

4 Conclusion

The ZnO prepared by the sonochemical method was successfully applied as an effective adsorbent for phosphate removal in wastewater. The suitable condition for the adsorption was found to be at pH 3–7 and ZnO amount of 0.6 g/L for 90 min. In the real treatment of Lam Thao Fertilizers and Chemicals Corporation's wastewater, the phosphorus removal efficiency achieved 88.75% and the outlet concentration of phosphate after adsorption was 5.05 mg/L.

References

1. Liu, Q., Wan, J., Wang, J., Li, S., Dagot, C., Wang, Y.: Recovery of phosphorus via harvesting phosphorus-accumulating granular sludge in sequencing batch airlift reactor. Biores. Technol. **224**, 87–93 (2017)
2. Djurisic, A.B., Chen, X., Leung, Y.H., Man Ching Ng, A.: ZnO nanostructures: growth, properties and applications. J. Mater. Chem. **22**(14), 6526–6535 (2012)
3. Yang, K., et al.: Adsorptive removal of phosphate by Mg–Al and Zn–Al layered double hydroxides: kinetics, isotherms and mechanisms. Sep. Purif. Technol. **124**, 36–42 (2014)
4. Luo, Z., Zhu, S., Liu, Z., Liu, J., Huo, M., Yang, W.: Study of phosphate removal from aqueous solution by zinc oxide. J. Water Health **13**(3), 704–713 (2015)
5. Huy, N.N., et al.: Facile one-step synthesis of zinc oxide nanoparticles by ultrasonic-assisted precipitation method and its application for H2S adsorption in air. J. Phys. Chem. Solids **132**, 99–103 (2019)
6. Lu, F., Cai, W., Zhang, Y.: ZnO hierarchical micro/nanoarchitectures: solvothermal synthesis and structurally enhanced photocatalytic performance. Adv. Func. Mater. **18**(7), 1047–1056 (2008)
7. Elkady, M.F., Shokry Hassan, H., Salama, E.: Sorption profile of phosphorus Ions onto ZnO nanorods synthesized via sonic technique. J. Eng. **2016**, 1–9 (2016)

Research and Test of the Self-designed and Manufactured Rotary Friction Welding Machine with CT3 Steel Samples

Tran Vinh Hung[1,2], Ta Van Ranh[3], Tran Thi Van Nga[4], and Le Hong Ky[3](✉)

[1] Faculty of Mechanical Engineering and Mechatronics, PHENIKAA University, Hanoi 12116, Vietnam
[2] PHENIKAA Research and Technology Institute (PRATI), A&A Green Phoenix Group JSC, No. 167 Hoang Ngan, Trung Hoa, Cau Giay, Hanoi 11313, Vietnam
[3] Vinh Long University of Technology Education, Vinh Long, Vietnam
kylh@vlute.edu.vn
[4] Faculty of Mechanical Engineering, University of Transport and Communications, Hanoi, Vietnam

Abstract. This paper presents the results of research and test of the self-designed and manufactured rotary friction welding machine. Tensile test results show that the tensile strength of the material after welding is satisfactory according to the standards of the material; the elongation is within the elongation limit of the welding specimen; the yield limit is greater than the minimum yield limit of the material. The parameters of the welding equipment are guaranteed according to the design requirements.

Keywords: Tensile strength · Elongation · Yield limit · Rotary friction welding · Welding material

1 Introduction

Friction welding is the welding method with the most stable and durable welded joint structure compared with the method of fusion welding of the material. In addition to the outstanding advantage in terms of ensuring the mechanical properties of the welded joint, an advantage that other welding methods cannot do is to connect different materials through the process of friction, the materials are diffused into each other. With rotary friction welding, when two parts rotate relative to each other, generating frictional heat that melts the material at the contact area, the two parts are pressed together to create a welded joint [1–3, 14–18], Fig. 1.

D. C. Nguyen et al. (Eds.): ICERA 2021, LNNS 366, pp. 799–812, 2022.
https://doi.org/10.1007/978-3-030-92574-1_82

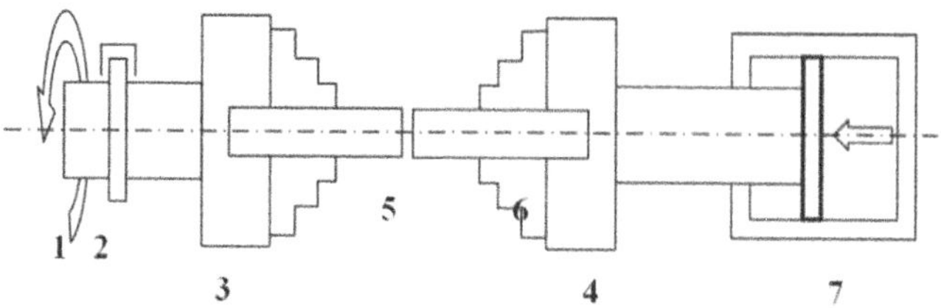

Fig. 1. Operating principle of rotary friction welding machine

In the 60s of the twentieth century, rotary friction welding developed strongly in the United States. Most of the equipment manufactured during this period was researched by AMF, Caterpillar, and Rockwell International companies [1–3]. Currently, rotary friction welding is strongly developed to meet the increasing needs of the manufacturing industry. At universities, a lot of research on rotary friction welding has been carried out in various aspects. Several in-depth studies on rotary friction have been carried out in the world. Researches on rotary friction welding machines: Gourav sardana and Ajay Lohan, have researched special fixture for friction welding on lathe machine [1]; Jagroop Singh, Karamdeep Singh, Department of Mechanical Engineering, ACEM, Kapur thala, Punjab, India have researched rotary friction welding on lathe machine with stailstock [2]; P. A. Thakare, Lt Randheer Singh designed and developed the micro rotary friction welding machine and investigated the friction welding parameters for similar materials [3]. Researches on welding of different materials include: H. Ochi, working at Welding Research Center, Osaka Technical Research Institute, Japan, together with K. Ogawa studied the properties of the welds of aluminum alloy and steel [25]; Z. Lindemann, K. Skalski, W. Wosin Ski, and J. Zimmerman studied the thermo-mechanical phenomenon in the process of friction welding of corundum ceramics and aluminium [14]; Eder Paduan Alves, Francisco Piorino Neto, Chen Ying An, Euclides Castorino da Silva conducted an experimental investigation of temperature during rotary friction welding of AA1050 aluminum with AISI 304 stainless steel [24]; M. B. Uday, M. N. Ahmad Fauzi, Zuhailawati H, A.B. Ismail of University of Materials and Mineral Resources Engineering, Malaysia, studied the effect of welding speed on the mechanical strength of friction welded joint of YSZ-Alumina composite and 6061 aluminum alloy [5]. Research on thermo-mechanical phenomena in the process of rotary friction welding includes authors with the following published work: Ruma Mohd Abdul Wahed, Mohammed Farhan studied the effect of external heating of the friction welded joint [17]; M. B. Uday, M. N. Ahmad Fauzi, Zuhailawati H, A.B. Ismail of University of Materials and Mineral Resources Engineering, Malaysia, studied the effect of welding speed on the mechanical strength of friction welded joint of YSZ-Alumina composite and 6061 aluminum alloy [5]. Jatinder Gill, Jagdev Singh experimentally studied the effect of heating time on the mechanical properties of nylon joints produced by friction welding [18]; Eder Paduan Alves, Chen Ying An conducted an experimental determination of temperature during rotary friction welding of dissimilar materials published in Scientific and Engineering Publishing Company [15]; Ali Moarrefzadeh studied the effect of heat in the friction welding process, published in Journal of Mechanical Engineering 2012 [16]. Evaluation of rotary friction welding includes the typical authors and works such as G. J. Baxter, M. Preuss and P.J. Withers at Manchester Institute of Natural Materials Research, UK, studied inertial friction welding of nickel base superalloys for aerospace applications [4]; Piaar Nagar, D. Ananthapadmanaban studied the friction weldability of low carbon steel with stainless steel and aluminum with copper [28]; P. Shiva Shankar, L. Suresh Kumar, N. Ravinder Reddy used Taguchi method to conduct an experimental investigation and analyze the friction welding parameters for Cu Zn28 copper alloy [6]; Amit Handa, Vikas Chawla made a review to evaluate friction welding [34]; Veerabhadrappa Algurl, Kabadi, Ganechari and Sharanabasappa conducted an experimental investigation on friction characteristics of modified ZA-27 alloy using taguchi technique [7]; Amit Handa Punjub, Vikas Chawla Dav conducted an experimental study on mechanical properties of friction welded AISI 1021 steels [8]; Mumin Sahin and Cenk Misirli studied the mechanical and metallurgical properties of Friction Welded Aluminium Joints [29]; Ali. Moarrefzadeh performed a Numerical Modeling of Friction Welding Process [35]; The research team of Baiju Sasidharan, Dr. K.P.Narayanan, R. Arivazhakan studied the influence of interface surface geometries in the tensile characteristics of friction welded joints from aluminium alloys [20].

Friction welding is one of the environmentally friendly welding methods. The main source of welding energy is the heat source due to friction between the contacting surfaces of the welding parts. This heat source causes the surface of the part to flow intensely in the structure of the material, a new feature in current biomechanical engineering. The welding principle is completely new, so it needs a lot of research from basic to in-depth. The material organization at the welded joints changes compared to the original material, and nanostructures can be achieved, which is very significant in the new materials industry. The requirement for experimental equipment is simple, and the theoretical basis is encapsulated in mechanical engineering, which is very suitable for current research conditions in universities.

2 Models of Rotary Friction Welding Machines Designed and Manufactured

Based on the reference to the studies [1–3, 6–8] and the actual working conditions of the research team, after analyzing the options, an operating principle of the rotary friction welding machine was selected as Fig. 1, 2D design model as Fig. 2. Main parameters typical for welding process include: spindle rotation speed, friction pressure and friction time, forging pressure and forging time, upset length.

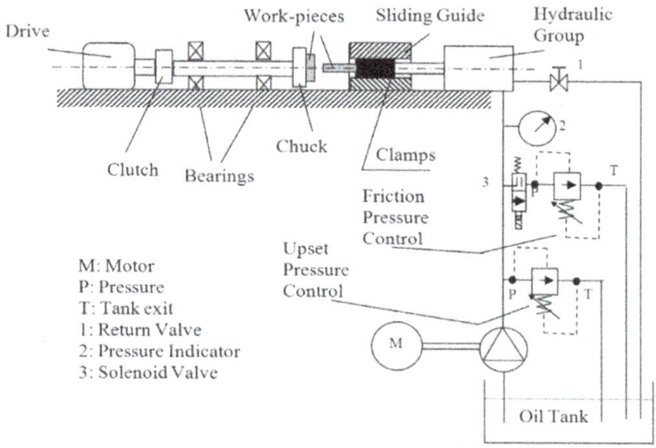

Fig. 2. 2D design model of rotary friction welding machine

Figure 2 and Fig. 3a show that when the rotation speed reaches a given speed *n*, hydraulic action causes two surfaces to rub against each other. Then the torque increases, and when the two frictional surfaces generate heat, the torque decreases. This stage is the friction phase. During the friction phase, the length of the workpiece is shortened by a small amount. When the friction surface has material overflow, the brake mechanism makes the rotation speed *n* stop suddenly. During braking the torque is maximum and the torque is zero when the rotation speed *n* is zero. This is the braking phase. The braking phase takes place as quickly as possible (if the braking time is large, the heat reduction of the welded joint is limited). When the braking phase is about to end, the pressure should be increased so that the two surfaces diffuse intensely into each other. This stage is the forging phase. The steps to perform the rotary friction welding process are shown in Fig. 3b.

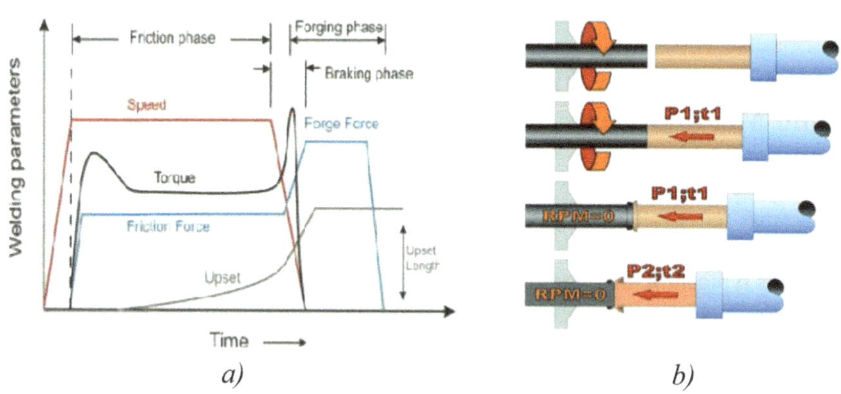

Fig. 3. Parameters and phases of rotary friction welding process

The rotary friction welding machine is manufactured (Fig. 4) fitted with a solenoid valve to control the hydraulic system (principle shown in Fig. 2).

Fig. 4. The rotary friction welding machine is manufactured

3 Experiments

3.1 Sample for Experiment

Test specimens of solid circular shape, with a diameter of 14 mm and a length of 100 mm, are made of CT3 steel with strength limits according to ГОСТ 380 - 89 standard as shown in Table 1.

Table 1. Mechanical properties of CT3 steel

Standard	Mechanical properties		
	Yield limit (N/mm^2)	Tensile strength (N/mm^2)	Elongation (%)
ГОСТ 380 - 89	225 min	373 ÷ 461	22 min

3.2 Measuring Instruments

In the study, the CHT4106 universal tensile strength testing machine was used to measure the mechanical properties of the samples after rotary friction welding (Fig. 5).

Fig. 5. HT4106 model universal compression tensile machine

HT4106 model universal compression tensile machine with load force of 1000 kN, can perform tensile, compression, bending and cutting tests. Test standards are in accordance with ISO, ASTM, etc. With PowerTest software of SANS, all data and management are in the database, easy to access and transfer to other softwares such as Excel. Parameters such as yield limit, strength limit, relative elongation, elastic modulus, etc. can be measured or calculated automatically. The machine also determines the bending resistance of metal and can display 3 curves at the same time: stress-deformation, force-time, force-elongation of the test specimen.

3.3 Experimental Design

Test parameters include: rotation speed of workpiece, friction force, friction time, forging time, forging force, upset length. In which, there are 2 fixed parameters (rotation speed of workpiece *n*, upset length *l*), 2 variable parameters that can be controlled (axial pressure in the friction stage *P*1, axial pressure in the forging stage *P*2) and 2 calculated dependent parameters: friction time *t*1, forging time *t*2 (Table 2).

Table 2. Input parameters of the rotary friction welding test specimen

No.	Axial pressure at stages	Symbol		Design variable value			Unit
		Nature	Encode	(−1)	(0)	(+1)	
1	*Friction*	***P*1**	A	50	70	90	N/mm^2
2	*Forging*	***P*2**	B	110	130	150	N/mm^2

The second-order symmetric mixed quadratic experimental planning of the form B with the polynomial regression equation has the general form as formula (1):

$$y = b_o + \sum_{i=1}^{k} b_i x_i + \sum_{i=1}^{k} b_{ii} x_i^2 \sum_{\substack{i<j \\ i,j=1}}^{k} b_{ij} x_i x_j \tag{1}$$

The study has chosen the second-order symmetric mixed planning of the form B, the experiment at the center of 2 to help keep the number of experiments to a minimum. Natural form planning matrix is shown in Table 3.

Table 3. Experimental planning matrix with input parameters

Test specimen	*n*(rpm)	Axial pressing force		Time		*l*(mm)
		*P*1 (N/mm^2)	*P*2 (N/mm^2)	*t*1 (s)	*t*2 (s)	
1	1.450	50	110	50	7	4
2	1.450	90	110	36	7	4
3	1.450	50	150	50	5	4
4	1.450	90	150	36	5	4
5	1.450	50	130	50	6	4
6	1.450	90	130	36	6	4
7	1.450	70	110	43	7	4
8	1.450	70	150	43	5	4
9	1.450	70	130	43	6	4
10	1.450	70	130	43	6	4

3.4 Conducting Experiments

The samples are cut to length and cleaned to remove oil, grease, etc. and ensure a flat, smooth, swarf-free work surface so that the two surfaces are in even contact, good friction, and faster heating time. 10 experiments with the same parameters are carried out as shown in Table 3 according to the basic stages: Friction process (Fig. 6) and Forging process (Fig. 7).

Fig. 6. Friction process

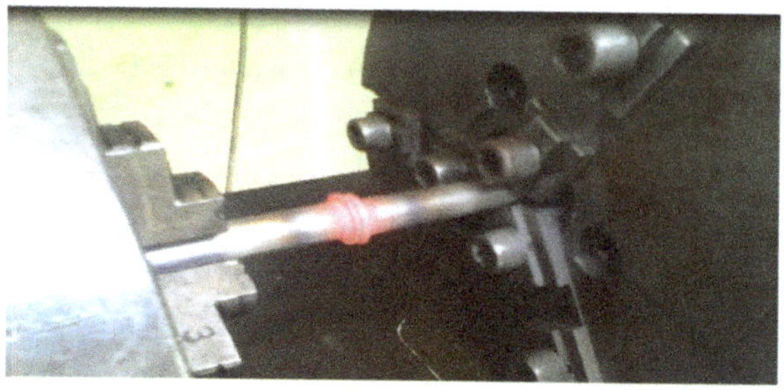

Fig. 7. Forging process (upset)

3.5 Results and Discussion

After the rotary friction welding equipment is manufactured and the above samples are tested, it shows that the friction welding equipment gets the set requirements: high rigidity, low vibration during operation, the main shaft motor ensuring the operating capacity and reaching the number of revolutions according to the design requirements, the sliding table mechanism operating smoothly and ensuring stability during the welding process, the hydraulic system ensuring the original design pressure.

According to the ASTM E8 standard, the samples after rotary friction welding (Fig. 8) are machined by turning method before the tensile test (Fig. 9). The sample after tensile test to determine mechanical properties is shown in Fig. 10.

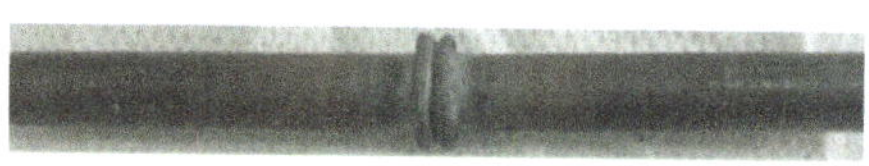

Fig. 8. CT3 steel welding product

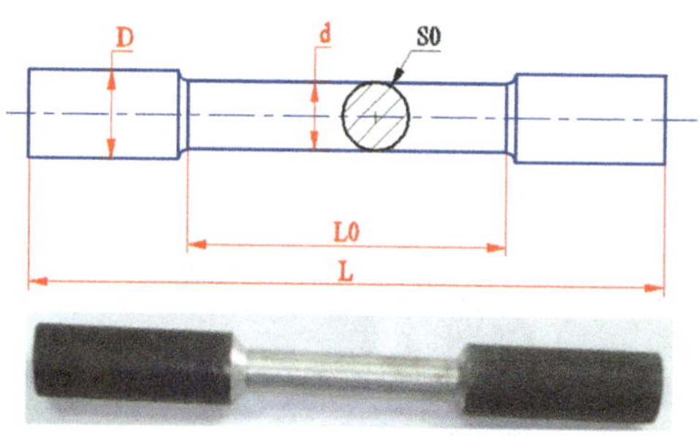

Fig. 9. Tensile test specimen according to ASTM E8 standard

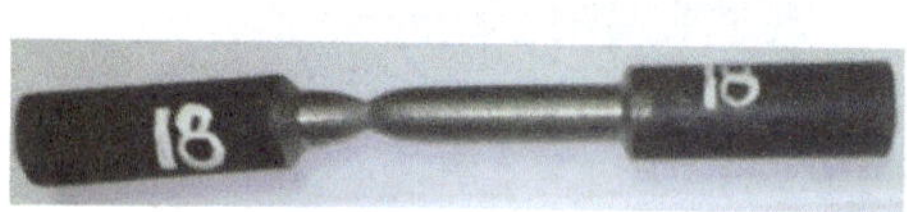

Fig. 10. Sample after tensile test

The number of repeated experiments of the planning was calculated and selected as 3. CHT4106 devices were used to measure the strength limit, yield limit and calculate the relative elongation with the results table as shown in Table 4.

Table 4. Tensile test results

Test specimen	Tensile strength, $\overline{\sigma_b}$ (N/mm^2)	Elongation, $\overline{\delta}$ (%)	Yield limit, $\overline{\sigma_{0.2}}$ (N/mm^2)
1	401,7	16,7	281,6
2	401,7	17,2	281,6
3	403,7	17,1	284,2
4	407,9	17,8	286,6
5	403,5	16,9	283,7
6	407,2	17,5	284,2
7	406,7	16,9	281,8
8	405,6	16,8	282,5
9	404,7	16,3	281,2
10	404,2	16,5	281,7

The relative elongation after break is calculated by the formula: $A = \left(\frac{L_u - L_0}{L_0}\right) \cdot 100.$

The relative contractility is calculated by the formula: $Z = \left(\frac{S_0 - S_u}{S_0}\right) \cdot 100.$

In which:

L_0: Initial session length.
L_u: Session length at the end.
L: Total length of the original sample.
L_c: Sample length at the end.
S_0: Original cross-sectional area of the parallel section.
L_u: Minimum cross-sectional area after cutting.

Tensile Strength σ_b
Synthesis and analysis of parameters affecting tensile strength are shown in Table 5.

Table 5. Parameters affecting tensile strength

Term	Coef	SE Coef	T-Value	P-Value
Constant	405.150	0.161	2511.87	0.000
P1	1.061	0.132	8.06	0.000
P2	0.867	0.132	6.58	0.000
P1 * P1	−1.189	0.208	−5.71	0.000
P1 * P2	0.692	0.161	4.29	0.000

The regression equation characterizing for the malfunction of the axial pressing force of friction and forging parts when welding friction can only be natural as follows:

$$\sigma_b = 396.97 + 0.2444\, P1 - 0.0777\, P2 - 0.002972\, P1 * P1 + 0.001729\, P1 * P2 \tag{2}$$

From here, graphs are built to show the influence of 2 input parameters on the tensile strength of the sample as shown in Fig. 11.

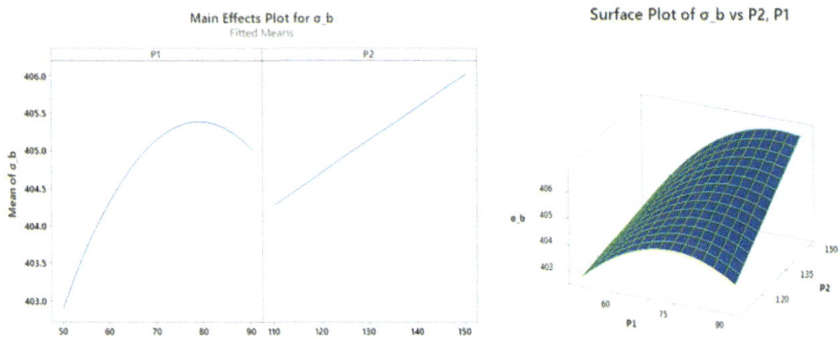

Fig. 11. Tensile strength analysis results

From the results of the analysis we realize that the higher the friction, the higher the strength, and when the maximum value is reached, it decreases. As the forging element increases, the tensile strength of the detail will increase steadily.

Elongation δ

Synthesis and analysis of parameters affecting elongation are shown in Table 6.

Table 6. Parameters affecting elongation

Term	Coef	SE Coef	T-Value	P-Value
Constant	16.6024	0.0548	303.02	0.000
P1	0.3222	0.0374	8.61	0.000
P2	0.1500	0.0374	4.01	0.000
P1 * P1	0.4119	0.0600	6.86	0.000
P2 * P2	0.1619	0.0600	2.70	0.012

The regression equation characterizing the influence of axial pressure of friction and forging parts when welding friction to the elongation of the sample of the natural form is as follows:

$$\delta = 26.39 - 0.1281\,P1 - 0.0977\,P2 + 0.001030\,P1 * P1 + 0.000405\,P2 * P2 \tag{3}$$

From here, graphs are built to show the influence of 2 input parameters on the elongation of the sample as shown in Fig. 12.

From the results of the analysis, it is noticed that the higher the friction, the longer the elongation will decrease to the point of urination will increase. As the Forging factor increases, the elongation of the parts tends to increase but will initially decrease when it comes to the pole, which will gradually increase.

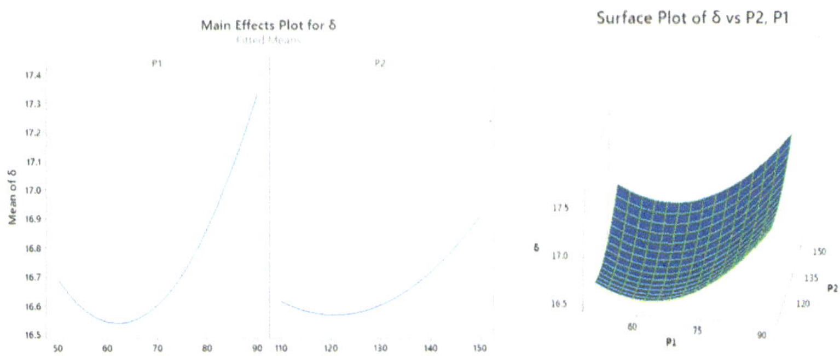

Fig. 12. Elongation analysis results

Yield Limit Analysis σ_{02}

Synthesis and analysis of parameters affecting yield limit analysis are shown in Table 7.

Table 7. Parameters affecting the yield limit

Term	Coef	SE Coef	T-Value	P-Value
Constant	281.558	0.083	3407.21	0.000
P1	0.6111	0.0675	9.06	0.000
P2	1.3500	0.0675	20.01	0.000
P1 * P1	1.942	0.107	18.20	0.000
P1 * P2	0.2667	0.0826	3.23	0.003

The regression equation characterizing for the influence of axial pressing force of friction and forging parts when friction welding arrives yield limit analysis samples of natural form is as follows:

$$\sigma_{0.2} = 300.50 - 0.7357\,P1 + 0.0208\,P2 + 0.004854\,P1 * P1 + 0.000667\,P1 * P2 \tag{4}$$

From here, graphs are built to show the influence of 2 input parameters on yield limit analysis of the sample as shown in Fig. 13.

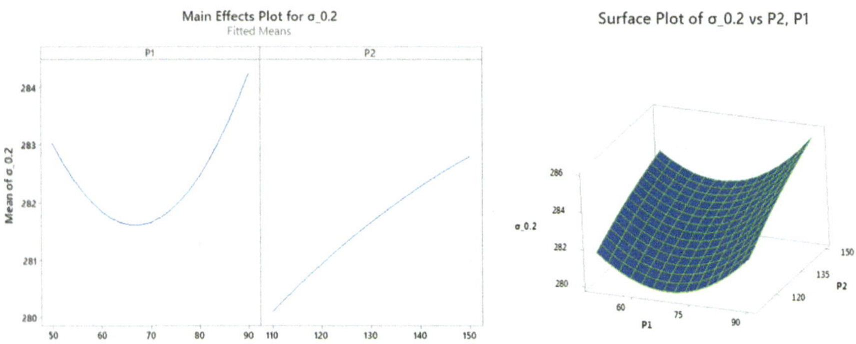

Fig. 13. Analysis results yield limit analysis

From the chart, it is noticed that the higher the friction, the lower yield limit. Especially, when it decreases to the minimized value, it tends to increase again. The more the forging factor increases, the yield limit increases.

Tensile test results with mechanical properties are synthesized and compared with the mechanical properties of the sample material (Table 1), with the table as shown in Table 8.

Table 8. Table comparing the properties of the welding specimen with that of the material

Tensile strength (N/mm^2)		Elongation (%)		Yield limit (N/mm^2)	
Standard	Experiment	Standard	Experiment	Standard	Experiment
373 ÷ 461	401.7 ÷ 407.9	22 min	16.3 ÷ 17.8	225 min	281.2 ÷ 286.6

4 Conclusion

The rotary friction welding machine has been successfully researched and manufactured, put to the test and achieved good results. The rotation speed of the part is stable and ensures the maximum number of revolutions required for friction welding is 1450 rpm. The hydraulic cylinder's propulsion mechanism is stable and generates pressure in accordance with the requirements of each material (160 Mpa). The structure of the welding machine is firm, ensuring stability during operation. Transmission sliding table mechanism is smooth and stable. Materials with the same properties can be welded, meeting the set requirements.

References

1. Sardana, G., Lohan, A.: Friction welding on lathe machine with special fixture. Int. J. Innov. Eng. Technol. **2**(3), 258–261 (2013)
2. Singh, J., Singh, K.: Fabrication of friction welding on centre lathe: a case study. Int. J. Res. Mech. Eng. Technol. **4**(2), 145–147 (2014)
3. Thakare, P.A., Randheer Singh, L.: Design and development of micro friction welding machine and investigation of welding parameters for similar materials. Int. J. Sci. Eng. Res. **5**(6), 1379 (2014)
4. Ahmet, C.A.N., Baxter, G.J., Preuss, M., Withers, P.J.: Research on the inertia friction welding of nickel base superalloys for aerospace applications studied by G.J. Baxter, M. Preuss and P.J. Withers. Manchester Institute of Natural Materials Research, UK
5. Uday, M.B., Ahmad Fauzi, M.N., Zuhailawati, H., Ismail, A.B.: Effect of welding speed on mechanical strength of friction welded joint of YSZ-Alumina composite and 6061 Aluminum Alloy. Mater. Sci. Eng. A **528**, 4753–4760 (2011)
6. Shiva Shankar, P., Suresh Kumar, L., Ravinder Reddy, N.: Experimental investigation and statistical analysis of the friction welding parameters for the copper alloy - Cu Zn28 using Taguchi method. Int. J. Res. Eng. Technol. eISSN: 2319–1163 pISSN: 2321–7308
7. Algur, V., Kabadi, V.R., Ganechari, S.M., Sharanabasappa, M.: Experimental investigation on friction characteristics of modified ZA-27 alloy using Taguchi technique. Int. J. Mech. Eng. Rob. Res. **3**(4), 24 (2014)
8. Handa, A., Chawla, V.: Experimental study of mechanical properties of friction welded AISI 1021 steels. Sadhana **38**(6), 1407–1419 (2013)
9. Kurt, A., Uygur, I., Paylasan, U.: Effect of friction welding parameters on mechanical and microstructural properties of dissimilar AISI 1010-ASTM B22 joints. Welding J. **90**(5), 102s–106s (2011)
10. Shiva Shankar, P., Suresh Kumar, L., Ravinder Reddy, N.: Experimental investigation and statistical analysis of the friction welding parameters for the copper alloy – Cu Zn28 using Taguchi method. Int. J. Res. Eng. Technol. eISSN: 2319–1163 | pISSN: 2321–7308
11. Sathiya, P., Aravindan, S., Noorulhaq, A., Panneerselvam, K.: Optimization of friction welding parameters using simulated annealing. Ind. J. Eng. Mater. Sci. **13**, 37–44 (2006)
12. Khany, S.E., Krishnan, K.N., Abdul Wahed, M.: Study of transient temperature distribution in a friction welding process and its effects on its joints. Int. J. Comput. Eng. Res. **2**(5), 1645 (2012)
13. Rashmikant Pandya, P., Bagesar, A., Patel, J.: Effect of welding parameters on burn-off length for friction welding of Inconel 718 and SS 304 for production of bimetal poppet exhaust valve. Int. J. Sci. Res. Dev. **2**(4), 1 (2014)
14. Lindemann, Z.Z., Skalski, K., Włosinski, W., Zimmerman, J.: Thermo-mechanical phenomena in the process of friction welding of corundum ceramics and aluminium. Bull. Polish Acad. Sci. Tech. Sci. **54**(1), 1 (2006)
15. Alves, E.P., An, C.Y., Neto, F.P., dos Ferro Santos, E.: Experimental Determination of Temperature During Rotary Friction Welding of Dissimilar Materials. Science and Engineering Publishing Company (2012)
16. Moarrefzadeh, A.: Study of heat affected zone (HAZ) in friction welding process. J. Mech. Eng. **1**(1), 11 (2012)
17. Ruma, Abdul Wahed, M., Farhan, M.: A study on the effect of external heating of the friction welded joint. Int. J. Emerg. Technol. Adv. Eng. **3**(5), 603 (2013)

18. Gill, J., Singh, J.: Experimental study on the effect of heating time on mechanical properties of nylon-6 joints produced by friction welding. Int. J. Adv. Eng. Res. Stud. E-ISSN 2249–8974
19. Can, A., Şahin, M., Küçük, M.: Modelling of friction welding. In: International Scientific Conference, 19–20 November 2010, Gabrovo
20. Sasidharan, B., Narayanan, K.P., Arivazhakan, R.: Influence of interface surface geometries in the tensile characteristics of friction welded joints from aluminium alloys. In: International Journal of Innovative Research in Science, Engineering and Technology. An ISO 3297:2007 Certified Organization, December 2013, vol. 2, no. 1 (2013)
21. Alves, E.P., An, C.Y., Neto, F.P., dos Santos, E.F.: Experimental determination of temperature during rotary friction welding of dissimilar materials. Front. Aerosp. Eng. **1**(1), 20 (2012)
22. Gill, J., Singh, J.: Experimental study on the effect of heating
23. Time on mechanical properties of nylon-6 joints produced by friction welding. Int. J. Adv. Eng. Res. Stud. E-ISSN2249–8974
24. Alves, E.P., Neto, F.P., An, C.Y., da Silva, E.C.: Experimental determination of temperature during rotary friction welding of AA1050 aluminum with AISI 304. Stainless Steel **4**(1), 61–67 (2012)
25. Ochi, H., Ogawa, K.: Friction welding of aluminum alloy and steel. Int. J. Offshore Polar Eng. **8**(2) (1998)
26. Şahin, M.: Friction welding of different materials. In: International Scientific Conference 19–20 November 2010, Gabrovo
27. Rombaut, P., De Waele, W., Faes, K.: Friction welding of steel to ceramic. Sustain. Construct. Design (2011)
28. Ananthapadmanaban, D.: Piaar Nagar, Chennai–119, Studies on Friction Weldability of Low Carbon Steel with Stainless Steel and Aluminium with Copper
29. Sahin, M., Misirli, C.: Mechanical and Metalurgical Properties of Friction Welded Aluminium Joints, 24 trang
30. Yuai, S., Yaoui, L., Zhu Xianong, Y., Siong, Z.: Strength distribution at interface of rotaryrictionelded aluminum to nodular cast iron. Trans. Nonferr. Metals Soc. China **18**, 14–18 (2008)
31. Boonseng, K., Chainarong, S., Meengam, C.: Microstructure and mechanical properties of friction welding in SSM356 aluminium alloys. Int. J. Emerg. Trends Eng. Res. **2**(4) (2014)
32. Rajamani, G.P., Shunmuga, M.S., Rao, K.P.: Parameter optimization and properties of friction welded quenched and tempered steel. Weld. Res. Suppl. 226-S (1992)
33. Gavade, J.P., Rajopadhye, R.D., Jadhav, N.D., Ramdasi, P.: A review on effects of processing parameters on mechanical behavior of different metals joined by continuous drive friction welding. IJESR **3**(4) (2013). 2777–2784, ISSN 2277–2685
34. Handa, A., Chawla, V.: Friction Welding - A Review. **1**(2), 34–38 (2014). http://asianonlinejournals.com/index.php/Mare
35. Moarrefzadeh, A.: Numerical modeling of friction welding process. Int. J. Multidiscip. Sci. Eng. **2**(8) (2011)
36. Narasimha Murthy, K., Raghupathy, V.P., Sethuram, D.: Two day workshop on friction welding & friction stir welding. In: International Center for Advancement of Manufacturing Technology, 24–25 November 2011 (2011)

Research Design and Experimental Manufacturing of Compound Non-circular Gear Train with an Improved Cycloid Profile of the Ellipse

Nguyen Hong Thai[1(✉)], Trinh Thi Khanh Ly[2], and Nguyen Thanh Trung[1,3]

[1] Department of Mechanical Design and Robotics, School of Mechanical Engineering, Hanoi University of Science and Technology, Hanoi, Vietnam
thai.nguyenhong@hust.edu.vn
[2] Faculty of Automation Technology, Electric Power University, Hanoi, Vietnam
[3] Vietnam National Research Institute of Mechanical Engineering, Hanoi, Vietnam

Abstract. This paper presents a method to design a compound non-circular gear (CNCG) train with an improved cycloid tooth profile. First, the research established a mathematical model describing the centrodes of the CNCG train to create a variable gear ratio function. Then, the tooth profile of non-circular gears in the CNCG trains is shaped by a rack cutter whose tooth profile is an improved cycloid. A CNCG train with a speed variation range from 1.4 to 5.8 has been experimentally designed and manufactured. A system of experimental equipment to determine the gear ratio function of the CNCG train based on the meshing between the gear pairs in the CNCG train has been manufactured. The experimental results show that the maximum error of the gear ratio function experimental compared to the theoretical one is 4.74%. It shows that the results of this research can be applied into practical systems that require speed variations.

Keywords: Non-circular gears · Oval gears · Compound non-circular gear train · Gear ratio function · Improved cycloid profile

1 Introduction

The first person to come up with the concept of the design and application of non-circular gears (NCG) in astronomical clock making was Giovanni Dondi (1330–1388), an Italian watchmaker [1, 2]. In the 15th century, NCG was studied and applied by Leonardo Da Vinchi (1452–1519) in designing watches, musical instruments, automatic key-making machines, pumps etc. These designs were also recorded in an official publication in 1493 [3]. However, the complexity of design and fabrication made the applications of NCG in the coming years face difficulty [4]. In the early 20th century, the development of science, especially information technology, encouraged researchers

D. C. Nguyen et al. (Eds.): ICERA 2021, LNNS 366, pp. 813–825, 2022.
https://doi.org/10.1007/978-3-030-92574-1_83

to develop NCG in various applications such as agricultural mechanisms [5], Vehicle step-less speed changes [6], Hydraulic machinery [7–9] or medical equipment [10] etc. But the above studies merely focused on researching each pair of non-circular gears. In recent years, the research of the compound non-circular gear (CNCG) train into practice has been developed such as Xu et al. [11] applied the CNCG train in combination with cylindrical gears in designing the gear drive of the new coal seam gas (CSG) drainage machine in CSG mining industry; Zhao et al. [12, 13] who improved the CNCG train of Guo's rice transplanter and developed it to transplant other leafy vegetables; Zhou et al. [14] applied a simple non-circular gear train to the design and manufacture of a rice potted seedling transplanting mechanism with a transplanting efficiency of up to 180 plants/min etc. But most of the researchers focus on using the involutes curve of a circle [10–14] or the Novikov type arc [15, 16] as the tooth profile. Applying other curves such as cycloid, improved cycloid [17–19] to the design of tooth profiles of NCG has not been studied much. Nevertheless, when using curves that are the involutes of a circle [15], the circular arc [16] and the involutes of an ellipse [20, 25] to design the tooth profile of the NCGs. Then at position radius small of the centrodes, the tooth is often small and tooth pointing. In addition, the disadvantage of these profiles is that the small modulus leads to small teeth, which reduces the load capacity of the gear drives. In recent years, the cycloid curve [26–28] and improved cycloid curve [22] have been applied in the design of tooth profiles for the NCGs with the advantage of more even teeth and designed with a large modulus leading to increased tooth size, improving the load capacity of the gear drives.

Therefore, in this research, the authors applied the improved cycloid curve from the generated ellipse to the tooth profile design of the CNCG train. To solve the above problem, the content of the paper includes: (i) Model the centrodes of the CNCG train; (ii) Design and manufacture the CNCG train with tooth profile is the improved cycloid curve and (iii) Experimentally verify the research theory.

2 The Design Method of the Compound Non-circular Gear Train with Tooth Profile is an Improved Cycloid Curve

2.1 Description of Compound Non-circular Gear Train

The CNCG train is formed from two pairs of NCGs as depicted in Fig. 1.

Fig. 1. The compound non-circular gear train

Wherein: pair gears 1–2 are eccentric cylindrical gear 1-NCG and pair gears 3–4 are two oval gears with the same size and geometric shape.

2.2 Mathematical Model of Improved Cycloid Curve

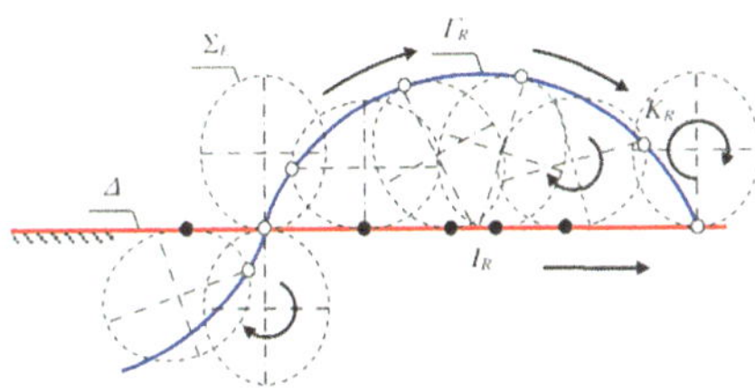

Fig. 2. The principle of generating an improved cycloid curve for the tooth profile

From Fig. 2, the improved cycloid curve Γ_R is the locus of a fixed point on the ellipse generating Σ_E. When Σ_E rolls without slipping on a fixed line Δ as depicted in Fig. 2 and the equation as follows:

$$\mathbf{r}_{K_R} = \begin{bmatrix} x_{K_R} \\ x_{K_R} \end{bmatrix} = \begin{bmatrix} s_2(\varphi) - a\sin(\psi) \\ (-1)^g s_3(\varphi) - a\cos(\psi) \end{bmatrix} \tag{1}$$

In the formula (1): $g = 0$ when Γ_R is the curve above Δ and $g = 1$ when Γ_R is the curve under Δ; $s_2(\phi) = \int_0^{\phi} \sqrt{r_E(\phi)^2 + \left(\frac{dr_E(\phi)}{d\phi}\right)^2} d\phi + r_E(\phi)\sin(\psi - \phi)$ with $r_E(\phi)$ is the polar radius of the generating ellipse Σ_E. According to [21] $r_E(\phi)$ is determined by: $r_E(\phi) = a\sqrt{\frac{1-\varepsilon^2}{1-\varepsilon^2\cos^2\phi}}$ wherein $\varepsilon = \frac{\sqrt{a^2-b^2}}{a}$ with a, b is the semi-major axis and semi-minor axis of Σ_E, respectively; $\phi \in [0 \div 2\pi]$; $s_3(\phi) = r_E(\phi)\cos(\psi - \phi)$ with: $\phi = \psi + arctg\left(\frac{r_E(\phi)}{dr_E/d\phi}\right) - \frac{\pi}{2}$.

2.3 Design Parameters of the Rack Cutter

From the improved cycloid curve Γ_R of the generating ellipse Σ_E presented in Sect. 2.2, we have the design of the rack cutter as shown in Fig. 3 and the design parameters of the rack cutter are given in Table 1 below.

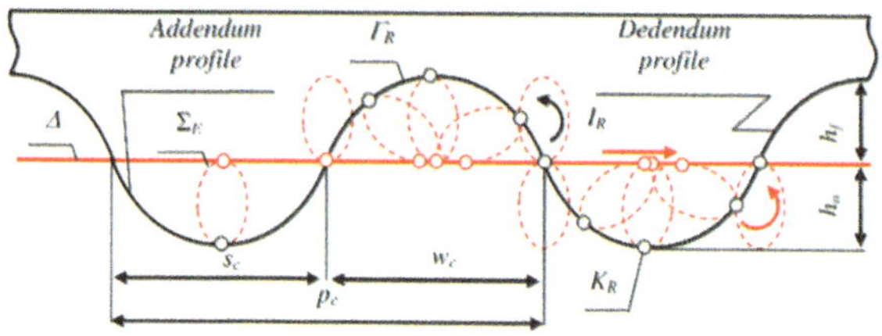

Fig. 3. The rack cutter

Table 1. Rack cutter design parameters

Design parameters	Notation	Unit	Equation
Tooth thickness on the datum line Δ	s_c	mm	$C_{\sum_E} = \int_0^{2\pi} \sqrt{r_E(\phi)^2 + \left(\frac{dr_E(\phi)}{d\phi}\right)^2} d\phi$
Width of space on the datum line Δ	w_c	mm	$C_{\sum_E} = \int_0^{2\pi} \sqrt{r_E(\phi)^2 + \left(\frac{dr_E(\phi)}{d\phi}\right)^2} d\phi$
Tooth pitch	p_c	mm	s + w
Tooth addendum	h_a	mm	$\frac{\lvert ab\cos\phi - ab\rvert}{\sqrt{(a\cos\phi)^2 + (b\sin\phi)^2}}$
Tooth dedendum	h_f	mm	$\frac{\lvert ab\cos\phi - ab\rvert}{\sqrt{(a\cos\phi)^2 + (b\sin\phi)^2}}$
Whole depth	H	mm	$h_f + h_a$

2.4 Kinematic Analysis of the Compound Non-circular Gear Train

2.4.1 Centrodes Design of the Compound Non-circular Gear Train

With the structure of the CNCG train shown in Fig. 1, we have the schematic diagram of the CNCG train described in Fig. 4.

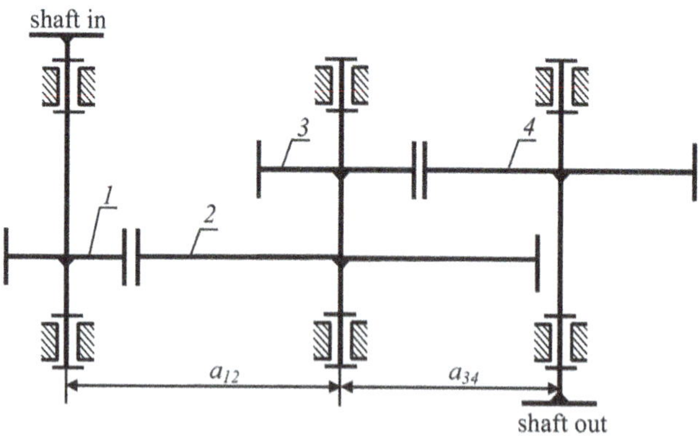

Fig. 4. Schematic diagram of the CNCG train

Wherein:

+ Gear 1 is an eccentric cylindrical gear whose centrode equation Σ_l is described by [29]:

$$\rho_1(\varphi_1) = \sqrt{R_1^2 - e_1^2 \sin^2 \varphi_1} + e_1 \cos \varphi_1 \tag{2}$$

Here: R_1 is the radius of the circle; e_1 is the eccentricity; $\varphi_1 \in [0 \div 2\pi]$ is the pole angle.

+ Gear 3 is an oval gear whose centrode equation Σ_3 is given by [23]:

$$\rho_3(\varphi_3) = \frac{2a_3b_3}{(a_3 + b_3) - (a_3 - b_3)\cos(2\varphi_3)} \tag{3}$$

Here: a_3, b_3 is the semi-major axis and semi-minor axis of Σ_3, respectively; $\varphi_3 \in [0 \div 2\pi]$.

Because gear pairs 1–2 and 3–4 are two pairs of external gears. According to [22, 24] we have a mathematical equation describing the centrode Σ_2 of the NCG 2 and the centrode Σ_4 of the NCG 4 as follows:

$$\{\Sigma_2\} : \begin{cases} \rho_2(\varphi_2(\varphi_1)) = a_{12} - \left(\sqrt{(R_1^2 - e_1^2 \sin^2 \varphi_1)} + e_1 \cos \varphi_1\right) \\ \varphi_2(\varphi_1) = \int\limits_0^{\varphi_1} \left(\dfrac{\sqrt{(R_1^2 - e_1^2 \sin^2 \varphi_1)} + e_1 \cos \varphi_1}{a_{12} - \left(\sqrt{(R_1^2 - e_1^2 \sin^2 \varphi_1)} + e_1 \cos \varphi_1\right)} \right) d\varphi_1 \end{cases} \tag{4}$$

$$\{\Sigma_4\} : \begin{cases} \rho_4(\varphi_4(\varphi_3)) = a_{34} - \left(\dfrac{2a_3b_3}{(a_3 + b_3) - (a_3 - b_3)\cos(2\varphi_3)} \right) \\ \varphi_4(\varphi_3) = \int\limits_0^{\varphi_4} \left(\dfrac{2a_3b_3}{a_{34}((a_3 + b_3) - (a_3 - b_3)\cos(2\varphi_3)) - 2a_3b_3} \right) d\varphi_3 \end{cases} \tag{5}$$

The shaft distance of the NCG pairs is determined by:

$$\begin{cases} a_{12}(n_1, e_1, R_1) = R_1(1 + n_1)\left(1 - \frac{(n_1 - 12)e_1^2}{4n_1 R_1^2}\right) \\ a_{34}(a_3, b_3, n_3) = \frac{(a_3 + b_3) + \sqrt{\left((a_3 + b_3)^2 - 4a_3b_3(1 - n_3^2)\right)}}{2} \end{cases} \tag{6}$$

Where: n_1 is the number of revolutions for NCG 1 to make one revolution of NCG 2 and n_3 is the number of revolutions for NCG 3 to make one revolution of NCG 4.

2.4.2 Kinematic Relationship Between NCG Shafts

Thus, we have the kinematic equation of the CNCG train:

$$\begin{cases} i_{12} = \frac{\omega_1}{\omega_2} = \frac{\rho_1(\varphi_1)}{\rho_2(\varphi_2(\varphi_1))} \\ i_{34} = \frac{\omega_4}{\omega_3} = \frac{\rho_4(\varphi_4(\varphi_3))}{\rho_3(\varphi_3)} \\ i_{14} = \frac{\omega_1}{\omega_4} = i_{12} i_{34} = \left(\frac{\rho_1(\varphi_1)}{\rho_2(\varphi_2(\varphi_1))}\right)\left(\frac{\rho_4(\varphi_4(\varphi_3))}{\rho_3(\varphi_3)}\right) \end{cases} \quad (7)$$

Since NCG 2 and NCG 3 are fixed on shaft 2 from Eq. (7), we have a kinematic relationship:

$$\begin{cases} \omega_2 = \frac{\rho_2(\varphi_2(\varphi_1))}{\rho_1(\varphi_1)} \omega_1 \\ \omega_4 = \left(\frac{\rho_2(\varphi_2(\varphi_1))}{\rho_1(\varphi_1)}\right)\left(\frac{\rho_3(\varphi_3)}{\rho_4(\varphi_4(\varphi_3))}\right) \omega_1 \end{cases} \quad (8)$$

3 Design and Manufacture of the Compound Non-circular Gear Train

The tooth profile of NCGs is shaped by having the datum line of the rack cutter rolls without slipping outside the centrodes of the NCG. In addition, in order for the NCG pairs after shaping to be able to mesh, each NCG pair must be shaped with a rack cutter.

By substitution centrode design parameter Σ_1: $R_1 = 25$ mm, $e_1 = 5$ mm, $n_1 = 3$ into Eqs. (4), (6) and substitution centrode design parameter Σ_3: $a_3 = 50$ mm, $b_3 = 30$ mm, $n_3 = 1$ into Eqs. (5), (6). After solving by numerical calculation on Matlab, we have the centrodes of the CNCG train as described in Fig. 5.

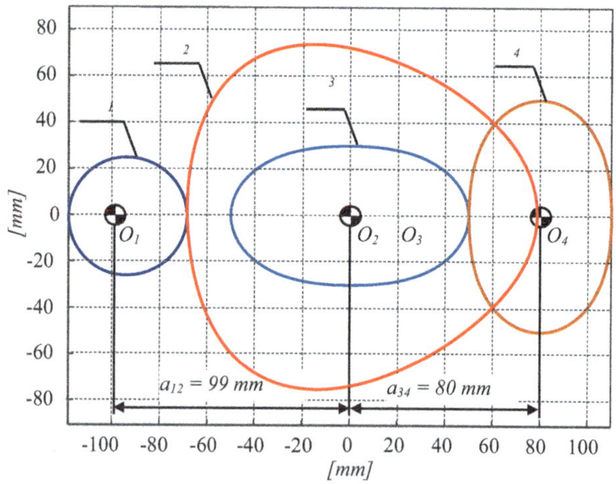

Fig. 5. Centrodes of the CNCG train

From Fig. 5, the parameters of Σ_{E1}: a_{E1} = 1.8 mm, b_{E1} = 1.4 mm are determined from the circumference of the centrodes (Σ_2, Σ_1) while the parameters of Σ_{E2}: a_{E1} = 1 mm, b_{E2} = 0.7 mm are determined from the circumference of the centrodes (Σ_3, Σ_4). After substituting the above parameters into Table 1, we have a set of design parameters of rack cutter 1 and rack cutter 2, described in Table 2 below.

Table 2. Design parameters of rack cutter

Parameter	Notation	Rack cutter 1	Rack cutter 2
Module (mm)	m	6.2	3.4
Tooth thickness (mm)	s_c	9.8	5.4
Width of space (mm)	w_c	9.8	5.4
Tooth pitch (mm)	p_c	19.6	10.8
Tooth addendum (mm)	h_a	3.5	2.0
Tooth dedendum (mm)	h_f	3.5	2.0
Whole depth (mm)	h	7.0	4.0

From Table 2, after shaping each pair of NCGs, we have the design of the CNCG train described in Fig. 6(a), and Fig. 6b is a photograph of the manufactured NCG pairs. The design parameters of NCGs in the CNCG train are given in Table 3.

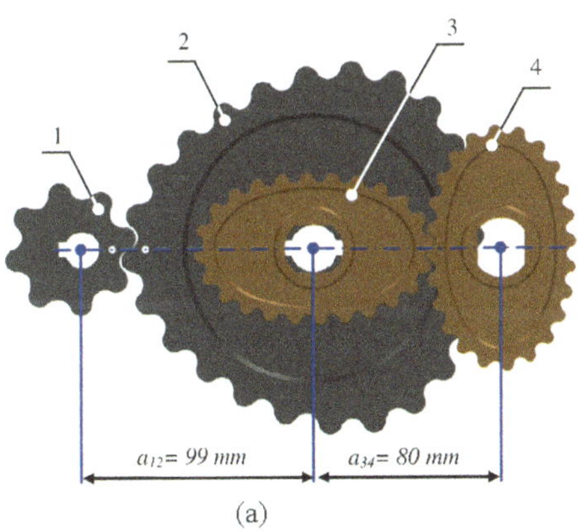

Fig. 6. The CNCG train after design and manufacture with (a) The design and (b) The photograph of NCG pairs

Table 3. Design parameters of NCG in the CNCG train

Parameter	Notation	Gears pair 1–2		Gears pair 3–4	
		NCG 1	NCG 2	Oval gear 3	Oval gear 4
Module	*m*	6.2	6.2	3.4	3.4
Circumference coefficient	*n*	3.0	–	1.0	–
Number of teeth	*z*	8	24	24	24
Tooth pitch *[mm]*	*p*	19.6	19.6	10.8	10.8
Tooth thickness *[mm]*	*s*	9.8	9.8	5.4	5.4
Width of space *[mm]*	*w*	9.8	9.8	5.4	5.4
Whole depth *[mm]*	*h*	7.0	7.0	4.0	4.0
Tooth addendum *[mm]*	*ha*	3.5	3.5	2.0	2.0
Tooth dedendum *[mm]*	*hf*	3.5	3.5	2.0	2.0

From the design parameters of the CNCG train in Table 3, we have a graph of the gear ratio function of the CNCG train described in Fig. 7.

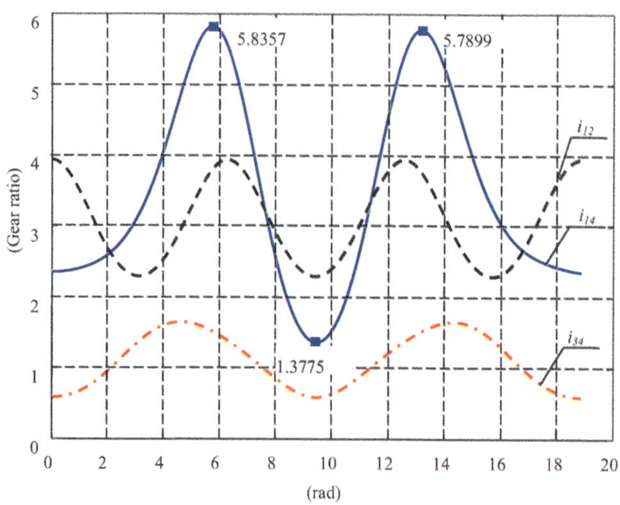

Fig. 7. The gear ratio of the CNCG train

4 Experimental Determination of Gear Ratio Function of Compound Non-circular Gear Train

The graph of the gear ratio function of the CNCG train, as shown in Fig. 7, is determined from the centrodes of the train according to Eq. (7). But in fact, the gear ratio function of the CNCG train is performed by the meshing process of pairs of NCGs in the system. Therefore, in this content, we manufacture an experiment system to determine the gear ratio function of the CNCG train from the meshing process of pairs of NCG as described in Fig. 8.

Fig. 8. Experimental system

4.1 Architecture and Hardware of the Experimental System

The experimental system consists of a CNCG train and hardware devices, as shown in Fig. 9. The rotation speed of the gear shafts is determined independently by encoders with a resolution of 400 pulses. Measurement data from the encoders is collected through the counter of the PLC (S7-1200, CPU 1214C). Data from PLC is sent to the industrial computer (BOXER-6640-A1-1010) for processing through software. The motor speed (1.5 Kw) is controlled from the computer to the PLC through the Delta inverter. To avoid overload, the speed from the motor shaft is transmitted to the CNCG train by a belt drive with a 1:1 speed ratio. The CNCG train is lubricated by shell grease Gadus s2 v2020-2.

System setting parameters: sampling interval T = 0.2 s.

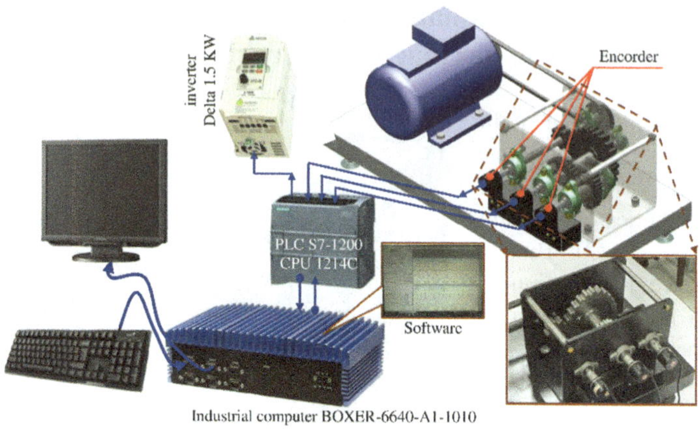

Fig. 9. The electrical configuration of an experimental system

4.2 Experimental Measurement Results and Discussion

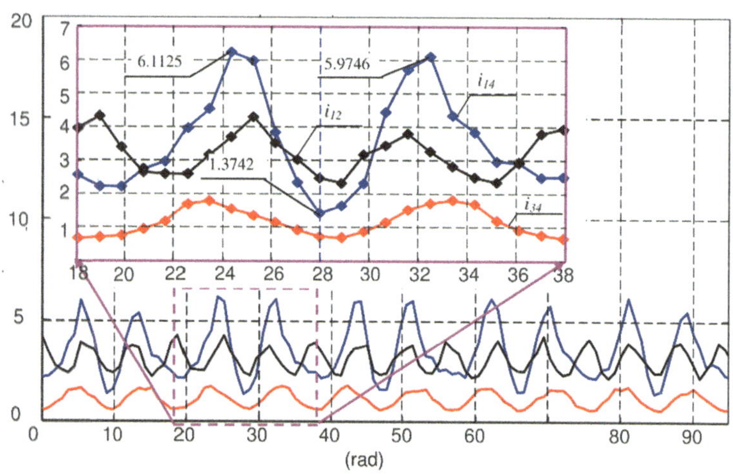

Fig. 10. Gear ratio measured experimentally from the meshing process of CNCG train

With the sampling interval setting to T = 0.2 s and the motor speed is set to 50 rpm corresponding to the frequency of the inverter f = 1.8 Hz. Figure 11 below is a graph of the gear ratio measured at 315 measuring points during the meshing process between pains of NCGs in the CNCG train. From Fig. 10, we can see that the geometry, values and rules are approximately identical to the theoretical gear ratio function determined from the centrodes in Fig. 7. After processing the data and set the standard point "**0**" of the graph, we have a comparison graph between theory and experimental measurement from the meshing process of the CNCG train as described in Fig. 11a.

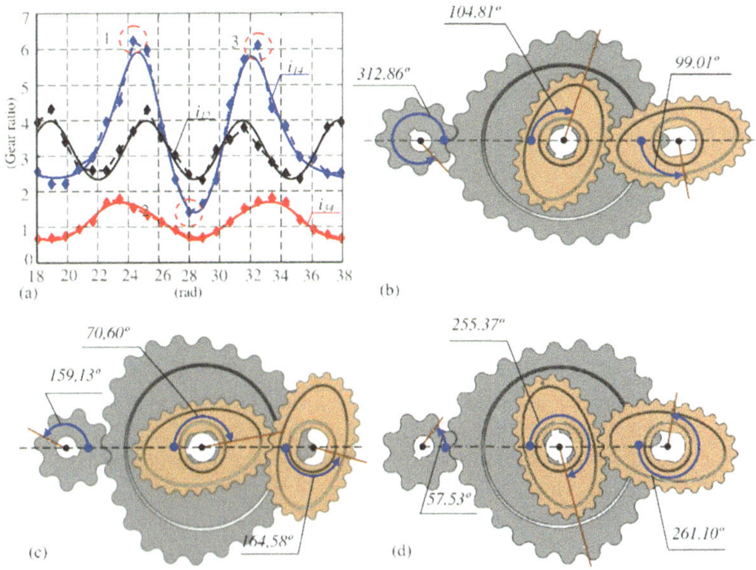

Fig. 11. The gear ratio of CNCG train between theory and experiment

From the graph of Fig. 11a, we can see that the error between theory and experiment is largest at three locations marked by red dashed circles and Fig. 11b, 11c, 11d are the meshing positions of NCG pairs in the CNCG train corresponding to positions 1, 2, 3 on Fig. 11a. These are exceptional positions, specifically: (*i*) Position ⬟, ⬢ has the maximum gear ratio i_{14} where gear 1 has the minimum radius, and gear 2 has the maximum radius (*see Fig. 11b and Fig. 11d*); (2) Position ⬣ has a minimum gear ratio i_{14} where gear 1 has the maximum radius, and gear 2 has the minimum radius (*see Fig. 11c*). The reason is that these positions are extreme points with sudden changes in speed and torque, and corresponding to positions ⬟, ⬣, ⬢, there are errors between theory and experiment: 4.74%, 3.19%, 0.24%, respectively. This error is entirely acceptable because it does not include assembly errors and manufacturing errors.

5 Conclusion

From the above theoretical and experimental research results, this study achieved the following results:

(i) A compound non-circular gear train has been designed and manufactured with a tooth profile of an improved cycloid curve. This gear train has a variable speed and torque range between the input and output shafts from 1.4 to 5.8.

(ii) Conducting experimental measurement to verify between experiment and theory, the results show that the max error is 4.74%.

The obtained results can be applied to design continuously variable transmissions for different application scenarios in production practice, such as CVTs in Automotive.

Acknowledgments. This work was supported by Project of Ministry of Education and Training, Vietnam. Under grant Number: B2019 - BKA – 09.

References

1. Poulle, E.: Équatoires et Horlogerie Planétaire du XIIIe au XVIe Siècle, Tome I, Librairie Droz, pp. 511–550 (1980)
2. Addomine, M., Figliolini, G., Pennestri, E.: A landmark in the history of non-circular gears design: the mechanical masterpiece of Dondi'sastrarium. Mech. Mach. Theory **122**, 219–232 (2018)
3. da Vinci, L.: Codex Madrid (1493)
4. Litvin, F.L., Gonzalez-Perez, I., Fuentes, A., Hayasaka, K.: Design and investigation of gear drives with non-circular gears applied for speed variation and generation of functions. Comput. Methods Appl. Mech. Eng **197**, 3783–3802 (2008)
5. Mundo, D., Gatti, G., Dooner, D.B.: Optimized five-bar linkages with non-circular gears for exact path generation. Mech. Mach. Theory **44**(4), 751–760 (2009)
6. Dooner, D.B.: Function generation utilizing an eight-link mechanism and optimized non-circular gear elements with application to automotive steering. In: Proceedings of the Institution of Mechanical Engineers, Part C: Journal of Mechanical Engineering Science, pp. 847–857 (2001)
7. Tien, T.N., Thai, N.H.: A novel design of the roots blower. Viet. J. Sci. Technol. **57**(2), 249–260 (2019)
8. Thai, N.H., Long, N.D.: A new design of the Lobe pump is based on the meshing principle of elliptical gear pairs. Sci. Technol. Dev. J. Eng. Technol. **4**(2), 861–871 (2021)
9. Kurasov, D.: Geometric calculation of planetary rotor hydraulic machines. IOP Conf. Ser. Mater. Sci. Eng. **862**, 1–6 (2020)
10. Ottaviano, E., Mundo, D., Danieli, G.A., Ceccarelli, M.: Numerical and experimental analysis of non-circular gears and cam-follower systems as function generators. Mech. Mach. Theory **43**(8), 996–1008 (2008)
11. Xu, G., Hua, D., Dai, W., Zhang, X.: Design and performance analysis of a coal bed gas drainage machine based on incomplete non-circular gears. Energies. 2–19 (2017)
12. Zhao, X., Chu, M., Ma, X., Dai, L., Ye, B., Chen, J.: Research on design method of non-circular planetary gear train transplanting mechanism based on precise poses and trajectory optimization. Adv. Mech. Eng. **10**(12), 1–12 (2018)
13. Zhao, X., Ye, J., Chu, M., Dai, L., Chen, J.: Automatic scallion seedling feeding mechanism with an asymmetrical high-order transmission gear train. Chinese J. Mech. Eng. **33**(1), 2–14 (2020)
14. Zhou, M., Yang, Y., Wei, M., Yin, D.: Method for generating non-circular gear with addendum modification and its application in transplanting mechanism. Int. J. Agricult. Biol. Eng. **13**(6), 68–75 (2020)
15. Bair, B.W.: Computer aided design of non-standard elliptical gear drives. Proc. Inst. Mech. Eng. Part C J. Mech. Eng. Sci. **216**(4), 473–482 (2001)
16. Bair, B.W., Sung, M.-H., Wang, J.-S., Chen, C.-F.: Tooth profile generation and analysis of oval gears with circular-arc teeth. Mech. Mach. Theory **44**, 1306–1317 (2009)
17. Thai, N.H., Trung, N.T.: Establishing formulas for design of Roots pump geometrical parameters with given specific flow rate. Viet. J. Sci. Technol. **53**(4), 533–542 (2015)
18. Thai, N.H., Giang, T.C.: Influence of geometrical dimensions on the profile slippage in the hypogerotor pump. Viet. J. Sci. Technol. **56**(4), 482–491 (2018)

19. Thai, N.H., Tien, T.N., Dung, P.T., Huy, N.Q.: Influence of the designing parameters on flow fluctuation and pressure of the improved roots blower. In: International Conference of Fluid Machinery and Automation Systems - ICFMAS2018, pp. 196–203 (2018)
20. Thai, N.H.: Shaping the tooth profile of elliptical gear with the involute ellipse curve. Sci. Technol. Dev. J. Eng. Technol. **24**(3), 1048–1056 (2021)
21. Litvin, F.L., Fuentes-Azna, A., Gonzalez-Perez, I., Hayasaka, K.: Noncircular Gears Design and Generation. Cambridge University Press, Cambridge (2009)
22. Thai, N.H., Trung, N.T., Viet, N.H.: Research and manufacture of external non-circular gear-pair with improved cycloid profile of the ellipse. Sci. Technol. Dev. J. Eng. Technol. **24**(2), 835–845 (2021)
23. Tien, T.N., Thai, N.H., Long, N.D.: Effects of head gaps and rotor gap on flow rate and hydraulic leakage of a novel non-contact rotor blower. Viet. J. Sci. Technol. **57**(4A), 125–140 (2019)
24. Litvin, F.L., Fuentes, A.: Gear Geometry and Applied Theory. Cambridge University Press, Cambridge (2004)
25. Thai, N.H., Thom, P.V., Lam, D.B.: Effects of pressure angle on uneven wear of a tooth profile of an elliptical gear generated by ellipse involute. Sci. Technol. Dev. J. Eng. Technol. **24**(3), 2031–2043 (2021)
26. Thai, N.H., Trung, N.T., Nghia, L.X., Duong, N.T.: Synthesis of the external non-circular gear-train with cycloid profile. J. Sci. Technol. Tech. Univ. **145**, 033–039 (2020)
27. Thai, N.H., Trung, N.T., Nghia, L.X., Duong, N.T.: Profile sliding phenomenon in the external non-circular gear-train with cycloidal profile. Eng. Technol. Sustain. Dev. **31**(2), 053–057 (2021)
28. Viet, N.H., Thai, N.H.: Geometric design and kinematics analysis of non - circular planetary gear train with cycloid profile. Eng. Technol. Sustain. Dev. **31**(3), 105–112 (2021)
29. Thai, N.H., Trung, N.T.: Pitch line synthesis of noncircular planetary gears. J. Sci. Technol. Tech. Univ. **140**, 05–010 (2020)

Research on Application of Machine Learning in Pedestrian Headform to Bonnet Top Tests

Van-Luc Ngo[1,2(✉)] and Van-Hai Nguyen[1,2]

[1] Faculty of Mechanical Engineering and Mechatronics, Phenikaa University, Hanoi 12116, Vietnam
luc.ngovan@phenikaa-uni.edu.vn

[2] Phenikaa Research and Technology Institute (PRATI), A&A Green Phoenix Group JSC, No. 167 Hoang Ngan, Trung Hoa, Cau Giay, Hanoi 11313, Vietnam

Abstract. Pedestrian vehicle safety accreditation is a mandatory requirement for all cars marketed in the European Union. Therefore, when researching and developing a new car model, all automakers must study to design pedestrian protection in collisions. In the process of research and development as well as pedestrian safety accreditation, there are a lot of impacting tests to evaluate pedestrian safety. As for bonnet top area, there are at least 18 tests of headform impact to bonnet top to evaluate pedestrian head safety. Since the number of positions selected for testing is finite and discrete, it is difficult to accurately conclude the pedestrian safety level for every position on the bonnet surface area. In this study, the four models of machine learning (ML) algorithm Linear regression (LIN), Support Vector Machine Regression (SVR), Extreme Gradient Boosting (XGB) and Artificial Neural Network (ANN) are applied to predict the safety level of every position on the bonnet top surface based on the data obtained from testing at several positions. In order to assess the predictive performance of ML model, the error metrics are used such as root mean squared error (RMSE), mean absolute error (MAE) and coefficient of determination (R^2). The GridSearchCV algorithm is also used to find the best hyperparameter of ML model. The results of this study will be very useful for the development of solutions to vehicle safety tests for pedestrians. LIN is the best model in predicting the HIC value of headform and bonnet collision.

Keywords: Pedestrian head safety · Machine learning · Prediction HIC value · Headform impactor to bonnet top tests · Pedestrian vehicle safety accreditation

1 Introduction

According to regulation [1], all cars sold on the market of the European Union are required to comply with pedestrian vehicle safety accreditation. In the process of accreditation, many tests of pedestrian impactor to frontal parts of a car are performed to evaluate the safety of these parts with pedestrian in collisions. After each test, the relevant parts and even the car have to be replaced in order to perform other tests. These tests are also carried out many times during the research and development to design a new model of vehicle for pedestrian safety. Performing these tests is time-consuming and expensive but there is still a possibility to exist positions that are

D. C. Nguyen et al. (Eds.): ICERA 2021, LNNS 366, pp. 826–833, 2022.
https://doi.org/10.1007/978-3-030-92574-1_84

dangerous for pedestrians in collisions which have not been tested, because the number of tests is finite and discrete while safety is regulated by area. As for bonnet top surface, the regulations require that 2/3 of the area of the bonnet top must give results at a safe threshold, the remaining 1/3 of the area can cause serious injury and there is no location result at a level that could endanger the life of a pedestrian [1, 2]. A minimum of 18 different positions on the bonnet surface are selected for impacting tests with headform to assess whether the bonnet top meets the regulatory requirements. It is very difficult to guarantee that from test results with several positions on the surface it can be concluded that the whole bonnet surface is safe for pedestrians. If the number of positions selected for testing increases, the more accurate the conclusion about the safety of the bonnet surface will be, the higher the cost will be. A question is given as to whether it is possible from experimental data with a number of positions to determine the pedestrian head safety level at every position on the bonnet top surface. There has been no study which performed the prediction of pedestrian head safety for untested positions until now, but this problem will be considered in this study.

Nowadays, with the development of computer science, Machine Learning (ML) has been applied to many fields of science and technology. In this study, the author will study to find an application of ML in predicting the safety of every position on the bonnet top surface for pedestrians based on data obtained from testing with several positions. The ML models are commonly sensitive to types of data and hyperparameter. Thus, selection and comparison of ML model is necessary to find out the best model for a specific situation [3]. This study aims to investigate the application of different advanced regression techniques such as LIN, SVR, XGB, and ANN to predict HIC value or level of pedestrian head injuries in headform impactors to bonnet top tests.

2 Description of Headform Impactors to Bonnet Top Tests

Headform impactors to bonnet top tests are performed to evaluate the level of pedestrian head safety in collisions with the vehicle. These tests are performed by using the child and adult headform impact to bonnet top with the procedure as brief description in Fig. 1 [4]. The child and adult headform impactors are respectively used to test positions in the forward and rearward area of the bonnet top surface. These regions are divided by WAD (Wrap Around Distance-Fig. 1) of 1500 mm. The impact angle for the child and adult headform is 500 and 650 respectively. The impact velocity for both child and adult headform impactors are 40 km/h.

The level of head injury is assessed through the HIC value calculated based on the resultant acceleration of the headform impactor center of gravity as the following equation:

$$\mathrm{HIC} = Max\left[\frac{1}{T_2 - T_1}\int_{T_1}^{T_2} \mathrm{AdT}\right]^{2.5}(\mathrm{T}_2 - T_1) \tag{1}$$

with condition $\Delta T = T_2 - T_1 \leq 15$ ms (mini-second).

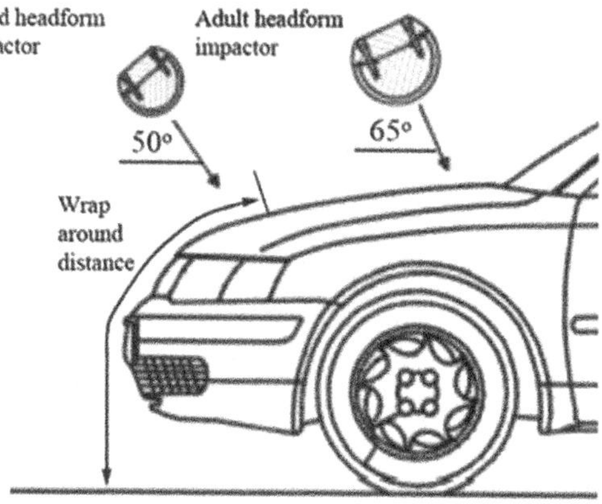

Fig. 1. Procedure of child and adult headform impactor to bonnet top tests

Where, T_1 and T_2 are the two instant times (unit of second) in the collision process and (A) is resultant acceleration measured during collision (unit of [g], 1[g] = 9.81 [m/s^2]). The HIC value reflects the level of pedestrian head injury. The higher the HIC value is, the greater level of pedestrian head injury will be. The HIC value of 1000 is pedestrian head safety threshold. The HIC value from 1000 to 2000 is the range of serious injury and the pedestrian can be killed if the HIC value is greater than 2000.

3 Methodology and Materials

3.1 Methodology and Research Objective

The equation for calculating shows that the HIC value depends on the variable characteristics of the resultant acceleration (A) and therefore will depend on the characteristics of the impacting position such as stiffness, structure, etc. However, the closer the two positions are to each other, the more similar the structural characteristics are, and therefore the HIC values as well as the variation of acceleration characteristics when impacting the two positions are more similar. It is assumed that the HIC value as well as the acceleration variation characteristics between different positions will change according to a definite rule which depends on the structure of the bonnet. Thus, if the rule of variation of HIC values between different positions can be found based on data obtained from impacting tests with different positions, it is possible to determine the HIC value of positions in the space between different positions. The objective of this study is to check the hypothesis mentioned above. If this hypothesis is correct, the HIC values can be found at every position on the bonnet surface by finding the rule of variation of the HIC value on the bonnet surface.

3.2 Materials and Methods

In this study, the tests of headform impactor to bonnet would be performed using a finite element model of an adult headform which is developed by Teng et al. [5], element model of Ford Taurus [6]. A total of 36 positions on the bonnet top surface are selected for testing collision with the adult headform impactor. These positions are selected equally spaced both horizontally and vertically on the test area with the adult headform. The data obtained from these tests include resultant acceleration (A), tangential acceleration (At), normal acceleration (An), HIC value, two instant times to calculate the HIC value. Table 1 shows the parameters that affect the HIC value. Where: $X_1 = T_2 - T_1$ duration for HIC calculating; $X_2 = Max(A)$ or the peak of resultant acceleration; X_3 = HIC value which is calculated based on only tangential acceleration (At); $X_4 = R$; $X_5 = (X4)^{2.5}$.

$$R = \frac{\overline{A}_n}{\overline{A}_t} \tag{2}$$

With: $\overline{A}_n = \frac{1}{T_2 - T_1}\int_{T_1}^{T_2} A_n dT$ and $\overline{A}_t = \frac{1}{T_2 - T_1}\int_{T_1}^{T_2} A_t dT$

Table 1. Database determined form results of simulations

No	X_1	X_2	X_3	X_4	X_5	HIC	No	X_1	X_2	X_3	X_4	X_5	HIC
1	15	199	360	1.6	5.45	**1962**	19	7	173	487	1.1	2.6	**1276**
2	8	183	378	1.7	5.45	**2061**	20	15	154	285	1.3	3.6	**1012**
3	15	172	279	1.9	7.34	**2050**	21	11	181	290	1.6	5.1	**1462**
4	6	165	384	1.3	3.29	**1264**	22	10	176	364	1.8	5.1	**1858**
5	5	189	664	0.8	1.93	**1279**	23	12	174	331	1.2	3.3	**1104**
6	6	158	444	1.0	2.28	**1011**	24	7	165	440	1.1	2.6	**1412**
7	10	185	568	1.1	3.09	**1755**	25	15	155	246	1.5	5.1	**1238**
8	10	185	363	1.6	5.24	**1905**	26	14	147	270	1.0	3.7	**989**
9	7	194	341	1.5	4.23	**1443**	27	12	153	297	1.3	3.8	**1136**
10	6	176	352	1.4	4.17	**1469**	28	8	162	357	1.3	3.4	**1229**
11	13	204	486	1.6	6.24	**3031**	29	7	153	375	1.2	3.2	**1214**
12	15	173	313	1.4	4.72	**1479**	30	14	149	398	1.4	3.8	**1521**
13	15	159	388	1.2	3.09	**1199**	31	10	161	333	1.4	4.1	**1351**
14	6	173	423	1.0	3.26	**998**	32	12	152	297	1.5	4.6	**1365**
15	14	373	305	3.0	19.48	**5951**	33	9	174	348	1.6	5.1	**1756**
16	15	172	460	1.3	2.85	**1311**	34	9	166	334	1.6	5.5	**1818**
17	15	231	312	1.7	8.38	**2613**	35	7	212	377	1.8	6.1	**2294**
18	15	139	239	1.5	4.41	**1054**	36	10	172	320	1.5	4.9	**1552**

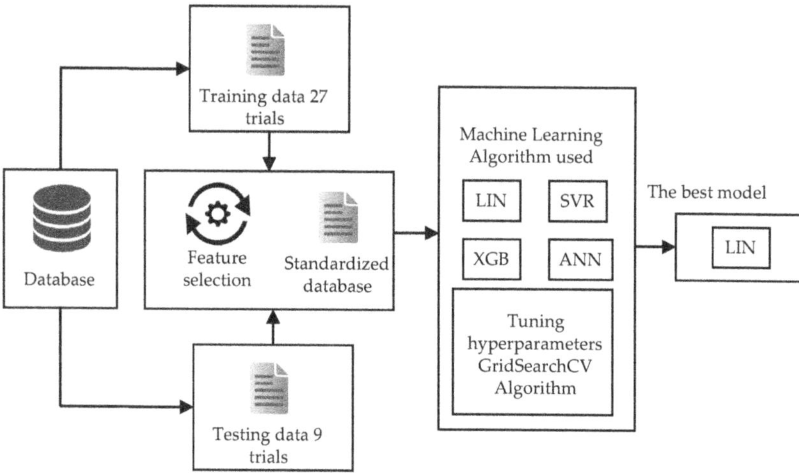

Fig. 2. Machine learning algorithm used in this study.

All parameters are calculated based on the results of simulations shown in Table 1. The parameters are split into 28 trials of training dataset and 8 trials of testing dataset.

$$RMSE = \sqrt{\frac{\sum_{i=1}^{N} (\widehat{y}_i - y_i)^2}{N}} \tag{3}$$

$$MAE = \frac{1}{N}\sum_{i=1}^{N} |\widehat{y}_i - y_i| \tag{4}$$

$$R^2 = 1 - \frac{\sum_{i=1}^{N} (\widehat{y}_i - y_i)^2}{\sum_{i=1}^{N} (\widehat{y}_i - \overline{y})^2} \tag{5}$$

The dataset has been normalized by the log1p(x) in Python function to ensure a distribution and a consistent scale for predictors. ML regression algorithms including LIN, SVR, XGB and ANN are used for predicting the HIC value. Various error metrics – RMSE, MAE and R2 were used to assess the performance of ML models. The metrics errors are described in Eqs. (3)–(5). Figure 2 presents the overall methodology used in this work: dataset extraction, feature selection, and various subdataset made to predict HIC values. Besides, the figure also describes how to fine-tune the model hyperparameter by using the GridSearchCV algorithm to find out the best parameter for each model with metric minimum of root mean squared errors.

4 Results and Discussions

By the analyzed performance of LIN, SVR, XGB and ANN regression models in respect of HIC value prediction, the accuracy metrics of model performance is shown in Table 2 with training and testing datasets. As indicated in this table, with training and testing dataset, LIN shows the best performance: this model provides the largest value of R2 and the lowest values of MAE and RMSE. When using the training dataset, the error metrics of RMSE and MAE are calculated in ascending order for model performance as LIN, SVR, XGB and ANN. This is the same ranking with testing dataset. On the contrary, ANN shows the worst performance, in respect of both testing and training datasets.

Table 2. Results of ML regression models.

Model	Training datasets			Testing dataset		
	RMSE	MAE	R^2	RMSE	MAE	R^2
LIN	91.8254	60.3961	0.9928	43.8023	31.3180	0.9943
SVR	126.9785	97.6998	0.9832	80.2660	70.2908	0.9262
XGB	672.0919	214.2872	0.7539	221.4571	190.8015	0.9451
ANN	743.5175	246.9982	0.9708	169.6440	122.6639	0.9827

Figure 3 shows the line scatter graphs among measured values in the experimental simulation and predicted HIC values for four model performances by using the training dataset, respectively. The various color of line scatter reflects the various ML models of predicted values. As indicated in this figure, the HIC values predicted by LIN model are the closest to the simulated measurement value. Moreover, the predicted HIC values of ANN model are the furthest simulated measurement value. The comparison of results of four ML models and simulated measurement by using the testing dataset is shown in Fig. 4 by line scatter plot with various colors and marks. The ranking of error metrics is the same by using the training dataset. The HIC values predicted by LIN model are the closest to the simulated measurement. This figure also indicates that the HIC values predicted by ANN model are not fitting to simulated measurement. The comparison in Fig. 3, Fig. 4 and the results in Table 2 show that LIN model exhibits the best predicting performance in respect to others model. This work proposed the explicit equation for calculating HIC values based on input independent values regarding to the following LIN model equation:

$$HIC = e^{[(-0.3989 + 0.048\ln(1+X1) - 0.0589\ln(1+X2) + 9.857\ln(1+X3) + 0.552\ln(1+X4) + 0.975\ln(1+X5)]} - 1$$

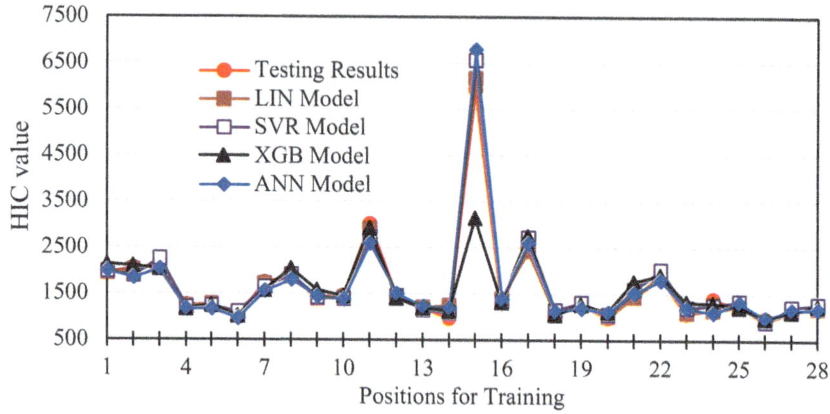

Fig. 3. Comparison between testing results and analysis using ML algorithm

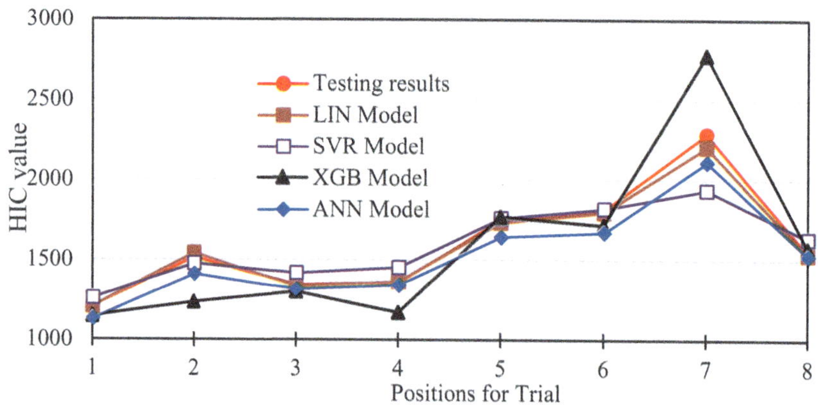

Fig. 4. Comparison between results of testing simulation and prediction using ML algorithm

The various color of line scatter reflects the various ML models of predicted values.

5 Conclusions and Recommendations

In this research, four machine learning models including LIN, SVR, XGB and ANN - have been applied to predict HIC values in adult headform to bonnet top tests. The various assessment metrics (RMSE, MAE, and R2) have been used to assess the performance of each model. Subsequently, hyperparameter optimization using Grid-SearchCV has been applied. By considering four ML model, the results indicate that LIN the best model using evaluation metrics of RMSE, MAE, and R2. Moreover, the explicit equation for determining HIC has been proposed. The result of this study shows that the hypothesis about the HIC value as well as the acceleration characteristics when colliding at positions between different locations will vary according to a definite rule. This result also shows that machine learning can be used in predicting the safety of pedestrians when having an impact to bonnet top and this direction of research should be developed in further studies.

References

1. Directive 2003/102/EC of the European Parliament and of the Council of 17 November 2003
2. Regulation (EC) No 78/2009 of the European Parliament and of the Council of 14 January 2009
3. Nguyen, V.-H., et al.: Applying Bayesian optimization for machine learning models in predicting the surface roughness in single-point diamond turning polycarbonate. Math. Prob. Eng. 2021 (2021)
4. EEVC/WG 17 Report. (1998, updated 2002). Improved Test Methods to Evaluate Pedestrian Protection Afforded by Passenger Cars, European Enhanced Vehicle-safety Committee
5. Information on www.ncac.gwu.edu/vml/models.html
6. Teng, T.L., Nguyen, T.H.: Development and validation of FE models of impactor for pedestrian testing. J. Mech. Sci. Technol. **22**(9), 1660–1667 (2008)

Robust Control Design for Wheeled Mobile Robotic Systems with Predictive Model

Dao Phuong Nam[1], Tran Nam Anh[1], and Nguyen Hong Quang[2](✉)

[1] Hanoi University of Science and Technology, 1 Dai Co Viet Road, Hanoi 100000, Vietnam
[2] Thai Nguyen University of Technology, 666, 3/2 Street, Tich Luong Ward, Thai Nguyen 251750, Vietnam
quang.nguyenhong@tnut.edu.vn

Abstract. In this article, robust control schemes are presented for time-varying trajectory tracking control of Wheeled Mobile Robots (WMRs) in the presence of disturbance. First, the robust model predictive control (RMPC) is implemented for the kinematic subsystem by solving the modified optimization problem with the fixed starting point. Second, the robust control is extended for the dynamic subsystem by the Backstepping technique and additional observer. This paper shows how to ensure the stability of the closed-loop tracking error system. The performance of the proposed robust control is verified through simulation studies using Casadi tool.

Keywords: Model predictive control (MPC) · Wheeled mobile robotic systems (WMRs) · Output feedback control · Terminal controller · Feasibility

1 Introduction

As an important branch of general robotics, wheeled mobile robots (WMRs) have been intensely studied with many remarkable results in many fields, such as industrial automation, space research, etc. Based on the property of non-holonomic constraint, it is necessary to separate the model of WMR into two subsystems, including the kinematic subsystem and dynamic subsystem [1–7]. This method can obtain an easier control design for each control loop because less difficulties than the kinematic subsystem influence each subsystem is an under-actuated system, and the dynamic subsystem is influenced by disturbance, uncertainties. Moreover, the Backstepping technique is usually investigated for this cascade model by considering the Lyapunov function candidate sequentially from the kinematic subsystem to the whole system [1–4, 7]. Several developments have been mentioned in investigating control law, such as using the modified transform matrix and observer in [1], disturbance observer [2], extending to WMR model under slip [4], etc.

Furthermore, it should be verified that the model-based control in [3] utilized the Kronecker product to achieve control structure without considering the cascade model. On the other hand, authors in [7] proposed the MPC law to design the outer loop with theoretical analysis to be discussed. As we all know, the difficulties of investigating MPC in mobile robotic systems are solving the optimization problem and guaranteeing

D. C. Nguyen et al. (Eds.): ICERA 2021, LNNS 366, pp. 834–842, 2022.
https://doi.org/10.1007/978-3-030-92574-1_85

the stability of the MPC method. In order to solve these challenges, Lyapunov stability theory is still employed with an appropriate Lyapunov function candidate [5–13]. Thanks to the optimization problem is given in the MPC method. We can utilize this optimal function to be the Lyapunov function candidate is considering the stability problem. However, comparing the Lyapunov function in two sampling times should be investigated using the terminal controller, feasible region, feasibility problem [5–13]. There are two main approaches to implement this comparison. First, the Linear Matrix Inequalities (LMIs) technique is used as in [5, 13]. It can be seen that the LMIs are achieved after implementing the comparison. Second, the intermediate function is pointed out by the intermediate control scheme using a terminal controller [6–12]. It is worth noting that MPC law is almost applied for the kinematic subsystem in robotics [6–12].

In this paper, several robust control schemes are proposed for a perturbed WMR system with non-holonomic constraints. With the proposed optimization problem in the MPC approach, the controller is easier implemented with the fixed starting point of an optimization problem. On the other hand, the observer is discussed in the second robust control method. Furthermore, the solution of a two-time scale-based filter is also discussed in estimating the disturbance. The rest of this paper is organized as follows. Section 2 introduced the mathematical model of WMRs and the control objective. Two control designs are presented in Sect. 3. In Sect. 4, the simulation studies are carried out to confirm the proposed control schemes. Finally, the conclusion is given in Sect. 5.

2 Problem Formulations

As we all have known in traditional Euler-Lagrange theory, the mathematical model of Wheeled Mobile Robotics (WMRs) is written by the following dynamic equations [1–4]:

$$\begin{cases} M(\eta)\ddot{\eta} + C(\eta, \dot{\eta})\dot{\eta} + B(\eta)(F(\dot{\eta}) + \tau_d) = B(\eta)\tau + J^T(\eta)\lambda \\ J(\eta)\dot{\eta} = 0 \end{cases} \tag{1}$$

$$\eta = [x \quad y \quad \theta]^T \tag{2}$$

where η is the joint variables vector with x and y, θ to be defined the position, and orientation of the WMR in term of the x axis. The matrices $M(q) \in \Re^{3\times3}; C(q, \dot{q}) \in \Re^{3\times3}; B(q) \in \Re^{3\times2}$ are the system inertia matrix, the centripetal, Corriolis torques vector, the input transformation matrix, respectively. It is similar to the work in [1–4], we achieve the two subsystems of WMRs to be described in (1), as follows:

$$\dot{\eta} = S(\eta)\sigma \tag{3}$$

$$D(\eta)S(\eta)\dot{\sigma} + C_1(\eta, \dot{\eta})\sigma = B(\eta)\tau \tag{4}$$

where $S(\eta)$ solutions in the equation $S^T(\eta)J^T(\eta) = 0$.

Remark 1: It is worth noting that the consideration of WMRs as a cascade system (3, 4) is able to achieve an easier controller. However, due to eliminating the constraint coefficient λ in the model (3, 4), it follows that the implementation of motion/force control is impossible.

The control task in this paper is to find the vector of torque τ with robust model predictive control (MPC) to ensure optimality principle and the tracking of trajectory. Moreover, this work is extended to output feedback control by adding more an appropriate observer. Thanks to the creation of separating this WMR model, we can establish the general cascade controller of WMR systems as described in Fig. 1.

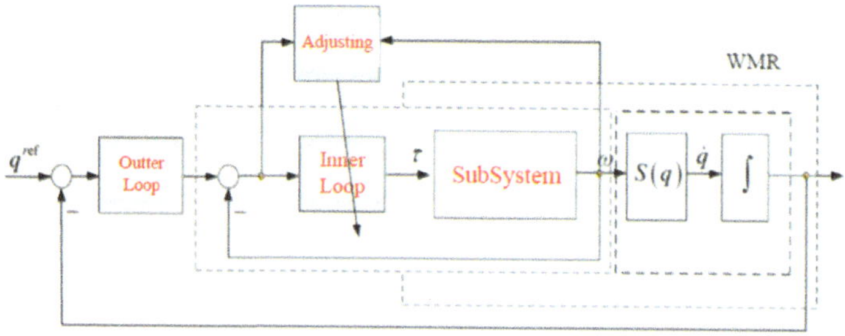

Fig. 1. The cascade control scheme of WMR systems

3 Robust Control Design for WMRs

As we all have known in traditional Euler-Lagrange theory, In this section, several approaches to implementing robust controllers are introduced to guarantee the trajectory tracking problem. First, we employ robust model predictive control (RMPC) for the kinematic model to eliminate disturbance. Therefore, the MPC algorithm is established easily with the fixed starting point to be described in the optimization problem. Second, we handle the dynamic subsystem continuously with the development of an observer to obtain an output feedback control scheme. Third, a solution of estimating the disturbance by a two-time scale filter [2] is also mentioned to compare.

3.1 Robust Model Predictive Control for WMRs

The Model Predictive Control (MPC) approach is developed for the tracking error model of the kinematic subsystem. Due to the effect of the trigonometric operator in (3) in establishing the control design, the transformation matrix is used to obtain the following model:

$$\begin{bmatrix} x_e(t) \\ y_e(t) \\ \theta_e(t) \end{bmatrix} = \begin{bmatrix} \cos\theta & \sin\theta & 0 \\ -\sin\theta & \cos\theta & 0 \\ 0 & 0 & 1 \end{bmatrix} \begin{bmatrix} x_r(t) - x(t) \\ y_r(t) - y(t) \\ \theta_r(t) - \theta(t) \end{bmatrix} \quad (5)$$

where $[x_r(t), y_r(t), \theta_r(t)]^T$ is the desired trajectory vector.

According to (5) and (3), the model can be written as:

$$\frac{d}{dt}\begin{bmatrix} x_e(t) \\ y_e(t) \end{bmatrix} = \begin{bmatrix} 0 & \omega \\ -\omega & 0 \end{bmatrix}\begin{bmatrix} x_e(t) \\ y_e(t) \end{bmatrix} + \begin{bmatrix} -\upsilon + (\upsilon_r)\cos\theta_e \\ \upsilon_r \sin\theta_e \end{bmatrix} \tag{6}$$

$$\frac{d}{dt}\begin{bmatrix} x_e(t) \\ y_e(t) \end{bmatrix} = \begin{bmatrix} 0 & \omega \\ -\omega & 0 \end{bmatrix}\begin{bmatrix} x_e(t) \\ y_e(t) \end{bmatrix} + u_e(t) \tag{7}$$

where $u_e(t)$ is given by

$$u_e(t) = \begin{bmatrix} -\upsilon + (\upsilon_r)\cos\theta_e \\ \upsilon_r \sin\theta_e \end{bmatrix} \tag{8}$$

For the purpose of obtaining tracking problem, an MPC algorithm with a fixed starting point in the optimization problem can be shown as follows:

Algorithm 1:

1. At time $t_k = k\delta$, collecting the actual state;
2. Solving the modified Optimization Problem to achieve the controller $u_e(t)$

$$\begin{aligned} \arg\min_{u(\tau|t_k)} J(x_e, y_e, u_e) = & \int_{t_k}^{t_k+T} L(x_e(\tau|t_k), y_e(\tau|t_k), u_e(\tau|t_k))d\tau \\ & + g(x_e(t_k+T|t_k), y_e(t_k+T|t_k)); T = N\delta \\ & \min_{u(\tau|t_k)} J(x_e, y_e, u_e) \\ & = \int_{t_k}^{t_k+T} \left\{ [x_e^T(\tau|t_k), y_e^T(\tau|t_k)]Q\begin{bmatrix} x_e(\tau|t_k) \\ y_e(\tau|t_k) \end{bmatrix} + u_e^T(\tau|t_k)Ru_e(\tau|t_k) \right\} d\tau \\ & + \frac{1}{2}[x_e^T(\tau|t_k), y_e^T(\tau|t_k)]\begin{bmatrix} x_e(\tau|t_k) \\ y_e(\tau|t_k) \end{bmatrix} \end{aligned} \tag{9}$$

$$\begin{aligned} u(\tau|t_k) \in U_{tube} = \left\{ [\upsilon, \omega]^T : \tfrac{|\upsilon|}{a} + \tfrac{|\omega|}{b} \le \lambda_{tube} = \tfrac{1}{\sqrt{2}} \right\} \\ \textit{Subject to } x_e(t_k+T|t_k) \& y_e(t_k+T|t_k) \\ \in \Omega_{tube} = \left\{ [x_e, y_e]^T : k_1|x_e| + k_2|y_e| \le a\lambda_{tube} \right\} \end{aligned} \tag{10}$$

where k_1, k_2 are selected from the matrix Q, R in cost function as described in Remark 3.

3. Apply this control input $[\upsilon, \omega]^T$ to WMR system during the sampling time interval $[t_k, t_{k+1})$.
4. Updating the time instant $t_k \to t_{k+1}$ and coming back step 1.

Theorem 1: The control input under Algorithm 1 guarantee the ISS stability in the outer loop error subsystem.

Remark 3: The key idea of Proof is utilizing the Lyapunov Candidate function being the minimum cost function in Algorithm 1. It enables us to compare Lyapunov function candidate in the sequent two steps with intermediate control input obtained from the terminal controller as above.

3.2 Robust Output Feedback Control for WMRs

In this section, the control design is realized from outer model to modified inner model with the first Lyapunov function being chosen as:

$$V_0 = \sqrt{1 + x_e^2 + y_e^2} + \sqrt{1 + \overline{\phi}_e^2} - 2. \tag{11}$$

The modified inner model is use a new variable as:

$$\overline{\phi}_e = \phi_e + \arcsin\left(\frac{k(t)y_e}{\sqrt{1 + x_e^2 + y_e^2}}\right). \tag{12}$$

Therefore, we obtain the modified inner model as:

$$\begin{bmatrix} \dot{x}_e \\ \dot{y}_e \\ \dot{\phi}_e \end{bmatrix} = \begin{bmatrix} \omega y_e + v + f_1 \\ -\omega x_e - \frac{k(t)v_r y_e}{\Gamma_1} + f_2 \\ \omega\left(\frac{k(t)x_e}{\Gamma_1}\right) + f_3 - v\frac{k(t)x_e y_e}{\Gamma_1^2\Gamma_2} \end{bmatrix} \tag{13}$$

By using the observer in [1], we obtain the derivative of Lyapunov function V_2 as:

$$\begin{aligned} \dot{V}_2 = & -k_1\frac{x_e^2}{\Gamma_1^2} - \frac{k(t)v_r y_e^2}{\Gamma_1^2} - k_2\frac{\overline{\phi}_e^2}{\Gamma_3^2} + \frac{y_e f_2}{\Gamma_1} - \tilde{\varpi}^T(D_R + K + C)\tilde{\varpi} - \alpha\tilde{m}\left(\dot{\hat{m}} + \hat{u}_1\breve{v}\right) \\ & - \beta\tilde{n}\left(\dot{\hat{n}} + \hat{u}_2\breve{\omega}\right) - \tilde{H}^T\Lambda\left(\dot{\hat{H}} + \Lambda^{-1}Y_c^T\breve{\varpi}\right) + \alpha\tilde{m}\left[\hat{u}_1\breve{v} - u_1\tilde{v}\right] + \beta\tilde{n}\left[\hat{u}_2\breve{\omega} - u_2\tilde{\omega}\right] \\ & + \tilde{H}^T\left[\hat{Y}_c^T\breve{\varpi} - Y_c\tilde{\varpi}\right] + \tilde{\varpi}^T(\hat{Y}_c - Y_c)\hat{H} + \tilde{\varpi}^T K\left(\tilde{\varpi} - \breve{\varpi}\right) + \alpha\tilde{m}(u_1 - \hat{u}_1)\tilde{v} \\ & + \beta\tilde{n}(u_2 - \hat{u}_2)\tilde{\omega} \end{aligned} \tag{14}$$

Remark 2: The computation in (14) is different from the work in [1] with three additional terms:

1) $-\tilde{\varpi}^T C\tilde{\varpi}$;
2) $+\tilde{\varpi}^T K\left(\tilde{\varpi} - \breve{\varpi}\right)$;
3) $+\alpha\tilde{m}(u_1 - \hat{u}_1)\tilde{v} + \beta\tilde{n}(u_2 - \hat{u}_2)\tilde{\omega}$;

The fact is the output feedback control input $\hat{U}(\hat{\varpi}) = -K\breve{\varpi} + \hat{Y}_c\hat{H} - \Omega$. The stability of the closed system is UUB because of the tracking effectiveness of the observer.

A. *Two Time-Scale Tracking Control for WMRs*

With the cascade controller (Fig. 1) and the transforming method (5) in tracking error model, the work in [2] proposed the control structure (15) with disturbance observer (16). It should be noted that, the proposed controller in section B is different from [2] considering the State observer to develop output feedback controller.

$$\begin{aligned} u &= u^* + l \\ u^* &= \begin{bmatrix} e_1 \cos\theta_c + e_2 \sin\theta_c - \dot{\upsilon}_r + \dot{\upsilon}_d - Q_2\tilde{\upsilon} \\ e_3 - \dot{\omega}_r + \dot{\omega}_d - Q_3\tilde{\omega} \end{bmatrix} \end{aligned} \tag{15}$$

$$\begin{cases} l = \frac{1}{\varepsilon}(\hat{z} - z) \\ \frac{d}{dt}\hat{z} = -\frac{1}{\varepsilon}(\hat{z} - z) + u + \dot{z}_d \end{cases} \tag{16}$$

4 Simulation Studies

The mathematical model of Wheeled Mobile Robotics (WMRs) is written by the following dynamic equations [1–4]:

The simulation studies is demonstrated by Casadi tool for Algorithm 1 with modified optimization in the kinematic subsystem of a WMR:

$v_r = 0.015(m/s); \omega_r = 0.04(rad/s); \xi_r(0) = [0 \quad 0 \quad \pi/3]^T;$
$\xi_f(0) = [0.4 \quad -0.2 \quad -\pi/2]^T;$
$T = 1; \delta = 0.1; P = diag\{\ 0.4, 0.4\}; R = diag\{0.2; 0.2\}; \tilde{k}_1 = \tilde{k}_2 = 1.2;$
$\lambda_{tube} = 0.6636,\ \ \Omega_{tube} = \{\tilde{p}_e : |\tilde{x}_e| + |\tilde{y}_e| < 0.0542\},\ \ P_{fe} = \{p_{fe}(t) : |p_{fe}| \leq [0.0017 \ 0.0017]^T\}.$

The results show good behavior as described in the following Figs. 2 and 3. Moreover, the simulations are developed by two-time scale based robust control in Section C with the parameters as follows:

r = 0.098 m, b = 0.2 m, $d = 0.3\,m,\ m_w = 22\,kg,\ m_M = 2\,kg,\ I_M = 0.02\,kg.m^2$, $I_D = 0.005\,kg.m^2$.

And the desired trajectory is chosen as

$x_r(t) = t;\ y_r(t) = sin\omega t$, with $\omega = \frac{\pi}{180};\ \theta_r(t) = tan^{-1}\left(\frac{\pi}{180}\left(cos\frac{\pi}{180}(t)\right)\right)$.

The disturbance is inserted into WMRs: $\tau_d = \left[\sin\left(\frac{1}{20}t\right); 2\sin\left(\frac{1}{20}t\right)\right]$

In this case, it can be seen that the good responses are shown in Figs. 4, 5, 6, 7, 8 and 9, with the improvement by a filter to be considered in two cases (Figs. 4 and 5). Finally, the proposed robust output feedback controller is implemented with high effectiveness in Figs. 10 and 11.

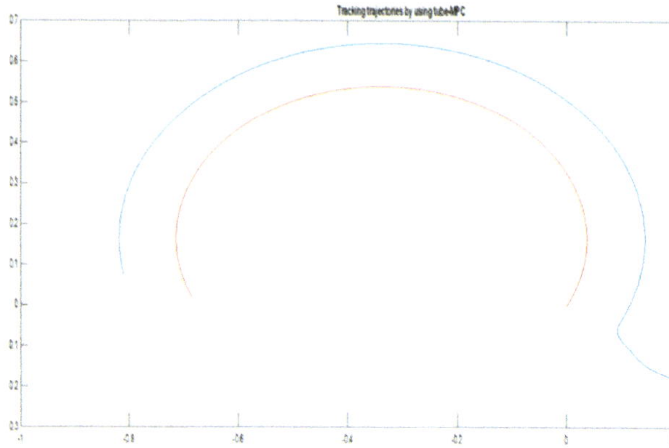

Fig. 2. The trajectory of a WMR using Robust MPC Law

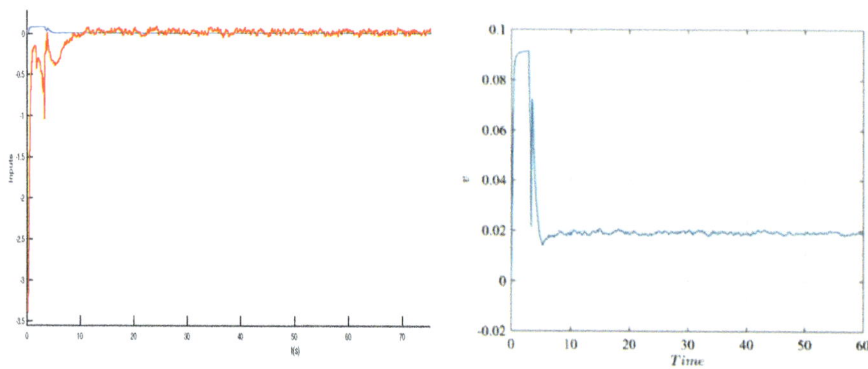

Fig. 3. The control inputs of a WMR

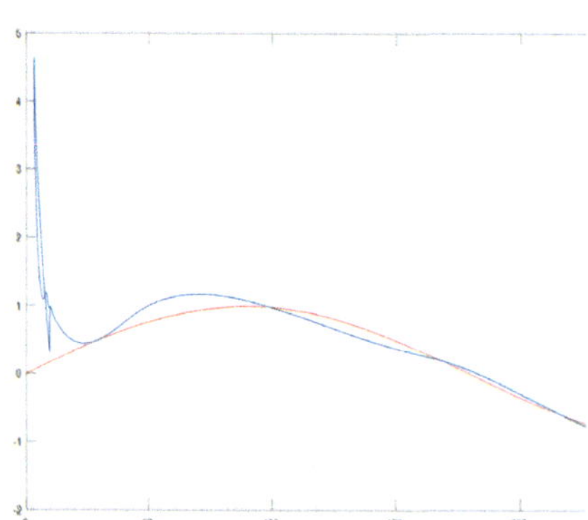

Fig. 4. The response by two-time scale based robust control (no filter)

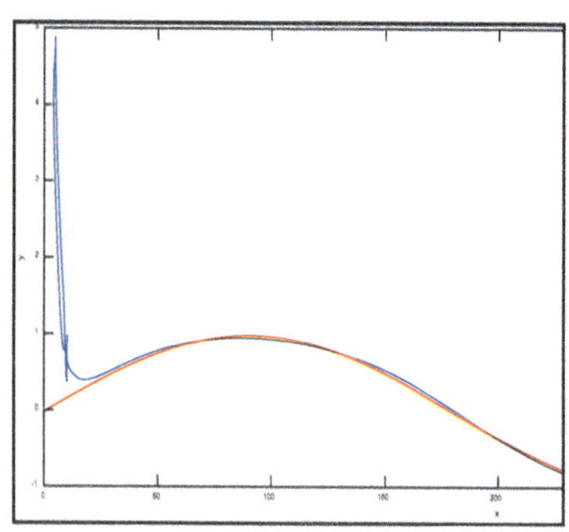

Fig. 5. The response by two-time scale based robust control (using filter)

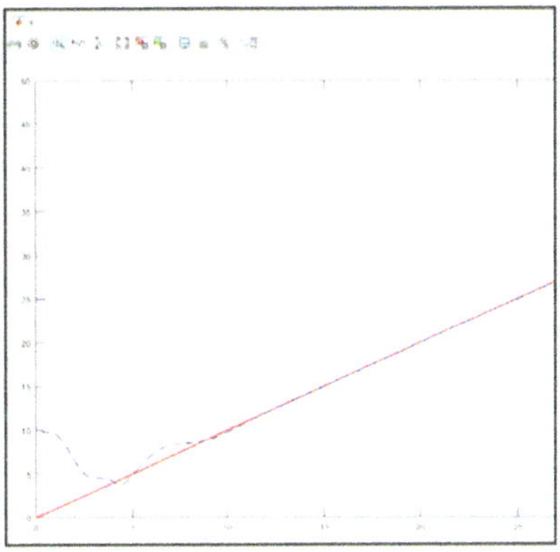

Fig. 6. The response of trajectory (axis X) by two-time scale based robust control

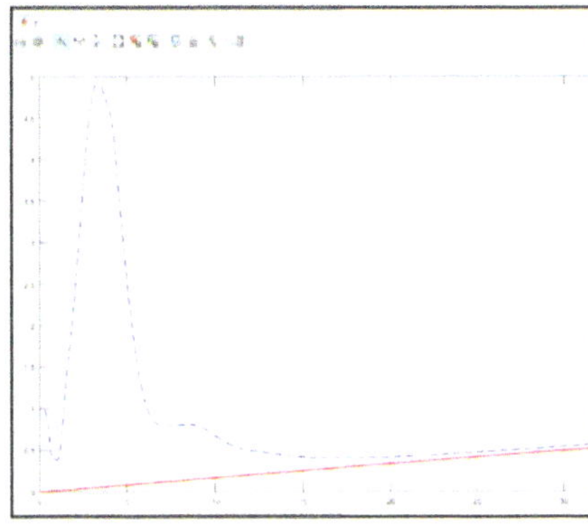

Fig. 7. The response of trajectory (axis Y) by two-time scale based robust control

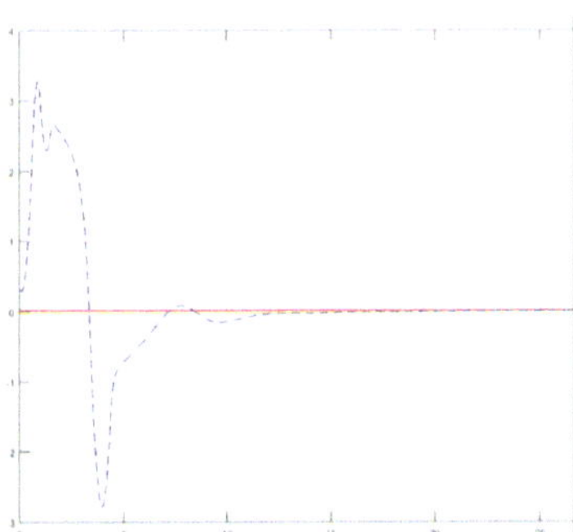

Fig. 8. The response of Trajectory (Direction Angle) by two-time scale based robust control

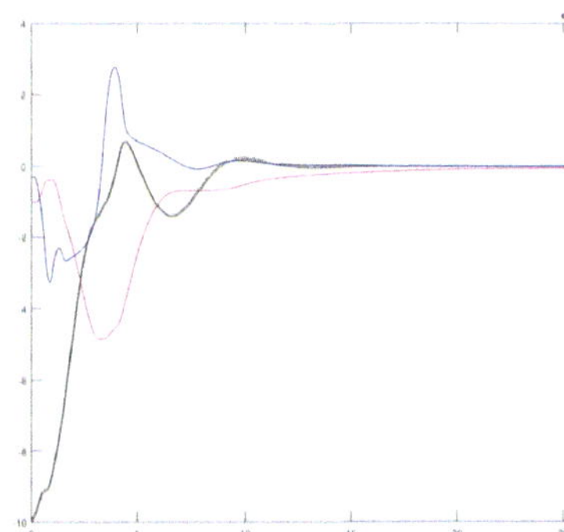

Fig. 9. The Tracking Error by two-time scale based robust control

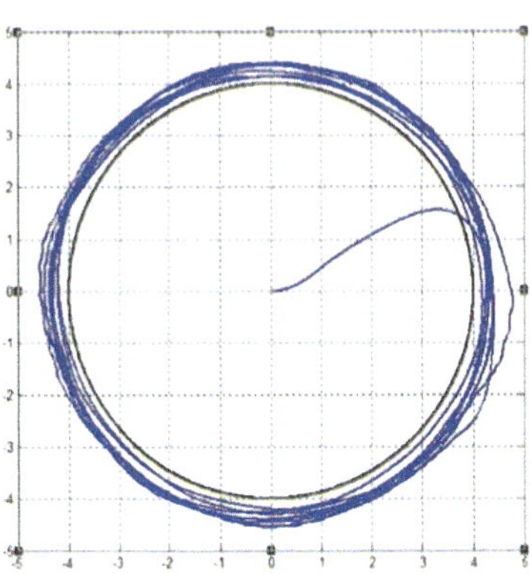

Fig. 10. The response of Trajectory in Cartesian Coordinate by two-time scale based robust control (Circle)

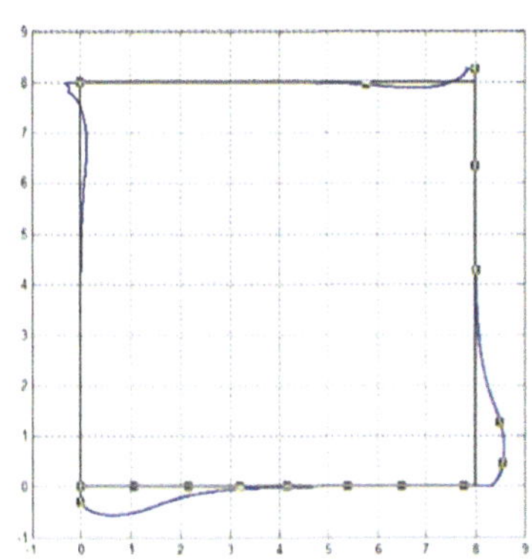

Fig. 11. The response of Trajectory in Cartesian Coordinate by two-time scale based robust control (Rectangle)

5 Conclusion

We have proposed robust control schemes for WMRs, including robust MPC for the kinematic subsystem and output feedback control for the cascade model with disturbance. The MPC algorithm is presented with a modified optimization problem, and the extension for cascade controller is considered with an additional observer. The theoretical analysis is carried out as well as simulation studies are verified to show the efficiency of the proposed algorithm.

Acknowledgments. The authors gratefully acknowledge the Thai Nguyen University of Technology for supporting this work.

References

1. Huang, J., Wen, C., Wang, W., Jiang, Z.P.: Adaptive output feedback tracking control of a nonholonomic mobile robot. Automatica **50**(3), 821–831 (2014)
2. Sun, W., Tang, S., Gao, H., Zhao, J.: Two time-scale tracking control of nonholonomic wheeled mobile robots. IEEE Trans. Control Syst. Technol. **24**(6), 2059–2069 (2016)
3. Chen, Z., Liu, Y., He, W., Qiao, H., Ji, H.: Adaptive-neural-network-based trajectory tracking control for a nonholonomic wheeled mobile robot with velocity constraints. IEEE Trans. Industr. Electron. **68**(6), 5057–5067 (2020)
4. Nguyena, T., Hoang, T., Pham, M., Dao, N.: A Gaussian wavelet network-based robust adaptive tracking controller for a wheeled mobile robot with unknown wheel slips. Int. J. Control **92**(11), 2681–2692 (2019)
5. Yang, W., Xu, D., Zhang, C., Yan, W.: A novel robust model predictive control approach with pseudo terminal designs. Inf. Sci. **1**(481), 128–140 (2019)
6. Wang, P., Feng, X., Li, W., Ping, X., Yu, W.: Robust RHC for wheeled vehicles with bounded disturbances. Int. J. Robust Nonlinear Control **29**(7), 2063–2081 (2019)
7. Dao, P.N., Nguyen, H.Q., Nguyen, T.L., Mai, X.S.: Finite horizon robust nonlinear model predictive control for wheeled mobile robots. Math. Probl. Eng. **18**, 2021 (2021)
8. Yang, H., Guo, M., Xia, Y., Cheng, L.: Trajectory tracking for wheeled mobile robots via model predictive control with softening constraints. IET Control Theory Appl. **12**(2), 206–214 (2018)
9. Sun, Z., Xia, Y.: Receding horizon tracking control of unicycle-type robots based on virtual structure. Int. J. Robust Nonlinear Control **26**(17), 3900–3918 (2016)
10. Nikou, A., Verginis, C.K., Heshmati-alamdari, S., Dimarogonas, D.V.: A robust non-linear MPC framework for control of underwater vehicle manipulator systems under high-level tasks. IET Control Theory Appl. (2020)
11. Dai, L., Lu, Y., Xie, H., Sun, Z., Xia, Y.: Robust tracking MPC with quadratic robustness constraint for mobile robots with incremental input constraints. IEEE Trans. Ind. Electron. (2020)
12. Dai, L., Yu, Y., Zhai, D.H., Huang, T., Xia, Y.: Robust model predictive tracking control for robot manipulators with disturbances. IEEE Trans. Industr. Electron. **68**(5), 4288–4297 (2020)
13. Xie, H., Dai, L., Luo, Y., Xia, Y.: Robust MPC for disturbed nonlinear discrete-time systems via a composite self-triggered scheme. Automatica **127**, 109499 (2021)

State Estimation of Discretely Controlled Continuous Systems Using Unscented Kalman Filter

Thanh Ha Tran[1,2](✉), Tuan Anh Nguyen[3], Lan Anh Dinh Thi[3], and Thu Ha Nguyen[3]

[1] ELANI Technology Solutions Corporation, Hanoi, Vietnam
ha.tt@elani.vn

[2] Institute of Science, Technology and Training OMEGA, Hanoi, Vietnam

[3] School of Electrical Engineering, Hanoi University of Science and Technology, Hanoi, Vietnam

Abstract. This paper presents a scheme to estimate continuous states of Discretely Controlled Continuous Systems where continuous plants with nonlinear dynamics are controlled by discrete controllers that generate discrete control signals depending on discrete measurement events occurred at an irregular time. The aim of the estimator is to compute the continuous states of such systems from these measurements in order to provide the full information of the continuous states so that the discrete controller can extract suitable inputs to issue to the plant. The proposed state estimator is a modified version of the Unscented Kalman Filter (UKF). Upon receiving discrete measurement events, the estimator computes the estimated state and the covariance of the estimation error from previous measurement events according to the known nonlinear dynamic, control signals, and transformed sigma points. The next step is to incorporate the measurement value at present discrete events into the estimated state and the estimation covariance. The approach is verified by applying it to the water three-tank system built in MATLAB Simscape Fluid Toolbox.

Keywords: Nscented Kalman Filter · Extended Kalman Filter · State estimation · Discrete measurement events · Hybrid dynamic systems

1 Introduction

This paper presents a scheme to estimate continuous states of Discretely Controlled Continuous Systems (DCCS). These systems are a class of hybrid dynamic systems where continuous plants with nonlinear dynamics are controlled by discrete controllers that generate discrete control signals depending on discrete measurement events [1, 3, 4]. Such an event is detected whenever a system trajectory reaches a threshold of discrete sensors configured for appropriate process variables [4, 5]. The values of the crossings and their time instants are the given information by these events. The time values are not equidistant because these events are autonomously generated and without predetermined sequences. Discrete sensors are widely applied in industrial applications because of theirs accurate and robust measurements [6]. As DCCS exhibit both the

D. C. Nguyen et al. (Eds.): ICERA 2021, LNNS 366, pp. 843–854, 2022.
https://doi.org/10.1007/978-3-030-92574-1_86

continuous and discrete dynamics as well as the complex interaction of continuous plants with discrete-event controllers, the design of such feedback controllers is a challenging mission [5]. From the given plant model and the specifications, discrete controllers are synthesized in a closed-loop system fashion. To solve this problem, a new approach employing dynamic programming for the design of discrete controllers was introduced in [1, 3]. For this approach, the feedback controller is computed for all the test points on the boundary regions of the cells resulted from the partitioning of the state space which is derived from the thresholds specified by the discrete sensors and specifications. The designed controller issues a new control signal every time the system trajectories reach a threshold and satisfy the specifications: steering the system trajectories from an initial set to a target set in the shortest time also staying away from forbidden sets.

In the design of the state feedback controller, it assumes that the full continuous state variables are available. Nevertheless, the information obtained at each occurrence time of discrete events is only the value of one of the state variables whose threshold has just been crossed. Moreover, under the effect of process noise and measurement noise of the plant, the result controller may give an inappropriate control signal and not be able to meet the requirements. Thus, the state estimation from the sequence of discrete measurement events is a crucial problem to implement the state feedback controller.

To handle this state estimation task, an event-based state estimation (EBSE) was presented in [2, 3] which is a modified version of the constrained extended Kalman filter (EKF) that considered mixed discrete-continuous or hybrid dynamics with asynchronous discrete measurements and constraints on continuous states. The EBSE operates recursively between prediction and correction stages. Between the occurrences of discrete measurement events, the prediction stage computes the estimated state according to the known nonlinear dynamic, discrete control signals, and its covariance matrix utilizing a simplified Riccati differential equation. Once a discrete measurement is received, the correction is triggered. It incorporates the measurement value at a present event into the state estimated state and its estimation covariance matrix. In comparison to the continuous-discrete or hybrid extended Kalman filter as presented in [10, 11], the main differences of EBSE are as follows: Firstly, the sampling time used for the prediction is not identical in each run; Secondly, the correction is executed which bases on the discrete events instead of time events; Thirdly, a constrained optimization problem considered state constraints is numerically solved to substitute for the calculation of the Kalman gain. Fourthly, the measured values at the original EKF are all values of a variable that have been determined since the beginning of the system implementation. In the case of DCCS, the measured value obtained when events occurred belongs to one of the state variables whose threshold has been reached. Therefore, the measurement matrix is different at each time events occur.

The implementation of the EBSE has shown that it is a practical and suitable technique for solving the state estimation of DCCS but there are two issues arisen. If the dynamic of the system is strongly nonlinear, the linearization will no longer be accurate enough which leads to the fact that the estimator might be unstable. In addition, computing Jacobian matrices are not always feasible in practice (either by hand or by computer). Although computer tools are supporting for calculating derivatives, the results are not simplified which limits the main advantage of this method that is simplicity. Meanwhile, the Unscented Kalman Filter (UKF) is originally considered for continuous-time dynamic systems with discrete-time measurements which uses a deterministic sampling technique known as the unscented transform to adapt the Kalman filter to nonlinear systems. The unscented transform works by passing a bunch of carefully chosen samples through a nonlinear system and computing the mean and covariance as the outputs [7, 10, 14, 15, 18–20]. As a result, it provides a better approximation of output distribution than analytical local linearization used by the EKF for similar computation costs. Also, this technique removes the requirement to explicitly calculating the Jacobian matrix which drastically reduces computation efforts.

This contribution presents a scheme that addresses the state estimation problem for DCCS. The estimator aims to compute the continuous states of such systems from these irregular measurements in order to provide the full information of the continuous states so that the discrete controller can extract suitable inputs to issue to the plant. The proposed state estimator is a modified version of the UKF. The formula of problem setting and the structure of the UKF in this paper is the same as the EBSE. The differences are the transformations of sigma points, the replacements of linearization in the prediction, and of Jacobian matrices computation the Kalman gain in the correction. Upon receiving discrete measurement events, the UKF computes the estimated state and the covariance of the estimation error from previous measurement events according to the known nonlinear dynamic, control signals, and transformed sigma points. The next step is to incorporate the measurement value at present discrete events into the estimated state and the estimation covariance.

The UKF is verified by applying it to the water three-tank system. This model is created with Simscape Fluid Toolbox [11] which is available in MathWorks. The dynamic of the model is nonlinear and the thresholds set are generated by the specified blocks in the Simulink library. The controller is responsible for bringing the model state trajectory from the initial set to the target set and avoiding the forbidden set. The UKF will calculate the estimated state each time a measurement event occurs to provide complete information about the system state for the state feedback controller. The simulation result will show the feasibility of this approach when it is applied to this model.

The rest of this paper is organized as follows. In Sect. 2, the problem formulation of DCCS and state feedback controller is presented. Section 3 describes the contents of the UKF scheme. The numerical simulation results when applying this approach are shown in Sect. 4. Section 5 provides conclusions and an outlook on future works.

2 Problem Formulation

2.1 Plant Model

Definition 1: A discretely controlled continuous system [1, 2] is given by $DCCS = (X, U, L, E, Y, f)$ with:

- $X = \left[x_1^{lb}, x_1^{ub}\right] \times \ldots \times \left[x_n^{lb}, x_n^{ub}\right] \subseteq \mathbb{R}^n$ is a continuous state space. Each x_i is defined on an interval $\left[x_i^{lb}, x_i^{ub}\right]$, $i \in \{1, \ldots, n\}$ with n is the dimension of X.

- $U = \{u_1, \ldots, u_m\} \subseteq \mathbb{R}^{n_p}$ is a set of discrete inputs. Each u_j is specified by $u_j = \left[u_j^1, \ldots, u_j^p\right]$ with $j \in \{1, \ldots, m\}$, where each value $u_j^k \in \mathbb{R}$ with $k \in \{1, \ldots, p\}$.
- $L = \{L_1, \ldots, L_n\}$ is an ordered set of thresholds. Each L_i is defined for each x_i such that $L_i = \{l_i^0, \ldots, l_i^{q_i}\}$ with $l_i^0 = x_i^{lb}$ and $l_i^{q_i} = x_i^{ub}$. The thresholds indicate those values of x_i at which the input can be set to a new valueu_k.
- n_c is the number of cells that result from a partition of X induced by L such that $X = \bigcup_{1 \le c \le n_c} X_c$. Each cell X_c is bounded by a set of boundary sections $\partial X_c = \left[l_1^{p_1}, l_1^{p_1+1}\right] \times \ldots \times \left[l_n^{p_n}, l_n^{p_n+1}\right]$, where $i \in \{1, \ldots, n\}$ and $p_j \in \{0, \ldots, q_i - 1\}$.
- $E = \{e_1, \ldots, e_{n_e}\} \subseteq \mathbb{R}^{n_e}$ is a set of discrete events. An event $e_k \in E$ occurs at a point of time $t = t_k$ at which a variable crosses a threshold: $x_i(t) = l_i^k$.
- $Y = \{y_1, \ldots, y_{n_e}\} \subseteq \mathbb{R}^{n_e}$ is a set of discrete measurements. y_k is the measurement at t_k which is determined from $y_k = C_k x_k + v_k$, $v_k \sim N(0, R_k)$. v_k is the measurement noise in discrete-time which assumes to be zero-mean Gaussian noise specified the covariance matrix $R_k \in \mathbb{R}^{n_k \times n_k}$. x_k is the value detected at t_k at which e_k occurs, i.e., $x_k = l_k^j$. C_k is the measurement matrix of dimension $n_k \times n$ with $rank(C_k) = n_k$, where $n_k \in \{1, \ldots, n\}$. Each row of C_k contains exactly one element of 1 at the ith position related to x_i and $(n-1)$ elements of 0 at other positions.
- $f : X \times U \times Q \to R^n$ is the stochastic nonlinear continuous dynamics $\dot{x} = f(x, u, t) + w(t)$ to each $x \in X$, $u \in U$ and $w \in Q$. $w \sim N(0, Q)$ is the process noise in continuous time and zero mean Gaussian noise specified by the covariance matrix $Q \in \mathbb{R}^{n \times n}$. $x(0) = x_0 + w_0$ is the initial condition of the system and $w_0 \sim N(0, P_0)$ is zero-mean Gaussian noise specified by the covariance matrix P_0. For constant u, f gives a continuous solution of $x(t)$ for $t \in [t_k, t_{k+1}]$ between events.

Let $T = \{t_0, t_1, \ldots, t_f\}, t_k \in \mathbb{R}^{\ge 0}$ be a sequence of time points, where t_0 is the initial time, t_k is the time at which an event occurs and t_f is the final time. $\phi_u = (u_0, u_1, \ldots, u_f)$ is an input sequence with $u_k = u(t_k)$ is piecewise constant for $t \in [t_k, t_{k+1}]$. An arbitrary input u_k is kept constant during the time interval $[t_k, t_{k+1}]$ when the system operates in X_i. A trajectory of DCCS is the sequence of state $\phi_x = (x_{t_0}, x_{t_1}, \ldots, x_{t_f})$ with $x_{t_k} = x(t_k)$ which is the evolution of the system under sequences of T and ϕ_u as well as Q.

If a continuous state $x(t)$ hits a specified threshold $l_{i,k}$, a corresponding event e_i is trigged. The input u_k remains constant until a new threshold is hit since the controller issues a new discrete input only when it receives a discrete event.

2.2 State Feedback Controller

Definition 2: A logic controller $C : E \times X_T \to U$ is a static mapping from E to U that is fixed for a set of target states X_T.

The logic controller generates a discrete input based upon current discrete events so that at each threshold crossing, there is one combination of discrete inputs that is optimal to drive the system to reach the target. Note that we do not consider any delay in the response of the controller in this setting. Therefore, it is assumed that the controller reacts immediately when it is triggered by discrete measurement events.

The task of the discrete controller is to optimally bring the system from a set of initial state X_0 to X_T while all unsafe sets of state X_F are avoided. The sets X_0, X_T, X_F are the control specifications that are specified as regions of the induced partition of X. As the solution to the task of discrete controller design has been presented in [1, 3], this paper only focuses on the task of estimator design.

3 Estimator Design

3.1 State Estimator

For the state estimation problem, the disturbed continuous plant with nonlinear dynamics and the measurement equation is given as defined in Sect. 2.1. A state estimator can be written as a dynamical system [2, 3]:

$$\begin{aligned} \dot{\bar{x}} &= \bar{f}(\bar{x}, u, y_k) \\ \bar{x}(0) &= \bar{x}_0 \\ \bar{y}_k &= C_k \bar{x}_k + v_k \end{aligned} \tag{1}$$

where $\bar{x} \in R^n$ is the estimated state, $\bar{f} \in F^n$ is the nonlinear dynamics of the state estimator, $\bar{y}_k \in R^{n_k}$ is the estimated measurement at event time t_k, $\bar{x}_0$ is the initial condition of the estimator.

Definition 3: The task of state estimator design is to compute the estimated state $\bar{x}(t)$ based on the sequences of discrete event measurements $y_t \in Y$ and discrete control inputs $u_t, t \in \{0, .., k\}$ such that $\bar{x}(t)$ is close to $x(t)$ and converges into $x(t)$ as $t \to \infty$.

3.2 Unscented Kalman Filter Design Scheme

In order to implement in this paper, the UKF estimator introduced in [10] needs some small modifications due to the special characteristic of the plant. The most important thing is the measurement function which is not the same in the whole procedure because the measurement value depends on the threshold which has been reached when the event occurs. The main steps of UKF used in this paper are described as below:

- Initialization:

$$k := 0; \bar{x}_{-1|-1} = \bar{x}_0; P_{-1|-1} = P_0 \tag{2}$$

- Prediction:

$$\begin{aligned} \overline{x}_{k-1|k-1}^{(i)} &= \overline{x}_{k|k-1} + \tilde{x}^{(i)} \quad i = 1, \ldots, 2n \\ \tilde{x}^{(i)} &= \left(\sqrt{nP_{k-1|k-1}}\right)_i^T \quad i = 1, \ldots, n \\ \tilde{x}^{(n+i)} &= -\left(\sqrt{nP_{k-1|k-1}}\right)_i^T \quad i = 1, \ldots, n \end{aligned} \tag{3}$$

$$\dot{\overline{x}}^{(i)}(t) = \overline{x}_{k|k-1}^{(i)} = \overline{f}\left(\overline{x}_{k-1|k-1}^{(i)}, u_{k-1}, t\right) \quad i = 1, \ldots, 2n \tag{4}$$

$$\overline{x}_{k|k-1} = \frac{1}{2n}\sum\nolimits_{i}^{2n} \overline{x}_{k|k-1}^{(i)} \tag{5}$$

$$\dot{P}(t) \approx P_{k|k-1} = \frac{1}{2n} \times \sum\nolimits_{i=1}^{2n} \left(\overline{x}_{k|k-1}^{(i)} - \overline{x}_{k|k-1}\right)\left(\overline{x}_{k|k-1}^{(i)} - \overline{x}_{k|k-1}\right)^T + Q \tag{6}$$

- Correction:

$$\begin{aligned} \overline{x}_{k|k}^{(i)} &= \overline{x}_{k|k-1} + \tilde{\underline{x}}^{(i)} \quad i = 1, \ldots, 2n \\ \tilde{\underline{x}}^{(i)} &= \left(\sqrt{nP_{k|k-1}}\right)_i^T \quad i = 1, \ldots, n \\ \tilde{\underline{x}}^{(n+i)} &= -\left(\sqrt{nP_{k|k-1}}\right)_i^T \quad i = 1, \ldots, n \end{aligned} \tag{7}$$

$$\overline{y}_k^{(i)} = C_k \overline{x}_{k|k}^{(i)} \quad i = 1, \ldots, 2n \tag{8}$$

$$C_k = \frac{\partial x_i}{\partial \overline{x}} \tag{9}$$

$$\overline{y}_k = \frac{1}{2n}\sum\nolimits_{i=1}^{2n} \overline{y}_k^{(i)} \tag{10}$$

$$P_{\overline{y}_k} = \frac{1}{2n} \times \sum\nolimits_{i}^{2n} \left(\overline{y}_k^{(i)} - \overline{y}_k\right)\left(\overline{y}_k^{(i)} - \overline{y}_k\right)^T + R_k \tag{11}$$

$$P_{\overline{x}_{k|k-1}, \overline{y}_k} = \frac{1}{2n}\sum\nolimits_{i}^{2n} \left(\overline{x}_{k|k}^{(i)} - \overline{x}_{k|k-1}\right)\left(\overline{y}_k^{(i)} - \overline{y}_k\right)^T \tag{12}$$

$$\begin{aligned} K_k &= P_{\overline{x}_{k|k-1}, \overline{y}_k} P_{\overline{y}_k}^{-1} \\ \overline{x}_{k|k} &= \overline{x}_{k|k-1} + K_k(y_k - \overline{y}_k) \\ P_{k|k} &= P_{k|k-1} - K_k P_{\overline{y}_k} K_k^T \end{aligned} \tag{13}$$

- Iteration:

$$k := k + 1;\ \bar{x}_{k-1|k-1} := \bar{x}_{k|k};\ P_{k-1|k-1} := P_{k|k} \tag{14}$$

The scheme starts with the initialization where the initial conditions $\bar{x}_0, P_0$ are given as in Eq. (2) and the input u_0 is assumed to be known. The prediction stage is carried out in between discrete events when no measurement is available. For this stage, the predicted state $\bar{x}_{k|k-1}$ and the predicted covariance matrix of the estimation error $P_{k|k-1}$ are propagated from the previous measurement event to the present one via Eqs. (3)–(6). $\bar{x}_{k|k-1}$ is computed from Eq. (5) by combining the sigma points $\bar{x}^{(i)}_{k|k-1}$ which are transformed from the integrations as given in Eq. (4). These integrations are performed with the known nonlinear dynamics which has no process noise and the discrete input u_{k-1} and $2n$ sigma points $\bar{x}^{(i)}_{k-1|k-1}$ derived from Eq. (3) where n is the number of states and $\sqrt{(\cdot)}$ means the matrix square root and sub-index i mean $i'th$ row. The integration time $t \in [t_{k-1}, t_k]$ in (4) is the interval from t_{k-1} of the previous measurement event to t_k of the latest one. $P_{k|k-1}$ is calculated from Eq. (6). The prediction continually executes until a discrete measurement y_k is detected at t_k.

Once the measurement value y_k is received, the correction stage is performed following Eqs. (7)–(13). This stage updates $\bar{x}_{k|k-1}$ and $P_{k|k-1}$ in order to get the corrected state $\bar{x}_{k|k}$ and the corrected covariance of the estimation error $P_{k|k}$ as indicated in Eq. (13). The calculations of $\bar{x}_{k|k}$ and $P_{k|k}$ are taken into account y_k, $P_{\bar{y}_k}$ and $P_{\bar{x}_{k|k-1}, \bar{y}_k}$ as in Eqs. (11) and (12). $P_{\bar{y}_k}$ is the covariance of the predicted measurement value $\bar{y}_k$. $P_{\bar{x}_{k|k-1}, \bar{y}_k}$ is the cross covariance between $\bar{x}_{k|k-1}$ and $\bar{y}_k$. According to Eq. (13), $\bar{y}_k$ is estimated by combining the sigma points $\bar{y}^{(i)}_k$ as in Eq. (10) which are determined from the transformation of $2n$ sigma points $\bar{x}^{(i)}_{k|k}$ resulting from Eqs. (7, 8) and the measurement matrix C_k as described in Eq. (9) where sub-index i means the $i'th$ order of state that has triggered threshold. The value of C_k depends on which threshold that has been reached when the event occurs.

The estimated state $\bar{x}_{k|k}$ is provided to the discrete controller to generate the next control signal u_k for implementation to the system. At the same time, the next iteration of the procedure is triggered with the prediction stage with the determined u_k and $\bar{x}_{k-1|k-1}$, $P_{k-1|k-1}$ as assigned in Eq. (14). Under the control of u_k, the continuous plant will evolve until the next discrete event occurs and then the correction stage is operated again. The design parameters of the described scheme are $\bar{x}_0$, P_0, Q, R_k.

From the algorithm above, it can be seen that the total computation cost is approximately n^3 with n is the number of the system state. Therefore, the computational complexity of this method is $\mathcal{O}(n^3)$. When n increases, the total computation cost does not increase too quickly, which makes the UKF more suitable in providing full-state information to the feedback controller.

4 Simulation

4.1 Simscape Water Three-Tank Model

The water three-tank system diagram set up in Simulink is exhibited in Fig. 1. This system consists of three tanks connected. These tanks are filled with water and the levels can be adjusted by logic control actuators. Tank 1 with level h_1 is filled with constant inflow F_{1in} and emptied with outflow F_{12}. The water level h_2 of tank 2 is determined by flow F_{12} connected to tank 1 and flow F_{23} connected to tank 3. Tank 3 has constant inflow F_{3in}, outflow F_{23} connected to tank 2, and another outflow F_{3out}. The flow rate of these flows is controlled by on/off valves $V_{1in}, V_{12}, V_{23}, V_{3in}, V_{3out}$. As a result, there are 32 different combinations of discrete inputs.

The dynamics of the tank block can be summarized as follows [11]:

$$
\begin{aligned}
F_i &= \sqrt{\frac{1}{K_i}} \times A_i \times \sqrt{\frac{2}{\rho}} \times \frac{p_{loss,i}}{\left(p_{loss,i}^2 + p_{cr,i}^2\right)^{1/4}} \\
p_{cr,i} &= K_i \times \frac{\rho}{2} \times \left(\frac{Re_{cr} \times \nu}{d_i}\right)^2
\end{aligned}
\qquad (15)
$$

where: i is one of three outlets A, B or C; F_i is the flow rate through outlet i of tank $[m^3/s]$; K_i is the pressure loss coefficient at outlet i; A_i is the connecting pipe area at outlet i $[m^2]$; ρ is the fluid density $[g/m^3]$; $p_{loss,i}$ is the pressure loss in the connecting pipe i $[Pa]$; $p_{cr,i}$ is the minimum pressure for the turbulent flow in the connecting pipe i $[Pa]$; Re_{cr} is the Critical Reynolds number; ν is the fluid kinematic viscosity $[m^2/s]$.

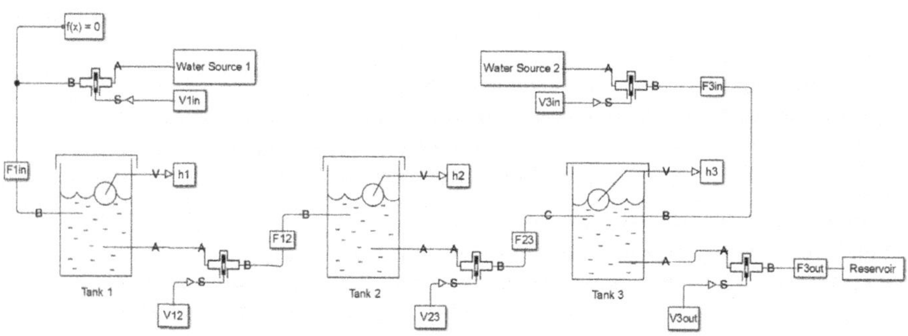

Fig. 1. Simulink water three-tank model

The dynamics of the valve block are briefly described as follows [11]:

$$\begin{aligned} F &= C_D \times A(h) \times \sqrt{\frac{2}{\rho}} \times \frac{p}{\left(p^2 + p_{cr}^2\right)^{1/4}} \\ p_{cr} &= \frac{\rho}{2} \times \left(\frac{Re_{cr} \times v}{C_D \times D_H}\right)^2 \\ A(h) &= \begin{cases} A_{leak} & h \leq 0 \ or \ h > 2D \\ D^2 \times \left[\frac{\alpha}{2} - \frac{\sin(2\alpha)}{4}\right], \alpha = a \times \cos\left|1 - \frac{h}{D}\right| & 0 < h \leq 2D \end{cases} \end{aligned} \tag{16}$$

where: F is the flow rate through the valve $[m^3/s]$; p is the pressure differential $[Pa]$; C_D is the flow discharge coefficient; $A(h)$ is the instantaneous orifice passage area $[m^2]$; h is the valve opening; D is the orifice diameter $[m]$; p_{cr} is the minimum pressure for the turbulent flow $[Pa]$; D_H is the valve instantaneous hydraulic diameter $[m]$.

4.2 Simulation Results

The aim of the discrete control is to lead the system trajectories from X_0 to X_T in minimum time while avoiding X_F. As a first step, a logic controller is synthesized by using the approach presented in [1, 3]. In the next step, the UKF introduced in Sect. 3 is applied to estimate the states in order to extract the control law of the controller. In addition, an EKF observer is also implemented for this system to compare its estimation performance with the UKF. For this aim, a simplified model was created from the test data of the model in Fig. 1 so that it can calculate the Jacobian matrix. The formula of the EKF used here is presented in [2, 3]. The main parameters of this Simulink model are given as follow: $A_1 = A_2 = A_3 = 0.1[m^2]$, $F_{1in} = F_{3in} = 1 \times 10^{-4}[m^3/sec]$. The constraints on the tank levels are $0 \leq h_1, h_2, h_3 \leq 0.6[m]$. The level in each tank is discretized by set of thresholds $L_i = \{0, 0.1, 0.2, 0.3, 0.4, 0.5, 0.6\}[m]$ with $i = 1, 2, 3$. The initial state is $\underline{x}_0 = [0.27 \quad 0.47 \quad 0.17]^T [m]$.

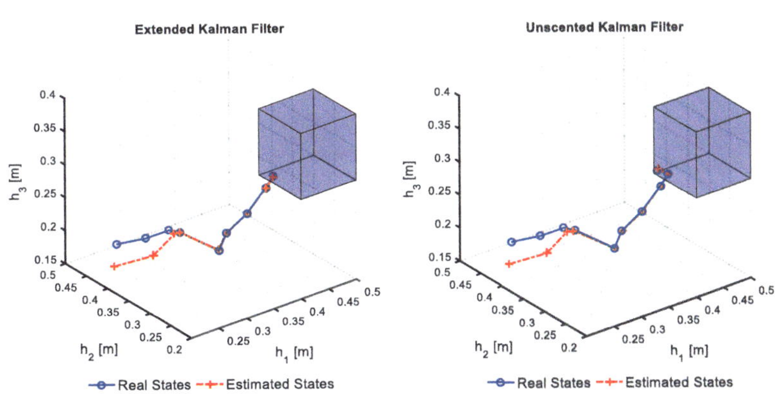

Fig. 2. The system trajectories

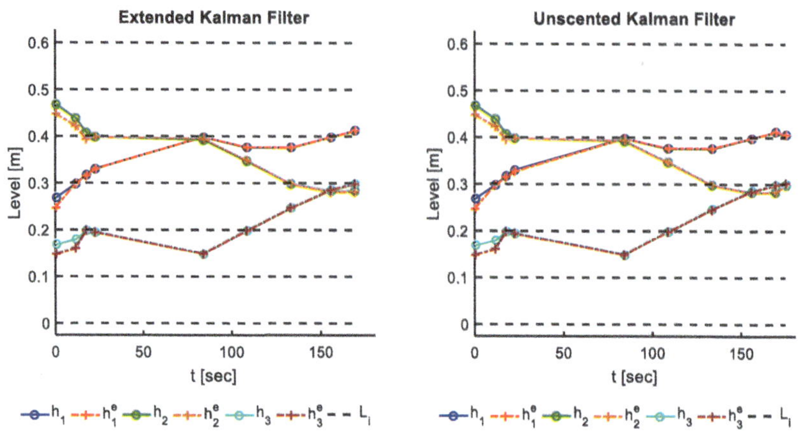

Fig. 3. The continuous states and the estimated states

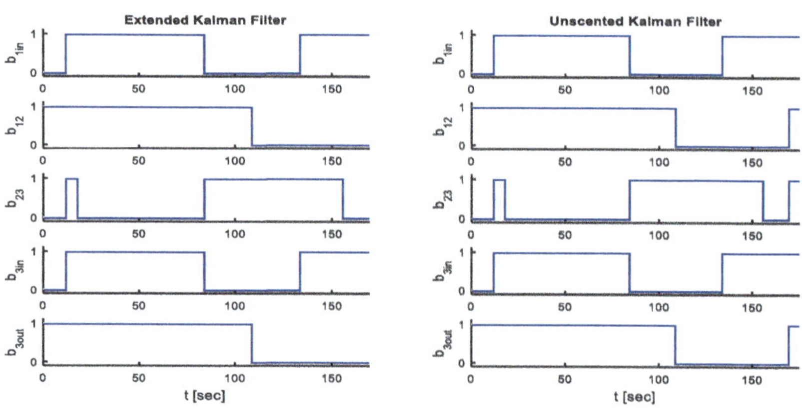

Fig. 4. The controller inputs

The control specifications are specified in term of:

- Initial set X_0: $0.2 \le h_1 \le 0.3[m]$, $0.4 \le h_2 \le 0.5[m]$, $0.1 \le h_3 \le 0.2[m]$.
- Target set X_T: $0.4 \le h_1 \le 0.5[m]$, $0.2 \le h_2 \le 0.3[m]$, $0.3 \le h_3 \le 0.4[m]$.
- Forbidden sets X_F: $0 \le h_1 \le 0.1[m]$, $0.5 \le h_1 \le 0.6[m]$, $0 \le h_2 \le 0.1[m]$, $0.5 \le h_2 \le 0.6[m]$, $0 \le h_3 \le 0.1[m]$, $0.5 \le h_3 \le 0.6[m]$.

The design parameters for the observer are specified as follow: $\underset{-0}{\hat{x}} = [0.25\ \ 0.45\ \ 0.15]^T[m]$, $P_0 = diag([0.05\ \ 0.05\ \ 0.05])$, $Q = diag([0.025^2\ \ 0.025^2\ \ 0.025^2])$, $R = 0.01^2$.

Figure 2 and Fig. 3 show the system trajectories and their estimated states, where the solid lines represent the real values and the dashed lines are the estimated values. Figure 4 shows the control signals extracted from the offline controller for deployment to the system. It can be seen that, under the influence of the controller and UKF observer, the state trajectory has been lead from the initial set to the target set and avoids the forbidden sets. At the same time, the estimated values are close to the real ones after some measurement events. Besides, although the measurements are quite close to the other from the beginning, the control signal does not change too fast, which shows the effectiveness of the offline controller. Comparing to the quality when implementing the EKF set with the same scheme and initialization parameters, the evolution of the system trajectory is almost the same (the trajectory applying UKF has an extra segment at the end to the implementation of EKF but it is insignificant compared to the whole system runtime). This is verified by the same control signals as shown in Fig. 4

5 Conclusions

The paper describes an approach of state estimation for DCCS where the measurements and input signals to plant are discrete values but the dynamics are continuous. The estimator proposed here is a modified version of the UKF. It estimates the continuous state of a system from the sequence of discrete measurements in order to provide a full feedback solution for the controller. This version of UKF uses the same formulas as the original one but the main difference herein is the information provided for the calculation. The numerical simulation results show the feasibility of applying this state estimation method to this class of hybrid dynamic systems. Because of basing on linearization, the EKF can give incorrect information about the estimated state if the dynamics of the system are strongly nonlinear. The UKF does not use linearization methods and has the same performance as the EKF but the design parameters of UKF need to be chosen more carefully. The performances of the EKF and UKF for DCCS depend on the asymptotic stability of the estimation error [8, 13, 16, 17]. Current work focuses on the investigation of this stability problem. Imposing constraints into the estimation schemes is a practical issue. The constrained EKF for DCCS described in [2, 3] is currently considered to extend it to the constrained Moving Horizon Estimation (MHE). The extension of the UKF to include state constraints is of interest [8]. Another state estimation method for DCCS using Particle Filter (PF) which follows the direction of applying numerical transformations to by-pass linearization is also studied [12].

References

1. Tran, T.H., Engell, S.: Design of logic feedback controllers for discretely controlled continuous systems. In: Proceedings of 3rd IFAC Conference on Analysis and Design of Hybrid Systems (ADHS'09), Zaragoza, Spain, 126–131 (2009)
2. Tran, T.H., Engell, S.: Design of state observer for discretely controlled continuous systems. In: Proceedings of UKACC International Conference on Control (Control 2010), Coventry, UK, 1113–1118 (2010)
3. Tran, T.H., Engell, S.: Estimator-based switching controllers for a class of hybrid systems. In: Proceedings of 18th IFAC World Congress (IFAC 2011), Milano, Italy, 6628–6633 (2011)

4. Stursberg, O., Kowalewski, S., Engell, S.: On the generation of timed discrete approximations for continuous systems. Math. Comput. Model. Dyn. Syst. **6**(1), 51–70 (2000)
5. Engell, S., et al.: Continuous-discrete interactions in chemical processing plants. Proc. IEEE **88**(7), 1050–1068 (2000)
6. Lewis, R. Lewis, R.W.: Programming industrial control systems using IEC 1131–3". No. 50. Iet (1998)
7. Julier, S.J., Uhlmann, J.K., Durrant-Whyte, H.F.: A new approach for filtering nonlinear systems. In: Proceedings of the 1995 American Control Conference (ACC 1995), Vol. 3. IEEE (1995)
8. Dymirkovski, G.M.: New results on stochastic stability of discrete-time unscented kalman filter. In: 7th IEEE Conference on Industrial Electronics and Applications (ICIEA) (2012)
9. Simon, D.: Optimal state estimation. In: Kalman, H., Infinity, and nonlinear approaches. John Wiley & Sons (2006)
10. Lewis, F.L., Xie, L., Popa, D.: Optimal and Robust Estimation: with an Introduction to Stochastic Control Theory. CRC Press (2017)
11. Simscape™ Fluids™ Reference, The MathWorks, Inc (2016). https://www.mathworks.com/help/releases/R2016a/pdf_doc/physmod/hydro/hydro_ref.pdf
12. Doucet, A., Johansen, A.M.: A tutorial on particle filtering and smoothing: fifteen years later. Handbook Nonlin. Filter. **12**(3), 656–704 (2009)
13. Daid, A., Busvelle, E., Aidene, M.: On the convergence of the unscented Kalmal filter. Eur. J. Control. **57**, 125–134 (2021)
14. Julier, S.J., Uhlmann, J.K., Durrant-Whyte, H.F.: A new method for the nonlinear transformation of means and covariances in filters and estimators. IEEE Trans. Autom. Control **45**(3), 477–482 (2000)
15. Julier, S.J., Uhlmann, J.K.: Unscented filtering and nonlinear estimation. Proc. IEEE **92**(3), 401–422 (2004)
16. Li, L., Xia, Y.: Stochastic stability of the unscented Kalman filter with intermittent observations. Automatica **48**(5), 978–981 (2012)
17. Reif, K., Unbehauen, R., Yaz, E., Guenther, S.: Stochastic stability of the continuous-time extended Kalman filter. IEEE Proc. Control Theory Appl. **147**(1), 45–52 (2000)
18. Särkkä, S.: On unscented Kalman filtering for state estimation of continuous-time nonlinear systems. IEEE Trans. Autom. Control **52**(9), 1631–1641 (2007)
19. Daid, A., Busvelle, E.,Aidene, M.: Unscented Kalman observer. In: Proceedings of the 8th International Conference on Systems and Control (2019)
20. Teixeira, B.O.S., Torres, L.A.B., Aguirre, L.A., Bernstein, D.S.: Unscented filtering for interval-constrained nonlinear systems. In: 47th IEEE Conference on Decision and Control (CDC), Cancun, Mexico, 5116–5121 (2008)

Stress Corrosion Cracking Behavior in API 5L Steels for Sour Service

A. Blanco[1], J. M. Hallen[1], and Tu Manh Le[2,3](✉)

[1] Departamento de Ingeniería en Metalurgia y Materiales, IPN-ESIQIE, Zacatenco, 07738 Ciudad de México, México
[2] Faculty of Materials Science and Engineering, Phenikaa Institute for Advanced Study (PIAS), Phenikaa University, Hanoi 100000, Vietnam
tu.lemanh@phenikaa-uni.edu.vn
[3] A&A Green Phoenix Group, Phenikaa Research and Technology Institute (PRATI), 167 Hoang Ngan, Hanoi 10000, Vietnam

Abstract. This work presents a study on the behavior of stress corrosion cracking (SCC) phenomenon in pipeline steels for sour service in near-neutral media. SCC tests under static stressing conditions were performed in planar steel samples using a Cortest proof-ring coupled with a fabricated cell containing near-neutral pH solution as a corrosive medium, which simulates the soil conditions, to observe the cracking initiation stage. Characterizations of the steel samples using surface techniques such as SEM, EDS, and Raman spectroscopy show that the presence of nonmetallic inclusions takes place as pitting sites (to form corrosion pits), which constitute the most susceptible sites for SCC crack initiation. SEM images taken on the steel sample surfaces exhibit both SCC transgranular and intergranular crack types in near-neutral media. Although the obtained results are interesting, further study is necessary to evaluate and confirm the role of inclusions in SCC initiation mechanisms.

Keywords: Pipeline steel · Stress corrosion cracking · Initiation stage

1 Introduction

Stress corrosion cracking (SCC) manifests in pipelines under specific conditions, the combined effects of susceptible materials in suitable corrosive environment under a specific stress, promoting the initiation and propagation of cracks through a material [1]. Buried pipelines could develop external SCC, which describes the action of the corrosive environment in the surrounding of the external wall of the pipe. Two types of the external SCC usually occur namely: intergranular cracking at high pH and transgranular cracking at near-neutral pH [2, 3]. The development of SCC arises in four stages: i) the first one is the cracking initiation in the conditions that allows to small cracks arise, ii) these cracks develop a certain size to be considered individual cracks; iii) the third stage is the continuous crack's growth and potential coalesce, and iv) finally, in the fourth stage the largest cracks coalesce, forward through the wall pipe until failure occurs [4]. To control the SCC, understanding the origin of the cracking initiation stage is very important.

D. C. Nguyen et al. (Eds.): ICERA 2021, LNNS 366, pp. 855–861, 2022.
https://doi.org/10.1007/978-3-030-92574-1_87

Anodic dissolution and hydrogen-based embrittlement are the main crack initiation mechanisms acting in SCC [5]. When the pipeline steel is exposured to the corrosive environment, local dissolution of the material occurs. At surface discontinuities such as grain boundaries, dissolved inclusions, and metallurgical defects, it is more successful to occur pitting corrosion, that later with the joint action of the exerted stress the surface arises a micro-strain at the pit bottom leading the pit-to-crack transition [6, 7]. In addition to the mechanism, the susceptibility of the pipeline to SCC is known to be dependent on both material and environment. The characteristics of the material (such as chemical composition, microstructure, the presence and distribution of inclusions and/or precipitates, crystallographic texture) strongly determine its susceptibility to developing SCC [5–9]. According to Asahi et al. [8], homogenous microstructures were more resistant SCC than mixed microstructures. Previous work [9] has shown that texture components such as {001}//ND texture fiber show higher susceptibility to develop SCC in API 5L X52 pipeline steel, while {111} planes (family) are lesser susceptible to form micro-cracks. Moreover, the crack nucleation could be due to high stress concentration regions produced by structural defects such as inclusions [9]. However, experimental evidence of the role that plays inclusions on the initial stage of SCC mechanism is still lacking, and therefore, this is the main purpose of the present work.

2 Materials and Methods

The API 5L X52 steel samples for sour service (SS), which are labeled as X52SS, were extracted from a new pipeline. Chemical composition (in wt.%) of these steels is given as follows: %C: 0.05%, %Mn: 0.63%, %Si: 0.22, %Cr: 0.02, %Ni: 0.02%, %Cu: 0.21, %Mo: 0.01, %Al: 0.04%, %Ca: 0.006, %Fe: Balance. Strain-stress curve was measured in a Shimadzu tensile testing machine, 200kN load. The surface sample for the energy dispersive X-ray (EDS) analysis at inclusions were prepared down to P4000 silica carbide abrasive paper and polished to 25 μm with diamond spray. ASTM E45 was used to determine inclusion content. EDS analysis was carried out in a scanning electron microscope (SEM) JEOL 6300. The near-neutral pH solution, which simulates the soil conditions, was NS4 as referred in literature [10]. The pH of the solution was adjusted near to 6.3, approximately, using deareated conditions by purging the solution with N_2 and CO_2 for 12 h prior to the test and the solution was remained purging through the entire SCC tests. The NS4 solution was prepared with Fermot chemical analytical grade reagents and with distilled water.

SCC tests were performed on a flat tensile (bone shape) sample with the dimensions as reported in [9]. A static load was applied to the steel samples with a Cortest proof-ring throughout all the duration of the experiments. For testing different stressing state, proof ring was calibrated using a strain gauge, to measure the applied stress ($\sigma_{applied}$)-yield stress ($\sigma_{0.2}$) ratio, $\sigma_{applied}/\sigma_{0.2}$ at different immersion time as shown in Fig. 1.

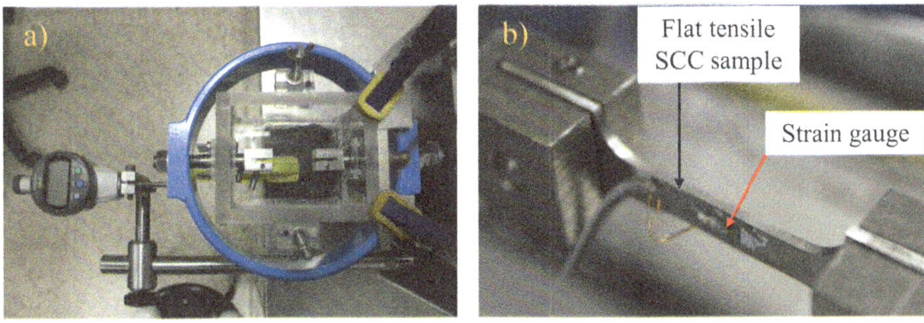

Fig. 1. Cortest proof-ring, a) SCC test and b) stress calibration with a strain gauge.

After testing, the steel samples were removed and cleaned with 0.5%wt ethylenediaminetetraacetic acid (EDTA) solution in an ultrasonic cleaner for 15 min (in three intervals of 5 min for each one), and then examined in a (high-resolution) field emission scanning electron microscope (FESEM), JEOL JSM-6701F. The corrosion products formed on the samples surface were examined by RAMAN spectroscopy, using a dispersive Bruker RAMAN spectrophotometer Senterra II.

3 Result and Discussion

3.1 Material Characterization

Figure 2a shows the microstructure of the steel constructed in three-dimensional (3D) image. It can be seen that the steel exhibits a relatively homogeneous distribution of ferrite and pearlite microstructure with uniaxial grains is shown in, grain size was 7.8 μm, 95% ferrite and 5% perlite. The material exhibits a good mechanical property as shown in Fig. 2b, stress-strain curve measured in a tensile testing machine of X52 SS pipeline steel, from which yield stress (YS) and ultimate tensile stress (UTS) are determined to be 68.6 ksi and 77.2 ksi, respectively. Similar behavior was observed using a SCC proof-ring in elastic region. According to the inclusion contents shown in Table 2, the studied steel mainly presents fine aluminates and fine/coarse globular oxides. Meanwhile, EDS analysis suggests the presence of Al-Ca-O and Al-Ca-O-S nonmetallic inclusions, Fig. 4a and 4b, respectively, in X52 SS steel. Wang *et al.* [10] have shown that complex oxide inclusions can be dissolved due to chemical dissolution, inducing the pit formation in the steel surface. The dissolution mode could have two opposite effects on the development of SCC; on the one hand, the circular shape of these inclusions prevents the cracking initiation, since no stress is concentrated on their walls. Besides, other works [8, 11] have reported that inclusions can act as hydrogen traps, due to their large diameters. In near-neutral pH environments, it generates hydrogen (H) due to water electrolysis, the accumulated hydrogen in pit cavity can promote pit-to-crack transition.

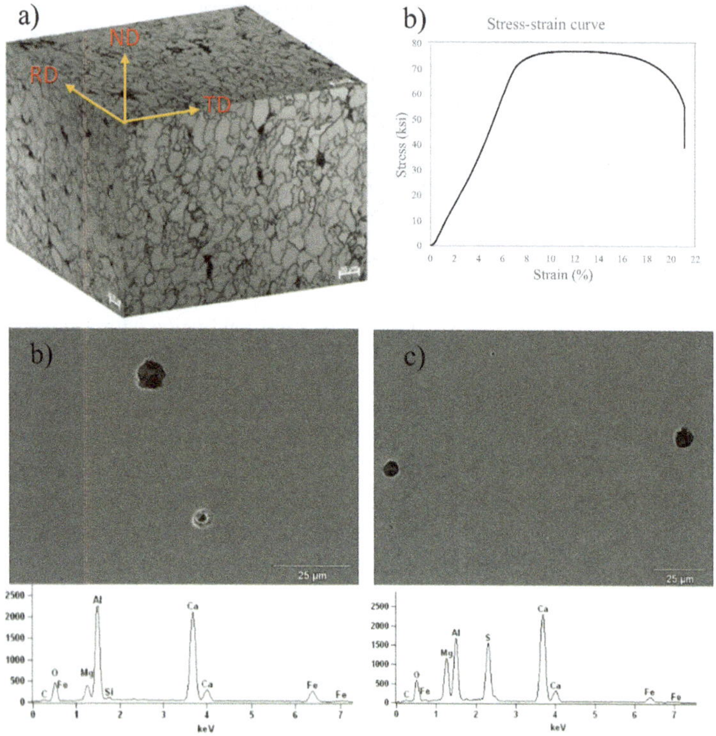

Fig. 2. a) 3D microstructure of API 5L X52 SS steel reconstructed by planes perpendicular to Rolling Direction (RD), Transversal Direction (TD), and Normal Direction (ND), b) Stress-Strain curve of X52 SS steel, c and d) SEM micrographs of inclusions and their respective EDS spectra showing c) Al-Ca-O and d) Al-Ca-O-S nonmetallic inclusions.

Table 1. API 5L SS pipeline steel inclusion contents.

Steel	Inclusion							
	Sulfides		Aluminates		Silicates		Globular oxides	
	Fine	Coarse	Fine	Coarse	Fine	Coarse	Fine	Coarse
X52 SS	0.2	0	0.4	0	0.2	0	1	0.3

3.2 Crack Initiation Stage

The results obtained from different working conditions used to evaluate SCC phenomena in the material are summarized in Table 2. It clearly indicates that the presence pitting corrosion (P) was observed in all the SCC tests and micro-cracks formation after 30 days of immersion and stress above 100% $\sigma 0.2$. These results are analogous to those reported by Zheng et al. [11], where most of SCC cracks in X80 pipeline steel were formed after 54 days of testing in NS4 solution with bubbling constant 5% CO_2

balanced with N_2. Table 2 also suggests that micro-crack initiation in X52 SS steel, at a static load of 116% $\sigma_{0.2}$, requires a minimum immersion time of 60 days.

Table 2. SCC test as a function of immersion time and applied stress.

Element	1	2	3	4	5	6	7
pH	6.48	7.03	6.86	6.38	6.37	6.86	6.71
Immersion time (days)	15	30	30	30	60	60	60
Applied stress ($\sigma_{applied}$/ $\sigma_{0.2}$)%	66.76	69.85	90.49	100.65	116.15	116.15	119.20
Type corrosion	P	P	P	P	P/SCC	P/SCC	P/SCC
Surface cracking	No	No	No	No	Yes	Yes	Yes

Figure 3 clearly shows experimental evidence that most of the observed cracks can be initiated from corrosion pits. According to their shapes, these pits are originated due to the presence of nonmetallic inclusions (see Figs. 2a and 2b) in the steel acting as the pitting sites (see Ref. [12]). This observation is consistent with the inclusion's analysis given in Table 1. From these pits, two cracking types can be easily identified as transgranular and intergranular (grain boundary) crack, since the grain boundaries are clearly defined in the steel's microstructure. Thus, if a crack is nucleated in a grain boundary, then the depth of the crack will increase as it advances. Therefore, the mechanical stress will gradually increase at the crack tip and the transition from intergranular to transgranular crack occurs.

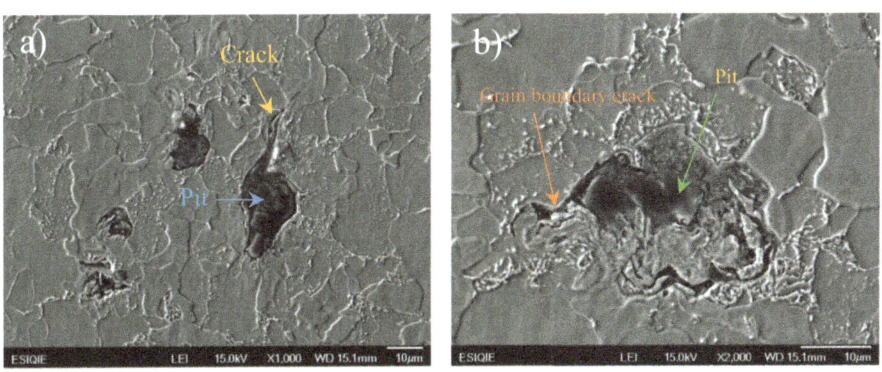

Fig. 3. Crack initiation from pits after 60 immersion days in NS4 solution, a) transgranular crack and b) grain boundary crack.

3.3 SCC Corrosion Products

Figure 4 shows a representative RAMAN spectrum of X52 SS steel corrosion products. Corrosion products are identified as α- Fe_2O_3 (218 cm^{-1}), α-FeOOH (397 cm^{-1}) and green rust (597 cm^{-1}) according with Colomban et al. [13]. Green rust reaffirms the presence of a compound of iron (II) or iron (III) anions and carbonate cation. The formation of iron (II) complex ion in bicarbonate solutions enhances pitting corrosion considered an aeriated solution, oxidation of iron (II) to iron (III), which covers the pits but does not prevent their growth [14]. Furthermore, iron (II) and iron (III) oxides, usually called mill scale, are related to pit formation on the steel surface, due to the galvanic pair between mill scale and bulk material [15]. The results suggest that mill scale formation in the steel generates the formation of pitting on the steel surface, and during the time of SCC process the corrosion reactions could generate carbonate complexes that are deposited later as a film on the pipeline surface. However, the pits continue their growth to a certain size that the action of the exerted stress can promote the pit-crack transition.

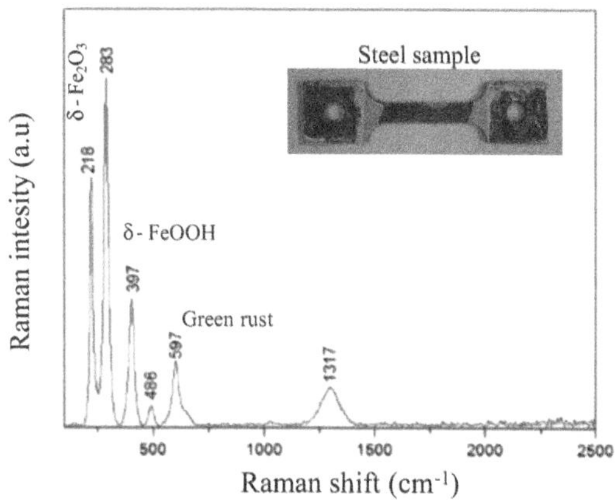

Fig. 4. Raman spectra of corrosion products at surface of X52 SS pipeline steel after 60 days immersed in NS4 solution.

4 Conclusions

SCC behavior of pipeline steel for sour service in near-neutral media has been studied. It was found that SCC crack initiation of the studied steel requires long immersion time in the corrosive environment. The applied stress does not have a direct influence in the cracking initiation, once 100% of the yield stress was exceeded, the same cracking behavior was observed at different immersion times. Microstructural parameters such as grain size, grain boundary and inclusion content play an important role in this first stage of SCC. Although X52 SS steel has a low content of non-metallic inclusions, and mainly of circular shapes, they tend to dissolve easily in the corrosive environment, being the most probable sites of crack initiation, including both transgranular and intergranular types. Furthermore, the formation of corrosion products at steel pipeline surface has a crucial play to SCC initiation process. At first mill scale promote pit corrosion, and then carbonate complexes film are deposited at surface, despite those pits still growth to certain size that allow the pit-crack transition. Finally, the obtained results have demonstrated that nonmetallic inclusions containing in the steel could provide more susceptible sites to SCC, however, further study is necessary to evaluate and confirm the role of inclusions in the SCC initiation mechanisms.

References

1. Zheng, W., Elboujdaini, M., Revie, R.W.: Stress Corrosion Cracking in Pipelines. In: Stress corrosion cracking, pp. 749–771. Elsevier (2011). https://doi.org/10.1533/9780857093769.4.749
2. Zhang, X.-Y., Lambert, S.B., Sutherby, R., Plumtree, A.: Transgranular stress corrosion cracking of X-60 pipeline steel in simulated ground water. Corrosion **55**(3), 297–305 (1999)
3. Ryakhovskikh, I.V., Bogdanov, R.I., Ignatenko, V.E.: Intergranular stress corrosion cracking of steel gas pipelines in weak alkaline soil electrolytes. Eng. Fail. Anal. **94**, 87–95 (2018)
4. Elboujdaini, M., Wang, Y.-Z., Revie, R. W.: Initiation of stress corrosion cracking on x-65 linepipe steels in near-neutral pH environment. (2000). http://www.asme.org/about-asme/terms-of-use
5. Mohtadi-Bonab, M.A., Ghesmati-Kucheki, H.: Important factors on the failure of pipeline steels with focus on hydrogen induced cracks and improvement of their resistance: review paper. Met. Mater. Int. **25**(5), 1109–1134 (2019). https://doi.org/10.1007/s12540-019-00266-7
6. Zhu, M., et al.: Effect of strength and microstructure on stress corrosion cracking behavior and mechanism of X80 pipeline steel in high pH carbonate/bicarbonate solution. J. Mater. Eng. Perform. **23**(4), 1358–1365 (2014). https://doi.org/10.1007/s11665-014-0880-4
7. Asahi, H., Kushida, T., Kimura, M., Fukai, H., Okano, S.: Role of microstructures on stress corrosion cracking of pipeline steels in carbonate-bicarbonate solution. Corrosion **55**(7), 644–652 (1999)
8. Wang, L., et al.: Influence of inclusions on initiation of pitting corrosion and stress corrosion cracking of X70 steel in near-neutral pH environment. Corros. Sci. **147**, 108–127 (2019). https://doi.org/10.1016/j.corsci.2018.11.007
9. Blanco, A., Hallen, J.M., Nguyen, T.S., Le Manh, T.: Influence of crystallographic texture on susceptibility to stress corrosion cracking mechanism of API 5L X52 steel for sour service. Eng. Fail. Anal. **119**, 105002 (2021)
10. Lizárraga, C.R., Pérez, T., Galván-Martinez, R., Quej-Aké, L.M., Contreras, A.: Stress-corrosion-cracking assessment of X52 pipeline steel exposed to mexican soils. Materiali in Tehnologije **53**(1), 81–86 (2019). https://doi.org/10.17222/MIT.2018.018
11. Zheng, W., et al.: Near-neutral pH SCC of two line pipe steels under quasi-static stressing conditions. (2006). http://www.asme.org/about-asme/terms-of-use
12. Madrigal-Cano, M., Hallen, J.M., Arce-Estrada, E.M.: TuLe Manh, "Effect of crystallographic texture and microstructure on pitting corrosion behavior of low carbon steels: a Monte Carlo model." Comput. Mater. Sci. **161**, 394–402 (2019). https://doi.org/10.1016/j.commatsci.2019.02.016
13. Colomban, P., Cherifi, S., Despert, G.: Raman identification of corrosion products on automotive galvanized steel sheets. J. Raman Spectrosc. **39**(7), 881–886 (2008). https://doi.org/10.1002/jrs.1927
14. Davies, D.H., Burstein, G.T.: The effects of bicarbonate on the corrosion and passivation of iron. Corrosion **36**(8), 416–422 (1980)
15. Wang, S., Lamborn, L., Chevil, K., Gamboa, E., Chen, W.: On the formation of stress corrosion crack colonies with different crack population. Corros. Sci. **168**, 108592 (2020). https://doi.org/10.1016/j.corsci.2020.108592

Study on Minimizing Cleaning Cost in Abrasive Blasting with Quartz Sand Using Aluminium Oxide Nozzle

Trinh Kieu Tuan[1], Do The Vinh[2], Tran Quoc Hoang[3], Nguyen Huu Quang[1], Le Hong Ky[4], Le Hoang Anh[4], Vu Ngoc Pi[2], and Nguyen Thanh Tu[2(✉)]

[1] Vinh Long University of Technology Education, Vĩnh Long, Vietnam
[2] Thai Nguyen University of Technology, Thai Nguyen, Vietnam
nguyenthanhtucnvl@tnut.edu.vn
[3] Nguyen Tat Thanh University, Ho Chi Minh City, Vietnam
[4] University of Economics - Technology for Industries, Hanoi, Vietnam

Abstract. In industry, sandblastinghas been used with significant demand for surface preparation for painting, engraving, etc. During sandblasting, the replacement of the nozzle is a regular practice. However, optimizing the nozzle replacement has not been given sufficient importance. In this work, the determination of optimum replaced aluminum oxide nozzle diameter for minimizing cleaning cost was carried out. Three main input parameters, including the time for changing a nozzle, the nozzle wear rate per hour, and the compressor power, were selected to determine their impact on the cleaning costs. Besides, the Taguchi method was used for the experimental design and results' analysis. The results show that the compressor power is the factor having the greatest impact on cleaning costs. Furthermore, the minimum cleaning cost were determined by the optimal parameters of the sandblasting process. The results of this study are a good experience for sandblasting processes using aluminum oxide nozzles.

Keywords: Abrasive blasting · Sand blasting · Replaced nozzle diameter · Aluminum oxide nozzle · Taguchi method

1 Introduction

Invented by Benjamin C. Tilghman in 1870 [1], sandblasting has become a very popular process for surface preparation of processes such as welding, painting, engraving, etc. [2]. Sand blasting technology adversely affects the health of workers because of the dust it produces. However, it is still widely used in industry because of its undisputed advantages [3, 4]. The demand for abrasives around the world is increasing year by year. In Global Abrasives Market Report 2020 [5], world demand for abrasives will increase from $8.6 billion in 2014 in 2014 to $11.4 billion in 2019. Within Europe alone, the demand for blasting and related abrasive materials is over 850,000 tonnes [6]. Certainly, the demand is increasing. Hence, the cost decrease of the abrasive process attracts not only manufacturers but also researchers.

D. C. Nguyen et al. (Eds.): ICERA 2021, LNNS 366, pp. 862–871, 2022.
https://doi.org/10.1007/978-3-030-92574-1_88

In [7], the optimization of replaced tungsten carbide nozzle diameter during abrasive blasting with quartz sand was carried out with the goal of reducing cleaning costs. The authors have found the minimum value of the cleaning cost obtained by determining the appropriate replaced nozzle diameter. In a study by Vu Ngoc Pi and A. M. Hoogstrate, the minimum value of the cost was determined by the optimization of the sandblasting parameters [8]. In another study by the same author, profit optimization of the sandblasting process was performed [2]. The influence of major factors such as the air pressure, the initial nozzle diameter, the nozzle wear, etc., on the outlet response has been introduced. Accordingly, the profit rate, as well as the blasting time, has been reduced significantly. A cost comparison study in sandblasting was conducted by James D. Hansink. In the study, the author concludes that the garnet was recycled three times after blasting, and the first recycled garnet used could receive the lowest cost [3]. Effects of four input parameters, including the nozzle diameter, the air pressure, the cost components, and the wear of the nozzle on the cleaning cost in the study of Vu et al. [4]. The cleaning time and the cleaning costs calculated by the proposed model have been significantly reduced.

The Taguchi method is a powerful yet simple statistical analysis method. Many researchers have applied Taguchi to optimize process parameters to reduce machining costs and time [5–7, 9]. In this method, orthogonal arrays are used for experimental design to obtain good results with the smallest number of test runs. The signal to noise (S/N) ratio is calculated to measure the performance characteristics and to give the contribution of the process parameters to the output response. Depending on the output responses, three types of S/N ratio, including the-bigger-is-the-better, the-smaller-is-the-better, and the-nominal-is-the-better, will be selected accordingly [6, 10].

In this study, the optimum replaced aluminum oxide nozzle diameter for minimizing cleaning cost was conducted. Three main input parameters, such as the nozzle wear rate per hour, the nozzle changing time, and the compressor power, were selected to determine their impact on the cleaning costs. The Taguchi method was applied for the experimental organization and statistical analysis.

2 Cost Analysis and Optimization

During the blasting process, the cleaning cost C_{cl} (USD/m^2) is determined by the Eq. (1):

$$C_{cl} = \left(C_{m,h} + C_{a,h} + C_{n,h}\right)/v_{cl} \tag{1}$$

Where, $C_{m,h}$, $C_{a,h}$ and $C_{n,h}$ are the cost of cleaning equipment, the cost of sands, and the nozzle cost per hour (USD/h), respectively; v_{cl} is cleaning rate (m^2/h).

Equation (2) provides a calculation for the abrasive cost:

$$C_{a,h} = 3600 \cdot \dot{m}_a \cdot C_{a,m} \tag{2}$$

In which, $C_{a,m}$ is the abrasive cost per kilogram (USD/kg); $\dot{m}_a$ is the abrasive mass flow rate (kg/h) determined by the Eq. (3) with the determination coefficient $R^2 = 0.998$:

$$\dot{m}_a = 15.14 \cdot d_n^{-0.0214} \cdot P^{1.0359} \quad (3)$$

The nozzle cost per hour is calculated by the following equation:

$$C_{n,h} = C_{n,p}/L_n \quad (4)$$

Where $C_{n,p}$ is the cost of a nozzle (USD/piece); L_n is the nozzle lifetime (h).

For aluminum oxide nozzle, the nozzle lifetime is from 10 to 30 (h) [11]. Also, a nozzle will be replaced when it wears 1/16-in. (or 1.59 mm) beyond its original size [12]. Therefore, the wear rate WR (mm/h) can be determined by (5):

$$WR = 1.59/L_n = 1.59/(10 \div 30) \approx 0.053 \div 0.159 \quad (5)$$

From the data in [11], the following regression equation (with $R^2 = 0.9984$) was used to calculate the cleaning rate v_{cl}:

$$v_{cl} = 6.6101 \cdot 10^{-6} \cdot d_n^{2.0121} \cdot p^{1.5988} \quad (6)$$

Where p is the air pressure (kPa) which can be determined by the following regression equation (with $R^2 = 0.9774$):

$$p = 3854.7 \cdot d_n^{-2.5783} \cdot P^{1.1801} \quad (7)$$

Wherein P is the power (kW); d_n is nozzle diameter (mm).

From Eq. (7), it was found that the air pressure depends strongly on the nozzle diameter (Fig. 1). Also, the air pressure depends inversely on the nozzle diameter (or the nozzle lifetime). From Eq. (6), it can be noticed that the cleaning rate depends on nozzle diameter and air pressure. When the nozzle diameter increases, the cleaning rate grows significantly (Fig. 2).

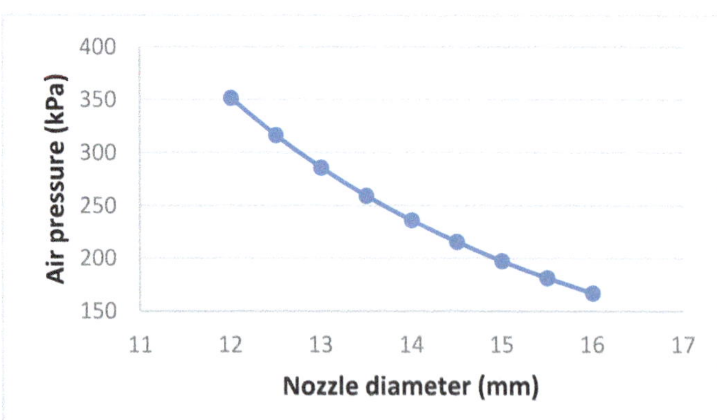

Fig. 1. Air pressure versus nozzle diameter

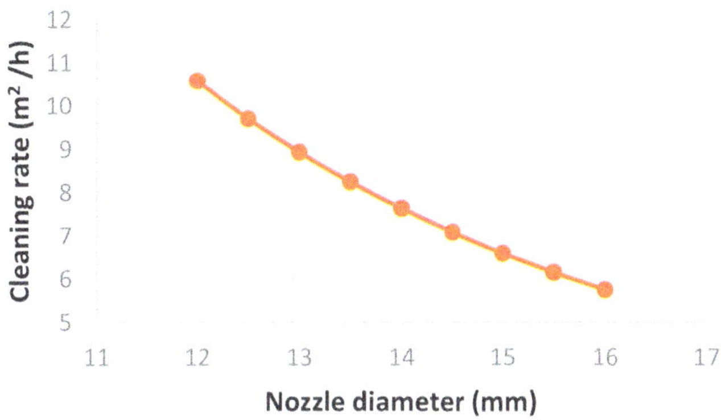

Fig. 2. Cleaning rate versus nozzle diameter

With $C_{m,h}$ = 10 (USD/h); $C_{a,h}$ = 1 (USD/kg); $C_{n,p}$ = 200 (USD/piece); WR = 0.002 (mm/h); P = 30 kW, the relationship between the cleaning cost (USD/m^2) and nozzle diameter are shown in Fig. 3. It is found that the cleaning cost depends powerfully on the replaced nozzle diameter. Besides, there is an optimum replaced nozzle diameter (d_{op}) at which the cleaning cost is minimal. If applying the optimum replaced nozzle diameter (in this case d_{op} = 12.2 mm), the cleaning cost is much smaller (22.56%) compared to replacing the nozzle in the normal way (normally replaced nozzle diameter is 13.9 mm).

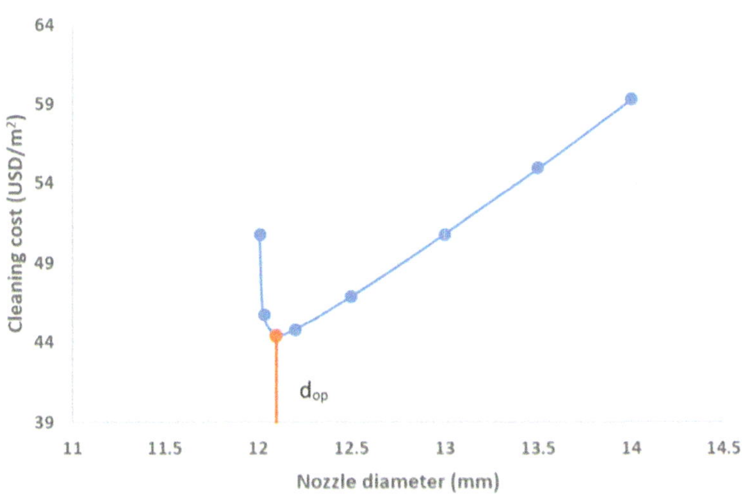

Fig. 3. Cleaning cost versus nozzle diameter

The optimization problem of cleaning cost is solved by programming. In this study, three input parameters, including the air pressure, the time for changing a nozzle, and the average nozzle wear rate were selected for determining the optimum replaced nozzle diameter. Besides, other input parameters are kept unchanged. These parameters include the initial nozzle diameter (D_0 = 12.5 mm), the machine cost, including the labor cost and the overhead cost ($C_{m,h}$ = 12 USD/h), the sand cost ($C_{a,h}$ = 0.2 USD/h), and the nozzle cost per piece ($C_{n,p}$ = 100 USD/h). The optimal problem is given as the following objective function:

$$\min C_l = f(d_n) \tag{8}$$

With the following constraints:

$$0.004 \leq WR \leq 0.0053 \tag{9}$$

$$4 \leq P_c \leq 80 \tag{10}$$

$$1 \leq t_{cn} \leq 15 \tag{11}$$

3 Experimental Details

In the study, sandblasting processes with the aluminum oxide nozzle were conducted to find out the effects of parameters details including nozzle wear rate per hour, time for changing a nozzle, and compressor power on the cleaning cost per square meter. Tests were carried out to find the smallest cost with an initial nozzle diameter d0 = 12.5 (mm). The parameters with levels are shown in Table 1.

Table 1. The parameters with levels

Factor	Code	Level 1	Level 2	Level 3
Nozzle wear rate	WR (mm/h)	0.053	0.106	0.159
Nozzle changing time	t_{cn}(min)	1	5	10
Compressor power	P (kW)	340	685	1030

The experimental design and data analysis were conducted by using the Minitab R18 software. In experimental design, the orthogonal array L9(3^^{3}) was performed. In all experiments, the initial nozzle diameter d_0 is 12.5 (mm). As mentioned above, other sandblasting parameters were fixed, including the machine cost of 12 (USD/h), the nozzle cost of 50 (USD/piece), and the cost of sand of 0.2 (USD/kg).

4 Experimental Results and Discussions

The optimum replaced aluminum oxide nozzle diameter (D_{op}) and the cleaning cost (Ccl) obtained from the experiments are shown in Table 2. The S/N is given by Minitab software. The purpose of this study is to reduce cleaning costs. Thus, the smaller - the – better type of S/N ratio was selected and calculated as the Eq. (12):

$$S/N = -10Log_{10}\left(\frac{1}{n}\sum_{i=1}^{n} y_{ij}^{2}\right) \tag{12}$$

Where n is the number of repetitions at each experiment, y_{ij} is the observed data.

Table 3 shows the response table for means. The ranked highest of each level of sandblasting parameters is the first level, the first level, and the third level for the nozzle wear rate per hour (WR), the nozzle changing time (t_{cn}), and the compressor power (P), respectively. Consequently, the optimal condition of the experiment will be (1-1-3). The rank of the compressor power is "1" meaning that compressor power (P) is the most influential factor on the cost for cleaning one square meter (C_{cl}). The second factor affecting cleaning cost is the nozzle wear rate per hour (WR).

The analysis of variance for means of cleaning cost is shown in Table 4. The compressor power is the most influential input factor on cleaning cost. It contributes 99.97% to the total effect. The nozzle wear rate per hour and the time for changing a nozzle have a little influence. The P values of three factors less than 0.05 mean that the effect of these parameters has statistical significance. The coefficient of determination R-squared of 100% means that 100% of the variability in the cleaning cost can be explained by the input factor in this study.

Table 2. The result of the experiment

Run	WR	t_{cn}	P	D_{op} (mm)	C_{cl} (USD/m2)	S/N
1	0.053	1	340	14.17	9.796	−19.8209
2	0.053	5	685	13.54	5.682	−15.0901
3	0.053	10	1030	13.55	4.183	−12.4304
4	0.106	1	685	13.66	5.691	−15.1034
5	0.106	5	1030	13.69	4.195	−12.4539
6	0.106	10	340	15.17	9.900	−19.9129
7	0.159	1	1030	13.6	4.184	−12.4306
8	0.159	5	340	15.28	9.906	−19.9183
9	0.159	10	685	14.6	5.790	−15.2543

Table 3. Response table for means

Response Table for Means

Level	WR	tcn	P
1	**6.554**	**6.557**	9.867
2	6.595	6.594	5.721
3	6.627	6.625	**4.187**
Delta	0.073	0.068	5.680
Rank	2	3	1

Figure 4 indicates the main effects plot for the mean of cleaning cost. Figure 4 shows that the optimal parameters of sandblasting process for minimum cleaning cost are the nozzle wear rate per hour of 0.053 (mm/h), the time for changing a nozzle of 1.0 (min), and the compressor power of 1030 (kW). In addition, the large slope in the plot of P shows that P is the factor having the most significant influence on the response.

Table 4. Analysis of Variance for Means.

Analysis of Variance for Means

Source	DF	Seq SS	Adj SS	Adj MS	F	P
WR	2	0.0080	0.0080	0.0040	157.97	0.006
tcn	2	0.0070	0.0070	0.0035	136.57	0.007
P	2	51.8122	51.8122	25.9061	1017862.31	0.000
Residual Error	2	0.0001	0.0001	0.0000		
Total	8	51.8272				

Model Summary

S	R-Sq	R-Sq(adj)
0.0050	100.00%	100.00%

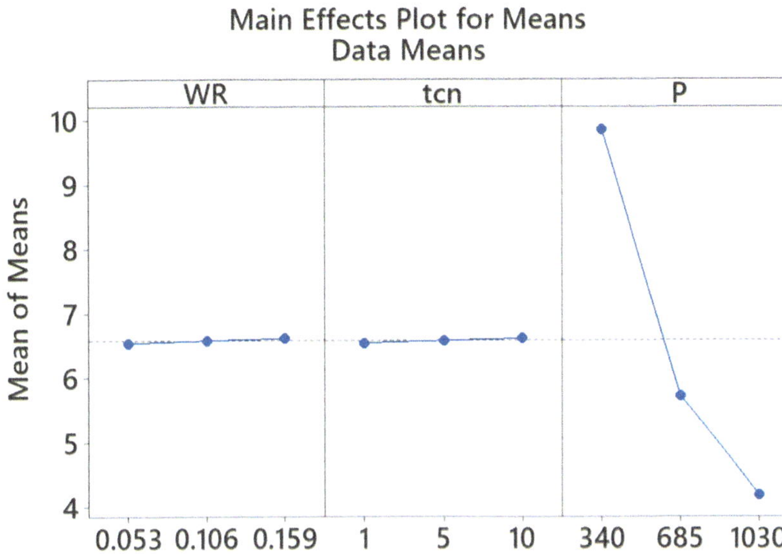

Fig. 4. Main effects plot for the mean of cleaning cost

A model was presented for calculating the optimum cost for cleaning one square meter as the following equation:

$$\overline{C_{clop}} = \overline{WR_1} + \overline{tcn_1} + \overline{P_3} - 2.\overline{T_{gg}} \tag{13}$$

Where $\overline{WR_1}$ is the average cleaning cost for WR at level 1, $\overline{tcn_1}$ is the average cleaning cost for tcn at level 1, $\overline{P_3}$ is the average cleaning cost for P at level 3, $\overline{T_{gg}}$ is the the average cleaning cost for all experiments.$\overline{T_{gg}}$ is determined by (14):

$$\overline{T_{gg}} = \frac{\sum_1^{18} C_{cl}}{9} = 6.592\ USD/m^2 \tag{14}$$

By (13):

$$\overline{C_{clop}} = 4.114\ USD/m^2$$

The optimum replaced nozzle diameter (D_{op}) with minimum cost for cleaning one square meter is calculated by the following equation:

$$\overline{D_{op}} = \overline{WR_1} + \overline{tcn_1} + \overline{P_3} - 2.\overline{T} \tag{15}$$

Where $\overline{WR_1}$ is the average cleaning cost for WR at level 1, $\overline{t_{cn1}}$ is the average cleaning cost for tcn at level 1, $\overline{P_3}$ is the average cleaning cost for P at level 3, $\overline{T}$ is the the average replaced nozzle diameter for all experiments. $\overline{T}$ is calculated by Eq. (16) as the following:

$$\overline{T} = \frac{\sum_{1}^{18} D_{op}}{9} = 14.14\ mm \tag{16}$$

By (15), the optimum replaced nozzle diameter (D_{op}) with minimum cost for cleaning one square meter is determined as the following:

$$D_{cp} = 12.89\,\text{mm}$$

A verification experiment was performed with the optimum input parameters. The values of the input factor were used in the experiment as follows: the nozzle wear rate per hour of 0.053 (mm/h); the time for changing a nozzle of 1.0 (min); and the compressor power of 1030 (kW). The experimental results are indicated in Table 5. The difference between the calculated value and the experimental value is 0.014% and 1.94% for C_{clop} and D_{op}, respectively. That evidences that the results of this study are highly reliable.

Table 5. The result of the verification experiment

Response	Result		
	Calculated	Experiment	% error
	$WR_1/t_{cn1}/P_3$	$WR_1/t_{cn1}/P_3$	
C_{clop}	4.114	4.11458	0.014
D_{op}	12.89	13.14	1.94

5 Conclusion

In the current study, a study on optimization of parameters of abrasive blasting with quartz sand using aluminum oxide nozzle for minimum cost for cleaning one square meter was conducted. Additionally, the optimum replaced nozzle diameter with minimum cleaning cost was also determined. These conclusions have been drawn:

- With a specific sandblasting process, an optimal replaced nozzle diameter value is confirmed, for which the cleaning cost is minimal.
- To minimize the cleaning cost, the optimal parameters of abrasive blasting applied are the nozzle wear rate per hour of 0.053 (mm/h), the time for changing a nozzle of 1.0 (min), and the compressor power of 1030 (kW).
- The compressor power is the most influential input factor on cleaning cost. It contributes 99.97% on the total effect. The influences of other factors are negligible.
- The verification experiment has proven that the suitability of the experimental data and the model is reliable.

Acknowledgment. This work was supported by Thai Nguyen University of Technology.

References

1. Tilghman, B.C.: Sandblasting. US Patent No. 108, 408 (1870)
2. Brantley, C.D., Reist, P.C.: Abrasive blasting with quartz sand: factors affecting the potential for incidential exposure to respirable silica. Am. Ind. Hyg. Assoc. J. **55**, 946–952 (1994)
3. Vu, N.-P., Le, X.-H., Nguyen, D.-N., Luu, A.-T., Nguyen, T.-T., Tran, N.-G., et al.: Optimizing replaced nozzle diameter of abrasive blasting systems using experiment technique design. Appl. Sci. **10**, 3920 (2020)
4. Kura, B., Kambham, K., Sangameswaran, S., Potana, S.: Atmospheric particulate emissions from dry abrasive blasting using coal slag. J. Air Waste Manag. Assoc. **56**, 1205–1215 (2006)
5. G. A. M. R.: 2020. (accessed on 15th January 2021). https://www.esherpamarketreports.com/reports/%EF%BB%BFglobal-abrasives-market-report-2020/
6. Briggs, J., and T. GIESE: The Abrasives Industry in Europe and North America: A Market/Technology Report, Abrasive Engineering Society (2009)
7. Tran, T.-H., Do, T.-V., Tran, N.-G., Vu, N.-P.: Optimization of replaced nozzle diameter for minimum cleaning cost during abrasive blasting with quartz sand. Int. J. Mech. Prod. Eng. Res. Dev. (IJMPERD) **10**, 813–822 (2021)
8. Tilghman, B.C.: Sandblasting, US Patent No. 108, 408 (1870)
9. Pi, V.N., Hoogstrate, A.: Cost optimization of abrasive blasting systems: A new and effective way for using blasting nozzles. In: Key Engineering Materials, pp. 323–328 (2007)
10. Pi, V.N.: Profit optimization of abrasive blasting systems. In: Key Engineering Materials, pp. 392–397 (2009)
11. Hansink, J.D.: Economics of abrasive selection for shipyard use. Protective Coatings Europe **3**, 25 (1998)
12. Pi, V.N., Chinh, P.C.: A study on cost optimization of steel shot blasting systems. In: Applied Mechanics and Materials, pp. 568–572 (2011)
13. Phadke, M.S.: Quality engineering using design of experiments. In: Quality control, robust design, and the Taguchi method, pp. 31–50. Springer (1989)
14. Do, T.-V., Hsu, Q.-C.: Optimization of minimum quantity lubricant conditions and cutting parameters in hard milling of AISI H13 steel. Appl. Sci. **6**, 83 (2016)
15. Do, T.-V., Vu, N.-C., Nguyen, Q.-M.: Optimization of cooling conditions and cutting parameters during hard milling of AISI H13 steel by using Taguchi method. In: 2018 IEEE International Conference on Advanced Manufacturing (ICAM), pp. 396–398 (2018)
16. Phan, T.-D., Do, T.-V., Pham, T.-L., Duong, H.-L.: Optimization of cutting parameters and nanoparticle concentration in hard milling for surface roughness of JIS SKD61 steel using linear regression and taguchi method. In: Sattler, K.-U., Nguyen, D.C., Vu, N.P., Long, B.T., Puta, H. (eds.) ICERA 2020. LNNS, vol. 178, pp. 628–635. Springer, Cham (2021). https://doi.org/10.1007/978-3-030-64719-3_69
17. Fratila, D., Caizar, C.: Application of Taguchi method to selection of optimal lubrication and cutting conditions in face milling of AlMg3. J. Clean. Prod. **19**, 640–645 (2011)
18. http://www.nortonsandblasting.com/nsbnozzles2.html. Accessed 15 Apr 2020
19. https://clemcoindustries.com/products/accessories/boron-carbide-lined-metal-jacketed-bsd-series/. Accessed 15 Apr 2020

Study on Productivity Improvement When Turning AISI 1045 Steel on Basis of Surface Roughness Assurance

Le Hoang Anh[1], Nguyen Hong Linh[2], Dang Quoc Cuong[3], Bui Thanh Danh[4], Nguyen Huu Quang[5], and Nguyen Anh Tuan[5](✉)

[1] Vinh Long University of Technology Education, Vinh Long, Vietnam
[2] Electric Power University, Hanoi, Vietnam
[3] Nguyen Tat Thanh University, Ho Chi Minh City, Vietnam
[4] University of Transport and Communications, Hanoi, Vietnam
[5] University of Economics - Technology for Industries, Hanoi, Vietnam
Natuan.ck@uneti.edu.vn

Abstract. AISI 1045 steel is considered the most popular steel type for machining mechanical products which is most often used to manufacture parts such as shafts, gears, forks, etc., … Therefore, the improvement of productivity and quality when machining this steel type is highly necessary to be studied. In this article, a study to improve the productivity when turning this steel type on the basis of ensuring the minimum surface roughness is presented. The experiments have been conducted in the order of Central Composite Design (*CCD*) matrix on CNC lathes and TiAlN coated inserts. From the experimental results, Response Surface Methodology (*RSM*) has been applied to build a regression model representing the relationship between surface roughness and cutting velocity, feed rate and cutting depth. Sub-regression models representing the relationship between surface roughness and each cutting parameter have also been set up. Since then, the charts showing the effect of each parameter on surface roughness have been built, as well as the values of cutting parameters have been determined to improve machining productivity while ensuring the minimum surface roughness. A number of verification experiments have been conducted and it is found that it is possible to increase the productivity by approximately 2 times, while the surface roughness is merely increased by 0.08 μm.

Keywords: Turning · AISI1045 steel · Surface roughness · Productivity improvement · RSM

1 Introduction

Turning is considered to be the most common method among cutting machining methods. This method accounts for about 40% the machining volume of mechanical products [1]. Therefore, the study for finding solutions to improve machining productivity and product quality when machining by means of turning method has been being carried out by numerous scientists. Among the parameters to evaluate the quality

D. C. Nguyen et al. (Eds.): ICERA 2021, LNNS 366, pp. 872–881, 2022.
https://doi.org/10.1007/978-3-030-92574-1_89

of product, surface roughness is the most common parameter chosen by the authors as the evaluation criterion. It is also easy to understand because the surface roughness has a great effect on workability and durability of the product, and the evaluation of surface roughness is not too difficult [2, 3].

Shihab et al. [4] have applied the RSM method to determine the values of cutting velocity, feed rate and cutting depth to ensure the minimum surface roughness when turning AISI 52100 steel. This work has also been performed by a number of authors in a similar manner when turning a number of different materials such as: Izelu et al. [5] when turning 41Cr4 steel; Xiao et al. [6] when turning 1Cr18Ni9Ti steel; Yusuf et al. [7] when turning 7050 aluminum alloy; Bhushan [8] when turning AA7075/Sic Composite; Laghari et al. [9] when turning Al/SiCp materials; Arfaoui et al. [10] when turning AISI 1040 steel; Nayak et al. [11] when dry turning AISI 203 steel; Sahoo et al. [12] when turning AISI 1040 steel; Tank et al. [13] when turning Glass Fiber Reinforced Plastic (GFRP); etc.

The RSM method has also been used to determine the values of other parameters. Makadia et al. [14] have determined the optimal values of tool nose radius, cutting velocity, feed rate and cutting depth to ensure the minimum surface roughness when turning AISI 140 steel; Kuntoglu et al. [15] have determined the values of cutting edge angle, cutting velocity and feed rate to ensure the minimum surface roughness when turning AISI 5140 steel, etc.

The RSM method has been applied with a number of other methods to determine the values of cutting velocity, feed rate and cutting depth in order to simultaneously ensure the minimum surface roughness and the maximum Material Removal Rate (MRR) in some studies such as: Jayaraman et al. [16] have combined RSM method and Genetic Algorithm (GA) when turning AA6351 materials; Nayak et al. [11] have also combined RSM method and Grey Relational Analysis (GRA) when turning AISI 304 steel; Dhabale et al. [17] have used a combination of RSM and Non-Dominated Sorted Genetic Algorithm (NDSGA) when turning AlMg1SiCu materials; etc.

A number of studies as mentioned above show that the RSM method has been successful in determining the values of the parameters of machining process to ensure the minimum surface roughness. The RSM has also been successful when being combined with some other methods to determine the values of the parameters of machining process to simultaneously ensure the minimum surface roughness and the maximum MRR. However, it can be seen that the methods (algorithms) having been combined with the RSM in those studies are quite complicated methods (algorithms) that are not familiar to all researches and especially, they are quite unfamiliar to the technicians at the factories. Therefore, it is necessary to find a simple method to combine with the RSM. In this study, on the basis of RSM method, building sub-regression models to determine the values of parameters has been performed. This method avoids using complicated algorithms as in published studies.

2 Turning Experiment

Leadwell TP5-220L CNC lathe has been used to conduct the experiments (Fig. 1). Mazatrol operating system has been installed in the software of machine to control the machine to comply with the requirements of the operator. The accuracy of movements of the machine reaches an accuracy of 0.015 μm.

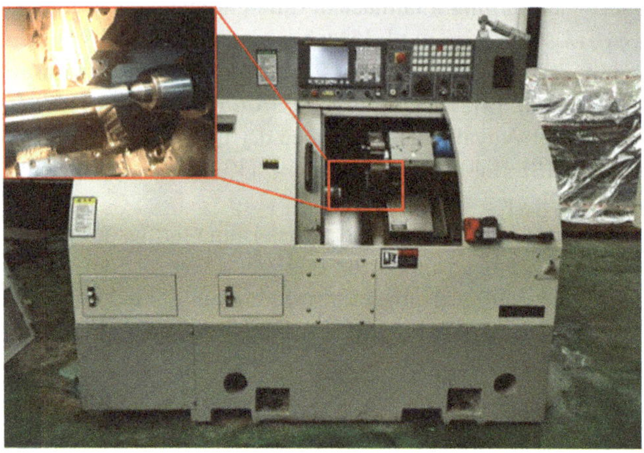

Fig. 1. TP5-200L CNC lathe

TiAlN coated inserts have been used during the experiment with a tool nose radius of 0.5 mm. In order to eliminate the effect of tool wear on experimental results, each insert is used only once. The surface roughness has been measured with SJ301 machine with the set standard length of 2.5 mm. During the measurement, the direction of displacement of the probe is parallel to the workpiece centerline, i. e. perpendicular to the cutting velocity vector. The roughness at each experiment is calculated as the average value of at least three consecutive measurements. Its purpose is to reduce the effect of random errors on measurement accuracy.

Cutting velocity, feed rate and cutting depth are the three parameters chosen as input parameters for the experiment. At each experiment, the values of these parameters will be adjusted in accordance with the intention of experimenter. The CCD design has been used for building the experimental matrix. According to this design, each input parameter will have five levels of value corresponding to the encoding values of $-\alpha$, $-1, 0, 1$ and α. In which, $\alpha = (2k)1/4$ where k is the number of input parameters. Thus, when $k = 3$, then $\alpha = 1.682$. The number of experiments of the matrix includes 2k experiments at the levels of $\pm$ 1, 2k experiments at the levels of $\pm$ α and 6 experiments at the levels of 0 [18]. The values of input parameters at the levels are presented in Table 1 [19]. The experimental matrix with a total of 20 experiments is presented in Table 2.

Table 1. Values of input parameters at the levels

Parameters	Actual symbol	Unit	Code	Value at levels				
				−1.682	−1	0	1	1.682
Cutting speed	v_c	m/min	x_1	66.96	80	105	130	147.04
Feed rate	f_d	mm/rev	x_2	0.056	0.07	0.09	0.11	0.124
Depth of cut	a_p	mm	x_3	0.332	0.4	0.5	0.6	0.668

Table 2. Experimental matrix and results

No.	*Code value*			*Actual value*			Ra (μm)
	x_1	x_2	x_3	v_c *(m/min*	f_d *(mm/rev)*	a_p *(mm)*	
1	1	−1	−1	130	0.07	0.4	0.242
2	1	1	−1	130	0.11	0.4	1.403
3	−1	−1	1	80	0.07	0.6	0.922
4	1	1	1	130	0.11	0.6	1.438
5	−1	1	1	80	0.11	0.6	2.017
6	0	0	0	105	0.09	0.5	1.163
7	1	−1	1	130	0.07	0.6	0.282
8	0	0	0	105	0.09	0.5	1.164
9	0	0	0	105	0.09	0.5	1.164
10	0	0	0	105	0.09	0.5	1.168
11	0	0	0	105	0.09	0.5	1.159
12	1.682	0	0	147.04	0.09	0.5	0.622
13	0	0	−1.682	105	0.09	0.332	1.121
14	0	0	1.682	105	0.09	0.668	1.203
15	−1	−1	−1	80	0.07	0.4	0.882
16	−1	1	−1	80	0.11	0.4	2.052
17	0	0	0	105	0.09	0.5	1.160
18	−1.682	0	0	62.96	0.09	0.5	1.698
19	0	−1.682	0	105	0.056	0.5	0.184
20	0	1.682	0	105	0.124	0.5	2.136

3 Results and Discussion

The experimental results have been included in Table 2. Formula (1) expresses a regression model of surface roughness. Building a regression model based on experimental data is known as the *RSM* method [18]. This model has a coefficient of determination $R^2 = 0.9997$ and an adjusted coefficient of determination $R^2(adj) = 0.9994$. Since both of these coefficients are very close to 1, it can be confirmed that this model has a very large compatibility (fitness) with the experimental data. Also from this model, it can be seen that the coefficients of squared quantities are very small, showing that the square of the cutting parameters has a very little effect on the surface roughness. However, the squared quantities should not be excluded from the regression model because the compatibility of model with experimental data will be weakened (the values of R^2 and $R^2(adj)$ will be reduced) [18].

$$Y = R_a = 1.1632 - 0,3161x_1 + 0,5759x_2 + 0,016x_3 - 0,0023x_1^2$$
$$-0,0023x_2^2 - 0,0016x_3^2 + 0,0065x_1x_2 + 0,0088x_1x_3 - 0,01x_2x_3 \quad (1)$$

In order to survey the influence of each cutting parameter on surface roughness, two of the three parameters are fixed at the levels of −1, 0 and 1, respectively. The remaining parameter will receive values from the smallest to the greatest [20, 21]:

Subsequently fixing two of the three parameters at −1, 0 and 1 and changing the remaining parameters, the regression equations for the effect of each parameter on surface roughness is presented in the formulas from (2, 3, 4, 5, 6 and 7), and (8, 9 and 10), respectively. The charts showing the effect of each parameter on the surface roughness is shown in Fig. 2, 3, and 4, respectively.

$$Y_{1(-1)} = 1,1333 - 0,3314x_1 - 0,0023x_1^2 \quad (2)$$

$$Y_{2(-1)} = 1,4682 + 0,5794x_2 - 0,0023x_2^2 \quad (3)$$

$$Y_{3(-1)} = 0,9053 + 0,0172x_3 - 0,0016x_3^2 \quad (4)$$

$$Y_{1(0)} = 1,1632 - 0,3161x_1 - 0,0023x_1^2 \quad (5)$$

$$Y_{2(0)} = 1,1632 + 0,5759x_2 - 0,0023x_2^2 \quad (6)$$

$$Y_{3(0)} = 1,1632 + 0,016x_3 - 0,0016x_3^2 \quad (7)$$

$$Y_{1(1)} = 1,7412 - 0,3008x_1 - 0,0023x_1^2 \quad (8)$$

$$Y_{2(1)} = 0,8464 + 0,576x_2 - 0,0023x_2^2 \quad (9)$$

$$Y_{3(1)} = 1,4249 + 0,0148x_3 - 0,0016x_3^2 \quad (10)$$

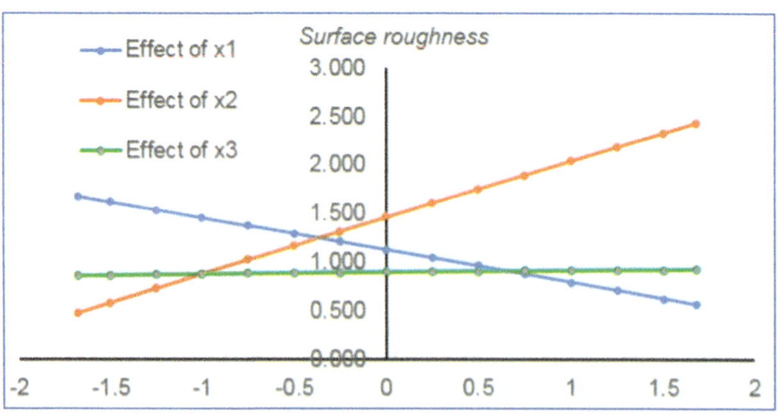

Fig. 2. The curve of the sub-regression model when 2 out of 3 parameters at the coding level -1

Fig. 3.The curve of the sub-regression model when 2 out of 3 parameters at the coding level 0

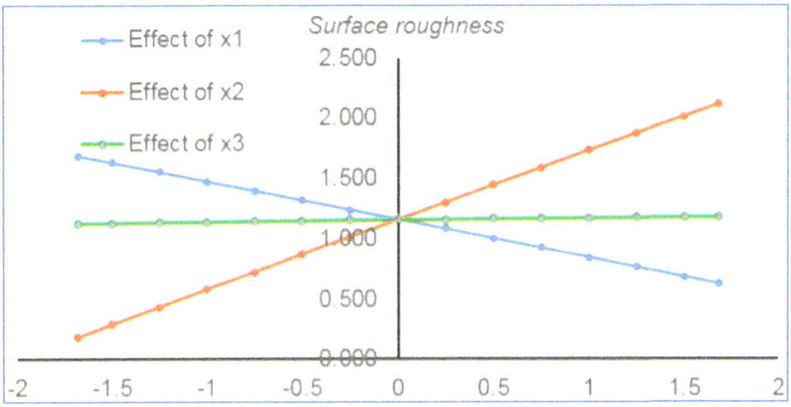

Fig. 3. The curve of the sub-regression model when 2 out of 3 parameters at the coding level 0

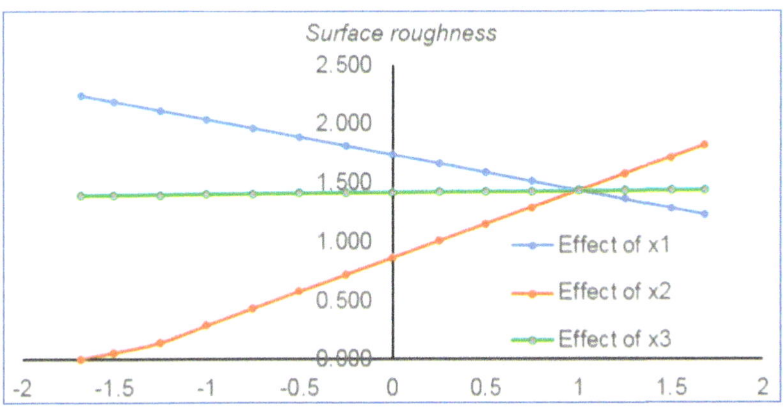

Fig. 4. The curve of the sub-regression model when 2 out of 3 parameters at the coding level 1

Observing Figures from 2, 3 and 4, it is shown that: In all three cases, the curve of feed rate (x2) has the greatest slope, followed by the slope of the curve of cutting velocity (x1). The slope of the curve of cutting depth (x3) is very small, this curve is almost the same as the horizontal axis of the graph. It shows that the feed rate is the parameter with the greatest effect on the surface roughness, followed by the effect of cutting velocity. The cutting depth has a little effect on the surface roughness. In order to reinforce this comment, the graph of the effect of cutting parameters on surface roughness can be observed in Fig. 5. This graph has been made by analyzing the data on the statistical software Minitab 16. Also from these graphs, it is shown that when increasing the cutting velocity, the surface roughness decreases, and when increasing the feed rate, the surface roughness increases. It means that to have a small surface roughness, it is necessary to choose the maximum value of cutting velocity and the minimum value of feed rate in the surveyed area.

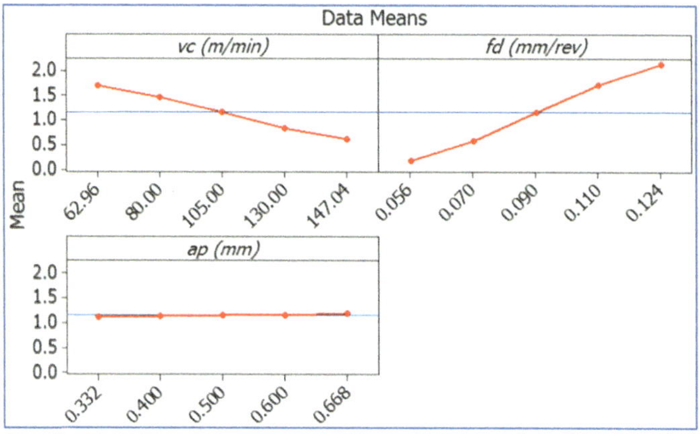

Fig. 5. Main effects plot for surface roughness

4 Determination of Cutting Parameters for Productivity Improvement on the Basis of Surface Roughness Assurance

As analyzed above, in order to ensure the small value of surface roughness, it is necessary to turn with the maximum value of cutting speed and the minimum value of feed rate. The change in cutting depth hardly changes the surface roughness. Therefore, the measure for machining productivity improvement while ensuring the minimum surface roughness is to choose the maximum cutting speed, the minimum feed rate and the maximum cutting depth. However, it is necessary to conduct some experiments to survey the surface roughness when changing the cutting depth. Carrying out the turning with 5 steel samples in accordance with some values of cutting depth has given the results as shown in Table 3.

Table 3. Milling results

No.	Code value			Actual value			*Ra (μm)*
	x_1	x_2	x_3	v_c *(m/min)*	f_d *(mm/rev)*	a_p *(mm)*	
1	1.682	−1.682	−1.682	147.04	0.056	0.332	0.461
2			−1			0.4	0.489
3			0			0.5	0.524
4			1			0.6	0.534
5			1.682			0.668	0.541

The results in Table 3 have shown that when increasing the cutting depth from 0.332 mm to 0.668 mm (about 2.01 times), the surface roughness also increases by only about 0.08 μm, and all five roughness values in Table 3 are within the level 8 of roughness. Thus, in this case, the cutting productivity has increased by approximately 2 times but the surface roughness is almost unchanged. Since then, it can be determined that in order to improve machining productivity while ensuring the "minimum" value of surface roughness, cutting speed, feed rate and cutting depth are 147.04 m/min, 0.056 mm/rev and 0.668 mm, respectively.

5 Conclusion

Experimental process of turning AISI 1045 steel on CNC lathe with TiAlN coated insert has been performed in this study. The *RSM* method has been applied to build the regression model of surface roughness. Sub-regression models showing the relationship between surface roughness and each cutting parameter have been set up. The effect of cutting parameters on surface roughness has been found. The set of values of cutting parameters to ensure the maximum cutting productivity while ensuring the minimum surface roughness has also been found. A number of conclusions are drawn as follows:

- The feed rate is the parameter having the greatest effect on surface roughness. When increasing feed rate, the surface roughness increases. The cutting speed is the parameter having the second greatest effect on surface roughness, when increasing cutting speed, the surface roughness decreases. The change in cutting depth has an insignificant effect on the surface roughness.
- In order to improve machining productivity while ensuring the small value of surface roughness, values of cutting speed, feed rate and cutting depth are 147.04 m/min, 0.056 mm/rev and 0.668 mm, respectively.
- The method of building sub-regression models presented in this study has opened up a new direction when analyzing experimental data. This method overcomes the necessity to learn about complicated algorithms in the research process, as well as can be easily applied in production units.

Acknowledgment. This work was supported by Thai Nguyen University of Technology.

References

1. Dich, T.V., Binh, N.T., Dat, N.T., Tiep, N.V., Viet, T.X.: Manufacturing Technology. Science and Technics Publishing House, Ha Noi (2003)
2. Kulshreshtha, M.: Analysis of the effect of machining parameters on surface roughness of EN 36 nickel steel. Int. J. Adv. Inf. Sci. Technol. **2**(8), 1–7 (2013). https://doi.org/10.15693/ijaist/2013.v2i8.9-15
3. Mohruni, A.S., Yanis, M., Kurniawan, E.: Development of surface roughness prediction model for hard turning on AISI D2 steel using Cubic Boron Nitride insert. Jurnal Teknologi **80**(1), 173–178 (2018). https://doi.org/10.11113/jt.v80.10492

4. Shihab, S., Khan, Z., Mohammad, A.A.S., Siddiquee, A.: Optimization of surface integrity in dry hard turning using RSM. Sadhana **39**(5), 1035–1053 (2014). https://doi.org/10.1007/s12046-014-0263-4
5. Izelu, C.O., Eze, S.C., Ashiedu, F.I.: Modeling and optimization of hard turning operation on 41Cr4 alloy steel using response surface methodology. Int. J. Mech. Eng. Appl. **4**(2), 88–102 (2016). https://doi.org/10.11648/j.ijmea.20160402.18
6. Xiao, M., et al.: Prediction of surface roughness and optimization of cutting parameters of stainless steel turning based on RSM. Math. Probl. Eng. **2018**, 1–15 (2018). https://doi.org/10.1155/2018/9051084
7. Yusuf, M., Ariffinm, M.K.A., Ismail, N., Sulaiman, S.: Optimization of cutting parameters on turning process based on surface roughness using response surface methodology. Appl. Mech. Mater. **117–119**, 1561–1565 (2012). https://doi.org/10.4028/www.scientific.net/AMM.117-119.1561
8. Bhushan, R.K.: Multi-response optimization of parameters during turning of AA7075/ SiC composite for minimum surface roughness and maximum tool life. SILICON **13**(9), 2845–2856 (2020). https://doi.org/10.1007/s12633-020-00640-w
9. Laghari, R.A., Li, J., Xie, Z., Wang, S.: Modeling and optimization of tool wear and surface roughness in turning of Al/SiCp using response surface methodology. 3D Res. **9**(4), 1–13 (2018). https://doi.org/10.1007/s13319-018-0199-2
10. Arfaoui, S., Zemzemi, F., Dakhli, M., Tourki, Z.: Optimization of hard turning process parameters using the response surface methodology and finite element simulations. Int. J. Adv. Manuf. Technol. **103**(1–4), 1279–1290 (2019). https://doi.org/10.1007/s00170-019-03535-2
11. Nayak, S.K., Patro, J.K., Dewangan, S., Gangopadhyay, S.: Multi-objective optimization of machining parameters during dry turning of AISI 304 austenitic stainless steel using grey relational analysis. Procedia Mat. Sci. **6**, 701–708 (2014). https://doi.org/10.1016/j.mspro.2014.07.086
12. Sahoo, P.: Optimization of turning parameters for surface roughness using RSM and GA. Adv. Prod. Eng. Manag. **6**(3), 197–208 (2011)
13. Tank, K., Shetty, N., Panchal, G., Tukrel, A.: Optimization of turning parameters for the finest surface roughnes characteristics using desirability function analysis coupled with fuzzy methodology and ANOVA. Mat. Today Proceed. **5**, 13015–13024 (2018)
14. Makadia, A.J., Nanavati, J.I.: Optimisation of machining parameters for turning operations based on response surface methodology. Measurement **46**, 1521–1529 (2013). https://doi.org/10.1016/j.measurement.2012.11.026
15. Kuntoglu, M., Aslan, A., Pimenov, D.Y., Giasin, K., Mikolajczyk, T., Sharma, S.: Modeling of cutting parameters and tool geometry for multi-criteria optimization of surface roughness and vibration via response surface methodology in turning of AISI 5140 steel. Materials **13** (4242), 1–21 (2020). https://doi.org/10.3390/ma13194242
16. Jayaraman, P., Kumar, L.M., Senthil kumar, V.S.: Optimization of cutting parameters in turning of AA6351 using response surface methodology and genetic algorithm. Int. J. Appl. Eng. Res. **10**(23), 43905–43911 (2015)
17. Dhabale, R., Jatti, V.K.S., Singh, T.P.: Multi-objective optimization of turning process during machining of AlMg1SiCu using non-dominated sorted genetic algorithm. Procedia Mat. Sci. **6**, 961–966 (2014)
18. Dean, A., Voss, D., Draguljić, D.: Design and Analysis of Experiments - Second Edition, Springer (2007).https://doi.org/10.1007/978-3-319-52250-0
19. Babu, V.S., Amuthakkannan, R., Kumar, S.S., Muruganandam, A.: Optimal cutting parameters estimation to improve surface finish in turning operation in AISI 1045 using Taguchi's robust design. Int. J. Ind. Syst. Eng. **15**(1), 19–36 (2013)

20. Trung, D.D., Son, N.H., Nguyen, N.-T.: A study on productivity improvement for the SKD11 steel grinding by using CBN grinding wheel - a new approach. Univ. J. Mech. Eng. **8**(1), 29–33 (2020). https://doi.org/10.13189/ujme.2020.080104
21. Trung, D.D., Son, N.H., Nguyen, N.-T.: Research on improving machining productivity when external cylindrical grinding of 160Cr12Mo steel. Univ. J. Mech. Eng. **8**(1), 34–40 (2020). https://doi.org/10.13189/ujme.2020.080105

Subordinate Control for Nonlinear Electric Drives of Exoskeleton with Compensator Based on Neural Network

Mikhail P. Belov[1], Dinh Dang Truong[1], and Thi Ha Nguyen[2(✉)]

[1] Department of Robotics and Industrial Automation, Saint Petersburg Electrotechnical University "LETI", Saint Petersburg, Russia
[2] Thai Nguyen University of Technology, Thai Nguyen, Vietnam
honghal983@tnut.edu.vn

Abstract. This paper presents a method of subordinate control with compensator based on a neural network for nonlinear electric drives of exoskeleton. The development of a dynamic model of an exoskeleton-human includes five links of two legs and a body, taking into account nonlinear electric drives of each joint. The neural network is used to approximate the nonlinear elements of the object model based on the dynamic error of the joint angles of the lower limb exoskeleton. Additionally, a robust element is used to correct the approximate error of the neural network in the control system. The convergence and asymptotic stability representations of the control system are proved on the basis of Lyapunov's theory. Also, the results of the experiment were obtained to illustrate the effectiveness of the proposed control strategy in the Matlab/Simulink program, including comparison with a subordinate controller without compensator.

Keywords: Exoskeleton · Neural network · Nonlinear electric drive · Subordinate control · Compensator · Exoskeleton electric drive

1 Introduction

The principle of subordinate control (SC) is widespread in the development of systems for regulating the speed and position of all kinds of mechanisms for general industrial and special purposes. The construction of more complex control systems (nonlinear, adaptive, intelligent, etc.) is carried out by including the method of subordinate regulation in them as a basic and unchanging structure, which is an integral part of the control object, considered as an initial description when synthesizing more complex systems [1]. Until now, some researches have used a neural network (NN) to control the lower limb exoskeleton, which exhibits asymptotic convergence, in contrast to traditional control systems that have tracking errors and lead to a relatively small field of application [2–6].

Neural network has great potential for improving movement tracking in the rehabilitation of the lower limb exoskeleton. In [2], a gravity compensator was used in the form of a PD-regulator for the reference phase of the exoskeleton. A simple PD

D. C. Nguyen et al. (Eds.): ICERA 2021, LNNS 366, pp. 882–890, 2022.
https://doi.org/10.1007/978-3-030-92574-1_90

controller has low accuracy, difficult to meet the control performance requirements of an exoskeleton.

This paper presents a method of subordinate control with loop control of position and speed, including a regulator with a neural network to compensate for uncertain changes in the forces of gravity, friction and elasticity of the dynamic model of the exoskeleton to obtain the desired trajectory. The neural network is trained in the off-line mode to approximate and adapt for uncertain changes in the forces of gravity, friction and elasticity of the dynamic model. This new control is natural combination of industrial linear PID control and neural compensation. The study results show that it is more effective than similar studies for single-joint or single-stage [7, 8]. The main content of this article is the semi global asymptotic stability of neural PID control and the local asymptotic stability of neural PID control.

2 Mathematical Model of Lower Limbs Human-Exoskeleton (LLHE)

Consider the simulation of the dynamics of a human - exoskeleton consists of a five-link bipedal system [9, 10] with a body in the sagittal plane based on Lagrange principles. The model of two-legged five-link is shown in Fig. 1.

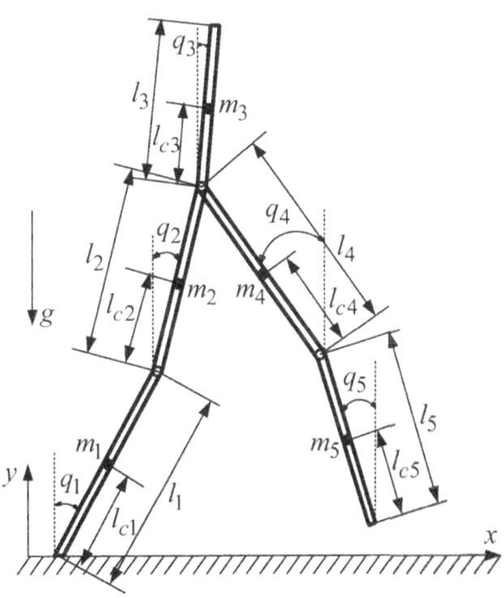

Fig. 1. Five-link exoskeleton model

Mathematical model of the dynamics of the lower extremities of a human exoskeleton with nonlinear electric drives can be written as:

$$\mathbf{A}(q)\ddot{q} + \mathbf{B}(q,\dot{q})\dot{q} + \mathbf{C}(q) + \mathbf{F}(q,\dot{q}) + \boldsymbol{\tau}_h = \mathbf{u} \qquad (1)$$

where $q,\ \dot{q},\ \ddot{q} \in \mathbb{R}^n$—the generalized coordinate of the position, speed, acceleration of the links; $\mathbf{q} = [q_1,\ q_2,\ q_3,\ q_4,\ q_5]^T$; $\dot{\mathbf{q}} = [\dot{q}_1,\ \dot{q}_2,\ \dot{q}_3,\ \dot{q}_4,\ \dot{q}_5]^T$; $\ddot{\mathbf{q}} = [\ddot{q}_1,\ \ddot{q}_2,\ \ddot{q}_3,\ \ddot{q}_4,\ \ddot{q}_5]^T$.

$$\mathbf{A}(q) = \begin{bmatrix} a_{11}+a_1 & a_{12} & a_{13} & a_{14} & a_{15} \\ a_{21} & a_{22}+a_2 & a_{23} & a_{24} & a_{25} \\ a_{31} & a_{32} & a_{33}+a_3 & a_{34} & a_{35} \\ a_{41} & a_{42} & a_{43} & a_{44}+a_4 & a_{45} \\ a_{51} & a_{52} & a_{53} & a_{54} & a_{55}+a_5 \end{bmatrix};$$

where $\mathbf{a} = \mathbf{diag}\{a_1,\ a_2,\ a_3,\ a_4,\ a_5\} = \mathbf{diag}\{r_i^2 J_{mi}\}$; a_{ij} - elements of inertia matrix $\mathbf{A}(q)$; r_i - gear ratio reducer of the joints; J_{mi} - inertia moment of motor.

$$\mathbf{B}(q,\dot{q}) = \begin{bmatrix} b_{11}+b_1 & b_{12} & b_{13} & b_{14} & b_{15} \\ b_{21} & b_{22}+b_2 & b_{23} & b_{24} & b_{25} \\ b_{31} & b_{32} & b_{33}+b_3 & b_{34} & b_{35} \\ b_{41} & b_{42} & b_{43} & b_{44}+b_4 & b_{45} \\ b_{51} & b_{52} & b_{53} & b_{54} & b_{55}+b_5 \end{bmatrix};$$

where $\mathbf{b} = \mathbf{diag}\{b_1,\ b_2,\ b_3,\ b_4,\ b_5\} = \mathbf{diag}\left\{r_i^2 b_{mi} + \frac{r_i^2 k_{ei} k_{ti}}{R_i}\right\}$; b_{ij} - elements of matrix $\mathbf{B}(q,\dot{q})$, containing centrifugal and Coriolis forces; b_{mi} - viscous friction coefficients of motor; k_e - electromotive force constant; k_t - motor torque constant; R - motor phase winding resistance; $\mathbf{C}(q) = [g_1,\ g_2,\ g_3,\ g_4,\ g_5]^T$ - vector, containing the moments of gravity; $\mathbf{F}(q,\dot{q}) = [f_1,\ f_2,\ f_3,\ f_4,\ f_5]^T$ - The function of the moment of elasticity and friction is represented by Coulomb friction, viscous friction; k_s - equivalent stiffness of the elastic connection of the drive; b_e - viscous friction coefficient of exoskeleton; α - dynamic friction coefficient of electric drives;
where:

$$f_i = b_e\dot{q}_i + \gamma sign(\dot{q}_i) + r_i k_s (r_i - 1) q_i + r_i \alpha sign(r_i \dot{q}_i)$$

τ_h—vector moment of the residual part human leg (considered external disturbance); $\mathbf{u} = \left[\frac{r_1 k_{t1}}{R_1} u_1,\ \frac{r_2 k_{t2}}{R_2} u_2,\ \frac{r_3 k_{t3}}{R_3} u_3,\ \frac{r_4 k_{t4}}{R_4} u_4,\ \frac{r_5 k_{t5}}{R_5} u_5\right]^T$ - vector control voltage.

3 Building the Structure of the Control System

The use of a subordinate structure does not require the division of the object into separate subsystems, which are evaluated online, as well as complete knowledge of the dynamics. Therefore, this method is easy to use and does not require complex calculations.

The block generality diagram of the exoskeleton electric drive control system based on subordinate regulation with a neural network compensator is shown in Fig. 2.

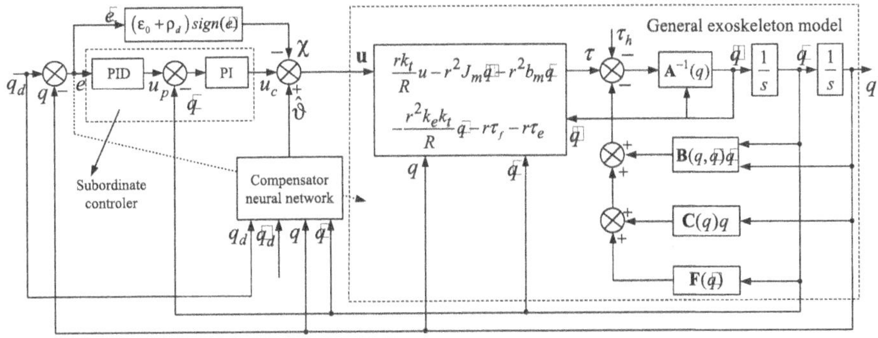

Fig. 2. The structure of the control system with a torque compensator neural network

where **u**—input voltages motor; $\tau_f(\dot{q}_m) = \alpha \text{sgn}(\dot{q}_m)$—frictional moment in the electric drive; $\tau_e = k_s(q_m - q) = k_s q(r - 1)$—elastic moment in the mechanical connection of the electric drive; $q_d = (q_{d1}, q_{d2}, q_{d3}, q_{d4}, q_{d5})$—given trajectory of the joint angles; $e_i = q_{di} - q_i$, $i = \overline{1, 5}$—control system error; $\chi = (\varepsilon_0 + \rho_d) sign(\dot{e})$—robust element is used to correct neural network approximation errors.

4 Synthesis of Control System

The subordinate controller is a compensator with a neural network based on a dynamic error (the actual trajectory and the specified trajectory of the exoskeleton) and is shown in Fig. 2.

To implement neural networks, two assumptions must be used. All sigmoidal functions used in neural networks usually satisfy the Livshitz condition, then [11]:

$$\tilde{\varphi} := \varphi(\mathbf{Z}^{*T}\mathrm{x}) - \hat{\varphi}(\mathbf{Z}^T\mathrm{x}) = P_\sigma \tilde{\mathbf{Z}}^T x + \upsilon_\sigma, \lim_{x \to \infty}$$

$$P_\sigma = \left.\frac{\partial \varphi^T(v)}{\partial v}\right|_{v = \hat{Z}^T x}, \|\upsilon_\sigma\|_\Lambda^2 \le l \left\|\tilde{\mathbf{Z}}^T x\right\|_\Lambda^2, l > 0$$

where $\tilde{\mathbf{W}} = \mathbf{W}^* - \hat{\mathbf{W}}$, $\tilde{\mathbf{Z}} = \mathbf{Z}^* - \hat{\mathbf{Z}}$, Λ—positive definite matrix; $\mathbf{W}^*$, $\mathbf{Z}^*$—fixed bounded weight matrices; $\hat{\mathbf{W}}$, $\hat{\mathbf{Z}}$—time-varying neural network weight matrices; υ_σ—found upper bound.

To implement a neural network with n outputs, k outputs and p hidden layers, use the sigmoidal nonlinearity activation function [12]:

$$\varphi_i(x) = \frac{1 - e^{-2x}}{1 + e^{-2x}} \tag{2}$$

$$\hat{\vartheta}_{ij} = \sum_{j=1}^{p} W_{ij} \varphi_j(Z_{ijk} x), i = 1, 2, ..., k, \tag{3}$$

The neural network is used to approximate the force of gravity, friction and elastic moment and is expressed as

$$\vartheta = \mathbf{C}(q) + \mathbf{F}(q, \dot{q}) = \mathbf{W}^{*T} \varphi(\mathbf{Z}^{*T} x) + \varepsilon(x) \tag{4}$$

where $\varepsilon(x)$—estimate error.

When considering nonlinear elements in the LLHE dynamics model (1), the control law can be written in the form:

$$\mathbf{u} = \mathbf{K}_p e + \mathbf{K}_d \dot{e} + \hat{\mathbf{W}}^T \varphi(\hat{\mathbf{Z}}^T x) + \xi - \chi \tag{5}$$

where; $\xi = \mathbf{k}_{ip}\mathbf{k}_{ic} \int_0^t e(t)dt - (\mathbf{k}_{pc} + \mathbf{k}_{ic})\dot{\mathbf{q}}$; $\mathbf{K}_p = \mathbf{k}_{pp}\mathbf{k}_{pc}$; $\mathbf{K}_d = \mathbf{k}_{dp}\mathbf{k}_{pc}$; $\mathbf{k}_{pp}$, $\mathbf{k}_{dp}$, $\mathbf{k}_{ip}$ - PID controller gains (position controller); $\mathbf{k}_{pc}$, $\mathbf{k}_{ic}$ - gains of PI regulator (speed regulator).

Substituting Eq. (4) and (5) into Eq. (1), we obtain:

$$\begin{aligned} &\mathbf{A}(q)\ddot{e} + \mathbf{B}(q, \dot{q})\dot{e} + \mathbf{K}_p e + \mathbf{K}_d \dot{e} + \hat{\mathbf{W}}^T \varphi(\hat{\mathbf{Z}}^T x) \\ &-\mathbf{W}^{*T} \varphi(\mathbf{Z}^{*T} x) - \varepsilon(x) - \tau_h + \xi - \chi = 0 \end{aligned}$$

$$\begin{aligned} &\mathbf{A}(q)\ddot{e} + \mathbf{B}(q, \dot{q})\dot{e} + \mathbf{K}_p e \\ &= -\mathbf{K}_d \dot{e} + \mathbf{W}^{*T} \varphi(\mathbf{Z}^{*T} x) - \hat{\mathbf{W}}^T \varphi(\hat{\mathbf{Z}}^T x) + \varepsilon(x) + \tau_h - \xi + \chi \end{aligned} \tag{6}$$

where

$$\begin{aligned} &-\left(\mathbf{W}^{*T} \varphi(\mathbf{Z}^{*T} x) - \hat{\mathbf{W}}^T \varphi(\hat{\mathbf{Z}}^T x)\right) \\ &= \mathbf{W}^{*T} \varphi(\mathbf{Z}^{*T} x) - \mathbf{W}^{*T} \varphi(\hat{\mathbf{Z}}^T x) + \mathbf{W}^{*T} \varphi(\hat{\mathbf{Z}}^T x) - \hat{\mathbf{W}}^T \varphi(\hat{\mathbf{Z}}^T x); \end{aligned} \tag{7}$$

$$\mathbf{W}^{*T} \varphi(\mathbf{Z}^{*T} x) - \hat{\mathbf{W}}^T \varphi(\hat{\mathbf{Z}}^T x) = -\mathbf{W}^{*T} \varphi(\tilde{\mathbf{Z}}^T x) - \tilde{\mathbf{W}}^T \varphi(\hat{\mathbf{Z}}^T x) \tag{8}$$

Substituting Eq. (8) into Eq. (6), we get:

$$\begin{aligned} &\mathbf{A}(q)\ddot{e} + \mathbf{B}(q, \dot{q})\dot{e} + \mathbf{K}_p e \\ &= -\mathbf{K}_d \dot{e} - \mathbf{W}^{*T} \varphi(\tilde{\mathbf{Z}}^T x) - \tilde{\mathbf{W}}^T \varphi(\hat{\mathbf{Z}}^T x) + \varepsilon(x) + \tau_h - \xi + \chi \end{aligned} \tag{9}$$

The adaptive law of neural network is written as:

$$\begin{cases} \dot{\hat{\mathbf{W}}} = -\mathbf{F} \varphi(\hat{\mathbf{Z}}^T x)\dot{\mathbf{e}}^T \\ \dot{\hat{\mathbf{Z}}} = -\mathbf{C}\mathbf{W}^{*T} \varphi(x)\dot{\mathbf{e}}^T \end{cases} \tag{10}$$

where $\mathbf{F}$ - friction force and elastic moment in a dynamic model of an exoskeleton with electric drives; $\mathbf{C}$ - moment of gravity.

Lyapunov Function. To analyze the stability of the control system, we define the Lyapunov function as follows:

$$V = \frac{1}{2}\dot{\mathbf{e}}^T\mathbf{A}(q)\dot{e} + \frac{1}{2}\mathbf{e}^T\mathbf{K}_p e + \frac{1}{2}tr\left(\tilde{\mathbf{W}}^T\mathbf{F}^{-1}\tilde{\mathbf{W}}\right) + \frac{1}{2}tr(\tilde{\mathbf{Z}}^T\mathbf{C}^{-1}\tilde{\mathbf{Z}}) \quad (11)$$

Taking the derivative of Eq. (11), we obtain:

$$\dot{V} = \dot{e}^T\mathbf{A}(q)\ddot{e} + \frac{1}{2}\dot{\mathbf{e}}^T\dot{\mathbf{A}}(q)\dot{e} + \dot{\mathbf{e}}^T\mathbf{K}_p e + tr\left(\tilde{\mathbf{W}}^T\mathbf{F}^{-1}\dot{\tilde{\mathbf{W}}}\right) + tr\left(\tilde{\mathbf{Z}}^T\mathbf{C}^{-1}\dot{\tilde{\mathbf{Z}}}\right) \quad (12)$$

Since $\dot{\mathbf{A}}(q) - 2\mathbf{B}(q,\dot{q})$ is a skew-symmetric matrix, it satisfies the following relation:

$$x^T\left[\dot{\mathbf{A}}(q) - 2\mathbf{B}(q,\dot{q})\right]x = 0,\ \forall x \in \mathbb{R}^n \quad (13)$$

Substituting $\dot{e}^T\mathbf{A}(q)\dot{e} = 2\dot{e}^T\mathbf{B}(q,\dot{q})\dot{e}$ in Eq. (12), we get:

$$\begin{aligned}\dot{V} &= \dot{\mathbf{e}}^T\mathbf{A}(q)\ddot{e} + \dot{\mathbf{e}}^T\mathbf{B}(q,\dot{q})\dot{e} + \dot{\mathbf{e}}^T\mathbf{K}_p e + tr\left(\tilde{\mathbf{W}}^T\mathbf{F}^{-1}\dot{\tilde{\mathbf{W}}}\right) + tr\left(\tilde{\mathbf{Z}}^T\mathbf{C}^{-1}\dot{\tilde{\mathbf{Z}}}\right)\\ \dot{V} &= \dot{\mathbf{e}}^T\left(\mathbf{A}(q)\ddot{e} + \mathbf{B}(q,\dot{q})\dot{e} + \mathbf{K}_p e\right) + tr\left(\tilde{\mathbf{W}}^T\mathbf{F}^{-1}\dot{\tilde{\mathbf{W}}}\right) + tr\left(\tilde{\mathbf{Z}}^T\mathbf{C}^{-1}\dot{\tilde{\mathbf{Z}}}\right)\end{aligned} \quad (14)$$

Substituting Eq. (6) into Eq. (14), we get:

$$\begin{aligned}\dot{V} = {} & \dot{\mathbf{e}}^T\left(-\mathbf{K}_d\dot{e} - \mathbf{W}^{*T}\varphi(\tilde{\mathbf{Z}}^T x) - \tilde{\mathbf{W}}^T\varphi(\hat{\mathbf{Z}}^T x) + \varepsilon(x) + \boldsymbol{\tau}_h - \xi + \chi\right)\\ & + tr\left(\tilde{\mathbf{W}}^T\mathbf{F}^{-1}\dot{\tilde{\mathbf{W}}}\right) + tr\left(\tilde{\mathbf{Z}}^T\mathbf{C}^{-1}\dot{\tilde{\mathbf{Z}}}\right)\end{aligned} \quad (15)$$

$$\begin{aligned}\dot{V} = {} & -\dot{\mathbf{e}}^T\mathbf{K}_d\dot{e} - \dot{\mathbf{e}}^T\mathbf{W}^{*T}\varphi(\tilde{\mathbf{Z}}^T x) + tr\left(\tilde{\mathbf{Z}}^T\mathbf{C}^{-1}\dot{\tilde{\mathbf{Z}}}\right)\\ & - \dot{\mathbf{e}}^T\tilde{\mathbf{W}}^T\varphi(\hat{\mathbf{Z}}^T x) + tr\left(\tilde{\mathbf{W}}^T\mathbf{F}^{-1}\dot{\tilde{\mathbf{W}}}\right) + \dot{\mathbf{e}}^T\left(\varepsilon(x) + \boldsymbol{\tau}_h - \xi + \chi\right)\end{aligned} \quad (16)$$

$$\begin{aligned}\dot{V} = {} & -\dot{\mathbf{e}}^T\mathbf{K}_d\dot{e} + tr\tilde{\mathbf{Z}}^T\left(\mathbf{C}^{-1}\dot{\tilde{\mathbf{Z}}} - \dot{e}^T\mathbf{W}^{*T}\varphi(x)\right)\\ & + tr\tilde{\mathbf{W}}^T\left(\mathbf{F}^{-1}\dot{\tilde{\mathbf{W}}} - \dot{e}^T\varphi(\hat{\mathbf{Z}}^T x)\right) + \dot{\mathbf{e}}^T\left(\varepsilon(x) + \boldsymbol{\tau}_h - \xi + \chi\right)\end{aligned} \quad (17)$$

Substituting adaptive law (10) into Eq. (17), we obtain:

$$\begin{cases} \dot{\tilde{\mathbf{W}}} = \dot{\mathbf{W}}^* - \dot{\hat{\mathbf{W}}} = 0 - \dot{\hat{\mathbf{W}}} = \mathbf{F}\varphi(\hat{\mathbf{Z}}^T x)\dot{\mathbf{e}}^T \\ \dot{\tilde{\mathbf{Z}}} = \dot{\mathbf{Z}}^* - \dot{\hat{\mathbf{Z}}} = 0 - \dot{\hat{\mathbf{Z}}} = \mathbf{C}\mathbf{W}^{*T}\varphi(x)\dot{\mathbf{e}}^T \end{cases} \tag{18}$$

Finally, we have:

$$\begin{aligned} \dot{V} &= -\dot{\mathbf{e}}^T\mathbf{K}_d\dot{e} + \dot{\mathbf{e}}^T(\varepsilon(x) + \tau_h - \xi + \chi) \\ &= -\dot{\mathbf{e}}^T\mathbf{K}_d\dot{e} + \dot{\mathbf{e}}^T(\varepsilon(x) + \tau_h + \chi) - \dot{\mathbf{e}}^T\xi \end{aligned} \tag{19}$$

Consider the expression:

$$\begin{aligned} \dot{\mathbf{e}}^T(\varepsilon(x) + \tau_h + \chi) &= \dot{\mathbf{e}}^T(\varepsilon(x) + \tau_h) + \dot{\mathbf{e}}^T\chi \\ &= \dot{\mathbf{e}}^T(\varepsilon(x) + \tau_h) - \|\dot{e}\|(\varepsilon_0 + \rho_d) \end{aligned} \tag{20}$$

Since $\begin{cases} \|\varepsilon(x)\| \le \varepsilon_0 \\ \|\tau_h\| \le \rho_d \end{cases}$, so we have $\dot{\mathbf{e}}^T(\varepsilon(x) + \tau_h + \chi) \le 0$.
From expression (20) we have:

$$\begin{cases} -\dot{\mathbf{e}}^T\mathbf{K}_d\dot{e} \le 0 \\ \dot{\mathbf{e}}^T(\varepsilon(x) + \tau_h + \chi) \le 0 \\ -\|\dot{\mathbf{e}}\|\xi \le 0 \end{cases}. \tag{21}$$

From the analysis we obtain $\dot{V} \le 0$, which means that the equilibrium state of the system is stable.

5 Simulation Results in Matlab Environment

The desired trajectories are set as follows:

$$q_{d1,5} = \sin(0.5t) + \sin(0.75t) + \cos(t) - 1; q_{d2,4} = 0.5\sin(t); q_{d3} = 0.25\sin(0.5t)$$

Calculate and select coefficients of PID controllers in control loops based on method Ziegler – Nichols with the same coefficients at each joint to ensure the motion kinematics of the exoskeleton when changing alternate walking phase of two legs. Position loop PID gain:

$$\begin{gathered} \mathbf{k}_{pp} = \mathbf{diag}\{420, 420, 420, 420, 420\}; \mathbf{k}_{ip} = \mathbf{diag}\{120, 120, 120, 120, 120\}; \\ \mathbf{k}_{dp} = \mathbf{diag}\{68, 68, 68, 68, 68\}; \mathbf{k}_{pc} = \mathbf{diag}\{60, 60, 60, 60, 60\}; \\ \mathbf{k}_{ic} = \mathbf{diag}\{12, 12, 12, 12, 12\} \end{gathered}$$

The simulation results are shown in Figs. 3, 4, 5, 6 and 7:

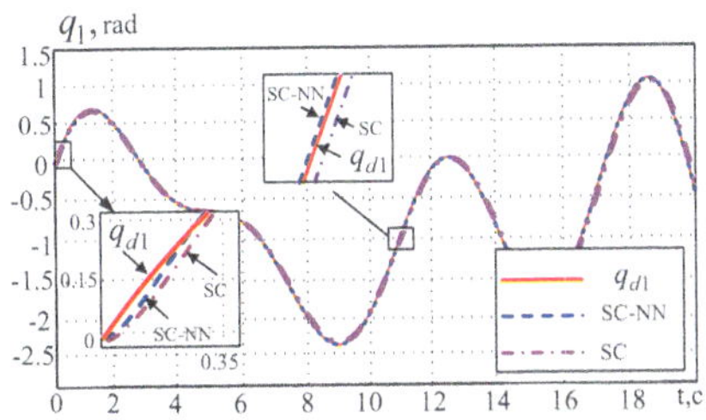

Fig. 3. Trajectories of movement of the knee joint q_1 (rad)

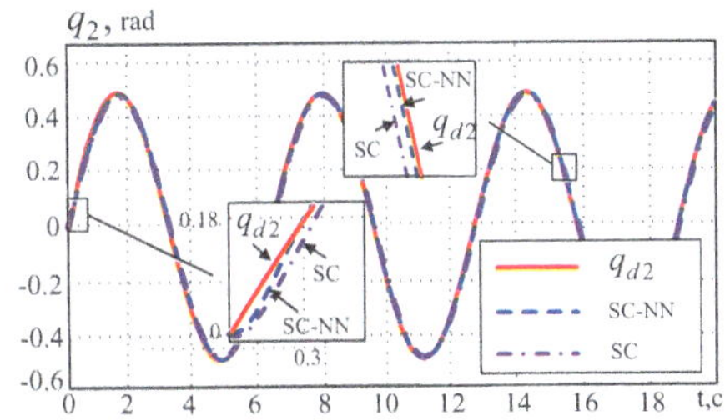

Fig. 4. Trajectories of movement of the hip joint q_2 (rad)

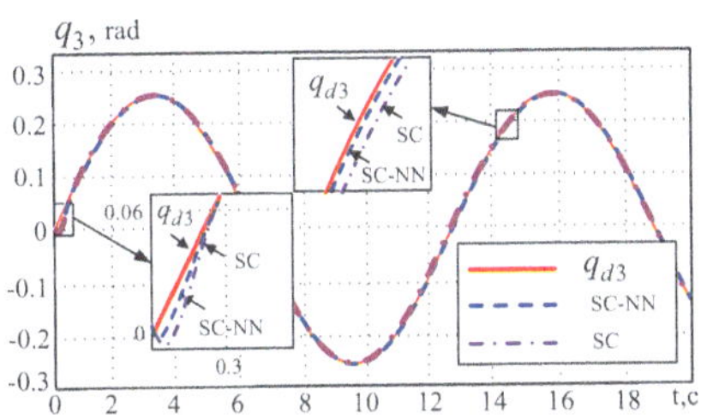

Fig. 5. Trajectories of movement of the joint of the body q_3 (rad)

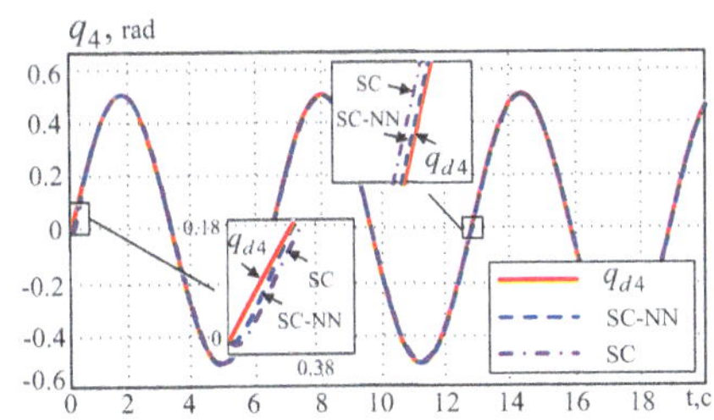

Fig. 6. Trajectories of movement of hip motion q_4 (rad)

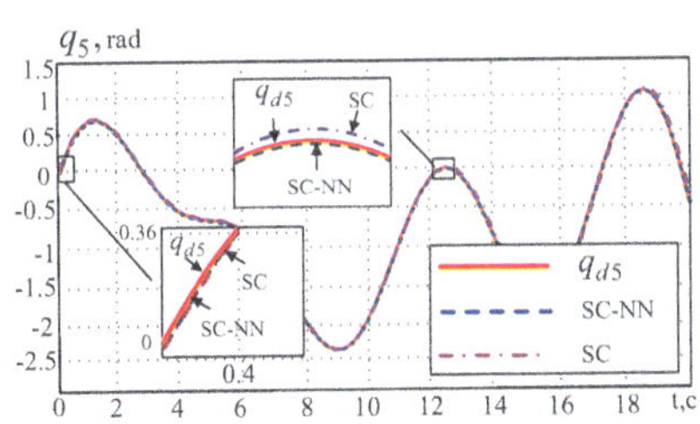

Fig. 7. Trajectories of movement of the joint of the knee q_5 (rad)

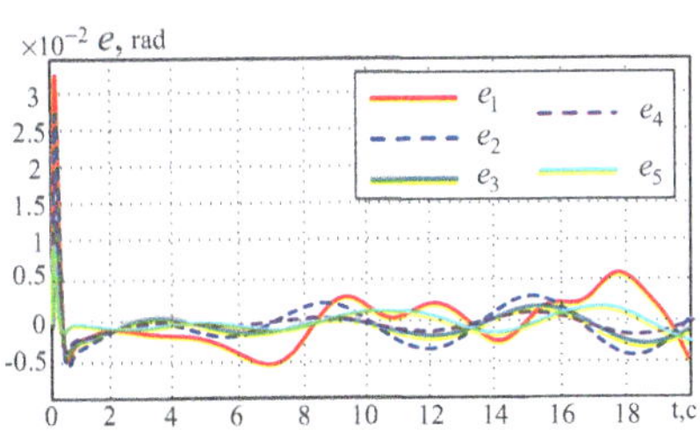

Fig. 8. Joint position error of SC-NN (rad)

It can be seen from the graphs (Figs. 3, 4, 5, 6 and 7) that the use of subordinate control with an NN compensator provides good tracking and stability of nonlinear electric drives of the exoskeleton. On the other hand, it has been proven that the proposed regulator is more efficient for controlling the electric drives of the exoskeleton (to follow the desired trajectory) in comparison with the SC method. On the basis of simulation results, it is shown that the control system provides accurate tracking of the trajectory of movement of the joints along the desired trajectory with a small error, shown in Fig. 8.

6 Conclusion

Mathematical model of the dynamics of the lower extremities of a human exoskeleton is developed taking into account the nonlinear elements of electric drives of each joint as a control object. The neural network allows to approximate the uncertain changes in gravity, friction and elasticity of the dynamics model. The proposed model makes it possible to solve the problems of the derivatives of the amplification coefficients in the case of subordinate regulation of the exoskeleton by electric drives. The control system of the subordinate regulation of the electric drives of the exoskeleton was developed without changing its structure and settings with the introduction of corrective elements in the form of an adaptive compensator with a neural network. Semi-global asymptotic stability of control with a neural slave controller is proved. Using simulation in Matlab/Simulink, the effectiveness of the neural network in compensating for gravity, friction and elasticity has been confirmed.

Acknowledgement. The work described in this paper was supported by Thai Nguyen University of Technology for a scientific project.

References

1. Putov, V.V., Sheludko, V.N.: Control of multi-stage mechanical objects: Publishing of Saint-Petersburg Electrotechnical University, 222 p. (2017)
2. Duong, M.K., Cheng, H., Tran, H.T., Jing, Q.: Minimizing human-exoskeleton interaction force using compensation for dynamic uncertainty error with adaptive RBF network. J. Intell. Robot. Syst. **82**(3–4), 413–433 (2016)
3. Mefoued, S.: A robust adaptive neural control scheme to drive an actuated orthosis for assistance of knee movements. Neurocomputing **140**, 27–40 (2014)
4. Lee, T., Kim, I., Lee, S.H.: Estimation of the continuous walking angle of knee and ankle (talocrural joint, subtalar joint) of a lower-limb exoskeleton robot using a neural network. Sensors **21**(8), 2807 (2021)
5. Jung, J.Y., et al.: A neural network-based gait phase classification method using sensors equipped on lower limb exoskeleton robots. Sensors **15**(11), 27738–27759 (2015)
6. Zhang, D., et al.: Cooperative control for a hybrid rehabilitation system combining functional electrical stimulation and robotic exoskeleton. Front. Neurosci. **11**, 725 (2017)
7. Xue, T., et al.: The control system for flexible hip assistive exoskeleton. In: 2018 IEEE International Conference on Robotics and Biomimetics (ROBIO), pp. 697–702. IEEE (2018)
8. Wu, J., et al.: The design and control of a 3DOF lower limb rehabilitation robot. Mechatronics **33**, 13–22 (2016)
9. Rigatos, G., et al.: A nonlinear optimal control approach for a lower-limb robotic exoskeleton. Int. J. Humanoid Robotics **17**(05), 2050018 (2020)
10. Ma, Y., et al.: Gait phase classification and assist torque prediction for a lower limb exoskeleton system using kernel recursive least-squares method. Sensors **19**(24), 5449 (2019)
11. Xie, X., et al.: Adaptive neural network control of a 5 DOF robot manipulator. In: 2010 International Conference on Intelligent Control and Information Processing, pp. 376–381. IEEE (2010)
12. Belov, M.P.: Modern Approaches to Structural-Parametric Synthesis, Research and Control of Automated Electric Drives, St. Petersburg, Publishing (2012)

Supercapacitor Electrode Based on Agricultural Waste Derived Biochar Materials

Nguyen Van-Truong[1], Phuoc-Anh Le[2], Pham Van Hao[3], Pham Minh Tan[1], Hoang Tien-Dat[4], Dang Van Thanh[5], and Le Dinh Trong[6(✉)]

[1] Faculty of Fundamental Sciences, Thai Nguyen University of Technology, Thai Nguyen, Vietnam

[2] Institute of Sustainability Science, Vietnam-Japan University, Vietnam National University, Hanoi, Vietnam

[3] Faculty of Basic Sciences, TNU-University of Information and Communication Technology, Thai Nguyen, Vietnam

[4] Faculty of International Training, Thai Nguyen University of Technology, Thai Nguyen, Vietnam

[5] Faculty of Basic Sciences, TNU-University of Medicine and Pharmacy, Thai Nguyen, Vietnam

[6] Hanoi Pedagogical University 2, Vinh Phuc, Vietnam

ledinhtrong@hpu2.edu.vn

Abstract. In this study, we report a fast and simple method for synthesis biochars from biomass sugarcane bagasse as a raw material. The biochars material possesses a good porous structure after one-step carbonization in a normal anaerobic furnace with an amorphous phase and a good specific surface area of 272.75 m^2 g^{-1}. The sugarcane bagasse-derived biochars show a harmonious agreement with the gel polymer electrolyte via electrochemical investigation. The electrical double layer supercapacitor cell demonstrates a good charge-discharge rate capability and also good specific capacitance. The porous biochars obtain electrode specific capacitance of 48 F g^{-1} at a current density of 2 A g^{-1} and the maximum energy density is about 7.71 W h kg^{-1} at a power density of 1028 W kg^{-1}. This study provides a promising route for recycling waste biomass in energy storage applications .

Keywords: Biochars · Biomass waste · Electrical double layer supercapacitor · Gel polymer electrolyte

1 Introduction

Recently, the biomass-derived carbon material is an interesting material for widely applications. The development of carbon material from biomass source really suits the development trend of energy technology with the excellent candidate replacing fossil material. Hydrothermal carbonization is one of most popular method for synthesis biomass-derived carbon due to low cost, easily control. Nowadays, biomass-derived

D. C. Nguyen et al. (Eds.): ICERA 2021, LNNS 366, pp. 891–896, 2022.
https://doi.org/10.1007/978-3-030-92574-1_91

carbon has been used widely in catalysis, lithium/sodium ion battery, supercapacitor, solar cell, fuel cell, etc. Currently, most of previous reports using carbonization method for synthesis activated carbon derived from biomass which were used base like KOH as active factor which increased the specific surface of biochars with small mesopores with highly porous structure [1–5]. However, the carbonization process with KOH active to make harmful gas at a high temperature. Furthermore, the waste water after washing step contain potassium which non-environmental friendly and hardly for treatment.

Herein, we propose a solution to carbonize raw material without KOH activation. Sugarcane bagasse is one of good candidate due to easy to carbonize, good specific surface area. In this report, we introduce one green-step for synthesis biochars from sugarcane bagasse by carbonization method without active material. This study provides a promising route for the recycling waste biomass in energy storage applications.

2 Experimental Section

2.1 Materials

Lithium sulfate tetrahydrate (Li_2SO_4) were from Sigma Aldrich. 1-methyl-2-pyrrolidinone (C_5H_9NO) and carbon black acetylene (100% compressed) was purchased from Alfa Aesar.

2.2 Preparation and Characterization of Porous Carbon

The porous carbons were prepared similar to our previous reported with a small modification [1]. For details, sugarcane bagasse (SB) heated under a pyrolysis temperature of 800 °C in oxygen limited conditions for 2 h. The as-prepared products then were sieved to obtain biochars flakes with size less than 0.5 mm and were stored in plastic bags for further use. The morphology, structure and the physical chemistry properties were studied by scanning electron microscopy (SEM, Hitachi S4800), X-ray diffraction spectroscopy (XRD, D2 Bruker), and Raman spectroscopy (Jobin won, Horiba).

2.3 Preparation of Biochars Electrode, and Electrolyte

The porous carbon electrode was prepared by solution casting method. The electrode slurry was obtained by mixing biochars (80 mg, 80 wt.%), carbon nanotube multiwall (10 mg, 10 wt.%) and polyvinylidene fluoride (PVDF; 10 mg, 10 wt.%) in solution of N-Methyl-2-Pyrrolidone (NMP) with constant stirring in 4 days. Then, this slurry was coated on nickel substrate (1×1 cm) and dried at 60 °C for 24 h to obtain the electrodes. The mass of volume for each electrode is 1 mg/cm^2.

2.4 Supercapacitor Fabrication and Characterization

The testing cell of supercapacitor was followed by sandwich structure, a layer gel polymer electrolyte between two layers of electrodes, and acrylic glass cover. The electrochemical measurement was characterized on an Autolab electrochemical workstation (PGSTAT302N) by cyclic voltammetry (CV), electrochemical impedance spectroscopy (EIS) ranging from 0.01 Hz to 100 kHz at potential amplitude of 10 mV, and Galvanostatic charge-discharge (GCD). The specific capacitance of the supercapacitor (C, F g^{-1}) and electrode specific supercapacitance (C_s, F g^{-1}) are calculated via charge-discharge curves by following equations [6]:

$$C = \frac{I \times \Delta t}{\Delta V \times m} \quad (1)$$

$$C_s = 4C \quad (2)$$

where I (A) is the discharge current, Δt is the discharge time, ΔV is the potential voltage, and m is the weight of biocharselectrode. The energy density (E, W h kg^{-1}) and the power density (P, W kg^{-1}) can calculate by the following equations:

$$E = \frac{C \times (\Delta V)^2}{2} \times \frac{1000}{3600} \quad (3)$$

$$P = \frac{E \times 3600}{\Delta t} \quad (4)$$

3 Results and Discussions

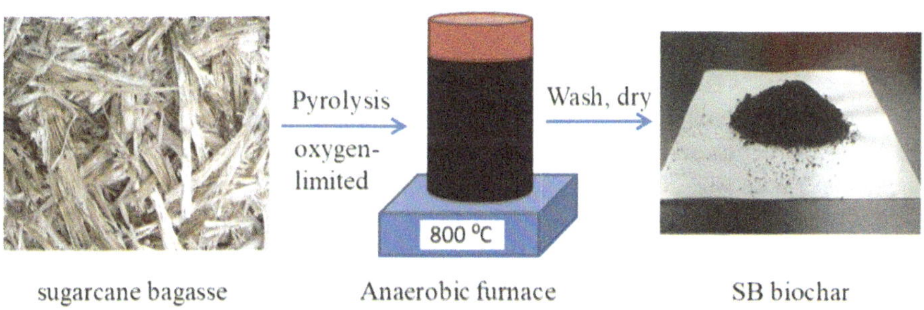

Fig. 1. The process for synthesis SB8 biochars.

The sintering process for preparing biochars from SB is demonstrated in Fig. 1. 100 g dried SB after washing and drying was put into anaerobic furnace at 800 °C for 1h. The biochars material after carbonization was still 70 wt.% in compare with raw SB material (~70 g). This weight ratio between raw material and biomass-carbon material is suitable with application which needs a large amount of biochars material.

Figure 2 shows the SEM imagines of biochars at low and high magnifications. The thick and large size around micro-meter with a porous surface can be seen. It is clearly confirmed that the carbonization at 800 °C for 2 h could transfer totally sugarcane bagasse to carbon material. Furthermore, at high resolution of SEM (Fig. 2b), the biochars shows a porous morphology on the surface. It is indicated that this structural porous carbon allows strongly a moving of ion from electrolyte to diffuse into electrode during charge time.

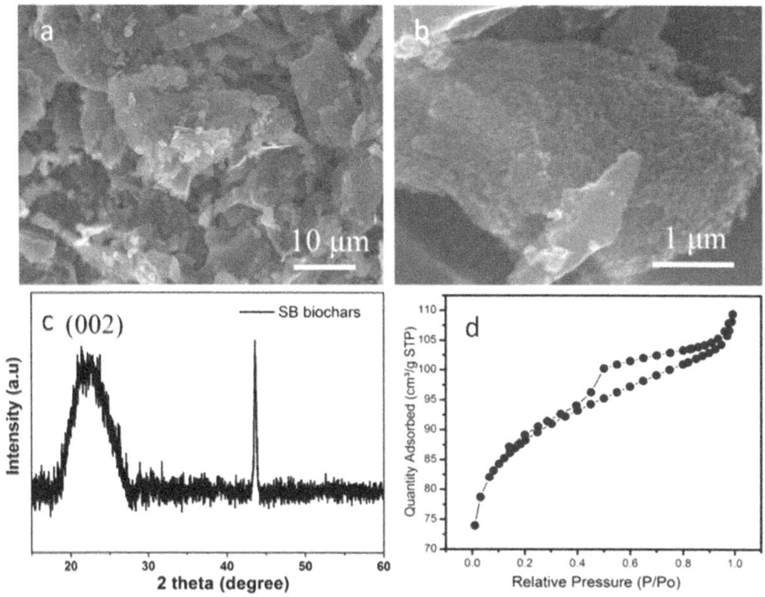

Fig. 2. SEM images of SB8 at (a) low magnification and (b) high magnification; (c) XRD pattern and (d) Adsorption and desorption isotherms of N_2.

Biochars structural was studied by XRD pattern in Fig. 2c. Two peaks two theta of the X-ray diffraction around 23^0 (002) and 43^0 (100) indicated that the carbonization at high temperature 800 °C was completed well. The large peak $2\theta = 23^0$ shows the construction of graphitic phase and the small peak of graphitized carbon around $2\theta = 43^0$ indicating the creation of carbon crystal from SB at high temperature [5]. The specific surface area of biochars is depicted via nitrogen adsorption-desorption isotherm measurements (Fig. 2d). The type of nitrogen sorption isotherm for the biochar sample suggests the micro pores which agree with SEM imagines. The (Brunauer, Emmett and Teller) BET surface area of biochar reaches 272.75 m^2 g^{-1}, the adsorption average pore width about 24.8 Angstrom. These results could be suitable for electrode material of supercapacitor.

The electrical double layer supercapacitor (EDLS) cell was fabricated following sandwich structure, a layer of gel polymer electrolyte (PVA-Li_2SO_4) between two layers activated carbon. The investigation of electrochemical study of EDLS cell depicts in Fig. 3. The CV curves at different at various sweep rate are shown in Fig. 3 (a). It is clearly seen that despite of high sweep rate of 200 mV s^{-1}, the CV curve still rectangular shape, indicating fast and stable charge-discharge in the positive and

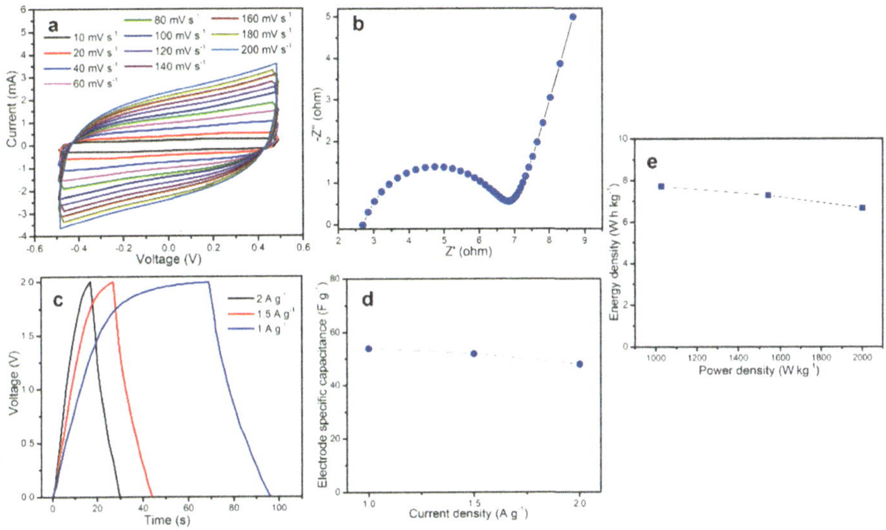

Fig. 3. Electrochemical properties of electrical double layer supercapacitor cell: (a) CV curves, (b) EIS plot, (c) GCD curves, (d) electrode specific capacitance as a function of current density, (e) Ragone plot.

negative electrode. Another side, at the low current 10 mV s^{-1}, the CV shape is almost rectangular to illustrate that the electrode has a high surface area and macrospores. Figure 3b shows the EIS analysis of EDLS cell, a low value around 2.7 Ω, indicating the excellent interface connection between a gel polymer electrolyte layer with two electrodes [7, 8]. This internal resistance includes resistance of gel polymer electrolyte, intrinsic resistance of electrodes, and interface resistance of electrode/electrolyte. Furthermore, the low value of resistance which indicates excellent and stable ion diffusion into electrode layer during charge-discharge process. The galvanostatic charge-discharge curves at various current densities 2, 1.5, 1 A g^{-1} (Fig. 3c). At 2 A g^{-1} the electrode specific capacitance is 48 F g^{-1}. At the lower current density of 1.5 and 1 A g^{-1}, the electrode specific capacitance obtains 51, and 54 F g^{-1}, which illustrates in Fig. 3d. The Ragone plot for the EDLS cell (Fig. 3e) shows that the maximum energy density is about 7.71 W h kg^{-1} at power density of 1028 W kg^{-1}. The good value of energy density of EDSL cell proofs that biomass SB-derived biochars material is a potential candidate material for supercapacitor.

4 Conclusions

The biomass sugarcane bagasse can be transformed into biochars material via one step carbonization without chemical activation. The convention of transformation from biomass raw material to carbon is quiet high of 70%. SB biochar reveals amorphous phase with good specific surface area of 272.75 m^2 g^{-1}. SB biochar further exhibited well electrode specific capacitance of 48 F g^{-1} at current density of 2 A g^{-1} and the maximum energy density of 7.71 W h kg^{-1} at power density of 1028 W kg^{-1}, which indicates that the biochars material from sugarcane bagasse can be used for energy storage application.

Acknowledgement. This research is funded by grant number B.2019-SP2-06.

References

1. Le, P.A., Nguyen, V.T., Sahoo, S.K., et al.: Porous carbon materials derived from areca palm leaves for high performance symmetrical solid-state supercapacitors. J. Mater. Sci. **55**(55), 10751–10764 (2020)
2. Gao, Z., Zhang, Y., Song, N., Li, X.: Biomass-derived renewable carbon materials for electrochemical energy storage. Mater. Res. Lett. **5**(2), 69–88 (2017)
3. Deng, J., Li, M., Wang, Y.: Biomass-derived carbon: synthesis and applications in energy storage and conversation. Green Chem. **18**, 4824–4854 (2016)
4. Wang, J., Kaskel, S.: KOH activation of carbon-based materials for energy storage. J. Mater. Chem. **22**, 23710 (2012)
5. Clemente, J.S., Beauchemin, S., Thibault, Y., MacKinnon, T., Smith, D.: Differentiating Inorganics in Biochars Produced at Commercial Scale Using Principal Component Analysis. ACS Omega **3**(6), 6931–6944 (2018)
6. Sudhan, N., Subramani, K., Karnan, M., Ilayaraja, N., Sathish, M.: Biomass-derived activated porous carbon from rice straw for a high-energy symmetric supercapacitor in aqueous and non-aqueous electrolytes. Energy Fuels **31**, 977–985 (2017)
7. Huang, C., Sun, T., Hulicova-Jurcakova, D.: Wide electrochemical window of supercapacitors from coffee bean-derived phosphorus-rich carbons. Chemsuschem **6**(12), 2330–2339 (2013)
8. Ma, F., Wan, J., Wu, G., Zhao, H.: Highly porous carbon micro flakes derived from catkins for high-performance supercapacitors. RSC Adv. **5**, 44416–44422 (2015)

Surface Roughness Model When Grinding 1066 Steel

Do Duc Trung[1], Trinh Kieu Tuan[2], Tran Quoc Hoang[3], Nguyen Van Tuan[1], and Luu Anh Tung[4](✉)

[1] Ha Noi University of Industry, Hanoi, Vietnam
[2] University of Economics - Technology for Industries, Hanoi, Vietnam
[3] Nguyen Tat Thanh University, Ho Chi Minh City, Vietnam
[4] Thai Nguyen University of Technology, Thai Nguyen, Vietnam
luuanhtung@tnut.edu.vn

Abstract. A steel cylindrical grinding process has been carried out in this study, in which 1066 steel was employed as the experimental material and an aluminum oxide grinding wheel was utilized. The experimental matrix has been designed in the form of Central Composite Designs (CCD) with four input parameters, including cutting velocity, workpiece velocity, feed rate, and cutting depth. The surface roughness has been selected as the output parameter of the grinding process. The analysis of the experimental results using Pareto chart has determined the effect of the input parameters on the surface roughness. A regression model showing the relationship between the surface roughness and input parameters has been set up. In order to improve the accuracy of the surface roughness model, two data transformations of Box-Cox and Johnson have been applied to build two more surface roughness models. These three roughness models have been used to predict the surface roughness and those predictions have been compared with experimental surface roughness. Through the comparison criteria including coefficient of determination R^2, adjusted coefficient of determination R^2(adj), and absolute percentage error (PAE), the roughness model with the highest accuracy has been determined. Finally, the direction for further studies has also been mentioned in this article.

Keywords: Cylindrical grinding · 1066 steel · Aluminum oxide grinding wheel · Box-Cox transformation · Johnson transformation

1 Introduction

The roughness of workpiece surface when grinding has always been a topic of interest to various scientists. It can be said that because surface roughness has a direct effect on the workability and service life of the product, and grinding is usually the final machining method requiring small surface roughness [1]. Cylindrical grinding is the most common among the grinding methods. It is able to machine many types of surfaces such as cylindrical, tapered, profiled surfaces or simultaneously machine many surfaces in a single attachment [2]. When studying the cylindrical grinding method, in

D. C. Nguyen et al. (Eds.): ICERA 2021, LNNS 366, pp. 897–908, 2022.
https://doi.org/10.1007/978-3-030-92574-1_92

order to have a basis for selecting the parameters of the machining process to ensure the surface roughness obtaining the desired value, there have been many experimental studies on the grinding process to build a relationship between surface roughness and parameters of the machining process. That relationship is known as surface roughness model or surface roughness regression model.

Building the relationship between surface roughness and workpiece velocity, feed rate and cutting depth when grinding X12M steel with aluminum oxide grinding wheel has been performed by N. V. Thien et al. [3]. C. Thiagarajan et al. [4] have built the relationship between surface roughness and cutting velocity, workpiece velocity, feed rate, and cutting depth when grinding Al/SiC Composites with an aluminum oxide grinding wheel. When using an aluminum oxide grinding wheel to grind ШХ15 steel, H. T. Dung et al. [5] have built the relationship between surface roughness and workpiece velocity, feed rate, and cutting depth. The relationship between surface roughness and cutting velocity, feed rate, and cutting depth when grinding EN8 steel with aluminum oxide grinding wheel has been performed by Aman et al. [6]. T. Singh et al. [7] have built the relationship between surface roughness and cutting velocity, workpiece velocity, cutting depth, abrasive grain size, the concentration of cutting fluid, and the number of passes when using aluminum oxide grinding wheel to grind AISI 4140 steel. T. Q. Hung et al. [8], V. N. Nguyet [9] have built the relationship between surface roughness and workpiece velocity, feed rate and cutting depth when using an aluminum oxide grinding wheel to grind SKD11 steel.

Through a number of the above stated studies, it is shown that the parameters of cutting parameters are often selected by the authors as the input parameters of the experimental process because the adjustment of values of these parameters will be easily carried out by the worker. In addition, an aluminum oxide grinding wheel has been used in many different cases due to a lower cost grinding wheel compared to other grinding wheels (for example, compared to CBN grinding wheel, diamond grinding wheel), on the other hand, this grinding wheel is also suitable for grinding many different types of materials [2, 10].

1066 Steel (equivalent to steel grades of 65G, 66Mn4, CK67, 080A67) is manganese steel with very high wear resistance. This steel is often used to manufacture parts and/or components with high requirements for wear resistance, such as pins, guide surfaces, etc. When manufacturing such workpieces, the grinding method is often chosen as the final machining method [11–14]. In manufacturing practice, an aluminum oxide grinding wheel has also been commonly used to grind this steel. However, in accordance with the research of the authors of this article, there has not been any published work on building a roughness model showing the published relationship between surface roughness, and parameters including cutting velocity, workpiece velocity, feed rate and cutting depth when cylindrically grinding 1066 steel so far.

From the above analysis, in this study, an experiment of grinding 1066 steel with aluminum oxide grinding wheel on the cylindrical grinder will be conducted. The parameters of cutting mode, including cutting velocity, workpiece velocity, feed rate, and cutting depth, will be selected as the input parameters of the experimental process.

The roughness model representing the relationship between surface roughness and input parameters will also be set up. Improvements in the accuracy of the surface roughness model will also be carried out.

2 Grinding Process Experiment

2.1 Experimental System

In this work, 1066 steel was used for the experimental process. The experimental sample has a diameter of 30 mm and a length of 340 mm. Before conducting the experiment, the steel sample has been turned and treated by heat to reach a hardness of 60–62 HRC.

The experiments have been conducted on the external cylindrical grinder with its serial number of SCHADT MAR 750. The grinder was fitted with two frequency converters so that the grinding wheel velocity and the workpiece velocity can be adjusted as desired. The cutting depth can be adjusted by means of the verner of the grinder with an accuracy of 5/1000. The feed rate is reckoned as the displacement of the platform after one minute. This parameter can be adjusted steplessly thanks to the hydraulic pump system. Aluminum oxide grinding wheel with the serial number of Cn80.TB1. G.V1. 320 × 28 × 160.35 m/s was used during the experiment. This is a type of grinding wheel manufactured by Hai Duong Grinding Wheel Factory (Vietnam). This type of grinding wheel is commonly used for grinding steel in general and alloy steel in particular [10]. The surface roughness is measured with SJ201 machine. In order to reduce the random error of the measurement, each experimental sample has been measured at least three times on three different sites. The roughness value at each experiment is the average value of consecutive measurements. In order to eliminate the effect of grinding wheel wear on the experimental results, before each experiment, the grinding wheel has been dressed with a 1-grain diamond dresser with the following dressing parameters: the cutting depth of 0.02 mm, the feed rate of 150 mm/min. In the grinding process, 10% emulsion solution has been used to irrigate the grinding area with a flow of about 16 L/min.

2.2 Experiment Design

The selected parameters are the input parameters of the experimental process, including cutting velocity, workpiece velocity, feed rate and cutting depth. The CCD design has been chosen to design the experimental matrix. In accordance with this design, each parameter will have 5 levels of values, encoded as $-\alpha$, −1, 0, 1 and $+\alpha$, of which $\alpha = (2k)1/4$, where k is the number of input parameters. In this study, the number of input parameters is 4, so α is 2. The values of parameters at all levels are shown in Table 1.

Table 1. Values of input parameters

Parameters	Code symbol	Actual symbol	Unit	Levels				
				−2	−1	0	1	2
Cutting velocity	x_1	v_c	m/s	18	22	26	30	34
Workpiece speed	x_2	v_w	m/min	35	40	45	50	55
Feed rate	x_3	f_d	m/min	0.2	0.3	0.4	0.5	0.6
Depth of cut	x_4	a_p	mm	0.005	0.01	0.015	0.02	0.025

The matrix includes 2k experiments at the base levels, 2k axial experiments (corresponding to the encoding level of ± α) and 5 selected experiments at the central levels [15, 16]. A total of 29 experiments of the matrix are shown in Table 2.

Table 2. Central composite designs matrix

No.	Code value				Actual value				Ra (μm)
	x_1	x_2	x_3	x_4	v_c (m/s)	V_w (m/min)	f_d (m/min)	a_p (mm)	
1	1	−1	−1	1	30	40	0.3	0.02	0.460
2	0	0	−2	0	26	45	0.2	0.015	0.102
3	−2	0	0	0	18	45	0.4	0.015	0.721
4	−1	1	−1	−1	22	50	0.3	0.01	0.514
5	−1	1	−1	1	22	50	0.3	0.02	0.736
6	0	0	0	0	26	45	0.4	0.015	0.428
7	−1	−1	1	−1	22	40	0.5	0.01	1.742
8	1	−1	1	1	30	40	0.5	0.02	0.926
9	−1	−1	−1	−1	22	40	0.3	0.01	0.736
10	1	1	1	1	30	50	0.5	0.02	0.654
11	0	0	0	−2	26	45	0.4	0.005	0.442
12	−1	−1	1	1	22	40	0.5	0.02	1.021
13	−1	1	1	−1	22	50	0.5	0.01	0.970
14	1	1	1	−1	30	50	0.5	0.01	0.490
15	0	0	0	0	26	45	0.4	0.015	0.440
16	0	0	0	0	26	45	0.4	0.015	0.432
17	0	0	0	0	26	45	0.4	0.015	0.436
18	0	0	0	2	26	45	0.4	0.025	0.461
19	1	1	−1	−1	30	40	0.3	0.01	0.422
20	1	−1	1	−1	30	40	0.5	0.01	0.674
21	1	1	−1	1	30	50	0.3	0.02	0.228
22	−1	−1	−1	1	22	40	0.3	0.02	1.030
23	−1	1	1	1	22	50	0.5	0.02	1.324
24	2	0	0	0	34	45	0.4	0.015	0.512
25	0	2	0	0	26	55	0.4	0.015	0.222
26	0	−2	0	0	26	35	0.4	0.015	0.219
27	1	1	−1	−1	30	50	0.3	0.01	0.054
28	0	0	2	0	26	45	0.6	0.015	0.918
29	0	0	0	0	26	45	0.4	0.015	0.452

3 Results and Discussion

The experiments were conducted in the same order as shown in Table 2, in which the experimental results have been included. In Fig. 1, the Pareto chart shows the effect of the input parameters on the surface roughness, in which the significance level is chosen as 0.05 [15, 16].

In Fig. 1, the curves of the feed rate and cutting velocity have exceeded the red limiting curve in the Pareto chart, proving that these two parameters have a great effect on the surface roughness. Moreover, the effect of the feed rate on the surface roughness is greater than that of the cutting velocity. The two parameters of the workpiece velocity and cutting depth have an insignificant effect on the surface roughness. However, if considered in detail, the effect of the workpiece velocity on the surface roughness is greater than the effect of the cutting depth.

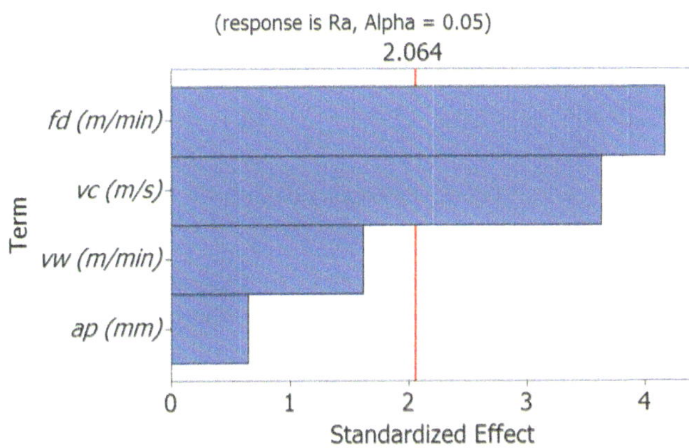

Fig. 1. Pareto chart of the standardized effects for Ra

From the experimental results in Table 2, the regression model showing the relationship between the surface roughness and the input parameters is shown in the formula (1). This model has a coefficient of determination R2 = 0.7089 and an adjusted coefficient of determination R2(adj) = 0.4177. The meaning and way to determine the values of these two coefficients have been presented in the documents on experimental planning and data processing [15, 16]. In this case, both coefficients, R2 and R2(adj), are much smaller than 1, so if the model (1) is used to predict the surface roughness, the predicted value will be different from the experimental results to a large extent [15]. Therefore, the problem here is to find a solution to improve the accuracy of the surface roughness model. This is the content that will be presented in the next part of this article.

$$\begin{aligned} R_a = {} & 0.4376 - 0.1910x_1 - 0.0848x_2 + 0.2189x_3 + 0.0340x_4 + 0.0946x_1^2 \\ & - 0.0044x_2^2 + 0.0680x_3^2 + 0.0533x_4^2 - 0.0044x_1x_2 \\ & - 0.0288x_1x_3 + 0.0299x_1x_4 + 0.0119x_2x_3 + 0.0657x_2x_4 - 0.0424x_3x_4 \end{aligned} \quad (1)$$

4 Improvement in the Accuracy of Surface Roughness Model

4.1 Distribution Law of Surface Roughness

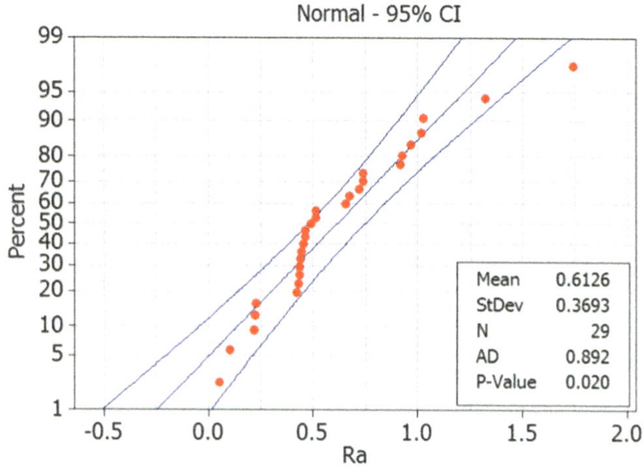

Fig. 2. Probability plot of Ra

To improve the accuracy of regression models, two known instruments are the Box-Cox transformation and the Johnson transformation. The Box-Cox transformation has been successfully applied to improve the accuracy of the roughness model when centreless grinding SCM435 steel [17], when milling EN353 steel [18], when milling AISI 1019 steel [19]. The Johnson transformation has also been successfully applied to improve the accuracy of the roughness model when milling AISI 1045 steel [20], when surface grinding 65G steel [21]. These two data transformations are two instruments to perform the transformation of a normally undistributed dataset into a normally distributed dataset. Thus, the condition to apply these data transformations is the normally undistributed initial data [15].

The distribution rule of the set of roughness values in the experiment is reported in Fig. 2, in which the red dots represent the surface roughness. Among these 29 red dots, there are also some red dots located relatively far from the central line (the middle straight line). On the other hand, the P-value of 0.02 is smaller than the significance level (normally, the significance level is 0.05 [15, 16]). Thus, it can be confirmed that the surface roughness dataset is normally undistributed. This is a necessary and sufficient condition to perform the Box-Cox and Johnson transformations [15].

4.2 Building the Surface Roughness Model Using the Box-Cox Transformation

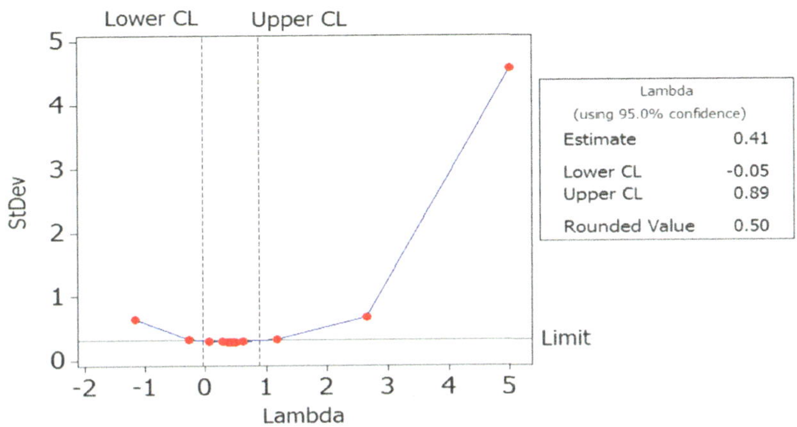

Fig. 3. Box-Cox transformation graph of surface roughness

The Box-Cox transformation: the data transformation is performed in accordance with the following formula [22]:

$$\begin{bmatrix} R_a' = R_a^{\lambda} \ when \ \lambda \neq 0 \\ R_a' = \ln(R_a) \ when \ \lambda = 0 \end{bmatrix} \tag{2}$$

Where, Ra' is the surface roughness value after the transformation, Ra is the surface roughness in the experiment, and λ is the exponent of transformation. Statistical software Minitab has been used to perform the transformation. The graph of Box-Cox transformation is shown in Fig. 3 with the λ value of 0.5. The surface roughness value after the transformation is shown in Table 3.

Figure 4 shows the distribution rule of the dataset after performing the Box-Cox transformation. The red dots represent data located close to the central line, and the P-value of 0.175 is much larger than the significance level. Thus, the dataset after the transformation has been normally distributed. Consequently, it is confirmed that the Box-Cox transformation has been successfully performed in this case. With the λ value of 0.5, the surface roughness model as in the formula (3) can be built. This model has the R^2 and R^2(adj) values of 0.7559 and 0.5119, respectively.

$$R_a = \begin{pmatrix} 0.6615 - 0.1188x_1 - 0.0575x_2 + 0.1479x_3 + 0.0294x_4 \\ +0.0614x_1^2 - 0.0168x_2^2 + 0.0255x_3^2 + 0.0338x_4^2 - 0.0250x_1x_2 \\ +0.0148x_1x_3 + 0.0231x_1x_4 + 0.0279x_2x_3 + 0.0402x_2x_4 \\ -0.0291x_3x_4 \end{pmatrix}^2 \tag{3}$$

Table 3. Value of surface roughness before and after transformations.

No	Surface roughness		
	Experimental value (μm)	Box-Cox transformation (dimensionless)	Johnson transformation (dimensionless)
1	0.460	0.67823	−0.27921
2	0.102	0.31937	−1.62842
3	0.721	0.84912	0.46161
4	0.514	0.71694	−0.10851
5	0.736	0.85790	0.49827
6	0.428	0.65422	−0.38498
7	1.742	1.31985	2.12721
8	0.926	0.96229	0.91786
9	0.736	0.85790	0.49827
10	0.654	0.80870	0.29069
11	0.442	0.66483	−0.33828
12	1.021	1.01045	1.10102
13	0.970	0.98489	1.00464
14	0.490	0.70000	−0.18317
15	0.440	0.66332	−0.34491
16	0.432	0.65727	−0.37157
17	0.436	0.66030	−0.35821
18	0.461	0.67897	−0.27596
19	0.422	0.64962	−0.40520
20	0.674	0.82098	0.34298
21	0.228	0.47749	−1.12017
22	1.030	1.01489	1.11758
23	1.324	1.15065	1.59680
24	0.512	0.71554	−0.11466
25	0.222	0.47117	−1.14389
26	0.219	0.46797	−1.15578
27	0.054	0.23238	−1.82340
28	0.918	0.95812	0.90170
29	0.452	0.67231	−0.30533

4.3 Building the Surface Roughness Model Using the Johnson Transformation

The statistical software Minitab has been used once again to perform the Johnson transformation. The dataset after the transformation is also included in Table 3. Figure 4 presents the graph of data transformation. The upper left part of this figure is the distribution rule of untransformed dataset, which has been analyzed in Sect. 4.1. The lower left part of this figure shows the distribution rule of the dataset after the transformation. The red dots have been close to the central line, the P-value of 0.219 is also much larger than the significance level. Thus, it can be confirmed that, the dataset after the Johnson transformation has also been normally distributed. When observing the upper right part of this figure, we also see that the curve of the dataset is bell-shaped, which further reinforces the idea that the dataset, after the transformation, has been normally distributed. The lower right partsupplies the information on transformation. Since then, a new model of surface roughness is shown as in the formula (4). This model has its coefficients of R2 and R2(adj) equal to 0.7592 and 0.5185, respectively.

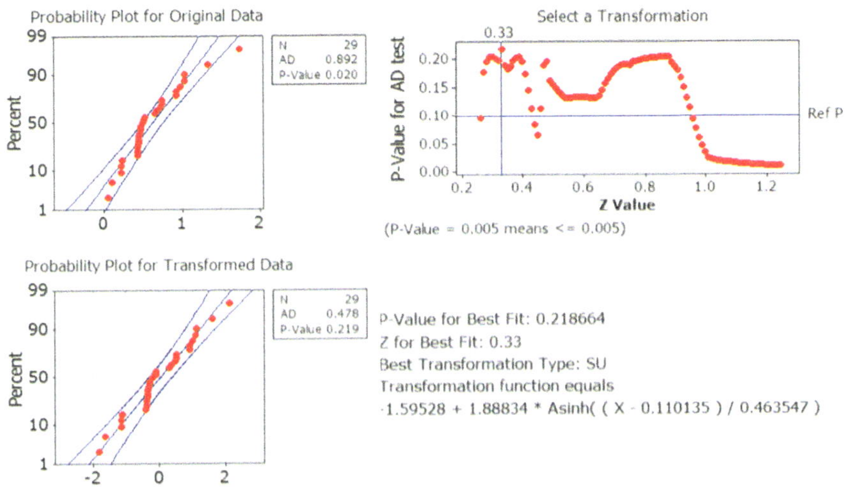

Fig. 4. Johnson transformation graph for *Ra*

$$R_a = 0.110135 + 0.463547 * Sinh(0.6579 - 0.2482x_1 - 0.1157x_2 + 0.3063x_3 + 0.0617x_4 + 0.1368x_1^2 - 0.0384x_2^2 + 0.0657x_3^2 + 0.0732x_4^2 - 0.0516x_1x_2 + 0.0388x_1x_3 + 0.0359x_1x_4 + 0.0564x_2x_3 + 0.0689x_2x_4 - 0.0477x_3x_4) \quad (4)$$

4.4 Comparison of Surface Roughness Models

The three surface roughness models as built above ((1), (3) and (4) are used to predict the surface roughness with the data of the input parameters as shown in Table 2. The results are shown in Table 4. Percentage absolute error (PAE) between the predicted results by the models and the experimental results has also been calculated and included in this table.

Table 5 exhibitsa summary of information on three built roughness models, inclusive of R2, R2(adj) and the average value of PAE.

Table 4. Values of surface roughness when predicting by models

No.	Experimental Ra (μm)	Predicted cutting force Ra (μm)			PAE		
		Without transformation	Box-Cox transformation	Johnson transformation	Without transformation	Box-Cox transformation	Johnson transformation
1	0.460	0.4097	0.4105	0.4064	10.93%	10.76%	11.66%
2	0.102	0.2718	0.2187	0.2552	166.47%	114.45%	150.22%
3	0.721	1.198	1.3103	1.3385	66.16%	81.74%	85.64%
4	0.514	0.3879	0.3845	0.4124	24.53%	25.19%	19.76%
5	0.736	0.6123	0.5949	0.5845	16.81%	19.17%	20.58%
6	0.428	0.4376	0.4376	0.4376	2.24%	2.24%	2.24%
7	1.742	1.2603	1.1879	1.1075	27.65%	31.81%	36.43%
8	0.926	0.6813	0.7261	0.7170	26.43%	21.59%	22.57%
9	0.736	0.7039	0.6745	0.6490	4.36%	8.35%	11.81%
10	0.654	0.6581	0.6778	0.6549	0.63%	3.64%	0.14%
11	0.442	0.5828	0.5744	0.5389	31.86%	29.96%	21.92%
12	1.021	1.0523	0.9291	0.9231	3.07%	9.00%	9.59%
13	0.970	0.9919	1.0006	0.9500	2.26%	3.15%	2.06%
14	0.490	0.4837	0.4846	0.4988	1.29%	1.11%	1.79%
15	0.440	0.4376	0.4376	0.4376	0.55%	0.55%	0.55%
16	0.432	0.4376	0.4376	0.4376	1.30%	1.29%	1.29%
17	0.436	0.4376	0.4376	0.4376	0.37%	0.36%	0.36%
18	0.461	0.7188	0.7665	0.7094	55.92%	66.27%	53.89%
19	0.422	0.3285	0.3113	0.3254	22.16%	26.24%	22.88%
20	0.674	0.7697	0.7845	0.7464	14.20%	16.39%	10.75%
21	0.228	0.3389	0.2503	0.2476	48.64%	9.78%	8.60%
22	1.030	0.6655	0.6589	0.6556	35.39%	36.03%	36.35%
23	1.324	1.0467	1.0714	1.0441	20.94%	19.08%	21.14%
24	0.512	0.434	0.4482	0.4668	15.23%	12.46%	8.82%
25	0.222	0.2504	0.2297	0.2382	12.79%	3.48%	7.30%
26	0.219	0.5896	0.5031	0.4828	169.22%	129.73%	120.44%
27	0.054	-0.0051	0.0659	0.0469	109.44%	22.03%	13.15%
28	0.918	1.1474	1.1221	1.1340	24.99%	22.23%	23.53%
29	0.452	0.4376	0.4376	0.4376	3.19%	3.19%	3.19%
Mean					31.69%	25.22%	25.13%

The data in Table 5 show that the values of both coefficients of R2 and R2(adj) of the model using the Johnson transformation have the largest value, followed by the model using the Box-Cox transformation. These two parameters of the model not using the data transformation are both smaller than those in the two models using the data transformation. Regarding PAE, the model using the Johnson transformation has the smallest value, followed by the model using the Box-Cox transformation, and for the model not using the data transformation, this parameter has the largest value. These comparison results have shown that the surface roughness model using the Johnson transformation has the highest accuracy, followed by the model using the Box-Cox transformation, and finally, the model not using the data transformation.

Table 5. Comparison of surface roughness models

Models	R^2	R^2(adj)	% PAE
Without using transformation	0.7089	0.4177	31.69
Using Box-Cox transformation	0.7559	0.5119	25.22
Using Johnson transformation	0.7592	0.5185	25.13

5 Conclusions

In this study, the grinding process of 1066 steel on the external cylindrical grinder has been performed. The effect of cutting velocity, workpiece velocity, feed rate, and cutting depth on surface roughness has been found. Three surface roughness models have been built. The results are drawn as follows:

- Feed rate is the parameter with the greatest effect on the surface roughness, followed by the effect of the cutting velocity. The workpiece velocity and cutting depth have an insignificant effect on the surface roughness.
- The roughness model using the Johnson transformation has the highest accuracy, followed by the mode using the Box-Cox transformation. The roughness model not using the data transmission has the lowest accuracy.
- The surface roughness model using the Johnson transformation will be used to optimize the grinding process, which is the direction for the next study of the authors of this article.

Acknowledgment. This work was supported by Thai Nguyen University of Technology.

References

1. Malkin, S., Guo, C.: Grinding Technology: Theory and Applications of Machining with Abrasives, 2nd edn. Industrial Press, New York (2008)
2. Dich, T.V., Binh, N.T., Dat, N.T., Tiep, N.V., Viet, T.X: Manufacturing Technology. Science and Technics Publishing House, Ha Noi (2003)
3. Thien, N.V., Dung, H.T., Trung, D.D., Cuong, N.: Optimization cutting parameters when grinding X12M steel using hai duong grinding wheel. Int. J. Eng. Technol. **7**(4), 6994–6996 (2018). https://doi.org/10.14419/ijet.v7i4.28526
4. Thiagarajan, C., Sivaramakrishnan, R., Somasundaram, S.: Modeling and optimization of cylindrical grinding of Al/SiC composites using genetic algorithms. J. Braz. Soc. Mech. Sci. Eng. **34**(1), 32–40 (2012)
5. Dung, H.T., Trung, D.D., Thien, N.V., Tung, N.N., Cuong, N.: Optimization cutting parameters when grinding шх15 steel using hai duong grinding wheel. Int. J. Mech. Prod. Eng. Res. Dev. **9**(4), 155–162 (2019). https://doi.org/10.24247/ijmperdaug201917
6. Thakur, A., Sahni, V.: Optimization of traverse cut cylindrical grinding process parameters of treated EN-8 steel. Int. J. Adv. Res. Sci. Eng. **6**(12), 1512–1523 (2017)

7. Singh, T., Kumar, P., Goyal, K.: Optimization of process parameters for minimum outof-roundness of cylindrical grinding of heat treated AISI 4140 steel. Am. J. Mech. Eng. **2**(2), 34–40 (2014). https://doi.org/10.12691/ajme-2-2-1
8. Hung, T.Q., Duc, D.V., Son, N.H.: Optimization of cutting parameters for minimum the surface roughness when grinding SKD11 steel on cylindrical grinder, In: Proceeding of the 5th National Conference on Mechanical Science & Technolog, pp. 560–565 (2018). In Vietnamese
9. Nguyet, V.N.: Effects of grinding parameters on surface roughness on when external cylindrical grinding SKD11 steel using Hai Duong grinding wheel. Int. J. Adv. Eng. Manage. Res. **6**(1), 9–18 (2021)
10. Long, B.T., Luc, T.T., Tuy, T.S.: Material Processing Principles. Science and Technics Publishing House, Hanoi (2013)
11. Liu, J.D., Wang, G.C., Wang, B.L., Chen, K.M.: Study on the formation of grind-hardening of steel AISI 1066. Key Eng. Mater. **329**, 57–62 (2017). https://doi.org/10.4028/www.scientific.net/KEM.329.57
12. Judong, L., Wei, Y., Songwei, H., Zhilong, X.: Experimental study on grinding-hardening of 1060 steel. Energy Procedia **16**, 103–108 (2012). https://doi.org/10.1016/j.egypro.2012.01.019
13. Zhuang, V.J.Z., Liu, J.D., Zhang, Y.Z.: Study on depth and its uniformity of 65Mn steel grindhardened layer. Key Eng. Mater. **487**, 94–98 (2011). https://doi.org/10.4028/www.scientific.net/KEM.487.94
14. Liu, J.D., Wang, G.C., Wang, Z., Fan, S.T.: Experimental Research on Grind-Hardening of 65Mn steel. Mater. Sci. Forum **505–507**, 787–792 (2006). https://doi.org/10.4028/www.scientific.net/MSF.505-507.787
15. Du, N.V., Binh, N.D.: Design of Experiment Techniques. Science and Technics Publishing House, Ho Chi Minh (2011)
16. Dean, A., Voss, D., Draguljić, D.: Design and Analysis of Experiments, 2nd (edn.). Springer, Cham (2007)
17. Trung, D.D., Nguyen, N.T.: Investigation of the surface roughness in infeed centerless grinding of SCM435 steel. Int. J. Autom. Technol. **15**(1), 123–130 (2021)
18. Bhardwaj, B., Kumar, R., Singh, P.K.: An improved surface roughness prediction model using box-cox transformation with RSM in end milling of EN 353. J. Mech. Sci. Technol. **28** (12), 5149–5157 (2014). https://doi.org/10.1007/s12206-014-0837-4
19. Bhardwaj, B., Kumar, R., Singh, P.K.: Effect of machining parameters on surface roughness in end milling of AISI 1019 steel. Proc. Inst. Mech. Eng. Part B J. Eng. Manuf. 1–11 (2013). https://doi.org/10.1177/0954405413506417
20. Trung, D.D: Influence of cutting parameters on surface roughness during milling AISI 1045 steel. Tribol. Ind. **42**(4), 658–665 (2020). https://doi.org/10.24874/ti.969.09.20.11
21. Trung, D.D.: Influence of cutting parameters on surface roughness in grinding of 65G steel. Tribol. Ind. **43**(1), 167–176 (2021). https://doi.org/10.24874/ti.1009.11.20.01
22. NCSS Statistical Software: Box-cox transformation. https://ncsswpengine.netdna-ssl.com/wpcontent/themes/ncss/pdf/Procedures/NCSS/Box-Cox_Transformation.pdf

The Effect of Bonnet Reinforcement Structure on Pedestrian Head Injuries in Collisions

Van-Luc Ngo[1,2](✉) and Van-Tuan Nguyen[1,2]

[1] Faculty of Mechanical Engineering and Mechatronics, Phenikaa University, Hanoi 12116, Vietnam
luc.ngovan@phenikaa-uni.edu.vn

[2] A&A Green Phoenix Group JSC, Phenikaa Research and Technology Institute (PRATI), No. 167 Hoang Ngan, Trung Hoa, Cau Giay, Hanoi 11313, Vietnam

Abstract. Pedestrian headform impactor on the bonnet top to assess the level of pedestrian head injury in a collision is one of the mandatory tests for all vehicles sold in the European Union market. Therefore, many automakers research how to design a bonnet reinforcement structure for pedestrian head protection when developing a new model of vehicle. Each automaker has their own designs and the reality shows that the different models of vehicles have different structures of bonnet reinforcement. Several structural models of the pedestrian-friendly bonnet reinforcement have been introduced by automakers, but the basis or design rules have never been published. Previous studies have focused on stiffness as it is believed that bonnet stiffness is the main cause of head injuries in collisions. This study was carried out with the aim of exploring how the structural properties of bonnet reinforcement affect on head injury in collision beside the factor of bonnet stiffness. The results of this study will contribute to the basis as well as the principles for the design of a bonnet reinforcement structure for pedestrian head safety.

Keywords: Pedestrian head safety · Bonnet reinforcement structure · Secondary collision · Headform impactor · HIC value

1 Introduction

Research into designing the frontal structure of vehicle for pedestrian protection in collisions is one of the solutions that is being carried out by automakers as well as many research centers. The structure of the bonnet interests a lot of researchers because it is related to the pedestrian headform impactor to bonnet top tests [1]. The structure of the bonnet includes the bonnet skin covering the top and the bonnet reinforcement that gives the bonnet stiffness. Bonnet skin is simply a cover sheet on top so the essence of the study of the bonnet structure is the study of bonnet reinforcement structure. There are many designs of bonnet structure for pedestrian protection introduced with different structures. Previous researchers believed that the stiffness of the bonnet is the main cause of head injuries in pedestrian head to bonnet top collisions. Therefore, they focused on the solutions to reduce the stiffness and mass of the bonnet by making more holes in the rib of reinforcement [2], reducing the thickness [3], or using composite

D. C. Nguyen et al. (Eds.): ICERA 2021, LNNS 366, pp. 909–916, 2022.
https://doi.org/10.1007/978-3-030-92574-1_93

materials [4]. Several patented models such as Ladder Honeycomb [5], or semi-spherical structures [6] are designed on the basis of the impact energy absorption and the flexibility of changing the stiffness.

This study aimed to find out other factors of bonnet reinforment structure that influence pedestrian head injury beyond factors such as stiffness, energy absorption, mass of bonnet. In this study, the pedestrian headform impacting on many positions on bonnet top with different structural characteristics will be simulated to analyze the effect of bonnet reinforcement structure on pedestrian head injuries. The results of this study will contribute to the development of a process for designing a bonnet structure for pedestrian protection in vehicle-pedestrian collisions.

2 Procedure of Pedestrian Headform Impacting on Bonnet Top Performance

The tests of pedestrian safety in pedestrian head to bonnet of vehicle collisions are performed base on a procedure proposed by the EEVC, which is briefly described in Fig. 1 [7]. Accordingly, the child and adult headform impactors are used to impact to the corresponding areas on the bonnet top area where often impact on the head of children and adults in vehicle-pedestrian collisions. The child and adult headform impactors are respectively used to test positions in the forward and rearward region of the bonnet, these regions are divided by WAD (Wrap Around Distance-Fig. 1) of 1500 mm. The impact angle for the child and adult headform is 500 and 650 respectively, the impact velocity for both child and adult headform impactors is 40 km/h.

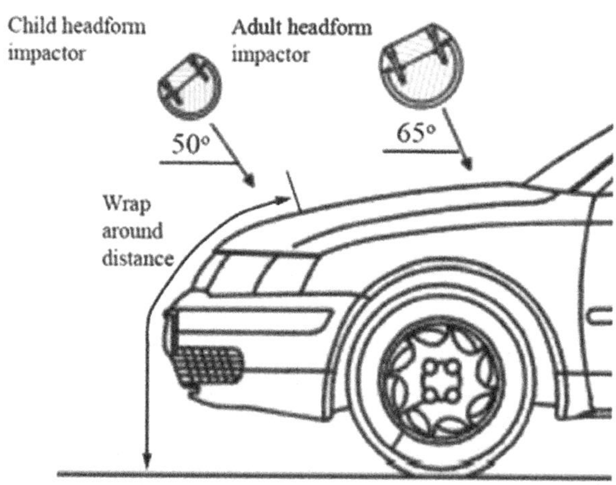

Fig. 1. Pedestrian headform to bonnet top tests

The child and adult headform impactors have a spherical structure with an aluminum core and covered by rubber outside. The child headform impactor has a diameter of 130 mm and weight of 2.5 kg, whereas these parameters of the adult headform are 165 mm and 4.8 kg, respectively. The child and adult headform impactors must pass dynamic tests according to EEVC regulations before using in the tests [7]. An accelerometer is placed at the center of gravity of the headforms to measure the acceleration during the collision. The level of head injury is assessed through the HIC value calculated based on the the headform impactor resultant acceleration of the center of gravity.

With T_1 and T_2 are the two instant times (*unit of second*) in the collision process and (A) is resultant acceleration measured during collision (*unit of [g], 1[g] = 9.81 [m/s^2]*), the HIC value is determined as follows equation:

$$\mathrm{HIC} = Max\left[\frac{1}{T_2 - T_1}\int_{T_1}^{T_2} \mathrm{AdT}\right]^{2.5} (\mathrm{T}_2 - T_1) \tag{1}$$

$$\text{With condition } \Delta T = T_2 - T_1 \leq 15\,\text{ms (mini - second)} \tag{2}$$

The higher HIC value, the greater level of pedestrian head injury will be. The HIC value of 1000 is pedestrian head safety threshold. The HIC value of 2000 is the limit for serious injury and pedestrian can be fatal if the HIC value is greater than 3000. The equation to calculate the HIC value shows that the HIC value not only depends on the maximum value of (A) but also depends on the variable characteristics of (A) during the processs of collision.

3 Simulations of Headform Impacting on the Bonnet Top

In this research, the finite element model of 1992 Ford Taurus, child and adult headform impactors as shown in Fig. 2 are used to simulate pedestrian head safety tests. These finite element models are developed and validated under conditions to ensure that these models are usable in simulations of pedestrian heaform impacting on bonnet top tests [8, 9].

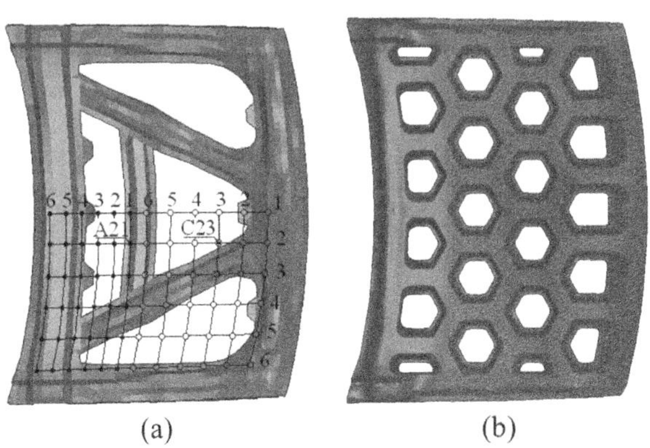

(a) (b)

Fig. 2. Positions on bonnet were selected for collisions by headform impactors

All parts in the engine room close to bonnet reinforcement structure such as oil cap, or battery will be moved down to avoid the contact during the collisions and affect the results of evaluation. The velocity and direction of the impact were established according to the procedures recommended by the EEVC as described briefly above.

The positions on the bonnet surface selected to perform the tests are described in Fig. 3a. Total of 72 positions were selected to test of collision, of which 36 positions would be used for tests with child headform impactor (marked as positions of Cij) and 36 positions for test with adult headform impactor (marked as positions of Aij). Because bonnet reinforcement structure is symmetric, to analyze the effect of bonnet reinforcement structure on head injury only positions on a half are selected. On each testing region with the child and adult headforms, the impact positions were selected equally spaced both horizontally and vertically. The total number of selected positions was large enough so that it is possible to collide positions with different reinforcement structure properties. After performing the tests with the original reinforcement structure as shown in Fig. 3a, the results will be analyzed to find out what factors of reinforcement structure influence on pedestrian head injuries. On the basis of analysis results, a new structure will be proposed as shown in Fig. 3b and continue to repeat the tests same selected positions. The results obtained when testing with the new bonnet reinforcement structure will be compared with the test results with the original one to evaluate and re-check the proposed argument analysis.

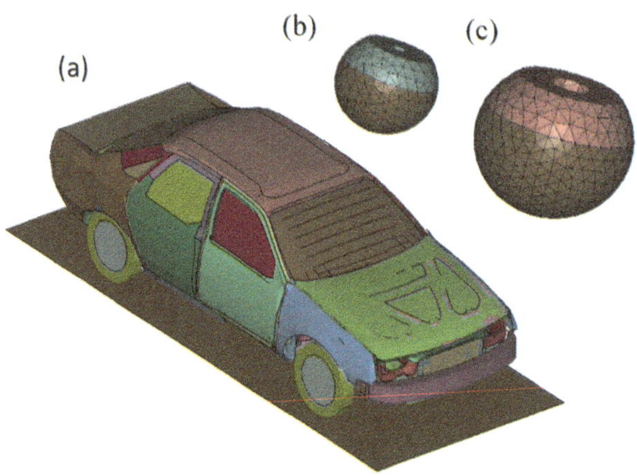

Fig. 3. Finite element models are used to simulation pedestrian headform impact to bonnet top

4 Results and Discussions

Figures 4 and 5 show the results of the collision tests with the original and new structure of reinforcements. Testing results with the original structure of reinforcement (*left* column) show that the HIC value is very high and uneven, the level of injuries for adults is higher than for children. Just intuitively one would think that position C14 is safer than position C16 because position C14 is in the middle of the hole of reinforcement where as position C16 is on the rib of reinforcement. But the testing results show that position C16 is in the safe range while position C14 is in the dangerous range. The testing results of impact to positon C25 is at the threshold of fatality even though position C25 is also on the hole of reinforcement. Positions on the rib of reinforcement are obviously harder than those on the hole of reinforcement but do not always give higher HIC values in impacting tests as shown in Figs. 3, 4 and 5.

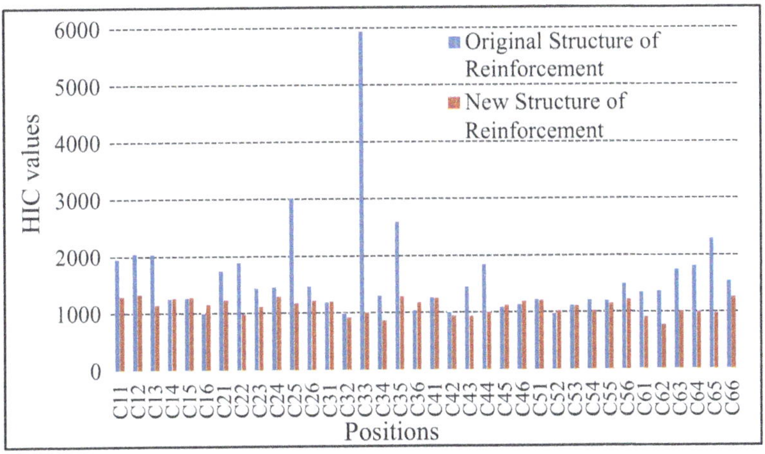

Fig. 4. The results of simulations child headform impactor collision to bonnet top

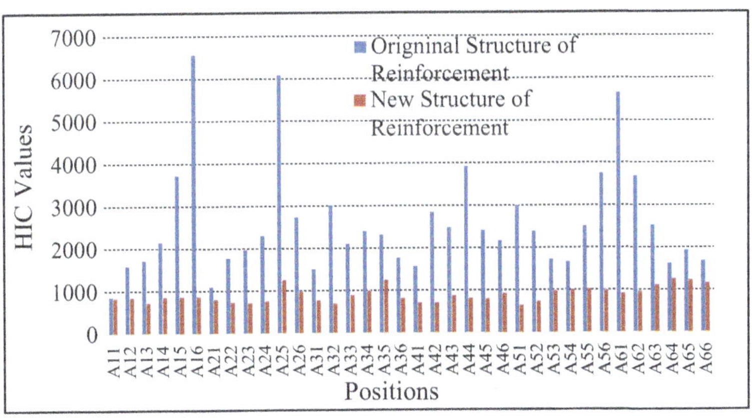

Fig. 5. The results of simulations adult headform impactor collision to bonnet top

These results demonstrate that the stiffness of the bonnet is only one of the factors affecting the HIC value, but not completely determining the HIC value. Kinetic analysis of headforms during impact time shows that, in addition to the stiffness at the impact position the uniformity of stiffness of the area around the impact position also has a great influence on the HIC value. Heaforms impact on the bonnet surface in an inclined direction, so the velocity can be divided into two components, perpendicular to the bonnet surface (Vn) and parallel to the bonnet surface (Vt) as shown in Fig. 6. The kinetic energy of (Vn) will have the effect on causing the bonnet surface to sag down and the kinetic energy of (Vt) will cause the headform impactor to slide on the surface of the bonnet. If the impact position and the surrounding area have uniform stiffness, the bonnet surface will sag evenly as the HA curve on the diagram. In this case, the headform will slide easily over the surface of the bonnet skin as described in Fig. 6a, and has little rotation during the sliding. If in the sliding direction of the headform during impact it is blocked by rib of reinforcement such as positions C14, C15 and C25, it will make the stiffness of the bonnet harder at the incoming sliding direction. In this case, the bonnet skin will not sag evenly, but following the HB curve as shown in the diagram in Fig. 6. During the sliding, the headform will be greatly hindered, causing the headform to rotate very strongly during the collision as shown in Fig. 6b. A strong rotation of headform around contact point will increase the normal acceleration (An) of the headform impactor center of gravity and therefore will increase the resultant acceleration (A). Figure 7 describes the changing characteristics of the components of acceleration getting from testing impact to position C25. The resultant acceleration (A) has two peaks, the first is decided by (At) and the second is decided by (An). If there is no tripping during the slide, causing the headform to rotate strongly around the contact point, there will be no sudden increase of (An) and therefore no second peak of (A). The occurrence of the second peak is equivalent to a secondary collision that occurs after the main collision. The combination of these two collisions is the cause of the sudden increase in the HIC value at position C25. An analysis of the variation characteristics of the acceleration components at positions with unusually high HIC values is all occurrence of secondary collisions. Position with characteristics that are not only higher stiffness at impact position, but also occurrence of secondary collisions, then the HIC value will be very high such as position C33 or some position

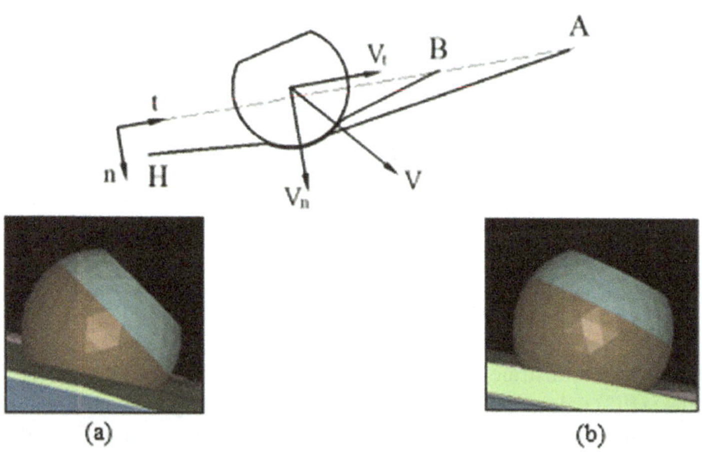

Fig. 6. Analysis of headform impactor kinetic characteristic

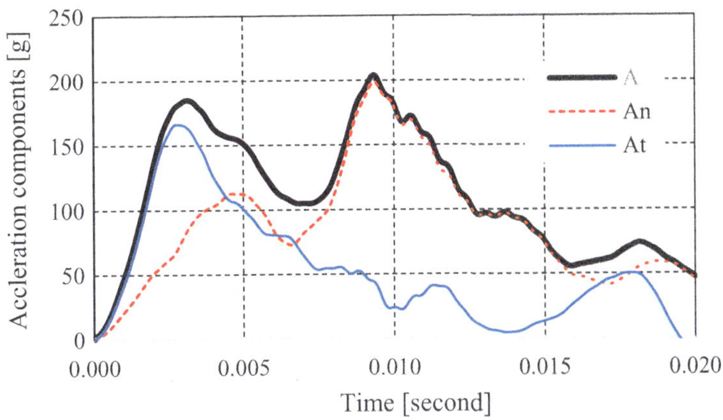

Fig. 7. The characteristic of acceleration components when impact to possion C25

tested with adult headform impactor. The testing results with the new structure of reinforcement (the right column in Figs. 4 and 5) with more uniform stiffness gives a more uniform HIC value and almost no difference between the child and adult headform and also not the valid positions HIC increased abnormally. This result proves that in addition to the stiffness, the uniform of stiffness also greatly affects the pedestrian head injuries in collisions to bonnet top.

5 Conclusions and Recomemdations

The results of this study have shown that besides the stiffness of the bonnet, the uniformity of the stiffness around the impact position also greatly affects on the level of pedestrian head injuries in headform to bonnet top tests. The obstruction during the headform impactor sliding over the bonnet surface in process of collision is equivalent to a second impact and contributes significantly to pedestrian head injury level. The uneven stiffness of the bonnet around the impact position is the main cause of the occurrence of obstruction during the slide of the headform impactor on bonnet surface. The design of reinforcement structure for pedestrian safety should first ensure uniformity of stiffness to avoid secondary collisions. It is the secondary collision that is really what causes the HIC value to reach dangerous levels.

References

1. Directive 2003/102/EC of the European Parliament and of the Council of 17 November 2003
2. Han, Y.H., Lee, Y.W.: Development of a vehicle structure with enhanced pedestrian safety. SAE World Congress, Detroit, Michigan, USA (2003)
3. Ingo, K., Flavio, F.: Improvements to pedestrian protection as exemplified on a standard-sized car. Report no. 283, Germany (2000)

4. Teng, T.-L., Liang, C.-C., Shih, C.-J., Nguyen, M.-T.: Design of car hood of sandwich structures for pedestrian safety. Int. J. Veh. Des. **63**(2/3), 185–198 (2013)
5. Ray, A.K.: Ladder honeycomb hood structure for a motor vehicle for pedestrian protection (2014). Pub. no.: US 2014/0110971 A1
6. Kaisare, S.K.: Vehicle bonnet structure for pedestrian protection (2014). Pub. no.: US 2014/013817.6 A1
7. EEVC/WG 17 Report: Improved test methods to evaluate pedestrian protection afforded by passenger cars. European Enhanced Vehicle-safety Committee (1998, updated 2002)
8. Information on www.ncac.gwu.edu/vml/models.html
9. Teng, T.L., Nguyen, T.H.: Development and validation of FE models of impactor for pedestrian testing. J. Mech. Sci. Technol. **22**(9), 1660–1667 (2008)

The Effects of High Tempering Parameters on the Relative Elongation and Yield Strength of 42CrMo Steel as the Turbine Shaft

Nguyen Ngoc Nhan, Nguyen Hoang Thien, and Le Hong Ky(✉)

Vinh Long University of Technology Education, Vinh Long, Vietnam
kylh@vlute.edu.vn

Abstract. The turbine shaft is an important part of the turbocharger structure. This part requires high mechanical properties, precision, and surface quality to ensure the working conditions with very high speed and temperature. Heat treatment of turbocharger turbine shafts is still a secret of the companies, and very few studies on this subject have been published. Automatic heat treatment system with turbine shafts is introduced. However, such an automated production system is not suitable for the actual conditions of the research facilities. This paper presents the research results of the high tempering process to give the optimal parameters of the relative elongation and yield strength of 42CrMo steel as the turbine shaft of the turbocharger structure.

Keywords: High tempering · Yield strength · Elongation · Tempering temperature · Tempering time · Turbocharger

1 Introduction

Engines using Turbocharger often have many times higher power than naturally loaded engines with the same cylinder capacity. Turbocharger works with very high temperature, and the working speed is usually 100,000 rpm (can be up to 250,000 rpm) [16]. The turbine shaft has a very special structure. This is an important part connecting the two turbine blades. Usually, the shaft part is manufactured in one piece with the flame blade, so in the process of using, the shaft is worn or when there is a problem with the turbine blade, it is required to replace both the shaft and the turbine blade. The manufacture of turbine blades is mainly by casting in metal molds [9, 10]. The heat treatment of turbocharger turbine shafts is still a secret of the companies, so there is almost no scientific publication. EMAG Corporation [13] introduces the automatic production of turbine shafts (Fig. 2). In which, in the heat treatment part, a robot picks it up from there and transfers it for heat treatment in a MIND-L 1000 from EMAG eldec, which very quickly and precisely heats the component to a specific temperature to relieve any possible stresses. However, such an automatic production system is not suitable for the actual conditions of the research facilities. Previously published studies [1–4] mentioned the design and manufacture of turbine blades; [11] published the research results on the effects of temperature and heat retention time at high tempering of steam turbine. High yield strength, tensile strength and percentage elongation are

D. C. Nguyen et al. (Eds.): ICERA 2021, LNNS 366, pp. 917–928, 2022.
https://doi.org/10.1007/978-3-030-92574-1_94

required along with high impact values at the same time. In addition, the material should exhibit good high temperature properties such as creep, elevated temperature strength and thermal stability. The material requires high tensile properties along with good impact strength for proper application. But it is generally observed that these two properties are inversely related to each other, i.e., the efforts to increase the tensile strength of materials results in decreasing in impact strength and vice versa. Therefore, one must judiciously select the heat treatment parameters to maintain a good combination of these properties [12]. The influence of sub-temperature quenching temperature on the tensile strength, hardness, and toughness of 35CrMo steel was studied [14]. The microstructure characteristics of this steel after sub-temperature quenching were analyzed. The results showed in the range of 780–810 °C, the strength and hardness of the 35CrMo steel sub-temperature quenched increased with the increasing quenching temperature, reached the highest strength and hardness at 800 °C quenching and then began to decrease, while its elongation decreased with the increasing quenching temperature. Effects of tempering temperature on the microstructures and low temperature impact on the toughness of 42CrMo4-V steel were investigated [15]. Microstructures of 42CrMo4-V steel after tempering at 570–720 °C for 3 h were experimentally investigated using scanning electron microscopy, electron back-scattered diffraction, transmission electron microscopy and X-ray diffraction.

This paper publishes the research results on the effects of temperature and heat retention time at high tempering on the relative elongation and yield strength of the turbine shaft body in the HX40W model turbocharger structure made of 42CrMo steel (Fig. 1). The study also determined the optimal technology parameters when highly tempering the turbine shaft.

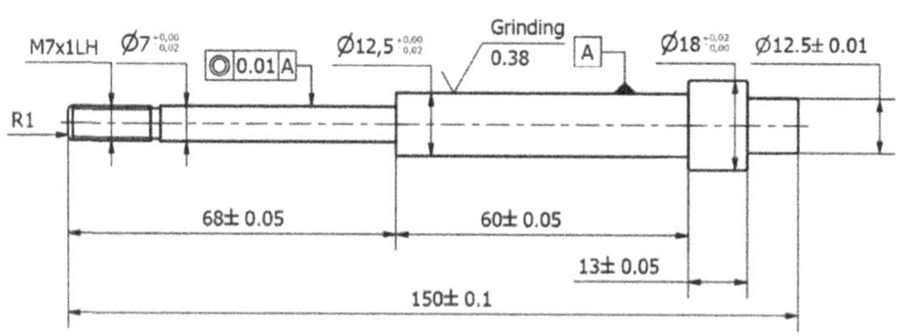

Fig. 1. Turbine shaft

Spectro Lab spectrum analyzer was used to determine the material composition of the Turbine shaft parts of the HX40W model Turbocharger applied to Dongfeng trucks. The microscopic organization of the turbine shaft was analyzed using an Optika metallographic microscope, thereby determining the organization of phases, the distribution of phases, the shape and size of phases, the thickness of the hardened layer, and the particle size in phases, etc. Analysis results show that the parts of the turbine shaft have a microscopic organization of sorbite, and the particle size is from 5 to 10 μm. In order to achieve this organization, the parts must go through the stages of quenching and high tempering in Fig. 2b. In the study, INSTRON's DX300 strength measuring device was used to determine the yield strength and relative elongation of the parts, respectively: $\sigma 0.2 = 1011\ N/mm^2$, $\delta = 14.3\%$.

2 Experiments

2.1 Sample for Experiment

From the above analysis, in this study, the material is 42CrMo low-alloy steel (Fig. 2a) with the chemical composition given in Table 1 to conduct heat treatment to achieve the microscopic organization of sorbite with the yield strength and low relative elongation mentioned above.

Table 1. Chemical composition of 42CrMo steel for turbine

Steel	Chemical composition						
	C	Si	Mn	Cr	Mo	P	S
42CrMo	0.38–0.45	≤ 0.40	0.60–0.90	0.90–1.20	0.15–0.30	≤ 0.035	≤ 0.035

40CrMo solid round workpiece has high strength, high hardness, toughness, small deformation at high temperature, high creep, and stable steel strength over time. This is a low alloy steel, with a Carbon content of about 0.42%, and the important chemical elements are Chromium (0.90–1.20%), Molybdenum (0.15–0.30%), Manganese (0.60–0.90%).

2.2 Experimental Equipment, Measuring Instruments

The purpose of high tempering is to reduce or lose internal stress, give the steel high mechanical properties, and improve machinability. The experiment used a volumetric furnace with LAM - 50 N2 shielding gas (see Fig. 2b) heated by a resistor, controlled by a pulse system to keep power supply so that the heating is always stable. In addition, the furnace also uses a protective nitrogen gas system to help the parts when the tempering organizations are transformed evenly. The method of single-stage calcination, sparse arrangement in furnace (see Fig. 2c) and cooling in SHL QUENCH 100B oil were used.

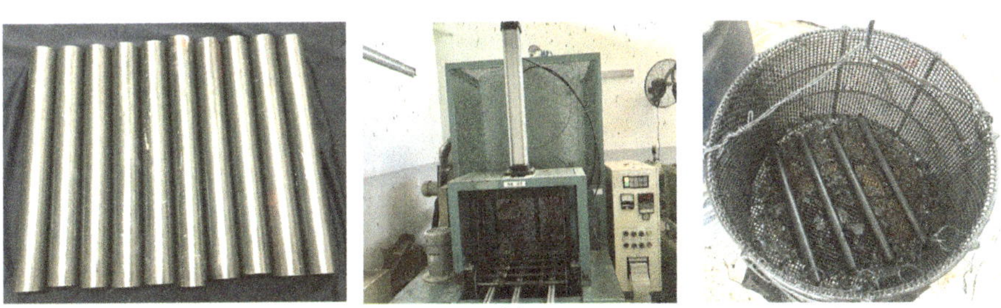

a) b) c)

Fig. 2. Workpieces, equipment, and tools of jig for placing workpieces in the experiment

To measure the experimental results, the INSTRON-300DX relative elongation and yield strength tester, (see Fig. 3) was used in the experiment.

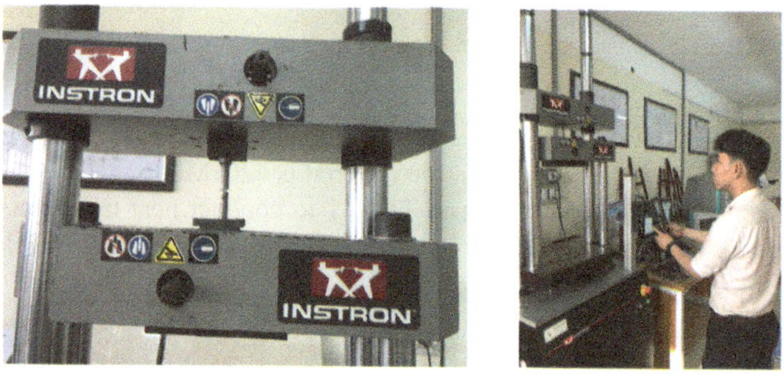

Fig. 3. The relative elongation and yield strength tester

2.3 Determining High Tempering Mode

With 42CrMo low-alloy steel workpiece, in order to have a microscopic organization of sorbite, with a particle size of 5–10 μm, and the desired yield strength and relative elongation, the part must go through the stages of quenching and high tempering. In which, according to [5] when quenching, the temperature is 850 °C; quenching time is 20 min, and the heat retention time is 55 min.

Selecting High Tempering Temperature (Temp R). The temperature range was selected to design experiments as two edges of high tempering mode of 550 °C (low temperature edge of high tempering) and 670 °C (high temperature edge of high tempering).

Selecting High Tempering Time (Time R). According to [5], the minimum effective time when high tempering is 120 min, here selected as the lower limit.

Heat Retention Time (T). According to [5] with 42CrMo steel, the heat retention time is up to 240 min to ensure the Martensite and Austenitic organizations to be residual after tempering turns into Sorbite. Natural design of experiments matrix is shown as in Table 2.

Table 2. High tempering experiment matrix

Experiment	*Temp R*	*Time R*
1	550	120
2	670	120
3	550	240
4	670	240
5	550	180
6	670	180
7	610	120
8	610	240
9	610	180
10	610	180

2.4 Experimentation [5]

The experiments were conducted according to the matrix in Table 2 with the LAM – 50 N2 furnace. The workpiece after high tempering is cooled in the air, Fig. 4a, and test sample according to ASTM E8, Fig. 4b.

a) *b)*

Fig. 4. Parts after high tempering and test samples

2.5 Carrying Out Mechanical Test of the Specimen

Before measuring the experiment results on the INSTRON-300DX device (Fig. 2), the test specimen for mechanical testing of the shafts must be machined according to ASTM E8 standard (American Society for Testing and Materials) [8] as in Fig. 3b. Tensile test results of specimens to determine yield limit of $\sigma0.2$, relative elongation δ are shown in Fig. 5 and Table 3.

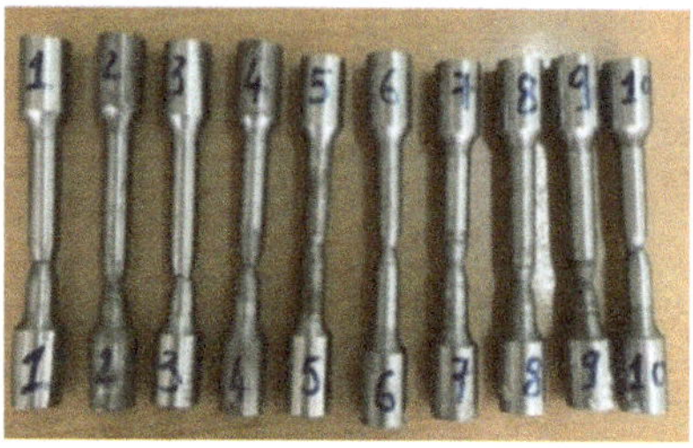

Fig. 5. Tensile test results of the specimen

Table 3. Experiment results

Experiment	Temp R	Time R	$\overline{\sigma_{0.2}}$	$s_{\sigma_{0.2}}$	$\overline{\delta}$	s_δ
1	550	120	1091	3	71.5	0.3
2	670	120	908	12	75	0.2
3	550	240	1080	8	72.1	0.1
4	670	240	868	12	75.6	0.2
5	550	180	1083	6	72.3	0.1
6	670	180	893	12	75.9	0.3
7	610	120	958	13	72.5	0.2
8	610	240	930	12	73.1	0.1
9	610	180	945	13	73	0.2
10	610	180	939	13	73.2	0.1

3 Analysis of Experimental Results [6]

3.1 Yield Strength $\sigma_{0.2}$

Minitab 19 software was used for preliminary experimental analysis of the experimental samples, removing the coefficients that do not affect the regression equation (with p-Value > 0.05) to get the results of experimental analysis as shown in Table 4, analysis of variance (ANOVA) for the regression equation just found as shown in Table 5.

Table 4. Analyzing experimental results

Term	Coef	SE Coef	T-Value	P-Value
Constant	942.92	2.82	334.20	0.000
Temp R	−97.61	2.30	−42.37	0.000
Time R	−13.22	2.30	−5.74	0.000
Temp R*Temp R	44.36	3.64	12.18	0.000
Temp R*time R	−7.25	2.82	−2.57	0.017

The analysis results show that the coefficients of the regression equation just calculated have p-Value < 0.05, which means that the coefficients have an influence on the regression equation.

Table 5. Analysis of variance (ANOVA)

Source	DF	Adj SS	Adj MS	F-Value	P-Value
Model	4	189449	47362	495.80	0.000
Linear	2	174650	87325	914.14	0.000
Temp R	1	171503	171503	1795.34	0.000
Time R	1	3147	3147	32.94	0.000
Square	1	14169	14169	148.32	0.000
Temp R*Temp R	1	14169	14169	148.32	0.000
2-Way Interaction	1	631	631	6.60	0.017
Temp R*time R	1	631	631	6.60	0.017
Error	25	2388	96		
Lack-of-Fit	4	84	21	0.19	0.940
Pure Error	21	2304	110		
Total	29	191837			

The non-conformity test of the regression equation and the experimental data (lack-of-fit test) was performed, showing that the p-Value of Lack-Of-Fit is 0.940 > 0.05. From that, it is concluded that there is no basis to prove that the regression equation found is not consistent with the experimental data.

The fit of the model was tested by the adjusted R^2 and predicted R^2 parameters, showing that these parameters are both $\geq 90\%$, which means that the model fits the data very well:

S	R-sq	R-sq(adj)	R-sq(pred)
9.77377	98.76%	98.56%	98.23%

The regression equation characterizing the effect of high tempering mode parameters on the yield limit of the test sample has the following natural form:

$$\sigma_{0.2} = 6339 - 16.30\,Temp\,R + 1.008\,time\,R + 0.01232\,Temp\,R * Temp\,R - 0.002014\,Temp\,R * Time\,R \quad (1)$$

The graphs show the influence of two input parameters, namely the tempering temperature, the tempering time on the yield limit of the test sample as shown in Fig. 6.

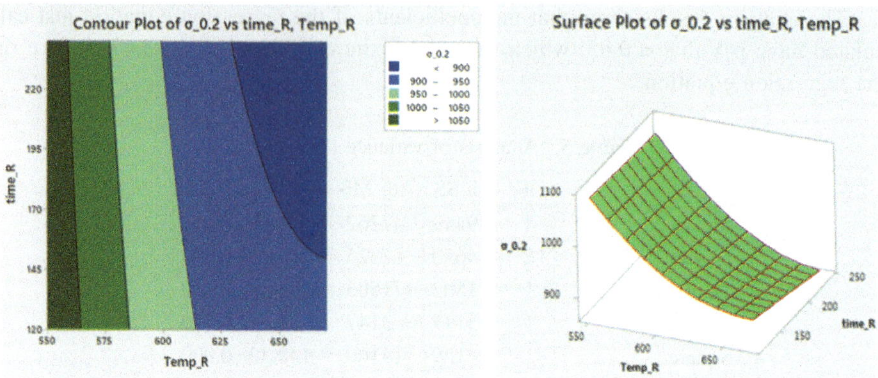

Fig. 6. Result of yield limit analysis $\sigma_{0.2}$

3.2 Relative Elongation δ

Table 6. Analyzing experimental results

Term	Coef	SE Coef	T-Value	P-Value
Constant	15.044	0.109	137.95	0.000
Temp R	2.8444	0.0745	38.18	0.000
Time R	0.4622	0.0745	6.20	0.000
Temp R*Temp R	1.379	0.119	11.54	0.000
Time R*time R	−0.648	0.119	−5.42	0.000
Temp R*time R	−0.0000	0.0912	−0.00	1.000

Similarly, Minitab 19 software was used for preliminary experimental analysis of the experimental samples, removing the coefficients that do not affect the regression equation to get the results of experimental analysis as shown in Table 6, analysis of variance (ANOVA) for the regression equation just found as shown in Table 7.

Table 7. Analysis of variance (ANOVA)

Source	DF	Adj SS	Adj MS	F-Value	P-Value
Model	4	164.050	41.013	427.67	0.000
Linear	2	149.481	74.741	779.37	0.000
Temp R	1	145.636	145.636	1518.64	0.000
Time R	1	3.846	3.846	40.10	0.000
Square	2	14.569	7.285	75.96	0.000
Temp R*Temp R	1	13.312	13.312	138.82	0.000
Time R*time R	1	2.936	2.936	30.61	0.000
Error	25	2.397	0.096		
Lack-of-Fit	4	0.469	0.117	1.28	0.311
Pure error	21	1.929	0.092		
Total	29	166.448			

The non-conformity test of the regression equation and the experimental data (lack-of-fit test) was performed, showing that the p-Value of Lack-Of-Fit is 0.311 > 0.05. From that, it is concluded that there is no basis to prove that the regression equation found is not consistent with the experimental data. The fit of the model was tested by the adjusted R^2 and predicted R^2 parameters, showing that these parameters are both $\geq 90\%$, which means that the model fits the data very well:

S	R-sq	R-sq(adj)	R-sq(pred)
0.309675	98.56%	98.33%	97.94%

The regression equation characterizing the effect of high tempering mode parameters on the relative elongation of the test sample has the following natural form:

$$\delta = 121.4 - 0.4199\, Temp\, R + 0.0725\, time\, R + 0.000383\, Temp\, R * Temp\, R - 0.000180\, time\, R * Time\, R \quad (2)$$

From the regression equation, the graphs show the influence of two input parameters, namely the tempering temperature, the tempering time on the relative elongation of the experimental sample, Fig. 7.

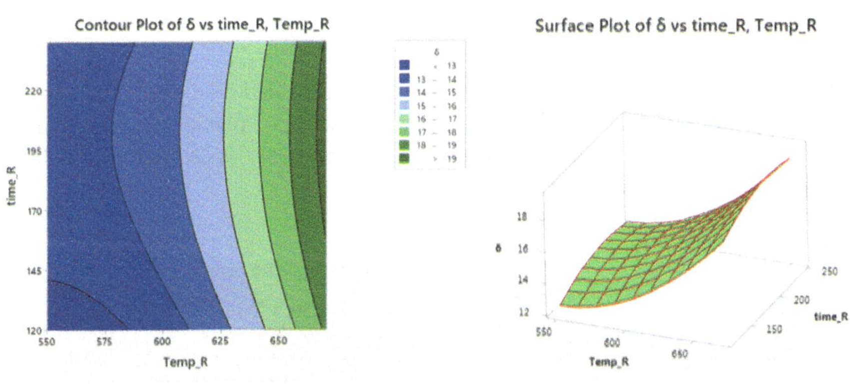

Fig. 7. Relative elongation analysis results δ

3.3 Optimization of High Tempering Technology Parameters

After having the regression equations, the two output parameters, namely yield strength and relative elongation were optimized. Optimizing these two parameters for the 10% error limit will have the allowable value of the result as shown in Table 8.

Table 8. Allowable value adjustment table of results

Response	Goal	Lower	Target	Upper	Weight	Importance
δ	Target	11.92	14.3	19.92	1	1
$\sigma\_{0.2}$	Target	857.00	1011.0	1095.00	1	1

Optimum value results:

Solution	Temp R	Time R	δFit	$\sigma_{0.2}$Fit	Composite desirability
1	575.455	198.788	13.9445	1010.99	0.922261

When there is an optimal result, this parameter is taken to perform high tempering to test the actual mechanical properties. Results after tensile test according to ASTM E8 standard:

$\sigma_{0.2} = 985$ N/mm^2

$\delta = 15.8\%$

Comparing the results with the sample shaft:

Sample parts	Parts with optimal parameters
$\sigma_{0.2} = 1011$ N/mm^2	$\sigma_{0.2} = 985$ N/mm^2
δ = 14.3%	$\delta = 15.8\%$

From the above comparison results, the yield strength and relative elongation of the optimal parts are satisfactory within the allowable range of planning calculations and experimental data processing. A review of the microscopic organization of the high tempering model was conducted (Fig. 8a).

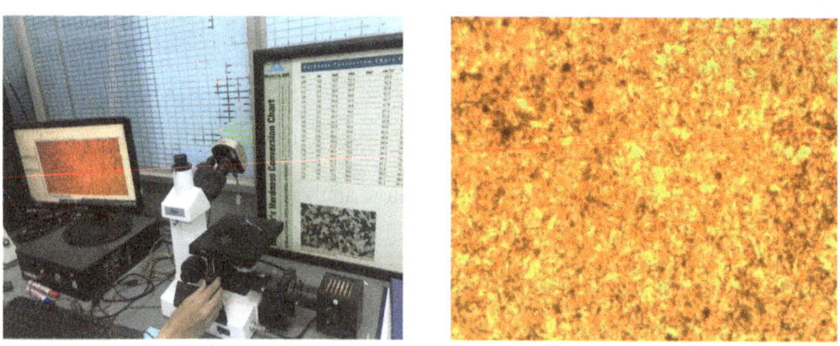

a) *b)*

Fig. 8. Perform a micro-organization review of the turbine

The results of microscopic organization review of the highly tempered shaft with optimal time and temperature show that this is a soorbite organization with a structure completely like that of the sample shaft (Fig. 8b).

4 Conclusion

This paper presents the research results on the effects of technological parameters on yield strength and relative elongation of 42CrMo steel as turbine shaft when high tempering in LAM - 50 N2 volume furnace using protected nitrogen gas system and being cooled in SHL QUENCH 100B oil. Research results have identified two regression equations and a set of technological parameters that have the greatest influence on the yield strength limit and relative elongation of turbine shaft parts, in which the optimum tempering temperature of *Temp R* is 575 °C, and the optimal heat retention time when high tempering *Time R* is 200 min.

References

1. Le, H.K., Dang, T.T.: Design analysis of turbine blade in the turbocharger. Vietnam J. Mech. (2020). ISSN 0866–7056
2. Dang, T.T., Tran, V.H., Le, H.K.: Evaluation on the accuracy of direct machining of turbine blade parts from inventor professional software on CNC milling machines. Vietnam J. Mech. (2020). ISSN 0866–7056
3. Le, H.K., Thong, N.C.: Research on turbo wing processing in turbocharger structure on CNC HAAS VF2 milling machine. Vietnam J. Mech. (2020). ISSN 0866–7056
4. Thong, N.C., Tran, V.H., Le, H.K.: Research on processing turbine blades in HX40W turbocharger structure on CNC milling machine. Vietnam J. Mech. (2020). ISSN 0866–7056
5. Nguyen, N.H.: Building heat treatment mode, HCMUT-VNU-HCM (2018)
6. Loc, N.H.: Planning and experimental analysis, HCMUT-VNU-HCM (2011)
7. Nguyen, C.C.: Handbook of Heat Treatment 1. Hanoi University of Science and Technology
8. Nguyen, D.T., Do, D.T.: Establishment of process and program for pure tensile fatigue testing of metallic materials on LFV 500-HH fatigue testing machine. Transp. Commun. Sci. J. **71**(2) (2020)
9. Takeuchi, Y., Watanabe, T.: Generation of 5-axis control collision-free tool path and post-processing for NC data. CIRP Ann. **41**, 539–542 (1992)
10. Bi, Q., Chen, H., Zou, X., Zhu, L., Ding, H.: Five-axis flank milling for design and manufacture of turbocharger compressor impeller. In: Proceedings of ASME Turbo Expo, Turbine Technical Conference and Exposition, pp. 1–11 (2014)
11. Wang, W.Z., Xuan, F.Z., Zhu, K.L., Tu, S.T.: Failure analysis of the final stage blade in steam turbine. School of Mechanical and Power Engineering, East China University of Science and Technology, Shanghai, PR China
12. Kindo, A.K., Goel, R.: Optimization of heat treatment of steel developed for turbine shafts. Department of Metallurgical and Materials Engineering National Institute of Technology, Rourkela (2012)
13. https://www.emag.com/blog/en/finish-machining-turbocharger-shafts/
14. Chen, R., Sung, W.P.: Effect of sub-temperature quenching temperature on microstructure and property of 35CrMo steel. Adv. Mater. Res. **490–495**, 3257–3261 (2012). https://doi.org/10.4028/www.scientific.net/AMR.490-495.3257

15. Sun, C., Pai-Xian, F., Liu, H.-W., Liu, H.-H., Ning-Yu, D.: Effect of tempering temperature on the low temperature impact toughness of 42CrMo4-V steel. Metals **8**(4), 232 (2018). https://doi.org/10.3390/met8040232
16. Seifoori, S., Parrany, A.M., Khodayari, M.: A high-cycle fatigue failure analysis for the turbocharger shaft of BELAZ 75131 mining dump truck. Mechanical Engineering Department, Faculty of Engineering, Vali-e-Asr University of Rafsanjan, Iran

Two Methods for Detecting the Linear Attack on SCADA Systems

Duc-Duong Nguyen[1,2](✉) and Manh-Hung Do[1]

[1] Hanoi University of Science and Technology, Hanoi, Vietnam
ndduong86.ddt@uneti.edu.vn, hung.domanh@hust.edu.vn
[2] University of Economics – Technology for Industries, Hanoi, Vietnam

Abstract. The linear attack is a new intrusion attack type that can change the normal operation of SCADA systems. It easily passes Kullback–Leibler (K-L) detection method. This paper presents the applied ability of the Cumulative Sum method (CUSUM) and Chi-squared (CHI2) to detect the linear attack in case the K-L method cannot detect. The object, which is attacked, is a wireless communication process from level sensor and temperature sensor to a controller in a thermal mixing tank system. By analyzing the tested results, which is implemented on simulation data, it shows that an appropriate range of threshold of both CUSUM and CHI2 methods can be chosen to detect the linear attack. In addition, the obtained results also show that the detection ability of the CUSUM method is better than that of the CHI2 method.

Keywords: Linear attack · CUSUM method · Chi-squared method · Detection threhold · Receiver operating characteristic – ROC

1 Introduction

Today, SCADA systems are universally used in industrial factories, and in power transformer stations. The task of data collection, data processing, and ensuring the integrity of the data is very important to keep the normal operation of such systems.

SCADA systems can be attacked at different points in different ways [1–8] (Fig. 1). There are two main directions to ensure information security in SCADA systems, that is, testing attack detection methods (through building and testing new attack types), and building methods to detect new attack types.

Recently, authors in [7] proposed a linear attack type that changed the data transmitted form sensors to controllers in a control system. The attack is proven to be extremely dangerous. It can pass the attack detection method K-L in any case [6].

In this paper, we build a mathematical model of a thermal mixing tank system as an object to test attack detection methods. Based on our published research results [9], we compare the linear attack detection ability of CUSUM and CHI2 methods on our built model, to confirm the effectiveness of two methods.

The following parts of the paper are organized as follows: the second section presents an overview of the linear attack and the K-L detection method; the third section introduces CHI2 and CUSUM detection algorithms; a model of thermal mixing process is presented in Sect. 4; application of CHI2, CUSUM and discussions are presented in Sect. 5; finally, some conclusions and our future works are presented in Sect. 6.

D. C. Nguyen et al. (Eds.): ICERA 2021, LNNS 366, pp. 929–940, 2022.
https://doi.org/10.1007/978-3-030-92574-1_95

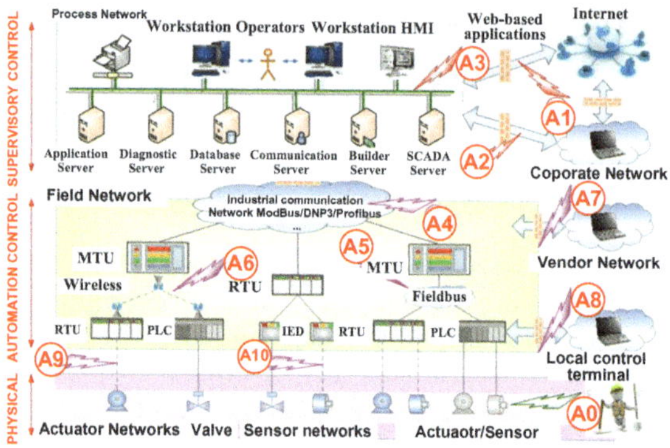

Fig. 1. Possible attack points to SCADA system [10]

2 Linear Attack and K-L Detection Algorithm

Consider an industrial control system with a linear attack point described in Fig. 2, which alters the wireless transmission data at the sensor's output [10].

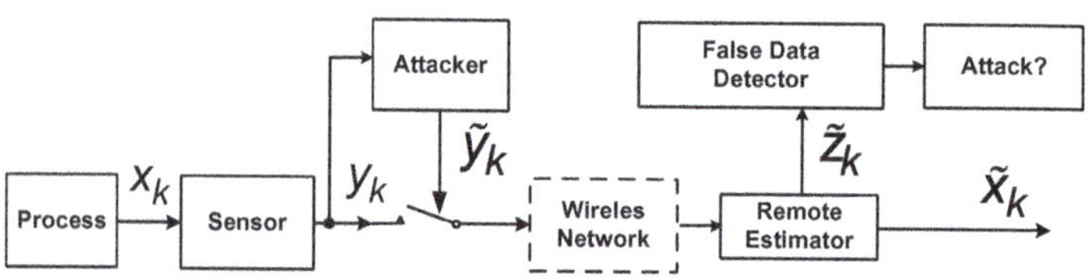

Fig. 2. Diagram illustrating the location of a linear attack

The equation describing the signal at the inputs and outputs of the sensors can be written as in (1) [6].

$$x_{k+1} = Ax_k + \omega_k;\ y_k = Cx_k + v_k \tag{1}$$

where: $k \in N$ - the time index; $x_k \in \mathbb{R}^n$ - the process states; $y_k \in \mathbb{R}^m$ - the sensor measurement; $v_k \in \mathbb{R}^m, v_k \sim N(0, R)$ - white noise effect on sensors (Gaussian noise); $\omega_k \in \mathbb{R}^n$, $\omega_k \sim N(0, Q)$ - white noise effect on state variable; $Q \geq 0; R > 0$ - covariance matrix of white noise; $\tilde{x}_k, \hat{x}_k^-$ - estimate of the state of the remote estimator when attacked, when not attacked; $A \in \mathbb{R}^{n \times n}, C \in \mathbb{R}^{m \times m}$ - system matrices; $\overline{P}$ - estimate of the covariance at steady state.

When not attacked, the sensor output bias estimate can be written as in (2) [5, 10] where $E\left[z_i, z_j^T\right]$ - the expected residual components z_k.

$$z_k = y_k - C\hat{x}_k^-; z_k \sim N(0; \Sigma);\ \Sigma = C\overline{P}C^T + R; E\left[z_i, z_j^T\right] = 0\,\forall \mathrm{i} \neq \mathrm{j} \tag{2}$$

When attacked, the output signals of the sensors are changed as described in (3):

$$\tilde{y}_k = \tilde{z}_k + C\tilde{x}_k^- \tag{3}$$

Authors in [6] have studied the possibility of a linear attack that passed the K-L detection method. This is a highly appreciated method of error detection, based on the principle of calculating the difference D between two series of random values $\tilde{z}_k$ and z_k which expressed in (4) [6, 11]:

$$D(\tilde{z}_k||z_k) = \int f_{\tilde{z}_k}(\chi)\log\frac{f_{\tilde{z}_k}(\chi)}{f_{z_k}(\chi)}d\chi \tag{4}$$

where $f_{\tilde{z}_k}(\chi)$ and $f_{z_k}(\chi)$ is the density function of $\tilde{z}_k$ and z_k.

When the difference exceeds a threshold, the data is evaluated as being subjected to a falsification attack and vice versa, as shown in (5):

$$D(\tilde{z}_k||z_k) \le \delta \rightarrow \text{Not attacked} \quad D(\tilde{z}_k||z_k) > \delta \rightarrow \text{Attacked} \tag{5}$$

where δ is the preset detection threshold of the K-L method.

According to the [6], under the action of linear attack, the sensor signals y_k are transformed into $\tilde{y}_k$, satisfied (3) and (6):

$$\tilde{z}_k = T_k z_k + b_k \tag{6}$$

where $T_k \in \mathbb{R}^{m\times m}$ - linear attack matrix; $b_k \sim N(0, \Gamma_k)$ - Gaussian random variable.

The linear attack can pass the K-L detection method when attack matrices T_k, Γ_k can be determined and satisfied (7) [6]:

$$\begin{cases} \min\limits_{(T_k)} Tr\left(C\overline{P}\overline{P}C^T\Sigma^{-1}T_k\right) \\ \begin{bmatrix} \tilde{\Sigma} & T_k \\ T_k^T & \Sigma^{-1} \end{bmatrix} \le 0 \end{cases} ;\ \Gamma_k = \tilde{\Sigma}_k - T_k\Sigma T_k^T \tag{7}$$

In which, according to the convex plan Karush Kuhn Tucker, the relationship between threshold δ and μ satisfies as the (8) [10]:

$$\mu\left(\frac{1}{2}Tr\left(\Sigma^{-1}\tilde{\Sigma}_k\right) - \frac{m}{2} + \frac{1}{2}\log\frac{|\Sigma|}{|\tilde{\Sigma}_k|} - \delta\right) = 0 \tag{8}$$

where $\mu > 2\min\limits_{1\le i\le m}\lambda_i$ and $\lambda_1, \lambda_2, \ldots, \lambda_m$ are the eigenvalues of $K^TK\Sigma$; K is the Kalman coefficient matrix, $Tr\left(\Sigma^{-1}\tilde{\Sigma}_k\right)$ is the trace of the matrix $\Sigma^{-1}\tilde{\Sigma}_k$.

Thus, for each threshold δ of the K-L method, it is always to find appropriate linear attack matrices T_k, Γ_k to pass the K-L.

3 CHI2 and CUSUM Detection Methods

The CHI2 method differs from the K-L method in that it uses the squared form of z_k value series to check the significant difference between the covariance and the desired value of the error between observed outputs and the estimated output as shown in formula (9) [6, 9, 12]:

$$\alpha_k = z_k^T \Sigma^{-1} z_k \tag{9}$$

The values of statistic decision g_k of the CHI2 method are determined according to (10):

$$g_k = \sum_{i=k-J+1}^{k} \alpha_k = \sum_{i=k-J+1}^{k} z_i^T \Sigma^{-1} z_i \tag{10}$$

When the statistic decision exceeds a threshold, the data is evaluated as being subjected to a falsification attack and vice versa, as shown in (11) and Fig. 3b.

$$g_k \leq h \rightarrow \text{not attacked: } g_k > h \rightarrow \text{attacked} \tag{11}$$

The attacked warning time T_a is determined from the condition:

$$T_a = \min(k : g_k \geq h) \tag{12}$$

where h is the attack detection threshold (preset).

The CUSUM method is different from the K-L and CHI2 methods in that it applies Wald's theory to analyze anomalies in tested data [13, 14]. Considering a random system $X = [x_1, x_2, \ldots x_k]^T \sim N(\mu, \Sigma)$ and assuming that $X \sim N(\mu, \Sigma_0)$ when non-attacked, and $X \sim N(\mu, \Sigma_1)$ when attacked. The CUSUM method considers the actual rate of change S_i^k (likelihood ratio – LLR), as illustrated in Fig. 3a, determined by formula (13) [9, 11, 13]. S_i^k tends to vary monotonically when there is no abnormal change in the signal, and reverses.

$$S_i^k = \sum_{t=i}^{k} \ln \frac{f_{\theta_1}(x_t)}{f_{\theta_0}(x_t)}; \; s_k = \frac{1}{2} \ln \frac{\det \Sigma_0}{\det \Sigma_1} - \frac{1}{2}(x-\mu)^T \left[\Sigma_1^{-1} - \Sigma_0^{-1}\right](x-\mu) \tag{13}$$

where $\det \Sigma_0, \det \Sigma_1$ are determinant of the Σ_0, Σ_1 matrix, calculated as (14):

$$\Sigma_0 = \Sigma = C\bar{P}C^T + R; \; \Sigma_1 = T_k \Sigma T_k^T + \Gamma_k \tag{14}$$

The statistic decision values g_k of the CUSUM method are determined according to (15) [9, 14]:

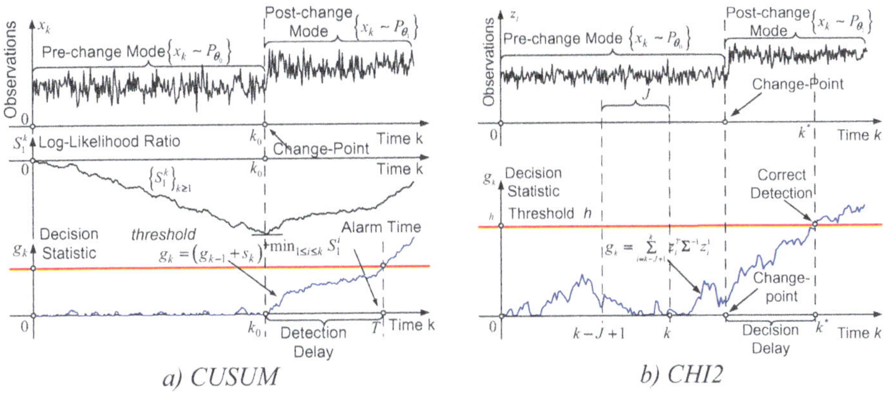

Fig. 3. Describe CUSUM(a) and CHI2(b) detection algorithm [13]

$$g_k = \begin{cases} g_{k-1} + s_k & \text{if } g_{k-1} + s_k > 0 \\ 0 & \text{if } g_{k-1} + s_k < 0 \end{cases} \tag{15}$$

The attack warning time T_a is determined from the condition (16) [9, 14]:

$$T_a = \min(k : g_k \geq h) \tag{16}$$

where h is the preset attack detection threshold.

Thus, the CHI2 method only uses the data before an anomal change, while the CUSUM uses the data both before and after an anomal change.

4 Model of a Thermal Mixing Process

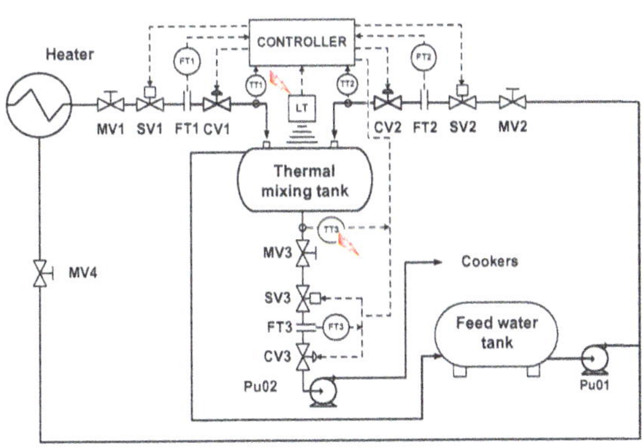

Fig. 4. P&ID of thermal mixing process

In this paper, we consider a model of thermal mixing process which widely used in food factories and brewers (Fig. 4).

In this model, F_1, F_2, F_3 are volume flow rate of hot water, additional water, output water (m^3/h); T_1, T_2, T_3 are temperature of hot water, additional water, tank's water (°C); H is water level in thermal mixing tank, with H_{max} = 2.8 (m); ρ is water's specific weight, with ρ = 1000 (kg/m^3); V is water volume in thermal mixing tank; L is length of thermal mixing tank, with L = 14.5 (m); R is radius of thermal mixing tank, with R = 1.4 (m).

According to the conservation law of mass and energy [9], we have:

$$\underbrace{\begin{bmatrix} \Delta \dot{H} \\ \Delta \dot{T}_3 \end{bmatrix}}_{\dot{x}(t)} = \underbrace{\begin{bmatrix} 0 & 0 \\ 0 & -\frac{\overline{F_1}+\overline{F_2}}{\rho \overline{V}} \end{bmatrix}}_{\tilde{A}} \underbrace{\begin{bmatrix} \Delta H \\ \Delta T_3 \end{bmatrix}}_{x(t)} + \underbrace{\begin{bmatrix} \frac{1}{\overline{A}} & \frac{1}{\overline{A}} \\ \frac{\overline{T_2}-\overline{T_3}}{\overline{V}} & \frac{\overline{T_1}-\overline{T_3}}{\overline{V}} \end{bmatrix}}_{\tilde{B}} \underbrace{\begin{bmatrix} \Delta F_2 \\ \Delta F_1 \end{bmatrix}}_{u(t)} + \begin{bmatrix} \frac{-1}{\overline{A}} & 0 & 0 \\ 0 & \frac{\overline{F_1}}{\rho \overline{V}} & \frac{\overline{F_2}}{\rho \overline{V}} \end{bmatrix} \begin{bmatrix} \Delta F_3 \\ \Delta T_1 \\ \Delta T_2 \end{bmatrix} \tag{17}$$

As flow rate F_3 and T_1,T_2 are process disturbance, it is assumed that these variables do not change in the whole mixing process ($\Delta F_3 = 0; \Delta T_1 = \Delta T_2 = 0$), we have:

$$\dot{x}(t) = \tilde{A}x(t) + \tilde{B}u(t) \tag{18}$$

Two sensors LT and TT3 are used to measure water level $H(t)$ and water temperature $T_3(t)$ in the thermal mixing tank. We have the measurement equation:

$$\underbrace{\begin{bmatrix} y_1(t) \\ y_2(t) \end{bmatrix}}_{y(t)} = \underbrace{\begin{bmatrix} K_1 & 0 \\ 0 & K_2 \end{bmatrix}}_{\tilde{C}} \underbrace{\begin{bmatrix} \Delta H \\ \Delta T_3 \end{bmatrix}}_{x(t)} \tag{19}$$

where $y_1(t)$ - level transmitter signal of mixing tank; $y_2(t)$ - temperature transmitter signal of mixing tank.

In this paper, we choose the following parameters for the thermal mixing tank:

$$F_{\max} = 160 m^3/h;\ \overline{H} = 0.5H_{\max};\ \overline{T_1} = 80\,°C;\ \overline{T_2} = 37\,°C, \overline{T_3} = 39\,°C, \overline{F_1} = \overline{F_2} = \overline{F_3} = 0.5F_{\max};$$
$$K_1 = 0.5;\ K_2 = 0.7;\ \textit{sampling cycle}\ T_S = 0.1\ \mathrm{s}$$

According to [5], we have a discontinuous state model of the object in Eq. (1) with parameters:

$$A = \begin{bmatrix} 1 & 0 \\ 0 & 1 \end{bmatrix};\ C = \begin{bmatrix} 0.5 & 0 \\ 0 & 0.7 \end{bmatrix};\ Q = \begin{bmatrix} 0.51 & 0 \\ 0 & 0.505 \end{bmatrix};\ R = \begin{bmatrix} 1 & 0 \\ 0 & 0.8 \end{bmatrix}$$

5 Application of CUSUM, CHI2 and Discussions

We perform simulations to evaluate the applicability of CHI2 and CUSUM methods to detect linear attacks in case the K-L method is passed. The formula (8) shows the relationship between δ and μ in the K-L method. Through choosing the thresholds δ we can calculate μ. Based on the analysis in [5], we select δ in range $\delta \in [0; 176]$.

$$\delta = \begin{cases} 0 \equiv \delta_0 \\ 0.1 \equiv \delta_1 \\ 1.0 \equiv \delta_2 \\ 1.5 \equiv \delta_3 \end{cases} \Rightarrow \mu = \begin{cases} \infty \equiv \mu_0 \\ 2.9813 \equiv \mu_1 \\ 1.4850 \equiv \mu_2 \\ 1.3005 \equiv \mu_3 \end{cases} \quad \delta = \begin{cases} 3.0 \equiv \delta_4 \\ 4.0 \equiv \delta_5 \\ 5.0 \equiv \delta_6 \\ 176 \equiv \delta_7 \end{cases} \Rightarrow \mu = \begin{cases} 1.2290 \equiv \mu_4 \\ 1.1862 \equiv \mu_5 \\ 1.1587 \equiv \mu_6 \\ 1.0235 \equiv \mu_7 \end{cases}$$

By solving the optimal system (7) using CVX toolbox in Matlab and from (13), we get parameter matrices of the linear attack $T_{k0} \div T_{k7}; \Gamma_{k0} \div \Gamma_{k7}$ so that it passes the K-L method. Then, we check the difference between attacked signals and non-attacked signals, through the traces of the covariance matrices $Tr(\tilde{P}_k)$ as shown in Fig. 5. A simulation data with the length of 25 s is used. There is not attack in the first 19 s. Attacks occurs in the ranges from 19 to 25 s [6].

The simulation results in Fig. 5 show that when not attacked the trace of the covariance matrix ($Tr(\tilde{P}_k)$) is approximately 2.1. When the linear attack appears, $Tr(\tilde{P}_k)$ of the system spiked, representing a large difference of the signal before and after attacked. Meanwhile, with selected matrices T_k, Γ_k, the linear attack is not detected by K-L method. For each given threshold δ, it is always possible to calculate a set of parameters T_k and Γ_k of the linear attack so that it passes the K-L detection method.

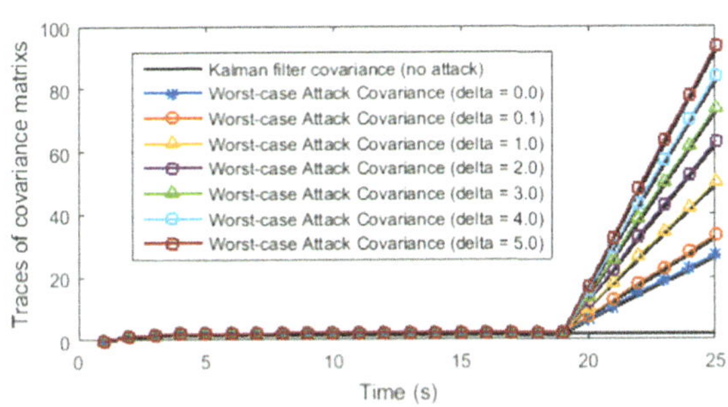

Fig. 5. Traces of covariance matrixs under linear attacks

In order to examine the **applicability** of CHI2 and CUSUM, we, first, consider the case of linear attack passing the K-L method at a low threshold $\delta = \delta_1 = 0.1$. A 1000 s length simulation dataset with linear attacks occurring in the range from 500 s to 700 s is constructed to test CHI2 and CUSUM At the detection threshold of the two methods CHI2 and CUSUM h = 3.4, applying the formulas (8, 10, 11, 12, 14, 15, 16) to calculate time of attack warning T_a, we obtianed results as shown in Fig. 6a. The result shows that the CUSUM method detected linear attack at $T_{a\_CUSUM} = 541s$, $nu = 1$, (correctly detected). However, the CHI2 method gave a false alarm, because $nu = 1$, $T_{a\_CHI2} = 4s$ which is not within the attacked period.

Similarly, with detection threshold $h = 16$, we obtain simulation results as shown in Fig. 6b. Simulation results show **that,** CUSUM method has detected linear attack at $T_{a\_CUSUM} = 647s$, $nu = 1$, (correct detection). However, the CHI2 method detected this attack at $T_{a\_CHI2} = 0s$, $nu = 0$ (false detection).

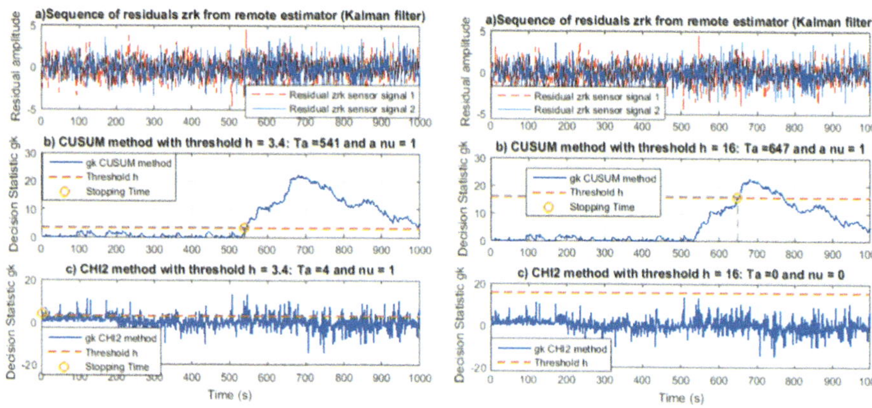

a) threshold h = 3.4 *b) threshold h = 16*

Fig. 6. Linear attack detection by CUSUM, CHI2 method with threshold $\delta = 0.1$

In general, to evaluate linear attack detection over time interval $[k_0\, k_0 + L)$ of each method, we use the correct detection probability P_D; probability of false alarm P_{FA} and the probability of wrong time detection P_{MD} (Fig. 7). Relationship between P_D and P_{FA} is represented on the ROC curve.

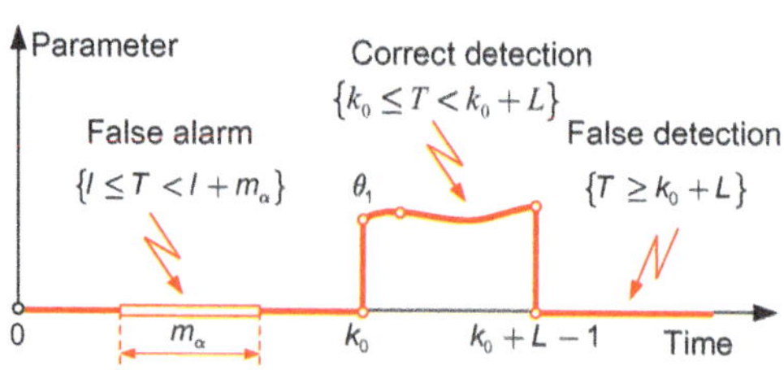

Fig. 7. Attack detection problem [13]

In which,

Correct detection is an abnormal change detection in the exact time it occurs. The probability of correct detection is illustrated in formula (20):

$$P_D = P(k_0 \le T \le k_0 + L - 1) \quad (20)$$

False warning: detect abnormal change before it happens $(T < k_0)$. The probability of false alarms can be determined in the corresponding time interval m_α with false alarm initial time l and threshold α illustrated in formula (21):

$$P_{FA} = P_{FA}(T, m_\alpha) = \sup_{l \ge 1} P_0[(l \le T \le l + m_\alpha) \le \alpha] \quad (21)$$

According to [15], the Monte-Carlo estimation of $\overline{P}_{FA}$ is calculated as the ratio of the number of false alarms to the total number of simulations nS; Monte-Carlo estimation of $\overline{P}_D$ is calculated as the ratio of the number of correct detections to the total number of simulations nS, illustrated in formula (22):

$$\overline{P}_{FA} = \frac{1}{nS}\sum_{k=1}^{nS} nu(k), k \le k_0; \overline{P}_D = \frac{1}{nS}\sum_{k=1}^{nS} nu(k); k_0 \le k \le k_0 + L - 1 \quad (22)$$

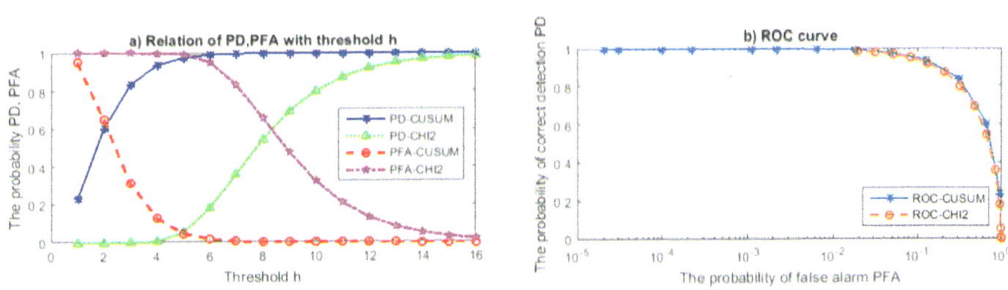

Fig. 8. Evaluation of linear attack detection by CUSUM, CHI2 with threshold h = 1 ÷ 16 when K-L is passed with threshold $\delta = 5$

Considering the case when K-L is passed at threshold $\delta = 5$, the detection thresholds of CHI2 and CUSUM are selected in the range $h = 1 \div 16$, and applying formula (22) to the simulated object, we obtain results as shown in Fig. 8. Figure 8a shows that with the same threshold of h, the CUSUM method has a higher probability of correct detection $\overline{P}_D$, and lower probability of false alarm $\overline{P}_{FA}$ than those of the CHI2 method. Results also show that when the threshold h increases, the probability of correct detection $\overline{P}_D$ of both methods increases, while the probability of false alarm $\overline{P}_{FA}$ decreases gradually. In other words, it can say that the detection ability of the CUSUM method is better than that of the CHI2 method.

For other thresholds δ (of the KL method), we obtain the characteristic curves similar to Fig. 8 and set of ROC curves of the CUSUM method (Fig. 9), and of the CHI2 method (Fig. 10).

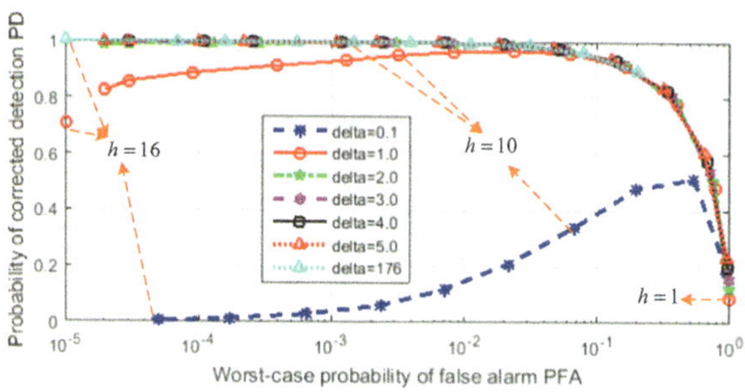

Fig. 9. Statistical performance to detect linear attack by CUSUM with several thresholds δ

Comparison of results in Figs. 9 and 10 shows that, under the same test conditions, the false alarm probability of the CHI2 method is higher than that of the CUSUM method. These figures also show that the linear attack which passes K-L method at high threshold δ is easily detected by the CHI2 and CUSUM methods (shown by the high correct detection probability and the low false detection probability in both CHI2 and CUSUM methods).

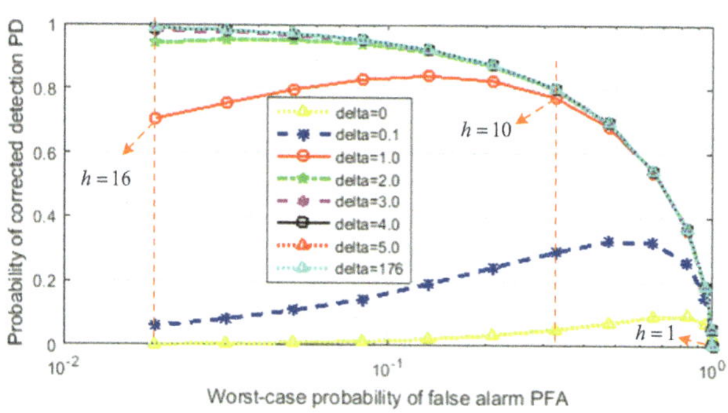

Fig. 10. Statistical performance to detect linear attack by CHI2 with several thresholds δ

Corresponding to each threshold value δ of the K-L method that the linear attack passed, we see the possibility that it exists a threshold h to detect linear attacks by the CUSUM or CHI2 method (Fig. 11) [9]. The results in Fig. 11 show the fact that it is possible to choose an appropriate threshold h for CUSUM or CHI2 method to detect a linear attack in all cases K-L is passed. The threshold range h of CUSUM method is wider than that of CHI2. It indicates the detection ability of CUSUM is better than that of CHI2.

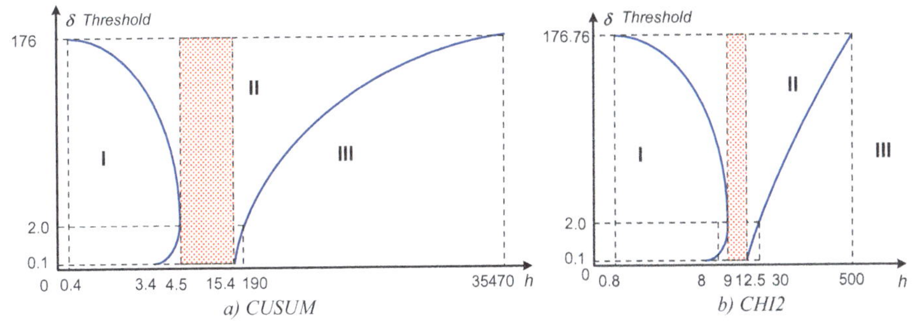

Fig. 11. Linear attack detection by CHI2 and CUSUM methods

6 Conclusions

In this paper, we use our model of thermal mixing process to test the ability to detect linear attack of CHI2 and CUSUM methods. Our analyses have shown that the range of threshold values h of CUSUM method is wider than that of CHI2 method. Through ROC curves of two methods on the model of thermal mixing process, it is easy to see that the detection ability of CUSUM is better and more effective than that of CHI2. For application, these methods can be used as a back-end detection layer in a series of techniques applied to ensure data integrity of SCADA systems from linear attack.

For our future works, we consider to apply the expanded Chi-squared, expanded CUSUM methods and other detection methods to evaluate their ability of linear attack detection.

References

1. Lehto, M., Neittaanmäki, P.: Cyber Security: Analytics, Technology and Automation, vol. 78. Springer, Cham (2015). https://doi.org/10.1007/978-3-319-18302-2
2. East, S., Butts, J., Papa, M., Shenoi, S.: A taxonomy of attacks on the DNP3 protocol. In: Palmer, C., Shenoi, S. (eds.) ICCIP 2009. IAICT, vol. 311, pp. 67–81. Springer, Heidelberg (2009). https://doi.org/10.1007/978-3-642-04798-5_5
3. Perez, R.L., Adamsky, F., Soua, R., Engel, T.: Machine learning for reliable network attack detection in SCADA systems. In: 2018 17th IEEE International Conference on Trust, Security and Privacy in Computing and Communications/12th IEEE International Conference on Big Data Science and Engineering (TrustCom/BigDataSE), pp. 633–638 (2018)
4. Hijazi, A., El Safadi, A., Flaus, J.M.: A deep learning approach for intrusion detection system in industry network. In: BDCSIntell, pp. 55–62 (2018)
5. Guo, Z., Shi, D., Johansson, K.H., Shi, L.: Optimal linear cyber-attack on remote state estimation. IEEE Trans. Control Netw. Syst. **4**(1), 4–13 (2017). https://doi.org/10.1109/TCNS.2016.2570003
6. Guo, Z., Shi, D., Johansson, K.H., Shi, L.: Worst-case stealthy innovation-based linear attack on remote state estimation. Automatica **89**, 117–124 (2018)

7. Wu, S., Guo, Z., Shi, D., Johansson, K.H., Shi, L.: Optimal innovation-based deception attack on remote state estimation. In: 2017 American Control Conference (ACC), pp. 3017–3022, May 2017. https://doi.org/10.23919/ACC.2017.7963410
8. Ding, D., Han, Q.-L., Xiang, Y., Ge, X., Zhang, X.-M.: A survey on security control and attack detection for industrial cyber-physical systems. Neurocomputing **275**, 1674–1683 (2018)
9. Duong, N.D., Le, M.T., Cung, T.L.: Ability to detect the linear attack in industrial control systems by CUSUM method. J. Sci. Techonol. Tech. Univ. 14–20 (2020)
10. Do, V.L., et al.: Sequential detection and isolation of cyber-physical attacks on SCADA systems, Troyes (2015)
11. Kullback, S.: Information Theory and statistics. Courier Corporation, Chelmsford (1997)
12. Nikulin, M.S.: Chi-squared test for normality. In: Proceedings of the International Vilnius Conference on Probability Theory and Mathematical Statistics, vol. 2, no. 1, pp. 119–122 (1973)
13. Basseville, M., Nikiforov, I.V.: Detection of Abrupt Changes: Theory and Application, vol. 104. Prentice Hall Englewood Cliffs, Hoboken (1993)
14. Woodward, R.H., Goldsmith, P.L.: Cumulative Sum Techniques. Imperial Chemical Industries Limited, London (1964)
15. Parloir, C., Kinnaert, M.: Performance evaluation of fault detection algorithms by monte carlo methods. IFAC Proc. Vol. **37**(21), 597–602 (2004)

Author Index

D. C. Nguyen et al. (Eds.): ICERA 2021, LNNS 366, pp. 941–944, 2022.
https://doi.org/10.1007/978-3-030-92574-1

The manufacturer's authorised representative in the EU is Springer Nature Customer Service Centre GmbH, Europaplatz 3, 69115 Heidelberg, Germany. If you have any concerns regarding our products, please contact ProductSafety@springernature.com

Printed and bound by CPI Group (UK) Ltd, Croydon, CR0 4YY
30/09/2026
02225255-0002